Numerical Recipes in C++

Second Edition

Numerical Recipes in C++

The Art of Scientific Computing

Second Edition

William H. Press

Los Alamos National Laboratory

Saul A. Teukolsky

Department of Physics, Cornell University

William T. Vetterling

Polaroid Corporation

Brian P. Flannery

EXXON Research and Engineering Company

CAMBRIDGE
UNIVERSITY PRESS

PUBLISHED BY THE PRESS SYNDICATE OF THE UNIVERSITY OF CAMBRIDGE
The Pitt Building, Trumpington Street, Cambridge, United Kingdom

CAMBRIDGE UNIVERSITY PRESS
The Edinburgh Building, Cambridge CB2 2RU, UK
40 West 20th Street, New York, NY 10011-4211, USA
477 Williamstown Road, Port Melbourne, VIC, 3207, Australia
Ruiz de Alarcón 13, 28014 Madrid, Spain
Dock House, The Waterfront, Cape Town 8001, South Africa

http://www.cambridge.org

Some sections of this book were originally published, in different form, in *Computers
in Physics* magazine, © American Institute of Physics, 1988–1992.

First Edition originally published 1988; Second Edition originally published 1992;
C++ edition originally published 2002.

This printing is corrected to software version 2.10

Printed in the United States of America
Typeface Times 10/12 pt.   *System* TEX   [AU]

Affiliations shown on title page are for purposes of identification only. No implication
that the works contained herein were created in the course of employment is intended,
nor is any knowledge of or endorsement of these works by the listed institutions to be inferred.

A catalog record for this book is available from the British Library.

Library of Congress Cataloging in Publication Data
Numerical recipes in C++ : the art of scientific computing / William H. Press
    ... [et al.]. – 2nd ed.
            p.   cm.
    Includes bibliographical references and index.
    ISBN 0-521-75033-4
    1. C++ (Computer program language) 2. Numerical analysis.   I. Press, William H.
QA76.73.C153 N85 2002
519.4′0285′5133–dc21   2001052699

ISBN   0 521 75033 4   Hardback
ISBN   0 521 75034 2   Example book in C++
ISBN   0 521 75037 7   C++/C CD-ROM (Windows/Macintosh)
ISBN   0 521 75035 0   Complete CD-ROM (Windows/Macintosh)
ISBN   0 521 75036 9   Complete CD-ROM (UNIX/Linux)

1002653863

Contents

Preface to the C++ Edition

C++ has gradually become the dominant language for computer programming, displacing C and Fortran even in many scientific and engineering applications. This version of *Numerical Recipes* contains the entire text of the Second Edition with all the programs presented in C++.

C++ poses special problems for numerical work. In particular, it is difficult to treat vectors and matrices in a manner that is simultaneously efficient and yet allows programming with high-level constructs. The fact that there is still no universally accepted standard library for doing this makes the problem even more difficult for authors of a book like this one. In Chapter 1 and the Appendices we describe how we have solved this problem. The default option is for you, the reader, to use a very simple class library that we provide. You can be up and running in a few minutes. We also show you how you can alternatively use any other matrix/vector class library of your choosing. This may take you a few minutes to set up the first time, but thereafter will provide transparent access to the Recipes with essentially no loss in efficiency.

We have taken this opportunity to respond to a clear consensus from our C readers, and converted all arrays and matrices to be "zero-based." We have also taken this opportunity to fix errors in the text and programs that have been reported to us by our readers. There are too many people to acknowledge individually, but to all who have written to us we are very grateful.

September 2001

William H. Press
Saul A. Teukolsky
William T. Vetterling
Brian P. Flannery

Preface to the Second Edition

Our aim in writing the original edition of *Numerical Recipes* was to provide a book that combined general discussion, analytical mathematics, algorithmics, and actual working programs. The success of the first edition puts us now in a difficult, though hardly unenviable, position. We wanted, then and now, to write a book that is informal, fearlessly editorial, unesoteric, and above all useful. There is a danger that, if we are not careful, we might produce a second edition that is weighty, balanced, scholarly, and boring.

It is a mixed blessing that we know more now than we did six years ago. Then, we were making educated guesses, based on existing literature and our own research, about which numerical techniques were the most important and robust. Now, we have the benefit of direct feedback from a large reader community. Letters to our alter-ego enterprise, Numerical Recipes Software, are in the thousands per year. (Please, *don't telephone* us.) Our post office box has become a magnet for letters pointing out that we have omitted some particular technique, well known to be important in a particular field of science or engineering. We value such letters, and digest them carefully, especially when they point us to specific references in the literature.

The inevitable result of this input is that this Second Edition of *Numerical Recipes* is substantially larger than its predecessor, in fact about 50% larger both in words and number of included programs (the latter now numbering well over 300). "Don't let the book grow in size," is the advice that we received from several wise colleagues. We have tried to follow the intended spirit of that advice, even as we violate the letter of it. We have not lengthened, or increased in difficulty, the book's principal discussions of mainstream topics. Many new topics are presented at this same accessible level. Some topics, both from the earlier edition and new to this one, are now set in smaller type that labels them as being "advanced." The reader who ignores such advanced sections completely will not, we think, find any lack of continuity in the shorter volume that results.

Here are some highlights of the new material in this Second Edition:

- a new chapter on integral equations and inverse methods
- a detailed treatment of multigrid methods for solving elliptic partial differential equations
- routines for band diagonal linear systems
- improved routines for linear algebra on sparse matrices
- Cholesky and QR decomposition
- orthogonal polynomials and Gaussian quadratures for arbitrary weight functions
- methods for calculating numerical derivatives
- Padé approximants, and rational Chebyshev approximation
- Bessel functions, and modified Bessel functions, of fractional order; and several other new special functions
- improved random number routines
- quasi-random sequences
- routines for adaptive and recursive Monte Carlo integration in high-dimensional spaces
- globally convergent methods for sets of nonlinear equations

- simulated annealing minimization for continuous control spaces
- fast Fourier transform (FFT) for real data in two and three dimensions
- fast Fourier transform (FFT) using external storage
- improved fast cosine transform routines
- wavelet transforms
- Fourier integrals with upper and lower limits
- spectral analysis on unevenly sampled data
- Savitzky-Golay smoothing filters
- fitting straight line data with errors in both coordinates
- a two-dimensional Kolmogorov-Smirnoff test
- the statistical bootstrap method
- embedded Runge-Kutta-Fehlberg methods for differential equations
- high-order methods for stiff differential equations
- a new chapter on "less-numerical" algorithms, including Huffman and arithmetic coding, arbitrary precision arithmetic, and several other topics.

Consult the Preface to the First Edition, following, or the table of Contents, for a list of the more "basic" subjects treated.

Acknowledgments

It is not possible for us to list by name here all the readers who have made useful suggestions; we are grateful for these. In the text, we attempt to give specific attribution for ideas that appear to be original, and are not known in the literature. We apologize in advance for any omissions.

Some readers and colleagues have been particularly generous in providing us with ideas, comments, suggestions, and programs for this Second Edition. We especially want to thank George Rybicki, Philip Pinto, Peter Lepage, Robert Lupton, Douglas Eardley, Ramesh Narayan, David Spergel, Alan Oppenheim, Sallie Baliunas, Scott Tremaine, Glennys Farrar, Steven Block, John Peacock, Thomas Loredo, Matthew Choptuik, Gregory Cook, L. Samuel Finn, P. Deuflhard, Harold Lewis, Peter Weinberger, David Syer, Richard Ferch, Steven Ebstein, Bradley Keister, and William Gould. We have been helped by Nancy Lee Snyder's mastery of a complicated TEX manuscript. We express appreciation to our editors Lauren Cowles and Alan Harvey at Cambridge University Press, and to our production editor Russell Hahn. We remain, of course, grateful to the individuals acknowledged in the Preface to the First Edition.

Special acknowledgment is due to programming consultant Seth Finkelstein, who wrote, rewrote, or influenced many of the routines in this book, as well as in its Fortran-language twin and the companion Example books. Our project has benefited enormously from Seth's talent for detecting, and following the trail of, even very slight anomalies (often compiler bugs, but occasionally our errors), and from his good programming sense. To the extent that this edition of *Numerical Recipes in C* has a more graceful and "C-like" programming style than its predecessor, most of the credit goes to Seth. (Of course, we accept the blame for the Fortranish lapses that still remain.)

We prepared this book for publication on DEC and Sun workstations running the UNIX operating system, and on a 486/33 PC compatible running MS-DOS 5.0/Windows 3.0. (See §1.0 for a list of additional computers used in program tests.) We enthusiastically recommend the principal software used: GNU Emacs, TEX, Perl, Adobe Illustrator, and PostScript. Also used were a variety of C compilers – too numerous (and sometimes too buggy) for individual acknowledgment. It is a sobering fact that our standard test suite (exercising all the routines in this book) has uncovered compiler bugs in many of the compilers tried. When possible, we work with developers to see that such bugs get fixed; we encourage interested compiler developers to contact us about such arrangements.

WHP and SAT acknowledge the continued support of the U.S. National Science Foundation for their research on computational methods. D.A.R.P.A. support is acknowledged for §13.10 on wavelets.

June 1992

William H. Press
Saul A. Teukolsky
William T. Vetterling
Brian P. Flannery

Preface to the First Edition

We call this book *Numerical Recipes* for several reasons. In one sense, this book is indeed a "cookbook" on numerical computation. However, there is an important distinction between a cookbook and a restaurant menu. The latter presents choices among complete dishes in each of which the individual flavors are blended and disguised. The former — and this book — reveals the individual ingredients and explains how they are prepared and combined.

Another purpose of the title is to connote an eclectic mixture of presentational techniques. This book is unique, we think, in offering, for each topic considered, a certain amount of general discussion, a certain amount of analytical mathematics, a certain amount of discussion of algorithmics, and (most important) actual implementations of these ideas in the form of working computer routines. Our task has been to find the right balance among these ingredients for each topic. You will find that for some topics we have tilted quite far to the analytic side; this where we have felt there to be gaps in the "standard" mathematical training. For other topics, where the mathematical prerequisites are universally held, we have tilted towards more in-depth discussion of the nature of the computational algorithms, or towards practical questions of implementation.

We admit, therefore, to some unevenness in the "level" of this book. About half of it is suitable for an advanced undergraduate course on numerical computation for science or engineering majors. The other half ranges from the level of a graduate course to that of a professional reference. Most cookbooks have, after all, recipes at varying levels of complexity. An attractive feature of this approach, we think, is that the reader can use the book at increasing levels of sophistication as his/her experience grows. Even inexperienced readers should be able to use our most advanced routines as black boxes. Having done so, we hope that these readers will subsequently go back and learn what secrets are inside.

If there is a single dominant theme in this book, it is that practical methods of numerical computation can be simultaneously efficient, clever, and — important — clear. The alternative viewpoint, that efficient computational methods must necessarily be so arcane and complex as to be useful only in "black box" form, we firmly reject.

Our purpose in this book is thus to open up a large number of computational black boxes to your scrutiny. We want to teach you to take apart these black boxes and to put them back together again, modifying them to suit your specific needs. We assume that you are mathematically literate, i.e., that you have the normal mathematical preparation associated with an undergraduate degree in a physical science, or engineering, or economics, or a quantitative social science. We assume that you know how to program a computer. We do not assume that you have any prior formal knowledge of numerical analysis or numerical methods.

The scope of *Numerical Recipes* is supposed to be "everything up to, but not including, partial differential equations." We honor this in the breach: First, we *do* have one introductory chapter on methods for partial differential equations (Chapter 19). Second, we obviously cannot include *everything* else. All the so-called "standard" topics of a numerical analysis course have been included in this book:

linear equations (Chapter 2), interpolation and extrapolation (Chaper 3), integration (Chaper 4), nonlinear root-finding (Chapter 9), eigensystems (Chapter 11), and ordinary differential equations (Chapter 16). Most of these topics have been taken beyond their standard treatments into some advanced material which we have felt to be particularly important or useful.

Some other subjects that we cover in detail are not usually found in the standard numerical analysis texts. These include the evaluation of functions and of particular special functions of higher mathematics (Chapters 5 and 6); random numbers and Monte Carlo methods (Chapter 7); sorting (Chapter 8); optimization, including multidimensional methods (Chapter 10); Fourier transform methods, including FFT methods and other spectral methods (Chapters 12 and 13); two chapters on the statistical description and modeling of data (Chapters 14 and 15); and two-point boundary value problems, both shooting and relaxation methods (Chapter 17).

The programs in this book are included in ANSI-standard C. Versions of the book in Fortran, Pascal, and BASIC are available separately. We have more to say about the C language, and the computational environment assumed by our routines, in §1.1 (Introduction).

Acknowledgments

Many colleagues have been generous in giving us the benefit of their numerical and computational experience, in providing us with programs, in commenting on the manuscript, or in general encouragement. We particularly wish to thank George Rybicki, Douglas Eardley, Philip Marcus, Stuart Shapiro, Paul Horowitz, Bruce Musicus, Irwin Shapiro, Stephen Wolfram, Henry Abarbanel, Larry Smarr, Richard Muller, John Bahcall, and A.G.W. Cameron.

We also wish to acknowledge two individuals whom we have never met: Forman Acton, whose 1970 textbook *Numerical Methods that Work* (New York: Harper and Row) has surely left its stylistic mark on us; and Donald Knuth, both for his series of books on *The Art of Computer Programming* (Reading, MA: Addison-Wesley), and for TEX, the computer typesetting language which immensely aided production of this book.

Research by the authors on computational methods was supported in part by the U.S. National Science Foundation.

October 1985

William H. Press
Brian P. Flannery
Saul A. Teukolsky
William T. Vetterling

License Information

Read this section if you want to use the programs in this book on a computer. You'll need to read the following Disclaimer of Warranty, get the programs onto your computer, and acquire a Numerical Recipes software license. (Without this license, which can be the free "immediate license" under terms described below, the book is intended as a text and reference book, for reading purposes only.)

Disclaimer of Warranty

We make no warranties, express or implied, that the programs contained in this volume are free of error, or are consistent with any particular standard of merchantability, or that they will meet your requirements for any particular application. They should not be relied on for solving a problem whose incorrect solution could result in injury to a person or loss of property. If you do use the programs in such a manner, it is at your own risk. The authors and publisher disclaim all liability for direct or consequential damages resulting from your use of the programs.

How to Get the Code onto Your Computer

Pick one of the following methods:

- You can type the programs from this book directly into your computer. In this case, the *only* kind of license available to you is the free "immediate license" (see below). You are not authorized to transfer or distribute a machine-readable copy to any other person, nor to have any other person type the programs into a computer on your behalf. We do not want to hear bug reports from you if you choose this option, because experience has shown that *virtually all* reported bugs in such cases are typing errors!

- You can download the Numerical Recipes programs electronically from the Numerical Recipes On-Line Software Store, located at our Web site http://www.nr.com. All the files (Recipes and demonstration programs) are packaged as a single compressed file. You'll need to purchase a license to download and unpack them. Any number of single-screen licenses can be purchased instantly (with discount for multiple screens) from the On-Line Store, with fees that depend on your operating system (Windows or Macintosh versus Linux or UNIX) and whether you are affiliated with an educational institution. Purchasing a single-screen license is also the way to start if you want to acquire a more general (site or corporate) license; your single-screen cost will be subtracted from the cost of any later license upgrade.

- You can purchase media containing the programs from Cambridge University Press. A CD-ROM version in ISO-9660 format for Windows and Macintosh systems contains the complete C++ software, and also the previously available C version. The CD-ROM is available with a single-screen license for Windows or Macintosh (order ISBN 0 521 750377). More extensive CD-ROMs in ISO-9660 format for Windows, Macintosh, and UNIX/Linux systems are also available; these include the C++, C, and Fortran versions on a single CD-ROM (as well as versions in Pascal and BASIC from the first edition). These CD-ROMs are available with a single-screen license for Windows or Macintosh (order ISBN 0 521 750350), or (at a slightly higher price) with a single-screen license for UNIX/Linux workstations (order ISBN 0 521 750369). Orders for media from Cambridge University Press can be placed at 800 872-7423 (North America only) or by email to orders@cup.org (North America) or directcustserv@cambridge.org (rest of world). Or, visit the Web site http://www.cambridge.org.

Types of License Offered

Here are the types of licenses that we offer. Note that some types are automatically acquired with the purchase of media from Cambridge University Press, or of an unlocking password from the Numerical Recipes On-Line Software Store, while other types of licenses require that you communicate specifically with Numerical Recipes Software (email: orders@nr.com or fax: 781 863-1739). Our Web site http://www.nr.com has additional information.

- ["Immediate License"] If you are the individual owner of a copy of this book and you type one or more of its routines into your computer, we authorize you to use them on that computer for your own personal and noncommercial purposes. You are not authorized to transfer or distribute machine-readable copies to any other person, or to use the routines on more than one machine, or to distribute executable programs containing our routines. This is the only free license.

- ["Single-Screen License"] This is the most common type of low-cost license, with terms governed by our Single-Screen (Shrinkwrap) License document (complete terms available through our Web site). Basically, this license lets you use *Numerical Recipes* routines on any one screen (laptop, workstation, X-terminal, etc.). You may also, under this license, transfer precompiled, executable programs incorporating our routines to other, unlicensed, screens or computers, providing that (i) your application is noncommercial (i.e., does not involve the selling of your program for a fee), (ii) the programs were first developed, compiled, and successfully run on a licensed screen, and (iii) our routines are bound into the programs in such a manner that they cannot be accessed as individual routines and cannot practicably be unbound and used in other programs. That is, under this license, your program user must not be able to use our programs as part of a program library or "mix-and-match" workbench. Conditions for other types of commercial or noncommercial distribution may be found on our Web site (`http://www.nr.com`).

- ["Multi-Screen, Server, Site, and Corporate Licenses"] The terms of the Single-Screen License can be extended to designated groups of machines, defined by number of screens, number of machines, locations, or ownership. Significant discounts from the corresponding single-screen prices are available when the estimated number of screens exceeds 40. Contact Numerical Recipes Software (email: orders@nr.com or fax: 781 863-1739) for details.

- ["Course Right-to-Copy License"] Instructors at accredited educational institutions who have adopted this book for a course, and who have already purchased a Single-Screen License (either acquired with the purchase of media, or from the Numerical Recipes On-Line Software Store), may license the programs for use in that course as follows: Mail your name, title, and address; the course name, number, dates, and estimated enrollment; and advance payment of $5 per (estimated) student to Numerical Recipes Software, at this address: P.O. Box 380243, Cambridge, MA 02238-0243 (USA). You will receive by return mail a license authorizing you to make copies of the programs for use by your students, and/or to transfer the programs to a machine accessible to your students (but only for the duration of the course).

About Copyrights on Computer Programs

Like artistic or literary compositions, computer programs are protected by copyright. Generally it is an infringement for you to copy into your computer a program from a copyrighted source. (It is also not a friendly thing to do, since it deprives the program's author of compensation for his or her creative effort.) Under copyright law, all "derivative works" (modified versions, or translations into another computer language) also come under the same copyright as the original work.

Copyright does not protect ideas, but only the expression of those ideas in a particular form. In the case of a computer program, the ideas consist of the program's methodology and algorithm, including the necessary sequence of steps adopted by the programmer. The expression of those ideas is the program source code (particularly any arbitrary or stylistic choices embodied in it), its derived object code, and any other derivative works.

If you analyze the ideas contained in a program, and then express those ideas in your own completely different implementation, then that new program implementation belongs to you. That is what we have done for those programs in this book that are not entirely of our own devising. When programs in this book are said to be "based" on programs published in copyright sources, we mean that the ideas are the same. The expression of these ideas as source code is our own. We believe that no material in this book infringes on an existing copyright.

Trademarks

Several registered trademarks appear within the text of this book: Sun, Solaris, Ultra, and WorkShop are trademarks of Sun Microsystems, Inc. Microsoft, Windows, and Visual C++ are trademarks of Microsoft Corporation. DEC and VMS are trademarks of Compaq Computer Corporation. IBM, AIX, and RS/6000 are trademarks of International Business Machines Corporation. Intel, Pentium and KAI C++ are trademarks of Intel Corporation. Linux is a trademark of Linus Torvalds. Apple and Macintosh are trademarks of Apple Computer, Inc. Borland C++ Builder is a trademark of Inprise Corporation. UNIX is a trademark of The Open Group. IMSL is a trademark of Visual Numerics, Inc. NAG refers to proprietary computer software of Numerical Algorithms Group (USA) Inc. PostScript and Adobe Illustrator are trademarks of Adobe Systems Incorporated. Last, and no doubt least, Numerical Recipes, NR, and nr.com (when identifying our products) are trademarks of Numerical Recipes Software.

Attributions

The fact that ideas are legally "free as air" in no way supersedes the ethical requirement that ideas be credited to their known originators. When programs in this book are based on known sources, whether copyrighted or in the public domain, published or "handed-down," we have attempted to give proper attribution. Unfortunately, the lineage of many programs in common circulation is often unclear. We would be grateful to readers for new or corrected information regarding attributions, which we will attempt to incorporate in subsequent printings.

Computer Programs
by Chapter and Section

Chapter 1.   Preliminaries

1.0 Introduction

This book, like its sibling versions in other computer languages, is supposed to teach you methods of numerical computing that are practical, efficient, and (insofar as possible) elegant. We presume throughout this book that you, the reader, have particular tasks that you want to get done. We view our job as educating you on how to proceed. Occasionally we may try to reroute you briefly onto a particularly beautiful side road; but by and large, we will guide you along main highways that lead to practical destinations.

Throughout this book, you will find us fearlessly editorializing, telling you what you should and shouldn't do. This prescriptive tone results from a conscious decision on our part, and we hope that you will not find it irritating. We do not claim that our advice is infallible! Rather, we are reacting against a tendency, in the textbook literature of computation, to discuss every possible method that has ever been invented, without ever offering a practical judgment on relative merit. We do, therefore, offer you our practical judgments whenever we can. As you gain experience, you will form your own opinion of how reliable our advice is.

We presume that you are able to read computer programs in C++, that being the language of this version of *Numerical Recipes*. The books *Numerical Recipes in Fortran 77*, *Numerical Recipes in Fortran 90*, and *Numerical Recipes in C* are separately available, if you prefer to program in one of those languages. Earlier editions of *Numerical Recipes in Pascal* and *Numerical Recipes Routines and Examples in BASIC* are also available; while not containing the additional material of the Second Edition versions, these versions are perfectly serviceable if Pascal or BASIC is your language of choice.

When we include programs in the text, they look like this:

```
#include <cmath>
#include "nr.h"
using namespace std;

void NR::flmoon(const int n, const int nph, int &jd, DP &frac)
```
Our programs begin with an introductory comment summarizing their purpose and explaining their calling sequence. This routine calculates the phases of the moon. Given an integer n and a code nph for the phase desired (nph = 0 for new moon, 1 for first quarter, 2 for full, 3 for last quarter), the routine returns the Julian Day Number jd, and the fractional part of a day frac to be added to it, of the nth such phase since January, 1900. Greenwich Mean Time is assumed.
```
{
    const DP RAD=3.141592653589793238/180.0;
    int i;
```

1

```
DP am,as,c,t,t2,xtra;

c=n+nph/4.0;
t=c/1236.85;
t2=t*t;
as=359.2242+29.105356*c;
am=306.0253+385.816918*c+0.010730*t2;
jd=2415020+28*n+7*nph;
xtra=0.75933+1.53058868*c+((1.178e-4)-(1.55e-7)*t)*t2;
if (nph == 0 || nph == 2)
    xtra += (0.1734-3.93e-4*t)*sin(RAD*as)-0.4068*sin(RAD*am);
else if (nph == 1 || nph == 3)
    xtra += (0.1721-4.0e-4*t)*sin(RAD*as)-0.6280*sin(RAD*am);
else nrerror("nph is unknown in flmoon");
i=int(xtra >= 0.0 ? floor(xtra) : ceil(xtra-1.0));
jd += i;
frac=xtra-i;
}
```

This is how we comment an individual line.

You aren't really intended to understand this algorithm, but it does work!

This is how we will indicate error conditions.

Note our convention of handling all errors and exceptional cases with a statement like nrerror("some error message");. The function nrerror() is part of a small file of utility programs, nrutil.h, listed in Appendix B at the back of the book. This Appendix includes a number of other utilities that we will describe later in this chapter. Function nrerror() prints the indicated error message to your stderr device (usually your terminal screen), and then invokes the function exit(), which terminates execution. You can modify nrerror() so that it does anything else that will halt execution. For example, you can have it pause for input from the keyboard, and then manually interrupt execution. In some applications, you will want to modify nrerror() to do more sophisticated error handling, for example to transfer control somewhere else by throwing a C++ exception.

We will have more to say about the C++ programming language, its conventions and style, in §1.1 and §1.2.

Quick Start: Using the C++ Numerical Recipes Routines

This section is for people who want to jump right in. We'll compute the mean and variance of the Julian Day numbers of the first 20 full moons after January 1900. (Now *there's* a useful pair of quantities!)

First, locate the important files nrtypes.h, nrutil.h, and nr.h, as listed in Appendices A and B. These contain the definitions of the various types used by our routines, the vector and matrix classes we use, various utility functions, and the function declarations for all the Recipe functions. (Actually, nrtypes.h includes by default the file nrtypes_nr.h, and nrutil.h includes by default the file nrutil_nr.h. This setup is to allow you to change the defaults easily, as will be discussed in §1.3.)

Second, create this main program file:

```
#include <iostream>
#include <iomanip>
#include "nr.h"
using namespace std;

int main(void)
{
```

```
      const int NTOT=20;
      int i,jd,nph=2;
      DP frac,ave,vrnce;
      Vec_DP data(NTOT);

      for (i=0;i<NTOT;i++) {
          NR::flmoon(i,nph,jd,frac);
          data[i]=jd;
      }
      NR::avevar(data,ave,vrnce);
      cout << "Average = " << setw(12) << ave;
      cout << " Variance = " << setw(13) << vrnce << endl;
      return 0;
  }
```

Here is a brief explanation of some elements of the above program:

You must always have the #include nr.h statement, which includes the function declarations. It also includes the files nrtypes.h and nrutil.h for you. The declaration Vec_DP data(NTOT) creates a double precision vector with NTOT elements. The types DP (double precision, i.e., double) and Vec_DP are defined in nrtypes.h, while the vector class is defined in nrutil.h. The scope resolution prefix NR:: makes accessible the Recipe functions flmoon and avevar, which are in the NR namespace defined in nr.h. We call flmoon, looping over 20 full moons, and store the Julian days it returns in data. Then we call the avevar routine, and print the answers.

Third, compile the main program file, and also the files flmoon.cpp and avevar.cpp. Link the resulting object files.

Fourth, run the resulting executable file. Typical output is:

```
Average = 2.41532e+06 Variance = 30480.7
```

The files nrtypes.h, nrutil.h, and nr.h and the concepts behind them will be discussed in detail in §1.2 and 1.3.

Computational Environment and Program Validation

Our goal is that the programs in this book be as portable as possible, across different platforms (models of computer), across different operating systems, and across different C++ compilers. C++ was designed with this type of portability in mind. Nevertheless, we have found that there is no substitute for actually checking all programs on a variety of compilers, in the process uncovering differences in library structure or contents, and even occasional differences in allowed syntax. As surrogates for the large number of hardware and software configurations, we have tested all the programs in this book on the combinations of machines, operating systems, and compilers shown on the accompanying table. More generally, the programs should run without modification on any compiler that implements the ANSI/ISO C++ standard, as described for example in Stroustrup's excellent book [1].

In validating the programs, we have taken the program source code directly from the machine-readable form of the book's manuscript, to decrease the chance of propagating typographical errors. "Driver" or demonstration programs that we used as part of our validations are available separately as the *Numerical Recipes Example Book (C++)*, as well as in machine-readable form. If you plan to use more than a few of the programs in this book, then you may find it useful to obtain

Tested Machines and Compilers	
O/S and Hardware	Compiler Version
Microsoft Windows / Intel	Microsoft Visual C++ 6.0
Microsoft Windows / Intel	Borland C++ 5.02
Linux / Intel	GNU C++ ("g++") 2.95.2
AIX 4.3 / IBM RS/6000	KAI C++ ("KCC") 4.0d
SunOS 5.7 (Solaris 7) / Sun Ultra 5	Sun Workshop Compiler C++ ("CC") 5.0
SunOS 5.7 (Solaris 7) / Sun Ultra 5	GNU C++ ("g++") 2.95.2

the machine-readable software distribution, which includes both the Recipes and the demonstration programs.

Of course we would be foolish to claim that there are no bugs in our programs, and we do not make such a claim. We have been very careful, and have benefitted from the experience of the many readers who have written to us. If you find a new bug, please document it and tell us! You can find contact information at http://www.nr.com.

Compatibility with the First Edition

If you are accustomed to the *Numerical Recipes* routines of the First Edition, rest assured: almost all of them are still here, with the same names and functionalities, often with major improvements in the code itself. In addition, we hope that you will soon become equally familiar with the added capabilities of the more than 100 routines that are new to this edition.

We have retired a small number of First Edition routines, those that we believe to be clearly dominated by better methods implemented in this edition. A table, following, lists the retired routines and suggests replacements.

Previous Routines Omitted from This Edition		
Name(s)	Replacement(s)	Comment
adi	mglin or mgfas	better method
cosft	cosft1 or cosft2	choice of boundary conditions
cel, el2	rf, rd, rj, rc	better algorithms
des, desks	ran4 now uses psdes	was too slow
iindexx	indexx	indexx overloaded for double and int
mdian1, mdian2	select, selip	more general
qcksrt	sort	name change (sort is now hpsort)
rkqc	rkqs	better method
smooft	use convlv with coefficients from savgol	
sparse	linbcg	more general

About References

You will find references, and suggestions for further reading, listed at the end of most sections of this book. References are cited in the text by bracketed numbers like this [2].

Because computer algorithms often circulate informally for quite some time before appearing in a published form, the task of uncovering "primary literature" is sometimes quite difficult. We have not attempted this, and we do not pretend to any degree of bibliographical completeness in this book. For topics where a substantial secondary literature exists (discussion in textbooks, reviews, etc.) we have consciously limited our references to a few of the more useful secondary sources, especially those with good references to the primary literature. Where the existing secondary literature is insufficient, we give references to a few primary sources that are intended to serve as starting points for further reading, not as complete bibliographies for the field.

The order in which references are listed is not necessarily significant. It reflects a compromise between listing cited references in the order cited, and listing suggestions for further reading in a roughly prioritized order, with the most useful ones first.

The remaining three sections of this chapter review some basic concepts of programming (control structures, etc.), discuss a set of conventions specific to C++ that we have adopted in this book, and introduce some fundamental concepts in numerical analysis (roundoff error, etc.). Thereafter, we plunge into the substantive material of the book.

CITED REFERENCES AND FURTHER READING:

Stroustrup, B. 1997, *The C++ Programming Language*, 3rd ed. (Reading, MA: Addison-Wesley). [1]

Meeus, J. 1982, *Astronomical Formulae for Calculators*, 2nd ed., revised and enlarged (Richmond, VA: Willmann-Bell). [2]

1.1 Program Organization and Control Structures

We sometimes like to point out the close analogies between computer programs, on the one hand, and written poetry or written musical scores, on the other. All three present themselves as visual media, symbols on a two-dimensional page or computer screen. Yet, in all three cases, the visual, two-dimensional, *frozen-in-time* representation communicates (or is supposed to communicate) something rather different, namely a process that *unfolds in time*. A poem is meant to be read; music, played; a program, executed as a sequential series of computer instructions.

In all three cases, the target of the communication, in its visual form, is a human being. The goal is to transfer to him/her, as efficiently as can be accomplished, the greatest degree of understanding, in advance, of how the process *will* unfold in time. In poetry, this human target is the reader. In music, it is the performer. In programming, it is the program user.

Now, you may object that the target of communication of a program is not a human but a computer, that the program user is only an irrelevant intermediary, a lackey who feeds the machine. This is perhaps the case in the situation where the naive user pops a CDROM into a desktop computer and feeds that computer a black-box program in binary executable form. The computer, in this case, doesn't much care whether that program was written with "good programming practice" or not.

We envision, however, that you, the readers of this book, are in quite a different situation. You need, or want, to know not just *what* a program does, but also *how* it does it, so that you can tinker with it and modify it to your particular application. You need others to be able to see what you have done, so that they can criticize or admire. In such cases, where the desired goal is *maintainable* or *reusable* code, the targets of a program's communication are surely human, not machine.

One key to achieving good programming practice is to recognize that programming, music, and poetry — all three being symbolic constructs of the human brain — are naturally structured into hierarchies that have many different nested levels. Sounds (phonemes) form small meaningful units (morphemes) which in turn form words; words group into phrases, which group into sentences; sentences make paragraphs, and these are organized into higher levels of meaning. Notes form musical phrases, which form themes, counterpoints, harmonies, etc.; which form movements, which form concertos, symphonies, and so on.

The structure in programs is equally hierarchical. Appropriately, good programming practice brings different techniques to bear on the different levels [1-3]. At a low level is the ascii character set. Then, constants, identifiers, operands, operators. Then program statements, like a[j+1]=b+c/3.0;. Here, the best programming advice is simply *be clear*, or (correspondingly) *don't be too tricky*. You might momentarily be proud of yourself at writing the single line

```
k=(2-j)*(1+3*j)/2;
```

if you want to permute cyclically one of the values $j = (0, 1, 2)$ into respectively $k = (1, 2, 0)$. You will regret it later, however, when you try to understand that line. Better, and likely also faster, is

```
k=j+1;
if (k == 3) k=0;
```

Many programming stylists would even argue for the ploddingly literal

```
switch (j) {
    case 0: k=1; break;
    case 1: k=2; break;
    case 2: k=0; break;
    default: {
        cerr << "unexpected value for j";
        exit(1);
    }
}
```

on the grounds that it is both clear and additionally safeguarded from wrong assumptions about the possible values of j. Our preference among the implementations is for the middle one.

In this simple example, we have in fact traversed several levels of hierarchy: Statements frequently come in "groups" or "blocks" which make sense only taken as a whole. The middle fragment above is one example. Another is

```
swap=a[j];
a[j]=b[j];
b[j]=swap;
```

which makes immediate sense to any programmer as the exchange of two variables, while

```
ans=sum=0.0;
n=1;
```

is very likely to be an initialization of variables prior to some iterative process. This level of hierarchy in a program is usually evident to the eye. It is good programming practice to put in comments at this level, e.g., "initialize" or "exchange variables."

The next level is that of *control structures*. These are things like the switch construction in the example above, for loops, and so on. This level is sufficiently important, and relevant to the hierarchical level of the routines in this book, that we will come back to it just below.

At still higher levels in the hierarchy, we have functions, classes, and methods, and the whole "global" organization of the computational task to be done. In the musical analogy, we are now at the level of movements and complete works. At these levels, *modularization* and *encapsulation* become important programming concepts, the general idea being that program units should interact with one another only through clearly defined and narrowly circumscribed interfaces. Good modularization practice is an essential prerequisite to the success of large, complicated software projects, especially those employing the efforts of more than one programmer. It is also good practice (if not quite as essential) in the less massive programming tasks that an individual scientist, or reader of this book, encounters.

Some computer languages, such as C++ and Modula-2, promote good modularization with higher-level language constructs. In C++ and Modula-2, functions, type definitions, and data structures can be encapsulated into "classes" or "modules" that communicate through declared public interfaces and whose internal workings are hidden from the rest of the program [4,5]. In C++, a class also serves as a user-definable generalization of data type that provides for data hiding, automatic initialization of data, memory management, dynamic typing, and operator overloading (i.e., the user-definable extension of operators like + and * so as to be appropriate to operands in any particular class) [5]. Properly used in defining the data structures that are passed between program units, classes can clarify and circumscribe these units' public interfaces, reducing the chances of programming error and also allowing a considerable degree of compile-time and run-time error checking.

Beyond modularization, though depending on it, lie the concepts of *object-oriented programming*. Here a programming language such as C++ allows a class's public interface to accept redefinitions of types or actions, and these redefinitions become shared all the way down through the class's hierarchy (so-called *polymorphism*). For example, a routine written to invert a matrix of real numbers could be made able to handle complex numbers by overloading complex data types and corresponding definitions of the arithmetic operations. Additional concepts of *inheritance* (the ability to define a data type that "inherits" all the structure of another type, plus additional structure of its own), and *object extensibility* (the ability to add functionality to a class without access to its source code), also come into play.

We have not attempted to modularize, or make objects out of, the routines in this book. The reason is that we envision that you, the reader, might want to incorporate the algorithms in this book, a few at a time, into modules or objects with a structure of your own choosing. There does not exist, at present, a standard or accepted set of classes for scientific object-oriented computing. While we might have tried to invent such a set, doing so would have inevitably tied the algorithmic content of the book (which is its *raison d'être*) to some rather specific, and perhaps haphazard, set of choices regarding class definitions.

On the other hand, we are not unfriendly to the goals of modular and object-oriented programming. We have therefore tried to structure our programs to be "object friendly." Also, in the implementation of our functions, we have paid particular attention to the practices of *structured programming*, as we now discuss.

Control Structures

An executing program unfolds in time, but not strictly in the linear order in which the statements are written. Program statements that affect the order in which statements are executed, or that affect whether statements are executed, are called *control statements*. Control statements never make useful sense by themselves. They make sense only in the context of the groups or blocks of statements that they in turn control. If you think of those blocks as paragraphs containing sentences, then the control statements are perhaps best thought of as the indentation of the paragraph and the punctuation between the sentences, not the words within the sentences.

We can now say what the goal of structured programming is. It is *to make program control manifestly apparent in the visual presentation of the program*. You see that this goal has nothing at all to do with how the computer sees the program. As already remarked, computers don't care whether you use structured programming or not. Human readers, however, *do* care. You yourself will also care, once you discover how much easier it is to perfect and debug a well-structured program than one whose control structure is obscure.

You accomplish the goals of structured programming in two complementary ways. First, you acquaint yourself with the small number of essential control structures that occur over and over again in programming, and that are therefore given convenient representations in most programming languages. You should learn to think about your programming tasks, insofar as possible, exclusively in terms of these standard control structures. In writing programs, you should get into the habit of representing these standard control structures in consistent, conventional ways.

"Doesn't this inhibit *creativity*?" our students sometimes ask. Yes, just as Mozart's creativity was inhibited by the sonata form, or Shakespeare's by the metrical requirements of the sonnet. The point is that creativity, when it is meant to communicate, does *well* under the inhibitions of appropriate restrictions on format.

Second, you *avoid*, insofar as possible, control statements whose controlled blocks or objects are difficult to discern at a glance. This means, in practice, that *you must avoid named labels on statements and* goto*'s*. It is not the goto's that are dangerous (although they do interrupt one's reading of a program); the named statement labels are the hazard. In fact, whenever you encounter a named statement label while reading a program, you will soon become conditioned to get a sinking feeling in the pit of your stomach. Why? Because the following questions will, by habit, immediately spring to mind: Where did control come *from* in a branch to this label? It could be anywhere in the function! What circumstances resulted in a branch to this label? They could be anything! Certainty becomes uncertainty, understanding dissolves into a morass of possibilities.

Some examples are now in order to make these considerations more concrete (see Figure 1.1.1).

Catalog of Standard Structures

Iteration.   In C++, simple iteration is performed with a for loop, for example

```
for (j=2;j<=1000;j++) {
    b[j]=a[j-1];
    a[j-1]=j;
}
```

Notice how we always indent the block of code that is acted upon by the control structure, leaving the structure itself unindented. Notice also our habit of putting the initial curly brace on the same line as the for statement, instead of on the next line. This saves a full line of white space, and our publisher loves us for it.

IF structure.   This structure in C++ is similar to that found in Pascal, Algol, Fortran and other languages, and typically looks like

```
if (...) {
    ...
}
else if (...) {
    ...
}
else {
    ...
}
```

Since compound-statement curly braces are required only when there is more than one statement in a block, however, C++'s if construction can be somewhat less explicit than that shown above. Some care must be exercised in constructing nested if clauses. For example, consider the following:

```
if (b > 3)
    if (a > 3) b += 1;
else b -= 1;                    /* questionable! */
```

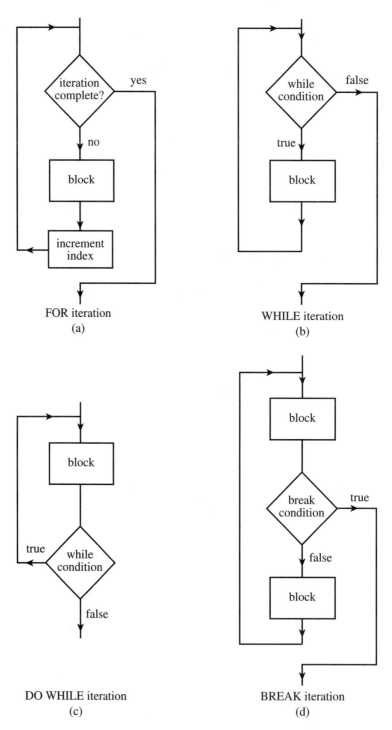

Figure 1.1.1.   Standard control structures used in structured programming: (a) `for` iteration; (b) `while` iteration; (c) `do while` iteration; (d) `break` iteration; (e) `if` structure; (f) `switch` structure

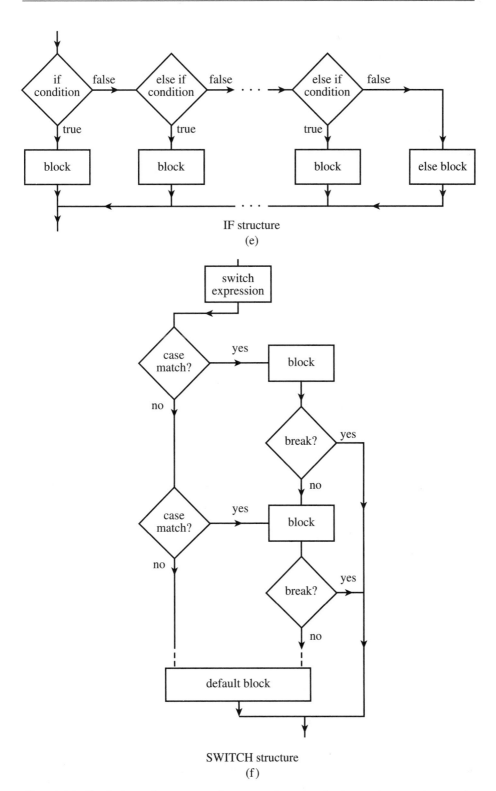

IF structure
(e)

SWITCH structure
(f)

Figure 1.1.1.   Standard control structures used in structured programming (see caption on previous page).

As judged by the indentation used on successive lines, the intent of the writer of this code is the following: 'If b is greater than 3 and a is greater than 3, then increment b. If b is not greater than 3, then decrement b.' According to the rules of C++, however, the actual meaning is 'If b is greater than 3, then evaluate a. If a is greater than 3, then increment b, and if a is less than or equal to 3, decrement b.' The point is that an `else` clause is associated with the most recent open `if` statement, no matter how you lay it out on the page. Such confusions in meaning are easily resolved by the inclusion of braces. They may in some instances be technically superfluous; nevertheless, they clarify your intent and improve the program. The above fragment should be written as

```
if (b > 3) {
    if (a > 3) b += 1;
} else {
    b -= 1;
}
```

Here is a working program that consists dominantly of `if` control statements:

```
#include <cmath>
#include "nr.h"
using namespace std;

int NR::julday(const int mm, const int id, const int iyyy)
```
In this routine julday returns the Julian Day Number that begins at noon of the calendar date specified by month mm, day id, and year iyyy, all integer variables. Positive year signifies A.D.; negative, B.C. Remember that the year after 1 B.C. was 1 A.D.
```
{
    const int IGREG=15+31*(10+12*1582);        Gregorian Calendar adopted Oct. 15, 1582.
    int ja,jul,jy=iyyy,jm;

    if (jy == 0) nrerror("julday: there is no year zero.");
    if (jy < 0) ++jy;
    if (mm > 2) {                              Here is an example of a block IF-structure.
        jm=mm+1;
    } else {
        --jy;
        jm=mm+13;
    }
    jul = int(floor(365.25*jy)+floor(30.6001*jm)+id+1720995);
    if (id+31*(mm+12*iyyy) >= IGREG) {         Test whether to change to Gregorian Cal-
        ja=int(0.01*jy);                       endar.
        jul += 2-ja+int(0.25*ja);
    }
    return jul;
}
```

(Astronomers number each 24-hour period, starting and ending at *noon*, with a unique integer, the Julian Day Number [6]. Julian Day Zero was a very long time ago; a convenient reference point is that Julian Day 2450000 began at noon of October 9, 1995. If you know the Julian Day Number that begins at noon of a given calendar date, then the day of the week of that date is obtained by adding 1 and taking the result modulo base 7; a zero answer corresponds to Sunday, 1 to Monday, ..., 6 to Saturday.)

While iteration.   Most languages (though not `Fortran`, incidentally) provide for structures like the following C++ example:

```
while (n < 1000) {
    n *= 2;
    j += 1;
}
```

It is the particular feature of this structure that the control-clause (in this case `n < 1000`) is evaluated *before* each iteration. If the clause is not true, the enclosed statements will not be executed. In particular, if this code is encountered at a time when `n` is greater than or equal to 1000, the statements will not even be executed once.

Do-While iteration.   Companion to the `while` iteration is a related control-structure that tests its control-clause at the *end* of each iteration. In C++, it looks like this:

```
do {
    n *= 2;
    j += 1;
} while (n < 1000);
```

In this case, the enclosed statements will be executed at least once, independent of the initial value of `n`.

Break.   In this case, you have a loop that is repeated indefinitely until some condition *tested somewhere in the middle of the loop* (and possibly tested in more than one place) becomes true. At that point you wish to exit the loop and proceed with what comes after it. In C++ the structure is implemented with the simple `break` statement, which terminates execution of the innermost `for`, `while`, `do`, or `switch` construction and proceeds to the next sequential instruction. (In `Pascal` and `Fortran` 77, this structure requires the use of statement labels, to the detriment of clear programming.) A typical usage of the break statement is:

```
for(;;) {
    [statements before the test]
    if (...) break;
    [statements after the test]
}
[next sequential instruction]
```

Here is a program that uses several different iteration structures. One of us was once asked, for a scavenger hunt, to find the date of a Friday the 13th on which the moon was full. This is a program which accomplishes that task, giving incidentally all other Fridays the 13th as a by-product.

```
#include <iostream>
#include <iomanip>
#include <cmath>
#include "nr.h"
using namespace std;

int main(void)     // Program badluk
```

```
{
    const int IYBEG=2000,IYEND=2100;          The range of dates to be searched.
    const DP ZON=-5.0;                        Time zone −5 is Eastern Standard Time.
    int ic,icon,idwk,im,iyyy,jd,jday,n;
    DP timzon=ZON/24.0,frac;

    cout << endl << "Full moons on Friday the 13th from ";
    cout << setw(5) << IYBEG << " to " << setw(5) << IYEND << endl;
    for (iyyy=IYBEG;iyyy<=IYEND;iyyy++) {     Loop over each year,
        for (im=1;im<=12;im++) {              and each month.
            jday=NR::julday(im,13,iyyy);      Is the 13th a Friday?
            idwk=int((jday+1) % 7);
            if (idwk == 5) {
                n=int(12.37*(iyyy-1900+(im-0.5)/12.0));
```
This value n is a first approximation to how many full moons have occurred
since 1900. We will feed it into the phase routine and adjust it up or down
until we determine that our desired 13th was or was not a full moon. The
variable icon signals the direction of adjustment.
```
                icon=0;
                for (;;) {
                    NR::flmoon(n,2,jd,frac);  Get date of full moon n.
                    frac=24.0*(frac+timzon);  Convert to hours in correct time zone.
                    if (frac < 0.0) {         Convert from Julian Days beginning at
                        --jd;                     noon to civil days beginning at mid-
                        frac += 24.0;             night.
                    }
                    if (frac > 12.0) {
                        ++jd;
                        frac -= 12.0;
                    } else
                        frac += 12.0;
                    if (jd == jday) {         Did we hit our target day?
                        cout << endl << setw(2) << im;
                        cout << "/13/" << setw(4) << iyyy << endl;
                        cout << fixed << setprecision(1);
                        cout << "Full moon" << setw(6) << frac;
                        cout << " hrs after midnight (EST)" << endl;
                        break;                Part of the break-structure, a match.
                    } else {                  Didn't hit it.
                        ic=(jday >= jd ? 1 : -1);
                        if (ic == (-icon)) break;   Another break, case of no match.
                        icon=ic;
                        n += ic;
                    }
                }
            }
        }
    }
    return 0;
}
```

If you are merely curious, there were (or will be) occurrences of a full moon
on Friday the 13th (time zone GMT−5) on: 3/13/1903, 10/13/1905, 6/13/1919,
1/13/1922, 11/13/1970, 2/13/1987, 10/13/2000, 9/13/2019, and 8/13/2049.

Other "standard" structures.   Our advice is generally to avoid them.
Every programming language has some number of "goodies" that the standards
committee just couldn't resist throwing in. They seemed like a good idea at the time.
Unfortunately they don't stand the *test* of time! Your program becomes difficult to
translate into other languages, and difficult to read (because rarely used structures

are unfamiliar to the reader). You can almost always accomplish the supposed conveniences of these structures in other ways.

About "Advanced Topics"

Material set in smaller type, like this, signals an "advanced topic," either one outside of the main argument of the chapter, or else one requiring of you more than the usual assumed mathematical background, or else (in a few cases) a discussion that is more speculative or an algorithm that is less well-tested. Nothing important will be lost if you skip the advanced topics on a first reading of the book.

You may have noticed that, by its looping over the months and years, the program badluk avoids using any algorithm for converting a Julian Day Number back into a calendar date. A routine for doing just this is not very interesting structurally, but it is occasionally useful:

```cpp
#include <cmath>
#include "nr.h"
using namespace std;

void NR::caldat(const int julian, int &mm, int &id, int &iyyy)
```
Inverse of the function julday given above. Here julian is input as a Julian Day Number, and the routine outputs mm,id, and iyyy as the month, day, and year on which the specified Julian Day started at noon.
```cpp
{
    const int IGREG=2299161;
    int ja,jalpha,jb,jc,jd,je;

    if (julian >= IGREG) {        Cross-over to Gregorian Calendar produces this correc-
        jalpha=int((DP(julian-1867216)-0.25)/36524.25);               tion.
        ja=julian+1+jalpha-int(0.25*jalpha);
    } else if (julian < 0) {    Make day number positive by adding integer number of
        ja=julian+36525*(1-julian/36525);       Julian centuries, then subtract them off
    } else                                      at the end.
        ja=julian;
    jb=ja+1524;
    jc=int(6680.0+(DP(jb-2439870)-122.1)/365.25);
    jd=int(365*jc+(0.25*jc));
    je=int((jb-jd)/30.6001);
    id=jb-jd-int(30.6001*je);
    mm=je-1;
    if (mm > 12) mm -= 12;
    iyyy=jc-4715;
    if (mm > 2) --iyyy;
    if (iyyy <= 0) --iyyy;
    if (julian < 0) iyyy -= 100*(1-julian/36525);
}
```

(For additional calendrical algorithms, applicable to various historical calendars, see [7].)

CITED REFERENCES AND FURTHER READING:

Harbison, S.P., and Steele, G.L., Jr. 1991, *C: A Reference Manual*, 3rd ed. (Englewood Cliffs, NJ: Prentice-Hall).

Kernighan, B.W. 1978, *The Elements of Programming Style* (New York: McGraw-Hill). [1]

Yourdon, E. 1975, *Techniques of Program Structure and Design* (Englewood Cliffs, NJ: Prentice-Hall). [2]

Jones, R., and Stewart, I. 1987, *The Art of C Programming* (New York: Springer-Verlag). [3]

Hoare, C.A.R. 1981, *Communications of the ACM*, vol. 24, pp. 75–83.

Wirth, N. 1983, *Programming in Modula-2*, 3rd ed. (New York: Springer-Verlag). [4]

Stroustrup, B. 1997, *The C++ Programming Language*, 3rd ed. (Reading, MA: Addison-Wesley). [5]

Meeus, J. 1982, *Astronomical Formulae for Calculators*, 2nd ed., revised and enlarged (Richmond, VA: Willmann-Bell). [6]

Hatcher, D.A. 1984, *Quarterly Journal of the Royal Astronomical Society*, vol. 25, pp. 53–55; see also *op. cit.* 1985, vol. 26, pp. 151–155, and 1986, vol. 27, pp. 506–507. [7]

1.2 Some C++ Conventions for Scientific Computing

The C++ language derives from C, which in turn was originally devised for systems programming work, not for scientific computing. Relative to other high-level programming languages, C puts the programmer "very close to the machine" in several respects. It is operator-rich, giving direct access to most capabilities of a machine-language instruction set. It has a large variety of intrinsic data types (short and long, signed and unsigned integers; floating and double-precision reals; pointer types; etc.), and a concise syntax for effecting conversions and indirections. It defines an arithmetic on pointers (addresses) that relates gracefully to array addressing and is highly compatible with the index register structure of many computers. All these features are present in C++ too.

Portability is another strong point of the C language, and so also of C++. C is the underlying language of the UNIX and Linux operating systems; both the language and these operating systems have by now been implemented on literally hundreds of different computer types. The language's universality, portability, and flexibility have attracted increasing numbers of scientists and engineers to it. It is commonly used for the real-time control of experimental hardware, often in spite of the fact that the standard UNIX or Linux kernel is less than ideal as an operating system for this purpose.

The use of C or C++ for higher level scientific calculations such as data analysis, modeling, and floating-point numerical work has generally been slower in developing. In part this is because of the entrenched position of Fortran as the mother-tongue of virtually all scientists and engineers born before 1960, and many born after. In part, also, the slowness of C's penetration into scientific computing has been due to deficiencies in the language that computer scientists have been (we think, stubbornly) slow to recognize. Examples are the lack of a good way to raise numbers to small integer powers, and the difficulty in getting compilers to emit optimized code when high-level constructs are used.

The C version of this book attempted to lay out by example a set of sensible, practical conventions for scientific C programming. In this version we will try to do the same for C++.

The need for programming conventions in C++ is very great. Far from the problem of overcoming constraints imposed by the language (our repeated experience with Pascal), the problem in C++ is to choose the best and most natural techniques from multiple opportunities — and then to use those techniques completely consistently from program to program. In the rest of this section, we set out some of the issues, and describe the adopted conventions that are used in all of the routines in this book. The best general reference for detailed information on the C++ language is Stroustrup's book [1].

Double Precision

When the C version of this book first came out, the default precision used by most scientists for most calculations was single precision. The reason was that double precision imposed a significant overhead both in execution speed and in memory requirements. (The fact that C automatically converts float variables to double in many situations was another insult to scientific programmers!)

Nowadays, the execution speed overhead has essentially disappeared. There are even machines where single precision is slower than double! And memory conservation is often not the concern it used to be. Accordingly, the default precision in this C++ edition is double precision. To make it easy to change to single precision (or quadruple precision!) if you want to, we have not hard coded the type double in the routines. Instead we have used the name DP, along with the typedef definition

```
typedef double DP;
```

This typedef occurs twice, once in nrtypes.h and once in nrutil.h. To change the default precision to single, just change the definition in both places to

```
typedef float DP;
```

You will also have to change the values of certain "accuracy parameters" in some routines. This is further described in Appendix C. For some of the Recipes roundoff error is a particular concern. In those cases we will explicitly warn you always to use double precision.

Function Declarations, Header Files, and Namespaces

In C++ a function cannot be called unless it has previously been *declared*. A function declaration gives the return type of the function, and the number and types of its arguments. (Function declarations are also called function prototypes or function headers.) For example,

```
int g(int x, int y, double z);        Function declaration.
```

A function must also be *defined* somewhere (and defined once only). A function definition consists of a function declaration plus the body of the function, the code that actually does the work. For example,

```
int g(int x, int y, double z)        Function definition.
{
    return x+int(z)/y;
}
```

If all your code is in one file, you can often arrange for each function definition to precede the functions that call it. Since the definition includes the declaration, the compiler can then check that a given function call invokes the function with the correct argument types. However, this setup is feasible only for small programs. In general, a C++ program consists of multiple source files that are separately compiled, and the compiler cannot check the consistency of each function call without some additional assistance. A simple and safe way to proceed is as follows [1]:

- Every external function should have a single declaration in a header (.h) file.
- The source file with the definition (body) of the function should also include the header file so that the compiler can check that the declaration and the definition match.
- Every source file that calls the function should include the same header file.

For the routines in this book, the header file containing all the declarations is nr.h, listed in Appendix A. You should put the statement #include "nr.h" at the top of every source file that invokes *Numerical Recipes* routines.

A *namespace* is a means of hiding the implementation of a program module, so that only its interface is accessible from the user program. This provides you with control over which variables and functions are "visible" in other parts of the program, and offers a way to group objects and functions of related purpose into separate modules.

All of the *Numerical Recipes* function declarations are in the NR namespace. Indeed, the file nr.h has the following structure:

```
namespace NR {
    void addint(Mat_O_DP &uf, Mat_I_DP &uc, Mat_O_DP &res);
        ...
    void zroots(Vec_I_CPLX_DP &a, Vec_O_CPLX_DP &roots, const bool &polish);
}
```

(The types Mat_O_DP, etc. will be explained below.) The *Numerical Recipes* function definitions have the following format:

```
#include "nr.h"

void NR::addint(Mat_O_DP &uf, Mat_I_DP &uc, Mat_O_DP &res)
    ...
```

Note that the code includes nr.h (since that is where the namespace NR is defined). Also, the function name must be prefixed with NR:: since NR is the scope in which the function is being defined. Note that if the function happens to call another *Numerical Recipes* function, no special declaration is required. Since all the Recipes are in namespace NR, they are all within each other's scope.

When you write a program that calls one of the Recipe functions, your program must include nr.h. You then have three ways of invoking a particular Recipe:

- Call the function with explicit scope resolution:

```
NR::addint(uf, uc, res);
```

- Precede the function call with a using declaration:

```
using NR::addint;
   ...
addint(uf, uc, res);
```

- Include a using directive in the program:

```
using namespace NR;
   ...
addint(uf, uc, res);
```

This directive makes *all* the functions in NR visible, something purists would regard as sloppy practice.

Some further details about the file nr.h are given in Appendix A.

We mention here another feature of namespaces that we make extensive use of in the Recipes. Often one wants to define an ancillary function that is used by another function. This adjunct function does some task that is of no general interest, so we want its definition to be local to the file in which the principal function is defined. In C you achieve this by defining the function as a static function. In C++ it is better to reserve static to declare objects inside functions and classes whose storage must be nonvolatile. Local functions hidden from the rest of your program can be put inside an *unnamed namespace*. Only functions within the same file can access objects inside an unnamed namespace (see, e.g., golden.cpp in §10.1).

Const Correctness

Few topics in discussions about C++ evoke more heat than questions about the keyword const. Here is our position: We are firm believers in using const wherever possible, to achieve what is called "const correctness." Many coding errors are automatically trapped by the compiler if you have qualified identifiers that should not change with const when they are declared. Also, using const makes your code much more readable: When you see const in front of an argument to a routine, you know immediately that the routine will not modify the object. Conversely, if const is absent you should be able to count on the object being changed somewhere.

We are such strong const believers that we insert const even where many people think it is redundant: If an argument is passed *by value* to a function, then the function makes a copy of it. Even if this copy is modified by the function, the original value is unchanged after the function exits. While this allows you to change, with impunity, the values of arguments that have been passed by value, we believe this usage is error-prone and hard to read. If your intention in passing something by value is that it is an input variable only, then make it clear. So we declare a function $f(x)$ as, for example,

```
double f(const double x);
```

If in the function you want to use a local variable that is initialized to x but then gets changed, define a new quantity — don't use x. If you put `const` in the declaration, the compiler will not let you get this wrong.

Some people think that using `const` on arguments makes your functions less general. Quite the opposite! Calling a function that expects a `const` argument with a non-`const` variable involves a "trivial" conversion. But trying to pass a `const` quantity to a non-`const` argument is an error.

The final reason for using `const` is that it allows certain user-defined conversions to be made. As we will see in §1.3, this is the key to writing our functions so that you can use them transparently with any matrix/vector class library.

Vectors and Matrices

Vectors and matrices are the fundamental building blocks of numerical programming. One-dimensional C-style arrays, declared for example as `double b[4]`, are part of C++ and may occasionally be used in numerical work. However, fixed size two-dimensional arrays with declarations like `double a[5][9]` are almost never desirable. Scientific programming requires some mechanism for implementing *variable dimension arrays*, which are passed to a function along with real-time information about their two-dimensional size.

There is no technical reason that a C or C++ compiler could not allow a syntax like

```
void someroutine(a,m,n)
double a[m][n];              /* ILLEGAL DECLARATION */
```

and emit code to evaluate the variable dimensions m and n (or any variable-dimension expression) each time `someroutine()` is entered. Alas! the above fragment is forbidden by the C++ language definition.

The natural way to deal with vectors and matrices in C++ is as classes. A matrix object can contain not only the values of the matrix elements, but also information on the matrix size. Moreover, the class can provide various overloaded operators and functions to facilitate high-level programming. The C++ standard already provides such a class for vectors (`valarray`), but not for matrices. Many people have written their own class libraries for one- and two-dimensional arrays. Most of these libraries are excellent in providing a suite of high-level constructs, but are terribly inefficient in execution.

When we began preparing this C++ version of *Numerical Recipes*, our original intention was to "do it right:" we would construct a class library with all the necessary high-level constructs that would also be efficient. This turns out to be a highly nontrivial task, especially since many compilers do not yet implement all the features of the C++ standard that are necessary to make such code efficient. Furthermore, we soon realized that this was exactly the wrong way to proceed. Why should you, the reader, adopt *our* class library when you have probably already invested a large effort in developing or using a different one? And what if you want to use our routines in a code written with another class library?

So, instead, we have gone to completely the opposite extreme. The Recipe functions in this book are written with vector and matrix classes called NRVec and NRMat. These classes are defined in `nrutil.h`, and provide a *minimal* implementation of vector and matrix operations. You can use our routines in either of two ways:

- Use our classes to handle vectors and matrices in all your programs. When you include nr.h to make the Recipes available, it automatically includes nrutil.h for you, so our vector and matrix classes will be accessible. The only disadvantage of this is that our classes do not provide many high-level constructs.
- Use any vector and matrix classes you like. In §1.3, we will show you how to set up a modified nrutil.h so that you will be able call our routines transparently, even though ours are declared with the Numerical Recipes classes NRVec and NRMat.

The machinery behind the NRVec and NRMat classes, and the way they can be used with other class libraries, is quite complicated. Most readers will not want to delve into this material. Accordingly, we defer its discussion to §1.3. All you really need to know to be able to understand how vectors and matrices are used in the Recipes is in the remainder of this section.

Vectors and Matrices are Zero-Based

Arrays in C and C++ are natively "zero-based" or "zero-offset." An array declared by the statement double a[3]; has the valid references a[0], a[1], and a[2], but *not* a[3]. However, many mathematical algorithms naturally like to go from 1 to n, not from 0 to $n - 1$. In the C version of this book we showed how to use the power of the C language to declare *unit-offset* arrays efficiently and elegantly, and how to use them alongside zero-offset arrays. However, one message that came through loud and clear from our readers' letters was that they didn't like unit-offset arrays. Accordingly, *all the routines in this book use zero-based arrays exclusively.* Even where a mathematical algorithm would be more naturally expressed with indices running from 1 to n, we have presented it in the text with indices from 0 to $n - 1$. Thus the indices in the code arrays are the same as those in the text.

Vector and Matrix Type Names in the Recipes: nrtypes.h

In §1.3 we describe the nitty-gritty of how our vector and matrix classes are implemented, and how you can alternatively use any other matrix/vector library instead. Many readers will find this tough going. Fortunately, it is possible to insulate you almost entirely from the details of the matrix/vector library. In fact, if you peruse the Recipe functions, you won't see the class names NRVec or NRMat appearing anywhere. Instead, you'll see names like Vec_I_DP and Mat_O_INT declaring vectors and matrices. These are the identifiers you should actually use, and it is the file nrtypes.h that encapsulates all these definitions in a set of typedef definitions.

The name we use for a typical type defined in nrtypes.h consists of three parts:

- Vec or Mat for vector or matrix.
- I, O, or IO for in, out, or in-out. These symbols happen to be based on the corresponding Fortran 90 names "intent in", "intent out", and "intent inout," but the concepts are universal. They describe whether the array being passed to the function supplies values to the function but does

not return values (intent in), returns values from the function but does not
supply any (intent out), or does both (intent inout).

- the scalar type that is the intended default. For example, INT for int,
DP (double precision) for double, CPLX_SP (complex single precision)
for complex<float>, and so on. Here is a complete list of all the types
we so denote:

```
BOOL            bool
CHR             char
UCHR            unsigned char
INT             int
UINT            unsigned int
LNG             long
ULNG            unsigned long
SP              float
DP              double
CPLX_SP         complex<float>
CPLX_DP         complex<double>
ULNG_p          unsigned long *
DP_p            double *
FSTREAM_p       fstream *
```

A sample entry in nrtypes.h for double precision vectors is

```
typedef const NRVec<DP> Vec_I_DP;
typedef NRVec<DP> Vec_DP, Vec_O_DP, Vec_IO_DP;
```

The first line says that intent in vectors will be const double precision. The second
line says that intent out and intent inout vectors will not be const. The identifier
Vec_DP is for vectors declared locally within a Recipe function, where the terms
"in" and "out" are not meaningful.

With these defined types, you can write programs completely without reference
to the implementation of the underlying vector and matrix classes. For example, you
might use statements like the following:

```
double func(Mat_I_DP &a, Vec_IO_DP &b);
{
    const int N=10;
    Mat_DP d(N,N);

    Vec_DP *e=new Vec_DP(N);
    ...

    delete e;
}
```

The complete set of these definitions is contained in the file nrtypes.h,
reproduced in Appendix B. But you'll find you seldom have to refer to it. In fact, if
you haven't already done so, now would be a good time to follow the instructions
in the Quick Start subsection in §1.0. You'll probably find you know enough to use
the Recipes without any problems.

A Few Wrinkles

We like to keep code compact, avoiding unnecessary spaces unless they add immediate clarity. We usually don't put space around the assignment operator "=". For historical reasons, you will see us write y= -10.0; or y=(-10.0);, and y= *a; or y=(*a);. This is just because there used to be some C compilers recognize the (nonexistent) operator "=-" as being equivalent to the subtractive assignment operator "-=", and "=*" as being the same as the multiplicative assignment operator "*=". We hope that this quirkiness has disappeared by now, but we still have lingering habits.

We have the same viewpoint regarding unnecessary parentheses. You can't write (or read) C++ effectively unless you memorize its operator precedence and associativity rules. Please study the accompanying table while you brush your teeth every night.

We never use the `register` storage class specifier. Good optimizing compilers are quite sophisticated in making their own decisions about what to keep in registers, and the best choices are sometimes rather counter-intuitive.

We like to use the C++ constructor for casting types: `int(x)` rather than the C-style `(int) x`. Similarly, if a pointer to a function is passed as an argument, we invoke the function with the C++ form `func(x)` rather than the C-style `*func(x)`.

Some of our routines need to define a global vector or matrix to communicate data between two routines, let's call them `one()` and `two()`, without using function arguments. Typically the size of the data set needs to be set dynamically at runtime. We handle this situation by making the global variable a pointer to the data, so that function `one` would look something like this:

```
Vec_DP *xvec_p;                 Definition of global pointer.

void one(...)
{
    ...
    xvec_p=new Vec_DP(n);       Allocate storage of size n.
    Vec_DP &xvec= *xvec_p;      Make alias to simplify subsequent coding.
    ...
    delete xvec_p;              Reclaim storage when done.
```

The reference variable `xvec` is defined only to make subsequent code easier to write and read. Instead of writing `(*xvec_p)[i]` we can write `xvec[i]`.

To use the global vector in function `two`, we use the following scheme:

```
extern Vec_DP *xvec_p;          Declaration of global pointer, defined elsewhere.

void two(...)
{
    ...
    Vec_DP &xvec= *xvec_p;      Make alias to simplify subsequent coding.
    ...
```

(Note for experts: An alternative scheme is to declare a zero-size vector globally, and then resize it appropriately inside function `one()`. However, we don't want to assume the existence of a resize function in the vector or matrix library.)

We have already alluded to the problem of computing small integer powers of numbers, most notably the square and cube. The omission of this operation

Operator Precedence and Associativity Rules in C++		
: :	scope resolution	left-to-right
() [] . -> ++ --	function call array element (subscripting) member selection member selection (by pointer) post increment post decrement	left-to-right **right-to-left**
! ~ - ++ -- & * new delete (type) sizeof	logical not bitwise complement unary minus pre increment pre decrement address of contents of (dereference) create destroy cast to type size in bytes	**right-to-left**
* / %	multiply divide remainder	left-to-right
+ -	add subtract	left-to-right
<< >>	bitwise left shift bitwise right shift	left-to-right
< > <= >=	arithmetic less than arithmetic greater than arithmetic less than or equal to arithmetic greater than or equal to	left-to-right
== !=	arithmetic equal arithmetic not equal	left-to-right
&	bitwise and	left-to-right
^	bitwise exclusive or	left-to-right
|	bitwise or	left-to-right
&&	logical and	left-to-right
||	logical or	left-to-right
? :	conditional expression	**right-to-left**
= also += -= *= /= %= <<= >>= &= ^= |=	assignment operator	**right-to-left**
,	sequential expression	left-to-right

from C++ is perhaps the language's most galling continuing insult to the scientific programmer. All good Fortran compilers recognize expressions like (a+b)**4 and produce in-line code, in this case with only *one* add and *two* multiplies. It is typical for constant integer powers up to 12 to be thus recognized.

In nrutil.h we provide an inline templated function to handle squaring. Its definition is

```
template<class T>
inline const T SQR(const T a) {return a*a;}
```

You're on your own for higher powers. We also provide a collection of similar functions for other simple operations: SQR, MAX, MIN, SIGN, and SWAP. These do the obvious things (SIGN(a,b) returns the magnitude of a times the sign of b.)

Scientific programming in C++ may someday become a bed of roses; for now, watch out for the thorns!

CITED REFERENCES AND FURTHER READING:

Stroustrup, B. 1997, *The C++ Programming Language*, 3rd ed. (Reading, MA: Addison-Wesley).
[1]

1.3 Implementation of the Vector and Matrix Classes

In this section we describe the details of the implementation of the vector and matrix classes we use. We start with the default classes, and then describe how you can instead use another matrix/vector library with our Recipe functions. Unless you are an experienced C++ programmer, you will probably not want to read beyond the first subsection. And remember: in practice you will actually never need to use the names NRVec or NRMat. Instead, you use identifiers like Vec_I_DP as described in the previous section.

The Default Vector and Matrix Classes

Here is the declaration of the NRVec class:

```
template <class T>
class NRVec {
    private:
        int nn;                                     Size of array, indices 0..nn-1.
        T *v;                                       Pointer to data array.
    public:
        NRVec();                                    Default constructor.
        explicit NRVec(int n);                      Construct vector of size n.
        NRVec(const T &a, int n);                   Initialize to constant value a.
        NRVec(const T *a, int n);                   Initialize to values in C-style array a.
        NRVec(const NRVec& rhs);                    Copy constructor.
        NRVec& operator=(const NRVec& rhs);         Assignment operator.
        NRVec& operator=(const T& a);               Assign a to every element.
```

```
        inline T & operator[](const int i);          Return element number i.
        inline const T & operator[](const int i) const;        const version.
        inline int size() const;                      Return size of vector.
        ~NRVec();                                      Destructor.
};
```

The private variables are discussed in Appendix B. Let's look at the public interface. The various constructors allow vectors to be declared as in the following examples:

```
    NRVec<double> v;              Zero-size array.
    NRVec<double> w(10);          w[0..9].
    NRVec<int> x(a,10);           x[0..9] = a.
    NRVec<int> y(b,10);           y[0..9] = b[0..9], b a C-style array.
    NRVec<int> z=y;               z[0..9] = y[0..9], y an NRVec.
```

The `explicit` keyword prevents the compiler from performing an implicit type conversion from an integer. Unless you know what you are doing, you almost certainly don't need to do such conversions, which are often a source of hard-to-diagnose errors. So using `explicit` is a good idea.

The overloaded assignment operators allow code like (assuming the declarations above)

```
    y=x;                 Copy values from x to y.
    x=1;                 Set all elements of x to 1.
```

The overloaded subscript operator `[]` is used repeatedly in the routines to access array elements. For example, `x[6]` is the seventh element in the vector. We will discuss the two forms of the subscript operator below in the subsection "More on `const` Correctness."

Finally, the `size()` function returns the number of elements in the vector:

```
    NRVec<double> w(10);
    int n = w.size();            Sets n to 10.
```

As we will discuss later, you can use any vector class you like with the *Numerical Recipes* functions as long as it provides the basic functions above. (The functions can be called something else; it's the *functionality* that must be the same.) In fact, you can get away with even less. As long as you provide the constructor for a vector of length n, the subscript operator, and the `size()` function, most of the Recipes will work. About ten will have to have some code replaced by explicit loops to handle initializing to values or arrays. (The *Numerical Recipes* example routines make more extensive use of the additional functions in the NRVec class.)

The matrix class NRMat is very similar to NRVec:

```
template <class T>
class NRMat {
    private:
        int nn;                              Number of rows and columns. Index
        int mm;                                  range is 0..nn-1, 0..mm-1.
        T **v;                               Storage for data.
    public:
        NRMat();                             Default constructor.
        NRMat(int n, int m);                 Construct n × m matrix.
        NRMat(const T& a, int n, int m);     Initialize to constant value a.
```

```
            NRMat(const T *a, int n, int m);           Initialize to values in C-style array a.
            NRMat(const NRMat& rhs);                   Copy constructor.
            NRMat& operator=(const NRMat& rhs);        Assignment operator.
            NRMat& operator=(const T& a);              Assign a to every element.
            inline T* operator[](const int i);         Subscripting: pointer to row i.
            inline const T* operator[](const int i) const;        const version.
            inline int nrows() const;                  Return number of rows.
            inline int ncols() const;                  Return number of columns.
            ~NRMat();                                   Destructor.
};
```

The only point to note is the return type of operator[]. If a is of type NRMat, we want a[i] to point to row i of the matrix so that a[i][j] will be the (i, j) matrix element. In our default class shown above, this means that a[i] must be of type T*. However, it might be something else in a different, more complicated, class library.

Just as in the case of vectors, you will find out below how to use any matrix class with the Recipes, as long as it provides the basic functions above. And if all you supply is the constructor for an $m \times n$ matrix, the subscript operator, and the functions for the number of rows and columns, only about five routines will need to be rewritten.

Only two of our routines use a *three*-dimensional array: rlft3 in §12.5 and solvde in §17.3. This data structure is provided by the class NRMat3d in nrutil.h. We have not made any special efforts for you to be able to use your own class instead of the one we provide, but if necessary you could follow the same technique we describe below for vectors and matrices.

The full implementation code for the NRVec and NRMat classes is in nrutil.h, which is reproduced in Appendix B. Note that you can easily use the standard library class valarray for vectors instead of NRVec: Simply replace the complete declaration and implementation of the NRVec class with the following:

```
    #define NRVec valarray
    #include <valarray>
```

Alas, the standard library doesn't provide anything comparable to valarray for matrices.

More on Const Correctness

At this point we need to elaborate on what exactly const does for a non-simple type such as a class that is an argument of a function. Basically it guarantees that the object is not modified by the function. In other words, the data members are unchanged. But note that if a data member is a *pointer* to some data, and the data itself is not a member variable, then *the data can be changed* even though the pointer cannot be.

Let's look at the implications of this for a function f that takes an NRVec argument a. To avoid unnecessary copying, we always pass arrays by reference. Consider the difference between declaring the argument of a function with and without const:

```
    void f(NRVec<double> &a)      OR      void f(const NRVec<double> &a)
```

The const version promises that f does not modify the data members of a. But a statement like

```
a[i] = 4;
```

inside the function declaration is in principle perfectly OK — you are modifying the data pointed to, not the pointer itself.

"Isn't there some way to protect the data?" you may ask. Yes, there is — you can declare the *return type* of operator[] to be const. This is why there are two versions of operator[] in the NRVec class,

```
T & operator[](const int i);
const T & operator[](const int i) const;
```

The first form returns a reference to an element of a modifiable vector, while the second is for a nonmodifiable vector. The way these work in our example is that if argument a of function f() above is declared as const reference, then when a is dereferenced the second form of operator[] will be invoked. (It is the trailing const in the declaration that promises not to modify the object being dereferenced, and which must agree in const'ness with the argument.) Then the return type const allows expressions like x = a[i] but forbids expressions like a[i] = 4 (a[i] is not an lvalue). By contrast, if a is declared in f() without the const qualifier, the first form of operator[] is invoked, which allows both kinds of statements (a[i] can be an lvalue).[1]

This is the kind of trickery you have to resort to to get const to protect the data. As judged from the large number of matrix/vector libraries that follow this scheme, many people feel that the payoff is worthwhile.

However, we must also recognize the existence of an alternative implementation of operator[], which is to stick with the basic C++ philosophy "const protects the container, not the contents." In this case you would want only *one* form of operator[], namely

```
T & operator[](const int i) const;
```

It would be invoked whether your vector was passed by const reference or not. In both cases the size and pointer of the vector are unchanged, and element i is returned as potentially modifiable.

While we do not use this second alternative in our default classes, since it is nice to be able to use const to protect the contents, we *must* use it if we replace our default classes with the classes described in the next subsection that allow you to use your own favorite matrix/vector library instead of our defaults. The reason is that with this alternative, all invocations of operator[] will enforce constness of the object so as to allow the possibility of automatic conversions. (When an object is passed to a function by reference, automatic type conversions can be made only if the object is const.) Although giving up the extra const checking is regrettable,

---

[1] "Wait!" you might object. "The first form of operator[] doesn't actually modify the object, only the data. Why can't I put a trailing const after it?" Answer: because then the compiler could not distinguish between the two forms. Overloaded functions and operators must be distinguishable by the argument types alone, not by the return types.

this nevertheless turns out to be the best route to allowing transparent use of other matrix/vector libraries, as we now explain.

Using Other Vector and Matrix Libraries

Suppose you have a vector class MyVec that you want to use instead of NRVec. The goal is to have automatic type conversions when a MyVec is passed to a Recipe function (and an NRVec is expected), and when an NRVec is returned by a Recipe function (and a MyVec is expected). C++ supplies us with the necessary tools: a "constructor conversion" to make an NRVec out of a MyVec, and an "operator conversion" to make a MyVec out of an NRVec.

The first key idea is to make NRVec be a "wrapper class" for MyVec, that is, an NRVec "holds-a" MyVec. The second key idea is to have the wrapper class hold both a MyVec *and* a reference to a MyVec. Then when we want to make an NRVec out of a MyVec, we can use the constructor initializer list to point the reference to the existing MyVec. No copying of data takes place at all.

The wrapper classes must provide all the functions of the default classes introduced earlier, and we shall describe them in the same order. So here is the start of the NRVec wrapper class:

```
template<class T>
class NRVec {
    private:
        MyVec<T> myvec;
        MyVec<T> &myref;
    public:
        NRVec<T>() : myvec(), myref(myvec) {}
            ...
```

The private variables are only a MyVec and a reference to one. The private variable myvec is used solely when creating a new NRVec. The default constructor just invokes the default constructor for a MyVec, and then points the reference accordingly. Similarly, the remaining constructors simply invoke the corresponding MyVec constructors:

```
explicit NRVec<T>(const int n) : myvec(n), myref(myvec) {}
NRVec<T>(const T &a, int n) : myvec(n,a), myref(myvec) {}
NRVec<T>(const T *a, int n) : myvec(n,a), myref(myvec) {}
```

Note we are assuming that the syntax of the MyVec class member functions is the same as that of the NRVec class, but it's easy to take care of different syntax. For example, some vector classes expect arguments in the opposite order: myvec(a,n) instead of myvec(n,a).

Next comes the conversion constructor. It makes a special NRVec that points to MyVec's data, taking care of MyVec actual argments passed as NRVec formal arguments in function calls:

```
NRVec<T>(MyVec<T> &rhs) : myref(rhs) {}
```

Since all the other functions in NRVec access the object only through myref, there is no need to initialize myvec.

The copy constructor and assignment operator simply call the MyVec copy constructor and assignment operator:

```
NRVec(const NRVec<T> &rhs) : myvec(rhs.myref), myref(myvec) {}
inline NRVec& operator=(const NRVec &rhs) { myref=rhs.myref; return *this;}
inline NRVec& operator=(const T &rhs) { myvec=rhs; return *this;}
```

(We assume here that the MyVec functions do the sensible thing of making a "deep" copy, that is, they copy the data, not just the reference. See the MTL example in Appendix B for a case where this is not true.)

Similarly, functions like size() are implemented by calling the corresponding MyVec functions:

```
inline int size() const {return myref.size();}
```

We pointed out earlier that for the subscript operator, only the form that guarantees constness of the container is allowed. We want automatic type conversion, and we get that only for objects passed by const reference. Accordingly, we cannot guarantee constness of the data (see the discussion in the previous subsection):

```
inline T & operator[](const int i) const {return myref[i];}
```

Next comes the conversion operator from NRVec to MyVec, which handles NRVec function return types when used in MyVec expressions:

```
inline operator MyVec<T>() const {return myref;}
```

Finally, the destructor is trivial: the MyVec destructor takes care of destroying the data.

```
~NRVec() {}
```

A wrapper class for NRMat follows exactly the same form as for NRVec. The only thing to watch out for is to make sure that the return type for operator[] is whatever a MyMat object returns for a single [] dereference. Then expressions like a[i][j] will work correctly.

Instead of listing the final form of the wrapper class here (pulling together all the lines above), we list in Appendix B two sample nrutil.h files for two popular matrix/vector libraries. They are for the Template Numerical Toolkit, or TNT [1], and the Matrix Template Library, or MTL [2]. We have found TNT to be particularly easy to use with the Recipe functions. All you need to do is use TNT::Vector for MyVec and TNT::Matrix for MyMat. If you are concerned about poor efficiency in using wrapper classes, our timing experiments show less than 10% overhead for TNT with most compilers, compared with using TNT directly.

One final instruction: the file nrtypes.h also needs to be changed to the "const protects the container, not the contents" convention for passing arguments by reference. To make this easy, we have supplied the file nrtypes_lib.h that correctly defines the constness of vector and matrix arguments in this way. So simply edit nrtypes.h to include this file instead of the default nrtypes_nr.h (see Appendix B).

A note for C++ aficionados: You can also implement the interface to other matrix/vector libraries by making NRVec a derived class of your vector class. However, this is not nearly as elegant as the wrapper class. In particular, it depends on the implementations inside of your vector class, while the wrapper class uses only the public interface and semantics of your vector class.

CITED REFERENCES AND FURTHER READING:

Pozo, R., *Template Numerical Toolkit*, http://math.nist.gov/tnt. [1]

Lumsdaine, A., and Siek, J. 1998, *The Matrix Template Library*, http://www.lsc.nd.edu/research/mtl. [2]

1.4 Error, Accuracy, and Stability

Although we assume no prior training of the reader in formal numerical analysis, we will need to presume a common understanding of a few key concepts. We will define these briefly in this section.

Computers store numbers not with infinite precision but rather in some approximation that can be packed into a fixed number of *bits* (binary digits) or *bytes* (groups of 8 bits). Almost all computers allow the programmer a choice among several different such *representations* or *data types*. Data types can differ in the number of bits utilized (the *wordlength*), but also in the more fundamental respect of whether the stored number is represented in *fixed-point* (`int` or `long`) or *floating-point* (`float` or `double`) format.

A number in integer representation is exact. Arithmetic between numbers in integer representation is also exact, with the provisos that (i) the answer is not outside the range of (usually, signed) integers that can be represented, and (ii) that division is interpreted as producing an integer result, throwing away any integer remainder.

In floating-point representation, a number is represented internally by a sign bit s (interpreted as plus or minus), an exact integer exponent e, and an exact positive integer mantissa M. Taken together these represent the number

$$s \times M \times B^{e-E} \tag{1.4.1}$$

where B is the base of the representation (usually $B = 2$, but sometimes $B = 16$), and E is the *bias* of the exponent, a fixed integer constant for any given machine and representation. An example is shown in Figure 1.4.1.

Several floating-point bit patterns can represent the same number. If $B = 2$, for example, a mantissa with leading (high-order) zero bits can be left-shifted, i.e., multiplied by a power of 2, if the exponent is decreased by a compensating amount. Bit patterns that are "as left-shifted as they can be" are termed *normalized*. Most computers always produce normalized results, since these don't waste any bits of the mantissa and thus allow a greater accuracy of the representation. Since the high-order bit of a properly normalized mantissa (when $B = 2$) is *always* one, some computers don't store this bit at all, giving one extra bit of significance.

Arithmetic among numbers in floating-point representation is not exact, even if the operands happen to be exactly represented (i.e., have exact values in the form of equation 1.4.1). For example, two floating numbers are added by first right-shifting (dividing by two) the mantissa of the smaller (in magnitude) one, simultaneously increasing its exponent, until the two operands have the same exponent. Low-order (least significant) bits of the smaller operand are lost by this shifting. If the two operands differ too greatly in magnitude, then the smaller operand is effectively replaced by zero, since it is right-shifted to oblivion.

The smallest (in magnitude) floating-point number which, when added to the floating-point number 1.0, produces a floating-point result different from 1.0 is termed the *machine accuracy* ϵ_m. A typical computer with $B = 2$ and a 32-bit wordlength has ϵ_m around 3×10^{-8}. (A more detailed discussion of machine characteristics, and a program to determine them, is given in §20.1.) Roughly speaking, the machine accuracy ϵ_m is the fractional accuracy to which floating-point

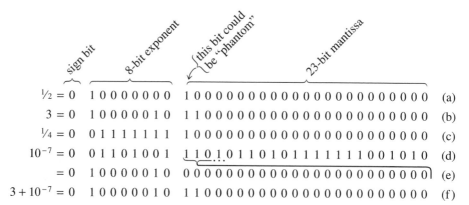

Figure 1.4.1.   Floating point representations of numbers in a typical 32-bit (4-byte) format. (a) The number $1/2$ (note the bias in the exponent); (b) the number 3; (c) the number $1/4$; (d) the number 10^{-7}, represented to machine accuracy; (e) the same number 10^{-7}, but shifted so as to have the same exponent as the number 3; with this shifting, all significance is lost and 10^{-7} becomes zero; shifting to a common exponent must occur before two numbers can be added; (f) sum of the numbers $3 + 10^{-7}$, which equals 3 to machine accuracy. Even though 10^{-7} can be represented accurately by itself, it cannot accurately be added to a much larger number.

numbers are represented, corresponding to a change of one in the least significant bit of the mantissa. Pretty much any arithmetic operation among floating numbers should be thought of as introducing an additional fractional error of at least ϵ_m. This type of error is called *roundoff error*.

It is important to understand that ϵ_m is not the smallest floating-point number that can be represented on a machine. *That* number depends on how many bits there are in the exponent, while ϵ_m depends on how many bits there are in the mantissa.

Roundoff errors accumulate with increasing amounts of calculation. If, in the course of obtaining a calculated value, you perform N such arithmetic operations, you *might* be so lucky as to have a total roundoff error on the order of $\sqrt{N}\epsilon_m$, if the roundoff errors come in randomly up or down. (The square root comes from a random-walk.) However, this estimate can be very badly off the mark for two reasons:

(i) It very frequently happens that the regularities of your calculation, or the peculiarities of your computer, cause the roundoff errors to accumulate preferentially in one direction. In this case the total will be of order $N\epsilon_m$.

(ii) Some especially unfavorable occurrences can vastly increase the roundoff error of single operations. Generally these can be traced to the subtraction of two very nearly equal numbers, giving a result whose only significant bits are those (few) low-order ones in which the operands differed. You might think that such a "coincidental" subtraction is unlikely to occur. Not always so. Some mathematical expressions magnify its probability of occurrence tremendously. For example, in the familiar formula for the solution of a quadratic equation,

$$x = \frac{-b + \sqrt{b^2 - 4ac}}{2a} \tag{1.4.2}$$

the addition becomes delicate and roundoff-prone whenever $ac \ll b^2$. (In §5.6 we will learn how to avoid the problem in this particular case.)

Roundoff error is a characteristic of computer hardware. There is another, different, kind of error that is a characteristic of the program or algorithm used, independent of the hardware on which the program is executed. Many numerical algorithms compute "discrete" approximations to some desired "continuous" quantity. For example, an integral is evaluated numerically by computing a function at a discrete set of points, rather than at "every" point. Or, a function may be evaluated by summing a finite number of leading terms in its infinite series, rather than all infinity terms. In cases like this, there is an adjustable parameter, e.g., the number of points or of terms, such that the "true" answer is obtained only when that parameter goes to infinity. Any practical calculation is done with a finite, but sufficiently large, choice of that parameter.

The discrepancy between the true answer and the answer obtained in a practical calculation is called the *truncation error*. Truncation error would persist even on a hypothetical, "perfect" computer that had an infinitely accurate representation and no roundoff error. As a general rule there is not much that a programmer can do about roundoff error, other than to choose algorithms that do not magnify it unnecessarily (see discussion of "stability" below). Truncation error, on the other hand, is entirely under the programmer's control. In fact, it is only a slight exaggeration to say that clever minimization of truncation error is practically the entire content of the field of numerical analysis!

Most of the time, truncation error and roundoff error do not strongly interact with one another. A calculation can be imagined as having, first, the truncation error that it would have if run on an infinite-precision computer, "plus" the roundoff error associated with the number of operations performed.

Sometimes, however, an otherwise attractive method can be *unstable*. This means that any roundoff error that becomes "mixed into" the calculation at an early stage is successively magnified until it comes to swamp the true answer. An unstable method would be useful on a hypothetical, perfect computer; but in this imperfect world it is necessary for us to require that algorithms be stable — or if unstable that we use them with great caution.

Here is a simple, if somewhat artificial, example of an unstable algorithm: Suppose that it is desired to calculate all integer powers of the so-called "Golden Mean," the number given by

$$\phi \equiv \frac{\sqrt{5} - 1}{2} \approx 0.61803398 \qquad (1.4.3)$$

It turns out (you can easily verify) that the powers ϕ^n satisfy a simple recursion relation,

$$\phi^{n+1} = \phi^{n-1} - \phi^n \qquad (1.4.4)$$

Thus, knowing the first two values $\phi^0 = 1$ and $\phi^1 = 0.61803398$, we can successively apply (1.4.4) performing only a single subtraction, rather than a slower multiplication by ϕ, at each stage.

Unfortunately, the recurrence (1.4.4) also has *another* solution, namely the value $-\frac{1}{2}(\sqrt{5} + 1)$. Since the recurrence is linear, and since this undesired solution has magnitude greater than unity, any small admixture of it introduced by roundoff errors will grow exponentially. On a typical machine with 32-bit wordlength, (1.4.4) starts to give completely wrong answers by about $n = 16$, at which point ϕ^n is down to only 10^{-4}. The recurrence (1.4.4) is *unstable*, and cannot be used for the purpose stated.

We will encounter the question of stability in many more sophisticated guises, later in this book.

CITED REFERENCES AND FURTHER READING:

Stoer, J., and Bulirsch, R. 1993, *Introduction to Numerical Analysis*, 2nd ed. (New York: Springer-Verlag), Chapter 1.

Kahaner, D., Moler, C., and Nash, S. 1989, *Numerical Methods and Software* (Englewood Cliffs, NJ: Prentice Hall), Chapter 2.

Johnson, L.W., and Riess, R.D. 1982, *Numerical Analysis*, 2nd ed. (Reading, MA: Addison-Wesley), §1.3.

Wilkinson, J.H. 1964, *Rounding Errors in Algebraic Processes* (Englewood Cliffs, NJ: Prentice-Hall).

Chapter 2.   Solution of Linear Algebraic Equations

2.0 Introduction

A set of linear algebraic equations looks like this:

$$a_{00}x_0 + a_{01}x_1 + a_{02}x_2 + \cdots + a_{0,N-1}x_{N-1} = b_0$$

$$a_{10}x_0 + a_{11}x_1 + a_{12}x_2 + \cdots + a_{1,N-1}x_{N-1} = b_1$$

$$a_{20}x_0 + a_{21}x_1 + a_{22}x_2 + \cdots + a_{2,N-1}x_{N-1} = b_2 \qquad (2.0.1)$$

$$\cdots \qquad\qquad \cdots$$

$$a_{M-1,0}x_0 + a_{M-1,1}x_1 + \cdots + a_{M-1,N-1}x_{N-1} = b_{M-1}$$

Here the N unknowns x_j, $j = 0, 1, \ldots, N - 1$ are related by M equations. The coefficients a_{ij} with $i = 0, 1, \ldots, M - 1$ and $j = 0, 1, \ldots, N - 1$ are known numbers, as are the *right-hand side* quantities b_i, $i = 0, 1, \ldots, M - 1$.

Nonsingular versus Singular Sets of Equations

If $N = M$ then there are as many equations as unknowns, and there is a good chance of solving for a unique solution set of x_j's. Analytically, there can fail to be a unique solution if one or more of the M equations is a linear combination of the others, a condition called *row degeneracy*, or if all equations contain certain variables only in exactly the same linear combination, called *column degeneracy*. (For square matrices, a row degeneracy implies a column degeneracy, and vice versa.) A set of equations that is degenerate is called *singular*. We will consider singular matrices in some detail in §2.6.

Numerically, at least two additional things can go wrong:

- While not exact linear combinations of each other, some of the equations may be so close to linearly dependent that roundoff errors in the machine render them linearly dependent at some stage in the solution process. In this case your numerical procedure will fail, and it can tell you that it has failed.

- Accumulated roundoff errors in the solution process can swamp the true solution. This problem particularly emerges if N is too large. The numerical procedure does not fail algorithmically. However, it returns a set of x's that are wrong, as can be discovered by direct substitution back into the original equations. The closer a set of equations is to being singular, the more likely this is to happen, since increasingly close cancellations will occur during the solution. In fact, the preceding item can be viewed as the special case where the loss of significance is unfortunately total.

Much of the sophistication of complicated "linear equation-solving packages" is devoted to the detection and/or correction of these two pathologies. As you work with large linear sets of equations, you will develop a feeling for when such sophistication is needed. It is difficult to give any firm guidelines, since there is no such thing as a "typical" linear problem. But here is a rough idea: Linear sets with N as large as 20 or 50 can be routinely solved in single precision (32 bit floating representations) without resorting to sophisticated methods, *if* the equations are not close to singular. With double precision (64 bits or more), this number can readily be extended to N as large as perhaps 1000, after which point the limiting factor is generally machine time, not accuracy.

Even larger linear sets, N in the thousands or greater, can be solved when the coefficients are sparse (that is, mostly zero), by methods that take advantage of the sparseness. We discuss this further in §2.7.

At the other end of the spectrum, one seems just as often to encounter linear problems which, by their underlying nature, are close to singular. In this case, you *might* need to resort to sophisticated methods even for the case of $N = 10$ (though rarely for $N = 5$). Singular value decomposition (§2.6) is a technique that can sometimes turn singular problems into nonsingular ones, in which case additional sophistication becomes unnecessary.

Matrices

Equation (2.0.1) can be written in matrix form as

$$\mathbf{A} \cdot \mathbf{x} = \mathbf{b} \qquad (2.0.2)$$

Here the raised dot denotes matrix multiplication, $\mathbf{A}$ is the matrix of coefficients, and $\mathbf{b}$ is the right-hand side written as a column vector,

$$\mathbf{A} = \begin{bmatrix} a_{00} & a_{01} & \cdots & a_{0,N-1} \\ a_{10} & a_{11} & \cdots & a_{1,N-1} \\ & \cdots & & \\ a_{M-1,0} & a_{M-1,1} & \cdots & a_{M-1,N-1} \end{bmatrix} \qquad \mathbf{b} = \begin{bmatrix} b_0 \\ b_1 \\ \cdots \\ b_{M-1} \end{bmatrix} \qquad (2.0.3)$$

By convention, the first index on an element a_{ij} denotes its row, the second index its column. For most purposes you don't need to know how a matrix is stored in a computer's physical memory; you simply reference matrix elements by their two-dimensional addresses, e.g., $a_{34} = $ a[3][4]. We have already seen, in §1.2 and

1.3, that this C++ notation can in fact hide a variety of subtle and versatile physical storage schemes. You might wish to review those sections at this point.

Tasks of Computational Linear Algebra

We will consider the following tasks as falling in the general purview of this chapter:

- Solution of the matrix equation $\mathbf{A} \cdot \mathbf{x} = \mathbf{b}$ for an unknown vector $\mathbf{x}$, where $\mathbf{A}$ is a square matrix of coefficients, raised dot denotes matrix multiplication, and $\mathbf{b}$ is a known right-hand side vector (§2.1–§2.10).
- Solution of more than one matrix equation $\mathbf{A} \cdot \mathbf{x}_j = \mathbf{b}_j$, for a set of vectors $\mathbf{x}_j, j = 0, 1, \ldots$, each corresponding to a different, known right-hand side vector $\mathbf{b}_j$. In this task the key simplification is that the matrix $\mathbf{A}$ is held constant, while the right-hand sides, the $\mathbf{b}$'s, are changed (§2.1–§2.10).
- Calculation of the matrix $\mathbf{A}^{-1}$ which is the matrix inverse of a square matrix $\mathbf{A}$, i.e., $\mathbf{A} \cdot \mathbf{A}^{-1} = \mathbf{A}^{-1} \cdot \mathbf{A} = \mathbf{1}$, where $\mathbf{1}$ is the identity matrix (all zeros except for ones on the diagonal). This task is equivalent, for an $N \times N$ matrix $\mathbf{A}$, to the previous task with N different $\mathbf{b}_j$'s $(j = 0, 1, \ldots, N-1)$, namely the unit vectors ($\mathbf{b}_j$ = all zero elements except for 1 in the jth component). The corresponding $\mathbf{x}$'s are then the columns of the matrix inverse of $\mathbf{A}$ (§2.1 and §2.3).
- Calculation of the determinant of a square matrix $\mathbf{A}$ (§2.3).

If $M < N$, or if $M = N$ but the equations are degenerate, then there are effectively fewer equations than unknowns. In this case there can be either no solution, or else more than one solution vector $\mathbf{x}$. In the latter event, the solution space consists of a particular solution $\mathbf{x}_p$ added to any linear combination of (typically) $N - M$ vectors (which are said to be in the nullspace of the matrix $\mathbf{A}$). The task of finding the solution space of $\mathbf{A}$ involves

- Singular value decomposition of a matrix $\mathbf{A}$.

This subject is treated in §2.6.

In the opposite case there are more equations than unknowns, $M > N$. When this occurs there is, in general, no solution vector $\mathbf{x}$ to equation (2.0.1), and the set of equations is said to be *overdetermined*. It happens frequently, however, that the best "compromise" solution is sought, the one that comes closest to satisfying all equations simultaneously. If closeness is defined in the least-squares sense, i.e., that the sum of the squares of the differences between the left- and right-hand sides of equation (2.0.1) be minimized, then the overdetermined linear problem reduces to a (usually) solvable linear problem, called the

- Linear least-squares problem.

The reduced set of equations to be solved can be written as the $N \times N$ set of equations

$$(\mathbf{A}^T \cdot \mathbf{A}) \cdot \mathbf{x} = (\mathbf{A}^T \cdot \mathbf{b}) \tag{2.0.4}$$

where $\mathbf{A}^T$ denotes the transpose of the matrix $\mathbf{A}$. Equations (2.0.4) are called the *normal equations* of the linear least-squares problem. There is a close connection between singular value decomposition and the linear least-squares problem, and the latter is also discussed in §2.6. You should be warned that direct solution of the normal equations (2.0.4) is not generally the best way to find least-squares solutions.

Some other topics in this chapter include

- Iterative improvement of a solution (§2.5)
- Various special forms: symmetric positive-definite (§2.9), tridiagonal (§2.4), band diagonal (§2.4), Toeplitz (§2.8), Vandermonde (§2.8), sparse (§2.7)
- Strassen's "fast matrix inversion" (§2.11).

Standard Subroutine Packages

We cannot hope, in this chapter or in this book, to tell you everything there is to know about the tasks that have been defined above. In many cases you will have no alternative but to use sophisticated black-box program packages. Several good ones are available, though not always in C++. LAPACK, a sucessor to the venerable LINPACK, was developed at Argonne National Laboratories and deserves particular mention because it is published, documented, and available for free use. ScaLAPACK is a version available for parallel architectures. Packages available commercially include those in the IMSL and NAG libraries.

You should keep in mind that the sophisticated packages are designed with very large linear systems in mind. They therefore go to great effort to minimize not only the number of operations, but also the required storage. Routines for the various tasks are usually provided in several versions, corresponding to several possible simplifications in the form of the input coefficient matrix: symmetric, triangular, banded, positive definite, etc. If you have a large matrix in one of these forms, you should certainly take advantage of the increased efficiency provided by these different routines, and not just use the form provided for general matrices.

There is also a great watershed dividing routines that are *direct* (i.e., execute in a predictable number of operations) from routines that are *iterative* (i.e., attempt to converge to the desired answer in however many steps are necessary). Iterative methods become preferable when the battle against loss of significance is in danger of being lost, either due to large N or because the problem is close to singular. We will treat iterative methods only incompletely in this book, in §2.7 and in Chapters 18 and 19. These methods are important, but mostly beyond our scope. We will, however, discuss in detail a technique that is on the borderline between direct and iterative methods, namely the iterative improvement of a solution that has been obtained by direct methods (§2.5).

CITED REFERENCES AND FURTHER READING:

Golub, G.H., and Van Loan, C.F. 1996, *Matrix Computations*, 3rd ed. (Baltimore: Johns Hopkins University Press).

Gill, P.E., Murray, W., and Wright, M.H. 1991, *Numerical Linear Algebra and Optimization*, vol. 1 (Redwood City, CA: Addison-Wesley).

Stoer, J., and Bulirsch, R. 1993, *Introduction to Numerical Analysis*, 2nd ed. (New York: Springer-Verlag), Chapter 4.

Anderson, E., et al. 2000, *LAPACK User's Guide*, 3rd ed. (Philadelphia: S.I.A.M.).

Coleman, T.F., and Van Loan, C. 1988, *Handbook for Matrix Computations* (Philadelphia: S.I.A.M.).

Forsythe, G.E., and Moler, C.B. 1967, *Computer Solution of Linear Algebraic Systems* (Englewood Cliffs, NJ: Prentice-Hall).

Wilkinson, J.H., and Reinsch, C. 1971, *Linear Algebra*, vol. II of *Handbook for Automatic Computation* (New York: Springer-Verlag).

Westlake, J.R. 1968, *A Handbook of Numerical Matrix Inversion and Solution of Linear Equations* (New York: Wiley).

Johnson, L.W., and Riess, R.D. 1982, *Numerical Analysis*, 2nd ed. (Reading, MA: Addison-Wesley), Chapter 2.

Ralston, A., and Rabinowitz, P. 1978, *A First Course in Numerical Analysis*, 2nd ed.; reprinted 2001 (New York: Dover), Chapter 9.

2.1 Gauss-Jordan Elimination

For inverting a matrix, *Gauss-Jordan elimination* is about as efficient as any other method. For solving sets of linear equations, Gauss-Jordan elimination produces *both* the solution of the equations for one or more right-hand side vectors **b**, and also the matrix inverse $\mathbf{A}^{-1}$. However, its principal weaknesses are (i) that it requires all the right-hand sides to be stored and manipulated at the same time, and (ii) that when the inverse matrix is *not* desired, Gauss-Jordan is three times slower than the best alternative technique for solving a single linear set (§2.3). The method's principal strength is that it is as stable as any other direct method, perhaps even a bit more stable when full pivoting is used (see below).

If you come along later with an additional right-hand side vector, you can multiply it by the inverse matrix, of course. This does give an answer, but one that is quite susceptible to roundoff error, not nearly as good as if the new vector had been included with the set of right-hand side vectors in the first instance.

For these reasons, Gauss-Jordan elimination should usually not be your method of first choice, either for solving linear equations or for matrix inversion. The decomposition methods in §2.3 are better. Why do we give you Gauss-Jordan at all? Because it is straightforward, understandable, solid as a rock, and an exceptionally good "psychological" backup for those times that something is going wrong and you think it *might* be your linear-equation solver.

Some people believe that the backup is more than psychological, that Gauss-Jordan elimination is an "independent" numerical method. This turns out to be mostly myth. Except for the relatively minor differences in pivoting, described below, the actual sequence of operations performed in Gauss-Jordan elimination is very closely related to that performed by the routines in the next two sections.

For clarity, and to avoid writing endless ellipses ($\cdots$) we will write out equations only for the case of four equations and four unknowns, and with three different right-hand side vectors that are known in advance. You can write bigger matrices and extend the equations to the case of $N \times N$ matrices, with M sets of right-hand side vectors, in completely analogous fashion. The routine implemented below is, of course, general.

Elimination on Column-Augmented Matrices

Consider the linear matrix equation

$$\begin{bmatrix} a_{00} & a_{01} & a_{02} & a_{03} \\ a_{10} & a_{11} & a_{12} & a_{13} \\ a_{20} & a_{21} & a_{22} & a_{23} \\ a_{30} & a_{31} & a_{32} & a_{33} \end{bmatrix} \cdot \left[\begin{pmatrix} x_{00} \\ x_{10} \\ x_{20} \\ x_{30} \end{pmatrix} \sqcup \begin{pmatrix} x_{01} \\ x_{11} \\ x_{21} \\ x_{31} \end{pmatrix} \sqcup \begin{pmatrix} x_{02} \\ x_{12} \\ x_{22} \\ x_{32} \end{pmatrix} \sqcup \begin{pmatrix} y_{00} & y_{01} & y_{02} & y_{03} \\ y_{10} & y_{11} & y_{12} & y_{13} \\ y_{20} & y_{21} & y_{22} & y_{23} \\ y_{30} & y_{31} & y_{32} & y_{33} \end{pmatrix} \right]$$

$$= \left[\begin{pmatrix} b_{00} \\ b_{10} \\ b_{20} \\ b_{30} \end{pmatrix} \sqcup \begin{pmatrix} b_{01} \\ b_{11} \\ b_{21} \\ b_{31} \end{pmatrix} \sqcup \begin{pmatrix} b_{02} \\ b_{12} \\ b_{22} \\ b_{32} \end{pmatrix} \sqcup \begin{pmatrix} 1 & 0 & 0 & 0 \\ 0 & 1 & 0 & 0 \\ 0 & 0 & 1 & 0 \\ 0 & 0 & 0 & 1 \end{pmatrix} \right] \qquad (2.1.1)$$

Here the raised dot ($\cdot$) signifies matrix multiplication, while the operator $\sqcup$ just signifies column augmentation, that is, removing the abutting parentheses and making a wider matrix out of the operands of the $\sqcup$ operator.

It should not take you long to write out equation (2.1.1) and to see that it simply states that x_{ij} is the ith component ($i = 0, 1, 2, 3$) of the vector solution of the jth right-hand side ($j = 0, 1, 2$), the one whose coefficients are $b_{ij}, i = 0, 1, 2, 3$; and that the matrix of unknown coefficients y_{ij} is the inverse matrix of a_{ij}. In other words, the matrix solution of

$$[\mathbf{A}] \cdot [\mathbf{x}_0 \sqcup \mathbf{x}_1 \sqcup \mathbf{x}_2 \sqcup \mathbf{Y}] = [\mathbf{b}_0 \sqcup \mathbf{b}_1 \sqcup \mathbf{b}_2 \sqcup \mathbf{1}] \qquad (2.1.2)$$

where $\mathbf{A}$ and $\mathbf{Y}$ are square matrices, the $\mathbf{b}_i$'s and $\mathbf{x}_i$'s are column vectors, and $\mathbf{1}$ is the identity matrix, simultaneously solves the linear sets

$$\mathbf{A} \cdot \mathbf{x}_0 = \mathbf{b}_0 \qquad \mathbf{A} \cdot \mathbf{x}_1 = \mathbf{b}_1 \qquad \mathbf{A} \cdot \mathbf{x}_2 = \mathbf{b}_2 \qquad (2.1.3)$$

and

$$\mathbf{A} \cdot \mathbf{Y} = \mathbf{1} \qquad (2.1.4)$$

Now it is also elementary to verify the following facts about (2.1.1):
- Interchanging any two *rows* of $\mathbf{A}$ and the corresponding *rows* of the $\mathbf{b}$'s and of $\mathbf{1}$, does not change (or scramble in any way) the solution $\mathbf{x}$'s and $\mathbf{Y}$. Rather, it just corresponds to writing the same set of linear equations in a different order.
- Likewise, the solution set is unchanged and in no way scrambled if we replace any row in $\mathbf{A}$ by a linear combination of itself and any other row, as long as we do the same linear combination of the rows of the $\mathbf{b}$'s and $\mathbf{1}$ (which then is no longer the identity matrix, of course).
- Interchanging any two *columns* of $\mathbf{A}$ gives the same solution set only if we simultaneously interchange corresponding *rows* of the $\mathbf{x}$'s and of $\mathbf{Y}$. In other words, this interchange scrambles the order of the rows in the solution. If we do this, we will need to unscramble the solution by restoring the rows to their original order.

Gauss-Jordan elimination uses one or more of the above operations to reduce the matrix $\mathbf{A}$ to the identity matrix. When this is accomplished, the right-hand side becomes the solution set, as one sees instantly from (2.1.2).

Pivoting

In "Gauss-Jordan elimination with no pivoting," only the second operation in the above list is used. The zeroth row is divided by the element a_{00} (this being a trivial linear combination of the zeroth row with any other row — zero coefficient for the other row). Then the right amount of the zeroth row is subtracted from each other row to make all the remaining a_{i0}'s zero. The zeroth column of **A** now agrees with the identity matrix. We move to column 1 and divide row 1 by a_{11}, then subtract the right amount of row 1 from rows 0, 2, and 3, so as to make their entries in column 1 zero. Column 1 is now reduced to the identity form. And so on for columns 2 and 3. As we do these operations to **A**, we of course also do the corresponding operations to the **b**'s and to **1** (which by now no longer resembles the identity matrix in any way!).

Obviously we will run into trouble if we ever encounter a zero element on the (then current) diagonal when we are going to divide by the diagonal element. (The element that we divide by, incidentally, is called the *pivot element* or *pivot*.) Not so obvious, but true, is the fact that Gauss-Jordan elimination with no pivoting (no use of the first or third procedures in the above list) is numerically unstable in the presence of any roundoff error, even when a zero pivot is not encountered. You must *never* do Gauss-Jordan elimination (or Gaussian elimination, see below) without pivoting!

So what *is* this magic pivoting? Nothing more than interchanging rows (*partial pivoting*) or rows and columns (*full pivoting*), so as to put a particularly desirable element in the diagonal position from which the pivot is about to be selected. Since we don't want to mess up the part of the identity matrix that we have already built up, we can choose among elements that are both (i) on rows below (or on) the one that is about to be normalized, and also (ii) on columns to the right (or on) the column we are about to eliminate. Partial pivoting is easier than full pivoting, because we don't have to keep track of the permutation of the solution vector. Partial pivoting makes available as pivots only the elements already in the correct column. It turns out that partial pivoting is "almost" as good as full pivoting, in a sense that can be made mathematically precise, but which need not concern us here (for discussion and references, see [1]). To show you both variants, we do full pivoting in the routine in this section, partial pivoting in §2.3.

We have to state how to recognize a particularly desirable pivot when we see one. The answer to this is not completely known theoretically. It is known, both theoretically and in practice, that simply picking the largest (in magnitude) available element as the pivot is a very good choice. A curiosity of this procedure, however, is that the choice of pivot will depend on the original scaling of the equations. If we take the third linear equation in our original set and multiply it by a factor of a million, it is almost guaranteed that it will contribute the first pivot; yet the underlying solution of the equations is not changed by this multiplication! One therefore sometimes sees routines which choose as pivot that element which *would* have been largest if the original equations had all been scaled to have their largest coefficient normalized to unity. This is called *implicit pivoting*. There is some extra bookkeeping to keep track of the scale factors by which the rows would have been multiplied. (The routines in §2.3 include implicit pivoting, but the routine in this section does not.)

Finally, let us consider the storage requirements of the method. With a little reflection you will see that at every stage of the algorithm, *either* an element of **A** is predictably a one or zero (if it is already in a part of the matrix that has been reduced to identity form) *or else* the exactly corresponding element of the matrix that started as **1** is predictably a one or zero (if its mate in **A** has not been reduced to the identity form). Therefore the matrix **1** does not have to exist as separate storage: The matrix inverse of **A** is gradually built up in **A** as the original **A** is destroyed. Likewise, the solution vectors **x** can gradually replace the right-hand side vectors **b** and share the same storage, since after each column in **A** is reduced, the corresponding row entry in the **b**'s is never again used.

Here is the routine for Gauss-Jordan elimination with full pivoting:

```
#include <cmath>
#include "nr.h"
using namespace std;

void NR::gaussj(Mat_IO_DP &a, Mat_IO_DP &b)
```
Linear equation solution by Gauss-Jordan elimination, equation (2.1.1) above. The input matrix is a[0..n-1][0..n-1]. b[0..n-1][0..m-1] is input containing the m right-hand side vectors. On output, a is replaced by its matrix inverse, and b is replaced by the corresponding set of solution vectors.
```
{
    int i,icol,irow,j,k,l,ll;
    DP big,dum,pivinv;

    int n=a.nrows();
    int m=b.ncols();
    Vec_INT indxc(n),indxr(n),ipiv(n);    These integer arrays are used for bookkeeping on
    for (j=0;j<n;j++) ipiv[j]=0;               the pivoting.
    for (i=0;i<n;i++) {                    This is the main loop over the columns to be
        big=0.0;                               reduced.
        for (j=0;j<n;j++)                  This is the outer loop of the search for a pivot
            if (ipiv[j] != 1)                  element.
                for (k=0;k<n;k++) {
                    if (ipiv[k] == 0) {
                        if (fabs(a[j][k]) >= big) {
                            big=fabs(a[j][k]);
                            irow=j;
                            icol=k;
                        }
                    }
                }
        ++(ipiv[icol]);
```
We now have the pivot element, so we interchange rows, if needed, to put the pivot element on the diagonal. The columns are not physically interchanged, only relabeled: indxc[i], the column of the $(i+1)$th pivot element, is the $(i+1)$th column that is reduced, while indxr[i] is the row in which that pivot element was originally located. If indxr[i] $\neq$ indxc[i] there is an implied column interchange. With this form of bookkeeping, the solution b's will end up in the correct order, and the inverse matrix will be scrambled by columns.
```
        if (irow != icol) {
            for (l=0;l<n;l++) SWAP(a[irow][l],a[icol][l]);
            for (l=0;l<m;l++) SWAP(b[irow][l],b[icol][l]);
        }
        indxr[i]=irow;                     We are now ready to divide the pivot row by the
        indxc[i]=icol;                         pivot element, located at irow and icol.
        if (a[icol][icol] == 0.0) nrerror("gaussj: Singular Matrix");
        pivinv=1.0/a[icol][icol];
        a[icol][icol]=1.0;
        for (l=0;l<n;l++) a[icol][l] *= pivinv;
        for (l=0;l<m;l++) b[icol][l] *= pivinv;
        for (ll=0;ll<n;ll++)               Next, we reduce the rows...
            if (ll != icol) {              ...except for the pivot one, of course.
```

```
            dum=a[ll][icol];
            a[ll][icol]=0.0;
            for (l=0;l<n;l++) a[ll][l] -= a[icol][l]*dum;
            for (l=0;l<m;l++) b[ll][l] -= b[icol][l]*dum;
        }
    }
```
This is the end of the main loop over columns of the reduction. It only remains to unscramble the solution in view of the column interchanges. We do this by interchanging pairs of columns in the reverse order that the permutation was built up.
```
    for (l=n-1;l>=0;l--) {
        if (indxr[l] != indxc[l])
            for (k=0;k<n;k++)
                SWAP(a[k][indxr[l]],a[k][indxc[l]]);
    }                                          And we are done.
}
```

Row versus Column Elimination Strategies

The above discussion can be amplified by a modest amount of formalism. Row operations on a matrix $\mathbf{A}$ correspond to pre- (that is, left-) multiplication by some simple matrix $\mathbf{R}$. For example, the matrix $\mathbf{R}$ with components

$$R_{ij} = \begin{cases} 1 & \text{if } i = j \text{ and } i \neq 2, 4 \\ 1 & \text{if } i = 2, j = 4 \\ 1 & \text{if } i = 4, j = 2 \\ 0 & \text{otherwise} \end{cases} \tag{2.1.5}$$

effects the interchange of rows 2 and 4. Gauss-Jordan elimination by row operations alone (including the possibility of *partial* pivoting) consists of a series of such left-multiplications, yielding successively

$$\mathbf{A} \cdot \mathbf{x} = \mathbf{b}$$

$$(\cdots \mathbf{R}_2 \cdot \mathbf{R}_1 \cdot \mathbf{R}_0 \cdot \mathbf{A}) \cdot \mathbf{x} = \cdots \mathbf{R}_2 \cdot \mathbf{R}_1 \cdot \mathbf{R}_0 \cdot \mathbf{b}$$

$$(\mathbf{1}) \cdot \mathbf{x} = \cdots \mathbf{R}_2 \cdot \mathbf{R}_1 \cdot \mathbf{R}_0 \cdot \mathbf{b} \tag{2.1.6}$$

$$\mathbf{x} = \cdots \mathbf{R}_2 \cdot \mathbf{R}_1 \cdot \mathbf{R}_0 \cdot \mathbf{b}$$

The key point is that since the $\mathbf{R}$'s build from right to left, the right-hand side is simply transformed at each stage from one vector to another.

Column operations, on the other hand, correspond to post-, or right-, multiplications by simple matrices, call them $\mathbf{C}$. The matrix in equation (2.1.5), if right-multiplied onto a matrix $\mathbf{A}$, will interchange *columns* 2 and 4 of $\mathbf{A}$. Elimination by column operations involves (conceptually) inserting a column operator, *and also its inverse,* between the matrix $\mathbf{A}$ and the unknown vector $\mathbf{x}$:

$$\mathbf{A} \cdot \mathbf{x} = \mathbf{b}$$

$$\mathbf{A} \cdot \mathbf{C}_0 \cdot \mathbf{C}_0^{-1} \cdot \mathbf{x} = \mathbf{b}$$

$$\mathbf{A} \cdot \mathbf{C}_0 \cdot \mathbf{C}_1 \cdot \mathbf{C}_1^{-1} \cdot \mathbf{C}_0^{-1} \cdot \mathbf{x} = \mathbf{b} \tag{2.1.7}$$

$$(\mathbf{A} \cdot \mathbf{C}_0 \cdot \mathbf{C}_1 \cdot \mathbf{C}_2 \cdots) \cdots \mathbf{C}_2^{-1} \cdot \mathbf{C}_1^{-1} \cdot \mathbf{C}_0^{-1} \cdot \mathbf{x} = \mathbf{b}$$

$$(\mathbf{1}) \cdots \mathbf{C}_2^{-1} \cdot \mathbf{C}_1^{-1} \cdot \mathbf{C}_0^{-1} \cdot \mathbf{x} = \mathbf{b}$$

which (peeling of the $\mathbf{C}^{-1}$'s one at a time) implies a solution

$$\mathbf{x} = \mathbf{C}_0 \cdot \mathbf{C}_1 \cdot \mathbf{C}_2 \cdots \mathbf{b} \tag{2.1.8}$$

Notice the essential difference between equation (2.1.8) and equation (2.1.6). In the latter case, the $\mathbf{C}$'s must be applied to $\mathbf{b}$ in the *reverse order* from that in which they become known. That is, they must all be stored along the way. This requirement greatly reduces the usefulness of column operations, generally restricting them to simple permutations, for example in support of full pivoting.

CITED REFERENCES AND FURTHER READING:

Wilkinson, J.H. 1965, *The Algebraic Eigenvalue Problem* (New York: Oxford University Press). [1]

Carnahan, B., Luther, H.A., and Wilkes, J.O. 1969, *Applied Numerical Methods* (New York: Wiley), Example 5.2, p. 282.

Bevington, P.R., and Robinson, D.K. 1992, *Data Reduction and Error Analysis for the Physical Sciences*, 2nd ed. (New York: McGraw-Hill), p. 247.

Westlake, J.R. 1968, *A Handbook of Numerical Matrix Inversion and Solution of Linear Equations* (New York: Wiley).

Ralston, A., and Rabinowitz, P. 1978, *A First Course in Numerical Analysis*, 2nd ed.; reprinted 2001 (New York: Dover), §9.3–1.

2.2 Gaussian Elimination with Backsubstitution

The usefulness of Gaussian elimination with backsubstitution is primarily pedagogical. It stands between full elimination schemes such as Gauss-Jordan, and triangular decomposition schemes such as will be discussed in the next section. Gaussian elimination reduces a matrix not all the way to the identity matrix, but only halfway, to a matrix whose components on the diagonal and above (say) remain nontrivial. Let us now see what advantages accrue.

Suppose that in doing Gauss-Jordan elimination, as described in §2.1, we at each stage subtract away rows only *below* the then-current pivot element. When a_{11} is the pivot element, for example, we divide the row 1 by its value (as before), but now use the pivot row to zero only a_{21} and a_{31}, not a_{01} (see equation 2.1.1). Suppose, also, that we do only partial pivoting, never interchanging columns, so that the order of the unknowns never needs to be modified.

Then, when we have done this for all the pivots, we will be left with a reduced equation that looks like this (in the case of a single right-hand side vector):

$$\begin{bmatrix} a'_{00} & a'_{01} & a'_{02} & a'_{03} \\ 0 & a'_{11} & a'_{12} & a'_{13} \\ 0 & 0 & a'_{22} & a'_{23} \\ 0 & 0 & 0 & a'_{33} \end{bmatrix} \cdot \begin{bmatrix} x_0 \\ x_1 \\ x_2 \\ x_3 \end{bmatrix} = \begin{bmatrix} b'_0 \\ b'_1 \\ b'_2 \\ b'_3 \end{bmatrix} \qquad (2.2.1)$$

Here the primes signify that the a's and b's do not have their original numerical values, but have been modified by all the row operations in the elimination to this point. The procedure up to this point is termed *Gaussian elimination*.

Backsubstitution

But how do we solve for the x's? The last x (x_3 in this example) is already isolated, namely

$$x_3 = b'_3 / a'_{33} \qquad (2.2.2)$$

With the last x known we can move to the penultimate x,

$$x_2 = \frac{1}{a'_{22}} [b'_2 - x_3 a'_{23}] \qquad (2.2.3)$$

and then proceed with the x before that one. The typical step is

$$x_i = \frac{1}{a'_{ii}} \left[b'_i - \sum_{j=i+1}^{N-1} a'_{ij} x_j \right] \qquad (2.2.4)$$

The procedure defined by equation (2.2.4) is called *backsubstitution*. The combination of Gaussian elimination and backsubstitution yields a solution to the set of equations.

The advantage of Gaussian elimination and backsubstitution over Gauss-Jordan elimination is simply that the former is faster in raw operations count: The innermost loops of Gauss-Jordan elimination, each containing one subtraction and one multiplication, are executed N^3 and $N^2 M$ times (where there are N equations and M unknowns). The corresponding loops in Gaussian elimination are executed only $\frac{1}{3} N^3$ times (only half the matrix is reduced, and the increasing numbers of predictable zeros reduce the count to one-third), and $\frac{1}{2} N^2 M$ times, respectively. Each backsubstitution of a right-hand side is $\frac{1}{2} N^2$ executions of a similar loop (one multiplication plus one subtraction). For $M \ll N$ (only a few right-hand sides) Gaussian elimination thus has about a factor three advantage over Gauss-Jordan. (We could reduce this advantage to a factor 1.5 by *not* computing the inverse matrix as part of the Gauss-Jordan scheme.)

For computing the inverse matrix (which we can view as the case of $M = N$ right-hand sides, namely the N unit vectors which are the columns of the identity matrix), Gaussian elimination and backsubstitution at first glance require $\frac{1}{3} N^3$ (matrix reduction) $+ \frac{1}{2} N^3$ (right-hand side manipulations) $+ \frac{1}{2} N^3$ (N backsubstitutions) $= \frac{4}{3} N^3$ loop executions, which is more than the N^3 for Gauss-Jordan. However, the unit vectors are quite special in containing all zeros except for one element. If this is taken into account, the right-side manipulations can be reduced to only $\frac{1}{6} N^3$ loop executions, and, for matrix inversion, the two methods have identical efficiencies.

Both Gaussian elimination and Gauss-Jordan elimination share the disadvantage that all right-hand sides must be known in advance. The LU decomposition method in the next section does not share that deficiency, and also has an equally small operations count, both for solution with any number of right-hand sides, and for matrix inversion. For this reason we will not implement the method of Gaussian elimination as a routine.

CITED REFERENCES AND FURTHER READING:

Ralston, A., and Rabinowitz, P. 1978, *A First Course in Numerical Analysis*, 2nd ed.; reprinted 2001 (New York: Dover), §9.3–1.

Isaacson, E., and Keller, H.B. 1966, *Analysis of Numerical Methods*; reprinted 1994 (New York: Dover), §2.1.

Johnson, L.W., and Riess, R.D. 1982, *Numerical Analysis*, 2nd ed. (Reading, MA: Addison-Wesley), §2.2.1.

Westlake, J.R. 1968, *A Handbook of Numerical Matrix Inversion and Solution of Linear Equations* (New York: Wiley).

2.3 LU Decomposition and Its Applications

Suppose we are able to write the matrix $\mathbf{A}$ as a product of two matrices,

$$\mathbf{L} \cdot \mathbf{U} = \mathbf{A} \tag{2.3.1}$$

where $\mathbf{L}$ is *lower triangular* (has elements only on the diagonal and below) and $\mathbf{U}$ is *upper triangular* (has elements only on the diagonal and above). For the case of a 4×4 matrix $\mathbf{A}$, for example, equation (2.3.1) would look like this:

$$\begin{bmatrix} \alpha_{00} & 0 & 0 & 0 \\ \alpha_{10} & \alpha_{11} & 0 & 0 \\ \alpha_{20} & \alpha_{21} & \alpha_{22} & 0 \\ \alpha_{30} & \alpha_{31} & \alpha_{32} & \alpha_{33} \end{bmatrix} \cdot \begin{bmatrix} \beta_{00} & \beta_{01} & \beta_{02} & \beta_{03} \\ 0 & \beta_{11} & \beta_{12} & \beta_{13} \\ 0 & 0 & \beta_{22} & \beta_{23} \\ 0 & 0 & 0 & \beta_{33} \end{bmatrix} = \begin{bmatrix} a_{00} & a_{01} & a_{02} & a_{03} \\ a_{10} & a_{11} & a_{12} & a_{13} \\ a_{20} & a_{21} & a_{22} & a_{23} \\ a_{30} & a_{31} & a_{32} & a_{33} \end{bmatrix}$$

$$\tag{2.3.2}$$

We can use a decomposition such as (2.3.1) to solve the linear set

$$\mathbf{A} \cdot \mathbf{x} = (\mathbf{L} \cdot \mathbf{U}) \cdot \mathbf{x} = \mathbf{L} \cdot (\mathbf{U} \cdot \mathbf{x}) = \mathbf{b} \tag{2.3.3}$$

by first solving for the vector $\mathbf{y}$ such that

$$\mathbf{L} \cdot \mathbf{y} = \mathbf{b} \tag{2.3.4}$$

and then solving

$$\mathbf{U} \cdot \mathbf{x} = \mathbf{y} \tag{2.3.5}$$

What is the advantage of breaking up one linear set into two successive ones? The advantage is that the solution of a triangular set of equations is quite trivial, as we have already seen in §2.2 (equation 2.2.4). Thus, equation (2.3.4) can be solved by *forward substitution* as follows,

$$
\begin{aligned}
y_0 &= \frac{b_0}{\alpha_{00}} \\
y_i &= \frac{1}{\alpha_{ii}} \left[b_i - \sum_{j=0}^{i-1} \alpha_{ij} y_j \right] \qquad i = 1, 2, \ldots, N-1
\end{aligned}
\tag{2.3.6}
$$

while (2.3.5) can then be solved by *backsubstitution* exactly as in equations (2.2.2)–(2.2.4),

$$
\begin{aligned}
x_{N-1} &= \frac{y_{N-1}}{\beta_{N-1,N-1}} \\
x_i &= \frac{1}{\beta_{ii}} \left[y_i - \sum_{j=i+1}^{N-1} \beta_{ij} x_j \right] \qquad i = N-2, N-3, \ldots, 0
\end{aligned}
\tag{2.3.7}
$$

Equations (2.3.6) and (2.3.7) total (for each right-hand side **b**) N^2 executions of an inner loop containing one multiply and one add. If we have N right-hand sides which are the unit column vectors (which is the case when we are inverting a matrix), then taking into account the leading zeros reduces the total execution count of (2.3.6) from $\frac{1}{2}N^3$ to $\frac{1}{6}N^3$, while (2.3.7) is unchanged at $\frac{1}{2}N^3$.

Notice that, once we have the *LU* decomposition of **A**, we can solve with as many right-hand sides as we then care to, one at a time. This is a distinct advantage over the methods of §2.1 and §2.2.

Performing the LU Decomposition

How then can we solve for **L** and **U**, given **A**? First, we write out the i,jth component of equation (2.3.1) or (2.3.2). That component always is a sum beginning with

$$\alpha_{i0}\beta_{0j} + \cdots = a_{ij}$$

The number of terms in the sum depends, however, on whether i or j is the smaller number. We have, in fact, the three cases,

$$i < j: \qquad \alpha_{i0}\beta_{0j} + \alpha_{i1}\beta_{1j} + \cdots + \alpha_{ii}\beta_{ij} = a_{ij} \qquad (2.3.8)$$

$$i = j: \qquad \alpha_{i0}\beta_{0j} + \alpha_{i1}\beta_{1j} + \cdots + \alpha_{ii}\beta_{jj} = a_{ij} \qquad (2.3.9)$$

$$i > j: \qquad \alpha_{i0}\beta_{0j} + \alpha_{i1}\beta_{1j} + \cdots + \alpha_{ij}\beta_{jj} = a_{ij} \qquad (2.3.10)$$

Equations (2.3.8)–(2.3.10) total N^2 equations for the $N^2 + N$ unknown α's and β's (the diagonal being represented twice). Since the number of unknowns is greater than the number of equations, we are invited to specify N of the unknowns arbitrarily and then try to solve for the others. In fact, as we shall see, it is always possible to take

$$\alpha_{ii} \equiv 1 \qquad i = 0, \ldots, N-1 \qquad (2.3.11)$$

A surprising procedure, now, is *Crout's algorithm*, which quite trivially solves the set of $N^2 + N$ equations (2.3.8)–(2.3.11) for all the α's and β's by just arranging the equations in a certain order! That order is as follows:

- Set $\alpha_{ii} = 1$, $i = 0, \ldots, N-1$ (equation 2.3.11).
- For each $j = 0, 1, 2, \ldots, N-1$ do these two procedures: First, for $i = 0, 1, \ldots, j$, use (2.3.8), (2.3.9), and (2.3.11) to solve for β_{ij}, namely

$$\beta_{ij} = a_{ij} - \sum_{k=0}^{i-1} \alpha_{ik}\beta_{kj}. \qquad (2.3.12)$$

(When $i = 0$ in 2.3.12 the summation term is taken to mean zero.) Second, for $i = j+1, j+2, \ldots, N-1$ use (2.3.10) to solve for α_{ij}, namely

$$\alpha_{ij} = \frac{1}{\beta_{jj}}\left(a_{ij} - \sum_{k=0}^{j-1} \alpha_{ik}\beta_{kj}\right). \qquad (2.3.13)$$

Be sure to do both procedures before going on to the next j.

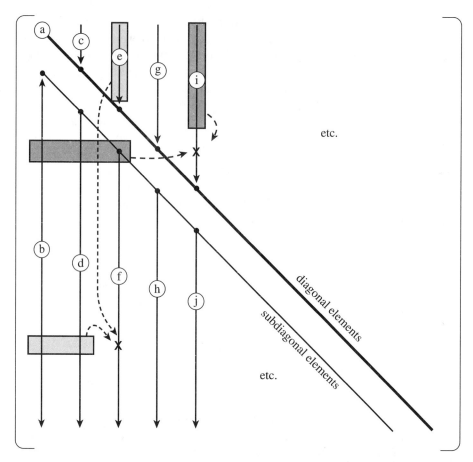

Figure 2.3.1.   Crout's algorithm for *LU* decomposition of a matrix. Elements of the original matrix are modified in the order indicated by lower case letters: a, b, c, etc. Shaded boxes show the previously modified elements that are used in modifying two typical elements, each indicated by an "x".

If you work through a few iterations of the above procedure, you will see that the α's and β's that occur on the right-hand side of equations (2.3.12) and (2.3.13) are already determined by the time they are needed. You will also see that every a_{ij} is used only once and never again. This means that the corresponding α_{ij} or β_{ij} can be stored in the location that the a used to occupy: the decomposition is "in place." [The diagonal unity elements α_{ii} (equation 2.3.11) are not stored at all.] In brief, Crout's method fills in the combined matrix of α's and β's,

$$\begin{bmatrix} \beta_{00} & \beta_{01} & \beta_{02} & \beta_{03} \\ \alpha_{10} & \beta_{11} & \beta_{12} & \beta_{13} \\ \alpha_{20} & \alpha_{21} & \beta_{22} & \beta_{23} \\ \alpha_{30} & \alpha_{31} & \alpha_{32} & \beta_{33} \end{bmatrix} \tag{2.3.14}$$

by columns from left to right, and within each column from top to bottom (see Figure 2.3.1).

What about pivoting? Pivoting (i.e., selection of a salubrious pivot element for the division in equation 2.3.13) is absolutely essential for the stability of Crout's method. Only partial pivoting (interchange of rows) can be implemented efficiently. However this is enough to make the method stable. This means, incidentally, that we don't actually decompose the matrix **A** into *LU* form, but rather we decompose a rowwise permutation of **A**. (If we keep track of what that permutation is, this decomposition is just as useful as the original one would have been.)

Pivoting is slightly subtle in Crout's algorithm. The key point to notice is that equation (2.3.12) in the case of $i = j$ (its final application) is *exactly the same* as equation (2.3.13) except for the division in the latter equation; in both cases the upper limit of the sum is $k = j - 1 (= i - 1)$. This means that we don't have to commit ourselves as to whether the diagonal element β_{jj} is the one that happens to fall on the diagonal in the first instance, or whether one of the (undivided) α_{ij}'s below it in the column, $i = j + 1, \ldots, N - 1$, is to be "promoted" to become the diagonal β. This can be decided after all the candidates in the column are in hand. As you should be able to guess by now, we will choose the largest one as the diagonal β (pivot element), then do all the divisions by that element *en masse*. This is *Crout's method with partial pivoting*. Our implementation has one additional wrinkle: It initially finds the largest element in each row, and subsequently (when it is looking for the maximal pivot element) scales the comparison *as if* we had initially scaled all the equations to make their maximum coefficient equal to unity; this is the *implicit pivoting* mentioned in §2.1.

```cpp
#include <cmath>
#include "nr.h"
using namespace std;

void NR::ludcmp(Mat_IO_DP &a, Vec_O_INT &indx, DP &d)
```
Given a matrix a[0..n-1][0..n-1], this routine replaces it by the *LU* decomposition of a rowwise permutation of itself. a is input. On output, it is arranged as in equation (2.3.14) above; indx[0..n-1] is an output vector that records the row permutation effected by the partial pivoting; d is output as ±1 depending on whether the number of row interchanges was even or odd, respectively. This routine is used in combination with lubksb to solve linear equations or invert a matrix.
```cpp
{
    const DP TINY=1.0e-20;              A small number.
    int i,imax,j,k;
    DP big,dum,sum,temp;

    int n=a.nrows();
    Vec_DP vv(n);                       vv stores the implicit scaling of each row.
    d=1.0;                              No row interchanges yet.
    for (i=0;i<n;i++) {                 Loop over rows to get the implicit scaling informa-
        big=0.0;                            tion.
        for (j=0;j<n;j++)
            if ((temp=fabs(a[i][j])) > big) big=temp;
        if (big == 0.0) nrerror("Singular matrix in routine ludcmp");
        No nonzero largest element.
        vv[i]=1.0/big;                  Save the scaling.
    }
    for (j=0;j<n;j++) {                 This is the loop over columns of Crout's method.
        for (i=0;i<j;i++) {            This is equation (2.3.12) except for i = j.
            sum=a[i][j];
            for (k=0;k<i;k++) sum -= a[i][k]*a[k][j];
            a[i][j]=sum;
        }
        big=0.0;                        Initialize for the search for largest pivot element.
        for (i=j;i<n;i++) {            This is i = j of equation (2.3.12) and i = j + 1...
            sum=a[i][j];                   N − 1 of equation (2.3.13).
```

```
            for (k=0;k<j;k++) sum -= a[i][k]*a[k][j];
            a[i][j]=sum;
            if ((dum=vv[i]*fabs(sum)) >= big) {
```
Is the figure of merit for the pivot better than the best so far?
```
                big=dum;
                imax=i;
            }
        }
        if (j != imax) {                    Do we need to interchange rows?
            for (k=0;k<n;k++) {             Yes, do so...
                dum=a[imax][k];
                a[imax][k]=a[j][k];
                a[j][k]=dum;
            }
            d = -d;                         ...and change the parity of d.
            vv[imax]=vv[j];                 Also interchange the scale factor.
        }
        indx[j]=imax;
        if (a[j][j] == 0.0) a[j][j]=TINY;
```
If the pivot element is zero the matrix is singular (at least to the precision of the algorithm). For some applications on singular matrices, it is desirable to substitute TINY for zero.
```
        if (j != n-1) {                     Now, finally, divide by the pivot element.
            dum=1.0/(a[j][j]);
            for (i=j+1;i<n;i++) a[i][j] *= dum;
        }
    }                                       Go back for the next column in the reduction.
}
```

Here is the routine for forward substitution and backsubstitution, implementing equations (2.3.6) and (2.3.7).

```
#include "nr.h"

void NR::lubksb(Mat_I_DP &a, Vec_I_INT &indx, Vec_IO_DP &b)
```
Solves the set of n linear equations $A \cdot X = B$. Here a[0..n-1][0..n-1] is input, not as the matrix A but rather as its LU decomposition, determined by the routine ludcmp. indx[0..n-1] is input as the permutation vector returned by ludcmp. b[0..n-1] is input as the right-hand side vector B, and returns with the solution vector X. a and indx are not modified by this routine and can be left in place for successive calls with different right-hand sides b. This routine takes into account the possibility that b will begin with many zero elements, so it is efficient for use in matrix inversion.
```
{
    int i,ii=0,ip,j;
    DP sum;

    int n=a.nrows();
    for (i=0;i<n;i++) {                     When ii is set to a positive value, it will become the
        ip=indx[i];                         index of the first nonvanishing element of b. We now
        sum=b[ip];                          do the forward substitution, equation (2.3.6). The
        b[ip]=b[i];                         only new wrinkle is to unscramble the permutation
        if (ii != 0)                        as we go.
            for (j=ii-1;j<i;j++) sum -= a[i][j]*b[j];
        else if (sum != 0.0)                A nonzero element was encountered, so from now on we
            ii=i+1;                         will have to do the sums in the loop above.
        b[i]=sum;
    }
    for (i=n-1;i>=0;i--) {                  Now we do the backsubstitution, equation (2.3.7).
        sum=b[i];
        for (j=i+1;j<n;j++) sum -= a[i][j]*b[j];
        b[i]=sum/a[i][i];                   Store a component of the solution vector X.
    }                                       All done!
}
```

The *LU* decomposition in `ludcmp` requires about $\frac{1}{3}N^3$ executions of the inner loops (each with one multiply and one add). This is thus the operation count for solving one (or a few) right-hand sides, and is a factor of 3 better than the Gauss-Jordan routine `gaussj` which was given in §2.1, and a factor of 1.5 better than a Gauss-Jordan routine (not given) that does not compute the inverse matrix. For inverting a matrix, the total count (including the forward and backsubstitution as discussed following equation 2.3.7 above) is $(\frac{1}{3} + \frac{1}{6} + \frac{1}{2})N^3 = N^3$, the same as `gaussj`.

To summarize, this is the preferred way to solve the linear set of equations $\mathbf{A} \cdot \mathbf{x} = \mathbf{b}$:

```
const int N = ...
Mat_DP a(N,N);
Vec_DP b(N);
Vec_INT indx(N);
DP d;
...
NR::ludcmp(a,indx,d);
NR::lubksb(a,indx,b);
```

The answer **x** will be given back in b. Your original matrix **A** will have been destroyed.

If you subsequently want to solve a set of equations with the same **A** but a different right-hand side **b**, you repeat *only*

```
NR::lubksb(a,indx,b);
```

not, of course, with the original matrix **A**, but with a and indx as were already set by `ludcmp`.

Inverse of a Matrix

Using the above *LU* decomposition and backsubstitution routines, it is completely straightforward to find the inverse of a matrix column by column.

```
const int N = ...
Mat_DP a(N,N),y(N,N);
Vec_DP col(N);
Vec_INT indx(N);
DP d;
int i,j;
...
NR::ludcmp(a,indx,d);            Decompose the matrix just once.
for(j=0;j<N;j++) {              Find inverse by columns.
    for(i=0;i<N;i++) col[i]=0.0;
    col[j]=1.0;
    NR::lubksb(a,indx,col);
    for(i=0;i<N;i++) y[i][j]=col[i];
}
```

The matrix y will now contain the inverse of the original matrix a, which will have been destroyed. Alternatively, there is nothing wrong with using a Gauss-Jordan routine like gaussj (§2.1) to invert a matrix in place, again destroying the original. Both methods have practically the same operations count.

Incidentally, if you ever have the need to compute $\mathbf{A}^{-1} \cdot \mathbf{B}$ from matrices $\mathbf{A}$ and $\mathbf{B}$, you should LU decompose $\mathbf{A}$ and then backsubstitute with the columns of $\mathbf{B}$ instead of with the unit vectors that would give $\mathbf{A}$'s inverse. This saves a whole matrix multiplication, and is also more accurate.

Determinant of a Matrix

The determinant of an LU decomposed matrix is just the product of the diagonal elements,

$$\det = \prod_{j=0}^{N-1} \beta_{jj} \tag{2.3.15}$$

We don't, recall, compute the decomposition of the original matrix, but rather a decomposition of a rowwise permutation of it. Luckily, we have kept track of whether the number of row interchanges was even or odd, so we just preface the product by the corresponding sign. (You now finally know the purpose of setting d in the routine ludcmp.)

Calculation of a determinant thus requires one call to ludcmp, with *no* subsequent backsubstitutions by lubksb.

```
const int N = ...
Mat_DP a(N,N);
Vec_INT indx(N);
DP d;
int j;
...
NR::ludcmp(a,indx,d);        This returns d as ±1.
for(j=0;j<N;j++) d *= a[j][j];
```

The variable d now contains the determinant of the original matrix a, which will have been destroyed.

For a matrix of any substantial size, it is quite likely that the determinant will overflow or underflow your computer's floating-point dynamic range. In this case you can modify the loop of the above fragment and (e.g.) divide by powers of ten, to keep track of the scale separately, or (e.g.) accumulate the sum of logarithms of the absolute values of the factors and the sign separately.

Complex Systems of Equations

If your matrix $\mathbf{A}$ is real, but the right-hand side vector is complex, say $\mathbf{b} + i\mathbf{d}$, then (i) LU decompose $\mathbf{A}$ in the usual way, (ii) backsubstitute $\mathbf{b}$ to get the real part of the solution vector, and (iii) backsubstitute $\mathbf{d}$ to get the imaginary part of the solution vector.

If the matrix itself is complex, so that you want to solve the system

$$(\mathbf{A} + i\mathbf{C}) \cdot (\mathbf{x} + i\mathbf{y}) = (\mathbf{b} + i\mathbf{d}) \tag{2.3.16}$$

then there are two possible ways to proceed. The best way is to rewrite `ludcmp` and `lubksb` as complex routines. Complex modulus substitutes for absolute value in the construction of the scaling vector vv and in the search for the largest pivot elements. Everything else goes through in the obvious way, with complex arithmetic used as needed.

A quick-and-dirty way to solve complex systems is to take the real and imaginary parts of (2.3.16), giving

$$\mathbf{A} \cdot \mathbf{x} - \mathbf{C} \cdot \mathbf{y} = \mathbf{b}$$
$$\mathbf{C} \cdot \mathbf{x} + \mathbf{A} \cdot \mathbf{y} = \mathbf{d} \tag{2.3.17}$$

which can be written as a $2N \times 2N$ set of *real* equations,

$$\begin{pmatrix} \mathbf{A} & -\mathbf{C} \\ \mathbf{C} & \mathbf{A} \end{pmatrix} \cdot \begin{pmatrix} \mathbf{x} \\ \mathbf{y} \end{pmatrix} = \begin{pmatrix} \mathbf{b} \\ \mathbf{d} \end{pmatrix} \tag{2.3.18}$$

and then solved with `ludcmp` and `lubksb` in their present forms. This scheme is a factor of 2 inefficient in storage, since $\mathbf{A}$ and $\mathbf{C}$ are stored twice. It is also a factor of 2 inefficient in time, since the complex multiplies in a complexified version of the routines would each use 4 real multiplies, while the solution of a $2N \times 2N$ problem involves 8 times the work of an $N \times N$ one. If you can tolerate these factor-of-two inefficiencies, then equation (2.3.18) is an easy way to proceed.

CITED REFERENCES AND FURTHER READING:

Golub, G.H., and Van Loan, C.F. 1996, *Matrix Computations*, 3rd ed. (Baltimore: Johns Hopkins University Press), Chapter 4.

Anderson, E., et al. 2000, LAPACK User's Guide, 3rd ed. (Philadelphia: S.I.A.M.).

Forsythe, G.E., Malcolm, M.A., and Moler, C.B. 1977, *Computer Methods for Mathematical Computations* (Englewood Cliffs, NJ: Prentice-Hall), §3.3, and p. 50.

Forsythe, G.E., and Moler, C.B. 1967, *Computer Solution of Linear Algebraic Systems* (Englewood Cliffs, NJ: Prentice-Hall), Chapters 9, 16, and 18.

Westlake, J.R. 1968, *A Handbook of Numerical Matrix Inversion and Solution of Linear Equations* (New York: Wiley).

Stoer, J., and Bulirsch, R. 1993, *Introduction to Numerical Analysis*, 2nd ed. (New York: Springer-Verlag), §4.2.

Ralston, A., and Rabinowitz, P. 1978, *A First Course in Numerical Analysis*, 2nd ed.; reprinted 2001 (New York: Dover), §9.11.

Horn, R.A., and Johnson, C.R. 1985, *Matrix Analysis* (Cambridge: Cambridge University Press).

2.4 Tridiagonal and Band Diagonal Systems of Equations

The special case of a system of linear equations that is *tridiagonal*, that is, has nonzero elements only on the diagonal plus or minus one column, is one that occurs frequently. Also common are systems that are *band diagonal*, with nonzero elements only along a few diagonal lines adjacent to the main diagonal (above and below).

For tridiagonal sets, the procedures of LU decomposition, forward- and backsubstitution each take only $O(N)$ operations, and the whole solution can be encoded very concisely. The resulting routine `tridag` is one that we will use in later chapters.

Naturally, one does not reserve storage for the full $N \times N$ matrix, but only for the nonzero components, stored as three vectors. The set of equations to be solved is

$$
\begin{bmatrix}
b_0 & c_0 & 0 & \cdots \\
a_1 & b_1 & c_1 & \cdots \\
& & \cdots \\
& \cdots & a_{N-2} & b_{N-2} & c_{N-2} \\
& \cdots & 0 & a_{N-1} & b_{N-1}
\end{bmatrix}
\cdot
\begin{bmatrix}
u_0 \\
u_1 \\
\cdots \\
u_{N-2} \\
u_{N-1}
\end{bmatrix}
=
\begin{bmatrix}
r_0 \\
r_1 \\
\cdots \\
r_{N-2} \\
r_{N-1}
\end{bmatrix}
\qquad (2.4.1)
$$

Notice that a_0 and c_{N-1} are undefined and are not referenced by the routine that follows.

```
#include "nr.h"

void NR::tridag(Vec_I_DP &a, Vec_I_DP &b, Vec_I_DP &c, Vec_I_DP &r, Vec_O_DP &u)
Solves for a vector u[0..n-1] the tridiagonal linear set given by equation (2.4.1). a[0..n-1],
b[0..n-1], c[0..n-1], and r[0..n-1] are input vectors and are not modified.
{
    int j;
    DP bet;

    int n=a.size();
    Vec_DP gam(n);                              One vector of workspace, gam is needed.
    if (b[0] == 0.0) nrerror("Error 1 in tridag");
    If this happens then you should rewrite your equations as a set of order N − 1, with u₁
    trivially eliminated.
    u[0]=r[0]/(bet=b[0]);
    for (j=1;j<n;j++) {                         Decomposition and forward substitution.
        gam[j]=c[j-1]/bet;
        bet=b[j]-a[j]*gam[j];
        if (bet == 0.0) nrerror("Error 2 in tridag");    Algorithm fails; see below.
        u[j]=(r[j]-a[j]*u[j-1])/bet;
    }
    for (j=(n-2);j>=0;j--)
        u[j] -= gam[j+1]*u[j+1];               Backsubstitution.
}
```

There is no pivoting in `tridag`. It is for this reason that `tridag` can fail even when the underlying matrix is nonsingular: A zero pivot can be encountered even for a nonsingular matrix. In practice, this is not something to lose sleep about. The kinds of problems that lead to tridiagonal linear sets usually have additional properties which guarantee that the algorithm in `tridag` will succeed. For example, if

$$
|b_j| > |a_j| + |c_j| \qquad j = 0, \ldots, N - 1 \qquad (2.4.2)
$$

(called *diagonal dominance*) then it can be shown that the algorithm cannot encounter a zero pivot.

It is possible to construct special examples in which the lack of pivoting in the algorithm causes numerical instability. In practice, however, such instability is almost never encountered — unlike the general matrix problem where pivoting is essential.

The tridiagonal algorithm is the rare case of an algorithm that, in practice, is more robust than theory says it should be. Of course, should you ever encounter a problem for which `tridag` fails, you can instead use the more general method for band diagonal systems, now described (routines `bandec` and `banbks`).

Some other matrix forms consisting of tridiagonal with a small number of additional elements (e.g., upper right and lower left corners) also allow rapid solution; see §2.7.

Band Diagonal Systems

Where tridiagonal systems have nonzero elements only on the diagonal plus or minus one, band diagonal systems are slightly more general and have (say) $m_1 \geq 0$ nonzero elements immediately to the left of (below) the diagonal and $m_2 \geq 0$ nonzero elements immediately to its right (above it). Of course, this is only a useful classification if m_1 and m_2 are both $\ll N$. In that case, the solution of the linear system by LU decomposition can be accomplished much faster, and in much less storage, than for the general $N \times N$ case.

The precise definition of a band diagonal matrix with elements a_{ij} is that

$$a_{ij} = 0 \quad \text{when} \quad j > i + m_2 \quad \text{or} \quad i > j + m_1 \qquad (2.4.3)$$

Band diagonal matrices are stored and manipulated in a so-called compact form, which results if the matrix is tilted $45°$ clockwise, so that its nonzero elements lie in a long, narrow matrix with $m_1 + 1 + m_2$ columns and N rows. This is best illustrated by an example: The band diagonal matrix

$$\begin{pmatrix} 3 & 1 & 0 & 0 & 0 & 0 & 0 \\ 4 & 1 & 5 & 0 & 0 & 0 & 0 \\ 9 & 2 & 6 & 5 & 0 & 0 & 0 \\ 0 & 3 & 5 & 8 & 9 & 0 & 0 \\ 0 & 0 & 7 & 9 & 3 & 2 & 0 \\ 0 & 0 & 0 & 3 & 8 & 4 & 6 \\ 0 & 0 & 0 & 0 & 2 & 4 & 4 \end{pmatrix} \qquad (2.4.4)$$

which has $N = 7$, $m_1 = 2$, and $m_2 = 1$, is stored compactly as the 7×4 matrix,

$$\begin{pmatrix} x & x & 3 & 1 \\ x & 4 & 1 & 5 \\ 9 & 2 & 6 & 5 \\ 3 & 5 & 8 & 9 \\ 7 & 9 & 3 & 2 \\ 3 & 8 & 4 & 6 \\ 2 & 4 & 4 & x \end{pmatrix} \qquad (2.4.5)$$

Here x denotes elements that are wasted space in the compact format; these will not be referenced by any manipulations and can have arbitrary values. Notice that the diagonal of the original matrix appears in column m_1, with subdiagonal elements to its left, superdiagonal elements to its right.

The simplest manipulation of a band diagonal matrix, stored compactly, is to multiply it by a vector to its right. Although this is algorithmically trivial, you might want to study the following routine carefully, as an example of how to pull nonzero elements a_{ij} out of the compact storage format in an orderly fashion.

```
#include "nr.h"

void NR::banmul(Mat_I_DP &a, const int m1, const int m2, Vec_I_DP &x,
    Vec_O_DP &b)
```
Matrix multiply $\mathbf{b} = \mathbf{A} \cdot \mathbf{x}$, where $\mathbf{A}$ is band diagonal with m1 rows below the diagonal and m2 rows above. The input vector $\mathbf{x}$ and output vector $\mathbf{b}$ are stored as x[0..n-1] and b[0..n-1], respectively. The array a[0..n-1][0..m1+m2] stores $\mathbf{A}$ as follows: The diagonal elements are in a[0..n-1][m1]. Subdiagonal elements are in a[j..n-1][0..m1-1] (with $j > 0$ appropriate to the number of elements on each subdiagonal). Superdiagonal elements are in a[0..j][m1+1..m1+m2] with $j < $ n-1 appropriate to the number of elements on each superdiagonal.
```
{
```

```
int i,j,k,tmploop;

int n=a.nrows();
for (i=0;i<n;i++) {
    k=i-m1;
    tmploop=MIN(m1+m2+1,int(n-k));
    b[i]=0.0;
    for (j=MAX(0,-k);j<tmploop;j++) b[i] += a[i][j]*x[j+k];
}
}
```

It is not possible to store the LU decomposition of a band diagonal matrix **A** quite as compactly as the compact form of **A** itself. The decomposition (essentially by Crout's method, see §2.3) produces additional nonzero "fill-ins." One straightforward storage scheme is to return the upper triangular factor (U) in the same space that **A** previously occupied, and to return the lower triangular factor (L) in a separate compact matrix of size $N \times m_1$. The diagonal elements of U (whose product, times $d = \pm 1$, gives the determinant) are returned in the first column of **A**'s storage space.

The following routine, bandec, is the band-diagonal analog of ludcmp in §2.3:

```
#include <cmath>
#include "nr.h"
using namespace std;

void NR::bandec(Mat_IO_DP &a, const int m1, const int m2, Mat_O_DP &al,
    Vec_O_INT &indx, DP &d)
```
Given an n × n band diagonal matrix **A** with m1 subdiagonal rows and m2 superdiagonal rows, compactly stored in the array a[0..n-1][0..m1+m2] as described in the comment for routine banmul, this routine constructs an LU decomposition of a rowwise permutation of **A**. The upper triangular matrix replaces a, while the lower triangular matrix is returned in al[0..n-1][0..m1-1]. indx[0..n-1] is an output vector that records the row permutation effected by the partial pivoting; d is output as ±1 depending on whether the number of row interchanges was even or odd, respectively. This routine is used in combination with banbks to solve band-diagonal sets of equations.
```
{
    const DP TINY=1.0e-20;
    int i,j,k,l,mm;
    DP dum;

    int n=a.nrows();
    mm=m1+m2+1;
    l=m1;
    for (i=0;i<m1;i++) {                    Rearrange the storage a bit.
        for (j=m1-i;j<mm;j++) a[i][j-l]=a[i][j];
        l--;
        for (j=mm-l-1;j<mm;j++) a[i][j]=0.0;
    }
    d=1.0;
    l=m1;
    for (k=0;k<n;k++) {                     For each row...
        dum=a[k][0];
        i=k;
        if (l<n) l++;
        for (j=k+1;j<l;j++) {               Find the pivot element.
            if (fabs(a[j][0]) > fabs(dum)) {
                dum=a[j][0];
                i=j;
            }
        }
        indx[k]=i+1;
        if (dum == 0.0) a[k][0]=TINY;
```

Matrix is algorithmically singular, but proceed anyway with TINY pivot (desirable in some applications).

```
    if (i != k) {                         Interchange rows.
        d = -d;
        for (j=0;j<mm;j++) SWAP(a[k][j],a[i][j]);
    }
    for (i=k+1;i<l;i++) {                 Do the elimination.
        dum=a[i][0]/a[k][0];
        al[k][i-k-1]=dum;
        for (j=1;j<mm;j++) a[i][j-1]=a[i][j]-dum*a[k][j];
        a[i][mm-1]=0.0;
    }
    }
}
```

Some pivoting is possible within the storage limitations of bandec, and the above routine does take advantage of the opportunity. In general, when TINY is returned as a diagonal element of U, then the original matrix (perhaps as modified by roundoff error) is in fact singular. In this regard, bandec is somewhat more robust than tridag above, which can fail algorithmically even for nonsingular matrices; bandec is thus also useful (with $m_1 = m_2 = 1$) for some ill-behaved tridiagonal systems.

Once the matrix **A** has been decomposed, any number of right-hand sides can be solved in turn by repeated calls to banbks, the backsubstitution routine whose analog in §2.3 is lubksb.

```
#include "nr.h"

void NR::banbks(Mat_I_DP &a, const int m1, const int m2, Mat_I_DP &al,
    Vec_I_INT &indx, Vec_IO_DP &b)
```
Given the arrays a, al, and indx as returned from bandec, and given a right-hand side vector b[0..n-1], solves the band diagonal linear equations $\mathbf{A} \cdot \mathbf{x} = \mathbf{b}$. The solution vector x overwrites b[0..n-1]. The other input arrays are not modified, and can be left in place for successive calls with different right-hand sides.
```
{
    int i,j,k,l,mm;
    DP dum;

    int n=a.nrows();
    mm=m1+m2+1;
    l=m1;
    for (k=0;k<n;k++) {                   Forward substitution, unscrambling the permuted rows
        j=indx[k]-1;                          as we go.
        if (j!=k) SWAP(b[k],b[j]);
        if (l<n) l++;
        for (j=k+1;j<l;j++) b[j] -= al[k][j-k-1]*b[k];
    }
    l=1;
    for (i=n-1;i>=0;i--) {                Backsubstitution.
        dum=b[i];
        for (k=1;k<l;k++) dum -= a[i][k]*b[k+i];
        b[i]=dum/a[i][0];
        if (l<mm) l++;
    }
}
```

The routines bandec and banbks are based on the Handbook routines *bandet1* and *bansol1* in [1].

CITED REFERENCES AND FURTHER READING:

Keller, H.B. 1968, *Numerical Methods for Two-Point Boundary-Value Problems*; reprinted 1991 (New York: Dover), p. 74.

Dahlquist, G., and Bjorck, A. 1974, *Numerical Methods* (Englewood Cliffs, NJ: Prentice-Hall), Example 5.4.3, p. 166.

Ralston, A., and Rabinowitz, P. 1978, *A First Course in Numerical Analysis*, 2nd ed.; reprinted 2001 (New York: Dover), §9.11.

Wilkinson, J.H., and Reinsch, C. 1971, *Linear Algebra*, vol. II of *Handbook for Automatic Computation* (New York: Springer-Verlag), Chapter I/6. [1]

Golub, G.H., and Van Loan, C.F. 1996, *Matrix Computations*, 3rd ed. (Baltimore: Johns Hopkins University Press), §4.3.

2.5 Iterative Improvement of a Solution to Linear Equations

Obviously it is not easy to obtain greater precision for the solution of a linear set than the precision of your computer's floating-point word. Unfortunately, for large sets of linear equations, it is not always easy to obtain precision equal to, or even comparable to, the computer's limit. In direct methods of solution, roundoff errors accumulate, and they are magnified to the extent that your matrix is close to singular. You can easily lose two or three significant figures for matrices which (you thought) were *far* from singular.

If this happens to you, there is a neat trick to restore the full machine precision, called *iterative improvement* of the solution. The theory is very straightforward (see Figure 2.5.1): Suppose that a vector $\mathbf{x}$ is the exact solution of the linear set

$$\mathbf{A} \cdot \mathbf{x} = \mathbf{b} \qquad (2.5.1)$$

You don't, however, know $\mathbf{x}$. You only know some slightly wrong solution $\mathbf{x} + \delta\mathbf{x}$, where $\delta\mathbf{x}$ is the unknown error. When multiplied by the matrix $\mathbf{A}$, your slightly wrong solution gives a product slightly discrepant from the desired right-hand side $\mathbf{b}$, namely

$$\mathbf{A} \cdot (\mathbf{x} + \delta\mathbf{x}) = \mathbf{b} + \delta\mathbf{b} \qquad (2.5.2)$$

Subtracting (2.5.1) from (2.5.2) gives

$$\mathbf{A} \cdot \delta\mathbf{x} = \delta\mathbf{b} \qquad (2.5.3)$$

But (2.5.2) can also be solved, trivially, for $\delta\mathbf{b}$. Substituting this into (2.5.3) gives

$$\mathbf{A} \cdot \delta\mathbf{x} = \mathbf{A} \cdot (\mathbf{x} + \delta\mathbf{x}) - \mathbf{b} \qquad (2.5.4)$$

In this equation, the whole right-hand side is known, since $\mathbf{x} + \delta\mathbf{x}$ is the wrong solution that you want to improve. It is essential to calculate the right-hand side in higher precision than the original solution, since there will be a lot of cancellation in the subtraction of $\mathbf{b}$. Then, we need only solve (2.5.4) for the error $\delta\mathbf{x}$, then subtract this from the wrong solution to get an improved solution.

An important extra benefit occurs if we obtained the original solution by *LU* decomposition. In this case we already have the *LU* decomposed form of $\mathbf{A}$, and all we need do to solve (2.5.4) is compute the right-hand side and backsubstitute!

The code to do all this is concise and straightforward:

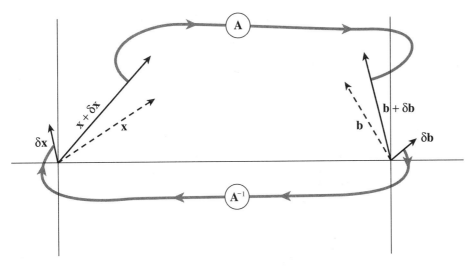

Figure 2.5.1. Iterative improvement of the solution to $\mathbf{A} \cdot \mathbf{x} = \mathbf{b}$. The first guess $\mathbf{x} + \delta\mathbf{x}$ is multiplied by $\mathbf{A}$ to produce $\mathbf{b} + \delta\mathbf{b}$. The known vector $\mathbf{b}$ is subtracted, giving $\delta\mathbf{b}$. The linear set with this right-hand side is inverted, giving $\delta\mathbf{x}$. This is subtracted from the first guess giving an improved solution $\mathbf{x}$.

```
#include "nr.h"

void NR::mprove(Mat_I_DP &a, Mat_I_DP &alud, Vec_I_INT &indx, Vec_I_DP &b,
    Vec_IO_DP &x)
```
Improves a solution vector x[0..n-1] of the linear set of equations $A \cdot X = B$. The matrix a[0..n-1][0..n-1], and the vectors b[0..n-1] and x[0..n-1] are input. Also input is alud[0..n-1][0..n-1], the *LU* decomposition of a as returned by ludcmp, and the vector indx[0..n-1] also returned by that routine. On output, only x[0..n-1] is modified, to an improved set of values.
```
{
    int i,j;

    int n=x.size();
    Vec_DP r(n);
    for (i=0;i<n;i++) {                      Calculate the right-hand side, accumulating
        long double sdp = -b[i];                 the residual in higher precision.
        for (j=0;j<n;j++)
            sdp += (long double) a[i][j]*(long double) x[j];
        r[i]=sdp;
    }
    lubksb(alud,indx,r);                     Solve for the error term,
    for (i=0;i<n;i++) x[i] -= r[i];          and subtract it from the old solution.
}
```

You should note that the routine ludcmp in §2.3 destroys the input matrix as it *LU* decomposes it. Since iterative improvement requires *both* the original matrix and its *LU* decomposition, you will need to copy $\mathbf{A}$ before calling ludcmp. Likewise lubksb destroys $\mathbf{b}$ in obtaining $\mathbf{x}$, so make a copy of $\mathbf{b}$ also. If you don't mind this extra storage, iterative improvement is *highly* recommended: It is a process of order only N^2 operations (multiply vector by matrix, and backsubstitute — see discussion following equation 2.3.7); it never hurts; and it can really give you your money's worth if it saves an otherwise ruined solution on which you have already spent of order N^3 operations.

You can call mprove several times in succession if you want. Unless you are starting quite far from the true solution, one call is generally enough; but a second call to verify convergence can be reassuring.

More on Iterative Improvement

It is illuminating (and will be useful later in the book) to give a somewhat more solid analytical foundation for equation (2.5.4), and also to give some additional results. Implicit in the previous discussion was the notion that the solution vector $\mathbf{x} + \delta\mathbf{x}$ has an error term; but we neglected the fact that the LU decomposition of $\mathbf{A}$ is itself not exact.

A different analytical approach starts with some matrix $\mathbf{B}_0$ that is assumed to be an *approximate* inverse of the matrix $\mathbf{A}$, so that $\mathbf{B}_0 \cdot \mathbf{A}$ is approximately the identity matrix $\mathbf{1}$. Define the *residual matrix* $\mathbf{R}$ of $\mathbf{B}_0$ as

$$\mathbf{R} \equiv \mathbf{1} - \mathbf{B}_0 \cdot \mathbf{A} \tag{2.5.5}$$

which is supposed to be "small" (we will be more precise below). Note that therefore

$$\mathbf{B}_0 \cdot \mathbf{A} = \mathbf{1} - \mathbf{R} \tag{2.5.6}$$

Next consider the following formal manipulation:

$$
\begin{aligned}
\mathbf{A}^{-1} &= \mathbf{A}^{-1} \cdot (\mathbf{B}_0^{-1} \cdot \mathbf{B}_0) = (\mathbf{A}^{-1} \cdot \mathbf{B}_0^{-1}) \cdot \mathbf{B}_0 = (\mathbf{B}_0 \cdot \mathbf{A})^{-1} \cdot \mathbf{B}_0 \\
&= (\mathbf{1} - \mathbf{R})^{-1} \cdot \mathbf{B}_0 = (\mathbf{1} + \mathbf{R} + \mathbf{R}^2 + \mathbf{R}^3 + \cdots) \cdot \mathbf{B}_0
\end{aligned}
\tag{2.5.7}
$$

We can define the nth partial sum of the last expression by

$$\mathbf{B}_n \equiv (\mathbf{1} + \mathbf{R} + \cdots + \mathbf{R}^n) \cdot \mathbf{B}_0 \tag{2.5.8}$$

so that $\mathbf{B}_\infty \to \mathbf{A}^{-1}$, if the limit exists.

It now is straightforward to verify that equation (2.5.8) satisfies some interesting recurrence relations. As regards solving $\mathbf{A} \cdot \mathbf{x} = \mathbf{b}$, where $\mathbf{x}$ and $\mathbf{b}$ are vectors, define

$$\mathbf{x}_n \equiv \mathbf{B}_n \cdot \mathbf{b} \tag{2.5.9}$$

Then it is easy to show that

$$\mathbf{x}_{n+1} = \mathbf{x}_n + \mathbf{B}_0 \cdot (\mathbf{b} - \mathbf{A} \cdot \mathbf{x}_n) \tag{2.5.10}$$

This is immediately recognizable as equation (2.5.4), with $-\delta\mathbf{x} = \mathbf{x}_{n+1} - \mathbf{x}_n$, and with $\mathbf{B}_0$ taking the role of $\mathbf{A}^{-1}$. We see, therefore, that equation (2.5.4) does not require that the LU decomposition of $\mathbf{A}$ be exact, but only that the implied residual $\mathbf{R}$ be small. In rough terms, if the residual is smaller than the square root of your computer's roundoff error, then after one application of equation (2.5.10) (that is, going from $\mathbf{x}_0 \equiv \mathbf{B}_0 \cdot \mathbf{b}$ to $\mathbf{x}_1$) the first neglected term, of order $\mathbf{R}^2$, will be smaller than the roundoff error. Equation (2.5.10), like equation (2.5.4), moreover, can be applied more than once, since it uses only $\mathbf{B}_0$, and not any of the higher $\mathbf{B}$'s.

A much more surprising recurrence which follows from equation (2.5.8) is one that more than *doubles* the order n at each stage:

$$\mathbf{B}_{2n+1} = 2\mathbf{B}_n - \mathbf{B}_n \cdot \mathbf{A} \cdot \mathbf{B}_n \qquad n = 0, 1, 3, 7, \ldots \tag{2.5.11}$$

Repeated application of equation (2.5.11), from a suitable starting matrix $\mathbf{B}_0$, converges *quadratically* to the unknown inverse matrix $\mathbf{A}^{-1}$ (see §9.4 for the definition of "quadratically"). Equation (2.5.11) goes by various names, including *Schultz's Method* and *Hotelling's Method*; see Pan and Reif [1] for references. In fact, equation (2.5.11) is simply the iterative Newton-Raphson method of root-finding (§9.4) applied to matrix inversion.

Before you get too excited about equation (2.5.11), however, you should notice that it involves two full matrix multiplications at each iteration. Each matrix multiplication involves N^3 adds and multiplies. But we already saw in §§2.1–2.3 that direct inversion of $\mathbf{A}$ requires only N^3 adds and N^3 multiplies *in toto*. Equation (2.5.11) is therefore practical only when special circumstances allow it to be evaluated much more rapidly than is the case for general matrices. We will meet such circumstances later, in §13.10.

In the spirit of delayed gratification, let us nevertheless pursue the two related issues: When does the series in equation (2.5.7) converge; and what is a suitable initial guess $\mathbf{B}_0$ (if, for example, an initial LU decomposition is not feasible)?

We can define the norm of a matrix as the largest amplification of length that it is able to induce on a vector,

$$\|\mathbf{R}\| \equiv \max_{\mathbf{v} \neq 0} \frac{|\mathbf{R} \cdot \mathbf{v}|}{|\mathbf{v}|} \tag{2.5.12}$$

If we let equation (2.5.7) act on some arbitrary right-hand side $\mathbf{b}$, as one wants a matrix inverse to do, it is obvious that a sufficient condition for convergence is

$$\|\mathbf{R}\| < 1 \tag{2.5.13}$$

Pan and Reif [1] point out that a suitable initial guess for $\mathbf{B}_0$ is any sufficiently small constant ϵ times the matrix transpose of $\mathbf{A}$, that is,

$$\mathbf{B}_0 = \epsilon \mathbf{A}^T \qquad \text{or} \qquad \mathbf{R} = 1 - \epsilon \mathbf{A}^T \cdot \mathbf{A} \tag{2.5.14}$$

To see why this is so involves concepts from Chapter 11; we give here only the briefest sketch: $\mathbf{A}^T \cdot \mathbf{A}$ is a symmetric, positive definite matrix, so it has real, positive eigenvalues. In its diagonal representation, $\mathbf{R}$ takes the form

$$\mathbf{R} = \mathrm{diag}(1 - \epsilon\lambda_0, 1 - \epsilon\lambda_1, \ldots, 1 - \epsilon\lambda_{N-1}) \tag{2.5.15}$$

where all the λ_i's are positive. Evidently any ϵ satisfying $0 < \epsilon < 2/(\max_i \lambda_i)$ will give $\|\mathbf{R}\| < 1$. It is not difficult to show that the optimal choice for ϵ, giving the most rapid convergence for equation (2.5.11), is

$$\epsilon = 2/(\max_i \lambda_i + \min_i \lambda_i) \tag{2.5.16}$$

Rarely does one know the eigenvalues of $\mathbf{A}^T \cdot \mathbf{A}$ in equation (2.5.16). Pan and Reif derive several interesting bounds, which are computable directly from $\mathbf{A}$. The following choices guarantee the convergence of $\mathbf{B}_n$ as $n \to \infty$,

$$\epsilon \leq 1 \Big/ \sum_{j,k} a_{jk}^2 \qquad \text{or} \qquad \epsilon \leq 1 \Big/ \left(\max_i \sum_j |a_{ij}| \times \max_j \sum_i |a_{ij}| \right) \tag{2.5.17}$$

The latter expression is truly a remarkable formula, which Pan and Reif derive by noting that the vector norm in equation (2.5.12) need not be the usual L_2 norm, but can instead be either the L_∞ (max) norm, or the L_1 (absolute value) norm. See their work for details.

Another approach, with which we have had some success, is to estimate the largest eigenvalue statistically, by calculating $s_i \equiv |\mathbf{A} \cdot \mathbf{v}_i|^2$ for several unit vector $\mathbf{v}_i$'s with randomly chosen directions in N-space. The largest eigenvalue λ can then be bounded by the maximum of $2\max s_i$ and $2N\mathrm{Var}(s_i)/\mu(s_i)$, where Var and μ denote the sample variance and mean, respectively.

CITED REFERENCES AND FURTHER READING:

Johnson, L.W., and Riess, R.D. 1982, *Numerical Analysis*, 2nd ed. (Reading, MA: Addison-Wesley), §2.3.4, p. 55.

Golub, G.H., and Van Loan, C.F. 1996, *Matrix Computations*, 3rd ed. (Baltimore: Johns Hopkins University Press), §3.5.3.

Dahlquist, G., and Bjorck, A. 1974, *Numerical Methods* (Englewood Cliffs, NJ: Prentice-Hall), §5.5.6, p. 183.

Forsythe, G.E., and Moler, C.B. 1967, *Computer Solution of Linear Algebraic Systems* (Englewood Cliffs, NJ: Prentice-Hall), Chapter 13.

Ralston, A., and Rabinowitz, P. 1978, *A First Course in Numerical Analysis*, 2nd ed.; reprinted 2001 (New York: Dover), §9.5, p. 437.

Pan, V., and Reif, J. 1985, in Proceedings of the Seventeenth Annual ACM Symposium on Theory of Computing (New York: Association for Computing Machinery). [1]

2.6 Singular Value Decomposition

There exists a very powerful set of techniques for dealing with sets of equations or matrices that are either singular or else numerically very close to singular. In many cases where Gaussian elimination and *LU* decomposition fail to give satisfactory results, this set of techniques, known as *singular value decomposition, or SVD*, will diagnose for you precisely what the problem is. In some cases, SVD will not only diagnose the problem, it will also solve it, in the sense of giving you a useful numerical answer, although, as we shall see, not necessarily "the" answer that you thought you should get.

SVD is also the method of choice for solving most *linear least-squares* problems. We will outline the relevant theory in this section, but defer detailed discussion of the use of SVD in this application to Chapter 15, whose subject is the parametric modeling of data.

SVD methods are based on the following theorem of linear algebra, whose proof is beyond our scope: Any $M \times N$ matrix $\mathbf{A}$ whose number of rows M is greater than or equal to its number of columns N, can be written as the product of an $M \times N$ column-orthogonal matrix $\mathbf{U}$, an $N \times N$ diagonal matrix $\mathbf{W}$ with positive or zero elements (the *singular values*), and the transpose of an $N \times N$ orthogonal matrix $\mathbf{V}$. The various shapes of these matrices will be made clearer by the following tableau:

$$
\begin{pmatrix} \\ \\ \mathbf{A} \\ \\ \\ \end{pmatrix} = \begin{pmatrix} \\ \\ \mathbf{U} \\ \\ \\ \end{pmatrix} \cdot \begin{pmatrix} w_0 & & & \\ & w_1 & & \\ & & \cdots & \\ & & & w_{N-1} \end{pmatrix} \cdot \begin{pmatrix} \\ \mathbf{V}^T \\ \\ \end{pmatrix}
$$

$$(2.6.1)$$

The matrices $\mathbf{U}$ and $\mathbf{V}$ are each orthogonal in the sense that their columns are orthonormal,

$$
\sum_{i=0}^{M-1} U_{ik}U_{in} = \delta_{kn} \qquad \begin{matrix} 0 \le k \le N-1 \\ 0 \le n \le N-1 \end{matrix} \qquad (2.6.2)
$$

$$
\sum_{j=0}^{N-1} V_{jk}V_{jn} = \delta_{kn} \qquad \begin{matrix} 0 \le k \le N-1 \\ 0 \le n \le N-1 \end{matrix} \qquad (2.6.3)
$$

or as a tableau,

$$
\left(\quad \mathbf{U}^T \quad\right) \cdot \left(\begin{array}{c} \\ \mathbf{U} \\ \\ \end{array}\right) = \left(\quad \mathbf{V}^T \quad\right) \cdot \left(\quad \mathbf{V} \quad\right)
$$

$$
= \left(\quad \mathbf{1} \quad\right)
$$

$$(2.6.4)$$

Since $\mathbf{V}$ is square, it is also row-orthonormal, $\mathbf{V} \cdot \mathbf{V}^T = 1$.

The SVD decomposition can also be carried out when $M < N$. In this case the singular values w_j for $j = M, \ldots, N-1$ are all zero, and the corresponding columns of $\mathbf{U}$ are also zero. Equation (2.6.2) then holds only for $k, n \leq M-1$.

The decomposition (2.6.1) can always be done, no matter how singular the matrix is, and it is "almost" unique. That is to say, it is unique up to (i) making the same permutation of the columns of $\mathbf{U}$, elements of $\mathbf{W}$, and columns of $\mathbf{V}$ (or rows of $\mathbf{V}^T$), or (ii) forming linear combinations of any columns of $\mathbf{U}$ and $\mathbf{V}$ whose corresponding elements of $\mathbf{W}$ happen to be exactly equal. An important consequence of the permutation freedom is that for the case $M < N$, a numerical algorithm for the decomposition need not return zero w_j's for $j = M, \ldots, N-1$; the $N-M$ zero singular values can be scattered among all positions $j = 0, 1, \ldots, N-1$.

At the end of this section, we give a routine, svdcmp, that performs SVD on an arbitrary matrix $\mathbf{A}$, replacing it by $\mathbf{U}$ (they are the same shape) and giving back $\mathbf{W}$ and $\mathbf{V}$ separately. The routine svdcmp is based on a routine by Forsythe et al. [1], which is in turn based on the original routine of Golub and Reinsch, found, in various forms, in [2-4] and elsewhere. These references include extensive discussion of the algorithm used. As much as we dislike the use of black-box routines, we are going to ask you to accept this one, since it would take us too far afield to cover its necessary background material here. Suffice it to say that the algorithm is very stable, and that it is very unusual for it ever to misbehave. Most of the concepts that enter the algorithm (Householder reduction to bidiagonal form, diagonalization by QR procedure with shifts) will be discussed further in Chapter 11.

If you are as suspicious of black boxes as we are, you will want to verify yourself that svdcmp does what we say it does. That is very easy to do: Generate an arbitrary matrix $\mathbf{A}$, call the routine, and then verify by matrix multiplication that (2.6.1) and (2.6.4) are satisfied. Since these two equations are the only defining requirements for SVD, this procedure is (for the chosen $\mathbf{A}$) a complete end-to-end check.

Now let us find out what SVD is good for.

SVD of a Square Matrix

If the matrix $\mathbf{A}$ is square, $N \times N$ say, then $\mathbf{U}$, $\mathbf{V}$, and $\mathbf{W}$ are all square matrices of the same size. Their inverses are also trivial to compute: $\mathbf{U}$ and $\mathbf{V}$ are orthogonal, so their inverses are equal to their transposes; $\mathbf{W}$ is diagonal, so its inverse is the diagonal matrix whose elements are the reciprocals of the elements w_j. From (2.6.1) it now follows immediately that the inverse of $\mathbf{A}$ is

$$\mathbf{A}^{-1} = \mathbf{V} \cdot [\text{diag} \ (1/w_j)] \cdot \mathbf{U}^T \qquad (2.6.5)$$

The only thing that can go wrong with this construction is for one of the w_j's to be zero, or (numerically) for it to be so small that its value is dominated by roundoff error and therefore unknowable. If more than one of the w_j's have this problem, then the matrix is even more singular. So, first of all, SVD gives you a clear diagnosis of the situation.

Formally, the *condition number* of a matrix is defined as the ratio of the largest (in magnitude) of the w_j's to the smallest of the w_j's. A matrix is singular if its condition number is infinite, and it is *ill-conditioned* if its condition number is too large, that is, if its reciprocal approaches the machine's floating-point precision (for example, less than 10^{-6} for single precision or 10^{-12} for double).

For singular matrices, the concepts of *nullspace* and *range* are important. Consider the familiar set of simultaneous equations

$$\mathbf{A} \cdot \mathbf{x} = \mathbf{b} \qquad (2.6.6)$$

where $\mathbf{A}$ is a square matrix, $\mathbf{b}$ and $\mathbf{x}$ are vectors. Equation (2.6.6) defines $\mathbf{A}$ as a linear mapping from the vector space $\mathbf{x}$ to the vector space $\mathbf{b}$. If $\mathbf{A}$ is singular, then there is some subspace of $\mathbf{x}$, called the nullspace, that is mapped to zero, $\mathbf{A} \cdot \mathbf{x} = 0$. The dimension of the nullspace (the number of linearly independent vectors $\mathbf{x}$ that can be found in it) is called the *nullity* of $\mathbf{A}$.

Now, there is also some subspace of $\mathbf{b}$ that can be "reached" by $\mathbf{A}$, in the sense that there exists some $\mathbf{x}$ which is mapped there. This subspace of $\mathbf{b}$ is called the range of $\mathbf{A}$. The dimension of the range is called the *rank* of $\mathbf{A}$. If $\mathbf{A}$ is nonsingular, then its range will be all of the vector space $\mathbf{b}$, so its rank is N. If $\mathbf{A}$ is singular, then the rank will be less than N. In fact, the relevant theorem is "rank plus nullity equals N."

What has this to do with SVD? SVD explicitly constructs orthonormal bases for the nullspace and range of a matrix. Specifically, the columns of $\mathbf{U}$ whose same-numbered elements w_j are *nonzero* are an orthonormal set of basis vectors that span the range; the columns of $\mathbf{V}$ whose same-numbered elements w_j are *zero* are an orthonormal basis for the nullspace.

Now let's have another look at solving the set of simultaneous linear equations (2.6.6) in the case that $\mathbf{A}$ is singular. First, the set of *homogeneous* equations, where $\mathbf{b} = 0$, is solved immediately by SVD: Any column of $\mathbf{V}$ whose corresponding w_j is zero yields a solution.

When the vector $\mathbf{b}$ on the right-hand side is not zero, the important question is whether it lies in the range of $\mathbf{A}$ or not. If it does, then the singular set of equations *does* have a solution $\mathbf{x}$; in fact it has more than one solution, since any vector in the nullspace (any column of $\mathbf{V}$ with a corresponding zero w_j) can be added to $\mathbf{x}$ in any linear combination.

If we want to single out one particular member of this solution-set of vectors as a representative, we might want to pick the one with the smallest length $|\mathbf{x}|^2$. Here is how to find that vector using SVD: Simply *replace $1/w_j$ by zero if $w_j = 0$.* (It is not very often that one gets to set $\infty = 0$!) Then compute (working from right to left)

$$\mathbf{x} = \mathbf{V} \cdot [\text{diag } (1/w_j)] \cdot (\mathbf{U}^T \cdot \mathbf{b}) \qquad (2.6.7)$$

This will be the solution vector of smallest length; the columns of $\mathbf{V}$ that are in the nullspace complete the specification of the solution set.

Proof: Consider $|\mathbf{x} + \mathbf{x}'|$, where $\mathbf{x}'$ lies in the nullspace. Then, if $\mathbf{W}^{-1}$ denotes the modified inverse of $\mathbf{W}$ with some elements zeroed,

$$\begin{aligned} |\mathbf{x} + \mathbf{x}'| &= \left| \mathbf{V} \cdot \mathbf{W}^{-1} \cdot \mathbf{U}^T \cdot \mathbf{b} + \mathbf{x}' \right| \\ &= \left| \mathbf{V} \cdot (\mathbf{W}^{-1} \cdot \mathbf{U}^T \cdot \mathbf{b} + \mathbf{V}^T \cdot \mathbf{x}') \right| \qquad (2.6.8) \\ &= \left| \mathbf{W}^{-1} \cdot \mathbf{U}^T \cdot \mathbf{b} + \mathbf{V}^T \cdot \mathbf{x}' \right| \end{aligned}$$

Here the first equality follows from (2.6.7), the second and third from the orthonormality of $\mathbf{V}$. If you now examine the two terms that make up the sum on the right-hand side, you will see that the first one has nonzero j components only where $w_j \neq 0$, while the second one, since $\mathbf{x}'$ is in the nullspace, has nonzero j components only where $w_j = 0$. Therefore the minimum length obtains for $\mathbf{x}' = 0$, q.e.d.

If $\mathbf{b}$ is not in the range of the singular matrix $\mathbf{A}$, then the set of equations (2.6.6) has no solution. But here is some good news: If $\mathbf{b}$ is not in the range of $\mathbf{A}$, then equation (2.6.7) can still be used to construct a "solution" vector $\mathbf{x}$. This vector $\mathbf{x}$ will not exactly solve $\mathbf{A} \cdot \mathbf{x} = \mathbf{b}$. But, among all possible vectors $\mathbf{x}$, it will do the closest possible job in the least squares sense. In other words (2.6.7) finds

$$\mathbf{x} \quad \text{which minimizes} \quad r \equiv |\mathbf{A} \cdot \mathbf{x} - \mathbf{b}| \qquad (2.6.9)$$

The number r is called the *residual* of the solution.

The proof is similar to (2.6.8): Suppose we modify $\mathbf{x}$ by adding some arbitrary $\mathbf{x}'$. Then $\mathbf{A} \cdot \mathbf{x} - \mathbf{b}$ is modified by adding some $\mathbf{b}' \equiv \mathbf{A} \cdot \mathbf{x}'$. Obviously $\mathbf{b}'$ is in the range of $\mathbf{A}$. We then have

$$\begin{aligned} \left| \mathbf{A} \cdot \mathbf{x} - \mathbf{b} + \mathbf{b}' \right| &= \left| (\mathbf{U} \cdot \mathbf{W} \cdot \mathbf{V}^T) \cdot (\mathbf{V} \cdot \mathbf{W}^{-1} \cdot \mathbf{U}^T \cdot \mathbf{b}) - \mathbf{b} + \mathbf{b}' \right| \\ &= \left| (\mathbf{U} \cdot \mathbf{W} \cdot \mathbf{W}^{-1} \cdot \mathbf{U}^T - 1) \cdot \mathbf{b} + \mathbf{b}' \right| \\ &= \left| \mathbf{U} \cdot \left[(\mathbf{W} \cdot \mathbf{W}^{-1} - 1) \cdot \mathbf{U}^T \cdot \mathbf{b} + \mathbf{U}^T \cdot \mathbf{b}' \right] \right| \qquad (2.6.10) \\ &= \left| (\mathbf{W} \cdot \mathbf{W}^{-1} - 1) \cdot \mathbf{U}^T \cdot \mathbf{b} + \mathbf{U}^T \cdot \mathbf{b}' \right| \end{aligned}$$

Now, $(\mathbf{W} \cdot \mathbf{W}^{-1} - 1)$ is a diagonal matrix which has nonzero j components only for $w_j = 0$, while $\mathbf{U}^T \mathbf{b}'$ has nonzero j components only for $w_j \neq 0$, since $\mathbf{b}'$ lies in the range of $\mathbf{A}$. Therefore the minimum obtains for $\mathbf{b}' = 0$, q.e.d.

Figure 2.6.1 summarizes our discussion of SVD thus far.

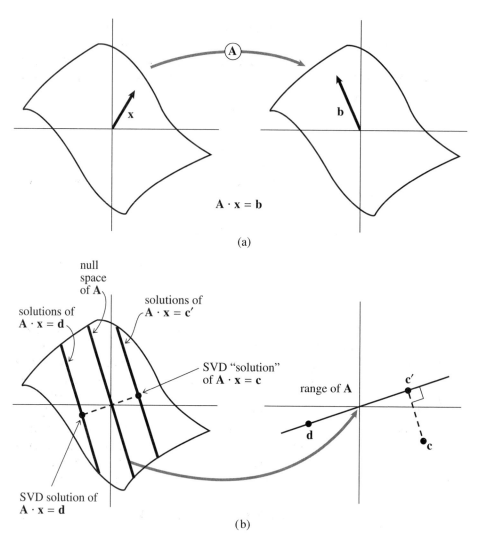

Figure 2.6.1.   (a) A nonsingular matrix **A** maps a vector space into one of the same dimension. The vector **x** is mapped into **b**, so that **x** satisfies the equation **A** · **x** = **b**. (b) A singular matrix **A** maps a vector space into one of lower dimensionality, here a plane into a line, called the "range" of **A**. The "nullspace" of **A** is mapped to zero. The solutions of **A** · **x** = **d** consist of any one particular solution plus any vector in the nullspace, here forming a line parallel to the nullspace. Singular value decomposition (SVD) selects the particular solution closest to zero, as shown. The point **c** lies outside of the range of **A**, so **A** · **x** = **c** has no solution. SVD finds the least-squares best compromise solution, namely a solution of **A** · **x** = **c**′, as shown.

In the discussion since equation (2.6.6), we have been pretending that a matrix either is singular or else isn't. That is of course true analytically. Numerically, however, the far more common situation is that some of the w_j's are very small but nonzero, so that the matrix is ill-conditioned. In that case, the direct solution methods of LU decomposition or Gaussian elimination may actually give a formal solution to the set of equations (that is, a zero pivot may not be encountered); but the solution vector may have wildly large components whose algebraic cancellation, when multiplying by the matrix **A**, may give a very poor approximation to the right-hand vector **b**. In such cases, the solution vector **x** obtained by *zeroing* the small w_j's and then using equation (2.6.7) is very often better (in the sense of the residual $|\mathbf{A} \cdot \mathbf{x} - \mathbf{b}|$ being smaller) than *both* the direct-method solution *and* the SVD solution where the small w_j's are left nonzero.

It may seem paradoxical that this can be so, since zeroing a singular value corresponds to throwing away one linear combination of the set of equations that we are trying to solve. The resolution of the paradox is that we are throwing away precisely a combination of equations that is so corrupted by roundoff error as to be at best useless; usually it is worse than useless since it "pulls" the solution vector way off towards infinity along some direction that is almost a nullspace vector. In doing this, it compounds the roundoff problem and makes the residual $|\mathbf{A} \cdot \mathbf{x} - \mathbf{b}|$ larger.

SVD cannot be applied blindly, then. You have to exercise some discretion in deciding at what threshold to zero the small w_j's, and/or you have to have some idea what size of computed residual $|\mathbf{A} \cdot \mathbf{x} - \mathbf{b}|$ is acceptable.

As an example, here is a "backsubstitution" routine svbksb for evaluating equation (2.6.7) and obtaining a solution vector $\mathbf{x}$ from a right-hand side $\mathbf{b}$, given that the SVD of a matrix $\mathbf{A}$ has already been calculated by a call to svdcmp. Note that this routine presumes that *you* have already zeroed the small w_j's. It does not do this for you. If you *haven't* zeroed the small w_j's, then this routine is just as ill-conditioned as any direct method, and you are misusing SVD.

```
#include "nr.h"

void NR::svbksb(Mat_I_DP &u, Vec_I_DP &w, Mat_I_DP &v, Vec_I_DP &b, Vec_O_DP &x)
Solves A · X = B for a vector X, where A is specified by the arrays u[0..m-1][0..n-1],
w[0..n-1], v[0..n-1][0..n-1] as returned by svdcmp. m and n will be equal for square
matrices. b[0..m-1] is the input right-hand side. x[0..n-1] is the output solution vector.
No input quantities are destroyed, so the routine may be called sequentially with different b's.
{
    int jj,j,i;
    DP s;

    int m=u.nrows();
    int n=u.ncols();
    Vec_DP tmp(n);
    for (j=0;j<n;j++) {                  Calculate U^T B.
        s=0.0;
        if (w[j] != 0.0) {              Nonzero result only if w_j is nonzero.
            for (i=0;i<m;i++) s += u[i][j]*b[i];
            s /= w[j];                  This is the divide by w_j.
        }
        tmp[j]=s;
    }
    for (j=0;j<n;j++) {                  Matrix multiply by V to get answer.
        s=0.0;
        for (jj=0;jj<n;jj++) s += v[j][jj]*tmp[jj];
        x[j]=s;
    }
}
```

Note that a typical use of svdcmp and svbksb superficially resembles the typical use of ludcmp and lubksb: In both cases, you decompose the left-hand matrix $\mathbf{A}$ just once, and then can use the decomposition either once or many times with different right-hand sides. The crucial difference is the "editing" of the singular values before svbksb is called:

```
const int N = ...
DP wmax,wmin;
Mat_DP a(N,N),u(N,N),v(N,N);
Vec_DP w(N),b(N),x(N);
int i,j;
...
for(i=0;i<N;i++)
    for j=0;j<N;j++)
        u[i][j]=a[i][j];
NR::svdcmp(u,w,v);
wmax=0.0;
for(j=0;j<N;j++) if (w[j] > wmax) wmax=w[j];
wmin=wmax*1.0e-12;
for(j=0;j<N;j++) if (w[j] < wmin) w[j]=0.0;
NR::svbksb(u,w,v,b,x);
```

Copy a into u if you don't want it to be destroyed.

SVD the square matrix a.
Will be the maximum singular value obtained.
This is where we set the threshold for singular values allowed to be nonzero. The constant is typical, but not universal. You have to experiment with your own application.

Now we can backsubstitute.

SVD for Fewer Equations than Unknowns

If you have fewer linear equations M than unknowns N, then you are not expecting a unique solution. Usually there will be an $N - M$ dimensional family of solutions. If you want to find this whole solution space, then SVD can readily do the job.

The SVD decomposition will yield $N - M$ zero or negligible w_j's, since $M < N$. There may be additional zero w_j's from any degeneracies in your M equations. Be sure that you find this many small w_j's, and zero them before calling svbksb, which will give you the particular solution vector $\mathbf{x}$. As before, the columns of $\mathbf{V}$ corresponding to zeroed w_j's are the basis vectors whose linear combinations, added to the particular solution, span the solution space.

SVD for More Equations than Unknowns

This situation will occur in Chapter 15, when we wish to find the least-squares solution to an overdetermined set of linear equations. In tableau, the equations to be solved are

$$\begin{pmatrix} & \\ & \mathbf{A} & \\ & \end{pmatrix} \cdot \begin{pmatrix} \mathbf{x} \end{pmatrix} = \begin{pmatrix} \mathbf{b} \end{pmatrix} \qquad (2.6.11)$$

The proofs that we gave above for the square case apply without modification to the case of more equations than unknowns. The least-squares solution vector $\mathbf{x}$ is given by (2.6.7), which, with nonsquare matrices, looks like this,

$$
\begin{pmatrix} \\ \mathbf{x} \\ \\ \end{pmatrix} = \begin{pmatrix} & & \\ & \mathbf{V} & \\ & & \end{pmatrix} \cdot \begin{pmatrix} \\ \mathrm{diag}(1/w_j) \\ \\ \end{pmatrix} \cdot \begin{pmatrix} & & \\ & \mathbf{U}^T & \\ & & \end{pmatrix} \cdot \begin{pmatrix} \\ \\ \mathbf{b} \\ \\ \\ \end{pmatrix}
$$

$$(2.6.12)$$

In general, the matrix $\mathbf{W}$ will not be singular, and no w_j's will need to be set to zero. Occasionally, however, there might be column degeneracies in $\mathbf{A}$. In this case you will need to zero some small w_j values after all. The corresponding column in $\mathbf{V}$ gives the linear combination of $\mathbf{x}$'s that is then ill-determined even by the supposedly overdetermined set.

Sometimes, although you do not need to zero any w_j's for *computational* reasons, you may nevertheless want to take note of any that are unusually small: Their corresponding columns in $\mathbf{V}$ are linear combinations of $\mathbf{x}$'s which are insensitive to your data. In fact, you may then wish to zero these w_j's, to reduce the number of free parameters in the fit. These matters are discussed more fully in Chapter 15.

Constructing an Orthonormal Basis

Suppose that you have N vectors in an M-dimensional vector space, with $N \leq M$. Then the N vectors span some subspace of the full vector space. Often you want to construct an orthonormal set of N vectors that span the same subspace. The textbook way to do this is by Gram-Schmidt orthogonalization, starting with one vector and then expanding the subspace one dimension at a time. Numerically, however, because of the build-up of roundoff errors, naive Gram-Schmidt orthogonalization is *terrible*.

The right way to construct an orthonormal basis for a subspace is by SVD: Form an $M \times N$ matrix $\mathbf{A}$ whose N columns are your vectors. Run the matrix through svdcmp. The columns of the matrix $\mathbf{U}$ (which in fact replaces $\mathbf{A}$ on output from svdcmp) are your desired orthonormal basis vectors.

You might also want to check the output w_j's for zero values. If any occur, then the spanned subspace was not, in fact, N dimensional; the columns of $\mathbf{U}$ corresponding to zero w_j's should be discarded from the orthonormal basis set.

(QR factorization, discussed in §2.10, also constructs an orthonormal basis, see [5].)

Approximation of Matrices

Note that equation (2.6.1) can be rewritten to express any matrix A_{ij} as a sum of outer products of columns of $\mathbf{U}$ and rows of $\mathbf{V}^T$, with the "weighting factors" being the singular values w_j,

$$ A_{ij} = \sum_{k=0}^{N-1} w_k \, U_{ik} V_{jk} \tag{2.6.13} $$

If you ever encounter a situation where *most* of the singular values w_j of a matrix **A** are very small, then **A** will be well-approximated by only a few terms in the sum (2.6.13). This means that you have to store only a few columns of **U** and **V** (the same k ones) and you will be able to recover, with good accuracy, the whole matrix.

Note also that it is very efficient to multiply such an approximated matrix by a vector **x**: You just dot **x** with each of the stored columns of **V**, multiply the resulting scalar by the corresponding w_k, and accumulate that multiple of the corresponding column of **U**. If your matrix is approximated by a small number K of singular values, then this computation of **A** · **x** takes only about $K(M+N)$ multiplications, instead of MN for the full matrix.

SVD Algorithm

Here is the algorithm for constructing the singular value decomposition of any matrix. See §11.2–§11.3, and also [4-5], for discussion relating to the underlying method.

```cpp
#include <cmath>
#include "nr.h"
using namespace std;

void NR::svdcmp(Mat_IO_DP &a, Vec_O_DP &w, Mat_O_DP &v)
```
Given a matrix a[0..m−1][0..n−1], this routine computes its singular value decomposition, $A = U \cdot W \cdot V^T$. The matrix U replaces a on output. The diagonal matrix of singular values W is output as a vector w[0..n−1]. The matrix V (not the transpose V^T) is output as v[0..n−1][0..n−1].
```cpp
{
    bool flag;
    int i,its,j,jj,k,l,nm;
    DP anorm,c,f,g,h,s,scale,x,y,z;

    int m=a.nrows();
    int n=a.ncols();
    Vec_DP rv1(n);
    g=scale=anorm=0.0;                      Householder reduction to bidiagonal form.
    for (i=0;i<n;i++) {
        l=i+2;
        rv1[i]=scale*g;
        g=s=scale=0.0;
        if (i < m) {
            for (k=i;k<m;k++) scale += fabs(a[k][i]);
            if (scale != 0.0) {
                for (k=i;k<m;k++) {
                    a[k][i] /= scale;
                    s += a[k][i]*a[k][i];
                }
                f=a[i][i];
                g = -SIGN(sqrt(s),f);
                h=f*g-s;
                a[i][i]=f-g;
                for (j=l-1;j<n;j++) {
                    for (s=0.0,k=i;k<m;k++) s += a[k][i]*a[k][j];
                    f=s/h;
                    for (k=i;k<m;k++) a[k][j] += f*a[k][i];
                }
                for (k=i;k<m;k++) a[k][i] *= scale;
            }
        }
        w[i]=scale *g;
```

```
            g=s=scale=0.0;
            if (i+1 <= m && i != n) {
                for (k=l-1;k<n;k++) scale += fabs(a[i][k]);
                if (scale != 0.0) {
                    for (k=l-1;k<n;k++) {
                        a[i][k] /= scale;
                        s += a[i][k]*a[i][k];
                    }
                    f=a[i][l-1];
                    g = -SIGN(sqrt(s),f);
                    h=f*g-s;
                    a[i][l-1]=f-g;
                    for (k=l-1;k<n;k++) rv1[k]=a[i][k]/h;
                    for (j=l-1;j<m;j++) {
                        for (s=0.0,k=l-1;k<n;k++) s += a[j][k]*a[i][k];
                        for (k=l-1;k<n;k++) a[j][k] += s*rv1[k];
                    }
                    for (k=l-1;k<n;k++) a[i][k] *= scale;
                }
            }
            anorm=MAX(anorm,(fabs(w[i])+fabs(rv1[i])));
        }
        for (i=n-1;i>=0;i--) {                  Accumulation of right-hand transformations.
            if (i < n-1) {
                if (g != 0.0) {
                    for (j=l;j<n;j++)           Double division to avoid possible underflow.
                        v[j][i]=(a[i][j]/a[i][l])/g;
                    for (j=l;j<n;j++) {
                        for (s=0.0,k=l;k<n;k++) s += a[i][k]*v[k][j];
                        for (k=l;k<n;k++) v[k][j] += s*v[k][i];
                    }
                }
                for (j=l;j<n;j++) v[i][j]=v[j][i]=0.0;
            }
            v[i][i]=1.0;
            g=rv1[i];
            l=i;
        }
        for (i=MIN(m,n)-1;i>=0;i--) {           Accumulation of left-hand transformations.
            l=i+1;
            g=w[i];
            for (j=l;j<n;j++) a[i][j]=0.0;
            if (g != 0.0) {
                g=1.0/g;
                for (j=l;j<n;j++) {
                    for (s=0.0,k=l;k<m;k++) s += a[k][i]*a[k][j];
                    f=(s/a[i][i])*g;
                    for (k=i;k<m;k++) a[k][j] += f*a[k][i];
                }
                for (j=i;j<m;j++) a[j][i] *= g;
            } else for (j=i;j<m;j++) a[j][i]=0.0;
            ++a[i][i];
        }
        for (k=n-1;k>=0;k--) {                  Diagonalization of the bidiagonal form: Loop over
            for (its=0;its<30;its++) {               singular values, and over allowed iterations.
                flag=true;
                for (l=k;l>=0;l--) {           Test for splitting.
                    nm=l-1;                    Note that rv1[0] is always zero.
                    if (fabs(rv1[l])+anorm == anorm) {
                        flag=false;
                        break;
                    }
                    if (fabs(w[nm])+anorm == anorm) break;
                }
```

```
if (flag) {
    c=0.0;                          Cancellation of rv1[1], if 1 > 0.
    s=1.0;
    for (i=l-1;i<k+1;i++) {
        f=s*rv1[i];
        rv1[i]=c*rv1[i];
        if (fabs(f)+anorm == anorm) break;
        g=w[i];
        h=pythag(f,g);
        w[i]=h;
        h=1.0/h;
        c=g*h;
        s = -f*h;
        for (j=0;j<m;j++) {
            y=a[j][nm];
            z=a[j][i];
            a[j][nm]=y*c+z*s;
            a[j][i]=z*c-y*s;
        }
    }
}
z=w[k];
if (l == k) {                      Convergence.
    if (z < 0.0) {                 Singular value is made nonnegative.
        w[k] = -z;
        for (j=0;j<n;j++) v[j][k] = -v[j][k];
    }
    break;
}
if (its == 29) nrerror("no convergence in 30 svdcmp iterations");
x=w[l];                            Shift from bottom 2-by-2 minor.
nm=k-1;
y=w[nm];
g=rv1[nm];
h=rv1[k];
f=((y-z)*(y+z)+(g-h)*(g+h))/(2.0*h*y);
g=pythag(f,1.0);
f=((x-z)*(x+z)+h*((y/(f+SIGN(g,f)))-h))/x;
c=s=1.0;                           Next QR transformation:
for (j=l;j<=nm;j++) {
    i=j+1;
    g=rv1[i];
    y=w[i];
    h=s*g;
    g=c*g;
    z=pythag(f,h);
    rv1[j]=z;
    c=f/z;
    s=h/z;
    f=x*c+g*s;
    g=g*c-x*s;
    h=y*s;
    y *= c;
    for (jj=0;jj<n;jj++) {
        x=v[jj][j];
        z=v[jj][i];
        v[jj][j]=x*c+z*s;
        v[jj][i]=z*c-x*s;
    }
    z=pythag(f,h);
    w[j]=z;                        Rotation can be arbitrary if z = 0.
    if (z) {
        z=1.0/z;
        c=f*z;
```

```
        s=h*z;
    }
    f=c*g+s*y;
    x=c*y-s*g;
    for (jj=0;jj<m;jj++) {
        y=a[jj][j];
        z=a[jj][i];
        a[jj][j]=y*c+z*s;
        a[jj][i]=z*c-y*s;
    }
}
rv1[l]=0.0;
rv1[k]=f;
w[k]=x;
        }
    }
}
```

```
#include <cmath>
#include "nr.h"
using namespace std;

DP NR::pythag(const DP a, const DP b)
```
Computes $(a^2 + b^2)^{1/2}$ without destructive underflow or overflow.
```
{
    DP absa,absb;

    absa=fabs(a);
    absb=fabs(b);
    if (absa > absb) return absa*sqrt(1.0+SQR(absb/absa));
    else return (absb == 0.0 ? 0.0 : absb*sqrt(1.0+SQR(absa/absb)));
}
```

CITED REFERENCES AND FURTHER READING:

Golub, G.H., and Van Loan, C.F. 1996, *Matrix Computations*, 3rd ed. (Baltimore: Johns Hopkins University Press), §8.6 and Chapter 12 (SVD). QR decomposition is discussed in §5.2.6. [5]

Lawson, C.L., and Hanson, R. 1974, *Solving Least Squares Problems* (Englewood Cliffs, NJ: Prentice-Hall), Chapter 18.

Forsythe, G.E., Malcolm, M.A., and Moler, C.B. 1977, *Computer Methods for Mathematical Computations* (Englewood Cliffs, NJ: Prentice-Hall), Chapter 9. [1]

Wilkinson, J.H., and Reinsch, C. 1971, *Linear Algebra*, vol. II of *Handbook for Automatic Computation* (New York: Springer-Verlag), Chapter I.10 by G.H. Golub and C. Reinsch. [2]

Anderson, E., et al. 2000, LAPACK User's Guide, 3rd ed. (Philadelphia: S.I.A.M.). [3]

Smith, B.T., et al. 1976, *Matrix Eigensystem Routines — EISPACK Guide*, 2nd ed., vol. 6 of Lecture Notes in Computer Science (New York: Springer-Verlag).

Stoer, J., and Bulirsch, R. 1993, *Introduction to Numerical Analysis*, 2nd ed. (New York: Springer-Verlag), §6.7. [4]

2.7 Sparse Linear Systems

A system of linear equations is called *sparse* if only a relatively small number of its matrix elements a_{ij} are nonzero. It is wasteful to use general methods of linear algebra on such problems, because most of the $O(N^3)$ arithmetic operations devoted to solving the set of equations or inverting the matrix involve zero operands. Furthermore, you might wish to work problems so large as to tax your available memory space, and it is wasteful to reserve storage for unfruitful zero elements. Note that there are two distinct (and not always compatible) goals for any sparse matrix method: saving time and/or saving space.

We have already considered one archetypal sparse form in §2.4, the band diagonal matrix. In the tridiagonal case, e.g., we saw that it was possible to save both time (order N instead of N^3) and space (order N instead of N^2). The method of solution was not different in principle from the general method of LU decomposition; it was just applied cleverly, and with due attention to the bookkeeping of zero elements. Many practical schemes for dealing with sparse problems have this same character. They are fundamentally decomposition schemes, or else elimination schemes akin to Gauss-Jordan, but carefully optimized so as to minimize the number of so-called *fill-ins*, initially zero elements which must become nonzero during the solution process, and for which storage must be reserved.

Direct methods for solving sparse equations, then, depend crucially on the precise pattern of sparsity of the matrix. Patterns that occur frequently, or that are useful as way-stations in the reduction of more general forms, already have special names and special methods of solution. We do not have space here for any detailed review of these. References listed at the end of this section will furnish you with an "in" to the specialized literature, and the following list of buzz words (and Figure 2.7.1) will at least let you hold your own at cocktail parties:

- tridiagonal
- band diagonal (or banded) with bandwidth M
- band triangular
- block diagonal
- block tridiagonal
- block triangular
- cyclic banded
- singly (or doubly) bordered block diagonal
- singly (or doubly) bordered block triangular
- singly (or doubly) bordered band diagonal
- singly (or doubly) bordered band triangular
- other (!)

You should also be aware of some of the special sparse forms that occur in the solution of partial differential equations in two or more dimensions. See Chapter 19.

If your particular pattern of sparsity is not a simple one, then you may wish to try an *analyze/factorize/operate* package, which automates the procedure of figuring out how fill-ins are to be minimized. The *analyze* stage is done once only for each pattern of sparsity. The *factorize* stage is done once for each particular matrix that fits the pattern. The *operate* stage is performed once for each right-hand side to

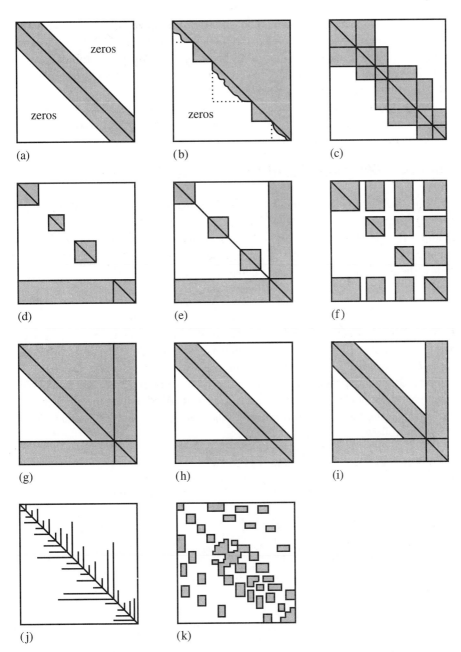

Figure 2.7.1. Some standard forms for sparse matrices. (a) Band diagonal; (b) block triangular; (c) block tridiagonal; (d) singly bordered block diagonal; (e) doubly bordered block diagonal; (f) singly bordered block triangular; (g) bordered band-triangular; (h) and (i) singly and doubly bordered band diagonal; (j) and (k) other!   (after Tewarson) [1].

be used with the particular matrix. Consult [2,3] for references on this. The NAG library [4] has an analyze/factorize/operate capability. A substantial collection of routines for sparse matrix calculation is also available from IMSL [5] as the *Yale Sparse Matrix Package* [6].

You should be aware that the special order of interchanges and eliminations, prescribed by a sparse matrix method so as to minimize fill-ins and arithmetic operations, generally acts to decrease the method's numerical stability as compared to, e.g., regular LU decomposition with pivoting. Scaling your problem so as to make its nonzero matrix elements have comparable magnitudes (if you can do it) will sometimes ameliorate this problem.

In the remainder of this section, we present some concepts which are applicable to some general classes of sparse matrices, and which do not necessarily depend on details of the pattern of sparsity.

Sherman-Morrison Formula

Suppose that you have already obtained, by herculean effort, the inverse matrix $\mathbf{A}^{-1}$ of a square matrix $\mathbf{A}$. Now you want to make a "small" change in $\mathbf{A}$, for example change one element a_{ij}, or a few elements, or one row, or one column. Is there any way of calculating the corresponding change in $\mathbf{A}^{-1}$ without repeating your difficult labors? Yes, if your change is of the form

$$\mathbf{A} \;\to\; (\mathbf{A} + \mathbf{u} \otimes \mathbf{v}) \tag{2.7.1}$$

for some vectors $\mathbf{u}$ and $\mathbf{v}$. If $\mathbf{u}$ is a unit vector $\mathbf{e}_i$, then (2.7.1) adds the components of $\mathbf{v}$ to the ith row. (Recall that $\mathbf{u} \otimes \mathbf{v}$ is a matrix whose i, jth element is the product of the ith component of $\mathbf{u}$ and the jth component of $\mathbf{v}$.) If $\mathbf{v}$ is a unit vector $\mathbf{e}_j$, then (2.7.1) adds the components of $\mathbf{u}$ to the jth column. If both $\mathbf{u}$ and $\mathbf{v}$ are proportional to unit vectors $\mathbf{e}_i$ and $\mathbf{e}_j$ respectively, then a term is added only to the element a_{ij}.

The *Sherman-Morrison* formula gives the inverse $(\mathbf{A} + \mathbf{u} \otimes \mathbf{v})^{-1}$, and is derived briefly as follows:

$$
\begin{aligned}
(\mathbf{A} + \mathbf{u} \otimes \mathbf{v})^{-1} &= (\mathbf{1} + \mathbf{A}^{-1} \cdot \mathbf{u} \otimes \mathbf{v})^{-1} \cdot \mathbf{A}^{-1} \\
&= (\mathbf{1} - \mathbf{A}^{-1} \cdot \mathbf{u} \otimes \mathbf{v} + \mathbf{A}^{-1} \cdot \mathbf{u} \otimes \mathbf{v} \cdot \mathbf{A}^{-1} \cdot \mathbf{u} \otimes \mathbf{v} - \ldots) \cdot \mathbf{A}^{-1} \\
&= \mathbf{A}^{-1} - \mathbf{A}^{-1} \cdot \mathbf{u} \otimes \mathbf{v} \cdot \mathbf{A}^{-1} \left(1 - \lambda + \lambda^2 - \ldots\right) \\
&= \mathbf{A}^{-1} - \frac{(\mathbf{A}^{-1} \cdot \mathbf{u}) \otimes (\mathbf{v} \cdot \mathbf{A}^{-1})}{1 + \lambda}
\end{aligned}
$$

$$\tag{2.7.2}$$

where

$$\lambda \equiv \mathbf{v} \cdot \mathbf{A}^{-1} \cdot \mathbf{u} \tag{2.7.3}$$

The second line of (2.7.2) is a formal power series expansion. In the third line, the associativity of outer and inner products is used to factor out the scalars λ.

The use of (2.7.2) is this: Given $\mathbf{A}^{-1}$ and the vectors $\mathbf{u}$ and $\mathbf{v}$, we need only perform two matrix multiplications and a vector dot product,

$$\mathbf{z} \equiv \mathbf{A}^{-1} \cdot \mathbf{u} \qquad \mathbf{w} \equiv (\mathbf{A}^{-1})^T \cdot \mathbf{v} \qquad \lambda = \mathbf{v} \cdot \mathbf{z} \tag{2.7.4}$$

to get the desired change in the inverse

$$\mathbf{A}^{-1} \;\to\; \mathbf{A}^{-1} - \frac{\mathbf{z} \otimes \mathbf{w}}{1 + \lambda} \tag{2.7.5}$$

The whole procedure requires only $3N^2$ multiplies and a like number of adds (an even smaller number if $\mathbf{u}$ or $\mathbf{v}$ is a unit vector).

The Sherman-Morrison formula can be directly applied to a class of sparse problems. If you already have a fast way of calculating the inverse of $\mathbf{A}$ (e.g., a tridiagonal matrix, or some other standard sparse form), then (2.7.4)–(2.7.5) allow you to build up to your related but more complicated form, adding for example a row or column at a time. Notice that you can apply the Sherman-Morrison formula more than once successively, using at each stage the most recent update of $\mathbf{A}^{-1}$ (equation 2.7.5). Of course, if you have to modify *every* row, then you are back to an N^3 method. The constant in front of the N^3 is only a few times worse than the better direct methods, but you have deprived yourself of the stabilizing advantages of pivoting — so be careful.

For some other sparse problems, the Sherman-Morrison formula cannot be directly applied for the simple reason that storage of the whole inverse matrix $\mathbf{A}^{-1}$ is not feasible. If you want to add only a single correction of the form $\mathbf{u} \otimes \mathbf{v}$, and solve the linear system

$$(\mathbf{A} + \mathbf{u} \otimes \mathbf{v}) \cdot \mathbf{x} = \mathbf{b} \tag{2.7.6}$$

then you proceed as follows. Using the fast method that is presumed available for the matrix $\mathbf{A}$, solve the two auxiliary problems

$$\mathbf{A} \cdot \mathbf{y} = \mathbf{b} \qquad \mathbf{A} \cdot \mathbf{z} = \mathbf{u} \tag{2.7.7}$$

for the vectors $\mathbf{y}$ and $\mathbf{z}$. In terms of these,

$$\mathbf{x} = \mathbf{y} - \left[\frac{\mathbf{v} \cdot \mathbf{y}}{1 + (\mathbf{v} \cdot \mathbf{z})} \right] \mathbf{z} \tag{2.7.8}$$

as we see by multiplying (2.7.2) on the right by $\mathbf{b}$.

Cyclic Tridiagonal Systems

So-called *cyclic tridiagonal systems* occur quite frequently, and are a good example of how to use the Sherman-Morrison formula in the manner just described. The equations have the form

$$\begin{bmatrix} b_0 & c_0 & 0 & \cdots & & & \beta \\ a_1 & b_1 & c_1 & \cdots & & & \\ & & \cdots & & & & \\ & & \cdots & a_{N-2} & b_{N-2} & c_{N-2} & \\ \alpha & & \cdots & 0 & a_{N-1} & b_{N-1} \end{bmatrix} \cdot \begin{bmatrix} x_0 \\ x_1 \\ \cdots \\ x_{N-2} \\ x_{N-1} \end{bmatrix} = \begin{bmatrix} r_0 \\ r_1 \\ \cdots \\ r_{N-2} \\ r_{N-1} \end{bmatrix} \tag{2.7.9}$$

This is a tridiagonal system, except for the matrix elements α and β in the corners. Forms like this are typically generated by finite-differencing differential equations with periodic boundary conditions (§19.4).

We use the Sherman-Morrison formula, treating the system as tridiagonal plus a correction. In the notation of equation (2.7.6), define vectors $\mathbf{u}$ and $\mathbf{v}$ to be

$$\mathbf{u} = \begin{bmatrix} \gamma \\ 0 \\ \vdots \\ 0 \\ \alpha \end{bmatrix} \qquad \mathbf{v} = \begin{bmatrix} 1 \\ 0 \\ \vdots \\ 0 \\ \beta/\gamma \end{bmatrix} \tag{2.7.10}$$

Here γ is arbitrary for the moment. Then the matrix $\mathbf{A}$ is the tridiagonal part of the matrix in (2.7.9), with two terms modified:

$$b_0' = b_0 - \gamma, \qquad b_{N-1}' = b_{N-1} - \alpha\beta/\gamma \qquad (2.7.11)$$

We now solve equations (2.7.7) with the standard tridiagonal algorithm, and then get the solution from equation (2.7.8).

The routine `cyclic` below implements this algorithm. We choose the arbitrary parameter $\gamma = -b_0$ to avoid loss of precision by subtraction in the first of equations (2.7.11). In the unlikely event that this causes loss of precision in the second of these equations, you can make a different choice.

```
#include "nr.h"

void NR::cyclic(Vec_I_DP &a, Vec_I_DP &b, Vec_I_DP &c, const DP alpha,
    const DP beta, Vec_I_DP &r, Vec_O_DP &x)
Solves for a vector x[0..n-1] the "cyclic" set of linear equations given by equation (2.7.9).
a, b, c, and r are input vectors, all dimensioned as [0..n-1], while alpha and beta are the
corner entries in the matrix. The input is not modified.
{
    int i;
    DP fact,gamma;

    int n=a.size();
    if (n <= 2) nrerror("n too small in cyclic");
    Vec_DP bb(n),u(n),z(n);
    gamma = -b[0];                        Avoid subtraction error in forming bb[0].
    bb[0]=b[0]-gamma;                     Set up the diagonal of the modified tridi-
    bb[n-1]=b[n-1]-alpha*beta/gamma;          agonal system.
    for (i=1;i<n-1;i++) bb[i]=b[i];
    tridag(a,bb,c,r,x);                   Solve A · x = r.
    u[0]=gamma;                           Set up the vector u.
    u[n-1]=alpha;
    for (i=1;i<n-1;i++) u[i]=0.0;
    tridag(a,bb,c,u,z);                   Solve A · z = u.
    fact=(x[0]+beta*x[n-1]/gamma)/        Form v · x/(1 + v · z).
        (1.0+z[0]+beta*z[n-1]/gamma);
    for (i=0;i<n;i++) x[i] -= fact*z[i];  Now get the solution vector x.
}
```

Woodbury Formula

If you want to add more than a single correction term, then you cannot use (2.7.8) repeatedly, since without storing a new $\mathbf{A}^{-1}$ you will not be able to solve the auxiliary problems (2.7.7) efficiently after the first step. Instead, you need the *Woodbury formula*, which is the block-matrix version of the Sherman-Morrison formula,

$$(\mathbf{A} + \mathbf{U} \cdot \mathbf{V}^T)^{-1}$$
$$= \mathbf{A}^{-1} - \left[\mathbf{A}^{-1} \cdot \mathbf{U} \cdot (1 + \mathbf{V}^T \cdot \mathbf{A}^{-1} \cdot \mathbf{U})^{-1} \cdot \mathbf{V}^T \cdot \mathbf{A}^{-1}\right] \qquad (2.7.12)$$

Here A is, as usual, an $N \times N$ matrix, while $\mathbf{U}$ and $\mathbf{V}$ are $N \times P$ matrices with $P < N$ and usually $P \ll N$. The inner piece of the correction term may become clearer if written as the tableau,

$$
\begin{bmatrix} & \\ & \\ \mathbf{U} & \\ & \\ & \end{bmatrix} \cdot \left[1 + \mathbf{V}^T \cdot \mathbf{A}^{-1} \cdot \mathbf{U} \right]^{-1} \cdot \begin{bmatrix} & \\ & \mathbf{V}^T & \\ & \end{bmatrix} \tag{2.7.13}
$$

where you can see that the matrix whose inverse is needed is only $P \times P$ rather than $N \times N$.

The relation between the Woodbury formula and successive applications of the Sherman-Morrison formula is now clarified by noting that, if $\mathbf{U}$ is the matrix formed by columns out of the P vectors $\mathbf{u}_0, \ldots, \mathbf{u}_{P-1}$, and $\mathbf{V}$ is the matrix formed by columns out of the P vectors $\mathbf{v}_0, \ldots, \mathbf{v}_{P-1}$,

$$
\mathbf{U} \equiv \begin{bmatrix} \mathbf{u}_0 \end{bmatrix} \cdots \begin{bmatrix} \mathbf{u}_{P-1} \end{bmatrix} \qquad \mathbf{V} \equiv \begin{bmatrix} \mathbf{v}_0 \end{bmatrix} \cdots \begin{bmatrix} \mathbf{v}_{P-1} \end{bmatrix} \tag{2.7.14}
$$

then two ways of expressing the same correction to $\mathbf{A}$ are

$$
\left(\mathbf{A} + \sum_{k=0}^{P-1} \mathbf{u}_k \otimes \mathbf{v}_k \right) = (\mathbf{A} + \mathbf{U} \cdot \mathbf{V}^T) \tag{2.7.15}
$$

(Note that the subscripts on $\mathbf{u}$ and $\mathbf{v}$ do *not* denote components, but rather distinguish the different column vectors.)

Equation (2.7.15) reveals that, if you have $\mathbf{A}^{-1}$ in storage, then you can either make the P corrections in one fell swoop by using (2.7.12), inverting a $P \times P$ matrix, or else make them by applying (2.7.5) P successive times.

If you don't have storage for $\mathbf{A}^{-1}$, then you *must* use (2.7.12) in the following way: To solve the linear equation

$$
\left(\mathbf{A} + \sum_{k=0}^{P-1} \mathbf{u}_k \otimes \mathbf{v}_k \right) \cdot \mathbf{x} = \mathbf{b} \tag{2.7.16}
$$

first solve the P auxiliary problems

$$
\mathbf{A} \cdot \mathbf{z}_0 = \mathbf{u}_0
$$
$$
\mathbf{A} \cdot \mathbf{z}_1 = \mathbf{u}_1
$$
$$
\cdots \tag{2.7.17}
$$
$$
\mathbf{A} \cdot \mathbf{z}_{P-1} = \mathbf{u}_{P-1}
$$

and construct the matrix $\mathbf{Z}$ by columns from the $\mathbf{z}$'s obtained,

$$
\mathbf{Z} \equiv \begin{bmatrix} \mathbf{z}_0 \end{bmatrix} \cdots \begin{bmatrix} \mathbf{z}_{P-1} \end{bmatrix} \tag{2.7.18}
$$

Next, do the $P \times P$ matrix inversion

$$
\mathbf{H} \equiv (1 + \mathbf{V}^T \cdot \mathbf{Z})^{-1} \tag{2.7.19}
$$

Finally, solve the one further auxiliary problem

$$\mathbf{A} \cdot \mathbf{y} = \mathbf{b} \tag{2.7.20}$$

In terms of these quantities, the solution is given by

$$\mathbf{x} = \mathbf{y} - \mathbf{Z} \cdot \left[\mathbf{H} \cdot (\mathbf{V}^T \cdot \mathbf{y}) \right] \tag{2.7.21}$$

Inversion by Partitioning

Once in a while, you will encounter a matrix (not even necessarily sparse) that can be inverted efficiently by partitioning. Suppose that the $N \times N$ matrix $\mathbf{A}$ is partitioned into

$$\mathbf{A} = \begin{bmatrix} \mathbf{P} & \mathbf{Q} \\ \mathbf{R} & \mathbf{S} \end{bmatrix} \tag{2.7.22}$$

where $\mathbf{P}$ and $\mathbf{S}$ are square matrices of size $p \times p$ and $s \times s$ respectively ($p + s = N$). The matrices $\mathbf{Q}$ and $\mathbf{R}$ are not necessarily square, and have sizes $p \times s$ and $s \times p$, respectively.

If the inverse of $\mathbf{A}$ is partitioned in the same manner,

$$\mathbf{A}^{-1} = \begin{bmatrix} \widetilde{\mathbf{P}} & \widetilde{\mathbf{Q}} \\ \widetilde{\mathbf{R}} & \widetilde{\mathbf{S}} \end{bmatrix} \tag{2.7.23}$$

then $\widetilde{\mathbf{P}}$, $\widetilde{\mathbf{Q}}$, $\widetilde{\mathbf{R}}$, $\widetilde{\mathbf{S}}$, which have the same sizes as $\mathbf{P}$, $\mathbf{Q}$, $\mathbf{R}$, $\mathbf{S}$, respectively, can be found by either the formulas

$$
\begin{aligned}
\widetilde{\mathbf{P}} &= (\mathbf{P} - \mathbf{Q} \cdot \mathbf{S}^{-1} \cdot \mathbf{R})^{-1} \\
\widetilde{\mathbf{Q}} &= -(\mathbf{P} - \mathbf{Q} \cdot \mathbf{S}^{-1} \cdot \mathbf{R})^{-1} \cdot (\mathbf{Q} \cdot \mathbf{S}^{-1}) \\
\widetilde{\mathbf{R}} &= -(\mathbf{S}^{-1} \cdot \mathbf{R}) \cdot (\mathbf{P} - \mathbf{Q} \cdot \mathbf{S}^{-1} \cdot \mathbf{R})^{-1} \\
\widetilde{\mathbf{S}} &= \mathbf{S}^{-1} + (\mathbf{S}^{-1} \cdot \mathbf{R}) \cdot (\mathbf{P} - \mathbf{Q} \cdot \mathbf{S}^{-1} \cdot \mathbf{R})^{-1} \cdot (\mathbf{Q} \cdot \mathbf{S}^{-1})
\end{aligned}
\tag{2.7.24}
$$

or else by the equivalent formulas

$$
\begin{aligned}
\widetilde{\mathbf{P}} &= \mathbf{P}^{-1} + (\mathbf{P}^{-1} \cdot \mathbf{Q}) \cdot (\mathbf{S} - \mathbf{R} \cdot \mathbf{P}^{-1} \cdot \mathbf{Q})^{-1} \cdot (\mathbf{R} \cdot \mathbf{P}^{-1}) \\
\widetilde{\mathbf{Q}} &= -(\mathbf{P}^{-1} \cdot \mathbf{Q}) \cdot (\mathbf{S} - \mathbf{R} \cdot \mathbf{P}^{-1} \cdot \mathbf{Q})^{-1} \\
\widetilde{\mathbf{R}} &= -(\mathbf{S} - \mathbf{R} \cdot \mathbf{P}^{-1} \cdot \mathbf{Q})^{-1} \cdot (\mathbf{R} \cdot \mathbf{P}^{-1}) \\
\widetilde{\mathbf{S}} &= (\mathbf{S} - \mathbf{R} \cdot \mathbf{P}^{-1} \cdot \mathbf{Q})^{-1}
\end{aligned}
\tag{2.7.25}
$$

The parentheses in equations (2.7.24) and (2.7.25) highlight repeated factors that you may wish to compute only once. (Of course, by associativity, you can instead do the matrix multiplications in any order you like.) The choice between using equation (2.7.24) and (2.7.25) depends on whether you want $\widetilde{\mathbf{P}}$ or $\widetilde{\mathbf{S}}$ to have the simpler formula; or on whether the repeated expression $(\mathbf{S} - \mathbf{R} \cdot \mathbf{P}^{-1} \cdot \mathbf{Q})^{-1}$ is easier to calculate than the expression $(\mathbf{P} - \mathbf{Q} \cdot \mathbf{S}^{-1} \cdot \mathbf{R})^{-1}$; or on the relative sizes of $\mathbf{P}$ and $\mathbf{S}$; or on whether $\mathbf{P}^{-1}$ or $\mathbf{S}^{-1}$ is already known.

Another sometimes useful formula is for the determinant of the partitioned matrix,

$$\det \mathbf{A} = \det \mathbf{P} \det(\mathbf{S} - \mathbf{R} \cdot \mathbf{P}^{-1} \cdot \mathbf{Q}) = \det \mathbf{S} \det(\mathbf{P} - \mathbf{Q} \cdot \mathbf{S}^{-1} \cdot \mathbf{R}) \qquad (2.7.26)$$

Indexed Storage of Sparse Matrices

We have already seen (§2.4) that tri- or band-diagonal matrices can be stored in a compact format that allocates storage only to elements which can be nonzero, plus perhaps a few wasted locations to make the bookkeeping easier. What about more general sparse matrices? When a sparse matrix of dimension $N \times N$ contains only a few times N nonzero elements (a typical case), it is surely inefficient — and often physically impossible — to allocate storage for all N^2 elements. Even if one did allocate such storage, it would be inefficient or prohibitive in machine time to loop over all of it in search of nonzero elements.

Obviously some kind of indexed storage scheme is required, one that stores only nonzero matrix elements, along with sufficient auxiliary information to determine where an element logically belongs and how the various elements can be looped over in common matrix operations. Unfortunately, there is no one standard scheme in general use. Knuth [7] describes one method. The Yale Sparse Matrix Package [6] and ITPACK [8] describe several other methods. For most applications, we favor the storage scheme used by PCGPACK [9], which is almost the same as that described by Bentley [10], and also similar to one of the Yale Sparse Matrix Package methods. The advantage of this scheme, which can be called *row-indexed sparse storage mode*, is that it requires storage of only about two times the number of nonzero matrix elements. (Other methods can require as much as three or five times.) For simplicity, we will treat only the case of square matrices, which occurs most frequently in practice.

To represent a matrix $\mathbf{A}$ of dimension $N \times N$, the row-indexed scheme sets up two one-dimensional arrays, call them sa and ija. The first of these stores matrix element values; the second stores integer values. The storage rules are:

- The first N locations of sa store $\mathbf{A}$'s diagonal matrix elements, in order. (Note that diagonal elements are stored even if they are zero; this is at most a slight storage inefficiency, since diagonal elements are nonzero in most realistic applications.)
- Each of the first N locations of ija stores the index of the array sa that contains the first *off-diagonal* element of the corresponding row of the matrix. (If there are no off-diagonal elements for that row, it is one greater than the index in sa of the most recently stored element of a previous row.)
- Location 0 of ija is always equal to $N + 1$. (It can be read to determine N.)
- Location N of ija is one greater than the index in sa of the last off-diagonal element of the last row. (It can be read to determine the number of nonzero elements in the matrix, or the number of elements in the arrays sa and ija.) Location N of sa is not used and can be set arbitrarily.
- Entries in sa at locations $\geq N + 1$ contain $\mathbf{A}$'s off-diagonal values, ordered by rows and, within each row, ordered by columns.
- Entries in ija at locations $\geq N+1$ contain the column number of the corresponding element in sa.

While these rules seem arbitrary at first sight, they result in a rather elegant storage scheme. As an example, consider the matrix

$$\begin{bmatrix} 3. & 0. & 1. & 0. & 0. \\ 0. & 4. & 0. & 0. & 0. \\ 0. & 7. & 5. & 9. & 0. \\ 0. & 0. & 0. & 0. & 2. \\ 0. & 0. & 0. & 6. & 5. \end{bmatrix} \qquad (2.7.27)$$

In row-indexed compact storage, matrix (2.7.27) is represented by the two arrays of length 11, as follows

index k	0	1	2	3	4	5	6	7	8	9	10
ija[k]	6	7	7	9	10	11	2	1	3	4	3
sa[k]	3.	4.	5.	0.	5.	x	1.	7.	9.	2.	6.

$$(2.7.28)$$

Here x is an arbitrary value. Notice that, according to the storage rules, the value of N (namely 5) is ija[0]-1, and the length of each array is ija[ija[0]-1], namely 11. The diagonal element in row i is sa[i], and the off-diagonal elements in that row are in sa[k] where k loops from ija[i] to ija[i+1]-1, if the upper limit is greater or equal to the lower one (as in C++'s for loops).

Here is a routine, sprsin, that converts a matrix from full storage mode into row-indexed sparse storage mode, throwing away any elements that are less than a specified threshold. Of course, the principal use of sparse storage mode is for matrices whose full storage mode won't fit into your machine at all; then you have to generate them directly into sparse format. Nevertheless sprsin is useful as a precise algorithmic definition of the storage scheme, for subscale testing of large problems, and for the case where execution time, rather than storage, furnishes the impetus to sparse storage.

```
#include <cmath>
#include "nr.h"
using namespace std;

void NR::sprsin(Mat_I_DP &a, const DP thresh, Vec_O_DP &sa, Vec_O_INT &ija)
Converts a square matrix a[0..n-1][0..n-1] into row-indexed sparse storage mode. Only
elements of a with magnitude ≥thresh are retained. Output is in two linear arrays: sa[0..]
contains array values, indexed by ija[0..]. These arrays should be supplied by the calling
routine with sizes at least sufficient to hold the compacted array. The number of elements filled
of sa and ija on output are both ija[ija[0]-1] (see text).
{
    int i,j,k;

    int n=a.nrows();
    int nmax=sa.size();
    for (j=0;j<n;j++) sa[j]=a[j][j];        Store diagonal elements.
    ija[0]=n+1;                             Index to row 0 off-diagonal element, if any.
    k=n;
    for (i=0;i<n;i++) {                      Loop over rows.
        for (j=0;j<n;j++) {                  Loop over columns.
            if (fabs(a[i][j]) >= thresh && i != j) {
                if (++k > nmax) nrerror("sprsin: sa and ija too small");
                sa[k]=a[i][j];              Store off-diagonal elements and their columns.
                ija[k]=j;
            }
        }
        ija[i+1]=k+1;                        As each row is completed, store index to
    }                                           next.
}
```

The single most important use of a matrix in row-indexed sparse storage mode is to multiply a vector to its right. In fact, the storage mode is optimized for just this purpose. The following routine is thus very simple.

```
#include "nr.h"

void NR::sprsax(Vec_I_DP &sa, Vec_I_INT &ija, Vec_I_DP &x, Vec_O_DP &b)
```
Multiply a matrix in row-index sparse storage arrays sa and ija by a vector x[0..n-1], giving a vector b[0..n-1].
```
{
    int i,k;

    int n=x.size();
    if (ija[0] != n+1)
        nrerror("sprsax: mismatched vector and matrix");
    for (i=0;i<n;i++) {
        b[i]=sa[i]*x[i];                        Start with diagonal term.
        for (k=ija[i];k<ija[i+1];k++) {         Loop over off-diagonal terms.
            b[i] += sa[k]*x[ija[k]];
        }
    }
}
```

It is also simple to multiply the *transpose* of a matrix by a vector to its right. (We will use this operation later in this section.) Note that the transpose matrix is not actually constructed.

```
#include "nr.h"

void NR::sprstx(Vec_I_DP &sa, Vec_I_INT &ija, Vec_I_DP &x, Vec_O_DP &b)
```
Multiply the transpose of a matrix in row-index sparse storage arrays sa and ija by a vector x[0..n-1], giving a vector b[0..n-1].
```
{
    int i,j,k;

    int n=x.size();
    if (ija[0] != (n+1))
        nrerror("mismatched vector and matrix in sprstx");
    for (i=0;i<n;i++) b[i]=sa[i]*x[i];          Start with diagonal terms.
    for (i=0;i<n;i++) {                         Loop over off-diagonal terms.
        for (k=ija[i];k<ija[i+1];k++) {
            j=ija[k];
            b[j] += sa[k]*x[i];
        }
    }
}
```

In fact, because the choice of row-indexed storage treats rows and columns quite differently, it is quite an involved operation to construct the transpose of a matrix, given the matrix itself in row-indexed sparse storage mode. When the operation cannot be avoided, it is done as follows: An index of all off-diagonal elements by their columns is constructed (see §8.4). The elements are then written to the output array in column order. As each element is written, its row is determined and stored. Finally, the elements in each column are sorted by row.

```
#include "nr.h"

void NR::sprstp(Vec_I_DP &sa, Vec_I_INT &ija, Vec_O_DP &sb, Vec_O_INT &ijb)
```
Construct the transpose of a sparse square matrix, from row-index sparse storage arrays sa and ija into arrays sb and ijb.
```
{
    int j,jl,jm,jp,ju,k,m,n2,noff,inc,iv;
    DP v;

    n2=ija[0];                                  Linear size of matrix plus 2.
    for (j=0;j<n2-1;j++) sb[j]=sa[j];           Diagonal elements.
    Vec_INT ija_vec((int *) &ija[n2],ija[n2-1]-ija[0]);
    Vec_INT ijb_vec(&ijb[n2],ija[n2-1]-ija[0]);
```

```
indexx(ija_vec,ijb_vec);
```
Overloaded version of indexx for long array. Index all off-diagonal elements by their columns.
```
for (j=n2,k=0;j < ija[n2-1];j++,k++) {
    ijb[j]=ijb_vec[k];
}
jp=0;
for (k=ija[0];k<ija[n2-1];k++) {                    Loop over output off-diagonal elements.
    m=ijb[k]+n2;                                    Use index table to store by (former) columns.
    sb[k]=sa[m];
    for (j=jp;j<ija[m]+1;j++)                       Fill in the index to any omitted rows.
        ijb[j]=k;
    jp=ija[m]+1;
    jl=0;                                           Use bisection to find which row element
    ju=n2-1;                                          m is in and put that into ijb[k].
    while (ju-jl > 1) {
        jm=(ju+jl)/2;
        if (ija[jm] > m) ju=jm; else jl=jm;
    }
    ijb[k]=jl;
}
for (j=jp;j<n2;j++) ijb[j]=ija[n2-1];
for (j=0;j<n2-1;j++) {                              Make a final pass to sort each row by
    jl=ijb[j+1]-ijb[j];                               Shell sort algorithm.
    noff=ijb[j];
    inc=1;
    do {
        inc *= 3;
        inc++;
    } while (inc <= jl);
    do {
        inc /= 3;
        for (k=noff+inc;k<noff+jl;k++) {
            iv=ijb[k];
            v=sb[k];
            m=k;
            while (ijb[m-inc] > iv) {
                ijb[m]=ijb[m-inc];
                sb[m]=sb[m-inc];
                m -= inc;
                if (m-noff+1 <= inc) break;
            }
            ijb[m]=iv;
            sb[m]=v;
        }
    } while (inc > 1);
}
}
```

The above routine embeds internally a sorting algorithm from §8.1, but calls the external routine indexx to construct the initial column index.

As final examples of the manipulation of sparse matrices, we give two routines for the multiplication of two sparse matrices. These are useful for techniques to be described in §13.10.

In general, the product of two sparse matrices is not itself sparse. One therefore wants to limit the size of the product matrix in one of two ways: either compute only those elements of the product that are specified in advance by a known pattern of sparsity, or else compute all nonzero elements, but store only those whose magnitude exceeds some threshold value. The former technique, when it can be used, is quite efficient. The pattern of sparsity is specified by furnishing an index array in row-index sparse storage format (e.g., ija). The program then constructs a corresponding value array (e.g., sa). The latter technique runs the danger of excessive compute times and unknown output sizes, so it must be used cautiously.

With row-index storage, it is much more natural to multiply a matrix (on the left) by the *transpose* of a matrix (on the right), so that one is crunching rows on rows, rather than rows on columns. Our routines therefore calculate $\mathbf{A} \cdot \mathbf{B}^T$, rather than $\mathbf{A} \cdot \mathbf{B}$. This means that you have to run your right-hand matrix through the transpose routine sprstp before sending it to the matrix multiply routine.

The two implementing routines, sprspm for "pattern multiply" and sprstm for "threshold multiply" are quite similar in structure. Both are complicated by the logic of the various combinations of diagonal or off-diagonal elements for the two input streams and output stream.

```
#include "nr.h"

void NR::sprspm(Vec_I_DP &sa, Vec_I_INT &ija, Vec_I_DP &sb, Vec_I_INT &ijb,
    Vec_O_DP &sc, Vec_I_INT &ijc)
```
Matrix multiply $\mathbf{A} \cdot \mathbf{B}^T$ where $\mathbf{A}$ and $\mathbf{B}$ are two sparse matrices in row-index storage mode, and $\mathbf{B}^T$ is the transpose of $\mathbf{B}$. Here, sa and ija store the matrix $\mathbf{A}$; sb and ijb store the matrix $\mathbf{B}$. This routine computes only those components of the matrix product that are *pre-specified* by the input index array ijc, which is not modified. On output, the arrays sc and ijc give the product matrix in row-index storage mode. For sparse matrix multiplication, this routine will often be preceded by a call to sprstp, so as to construct the transpose of a known matrix into sb, ijb.
```
{
    int i,ijma,ijmb,j,m,ma,mb,mbb,mn;
    DP sum;

    if (ija[0] != ijb[0] || ija[0] != ijc[0])
        nrerror("sprspm: sizes do not match");
    for (i=0;i<ijc[0]-1;i++) {          Loop over rows.
        m=i+1;
        j=m-1;                          Set up so that first pass through loop does the
        mn=ijc[i];                          diagonal component.
        sum=sa[i]*sb[i];
        for (;;) {                      Main loop over each component to be output.
            mb=ijb[j];
            for (ma=ija[i];ma<ija[i+1];ma++) {
            Loop through elements in A's row. Convoluted logic, following, accounts for the
            various combinations of diagonal and off-diagonal elements.
                ijma=ija[ma];
                if (ijma == j) sum += sa[ma]*sb[j];
                else {
                    while (mb < ijb[j+1]) {
                        ijmb=ijb[mb];
                        if (ijmb == i) {
                            sum += sa[i]*sb[mb++];
                            continue;
                        } else if (ijmb < ijma) {
                            mb++;
                            continue;
                        } else if (ijmb == ijma) {
                            sum += sa[ma]*sb[mb++];
                            continue;
                        }
                        break;
                    }
                }
            }
            for (mbb=mb;mbb<ijb[j+1];mbb++) {        Exhaust the remainder of B's row.
                if (ijb[mbb] == i) sum += sa[i]*sb[mbb];
            }
            sc[m-1]=sum;
            sum=0.0;                     Reset indices for next pass through loop.
            if (mn >= ijc[i+1]) break;
            j=ijc[(m= ++mn)-1];
        }
    }
}
```

```
}
```

```
#include <cmath>
#include "nr.h"
using namespace std;

void NR::sprstm(Vec_I_DP &sa, Vec_I_INT &ija, Vec_I_DP &sb, Vec_I_INT &ijb,
    const DP thresh, Vec_O_DP &sc, Vec_O_INT &ijc)
```
Matrix multiply $A \cdot B^T$ where A and B are two sparse matrices in row-index storage mode, and B^T is the transpose of B. Here, sa and ija store the matrix A; sb and ijb store the matrix B. This routine computes all components of the matrix product (which may be non-sparse!), but stores only those whose magnitude exceeds thresh. On output, the arrays sc and ijc give the product matrix in row-index storage mode. These arrays should be supplied by the calling routine with sizes at least sufficient to hold the product array. For sparse matrix multiplication, this routine will often be preceded by a call to sprstp, so as to construct the transpose of a known matrix into sb, ijb.
```
{
    int i,ijma,ijmb,j,k,ma,mb,mbb;
    DP sum;

    if (ija[0] != ijb[0]) nrerror("sprstm: sizes do not match");
    int nmax=sc.size();
    ijc[0]=k=ija[0];
    for (i=0;i<ija[0]-1;i++) {                    Loop over rows of A,
        for (j=0;j<ijb[0]-1;j++) {                and rows of B.
            if (i == j) sum=sa[i]*sb[j]; else sum=0.0e0;
            mb=ijb[j];
            for (ma=ija[i];ma<ija[i+1];ma++) {
            Loop through elements in A's row. Convoluted logic, following, accounts for the
            various combinations of diagonal and off-diagonal elements.
                ijma=ija[ma];
                if (ijma == j) sum += sa[ma]*sb[j];
                else {
                    while (mb < ijb[j+1]) {
                        ijmb=ijb[mb];
                        if (ijmb == i) {
                            sum += sa[i]*sb[mb++];
                            continue;
                        } else if (ijmb < ijma) {
                            mb++;
                            continue;
                        } else if (ijmb == ijma) {
                            sum += sa[ma]*sb[mb++];
                            continue;
                        }
                        break;
                    }
                }
            }
            for (mbb=mb;mbb<ijb[j+1];mbb++) {        Exhaust the remainder of B's row.
                if (ijb[mbb] == i) sum += sa[i]*sb[mbb];
            }
            if (i == j) sc[i]=sum;                  Where to put the answer...
            else if (fabs(sum) > thresh) {
                if (k > nmax-1) nrerror("sprstm: sc and ijc too small");
                sc[k]=sum;
                ijc[k++]=j;
            }
        }
        ijc[i+1]=k;
    }
}
```

Conjugate Gradient Method for a Sparse System

So-called *conjugate gradient methods* provide a quite general means for solving the $N \times N$ linear system

$$\mathbf{A} \cdot \mathbf{x} = \mathbf{b} \qquad (2.7.29)$$

The attractiveness of these methods for large sparse systems is that they reference $\mathbf{A}$ only through its multiplication of a vector, or the multiplication of its transpose and a vector. As we have seen, these operations can be very efficient for a properly stored sparse matrix. You, the "owner" of the matrix $\mathbf{A}$, can be asked to provide functions that perform these sparse matrix multiplications as efficiently as possible. We, the "grand strategists" supply the general routine, linbcg below, that solves the set of linear equations, (2.7.29), using your functions.

The simplest, "ordinary" conjugate gradient algorithm [11-13] solves (2.7.29) only in the case that $\mathbf{A}$ is symmetric and positive definite. It is based on the idea of minimizing the function

$$f(\mathbf{x}) = \frac{1}{2} \mathbf{x} \cdot \mathbf{A} \cdot \mathbf{x} - \mathbf{b} \cdot \mathbf{x} \qquad (2.7.30)$$

This function is minimized when its gradient

$$\nabla f = \mathbf{A} \cdot \mathbf{x} - \mathbf{b} \qquad (2.7.31)$$

is zero, which is equivalent to (2.7.29). The minimization is carried out by generating a succession of search directions $\mathbf{p}_k$ and improved minimizers $\mathbf{x}_k$. At each stage a quantity α_k is found that minimizes $f(\mathbf{x}_k + \alpha_k \mathbf{p}_k)$, and $\mathbf{x}_{k+1}$ is set equal to the new point $\mathbf{x}_k + \alpha_k \mathbf{p}_k$. The $\mathbf{p}_k$ and $\mathbf{x}_k$ are built up in such a way that $\mathbf{x}_{k+1}$ is also the minimizer of f over the whole vector space of directions already taken, $\{\mathbf{p}_0, \mathbf{p}_1, \ldots, \mathbf{p}_{k-1}\}$. After N iterations you arrive at the minimizer over the entire vector space, i.e., the solution to (2.7.29).

Later, in §10.6, we will generalize this "ordinary" conjugate gradient algorithm to the minimization of arbitrary nonlinear functions. Here, where our interest is in solving linear, but not necessarily positive definite or symmetric, equations, a different generalization is important, the *biconjugate gradient method*. This method does not, in general, have a simple connection with function minimization. It constructs four sequences of vectors, $\mathbf{r}_k$, $\bar{\mathbf{r}}_k$, $\mathbf{p}_k$, $\bar{\mathbf{p}}_k$, $k = 0, 1, \ldots$. You supply the initial vectors $\mathbf{r}_0$ and $\bar{\mathbf{r}}_0$, and set $\mathbf{p}_0 = \mathbf{r}_0$, $\bar{\mathbf{p}}_0 = \bar{\mathbf{r}}_0$. Then you carry out the following recurrence:

$$\alpha_k = \frac{\bar{\mathbf{r}}_k \cdot \mathbf{r}_k}{\bar{\mathbf{p}}_k \cdot \mathbf{A} \cdot \mathbf{p}_k}$$

$$\mathbf{r}_{k+1} = \mathbf{r}_k - \alpha_k \mathbf{A} \cdot \mathbf{p}_k$$

$$\bar{\mathbf{r}}_{k+1} = \bar{\mathbf{r}}_k - \alpha_k \mathbf{A}^T \cdot \bar{\mathbf{p}}_k$$

$$\beta_k = \frac{\bar{\mathbf{r}}_{k+1} \cdot \mathbf{r}_{k+1}}{\bar{\mathbf{r}}_k \cdot \mathbf{r}_k} \qquad (2.7.32)$$

$$\mathbf{p}_{k+1} = \mathbf{r}_{k+1} + \beta_k \mathbf{p}_k$$

$$\bar{\mathbf{p}}_{k+1} = \bar{\mathbf{r}}_{k+1} + \beta_k \bar{\mathbf{p}}_k$$

This sequence of vectors satisfies the *biorthogonality* condition

$$\bar{\mathbf{r}}_i \cdot \mathbf{r}_j = \mathbf{r}_i \cdot \bar{\mathbf{r}}_j = 0, \qquad j < i \qquad (2.7.33)$$

and the *biconjugacy* condition

$$\bar{\mathbf{p}}_i \cdot \mathbf{A} \cdot \mathbf{p}_j = \mathbf{p}_i \cdot \mathbf{A}^T \cdot \bar{\mathbf{p}}_j = 0, \qquad j < i \qquad (2.7.34)$$

There is also a mutual orthogonality,

$$\bar{\mathbf{r}}_i \cdot \mathbf{p}_j = \mathbf{r}_i \cdot \bar{\mathbf{p}}_j = 0, \qquad j < i \qquad (2.7.35)$$

The proof of these properties proceeds by straightforward induction [14]. As long as the recurrence does not break down earlier because one of the denominators is zero, it must terminate after $m \leq N$ steps with $\mathbf{r}_m = \bar{\mathbf{r}}_m = 0$. This is basically because after at most N steps you run out of new orthogonal directions to the vectors you've already constructed.

To use the algorithm to solve the system (2.7.29), make an initial guess $\mathbf{x}_0$ for the solution. Choose $\mathbf{r}_0$ to be the *residual*

$$\mathbf{r}_0 = \mathbf{b} - \mathbf{A} \cdot \mathbf{x}_0 \tag{2.7.36}$$

and choose $\bar{\mathbf{r}}_0 = \mathbf{r}_0$. Then form the sequence of improved estimates

$$\mathbf{x}_{k+1} = \mathbf{x}_k + \alpha_k \mathbf{p}_k \tag{2.7.37}$$

while carrying out the recurrence (2.7.32). Equation (2.7.37) guarantees that $\mathbf{r}_{k+1}$ from the recurrence is in fact the residual $\mathbf{b} - \mathbf{A} \cdot \mathbf{x}_{k+1}$ corresponding to $\mathbf{x}_{k+1}$. Since $\mathbf{r}_m = 0$, $\mathbf{x}_m$ is the solution to equation (2.7.29).

While there is no guarantee that this whole procedure will not break down or become unstable for general $\mathbf{A}$, in practice this is rare. More importantly, the exact termination in at most N iterations occurs only with exact arithmetic. Roundoff error means that you should regard the process as a genuinely iterative procedure, to be halted when some appropriate error criterion is met.

The ordinary conjugate gradient algorithm is the special case of the biconjugate gradient algorithm when $\mathbf{A}$ is symmetric, and we choose $\bar{\mathbf{r}}_0 = \mathbf{r}_0$. Then $\bar{\mathbf{r}}_k = \mathbf{r}_k$ and $\bar{\mathbf{p}}_k = \mathbf{p}_k$ for all k; you can omit computing them and halve the work of the algorithm. This conjugate gradient version has the interpretation of minimizing equation (2.7.30). If $\mathbf{A}$ is positive definite as well as symmetric, the algorithm cannot break down (in theory!). The routine `linbcg` below indeed reduces to the ordinary conjugate gradient method if you input a symmetric $\mathbf{A}$, but it does all the redundant computations.

Another variant of the general algorithm corresponds to a symmetric but non-positive definite $\mathbf{A}$, with the choice $\bar{\mathbf{r}}_0 = \mathbf{A} \cdot \mathbf{r}_0$ instead of $\bar{\mathbf{r}}_0 = \mathbf{r}_0$. In this case $\bar{\mathbf{r}}_k = \mathbf{A} \cdot \mathbf{r}_k$ and $\bar{\mathbf{p}}_k = \mathbf{A} \cdot \mathbf{p}_k$ for all k. This algorithm is thus equivalent to the ordinary conjugate gradient algorithm, but with all dot products $\mathbf{a} \cdot \mathbf{b}$ replaced by $\mathbf{a} \cdot \mathbf{A} \cdot \mathbf{b}$. It is called the *minimum residual* algorithm, because it corresponds to successive minimizations of the function

$$\Phi(\mathbf{x}) = \frac{1}{2}\, \mathbf{r} \cdot \mathbf{r} = \frac{1}{2}\, |\mathbf{A} \cdot \mathbf{x} - \mathbf{b}|^2 \tag{2.7.38}$$

where the successive iterates $\mathbf{x}_k$ minimize Φ over the same set of search directions $\mathbf{p}_k$ generated in the conjugate gradient method. This algorithm has been generalized in various ways for unsymmetric matrices. The *generalized minimum residual* method (GMRES; see [9,15]) is probably the most robust of these methods.

Note that equation (2.7.38) gives

$$\nabla\Phi(\mathbf{x}) = \mathbf{A}^T \cdot (\mathbf{A} \cdot \mathbf{x} - \mathbf{b}) \tag{2.7.39}$$

For any nonsingular matrix $\mathbf{A}$, $\mathbf{A}^T \cdot \mathbf{A}$ is symmetric and positive definite. You might therefore be tempted to solve equation (2.7.29) by applying the ordinary conjugate gradient algorithm to the problem

$$(\mathbf{A}^T \cdot \mathbf{A}) \cdot \mathbf{x} = \mathbf{A}^T \cdot \mathbf{b} \tag{2.7.40}$$

Don't! The condition number of the matrix $\mathbf{A}^T \cdot \mathbf{A}$ is the square of the condition number of $\mathbf{A}$ (see §2.6 for definition of condition number). A large condition number both increases the number of iterations required, and limits the accuracy to which a solution can be obtained. It is almost always better to apply the biconjugate gradient method to the original matrix $\mathbf{A}$.

So far we have said nothing about the *rate* of convergence of these methods. The ordinary conjugate gradient method works well for matrices that are well-conditioned, i.e., "close" to the identity matrix. This suggests applying these methods to the *preconditioned* form of equation (2.7.29),

$$(\widetilde{\mathbf{A}}^{-1} \cdot \mathbf{A}) \cdot \mathbf{x} = \widetilde{\mathbf{A}}^{-1} \cdot \mathbf{b} \tag{2.7.41}$$

The idea is that you might already be able to solve your linear system easily for some $\widetilde{\mathbf{A}}$ close to $\mathbf{A}$, in which case $\widetilde{\mathbf{A}}^{-1} \cdot \mathbf{A} \approx \mathbf{1}$, allowing the algorithm to converge in fewer steps. The matrix $\widetilde{\mathbf{A}}$ is called a *preconditioner* [11], and the overall scheme given here is known as the *preconditioned biconjugate gradient method* or *PBCG*.

For efficient implementation, the PBCG algorithm introduces an additional set of vectors $\mathbf{z}_k$ and $\overline{\mathbf{z}}_k$ defined by

$$\widetilde{\mathbf{A}} \cdot \mathbf{z}_k = \mathbf{r}_k \qquad \text{and} \qquad \widetilde{\mathbf{A}}^T \cdot \overline{\mathbf{z}}_k = \overline{\mathbf{r}}_k \tag{2.7.42}$$

and modifies the definitions of α_k, β_k, $\mathbf{p}_k$, and $\overline{\mathbf{p}}_k$ in equation (2.7.32):

$$\alpha_k = \frac{\overline{\mathbf{r}}_k \cdot \mathbf{z}_k}{\overline{\mathbf{p}}_k \cdot \mathbf{A} \cdot \mathbf{p}_k}$$

$$\beta_k = \frac{\overline{\mathbf{r}}_{k+1} \cdot \mathbf{z}_{k+1}}{\overline{\mathbf{r}}_k \cdot \mathbf{z}_k} \tag{2.7.43}$$

$$\mathbf{p}_{k+1} = \mathbf{z}_{k+1} + \beta_k \mathbf{p}_k$$

$$\overline{\mathbf{p}}_{k+1} = \overline{\mathbf{z}}_{k+1} + \beta_k \overline{\mathbf{p}}_k$$

For `linbcg`, below, we will ask you to supply routines that solve the auxiliary linear systems (2.7.42). If you have no idea what to use for the preconditioner $\widetilde{\mathbf{A}}$, then use the diagonal part of $\mathbf{A}$, or even the identity matrix, in which case the burden of convergence will be entirely on the biconjugate gradient method itself.

The routine `linbcg`, below, is based on a program originally written by Anne Greenbaum. (See [13] for a different, less sophisticated, implementation.) There are a few wrinkles you should know about.

What constitutes "good" convergence is rather application dependent. The routine `linbcg` therefore provides for four possibilities, selected by setting the flag `itol` on input. If `itol=1`, iteration stops when the quantity $|\mathbf{A} \cdot \mathbf{x} - \mathbf{b}|/|\mathbf{b}|$ is less than the input quantity `tol`. If `itol=2`, the required criterion is

$$|\widetilde{\mathbf{A}}^{-1} \cdot (\mathbf{A} \cdot \mathbf{x} - \mathbf{b})|/|\widetilde{\mathbf{A}}^{-1} \cdot \mathbf{b}| < \texttt{tol} \tag{2.7.44}$$

If `itol=3`, the routine uses its own estimate of the error in $\mathbf{x}$, and requires its magnitude, divided by the magnitude of $\mathbf{x}$, to be less than `tol`. The setting `itol=4` is the same as `itol=3`, except that the largest (in absolute value) component of the error and largest component of $\mathbf{x}$ are used instead of the vector magnitude (that is, the L_∞ norm instead of the L_2 norm). You may need to experiment to find which of these convergence criteria is best for your problem.

On output, `err` is the tolerance actually achieved. If the returned count `iter` does not indicate that the maximum number of allowed iterations `itmax` was exceeded, then `err` should be less than `tol`. If you want to do further iterations, leave all returned quantities as they are and call the routine again. The routine loses its memory of the spanned conjugate gradient subspace between calls, however, so you should not force it to return more often than about every N iterations.

```
#include <iostream>
#include <iomanip>
#include <cmath>
#include "nr.h"
using namespace std;
```

```
void NR::linbcg(Vec_I_DP &b, Vec_IO_DP &x, const int itol, const DP tol,
    const int itmax, int &iter, DP &err)
```
Solves $\mathbf{A} \cdot \mathbf{x} = \mathbf{b}$ for $\mathbf{x}$`[0..n-1]`, given $\mathbf{b}$`[0..n-1]`, by the iterative biconjugate gradient method. On input $\mathbf{x}$`[0..n-1]` should be set to an initial guess of the solution (or all zeros); `itol` is 1,2,3, or 4, specifying which convergence test is applied (see text); `itmax` is the maximum number of allowed iterations; and `tol` is the desired convergence tolerance. On output, $\mathbf{x}$`[0..n-1]` is reset to the improved solution, `iter` is the number of iterations actually taken, and `err` is the estimated error. The matrix $\mathbf{A}$ is referenced only through the user-supplied routines `atimes`, which computes the product of either $\mathbf{A}$ or its transpose on a vector; and `asolve`, which solves $\widetilde{\mathbf{A}} \cdot \mathbf{x} = \mathbf{b}$ or $\widetilde{\mathbf{A}}^T \cdot \mathbf{x} = \mathbf{b}$ for some preconditioner matrix $\widetilde{\mathbf{A}}$ (possibly the trivial diagonal part of $\mathbf{A}$).
```
{
```

```
DP ak,akden,bk,bkden=1.0,bknum,bnrm,dxnrm,xnrm,zm1nrm,znrm;
Double precision is a good idea in this routine.
const DP EPS=1.0e-14;
int j;

int n=b.size();
Vec_DP p(n),pp(n),r(n),rr(n),z(n),zz(n);
iter=0;                                   Calculate initial residual.
atimes(x,r,0);                            Input to atimes is x[0..n-1], output is r[0..n-1];
for (j=0;j<n;j++) {                            the final 0 indicates that the matrix (not its
    r[j]=b[j]-r[j];                            transpose) is to be used.
    rr[j]=r[j];
}
//atimes(r,rr,0);                         Uncomment this line to get the "minimum resid-
if (itol == 1) {                              ual" variant of the algorithm.
    bnrm=snrm(b,itol);
    asolve(r,z,0);                        Input to asolve is r[0..n-1], output is z[0..n-1];
}                                             the final 0 indicates that the matrix Ã (not
else if (itol == 2) {                         its transpose) is to be used.
    asolve(b,z,0);
    bnrm=snrm(z,itol);
    asolve(r,z,0);
}
else if (itol == 3 || itol == 4) {
    asolve(b,z,0);
    bnrm=snrm(z,itol);
    asolve(r,z,0);
    znrm=snrm(z,itol);
} else nrerror("illegal itol in linbcg");
cout << fixed << setprecision(6);
while (iter < itmax) {                    Main loop.
    ++iter;
    asolve(rr,zz,1);                      Final 1 indicates use of transpose matrix Ã^T.
    for (bknum=0.0,j=0;j<n;j++) bknum += z[j]*rr[j];
    Calculate coefficient bk and direction vectors p and pp.
    if (iter == 1) {
        for (j=0;j<n;j++) {
            p[j]=z[j];
            pp[j]=zz[j];
        }
    } else {
        bk=bknum/bkden;
        for (j=0;j<n;j++) {
            p[j]=bk*p[j]+z[j];
            pp[j]=bk*pp[j]+zz[j];
        }
    }
    bkden=bknum;                          Calculate coefficient ak, new iterate x, and new
    atimes(p,z,0);                                  residuals r and rr.
    for (akden=0.0,j=0;j<n;j++) akden += z[j]*pp[j];
    ak=bknum/akden;
    atimes(pp,zz,1);
    for (j=0;j<n;j++) {
        x[j] += ak*p[j];
        r[j] -= ak*z[j];
        rr[j] -= ak*zz[j];
    }
    asolve(r,z,0);                        Solve Ã · z = r and check stopping criterion.
    if (itol == 1)
        err=snrm(r,itol)/bnrm;
    else if (itol == 2)
        err=snrm(z,itol)/bnrm;
    else if (itol == 3 || itol == 4) {
        zm1nrm=znrm;
```

```
            znrm=snrm(z,itol);
            if (fabs(zm1nrm-znrm) > EPS*znrm) {
                dxnrm=fabs(ak)*snrm(p,itol);
                err=znrm/fabs(zm1nrm-znrm)*dxnrm;
            } else {
                err=znrm/bnrm;           Error may not be accurate, so loop again.
                continue;
            }
            xnrm=snrm(x,itol);
            if (err <= 0.5*xnrm) err /= xnrm;
            else {
                err=znrm/bnrm;           Error may not be accurate, so loop again.
                continue;
            }
        }
        cout << "iter=" << setw(4) << iter+1 << setw(12) << err << endl;
        if (err <= tol) break;
    }
}
```

The routine `linbcg` uses this short utility for computing vector norms:

```
#include <cmath>
#include "nr.h"
using namespace std;

DP NR::snrm(Vec_I_DP &sx, const int itol)
Compute one of two norms for a vector sx[0..n-1], as signaled by itol. Used by linbcg.
{
    int i,isamax;
    DP ans;

    int n=sx.size();
    if (itol <= 3) {
        ans = 0.0;
        for (i=0;i<n;i++) ans += sx[i]*sx[i];        Vector magnitude norm.
        return sqrt(ans);
    } else {
        isamax=0;
        for (i=0;i<n;i++) {                          Largest component norm.
            if (fabs(sx[i]) > fabs(sx[isamax])) isamax=i;
        }
        return fabs(sx[isamax]);
    }
}
```

So that the specifications for the routines `atimes` and `asolve` are clear, we list here simple versions that assume a matrix **A** stored somewhere in row-index sparse format.

```
#include "nr.h"

extern Vec_INT *ija_p;
extern Vec_DP *sa_p;                 The matrix is stored somewhere.

void NR::atimes(Vec_I_DP &x, Vec_O_DP &r, const int itrnsp)
{
    if (itrnsp) sprstx(*sa_p,*ija_p,x,r);
    else sprsax(*sa_p,*ija_p,x,r);
}
```

```
#include "nr.h"

extern Vec_INT *ija_p;
extern Vec_DP *sa_p;                    The matrix is stored somewhere.

void NR::asolve(Vec_I_DP &b, Vec_O_DP &x, const int itrnsp)
{
    int i;

    int n=b.size();
    for(i=0;i<n;i++) x[i]=((*sa_p)[i] != 0.0 ? b[i]/(*sa_p)[i] : b[i]);
```
The matrix $\tilde{\mathbf{A}}$ is the diagonal part of $\mathbf{A}$, stored in the first n elements of sa. Since the transpose matrix has the same diagonal, the flag itrnsp is not used in this example.
```
}
```

CITED REFERENCES AND FURTHER READING:

Tewarson, R.P. 1973, *Sparse Matrices* (New York: Academic Press). [1]

Jacobs, D.A.H. (ed.) 1977, *The State of the Art in Numerical Analysis* (London: Academic Press), Chapter I.3 (by J.K. Reid). [2]

George, A., and Liu, J.W.H. 1981, *Computer Solution of Large Sparse Positive Definite Systems* (Englewood Cliffs, NJ: Prentice-Hall). [3]

NAG Fortran Library (Numerical Algorithms Group, 256 Banbury Road, Oxford OX27DE, U.K.). [4]

IMSL Math/Library Users Manual (IMSL Inc., 2500 CityWest Boulevard, Houston TX 77042). [5]

Eisenstat, S.C., Gursky, M.C., Schultz, M.H., and Sherman, A.H. 1977, *Yale Sparse Matrix Package*, Technical Reports 112 and 114 (Yale University Department of Computer Science). [6]

Knuth, D.E. 1997, *Fundamental Algorithms*, 3rd ed., vol. 1 of *The Art of Computer Programming* (Reading, MA: Addison-Wesley), §2.2.6. [7]

Kincaid, D.R., Respess, J.R., Young, D.M., and Grimes, R.G. 1982, *ACM Transactions on Mathematical Software*, vol. 8, pp. 302–322. [8]

PCGPAK User's Guide (New Haven: Scientific Computing Associates, Inc.). [9]

Bentley, J. 1986, *Programming Pearls* (Reading, MA: Addison-Wesley), §9. [10]

Golub, G.H., and Van Loan, C.F. 1996, *Matrix Computations*, 3rd ed. (Baltimore: Johns Hopkins University Press), Chapters 4 and 10, particularly §§10.2–10.3. [11]

Stoer, J., and Bulirsch, R. 1993, *Introduction to Numerical Analysis*, 2nd ed. (New York: Springer-Verlag), Chapter 8. [12]

Baker, L. 1991, *More C Tools for Scientists and Engineers* (New York: McGraw-Hill). [13]

Fletcher, R. 1976, in *Numerical Analysis Dundee 1975*, Lecture Notes in Mathematics, vol. 506, A. Dold and B Eckmann, eds. (Berlin: Springer-Verlag), pp. 73–89. [14]

Saad, Y., and Schulz, M. 1986, *SIAM Journal on Scientific and Statistical Computing*, vol. 7, pp. 856–869. [15]

Bunch, J.R., and Rose, D.J. (eds.) 1976, *Sparse Matrix Computations* (New York: Academic Press).

Duff, I.S., and Stewart, G.W. (eds.) 1979, *Sparse Matrix Proceedings 1978* (Philadelphia: S.I.A.M.).

2.8 Vandermonde Matrices and Toeplitz Matrices

In §2.4 the case of a tridiagonal matrix was treated specially, because that particular type of linear system admits a solution in only of order N operations, rather than of order N^3 for the general linear problem. When such particular types exist, it is important to know about them. Your computational savings, should you ever happen to be working on a problem that involves the right kind of particular type, can be enormous.

This section treats two special types of matrices that can be solved in of order N^2 operations, not as good as tridiagonal, but a lot better than the general case. (Other than the operations count, these two types having nothing in common.) Matrices of the first type, termed *Vandermonde matrices*, occur in some problems having to do with the fitting of polynomials, the reconstruction of distributions from their moments, and also other contexts. In this book, for example, a Vandermonde problem crops up in §3.5. Matrices of the second type, termed *Toeplitz matrices*, tend to occur in problems involving deconvolution and signal processing. In this book, a Toeplitz problem is encountered in §13.7.

These are not the *only* special types of matrices worth knowing about. The *Hilbert matrices*, whose components are of the form $a_{ij} = 1/(i+j+1)$, $i,j = 0,\ldots,N-1$ can be inverted by an exact integer algorithm, and are very *difficult* to invert in any other way, since they are notoriously ill-conditioned (see [1] for details). The Sherman-Morrison and Woodbury formulas, discussed in §2.7, can sometimes be used to convert new special forms into old ones. Reference [2] gives some other special forms. We have not found these additional forms to arise as frequently as the two that we now discuss.

Vandermonde Matrices

A Vandermonde matrix of size $N \times N$ is completely determined by N arbitrary numbers $x_0, x_1, \ldots, x_{N-1}$, in terms of which its N^2 components are the integer powers x_i^j, $i,j = 0,\ldots,N-1$. Evidently there are two possible such forms, depending on whether we view the i's as rows, j's as columns, or vice versa. In the former case, we get a linear system of equations that looks like this,

$$
\begin{bmatrix}
1 & x_0 & x_0^2 & \cdots & x_0^{N-1} \\
1 & x_1 & x_1^2 & \cdots & x_1^{N-1} \\
\vdots & \vdots & \vdots & & \vdots \\
1 & x_{N-1} & x_{N-1}^2 & \cdots & x_{N-1}^{N-1}
\end{bmatrix}
\cdot
\begin{bmatrix}
c_0 \\
c_1 \\
\vdots \\
c_{N-1}
\end{bmatrix}
=
\begin{bmatrix}
y_0 \\
y_1 \\
\vdots \\
y_{N-1}
\end{bmatrix}
\tag{2.8.1}
$$

Performing the matrix multiplication, you will see that this equation solves for the unknown coefficients c_i which fit a polynomial to the N pairs of abscissas and ordinates (x_j, y_j). Precisely this problem will arise in §3.5, and the routine given there will solve (2.8.1) by the method that we are about to describe.

The alternative identification of rows and columns leads to the set of equations

$$
\begin{bmatrix}
1 & 1 & \cdots & 1 \\
x_0 & x_1 & \cdots & x_{N-1} \\
x_0^2 & x_1^2 & \cdots & x_{N-1}^2 \\
& & \cdots & \\
x_0^{N-1} & x_1^{N-1} & \cdots & x_{N-1}^{N-1}
\end{bmatrix}
\cdot
\begin{bmatrix}
w_0 \\ w_1 \\ w_2 \\ \cdots \\ w_{N-1}
\end{bmatrix}
=
\begin{bmatrix}
q_0 \\ q_1 \\ q_2 \\ \cdots \\ q_{N-1}
\end{bmatrix}
\tag{2.8.2}
$$

Write this out and you will see that it relates to the *problem of moments*: Given the values of N points x_i, find the unknown weights w_i, assigned so as to match the given values q_j of the first N moments. (For more on this problem, consult [3].) The routine given in this section solves (2.8.2).

The method of solution of both (2.8.1) and (2.8.2) is closely related to Lagrange's polynomial interpolation formula, which we will not formally meet until §3.1 below. Notwithstanding, the following derivation should be comprehensible:
Let $P_j(x)$ be the polynomial of degree $N-1$ defined by

$$
P_j(x) = \prod_{\substack{n=0 \\ (n \neq j)}}^{N-1} \frac{x - x_n}{x_j - x_n} = \sum_{k=0}^{N-1} A_{jk} x^k
\tag{2.8.3}
$$

Here the meaning of the last equality is to define the components of the matrix A_{ij} as the coefficients that arise when the product is multiplied out and like terms collected.
The polynomial $P_j(x)$ is a function of x generally. But you will notice that it is specifically designed so that it takes on a value of zero at all x_i with $i \neq j$, and has a value of unity at $x = x_j$. In other words,

$$
P_j(x_i) = \delta_{ij} = \sum_{k=0}^{N-1} A_{jk} x_i^k
\tag{2.8.4}
$$

But (2.8.4) says that A_{jk} is exactly the inverse of the matrix of components x_i^k, which appears in (2.8.2), with the subscript as the column index. Therefore the solution of (2.8.2) is just that matrix inverse times the right-hand side,

$$
w_j = \sum_{k=0}^{N-1} A_{jk} q_k
\tag{2.8.5}
$$

As for the transpose problem (2.8.1), we can use the fact that the inverse of the transpose is the transpose of the inverse, so

$$
c_j = \sum_{k=0}^{N-1} A_{kj} y_k
\tag{2.8.6}
$$

The routine in §3.5 implements this.
It remains to find a good way of multiplying out the monomial terms in (2.8.3), in order to get the components of A_{jk}. This is essentially a bookkeeping problem, and we will let you read the routine itself to see how it can be solved. One trick is to define a master $P(x)$ by

$$
P(x) \equiv \prod_{n=0}^{N-1} (x - x_n)
\tag{2.8.7}
$$

work out its coefficients, and then obtain the numerators and denominators of the specific P_j's via synthetic division by the one supernumerary term. (See §5.3 for more on synthetic division.) Since each such division is only a process of order N, the total procedure is of order N^2.
You should be warned that Vandermonde systems are notoriously ill-conditioned, by their very nature. (As an aside anticipating §5.8, the reason is the same as that which makes Chebyshev fitting so impressively accurate: there exist high-order polynomials that are very good uniform fits to zero. Hence roundoff error can introduce rather substantial coefficients of the leading terms of these polynomials.) It is a good idea always to compute Vandermonde problems in double precision.

The routine for (2.8.2) which follows is due to G.B. Rybicki.

```
#include "nr.h"

void NR::vander(Vec_I_DP &x, Vec_O_DP &w, Vec_I_DP &q)
```
Solves the Vandermonde linear system $\sum_{i=0}^{N-1} x_i^k w_i = q_k$ $(k = 0, \dots, N-1)$. Input consists of the vectors x[0..n-1] and q[0..n-1]; the vector w[0..n-1] is output.
```
{
    int i,j,k;
    DP b,s,t,xx;

    int n=q.size();
    Vec_DP c(n);
    if (n == 1) w[0]=q[0];
    else {
        for (i=0;i<n;i++) c[i]=0.0;          Initialize array.
        c[n-1] = -x[0];                      Coefficients of the master polynomial are found
        for (i=1;i<n;i++) {                      by recursion.
            xx = -x[i];
            for (j=(n-1-i);j<(n-1);j++) c[j] += xx*c[j+1];
            c[n-1] += xx;
        }
        for (i=0;i<n;i++) {                  Each subfactor in turn
            xx=x[i];
            t=b=1.0;
            s=q[n-1];
            for (k=n-1;k>0;k--) {            is synthetically divided,
                b=c[k]+xx*b;
                s += q[k-1]*b;               matrix-multiplied by the right-hand side,
                t=xx*t+b;
            }
            w[i]=s/t;                        and supplied with a denominator.
        }
    }
}
```

Toeplitz Matrices

An $N \times N$ Toeplitz matrix is specified by giving $2N - 1$ numbers R_k, where the index k ranges over $k = -N + 1, \dots, -1, 0, 1, \dots, N - 1$. Those numbers are then emplaced as matrix elements constant along the (upper-left to lower-right) diagonals of the matrix:

$$
\begin{bmatrix}
R_0 & R_{-1} & R_{-2} & \cdots & R_{-(N-2)} & R_{-(N-1)} \\
R_1 & R_0 & R_{-1} & \cdots & R_{-(N-3)} & R_{-(N-2)} \\
R_2 & R_1 & R_0 & \cdots & R_{-(N-4)} & R_{-(N-3)} \\
\cdots & & & \cdots & & \\
R_{N-2} & R_{N-3} & R_{N-4} & \cdots & R_0 & R_{-1} \\
R_{N-1} & R_{N-2} & R_{N-3} & \cdots & R_1 & R_0
\end{bmatrix}
\tag{2.8.8}
$$

The linear Toeplitz problem can thus be written as

$$
\sum_{j=0}^{N-1} R_{i-j} x_j = y_i \qquad (i = 0, \dots, N - 1)
\tag{2.8.9}
$$

where the x_j's, $j = 0, \dots, N - 1$, are the unknowns to be solved for.

The Toeplitz matrix is symmetric if $R_k = R_{-k}$ for all k. Levinson [4] developed an algorithm for fast solution of the symmetric Toeplitz problem, by a *bordering method*, that is, a recursive procedure that solves the $(M + 1)$-dimensional Toeplitz problem

$$
\sum_{j=0}^{M} R_{i-j} x_j^{(M)} = y_i \qquad (i = 0, \dots, M)
\tag{2.8.10}
$$

in turn for $M = 0, 1, \ldots$ until $M = N - 1$, the desired result, is finally reached. The vector $x_j^{(M)}$ is the result at the Mth stage, and becomes the desired answer only when $N - 1$ is reached.

Levinson's method is well documented in standard texts (e.g., [5]). The useful fact that the method generalizes to the *nonsymmetric* case seems to be less well known. At some risk of excessive detail, we therefore give a derivation here, due to G.B. Rybicki.

In following a recursion from step M to step $M + 1$ we find that our developing solution $x^{(M)}$ changes in this way:

$$\sum_{j=0}^{M} R_{i-j} x_j^{(M)} = y_i \qquad i = 0, \ldots, M \tag{2.8.11}$$

becomes

$$\sum_{j=0}^{M} R_{i-j} x_j^{(M+1)} + R_{i-(M+1)} x_{M+1}^{(M+1)} = y_i \qquad i = 0, \ldots, M + 1 \tag{2.8.12}$$

By eliminating y_i we find

$$\sum_{j=0}^{M} R_{i-j} \left(\frac{x_j^{(M)} - x_j^{(M+1)}}{x_{M+1}^{(M+1)}} \right) = R_{i-(M+1)} \qquad i = 0, \ldots, M \tag{2.8.13}$$

or by letting $i \to M - i$ and $j \to M - j$,

$$\sum_{j=0}^{M} R_{j-i} G_j^{(M)} = R_{-(i+1)} \tag{2.8.14}$$

where

$$G_j^{(M)} \equiv \frac{x_{M-j}^{(M)} - x_{M-j}^{(M+1)}}{x_{M+1}^{(M+1)}} \tag{2.8.15}$$

To put this another way,

$$x_{M-j}^{(M+1)} = x_{M-j}^{(M)} - x_{M+1}^{(M+1)} G_j^{(M)} \qquad j = 0, \ldots, M \tag{2.8.16}$$

Thus, if we can use recursion to find the order M quantities $x^{(M)}$ and $G^{(M)}$ *and* the single order $M + 1$ quantity $x_{M+1}^{(M+1)}$, then all of the other $x_j^{(M+1)}$ will follow. Fortunately, the quantity $x_{M+1}^{(M+1)}$ follows from equation (2.8.12) with $i = M + 1$,

$$\sum_{j=0}^{M} R_{M+1-j} x_j^{(M+1)} + R_0 x_{M+1}^{(M+1)} = y_{M+1} \tag{2.8.17}$$

For the unknown order $M + 1$ quantities $x_j^{(M+1)}$ we can substitute the previous order quantities in G since

$$G_{M-j}^{(M)} = \frac{x_j^{(M)} - x_j^{(M+1)}}{x_{M+1}^{(M+1)}} \tag{2.8.18}$$

The result of this operation is

$$x_{M+1}^{(M+1)} = \frac{\sum_{j=0}^{M} R_{M+1-j} x_j^{(M)} - y_{M+1}}{\sum_{j=0}^{M} R_{M+1-j} G_{M-j}^{(M)} - R_0} \tag{2.8.19}$$

The only remaining problem is to develop a recursion relation for G. Before we do that, however, we should point out that there are actually two distinct sets of solutions to the original linear problem for a nonsymmetric matrix, namely right-hand solutions (which we have been discussing) and left-hand solutions z_i. The formalism for the left-hand solutions differs only in that we deal with the equations

$$\sum_{j=0}^{M} R_{j-i} z_j^{(M)} = y_i \qquad i = 0, \ldots, M \tag{2.8.20}$$

Then, the same sequence of operations on this set leads to

$$\sum_{j=0}^{M} R_{i-j} H_j^{(M)} = R_{i+1} \tag{2.8.21}$$

where

$$H_j^{(M)} \equiv \frac{z_{M-j}^{(M)} - z_{M-j}^{(M+1)}}{z_{M+1}^{(M+1)}} \tag{2.8.22}$$

(compare with 2.8.14 – 2.8.15). The reason for mentioning the left-hand solutions now is that, by equation (2.8.21), the H_j satisfy exactly the same equation as the x_j except for the substitution $y_i \to R_{i+1}$ on the right-hand side. Therefore we can quickly deduce from equation (2.8.19) that

$$H_{M+1}^{(M+1)} = \frac{\sum_{j=0}^{M} R_{M+1-j} H_j^{(M)} - R_{M+2}}{\sum_{j=0}^{M} R_{M+1-j} G_{M-j}^{(M)} - R_0} \tag{2.8.23}$$

By the same token, G satisfies the same equation as z, except for the substitution $y_i \to R_{-(i+1)}$. This gives

$$G_{M+1}^{(M+1)} = \frac{\sum_{j=0}^{M} R_{j-M-1} G_j^{(M)} - R_{-M-2}}{\sum_{j=0}^{M} R_{j-M-1} H_{M-j}^{(M)} - R_0} \tag{2.8.24}$$

The same "morphism" also turns equation (2.8.16), and its partner for z, into the final equations

$$\begin{aligned} G_j^{(M+1)} &= G_j^{(M)} - G_{M+1}^{(M+1)} H_{M-j}^{(M)} \\ H_j^{(M+1)} &= H_j^{(M)} - H_{M+1}^{(M+1)} G_{M-j}^{(M)} \end{aligned} \tag{2.8.25}$$

Now, starting with the initial values

$$x_0^{(0)} = y_0/R_0 \qquad G_0^{(0)} = R_{-1}/R_0 \qquad H_0^{(0)} = R_1/R_0 \tag{2.8.26}$$

we can recurse away. At each stage M we use equations (2.8.23) and (2.8.24) to find $H_{M+1}^{(M+1)}, G_{M+1}^{(M+1)}$, and then equation (2.8.25) to find the other components of $H^{(M+1)}, G^{(M+1)}$. From there the vectors $x^{(M+1)}$ and/or $z^{(M+1)}$ are easily calculated.

The program below does this. It incorporates the second equation in (2.8.25) in the form

$$H_{M-j}^{(M+1)} = H_{M-j}^{(M)} - H_{M+1}^{(M+1)} G_j^{(M)} \tag{2.8.27}$$

so that the computation can be done "in place."

Notice that the above algorithm fails if $R_0 = 0$. In fact, because the bordering method does not allow pivoting, the algorithm will fail if any of the diagonal principal minors of the original Toeplitz matrix vanish. (Compare with discussion of the tridiagonal algorithm in §2.4.) If the algorithm fails, your matrix is not necessarily singular — you might just have to solve your problem by a slower and more general algorithm such as *LU* decomposition with pivoting.

The routine that implements equations (2.8.23)–(2.8.27) is also due to Rybicki. Note that the routine's r[n-1+j] is equal to R_j above, so that subscripts on the r array vary from 0 to $2N - 2$.

```
#include "nr.h"

void NR::toeplz(Vec_I_DP &r, Vec_O_DP &x, Vec_I_DP &y)
```
Solves the Toeplitz system $\sum_{j=0}^{N-1} R_{(N-1+i-j)}x_j = y_i \ (i = 0, \ldots, N-1)$. The Toeplitz matrix
need not be symmetric. y[0..n-1] and r[0..2*n-2] are input arrays; x[0..n-1] is the
output array.
```
{
    int j,k,m,m1,m2,n,n1;
    DP pp,pt1,pt2,qq,qt1,qt2,sd,sgd,sgn,shn,sxn;

    n=y.size();
    n1=n-1;
    if (r[n1] == 0.0) nrerror("toeplz-1 singular principal minor");
    x[0]=y[0]/r[n1];                       Initialize for the recursion.
    if (n1 == 0) return;
    Vec_DP g(n1),h(n1);
    g[0]=r[n1-1]/r[n1];
    h[0]=r[n1+1]/r[n1];
    for (m=0;m<n;m++) {                     Main loop over the recursion.
        m1=m+1;
        sxn = -y[m1];                       Compute numerator and denominator for x from eq.
        sd = -r[n1];                          (2.8.19),
        for (j=0;j<m+1;j++) {
            sxn += r[n1+m1-j]*x[j];
            sd += r[n1+m1-j]*g[m-j];
        }
        if (sd == 0.0) nrerror("toeplz-2 singular principal minor");
        x[m1]=sxn/sd;                       whence x.
        for (j=0;j<m+1;j++)                 Eq. (2.8.16).
            x[j] -= x[m1]*g[m-j];
        if (m1 == n1) return;
        sgn = -r[n1-m1-1];                  Compute numerator and denominator for G and H,
        shn = -r[n1+m1+1];                     eqs. (2.8.24) and (2.8.23),
        sgd = -r[n1];
        for (j=0;j<m+1;j++) {
            sgn += r[n1+j-m1]*g[j];
            shn += r[n1+m1-j]*h[j];
            sgd += r[n1+j-m1]*h[m-j];
        }
        if (sgd == 0.0) nrerror("toeplz-3 singular principal minor");
        g[m1]=sgn/sgd;                      whence G and H.
        h[m1]=shn/sd;
        k=m;
        m2=(m+2) >> 1;
        pp=g[m1];
        qq=h[m1];
        for (j=0;j<m2;j++) {
            pt1=g[j];
            pt2=g[k];
            qt1=h[j];
            qt2=h[k];
            g[j]=pt1-pp*qt2;
            g[k]=pt2-pp*qt1;
            h[j]=qt1-qq*pt2;
            h[k--]=qt2-qq*pt1;
        }
    }                                       Back for another recurrence.
    nrerror("toeplz - should not arrive here!");
}
```

If you are in the business of solving *very* large Toeplitz systems, you should find out about
so-called "new, fast" algorithms, which require only on the order of $N(\log N)^2$ operations,
compared to N^2 for Levinson's method. These methods are too complicated to include here.

Papers by Bunch [6] and de Hoog [7] will give entry to the literature.

CITED REFERENCES AND FURTHER READING:

Golub, G.H., and Van Loan, C.F. 1996, *Matrix Computations*, 3rd ed. (Baltimore: Johns Hopkins University Press), Chapter 5 [also treats some other special forms].

Forsythe, G.E., and Moler, C.B. 1967, *Computer Solution of Linear Algebraic Systems* (Englewood Cliffs, NJ: Prentice-Hall), §19. [1]

Westlake, J.R. 1968, *A Handbook of Numerical Matrix Inversion and Solution of Linear Equations* (New York: Wiley). [2]

von Mises, R. 1964, *Mathematical Theory of Probability and Statistics* (New York: Academic Press), pp. 394ff. [3]

Levinson, N., Appendix B of N. Wiener, 1949, *Extrapolation, Interpolation and Smoothing of Stationary Time Series* (New York: Wiley). [4]

Robinson, E.A., and Treitel, S. 1980, *Geophysical Signal Analysis* (Englewood Cliffs, NJ: Prentice-Hall), pp. 163ff. [5]

Bunch, J.R. 1985, *SIAM Journal on Scientific and Statistical Computing*, vol. 6, pp. 349–364. [6]

de Hoog, F. 1987, *Linear Algebra and Its Applications*, vol. 88/89, pp. 123–138. [7]

2.9 Cholesky Decomposition

If a square matrix $\mathbf{A}$ happens to be symmetric and positive definite, then it has a special, more efficient, triangular decomposition. *Symmetric* means that $a_{ij} = a_{ji}$ for $i, j = 0, \ldots, N - 1$, while *positive definite* means that

$$\mathbf{v} \cdot \mathbf{A} \cdot \mathbf{v} > 0 \quad \text{for all vectors } \mathbf{v} \tag{2.9.1}$$

(In Chapter 11 we will see that positive definite has the equivalent interpretation that $\mathbf{A}$ has all positive eigenvalues.) While symmetric, positive definite matrices are rather special, they occur quite frequently in some applications, so their special factorization, called *Cholesky decomposition*, is good to know about. When you can use it, Cholesky decomposition is about a factor of two faster than alternative methods for solving linear equations.

Instead of seeking arbitrary lower and upper triangular factors $\mathbf{L}$ and $\mathbf{U}$, Cholesky decomposition constructs a lower triangular matrix $\mathbf{L}$ whose transpose $\mathbf{L}^T$ can itself serve as the upper triangular part. In other words we replace equation (2.3.1) by

$$\mathbf{L} \cdot \mathbf{L}^T = \mathbf{A} \tag{2.9.2}$$

This factorization is sometimes referred to as "taking the square root" of the matrix $\mathbf{A}$. The components of $\mathbf{L}^T$ are of course related to those of $\mathbf{L}$ by

$$L_{ij}^T = L_{ji} \tag{2.9.3}$$

Writing out equation (2.9.2) in components, one readily obtains the analogs of equations (2.3.12)–(2.3.13),

$$L_{ii} = \left(a_{ii} - \sum_{k=0}^{i-1} L_{ik}^2 \right)^{1/2} \tag{2.9.4}$$

and

$$L_{ji} = \frac{1}{L_{ii}} \left(a_{ij} - \sum_{k=0}^{i-1} L_{ik} L_{jk} \right) \qquad j = i+1, i+2, \ldots, N-1 \tag{2.9.5}$$

If you apply equations (2.9.4) and (2.9.5) in the order $i = 0, 1, \ldots, N - 1$, you will see that the L's that occur on the right-hand side are already determined by the time they are needed. Also, only components a_{ij} with $j \geq i$ are referenced. (Since $\mathbf{A}$ is symmetric, these have complete information.) It is convenient, then, to have the factor $\mathbf{L}$ overwrite the subdiagonal (lower triangular but not including the diagonal) part of $\mathbf{A}$, preserving the input upper triangular values of $\mathbf{A}$. Only one extra vector of length N is needed to store the diagonal part of $\mathbf{L}$. The operations count is $N^3/6$ executions of the inner loop (consisting of one multiply and one subtract), with also N square roots. As already mentioned, this is about a factor 2 better than LU decomposition of $\mathbf{A}$ (where its symmetry would be ignored).

A straightforward implementation is

```
#include <cmath>
#include "nr.h"
using namespace std;

void NR::choldc(Mat_IO_DP &a, Vec_O_DP &p)
```
Given a positive-definite symmetric matrix a[0..n-1][0..n-1], this routine constructs its Cholesky decomposition, $\mathbf{A} = \mathbf{L} \cdot \mathbf{L}^T$. On input, only the upper triangle of a need be given; it is not modified. The Cholesky factor $\mathbf{L}$ is returned in the lower triangle of a, except for its diagonal elements which are returned in p[0..n-1].
```
{
    int i,j,k;
    DP sum;

    int n=a.nrows();
    for (i=0;i<n;i++) {
        for (j=i;j<n;j++) {
            for (sum=a[i][j],k=i-1;k>=0;k--) sum -= a[i][k]*a[j][k];
            if (i == j) {
                if (sum <= 0.0)          a, with rounding errors, is not positive definite.
                    nrerror("choldc failed");
                p[i]=sqrt(sum);
            } else a[j][i]=sum/p[i];
        }
    }
}
```

You might at this point wonder about pivoting. The pleasant answer is that Cholesky decomposition is extremely stable numerically, without any pivoting at all. Failure of choldc simply indicates that the matrix $\mathbf{A}$ (or, with roundoff error, another very nearby matrix) is not positive definite. In fact, choldc is an efficient way to test *whether* a symmetric matrix is positive definite. (In this application, you will want to replace the call to nrerror with some less drastic signaling method.)

Once your matrix is decomposed, the triangular factor can be used to solve a linear equation by backsubstitution. The straightforward implementation of this is

```
#include "nr.h"

void NR::cholsl(Mat_I_DP &a, Vec_I_DP &p, Vec_I_DP &b, Vec_O_DP &x)
```
Solves the set of n linear equations $\mathbf{A} \cdot \mathbf{x} = \mathbf{b}$, where a is a positive-definite symmetric matrix. a[0..n-1][0..n-1] and p[0..n-1] are input as the output of the routine choldc. Only the lower subdiagonal portion of a is accessed. b[0..n-1] is input as the right-hand side vector. The solution vector is returned in x[0..n-1]. a, n, and p are not modified and can be left in place for successive calls with different right-hand sides b. b is not modified.
```
{
    int i,k;
    DP sum;

    int n=a.nrows();
    for (i=0;i<n;i++) {                    Solve L · y = b, storing y in x.
        for (sum=b[i],k=i-1;k>=0;k--) sum -= a[i][k]*x[k];
```

```
        x[i]=sum/p[i];
    }
    for (i=n-1;i>=0;i--) {              Solve L^T · x = y.
        for (sum=x[i],k=i+1;k<n;k++) sum -= a[k][i]*x[k];
        x[i]=sum/p[i];
    }
}
```

A typical use of choldc and cholsl is in the inversion of covariance matrices describing the fit of data to a model; see, e.g., §15.6. In this, and many other applications, one often needs $\mathbf{L}^{-1}$. The lower triangle of this matrix can be efficiently found from the output of choldc:

```
for (i=0;i<n;i++) {
    a[i][i]=1.0/p[i];
    for (j=i+1;j<n;j++) {
        sum=0.0;
        for (k=i;k<j;k++) sum -= a[j][k]*a[k][i];
        a[j][i]=sum/p[j];
    }
}
```

CITED REFERENCES AND FURTHER READING:

Wilkinson, J.H., and Reinsch, C. 1971, *Linear Algebra*, vol. II of *Handbook for Automatic Computation* (New York: Springer-Verlag), Chapter I/1.

Gill, P.E., Murray, W., and Wright, M.H. 1991, *Numerical Linear Algebra and Optimization*, vol. 1 (Redwood City, CA: Addison-Wesley), §4.9.2.

Dahlquist, G., and Bjorck, A. 1974, *Numerical Methods* (Englewood Cliffs, NJ: Prentice-Hall), §5.3.5.

Golub, G.H., and Van Loan, C.F. 1996, *Matrix Computations*, 3rd ed. (Baltimore: Johns Hopkins University Press), §4.2.

2.10 QR Decomposition

There is another matrix factorization that is sometimes very useful, the so-called QR *decomposition*,

$$\mathbf{A} = \mathbf{Q} \cdot \mathbf{R} \tag{2.10.1}$$

Here $\mathbf{R}$ is upper triangular, while $\mathbf{Q}$ is orthogonal, that is,

$$\mathbf{Q}^T \cdot \mathbf{Q} = \mathbf{1} \tag{2.10.2}$$

where $\mathbf{Q}^T$ is the transpose matrix of $\mathbf{Q}$. Although the decomposition exists for a general rectangular matrix, we shall restrict our treatment to the case when all the matrices are square, with dimensions $N \times N$.

Like the other matrix factorizations we have met (LU, SVD, Cholesky), QR decomposition can be used to solve systems of linear equations. To solve

$$\mathbf{A} \cdot \mathbf{x} = \mathbf{b} \tag{2.10.3}$$

first form $\mathbf{Q}^T \cdot \mathbf{b}$ and then solve

$$\mathbf{R} \cdot \mathbf{x} = \mathbf{Q}^T \cdot \mathbf{b} \tag{2.10.4}$$

by backsubstitution. Since QR decomposition involves about twice as many operations as LU decomposition, it is not used for typical systems of linear equations. However, we will meet special cases where QR is the method of choice.

The standard algorithm for the QR decomposition involves successive Householder transformations (to be discussed later in §11.2). We write a Householder matrix in the form $\mathbf{1} - \mathbf{u} \otimes \mathbf{u}/c$ where $c = \frac{1}{2}\mathbf{u} \cdot \mathbf{u}$. An appropriate Householder matrix applied to a given matrix can zero all elements in a column of the matrix situated below a chosen element. Thus we arrange for the first Householder matrix $\mathbf{Q}_0$ to zero all elements in column 0 of $\mathbf{A}$ below the zeroth element. Similarly $\mathbf{Q}_1$ zeroes all elements in column 1 below element 1, and so on up to $\mathbf{Q}_{n-2}$. Thus

$$\mathbf{R} = \mathbf{Q}_{n-2} \cdots \mathbf{Q}_0 \cdot \mathbf{A} \qquad (2.10.5)$$

Since the Householder matrices are orthogonal,

$$\mathbf{Q} = (\mathbf{Q}_{n-2} \cdots \mathbf{Q}_0)^{-1} = \mathbf{Q}_0 \cdots \mathbf{Q}_{n-2} \qquad (2.10.6)$$

In most applications we don't need to form $\mathbf{Q}$ explicitly; we instead store it in the factored form (2.10.6). Pivoting is not usually necessary unless the matrix $\mathbf{A}$ is very close to singular. A general QR algorithm for rectangular matrices including pivoting is given in [1]. For square matrices, an implementation is the following:

```cpp
#include <cmath>
#include "nr.h"
using namespace std;

void NR::qrdcmp(Mat_IO_DP &a, Vec_O_DP &c, Vec_O_DP &d, bool &sing)
```
Constructs the QR decomposition of a[0..n-1][0..n-1]. The upper triangular matrix $\mathbf{R}$ is returned in the upper triangle of a, except for the diagonal elements of $\mathbf{R}$ which are returned in d[0..n-1]. The orthogonal matrix $\mathbf{Q}$ is represented as a product of $n-1$ Householder matrices $\mathbf{Q}_0 \cdots \mathbf{Q}_{n-2}$, where $\mathbf{Q}_j = \mathbf{1} - \mathbf{u}_j \otimes \mathbf{u}_j/c_j$. The ith component of $\mathbf{u}_j$ is zero for $i = 0, \ldots, j-1$ while the nonzero components are returned in a[i][j] for $i = j, \ldots, n-1$. sing returns as true if singularity is encountered during the decomposition, but the decomposition is still completed in this case; otherwise it returns false.
```cpp
{
    int i,j,k;
    DP scale,sigma,sum,tau;

    int n=a.nrows();
    sing=false;
    for (k=0;k<n-1;k++) {
        scale=0.0;
        for (i=k;i<n;i++) scale=MAX(scale,fabs(a[i][k]));
        if (scale == 0.0) {                        Singular case.
            sing=true;
            c[k]=d[k]=0.0;
        } else {                                   Form Qk and Qk · A.
            for (i=k;i<n;i++) a[i][k] /= scale;
            for (sum=0.0,i=k;i<n;i++) sum += SQR(a[i][k]);
            sigma=SIGN(sqrt(sum),a[k][k]);
            a[k][k] += sigma;
            c[k]=sigma*a[k][k];
            d[k] = -scale*sigma;
            for (j=k+1;j<n;j++) {
                for (sum=0.0,i=k;i<n;i++) sum += a[i][k]*a[i][j];
                tau=sum/c[k];
                for (i=k;i<n;i++) a[i][j] -= tau*a[i][k];
            }
        }
    }
    d[n-1]=a[n-1][n-1];
    if (d[n-1] == 0.0) sing=true;
}
```

The next routine, qrsolv, is used to solve linear systems. In many applications only the part (2.10.4) of the algorithm is needed, so we separate it off into its own routine rsolv.

```
#include "nr.h"

void NR::qrsolv(Mat_I_DP &a, Vec_I_DP &c, Vec_I_DP &d, Vec_IO_DP &b)
Solves the set of n linear equations A · x = b. a[0..n-1][0..n-1], c[0..n-1], and
d[0..n-1] are input as the output of the routine qrdcmp and are not modified. b[0..n-1]
is input as the right-hand side vector, and is overwritten with the solution vector on output.
{
    int i,j;
    DP sum,tau;

    int n=a.nrows();
    for (j=0;j<n-1;j++) {          Form Qᵀ · b.
        for (sum=0.0,i=j;i<n;i++) sum += a[i][j]*b[i];
        tau=sum/c[j];
        for (i=j;i<n;i++) b[i] -= tau*a[i][j];
    }
    rsolv(a,d,b);                  Solve R · x = Qᵀ · b.
}
```

```
#include "nr.h"

void NR::rsolv(Mat_I_DP &a, Vec_I_DP &d, Vec_IO_DP &b)
Solves the set of n linear equations R · x = b, where R is an upper triangular matrix stored in a
and d. a[0..n-1][0..n-1] and d[0..n-1] are input as the output of the routine qrdcmp
and are not modified. b[0..n-1] is input as the right-hand side vector, and is overwritten
with the solution vector on output.
{
    int i,j;
    DP sum;

    int n=a.nrows();
    b[n-1] /= d[n-1];
    for (i=n-2;i>=0;i--) {
        for (sum=0.0,j=i+1;j<n;j++) sum += a[i][j]*b[j];
        b[i]=(b[i]-sum)/d[i];
    }
}
```

See [2] for details on how to use QR decomposition for constructing orthogonal bases, and for solving least-squares problems. (We prefer to use SVD, §2.6, for these purposes, because of its greater diagnostic capability in pathological cases.)

Updating a QR decomposition

Some numerical algorithms involve solving a succession of linear systems each of which differs only slightly from its predecessor. Instead of doing $O(N^3)$ operations each time to solve the equations from scratch, one can often update a matrix factorization in $O(N^2)$ operations and use the new factorization to solve the next set of linear equations. The LU decomposition is complicated to update because of pivoting. However, QR turns out to be quite simple for a very common kind of update,

$$\mathbf{A} \rightarrow \mathbf{A} + \mathbf{s} \otimes \mathbf{t} \tag{2.10.7}$$

(compare equation 2.7.1). In practice it is more convenient to work with the equivalent form

$$\mathbf{A} = \mathbf{Q} \cdot \mathbf{R} \quad \rightarrow \quad \mathbf{A}' = \mathbf{Q}' \cdot \mathbf{R}' = \mathbf{Q} \cdot (\mathbf{R} + \mathbf{u} \otimes \mathbf{v}) \tag{2.10.8}$$

One can go back and forth between equations (2.10.7) and (2.10.8) using the fact that $\mathbf{Q}$ is orthogonal, giving

$$\mathbf{t} = \mathbf{v} \quad \text{and either} \quad \mathbf{s} = \mathbf{Q} \cdot \mathbf{u} \quad \text{or} \quad \mathbf{u} = \mathbf{Q}^T \cdot \mathbf{s} \qquad (2.10.9)$$

The algorithm [2] has two phases. In the first we apply $N - 1$ Jacobi rotations (§11.1) to reduce $\mathbf{R} + \mathbf{u} \otimes \mathbf{v}$ to upper Hessenberg form. Another $N - 1$ Jacobi rotations transform this upper Hessenberg matrix to the new upper triangular matrix $\mathbf{R}'$. The matrix $\mathbf{Q}'$ is simply the product of $\mathbf{Q}$ with the $2(N - 1)$ Jacobi rotations. In applications we usually want $\mathbf{Q}^T$, and the algorithm can easily be rearranged to work with this matrix instead of with $\mathbf{Q}$.

```
#include <cmath>
#include "nr.h"
using namespace std;

void NR::qrupdt(Mat_IO_DP &r, Mat_IO_DP &qt, Vec_IO_DP &u, Vec_I_DP &v)
Given the QR decomposition of some n × n matrix, calculates the QR decomposition of the ma-
trix Q·(R+u⊗v). Quantities are dimensioned as r[0..n-1][0..n-1], qt[0..n-1][0..n-1],
u[0..n-1], and v[0..n-1]. Note that Qᵀ is input and returned in qt.
{
    int i,k;

    int n=u.size();
    for (k=n-1;k>=0;k--)                    Find largest k such that u[k] ≠ 0.
        if (u[k] != 0.0) break;
    if (k < 0) k=0;
    for (i=k-1;i>=0;i--) {                  Transform R + u ⊗ v to upper Hessenberg.
        rotate(r,qt,i,u[i],-u[i+1]);
        if (u[i] == 0.0)
            u[i]=fabs(u[i+1]);
        else if (fabs(u[i]) > fabs(u[i+1]))
            u[i]=fabs(u[i])*sqrt(1.0+SQR(u[i+1]/u[i]));
        else u[i]=fabs(u[i+1])*sqrt(1.0+SQR(u[i]/u[i+1]));
    }
    for (i=0;i<n;i++) r[0][i] += u[0]*v[i];
    for (i=0;i<k;i++)                       Transform upper Hessenberg matrix to upper tri-
        rotate(r,qt,i,r[i][i],-r[i+1][i]);         angular.
}
```

```
#include <cmath>
#include "nr.h"
using namespace std;

void NR::rotate(Mat_IO_DP &r, Mat_IO_DP &qt, const int i, const DP a,
    const DP b)
Given matrices r[0..n-1][0..n-1] and qt[0..n-1][0..n-1], carry out a Jacobi rotation
on rows i and i + 1 of each matrix. a and b are the parameters of the rotation: cos θ =
a/√(a² + b²), sin θ = b/√(a² + b²).
{
    int j;
    DP c,fact,s,w,y;

    int n=r.nrows();
    if (a == 0.0) {                        Avoid unnecessary overflow or underflow.
        c=0.0;
        s=(b >= 0.0 ? 1.0 : -1.0);
    } else if (fabs(a) > fabs(b)) {
        fact=b/a;
        c=SIGN(1.0/sqrt(1.0+(fact*fact)),a);
        s=fact*c;
    } else {
```

```
      fact=a/b;
      s=SIGN(1.0/sqrt(1.0+(fact*fact)),b);
      c=fact*s;
   }
   for (j=i;j<n;j++) {            Premultiply r by Jacobi rotation.
      y=r[i][j];
      w=r[i+1][j];
      r[i][j]=c*y-s*w;
      r[i+1][j]=s*y+c*w;
   }
   for (j=0;j<n;j++) {            Premultiply qt by Jacobi rotation.
      y=qt[i][j];
      w=qt[i+1][j];
      qt[i][j]=c*y-s*w;
      qt[i+1][j]=s*y+c*w;
   }
}
```

We will make use of QR decomposition, and its updating, in §9.7.

CITED REFERENCES AND FURTHER READING:

Wilkinson, J.H., and Reinsch, C. 1971, *Linear Algebra*, vol. II of *Handbook for Automatic Computation* (New York: Springer-Verlag), Chapter I/8. [1]

Golub, G.H., and Van Loan, C.F. 1996, *Matrix Computations*, 3rd ed. (Baltimore: Johns Hopkins University Press), §§5.2, 5.3, 12.5. [2]

2.11 Is Matrix Inversion an N^3 Process?

We close this chapter with a little entertainment, a bit of algorithmic prestidigitation which probes more deeply into the subject of matrix inversion. We start with a seemingly simple question:

How many individual multiplications does it take to perform the matrix multiplication of two 2×2 matrices,

$$\begin{pmatrix} a_{00} & a_{01} \\ a_{10} & a_{11} \end{pmatrix} \cdot \begin{pmatrix} b_{00} & b_{01} \\ b_{10} & b_{11} \end{pmatrix} = \begin{pmatrix} c_{00} & c_{01} \\ c_{10} & c_{11} \end{pmatrix} \qquad (2.11.1)$$

Eight, right? Here they are written explicitly:

$$c_{00} = a_{00} \times b_{00} + a_{01} \times b_{10}$$
$$c_{01} = a_{00} \times b_{01} + a_{01} \times b_{11}$$
$$c_{10} = a_{10} \times b_{00} + a_{11} \times b_{10} \qquad (2.11.2)$$
$$c_{11} = a_{10} \times b_{01} + a_{11} \times b_{11}$$

Do you think that one can write formulas for the c's that involve only *seven* multiplications? (Try it yourself, before reading on.)

Such a set of formulas was, in fact, discovered by Strassen [1]. The formulas are:

$$Q_0 \equiv (a_{00} + a_{11}) \times (b_{00} + b_{11})$$
$$Q_1 \equiv (a_{10} + a_{11}) \times b_{00}$$
$$Q_2 \equiv a_{00} \times (b_{01} - b_{11})$$
$$Q_3 \equiv a_{11} \times (-b_{00} + b_{10}) \tag{2.11.3}$$
$$Q_4 \equiv (a_{00} + a_{01}) \times b_{11}$$
$$Q_5 \equiv (-a_{00} + a_{10}) \times (b_{00} + b_{01})$$
$$Q_6 \equiv (a_{01} - a_{11}) \times (b_{10} + b_{11})$$

in terms of which

$$c_{00} = Q_0 + Q_3 - Q_4 + Q_6$$
$$c_{10} = Q_1 + Q_3$$
$$c_{01} = Q_2 + Q_4 \tag{2.11.4}$$
$$c_{11} = Q_0 + Q_2 - Q_1 + Q_5$$

What's the use of this? There is one fewer multiplication than in equation (2.11.2), but *many more* additions and subtractions. It is not clear that anything has been gained. But notice that in (2.11.3) the a's and b's are never commuted. Therefore (2.11.3) and (2.11.4) are valid when the a's and b's are themselves matrices. The problem of multiplying two very large matrices (of order $N = 2^m$ for some integer m) can now be broken down recursively by partitioning the matrices into quarters, sixteenths, etc. And note the key point: The savings is not just a factor "7/8"; it is that factor at *each* hierarchical level of the recursion. In total it reduces the process of matrix multiplication to order $N^{\log_2 7}$ instead of N^3.

What about all the extra additions in (2.11.3)–(2.11.4)? Don't they outweigh the advantage of the fewer multiplications? For large N, it turns out that there are six times as many additions as multiplications implied by (2.11.3)–(2.11.4). But, if N is very large, this constant factor is no match for the change in the *exponent* from N^3 to $N^{\log_2 7}$.

With this "fast" matrix multiplication, Strassen also obtained a surprising result for matrix inversion [1]. Suppose that the matrices

$$\begin{pmatrix} a_{00} & a_{01} \\ a_{10} & a_{11} \end{pmatrix} \quad \text{and} \quad \begin{pmatrix} c_{00} & c_{01} \\ c_{10} & c_{11} \end{pmatrix} \tag{2.11.5}$$

are inverses of each other. Then the c's can be obtained from the a's by the following operations (compare equations 2.7.11 and 2.7.25):

$$R_0 = \text{Inverse}(a_{00})$$

$$R_1 = a_{10} \times R_0$$

$$R_2 = R_0 \times a_{01}$$

$$R_3 = a_{10} \times R_2$$

$$R_4 = R_3 - a_{11}$$

$$R_5 = \text{Inverse}(R_4) \qquad\qquad (2.11.6)$$

$$c_{01} = R_2 \times R_5$$

$$c_{10} = R_5 \times R_1$$

$$R_6 = R_2 \times c_{10}$$

$$c_{00} = R_0 - R_6$$

$$c_{11} = -R_5$$

In (2.11.6) the "inverse" operator occurs just twice. It is to be interpreted as the reciprocal if the a's and c's are scalars, but as matrix inversion if the a's and c's are themselves submatrices. Imagine doing the inversion of a very large matrix, of order $N = 2^m$, recursively by partitions in half. At each step, halving the order *doubles* the number of inverse operations. But this means that there are only N divisions in all! So divisions don't dominate in the recursive use of (2.11.6). Equation (2.11.6) is dominated, in fact, by its 6 multiplications. Since these can be done by an $N^{\log_2 7}$ algorithm, so can the matrix inversion!

This is fun, but let's look at practicalities: If you estimate how large N has to be before the difference between exponent 3 and exponent $\log_2 7 = 2.807$ is substantial enough to outweigh the bookkeeping overhead, arising from the complicated nature of the recursive Strassen algorithm, you will find that LU decomposition is in no immediate danger of becoming obsolete.

If, on the other hand, you like this kind of fun, then try these: (1) Can you multiply the complex numbers $(a+ib)$ and $(c+id)$ in only *three* real multiplications? [Answer: see §5.4.] (2) Can you evaluate a general fourth-degree polynomial in x for many different values of x with only *three* multiplications per evaluation? [Answer: see §5.3.]

CITED REFERENCES AND FURTHER READING:

Strassen, V. 1969, *Numerische Mathematik*, vol. 13, pp. 354–356. [1]

Kronsjö, L. 1987, *Algorithms: Their Complexity and Efficiency*, 2nd ed. (New York: Wiley).

Winograd, S. 1971, *Linear Algebra and Its Applications*, vol. 4, pp. 381–388.

Pan, V. Ya. 1980, *SIAM Journal on Computing*, vol. 9, pp. 321–342.

Pan, V. 1984, *How to Multiply Matrices Faster*, Lecture Notes in Computer Science, vol. 179 (New York: Springer-Verlag)

Pan, V. 1984, *SIAM Review*, vol. 26, pp. 393–415. [More recent results that show that an exponent of 2.496 can be achieved — theoretically!]

Chapter 3. Interpolation and Extrapolation

3.0 Introduction

We sometimes know the value of a function $f(x)$ at a set of points $x_0, x_1, \ldots,$ x_{N-1} (say, with $x_0 < \ldots < x_{N-1}$), but we don't have an analytic expression for $f(x)$ that lets us calculate its value at an arbitrary point. For example, the $f(x_i)$'s might result from some physical measurement or from long numerical calculation that cannot be cast into a simple functional form. Often the x_i's are equally spaced, but not necessarily.

The task now is to estimate $f(x)$ for arbitrary x by, in some sense, drawing a smooth curve through (and perhaps beyond) the x_i. If the desired x is in between the largest and smallest of the x_i's, the problem is called *interpolation*; if x is outside that range, it is called *extrapolation*, which is considerably more hazardous (as many former stock-market analysts can attest).

Interpolation and extrapolation schemes must model the function, between or beyond the known points, by some plausible functional form. The form should be sufficiently general so as to be able to approximate large classes of functions which might arise in practice. By far most common among the functional forms used are polynomials (§3.1). Rational functions (quotients of polynomials) also turn out to be extremely useful (§3.2). Trigonometric functions, sines and cosines, give rise to *trigonometric interpolation* and related Fourier methods, which we defer to Chapters 12 and 13.

There is an extensive mathematical literature devoted to theorems about what sort of functions can be well approximated by which interpolating functions. These theorems are, alas, almost completely useless in day-to-day work: If we know enough about our function to apply a theorem of any power, we are usually not in the pitiful state of having to interpolate on a table of its values!

Interpolation is related to, but distinct from, *function approximation*. That task consists of finding an approximate (but easily computable) function to use in place of a more complicated one. In the case of interpolation, you are given the function f at points *not of your own choosing*. For the case of function approximation, you are allowed to compute the function f at *any* desired points for the purpose of developing your approximation. We deal with function approximation in Chapter 5.

One can easily find pathological functions that make a mockery of any interpolation scheme. Consider, for example, the function

$$f(x) = 3x^2 + \frac{1}{\pi^4} \ln\left[(\pi - x)^2\right] + 1 \qquad (3.0.1)$$

which is well-behaved everywhere except at $x = \pi$, very mildly singular at $x = \pi$, and otherwise takes on all positive and negative values. Any interpolation based on the values $x = 3.13, 3.14, 3.15, 3.16$, will assuredly get a very wrong answer for the value $x = 3.1416$, even though a graph plotting those five points looks really quite smooth! (Try it on your calculator.)

Because pathologies can lurk anywhere, it is highly desirable that an interpolation and extrapolation routine should provide an estimate of its own error. Such an error estimate can never be foolproof, of course. We could have a function that, for reasons known only to its maker, takes off wildly and unexpectedly between two tabulated points. Interpolation always presumes some degree of smoothness for the function interpolated, but within this framework of presumption, deviations from smoothness can be detected.

Conceptually, the interpolation process has two stages: (1) Fit an interpolating function to the data points provided. (2) Evaluate that interpolating function at the target point x.

However, this two-stage method is generally not the best way to proceed in practice. Typically it is computationally less efficient, and more susceptible to roundoff error, than methods which construct a functional estimate $f(x)$ directly from the N tabulated values every time one is desired. Most practical schemes start at a nearby point $f(x_i)$, then add a sequence of (hopefully) decreasing corrections, as information from other $f(x_i)$'s is incorporated. The procedure typically takes $O(N^2)$ operations. If everything is well behaved, the last correction will be the smallest, and it can be used as an informal (though not rigorous) bound on the error.

In the case of polynomial interpolation, it sometimes does happen that the coefficients of the interpolating polynomial are of interest, even though their use in *evaluating* the interpolating function should be frowned on. We deal with this eventuality in §3.5.

Local interpolation, using a finite number of "nearest-neighbor" points, gives interpolated values $f(x)$ that do not, in general, have continuous first or higher derivatives. That happens because, as x crosses the tabulated values x_i, the interpolation scheme switches which tabulated points are the "local" ones. (If such a switch is allowed to occur anywhere *else*, then there will be a discontinuity in the interpolated function itself at that point. Bad idea!)

In situations where continuity of derivatives is a concern, one must use the "stiffer" interpolation provided by a so-called *spline* function. A spline is a polynomial between each pair of table points, but one whose coefficients are determined "slightly" nonlocally. The nonlocality is designed to guarantee global smoothness in the interpolated function up to some order of derivative. Cubic splines (§3.3) are the most popular. They produce an interpolated function that is continuous through the second derivative. Splines tend to be stabler than polynomials, with less possibility of wild oscillation between the tabulated points.

The number of points (minus one) used in an interpolation scheme is called the *order* of the interpolation. Increasing the order does not necessarily increase

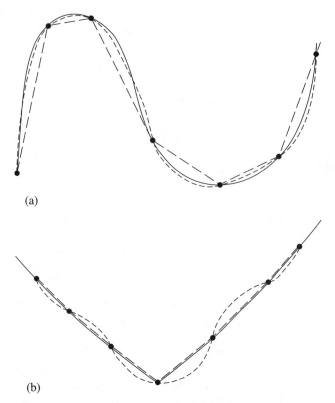

(a)

(b)

Figure 3.0.1.   (a) A smooth function (solid line) is more accurately interpolated by a high-order polynomial (shown schematically as dotted line) than by a low-order polynomial (shown as a piecewise linear dashed line). (b) A function with sharp corners or rapidly changing higher derivatives is *less* accurately approximated by a high-order polynomial (dotted line), which is too "stiff," than by a low-order polynomial (dashed lines). Even some smooth functions, such as exponentials or rational functions, can be badly approximated by high-order polynomials.

the accuracy, especially in polynomial interpolation. If the added points are distant from the point of interest x, the resulting higher-order polynomial, with its additional constrained points, tends to oscillate wildly between the tabulated values. This oscillation may have no relation at all to the behavior of the "true" function (see Figure 3.0.1). Of course, adding points *close* to the desired point usually does help, but a finer mesh implies a larger table of values, not always available.

Unless there is solid evidence that the interpolating function is close in form to the true function f, it is a good idea to be cautious about high-order interpolation. We enthusiastically endorse interpolations with 3 or 4 points, we are perhaps tolerant of 5 or 6; but we rarely go higher than that unless there is quite rigorous monitoring of estimated errors.

When your table of values contains many more points than the desirable order of interpolation, you must begin each interpolation with a search for the right "local" place in the table. While not strictly a part of the subject of interpolation, this task is important enough (and often enough botched) that we devote §3.4 to its discussion.

The routines given for interpolation are also routines for extrapolation. An important application, in Chapter 16, is their use in the integration of ordinary differential equations. There, considerable care *is* taken with the monitoring of errors. Otherwise, the dangers of extrapolation cannot be overemphasized: An interpolating function, which is perforce an extrapolating function, will typically go berserk when the argument x is outside the range of tabulated values by more than the typical spacing of tabulated points.

Interpolation can be done in more than one dimension, e.g., for a function $f(x, y, z)$. Multidimensional interpolation is often accomplished by a sequence of one-dimensional interpolations. We discuss this in §3.6.

CITED REFERENCES AND FURTHER READING:

Abramowitz, M., and Stegun, I.A. 1964, *Handbook of Mathematical Functions*, Applied Mathematics Series, Volume 55 (Washington: National Bureau of Standards; reprinted 1968 by Dover Publications, New York), §25.2.

Stoer, J., and Bulirsch, R. 1993, *Introduction to Numerical Analysis*, 2nd ed. (New York: Springer-Verlag), Chapter 2.

Acton, F.S. 1970, *Numerical Methods That Work*; 1990, corrected edition (Washington: Mathematical Association of America), Chapter 3.

Kahaner, D., Moler, C., and Nash, S. 1989, *Numerical Methods and Software* (Englewood Cliffs, NJ: Prentice Hall), Chapter 4.

Johnson, L.W., and Riess, R.D. 1982, *Numerical Analysis*, 2nd ed. (Reading, MA: Addison-Wesley), Chapter 5.

Ralston, A., and Rabinowitz, P. 1978, *A First Course in Numerical Analysis*, 2nd ed.; reprinted 2001 (New York: Dover), Chapter 3.

Isaacson, E., and Keller, H.B. 1966, *Analysis of Numerical Methods*; reprinted 1994 (New York: Dover), Chapter 6.

3.1 Polynomial Interpolation and Extrapolation

Through any two points there is a unique line. Through any three points, a unique quadratic. Et cetera. The interpolating polynomial of degree $N - 1$ through the N points $y_0 = f(x_0), y_1 = f(x_1), \ldots, y_{N-1} = f(x_{N-1})$ is given explicitly by Lagrange's classical formula,

$$
\begin{aligned}
P(x) = {} & \frac{(x - x_1)(x - x_2)...(x - x_{N-1})}{(x_0 - x_1)(x_0 - x_2)...(x_0 - x_{N-1})} y_0 \\
& + \frac{(x - x_0)(x - x_2)...(x - x_{N-1})}{(x_1 - x_0)(x_1 - x_2)...(x_1 - x_{N-1})} y_1 + \cdots \\
& + \frac{(x - x_0)(x - x_1)...(x - x_{N-2})}{(x_{N-1} - x_0)(x_{N-1} - x_1)...(x_{N-1} - x_{N-2})} y_{N-1}
\end{aligned}
\tag{3.1.1}
$$

There are N terms, each a polynomial of degree $N - 1$ and each constructed to be zero at all of the x_i except one, at which it is constructed to be y_i.

It is not terribly wrong to implement the Lagrange formula straightforwardly, but it is not terribly right either. The resulting algorithm gives no error estimate, and it is also somewhat awkward to program. A much better algorithm (for constructing the same, unique, interpolating polynomial) is *Neville's algorithm*, closely related to and sometimes confused with *Aitken's algorithm*, the latter now considered obsolete.

Let P_0 be the value at x of the unique polynomial of degree zero (i.e., a constant) passing through the point (x_0, y_0); so $P_0 = y_0$. Likewise define $P_1, P_2, \ldots, P_{N-1}$. Now let P_{01} be the value at x of the unique polynomial of degree one passing through both (x_0, y_0) and (x_1, y_1). Likewise $P_{12}, P_{23}, \ldots, P_{(N-2)(N-1)}$. Similarly, for higher-order polynomials, up to $P_{012\ldots(N-1)}$, which is the value of the unique interpolating polynomial through all N points, i.e., the desired answer. The various P's form a "tableau" with "ancestors" on the left leading to a single "descendant" at the extreme right. For example, with $N = 4$,

$$
\begin{array}{cccc}
x_0: & y_0 = P_0 & & \\
& & P_{01} & \\
x_1: & y_1 = P_1 & & P_{012} \\
& & P_{12} & & P_{0123} \\
x_2: & y_2 = P_2 & & P_{123} \\
& & P_{23} & \\
x_3: & y_3 = P_3 & &
\end{array}
\qquad (3.1.2)
$$

Neville's algorithm is a recursive way of filling in the numbers in the tableau a column at a time, from left to right. It is based on the relationship between a "daughter" P and its two "parents,"

$$
P_{i(i+1)\ldots(i+m)} = \frac{(x - x_{i+m})P_{i(i+1)\ldots(i+m-1)} + (x_i - x)P_{(i+1)(i+2)\ldots(i+m)}}{x_i - x_{i+m}}
$$
$$(3.1.3)$$

This recurrence works because the two parents already agree at points $x_{i+1} \ldots x_{i+m-1}$.

An improvement on the recurrence (3.1.3) is to keep track of the small *differences* between parents and daughters, namely to define (for $m = 1, 2, \ldots, N-1$),

$$C_{m,i} \equiv P_{i\ldots(i+m)} - P_{i\ldots(i+m-1)}$$
$$D_{m,i} \equiv P_{i\ldots(i+m)} - P_{(i+1)\ldots(i+m)}. \qquad (3.1.4)$$

Then one can easily derive from (3.1.3) the relations

$$
D_{m+1,i} = \frac{(x_{i+m+1} - x)(C_{m,i+1} - D_{m,i})}{x_i - x_{i+m+1}}
$$
$$
C_{m+1,i} = \frac{(x_i - x)(C_{m,i+1} - D_{m,i})}{x_i - x_{i+m+1}}
$$
$$(3.1.5)$$

At each level m, the C's and D's are the corrections that make the interpolation one order higher. The final answer $P_{0\ldots(N-1)}$ is equal to the sum of *any* y_i plus a set of C's and/or D's that form a path through the family tree to the rightmost daughter.

Here is a routine for polynomial interpolation or extrapolation:

```
#include <cmath>
#include "nr.h"
using namespace std;

void NR::polint(Vec_I_DP &xa, Vec_I_DP &ya, const DP x, DP &y, DP &dy)
Given arrays xa[0..n-1] and ya[0..n-1], and given a value x, this routine returns a value
y, and an error estimate dy. If P(x) is the polynomial of degree N - 1 such that P(xa_i) =
ya_i, i = 0,...,n-1, then the returned value y = P(x).
{
    int i,m,ns=0;
    DP den,dif,dift,ho,hp,w;

    int n=xa.size();
    Vec_DP c(n),d(n);
    dif=fabs(x-xa[0]);
    for (i=0;i<n;i++) {                    Here we find the index ns of the closest table entry,
        if ((dift=fabs(x-xa[i])) < dif) {
            ns=i;
            dif=dift;
        }
        c[i]=ya[i];                        and initialize the tableau of c's and d's.
        d[i]=ya[i];
    }
    y=ya[ns--];                            This is the initial approximation to y.
    for (m=1;m<n;m++) {                    For each column of the tableau,
        for (i=0;i<n-m;i++) {             we loop over the current c's and d's and update
            ho=xa[i]-x;                        them.
            hp=xa[i+m]-x;
            w=c[i+1]-d[i];
            if ((den=ho-hp) == 0.0) nrerror("Error in routine polint");
            This error can occur only if two input xa's are (to within roundoff) identical.
            den=w/den;
            d[i]=hp*den;                   Here the c's and d's are updated.
            c[i]=ho*den;
        }
        y += (dy=(2*(ns+1) < (n-m) ? c[ns+1] : d[ns--]));
        After each column in the tableau is completed, we decide which correction, c or d,
        we want to add to our accumulating value of y, i.e., which path to take through the
        tableau—forking up or down. We do this in such a way as to take the most "straight
        line" route through the tableau to its apex, updating ns accordingly to keep track of
        where we are. This route keeps the partial approximations centered (insofar as possible)
        on the target x. The last dy added is thus the error indication.
    }
}
```

Quite often you will want to call `polint` with the dummy arguments xa and ya referencing a sub-range of an array, e.g., x[5]...x[8]. However, since polint takes arguments of type NRVec, and NRVec has no facilities for specifying such subarrays, you have to copy the desired points into temporary arrays of the appropriate length. For more on this, see the end of §3.4.

CITED REFERENCES AND FURTHER READING:

Abramowitz, M., and Stegun, I.A. 1964, *Handbook of Mathematical Functions*, Applied Mathematics Series, Volume 55 (Washington: National Bureau of Standards; reprinted 1968 by Dover Publications, New York), §25.2.

Stoer, J., and Bulirsch, R. 1993, *Introduction to Numerical Analysis*, 2nd ed. (New York: Springer-Verlag), §2.1.

Gear, C.W. 1971, *Numerical Initial Value Problems in Ordinary Differential Equations* (Englewood Cliffs, NJ: Prentice-Hall), §6.1.

3.2 Rational Function Interpolation and Extrapolation

Some functions are not well approximated by polynomials, but *are* well approximated by rational functions, that is quotients of polynomials. We denote by $R_{i(i+1)...(i+m)}$ a rational function passing through the $m+1$ points $(x_i, y_i) \ldots (x_{i+m}, y_{i+m})$. More explicitly, suppose

$$R_{i(i+1)...(i+m)} = \frac{P_\mu(x)}{Q_\nu(x)} = \frac{p_0 + p_1 x + \cdots + p_\mu x^\mu}{q_0 + q_1 x + \cdots + q_\nu x^\nu} \qquad (3.2.1)$$

Since there are $\mu + \nu + 1$ unknown p's and q's (q_0 being arbitrary), we must have

$$m + 1 = \mu + \nu + 1 \qquad (3.2.2)$$

In specifying a rational function interpolating function, you must give the desired order of both the numerator and the denominator.

Rational functions are sometimes superior to polynomials, roughly speaking, because of their ability to model functions with poles, that is, zeros of the denominator of equation (3.2.1). These poles might occur for real values of x, if the function to be interpolated itself has poles. More often, the function $f(x)$ is finite for all finite *real* x, but has an analytic continuation with poles in the complex x-plane. Such poles can themselves ruin a polynomial approximation, even one restricted to real values of x, just as they can ruin the convergence of an infinite power series in x. If you draw a circle in the complex plane around your m tabulated points, then you should not expect polynomial interpolation to be good unless the nearest pole is rather far outside the circle. A rational function approximation, by contrast, will stay "good" as long as it has enough powers of x in its denominator to account for (cancel) any nearby poles.

For the interpolation problem, a rational function is constructed so as to go through a chosen set of tabulated functional values. However, we should also mention in passing that rational function approximations can be used in analytic work. One sometimes constructs a rational function approximation by the criterion that the rational function of equation (3.2.1) itself have a power series expansion that agrees with the first $m+1$ terms of the power series expansion of the desired function $f(x)$. This is called *Padé approximation*, and is discussed in §5.12.

Bulirsch and Stoer found an algorithm of the Neville type which performs rational function extrapolation on tabulated data. A tableau like that of equation (3.1.2) is constructed column by column, leading to a result and an error estimate. The Bulirsch-Stoer algorithm produces the so-called *diagonal* rational function, with the degrees of numerator and denominator equal (if m is even) or with the degree of the denominator larger by one (if m is odd, cf. equation 3.2.2 above). For the derivation of the algorithm, refer to [1]. The algorithm is summarized by a recurrence relation exactly analogous to equation (3.1.3) for polynomial approximation:

$$R_{i(i+1)...(i+m)} = R_{(i+1)...(i+m)}$$
$$+ \frac{R_{(i+1)...(i+m)} - R_{i...(i+m-1)}}{\left(\frac{x-x_i}{x-x_{i+m}}\right)\left(1 - \frac{R_{(i+1)...(i+m)} - R_{i...(i+m-1)}}{R_{(i+1)...(i+m)} - R_{(i+1)...(i+m-1)}}\right) - 1}$$

(3.2.3)

This recurrence generates the rational functions through $m + 1$ points from the ones through m and (the term $R_{(i+1)...(i+m-1)}$ in equation 3.2.3) $m - 1$ points. It is started with

$$R_i = y_i \tag{3.2.4}$$

and with

$$R \equiv [R_{i(i+1)...(i+m)} \quad \text{with} \quad m = -1] = 0 \tag{3.2.5}$$

Now, exactly as in equations (3.1.4) and (3.1.5) above, we can convert the recurrence (3.2.3) to one involving only the small differences

$$
\begin{aligned}
C_{m,i} &\equiv R_{i...(i+m)} - R_{i...(i+m-1)} \\
D_{m,i} &\equiv R_{i...(i+m)} - R_{(i+1)...(i+m)}
\end{aligned}
\tag{3.2.6}
$$

Note that these satisfy the relation

$$C_{m+1,i} - D_{m+1,i} = C_{m,i+1} - D_{m,i} \tag{3.2.7}$$

which is useful in proving the recurrences

$$
D_{m+1,i} = \frac{C_{m,i+1}(C_{m,i+1} - D_{m,i})}{\left(\frac{x-x_i}{x-x_{i+m+1}}\right)D_{m,i} - C_{m,i+1}}
$$

$$
C_{m+1,i} = \frac{\left(\frac{x-x_i}{x-x_{i+m+1}}\right)D_{m,i}(C_{m,i+1} - D_{m,i})}{\left(\frac{x-x_i}{x-x_{i+m+1}}\right)D_{m,i} - C_{m,i+1}}
$$

(3.2.8)

This recurrence is implemented in the following function, whose use is analogous in every way to `polint` in §3.1.

```
#include <cmath>
#include "nr.h"
using namespace std;

void NR::ratint(Vec_I_DP &xa, Vec_I_DP &ya, const DP x, DP &y, DP &dy)
Given arrays xa[0..n-1] and ya[0..n-1], and given a value of x, this routine returns a value
of y and an accuracy estimate dy. The value returned is that of the diagonal rational function,
evaluated at x, that passes through the n points (xa_i, ya_i), i = 0, ..., n-1.
{
    const DP TINY=1.0e-25;          A small number.
    int m,i,ns=0;
    DP w,t,hh,h,dd;
```

```
int n=xa.size();
Vec_DP c(n),d(n);
hh=fabs(x-xa[0]);
for (i=0;i<n;i++) {
    h=fabs(x-xa[i]);
    if (h == 0.0) {
        y=ya[i];
        dy=0.0;
        return;
    } else if (h < hh) {
        ns=i;
        hh=h;
    }
    c[i]=ya[i];
    d[i]=ya[i]+TINY;        The TINY part is needed to prevent a rare zero-over-zero
}                           condition.
y=ya[ns--];
for (m=1;m<n;m++) {
    for (i=0;i<n-m;i++) {
        w=c[i+1]-d[i];
        h=xa[i+m]-x;        h will never be zero, since this was tested in the initial-
        t=(xa[i]-x)*d[i]/h;      izing loop.
        dd=t-c[i+1];
        if (dd == 0.0) nrerror("Error in routine ratint");
        This error condition indicates that the interpolating function has a pole at the
        requested value of x.
        dd=w/dd;
        d[i]=c[i+1]*dd;
        c[i]=t*dd;
    }
    y += (dy=(2*(ns+1) < (n-m) ? c[ns+1] : d[ns--]));
}
}
```

CITED REFERENCES AND FURTHER READING:

Stoer, J., and Bulirsch, R. 1993, *Introduction to Numerical Analysis*, 2nd ed. (New York: Springer-Verlag), §2.2. [1]

Gear, C.W. 1971, *Numerical Initial Value Problems in Ordinary Differential Equations* (Englewood Cliffs, NJ: Prentice-Hall), §6.2.

Cuyt, A., and Wuytack, L. 1987, *Nonlinear Methods in Numerical Analysis* (Amsterdam: North-Holland), Chapter 3.

3.3 Cubic Spline Interpolation

Given a tabulated function $y_i = y(x_i)$, $i = 0...N - 1$, focus attention on one particular interval, between x_j and x_{j+1}. Linear interpolation in that interval gives the interpolation formula

$$y = Ay_j + By_{j+1} \tag{3.3.1}$$

where

$$A \equiv \frac{x_{j+1} - x}{x_{j+1} - x_j} \qquad B \equiv 1 - A = \frac{x - x_j}{x_{j+1} - x_j} \tag{3.3.2}$$

Equations (3.3.1) and (3.3.2) are a special case of the general Lagrange interpolation formula (3.1.1).

Since it is (piecewise) linear, equation (3.3.1) has zero second derivative in the interior of each interval, and an undefined, or infinite, second derivative at the abscissas x_j. The goal of cubic spline interpolation is to get an interpolation formula that is smooth in the first derivative, and continuous in the second derivative, both within an interval and at its boundaries.

Suppose, contrary to fact, that in addition to the tabulated values of y_i, we also have tabulated values for the function's second derivatives, y'', that is, a set of numbers y_i''. Then, within each interval, we can add to the right-hand side of equation (3.3.1) a cubic polynomial whose second derivative varies linearly from a value y_j'' on the left to a value y_{j+1}'' on the right. Doing so, we will have the desired continuous second derivative. If we also construct the cubic polynomial to have zero *values* at x_j and x_{j+1}, then adding it in will not spoil the agreement with the tabulated functional values y_j and y_{j+1} at the endpoints x_j and x_{j+1}.

A little side calculation shows that there is only one way to arrange this construction, namely replacing (3.3.1) by

$$y = Ay_j + By_{j+1} + Cy_j'' + Dy_{j+1}'' \qquad (3.3.3)$$

where A and B are defined in (3.3.2) and

$$C \equiv \frac{1}{6}(A^3 - A)(x_{j+1} - x_j)^2 \qquad D \equiv \frac{1}{6}(B^3 - B)(x_{j+1} - x_j)^2 \qquad (3.3.4)$$

Notice that the dependence on the independent variable x in equations (3.3.3) and (3.3.4) is entirely through the linear x-dependence of A and B, and (through A and B) the cubic x-dependence of C and D.

We can readily check that y'' is in fact the second derivative of the new interpolating polynomial. We take derivatives of equation (3.3.3) with respect to x, using the definitions of A, B, C, D to compute $dA/dx, dB/dx, dC/dx$, and dD/dx. The result is

$$\frac{dy}{dx} = \frac{y_{j+1} - y_j}{x_{j+1} - x_j} - \frac{3A^2 - 1}{6}(x_{j+1} - x_j)y_j'' + \frac{3B^2 - 1}{6}(x_{j+1} - x_j)y_{j+1}'' \quad (3.3.5)$$

for the first derivative, and

$$\frac{d^2y}{dx^2} = Ay_j'' + By_{j+1}'' \qquad (3.3.6)$$

for the second derivative. Since $A = 1$ at x_j, $A = 0$ at x_{j+1}, while B is just the other way around, (3.3.6) shows that y'' is just the tabulated second derivative, and also that the second derivative will be continuous across (e.g.) the boundary between the two intervals (x_{j-1}, x_j) and (x_j, x_{j+1}).

The only problem now is that we supposed the y_i'''s to be known, when, actually, they are not. However, we have not yet required that the *first* derivative, computed from equation (3.3.5), be continuous across the boundary between two intervals. The key idea of a cubic spline is to require this continuity and to use it to get equations for the second derivatives y_i''.

The required equations are obtained by setting equation (3.3.5) evaluated for $x = x_j$ in the interval (x_{j-1}, x_j) equal to the same equation evaluated for $x = x_j$ but in the interval (x_j, x_{j+1}). With some rearrangement, this gives (for $j = 1, \ldots, N-2$)

$$\frac{x_j - x_{j-1}}{6} y''_{j-1} + \frac{x_{j+1} - x_{j-1}}{3} y''_j + \frac{x_{j+1} - x_j}{6} y''_{j+1} = \frac{y_{j+1} - y_j}{x_{j+1} - x_j} - \frac{y_j - y_{j-1}}{x_j - x_{j-1}}$$

$$(3.3.7)$$

These are $N-2$ linear equations in the N unknowns y''_i, $i = 0, \ldots, N-1$. Therefore there is a two-parameter family of possible solutions.

For a unique solution, we need to specify two further conditions, typically taken as boundary conditions at x_0 and x_{N-1}. The most common ways of doing this are either

- set one or both of y''_0 and y''_{N-1} equal to zero, giving the so-called *natural cubic spline*, which has zero second derivative on one or both of its boundaries, or
- set either of y''_0 and y''_{N-1} to values calculated from equation (3.3.5) so as to make the first derivative of the interpolating function have a specified value on either or both boundaries.

One reason that cubic splines are especially practical is that the set of equations (3.3.7), along with the two additional boundary conditions, are not only linear, but also *tridiagonal*. Each y''_j is coupled only to its nearest neighbors at $j \pm 1$. Therefore, the equations can be solved in $O(N)$ operations by the tridiagonal algorithm (§2.4). That algorithm is concise enough to build right into the spline calculational routine. This makes the routine not completely transparent as an implementation of (3.3.7), so we encourage you to study it carefully, comparing with `tridag` (§2.4).

```
#include "nr.h"

void NR::spline(Vec_I_DP &x, Vec_I_DP &y, const DP yp1, const DP ypn,
    Vec_O_DP &y2)
```
Given arrays `x[0..n-1]` and `y[0..n-1]` containing a tabulated function, i.e., $y_i = f(x_i)$, with $x_0 < x_1 < \ldots < x_{N-1}$, and given values `yp1` and `ypn` for the first derivative of the interpolating function at points 0 and n−1, respectively, this routine returns an array `y2[0..n-1]` that contains the second derivatives of the interpolating function at the tabulated points x_i. If `yp1` and/or `ypn` are equal to 1×10^{30} or larger, the routine is signaled to set the corresponding boundary condition for a natural spline, with zero second derivative on that boundary.

```
{
    int i,k;
    DP p,qn,sig,un;

    int n=y2.size();
    Vec_DP u(n-1);
    if (yp1 > 0.99e30)                  The lower boundary condition is set either to be "nat-
        y2[0]=u[0]=0.0;                    ural"
    else {                              or else to have a specified first derivative.
        y2[0] = -0.5;
        u[0]=(3.0/(x[1]-x[0]))*((y[1]-y[0])/(x[1]-x[0])-yp1);
    }
    for (i=1;i<n-1;i++) {               This is the decomposition loop of the tridiagonal al-
        sig=(x[i]-x[i-1])/(x[i+1]-x[i-1]);   gorithm. y2 and u are used for tem-
        p=sig*y2[i-1]+2.0;                 porary storage of the decomposed
        y2[i]=(sig-1.0)/p;                 factors.
        u[i]=(y[i+1]-y[i])/(x[i+1]-x[i]) - (y[i]-y[i-1])/(x[i]-x[i-1]);
        u[i]=(6.0*u[i]/(x[i+1]-x[i-1])-sig*u[i-1])/p;
```

```
    }
    if (ypn > 0.99e30)                    The upper boundary condition is set either to be
        qn=un=0.0;                        "natural"
    else {                                or else to have a specified first derivative.
        qn=0.5;
        un=(3.0/(x[n-1]-x[n-2]))*(ypn-(y[n-1]-y[n-2])/(x[n-1]-x[n-2]));
    }
    y2[n-1]=(un-qn*u[n-2])/(qn*y2[n-2]+1.0);
    for (k=n-2;k>=0;k--)                   This is the backsubstitution loop of the tridiagonal
        y2[k]=y2[k]*y2[k+1]+u[k];          algorithm.
}
```

It is important to understand that the program spline is called only *once* to process an entire tabulated function in arrays x_i and y_i. Once this has been done, values of the interpolated function for any value of x are obtained by calls (as many as desired) to a separate routine splint (for "*spl*ine *int*erpolation"):

```
#include "nr.h"

void NR::splint(Vec_I_DP &xa, Vec_I_DP &ya, Vec_I_DP &y2a, const DP x, DP &y)
Given the arrays xa[0..n-1] and ya[0..n-1], which tabulate a function (with the xa_i's in
order), and given the array y2a[0..n-1], which is the output from spline above, and given
a value of x, this routine returns a cubic-spline interpolated value y.
{
    int k;
    DP h,b,a;

    int n=xa.size();
    int klo=0;                            We will find the right place in the table by means of
    int khi=n-1;                          bisection. This is optimal if sequential calls to this
    while (khi-klo > 1) {                 routine are at random values of x. If sequential calls
        k=(khi+klo) >> 1;                 are in order, and closely spaced, one would do better
        if (xa[k] > x) khi=k;             to store previous values of klo and khi and test if
        else klo=k;                       they remain appropriate on the next call.
    }                                     klo and khi now bracket the input value of x.
    h=xa[khi]-xa[klo];
    if (h == 0.0) nrerror("Bad xa input to routine splint");   The xa's must be dis-
    a=(xa[khi]-x)/h;                                           tinct.
    b=(x-xa[klo])/h;                      Cubic spline polynomial is now evaluated.
    y=a*ya[klo]+b*ya[khi]+((a*a*a-a)*y2a[klo]
        +(b*b*b-b)*y2a[khi])*(h*h)/6.0;
}
```

CITED REFERENCES AND FURTHER READING:

De Boor, C. 1978, *A Practical Guide to Splines* (New York: Springer-Verlag).

Forsythe, G.E., Malcolm, M.A., and Moler, C.B. 1977, *Computer Methods for Mathematical Computations* (Englewood Cliffs, NJ: Prentice-Hall), §§4.4–4.5.

Stoer, J., and Bulirsch, R. 1993, *Introduction to Numerical Analysis*, 2nd ed. (New York: Springer-Verlag), §2.4.

Ralston, A., and Rabinowitz, P. 1978, *A First Course in Numerical Analysis*, 2nd ed.; reprinted 2001 (New York: Dover), §3.8.

3.4 How to Search an Ordered Table

Suppose that you have decided to use some particular interpolation scheme, such as fourth-order polynomial interpolation, to compute a function $f(x)$ from a set of tabulated x_i's and f_i's. Then you will need a fast way of finding your place in the table of x_i's, given some particular value x at which the function evaluation is desired. This problem is not properly one of numerical analysis, but it occurs so often in practice that it would be negligent of us to ignore it.

Formally, the problem is this: Given an array of abscissas xx[j], j=0, 1,...,n-1, with the elements either monotonically increasing or monotonically decreasing, and given a number x, find an integer j such that x lies between xx[j] and xx[j+1]. For this task, let us define fictitious array elements xx[-1] and xx[n] equal to plus or minus infinity (in whichever order is consistent with the monotonicity of the table). Then j will always be between -1 and n-1, inclusive; a value of -1 indicates "off-scale" at one end of the table, n-1 indicates off-scale at the other end.

In most cases, when all is said and done, it is hard to do better than *bisection*, which will find the right place in the table in about $\log_2 n$ tries. We already did use bisection in the spline evaluation routine splint of the preceding section, so you might glance back at that. Standing by itself, a bisection routine looks like this:

```
#include "nr.h"

void NR::locate(Vec_I_DP &xx, const DP x, int &j)
Given an array xx[0..n-1], and given a value x, returns a value j such that x is between
xx[j] and xx[j+1]. xx must be monotonic, either increasing or decreasing. j=-1 or j=n-1
is returned to indicate that x is out of range.
{
    int ju,jm,jl;
    bool ascnd;

    int n=xx.size();
    jl=-1;                              Initialize lower
    ju=n;                              and upper limits.
    ascnd=(xx[n-1] >= xx[0]);          True if ascending order of table, false otherwise.
    while (ju-jl > 1) {                If we are not yet done,
        jm=(ju+jl) >> 1;              compute a midpoint,
        if (x >= xx[jm] == ascnd)
            jl=jm;                    and replace either the lower limit
        else
            ju=jm;                    or the upper limit, as appropriate.
    }                                  Repeat until the test condition is satisfied.
    if (x == xx[0]) j=0;              Then set the output
    else if (x == xx[n-1]) j=n-2;
    else j=jl;
}                                      and return.
```

Search with Correlated Values

Sometimes you will be in the situation of searching a large table many times, and with nearly identical abscissas on consecutive searches. For example, you may be generating a function that is used on the right-hand side of a differential equation: Most differential-equation integrators, as we shall see in Chapter 16, call

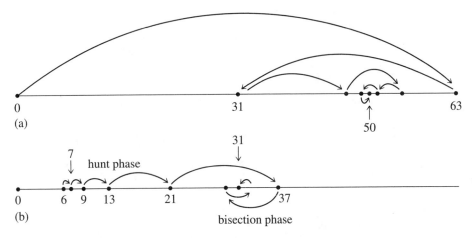

Figure 3.4.1.   (a) The routine `locate` finds a table entry by bisection. Shown here is the sequence of steps that converge to element 50 in a table of length 64. (b) The routine `hunt` searches from a previous known position in the table by increasing steps, then converges by bisection. Shown here is a particularly unfavorable example, converging to element 31 from element 6. A favorable example would be convergence to an element near 6, such as 8, which would require just three "hops."

for right-hand side evaluations at points that hop back and forth a bit, but whose trend moves slowly in the direction of the integration.

In such cases it is wasteful to do a full bisection, *ab initio*, on each call. The following routine instead starts with a guessed position in the table. It first "hunts," either up or down, in increments of 1, then 2, then 4, etc., until the desired value is bracketed. Second, it then bisects in the bracketed interval. At worst, this routine is about a factor of 2 slower than `locate` above (if the hunt phase expands to include the whole table). At best, it can be a factor of $\log_2 n$ faster than `locate`, if the desired point is usually quite close to the input guess. Figure 3.4.1 compares the two routines.

```
#include "nr.h"

void NR::hunt(Vec_I_DP &xx, const DP x, int &jlo)
Given an array xx[0..n-1], and given a value x, returns a value jlo such that x is between
xx[jlo] and xx[jlo+1]. xx[0..n-1] must be monotonic, either increasing or decreasing.
jlo=-1 or jlo=n-1 is returned to indicate that x is out of range. jlo on input is taken as
the initial guess for jlo on output.
{
    int jm,jhi,inc;
    bool ascnd;

    int n=xx.size();
    ascnd=(xx[n-1] >= xx[0]);          True if ascending order of table, false otherwise.
    if (jlo < 0 || jlo > n-1) {        Input guess not useful. Go immediately to bisec-
        jlo=-1;                            tion.
        jhi=n;
    } else {
        inc=1;                         Set the hunting increment.
        if (x >= xx[jlo] == ascnd) {   Hunt up:
            if (jlo == n-1) return;
            jhi=jlo+1;
            while (x >= xx[jhi] == ascnd) {     Not done hunting,
                jlo=jhi;
                inc += inc;            so double the increment
                jhi=jlo+inc;
                if (jhi > n-1) {       Done hunting, since off end of table.
```

```
                    jhi=n;
                    break;
                }                             Try again.
            }                                 Done hunting, value bracketed.
        } else {                              Hunt down:
            if (jlo == 0) {
                jlo=-1;
                return;
            }
            jhi=jlo--;
            while (x < xx[jlo] == ascnd) {            Not done hunting,
                jhi=jlo;
                inc <<= 1;                    so double the increment
                if (inc >= jhi) {             Done hunting, since off end of table.
                    jlo=-1;
                    break;
                }
                else jlo=jhi-inc;
            }                                 and try again.
        }                                     Done hunting, value bracketed.
    }                                         Hunt is done, so begin the final bisection phase:
    while (jhi-jlo != 1) {
        jm=(jhi+jlo) >> 1;
        if (x >= xx[jm] == ascnd)
            jlo=jm;
        else
            jhi=jm;
    }
    if (x == xx[n-1]) jlo=n-2;
    if (x == xx[0]) jlo=0;
}
```

After the Hunt

The problem: Routines `locate` and `hunt` return an index j such that your desired value lies between table entries xx[j] and xx[j+1], where xx[0..n-1] is the full length of the table. But, to obtain an m-point interpolated value using a routine like `polint` (§3.1) or `ratint` (§3.2), you need to supply shorter subarrays of xx and yy, of length m. How should the indices of these subarrays be specified?

The solution: Calculate

$$k = \mathtt{MIN}(\mathtt{MAX}(\mathtt{j-(m-1)/2,0}),\mathtt{n-m})$$

(The functions MIN and MAX give the minimum and maximum of two integer arguments; see §1.2 and Appendix B.) This expression produces the index of the leftmost member of an m-point set of points centered (insofar as possible) between j and j+1, but bounded by 0 at the left and n-1 at the right. You then need to copy the values xx[k..k+m-1] and yy[k..k+m-1] into temporary NRVecs to pass to `polint`.

CITED REFERENCES AND FURTHER READING:

Knuth, D.E. 1997, *Sorting and Searching*, 3rd ed., vol. 3 of *The Art of Computer Programming* (Reading, MA: Addison-Wesley), §6.2.1.

3.5 Coefficients of the Interpolating Polynomial

Occasionally you may wish to know not the value of the interpolating polynomial that passes through a (small!) number of points, but the coefficients of that polynomial. A valid use of the coefficients might be, for example, to compute simultaneous interpolated values of the function and of several of its derivatives (see §5.3), or to convolve a segment of the tabulated function with some other function, where the moments of that other function (i.e., its convolution with powers of x) are known analytically.

However, please be certain that the coefficients are what you need. Generally the coefficients of the interpolating polynomial can be determined much less accurately than its value at a desired abscissa. Therefore it is not a good idea to determine the coefficients only for use in calculating interpolating values. Values thus calculated will not pass exactly through the tabulated points, for example, while values computed by the routines in §3.1–§3.3 will pass exactly through such points.

Also, you should not mistake the interpolating polynomial (and its coefficients) for its cousin, the *best fit* polynomial through a data set. Fitting is a *smoothing* process, since the number of fitted coefficients is typically much less than the number of data points. Therefore, fitted coefficients can be accurately and stably determined even in the presence of statistical errors in the tabulated values. (See §14.8.) Interpolation, where the number of coefficients and number of tabulated points are equal, takes the tabulated values as perfect. If they in fact contain statistical errors, these can be magnified into oscillations of the interpolating polynomial in between the tabulated points.

As before, we take the tabulated points to be $y_i \equiv y(x_i)$. If the interpolating polynomial is written as

$$y = c_0 + c_1 x + c_2 x^2 + \cdots + c_{N-1} x^{N-1} \tag{3.5.1}$$

then the c_i's are required to satisfy the linear equation

$$
\begin{bmatrix}
1 & x_0 & x_0^2 & \cdots & x_0^{N-1} \\
1 & x_1 & x_1^2 & \cdots & x_1^{N-1} \\
\vdots & \vdots & \vdots & & \vdots \\
1 & x_{N-1} & x_{N-1}^2 & \cdots & x_{N-1}^{N-1}
\end{bmatrix}
\cdot
\begin{bmatrix}
c_0 \\
c_1 \\
\vdots \\
c_{N-1}
\end{bmatrix}
=
\begin{bmatrix}
y_0 \\
y_1 \\
\vdots \\
y_{N-1}
\end{bmatrix}
\tag{3.5.2}
$$

This is a *Vandermonde matrix*, as described in §2.8. One could in principle solve equation (3.5.2) by standard techniques for linear equations generally (§2.3); however the special method that was derived in §2.8 is more efficient by a large factor, of order N, so it is much better.

Remember that Vandermonde systems can be quite ill-conditioned. In such a case, *no* numerical method is going to give a very accurate answer. Such cases do not, please note, imply any difficulty in finding interpolated *values* by the methods of §3.1, but only difficulty in finding *coefficients*.

Like the routine in §2.8, the following is due to G.B. Rybicki.

```
#include "nr.h"
```

```
void NR::polcoe(Vec_I_DP &x, Vec_I_DP &y, Vec_O_DP &cof)
```
Given arrays $x[0..n-1]$ and $y[0..n-1]$ containing a tabulated function $y_i = f(x_i)$, this
routine returns an array of coefficients $\text{cof}[0..n-1]$, such that $y_i = \sum_{j=0}^{n-1} \text{cof}_j x_i^j$.
```
{
    int k,j,i;
    DP phi,ff,b;

    int n=x.size();
    Vec_DP s(n);
    for (i=0;i<n;i++) s[i]=cof[i]=0.0;
    s[n-1]= -x[0];
    for (i=1;i<n;i++) {                    Coefficients s_i of the master polynomial P(x) are
        for (j=n-1-i;j<n-1;j++)                found by recurrence.
            s[j] -= x[i]*s[j+1];
        s[n-1] -= x[i];
    }
    for (j=0;j<n;j++) {
        phi=n;
        for (k=n-1;k>0;k--)                The quantity phi = ∏_{j≠k}(x_j − x_k) is found as a
            phi=k*s[k]+x[j]*phi;              derivative of P(x_j).
        ff=y[j]/phi;
        b=1.0;                            Coefficients of polynomials in each term of the La-
        for (k=n-1;k>=0;k--) {                grange formula are found by synthetic division of
            cof[k] += b*ff;                    P(x) by (x − x_j). The solution c_k is accumu-
            b=s[k]+x[j]*b;                     lated.
        }
    }
}
```

The quantity $\text{phi} = \prod_{j \neq k}(x_j - x_k)$ is found as a derivative of $P(x_j)$.

Coefficients of polynomials in each term of the Lagrange formula are found by synthetic division of $P(x)$ by $(x - x_j)$. The solution c_k is accumulated.

Another Method

Another technique is to make use of the function value interpolation routine already given (`polint` §3.1). If we interpolate (or extrapolate) to find the value of the interpolating polynomial at $x = 0$, then this value will evidently be c_0. Now we can subtract c_0 from the y_i's and divide each by its corresponding x_i. Throwing out one point (the one with smallest x_i is a good candidate), we can repeat the procedure to find c_1, and so on.

It is not instantly obvious that this procedure is stable, but we have generally found it to be somewhat *more* stable than the routine immediately preceding. This method is of order N^3, while the preceding one was of order N^2. You will find, however, that neither works very well for large N, because of the intrinsic ill-condition of the Vandermonde problem. In single precision, N up to 8 or 10 is satisfactory; about double this in double precision.

```
#include <cmath>
#include "nr.h"
using namespace std;
```

```
void NR::polcof(Vec_I_DP &xa, Vec_I_DP &ya, Vec_O_DP &cof)
```
Given arrays $xa[0..n-1]$ and $ya[0..n-1]$ containing a tabulated function $ya_i = f(xa_i)$,
this routine returns an array of coefficients $\text{cof}[0..n-1]$ such that $ya_i = \sum_{j=0}^{n-1} \text{cof}_j xa_i^j$.
```
{
    int k,j,i;
    DP xmin,dy;
```

```
int n=xa.size();
Vec_DP x(n),y(n);
for (j=0;j<n;j++) {
    x[j]=xa[j];
    y[j]=ya[j];
}
for (j=0;j<n;j++) {
    Vec_DP x_t(n-j),y_t(n-j);
    for (k=0;k<n-j;k++) {
        x_t[k]=x[k];
        y_t[k]=y[k];
    }
    polint(x_t,y_t,0.0,cof[j],dy);          Extrapolate to x = 0.
    xmin=1.0e38;
    k = -1;
    for (i=0;i<n-j;i++) {                    Find the remaining x_i of smallest
        if (fabs(x[i]) < xmin) {                absolute value,
            xmin=fabs(x[i]);
            k=i;
        }
        if (x[i] != 0.0)                     (meanwhile reducing all the terms)
            y[i]=(y[i]-cof[j])/x[i];
    }
    for (i=k+1;i<n-j;i++) {                  and eliminate it.
        y[i-1]=y[i];
        x[i-1]=x[i];
    }
}
}
```

If the point $x = 0$ is not in (or at least close to) the range of the tabulated x_i's, then the coefficients of the interpolating polynomial will in general become very large. However, the real "information content" of the coefficients is in small differences from the "translation-induced" large values. This is one cause of ill-conditioning, resulting in loss of significance and poorly determined coefficients. You should consider redefining the origin of the problem, to put $x = 0$ in a sensible place.

Another pathology is that, if too high a degree of interpolation is attempted on a smooth function, the interpolating polynomial will attempt to use its high-degree coefficients, in combinations with large and almost precisely canceling combinations, to match the tabulated values down to the last possible epsilon of accuracy. This effect is the same as the intrinsic tendency of the interpolating polynomial values to oscillate (wildly) between its constrained points, and would be present even if the machine's floating precision were infinitely good. The above routines polcoe and polcof have slightly different sensitivities to the pathologies that can occur.

Are you still quite certain that using the *coefficients* is a good idea?

CITED REFERENCES AND FURTHER READING:

Isaacson, E., and Keller, H.B. 1966, *Analysis of Numerical Methods*; reprinted 1994 (New York: Dover), §5.2.

3.6 Interpolation in Two or More Dimensions

In multidimensional interpolation, we seek an estimate of $y(x_1, x_2, \ldots, x_n)$ from an n-dimensional grid of tabulated values y and n one-dimensional vectors giving the tabulated values of each of the independent variables $x_1, x_2, \ldots, x_n$. We will not here consider the problem of interpolating on a mesh that is not Cartesian, i.e., has tabulated function values at "random" points in n-dimensional space rather than at the vertices of a rectangular array. For clarity, we will consider explicitly only the case of two dimensions, the cases of three or more dimensions being analogous in every way.

In two dimensions, we imagine that we are given a matrix of functional values ya[0..m-1][0..n-1]. We are also given an array x1a[0..m-1], and an array x2a[0..n-1]. The relation of these input quantities to an underlying function $y(x_1, x_2)$ is

$$\texttt{ya[j][k]} = y(\texttt{x1a[j]}, \texttt{x2a[k]}) \qquad (3.6.1)$$

We want to estimate, by interpolation, the function y at some untabulated point (x_1, x_2).

An important concept is that of the *grid square* in which the point (x_1, x_2) falls, that is, the four tabulated points that surround the desired interior point. For convenience, we will number these points from 0 to 3, counterclockwise starting from the lower left (see Figure 3.6.1). More precisely, if

$$
\begin{aligned}
\texttt{x1a[j]} &\leq x_1 \leq \texttt{x1a[j+1]} \\
\texttt{x2a[k]} &\leq x_2 \leq \texttt{x2a[k+1]}
\end{aligned}
\qquad (3.6.2)
$$

defines j and k, then

$$
\begin{aligned}
y_0 &\equiv \texttt{ya[j][k]} \\
y_1 &\equiv \texttt{ya[j+1][k]} \\
y_2 &\equiv \texttt{ya[j+1][k+1]} \\
y_3 &\equiv \texttt{ya[j][k+1]}
\end{aligned}
\qquad (3.6.3)
$$

The simplest interpolation in two dimensions is *bilinear interpolation* on the grid square. Its formulas are:

$$
\begin{aligned}
t &\equiv (x_1 - \texttt{x1a[j]})/(\texttt{x1a[j+1]} - \texttt{x1a[j]}) \\
u &\equiv (x_2 - \texttt{x2a[k]})/(\texttt{x2a[k+1]} - \texttt{x2a[k]})
\end{aligned}
\qquad (3.6.4)
$$

(so that t and u each lie between 0 and 1), and

$$y(x_1, x_2) = (1-t)(1-u)y_0 + t(1-u)y_1 + tuy_2 + (1-t)uy_3 \qquad (3.6.5)$$

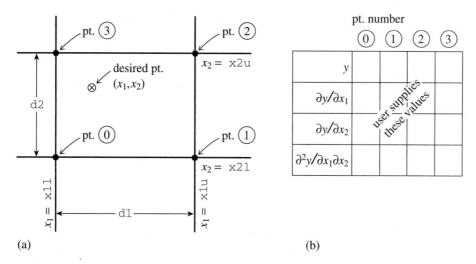

Figure 3.6.1.   (a) Labeling of points used in the two-dimensional interpolation routines bcuint and bcucof. (b) For each of the four points in (a), the user supplies one function value, two first derivatives, and one cross-derivative, a total of 16 numbers.

Bilinear interpolation is frequently "close enough for government work." As the interpolating point wanders from grid square to grid square, the interpolated function value changes continuously. However, the gradient of the interpolated function changes discontinuously at the boundaries of each grid square.

There are two distinctly different directions that one can take in going beyond bilinear interpolation to higher-order methods: One can use higher order to obtain increased accuracy for the interpolated function (for sufficiently smooth functions!), without necessarily trying to fix up the continuity of the gradient and higher derivatives. Or, one can make use of higher order to enforce smoothness of some of these derivatives as the interpolating point crosses grid-square boundaries. We will now consider each of these two directions in turn.

Higher Order for Accuracy

The basic idea is to break up the problem into a succession of one-dimensional interpolations. If we want to do $m-1$ order interpolation in the x_1 direction, and $n-1$ order in the x_2 direction, we first locate an $m \times n$ sub-block of the tabulated function matrix that contains our desired point (x_1, x_2). We then do m one-dimensional interpolations in the x_2 direction, i.e., on the rows of the sub-block, to get function values at the points $(x1a[j], x_2)$, $j = 0, \ldots, m-1$. Finally, we do a last interpolation in the x_1 direction to get the answer. If we use the polynomial interpolation routine polint of §3.1, and a sub-block which is presumed to be already located (and stored in the matrix ya), the procedure looks like this:

```
#include "nr.h"

void NR::polin2(Vec_I_DP &x1a, Vec_I_DP &x2a, Mat_I_DP &ya, const DP x1,
    const DP x2, DP &y, DP &dy)
```
Given arrays x1a[0..m−1] and x2a[0..n−1] of independent variables, and a submatrix of function values ya[0..m−1][0..n−1], tabulated at the grid points defined by x1a and x2a; and given values x1 and x2 of the independent variables; this routine returns an interpolated

function value y, and an accuracy indication dy (based only on the interpolation in the x1 direction, however).

```
{
    int j,k;

    int m=x1a.size();
    int n=x2a.size();
    Vec_DP ymtmp(m),ya_t(n);
    for (j=0;j<m;j++) {                          Loop over rows.
        for (k=0;k<n;k++) ya_t[k]=ya[j][k];
        polint(x2a,ya_t,x2,ymtmp[j],dy);         Interpolate answer into temporary stor-
    }                                            age.
    polint(x1a,ymtmp,x1,y,dy);                   Do the final interpolation.
}
```

Higher Order for Smoothness: Bicubic Interpolation

We will give two methods that are in common use, and which are themselves not unrelated. The first is usually called *bicubic interpolation*.

Bicubic interpolation requires the user to specify at each grid point not just the function $y(x_1, x_2)$, but also the gradients $\partial y/\partial x_1 \equiv y_{,1}$, $\partial y/\partial x_2 \equiv y_{,2}$ and the cross derivative $\partial^2 y/\partial x_1 \partial x_2 \equiv y_{,12}$. Then an interpolating function that is *cubic* in the scaled coordinates t and u (equation 3.6.4) can be found, with the following properties: (i) The values of the function and the specified derivatives are reproduced exactly on the grid points, and (ii) the values of the function and the specified derivatives change continuously as the interpolating point crosses from one grid square to another.

It is important to understand that nothing in the equations of bicubic interpolation requires you to specify the extra derivatives *correctly*! The smoothness properties are tautologically "forced," and have nothing to do with the "accuracy" of the specified derivatives. It is a separate problem for you to decide how to obtain the values that are specified. The better you do, the more *accurate* the interpolation will be. But it will be *smooth* no matter what you do.

Best of all is to know the derivatives analytically, or to be able to compute them accurately by numerical means, at the grid points. Next best is to determine them by numerical differencing from the functional values already tabulated on the grid. The relevant code would be something like this (using centered differencing):

```
y1a[j][k]=(ya[j+1][k]-ya[j-1][k])/(x1a[j+1]-x1a[j-1]);
y2a[j][k]=(ya[j][k+1]-ya[j][k-1])/(x2a[k+1]-x2a[k-1]);
y12a[j][k]=(ya[j+1][k+1]-ya[j+1][k-1]-ya[j-1][k+1]+ya[j-1][k-1])
        /((x1a[j+1]-x1a[j-1])*(x2a[k+1]-x2a[k-1]));
```

To do a bicubic interpolation within a grid square, given the function y and the derivatives y1, y2, y12 at each of the four corners of the square, there are two steps: First obtain the sixteen quantities c_{ij}, $i, j = 0, \ldots, 3$ using the routine bcucof below. (The formulas that obtain the c's from the function and derivative values are just a complicated linear transformation, with coefficients which, having been determined once in the mists of numerical history, can be tabulated and forgotten.) Next, substitute the c's into any or all of the following bicubic formulas for function and derivatives, as desired:

$$y(x_1, x_2) = \sum_{i=0}^{3} \sum_{j=0}^{3} c_{ij} t^i u^j$$

$$y_{,1}(x_1, x_2) = \sum_{i=0}^{3} \sum_{j=0}^{3} i c_{ij} t^{i-1} u^j \, (dt/dx_1)$$

$$y_{,2}(x_1, x_2) = \sum_{i=0}^{3} \sum_{j=0}^{3} j c_{ij} t^i u^{j-1} (du/dx_2)$$ (3.6.6)

$$y_{,12}(x_1, x_2) = \sum_{i=0}^{3} \sum_{j=0}^{3} i j c_{ij} t^{i-1} u^{j-1} (dt/dx_1)(du/dx_2)$$

where t and u are again given by equation (3.6.4).

```
#include "nr.h"

void NR::bcucof(Vec_I_DP &y, Vec_I_DP &y1, Vec_I_DP &y2, Vec_I_DP &y12,
    const DP d1, const DP d2, Mat_O_DP &c)
```
Given arrays y[0..3], y1[0..3], y2[0..3], and y12[0..3], containing the function, gra-
dients, and cross derivative at the four grid points of a rectangular grid cell (numbered counterclockwise from the lower left), and given d1 and d2, the length of the grid cell in the 1- and
2-directions, this routine returns the table c[0..3][0..3] that is used by routine bcuint
for bicubic interpolation.
```
{
    static int wt_d[16*16]=
        {1, 0, 0, 0, 0, 0, 0, 0, 0, 0, 0, 0, 0, 0, 0, 0,
        0, 0, 0, 0, 0, 0, 0, 0, 1, 0, 0, 0, 0, 0, 0, 0,
        -3, 0, 0, 3, 0, 0, 0, 0,-2, 0, 0,-1, 0, 0, 0, 0,
        2, 0, 0,-2, 0, 0, 0, 0, 1, 0, 0, 1, 0, 0, 0, 0,
        0, 0, 0, 0, 1, 0, 0, 0, 0, 0, 0, 0, 0, 0, 0, 0,
        0, 0, 0, 0, 0, 0, 0, 0, 0, 0, 0, 0, 1, 0, 0, 0,
        0, 0, 0, 0,-3, 0, 0, 3, 0, 0, 0, 0,-2, 0, 0,-1,
        0, 0, 0, 0, 2, 0, 0,-2, 0, 0, 0, 0, 1, 0, 0, 1,
        -3, 3, 0, 0,-2,-1, 0, 0, 0, 0, 0, 0, 0, 0, 0, 0,
        0, 0, 0, 0, 0, 0, 0, 0,-3, 3, 0, 0,-2,-1, 0, 0,
        9,-9, 9,-9, 6, 3,-3,-6, 6,-6,-3, 3, 4, 2, 1, 2,
        -6, 6,-6, 6,-4,-2, 2, 4,-3, 3, 3,-3,-2,-1,-1,-2,
        2,-2, 0, 0, 1, 1, 0, 0, 0, 0, 0, 0, 0, 0, 0, 0,
        0, 0, 0, 0, 0, 0, 0, 0, 2,-2, 0, 0, 1, 1, 0, 0,
        -6, 6,-6, 6,-3,-3, 3, 3,-4, 4, 2,-2,-2,-2,-1,-1,
        4,-4, 4,-4, 2, 2,-2, 2,-2,-2, 2, 1, 1, 1, 1};
    int l,k,j,i;
    DP xx,d1d2;
    Vec_DP cl(16),x(16);
    static Mat_INT wt(wt_d,16,16);

    d1d2=d1*d2;
    for (i=0;i<4;i++) {             Pack a temporary vector x.
        x[i]=y[i];
        x[i+4]=y1[i]*d1;
        x[i+8]=y2[i]*d2;
        x[i+12]=y12[i]*d1d2;
    }
    for (i=0;i<16;i++) {            Matrix multiply by the stored table.
        xx=0.0;
        for (k=0;k<16;k++) xx += wt[i][k]*x[k];
        cl[i]=xx;
    }
    l=0;
```

```
for (i=0;i<4;i++)                Unpack the result into the output table.
    for (j=0;j<4;j++) c[i][j]=cl[l++];
}
```

The implementation of equation (3.6.6), which performs a bicubic interpolation, gives back the interpolated function value and the two gradient values, and uses the above routine bcucof, is simply:

```
#include "nr.h"

void NR::bcuint(Vec_I_DP &y, Vec_I_DP &y1, Vec_I_DP &y2, Vec_I_DP &y12,
    const DP x1l, const DP x1u, const DP x2l, const DP x2u,
    const DP x1, const DP x2, DP &ansy, DP &ansy1, DP &ansy2)
Bicubic interpolation within a grid square. Input quantities are y,y1,y2,y12 (as described in
bcucof); x1l and x1u, the lower and upper coordinates of the grid square in the 1-direction;
x2l and x2u likewise for the 2-direction; and x1,x2, the coordinates of the desired point for
the interpolation. The interpolated function value is returned as ansy, and the interpolated
gradient values as ansy1 and ansy2. This routine calls bcucof.
{
    int i;
    DP t,u,d1,d2;
    Mat_DP c(4,4);

    d1=x1u-x1l;
    d2=x2u-x2l;
    bcucof(y,y1,y2,y12,d1,d2,c);        Get the c's.
    if (x1u == x1l || x2u == x2l)
        nrerror("Bad input in routine bcuint");
    t=(x1-x1l)/d1;                       Equation (3.6.4).
    u=(x2-x2l)/d2;
    ansy=ansy2=ansy1=0.0;
    for (i=3;i>=0;i--) {                 Equation (3.6.6).
        ansy=t*ansy+((c[i][3]*u+c[i][2])*u+c[i][1])*u+c[i][0];
        ansy2=t*ansy2+(3.0*c[i][3]*u+2.0*c[i][2])*u+c[i][1];
        ansy1=u*ansy1+(3.0*c[3][i]*t+2.0*c[2][i])*t+c[1][i];
    }
    ansy1 /= d1;
    ansy2 /= d2;
}
```

Higher Order for Smoothness: Bicubic Spline

The other common technique for obtaining smoothness in two-dimensional interpolation is the *bicubic spline*. Actually, this is equivalent to a special case of bicubic interpolation: The interpolating function is of the same functional form as equation (3.6.6); the values of the derivatives at the grid points are, however, determined "globally" by one-dimensional splines. However, bicubic splines are usually implemented in a form that looks rather different from the above bicubic interpolation routines, instead looking much closer in form to the routine polin2 above: To interpolate one functional value, one performs m one-dimensional splines across the rows of the table, followed by one additional one-dimensional spline down the newly created column. It is a matter of taste (and trade-off between time and memory) as to how much of this process one wants to precompute and store. Instead of precomputing and storing all the derivative information (as in bicubic interpolation), spline users typically precompute and store only one auxiliary table, of second derivatives in one direction only. Then one need only do spline *evaluations* (not constructions) for the m row splines; one must still do a construction *and* an evaluation for the final column spline. (Recall that a spline construction is a process of order N, while a spline evaluation is only of order $\log N$ — and that is just to find the place in the table!)

Here is a routine to precompute the auxiliary second-derivative table:

```
#include "nr.h"

void NR::splie2(Vec_I_DP &x1a, Vec_I_DP &x2a, Mat_I_DP &ya, Mat_O_DP &y2a)
Given an m by n tabulated function ya[0..m-1][0..n-1], and tabulated independent vari-
ables x2a[0..n-1], this routine constructs one-dimensional natural cubic splines of the rows
of ya and returns the second-derivatives in the array y2a[0..m-1][0..n-1]. (The array
x1a[0..m-1] is included in the argument list merely for consistency with routine splin2.)
{
    int m,n,j,k;

    m=x1a.size();
    n=x2a.size();
    Vec_DP ya_t(n),y2a_t(n);
    for (j=0;j<m;j++) {
        for (k=0;k<n;k++) ya_t[k]=ya[j][k];
        spline(x2a,ya_t,1.0e30,1.0e30,y2a_t);        Values 1 × 10^30 signal a natural
        for (k=0;k<n;k++) y2a[j][k]=y2a_t[k];            spline.
    }
}
```

After the above routine has been executed once, any number of bicubic spline interpolations can be performed by successive calls of the following routine:

```
#include "nr.h"

void NR::splin2(Vec_I_DP &x1a, Vec_I_DP &x2a, Mat_I_DP &ya, Mat_I_DP &y2a,
    const DP x1, const DP x2, DP &y)
Given x1a, x2a, ya, m, n as described in splie2 and y2a as produced by that routine; and
given a desired interpolating point x1,x2; this routine returns an interpolated function value y
by bicubic spline interpolation.
{
    int j,k;

    int m=x1a.size();
    int n=x2a.size();
    Vec_DP ya_t(n),y2a_t(n),yytmp(m),ytmp(m);
    for (j=0;j<m;j++) {                          Perform m evaluations of the row splines constructed
        for (k=0;k<n;k++) {                          by splie2, using the one-dimensional spline eval-
            ya_t[k]=ya[j][k];                        uator splint.
            y2a_t[k]=y2a[j][k];
        }
        splint(x2a,ya_t,y2a_t,x2,yytmp[j]);
    }
    spline(x1a,yytmp,1.0e30,1.0e30,ytmp);       Construct the one-dimensional column
    splint(x1a,yytmp,ytmp,x1,y);                    spline and evaluate it.
}
```

CITED REFERENCES AND FURTHER READING:

Abramowitz, M., and Stegun, I.A. 1964, *Handbook of Mathematical Functions*, Applied Mathematics Series, Volume 55 (Washington: National Bureau of Standards; reprinted 1968 by Dover Publications, New York), §25.2.

Kinahan, B.F., and Harm, R. 1975, *Astrophysical Journal*, vol. 200, pp. 330–335.

Johnson, L.W., and Riess, R.D. 1982, *Numerical Analysis*, 2nd ed. (Reading, MA: Addison-Wesley), §5.2.7.

Dahlquist, G., and Bjorck, A. 1974, *Numerical Methods* (Englewood Cliffs, NJ: Prentice-Hall), §7.7.

Chapter 4.   Integration of Functions

4.0 Introduction

Numerical integration, which is also called *quadrature*, has a history extending back to the invention of calculus and before. The fact that integrals of elementary functions could not, in general, be computed analytically, while derivatives *could* be, served to give the field a certain panache, and to set it a cut above the arithmetic drudgery of numerical analysis during the whole of the 18th and 19th centuries.

With the invention of automatic computing, quadrature became just one numerical task among many, and not a very interesting one at that. Automatic computing, even the most primitive sort involving desk calculators and rooms full of "computers" (that were, until the 1950s, people rather than machines), opened to feasibility the much richer field of numerical integration of differential equations. Quadrature is merely the simplest special case: The evaluation of the integral

$$I = \int_a^b f(x)dx \tag{4.0.1}$$

is precisely equivalent to solving for the value $I \equiv y(b)$ the differential equation

$$\frac{dy}{dx} = f(x) \tag{4.0.2}$$

with the boundary condition

$$y(a) = 0 \tag{4.0.3}$$

Chapter 16 of this book deals with the numerical integration of differential equations. In that chapter, much emphasis is given to the concept of "variable" or "adaptive" choices of stepsize. We will not, therefore, develop that material here. If the function that you propose to integrate is sharply concentrated in one or more peaks, or if its shape is not readily characterized by a single length-scale, then it is likely that you should cast the problem in the form of (4.0.2)–(4.0.3) and use the methods of Chapter 16.

The quadrature methods in this chapter are based, in one way or another, on the obvious device of adding up the value of the integrand at a sequence of abscissas within the range of integration. The game is to obtain the integral as accurately as possible with the smallest number of function evaluations of the integrand. Just as in the case of interpolation (Chapter 3), one has the freedom to choose methods

133

of various *orders*, with higher order sometimes, but not always, giving higher accuracy. "Romberg integration," which is discussed in §4.3, is a general formalism for making use of integration methods of a variety of different orders, and we recommend it highly.

Apart from the methods of this chapter and of Chapter 16, there are yet other methods for obtaining integrals. One important class is based on function approximation. We discuss explicitly the integration of functions by Chebyshev approximation ("Clenshaw-Curtis" quadrature) in §5.9. Although not explicitly discussed here, you ought to be able to figure out how to do *cubic spline quadrature* using the output of the routine spline in §3.3. (Hint: Integrate equation 3.3.3 over x analytically. See [1].)

Some integrals related to Fourier transforms can be calculated using the fast Fourier transform (FFT) algorithm. This is discussed in §13.9.

Multidimensional integrals are another whole multidimensional bag of worms. Section 4.6 is an introductory discussion in this chapter; the important technique of *Monte-Carlo integration* is treated in Chapter 7.

CITED REFERENCES AND FURTHER READING:

Carnahan, B., Luther, H.A., and Wilkes, J.O. 1969, *Applied Numerical Methods* (New York: Wiley), Chapter 2.

Isaacson, E., and Keller, H.B. 1966, *Analysis of Numerical Methods*; reprinted 1994 (New York: Dover), Chapter 7.

Acton, F.S. 1970, *Numerical Methods That Work*; 1990, corrected edition (Washington: Mathematical Association of America), Chapter 4.

Stoer, J., and Bulirsch, R. 1993, *Introduction to Numerical Analysis*, 2nd ed. (New York: Springer-Verlag), Chapter 3.

Ralston, A., and Rabinowitz, P. 1978, *A First Course in Numerical Analysis*, 2nd ed.; reprinted 2001 (New York: Dover), Chapter 4.

Dahlquist, G., and Bjorck, A. 1974, *Numerical Methods* (Englewood Cliffs, NJ: Prentice-Hall), §7.4.

Kahaner, D., Moler, C., and Nash, S. 1989, *Numerical Methods and Software* (Englewood Cliffs, NJ: Prentice Hall), Chapter 5.

Forsythe, G.E., Malcolm, M.A., and Moler, C.B. 1977, *Computer Methods for Mathematical Computations* (Englewood Cliffs, NJ: Prentice-Hall), §5.2, p. 89. [1]

Davis, P., and Rabinowitz, P. 1984, *Methods of Numerical Integration*, 2nd ed. (Orlando, FL: Academic Press).

4.1 Classical Formulas for Equally Spaced Abscissas

Where would any book on numerical analysis be without Mr. Simpson and his "rule"? The classical formulas for integrating a function whose value is known at equally spaced steps have a certain elegance about them, and they are redolent with historical association. Through them, the modern numerical analyst communes with the spirits of his or her predecessors back across the centuries, as far as the time of Newton, if not farther. Alas, times *do* change; with the exception of two of the

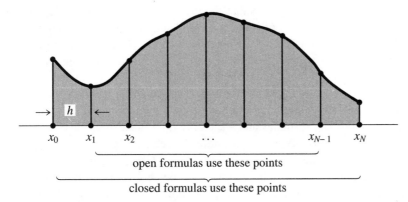

open formulas use these points

closed formulas use these points

Figure 4.1.1.   Quadrature formulas with equally spaced abscissas compute the integral of a function between x_0 and x_N. Closed formulas evaluate the function on the boundary points, while open formulas refrain from doing so (useful if the evaluation algorithm breaks down on the boundary points).

most modest formulas ("extended trapezoidal rule," equation 4.1.11, and "extended midpoint rule," equation 4.1.19, see §4.2), the classical formulas are almost entirely useless. They are museum pieces, but beautiful ones.

Some notation: We have a sequence of abscissas, denoted $x_0, x_1, \ldots, x_{N-1}$, x_N which are spaced apart by a constant step h,

$$x_i = x_0 + ih \qquad i = 0, 1, \ldots, N \tag{4.1.1}$$

A function $f(x)$ has known values at the x_i's,

$$f(x_i) \equiv f_i \tag{4.1.2}$$

We want to integrate the function $f(x)$ between a lower limit a and an upper limit b, where a and b are each equal to one or the other of the x_i's. An integration formula that uses the value of the function at the endpoints, $f(a)$ or $f(b)$, is called a *closed* formula. Occasionally, we want to integrate a function whose value at one or both endpoints is difficult to compute (e.g., the computation of f goes to a limit of zero over zero there, or worse yet has an integrable singularity there). In this case we want an *open* formula, which estimates the integral using only x_i's strictly *between* a and b (see Figure 4.1.1).

The basic building blocks of the classical formulas are rules for integrating a function over a small number of intervals. As that number increases, we can find rules that are exact for polynomials of increasingly high order. (Keep in mind that higher order does not always imply higher accuracy in real cases.) A sequence of such closed formulas is now given.

Closed Newton-Cotes Formulas

Trapezoidal rule:

$$\int_{x_0}^{x_1} f(x)dx = h\left[\frac{1}{2}f_0 + \frac{1}{2}f_1\right] + O(h^3 f'') \tag{4.1.3}$$

Here the error term $O(\ )$ signifies that the true answer differs from the estimate by an amount that is the product of some numerical coefficient times h^3 times the value of the function's second derivative somewhere in the interval of integration. The coefficient is knowable, and it can be found in all the standard references on this subject. The point at which the second derivative is to be evaluated is, however, unknowable. If we knew it, we could evaluate the function there and have a higher-order method! Since the product of a knowable and an unknowable is unknowable, we will streamline our formulas and write only $O(\ )$, instead of the coefficient.

Equation (4.1.3) is a two-point formula (x_0 and x_1). It is exact for polynomials up to and including degree 1, i.e., $f(x) = x$. One anticipates that there is a three-point formula exact up to polynomials of degree 2. This is true; moreover, by a cancellation of coefficients due to left-right symmetry of the formula, the three-point formula is exact for polynomials up to and including degree 3, i.e., $f(x) = x^3$:

Simpson's rule:

$$\int_{x_0}^{x_2} f(x)dx = h\left[\frac{1}{3}f_0 + \frac{4}{3}f_1 + \frac{1}{3}f_2\right] + O(h^5 f^{(4)}) \qquad (4.1.4)$$

Here $f^{(4)}$ means the fourth derivative of the function f evaluated at an unknown place in the interval. Note also that the formula gives the integral over an interval of size $2h$, so the coefficients add up to 2.

There is no lucky cancellation in the four-point formula, so it is also exact for polynomials up to and including degree 3.

Simpson's $\frac{3}{8}$ rule:

$$\int_{x_0}^{x_3} f(x)dx = h\left[\frac{3}{8}f_0 + \frac{9}{8}f_1 + \frac{9}{8}f_2 + \frac{3}{8}f_3\right] + O(h^5 f^{(4)}) \qquad (4.1.5)$$

The five-point formula again benefits from a cancellation:

Bode's rule:

$$\int_{x_0}^{x_4} f(x)dx = h\left[\frac{14}{45}f_0 + \frac{64}{45}f_1 + \frac{24}{45}f_2 + \frac{64}{45}f_3 + \frac{14}{45}f_4\right] + O(h^7 f^{(6)}) \quad (4.1.6)$$

This is exact for polynomials up to and including degree 5.

At this point the formulas stop being named after famous personages, so we will not go any further. Consult [1] for additional formulas in the sequence.

Extrapolative Formulas for a Single Interval

We are going to depart from historical practice for a moment. Many texts would give, at this point, a sequence of "Newton-Cotes Formulas of Open Type." Here is an example:

$$\int_{x_0}^{x_5} f(x)dx = h\left[\frac{55}{24}f_1 + \frac{5}{24}f_2 + \frac{5}{24}f_3 + \frac{55}{24}f_4\right] + O(h^5 f^{(4)})$$

Notice that the integral from $a = x_0$ to $b = x_5$ is estimated, using only the interior points x_1, x_2, x_3, x_4. In our opinion, formulas of this type are not useful for the reasons that (i) they cannot usefully be strung together to get "extended" rules, as we are about to do with the closed formulas, and (ii) for all other possible uses they are dominated by the Gaussian integration formulas which we will introduce in §4.5.

Instead of the Newton-Cotes open formulas, let us set out the formulas for estimating the integral in the single interval from x_0 to x_1, using values of the function f at $x_1, x_2, \ldots$. These will be useful building blocks for the "extended" open formulas.

$$\int_{x_0}^{x_1} f(x)dx = h[f_1] + O(h^2 f') \tag{4.1.7}$$

$$\int_{x_0}^{x_1} f(x)dx = h\left[\frac{3}{2}f_1 - \frac{1}{2}f_2\right] + O(h^3 f'') \tag{4.1.8}$$

$$\int_{x_0}^{x_1} f(x)dx = h\left[\frac{23}{12}f_1 - \frac{16}{12}f_2 + \frac{5}{12}f_3\right] + O(h^4 f^{(3)}) \tag{4.1.9}$$

$$\int_{x_0}^{x_1} f(x)dx = h\left[\frac{55}{24}f_1 - \frac{59}{24}f_2 + \frac{37}{24}f_3 - \frac{9}{24}f_4\right] + O(h^5 f^{(4)}) \tag{4.1.10}$$

Perhaps a word here would be in order about how formulas like the above can be derived. There are elegant ways, but the most straightforward is to write down the basic form of the formula, replacing the numerical coefficients with unknowns, say p, q, r, s. Without loss of generality take $x_0 = 0$ and $x_1 = 1$, so $h = 1$. Substitute in turn for $f(x)$ (and for f_1, f_2, f_3, f_4) the functions $f(x) = 1$, $f(x) = x$, $f(x) = x^2$, and $f(x) = x^3$. Doing the integral in each case reduces the left-hand side to a number, and the right-hand side to a linear equation for the unknowns p, q, r, s. Solving the four equations produced in this way gives the coefficients.

Extended Formulas (Closed)

If we use equation (4.1.3) $N - 1$ times, to do the integration in the intervals $(x_0, x_1), (x_1, x_2), \ldots, (x_{N-2}, x_{N-1})$, and then add the results, we obtain an "extended" or "composite" formula for the integral from x_0 to x_{N-1}.

Extended trapezoidal rule:

$$\int_{x_0}^{x_{N-1}} f(x)dx = h\left[\frac{1}{2}f_0 + f_1 + f_2 + \right.$$

$$\left. \cdots + f_{N-2} + \frac{1}{2}f_{N-1}\right] \quad + O\left(\frac{(b-a)^3 f''}{N^2}\right) \qquad (4.1.11)$$

Here we have written the error estimate in terms of the interval $b - a$ and the number of points N instead of in terms of h. This is clearer, since one is usually holding a and b fixed and wanting to know (e.g.) how much the error will be decreased by taking twice as many steps (in this case, it is by a factor of 4). In subsequent equations we will show *only* the scaling of the error term with the number of steps.

For reasons that will not become clear until §4.2, equation (4.1.11) is in fact the most important equation in this section, the basis for most practical quadrature schemes.

The *extended formula of order* $1/N^3$ is:

$$\int_{x_0}^{x_{N-1}} f(x)dx = h\left[\frac{5}{12}f_0 + \frac{13}{12}f_1 + f_2 + f_3 + \right.$$

$$\left. \cdots + f_{N-3} + \frac{13}{12}f_{N-2} + \frac{5}{12}f_{N-1}\right] \quad + O\left(\frac{1}{N^3}\right)$$

$$(4.1.12)$$

(We will see in a moment where this comes from.)

If we apply equation (4.1.4) to successive, nonoverlapping *pairs* of intervals, we get the *extended Simpson's rule:*

$$\int_{x_0}^{x_{N-1}} f(x)dx = h\left[\frac{1}{3}f_0 + \frac{4}{3}f_1 + \frac{2}{3}f_2 + \frac{4}{3}f_3 + \right.$$

$$\left. \cdots + \frac{2}{3}f_{N-3} + \frac{4}{3}f_{N-2} + \frac{1}{3}f_{N-1}\right] \quad + O\left(\frac{1}{N^4}\right)$$

$$(4.1.13)$$

Notice that the 2/3, 4/3 alternation continues throughout the interior of the evaluation. Many people believe that the wobbling alternation somehow contains deep information about the integral of their function that is not apparent to mortal eyes. In fact, the alternation is an artifact of using the building block (4.1.4). Another extended formula with the same order as Simpson's rule is

$$\int_{x_0}^{x_{N-1}} f(x)dx = h\left[\frac{3}{8}f_0 + \frac{7}{6}f_1 + \frac{23}{24}f_2 + f_3 + f_4 + \right.$$

$$\left. \cdots + f_{N-5} + f_{N-4} + \frac{23}{24}f_{N-3} + \frac{7}{6}f_{N-2} + \frac{3}{8}f_{N-1}\right]$$

$$+ O\left(\frac{1}{N^4}\right)$$

$$(4.1.14)$$

This equation is constructed by fitting cubic polynomials through successive groups of four points; we defer details to §18.3, where a similar technique is used in the solution of integral equations. We can, however, tell you where equation (4.1.12) came from. It is Simpson's extended rule, averaged with a modified version of itself in which the first and last step are done with the trapezoidal rule (4.1.3). The trapezoidal step is *two* orders lower than Simpson's rule; however, its contribution to the integral goes down as an additional power of N (since it is used only twice, not N times). This makes the resulting formula of degree *one* less than Simpson.

Extended Formulas (Open and Semi-open)

We can construct open and semi-open extended formulas by adding the closed formulas (4.1.11)–(4.1.14), evaluated for the second and subsequent steps, to the extrapolative open formulas for the first step, (4.1.7)–(4.1.10). As discussed immediately above, it is consistent to use an end step that is of one order lower than the (repeated) interior step. The resulting formulas for an interval open at both ends are as follows:

Equations (4.1.7) and (4.1.11) give

$$\int_{x_0}^{x_{N-1}} f(x)dx = h\left[\frac{3}{2}f_1 + f_2 + f_3 + \cdots + f_{N-3} + \frac{3}{2}f_{N-2}\right] + O\left(\frac{1}{N^2}\right) \quad (4.1.15)$$

Equations (4.1.8) and (4.1.12) give

$$\int_{x_0}^{x_{N-1}} f(x)dx = h\left[\frac{23}{12}f_1 + \frac{7}{12}f_2 + f_3 + f_4 + \right.$$
$$\left. \cdots + f_{N-4} + \frac{7}{12}f_{N-3} + \frac{23}{12}f_{N-2}\right] \quad (4.1.16)$$
$$+ O\left(\frac{1}{N^3}\right)$$

Equations (4.1.9) and (4.1.13) give

$$\int_{x_0}^{x_{N-1}} f(x)dx = h\left[\frac{27}{12}f_1 + 0 + \frac{13}{12}f_3 + \frac{4}{3}f_4 + \right.$$
$$\left. \cdots + \frac{4}{3}f_{N-5} + \frac{13}{12}f_{N-4} + 0 + \frac{27}{12}f_{N-2}\right] \quad (4.1.17)$$
$$+ O\left(\frac{1}{N^4}\right)$$

The interior points alternate 4/3 and 2/3. If we want to avoid this alternation, we can combine equations (4.1.9) and (4.1.14), giving

$$\int_{x_0}^{x_{N-1}} f(x)dx = h\left[\frac{55}{24}f_1 - \frac{1}{6}f_2 + \frac{11}{8}f_3 + f_4 + f_5 + f_6 + \right.$$

$$\left. \cdots + f_{N-6} + f_{N-5} + \frac{11}{8}f_{N-4} - \frac{1}{6}f_{N-3} + \frac{55}{24}f_{N-2}\right]$$

$$+ O\left(\frac{1}{N^4}\right)$$

$$(4.1.18)$$

We should mention in passing another extended open formula, for use where the limits of integration are located halfway between tabulated abscissas. This one is known as the *extended midpoint rule*, and is accurate to the same order as (4.1.15):

$$\int_{x_0}^{x_{N-1}} f(x)dx = h[f_{1/2} + f_{3/2} + f_{5/2} + $$

$$\cdots + f_{N-5/2} + f_{N-3/2}] \quad + O\left(\frac{1}{N^2}\right)$$

$$(4.1.19)$$

There are also formulas of higher order for this situation, but we will refrain from giving them.

The *semi-open formulas* are just the obvious combinations of equations (4.1.11)–(4.1.14) with (4.1.15)–(4.1.18), respectively. At the closed end of the integration, use the weights from the former equations; at the open end use the weights from the latter equations. One example should give the idea, the formula with error term decreasing as $1/N^3$ which is closed on the right and open on the left:

$$\int_{x_0}^{x_{N-1}} f(x)dx = h\left[\frac{23}{12}f_1 + \frac{7}{12}f_2 + f_3 + f_4 + \right.$$

$$\left. \cdots + f_{N-3} + \frac{13}{12}f_{N-2} + \frac{5}{12}f_{N-1}\right] \quad + O\left(\frac{1}{N^3}\right)$$

$$(4.1.20)$$

CITED REFERENCES AND FURTHER READING:

Abramowitz, M., and Stegun, I.A. 1964, *Handbook of Mathematical Functions*, Applied Mathematics Series, Volume 55 (Washington: National Bureau of Standards; reprinted 1968 by Dover Publications, New York), §25.4. [1]

Isaacson, E., and Keller, H.B. 1966, *Analysis of Numerical Methods*; reprinted 1994 (New York: Dover), §7.1.

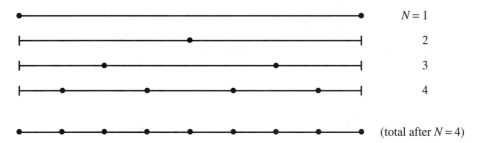

Figure 4.2.1.   Sequential calls to the routine `trapzd` incorporate the information from previous calls and evaluate the integrand only at those new points necessary to refine the grid. The bottom line shows the totality of function evaluations after the fourth call. The routine `qsimp`, by weighting the intermediate results, transforms the trapezoid rule into Simpson's rule with essentially no additional overhead.

4.2 Elementary Algorithms

Our starting point is equation (4.1.11), the extended trapezoidal rule. There are two facts about the trapezoidal rule which make it the starting point for a variety of algorithms. One fact is rather obvious, while the second is rather "deep."

The obvious fact is that, for a fixed function $f(x)$ to be integrated between fixed limits a and b, one can double the number of intervals in the extended trapezoidal rule without losing the benefit of previous work. The coarsest implementation of the trapezoidal rule is to average the function at its endpoints a and b. The first stage of refinement is to add to this average the value of the function at the halfway point. The second stage of refinement is to add the values at the 1/4 and 3/4 points. And so on (see Figure 4.2.1).

Without further ado we can write a routine with this kind of logic to it:

```
#include "nr.h"

DP NR::trapzd(DP func(const DP), const DP a, const DP b, const int n)
```
This routine computes the nth stage of refinement of an extended trapezoidal rule. `func` is input as the function to be integrated between limits a and b, also input. When called with n=1, the routine returns the crudest estimate of $\int_a^b f(x)dx$. Subsequent calls with n=2,3,... (in that sequential order) will improve the accuracy by adding 2^{n-2} additional interior points.
```
{
    DP x,tnm,sum,del;
    static DP s;
    int it,j;

    if (n == 1) {
        return (s=0.5*(b-a)*(func(a)+func(b)));
    } else {
        for (it=1,j=1;j<n-1;j++) it <<= 1;
        tnm=it;
        del=(b-a)/tnm;                  This is the spacing of the points to be added.
        x=a+0.5*del;
        for (sum=0.0,j=0;j<it;j++,x+=del) sum += func(x);
        s=0.5*(s+(b-a)*sum/tnm);        This replaces s by its refined value.
        return s;
    }
}
```

The above routine (`trapzd`) is a workhorse that can be harnessed in several ways. The simplest and crudest is to integrate a function by the extended trapezoidal rule where you know in advance (we can't imagine how!) the number of steps you want. If you want $2^M + 1$, you can accomplish this by the fragment

```
for(j=1;j<=m+1;j++) s=NR::trapzd(func,a,b,j);
```

with the answer returned as `s`.

Much better, of course, is to refine the trapezoidal rule until some specified degree of accuracy has been achieved:

```
#include <cmath>
#include "nr.h"
using namespace std;

DP NR::qtrap(DP func(const DP), const DP a, const DP b)
Returns the integral of the function func from a to b. The constants EPS can be set to the
desired fractional accuracy and JMAX so that 2 to the power JMAX-1 is the maximum allowed
number of steps. Integration is performed by the trapezoidal rule.
{
    const int JMAX=20;
    const DP EPS=1.0e-10;
    int j;
    DP s,olds=0.0;                    Initial value of olds is arbitrary.

    for (j=0;j<JMAX;j++) {
        s=trapzd(func,a,b,j+1);
        if (j > 5)                    Avoid spurious early convergence.
            if (fabs(s-olds) < EPS*fabs(olds) ||
                (s == 0.0 && olds == 0.0)) return s;
        olds=s;
    }
    nrerror("Too many steps in routine qtrap");
    return 0.0;                       Never get here.
}
```

Unsophisticated as it is, routine `qtrap` is in fact a fairly robust way of doing integrals of functions that are not very smooth. Increased sophistication will usually translate into a higher-order method whose efficiency will be greater only for sufficiently smooth integrands. `qtrap` is the method of choice, e.g., for an integrand which is a function of a variable that is linearly interpolated between measured data points. Be sure that you do not require too stringent an EPS, however: If `qtrap` takes too many steps in trying to achieve your required accuracy, accumulated roundoff errors may start increasing, and the routine may never converge. A value 10^{-10} or even smaller is usually no problem in double precision when the convergence is moderately rapid, but not otherwise. In single precision, accuracy is limited to about 10^{-6} at best.

We come now to the "deep" fact about the extended trapezoidal rule, equation (4.1.11). It is this: The error of the approximation, which begins with a term of order $1/N^2$, is in fact *entirely even* when expressed in powers of $1/N$. This follows directly from the *Euler-Maclaurin Summation Formula*,

$$\int_{x_0}^{x_{N-1}} f(x)dx = h\left[\frac{1}{2}f_0 + f_1 + f_2 + \cdots + f_{N-2} + \frac{1}{2}f_{N-1}\right]$$

$$- \frac{B_2 h^2}{2!}(f'_{N-1} - f'_0) - \cdots - \frac{B_{2k} h^{2k}}{(2k)!}(f_{N-1}^{(2k-1)} - f_0^{(2k-1)}) - \cdots$$

$$(4.2.1)$$

Here B_{2k} is a *Bernoulli number*, defined by the generating function

$$\frac{t}{e^t - 1} = \sum_{n=0}^{\infty} B_n \frac{t^n}{n!} \tag{4.2.2}$$

with the first few even values (odd values vanish except for $B_1 = -1/2$)

$$B_0 = 1 \quad B_2 = \frac{1}{6} \quad B_4 = -\frac{1}{30} \quad B_6 = \frac{1}{42}$$

$$B_8 = -\frac{1}{30} \quad B_{10} = \frac{5}{66} \quad B_{12} = -\frac{691}{2730} \tag{4.2.3}$$

Equation (4.2.1) is not a convergent expansion, but rather only an asymptotic expansion whose error when truncated at any point is always less than twice the magnitude of the first neglected term. The reason that it is not convergent is that the Bernoulli numbers become very large, e.g.,

$$B_{50} = \frac{495057205241079648212477525}{66}$$

The key point is that only even powers of h occur in the error series of (4.2.1). This fact is not, in general, shared by the higher-order quadrature rules in §4.1. For example, equation (4.1.12) has an error series beginning with $O(1/N^3)$, but continuing with all subsequent powers of N: $1/N^4$, $1/N^5$, etc.

Suppose we evaluate (4.1.11) with N steps, getting a result S_N, and then again with $2N$ steps, getting a result S_{2N}. (This is done by any two consecutive calls of trapzd.) The leading error term in the second evaluation will be 1/4 the size of the error in the first evaluation. Therefore the combination

$$S = \frac{4}{3}S_{2N} - \frac{1}{3}S_N \tag{4.2.4}$$

will cancel out the leading order error term. But there *is* no error term of order $1/N^3$, by (4.2.1). The surviving error is of order $1/N^4$, the same as Simpson's rule. In fact, it should not take long for you to see that (4.2.4) is *exactly* Simpson's rule (4.1.13), alternating 2/3's, 4/3's, and all. This is the preferred method for evaluating that rule, and we can write it as a routine exactly analogous to qtrap above:

```
#include <cmath>
#include "nr.h"
using namespace std;

DP NR::qsimp(DP func(const DP), const DP a, const DP b)
```
Returns the integral of the function func from a to b. The constants EPS can be set to the
desired fractional accuracy and JMAX so that 2 to the power JMAX-1 is the maximum allowed
number of steps. Integration is performed by Simpson's rule.
```
{
    const int JMAX=20;
    const DP EPS=1.0e-10;
    int j;
    DP s,st,ost=0.0,os=0.0;

    for (j=0;j<JMAX;j++) {
        st=trapzd(func,a,b,j+1);
        s=(4.0*st-ost)/3.0;             Compare equation (4.2.4), above.
        if (j > 5)                      Avoid spurious early convergence.
            if (fabs(s-os) < EPS*fabs(os) ||
                (s == 0.0 && os == 0.0)) return s;
        os=s;
        ost=st;
    }
    nrerror("Too many steps in routine qsimp");
    return 0.0;                         Never get here.
}
```

The routine qsimp will in general be more efficient than qtrap (i.e., require
fewer function evaluations) when the function to be integrated has a finite 4th
derivative (i.e., a continuous 3rd derivative). The combination of qsimp and its
necessary workhorse trapzd is a good one for light-duty work.

CITED REFERENCES AND FURTHER READING:

Stoer, J., and Bulirsch, R. 1993, *Introduction to Numerical Analysis*, 2nd ed. (New York: Springer-
Verlag), §3.3.

Dahlquist, G., and Bjorck, A. 1974, *Numerical Methods* (Englewood Cliffs, NJ: Prentice-Hall),
§§7.4.1–7.4.2.

Forsythe, G.E., Malcolm, M.A., and Moler, C.B. 1977, *Computer Methods for Mathematical
Computations* (Englewood Cliffs, NJ: Prentice-Hall), §5.3.

4.3 Romberg Integration

We can view Romberg's method as the natural generalization of the routine
qsimp in the last section to integration schemes that are of higher order than
Simpson's rule. The basic idea is to use the results from k successive refinements
of the extended trapezoidal rule (implemented in trapzd) to remove all terms in
the error series up to but not including $O(1/N^{2k})$. The routine qsimp is the case
of $k = 2$. This is one example of a very general idea that goes by the name of
Richardson's deferred approach to the limit: Perform some numerical algorithm for
various values of a parameter h, and then extrapolate the result to the continuum
limit $h = 0$.

Equation (4.2.4), which subtracts off the leading error term, is a special case of polynomial extrapolation. In the more general Romberg case, we can use Neville's algorithm (see §3.1) to extrapolate the successive refinements to zero stepsize. Neville's algorithm can in fact be coded very concisely within a Romberg integration routine. For clarity of the program, however, it seems better to do the extrapolation by function call to `polint`, already given in §3.1.

```cpp
#include <cmath>
#include "nr.h"
using namespace std;

DP NR::qromb(DP func(const DP), DP a, DP b)
```
Returns the integral of the function `func` from `a` to `b`. Integration is performed by Romberg's method of order 2K, where, e.g., K=2 is Simpson's rule.
```cpp
{
    const int JMAX=20, JMAXP=JMAX+1, K=5;
    const DP EPS=1.0e-10;
```
Here EPS is the fractional accuracy desired, as determined by the extrapolation error estimate; JMAX limits the total number of steps; K is the number of points used in the extrapolation.
```cpp
    DP ss,dss;
    Vec_DP s(JMAX),h(JMAXP),s_t(K),h_t(K);    These store the successive trapezoidal
    int i,j;                                  approximations and their relative
                                              stepsizes.
    h[0]=1.0;
    for (j=1;j<=JMAX;j++) {
        s[j-1]=trapzd(func,a,b,j);
        if (j >= K) {
            for (i=0;i<K;i++) {
                h_t[i]=h[j-K+i];
                s_t[i]=s[j-K+i];
            }
            polint(h_t,s_t,0.0,ss,dss);
            if (fabs(dss) <= EPS*fabs(ss)) return ss;
        }
        h[j]=0.25*h[j-1];
```
This is a key step: The factor is 0.25 even though the stepsize is decreased by only 0.5. This makes the extrapolation a polynomial in h^2 as allowed by equation (4.2.1), not just a polynomial in h.
```cpp
    }
    nrerror("Too many steps in routine qromb");
    return 0.0;                        Never get here.
}
```

The routine `qromb`, along with its required `trapzd` and `polint`, is quite powerful for sufficiently smooth (e.g., analytic) integrands, integrated over intervals which contain no singularities, and where the endpoints are also nonsingular. `qromb`, in such circumstances, takes many, *many* fewer function evaluations than either of the routines in §4.2. For example, the integral

$$\int_0^2 x^4 \log(x + \sqrt{x^2 + 1}) dx$$

converges (with parameters as shown above) on the second extrapolation, after just 6 calls to `trapzd`, while `qsimp` requires 11 calls (32 times as many evaluations of the integrand) and `qtrap` requires 19 calls (making 8192 times as many evaluations of the integrand).

CITED REFERENCES AND FURTHER READING:

Stoer, J., and Bulirsch, R. 1993, *Introduction to Numerical Analysis*, 2nd ed. (New York: Springer-Verlag), §§3.4–3.5.

Dahlquist, G., and Bjorck, A. 1974, *Numerical Methods* (Englewood Cliffs, NJ: Prentice-Hall), §§7.4.1–7.4.2.

Ralston, A., and Rabinowitz, P. 1978, *A First Course in Numerical Analysis*, 2nd ed.; reprinted 2001 (New York: Dover), §4.10–2.

4.4 Improper Integrals

For our present purposes, an integral will be "improper" if it has any of the following problems:

- its integrand goes to a finite limiting value at finite upper and lower limits, but cannot be evaluated *right on* one of those limits (e.g., $\sin x / x$ at $x = 0$)
- its upper limit is ∞, or its lower limit is $-\infty$
- it has an integrable singularity at either limit (e.g., $x^{-1/2}$ at $x = 0$)
- it has an integrable singularity at a known place between its upper and lower limits
- it has an integrable singularity at an unknown place between its upper and lower limits

If an integral is infinite (e.g., $\int_1^\infty x^{-1} dx$), or does not exist in a limiting sense (e.g., $\int_{-\infty}^\infty \cos x dx$), we do not call it improper; we call it impossible. No amount of clever algorithmics will return a meaningful answer to an ill-posed problem.

In this section we will generalize the techniques of the preceding two sections to cover the first four problems on the above list. A more advanced discussion of quadrature with integrable singularities occurs in Chapter 18, notably §18.3. The fifth problem, singularity at unknown location, can really only be handled by the use of a variable stepsize differential equation integration routine, as will be given in Chapter 16.

We need a workhorse like the extended trapezoidal rule (equation 4.1.11), but one which is an *open* formula in the sense of §4.1, i.e., does not require the integrand to be evaluated at the endpoints. Equation (4.1.19), the extended midpoint rule, is the best choice. The reason is that (4.1.19) shares with (4.1.11) the "deep" property of having an error series that is entirely even in h. Indeed there is a formula, not as well known as it ought to be, called the *Second Euler-Maclaurin summation formula*,

$$
\int_{x_0}^{x_{N-1}} f(x)dx = h[f_{1/2} + f_{3/2} + f_{5/2} + \cdots + f_{N-5/2} + f_{N-3/2}]
$$

$$
+ \frac{B_2 h^2}{4}(f'_{N-1} - f'_0) + \cdots \tag{4.4.1}
$$

$$
+ \frac{B_{2k} h^{2k}}{(2k)!}(1 - 2^{-2k+1})(f_{N-1}^{(2k-1)} - f_0^{(2k-1)}) + \cdots
$$

This equation can be derived by writing out (4.2.1) with stepsize h, then writing it out again with stepsize $h/2$, then subtracting the first from twice the second.

It is not possible to double the number of steps in the extended midpoint rule and still have the benefit of previous function evaluations (try it!). However, it is possible to *triple* the number of steps and do so. Shall we do this, or double and accept the loss? On the average, tripling does a factor $\sqrt{3}$ of unnecessary work, since the "right" number of steps for a desired accuracy criterion may in fact fall anywhere in the logarithmic interval implied by tripling. For doubling, the factor is only $\sqrt{2}$, but we lose an extra factor of 2 in being unable to use all the previous evaluations. Since $1.732 < 2 \times 1.414$, it is better to triple.

Here is the resulting routine, which is directly comparable to trapzd.

```
#include "nr.h"

DP NR::midpnt(DP func(const DP), const DP a, const DP b, const int n)
```
This routine computes the nth stage of refinement of an extended midpoint rule. func is input as the function to be integrated between limits a and b, also input. When called with n=1, the routine returns the crudest estimate of $\int_a^b f(x)dx$. Subsequent calls with n=2,3,... (in that sequential order) will improve the accuracy of s by adding $(2/3) \times 3^{n-1}$ additional interior points. s should not be modified between sequential calls.
```
{
    int it,j;
    DP x,tnm,sum,del,ddel;
    static DP s;

    if (n == 1) {
        return (s=(b-a)*func(0.5*(a+b)));
    } else {
        for(it=1,j=1;j<n-1;j++) it *= 3;
        tnm=it;
        del=(b-a)/(3.0*tnm);
        ddel=del+del;                   The added points alternate in spacing between
        x=a+0.5*del;                        del and ddel.
        sum=0.0;
        for (j=0;j<it;j++) {
            sum += func(x);
            x += ddel;
            sum += func(x);
            x += del;
        }
        s=(s+(b-a)*sum/tnm)/3.0;        The new sum is combined with the old integral
        return s;                           to give a refined integral.
    }
}
```

The routine midpnt can exactly replace trapzd in a driver routine like qtrap (§4.2); one simply changes trapzd(func,a,b,j) to midpnt(func,a,b, j), and perhaps also decreases the parameter JMAX since $3^{\text{JMAX}-1}$ (from step tripling) is a much larger number than $2^{\text{JMAX}-1}$ (step doubling).

The open formula implementation analogous to Simpson's rule (qsimp in §4.2) substitutes midpnt for trapzd and decreases JMAX as above, but now also changes the extrapolation step to be

```
s=(9.0*st-ost)/8.0;
```

since, when the number of steps is tripled, the error decreases to 1/9th its size, not 1/4th as with step doubling.

Either the modified qtrap or the modified qsimp will fix the first problem on the list at the beginning of this section. Yet more sophisticated is to generalize Romberg integration in like manner:

```
#include <cmath>
#include "nr.h"
using namespace std;

DP NR::qromo(DP func(const DP), const DP a, const DP b,
    DP choose(DP (*)(const DP), const DP, const DP, const int))
```
Romberg integration on an open interval. Returns the integral of the function `func` from a to b, using any specified integrating function `choose` and Romberg's method. Normally `choose` will be an open formula, not evaluating the function at the endpoints. It is assumed that `choose` triples the number of steps on each call, and that its error series contains only even powers of the number of steps. The routines `midpnt`, `midinf`, `midsql`, `midsqu`, `midexp`, are possible choices for `choose`. The constants below have the same meaning as in `qromb`.
```
{
    const int JMAX=14, JMAXP=JMAX+1, K=5;
    const DP EPS=3.0e-9;
    int i,j;
    DP ss,dss;
    Vec_DP h(JMAXP),s(JMAX),h_t(K),s_t(K);

    h[0]=1.0;
    for (j=1;j<=JMAX;j++) {
        s[j-1]=choose(func,a,b,j);
        if (j >= K) {
            for (i=0;i<K;i++) {
                h_t[i]=h[j-K+i];
                s_t[i]=s[j-K+i];
            }
            polint(h_t,s_t,0.0,ss,dss);
            if (fabs(dss) <= EPS*fabs(ss)) return ss;
        }
        h[j]=h[j-1]/9.0;              This is where the assumption of step tripling and an even
    }                                         error series is used.
    nrerror("Too many steps in routine qromo");
    return 0.0;                   Never get here.
}
```

Don't be put off by `qromo`'s complicated declaration. A typical invocation (integrating the Bessel function $Y_0(x)$ from 0 to 2) is simply

```
#include "nr.h"
using namespace NR;
DP answer;
...
answer=qromo(bessy0,0.0,2.0,midpnt);
```

The differences between `qromo` and `qromb` (§4.3) are so slight that it is perhaps gratuitous to list `qromo` in full. It, however, is an excellent driver routine for solving all the other problems of improper integrals in our first list (except the intractable fifth), as we shall now see.

The basic trick for improper integrals is to make a change of variables to eliminate the singularity, or to map an infinite range of integration to a finite one. For example, the identity

$$\int_a^b f(x)dx = \int_{1/b}^{1/a} \frac{1}{t^2} f\left(\frac{1}{t}\right) dt \qquad ab > 0 \tag{4.4.2}$$

can be used with *either* $b \to \infty$ and a positive, *or* with $a \to -\infty$ and b negative, and works for any function which decreases towards infinity faster than $1/x^2$.

You can make the change of variable implied by (4.4.2) either analytically and then use (e.g.) qromo and midpnt to do the numerical evaluation, *or* you can let the numerical algorithm make the change of variable for you. We prefer the latter method as being more transparent to the user. To implement equation (4.4.2) we simply write a modified version of midpnt, called midinf, which allows b to be infinite (or, more precisely, a very large number on your particular machine, such as 1×10^{30}), or a to be negative and infinite.

```
#include "nr.h"

namespace {
    DP func(DP funk(const DP), const DP x)
    {
        return funk(1.0/x)/(x*x);              Effects the change of variable.
    }
}
```

```
DP NR::midinf(DP funk(const DP), const DP aa, const DP bb, const int n)
This routine is an exact replacement for midpnt, i.e., returns the nth stage of refinement of
the integral of funk from aa to bb, except that the function is evaluated at evenly spaced
points in 1/x rather than in x. This allows the upper limit bb to be as large and positive as
the computer allows, or the lower limit aa to be as large and negative, but not both. aa and
bb must have the same sign.
{
    DP x,tnm,sum,del,ddel,b,a;
    static DP s;
    int it,j;

    b=1.0/aa;                      These two statements change the limits of integration.
    a=1.0/bb;
    if (n == 1) {                  From this point on, the routine is identical to midpnt.
        return (s=(b-a)*func(funk,0.5*(a+b)));
    } else {
        for(it=1,j=1;j<n-1;j++) it *= 3;
        tnm=it;
        del=(b-a)/(3.0*tnm);
        ddel=del+del;
        x=a+0.5*del;
        sum=0.0;
        for (j=0;j<it;j++) {
            sum += func(funk,x);
            x += ddel;
            sum += func(funk,x);
            x += del;
        }
        return (s=(s+(b-a)*sum/tnm)/3.0);
    }
}
```

If you need to integrate from a negative lower limit to positive infinity, you do this by breaking the integral into two pieces at some positive value, for example,

```
answer=qromo(funk,-5.0,2.0,midpnt)+qromo(funk,2.0,1.0e30,midinf);
```

Where should you choose the breakpoint? At a sufficiently large positive value so that the function funk is at least beginning to approach its asymptotic decrease to zero value at infinity. The polynomial extrapolation implicit in the second call to qromo deals with a polynomial in $1/x$, not in x.

To deal with an integral that has an integrable power-law singularity at its lower limit, one also makes a change of variable. If the integrand diverges as $(x-a)^{-\gamma}$, $0 \le \gamma < 1$, near $x = a$, use the identity

$$\int_a^b f(x)dx = \frac{1}{1-\gamma}\int_0^{(b-a)^{1-\gamma}} t^{\frac{\gamma}{1-\gamma}}f(t^{\frac{1}{1-\gamma}}+a)dt \qquad (b>a) \qquad (4.4.3)$$

If the singularity is at the upper limit, use the identity

$$\int_a^b f(x)dx = \frac{1}{1-\gamma}\int_0^{(b-a)^{1-\gamma}} t^{\frac{\gamma}{1-\gamma}}f(b-t^{\frac{1}{1-\gamma}})dt \qquad (b>a) \qquad (4.4.4)$$

If there is a singularity at both limits, divide the integral at an interior breakpoint as in the example above.

Equations (4.4.3) and (4.4.4) are particularly simple in the case of inverse square-root singularities, a case that occurs frequently in practice:

$$\int_a^b f(x)dx = \int_0^{\sqrt{b-a}} 2tf(a+t^2)dt \qquad (b>a) \qquad (4.4.5)$$

for a singularity at a, and

$$\int_a^b f(x)dx = \int_0^{\sqrt{b-a}} 2tf(b-t^2)dt \qquad (b>a) \qquad (4.4.6)$$

for a singularity at b. Once again, we can implement these changes of variable transparently to the user by defining substitute routines for `midpnt` which make the change of variable automatically:

```
#include <cmath>
#include "nr.h"
using namespace std;

namespace {
    DP func(DP funk(const DP), const DP aa, const DP x)
    {
        return 2.0*x*funk(aa+x*x);
    }
}
```

```
DP NR::midsql(DP funk(const DP), const DP aa, const DP bb, const int n)
```
This routine is an exact replacement for `midpnt`, except that it allows for an inverse square-root singularity in the integrand at the lower limit aa.
```
{
    DP x,tnm,sum,del,ddel,a,b;
    static DP s;
    int it,j;

    b=sqrt(bb-aa);
    a=0.0;
    if (n == 1) {
```
The rest of the routine is omitted since it is identical to `midinf`, except for the calling sequence of function func.

Similarly,

```
#include <cmath>
#include "nr.h"
using namespace std;

namespace {
    DP func(DP funk(const DP), const DP bb, const DP x)
    {
        return 2.0*x*funk(bb-x*x);
    }
}
```

DP NR::midsqu(DP funk(const DP), const DP aa, const DP bb, const int n)
This routine is an exact replacement for midpnt, except that it allows for an inverse square-root singularity in the integrand at the upper limit bb.
```
{
    DP x,tnm,sum,del,ddel,a,b;
    static DP s;
    int it,j;

    b=sqrt(bb-aa);
    a=0.0;
    if (n == 1) {
```
The rest of the routine is omitted since it is identical to midinf, except for the calling sequence of function func.

One last example should suffice to show how these formulas are derived in general. Suppose the upper limit of integration is infinite, and the integrand falls off exponentially. Then we want a change of variable that maps $e^{-x}dx$ into $(\pm)dt$ (with the sign chosen to keep the upper limit of the new variable larger than the lower limit). Doing the integration gives by inspection

$$t = e^{-x} \qquad \text{or} \qquad x = -\log t \tag{4.4.7}$$

so that

$$\int_{x=a}^{x=\infty} f(x)dx = \int_{t=0}^{t=e^{-a}} f(-\log t)\frac{dt}{t} \tag{4.4.8}$$

The user-transparent implementation would be

```
#include <cmath>
#include "nr.h"
using namespace std;

namespace {
    DP func(DP funk(const DP), const DP x)
    {
        return funk(-log(x))/x;
    }
}
```

DP NR::midexp(DP funk(const DP), const DP aa, const DP bb, const int n)
This routine is an exact replacement for midpnt, except that bb is assumed to be infinite (value passed not actually used). It is assumed that the function funk decreases exponentially rapidly at infinity.
```
{
    DP x,tnm,sum,del,ddel,a,b;
```

```
static DP s;
int it,j;

b=exp(-aa);
a=0.0;
if (n == 1) {
The rest of the routine is omitted since it is identical to midinf.
```

CITED REFERENCES AND FURTHER READING:

Acton, F.S. 1970, *Numerical Methods That Work*; 1990, corrected edition (Washington: Mathematical Association of America), Chapter 4.

Dahlquist, G., and Bjorck, A. 1974, *Numerical Methods* (Englewood Cliffs, NJ: Prentice-Hall), §7.4.3, p. 294.

Stoer, J., and Bulirsch, R. 1993, *Introduction to Numerical Analysis*, 2nd ed. (New York: Springer-Verlag), §3.7.

4.5 Gaussian Quadratures and Orthogonal Polynomials

In the formulas of §4.1, the integral of a function was approximated by the sum of its functional values at a set of equally spaced points, multiplied by certain aptly chosen weighting coefficients. We saw that as we allowed ourselves more freedom in choosing the coefficients, we could achieve integration formulas of higher and higher order. The idea of *Gaussian quadratures* is to give ourselves the freedom to choose not only the weighting coefficients, but also the location of the abscissas at which the function is to be evaluated: They will no longer be equally spaced. Thus, we will have *twice* the number of degrees of freedom at our disposal; it will turn out that we can achieve Gaussian quadrature formulas whose order is, essentially, twice that of the Newton-Cotes formula with the same number of function evaluations.

Does this sound too good to be true? Well, in a sense it is. The catch is a familiar one, which cannot be overemphasized: High order is not the same as high accuracy. High order translates to high accuracy only when the integrand is very smooth, in the sense of being "well-approximated by a polynomial."

There is, however, one additional feature of Gaussian quadrature formulas that adds to their usefulness: We can arrange the choice of weights and abscissas to make the integral exact for a class of integrands "polynomials times some known function $W(x)$" rather than for the usual class of integrands "polynomials." The function $W(x)$ can then be chosen to remove integrable singularities from the desired integral. Given $W(x)$, in other words, and given an integer N, we can find a set of weights w_j and abscissas x_j such that the approximation

$$\int_a^b W(x)f(x)dx \approx \sum_{j=0}^{N-1} w_j f(x_j) \tag{4.5.1}$$

is exact if $f(x)$ is a polynomial. For example, to do the integral

$$\int_{-1}^{1} \frac{\exp(-\cos^2 x)}{\sqrt{1-x^2}} dx \tag{4.5.2}$$

(not a very natural looking integral, it must be admitted), we might well be interested in a Gaussian quadrature formula based on the choice

$$W(x) = \frac{1}{\sqrt{1-x^2}} \tag{4.5.3}$$

in the interval $(-1, 1)$. (This particular choice is called *Gauss-Chebyshev integration*, for reasons that will become clear shortly.)

Notice that the integration formula (4.5.1) can also be written with the weight function $W(x)$ not overtly visible: Define $g(x) \equiv W(x)f(x)$ and $v_j \equiv w_j/W(x_j)$. Then (4.5.1) becomes

$$\int_a^b g(x)dx \approx \sum_{j=0}^{N-1} v_j g(x_j) \tag{4.5.4}$$

Where did the function $W(x)$ go? It is lurking there, ready to give high-order accuracy to integrands of the form polynomials times $W(x)$, and ready to *deny* high-order accuracy to integrands that are otherwise perfectly smooth and well-behaved. When you find tabulations of the weights and abscissas for a given $W(x)$, you have to determine carefully whether they are to be used with a formula in the form of (4.5.1), or like (4.5.4).

Here is an example of a quadrature routine that contains the tabulated abscissas and weights for the case $W(x) = 1$ and $N = 10$. Since the weights and abscissas are, in this case, symmetric around the midpoint of the range of integration, there are actually only five distinct values of each:

```
#include "nr.h"

DP NR::qgaus(DP func(const DP), const DP a, const DP b)
Returns the integral of the function func between a and b, by ten-point Gauss-Legendre inte-
gration: the function is evaluated exactly ten times at interior points in the range of integration.
{
    Here are the abscissas and weights:
    static const DP x[]={0.1488743389816312,0.4333953941292472,
        0.6794095682990244,0.8650633666889845,0.9739065285171717};
    static const DP w[]={0.2955242247147529,0.2692667193099963,
        0.2190863625159821,0.1494513491505806,0.0666713443086881};
    int j;
    DP xr,xm,dx,s;

    xm=0.5*(b+a);
    xr=0.5*(b-a);
    s=0;                          Will be twice the average value of the function, since the
    for (j=0;j<5;j++) {               ten weights (five numbers above each used twice)
        dx=xr*x[j];                   sum to 2.
        s += w[j]*(func(xm+dx)+func(xm-dx));
    }
    return s *= xr;               Scale the answer to the range of integration.
}
```

The above routine illustrates that one can use Gaussian quadratures without necessarily understanding the theory behind them: One just locates tabulated weights and abscissas in a book (e.g., [1] or [2]). However, the theory is very pretty, and it will come in handy if you ever need to construct your own tabulation of weights and abscissas for an unusual choice of $W(x)$. We will therefore give, without any proofs, some useful results that will enable you to do this. Several of the results assume that $W(x)$ does not change sign inside (a, b), which is usually the case in practice.

The theory behind Gaussian quadratures goes back to Gauss in 1814, who used continued fractions to develop the subject. In 1826 Jacobi rederived Gauss's results by means of orthogonal polynomials. The systematic treatment of arbitrary weight functions $W(x)$ using orthogonal polynomials is largely due to Christoffel in 1877. To introduce these orthogonal polynomials, let us fix the interval of interest to be (a, b). We can define the "scalar product of two functions f and g over a weight function W" as

$$\langle f|g \rangle \equiv \int_a^b W(x) f(x) g(x) dx \tag{4.5.5}$$

The scalar product is a number, not a function of x. Two functions are said to be *orthogonal* if their scalar product is zero. A function is said to be *normalized* if its scalar product with itself is unity. A set of functions that are all mutually orthogonal and also all individually normalized is called an *orthonormal* set.

We can find a set of polynomials (i) that includes exactly one polynomial of order j, called $p_j(x)$, for each $j = 0, 1, 2, \ldots$, and (ii) all of which are mutually orthogonal over the specified weight function $W(x)$. A constructive procedure for finding such a set is the recurrence relation

$$p_{-1}(x) \equiv 0$$
$$p_0(x) \equiv 1 \tag{4.5.6}$$
$$p_{j+1}(x) = (x - a_j)p_j(x) - b_j p_{j-1}(x) \qquad j = 0, 1, 2, \ldots$$

where

$$a_j = \frac{\langle xp_j|p_j \rangle}{\langle p_j|p_j \rangle} \qquad j = 0, 1, \ldots$$
$$b_j = \frac{\langle p_j|p_j \rangle}{\langle p_{j-1}|p_{j-1} \rangle} \qquad j = 1, 2, \ldots \tag{4.5.7}$$

The coefficient b_0 is arbitrary; we can take it to be zero.

The polynomials defined by (4.5.6) are *monic*, i.e., the coefficient of their leading term [x^j for $p_j(x)$] is unity. If we divide each $p_j(x)$ by the constant $[\langle p_j|p_j \rangle]^{1/2}$ we can render the set of polynomials orthonormal. One also encounters orthogonal polynomials with various other normalizations. You can convert from a given normalization to monic polynomials if you know that the coefficient of x^j in p_j is λ_j, say; then the monic polynomials are obtained by dividing each p_j by λ_j. Note that the coefficients in the recurrence relation (4.5.6) depend on the adopted normalization.

The polynomial $p_j(x)$ can be shown to have exactly j distinct roots in the interval (a, b). Moreover, it can be shown that the roots of $p_j(x)$ "interleave" the $j - 1$ roots of $p_{j-1}(x)$, i.e., there is exactly one root of the former in between each two adjacent roots of the latter. This fact comes in handy if you need to find all the roots: You can start with the one root of $p_1(x)$ and then, in turn, bracket the roots of each higher j, pinning them down at each stage more precisely by Newton's rule or some other root-finding scheme (see Chapter 9).

Why would you ever want to find all the roots of an orthogonal polynomial $p_j(x)$? Because the abscissas of the N-point Gaussian quadrature formulas (4.5.1) and (4.5.4) with weighting function $W(x)$ in the interval (a, b) are precisely the roots of the orthogonal polynomial $p_N(x)$ for the same interval and weighting function. This is the fundamental theorem of Gaussian quadratures, and lets you find the abscissas for any particular case.

Once you know the abscissas $x_0, \ldots, x_{N-1}$, you need to find the weights w_j, $j = 0, \ldots, N - 1$. One way to do this (not the most efficient) is to solve the set of linear equations

$$
\begin{bmatrix}
p_0(x_0) & \cdots & p_0(x_{N-1}) \\
p_1(x_0) & \cdots & p_1(x_{N-1}) \\
\vdots & & \vdots \\
p_{N-1}(x_0) & \cdots & p_{N-1}(x_{N-1})
\end{bmatrix}
\begin{bmatrix}
w_0 \\
w_1 \\
\vdots \\
w_{N-1}
\end{bmatrix}
=
\begin{bmatrix}
\int_a^b W(x)p_0(x)dx \\
0 \\
\vdots \\
0
\end{bmatrix}
\tag{4.5.8}
$$

Equation (4.5.8) simply solves for those weights such that the quadrature (4.5.1) gives the correct answer for the integral of the first N orthogonal polynomials. Note that the zeros on the right-hand side of (4.5.8) appear because $p_1(x), \ldots, p_{N-1}(x)$ are all orthogonal to $p_0(x)$, which is a constant. It can be shown that, with those weights, the integral of the *next* $N - 1$ polynomials is also exact, so that the quadrature is exact for all polynomials of degree $2N - 1$ or less. Another way to evaluate the weights (though one whose proof is beyond our scope) is by the formula

$$
w_j = \frac{\langle p_{N-1} | p_{N-1} \rangle}{p_{N-1}(x_j) p_N'(x_j)}
\tag{4.5.9}
$$

where $p_N'(x_j)$ is the derivative of the orthogonal polynomial at its zero x_j.

The computation of Gaussian quadrature rules thus involves two distinct phases: (i) the generation of the orthogonal polynomials $p_0, \ldots, p_N$, i.e., the computation of the coefficients a_j, b_j in (4.5.6); (ii) the determination of the zeros of $p_N(x)$, and the computation of the associated weights. For the case of the "classical" orthogonal polynomials, the coefficients a_j and b_j are explicitly known (equations 4.5.10 – 4.5.14 below) and phase (i) can be omitted. However, if you are confronted with a "nonclassical" weight function $W(x)$, and you don't know the coefficients a_j and b_j, the construction of the associated set of orthogonal polynomials is not trivial. We discuss it at the end of this section.

Computation of the Abscissas and Weights

This task can range from easy to difficult, depending on how much you already know about your weight function and its associated polynomials. In the case of classical, well-studied, orthogonal polynomials, practically everything is known, including good approximations for their zeros. These can be used as starting guesses, enabling Newton's method (to be discussed in §9.4) to converge very rapidly. Newton's method requires the derivative $p'_N(x)$, which is evaluated by standard relations in terms of p_N and p_{N-1}. The weights are then conveniently evaluated by equation (4.5.9). For the following named cases, this direct root-finding is faster, by a factor of 3 to 5, than any other method.

Here are the weight functions, intervals, and recurrence relations that generate the most commonly used orthogonal polynomials and their corresponding Gaussian quadrature formulas.

Gauss-Legendre:

$$W(x) = 1 \qquad -1 < x < 1$$

$$(j+1)P_{j+1} = (2j+1)xP_j - jP_{j-1} \tag{4.5.10}$$

Gauss-Chebyshev:

$$W(x) = (1 - x^2)^{-1/2} \qquad -1 < x < 1$$

$$T_{j+1} = 2xT_j - T_{j-1} \tag{4.5.11}$$

Gauss-Laguerre:

$$W(x) = x^\alpha e^{-x} \qquad 0 < x < \infty$$

$$(j+1)L^\alpha_{j+1} = (-x + 2j + \alpha + 1)L^\alpha_j - (j + \alpha)L^\alpha_{j-1} \tag{4.5.12}$$

Gauss-Hermite:

$$W(x) = e^{-x^2} \qquad -\infty < x < \infty$$

$$H_{j+1} = 2xH_j - 2jH_{j-1} \tag{4.5.13}$$

Gauss-Jacobi:

$$W(x) = (1-x)^\alpha(1+x)^\beta \qquad -1 < x < 1$$

$$c_j P^{(\alpha,\beta)}_{j+1} = (d_j + e_j x)P^{(\alpha,\beta)}_j - f_j P^{(\alpha,\beta)}_{j-1} \tag{4.5.14}$$

where the coefficients c_j, d_j, e_j, and f_j are given by

$$
\begin{aligned}
c_j &= 2(j+1)(j+\alpha+\beta+1)(2j+\alpha+\beta) \\
d_j &= (2j+\alpha+\beta+1)(\alpha^2 - \beta^2) \\
e_j &= (2j+\alpha+\beta)(2j+\alpha+\beta+1)(2j+\alpha+\beta+2) \\
f_j &= 2(j+\alpha)(j+\beta)(2j+\alpha+\beta+2)
\end{aligned}
\tag{4.5.15}
$$

We now give individual routines that calculate the abscissas and weights for these cases. First comes the most common set of abscissas and weights, those of Gauss-Legendre. The routine, due to G.B. Rybicki, uses equation (4.5.9) in the special form for the Gauss-Legendre case,

$$w_j = \frac{2}{(1 - x_j^2)[P_N'(x_j)]^2} \tag{4.5.16}$$

The routine also scales the range of integration from (x_1, x_2) to $(-1, 1)$, and provides abscissas x_j and weights w_j for the Gaussian formula

$$\int_{x_1}^{x_2} f(x)dx = \sum_{j=0}^{N-1} w_j f(x_j) \tag{4.5.17}$$

```cpp
#include <cmath>
#include "nr.h"
using namespace std;

void NR::gauleg(const DP x1, const DP x2, Vec_O_DP &x, Vec_O_DP &w)
```
Given the lower and upper limits of integration x1 and x2, this routine returns arrays x[0..n-1] and w[0..n-1] of length n, containing the abscissas and weights of the Gauss-Legendre n-point quadrature formula.
```cpp
{
    const DP EPS=1.0e-14;              // EPS is the relative precision.
    int m,j,i;
    DP z1,z,xm,xl,pp,p3,p2,p1;         // High precision is a good idea for this rou-
                                       // tine.
    int n=x.size();
    m=(n+1)/2;                         // The roots are symmetric in the interval, so
    xm=0.5*(x2+x1);                    // we only have to find half of them.
    xl=0.5*(x2-x1);
    for (i=0;i<m;i++) {                // Loop over the desired roots.
        z=cos(3.141592654*(i+0.75)/(n+0.5));
```
Starting with this approximation to the ith root, we enter the main loop of refinement by Newton's method.
```cpp
        do {
            p1=1.0;
            p2=0.0;
            for (j=0;j<n;j++) {        // Loop up the recurrence relation to get the
                p3=p2;                 // Legendre polynomial evaluated at z.
                p2=p1;
                p1=((2.0*j+1.0)*z*p2-j*p3)/(j+1);
            }
```
p1 is now the desired Legendre polynomial. We next compute pp, its derivative, by a standard relation involving also p2, the polynomial of one lower order.
```cpp
            pp=n*(z*p1-p2)/(z*z-1.0);
            z1=z;
            z=z1-p1/pp;                // Newton's method.
        } while (fabs(z-z1) > EPS);
        x[i]=xm-xl*z;                  // Scale the root to the desired interval,
        x[n-1-i]=xm+xl*z;             // and put in its symmetric counterpart.
        w[i]=2.0*xl/((1.0-z*z)*pp*pp); // Compute the weight
        w[n-1-i]=w[i];                 // and its symmetric counterpart.
    }
}
```

Next we give three routines that use initial approximations for the roots given by Stroud and Secrest [2]. The first is for Gauss-Laguerre abscissas and weights, to be used with the integration formula

$$\int_0^\infty x^\alpha e^{-x} f(x)dx = \sum_{j=0}^{N-1} w_j f(x_j) \qquad (4.5.18)$$

```
#include <cmath>
#include "nr.h"
using namespace std;

void NR::gaulag(Vec_O_DP &x, Vec_O_DP &w, const DP alf)
```
Given `alf`, the parameter α of the Laguerre polynomials, this routine returns arrays `x[0..n-1]` and `w[0..n-1]` containing the abscissas and weights of the n-point Gauss-Laguerre quadrature formula. The smallest abscissa is returned in `x[0]`, the largest in `x[n-1]`.
```
{
    const int MAXIT=10;
    const DP EPS=1.0e-14;            EPS is the relative precision.
    int i,its,j;
    DP ai,p1,p2,p3,pp,z,z1;          High precision is a good idea for this rou-
                                     tine.
    int n=x.size();
    for (i=0;i<n;i++) {              Loop over the desired roots.
        if (i == 0) {               Initial guess for the smallest root.
            z=(1.0+alf)*(3.0+0.92*alf)/(1.0+2.4*n+1.8*alf);
        } else if (i == 1) {        Initial guess for the second root.
            z += (15.0+6.25*alf)/(1.0+0.9*alf+2.5*n);
        } else {                    Initial guess for the other roots.
            ai=i-1;
            z += ((1.0+2.55*ai)/(1.9*ai)+1.26*ai*alf/
                (1.0+3.5*ai))*(z-x[i-2])/(1.0+0.3*alf);
        }
        for (its=0;its<MAXIT;its++) {    Refinement by Newton's method.
            p1=1.0;
            p2=0.0;
            for (j=0;j<n;j++) {          Loop up the recurrence relation to get the
                p3=p2;                   Laguerre polynomial evaluated at z.
                p2=p1;
                p1=((2*j+1+alf-z)*p2-(j+alf)*p3)/(j+1);
            }
            p1 is now the desired Laguerre polynomial. We next compute pp, its derivative,
            by a standard relation involving also p2, the polynomial of one lower order.
            pp=(n*p1-(n+alf)*p2)/z;
            z1=z;
            z=z1-p1/pp;                  Newton's formula.
            if (fabs(z-z1) <= EPS) break;
        }
        if (its >= MAXIT) nrerror("too many iterations in gaulag");
        x[i]=z;                          Store the root and the weight.
        w[i] = -exp(gammln(alf+n)-gammln(DP(n)))/(pp*n*p2);
    }
}
```

Next is a routine for Gauss-Hermite abscissas and weights. If we use the "standard" normalization of these functions, as given in equation (4.5.13), we find that the computations overflow for large N because of various factorials that occur. We can avoid this by using instead the orthonormal set of polynomials $\widetilde{H}_j$. They are generated by the recurrence

$$\widetilde{H}_{-1} = 0, \quad \widetilde{H}_0 = \frac{1}{\pi^{1/4}}, \quad \widetilde{H}_{j+1} = x\sqrt{\frac{2}{j+1}}\widetilde{H}_j - \sqrt{\frac{j}{j+1}}\widetilde{H}_{j-1} \quad (4.5.19)$$

The formula for the weights becomes

$$w_j = \frac{2}{[\widetilde{H}'_N(x_j)]^2} \tag{4.5.20}$$

while the formula for the derivative with this normalization is

$$\widetilde{H}'_j = \sqrt{2j}\widetilde{H}_{j-1} \tag{4.5.21}$$

The abscissas and weights returned by gauher are used with the integration formula

$$\int_{-\infty}^{\infty} e^{-x^2} f(x)dx = \sum_{j=0}^{N-1} w_j f(x_j) \tag{4.5.22}$$

```
#include <cmath>
#include "nr.h"
using namespace std;

void NR::gauher(Vec_O_DP &x, Vec_O_DP &w)
```
This routine returns arrays x[0..n-1] and w[0..n-1] containing the abscissas and weights of the n-point Gauss-Hermite quadrature formula. The largest abscissa is returned in x[0], the most negative in x[n-1].
```
{
    const DP EPS=1.0e-14,PIM4=0.7511255444649425;
```
Relative precision and $1/\pi^{1/4}$.
```
    const int MAXIT=10;                         Maximum iterations.
    int i,its,j,m;
    DP p1,p2,p3,pp,z,z1;                        High precision is a good idea for this rou-
                                                    tine.
    int n=x.size();
    m=(n+1)/2;
```
The roots are symmetric about the origin, so we have to find only half of them.
```
    for (i=0;i<m;i++) {                         Loop over the desired roots.
        if (i == 0) {                           Initial guess for the largest root.
            z=sqrt(DP(2*n+1))-1.85575*pow(DP(2*n+1),-0.16667);
        } else if (i == 1) {                    Initial guess for the second largest root.
            z -= 1.14*pow(DP(n),0.426)/z;
        } else if (i == 2) {                    Initial guess for the third largest root.
            z=1.86*z-0.86*x[0];
        } else if (i == 3) {                    Initial guess for the fourth largest root.
            z=1.91*z-0.91*x[1];
        } else {                                Initial guess for the other roots.
            z=2.0*z-x[i-2];
        }
        for (its=0;its<MAXIT;its++) {           Refinement by Newton's method.
            p1=PIM4;
            p2=0.0;
            for (j=0;j<n;j++) {                 Loop up the recurrence relation to get
                p3=p2;                              the Hermite polynomial evaluated at
                p2=p1;                              z.
                p1=z*sqrt(2.0/(j+1))*p2-sqrt(DP(j)/(j+1))*p3;
            }
```

p1 is now the desired Hermite polynomial. We next compute pp, its derivative, by
the relation (4.5.21) using p2, the polynomial of one lower order.

```
        pp=sqrt(DP(2*n))*p2;
        z1=z;
        z=z1-p1/pp;                          Newton's formula.
        if (fabs(z-z1) <= EPS) break;
    }
    if (its >= MAXIT) nrerror("too many iterations in gauher");
    x[i]=z;                                  Store the root
    x[n-1-i] = -z;                           and its symmetric counterpart.
    w[i]=2.0/(pp*pp);                        Compute the weight
    w[n-1-i]=w[i];                           and its symmetric counterpart.
    }
}
```

Finally, here is a routine for Gauss-Jacobi abscissas and weights, which
implement the integration formula

$$\int_{-1}^{1} (1-x)^\alpha (1+x)^\beta f(x)dx = \sum_{j=0}^{N-1} w_j f(x_j) \qquad (4.5.23)$$

```
#include <cmath>
#include "nr.h"
using namespace std;

void NR::gaujac(Vec_O_DP &x, Vec_O_DP &w, const DP alf, const DP bet)
```
Given alf and bet, the parameters α and β of the Jacobi polynomials, this routine returns
arrays x[0..n-1] and w[0..n-1] containing the abscissas and weights of the n-point Gauss-
Jacobi quadrature formula. The largest abscissa is returned in x[0], the smallest in x[n-1].
```
{
    const int MAXIT=10;
    const DP EPS=1.0e-14;                     EPS is the relative precision.
    int i,its,j;
    DP alfbet,an,bn,r1,r2,r3;
    DP a,b,c,p1,p2,p3,pp,temp,z,z1;           High precision is a good idea for this rou-
                                              tine.
    int n=x.size();
    for (i=0;i<n;i++) {                        Loop over the desired roots.
        if (i == 0) {                         Initial guess for the largest root.
            an=alf/n;
            bn=bet/n;
            r1=(1.0+alf)*(2.78/(4.0+n*n)+0.768*an/n);
            r2=1.0+1.48*an+0.96*bn+0.452*an*an+0.83*an*bn;
            z=1.0-r1/r2;
        } else if (i == 1) {                  Initial guess for the second largest root.
            r1=(4.1+alf)/((1.0+alf)*(1.0+0.156*alf));
            r2=1.0+0.06*(n-8.0)*(1.0+0.12*alf)/n;
            r3=1.0+0.012*bet*(1.0+0.25*fabs(alf))/n;
            z -= (1.0-z)*r1*r2*r3;
        } else if (i == 2) {                  Initial guess for the third largest root.
            r1=(1.67+0.28*alf)/(1.0+0.37*alf);
            r2=1.0+0.22*(n-8.0)/n;
            r3=1.0+8.0*bet/((6.28+bet)*n*n);
            z -= (x[0]-z)*r1*r2*r3;
        } else if (i == n-2) {                Initial guess for the second smallest root.
            r1=(1.0+0.235*bet)/(0.766+0.119*bet);
            r2=1.0/(1.0+0.639*(n-4.0)/(1.0+0.71*(n-4.0)));
            r3=1.0/(1.0+20.0*alf/((7.5+alf)*n*n));
            z += (z-x[n-4])*r1*r2*r3;
        } else if (i == n-1) {                Initial guess for the smallest root.
```

```
        r1=(1.0+0.37*bet)/(1.67+0.28*bet);
        r2=1.0/(1.0+0.22*(n-8.0)/n);
        r3=1.0/(1.0+8.0*alf/((6.28+alf)*n*n));
        z += (z-x[n-3])*r1*r2*r3;
    } else {                          Initial guess for the other roots.
        z=3.0*x[i-1]-3.0*x[i-2]+x[i-3];
    }
    alfbet=alf+bet;
    for (its=1;its<=MAXIT;its++) {      Refinement by Newton's method.
        temp=2.0+alfbet;                Start the recurrence with P_0 and P_1 to avoid
        p1=(alf-bet+temp*z)/2.0;          a division by zero when α + β = 0 or
        p2=1.0;                           −1.
        for (j=2;j<=n;j++) {            Loop up the recurrence relation to get the
            p3=p2;                        Jacobi polynomial evaluated at z.
            p2=p1;
            temp=2*j+alfbet;
            a=2*j*(j+alfbet)*(temp-2.0);
            b=(temp-1.0)*(alf*alf-bet*bet+temp*(temp-2.0)*z);
            c=2.0*(j-1+alf)*(j-1+bet)*temp;
            p1=(b*p2-c*p3)/a;
        }
        pp=(n*(alf-bet-temp*z)*p1+2.0*(n+alf)*(n+bet)*p2)/(temp*(1.0-z*z));
        p1 is now the desired Jacobi polynomial. We next compute pp, its derivative, by
        a standard relation involving also p2, the polynomial of one lower order.
        z1=z;
        z=z1-p1/pp;                     Newton's formula.
        if (fabs(z-z1) <= EPS) break;
    }
    if (its > MAXIT) nrerror("too many iterations in gaujac");
    x[i]=z;                             Store the root and the weight.
    w[i]=exp(gammln(alf+n)+gammln(bet+n)-gammln(n+1.0)-
        gammln(n+alfbet+1.0))*temp*pow(2.0,alfbet)/(pp*p2);
    }
}
```

Legendre polynomials are special cases of Jacobi polynomials with $\alpha = \beta = 0$, but it is worth having the separate routine for them, gauleg, given above. Chebyshev polynomials correspond to $\alpha = \beta = -1/2$ (see §5.8). They have analytic abscissas and weights:

$$
\begin{aligned}
x_j &= \cos\left(\frac{\pi(j + \frac{1}{2})}{N}\right) \\
w_j &= \frac{\pi}{N}
\end{aligned}
\tag{4.5.24}
$$

Case of Known Recurrences

Turn now to the case where you do not know good initial guesses for the zeros of your orthogonal polynomials, but you do have available the coefficients a_j and b_j that generate them. As we have seen, the zeros of $p_N(x)$ are the abscissas for the N-point Gaussian quadrature formula. The most useful computational formula for the weights is equation (4.5.9) above, since the derivative p'_N can be efficiently computed by the derivative of (4.5.6) in the general case, or by special relations for the classical polynomials. Note that (4.5.9) is valid as written only for monic polynomials; for other normalizations, there is an extra factor of λ_N/λ_{N-1}, where λ_N is the coefficient of x^N in p_N.

Except in those special cases already discussed, the best way to find the abscissas is *not* to use a root-finding method like Newton's method on $p_N(x)$. Rather, it is generally faster to use the Golub-Welsch [3] algorithm, which is based on a result of Wilf [4]. This algorithm notes that if you bring the term xp_j to the left-hand side of (4.5.6) and the term p_{j+1} to the right-hand side, the recurrence relation can be written in matrix form as

$$
x \begin{bmatrix} p_0 \\ p_1 \\ \vdots \\ p_{N-2} \\ p_{N-1} \end{bmatrix} = \begin{bmatrix} a_0 & 1 & & & \\ b_1 & a_1 & 1 & & \\ & \vdots & \vdots & & \\ & & b_{N-2} & a_{N-2} & 1 \\ & & & b_{N-1} & a_{N-1} \end{bmatrix} \cdot \begin{bmatrix} p_0 \\ p_1 \\ \vdots \\ p_{N-2} \\ p_{N-1} \end{bmatrix} + \begin{bmatrix} 0 \\ 0 \\ \vdots \\ 0 \\ p_N \end{bmatrix}
$$

or

$$x\mathbf{p} = \mathbf{T} \cdot \mathbf{p} + p_N \mathbf{e}_{N-1} \tag{4.5.25}$$

Here $\mathbf{T}$ is a tridiagonal matrix, $\mathbf{p}$ is a column vector of $p_0, p_1, \ldots, p_{N-1}$, and $\mathbf{e}_{N-1}$ is a unit vector with a 1 in the $(N-1)$st (last) position and zeros elsewhere. The matrix $\mathbf{T}$ can be symmetrized by a diagonal similarity transformation $\mathbf{D}$ to give

$$
\mathbf{J} = \mathbf{DTD}^{-1} = \begin{bmatrix} a_0 & \sqrt{b_1} & & & \\ \sqrt{b_1} & a_1 & \sqrt{b_2} & & \\ & \vdots & \vdots & & \\ & & \sqrt{b_{N-2}} & a_{N-2} & \sqrt{b_{N-1}} \\ & & & \sqrt{b_{N-1}} & a_{N-1} \end{bmatrix} \tag{4.5.26}
$$

The matrix $\mathbf{J}$ is called the *Jacobi matrix* (not to be confused with other matrices named after Jacobi that arise in completely different problems!). Now we see from (4.5.25) that $p_N(x_j) = 0$ is equivalent to x_j being an eigenvalue of $\mathbf{T}$. Since eigenvalues are preserved by a similarity transformation, x_j is an eigenvalue of the symmetric tridiagonal matrix $\mathbf{J}$. Moreover, Wilf [4] shows that if $\mathbf{v}_j$ is the eigenvector corresponding to the eigenvalue x_j, normalized so that $\mathbf{v} \cdot \mathbf{v} = 1$, then

$$w_j = \mu_0 v_{j,0}^2 \tag{4.5.27}$$

where

$$\mu_0 = \int_a^b W(x)\, dx \tag{4.5.28}$$

and where $v_{j,0}$ is the zeroth component of $\mathbf{v}$. As we shall see in Chapter 11, finding all eigenvalues and eigenvectors of a symmetric tridiagonal matrix is a relatively efficient and well-conditioned procedure. We accordingly give a routine, gaucof, for finding the abscissas and weights, given the coefficients a_j and b_j. Remember that if you know the recurrence relation for orthogonal polynomials that are not normalized to be monic, you can easily convert it to monic form by means of the quantities λ_j.

```
#include <cmath>
#include "nr.h"
using namespace std;

void NR::gaucof(Vec_IO_DP &a, Vec_IO_DP &b, const DP amu0, Vec_O_DP &x,
    Vec_O_DP &w)
```
Computes the abscissas and weights for a Gaussian quadrature formula from the Jacobi matrix. On input, a[0..n-1] and b[0..n-1] are the coefficients of the recurrence relation for the set of monic orthogonal polynomials. The quantity $\mu_0 \equiv \int_a^b W(x)\,dx$ is input as amu0. The abscissas x[0..n-1] are returned in descending order, with the corresponding weights in w[0..n-1]. The arrays a and b are modified. Execution can be speeded up by modifying tqli and eigsrt to compute only the zeroth component of each eigenvector.
```
{
    int i,j;

    int n=a.size();
    Mat_DP z(n,n);
```

```
    for (i=0;i<n;i++) {
        if (i != 0) b[i]=sqrt(b[i]);        Set up superdiagonal of Jacobi matrix.
        for (j=0;j<n;j++) z[i][j]=DP(i == j);
        Set up identity matrix for tqli to compute eigenvectors.
    }
    tqli(a,b,z);
    eigsrt(a,z);                            Sort eigenvalues into descending order.
    for (i=0;i<n;i++) {
        x[i]=a[i];
        w[i]=amu0*z[0][i]*z[0][i];          Equation (4.5.27).
    }
}
```

Orthogonal Polynomials with Nonclassical Weights

This somewhat specialized subsection will tell you what to do if your weight function is not one of the classical ones dealt with above and you do not know the a_j's and b_j's of the recurrence relation (4.5.6) to use in gaucof. Then, a method of finding the a_j's and b_j's is needed.

The *procedure of Stieltjes* is to compute a_0 from (4.5.7), then $p_1(x)$ from (4.5.6). Knowing p_0 and p_1, we can compute a_1 and b_1 from (4.5.7), and so on. But how are we to compute the inner products in (4.5.7)?

The textbook approach is to represent each $p_j(x)$ explicitly as a polynomial in x and to compute the inner products by multiplying out term by term. This will be feasible if we know the first $2N$ moments of the weight function,

$$\mu_j = \int_a^b x^j W(x) dx \qquad j = 0, 1, \ldots, 2N - 1 \tag{4.5.29}$$

However, the solution of the resulting set of algebraic equations for the coefficients a_j and b_j in terms of the moments μ_j is in general *extremely* ill-conditioned. Even in double precision, it is not unusual to lose all accuracy by the time $N = 12$. We thus reject any procedure based on the moments (4.5.29).

Sack and Donovan [5] discovered that the numerical stability is greatly improved if, instead of using powers of x as a set of basis functions to represent the p_j's, one uses some other known set of orthogonal polynomials $\pi_j(x)$, say. Roughly speaking, the improved stability occurs because the polynomial basis "samples" the interval (a, b) better than the power basis when the inner product integrals are evaluated, especially if its weight function resembles $W(x)$.

So assume that we know the *modified moments*

$$\nu_j = \int_a^b \pi_j(x) W(x) dx \qquad j = 0, 1, \ldots, 2N - 1 \tag{4.5.30}$$

where the π_j's satisfy a recurrence relation analogous to (4.5.6),

$$\pi_{-1}(x) \equiv 0$$

$$\pi_0(x) \equiv 1 \tag{4.5.31}$$

$$\pi_{j+1}(x) = (x - \alpha_j)\pi_j(x) - \beta_j \pi_{j-1}(x) \qquad j = 0, 1, 2, \ldots$$

and the coefficients α_j, β_j are known explicitly. Then Wheeler [6] has given an efficient $O(N^2)$ algorithm equivalent to that of Sack and Donovan for finding a_j and b_j via a set of intermediate quantities

$$\sigma_{k,l} = \langle p_k | \pi_l \rangle \qquad k, l \geq -1 \tag{4.5.32}$$

Initialize

$$\sigma_{-1,l} = 0 \qquad\qquad\qquad l = 1, 2, \ldots, 2N - 2$$
$$\sigma_{0,l} = \nu_l \qquad\qquad\qquad l = 0, 1, \ldots, 2N - 1$$
$$a_0 = \alpha_0 + \frac{\nu_1}{\nu_0} \qquad\qquad\qquad\qquad (4.5.33)$$
$$b_0 = 0$$

Then, for $k = 1, 2, \ldots, N - 1$, compute

$$\sigma_{k,l} = \sigma_{k-1,l+1} - (a_{k-1} - \alpha_l)\sigma_{k-1,l} - b_{k-1}\sigma_{k-2,l} + \beta_l\sigma_{k-1,l-1}$$

$$l = k, k+1, \ldots, 2N - k - 1$$

$$a_k = \alpha_k - \frac{\sigma_{k-1,k}}{\sigma_{k-1,k-1}} + \frac{\sigma_{k,k+1}}{\sigma_{k,k}}$$

$$b_k = \frac{\sigma_{k,k}}{\sigma_{k-1,k-1}}$$

$$(4.5.34)$$

Note that the normalization factors can also easily be computed if needed:

$$\langle p_0 | p_0 \rangle = \nu_0$$
$$\langle p_j | p_j \rangle = b_j \langle p_{j-1} | p_{j-1} \rangle \qquad j = 1, 2, \ldots \qquad (4.5.35)$$

You can find a derivation of the above algorithm in Ref. [7].

Wheeler's algorithm requires that the modified moments (4.5.30) be accurately computed. In practical cases there is often a closed form, or else recurrence relations can be used. The algorithm is extremely successful for *finite* intervals (a, b). For infinite intervals, the algorithm does not completely remove the ill-conditioning. In this case, Gautschi [8,9] recommends reducing the interval to a finite interval by a change of variable, and then using a suitable discretization procedure to compute the inner products. You will have to consult the references for details.

We give the routine orthog for generating the coefficients a_j and b_j by Wheeler's algorithm, given the coefficients α_j and β_j, and the modified moments ν_j. Note that in the routine, sig[k][l] $= \sigma_{k-1,l-1}$.

```
#include "nr.h"

void NR::orthog(Vec_I_DP &anu, Vec_I_DP &alpha, Vec_I_DP &beta, Vec_O_DP &a,
    Vec_O_DP &b)
```
Computes the coefficients a_j and b_j, $j = 0, \ldots N - 1$, of the recurrence relation for monic orthogonal polynomials with weight function $W(x)$ by Wheeler's algorithm. On input, the arrays alpha[0..2*n-2] and beta[0..2*n-2] are the coefficients α_j and β_j, $j = 0, \ldots 2N - 2$, of the recurrence relation for the chosen basis of orthogonal polynomials. The modified moments ν_j are input in anu[0..2*n-1]. The first n coefficients are returned in a[0..n-1] and b[0..n-1].
```
{
    int k,l,looptmp;

    int n=a.size();
    Mat_DP sig(2*n+1,2*n+1);
    looptmp=2*n;
    for (l=2;l<looptmp;l++) sig[0][l]=0.0;        Initialization, Equation (4.5.33).
    looptmp++;
    for (l=1;l<looptmp;l++) sig[1][l]=anu[l-1];
    a[0]=alpha[0]+anu[1]/anu[0];
    b[0]=0.0;
    for (k=2;k<n+1;k++) {                          Equation (4.5.34).
        looptmp=2*n-k+2;
        for (l=k;l<looptmp;l++) {
            sig[k][l]=sig[k-1][l+1]+(alpha[l-1]-a[k-2])*sig[k-1][l]
```

```
        -b[k-2]*sig[k-2][1]+beta[1-1]*sig[k-1][1-1];
    }
    a[k-1]=alpha[k-1]+sig[k][k+1]/sig[k][k]-sig[k-1][k]/sig[k-1][k-1];
    b[k-1]=sig[k][k]/sig[k-1][k-1];
    }
}
```

As an example of the use of `orthog`, consider the problem [7] of generating orthogonal polynomials with the weight function $W(x) = -\log x$ on the interval $(0, 1)$. A suitable set of π_j's is the shifted Legendre polynomials

$$\pi_j = \frac{(j!)^2}{(2j)!} P_j(2x - 1) \tag{4.5.36}$$

The factor in front of P_j makes the polynomials monic. The coefficients in the recurrence relation (4.5.31) are

$$\alpha_j = \frac{1}{2} \qquad\qquad j = 0, 1, \ldots$$
$$\beta_j = \frac{1}{4(4 - j^{-2})} \qquad j = 1, 2, \ldots \tag{4.5.37}$$

while the modified moments are

$$\nu_j = \begin{cases} 1 & j = 0 \\ \dfrac{(-1)^j (j!)^2}{j(j+1)(2j)!} & j \geq 1 \end{cases} \tag{4.5.38}$$

A call to `orthog` with this input allows one to generate the required polynomials to machine accuracy for very large N, and hence do Gaussian quadrature with this weight function. Before Sack and Donovan's observation, this seemingly simple problem was essentially intractable.

Extensions of Gaussian Quadrature

There are many different ways in which the ideas of Gaussian quadrature have been extended. One important extension is the case of *preassigned nodes*: Some points are required to be included in the set of abscissas, and the problem is to choose the weights and the remaining abscissas to maximize the degree of exactness of the the quadrature rule. The most common cases are *Gauss-Radau* quadrature, where one of the nodes is an endpoint of the interval, either a or b, and *Gauss-Lobatto* quadrature, where both a and b are nodes. Golub [10] has given an algorithm similar to `gaucof` for these cases.

The second important extension is the *Gauss-Kronrod* formulas. For ordinary Gaussian quadrature formulas, as N increases the sets of abscissas have no points in common. This means that if you compare results with increasing N as a way of estimating the quadrature error, you cannot reuse the previous function evaluations. Kronrod [11] posed the problem of searching for optimal sequences of rules, each of which reuses all abscissas of its predecessor. If one starts with $N = m$, say, and then adds n new points, one has $2n + m$ free parameters: the n new abscissas and weights, and m new weights for the fixed previous abscissas. The maximum degree of exactness one would expect to achieve would therefore be $2n + m - 1$. The question is whether this maximum degree of exactness can actually be achieved in practice, when the abscissas are required to all lie inside (a, b). The answer to this question is not known in general.

Kronrod showed that if you choose $n = m + 1$, an optimal extension can be found for Gauss-Legendre quadrature. Patterson [12] showed how to compute continued extensions of this kind. Sequences such as $N = 10, 21, 43, 87, \ldots$ are popular in automatic quadrature routines [13] that attempt to integrate a function until some specified accuracy has been achieved.

CITED REFERENCES AND FURTHER READING:

Abramowitz, M., and Stegun, I.A. 1964, *Handbook of Mathematical Functions*, Applied Mathematics Series, Volume 55 (Washington: National Bureau of Standards; reprinted 1968 by Dover Publications, New York), §25.4. [1]

Stroud, A.H., and Secrest, D. 1966, *Gaussian Quadrature Formulas* (Englewood Cliffs, NJ: Prentice-Hall). [2]

Golub, G.H., and Welsch, J.H. 1969, *Mathematics of Computation*, vol. 23, pp. 221–230 and A1–A10. [3]

Wilf, H.S. 1962, *Mathematics for the Physical Sciences* (New York: Wiley), Problem 9, p. 80. [4]

Sack, R.A., and Donovan, A.F. 1971/72, *Numerische Mathematik*, vol. 18, pp. 465–478. [5]

Wheeler, J.C. 1974, *Rocky Mountain Journal of Mathematics*, vol. 4, pp. 287–296. [6]

Gautschi, W. 1978, in *Recent Advances in Numerical Analysis*, C. de Boor and G.H. Golub, eds. (New York: Academic Press), pp. 45–72. [7]

Gautschi, W. 1981, in *E.B. Christoffel*, P.L. Butzer and F. Fehér, eds. (Basel: Birkhauser Verlag), pp. 72–147. [8]

Gautschi, W. 1990, in *Orthogonal Polynomials*, P. Nevai, ed. (Dordrecht: Kluwer Academic Publishers), pp. 181–216. [9]

Golub, G.H. 1973, *SIAM Review*, vol. 15, pp. 318–334. [10]

Kronrod, A.S. 1964, *Doklady Akademii Nauk SSSR*, vol. 154, pp. 283–286 (in Russian). [11]

Patterson, T.N.L. 1968, *Mathematics of Computation*, vol. 22, pp. 847–856 and C1–C11; 1969, *op. cit.*, vol. 23, p. 892. [12]

Piessens, R., de Doncker, E., Uberhuber, C.W., and Kahaner, D.K. 1983, *QUADPACK: A Subroutine Package for Automatic Integration* (New York: Springer-Verlag). [13]

Stoer, J., and Bulirsch, R. 1993, *Introduction to Numerical Analysis*, 2nd ed. (New York: Springer-Verlag), §3.6.

Johnson, L.W., and Riess, R.D. 1982, *Numerical Analysis*, 2nd ed. (Reading, MA: Addison-Wesley), §6.5.

Carnahan, B., Luther, H.A., and Wilkes, J.O. 1969, *Applied Numerical Methods* (New York: Wiley), §§2.9–2.10.

Ralston, A., and Rabinowitz, P. 1978, *A First Course in Numerical Analysis*, 2nd ed.; reprinted 2001 (New York: Dover), §§4.4–4.8.

4.6 Multidimensional Integrals

Integrals of functions of several variables, over regions with dimension greater than one, are *not easy*. There are two reasons for this. First, the number of function evaluations needed to sample an N-dimensional space increases as the Nth power of the number needed to do a one-dimensional integral. If you need 30 function evaluations to do a one-dimensional integral crudely, then you will likely need on the order of 30000 evaluations to reach the same crude level for a three-dimensional integral. Second, the region of integration in N-dimensional space is defined by an $N - 1$ dimensional boundary which can itself be terribly complicated: It need not be convex or simply connected, for example. By contrast, the boundary of a one-dimensional integral consists of two numbers, its upper and lower limits.

The first question to be asked, when faced with a multidimensional integral, is, "can it be reduced analytically to a lower dimensionality?" For example, so-called *iterated integrals* of a function of one variable $f(t)$ can be reduced to one-dimensional integrals by the formula

$$
\int_0^x dt_n \int_0^{t_n} dt_{n-1} \cdots \int_0^{t_3} dt_2 \int_0^{t_2} f(t_1)dt_1
$$
$$
= \frac{1}{(n-1)!} \int_0^x (x-t)^{n-1}f(t)dt
$$
(4.6.1)

Alternatively, the function may have some special symmetry in the way it depends on its independent variables. If the boundary also has this symmetry, then the dimension can be reduced. In three dimensions, for example, the integration of a spherically symmetric function over a spherical region reduces, in polar coordinates, to a one-dimensional integral.

The next questions to be asked will guide your choice between two entirely different approaches to doing the problem. The questions are: Is the shape of the boundary of the region of integration simple or complicated? Inside the region, is the integrand smooth and simple, or complicated, or locally strongly peaked? Does the problem require high accuracy, or does it require an answer accurate only to a percent, or a few percent?

If your answers are that the boundary is complicated, the integrand is *not* strongly peaked in very small regions, and relatively low accuracy is tolerable, then your problem is a good candidate for *Monte Carlo integration*. This method is very straightforward to program, in its cruder forms. One needs only to know a region with simple boundaries that *includes* the complicated region of integration, plus a method of determining whether a random point is inside or outside the region of integration. Monte Carlo integration evaluates the function at a random sample of points, and estimates its integral based on that random sample. We will discuss it in more detail, and with more sophistication, in Chapter 7.

If the boundary is simple, and the function is very smooth, then the remaining approaches, breaking up the problem into repeated one-dimensional integrals, or multidimensional Gaussian quadratures, will be effective and relatively fast [1]. If you require high accuracy, these approaches are in any case the *only* ones available to you, since Monte Carlo methods are by nature asymptotically slow to converge.

For low accuracy, use repeated one-dimensional integration or multidimensional Gaussian quadratures when the integrand is slowly varying and smooth in the region of integration, Monte Carlo when the integrand is oscillatory or discontinuous, but not strongly peaked in small regions.

If the integrand *is* strongly peaked in small regions, and you know where those regions are, break the integral up into several regions so that the integrand is smooth in each, and do each separately. If you don't know where the strongly peaked regions are, you might as well (at the level of sophistication of this book) quit: It is hopeless to expect an integration routine to search out unknown pockets of large contribution in a huge N-dimensional space. (But see §7.8.)

If, on the basis of the above guidelines, you decide to pursue the repeated one-dimensional integration approach, here is how it works. For definiteness, we will consider the case of a three-dimensional integral in x, y, z-space. Two dimensions, or more than three dimensions, are entirely analogous.

The first step is to specify the region of integration by (i) its lower and upper limits in x, which we will denote x_1 and x_2; (ii) its lower and upper limits in y at a specified value of x, denoted $y_1(x)$ and $y_2(x)$; and (iii) its lower and upper limits in z at specified x and y, denoted $z_1(x, y)$ and $z_2(x, y)$. In other words, find the numbers x_1 and x_2, and the functions $y_1(x), y_2(x), z_1(x, y)$, and $z_2(x, y)$ such that

$$
\begin{aligned}
I &\equiv \int \int \int dx \, dy \, dz f(x, y, z) \\
&= \int_{x_1}^{x_2} dx \int_{y_1(x)}^{y_2(x)} dy \int_{z_1(x,y)}^{z_2(x,y)} dz \; f(x, y, z)
\end{aligned}
\tag{4.6.2}
$$

For example, a two-dimensional integral over a circle of radius one centered on the origin becomes

$$
\int_{-1}^{1} dx \int_{-\sqrt{1-x^2}}^{\sqrt{1-x^2}} dy \; f(x, y)
\tag{4.6.3}
$$

Now we can define a function $G(x, y)$ that does the innermost integral,

$$
G(x, y) \equiv \int_{z_1(x,y)}^{z_2(x,y)} f(x, y, z) dz
\tag{4.6.4}
$$

and a function $H(x)$ that does the integral of $G(x, y)$,

$$
H(x) \equiv \int_{y_1(x)}^{y_2(x)} G(x, y) dy
\tag{4.6.5}
$$

and finally our answer as an integral over $H(x)$

$$
I = \int_{x_1}^{x_2} H(x) dx
\tag{4.6.6}
$$

In an implementation of equations (4.6.4)–(4.6.6), some basic one-dimensional integration routine (e.g., qgaus in the program following) gets called recursively: once to evaluate the outer integral I, then many times to evaluate the middle integral H, then even more times to evaluate the inner integral G (see Figure 4.6.1). Current values of x and y, and the pointer to your function func, are passed "over the head" of the intermediate calls through global variables defined in the namespace NRquad3d.

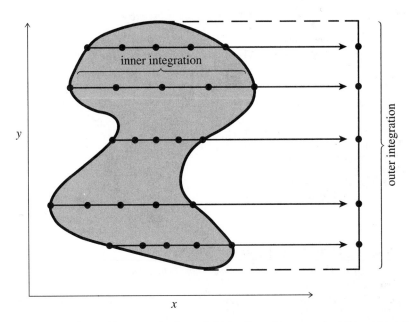

Figure 4.6.1.   Function evaluations for a two-dimensional integral over an irregular region, shown schematically. The outer integration routine, in y, requests values of the inner, x, integral at locations along the y axis of its own choosing. The inner integration routine then evaluates the function at x locations suitable to *it*. This is more accurate in general than, e.g., evaluating the function on a Cartesian mesh of points.

```
#include "nr.h"

extern DP yy1(const DP),yy2(const DP);
extern DP z1(const DP, const DP);
extern DP z2(const DP, const DP);

namespace NRquad3d {
    DP xsav,ysav;
    DP (*nrfunc)(const DP, const DP, const DP);

    DP f3(const DP z)                 The integrand f(x,y,z) evaluated at fixed x and y.
    {
        return nrfunc(xsav,ysav,z);
    }

    DP f2(const DP y)                 This is G of eq. (4.6.4).
    {
        ysav=y;
        return NR::qgaus(f3,z1(xsav,y),z2(xsav,y));
    }

    DP f1(const DP x)                 This is H of eq. (4.6.5).
    {
        xsav=x;
        return NR::qgaus(f2,yy1(x),yy2(x));
    }
}
```

DP NR::quad3d(DP func(const DP, const DP, const DP), const DP x1, const DP x2)
Returns the integral of a user-supplied function func over a three-dimensional region specified by the limits x1, x2, and by the user-supplied functions yy1, yy2, z1, and z2, as defined in (4.6.2). (The functions y_1 and y_2 are here called yy1 and yy2 to avoid conflict with the names

of Bessel functions in some C libraries). Integration is performed by calling qgaus recursively.

```
{
    NRquad3d::nrfunc=func;
    return qgaus(NRquad3d::f1,x1,x2);
}
```

The necessary user-supplied functions have the following prototypes:

```
DP func(const DP x, const DP y, const DP z);
The 3-dimensional function to be integrated.
DP yy1(const DP x);
DP yy2(const DP x);
DP z1(const DP x, const DP y);
DP z2(const DP x, const DP y);
```

CITED REFERENCES AND FURTHER READING:

Stroud, A.H. 1971, *Approximate Calculation of Multiple Integrals* (Englewood Cliffs, NJ: Prentice-Hall). [1]

Dahlquist, G., and Bjorck, A. 1974, *Numerical Methods* (Englewood Cliffs, NJ: Prentice-Hall), §7.7, p. 318.

Johnson, L.W., and Riess, R.D. 1982, *Numerical Analysis*, 2nd ed. (Reading, MA: Addison-Wesley), §6.2.5, p. 307.

Abramowitz, M., and Stegun, I.A. 1964, *Handbook of Mathematical Functions*, Applied Mathematics Series, Volume 55 (Washington: National Bureau of Standards; reprinted 1968 by Dover Publications, New York), equations 25.4.58ff.

Chapter 5.   Evaluation of Functions

5.0 Introduction

The purpose of this chapter is to acquaint you with a selection of the techniques that are frequently used in evaluating functions. In Chapter 6, we will apply and illustrate these techniques by giving routines for a variety of specific functions. The purposes of this chapter and the next are thus mostly in harmony, but there is nevertheless some tension between them: Routines that are clearest and most illustrative of the general techniques of this chapter are not always the methods of choice for a particular special function. By comparing this chapter to the next one, you should get some idea of the balance between "general" and "special" methods that occurs in practice.

Insofar as that balance favors general methods, this chapter should give you ideas about how to write your own routine for the evaluation of a function which, while "special" to you, is not so special as to be included in Chapter 6 or the standard program libraries.

CITED REFERENCES AND FURTHER READING:

Fike, C.T. 1968, *Computer Evaluation of Mathematical Functions* (Englewood Cliffs, NJ: Prentice-Hall).

Lanczos, C. 1956, *Applied Analysis*; reprinted 1988 (New York: Dover), Chapter 7.

5.1 Series and Their Convergence

Everybody knows that an analytic function can be expanded in the neighborhood of a point x_0 in a power series,

$$f(x) = \sum_{k=0}^{\infty} a_k (x - x_0)^k \tag{5.1.1}$$

Such series are straightforward to evaluate. You don't, of course, evaluate the kth power of $x - x_0$ *ab initio* for each term; rather you keep the $k-1$st power and update it with a multiply. Similarly, the form of the coefficients a is often such as to make use of previous work: Terms like $k!$ or $(2k)!$ can be updated in a multiply or two.

How do you know when you have summed enough terms? In practice, the terms had better be getting small fast, otherwise the series is not a good technique to use in the first place. While not mathematically rigorous in all cases, standard practice is to quit when the term you have just added is smaller in magnitude than some small ϵ times the magnitude of the sum thus far accumulated. (But watch out if isolated instances of $a_k = 0$ are possible!).

A weakness of a power series representation is that it is guaranteed *not* to converge farther than that distance from x_0 at which a singularity is encountered *in the complex plane*. This catastrophe is not usually unexpected: When you find a power series in a book (or when you work one out yourself), you will generally also know the radius of convergence. An insidious problem occurs with series that converge everywhere (in the mathematical sense), but almost nowhere fast enough to be useful in a numerical method. Two familiar examples are the sine function and the Bessel function of the first kind,

$$\sin x = \sum_{k=0}^{\infty} \frac{(-1)^k}{(2k+1)!} x^{2k+1} \tag{5.1.2}$$

$$J_n(x) = \left(\frac{x}{2}\right)^n \sum_{k=0}^{\infty} \frac{(-\frac{1}{4}x^2)^k}{k!(k+n)!} \tag{5.1.3}$$

Both of these series converge for all x. But both don't even start to converge until $k \gg |x|$; before this, their terms are increasing. This makes these series useless for large x.

Accelerating the Convergence of Series

There are several tricks for accelerating the rate of convergence of a series (or, equivalently, of a sequence of partial sums). These tricks will *not* generally help in cases like (5.1.2) or (5.1.3) while the size of the terms is still increasing. For series with terms of decreasing magnitude, however, some accelerating methods can be startlingly good. *Aitken's δ^2-process* is simply a formula for extrapolating the partial sums of a series whose convergence is approximately geometric. If S_{n-1}, S_n, S_{n+1} are three successive partial sums, then an improved estimate is

$$S_n' \equiv S_{n+1} - \frac{(S_{n+1} - S_n)^2}{S_{n+1} - 2S_n + S_{n-1}} \tag{5.1.4}$$

You can also use (5.1.4) with $n+1$ and $n-1$ replaced by $n+p$ and $n-p$ respectively, for any integer p. If you form the sequence of S_i''s, you can apply (5.1.4) a second time to *that* sequence, and so on. (In practice, this iteration will only rarely do much for you after the first stage.) Note that equation (5.1.4) should be computed as written; there exist algebraically equivalent forms that are much more susceptible to roundoff error.

For *alternating series* (where the terms in the sum alternate in sign), *Euler's transformation* can be a powerful tool. Generally it is advisable to do a small number of terms directly, through term $n - 1$ say, then apply the transformation to the rest of the series beginning with term n. The formula (for n even) is

$$\sum_{s=0}^{\infty} (-1)^s u_s = u_0 - u_1 + u_2 \ldots - u_{n-1} + \sum_{s=0}^{\infty} \frac{(-1)^s}{2^{s+1}} [\Delta^s u_n] \qquad (5.1.5)$$

Here Δ is the *forward difference operator*, i.e.,

$$\Delta u_n \equiv u_{n+1} - u_n$$

$$\Delta^2 u_n \equiv u_{n+2} - 2u_{n+1} + u_n \qquad\qquad (5.1.6)$$

$$\Delta^3 u_n \equiv u_{n+3} - 3u_{n+2} + 3u_{n+1} - u_n \qquad \text{etc.}$$

Of course you don't actually do the infinite sum on the right-hand side of (5.1.5), but only the first, say, p terms, thus requiring the first p differences (5.1.6) obtained from the terms starting at u_n.

Euler's transformation can be applied not only to convergent series. In some cases it will produce accurate answers from the first terms of a series that is formally divergent. It is widely used in the summation of asymptotic series. In this case it is generally wise not to sum farther than where the terms start increasing in magnitude; and you should devise some independent numerical check that the results are meaningful.

There is an elegant and subtle implementation of Euler's transformation due to van Wijngaarden [1]: It incorporates the terms of the original alternating series one at a time, in order. For each incorporation it *either* increases p by 1, equivalent to computing one further difference (5.1.6), or else *retroactively* increases n by 1, without having to redo all the difference calculations based on the old n value! The decision as to which to increase, n or p, is taken in such a way as to make the convergence most rapid. Van Wijngaarden's technique requires only one vector of saved partial differences. Here is the algorithm:

```cpp
#include <cmath>
#include "nr.h"
using namespace std;

void NR::eulsum(DP &sum, const DP term, const int jterm, Vec_IO_DP &wksp)
Incorporates into sum the jterm'th term, with value term, of an alternating series. sum is
input as the previous partial sum, and is output as the new partial sum. The first call to this
routine, with the first term in the series, should be with jterm=0. On the second call, term
should be set to the second term of the series, with sign opposite to that of the first call, and
jterm should be 1. And so on. wksp is a workspace array provided by the calling program,
dimensioned at least as large as the maximum number of terms to be incorporated.
{
    int j;
    static int nterm;
    DP tmp,dum;

    if (jterm == 0) {                           Initialize:
        nterm=1;                                Number of saved differences in wksp.
        sum=0.5*(wksp[0]=term);                 Return first estimate.
    } else {
        if (nterm+1 > wksp.size()) nrerror("wksp too small in euler");
        tmp=wksp[0];
```

```
wksp[0]=term;
for (j=1;j<nterm;j++) {                 Update saved quantities by van Wijn-
    dum=wksp[j];                         gaarden's algorithm.
    wksp[j]=0.5*(wksp[j-1]+tmp);
    tmp=dum;
}
wksp[nterm]=0.5*(wksp[nterm-1]+tmp);
if (fabs(wksp[nterm]) <= fabs(wksp[nterm-1]))      Favorable to increase p,
    sum += (0.5*wksp[nterm++]);          and the table becomes longer.
else                                     Favorable to increase n,
    sum += wksp[nterm];                  the table doesn't become longer.
    }
}
```

The powerful Euler technique is not directly applicable to a series of positive terms. Occasionally it is useful to convert a series of positive terms into an alternating series, just so that the Euler transformation can be used! Van Wijngaarden has given a transformation for accomplishing this [1]:

$$\sum_{r=1}^{\infty} v_r = \sum_{r=1}^{\infty} (-1)^{r-1} w_r \qquad (5.1.7)$$

where

$$w_r \equiv v_r + 2v_{2r} + 4v_{4r} + 8v_{8r} + \cdots \qquad (5.1.8)$$

Equations (5.1.7) and (5.1.8) replace a simple sum by a two-dimensional sum, each term in (5.1.7) being itself an infinite sum (5.1.8). This may seem a strange way to save on work! Since, however, the indices in (5.1.8) increase tremendously rapidly, as powers of 2, it often requires only a few terms to converge (5.1.8) to extraordinary accuracy. You do, however, need to be able to compute the v_r's efficiently for "random" values r. The standard "updating" tricks for sequential r's, mentioned above following equation (5.1.1), can't be used.

Actually, Euler's transformation is a special case of a more general transformation of power series. Suppose that some known function $g(z)$ has the series

$$g(z) = \sum_{n=0}^{\infty} b_n z^n \qquad (5.1.9)$$

and that you want to sum the new, unknown, series

$$f(z) = \sum_{n=0}^{\infty} c_n b_n z^n \qquad (5.1.10)$$

Then it is not hard to show (see [2]) that equation (5.1.10) can be written as

$$f(z) = \sum_{n=0}^{\infty} [\Delta^{(n)} c_0] \frac{g^{(n)}}{n!} z^n \qquad (5.1.11)$$

which often converges much more rapidly. Here $\Delta^{(n)} c_0$ is the nth finite-difference operator (equation 5.1.6), with $\Delta^{(0)} c_0 \equiv c_0$, and $g^{(n)}$ is the nth derivative of $g(z)$.

The usual Euler transformation (equation 5.1.5 with $n = 0$) can be obtained, for example, by substituting

$$g(z) = \frac{1}{1+z} = 1 - z + z^2 - z^3 + \cdots \tag{5.1.12}$$

into equation (5.1.11), and then setting $z = 1$.

Sometimes you will want to compute a function from a series representation even when the computation is *not* efficient. For example, you may be using the values obtained to fit the function to an approximating form that you will use subsequently (cf. §5.8). If you are summing very large numbers of slowly convergent terms, pay attention to roundoff errors! In floating-point representation it is more accurate to sum a list of numbers in the order starting with the smallest one, rather than starting with the largest one. It is even better to group terms pairwise, then in pairs of pairs, etc., so that all additions involve operands of comparable magnitude.

CITED REFERENCES AND FURTHER READING:

Goodwin, E.T. (ed.) 1961, *Modern Computing Methods*, 2nd ed. (New York: Philosophical Library), Chapter 13 [van Wijngaarden's transformations]. [1]

Dahlquist, G., and Bjorck, A. 1974, *Numerical Methods* (Englewood Cliffs, NJ: Prentice-Hall), Chapter 3.

Abramowitz, M., and Stegun, I.A. 1964, *Handbook of Mathematical Functions*, Applied Mathematics Series, Volume 55 (Washington: National Bureau of Standards; reprinted 1968 by Dover Publications, New York), §3.6.

Mathews, J., and Walker, R.L. 1970, *Mathematical Methods of Physics*, 2nd ed. (Reading, MA: W.A. Benjamin/Addison-Wesley), §2.3. [2]

5.2 Evaluation of Continued Fractions

Continued fractions are often powerful ways of evaluating functions that occur in scientific applications. A continued fraction looks like this:

$$f(x) = b_0 + \cfrac{a_1}{b_1 + \cfrac{a_2}{b_2 + \cfrac{a_3}{b_3 + \cfrac{a_4}{b_4 + \cfrac{a_5}{b_5 + \cdots}}}}} \tag{5.2.1}$$

Printers prefer to write this as

$$f(x) = b_0 + \frac{a_1}{b_1 +} \frac{a_2}{b_2 +} \frac{a_3}{b_3 +} \frac{a_4}{b_4 +} \frac{a_5}{b_5 +} \cdots \tag{5.2.2}$$

In either (5.2.1) or (5.2.2), the a's and b's can themselves be functions of x, usually linear or quadratic monomials at worst (i.e., constants times x or times x^2). For example, the continued fraction representation of the tangent function is

$$\tan x = \frac{x}{1 -} \frac{x^2}{3 -} \frac{x^2}{5 -} \frac{x^2}{7 -} \cdots \tag{5.2.3}$$

Continued fractions frequently converge much more rapidly than power series expansions, and in a much larger domain in the complex plane (not necessarily including the domain of convergence of the series, however). Sometimes the continued fraction converges best where the series does worst, although this is not a general rule. Blanch [1] gives a good review of the most useful convergence tests for continued fractions.

There are standard techniques, including the important *quotient-difference algorithm*, for going back and forth between continued fraction approximations, power series approximations, and rational function approximations. Consult Acton [2] for an introduction to this subject, and Fike [3] for further details and references.

How do you tell how far to go when evaluating a continued fraction? Unlike a series, you can't just evaluate equation (5.2.1) from left to right, stopping when the change is small. Written in the form of (5.2.1), the only way to evaluate the continued fraction is from right to left, first (blindly!) guessing how far out to start. This is not the right way.

The right way is to use a result that relates continued fractions to rational approximations, and that gives a means of evaluating (5.2.1) or (5.2.2) from left to right. Let f_n denote the result of evaluating (5.2.2) with coefficients through a_n and b_n. Then

$$f_n = \frac{A_n}{B_n} \tag{5.2.4}$$

where A_n and B_n are given by the following recurrence:

$$A_{-1} \equiv 1 \qquad B_{-1} \equiv 0$$

$$A_0 \equiv b_0 \qquad B_0 \equiv 1$$

$$A_j = b_j A_{j-1} + a_j A_{j-2} \qquad B_j = b_j B_{j-1} + a_j B_{j-2} \qquad j = 1, 2, \ldots, n \tag{5.2.5}$$

This method was invented by J. Wallis in 1655 (!), and is discussed in his *Arithmetica Infinitorum* [4]. You can easily prove it by induction.

In practice, this algorithm has some unattractive features: The recurrence (5.2.5) frequently generates very large or very small values for the partial numerators and denominators A_j and B_j. There is thus the danger of overflow or underflow of the floating-point representation. However, the recurrence (5.2.5) is linear in the A's and B's. At any point you can rescale the currently saved two levels of the recurrence, e.g., divide A_j, B_j, A_{j-1}, and B_{j-1} all by B_j. This incidentally makes $A_j = f_j$ and is convenient for testing whether you have gone far enough: See if f_j and f_{j-1} from the last iteration are as close as you would like them to be. (If B_j happens to be zero, which can happen, just skip the renormalization for this cycle. A fancier level of optimization is to renormalize only when an overflow is imminent, saving the unnecessary divides. All this complicates the program logic.)

Two newer algorithms have been proposed for evaluating continued fractions. *Steed's method* does not use A_j and B_j explicitly, but only the *ratio* $D_j = B_{j-1}/B_j$. One calculates D_j and $\Delta f_j = f_j - f_{j-1}$ recursively using

$$D_j = 1/(b_j + a_j D_{j-1}) \tag{5.2.6}$$

$$\Delta f_j = (b_j D_j - 1) \Delta f_{j-1} \tag{5.2.7}$$

Steed's method (see, e.g., [5]) avoids the need for rescaling of intermediate results. However, for certain continued fractions you can occasionally run into a situation where the denominator in (5.2.6) approaches zero, so that D_j and Δf_j are very large. The next Δf_{j+1} will typically cancel this large change, but with loss of accuracy in the numerical running sum of the f_j's. It is awkward to program around this, so Steed's method can be recommended only for cases where you know in advance that no denominator can vanish. We will use it for a special purpose in the routine bessik (§6.7).

The best general method for evaluating continued fractions seems to be the *modified Lentz's method* [6]. The need for rescaling intermediate results is avoided by using *both* the ratios

$$C_j = A_j/A_{j-1}, \qquad D_j = B_{j-1}/B_j \qquad (5.2.8)$$

and calculating f_j by

$$f_j = f_{j-1}C_jD_j \qquad (5.2.9)$$

From equation (5.2.5), one easily shows that the ratios satisfy the recurrence relations

$$D_j = 1/(b_j + a_jD_{j-1}), \qquad C_j = b_j + a_j/C_{j-1} \qquad (5.2.10)$$

In this algorithm there is the danger that the denominator in the expression for D_j, or the quantity C_j itself, might approach zero. Either of these conditions invalidates (5.2.10). However, Thompson and Barnett [5] show how to modify Lentz's algorithm to fix this: Just shift the offending term by a small amount, e.g., 10^{-30}. If you work through a cycle of the algorithm with this prescription, you will see that f_{j+1} is accurately calculated.

In detail, the modified Lentz's algorithm is this:

- Set $f_0 = b_0$; if $b_0 = 0$ set $f_0 = tiny$.
- Set $C_0 = f_0$.
- Set $D_0 = 0$.
- For $j = 1, 2, \ldots$
  Set $D_j = b_j + a_jD_{j-1}$.
  If $D_j = 0$, set $D_j = tiny$.
  Set $C_j = b_j + a_j/C_{j-1}$.
  If $C_j = 0$ set $C_j = tiny$.
  Set $D_j = 1/D_j$.
  Set $\Delta_j = C_jD_j$.
  Set $f_j = f_{j-1}\Delta_j$.
  If $|\Delta_j - 1| < eps$ then exit.

Here *eps* is your floating-point precision, say 10^{-7} or 10^{-15}. The parameter *tiny* should be less than typical values of $eps|b_j|$, say 10^{-30}.

The above algorithm assumes that you can terminate the evaluation of the continued fraction when $|f_j - f_{j-1}|$ is sufficiently small. This is usually the case, but by no means guaranteed. Jones [7] gives a list of theorems that can be used to justify this termination criterion for various kinds of continued fractions.

There is at present no rigorous analysis of error propagation in Lentz's algorithm. However, empirical tests suggest that it is at least as good as other methods.

Manipulating Continued Fractions

Several important properties of continued fractions can be used to rewrite them in forms that can speed up numerical computation. An *equivalence transformation*

$$a_n \to \lambda a_n, \quad b_n \to \lambda b_n, \quad a_{n+1} \to \lambda a_{n+1} \qquad (5.2.11)$$

leaves the value of a continued fraction unchanged. By a suitable choice of the scale factor λ you can often simplify the form of the a's and the b's. Of course, you can carry out successive equivalence transformations, possibly with different λ's, on successive terms of the continued fraction.

The *even* and *odd* parts of a continued fraction are continued fractions whose successive convergents are f_{2n} and f_{2n+1}, respectively. Their main use is that they converge twice as fast as the original continued fraction, and so if their terms are not much more complicated than the terms in the original there can be a big savings in computation. The formula for the even part of (5.2.2) is

$$f_{\text{even}} = d_0 + \cfrac{c_1}{d_1 +} \; \cfrac{c_2}{d_2 +} \; \cdots \qquad (5.2.12)$$

where in terms of intermediate variables

$$\begin{aligned}
\alpha_1 &= \frac{a_1}{b_1} \\
\alpha_n &= \frac{a_n}{b_n b_{n-1}}, \qquad n \geq 2
\end{aligned} \qquad (5.2.13)$$

we have

$$d_0 = b_0, \quad c_1 = \alpha_1, \quad d_1 = 1 + \alpha_2$$

$$c_n = -\alpha_{2n-1}\alpha_{2n-2}, \quad d_n = 1 + \alpha_{2n-1} + \alpha_{2n}, \qquad n \geq 2 \qquad (5.2.14)$$

You can find the similar formula for the odd part in the review by Blanch [1]. Often a combination of the transformations (5.2.14) and (5.2.11) is used to get the best form for numerical work.

We will make frequent use of continued fractions in the next chapter.

CITED REFERENCES AND FURTHER READING:

Abramowitz, M., and Stegun, I.A. 1964, *Handbook of Mathematical Functions*, Applied Mathematics Series, Volume 55 (Washington: National Bureau of Standards; reprinted 1968 by Dover Publications, New York), §3.10.

Blanch, G. 1964, *SIAM Review*, vol. 6, pp. 383–421. [1]

Acton, F.S. 1970, *Numerical Methods That Work*; 1990, corrected edition (Washington: Mathematical Association of America), Chapter 11. [2]

Cuyt, A., and Wuytack, L. 1987, *Nonlinear Methods in Numerical Analysis* (Amsterdam: North-Holland), Chapter 1.

Fike, C.T. 1968, *Computer Evaluation of Mathematical Functions* (Englewood Cliffs, NJ: Prentice-Hall), §§8.2, 10.4, and 10.5. [3]

Wallis, J. 1695, in *Opera Mathematica*, vol. 1, p. 355, Oxoniae e Theatro Shedoniano. Reprinted by Georg Olms Verlag, Hildeshein, New York (1972). [4]

Thompson, I.J., and Barnett, A.R. 1986, *Journal of Computational Physics*, vol. 64, pp. 490–509.
  [5]

Lentz, W.J. 1976, *Applied Optics*, vol. 15, pp. 668–671. [6]

Jones, W.B. 1973, in *Padé Approximants and Their Applications*, P.R. Graves-Morris, ed. (London: Academic Press), p. 125. [7]

5.3 Polynomials and Rational Functions

A polynomial of degree N is represented numerically as a stored array of coefficients, c[j] with j$= 0, \ldots, N$. We will always take c[0] to be the constant term in the polynomial, c[N] the coefficient of x^N; but of course other conventions are possible. There are two kinds of manipulations that you can do with a polynomial: *numerical* manipulations (such as evaluation), where you are given the numerical value of its argument, or *algebraic* manipulations, where you want to transform the coefficient array in some way without choosing any particular argument. Let's start with the numerical.

We assume that you know enough *never* to evaluate a polynomial this way:

```
p=c[0]+c[1]*x+c[2]*x*x+c[3]*x*x*x+c[4]*x*x*x*x;
```

or (even worse!),

```
p=c[0]+c[1]*x+c[2]*pow(x,2.0)+c[3]*pow(x,3.0)+c[4]*pow(x,4.0);
```

Come the (computer) revolution, all persons found guilty of such criminal behavior will be summarily executed, and their programs won't be! It is a matter of taste, however, whether to write

```
p=c[0]+x*(c[1]+x*(c[2]+x*(c[3]+x*c[4])));
```

or

```
p=(((c[4]*x+c[3])*x+c[2])*x+c[1])*x+c[0];
```

If the number of coefficients c[0..n] is large, one writes

```
p=c[n];
for(j=n-1;j>=0;j--) p=p*x+c[j];
```

or

```
p=c[j=n];
while (j>0) p=p*x+c[--j];
```

Another useful trick is for evaluating a polynomial $P(x)$ and its derivative $dP(x)/dx$ simultaneously:

```
p=c[n];
dp=0.0;
for(j=n-1;j>=0;j--) {dp=dp*x+p; p=p*x+c[j];}
```

or

```
p=c[j=n];
dp=0.0;
while (j>0) {dp=dp*x+p; p=p*x+c[--j];}
```

which yields the polynomial as p and its derivative as dp.

The above trick, which is basically *synthetic division* [1,2], generalizes to the evaluation of the polynomial and nd of its derivatives simultaneously:

```
#include "nr.h"

void NR::ddpoly(Vec_I_DP &c, const DP x, Vec_O_DP &pd)
Given the nc+1 coefficients of a polynomial of degree nc as an array c[0..nc] with c[0]
being the constant term, and given a value x, this routine returns the output array pd[0..nd].
The value of the polynomial evaluated at x is returned as pd[0], and the first nd derivatives
at x are returned in pd[1..nd].
{
    int nnd,j,i;
    DP cnst=1.0;

    int nc=c.size()-1;
    int nd=pd.size()-1;
    pd[0]=c[nc];
    for (j=1;j<nd+1;j++) pd[j]=0.0;
    for (i=nc-1;i>=0;i--) {
        nnd=(nd < (nc-i) ? nd : nc-i);
        for (j=nnd;j>0;j--)
            pd[j]=pd[j]*x+pd[j-1];
        pd[0]=pd[0]*x+c[i];
    }
    for (i=2;i<nd+1;i++) {          After the first derivative, factorial constants come in.
        cnst *= i;
        pd[i] *= cnst;
    }
}
```

As a curiosity, you might be interested to know that polynomials of degree $n > 3$ can be evaluated in *fewer* than n multiplications, at least if you are willing to precompute some auxiliary coefficients and, in some cases, do an extra addition. For example, the polynomial

$$P(x) = a_0 + a_1 x + a_2 x^2 + a_3 x^3 + a_4 x^4 \qquad (5.3.1)$$

where $a_4 > 0$, can be evaluated with 3 multiplications and 5 additions as follows:

$$P(x) = [(Ax + B)^2 + Ax + C][(Ax + B)^2 + D] + E \qquad (5.3.2)$$

where A, B, C, D, and E are to be precomputed by

$$A = (a_4)^{1/4}$$

$$B = \frac{a_3 - A^3}{4A^3}$$

$$D = 3B^2 + 8B^3 + \frac{a_1 A - 2a_2 B}{A^2} \qquad (5.3.3)$$

$$C = \frac{a_2}{A^2} - 2B - 6B^2 - D$$

$$E = a_0 - B^4 - B^2(C + D) - CD$$

Fifth degree polynomials can be evaluated in 4 multiplies and 5 adds; sixth degree polynomials can be evaluated in 4 multiplies and 7 adds; if any of this strikes you as interesting, consult references [3-5]. The subject has something of the same entertaining, if impractical, flavor as that of fast matrix multiplication, discussed in §2.11.

Turn now to algebraic manipulations. You multiply a polynomial of degree $n - 1$ (array of range [0..n-1]) by a monomial factor $x - a$ by a bit of code like the following,

```
c[n]=c[n-1];
for (j=n-1;j>=1;j--) c[j]=c[j-1]-c[j]*a;
c[0] *= (-a);
```

Likewise, you divide a polynomial of degree n by a monomial factor $x - a$ (synthetic division again) using

```
rem=c[n];
c[n]=0.0;
for(i=n-1;i>=0;i--) {
    swap=c[i];
    c[i]=rem;
    rem=swap+rem*a;
}
```

which leaves you with a new polynomial array and a numerical remainder `rem`.

Multiplication of two general polynomials involves straightforward summing of the products, each involving one coefficient from each polynomial. Division of two general polynomials, while it can be done awkwardly in the fashion taught using pencil and paper, is susceptible to a good deal of streamlining. Witness the following routine based on the algorithm in [3].

```
#include "nr.h"

void NR::poldiv(Vec_I_DP &u, Vec_I_DP &v, Vec_O_DP &q, Vec_O_DP &r)
```
Given the `n+1` coefficients of a polynomial of degree n in `u[0..n]`, and the `nv+1` coefficients of another polynomial of degree nv in `v[0..nv]`, divide the polynomial u by the polynomial v ("u"/"v") giving a quotient polynomial whose coefficents are returned in `q[0..n]`, and a remainder polynomial whose coefficients are returned in `r[0..n]`. The elements `r[nv..n]` and `q[n-nv+1..n]` are returned as zero.
```
{
    int k,j;

    int n=u.size()-1;
    int nv=v.size()-1;
    for (j=0;j<=n;j++) {
        r[j]=u[j];
        q[j]=0.0;
    }
    for (k=n-nv;k>=0;k--) {
        q[k]=r[nv+k]/v[nv];
        for (j=nv+k-1;j>=k;j--) r[j] -= q[k]*v[j-k];
    }
    for (j=nv;j<=n;j++) r[j]=0.0;
}
```

Rational Functions

You evaluate a rational function like

$$R(x) = \frac{P_\mu(x)}{Q_\nu(x)} = \frac{p_0 + p_1 x + \cdots + p_\mu x^\mu}{q_0 + q_1 x + \cdots + q_\nu x^\nu} \tag{5.3.4}$$

in the obvious way, namely as two separate polynomials followed by a divide. As a matter of convention one usually chooses $q_0 = 1$, obtained by dividing numerator and denominator by any other q_0. It is often convenient to have both sets of coefficients stored in a single array, and to have a standard function available for doing the evaluation:

```
#include "nr.h"

DP NR::ratval(const DP x, Vec_I_DP &cof, const int mm, const int kk)
```
Given mm, kk, and cof[0..mm+kk], evaluate and return the rational function (cof[0] + cof[1]x + $\cdots$ + cof[mm]x^{mm})/(1 + cof[mm+1]x + $\cdots$ + cof[mm+kk]x^{kk}).
```
{
    int j;
    DP sumd,sumn;

    for (sumn=cof[mm],j=mm-1;j>=0;j--) sumn=sumn*x+cof[j];
    for (sumd=0.0,j=mm+kk;j>mm;j--) sumd=(sumd+cof[j])*x;
    return sumn/(1.0+sumd);
}
```

CITED REFERENCES AND FURTHER READING:

Acton, F.S. 1970, *Numerical Methods That Work*; 1990, corrected edition (Washington: Mathematical Association of America), pp. 183, 190. [1]

Mathews, J., and Walker, R.L. 1970, *Mathematical Methods of Physics*, 2nd ed. (Reading, MA: W.A. Benjamin/Addison-Wesley), pp. 361–363. [2]

Knuth, D.E. 1997, *Seminumerical Algorithms*, 3rd ed., vol. 2 of *The Art of Computer Programming* (Reading, MA: Addison-Wesley), §4.6. [3]

Fike, C.T. 1968, *Computer Evaluation of Mathematical Functions* (Englewood Cliffs, NJ: Prentice-Hall), Chapter 4.

Winograd, S. 1970, *Communications on Pure and Applied Mathematics*, vol. 23, pp. 165–179. [4]

Kronsjö, L. 1987, *Algorithms: Their Complexity and Efficiency*, 2nd ed. (New York: Wiley). [5]

5.4 Complex Arithmetic

Since C++ has a built-in class complex, you can generally let the compiler and the class library take care of complex arithmetic for you. Generally, but not always. For a program with only a small number of complex operations, you may want to code these yourself, in-line. Or, you may find that your compiler is not up to snuff: It is disconcertingly common to encounter complex operations that produce overflows or underflows when both the complex operands and the complex result are perfectly representable. This occurs, we think, because software companies mistake the implementation of complex arithmetic for a completely trivial task, not requiring any particular finesse.

Actually, complex arithmetic is not *quite* trivial. Addition and subtraction are done in the obvious way, performing the operation separately on the real and imaginary parts of the operands. Multiplication can also be done in the obvious way, with 4 multiplications, one addition, and one subtraction,

$$(a + ib)(c + id) = (ac - bd) + i(bc + ad) \tag{5.4.1}$$

(the addition sign before the i doesn't count; it just separates the real and imaginary parts notationally). But it is sometimes faster to multiply via

$$(a + ib)(c + id) = (ac - bd) + i[(a + b)(c + d) - ac - bd] \tag{5.4.2}$$

which has only three multiplications (ac, bd, $(a + b)(c + d)$), plus two additions and three subtractions. The total operations count is higher by two, but multiplication is a slow operation on some machines.

While it is true that intermediate results in equations (5.4.1) and (5.4.2) can overflow even when the final result is representable, this happens only when the final answer is on the edge of representability. Not so for the complex modulus, if you or your compiler are misguided enough to compute it as

$$|a + ib| = \sqrt{a^2 + b^2} \qquad \text{(bad!)} \tag{5.4.3}$$

whose intermediate result will overflow if either a or b is as large as the square root of the largest representable number (e.g., 10^{19} as compared to 10^{38}). The right way to do the calculation is

$$|a + ib| = \begin{cases} |a|\sqrt{1 + (b/a)^2} & |a| \geq |b| \\ |b|\sqrt{1 + (a/b)^2} & |a| < |b| \end{cases} \tag{5.4.4}$$

Complex division should use a similar trick to prevent avoidable overflows, underflow, or loss of precision,

$$\frac{a + ib}{c + id} = \begin{cases} \dfrac{[a + b(d/c)] + i[b - a(d/c)]}{c + d(d/c)} & |c| \geq |d| \\[3ex] \dfrac{[a(c/d) + b] + i[b(c/d) - a]}{c(c/d) + d} & |c| < |d| \end{cases} \tag{5.4.5}$$

Of course you should calculate repeated subexpressions, like c/d or d/c, only once.

Complex square root is even more complicated, since we must both guard intermediate results, and also enforce a chosen branch cut (here taken to be the negative real axis). To take the square root of $c + id$, first compute

$$w \equiv \begin{cases} 0 & c = d = 0 \\[2ex] \sqrt{|c|}\sqrt{\dfrac{1 + \sqrt{1 + (d/c)^2}}{2}} & |c| \geq |d| \\[3ex] \sqrt{|d|}\sqrt{\dfrac{|c/d| + \sqrt{1 + (c/d)^2}}{2}} & |c| < |d| \end{cases} \tag{5.4.6}$$

Then the answer is

$$\sqrt{c+id} = \begin{cases} 0 & w = 0 \\ w + i\left(\dfrac{d}{2w}\right) & w \neq 0, c \geq 0 \\ \dfrac{|d|}{2w} + iw & w \neq 0, c < 0, d \geq 0 \\ \dfrac{|d|}{2w} - iw & w \neq 0, c < 0, d < 0 \end{cases} \tag{5.4.7}$$

CITED REFERENCES AND FURTHER READING:

Midy, P., and Yakovlev, Y. 1991, *Mathematics and Computers in Simulation*, vol. 33, pp. 33–49.

Knuth, D.E. 1997, *Seminumerical Algorithms*, 3rd ed., vol. 2 of *The Art of Computer Programming* (Reading, MA: Addison-Wesley) [see solutions to exercises 4.2.1.16 and 4.6.4.41].

5.5 Recurrence Relations and Clenshaw's Recurrence Formula

Many useful functions satisfy recurrence relations, e.g.,

$$(n+1)P_{n+1}(x) = (2n+1)xP_n(x) - nP_{n-1}(x) \tag{5.5.1}$$

$$J_{n+1}(x) = \frac{2n}{x}J_n(x) - J_{n-1}(x) \tag{5.5.2}$$

$$nE_{n+1}(x) = e^{-x} - xE_n(x) \tag{5.5.3}$$

$$\cos n\theta = 2\cos\theta\cos(n-1)\theta - \cos(n-2)\theta \tag{5.5.4}$$

$$\sin n\theta = 2\cos\theta\sin(n-1)\theta - \sin(n-2)\theta \tag{5.5.5}$$

where the first three functions are Legendre polynomials, Bessel functions of the first kind, and exponential integrals, respectively. (For notation see [1].) These relations are useful for extending computational methods from two successive values of n to other values, either larger or smaller.

Equations (5.5.4) and (5.5.5) motivate us to say a few words about trigonometric functions. If your program's running time is dominated by evaluating trigonometric functions, you are probably doing something wrong. Trig functions whose arguments form a linear sequence $\theta = \theta_0 + n\delta$, $n = 0, 1, 2, \ldots$, are efficiently calculated by the following recurrence,

$$\cos(\theta + \delta) = \cos\theta - [\alpha\cos\theta + \beta\sin\theta]$$
$$\sin(\theta + \delta) = \sin\theta - [\alpha\sin\theta - \beta\cos\theta] \tag{5.5.6}$$

where α and β are the precomputed coefficients

$$\alpha \equiv 2\sin^2\left(\frac{\delta}{2}\right) \qquad \beta \equiv \sin\delta \qquad (5.5.7)$$

The reason for doing things this way, rather than with the standard (and equivalent) identities for sums of angles, is that here α and β do not lose significance if the incremental δ is small. Likewise, the adds in equation (5.5.6) should be done in the order indicated by square brackets. We will use (5.5.6) repeatedly in Chapter 12, when we deal with Fourier transforms.

Another trick, occasionally useful, is to note that both $\sin\theta$ and $\cos\theta$ can be calculated via a single call to tan:

$$t \equiv \tan\left(\frac{\theta}{2}\right) \qquad \cos\theta = \frac{1-t^2}{1+t^2} \qquad \sin\theta = \frac{2t}{1+t^2} \qquad (5.5.8)$$

The cost of getting both sin and cos, if you need them, is thus the cost of tan plus 2 multiplies, 2 divides, and 2 adds. On machines with slow trig functions, this can be a savings. *However*, note that special treatment is required if $\theta \to \pm\pi$. And also note that many modern machines have *very fast* trig functions; so you should not assume that equation (5.5.8) is faster without testing.

Stability of Recurrences

You need to be aware that recurrence relations are not necessarily *stable* against roundoff error in the direction that you propose to go (either increasing n or decreasing n). A three-term linear recurrence relation

$$y_{n+1} + a_n y_n + b_n y_{n-1} = 0, \qquad n = 1, 2, \ldots \qquad (5.5.9)$$

has two linearly independent solutions, f_n and g_n say. Only one of these corresponds to the sequence of functions f_n that you are trying to generate. The other one g_n *may* be exponentially growing in the direction that you want to go, or exponentially damped, or exponentially neutral (growing or dying as some power law, for example). If it is exponentially growing, then the recurrence relation is of little or no practical use in that direction. This is the case, e.g., for (5.5.2) in the direction of increasing n, when $x < n$. You cannot generate Bessel functions of high n by forward recurrence on (5.5.2).

To state things a bit more formally, if

$$f_n/g_n \to 0 \quad \text{as} \quad n \to \infty \qquad (5.5.10)$$

then f_n is called the *minimal* solution of the recurrence relation (5.5.9). Nonminimal solutions like g_n are called *dominant* solutions. The minimal solution is unique, if it exists, but dominant solutions are not — you can add an arbitrary multiple of f_n to a given g_n. You can evaluate any dominant solution by forward recurrence, *but not the minimal solution*. (Unfortunately it is sometimes the one you want.)

Abramowitz and Stegun (in their Introduction) [1] give a list of recurrences that are stable in the increasing or decreasing directions. That list does not contain all possible formulas, of course. Given a recurrence relation for some function $f_n(x)$ you can test it yourself with about five minutes of (human) labor: For a fixed x in your range of interest, start the recurrence not with true values of $f_j(x)$ and $f_{j+1}(x)$, but (first) with the values 1 and 0, respectively, and then (second) with 0 and 1, respectively. Generate 10 or 20 terms of the recursive sequences in the direction that you want to go (increasing or decreasing from j), for each of the two starting conditions. Look at the difference between the corresponding members of the two sequences. If the differences stay of order unity (absolute value less than 10, say), then the recurrence is stable. If they increase slowly, then the recurrence may be mildly unstable but quite tolerably so. If they increase catastrophically, then there is an exponentially growing solution of the recurrence. If you know that the function that you want actually corresponds to the growing solution, then you can keep the recurrence formula anyway e.g., the case of the Bessel function $Y_n(x)$ for increasing n, see §6.5; if you don't know which solution your function corresponds to, you must at this point reject the recurrence formula. Notice that you can do this test *before* you go to the trouble of finding a numerical method for computing the two starting functions $f_j(x)$ and $f_{j+1}(x)$: stability is a property of the recurrence, not of the starting values.

An alternative heuristic procedure for testing stability is to replace the recurrence relation by a similar one that is linear with constant coefficients. For example, the relation (5.5.2) becomes

$$y_{n+1} - 2\gamma y_n + y_{n-1} = 0 \qquad\qquad (5.5.11)$$

where $\gamma \equiv n/x$ is treated as a constant. You solve such recurrence relations by trying solutions of the form $y_n = a^n$. Substituting into the above recurrence gives

$$a^2 - 2\gamma a + 1 = 0 \qquad \text{or} \qquad a = \gamma \pm \sqrt{\gamma^2 - 1} \qquad (5.5.12)$$

The recurrence is stable if $|a| \leq 1$ for all solutions a. This holds (as you can verify) if $|\gamma| \leq 1$ or $n \leq x$. The recurrence (5.5.2) thus cannot be used, starting with $J_0(x)$ and $J_1(x)$, to compute $J_n(x)$ for large n.

Possibly you would at this point like the security of some real theorems on this subject (although we ourselves always follow one of the heuristic procedures). Here are two theorems, due to Perron [2]:

Theorem A.   If in (5.5.9) $a_n \sim an^\alpha$, $b_n \sim bn^\beta$ as $n \to \infty$, and $\beta < 2\alpha$, then

$$g_{n+1}/g_n \sim -an^\alpha, \qquad f_{n+1}/f_n \sim -(b/a)n^{\beta-\alpha} \qquad (5.5.13)$$

and f_n is the minimal solution to (5.5.9).

Theorem B.   Under the same conditions as Theorem A, but with $\beta = 2\alpha$, consider the *characteristic polynomial*

$$t^2 + at + b = 0 \qquad\qquad (5.5.14)$$

If the roots t_1 and t_2 of (5.5.14) have distinct moduli, $|t_1| > |t_2|$ say, then

$$g_{n+1}/g_n \sim t_1 n^\alpha, \qquad f_{n+1}/f_n \sim t_2 n^\alpha \qquad (5.5.15)$$

and f_n is again the minimal solution to (5.5.9). Cases other than those in these two theorems are inconclusive for the existence of minimal solutions. (For more on the stability of recurrences, see [3].)

How do you proceed if the solution that you desire *is* the minimal solution? The answer lies in that old aphorism, that every cloud has a silver lining: If a recurrence relation is catastrophically unstable in one direction, then that (undesired) solution will decrease very rapidly in the reverse direction. This means that you can start with *any* seed values for the consecutive f_j and f_{j+1} and (when you have gone enough steps in the stable direction) you will converge to the sequence of functions that you want, times an unknown normalization factor. If there is some other way to normalize the sequence (e.g., by a formula for the sum of the f_n's), then this can be a practical means of function evaluation. The method is called *Miller's algorithm*. An example often given [1,4] uses equation (5.5.2) in just this way, along with the normalization formula

$$1 = J_0(x) + 2J_2(x) + 2J_4(x) + 2J_6(x) + \cdots \qquad (5.5.16)$$

Incidentally, there is an important relation between three-term recurrence relations and *continued fractions*. Rewrite the recurrence relation (5.5.9) as

$$\frac{y_n}{y_{n-1}} = -\frac{b_n}{a_n + y_{n+1}/y_n} \qquad (5.5.17)$$

Iterating this equation, starting with n, gives

$$\frac{y_n}{y_{n-1}} = -\frac{b_n}{a_n -} \frac{b_{n+1}}{a_{n+1} -} \cdots \qquad (5.5.18)$$

Pincherle's Theorem [2] tells us that (5.5.18) converges if and only if (5.5.9) has a minimal solution f_n, in which case it converges to f_n/f_{n-1}. This result, usually for the case $n = 1$ and combined with some way to determine f_0, underlies many of the practical methods for computing special functions that we give in the next chapter.

Clenshaw's Recurrence Formula

Clenshaw's recurrence formula [5] is an elegant and efficient way to evaluate a sum of coefficients times functions that obey a recurrence formula, e.g.,

$$f(\theta) = \sum_{k=0}^{N} c_k \cos k\theta \qquad \text{or} \qquad f(x) = \sum_{k=0}^{N} c_k P_k(x)$$

Here is how it works: Suppose that the desired sum is

$$f(x) = \sum_{k=0}^{N} c_k F_k(x) \qquad (5.5.19)$$

and that F_k obeys the recurrence relation

$$F_{n+1}(x) = \alpha(n, x) F_n(x) + \beta(n, x) F_{n-1}(x) \qquad (5.5.20)$$

for some functions $\alpha(n, x)$ and $\beta(n, x)$. Now define the quantities y_k ($k = N, N-1, \ldots, 1$) by the following recurrence:

$$y_{N+2} = y_{N+1} = 0$$
$$y_k = \alpha(k, x)y_{k+1} + \beta(k+1, x)y_{k+2} + c_k \quad (k = N, N-1, \ldots, 1) \tag{5.5.21}$$

If you solve equation (5.5.21) for c_k on the left, and then write out explicitly the sum (5.5.19), it will look (in part) like this:

$$f(x) = \cdots$$
$$+ [y_8 - \alpha(8, x)y_9 - \beta(9, x)y_{10}]F_8(x)$$
$$+ [y_7 - \alpha(7, x)y_8 - \beta(8, x)y_9]F_7(x)$$
$$+ [y_6 - \alpha(6, x)y_7 - \beta(7, x)y_8]F_6(x)$$
$$+ [y_5 - \alpha(5, x)y_6 - \beta(6, x)y_7]F_5(x) \tag{5.5.22}$$
$$+ \cdots$$
$$+ [y_2 - \alpha(2, x)y_3 - \beta(3, x)y_4]F_2(x)$$
$$+ [y_1 - \alpha(1, x)y_2 - \beta(2, x)y_3]F_1(x)$$
$$+ [c_0 + \beta(1, x)y_2 - \beta(1, x)y_2]F_0(x)$$

Notice that we have added and subtracted $\beta(1, x)y_2$ in the last line. If you examine the terms containing a factor of y_8 in (5.5.22), you will find that they sum to zero as a consequence of the recurrence relation (5.5.20); similarly all the other y_k's down through y_2. The only surviving terms in (5.5.22) are

$$f(x) = \beta(1, x)F_0(x)y_2 + F_1(x)y_1 + F_0(x)c_0 \tag{5.5.23}$$

Equations (5.5.21) and (5.5.23) are *Clenshaw's recurrence formula* for doing the sum (5.5.19): You make one pass down through the y_k's using (5.5.21); when you have reached y_2 and y_1 you apply (5.5.23) to get the desired answer.

Clenshaw's recurrence as written above incorporates the coefficients c_k in a downward order, with k decreasing. At each stage, the effect of all previous c_k's is "remembered" as two coefficients which multiply the functions F_{k+1} and F_k (ultimately F_0 and F_1). If the functions F_k are small when k is large, *and* if the coefficients c_k are small when k is *small*, then the sum can be dominated by small F_k's. In this case the remembered coefficients will involve a delicate cancellation and there can be a catastrophic loss of significance. An example would be to sum the trivial series

$$J_{15}(1) = 0 \times J_0(1) + 0 \times J_1(1) + \ldots + 0 \times J_{14}(1) + 1 \times J_{15}(1) \tag{5.5.24}$$

Here J_{15}, which is tiny, ends up represented as a canceling linear combination of J_0 and J_1, which are of order unity.

The solution in such cases is to use an alternative Clenshaw recurrence that incorporates c_k's in an upward direction. The relevant equations are

$$y_{-2} = y_{-1} = 0 \qquad (5.5.25)$$

$$y_k = \frac{1}{\beta(k+1,x)}[y_{k-2} - \alpha(k,x)y_{k-1} - c_k],$$
$$(k = 0, 1, \ldots, N-1) \qquad (5.5.26)$$

$$f(x) = c_N F_N(x) - \beta(N,x)F_{N-1}(x)y_{N-1} - F_N(x)y_{N-2} \qquad (5.5.27)$$

The rare case where equations (5.5.25)–(5.5.27) should be used instead of equations (5.5.21) and (5.5.23) can be detected automatically by testing whether the operands in the first sum in (5.5.23) are opposite in sign and nearly equal in magnitude. Other than in this special case, Clenshaw's recurrence is always stable, independent of whether the recurrence for the functions F_k is stable in the upward or downward direction.

CITED REFERENCES AND FURTHER READING:

Abramowitz, M., and Stegun, I.A. 1964, *Handbook of Mathematical Functions*, Applied Mathematics Series, Volume 55 (Washington: National Bureau of Standards; reprinted 1968 by Dover Publications, New York), pp. xiii, 697. [1]

Gautschi, W. 1967, *SIAM Review*, vol. 9, pp. 24–82. [2]

Lakshmikantham, V., and Trigiante, D. 1988, *Theory of Difference Equations: Numerical Methods and Applications* (San Diego: Academic Press). [3]

Acton, F.S. 1970, *Numerical Methods That Work*; 1990, corrected edition (Washington: Mathematical Association of America), pp. 20ff. [4]

Clenshaw, C.W. 1962, *Mathematical Tables*, vol. 5, National Physical Laboratory (London: H.M. Stationery Office). [5]

Dahlquist, G., and Bjorck, A. 1974, *Numerical Methods* (Englewood Cliffs, NJ: Prentice-Hall), §4.4.3, p. 111.

Goodwin, E.T. (ed.) 1961, *Modern Computing Methods*, 2nd ed. (New York: Philosophical Library), p. 76.

5.6 Quadratic and Cubic Equations

The roots of simple algebraic equations can be viewed as being functions of the equations' coefficients. We are taught these functions in elementary algebra. Yet, surprisingly many people don't know the right way to solve a quadratic equation with two real roots, or to obtain the roots of a cubic equation.

There are two ways to write the solution of the *quadratic equation*

$$ax^2 + bx + c = 0 \qquad (5.6.1)$$

with real coefficients a, b, c, namely

$$x = \frac{-b \pm \sqrt{b^2 - 4ac}}{2a} \qquad (5.6.2)$$

and

$$x = \frac{2c}{-b \pm \sqrt{b^2 - 4ac}} \qquad (5.6.3)$$

If you use *either* (5.6.2) *or* (5.6.3) to get the two roots, you are asking for trouble: If either a or c (or both) are small, then one of the roots will involve the subtraction of b from a very nearly equal quantity (the discriminant); you will get that root very inaccurately. The correct way to compute the roots is

$$q \equiv -\frac{1}{2}\left[b + \operatorname{sgn}(b)\sqrt{b^2 - 4ac}\right] \qquad (5.6.4)$$

Then the two roots are

$$x_1 = \frac{q}{a} \qquad \text{and} \qquad x_2 = \frac{c}{q} \qquad (5.6.5)$$

If the coefficients a, b, c, are complex rather than real, then the above formulas still hold, except that in equation (5.6.4) the sign of the square root should be chosen so as to make

$$\operatorname{Re}(b^*\sqrt{b^2 - 4ac}) \geq 0 \qquad (5.6.6)$$

where Re denotes the real part and asterisk denotes complex conjugation.

Apropos of quadratic equations, this seems a convenient place to recall that the inverse hyperbolic functions $\sinh^{-1}$ and $\cosh^{-1}$ are in fact just logarithms of solutions to such equations,

$$\sinh^{-1}(x) = \quad \ln\left(x + \sqrt{x^2 + 1}\right) \qquad (5.6.7)$$
$$\cosh^{-1}(x) = \pm\ln\left(x + \sqrt{x^2 - 1}\right) \qquad (5.6.8)$$

Equation (5.6.7) is numerically robust for $x \geq 0$. For negative x, use the symmetry $\sinh^{-1}(-x) = -\sinh^{-1}(x)$. Equation (5.6.8) is of course valid only for $x \geq 1$.

For the *cubic equation*

$$x^3 + ax^2 + bx + c = 0 \qquad (5.6.9)$$

with real or complex coefficients a, b, c, first compute

$$Q \equiv \frac{a^2 - 3b}{9} \qquad \text{and} \qquad R \equiv \frac{2a^3 - 9ab + 27c}{54} \qquad (5.6.10)$$

If Q and R are real (always true when a, b, c are real) *and* $R^2 < Q^3$, then the cubic equation has three real roots. Find them by computing

$$\theta = \arccos(R/\sqrt{Q^3}) \qquad (5.6.11)$$

in terms of which the three roots are

$$x_1 = -2\sqrt{Q}\cos\left(\frac{\theta}{3}\right) - \frac{a}{3}$$

$$x_2 = -2\sqrt{Q}\cos\left(\frac{\theta + 2\pi}{3}\right) - \frac{a}{3} \qquad (5.6.12)$$

$$x_3 = -2\sqrt{Q}\cos\left(\frac{\theta - 2\pi}{3}\right) - \frac{a}{3}$$

(This equation first appears in Chapter VI of François Viète's treatise "De emendatione," published in 1615!)

Otherwise, compute

$$A = -\left[R + \sqrt{R^2 - Q^3}\right]^{1/3} \qquad (5.6.13)$$

where the sign of the square root is chosen to make

$$\mathrm{Re}(R^*\sqrt{R^2 - Q^3}) \geq 0 \qquad (5.6.14)$$

(asterisk again denoting complex conjugation). If Q and R are both real, equations (5.6.13)–(5.6.14) are equivalent to

$$A = -\mathrm{sgn}(R)\left[|R| + \sqrt{R^2 - Q^3}\right]^{1/3} \qquad (5.6.15)$$

where the positive square root is assumed. Next compute

$$B = \begin{cases} Q/A & (A \neq 0) \\ 0 & (A = 0) \end{cases} \qquad (5.6.16)$$

in terms of which the three roots are

$$x_1 = (A + B) - \frac{a}{3} \qquad (5.6.17)$$

(the single real root when a, b, c are real) and

$$x_2 = -\frac{1}{2}(A + B) - \frac{a}{3} + i\frac{\sqrt{3}}{2}(A - B)$$

$$\qquad\qquad\qquad\qquad\qquad\qquad\qquad\qquad\qquad (5.6.18)$$

$$x_3 = -\frac{1}{2}(A + B) - \frac{a}{3} - i\frac{\sqrt{3}}{2}(A - B)$$

(in that same case, a complex conjugate pair). Equations (5.6.13)–(5.6.16) are arranged both to minimize roundoff error, and also (as pointed out by A.J. Glassman) to ensure that no choice of branch for the complex cube root can result in the spurious loss of a distinct root.

If you need to solve many cubic equations with only slightly different coefficients, it is more efficient to use Newton's method (§9.4).

CITED REFERENCES AND FURTHER READING:

Weast, R.C. (ed.) 1967, *Handbook of Tables for Mathematics*, 3rd ed. (Cleveland: The Chemical Rubber Co.), pp. 130–133.

Pachner, J. 1983, *Handbook of Numerical Analysis Applications* (New York: McGraw-Hill), §6.1.

McKelvey, J.P. 1984, *American Journal of Physics*, vol. 52, pp. 269–270; see also vol. 53, p. 775, and vol. 55, pp. 374–375.

5.7 Numerical Derivatives

Imagine that you have a procedure which computes a function $f(x)$, and now you want to compute its derivative $f'(x)$. Easy, right? The definition of the derivative, the limit as $h \to 0$ of

$$f'(x) \approx \frac{f(x+h) - f(x)}{h} \tag{5.7.1}$$

practically suggests the program: Pick a small value h; evaluate $f(x+h)$; you probably have $f(x)$ already evaluated, but if not, do it too; finally apply equation (5.7.1). What more needs to be said?

Quite a lot, actually. Applied uncritically, the above procedure is almost guaranteed to produce inaccurate results. Applied properly, it can be the right way to compute a derivative only when the function f is *fiercely* expensive to compute, when you already have invested in computing $f(x)$, and when, therefore, you want to get the derivative in no more than a single additional function evaluation. In such a situation, the remaining issue is to choose h properly, an issue we now discuss:

There are two sources of error in equation (5.7.1), truncation error and roundoff error. The truncation error comes from higher terms in the Taylor series expansion,

$$f(x+h) = f(x) + hf'(x) + \frac{1}{2}h^2 f''(x) + \frac{1}{6}h^3 f'''(x) + \cdots \tag{5.7.2}$$

whence

$$\frac{f(x+h) - f(x)}{h} = f' + \frac{1}{2}hf'' + \cdots \tag{5.7.3}$$

The roundoff error has various contributions. First there is roundoff error in h: Suppose, by way of an example, that you are at a point $x = 10.3$ and you blindly choose $h = 0.0001$. Neither $x = 10.3$ nor $x + h = 10.30001$ is a number with an exact representation in binary; each is therefore represented with some fractional error characteristic of the machine's floating-point format, ϵ_m, whose value in single precision may be $\sim 10^{-7}$. The error in the *effective* value of h, namely the difference between $x + h$ and x as represented in the machine, is therefore on the order of $\epsilon_m x$, which implies a fractional error in h of order $\sim \epsilon_m x/h \sim 10^{-2}$! By equation (5.7.1) this immediately implies at least the same large fractional error in the derivative.

We arrive at Lesson 1: Always choose h so that $x + h$ and x differ by an exactly representable number. This can usually be accomplished by the program steps

$$\begin{aligned} \texttt{temp} &= x + h \\ h &= \texttt{temp} - x \end{aligned} \tag{5.7.4}$$

Some optimizing compilers, and some computers whose floating-point chips have higher internal accuracy than is stored externally, can foil this trick; if so, it is usually enough to declare temp as volatile, or else to call a dummy function donothing(temp) between the two equations (5.7.4). This forces temp into and out of addressable memory.

With h an "exact" number, the roundoff error in equation (5.7.1) is $e_r \sim \epsilon_f |f(x)/h|$. Here ϵ_f is the fractional accuracy with which f is computed; for a simple function this may be comparable to the machine accuracy, $\epsilon_f \approx \epsilon_m$, but for a complicated calculation with additional sources of inaccuracy it may be larger. The truncation error in equation (5.7.3) is on the order of $e_t \sim |hf''(x)|$. Varying h to minimize the sum $e_r + e_t$ gives the optimal choice of h,

$$h \sim \sqrt{\frac{\epsilon_f f}{f''}} \approx \sqrt{\epsilon_f} x_c \qquad (5.7.5)$$

where $x_c \equiv (f/f'')^{1/2}$ is the "curvature scale" of the function f, or "characteristic scale" over which it changes. In the absence of any other information, one often assumes $x_c = x$ (except near $x = 0$ where some other estimate of the typical x scale should be used).

With the choice of equation (5.7.5), the fractional accuracy of the computed derivative is

$$(e_r + e_t)/|f'| \sim \sqrt{\epsilon_f}(ff''/f'^2)^{1/2} \sim \sqrt{\epsilon_f} \qquad (5.7.6)$$

Here the last order-of-magnitude equality assumes that f, f', and f'' all share the same characteristic length scale, usually the case. One sees that the simple finite-difference equation (5.7.1) gives *at best* only the square root of the machine accuracy ϵ_m.

If you can afford two function evaluations for each derivative calculation, then it is significantly better to use the symmetrized form

$$f'(x) \approx \frac{f(x+h) - f(x-h)}{2h} \qquad (5.7.7)$$

In this case, by equation (5.7.2), the truncation error is $e_t \sim h^2 f'''$. The roundoff error e_r is about the same as before. The optimal choice of h, by a short calculation analogous to the one above, is now

$$h \sim \left(\frac{\epsilon_f f}{f'''}\right)^{1/3} \sim (\epsilon_f)^{1/3} x_c \qquad (5.7.8)$$

and the fractional error is

$$(e_r + e_t)/|f'| \sim (\epsilon_f)^{2/3} f^{2/3} (f''')^{1/3}/f' \sim (\epsilon_f)^{2/3} \qquad (5.7.9)$$

which will typically be an order of magnitude (single precision) or two orders of magnitude (double precision) *better* than equation (5.7.6). We have arrived at Lesson 2: Choose h to be *the correct* power of ϵ_f or ϵ_m times a characteristic scale x_c.

You can easily derive the correct powers for other cases [1]. For a function of two dimensions, for example, and the mixed derivative formula

$$\frac{\partial^2 f}{\partial x \partial y} = \frac{[f(x+h, y+h) - f(x+h, y-h)] - [f(x-h, y+h) - f(x-h, y-h)]}{4h^2}$$

$$(5.7.10)$$

the correct scaling is $h \sim \epsilon_f^{1/4} x_c$.

It is disappointing, certainly, that no simple finite-difference formula like equation (5.7.1) or (5.7.7) gives an accuracy comparable to the machine accuracy ϵ_m, or even the lower accuracy to which f is evaluated, ϵ_f. Are there no better methods?

Yes, there are. All, however, involve exploration of the function's behavior over scales comparable to x_c, plus some assumption of smoothness, or analyticity, so that the high-order terms in a Taylor expansion like equation (5.7.2) have some meaning. Such methods also involve multiple evaluations of the function f, so their increased accuracy must be weighed against increased cost.

The general idea of "Richardson's deferred approach to the limit" is particularly attractive. For numerical integrals, that idea leads to so-called Romberg integration (for review, see §4.3). For derivatives, one seeks to extrapolate, to $h \to 0$, the result of finite-difference calculations with smaller and smaller finite values of h. By the use of Neville's algorithm (§3.1), one uses each new finite-difference calculation to produce both an extrapolation of higher order, and also extrapolations of previous, lower, orders but with smaller scales h. Ridders [2] has given a nice implementation of this idea; the following program, dfridr, is based on his algorithm, modified by an improved termination criterion. Input to the routine is a function f (called func), a position x, and a *largest* stepsize h (more analogous to what we have called x_c above than to what we have called h). Output is the returned value of the derivative, and an estimate of its error, err.

```
#include <cmath>
#include <limits>
#include "nr.h"
using namespace std;

DP NR::dfridr(DP func(const DP), const DP x, const DP h, DP &err)
```
Returns the derivative of a function func at a point x by Ridders' method of polynomial extrapolation. The value h is input as an estimated initial stepsize; it need not be small, but rather should be an increment in x over which func changes *substantially*. An estimate of the error in the derivative is returned as err.
```
{
      const int NTAB=10;                       Sets maximum size of tableau.
      const DP CON=1.4, CON2=(CON*CON);        Stepsize decreased by CON at each iteration.
      const DP BIG=numeric_limits<DP>::max();
      const DP SAFE=2.0;                       Return when error is SAFE worse than the
      int i,j;                                        best so far.
      DP errt,fac,hh,ans;
      Mat_DP a(NTAB,NTAB);

      if (h == 0.0) nrerror("h must be nonzero in dfridr.");
      hh=h;
      a[0][0]=(func(x+hh)-func(x-hh))/(2.0*hh);
      err=BIG;
      for (i=1;i<NTAB;i++) {
```
Successive columns in the Neville tableau will go to smaller stepsizes and higher orders of extrapolation.
```
            hh /= CON;
            a[0][i]=(func(x+hh)-func(x-hh))/(2.0*hh);      Try new, smaller stepsize.
            fac=CON2;
            for (j=1;j<=i;j++) {              Compute extrapolations of various orders, requiring
                  a[j][i]=(a[j-1][i]*fac-a[j-1][i-1])/(fac-1.0);      no new function eval-
                  fac=CON2*fac;                                        uations.
                  errt=MAX(fabs(a[j][i]-a[j-1][i]),fabs(a[j][i]-a[j-1][i-1]));
```

The error strategy is to compare each new extrapolation to one order lower, both at the present stepsize and the previous one.

```
        if (errt <= err) {          If error is decreased, save the improved answer.
            err=errt;
            ans=a[j][i];
        }
    }
    if (fabs(a[i][i]-a[i-1][i-1]) >= SAFE*err) break;
    If higher order is worse by a significant factor SAFE, then quit early.
}
return ans;
}
```

In dfridr, the number of evaluations of func is typically 6 to 12, but is allowed to be as great as $2\times$NTAB. As a function of input h, it is typical for the accuracy to get *better* as h is made larger, until a sudden point is reached where nonsensical extrapolation produces early return with a large error. You should therefore choose a fairly large value for h, but monitor the returned value err, decreasing h if it is not small. For functions whose characteristic x scale is of order unity, we typically take h to be a few tenths.

Besides Ridders' method, there are other possible techniques. If your function is fairly smooth, and you know that you will want to evaluate its derivative many times at arbitrary points in some interval, then it makes sense to construct a Chebyshev polynomial approximation to the function in that interval, and to evaluate the derivative directly from the resulting Chebyshev coefficients. This method is described in §§5.8–5.9, following.

Another technique applies when the function consists of data that is tabulated at equally spaced intervals, and perhaps also noisy. One might then want, at each point, to least-squares *fit* a polynomial of some degree M, using an additional number n_L of points to the left and some number n_R of points to the right of each desired x value. The estimated derivative is then the derivative of the resulting fitted polynomial. A very efficient way to do this construction is via Savitzky-Golay smoothing filters, which will be discussed later, in §14.8. There we will give a routine for getting filter coefficients that not only construct the fitting polynomial but, in the accumulation of a single sum of data points times filter coefficients, evaluate it as well. In fact, the routine given, savgol, has an argument ld that determines which derivative of the fitted polynomial is evaluated. For the first derivative, the appropriate setting is ld=1, and the value of the derivative is the accumulated sum divided by the sampling interval h.

CITED REFERENCES AND FURTHER READING:

Dennis, J.E., and Schnabel, R.B. 1983, *Numerical Methods for Unconstrained Optimization and Nonlinear Equations*; reprinted 1996 (Philadelphia: S.I.A.M.), §§5.4–5.6. [1]

Ridders, C.J.F. 1982, *Advances in Engineering Software*, vol. 4, no. 2, pp. 75–76. [2]

5.8 Chebyshev Approximation

The Chebyshev polynomial of degree n is denoted $T_n(x)$, and is given by the explicit formula

$$T_n(x) = \cos(n \arccos x) \qquad (5.8.1)$$

This may look trigonometric at first glance (and there is in fact a close relation between the Chebyshev polynomials and the discrete Fourier transform); however (5.8.1) can be combined with trigonometric identities to yield explicit expressions for $T_n(x)$ (see Figure 5.8.1),

$$
\begin{aligned}
T_0(x) &= 1 \\
T_1(x) &= x \\
T_2(x) &= 2x^2 - 1 \\
T_3(x) &= 4x^3 - 3x \\
T_4(x) &= 8x^4 - 8x^2 + 1
\end{aligned}
\qquad (5.8.2)
$$

$$\cdots$$

$$T_{n+1}(x) = 2xT_n(x) - T_{n-1}(x) \quad n \geq 1.$$

(There also exist inverse formulas for the powers of x in terms of the T_n's — see equations 5.11.2-5.11.3.)

The Chebyshev polynomials are orthogonal in the interval $[-1, 1]$ over a weight $(1 - x^2)^{-1/2}$. In particular,

$$\int_{-1}^{1} \frac{T_i(x)T_j(x)}{\sqrt{1-x^2}}dx = \begin{cases} 0 & i \neq j \\ \pi/2 & i = j \neq 0 \\ \pi & i = j = 0 \end{cases} \qquad (5.8.3)$$

The polynomial $T_n(x)$ has n zeros in the interval $[-1, 1]$, and they are located at the points

$$x = \cos\left(\frac{\pi(k + \frac{1}{2})}{n}\right) \qquad k = 0, 1, \ldots, n-1 \qquad (5.8.4)$$

In this same interval there are $n + 1$ extrema (maxima and minima), located at

$$x = \cos\left(\frac{\pi k}{n}\right) \qquad k = 0, 1, \ldots, n \qquad (5.8.5)$$

At all of the maxima $T_n(x) = 1$, while at all of the minima $T_n(x) = -1$; it is precisely this property that makes the Chebyshev polynomials so useful in polynomial approximation of functions.

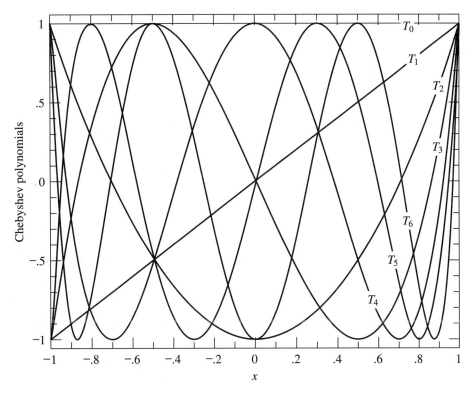

Figure 5.8.1.   Chebyshev polynomials $T_0(x)$ through $T_6(x)$. Note that T_j has j roots in the interval $(-1, 1)$ and that all the polynomials are bounded between ± 1.

The Chebyshev polynomials satisfy a discrete orthogonality relation as well as the continuous one (5.8.3): If x_k $(k = 0, \ldots, m - 1)$ are the m zeros of $T_m(x)$ given by (5.8.4), and if $i, j < m$, then

$$\sum_{k=0}^{m-1} T_i(x_k)T_j(x_k) = \begin{cases} 0 & i \neq j \\ m/2 & i = j \neq 0 \\ m & i = j = 0 \end{cases} \tag{5.8.6}$$

It is not too difficult to combine equations (5.8.1), (5.8.4), and (5.8.6) to prove the following theorem: If $f(x)$ is an arbitrary function in the interval $[-1, 1]$, and if N coefficients $c_j, j = 0, \ldots, N - 1$, are defined by

$$
\begin{aligned}
c_j &= \frac{2}{N} \sum_{k=0}^{N-1} f(x_k)T_j(x_k) \\
&= \frac{2}{N} \sum_{k=0}^{N-1} f\left[\cos\left(\frac{\pi(k + \frac{1}{2})}{N}\right)\right] \cos\left(\frac{\pi j(k + \frac{1}{2})}{N}\right)
\end{aligned}
\tag{5.8.7}
$$

then the approximation formula

$$f(x) \approx \left[\sum_{k=0}^{N-1} c_k T_k(x)\right] - \frac{1}{2}c_0 \tag{5.8.8}$$

is *exact* for x equal to all of the N zeros of $T_N(x)$.

For a fixed N, equation (5.8.8) is a polynomial in x which approximates the function $f(x)$ in the interval $[-1, 1]$ (where all the zeros of $T_N(x)$ are located). Why is this particular approximating polynomial better than any other one, exact on some other set of N points? The answer is *not* that (5.8.8) is necessarily more accurate than some other approximating polynomial of the same order N (for some specified definition of "accurate"), but rather that (5.8.8) can be truncated to a polynomial of *lower* degree $m \ll N$ in a very graceful way, one that *does* yield the "most accurate" approximation of degree m (in a sense that can be made precise). Suppose N is so large that (5.8.8) is virtually a perfect approximation of $f(x)$. Now consider the truncated approximation

$$f(x) \approx \left[\sum_{k=0}^{m-1} c_k T_k(x) \right] - \frac{1}{2} c_0 \qquad (5.8.9)$$

with the same c_j's, computed from (5.8.7). Since the $T_k(x)$'s are all bounded between ± 1, the difference between (5.8.9) and (5.8.8) can be no larger than the sum of the neglected c_k's ($k = m, \ldots, N - 1$). In fact, if the c_k's are rapidly decreasing (which is the typical case), then the error is dominated by $c_m T_m(x)$, an oscillatory function with $m + 1$ equal extrema distributed smoothly over the interval $[-1, 1]$. This smooth spreading out of the error is a very important property: The Chebyshev approximation (5.8.9) is very nearly the same polynomial as that holy grail of approximating polynomials the *minimax polynomial*, which (among all polynomials of the same degree) has the smallest maximum deviation from the true function $f(x)$. The minimax polynomial is very difficult to find; the Chebyshev approximating polynomial is almost identical and is very easy to compute!

So, given some (perhaps difficult) means of computing the function $f(x)$, we now need algorithms for implementing (5.8.7) and (after inspection of the resulting c_k's and choice of a truncating value m) evaluating (5.8.9). The latter equation then becomes an easy way of computing $f(x)$ for all subsequent time.

The first of these tasks is straightforward. A generalization of equation (5.8.7) that is here implemented is to allow the range of approximation to be between two arbitrary limits a and b, instead of just -1 to 1. This is effected by a change of variable

$$y \equiv \frac{x - \frac{1}{2}(b + a)}{\frac{1}{2}(b - a)} \qquad (5.8.10)$$

and by the approximation of $f(x)$ by a Chebyshev polynomial in y.

```
#include <cmath>
#include "nr.h"
using namespace std;

void NR::chebft(const DP a, const DP b, Vec_O_DP &c, DP func(const DP))
```
Chebyshev fit: Given a function func, lower and upper limits of the interval [a,b], and an output array c[0..n-1] of dimension n, this routine computes the n coefficients of the Chebyshev approximation such that func$(x) \approx [\sum_{k=0}^{n-1} c_k T_k(y)] - c_0/2$, where y and x are related by (5.8.10). This routine is to be used with moderately large n (e.g., 30 or 50), the array of c's subsequently to be truncated at the smaller value m such that c_m and subsequent elements are negligible.
```
{
    const DP PI=3.141592653589793;
```

```
    int k,j;
    DP fac,bpa,bma,y,sum;

    int n=c.size();
    Vec_DP f(n);
    bma=0.5*(b-a);
    bpa=0.5*(b+a);
    for (k=0;k<n;k++) {                      We evaluate the function at the n points required
        y=cos(PI*(k+0.5)/n);                     by (5.8.7).
        f[k]=func(y*bma+bpa);
    }
    fac=2.0/n;
    for (j=0;j<n;j++) {                      Now evaluate (5.8.7).
        sum=0.0;
        for (k=0;k<n;k++)
            sum += f[k]*cos(PI*j*(k+0.5)/n);
        c[j]=fac*sum;
    }
}
```

(If you find that the execution time of `chebft` is dominated by the calculation of N^2 cosines, rather than by the N evaluations of your function, then you should look ahead to §12.3, especially equation 12.3.22, which shows how fast cosine transform methods can be used to evaluate equation 5.8.7.)

Now that we have the Chebyshev coefficients, how do we evaluate the approximation? One could use the recurrence relation of equation (5.8.2) to generate values for $T_k(x)$ from $T_0 = 1, T_1 = x$, while also accumulating the sum of (5.8.9). It is better to use Clenshaw's recurrence formula (§5.5), effecting the two processes simultaneously. Applied to the Chebyshev series (5.8.9), the recurrence is

$$d_{m+1} \equiv d_m \equiv 0$$

$$d_j = 2xd_{j+1} - d_{j+2} + c_j \qquad j = m-1, m-2, \ldots, 1 \qquad (5.8.11)$$

$$f(x) \equiv d_0 = xd_1 - d_2 + \frac{1}{2}c_0$$

```
#include "nr.h"

DP NR::chebev(const DP a, const DP b, Vec_I_DP &c, const int m, const DP x)
```
Chebyshev evaluation: All arguments are input. `c[0..m-1]` is an array of Chebyshev coefficients, the first m elements of c output from `chebft` (which must have been called with the same a and b). The Chebyshev polynomial $\sum_{k=0}^{m-1} c_k T_k(y) - c_0/2$ is evaluated at a point $y = [x - (b+a)/2]/[(b-a)/2]$, and the result is returned as the function value.
```
{
    DP d=0.0,dd=0.0,sv,y,y2;
    int j;

    if ((x-a)*(x-b) > 0.0)
        nrerror("x not in range in routine chebev");
    y2=2.0*(y=(2.0*x-a-b)/(b-a));            Change of variable.
    for (j=m-1;j>0;j--) {                    Clenshaw's recurrence.
        sv=d;
        d=y2*d-dd+c[j];
        dd=sv;
    }
    return y*d-dd+0.5*c[0];                   Last step is different.
}
```

If we are approximating an *even* function on the interval $[-1, 1]$, its expansion will involve only even Chebyshev polynomials. It is wasteful to call chebev with all the odd coefficients zero [1]. Instead, using the half-angle identity for the cosine in equation (5.8.1), we get the relation

$$T_{2n}(x) = T_n(2x^2 - 1) \qquad (5.8.12)$$

Thus we can evaluate a series of even Chebyshev polynomials by calling chebev with the even coefficients stored consecutively in the array c, but with the argument x replaced by $2x^2 - 1$.

An odd function will have an expansion involving only odd Chebyshev polynomials. It is best to rewrite it as an expansion for the function $f(x)/x$, which involves only even Chebyshev polynomials. This will give accurate values for $f(x)/x$ near $x = 0$. The coefficients c_n' for $f(x)/x$ can be found from those for $f(x)$ by recurrence:

$$
\begin{aligned}
c_{N+1}' &= 0 \\
c_{n-1}' &= 2c_n - c_{n+1}', \qquad n = N-1, N-3, \ldots
\end{aligned}
\qquad (5.8.13)
$$

Equation (5.8.13) follows from the recurrence relation in equation (5.8.2).

If you insist on evaluating an odd Chebyshev series, the efficient way is to once again use chebev with x replaced by $y = 2x^2 - 1$, and with the odd coefficients stored consecutively in the array c. Now, however, you must also change the last formula in equation (5.8.11) to be

$$f(x) = x[(2y - 1)d_1 - d_2 + c_0] \qquad (5.8.14)$$

and change the corresponding line in chebev.

CITED REFERENCES AND FURTHER READING:

Clenshaw, C.W. 1962, *Mathematical Tables*, vol. 5, National Physical Laboratory, (London: H.M. Stationery Office). [1]

Goodwin, E.T. (ed.) 1961, *Modern Computing Methods*, 2nd ed. (New York: Philosophical Library), Chapter 8.

Dahlquist, G., and Bjorck, A. 1974, *Numerical Methods* (Englewood Cliffs, NJ: Prentice-Hall), §4.4.1, p. 104.

Johnson, L.W., and Riess, R.D. 1982, *Numerical Analysis*, 2nd ed. (Reading, MA: Addison-Wesley), §6.5.2, p. 334.

Carnahan, B., Luther, H.A., and Wilkes, J.O. 1969, *Applied Numerical Methods* (New York: Wiley), §1.10, p. 39.

5.9 Derivatives or Integrals of a Chebyshev-approximated Function

If you have obtained the Chebyshev coefficients that approximate a function in a certain range (e.g., from chebft in §5.8), then it is a simple matter to transform them to Chebyshev coefficients corresponding to the derivative or integral of the function. Having done this, you can evaluate the derivative or integral just as if it were a function that you had Chebyshev-fitted *ab initio*.

The relevant formulas are these: If c_i, $i = 0, \ldots, m - 1$ are the coefficients that approximate a function f in equation (5.8.9), C_i are the coefficients that approximate the indefinite integral of f, and c_i' are the coefficients that approximate the derivative of f, then

$$C_i = \frac{c_{i-1} - c_{i+1}}{2i} \qquad (i > 0) \tag{5.9.1}$$

$$c_{i-1}' = c_{i+1}' + 2ic_i \qquad (i = m - 1, m - 2, \ldots, 1) \tag{5.9.2}$$

Equation (5.9.1) is augmented by an arbitrary choice of C_0, corresponding to an arbitrary constant of integration. Equation (5.9.2), which is a recurrence, is started with the values $c_m' = c_{m-1}' = 0$, corresponding to no information about the $m + 1$st Chebyshev coefficient of the original function f.

Here are routines for implementing equations (5.9.1) and (5.9.2).

```
#include "nr.h"

void NR::chder(const DP a, const DP b, Vec_I_DP &c, Vec_O_DP &cder, const int n)
Given a,b,c[0..n-1], as output from routine chebft §5.8, and given n, the desired degree
of approximation (length of c to be used), this routine returns the array cder[0..n-1], the
Chebyshev coefficients of the derivative of the function whose coefficients are c.
{
    int j;
    DP con;

    cder[n-1]=0.0;                              n-1 and n-2 are special cases.
    cder[n-2]=2*(n-1)*c[n-1];
    for (j=n-2;j>0;j--)
        cder[j-1]=cder[j+1]+2*j*c[j];           Equation (5.9.2).
    con=2.0/(b-a);
    for (j=0;j<n;j++)                           Normalize to the interval b-a.
        cder[j] *= con;
}
```

```
#include "nr.h"

void NR::chint(const DP a, const DP b, Vec_I_DP &c, Vec_O_DP &cint, const int n)
Given a,b,c[0..n-1], as output from routine chebft §5.8, and given n, the desired degree
of approximation (length of c to be used), this routine returns the array cint[0..n-1], the
Chebyshev coefficients of the integral of the function whose coefficients are c. The constant of
integration is set so that the integral vanishes at a.
{
```

```
int j;
DP sum=0.0,fac=1.0,con;
```

```
con=0.25*(b-a);                         Factor that normalizes to the interval b-a.
for (j=1;j<n-1;j++) {
    cint[j]=con*(c[j-1]-c[j+1])/j;      Equation (5.9.1).
    sum += fac*cint[j];                 Accumulates the constant of integration.
    fac = -fac;                         Will equal ±1.
}
cint[n-1]=con*c[n-2]/(n-1);             Special case of (5.9.1) for n-1.
sum += fac*cint[n-1];
cint[0]=2.0*sum;                        Set the constant of integration.
}
```

Clenshaw-Curtis Quadrature

Since a smooth function's Chebyshev coefficients c_i decrease rapidly, generally exponentially, equation (5.9.1) is often quite efficient as the basis for a quadrature scheme. The routines chebft and chint, used in that order, can be followed by repeated calls to chebev if $\int_a^x f(x)dx$ is required for many different values of x in the range $a \le x \le b$.

If only the single definite integral $\int_a^b f(x)dx$ is required, then chint and chebev are replaced by the simpler formula, derived from equation (5.9.1),

$$\int_a^b f(x)dx = (b-a)\left[\frac{1}{2}c_0 - \frac{1}{3}c_2 - \frac{1}{15}c_4 - \cdots - \frac{1}{(2k+1)(2k-1)}c_{2k} - \cdots\right]$$

(5.9.3)

where the c_i's are as returned by chebft. The series can be truncated when c_{2k} becomes negligible, and the first neglected term gives an error estimate.

This scheme is known as *Clenshaw-Curtis quadrature* [1]. It is often combined with an adaptive choice of N, the number of Chebyshev coefficients calculated via equation (5.8.7), which is also the number of function evaluations of $f(x)$. If a modest choice of N does not give a sufficiently small c_{2k} in equation (5.9.3), then a larger value is tried. In this adaptive case, it is even better to replace equation (5.8.7) by the so-called "trapezoidal" or Gauss-Lobatto (§4.5) variant,

$$c_j = \frac{2}{N}\sum_{k=0}^{N}{}'' f\left[\cos\left(\frac{\pi k}{N}\right)\right]\cos\left(\frac{\pi jk}{N}\right) \qquad j = 0,\ldots,N-1$$

(5.9.4)

where (N.B.!) the two primes signify that the first and last terms in the sum are to be multiplied by $1/2$. If N is doubled in equation (5.9.4), then half of the new function evaluation points are identical to the old ones, allowing the previous function evaluations to be reused. This feature, plus the analytic weights and abscissas (cosine functions in 5.9.4), give Clenshaw-Curtis quadrature an edge over high-order adaptive Gaussian quadrature (cf. §4.5), which the method otherwise resembles.

If your problem forces you to large values of N, you should be aware that equation (5.9.4) can be evaluated rapidly, and simultaneously for all the values of j, by a fast cosine transform. (See §12.3, especially equation 12.3.17.) (We already remarked that the nontrapezoidal form (5.8.7) can also be done by fast cosine methods, cf. equation 12.3.22.)

CITED REFERENCES AND FURTHER READING:

Goodwin, E.T. (ed.) 1961, *Modern Computing Methods*, 2nd ed. (New York: Philosophical Library), pp. 78–79.

Clenshaw, C.W., and Curtis, A.R. 1960, *Numerische Mathematik*, vol. 2, pp. 197–205. [1]

5.10 Polynomial Approximation from Chebyshev Coefficients

You may well ask after reading the preceding two sections, "Must I store and evaluate my Chebyshev approximation as an array of Chebyshev coefficients for a transformed variable y? Can't I convert the c_k's into actual polynomial coefficients in the original variable x and have an approximation of the following form?"

$$f(x) \approx \sum_{k=0}^{m-1} g_k x^k \qquad (5.10.1)$$

Yes, you can do this (and we will give you the algorithm to do it), but we caution you against it: Evaluating equation (5.10.1), where the coefficient g's reflect an underlying Chebyshev approximation, usually requires more significant figures than evaluation of the Chebyshev sum directly (as by chebev). This is because the Chebyshev polynomials themselves exhibit a rather delicate cancellation: The leading coefficient of $T_n(x)$, for example, is 2^{n-1}; other coefficients of $T_n(x)$ are even bigger; yet they all manage to combine into a polynomial that lies between ± 1. *Only* when m is no larger than 7 or 8 should you contemplate writing a Chebyshev fit as a direct polynomial, and even in those cases you should be willing to tolerate two or so significant figures less accuracy than the roundoff limit of your machine.

You get the g's in equation (5.10.1) from the c's output from chebft (suitably truncated at a modest value of m) by calling in sequence the following two procedures:

```
#include "nr.h"

void NR::chebpc(Vec_I_DP &c, Vec_O_DP &d)
```
Chebyshev polynomial coefficients. Given a coefficient array c[0..n-1], this routine generates a coefficient array d[0..n-1] such that $\sum_{k=0}^{n-1} d_k y^k = \sum_{k=0}^{n-1} c_k T_k(y) - c_0/2$. The method is Clenshaw's recurrence (5.8.11), but now applied algebraically rather than arithmetically.
```
{
    int k,j;
    DP sv;

    int n=c.size();
    Vec_DP dd(n);
    for (j=0;j<n;j++) d[j]=dd[j]=0.0;
    d[0]=c[n-1];
    for (j=n-2;j>0;j--) {
        for (k=n-j;k>0;k--) {
            sv=d[k];
            d[k]=2.0*d[k-1]-dd[k];
            dd[k]=sv;
        }
        sv=d[0];
        d[0] = -dd[0]+c[j];
        dd[0]=sv;
    }
    for (j=n-1;j>0;j--)
        d[j]=d[j-1]-dd[j];
    d[0] = -dd[0]+0.5*c[0];
}
```

```
#include "nr.h"

void NR::pcshft(const DP a, const DP b, Vec_IO_DP &d)
```
Polynomial coefficient shift. Given a coefficient array d[0..n-1], this routine generates a coefficient array g[0..n-1] such that $\sum_{k=0}^{n-1} d_k y^k = \sum_{k=0}^{n-1} g_k x^k$, where x and y are related by (5.8.10), i.e., the interval $-1 < y < 1$ is mapped to the interval a < x < b. The array g is returned in d.
```
{
    int k,j;
    DP fac,cnst;

    int n=d.size();
    cnst=2.0/(b-a);
    fac=cnst;
    for (j=1;j<n;j++) {              First we rescale by the factor const...
        d[j] *= fac;
        fac *= cnst;
    }
    cnst=0.5*(a+b);                 ...which is then redefined as the desired shift.
    for (j=0;j<=n-2;j++)            We accomplish the shift by synthetic division. Synthetic
        for (k=n-2;k>=j;k--)           division is a miracle of high-school algebra. If you
            d[k] -= cnst*d[k+1];       never learned it, go do so. You won't be sorry.
}
```

CITED REFERENCES AND FURTHER READING:

Acton, F.S. 1970, *Numerical Methods That Work*; 1990, corrected edition (Washington: Mathematical Association of America), pp. 59, 182–183 [synthetic division].

5.11 Economization of Power Series

One particular application of Chebyshev methods, the *economization of power series*, is an occasionally useful technique, with a flavor of getting something for nothing.

Suppose that you are already computing a function by the use of a convergent power series, for example

$$f(x) \equiv 1 - \frac{x}{3!} + \frac{x^2}{5!} - \frac{x^3}{7!} + \cdots \tag{5.11.1}$$

(This function is actually $\sin(\sqrt{x})/\sqrt{x}$, but pretend you don't know that.) You might be doing a problem that requires evaluating the series many times in some particular interval, say $[0, (2\pi)^2]$. Everything is fine, except that the series requires a large number of terms before its error (approximated by the first neglected term, say) is tolerable. In our example, with $x = (2\pi)^2$, the first term smaller than 10^{-7} is $x^{13}/(27!)$. This then approximates the error of the finite series whose last term is $x^{12}/(25!)$.

Notice that because of the large exponent in x^{13}, the error is *much smaller* than 10^{-7} everywhere in the interval except at the very largest values of x. This is the feature that allows "economization": if we are willing to let the error elsewhere in the interval rise to about the same value that the first neglected term has at the extreme end of the interval, then we can replace the 13-term series by one that is significantly shorter.

Here are the steps for doing so:

1. Change variables from x to y, as in equation (5.8.10), to map the x interval into $-1 \le y \le 1$.

2. Find the coefficients of the Chebyshev sum (like equation 5.8.8) that exactly equals your truncated power series (the one with enough terms for accuracy).

3. Truncate this Chebyshev series to a smaller number of terms, using the coefficient of the first neglected Chebyshev polynomial as an estimate of the error.
4. Convert back to a polynomial in y.
5. Change variables back to x.

All of these steps can be done numerically, given the coefficients of the original power series expansion. The first step is exactly the inverse of the routine pcshft (§5.10), which mapped a polynomial from y (in the interval $[-1, 1]$) to x (in the interval $[a, b]$). But since equation (5.8.10) is a linear relation between x and y, one can also use pcshft for the inverse. The inverse of

$$\texttt{pcshft}(a, b, \texttt{d}, \texttt{n})$$

turns out to be (you can check this)

$$\texttt{pcshft}\left(\frac{-2 - b - a}{b - a}, \frac{2 - b - a}{b - a}, \texttt{d}, \texttt{n}\right)$$

The second step requires the inverse operation to that done by the routine chebpc (which took Chebyshev coefficients into polynomial coefficients). The following routine, pccheb, accomplishes this, using the formula [1]

$$x^k = \frac{1}{2^{k-1}}\left[T_k(x) + \binom{k}{1}T_{k-2}(x) + \binom{k}{2}T_{k-4}(x) + \cdots\right] \tag{5.11.2}$$

where the last term depends on whether k is even or odd,

$$\cdots + \binom{k}{(k-1)/2}T_1(x) \quad (k \text{ odd}), \qquad \cdots + \frac{1}{2}\binom{k}{k/2}T_0(x) \quad (k \text{ even}). \tag{5.11.3}$$

```
#include "nr.h"

void NR::pccheb(Vec_I_DP &d, Vec_O_DP &c)
Inverse of routine chebpc: given an array of polynomial coefficients d[0..n-1], returns an
equivalent array of Chebyshev coefficients c[0..n-1].
{
    int j,jm,jp,k;
    DP fac,pow;

    int n=d.size();
    pow=1.0;                        Will be powers of 2.
    c[0]=2.0*d[0];
    for (k=1;k<n;k++) {             Loop over orders of x in the polynomial.
        c[k]=0.0;                   Zero corresponding order of Chebyshev.
        fac=d[k]/pow;
        jm=k;
        jp=1;
        for (j=k;j>=0;j-=2,jm--,jp++) {
        Increment this and lower orders of Chebyshev with the combinatorial coefficent times
        d[k]; see text for formula.
            c[j] += fac;
            fac *= DP(jm)/DP(jp);
        }
        pow += pow;
    }
}
```

The fourth and fifth steps are accomplished by the routines chebpc and pcshft, respectively. Here is how the procedure looks all together:

```
const int NFEW = ... NMANY = ...
DP a,b;
Vec_DP c(NMANY),d(NFEW),e(NMANY);
```
Economize NMANY power series coefficients e[0..NMANY-1] in the range (a,b) into NFEW coefficients d[0..NFEW-1].

```
NR::pcshft((-2.0-b-a)/(b-a),(2.0-b-a)/(b-a),e);
NR::pccheb(e,c);
...
```
Here one would normally examine the Chebyshev coefficients c[0..NMANY-1] to decide how small NFEW can be.
```
NR::chebpc(c,d);
NR::pcshft(a,b,d);
```

In our example, by the way, the 8th through 10th Chebyshev coefficients turn out to be on the order of -7×10^{-6}, 3×10^{-7}, and -9×10^{-9}, so reasonable truncations (for single precision calculations) are somewhere in this range, yielding a polynomial with 8 – 10 terms instead of the original 13.

Replacing a 13-term polynomial with a (say) 10-term polynomial without any loss of accuracy — that does seem to be getting something for nothing. Is there some magic in this technique? Not really. The 13-term polynomial defined a function $f(x)$. Equivalent to economizing the series, we could instead have evaluated $f(x)$ at enough points to construct its Chebyshev approximation in the interval of interest, by the methods of §5.8. We would have obtained just the same lower-order polynomial. The principal lesson is that the rate of convergence of Chebyshev coefficients has nothing to do with the rate of convergence of power series coefficients; and it is the *former* that dictates the number of terms needed in a polynomial approximation. A function might have a *divergent* power series in some region of interest, but if the function itself is well-behaved, it will have perfectly good polynomial approximations. These can be found by the methods of §5.8, but *not* by economization of series. There is slightly less to economization of series than meets the eye.

CITED REFERENCES AND FURTHER READING:

Acton, F.S. 1970, *Numerical Methods That Work*; 1990, corrected edition (Washington: Mathematical Association of America), Chapter 12.

Arfken, G. 1970, *Mathematical Methods for Physicists*, 2nd ed. (New York: Academic Press), p. 631. [1]

5.12 Padé Approximants

A *Padé approximant*, so called, is that rational function (of a specified order) whose power series expansion agrees with a given power series to the highest possible order. If the rational function is

$$R(x) \equiv \frac{\displaystyle\sum_{k=0}^{M} a_k x^k}{1 + \displaystyle\sum_{k=1}^{N} b_k x^k} \tag{5.12.1}$$

then $R(x)$ is said to be a Padé approximant to the series

$$f(x) \equiv \sum_{k=0}^{\infty} c_k x^k \tag{5.12.2}$$

if

$$R(0) = f(0) \tag{5.12.3}$$

and also

$$\frac{d^k}{dx^k} R(x) \bigg|_{x=0} = \frac{d^k}{dx^k} f(x) \bigg|_{x=0} , \qquad k = 1, 2, \ldots, M + N \tag{5.12.4}$$

Equations (5.12.3) and (5.12.4) furnish $M + N + 1$ equations for the unknowns $a_0, \ldots, a_M$ and $b_1, \ldots, b_N$. The easiest way to see what these equations are is to equate (5.12.1) and (5.12.2), multiply both by the denominator of equation (5.12.1), and equate all powers of x that have either a's or b's in their coefficients. If we consider only the special case of a diagonal rational approximation, $M = N$ (cf. §3.2), then we have $a_0 = c_0$, with the remaining a's and b's satisfying

$$\sum_{m=1}^{N} b_m c_{N-m+k} = -c_{N+k}, \qquad k = 1, \ldots, N \tag{5.12.5}$$

$$\sum_{m=0}^{k} b_m c_{k-m} = a_k, \qquad k = 1, \ldots, N \tag{5.12.6}$$

(note, in equation 5.12.1, that $b_0 = 1$). To solve these, start with equations (5.12.5), which are a set of linear equations for all the unknown b's. Although the set is in the form of a Toeplitz matrix (compare equation 2.8.8), experience shows that the equations are frequently close to singular, so that one should not solve them by the methods of §2.8, but rather by full LU decomposition. Additionally, it is a good idea to refine the solution by iterative improvement (routine mprove in §2.5) [1].

Once the b's are known, then equation (5.12.6) gives an explicit formula for the unknown a's, completing the solution.

Padé approximants are typically used when there is some unknown underlying function $f(x)$. We suppose that you are able somehow to compute, perhaps by laborious analytic expansions, the values of $f(x)$ and a few of its derivatives at $x = 0$: $f(0)$, $f'(0)$, $f''(0)$, and so on. These are of course the first few coefficients in the power series expansion of $f(x)$; but they are not necessarily getting small, and you have no idea where (or whether) the power series is convergent.

By contrast with techniques like Chebyshev approximation (§5.8) or economization of power series (§5.11) that only condense the information that you already know about a function, Padé approximants can give you genuinely new information about your function's values. It is sometimes quite mysterious how well this can work. (Like other mysteries in mathematics, it relates to *analyticity*.) An example will illustrate.

Imagine that, by extraordinary labors, you have ground out the first five terms in the power series expansion of an unknown function $f(x)$,

$$f(x) \approx 2 + \frac{1}{9}x + \frac{1}{81}x^2 - \frac{49}{8748}x^3 + \frac{175}{78732}x^4 + \cdots \tag{5.12.7}$$

(It is not really necessary that you know the coefficients in exact rational form — numerical values are just as good. We here write them as rationals to give you the impression that they derive from some side analytic calculation.) Equation (5.12.7) is plotted as the curve labeled "power series" in Figure 5.12.1. One sees that for $x \gtrsim 4$ it is dominated by its largest, quartic, term.

We now take the five coefficients in equation (5.12.7) and run them through the routine pade listed below. It returns five rational coefficients, three a's and two b's, for use in equation (5.12.1) with $M = N = 2$. The curve in the figure labeled "Padé" plots the resulting rational function. Note that both solid curves derive from the *same* five original coefficient values.

To evaluate the results, we need *Deus ex machina* (a useful fellow, when he is available) to tell us that equation (5.12.7) is in fact the power series expansion of the function

$$f(x) = [7 + (1 + x)^{4/3}]^{1/3} \tag{5.12.8}$$

which is plotted as the dotted curve in the figure. This function has a branch point at $x = -1$, so its power series is convergent only in the range $-1 < x < 1$. In most of the range

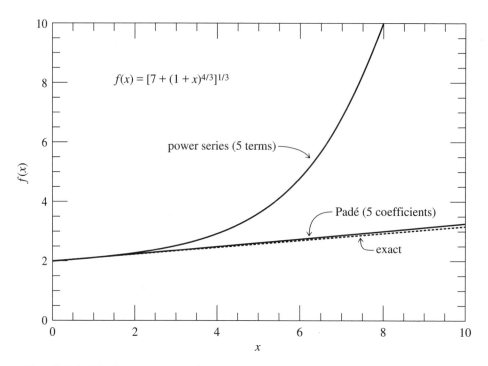

Figure 5.12.1.   The five-term power series expansion and the derived five-coefficient Padé approximant for a sample function $f(x)$. The full power series converges only for $x < 1$. Note that the Padé approximant maintains accuracy far outside the radius of convergence of the series.

shown in the figure, the series is divergent, and the value of its truncation to five terms is rather meaningless. Nevertheless, those five terms, converted to a Padé approximant, give a remarkably good representation of the function up to at least $x \sim 10$.

   Why does this work? Are there not other functions with the same first five terms in their power series, but completely different behavior in the range (say) $2 < x < 10$? Indeed there are. Padé approximation has the uncanny knack of picking the function *you had in mind* from among all the possibilities. *Except when it doesn't!* That is the downside of Padé approximation: it is uncontrolled. There is, in general, no way to tell how accurate it is, or how far out in x it can usefully be extended. It is a powerful, but in the end still mysterious, technique.

   Here is the routine that gets a's and b's from your c's. Note that the routine is specialized to the case $M = N$, and also that, on output, the rational coefficients are arranged in a format for use with the evaluation routine `ratval` (§5.3).

```
#include <cmath>
#include "nr.h"
using namespace std;
```

```
void NR::pade(Vec_IO_DP &cof, DP &resid)
```
Given `cof[0..2*n]`, the leading terms in the power series expansion of a function, solve the linear Padé equations to return the coefficients of a diagonal rational function approximation to the same function, namely $(\text{cof}[0] + \text{cof}[1]x + \cdots + \text{cof}[n]x^N)/(1 + \text{cof}[n+1]x + \cdots + \text{cof}[2*n]x^N)$. The value `resid` is the norm of the residual vector; a small value indicates a well-converged solution.
```
{
    const DP BIG=1.0e30;
    int j,k;
    DP d,rr,rrold,sum;

    int n=(cof.size()-1)/2;
```

```
Mat_DP q(n,n),qlu(n,n);
Vec_INT indx(n);
Vec_DP x(n),y(n),z(n);
for (j=0;j<n;j++) {                          Set up matrix for solving.
    y[j]=x[j]=cof[n+j+1];
    for (k=0;k<n;k++) {
        q[j][k]=cof[j-k+n];
        qlu[j][k]=q[j][k];
    }
}
ludcmp(qlu,indx,d);                          Solve by LU decomposition and backsubstitu-
lubksb(qlu,indx,x);                              tion.
rr=BIG;
do {                                         Important to use iterative improvement, since
    rrold=rr;                                    the Padé equations tend to be ill-conditioned.
    for (j=0;j<n;j++) z[j]=x[j];
    mprove(q,qlu,indx,y,x);
    for (rr=0.0,j=0;j<n;j++)                 Calculate residual.
        rr += SQR(z[j]-x[j]);
} while (rr < rrold);                        If it is no longer improving, call it quits.
resid=sqrt(rrold);
for (k=0;k<n;k++) {                          Calculate the remaining coefficients.
    for (sum=cof[k+1],j=0;j<=k;j++)
        sum -= z[j]*cof[k-j];
    y[k]=sum;
}                                            Copy answers to output.
for (j=0;j<n;j++) {
    cof[j+1]=y[j];
    cof[j+1+n] = -z[j];
}
}
```

CITED REFERENCES AND FURTHER READING:

Ralston, A. and Wilf, H.S. 1960, *Mathematical Methods for Digital Computers* (New York: Wiley),
    p. 14.

Cuyt, A., and Wuytack, L. 1987, *Nonlinear Methods in Numerical Analysis* (Amsterdam: North-
    Holland), Chapter 2.

Graves-Morris, P.R. 1979, in *Padé Approximation and Its Applications*, Lecture Notes in Mathe-
    matics, vol. 765, L. Wuytack, ed. (Berlin: Springer-Verlag). [1]

5.13 Rational Chebyshev Approximation

In §5.8 and §5.10 we learned how to find good polynomial approximations to a given function $f(x)$ in a given interval $a \leq x \leq b$. Here, we want to generalize the task to find good approximations that are rational functions (see §5.3). The reason for doing so is that, for some functions and some intervals, the optimal rational function approximation is able to achieve substantially higher accuracy than the optimal polynomial approximation with the same number of coefficients. This must be weighed against the fact that finding a rational function approximation is not as straightforward as finding a polynomial approximation, which, as we saw, could be done elegantly via Chebyshev polynomials.

Let the desired rational function $R(x)$ have numerator of degree m and denominator of degree k. Then we have

$$R(x) \equiv \frac{p_0 + p_1 x + \cdots + p_m x^m}{1 + q_1 x + \cdots + q_k x^k} \approx f(x) \qquad \text{for } a \leq x \leq b \qquad (5.13.1)$$

The unknown quantities that we need to find are $p_0, \ldots, p_m$ and $q_1, \ldots, q_k$, that is, $m + k + 1$ quantities in all. Let $r(x)$ denote the deviation of $R(x)$ from $f(x)$, and let r denote its maximum absolute value,

$$r(x) \equiv R(x) - f(x) \qquad r \equiv \max_{a \le x \le b} |r(x)| \qquad (5.13.2)$$

The ideal *minimax* solution would be that choice of p's and q's that minimizes r. Obviously there is *some* minimax solution, since r is bounded below by zero. How can we find it, or a reasonable approximation to it?

A first hint is furnished by the following fundamental theorem: If $R(x)$ is nondegenerate (has no common polynomial factors in numerator and denominator), then there is a unique choice of p's and q's that minimizes r; for this choice, $r(x)$ has $m + k + 2$ extrema in $a \le x \le b$, *all of magnitude r and with alternating sign*. (We have omitted some technical assumptions in this theorem. See Ralston [1] for a precise statement.) We thus learn that the situation with rational functions is quite analogous to that for minimax polynomials: In §5.8 we saw that the error term of an nth order approximation, with $n + 1$ Chebyshev coefficients, was generally dominated by the first neglected Chebyshev term, namely T_{n+1}, which itself has $n + 2$ extrema of equal magnitude and alternating sign. So, here, the number of rational coefficients, $m + k + 1$, plays the same role of the number of polynomial coefficients, $n + 1$.

A different way to see why $r(x)$ should have $m + k + 2$ extrema is to note that $R(x)$ can be made exactly equal to $f(x)$ at any $m + k + 1$ points x_i. Multiplying equation (5.13.1) by its denominator gives the equations

$$p_0 + p_1 x_i + \cdots + p_m x_i^m = f(x_i)(1 + q_1 x_i + \cdots + q_k x_i^k)$$
$$i = 0, 1, \ldots, m + k \qquad (5.13.3)$$

This is a set of $m + k + 1$ linear equations for the unknown p's and q's, which can be solved by standard methods (e.g., LU decomposition). If we choose the x_i's to all be in the interval (a, b), then there will generically be an extremum between each chosen x_i and x_{i+1}, plus also extrema where the function goes out of the interval at a and b, for a total of $m + k + 2$ extrema. For arbitrary x_i's, the extrema will not have the same magnitude. The theorem says that, for one particular choice of x_i's, the magnitudes can be beaten down to the identical, minimal, value of r.

Instead of making $f(x_i)$ and $R(x_i)$ equal at the points x_i, one can instead force the residual $r(x_i)$ to any desired values y_i by solving the linear equations

$$p_0 + p_1 x_i + \cdots + p_m x_i^m = [f(x_i) - y_i](1 + q_1 x_i + \cdots + q_k x_i^k)$$
$$i = 0, 1, \ldots, m + k \qquad (5.13.4)$$

In fact, if the x_i's are chosen to be the extrema (not the zeros) of the minimax solution, then the equations satisfied will be

$$p_0 + p_1 x_i + \cdots + p_m x_i^m = [f(x_i) \pm r](1 + q_1 x_i + \cdots + q_k x_i^k)$$
$$i = 0, 1, \ldots, m + k + 1 \qquad (5.13.5)$$

where the $\pm$ alternates for the alternating extrema. Notice that equation (5.13.5) is satisfied at $m + k + 2$ extrema, while equation (5.13.4) was satisfied only at $m + k + 1$ arbitrary points. How can this be? The answer is that r in equation (5.13.5) is an additional unknown, so that the number of both equations and unknowns is $m + k + 2$. True, the set is mildly nonlinear (in r), but in general it is still perfectly soluble by methods that we will develop in Chapter 9.

We thus see that, given only the *locations* of the extrema of the minimax rational function, we can solve for its coefficients and maximum deviation. Additional theorems, leading up to the so-called *Remes algorithms* [1], tell how to converge to these locations by an iterative process. For example, here is a (slightly simplified) statement of *Remes' Second Algorithm*: (1) Find an initial rational function with $m + k + 2$ extrema x_i (not having equal deviation). (2) Solve equation (5.13.5) for new rational coefficients and r. (3) Evaluate the resulting $R(x)$ to find its actual extrema (which will not be the same as the guessed values).

(4) Replace each guessed value with the nearest actual extremum of the same sign. (5) Go back to step 2 and iterate to convergence. Under a broad set of assumptions, this method will converge. Ralston [1] fills in the necessary details, including how to find the initial set of x_i's.

Up to this point, our discussion has been textbook-standard. We now reveal ourselves as heretics. We don't much like the elegant Remes algorithm. Its two nested iterations (on r in the nonlinear set 5.13.5, and on the new sets of x_i's) are finicky and require a lot of special logic for degenerate cases. Even more heretical, we doubt that compulsive searching for the *exactly best*, equal deviation, approximation is worth the effort — except perhaps for those few people in the world whose business it is to find optimal approximations that get built into compilers and microchips.

When we use rational function approximation, the goal is usually much more pragmatic: Inside some inner loop we are evaluating some function a zillion times, and we want to speed up its evaluation. Almost never do we need this function to the last bit of machine accuracy. Suppose (heresy!) we use an approximation whose error has $m + k + 2$ extrema whose deviations differ by a factor of 2. The theorems on which the Remes algorithms are based guarantee that the perfect minimax solution will have extrema somewhere within this factor of 2 range – forcing down the higher extrema will cause the lower ones to rise, until all are equal. So our "sloppy" approximation is in fact within a fraction of a least significant bit of the minimax one.

That is good enough for us, especially when we have available a very robust method for finding the so-called "sloppy" approximation. Such a method is the least-squares solution of overdetermined linear equations by singular value decomposition (§2.6 and §15.4). We proceed as follows: First, solve (in the least-squares sense) equation (5.13.3), not just for $m + k + 1$ values of x_i, but for a significantly larger number of x_i's, spaced approximately like the zeros of a high-order Chebyshev polynomial. This gives an initial guess for $R(x)$. Second, tabulate the resulting deviations, find the mean absolute deviation, call it r, and then solve (again in the least-squares sense) equation (5.13.5) with r fixed and the $\pm$ chosen to be the sign of the observed deviation at each point x_i. Third, repeat the second step a few times.

You can spot some Remes orthodoxy lurking in our algorithm: The equations we solve are trying to bring the deviations not to zero, but rather to plus-or-minus some consistent value. However, we dispense with keeping track of actual extrema; and we solve only linear equations at each stage. One additional trick is to solve a *weighted* least-squares problem, where the weights are chosen to beat down the largest deviations fastest.

Here is a program implementing these ideas. Notice that the only calls to the function fn occur in the initial filling of the table fs. You could easily modify the code to do this filling outside of the routine. It is not even necessary that your abscissas xs be exactly the ones that we use, though the quality of the fit will deteriorate if you do not have several abscissas between each extremum of the (underlying) minimax solution. Notice that the rational coefficients are output in a format suitable for evaluation by the routine ratval in §5.3.

```cpp
#include <iostream>
#include <iomanip>
#include <cmath>
#include "nr.h"
using namespace std;

void NR::ratlsq(DP fn(const DP), const DP a, const DP b, const int mm,
    const int kk, Vec_O_DP &cof, DP &dev)
```
Returns in cof [0..mm+kk] the coefficients of a rational function approximation to the function fn in the interval (a, b). Input quantities mm and kk specify the order of the numerator and denominator, respectively. The maximum absolute deviation of the approximation (insofar as is known) is returned as dev.
```cpp
{
    const int NPFAC=8,MAXIT=5;
    const DP BIG=1.0e30,PIO2=1.570796326794896619;
    int i,it,j,ncof,npt;
    DP devmax,e,hth,power,sum;

    ncof=mm+kk+1;
```

```
npt=NPFAC*ncof;
Number of points where function is evaluated, i.e., fineness of the mesh.
Vec_DP bb(npt),coff(ncof),ee(npt),fs(npt),w(ncof),wt(npt),xs(npt);
Mat_DP u(npt,ncof),v(ncof,ncof);
dev=BIG;
for (i=0;i<npt;i++) {                    Fill arrays with mesh abscissas and function val-
    if (i < (npt/2)-1) {                     ues.
        hth=PIO2*i/(npt-1.0);            At each end, use formula that minimizes round-
        xs[i]=a+(b-a)*SQR(sin(hth));      off sensitivity.
    } else {
        hth=PIO2*(npt-i)/(npt-1.0);
        xs[i]=b-(b-a)*SQR(sin(hth));
    }
    fs[i]=fn(xs[i]);
    wt[i]=1.0;                           In later iterations we will adjust these weights to
    ee[i]=1.0;                             combat the largest deviations.
}
e=0.0;
for (it=0;it<MAXIT;it++) {               Loop over iterations.
    for (i=0;i<npt;i++) {               Set up the "design matrix" for the least-squares
        power=wt[i];                      fit.
        bb[i]=power*(fs[i]+SIGN(e,ee[i]));
        Key idea here: Fit to fn(x)+e where the deviation is positive, to fn(x)-e where
        it is negative. Then e is supposed to become an approximation to the equal-ripple
        deviation.
        for (j=0;j<mm+1;j++) {
            u[i][j]=power;
            power *= xs[i];
        }
        power = -bb[i];
        for (j=mm+1;j<ncof;j++) {
            power *= xs[i];
            u[i][j]=power;
        }
    }
    svdcmp(u,w,v);                       Singular Value Decomposition.
    In especially singular or difficult cases, one might here edit the singular values, replacing
    small values by zero in w[0..ncof-1].
    svbksb(u,w,v,bb,coff);
    devmax=sum=0.0;
    for (j=0;j<npt;j++) {               Tabulate the deviations and revise the weights.
        ee[j]=ratval(xs[j],coff,mm,kk)-fs[j];
        wt[j]=fabs(ee[j]);              Use weighting to emphasize most deviant points.
        sum += wt[j];
        if (wt[j] > devmax) devmax=wt[j];
    }
    e=sum/npt;                          Update e to be the mean absolute deviation.
    if (devmax <= dev) {                Save only the best coefficient set found.
        for (j=0;j<ncof;j++) cof[j]=coff[j];
        dev=devmax;
    }
    cout << " ratlsq iteration= " << it;
    cout << " max error= " << setw(10) << devmax << endl;
}
}
```

Figure 5.13.1 shows the discrepancies for the first five iterations of `ratlsq` when it is applied to find the $m = k = 4$ rational fit to the function $f(x) = \cos x/(1 + e^x)$ in the interval $(0, \pi)$. One sees that after the first iteration, the results are virtually as good as the minimax solution. The iterations do not converge in the order that the figure suggests: In fact, it is the second iteration that is best (has smallest maximum deviation). The routine `ratlsq` accordingly returns the best of its iterations, not necessarily the last one; there is no advantage in doing more than five iterations.

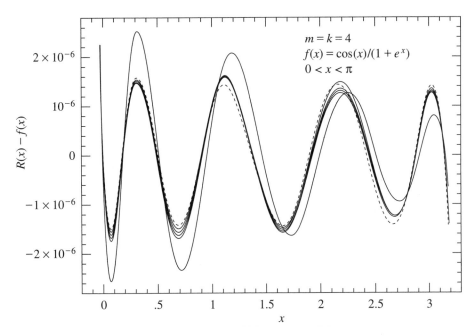

Figure 5.13.1.   Solid curves show deviations $r(x)$ for five successive iterations of the routine `ratlsq` for an arbitrary test problem. The algorithm does not converge to exactly the minimax solution (shown as the dotted curve). But, after one iteration, the discrepancy is a small fraction of the last significant bit of accuracy.

CITED REFERENCES AND FURTHER READING:

Ralston, A. and Wilf, H.S. 1960, *Mathematical Methods for Digital Computers* (New York: Wiley), Chapter 13. [1]

5.14 Evaluation of Functions by Path Integration

In computer programming, the technique of choice is not necessarily the most efficient, or elegant, or fastest executing one. Instead, it may be the one that is quick to implement, general, and easy to check.

One sometimes needs only a few, or a few thousand, evaluations of a special function, perhaps a complex valued function of a complex variable, that has many different parameters, or asymptotic regimes, or both. Use of the usual tricks (series, continued fractions, rational function approximations, recurrence relations, and so forth) may result in a patchwork program with tests and branches to different formulas. While such a program may be highly efficient in execution, it is often not the shortest way to the answer from a standing start.

A different technique of considerable generality is direct integration of a function's defining differential equation – an ab initio integration for each desired function value — along a path in the complex plane if necessary. While this may at first seem like swatting a fly with a golden brick, it turns out that when you already have the brick, and the fly is asleep right under it, all you have to do is let it fall!

As a specific example, let us consider the complex hypergeometric function $_2F_1(a, b, c; z)$, which is defined as the analytic continuation of the so-called hypergeometric series,

$$_2F_1(a, b, c; z) = 1 + \frac{ab}{c}\frac{z}{1!} + \frac{a(a+1)b(b+1)}{c(c+1)}\frac{z^2}{2!} + \cdots$$
$$+ \frac{a(a+1)\ldots(a+j-1)b(b+1)\ldots(b+j-1)}{c(c+1)\ldots(c+j-1)}\frac{z^j}{j!} + \cdots$$

$$(5.14.1)$$

The series converges only within the unit circle $|z| < 1$ (see [1]), but one's interest in the function is often not confined to this region.

The hypergeometric function $_2F_1$ is a solution (in fact *the* solution that is regular at the origin) of the hypergeometric differential equation, which we can write as

$$z(1-z)F'' = abF - [c - (a+b+1)z]F' \qquad (5.14.2)$$

Here prime denotes d/dz. One can see that the equation has regular singular points at $z = 0, 1$, and ∞. Since the desired solution is regular at $z = 0$, the values 1 and ∞ will in general be branch points. If we want $_2F_1$ to be a single valued function, we must have a branch cut connecting these two points. A conventional position for this cut is along the positive real axis from 1 to ∞, though we may wish to keep open the possibility of altering this choice for some applications.

Our golden brick consists of a collection of routines for the integration of sets of ordinary differential equations, which we will develop in detail later, in Chapter 16. For now, we need only a high-level, "black-box" routine that integrates such a set from initial conditions at one value of a (real) independent variable to final conditions at some other value of the independent variable, while automatically adjusting its internal stepsize to maintain some specified accuracy. That routine is called odeint and, in one particular invocation, calculates its individual steps with a sophisticated Bulirsch-Stoer technique.

Suppose that we know values for F and its derivative F' at some value z_0, and that we want to find F at some other point z_1 in the complex plane. The straight-line path connecting these two points is parametrized by

$$z(s) = z_0 + s(z_1 - z_0) \qquad (5.14.3)$$

with s a real parameter. The differential equation (5.14.2) can now be written as a set of two first-order equations,

$$\frac{dF}{ds} = (z_1 - z_0)F'$$
$$\frac{dF'}{ds} = (z_1 - z_0)\left(\frac{abF - [c - (a+b+1)z]F'}{z(1-z)}\right) \qquad (5.14.4)$$

to be integrated from $s = 0$ to $s = 1$. Here F and F' are to be viewed as two independent complex variables. The fact that prime means d/dz can be ignored; it will emerge as a consequence of the first equation in (5.14.4). Moreover, the real and imaginary parts of equation (5.14.4) define a set of four *real* differential equations,

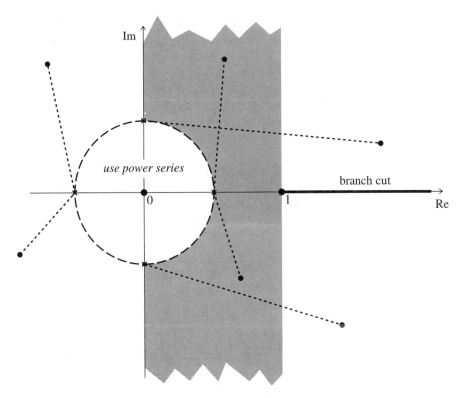

Figure 5.14.1. Complex plane showing the singular points of the hypergeometric function, its branch cut, and some integration paths from the circle $|z| = 1/2$ (where the power series converges rapidly) to other points in the plane.

with independent variable s. The complex arithmetic on the right-hand side can be viewed as mere shorthand for how the four components are to be coupled. It is precisely this point of view that gets passed to the routine odeint, since it knows nothing of either complex functions or complex independent variables.

It remains only to decide where to start, and what path to take in the complex plane, to get to an arbitrary point z. This is where consideration of the function's singularities, and the adopted branch cut, enter. Figure 5.14.1 shows the strategy that we adopt. For $|z| \leq 1/2$, the series in equation (5.14.1) will in general converge rapidly, and it makes sense to use it directly. Otherwise, we integrate along a straight line path from one of the starting points $(\pm 1/2, 0)$ or $(0, \pm 1/2)$. The former choices are natural for $0 < \mathrm{Re}(z) < 1$ and $\mathrm{Re}(z) < 0$, respectively. The latter choices are used for $\mathrm{Re}(z) > 1$, above and below the branch cut; the purpose of starting away from the real axis in these cases is to avoid passing too close to the singularity at $z = 1$ (see Figure 5.14.1). The location of the branch cut is *defined* by the fact that our adopted strategy never integrates across the real axis for $\mathrm{Re}(z) > 1$.

An implementation of this algorithm is given in §6.12 as the routine hypgeo.

A number of variants on the procedure described thus far are possible, and easy to program. If successively called values of z are close together (with identical values of a, b, and c), then you can save the state vector (F, F') and the corresponding value of z on each call, and use these as starting values for the next call. The incremental integration may then take only one or two steps. Avoid integrating across the branch cut unintentionally: the function value will be "correct," but not the one you want.

Alternatively, you may wish to integrate to some position z by a dog-leg path that *does* cross the real axis Re $z > 1$, as a means of *moving* the branch cut. For example, in some cases you might want to integrate from $(0, 1/2)$ to $(3/2, 1/2)$, and go from there to any point with Re $z > 1$ — with either sign of Im z. (If you are, for example, finding roots of a function by an iterative method, you do not want the integration for nearby values to take different paths around a branch point. If it does, your root-finder will see discontinuous function values, and will likely not converge correctly!)

In any case, be aware that a loss of numerical accuracy can result if you integrate through a region of large function value on your way to a final answer where the function value is small. (For the hypergeometric function, a particular case of this is when a and b are both large and positive, with c and $x \gtrsim 1$.) In such cases, you'll need to find a better dog-leg path.

The general technique of evaluating a function by integrating its differential equation in the complex plane can also be applied to other special functions. For example, the complex Bessel function, Airy function, Coulomb wave function, and Weber function are all special cases of the *confluent hypergeometric function*, with a differential equation similar to the one used above (see, e.g., [1] §13.6, for a table of special cases). The confluent hypergeometric function has no singularities at finite z: That makes it easy to integrate. However, its essential singularity at infinity means that it can have, along some paths and for some parameters, highly oscillatory or exponentially decreasing behavior: That makes it hard to integrate. Some case by case judgment (or experimentation) is therefore required.

CITED REFERENCES AND FURTHER READING:

Abramowitz, M., and Stegun, I.A. 1964, *Handbook of Mathematical Functions*, Applied Mathematics Series, Volume 55 (Washington: National Bureau of Standards; reprinted 1968 by Dover Publications, New York). [1]

Chapter 6.   Special Functions

6.0 Introduction

There is nothing particularly special about a *special function*, except that some person in authority or textbook writer (not the same thing!) has decided to bestow the moniker. Special functions are sometimes called *higher transcendental functions* (higher than what?) or *functions of mathematical physics* (but they occur in other fields also) or *functions that satisfy certain frequently occurring second-order differential equations* (but not all special functions do). One might simply call them "useful functions" and let it go at that; it is surely only a matter of taste which functions we have chosen to include in this chapter.

Good commercially available program libraries, such as NAG or IMSL, contain routines for a number of special functions. These routines are intended for users who will have no idea what goes on inside them. Such state of the art "black boxes" are often very messy things, full of branches to completely different methods depending on the value of the calling arguments. Black boxes have, or should have, careful control of accuracy, to some stated uniform precision in all regimes.

We will not be quite so fastidious in our examples, in part because we want to illustrate techniques from Chapter 5, and in part because we *want* you to understand what goes on in the routines presented. Some of our routines have an accuracy parameter that can be made as small as desired, while others (especially those involving polynomial fits) give only a certain accuracy, one that we believe serviceable (typically six significant figures or more). We do *not* certify that the routines are perfect black boxes. We do hope that, if you ever encounter trouble in a routine, you will be able to diagnose and correct the problem on the basis of the information that we have given.

In short, the special function routines of this chapter are meant to be used — we use them all the time — but we also want you to be prepared to understand their inner workings.

CITED REFERENCES AND FURTHER READING:

Abramowitz, M., and Stegun, I.A. 1964, *Handbook of Mathematical Functions*, Applied Mathematics Series, Volume 55 (Washington: National Bureau of Standards; reprinted 1968 by Dover Publications, New York) [full of useful numerical approximations to a great variety of functions].

IMSL Sfun/Library Users Manual (IMSL Inc., 2500 CityWest Boulevard, Houston TX 77042).

NAG Fortran Library (Numerical Algorithms Group, 256 Banbury Road, Oxford OX27DE, U.K.), Chapter S.

Hart, J.F., et al. 1968, *Computer Approximations* (New York: Wiley).

Hastings, C. 1955, *Approximations for Digital Computers* (Princeton: Princeton University Press).

Luke, Y.L. 1975, *Mathematical Functions and Their Approximations* (New York: Academic Press).

6.1 Gamma Function, Beta Function, Factorials, Binomial Coefficients

The gamma function is defined by the integral

$$\Gamma(z) = \int_0^\infty t^{z-1} e^{-t} dt \qquad (6.1.1)$$

When the argument z is an integer, the gamma function is just the familiar factorial function, but offset by one,

$$n! = \Gamma(n+1) \qquad (6.1.2)$$

The gamma function satisfies the recurrence relation

$$\Gamma(z+1) = z\Gamma(z) \qquad (6.1.3)$$

If the function is known for arguments $z > 1$ or, more generally, in the half complex plane $\text{Re}(z) > 1$ it can be obtained for $z < 1$ or $\text{Re}(z) < 1$ by the reflection formula

$$\Gamma(1-z) = \frac{\pi}{\Gamma(z)\sin(\pi z)} = \frac{\pi z}{\Gamma(1+z)\sin(\pi z)} \qquad (6.1.4)$$

Notice that $\Gamma(z)$ has a pole at $z = 0$, and at all negative integer values of z.

There are a variety of methods in use for calculating the function $\Gamma(z)$ numerically, but none is quite as neat as the approximation derived by Lanczos [1]. This scheme is entirely specific to the gamma function, seemingly plucked from thin air. We will not attempt to derive the approximation, but only state the resulting formula: For certain integer choices of γ and N, and for certain coefficients $c_1, c_2, \ldots, c_N$, the gamma function is given by

$$\Gamma(z+1) = (z+\gamma+\tfrac{1}{2})^{z+\frac{1}{2}} e^{-(z+\gamma+\frac{1}{2})}$$
$$\times \sqrt{2\pi} \left[c_0 + \frac{c_1}{z+1} + \frac{c_2}{z+2} + \cdots + \frac{c_N}{z+N} + \epsilon \right] \quad (z>0) \qquad (6.1.5)$$

You can see that this is a sort of take-off on Stirling's approximation, but with a series of corrections that take into account the first few poles in the left complex plane. The constant c_0 is very nearly equal to 1. The error term is parametrized by ϵ. For $\gamma = 5$, $N = 6$, and a certain set of c's, the error is smaller than $|\epsilon| < 2 \times 10^{-10}$. Impressed? If not, then perhaps you will be impressed by the fact that (with these same parameters) the formula (6.1.5) and bound on ϵ apply for the *complex* gamma function, *everywhere in the half complex plane Re $z > 0$.*

It is better to implement $\ln \Gamma(x)$ than $\Gamma(x)$, since the latter will overflow many computers' floating-point representation at quite modest values of x. Often the gamma function is used in calculations where the large values of $\Gamma(x)$ are divided by other large numbers, with the result being a perfectly ordinary value. Such operations would normally be coded as subtraction of logarithms. With (6.1.5) in hand, we can compute the logarithm of the gamma function with two calls to a logarithm and 25 or so arithmetic operations. This makes it not much more difficult than other built-in functions that we take for granted, such as $\sin x$ or e^x:

```cpp
#include <cmath>
#include "nr.h"
using namespace std;

DP NR::gammln(const DP xx)
Returns the value ln[Γ(xx)] for xx > 0.
{
    int j;
    DP x,y,tmp,ser;
    static const DP cof[6]={76.18009172947146,-86.50532032941677,
        24.01409824083091,-1.231739572450155,0.1208650973866179e-2,
        -0.5395239384953e-5};

    y=x=xx;
    tmp=x+5.5;
    tmp -= (x+0.5)*log(tmp);
    ser=1.000000000190015;
    for (j=0;j<6;j++) ser += cof[j]/++y;
    return -tmp+log(2.5066282746310005*ser/x);
}
```

How shall we write a routine for the factorial function $n!$? Generally the factorial function will be called for small integer values (for large values it will overflow anyway!), and in most applications the same integer value will be called for many times. It is a profligate waste of computer time to call `exp(gammln(n+1.0))` for each required factorial. Better to go back to basics, holding `gammln` in reserve for unlikely calls:

```cpp
#include <cmath>
#include "nr.h"
using namespace std;

DP NR::factrl(const int n)
Returns the value n! as a floating-point number.
{
    static int ntop=4;
    static DP a[33]={1.0,1.0,2.0,6.0,24.0};        Fill in table only as required.
    int j;

    if (n < 0) nrerror("Negative factorial in routine factrl");
    if (n > 32) return exp(gammln(n+1.0));
    Larger value than size of table is required. Actually, this big a value is going to overflow
    on many computers, but no harm in trying.
    while (ntop<n) {                               Fill in table up to desired value.
        j=ntop++;
        a[ntop]=a[j]*ntop;
    }
    return a[n];
}
```

A useful point is that `factrl` will be *exact* for the smaller values of n, since floating-point multiplies on small integers are exact on all computers. This exactness will not hold if we turn to the logarithm of the factorials. For binomial coefficients, however, we must do exactly this, since the individual factorials in a binomial coefficient will overflow long before the coefficient itself will.

The binomial coefficient is defined by

$$\binom{n}{k} = \frac{n!}{k!(n-k)!} \quad 0 \le k \le n \tag{6.1.6}$$

```
#include <cmath>
#include "nr.h"
using namespace std;

DP NR::bico(const int n, const int k)
```
Returns the binomial coefficient $\binom{n}{k}$ as a floating-point number.
```
{
    return floor(0.5+exp(factln(n)-factln(k)-factln(n-k)));
    The floor function cleans up roundoff error for smaller values of n and k.
}
```

which uses

```
#include "nr.h"

DP NR::factln(const int n)
```
Returns $\ln(n!)$.
```
{
    static DP a[101];                    A static array is automatically initialized to zero.

    if (n < 0) nrerror("Negative factorial in routine factln");
    if (n <= 1) return 0.0;
    if (n <= 100)                        In range of table.
        return (a[n] != 0.0 ? a[n] : (a[n]=gammln(n+1.0)));
    else return gammln(n+1.0);           Out of range of table.
}
```

If your problem requires a series of related binomial coefficients, a good idea is to use recurrence relations, for example

$$\binom{n+1}{k} = \frac{n+1}{n-k+1}\binom{n}{k} = \binom{n}{k} + \binom{n}{k-1}$$

$$\binom{n}{k+1} = \frac{n-k}{k+1}\binom{n}{k} \tag{6.1.7}$$

Finally, turning away from the combinatorial functions with integer valued arguments, we come to the beta function,

$$B(z,w) = B(w,z) = \int_0^1 t^{z-1}(1-t)^{w-1}dt \tag{6.1.8}$$

which is related to the gamma function by

$$B(z,w) = \frac{\Gamma(z)\Gamma(w)}{\Gamma(z+w)} \tag{6.1.9}$$

hence

```cpp
#include <cmath>
#include "nr.h"
using namespace std;

DP NR::beta(const DP z, const DP w)
Returns the value of the beta function B(z, w).
{
    return exp(gammln(z)+gammln(w)-gammln(z+w));
}
```

CITED REFERENCES AND FURTHER READING:

Abramowitz, M., and Stegun, I.A. 1964, *Handbook of Mathematical Functions*, Applied Mathematics Series, Volume 55 (Washington: National Bureau of Standards; reprinted 1968 by Dover Publications, New York), Chapter 6.

Lanczos, C. 1964, *SIAM Journal on Numerical Analysis*, ser. B, vol. 1, pp. 86–96. [1]

6.2 Incomplete Gamma Function, Error Function, Chi-Square Probability Function, Cumulative Poisson Function

The incomplete gamma function is defined by

$$P(a, x) \equiv \frac{\gamma(a, x)}{\Gamma(a)} \equiv \frac{1}{\Gamma(a)} \int_0^x e^{-t} t^{a-1} dt \qquad (a > 0) \qquad (6.2.1)$$

It has the limiting values

$$P(a, 0) = 0 \qquad \text{and} \qquad P(a, \infty) = 1 \qquad (6.2.2)$$

The incomplete gamma function $P(a, x)$ is monotonic and (for a greater than one or so) rises from "near-zero" to "near-unity" in a range of x centered on about $a - 1$, and of width about $\sqrt{a}$ (see Figure 6.2.1).

The complement of $P(a, x)$ is also confusingly called an incomplete gamma function,

$$Q(a, x) \equiv 1 - P(a, x) \equiv \frac{\Gamma(a, x)}{\Gamma(a)} \equiv \frac{1}{\Gamma(a)} \int_x^\infty e^{-t} t^{a-1} dt \qquad (a > 0) \quad (6.2.3)$$

It has the limiting values

$$Q(a, 0) = 1 \qquad \text{and} \qquad Q(a, \infty) = 0 \qquad (6.2.4)$$

The notations $P(a, x), \gamma(a, x)$, and $\Gamma(a, x)$ are standard; the notation $Q(a, x)$ is specific to this book.

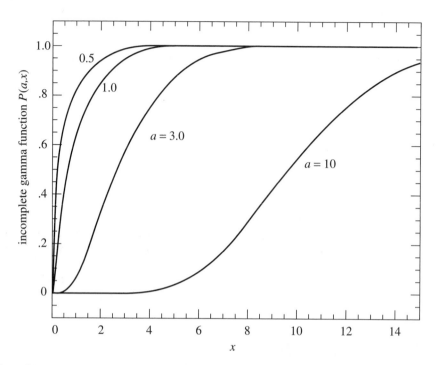

Figure 6.2.1.   The incomplete gamma function $P(a, x)$ for four values of a.

There is a series development for $\gamma(a, x)$ as follows:

$$\gamma(a, x) = e^{-x} x^a \sum_{n=0}^{\infty} \frac{\Gamma(a)}{\Gamma(a + 1 + n)} x^n \qquad (6.2.5)$$

One does not actually need to compute a new $\Gamma(a + 1 + n)$ for each n; one rather uses equation (6.1.3) and the previous coefficient.

A continued fraction development for $\Gamma(a, x)$ is

$$\Gamma(a, x) = e^{-x} x^a \left(\frac{1}{x+} \frac{1-a}{1+} \frac{1}{x+} \frac{2-a}{1+} \frac{2}{x+} \cdots \right) \qquad (x > 0) \qquad (6.2.6)$$

It is computationally better to use the even part of (6.2.6), which converges twice as fast (see §5.2):

$$\Gamma(a, x) = e^{-x} x^a \left(\frac{1}{x+1-a-} \frac{1 \cdot (1-a)}{x+3-a-} \frac{2 \cdot (2-a)}{x+5-a-} \cdots \right) \qquad (x > 0)$$
$$(6.2.7)$$

It turns out that (6.2.5) converges rapidly for x less than about $a + 1$, while (6.2.6) or (6.2.7) converges rapidly for x greater than about $a + 1$. In these respective regimes each requires at most a few times $\sqrt{a}$ terms to converge, and this many only near $x = a$, where the incomplete gamma functions are varying most rapidly. Thus (6.2.5) and (6.2.7) together allow evaluation of the function for all positive a and x. An extra dividend is that we never need compute a function value near zero by subtracting two nearly equal numbers. The higher-level functions that return $P(a, x)$ and $Q(a, x)$ are

```
#include "nr.h"

DP NR::gammp(const DP a, const DP x)
Returns the incomplete gamma function P(a,x).
{
    DP gamser,gammcf,gln;

    if (x < 0.0 || a <= 0.0)
        nrerror("Invalid arguments in routine gammp");
    if (x < a+1.0) {                    Use the series representation.
        gser(gamser,a,x,gln);
        return gamser;
    } else {                            Use the continued fraction representation
        gcf(gammcf,a,x,gln);
        return 1.0-gammcf;              and take its complement.
    }
}
```

```
#include "nr.h"

DP NR::gammq(const DP a, const DP x)
Returns the incomplete gamma function Q(a,x) ≡ 1 − P(a,x).
{
    DP gamser,gammcf,gln;

    if (x < 0.0 || a <= 0.0)
        nrerror("Invalid arguments in routine gammq");
    if (x < a+1.0) {                    Use the series representation
        gser(gamser,a,x,gln);
        return 1.0-gamser;              and take its complement.
    } else {                            Use the continued fraction representation.
        gcf(gammcf,a,x,gln);
        return gammcf;
    }
}
```

The argument gln is set by both the series and continued fraction procedures to the value $\ln\Gamma(a)$; the reason for this is so that it is available to you if you want to modify the above two procedures to give $\gamma(a,x)$ and $\Gamma(a,x)$, in addition to $P(a,x)$ and $Q(a,x)$ (cf. equations 6.2.1 and 6.2.3).

The functions gser and gcf which implement (6.2.5) and (6.2.7) are

```
#include <cmath>
#include <limits>
#include "nr.h"
using namespace std;

void NR::gser(DP &gamser, const DP a, const DP x, DP &gln)
Returns the incomplete gamma function P(a,x) evaluated by its series representation as gamser.
Also returns ln Γ(a) as gln.
{
    const int ITMAX=100;
    const DP EPS=numeric_limits<DP>::epsilon();
    int n;
    DP sum,del,ap;

    gln=gammln(a);
    if (x <= 0.0) {
        if (x < 0.0) nrerror("x less than 0 in routine gser");
```

```
        gamser=0.0;
        return;
    } else {
        ap=a;
        del=sum=1.0/a;
        for (n=0;n<ITMAX;n++) {
            ++ap;
            del *= x/ap;
            sum += del;
            if (fabs(del) < fabs(sum)*EPS) {
                gamser=sum*exp(-x+a*log(x)-gln);
                return;
            }
        }
        nrerror("a too large, ITMAX too small in routine gser");
        return;
    }
}
```

```
#include <cmath>
#include <limits>
#include "nr.h"
using namespace std;
```

void NR::gcf(DP &gammcf, const DP a, const DP x, DP &gln)
Returns the incomplete gamma function $Q(a,x)$ evaluated by its continued fraction representation as gammcf. Also returns $\ln\Gamma(a)$ as gln.
```
{
    const int ITMAX=100;
    const DP EPS=numeric_limits<DP>::epsilon();
    const DP FPMIN=numeric_limits<DP>::min()/EPS;
```
ITMAX is the maximum allowed number of iterations; EPS is the relative accuracy; FPMIN is a number near the smallest representable floating-point number.
```
    int i;
    DP an,b,c,d,del,h;

    gln=gammln(a);
    b=x+1.0-a;                                  Set up for evaluating continued fraction
    c=1.0/FPMIN;                                    by modified Lentz's method (§5.2)
    d=1.0/b;                                        with $b_0 = 0$.
    h=d;
    for (i=1;i<=ITMAX;i++) {                     Iterate to convergence.
        an = -i*(i-a);
        b += 2.0;
        d=an*d+b;
        if (fabs(d) < FPMIN) d=FPMIN;
        c=b+an/c;
        if (fabs(c) < FPMIN) c=FPMIN;
        d=1.0/d;
        del=d*c;
        h *= del;
        if (fabs(del-1.0) <= EPS) break;
    }
    if (i > ITMAX) nrerror("a too large, ITMAX too small in gcf");
    gammcf=exp(-x+a*log(x)-gln)*h;              Put factors in front.
}
```

Error Function

The error function and complementary error function are special cases of the incomplete gamma function, and are obtained moderately efficiently by the above procedures. Their definitions are

$$\text{erf}(x) = \frac{2}{\sqrt{\pi}} \int_0^x e^{-t^2} dt \qquad (6.2.8)$$

and

$$\text{erfc}(x) \equiv 1 - \text{erf}(x) = \frac{2}{\sqrt{\pi}} \int_x^\infty e^{-t^2} dt \qquad (6.2.9)$$

The functions have the following limiting values and symmetries:

$$\text{erf}(0) = 0 \qquad \text{erf}(\infty) = 1 \qquad \text{erf}(-x) = -\text{erf}(x) \qquad (6.2.10)$$

$$\text{erfc}(0) = 1 \qquad \text{erfc}(\infty) = 0 \qquad \text{erfc}(-x) = 2 - \text{erfc}(x) \qquad (6.2.11)$$

They are related to the incomplete gamma functions by

$$\text{erf}(x) = P\left(\frac{1}{2}, x^2\right) \qquad (x \geq 0) \qquad (6.2.12)$$

and

$$\text{erfc}(x) = Q\left(\frac{1}{2}, x^2\right) \qquad (x \geq 0) \qquad (6.2.13)$$

We'll put an extra "f" into our routine names to avoid conflicts with names already in some C libraries, even though the NR namespace should provide sufficient protection:

```
#include "nr.h"

DP NR::erff(const DP x)
Returns the error function erf(x).
{
    return x < 0.0 ? -gammp(0.5,x*x) : gammp(0.5,x*x);
}
```

```
#include "nr.h"

DP NR::erffc(const DP x)
Returns the complementary error function erfc(x).
{
    return x < 0.0 ? 1.0+gammp(0.5,x*x) : gammq(0.5,x*x);
}
```

If you care to do so, you can easily remedy the minor inefficiency in `erff` and `erffc`, namely that $\Gamma(0.5) = \sqrt{\pi}$ is computed unnecessarily when `gammp` or `gammq` is called. Before you do that, however, you might wish to consider the following routine, based on Chebyshev fitting to an inspired guess as to the functional form:

```
#include <cmath>
#include "nr.h"
using namespace std;

DP NR::erfcc(const DP x)
Returns the complementary error function erfc(x) with fractional error everywhere less than
1.2 × 10⁻⁷.
{
    DP t,z,ans;

    z=fabs(x);
    t=1.0/(1.0+0.5*z);
    ans=t*exp(-z*z-1.26551223+t*(1.00002368+t*(0.37409196+t*(0.09678418+
        t*(-0.18628806+t*(0.27886807+t*(-1.13520398+t*(1.48851587+
        t*(-0.82215223+t*0.17087277)))))))));
    return (x >= 0.0 ? ans : 2.0-ans);
}
```

There are also some functions of *two* variables that are special cases of the incomplete gamma function:

Cumulative Poisson Probability Function

$P_x(< k)$, for positive x and integer $k \geq 1$, denotes the *cumulative Poisson probability* function. It is defined as the probability that the number of Poisson random events occurring will be between 0 and $k - 1$ *inclusive*, if the expected mean number is x. It has the limiting values

$$P_x(< 1) = e^{-x} \qquad P_x(< \infty) = 1 \qquad (6.2.14)$$

Its relation to the incomplete gamma function is simply

$$P_x(< k) = Q(k, x) = \text{gammq}\,(k, x) \qquad (6.2.15)$$

Chi-Square Probability Function

$P(\chi^2|\nu)$ is defined as the probability that the observed chi-square for a correct model should be less than a value χ^2. (We will discuss the use of this function in Chapter 15.) Its complement $Q(\chi^2|\nu)$ is the probability that the observed chi-square will exceed the value χ^2 by chance *even* for a correct model. In both cases ν is an integer, the number of degrees of freedom. The functions have the limiting values

$$P(0|\nu) = 0 \qquad P(\infty|\nu) = 1 \qquad (6.2.16)$$
$$Q(0|\nu) = 1 \qquad Q(\infty|\nu) = 0 \qquad (6.2.17)$$

and the following relation to the incomplete gamma functions,

$$P(\chi^2|\nu) = P\left(\frac{\nu}{2}, \frac{\chi^2}{2}\right) = \text{gammp}\left(\frac{\nu}{2}, \frac{\chi^2}{2}\right) \qquad (6.2.18)$$

$$Q(\chi^2|\nu) = Q\left(\frac{\nu}{2}, \frac{\chi^2}{2}\right) = \text{gammq}\left(\frac{\nu}{2}, \frac{\chi^2}{2}\right) \qquad (6.2.19)$$

CITED REFERENCES AND FURTHER READING:

Abramowitz, M., and Stegun, I.A. 1964, *Handbook of Mathematical Functions*, Applied Mathematics Series, Volume 55 (Washington: National Bureau of Standards; reprinted 1968 by Dover Publications, New York), Chapters 6, 7, and 26.

Pearson, K. (ed.) 1951, *Tables of the Incomplete Gamma Function* (Cambridge: Cambridge University Press).

6.3 Exponential Integrals

The standard definition of the exponential integral is

$$E_n(x) = \int_1^\infty \frac{e^{-xt}}{t^n} dt, \qquad x > 0, \quad n = 0, 1, \dots \qquad (6.3.1)$$

The function defined by the principal value of the integral

$$\mathrm{Ei}(x) = -\int_{-x}^\infty \frac{e^{-t}}{t} dt = \int_{-\infty}^x \frac{e^t}{t} dt, \qquad x > 0 \qquad (6.3.2)$$

is also called an exponential integral. Note that $\mathrm{Ei}(-x)$ is related to $-E_1(x)$ by analytic continuation.

The function $E_n(x)$ is a special case of the incomplete gamma function

$$E_n(x) = x^{n-1}\Gamma(1-n, x) \qquad (6.3.3)$$

We can therefore use a similar strategy for evaluating it. The continued fraction — just equation (6.2.6) rewritten — converges for all $x > 0$:

$$E_n(x) = e^{-x}\left(\frac{1}{x+}\ \frac{n}{1+}\ \frac{1}{x+}\ \frac{n+1}{1+}\ \frac{2}{x+}\cdots\right) \qquad (6.3.4)$$

We use it in its more rapidly converging even form,

$$E_n(x) = e^{-x}\left(\frac{1}{x+n-}\ \frac{1\cdot n}{x+n+2-}\ \frac{2(n+1)}{x+n+4-}\cdots\right) \qquad (6.3.5)$$

The continued fraction only really converges fast enough to be useful for $x \gtrsim 1$. For $0 < x \lesssim 1$, we can use the series representation

$$E_n(x) = \frac{(-x)^{n-1}}{(n-1)!}[-\ln x + \psi(n)] - \sum_{\substack{m=0 \\ m \neq n-1}}^\infty \frac{(-x)^m}{(m-n+1)m!} \qquad (6.3.6)$$

The quantity $\psi(n)$ here is the digamma function, given for integer arguments by

$$\psi(1) = -\gamma, \qquad \psi(n) = -\gamma + \sum_{m=1}^{n-1} \frac{1}{m} \qquad (6.3.7)$$

where $\gamma = 0.5772156649\ldots$ is Euler's constant. We evaluate the expression (6.3.6) in order of ascending powers of x:

$$E_n(x) = -\left[\frac{1}{(1-n)} - \frac{x}{(2-n)\cdot 1} + \frac{x^2}{(3-n)(1\cdot 2)} - \cdots + \frac{(-x)^{n-2}}{(-1)(n-2)!}\right]$$
$$+ \frac{(-x)^{n-1}}{(n-1)!}[-\ln x + \psi(n)] - \left[\frac{(-x)^n}{1\cdot n!} + \frac{(-x)^{n+1}}{2\cdot (n+1)!} + \cdots\right]$$

$$(6.3.8)$$

The first square bracket is omitted when $n = 1$. This method of evaluation has the advantage that for large n the series converges before reaching the term containing $\psi(n)$. Accordingly, one needs an algorithm for evaluating $\psi(n)$ only for small n, $n \lesssim 20 - 40$. We use equation (6.3.7), although a table look-up would improve efficiency slightly.

Amos [1] presents a careful discussion of the truncation error in evaluating equation (6.3.8), and gives a fairly elaborate termination criterion. We have found that simply stopping when the last term added is smaller than the required tolerance works about as well.

Two special cases have to be handled separately:

$$E_0(x) = \frac{e^{-x}}{x}$$
$$E_n(0) = \frac{1}{n-1}, \qquad n > 1 \tag{6.3.9}$$

The routine expint allows fast evaluation of $E_n(x)$ to any accuracy EPS within the reach of your machine's word length for floating-point numbers. The only modification required for increased accuracy is to supply Euler's constant with enough significant digits. Wrench [2] can provide you with the first 328 digits if necessary!

```
#include <cmath>
#include <limits>
#include "nr.h"
using namespace std;

DP NR::expint(const int n, const DP x)
Evaluates the exponential integral En(x).
{
    const int MAXIT=100;
    const DP EULER=0.577215664901533;
    const DP EPS=numeric_limits<DP>::epsilon();
    const DP BIG=numeric_limits<DP>::max()*EPS;
    Here MAXIT is the maximum allowed number of iterations; EULER is Euler's constant γ; EPS
    is the desired relative error, not smaller than the machine precision; BIG is a number near
    the largest representable floating-point number.
    int i,ii,nm1;
    DP a,b,c,d,del,fact,h,psi,ans;

    nm1=n-1;
    if (n < 0 || x < 0.0 || (x==0.0 && (n==0 || n==1)))
    nrerror("bad arguments in expint");
    else {
```

```
        if (n == 0) ans=exp(-x)/x;                    Special case.
        else {
            if (x == 0.0) ans=1.0/nm1;                Another special case.
            else {
                if (x > 1.0) {                        Lentz's algorithm (§5.2).
                    b=x+n;
                    c=BIG;
                    d=1.0/b;
                    h=d;
                    for (i=1;i<=MAXIT;i++) {
                        a = -i*(nm1+i);
                        b += 2.0;
                        d=1.0/(a*d+b);                 Denominators cannot be zero.
                        c=b+a/c;
                        del=c*d;
                        h *= del;
                        if (fabs(del-1.0) <= EPS) {
                            ans=h*exp(-x);
                            return ans;
                        }
                    }
                    nrerror("continued fraction failed in expint");
                } else {                               Evaluate series.
                    ans = (nm1!=0 ? 1.0/nm1 : -log(x)-EULER);    Set first term.
                    fact=1.0;
                    for (i=1;i<=MAXIT;i++) {
                        fact *= -x/i;
                        if (i != nm1) del = -fact/(i-nm1);
                        else {
                            psi = -EULER;             Compute ψ(n).
                            for (ii=1;ii<=nm1;ii++) psi += 1.0/ii;
                            del=fact*(-log(x)+psi);
                        }
                        ans += del;
                        if (fabs(del) < fabs(ans)*EPS) return ans;
                    }
                    nrerror("series failed in expint");
                }
            }
        }
    }
    return ans;
}
```

A good algorithm for evaluating Ei is to use the power series for small x and the asymptotic series for large x. The power series is

$$\mathrm{Ei}(x) = \gamma + \ln x + \frac{x}{1 \cdot 1!} + \frac{x^2}{2 \cdot 2!} + \cdots \qquad (6.3.10)$$

where γ is Euler's constant. The asymptotic expansion is

$$\mathrm{Ei}(x) \sim \frac{e^x}{x}\left(1 + \frac{1!}{x} + \frac{2!}{x^2} + \cdots\right) \qquad (6.3.11)$$

The lower limit for the use of the asymptotic expansion is approximately $|\ln \mathrm{EPS}|$, where EPS is the required relative error.

```
#include <cmath>
#include <limits>
#include "nr.h"
using namespace std;

DP NR::ei(const DP x)
Computes the exponential integral Ei(x) for x > 0.
{
    const int MAXIT=100;
    const DP EULER=0.577215664901533;
    const DP EPS=numeric_limits<DP>::epsilon();
    const DP FPMIN=numeric_limits<DP>::min()/EPS;
    Here MAXIT is the maximum number of iterations allowed; EULER is Euler's constant γ; EPS
    is the relative error, or absolute error near the zero of Ei at x = 0.3725; FPMIN is a number
    close to the smallest representable floating-point number.
    int k;
    DP fact,prev,sum,term;

    if (x <= 0.0) nrerror("Bad argument in ei");
    if (x < FPMIN) return log(x)+EULER;         Special case: avoid failure of convergence
    if (x <= -log(EPS)) {                             test because of underflow.
        sum=0.0;                                Use power series.
        fact=1.0;
        for (k=1;k<=MAXIT;k++) {
            fact *= x/k;
            term=fact/k;
            sum += term;
            if (term < EPS*sum) break;
        }
        if (k > MAXIT) nrerror("Series failed in ei");
        return sum+log(x)+EULER;
    } else {                                    Use asymptotic series.
        sum=0.0;                                Start with second term.
        term=1.0;
        for (k=1;k<=MAXIT;k++) {
            prev=term;
            term *= k/x;
            if (term < EPS) break;
            Since final sum is greater than one, term itself approximates the relative error.
            if (term < prev) sum += term;       Still converging: add new term.
            else {
                sum -= prev;                    Diverging: subtract previous term and
                break;                              exit.
            }
        }
        return exp(x)*(1.0+sum)/x;
    }
}
```

CITED REFERENCES AND FURTHER READING:

Stegun, I.A., and Zucker, R. 1974, *Journal of Research of the National Bureau of Standards*, vol. 78B, pp. 199–216; 1976, *op. cit.*, vol. 80B, pp. 291–311.

Amos D.E. 1980, *ACM Transactions on Mathematical Software*, vol. 6, pp. 365–377 [1]; also vol. 6, pp. 420–428.

Abramowitz, M., and Stegun, I.A. 1964, *Handbook of Mathematical Functions*, Applied Mathematics Series, Volume 55 (Washington: National Bureau of Standards; reprinted 1968 by Dover Publications, New York), Chapter 5.

Wrench J.W. 1952, *Mathematical Tables and Other Aids to Computation*, vol. 6, p. 255. [2]

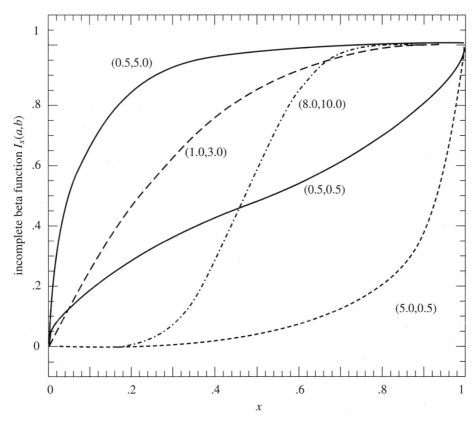

Figure 6.4.1.   The incomplete beta function $I_x(a, b)$ for five different pairs of (a, b). Notice that the pairs $(0.5, 5.0)$ and $(5.0, 0.5)$ are symmetrically related as indicated in equation (6.4.3).

6.4 Incomplete Beta Function, Student's Distribution, F-Distribution, Cumulative Binomial Distribution

The incomplete beta function is defined by

$$I_x(a, b) \equiv \frac{B_x(a, b)}{B(a, b)} \equiv \frac{1}{B(a, b)} \int_0^x t^{a-1}(1 - t)^{b-1} dt \qquad (a, b > 0) \qquad (6.4.1)$$

It has the limiting values

$$I_0(a, b) = 0 \qquad I_1(a, b) = 1 \qquad (6.4.2)$$

and the symmetry relation

$$I_x(a, b) = 1 - I_{1-x}(b, a) \qquad (6.4.3)$$

If a and b are both rather greater than one, then $I_x(a, b)$ rises from "near-zero" to "near-unity" quite sharply at about $x = a/(a + b)$. Figure 6.4.1 plots the function for several pairs (a, b).

The incomplete beta function has a series expansion

$$I_x(a,b) = \frac{x^a(1-x)^b}{aB(a,b)}\left[1 + \sum_{n=0}^{\infty}\frac{B(a+1,n+1)}{B(a+b,n+1)}x^{n+1}\right], \qquad (6.4.4)$$

but this does not prove to be very useful in its numerical evaluation. (Note, however, that the beta functions in the coefficients can be evaluated for each value of n with just the previous value and a few multiplies, using equations 6.1.9 and 6.1.3.)

The continued fraction representation proves to be much more useful,

$$I_x(a,b) = \frac{x^a(1-x)^b}{aB(a,b)}\left[\frac{1}{1+}\frac{d_1}{1+}\frac{d_2}{1+}\cdots\right] \qquad (6.4.5)$$

where

$$d_{2m+1} = -\frac{(a+m)(a+b+m)x}{(a+2m)(a+2m+1)}$$

$$d_{2m} = \frac{m(b-m)x}{(a+2m-1)(a+2m)} \qquad (6.4.6)$$

This continued fraction converges rapidly for $x < (a+1)/(a+b+2)$, taking in the worst case $O(\sqrt{\max(a,b)})$ iterations. But for $x > (a+1)/(a+b+2)$ we can just use the symmetry relation (6.4.3) to obtain an equivalent computation where the continued fraction will also converge rapidly. Hence we have

```
#include <cmath>
#include "nr.h"
using namespace std;

DP NR::betai(const DP a, const DP b, const DP x)
Returns the incomplete beta function Ix(a,b).
{
    DP bt;

    if (x < 0.0 || x > 1.0) nrerror("Bad x in routine betai");
    if (x == 0.0 || x == 1.0) bt=0.0;
    else                                        Factors in front of the continued fraction.
        bt=exp(gammln(a+b)-gammln(a)-gammln(b)+a*log(x)+b*log(1.0-x));
    if (x < (a+1.0)/(a+b+2.0))                  Use continued fraction directly.
        return bt*betacf(a,b,x)/a;
    else                                        Use continued fraction after making the sym-
        return 1.0-bt*betacf(b,a,1.0-x)/b;      metry transformation.
}
```

which utilizes the continued fraction evaluation routine

```
#include <cmath>
#include <limits>
#include "nr.h"
using namespace std;

DP NR::betacf(const DP a, const DP b, const DP x)
Used by betai: Evaluates continued fraction for incomplete beta function by modified Lentz's
method (§5.2).
{
    const int MAXIT=100;
    const DP EPS=numeric_limits<DP>::epsilon();
```

```
const DP FPMIN=numeric_limits<DP>::min()/EPS;
int m,m2;
DP aa,c,d,del,h,qab,qam,qap;

qab=a+b;                                 These q's will be used in factors that occur
qap=a+1.0;                                  in the coefficients (6.4.6).
qam=a-1.0;
c=1.0;                                   First step of Lentz's method.
d=1.0-qab*x/qap;
if (fabs(d) < FPMIN) d=FPMIN;
d=1.0/d;
h=d;
for (m=1;m<=MAXIT;m++) {
    m2=2*m;
    aa=m*(b-m)*x/((qam+m2)*(a+m2));
    d=1.0+aa*d;                          One step (the even one) of the recurrence.
    if (fabs(d) < FPMIN) d=FPMIN;
    c=1.0+aa/c;
    if (fabs(c) < FPMIN) c=FPMIN;
    d=1.0/d;
    h *= d*c;
    aa = -(a+m)*(qab+m)*x/((a+m2)*(qap+m2));
    d=1.0+aa*d;                          Next step of the recurrence (the odd one).
    if (fabs(d) < FPMIN) d=FPMIN;
    c=1.0+aa/c;
    if (fabs(c) < FPMIN) c=FPMIN;
    d=1.0/d;
    del=d*c;
    h *= del;
    if (fabs(del-1.0) <= EPS) break;    Are we done?
}
if (m > MAXIT) nrerror("a or b too big, or MAXIT too small in betacf");
return h;
}
```

Student's Distribution Probability Function

Student's distribution, denoted $A(t|\nu)$, is useful in several statistical contexts, notably in the test of whether two observed distributions have the same mean. $A(t|\nu)$ is the probability, for ν degrees of freedom, that a certain statistic t (measuring the observed difference of means) would be smaller than the observed value if the means were in fact the same. (See Chapter 14 for further details.) Two means are significantly different if, e.g., $A(t|\nu) > 0.99$. In other words, $1 - A(t|\nu)$ is the significance level at which the hypothesis that the means are equal is disproved.

The mathematical definition of the function is

$$A(t|\nu) = \frac{1}{\nu^{1/2} B(\frac{1}{2}, \frac{\nu}{2})} \int_{-t}^{t} \left(1 + \frac{x^2}{\nu}\right)^{-\frac{\nu+1}{2}} dx \qquad (6.4.7)$$

Limiting values are

$$A(0|\nu) = 0 \qquad A(\infty|\nu) = 1 \qquad (6.4.8)$$

$A(t|\nu)$ is related to the incomplete beta function $I_x(a,b)$ by

$$A(t|\nu) = 1 - I_{\frac{\nu}{\nu+t^2}}\left(\frac{\nu}{2}, \frac{1}{2}\right) \qquad (6.4.9)$$

So, you can use (6.4.9) and the above routine `betai` to evaluate the function.

F-Distribution Probability Function

This function occurs in the statistical test of whether two observed samples have the same variance. A certain statistic F, essentially the ratio of the observed dispersion of the first sample to that of the second one, is calculated. (For further details, see Chapter 14.) The probability that F would be as *large* as it is if the first sample's underlying distribution actually has *smaller* variance than the second's is denoted $Q(F|\nu_1, \nu_2)$, where ν_1 and ν_2 are the number of degrees of freedom in the first and second samples, respectively. In other words, $Q(F|\nu_1, \nu_2)$ is the significance level at which the hypothesis "1 has smaller variance than 2" can be rejected. A small numerical value implies a very significant rejection, in turn implying high confidence in the hypothesis "1 has variance greater or equal to 2."

$Q(F|\nu_1, \nu_2)$ has the limiting values

$$Q(0|\nu_1, \nu_2) = 1 \qquad Q(\infty|\nu_1, \nu_2) = 0 \qquad (6.4.10)$$

Its relation to the incomplete beta function $I_x(a, b)$ as evaluated by `betai` above is

$$Q(F|\nu_1, \nu_2) = I_{\frac{\nu_2}{\nu_2 + \nu_1 F}}\left(\frac{\nu_2}{2}, \frac{\nu_1}{2}\right) \qquad (6.4.11)$$

Cumulative Binomial Probability Distribution

Suppose an event occurs with probability p per trial. Then the probability P of its occurring k *or more* times in n trials is termed a *cumulative binomial probability*, and is related to the incomplete beta function $I_x(a, b)$ as follows:

$$P \equiv \sum_{j=k}^{n} \binom{n}{j} p^j (1-p)^{n-j} = I_p(k, n-k+1) \qquad (6.4.12)$$

For n larger than a dozen or so, `betai` is a much better way to evaluate the sum in (6.4.12) than would be the straightforward sum with concurrent computation of the binomial coefficients. (For n smaller than a dozen, either method is acceptable.)

CITED REFERENCES AND FURTHER READING:

Abramowitz, M., and Stegun, I.A. 1964, *Handbook of Mathematical Functions*, Applied Mathematics Series, Volume 55 (Washington: National Bureau of Standards; reprinted 1968 by Dover Publications, New York), Chapters 6 and 26.

Pearson, E., and Johnson, N. 1968, *Tables of the Incomplete Beta Function* (Cambridge: Cambridge University Press).

6.5 Bessel Functions of Integer Order

This section and the next one present practical algorithms for computing various kinds of Bessel functions of integer order. In §6.7 we deal with fractional order. In fact, the more complicated routines for fractional order work fine for integer order too. For integer order, however, the routines in this section (and §6.6) are simpler and faster. Their only drawback is that they are limited by the precision of the underlying rational approximations. For full double precision, it is best to work with the routines for fractional order in §6.7.

For any real ν, the Bessel function $J_\nu(x)$ can be defined by the series representation

$$J_\nu(x) = \left(\frac{1}{2}x\right)^\nu \sum_{k=0}^{\infty} \frac{(-\frac{1}{4}x^2)^k}{k!\,\Gamma(\nu + k + 1)} \tag{6.5.1}$$

The series converges for all x, but it is not computationally very useful for $x \gg 1$.

For ν *not* an integer the Bessel function $Y_\nu(x)$ is given by

$$Y_\nu(x) = \frac{J_\nu(x)\cos(\nu\pi) - J_{-\nu}(x)}{\sin(\nu\pi)} \tag{6.5.2}$$

The right-hand side goes to the correct limiting value $Y_n(x)$ as ν goes to some integer n, but this is also not computationally useful.

For arguments $x < \nu$, both Bessel functions look qualitatively like simple power laws, with the asymptotic forms for $0 < x \ll \nu$

$$J_\nu(x) \sim \frac{1}{\Gamma(\nu + 1)}\left(\frac{1}{2}x\right)^\nu \qquad \nu \geq 0$$

$$Y_0(x) \sim \frac{2}{\pi}\ln(x) \tag{6.5.3}$$

$$Y_\nu(x) \sim -\frac{\Gamma(\nu)}{\pi}\left(\frac{1}{2}x\right)^{-\nu} \qquad \nu > 0$$

For $x > \nu$, both Bessel functions look qualitatively like sine or cosine waves whose amplitude decays as $x^{-1/2}$. The asymptotic forms for $x \gg \nu$ are

$$J_\nu(x) \sim \sqrt{\frac{2}{\pi x}}\cos\left(x - \frac{1}{2}\nu\pi - \frac{1}{4}\pi\right)$$

$$Y_\nu(x) \sim \sqrt{\frac{2}{\pi x}}\sin\left(x - \frac{1}{2}\nu\pi - \frac{1}{4}\pi\right) \tag{6.5.4}$$

In the transition region where $x \sim \nu$, the typical amplitudes of the Bessel functions are on the order

$$J_\nu(\nu) \sim \frac{2^{1/3}}{3^{2/3}\Gamma(\frac{2}{3})}\frac{1}{\nu^{1/3}} \sim \frac{0.4473}{\nu^{1/3}}$$

$$Y_\nu(\nu) \sim -\frac{2^{1/3}}{3^{1/6}\Gamma(\frac{2}{3})}\frac{1}{\nu^{1/3}} \sim -\frac{0.7748}{\nu^{1/3}} \tag{6.5.5}$$

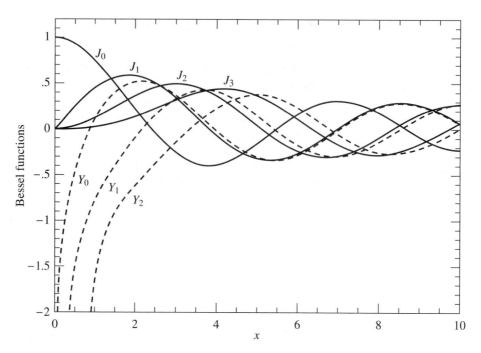

Figure 6.5.1.   Bessel functions $J_0(x)$ through $J_3(x)$ and $Y_0(x)$ through $Y_2(x)$.

which holds asymptotically for large ν. Figure 6.5.1 plots the first few Bessel functions of each kind.

The Bessel functions satisfy the recurrence relations

$$J_{n+1}(x) = \frac{2n}{x} J_n(x) - J_{n-1}(x) \tag{6.5.6}$$

and

$$Y_{n+1}(x) = \frac{2n}{x} Y_n(x) - Y_{n-1}(x) \tag{6.5.7}$$

As already mentioned in §5.5, only the second of these (6.5.7) is stable in the direction of increasing n for $x < n$. The reason that (6.5.6) is unstable in the direction of increasing n is simply that it is *the same recurrence* as (6.5.7): A small amount of "polluting" Y_n introduced by roundoff error will quickly come to swamp the desired J_n, according to equation (6.5.3).

A practical strategy for computing the Bessel functions of integer order divides into two tasks: first, how to compute J_0, J_1, Y_0, and Y_1, and second, how to use the recurrence relations stably to find other J's and Y's. We treat the first task first:

For x between zero and some arbitrary value (we will use the value 8), approximate $J_0(x)$ and $J_1(x)$ by rational functions in x. Likewise approximate by rational functions the "regular part" of $Y_0(x)$ and $Y_1(x)$, defined as

$$Y_0(x) - \frac{2}{\pi} J_0(x) \ln(x) \qquad \text{and} \qquad Y_1(x) - \frac{2}{\pi}\left[J_1(x)\ln(x) - \frac{1}{x}\right] \tag{6.5.8}$$

For $8 < x < \infty$, use the approximating forms ($n = 0, 1$)

$$J_n(x) = \sqrt{\frac{2}{\pi x}}\left[P_n\left(\frac{8}{x}\right)\cos(X_n) - Q_n\left(\frac{8}{x}\right)\sin(X_n)\right] \tag{6.5.9}$$

$$Y_n(x) = \sqrt{\frac{2}{\pi x}} \left[P_n\left(\frac{8}{x}\right) \sin(X_n) + Q_n\left(\frac{8}{x}\right) \cos(X_n) \right] \qquad (6.5.10)$$

where

$$X_n \equiv x - \frac{2n+1}{4}\pi \qquad (6.5.11)$$

and where P_0, P_1, Q_0, and Q_1 are each polynomials in their arguments, for $0 < 8/x < 1$. The P's are even polynomials, the Q's odd.

Coefficients of the various rational functions and polynomials are given by Hart [1], for various levels of desired accuracy. A straightforward implementation is

```
#include <cmath>
#include "nr.h"
using namespace std;

DP NR::bessj0(const DP x)
Returns the Bessel function J_0(x) for any real x.
{
    DP ax,z,xx,y,ans,ans1,ans2;

    if ((ax=fabs(x)) < 8.0) {               Direct rational function fit.
        y=x*x;
        ans1=57568490574.0+y*(-13362590354.0+y*(651619640.7
            +y*(-11214424.18+y*(77392.33017+y*(-184.9052456)))));
        ans2=57568490411.0+y*(1029532985.0+y*(9494680.718
            +y*(59272.64853+y*(267.8532712+y*1.0))));
        ans=ans1/ans2;
    } else {                                Fitting function (6.5.9).
        z=8.0/ax;
        y=z*z;
        xx=ax-0.785398164;
        ans1=1.0+y*(-0.1098628627e-2+y*(0.2734510407e-4
            +y*(-0.2073370639e-5+y*0.2093887211e-6)));
        ans2 = -0.1562499995e-1+y*(0.1430488765e-3
            +y*(-0.6911147651e-5+y*(0.7621095161e-6
            -y*0.934945152e-7)));
        ans=sqrt(0.636619772/ax)*(cos(xx)*ans1-z*sin(xx)*ans2);
    }
    return ans;
}
```

```
#include <cmath>
#include "nr.h"
using namespace std;

DP NR::bessy0(const DP x)
Returns the Bessel function Y_0(x) for positive x.
{
    DP z,xx,y,ans,ans1,ans2;

    if (x < 8.0) {                          Rational function approximation of (6.5.8).
        y=x*x;
        ans1 = -2957821389.0+y*(7062834065.0+y*(-512359803.6
            +y*(10879881.29+y*(-86327.92757+y*228.4622733))));
        ans2=40076544269.0+y*(745249964.8+y*(7189466.438
            +y*(47447.26470+y*(226.1030244+y*1.0))));
        ans=(ans1/ans2)+0.636619772*bessj0(x)*log(x);
    } else {                                Fitting function (6.5.10).
        z=8.0/x;
```

```
        y=z*z;
        xx=x-0.785398164;
        ans1=1.0+y*(-0.1098628627e-2+y*(0.2734510407e-4
            +y*(-0.2073370639e-5+y*0.2093887211e-6)));
        ans2 = -0.1562499995e-1+y*(0.1430488765e-3
            +y*(-0.6911147651e-5+y*(0.7621095161e-6
            +y*(-0.934945152e-7))));
        ans=sqrt(0.636619772/x)*(sin(xx)*ans1+z*cos(xx)*ans2);
    }
    return ans;
}


#include <cmath>
#include "nr.h"
using namespace std;

DP NR::bessj1(const DP x)
```
Returns the Bessel function $J_1(x)$ for any real x.
```
{
    DP ax,z,xx,y,ans,ans1,ans2;

    if ((ax=fabs(x)) < 8.0) {            Direct rational approximation.
        y=x*x;
        ans1=x*(72362614232.0+y*(-7895059235.0+y*(242396853.1
            +y*(-2972611.439+y*(15704.48260+y*(-30.16036606))))));
        ans2=144725228442.0+y*(2300535178.0+y*(18583304.74
            +y*(99447.43394+y*(376.9991397+y*1.0))));
        ans=ans1/ans2;
    } else {                             Fitting function (6.5.9).
        z=8.0/ax;
        y=z*z;
        xx=ax-2.356194491;
        ans1=1.0+y*(0.183105e-2+y*(-0.3516396496e-4
            +y*(0.2457520174e-5+y*(-0.240337019e-6))));
        ans2=0.04687499995+y*(-0.2002690873e-3
            +y*(0.8449199096e-5+y*(-0.88228987e-6
            +y*0.105787412e-6)));
        ans=sqrt(0.636619772/ax)*(cos(xx)*ans1-z*sin(xx)*ans2);
        if (x < 0.0) ans = -ans;
    }
    return ans;
}


#include <cmath>
#include "nr.h"
using namespace std;

DP NR::bessy1(const DP x)
```
Returns the Bessel function $Y_1(x)$ for positive x.
```
{
    DP z,xx,y,ans,ans1,ans2;

    if (x < 8.0) {                       Rational function approximation of (6.5.8).
        y=x*x;
        ans1=x*(-0.4900604943e13+y*(0.1275274390e13
            +y*(-0.5153438139e11+y*(0.7349264551e9
            +y*(-0.4237922726e7+y*0.8511937935e4)))));
        ans2=0.2499580570e14+y*(0.4244419664e12
            +y*(0.3733650367e10+y*(0.2245904002e8
            +y*(0.1020426050e6+y*(0.3549632885e3+y)))));
        ans=(ans1/ans2)+0.636619772*(bessj1(x)*log(x)-1.0/x);
    } else {                             Fitting function (6.5.10).
```

```
    z=8.0/x;
    y=z*z;
    xx=x-2.356194491;
    ans1=1.0+y*(0.183105e-2+y*(-0.3516396496e-4
        +y*(0.2457520174e-5+y*(-0.240337019e-6))));
    ans2=0.04687499995+y*(-0.2002690873e-3
        +y*(0.8449199096e-5+y*(-0.88228987e-6
        +y*0.105787412e-6)));
    ans=sqrt(0.636619772/x)*(sin(xx)*ans1+z*cos(xx)*ans2);
}
    return ans;
}
```

We now turn to the second task, namely how to use the recurrence formulas (6.5.6) and (6.5.7) to get the Bessel functions $J_n(x)$ and $Y_n(x)$ for $n \geq 2$. The latter of these is straightforward, since its upward recurrence is always stable:

```
#include "nr.h"

DP NR::bessy(const int n, const DP x)
Returns the Bessel function Yn(x) for positive x and n ≥ 2.
{
    int j;
    DP by,bym,byp,tox;

    if (n < 2) nrerror("Index n less than 2 in bessy");
    tox=2.0/x;
    by=bessy1(x);                   Starting values for the recurrence.
    bym=bessy0(x);
    for (j=1;j<n;j++) {             Recurrence (6.5.7).
        byp=j*tox*by-bym;
        bym=by;
        by=byp;
    }
    return by;
}
```

The cost of this algorithm is the call to bessy1 and bessy0 (which generate a call to each of bessj1 and bessj0), plus $O(n)$ operations in the recurrence.

As for $J_n(x)$, things are a bit more complicated. We can start the recurrence upward on n from J_0 and J_1, but it will remain stable only while n does not exceed x. This is, however, just fine for calls with large x and small n, a case which occurs frequently in practice.

The harder case to provide for is that with $x < n$. The best thing to do here is to use Miller's algorithm (see discussion preceding equation 5.5.16), applying the recurrence *downward* from some arbitrary starting value and making use of the upward-unstable nature of the recurrence to put us *onto* the correct solution. When we finally arrive at J_0 or J_1 we are able to normalize the solution with the sum (5.5.16) accumulated along the way.

The only subtlety is in deciding at how large an n we need start the downward recurrence so as to obtain a desired accuracy by the time we reach the n that we really want. If you play with the asymptotic forms (6.5.3) and (6.5.5), you should be able to convince yourself that the answer is to start larger than the desired n by an additive amount of order $[\text{constant} \times n]^{1/2}$, where the square root of the constant is, very roughly, the number of significant figures of accuracy.

The above considerations lead to the following function.

```cpp
#include <cmath>
#include <limits>
#include "nr.h"
using namespace std;

DP NR::bessj(const int n, const DP x)
Returns the Bessel function Jn(x) for any real x and n ≥ 2.
{
    const DP ACC=160.0;                          Make ACC larger to increase accuracy.
    const int IEXP=numeric_limits<DP>::max_exponent/2;
    bool jsum;
    int j,k,m;
    DP ax,bj,bjm,bjp,dum,sum,tox,ans;

    if (n < 2) nrerror("Index n less than 2 in bessj");
    ax=fabs(x);
    if (ax*ax <= 8.0*numeric_limits<DP>::min()) return 0.0;
    else if (ax > DP(n)) {                       Upwards recurrence from J0 and J1.
        tox=2.0/ax;
        bjm=bessj0(ax);
        bj=bessj1(ax);
        for (j=1;j<n;j++) {
            bjp=j*tox*bj-bjm;
            bjm=bj;
            bj=bjp;
        }
        ans=bj;
    } else {                                     Downwards recurrence from an even m here
        tox=2.0/ax;                                 computed.
        m=2*((n+int(sqrt(ACC*n)))/2);
        jsum=false;                              jsum will alternate between false and true;
        bjp=ans=sum=0.0;                            when it is true, we accumulate in sum
        bj=1.0;                                     the even terms in (5.5.16).
        for (j=m;j>0;j--) {                       The downward recurrence.
            bjm=j*tox*bj-bjp;
            bjp=bj;
            bj=bjm;
            dum=frexp(bj,&k);
            if (k > IEXP) {                       Renormalize to prevent overflows.
                bj=ldexp(bj,-IEXP);
                bjp=ldexp(bjp,-IEXP);
                ans=ldexp(ans,-IEXP);
                sum=ldexp(sum,-IEXP);
            }
            if (jsum) sum += bj;                  Accumulate the sum.
            jsum=!jsum;                           Change false to true or vice versa.
            if (j == n) ans=bjp;                  Save the unnormalized answer.
        }
        sum=2.0*sum-bj;                           Compute (5.5.16)
        ans /= sum;                               and use it to normalize the answer.
    }
    return x < 0.0 && (n & 1) ? -ans : ans;
}
```

CITED REFERENCES AND FURTHER READING:

Abramowitz, M., and Stegun, I.A. 1964, *Handbook of Mathematical Functions*, Applied Mathematics Series, Volume 55 (Washington: National Bureau of Standards; reprinted 1968 by Dover Publications, New York), Chapter 9.

Hart, J.F., et al. 1968, *Computer Approximations* (New York: Wiley), §6.8, p. 141. [1]

6.6 Modified Bessel Functions of Integer Order

The modified Bessel functions $I_n(x)$ and $K_n(x)$ are equivalent to the usual Bessel functions J_n and Y_n evaluated for purely imaginary arguments. In detail, the relationship is

$$I_n(x) = (-i)^n J_n(ix)$$
$$K_n(x) = \frac{\pi}{2} i^{n+1} [J_n(ix) + iY_n(ix)]$$

(6.6.1)

The particular choice of prefactor and of the linear combination of J_n and Y_n to form K_n are simply choices that make the functions real-valued for real arguments x.

For small arguments $x \ll n$, both $I_n(x)$ and $K_n(x)$ become, asymptotically, simple powers of their argument

$$I_n(x) \approx \frac{1}{n!} \left(\frac{x}{2}\right)^n \qquad n \geq 0$$

$$K_0(x) \approx -\ln(x)$$

(6.6.2)

$$K_n(x) \approx \frac{(n-1)!}{2} \left(\frac{x}{2}\right)^{-n} \qquad n > 0$$

These expressions are virtually identical to those for $J_n(x)$ and $Y_n(x)$ in this region, except for the factor of $-2/\pi$ difference between $Y_n(x)$ and $K_n(x)$. In the region $x \gg n$, however, the modified functions have quite different behavior than the Bessel functions,

$$I_n(x) \approx \frac{1}{\sqrt{2\pi x}} \exp(x)$$
$$K_n(x) \approx \frac{\pi}{\sqrt{2\pi x}} \exp(-x)$$

(6.6.3)

The modified functions evidently have exponential rather than sinusoidal behavior for large arguments (see Figure 6.6.1). The smoothness of the modified Bessel functions, once the exponential factor is removed, makes a simple polynomial approximation of a few terms quite suitable for the functions I_0, I_1, K_0, and K_1. The following routines, based on polynomial coefficients given by Abramowitz and Stegun [1], evaluate these four functions, and will provide the basis for upward recursion for $n > 1$ when $x > n$.

```
#include <cmath>
#include "nr.h"
using namespace std;

DP NR::bessi0(const DP x)
Returns the modified Bessel function I_0(x) for any real x.
{
    DP ax,ans,y;

    if ((ax=fabs(x)) < 3.75) {        Polynomial fit.
        y=x/3.75;
```

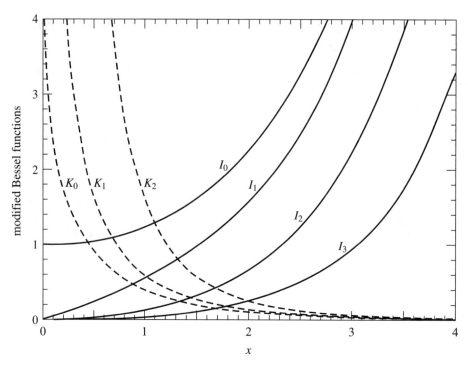

Figure 6.6.1.   Modified Bessel functions $I_0(x)$ through $I_3(x)$, $K_0(x)$ through $K_2(x)$.

```
        y*=y;
        ans=1.0+y*(3.5156229+y*(3.0899424+y*(1.2067492
            +y*(0.2659732+y*(0.360768e-1+y*0.45813e-2)))));
    } else {
        y=3.75/ax;
        ans=(exp(ax)/sqrt(ax))*(0.39894228+y*(0.1328592e-1
            +y*(0.225319e-2+y*(-0.157565e-2+y*(0.916281e-2
            +y*(-0.2057706e-1+y*(0.2635537e-1+y*(-0.1647633e-1
            +y*0.392377e-2))))))));
    }
    return ans;
}




#include <cmath>
#include "nr.h"
using namespace std;

DP NR::bessk0(const DP x)
Returns the modified Bessel function K_0(x) for positive real x.
{
    DP y,ans;

    if (x <= 2.0) {                      Polynomial fit.
        y=x*x/4.0;
        ans=(-log(x/2.0)*bessi0(x))+(-0.57721566+y*(0.42278420
            +y*(0.23069756+y*(0.3488590e-1+y*(0.262698e-2
            +y*(0.10750e-3+y*0.74e-5))))));
    } else {
        y=2.0/x;
        ans=(exp(-x)/sqrt(x))*(1.25331414+y*(-0.7832358e-1
            +y*(0.2189568e-1+y*(-0.1062446e-1+y*(0.587872e-2
```

```
                    +y*(-0.251540e-2+y*0.53208e-3))))));
    }
    return ans;
}
```

```cpp
#include <cmath>
#include "nr.h"
using namespace std;

DP NR::bessi1(const DP x)
Returns the modified Bessel function I₁(x) for any real x.
{
    DP ax,ans,y;

    if ((ax=fabs(x)) < 3.75) {            Polynomial fit.
        y=x/3.75;
        y*=y;
        ans=ax*(0.5+y*(0.87890594+y*(0.51498869+y*(0.15084934
            +y*(0.2658733e-1+y*(0.301532e-2+y*0.32411e-3))))));
    } else {
        y=3.75/ax;
        ans=0.2282967e-1+y*(-0.2895312e-1+y*(0.1787654e-1
            -y*0.420059e-2));
        ans=0.39894228+y*(-0.3988024e-1+y*(-0.362018e-2
            +y*(0.163801e-2+y*(-0.1031555e-1+y*ans))));
        ans *= (exp(ax)/sqrt(ax));
    }
    return x < 0.0 ? -ans : ans;
}
```

```cpp
#include <cmath>
#include "nr.h"
using namespace std;

DP NR::bessk1(const DP x)
Returns the modified Bessel function K₁(x) for positive real x.
{
    DP y,ans;

    if (x <= 2.0) {                   Polynomial fit.
        y=x*x/4.0;
        ans=(log(x/2.0)*bessi1(x))+(1.0/x)*(1.0+y*(0.15443144
            +y*(-0.67278579+y*(-0.18156897+y*(-0.1919402e-1
            +y*(-0.110404e-2+y*(-0.4686e-4)))))));
    } else {
        y=2.0/x;
        ans=(exp(-x)/sqrt(x))*(1.25331414+y*(0.23498619
            +y*(-0.3655620e-1+y*(0.1504268e-1+y*(-0.780353e-2
            +y*(0.325614e-2+y*(-0.68245e-3)))))));
    }
    return ans;
}
```

The recurrence relation for $I_n(x)$ and $K_n(x)$ is the same as that for $J_n(x)$ and $Y_n(x)$ provided that ix is substituted for x. This has the effect of changing a sign in the relation,

$$I_{n+1}(x) = -\left(\frac{2n}{x}\right) I_n(x) + I_{n-1}(x)$$

$$K_{n+1}(x) = +\left(\frac{2n}{x}\right) K_n(x) + K_{n-1}(x)$$

(6.6.4)

These relations are always *unstable* for upward recurrence. For K_n, itself growing, this presents no problem. For I_n, however, the strategy of downward recursion is therefore required once again, and the starting point for the recursion may be chosen in the same manner as for the routine bessj. The only fundamental difference is that the normalization formula for $I_n(x)$ has an alternating minus sign in successive terms, which again arises from the substitution of ix for x in the formula used previously for J_n

$$1 = I_0(x) - 2I_2(x) + 2I_4(x) - 2I_6(x) + \cdots$$

(6.6.5)

In fact, we prefer simply to normalize with a call to bessi0.

With this simple modification, the recursion routines bessj and bessy become the new routines bessi and bessk:

```
#include "nr.h"

DP NR::bessk(const int n, const DP x)
Returns the modified Bessel function Kn(x) for positive x and n ≥ 2.
{
    int j;
    DP bk,bkm,bkp,tox;

    if (n < 2) nrerror("Index n less than 2 in bessk");
    tox=2.0/x;
    bkm=bessk0(x);                        Upward recurrence for all x...
    bk=bessk1(x);
    for (j=1;j<n;j++) {                   ...and here it is.
        bkp=bkm+j*tox*bk;
        bkm=bk;
        bk=bkp;
    }
    return bk;
}
```

```
#include <cmath>
#include <limits>
#include "nr.h"
using namespace std;

DP NR::bessi(const int n, const DP x)
Returns the modified Bessel function In(x) for any real x and n ≥ 2.
{
    const DP ACC=200.0;                    Make ACC larger to increase accuracy.
    const int IEXP=numeric_limits<DP>::max_exponent/2;
    int j,k;
    DP bi,bim,bip,dum,tox,ans;

    if (n < 2) nrerror("Index n less than 2 in bessi");
```

```
    if (x*x <= 8.0*numeric_limits<DP>::min()) return 0.0;
    else {
        tox=2.0/fabs(x);
        bip=ans=0.0;
        bi=1.0;
        for (j=2*(n+int(sqrt(ACC*n)));j>0;j--) {        Downward recurrence from even
            bim=bip+j*tox*bi;                           m.
            bip=bi;
            bi=bim;
            dum=frexp(bi,&k);
            if (k > IEXP) {                             Renormalize to prevent overflows.
                ans=ldexp(ans,-IEXP);
                bi=ldexp(bi,-IEXP);
                bip=ldexp(bip,-IEXP);
            }
            if (j == n) ans=bip;
        }
        ans *= bessi0(x)/bi;                            Normalize with bessi0.
        return x < 0.0 && (n & 1) ? -ans : ans;
    }
}
```

CITED REFERENCES AND FURTHER READING:

Abramowitz, M., and Stegun, I.A. 1964, *Handbook of Mathematical Functions*, Applied Mathematics Series, Volume 55 (Washington: National Bureau of Standards; reprinted 1968 by Dover Publications, New York), §9.8. [1]

Carrier, G.F., Krook, M. and Pearson, C.E. 1966, *Functions of a Complex Variable* (New York: McGraw-Hill), pp. 220ff.

6.7 Bessel Functions of Fractional Order, Airy Functions, Spherical Bessel Functions

Many algorithms have been proposed for computing Bessel functions of fractional order numerically. Most of them are, in fact, not very good in practice. The routines given here are rather complicated, but they can be recommended wholeheartedly.

Ordinary Bessel Functions

The basic idea is *Steed's method*, which was originally developed [1] for Coulomb wave functions. The method calculates J_ν, J_ν', Y_ν, and Y_ν' simultaneously, and so involves four relations among these functions. Three of the relations come from two continued fractions, one of which is complex. The fourth is provided by the Wronskian relation

$$W \equiv J_\nu Y_\nu' - Y_\nu J_\nu' = \frac{2}{\pi x} \qquad (6.7.1)$$

The first continued fraction, CF1, is defined by

$$f_\nu \equiv \frac{J_\nu'}{J_\nu} = \frac{\nu}{x} - \frac{J_{\nu+1}}{J_\nu}$$

$$= \frac{\nu}{x} - \frac{1}{2(\nu+1)/x -} \; \frac{1}{2(\nu+2)/x -} \cdots \qquad (6.7.2)$$

You can easily derive it from the three-term recurrence relation for Bessel functions: Start with equation (6.5.6) and use equation (5.5.18). Forward evaluation of the continued fraction by one of the methods of §5.2 is essentially equivalent to backward recurrence of the recurrence relation. The rate of convergence of CF1 is determined by the position of the *turning point* $x_{\rm tp} = \sqrt{\nu(\nu+1)} \approx \nu$, beyond which the Bessel functions become oscillatory. If $x \lesssim x_{\rm tp}$, convergence is very rapid. If $x \gtrsim x_{\rm tp}$, then each iteration of the continued fraction effectively increases ν by one until $x \lesssim x_{\rm tp}$; thereafter rapid convergence sets in. Thus the number of iterations of CF1 is of order x for large x. In the routine bessjy we set the maximum allowed number of iterations to 10,000. For larger x, you can use the usual asymptotic expressions for Bessel functions.

One can show that the sign of J_ν is the same as the sign of the denominator of CF1 once it has converged.

The complex continued fraction CF2 is defined by

$$p + iq \equiv \frac{J_\nu' + iY_\nu'}{J_\nu + iY_\nu} = -\frac{1}{2x} + i + \frac{i}{x}\,\frac{(1/2)^2 - \nu^2}{2(x+i)+}\,\frac{(3/2)^2 - \nu^2}{2(x+2i)+}\cdots \qquad (6.7.3)$$

(We sketch the derivation of CF2 in the analogous case of modified Bessel functions in the next subsection.) This continued fraction converges rapidly for $x \gtrsim x_{\rm tp}$, while convergence fails as $x \to 0$. We have to adopt a special method for small x, which we describe below. For x not too small, we can ensure that $x \gtrsim x_{\rm tp}$ by a stable recurrence of J_ν and J_ν' downwards to a value $\nu = \mu \lesssim x$, thus yielding the ratio f_μ at this lower value of ν. This is the stable direction for the recurrence relation. The initial values for the recurrence are

$$J_\nu = \text{arbitrary}, \qquad J_\nu' = f_\nu J_\nu, \qquad (6.7.4)$$

with the sign of the arbitrary initial value of J_ν chosen to be the sign of the denominator of CF1. Choosing the initial value of J_ν very small minimizes the possibility of overflow during the recurrence. The recurrence relations are

$$\begin{aligned} J_{\nu-1} &= \frac{\nu}{x}J_\nu + J_\nu' \\ J_{\nu-1}' &= \frac{\nu-1}{x}J_{\nu-1} - J_\nu \end{aligned} \qquad (6.7.5)$$

Once CF2 has been evaluated at $\nu = \mu$, then with the Wronskian (6.7.1) we have enough relations to solve for all four quantities. The formulas are simplified by introducing the quantity

$$\gamma \equiv \frac{p - f_\mu}{q} \qquad (6.7.6)$$

Then

$$J_\mu = \pm\left(\frac{W}{q + \gamma(p - f_\mu)}\right)^{1/2} \qquad (6.7.7)$$

$$J_\mu' = f_\mu J_\mu \qquad (6.7.8)$$

$$Y_\mu = \gamma J_\mu \qquad (6.7.9)$$

$$Y_\mu' = Y_\mu\left(p + \frac{q}{\gamma}\right) \qquad (6.7.10)$$

The sign of J_μ in (6.7.7) is chosen to be the same as the sign of the initial J_ν in (6.7.4).

Once all four functions have been determined at the value $\nu = \mu$, we can find them at the original value of ν. For J_ν and J_ν', simply scale the values in (6.7.4) by the ratio of (6.7.7) to the value found after applying the recurrence (6.7.5). The quantities Y_ν and Y_ν' can be found by starting with the values in (6.7.9) and (6.7.10) and using the stable upwards recurrence

$$Y_{\nu+1} = \frac{2\nu}{x}Y_\nu - Y_{\nu-1} \qquad (6.7.11)$$

together with the relation

$$Y_\nu' = \frac{\nu}{x}Y_\nu - Y_{\nu+1} \qquad (6.7.12)$$

Now turn to the case of small x, when CF2 is not suitable. Temme [2] has given a good method of evaluating Y_ν and $Y_{\nu+1}$, and hence Y_ν' from (6.7.12), by series expansions that accurately handle the singularity as $x \to 0$. The expansions work only for $|\nu| \le 1/2$, and so now the recurrence (6.7.5) is used to evaluate f_ν at a value $\nu = \mu$ in this interval. Then one calculates J_μ from

$$J_\mu = \frac{W}{Y_\mu' - Y_\mu f_\mu} \tag{6.7.13}$$

and J_μ' from (6.7.8). The values at the original value of ν are determined by scaling as before, and the Y's are recurred up as before.

Temme's series are

$$Y_\nu = -\sum_{k=0}^{\infty} c_k g_k \qquad Y_{\nu+1} = -\frac{2}{x} \sum_{k=0}^{\infty} c_k h_k \tag{6.7.14}$$

Here

$$c_k = \frac{(-x^2/4)^k}{k!} \tag{6.7.15}$$

while the coefficients g_k and h_k are defined in terms of quantities p_k, q_k, and f_k that can be found by recursion:

$$g_k = f_k + \frac{2}{\nu} \sin^2\left(\frac{\nu\pi}{2}\right) q_k$$
$$h_k = -k g_k + p_k$$
$$p_k = \frac{p_{k-1}}{k - \nu} \tag{6.7.16}$$
$$q_k = \frac{q_{k-1}}{k + \nu}$$
$$f_k = \frac{k f_{k-1} + p_{k-1} + q_{k-1}}{k^2 - \nu^2}$$

The initial values for the recurrences are

$$p_0 = \frac{1}{\pi} \left(\frac{x}{2}\right)^{-\nu} \Gamma(1+\nu)$$
$$q_0 = \frac{1}{\pi} \left(\frac{x}{2}\right)^{\nu} \Gamma(1-\nu) \tag{6.7.17}$$
$$f_0 = \frac{2}{\pi} \frac{\nu\pi}{\sin\nu\pi} \left[\cosh\sigma\, \Gamma_1(\nu) + \frac{\sinh\sigma}{\sigma} \ln\left(\frac{2}{x}\right) \Gamma_2(\nu)\right]$$

with

$$\sigma = \nu \ln\left(\frac{2}{x}\right)$$
$$\Gamma_1(\nu) = \frac{1}{2\nu} \left[\frac{1}{\Gamma(1-\nu)} - \frac{1}{\Gamma(1+\nu)}\right] \tag{6.7.18}$$
$$\Gamma_2(\nu) = \frac{1}{2} \left[\frac{1}{\Gamma(1-\nu)} + \frac{1}{\Gamma(1+\nu)}\right]$$

The whole point of writing the formulas in this way is that the potential problems as $\nu \to 0$ can be controlled by evaluating $\nu\pi/\sin\nu\pi$, $\sinh\sigma/\sigma$, and Γ_1 carefully. In particular, Temme gives Chebyshev expansions for $\Gamma_1(\nu)$ and $\Gamma_2(\nu)$. We have rearranged his expansion for Γ_1 to be explicitly an even series in ν so that we can use our routine chebev as explained in §5.8.

The routine assumes $\nu \ge 0$. For negative ν you can use the reflection formulas

$$J_{-\nu} = \cos\nu\pi\, J_\nu - \sin\nu\pi\, Y_\nu$$
$$Y_{-\nu} = \sin\nu\pi\, J_\nu + \cos\nu\pi\, Y_\nu \tag{6.7.19}$$

The routine also assumes $x > 0$. For $x < 0$ the functions are in general complex, but expressible in terms of functions with $x > 0$. For $x = 0$, Y_ν is singular.

Internal arithmetic in the routine should be carried out in double precision, even if only single precision results are wanted. The complex arithmetic is carried out explicitly with real variables.

```
#include <cmath>
#include <limits>
#include "nr.h"
using namespace std;
```

void NR::bessjy(const DP x, const DP xnu, DP &rj, DP &ry, DP &rjp, DP &ryp)
Returns the Bessel functions $rj = J_\nu$, $ry = Y_\nu$ and their derivatives $rjp = J'_\nu$, $ryp = Y'_\nu$, for positive x and for $xnu = \nu \ge 0$. The relative accuracy is within one or two significant digits of EPS, except near a zero of one of the functions, where EPS controls its absolute accuracy. FPMIN is a number close to the machine's smallest floating-point number.
```
{
    const int MAXIT=10000;
    const DP EPS=numeric_limits<DP>::epsilon();
    const DP FPMIN=numeric_limits<DP>::min()/EPS;
    const DP XMIN=2.0, PI=3.141592653589793;
    DP a,b,br,bi,c,cr,ci,d,del,del1,den,di,dlr,dli,dr,e,f,fact,fact2,
        fact3,ff,gam,gam1,gam2,gammi,gampl,h,p,pimu,pimu2,q,r,rjl,
        rjl1,rjmu,rjp1,rjpl,rjtemp,ry1,rymu,rymup,rytemp,sum,sum1,
        temp,w,x2,xi,xi2,xmu,xmu2;
    int i,isign,l,nl;

    if (x <= 0.0 || xnu < 0.0)
        nrerror("bad arguments in bessjy");
    nl=(x < XMIN ? int(xnu+0.5) : MAX(0,int(xnu-x+1.5)));
```
    nl is the number of downward recurrences of the J's and upward recurrences of Y's. xmu lies between $-1/2$ and $1/2$ for x < XMIN, while it is chosen so that x is greater than the turning point for $x \ge$ XMIN.
```
    xmu=xnu-nl;
    xmu2=xmu*xmu;
    xi=1.0/x;
    xi2=2.0*xi;
    w=xi2/PI;                        The Wronskian.
    isign=1;                         Evaluate CF1 by modified Lentz's method (§5.2).
    h=xnu*xi;                            isign keeps track of sign changes in the de-
    if (h < FPMIN) h=FPMIN;              nominator.
    b=xi2*xnu;
    d=0.0;
    c=h;
    for (i=0;i<MAXIT;i++) {
        b += xi2;
        d=b-d;
        if (fabs(d) < FPMIN) d=FPMIN;
        c=b-1.0/c;
        if (fabs(c) < FPMIN) c=FPMIN;
        d=1.0/d;
        del=c*d;
        h=del*h;
        if (d < 0.0) isign = -isign;
        if (fabs(del-1.0) <= EPS) break;
    }
    if (i >= MAXIT)
        nrerror("x too large in bessjy; try asymptotic expansion");
    rjl=isign*FPMIN;                 Initialize $J_\nu$ and $J'_\nu$ for downward recurrence.
    rjpl=h*rjl;
    rjl1=rjl;                        Store values for later rescaling.
    rjp1=rjpl;
    fact=xnu*xi;
    for (l=nl-1;l>=0;l--) {
        rjtemp=fact*rjl+rjpl;
```

```
        fact -= xi;
        rjpl=fact*rjtemp-rjl;
        rjl=rjtemp;
    }
    if (rjl == 0.0) rjl=EPS;
    f=rjpl/rjl;                       Now have unnormalized Jμ and J'μ.
    if (x < XMIN) {                   Use series.
        x2=0.5*x;
        pimu=PI*xmu;
        fact = (fabs(pimu) < EPS ? 1.0 : pimu/sin(pimu));
        d = -log(x2);
        e=xmu*d;
        fact2 = (fabs(e) < EPS ? 1.0 : sinh(e)/e);
        beschb(xmu,gam1,gam2,gampl,gammi);      Chebyshev evaluation of Γ1 and Γ2.
        ff=2.0/PI*fact*(gam1*cosh(e)+gam2*fact2*d);      f0.
        e=exp(e);
        p=e/(gampl*PI);                   p0.
        q=1.0/(e*PI*gammi);               q0.
        pimu2=0.5*pimu;
        fact3 = (fabs(pimu2) < EPS ? 1.0 : sin(pimu2)/pimu2);
        r=PI*pimu2*fact3*fact3;
        c=1.0;
        d = -x2*x2;
        sum=ff+r*q;
        sum1=p;
        for (i=1;i<=MAXIT;i++) {
            ff=(i*ff+p+q)/(i*i-xmu2);
            c *= (d/i);
            p /= (i-xmu);
            q /= (i+xmu);
            del=c*(ff+r*q);
            sum += del;
            del1=c*p-i*del;
            sum1 += del1;
            if (fabs(del) < (1.0+fabs(sum))*EPS) break;
        }
        if (i > MAXIT)
            nrerror("bessy series failed to converge");
        rymu = -sum;
        ry1 = -sum1*xi2;
        rymup=xmu*xi*rymu-ry1;
        rjmu=w/(rymup-f*rymu);            Equation (6.7.13).
    } else {                             Evaluate CF2 by modified Lentz's method (§5.2).
        a=0.25-xmu2;
        p = -0.5*xi;
        q=1.0;
        br=2.0*x;
        bi=2.0;
        fact=a*xi/(p*p+q*q);
        cr=br+q*fact;
        ci=bi+p*fact;
        den=br*br+bi*bi;
        dr=br/den;
        di = -bi/den;
        dlr=cr*dr-ci*di;
        dli=cr*di+ci*dr;
        temp=p*dlr-q*dli;
        q=p*dli+q*dlr;
        p=temp;
        for (i=1;i<MAXIT;i++) {
            a += 2*i;
            bi += 2.0;
            dr=a*dr+br;
            di=a*di+bi;
```

```
                    if (fabs(dr)+fabs(di) < FPMIN) dr=FPMIN;
                    fact=a/(cr*cr+ci*ci);
                    cr=br+cr*fact;
                    ci=bi-ci*fact;
                    if (fabs(cr)+fabs(ci) < FPMIN) cr=FPMIN;
                    den=dr*dr+di*di;
                    dr /= den;
                    di /= -den;
                    dlr=cr*dr-ci*di;
                    dli=cr*di+ci*dr;
                    temp=p*dlr-q*dli;
                    q=p*dli+q*dlr;
                    p=temp;
                    if (fabs(dlr-1.0)+fabs(dli) <= EPS) break;
                }
                if (i >= MAXIT) nrerror("cf2 failed in bessjy");
                gam=(p-f)/q;                         Equations (6.7.6) – (6.7.10).
                rjmu=sqrt(w/((p-f)*gam+q));
                rjmu=SIGN(rjmu,rjl);
                rymu=rjmu*gam;
                rymup=rymu*(p+q/gam);
                ry1=xmu*xi*rymu-rymup;
            }
            fact=rjmu/rjl;
            rj=rjl1*fact;                            Scale original Jν and J'ν.
            rjp=rjp1*fact;
            for (i=1;i<=nl;i++) {                    Upward recurrence of Yν.
                rytemp=(xmu+i)*xi2*ry1-rymu;
                rymu=ry1;
                ry1=rytemp;
            }
            ry=rymu;
            ryp=xnu*xi*rymu-ry1;
        }
```

```
#include "nr.h"

void NR::beschb(const DP x, DP &gam1, DP &gam2, DP &gampl, DP &gammi)
```
Evaluates Γ_1 and Γ_2 by Chebyshev expansion for $|x| \leq 1/2$. Also returns $1/\Gamma(1 + x)$ and $1/\Gamma(1 - x)$.
```
{
    const int NUSE1=7, NUSE2=8;
    static const DP c1_d[7] = {
        -1.142022680371168e0,6.5165112670737e-3,
        3.087090173086e-4,-3.4706269649e-6,6.9437664e-9,
        3.67795e-11,-1.356e-13};
    static const DP c2_d[8] = {
        1.843740587300905e0,-7.68528408447867e-2,
        1.2719271366546e-3,-4.9717367042e-6,-3.31261198e-8,
        2.423096e-10,-1.702e-13,-1.49e-15};
    DP xx;
    static Vec_DP c1(c1_d,7),c2(c2_d,8);

    xx=8.0*x*x-1.0;                          Multiply x by 2 to make range be −1 to 1,
    gam1=chebev(-1.0,1.0,c1,NUSE1,xx);           and then apply transformation for eval-
    gam2=chebev(-1.0,1.0,c2,NUSE2,xx);           uating even Chebyshev series.
    gampl= gam2-x*gam1;
    gammi= gam2+x*gam1;
}
```

Modified Bessel Functions

Steed's method does not work for modified Bessel functions because in this case CF2 is purely imaginary and we have only three relations among the four functions. Temme [3] has given a normalization condition that provides the fourth relation.

The Wronskian relation is

$$W \equiv I_\nu K'_\nu - K_\nu I'_\nu = -\frac{1}{x} \qquad (6.7.20)$$

The continued fraction CF1 becomes

$$f_\nu \equiv \frac{I'_\nu}{I_\nu} = \frac{\nu}{x} + \frac{1}{2(\nu+1)/x +} \; \frac{1}{2(\nu+2)/x +} \cdots \qquad (6.7.21)$$

To get CF2 and the normalization condition in a convenient form, consider the sequence of confluent hypergeometric functions

$$z_n(x) = U(\nu + 1/2 + n, 2\nu + 1, 2x) \qquad (6.7.22)$$

for fixed ν. Then

$$K_\nu(x) = \pi^{1/2}(2x)^\nu e^{-x} z_0(x) \qquad (6.7.23)$$

$$\frac{K_{\nu+1}(x)}{K_\nu(x)} = \frac{1}{x} \left[\nu + \frac{1}{2} + x + \left(\nu^2 - \frac{1}{4} \right) \frac{z_1}{z_0} \right] \qquad (6.7.24)$$

Equation (6.7.23) is the standard expression for K_ν in terms of a confluent hypergeometric function, while equation (6.7.24) follows from relations between contiguous confluent hypergeometric functions (equations 13.4.16 and 13.4.18 in Abramowitz and Stegun). Now the functions z_n satisfy the three-term recurrence relation (equation 13.4.15 in Abramowitz and Stegun)

$$z_{n-1}(x) = b_n z_n(x) + a_{n+1} z_{n+1} \qquad (6.7.25)$$

with

$$b_n = 2(n+x)$$
$$a_{n+1} = -[(n+1/2)^2 - \nu^2] \qquad (6.7.26)$$

Following the steps leading to equation (5.5.18), we get the continued fraction CF2

$$\frac{z_1}{z_0} = \frac{1}{b_1 +} \; \frac{a_2}{b_2 +} \cdots \qquad (6.7.27)$$

from which (6.7.24) gives $K_{\nu+1}/K_\nu$ and thus K'_ν/K_ν.

Temme's normalization condition is that

$$\sum_{n=0}^{\infty} C_n z_n = \left(\frac{1}{2x} \right)^{\nu+1/2} \qquad (6.7.28)$$

where

$$C_n = \frac{(-1)^n}{n!} \frac{\Gamma(\nu + 1/2 + n)}{\Gamma(\nu + 1/2 - n)} \qquad (6.7.29)$$

Note that the C_n's can be determined by recursion:

$$C_0 = 1, \qquad C_{n+1} = -\frac{a_{n+1}}{n+1} C_n \qquad (6.7.30)$$

We use the condition (6.7.28) by finding

$$S = \sum_{n=1}^{\infty} C_n \frac{z_n}{z_0} \qquad (6.7.31)$$

Then

$$z_0 = \left(\frac{1}{2x} \right)^{\nu+1/2} \frac{1}{1+S} \qquad (6.7.32)$$

and (6.7.23) gives K_ν.

Thompson and Barnett [4] have given a clever method of doing the sum (6.7.31) simultaneously with the forward evaluation of the continued fraction CF2. Suppose the continued fraction is being evaluated as

$$\frac{z_1}{z_0} = \sum_{n=0}^{\infty} \Delta h_n \tag{6.7.33}$$

where the increments Δh_n are being found by, e.g., Steed's algorithm or the modified Lentz's algorithm of §5.2. Then the approximation to S keeping the first N terms can be found as

$$S_N = \sum_{n=1}^{N} Q_n \Delta h_n \tag{6.7.34}$$

Here

$$Q_n = \sum_{k=1}^{n} C_k q_k \tag{6.7.35}$$

and q_k is found by recursion from

$$q_{k+1} = (q_{k-1} - b_k q_k)/a_{k+1} \tag{6.7.36}$$

starting with $q_0 = 0$, $q_1 = 1$. For the case at hand, approximately three times as many terms are needed to get S to converge as are needed simply for CF2 to converge.

To find K_ν and $K_{\nu+1}$ for small x we use series analogous to (6.7.14):

$$K_\nu = \sum_{k=0}^{\infty} c_k f_k \qquad K_{\nu+1} = \frac{2}{x} \sum_{k=0}^{\infty} c_k h_k \tag{6.7.37}$$

Here

$$\begin{aligned}
c_k &= \frac{(x^2/4)^k}{k!} \\
h_k &= -k f_k + p_k \\
p_k &= \frac{p_{k-1}}{k - \nu} \\
q_k &= \frac{q_{k-1}}{k + \nu} \\
f_k &= \frac{k f_{k-1} + p_{k-1} + q_{k-1}}{k^2 - \nu^2}
\end{aligned} \tag{6.7.38}$$

The initial values for the recurrences are

$$\begin{aligned}
p_0 &= \frac{1}{2} \left(\frac{x}{2}\right)^{-\nu} \Gamma(1+\nu) \\
q_0 &= \frac{1}{2} \left(\frac{x}{2}\right)^{\nu} \Gamma(1-\nu) \\
f_0 &= \frac{\nu\pi}{\sin\nu\pi} \left[\cosh\sigma\,\Gamma_1(\nu) + \frac{\sinh\sigma}{\sigma} \ln\left(\frac{2}{x}\right) \Gamma_2(\nu)\right]
\end{aligned} \tag{6.7.39}$$

Both the series for small x, and CF2 and the normalization relation (6.7.28) require $|\nu| \le 1/2$. In both cases, therefore, we recurse I_ν down to a value $\nu = \mu$ in this interval, find K_μ there, and recurse K_ν back up to the original value of ν.

The routine assumes $\nu \ge 0$. For negative ν use the reflection formulas

$$\begin{aligned}
I_{-\nu} &= I_\nu + \frac{2}{\pi} \sin(\nu\pi)\, K_\nu \\
K_{-\nu} &= K_\nu
\end{aligned} \tag{6.7.40}$$

Note that for large x, $I_\nu \sim e^x$, $K_\nu \sim e^{-x}$, and so these functions will overflow or underflow. It is often desirable to be able to compute the scaled quantities $e^{-x}I_\nu$ and $e^x K_\nu$. Simply omitting the factor e^{-x} in equation (6.7.23) will ensure that all four quantities will have the appropriate scaling. If you also want to scale the four quantities for small x when the series in equation (6.7.37) are used, you must multiply each series by e^x.

```
#include <cmath>
#include <limits>
#include "nr.h"
using namespace std;
```

void NR::bessik(const DP x, const DP xnu, DP &ri, DP &rk, DP &rip, DP &rkp)
Returns the modified Bessel functions $ri = I_\nu$, $rk = K_\nu$ and their derivatives $rip = I'_\nu$,
$rkp = K'_\nu$, for positive x and for $xnu = \nu \geq 0$. The relative accuracy is within one or two
significant digits of EPS. FPMIN is a number close to the machine's smallest floating-point
number.

```
{
    const int MAXIT=10000;
    const DP EPS=numeric_limits<DP>::epsilon();
    const DP FPMIN=numeric_limits<DP>::min()/EPS;
    const DP XMIN=2.0, PI=3.141592653589793;
    DP a,a1,b,c,d,del,del1,delh,dels,e,f,fact,fact2,ff,gam1,gam2,
        gammi,gampl,h,p,pimu,q,q1,q2,qnew,ril,ril1,rimu,rip1,ripl,
        ritemp,rk1,rkmu,rkmup,rktemp,s,sum,sum1,x2,xi,xi2,xmu,xmu2;
    int i,l,nl;

    if (x <= 0.0 || xnu < 0.0) nrerror("bad arguments in bessik");
    nl=int(xnu+0.5);                          nl is the number of downward re-
    xmu=xnu-nl;                               currences of the I's and upward
    xmu2=xmu*xmu;                             recurrences of K's. xmu lies be-
    xi=1.0/x;                                 tween −1/2 and 1/2.
    xi2=2.0*xi;
    h=xnu*xi;                                 Evaluate CF1 by modified Lentz's
    if (h < FPMIN) h=FPMIN;                   method (§5.2).
    b=xi2*xnu;
    d=0.0;
    c=h;
    for (i=0;i<MAXIT;i++) {
        b += xi2;
        d=1.0/(b+d);                          Denominators cannot be zero here,
        c=b+1.0/c;                            so no need for special precau-
        del=c*d;                              tions.
        h=del*h;
        if (fabs(del-1.0) <= EPS) break;
    }
    if (i >= MAXIT)
        nrerror("x too large in bessik; try asymptotic expansion");
    ril=FPMIN;                                Initialize $I_\nu$ and $I'_\nu$ for downward re-
    ripl=h*ril;                               currence.
    ril1=ril;                                 Store values for later rescaling.
    rip1=ripl;
    fact=xnu*xi;
    for (l=nl-1;l >= 0;l--) {
        ritemp=fact*ril+ripl;
        fact -= xi;
        ripl=fact*ritemp+ril;
        ril=ritemp;
    }
    f=ripl/ril;                               Now have unnormalized $I_\mu$ and $I'_\mu$.
    if (x < XMIN) {                           Use series.
        x2=0.5*x;
        pimu=PI*xmu;
        fact = (fabs(pimu) < EPS ? 1.0 : pimu/sin(pimu));
        d = -log(x2);
        e=xmu*d;
        fact2 = (fabs(e) < EPS ? 1.0 : sinh(e)/e);
        beschb(xmu,gam1,gam2,gampl,gammi);    Chebyshev evaluation of $\Gamma_1$ and $\Gamma_2$.
        ff=fact*(gam1*cosh(e)+gam2*fact2*d);  $f_0$.
        sum=ff;
        e=exp(e);
```

```
        p=0.5*e/gamp1;                                  p0.
        q=0.5/(e*gammi);                                q0.
        c=1.0;
        d=x2*x2;
        sum1=p;
        for (i=1;i<=MAXIT;i++) {
            ff=(i*ff+p+q)/(i*i-xmu2);
            c *= (d/i);
            p /= (i-xmu);
            q /= (i+xmu);
            del=c*ff;
            sum += del;
            del1=c*(p-i*ff);
            sum1 += del1;
            if (fabs(del) < fabs(sum)*EPS) break;
        }
        if (i > MAXIT) nrerror("bessk series failed to converge");
        rkmu=sum;
        rk1=sum1*xi2;
    } else {                                     Evaluate CF2 by Steed's algorithm
        b=2.0*(1.0+x);                              (§5.2), which is OK because there
        d=1.0/b;                                    can be no zero denominators.
        h=delh=d;
        q1=0.0;                                  Initializations for recurrence (6.7.35).
        q2=1.0;
        a1=0.25-xmu2;
        q=c=a1;                                  First term in equation (6.7.34).
        a = -a1;
        s=1.0+q*delh;
        for (i=1;i<MAXIT;i++) {
            a -= 2*i;
            c = -a*c/(i+1.0);
            qnew=(q1-b*q2)/a;
            q1=q2;
            q2=qnew;
            q += c*qnew;
            b += 2.0;
            d=1.0/(b+a*d);
            delh=(b*d-1.0)*delh;
            h += delh;
            dels=q*delh;
            s += dels;
            if (fabs(dels/s) <= EPS) break;
            Need only test convergence of sum since CF2 itself converges more quickly.
        }
        if (i >= MAXIT) nrerror("bessik: failure to converge in cf2");
        h=a1*h;
        rkmu=sqrt(PI/(2.0*x))*exp(-x)/s;            Omit the factor $\exp(-x)$ to scale
        rk1=rkmu*(xmu+x+0.5-h)*xi;                     all the returned functions by $\exp(x)$
    }                                                  for $x \geq$ XMIN.
    rkmup=xmu*xi*rkmu-rk1;
    rimu=xi/(f*rkmu-rkmup);                       Get $I_\mu$ from Wronskian.
    ri=(rimu*ril1)/ril;                           Scale original $I_\nu$ and $I_\nu'$.
    rip=(rimu*rip1)/ril;
    for (i=1;i <= nl;i++) {                       Upward recurrence of $K_\nu$.
        rktemp=(xmu+i)*xi2*rk1+rkmu;
        rkmu=rk1;
        rk1=rktemp;
    }
    rk=rkmu;
    rkp=xnu*xi*rkmu-rk1;
}
```

Airy Functions

For positive x, the Airy functions are defined by

$$\mathrm{Ai}(x) = \frac{1}{\pi}\sqrt{\frac{x}{3}}K_{1/3}(z) \tag{6.7.41}$$

$$\mathrm{Bi}(x) = \sqrt{\frac{x}{3}}[I_{1/3}(z) + I_{-1/3}(z)] \tag{6.7.42}$$

where

$$z = \frac{2}{3}x^{3/2} \tag{6.7.43}$$

By using the reflection formula (6.7.40), we can convert (6.7.42) into the computationally more useful form

$$\mathrm{Bi}(x) = \sqrt{x}\left[\frac{2}{\sqrt{3}}I_{1/3}(z) + \frac{1}{\pi}K_{1/3}(z)\right] \tag{6.7.44}$$

so that Ai and Bi can be evaluated with a single call to `bessik`.

The derivatives should not be evaluated by simply differentiating the above expressions because of possible subtraction errors near $x = 0$. Instead, use the equivalent expressions

$$\mathrm{Ai}'(x) = -\frac{x}{\pi\sqrt{3}}K_{2/3}(z)$$
$$\mathrm{Bi}'(x) = x\left[\frac{2}{\sqrt{3}}I_{2/3}(z) + \frac{1}{\pi}K_{2/3}(z)\right] \tag{6.7.45}$$

The corresponding formulas for negative arguments are

$$\mathrm{Ai}(-x) = \frac{\sqrt{x}}{2}\left[J_{1/3}(z) - \frac{1}{\sqrt{3}}Y_{1/3}(z)\right]$$
$$\mathrm{Bi}(-x) = -\frac{\sqrt{x}}{2}\left[\frac{1}{\sqrt{3}}J_{1/3}(z) + Y_{1/3}(z)\right]$$
$$\mathrm{Ai}'(-x) = \frac{x}{2}\left[J_{2/3}(z) + \frac{1}{\sqrt{3}}Y_{2/3}(z)\right] \tag{6.7.46}$$
$$\mathrm{Bi}'(-x) = \frac{x}{2}\left[\frac{1}{\sqrt{3}}J_{2/3}(z) - Y_{2/3}(z)\right]$$

```
#include <cmath>
#include "nr.h"
using namespace std;

void NR::airy(const DP x, DP &ai, DP &bi, DP &aip, DP &bip)
Returns Airy functions Ai(x), Bi(x), and their derivatives Ai'(x), Bi'(x).
{
    const DP PI=3.141592653589793238, ONOVRT=0.577350269189626;
    const DP THIRD=(1.0/3.0), TWOTHR=2.0*THIRD;
    DP absx,ri,rip,rj,rjp,rk,rkp,rootx,ry,ryp,z;

    absx=fabs(x);
    rootx=sqrt(absx);
    z=TWOTHR*absx*rootx;
    if (x > 0.0) {
        bessik(z,THIRD,ri,rk,rip,rkp);
        ai=rootx*ONOVRT*rk/PI;
        bi=rootx*(rk/PI+2.0*ONOVRT*ri);
        bessik(z,TWOTHR,ri,rk,rip,rkp);
        aip = -x*ONOVRT*rk/PI;
        bip=x*(rk/PI+2.0*ONOVRT*ri);
```

```
    } else if (x < 0.0) {
        bessjy(z,THIRD,rj,ry,rjp,ryp);
        ai=0.5*rootx*(rj-ONOVRT*ry);
        bi = -0.5*rootx*(ry+ONOVRT*rj);
        bessjy(z,TWOTHR,rj,ry,rjp,ryp);
        aip=0.5*absx*(ONOVRT*ry+rj);
        bip=0.5*absx*(ONOVRT*rj-ry);
    } else {                        Case x = 0.
        ai=0.355028053887817;
        bi=ai/ONOVRT;
        aip = -0.258819403792807;
        bip = -aip/ONOVRT;
    }
}
```

Spherical Bessel Functions

For integer n, spherical Bessel functions are defined by

$$j_n(x) = \sqrt{\frac{\pi}{2x}} J_{n+(1/2)}(x)$$

$$y_n(x) = \sqrt{\frac{\pi}{2x}} Y_{n+(1/2)}(x)$$

(6.7.47)

They can be evaluated by a call to bessjy, and the derivatives can safely be found from the derivatives of equation (6.7.47).

Note that in the continued fraction CF2 in (6.7.3) just the first term survives for $\nu = 1/2$. Thus one can make a very simple algorithm for spherical Bessel functions along the lines of bessjy by always recursing j_n down to $n = 0$, setting p and q from the first term in CF2, and then recursing y_n up. No special series is required near $x = 0$. However, bessjy is already so efficient that we have not bothered to provide an independent routine for spherical Bessels.

```
#include <cmath>
#include "nr.h"
using namespace std;

void NR::sphbes(const int n, const DP x, DP &sj, DP &sy, DP &sjp, DP &syp)
Returns spherical Bessel functions jₙ(x), yₙ(x), and their derivatives jₙ'(x), yₙ'(x) for integer n.
{
    const DP RTPIO2=1.253314137315500251;
    DP factor,order,rj,rjp,ry,ryp;

    if (n < 0 || x <= 0.0) nrerror("bad arguments in sphbes");
    order=n+0.5;
    bessjy(x,order,rj,ry,rjp,ryp);
    factor=RTPIO2/sqrt(x);
    sj=factor*rj;
    sy=factor*ry;
    sjp=factor*rjp-sj/(2.0*x);
    syp=factor*ryp-sy/(2.0*x);
}
```

CITED REFERENCES AND FURTHER READING:

Barnett, A.R., Feng, D.H., Steed, J.W., and Goldfarb, L.J.B. 1974, *Computer Physics Communications*, vol. 8, pp. 377–395. [1]

Temme, N.M. 1976, *Journal of Computational Physics*, vol. 21, pp. 343–350 [2]; 1975, *op. cit.*, vol. 19, pp. 324–337. [3]

Thompson, I.J., and Barnett, A.R. 1987, *Computer Physics Communications*, vol. 47, pp. 245–257. [4]

Barnett, A.R. 1981, *Computer Physics Communications*, vol. 21, pp. 297–314.

Thompson, I.J., and Barnett, A.R. 1986, *Journal of Computational Physics*, vol. 64, pp. 490–509.

Abramowitz, M., and Stegun, I.A. 1964, *Handbook of Mathematical Functions*, Applied Mathematics Series, Volume 55 (Washington: National Bureau of Standards; reprinted 1968 by Dover Publications, New York), Chapter 10.

6.8 Spherical Harmonics

Spherical harmonics occur in a large variety of physical problems, for example, whenever a wave equation, or Laplace's equation, is solved by separation of variables in spherical coordinates. The spherical harmonic $Y_{lm}(\theta, \phi)$, $-l \leq m \leq l$, is a function of the two coordinates θ, ϕ on the surface of a sphere.

The spherical harmonics are orthogonal for different l and m, and they are normalized so that their integrated square over the sphere is unity:

$$\int_0^{2\pi} d\phi \int_{-1}^1 d(\cos \theta) Y_{l'm'}{}^*(\theta, \phi) Y_{lm}(\theta, \phi) = \delta_{l'l}\delta_{m'm} \qquad (6.8.1)$$

Here asterisk denotes complex conjugation.

Mathematically, the spherical harmonics are related to *associated Legendre polynomials* by the equation

$$Y_{lm}(\theta, \phi) = \sqrt{\frac{2l+1}{4\pi}\frac{(l-m)!}{(l+m)!}} P_l^m(\cos \theta) e^{im\phi} \qquad (6.8.2)$$

By using the relation

$$Y_{l,-m}(\theta, \phi) = (-1)^m Y_{lm}{}^*(\theta, \phi) \qquad (6.8.3)$$

we can always relate a spherical harmonic to an associated Legendre polynomial with $m \geq 0$. With $x \equiv \cos \theta$, these are defined in terms of the ordinary Legendre polynomials (cf. §4.5 and §5.5) by

$$P_l^m(x) = (-1)^m (1-x^2)^{m/2} \frac{d^m}{dx^m} P_l(x) \qquad (6.8.4)$$

The first few associated Legendre polynomials, and their corresponding normalized spherical harmonics, are

$P_0^0(x) = \quad 1$	$Y_{00} = \quad \sqrt{\frac{1}{4\pi}}$
$P_1^1(x) = \quad -(1-x^2)^{1/2}$	$Y_{11} = -\sqrt{\frac{3}{8\pi}} \sin \theta e^{i\phi}$
$P_1^0(x) = \quad x$	$Y_{10} = \quad \sqrt{\frac{3}{4\pi}} \cos \theta$
$P_2^2(x) = \quad 3(1-x^2)$	$Y_{22} = \frac{1}{4}\sqrt{\frac{15}{2\pi}} \sin^2 \theta e^{2i\phi}$
$P_2^1(x) = -3(1-x^2)^{1/2}x$	$Y_{21} = -\sqrt{\frac{15}{8\pi}} \sin \theta \cos \theta e^{i\phi}$
$P_2^0(x) = \quad \frac{1}{2}(3x^2 - 1)$	$Y_{20} = \quad \sqrt{\frac{5}{4\pi}}\left(\frac{3}{2}\cos^2 \theta - \frac{1}{2}\right)$

$$(6.8.5)$$

There are many bad ways to evaluate associated Legendre polynomials numerically. For example, there are explicit expressions, such as

$$P_l^m(x) = \frac{(-1)^m(l+m)!}{2^m m!(l-m)!}(1-x^2)^{m/2}\left[1 - \frac{(l-m)(m+l+1)}{1!(m+1)}\left(\frac{1-x}{2}\right)\right.$$
$$\left. + \frac{(l-m)(l-m-1)(m+l+1)(m+l+2)}{2!(m+1)(m+2)}\left(\frac{1-x}{2}\right)^2 - \cdots\right]$$

(6.8.6)

where the polynomial continues up through the term in $(1-x)^{l-m}$. (See [1] for this and related formulas.) This is not a satisfactory method because evaluation of the polynomial involves delicate cancellations between successive terms, which alternate in sign. For large l, the individual terms in the polynomial become very much larger than their sum, and all accuracy is lost.

In practice, (6.8.6) can be used only in single precision (32-bit) for l up to 6 or 8, and in double precision (64-bit) for l up to 15 or 18, depending on the precision required for the answer. A more robust computational procedure is therefore desirable, as follows:

The associated Legendre functions satisfy numerous recurrence relations, tabulated in [1-2]. These are recurrences on l alone, on m alone, and on both l and m simultaneously. Most of the recurrences involving m are unstable, and so dangerous for numerical work. The following recurrence on l is, however, stable (compare 5.5.1):

$$(l-m)P_l^m = x(2l-1)P_{l-1}^m - (l+m-1)P_{l-2}^m \qquad (6.8.7)$$

It is useful because there is a closed-form expression for the starting value,

$$P_m^m = (-1)^m(2m-1)!!(1-x^2)^{m/2} \qquad (6.8.8)$$

(The notation $n!!$ denotes the product of all *odd* integers less than or equal to n.) Using (6.8.7) with $l = m+1$, and setting $P_{m-1}^m = 0$, we find

$$P_{m+1}^m = x(2m+1)P_m^m \qquad (6.8.9)$$

Equations (6.8.8) and (6.8.9) provide the two starting values required for (6.8.7) for general l.

The function that implements this is

```
#include <cmath>
#include "nr.h"
using namespace std;

DP NR::plgndr(const int l, const int m, const DP x)
```
Computes the associated Legendre polynomial $P_l^m(x)$. Here m and l are integers satisfying $0 \le m \le l$, while x lies in the range $-1 \le x \le 1$.
```
{
    int i,ll;
    DP fact,pll,pmm,pmmp1,somx2;

    if (m < 0 || m > l || fabs(x) > 1.0)
```

```
        nrerror("Bad arguments in routine plgndr");
    pmm=1.0;                        Compute Pᵐₘ.
    if (m > 0) {
        somx2=sqrt((1.0-x)*(1.0+x));
        fact=1.0;
        for (i=1;i<=m;i++) {
            pmm *= -fact*somx2;
            fact += 2.0;
        }
    }
    if (l == m)
        return pmm;
    else {                          Compute Pᵐₘ₊₁.
        pmmp1=x*(2*m+1)*pmm;
        if (l == (m+1))
            return pmmp1;
        else {                      Compute Pₗᵐ, l > m + 1.
            for (ll=m+2;ll<=l;ll++) {
                pll=(x*(2*ll-1)*pmmp1-(ll+m-1)*pmm)/(ll-m);
                pmm=pmmp1;
                pmmp1=pll;
            }
            return pll;
        }
    }
}
```

CITED REFERENCES AND FURTHER READING:

Magnus, W., and Oberhettinger, F. 1949, *Formulas and Theorems for the Functions of Mathematical Physics* (New York: Chelsea), pp. 54ff. [1]

Abramowitz, M., and Stegun, I.A. 1964, *Handbook of Mathematical Functions*, Applied Mathematics Series, Volume 55 (Washington: National Bureau of Standards; reprinted 1968 by Dover Publications, New York), Chapter 8. [2]

6.9 Fresnel Integrals, Cosine and Sine Integrals

Fresnel Integrals

The two Fresnel integrals are defined by

$$C(x) = \int_0^x \cos\left(\frac{\pi}{2}t^2\right) dt, \qquad S(x) = \int_0^x \sin\left(\frac{\pi}{2}t^2\right) dt \qquad (6.9.1)$$

The most convenient way of evaluating these functions to arbitrary precision is to use power series for small x and a continued fraction for large x. The series are

$$C(x) = x - \left(\frac{\pi}{2}\right)^2 \frac{x^5}{5 \cdot 2!} + \left(\frac{\pi}{2}\right)^4 \frac{x^9}{9 \cdot 4!} - \cdots$$

$$S(x) = \left(\frac{\pi}{2}\right) \frac{x^3}{3 \cdot 1!} - \left(\frac{\pi}{2}\right)^3 \frac{x^7}{7 \cdot 3!} + \left(\frac{\pi}{2}\right)^5 \frac{x^{11}}{11 \cdot 5!} - \cdots \qquad (6.9.2)$$

There is a complex continued fraction that yields both $S(x)$ and $C(x)$ simultaneously:

$$C(x) + iS(x) = \frac{1+i}{2} \operatorname{erf} z, \qquad z = \frac{\sqrt{\pi}}{2}(1-i)x \qquad (6.9.3)$$

where

$$
\begin{aligned}
e^{z^2} \operatorname{erfc} z &= \frac{1}{\sqrt{\pi}} \left(\frac{1}{z+} \frac{1/2}{z+} \frac{1}{z+} \frac{3/2}{z+} \frac{2}{z+} \cdots \right) \\
&= \frac{2z}{\sqrt{\pi}} \left(\frac{1}{2z^2+1-} \frac{1\cdot 2}{2z^2+5-} \frac{3\cdot 4}{2z^2+9-} \cdots \right)
\end{aligned}
\qquad (6.9.4)
$$

In the last line we have converted the "standard" form of the continued fraction to its "even" form (see §5.2), which converges twice as fast. We must be careful not to evaluate the alternating series (6.9.2) at too large a value of x; inspection of the terms shows that $x = 1.5$ is a good point to switch over to the continued fraction.

Note that for large x

$$C(x) \sim \frac{1}{2} + \frac{1}{\pi x} \sin\left(\frac{\pi}{2}x^2\right), \qquad S(x) \sim \frac{1}{2} - \frac{1}{\pi x} \cos\left(\frac{\pi}{2}x^2\right) \qquad (6.9.5)$$

Thus the precision of the routine `frenel` may be limited by the precision of the library routines for sine and cosine for large x.

```
#include <cmath>
#include <complex>
#include <limits>
#include "nr.h"
using namespace std;

void NR::frenel(const DP x, complex<DP> &cs)
```
Computes the Fresnel integrals $S(x)$ and $C(x)$ for all real x. $C(x)$ is returned as the real part of cs and $S(x)$ as the imaginary part.
```
{
    const int MAXIT=100;
    const DP EPS=numeric_limits<DP>::epsilon();
    const DP FPMIN=numeric_limits<DP>::min();
    const DP BIG=numeric_limits<DP>::max()*EPS;
    const DP PI=3.141592653589793238, PIBY2=(PI/2.0), XMIN=1.5;
```
Here MAXIT is the maximum number of iterations allowed; EPS is the relative error; FPMIN is a number near the smallest representable floating-point number; BIG is a number near the machine overflow limit; XMIN is the dividing line between using the series and continued fraction.
```
    bool odd;
    int k,n;
    DP a,ax,fact,pix2,sign,sum,sumc,sums,term,test;
    complex<DP> b,cc,d,h,del;

    ax=fabs(x);
    if (ax < sqrt(FPMIN)) {              Special case: avoid failure of convergence
        cs=ax;                          test because of underflow.
    } else if (ax <= XMIN) {            Evaluate both series simultaneously.
        sum=sums=0.0;
        sumc=ax;
        sign=1.0;
```

```
    fact=PIBY2*ax*ax;
    odd=true;
    term=ax;
    n=3;
    for (k=1;k<=MAXIT;k++) {
        term *= fact/k;
        sum += sign*term/n;
        test=fabs(sum)*EPS;
        if (odd) {
            sign = -sign;
            sums=sum;
            sum=sumc;
        } else {
            sumc=sum;
            sum=sums;
        }
        if (term < test) break;
        odd=!odd;
        n += 2;
    }
    if (k > MAXIT) nrerror("series failed in frenel");
    cs=complex<DP>(sumc,sums);
} else {                                 Evaluate continued fraction by modified
    pix2=PI*ax*ax;                         Lentz's method (§5.2).
    b=complex<DP>(1.0,-pix2);
    cc=BIG;
    d=h=1.0/b;
    n = -1;
    for (k=2;k<=MAXIT;k++) {
        n += 2;
        a = -n*(n+1);
        b += 4.0;
        d=1.0/(a*d+b);                   Denominators cannot be zero.
        cc=b+a/cc;
        del=cc*d;
        h *= del;
        if (fabs(real(del)-1.0)+fabs(imag(del)) <= EPS) break;
    }
    if (k > MAXIT) nrerror("cf failed in frenel");
    h *= complex<DP>(ax,-ax);
    cs=complex<DP>(0.5,0.5)
        *(1.0-complex<DP>(cos(0.5*pix2),sin(0.5*pix2))*h);
}
if (x < 0.0) {                           Use antisymmetry.
    cs = -cs;
}
return;
}
```

Cosine and Sine Integrals

The cosine and sine integrals are defined by

$$\mathrm{Ci}(x) = \gamma + \ln x + \int_0^x \frac{\cos t - 1}{t}\, dt$$

$$\mathrm{Si}(x) = \int_0^x \frac{\sin t}{t}\, dt$$

(6.9.6)

Here $\gamma \approx 0.5772\dots$ is Euler's constant. We only need a way to calculate the functions for $x > 0$, because

$$\mathrm{Si}(-x) = -\,\mathrm{Si}(x), \qquad \mathrm{Ci}(-x) = \mathrm{Ci}(x) - i\pi \tag{6.9.7}$$

Once again we can evaluate these functions by a judicious combination of power series and complex continued fraction. The series are

$$\mathrm{Si}(x) = x - \frac{x^3}{3\cdot 3!} + \frac{x^5}{5\cdot 5!} - \cdots$$
$$\mathrm{Ci}(x) = \gamma + \ln x + \left(-\frac{x^2}{2\cdot 2!} + \frac{x^4}{4\cdot 4!} - \cdots\right) \tag{6.9.8}$$

The continued fraction for the exponential integral $E_1(ix)$ is

$$E_1(ix) = -\,\mathrm{Ci}(x) + i[\mathrm{Si}(x) - \pi/2]$$
$$= e^{-ix}\left(\frac{1}{ix+}\,\frac{1}{1+}\,\frac{1}{ix+}\,\frac{2}{1+}\,\frac{2}{ix+}\cdots\right) \tag{6.9.9}$$
$$= e^{-ix}\left(\frac{1}{1+ix-}\,\frac{1^2}{3+ix-}\,\frac{2^2}{5+ix-}\cdots\right)$$

The "even" form of the continued fraction is given in the last line and converges twice as fast for about the same amount of computation. A good crossover point from the alternating series to the continued fraction is $x = 2$ in this case. As for the Fresnel integrals, for large x the precision may be limited by the precision of the sine and cosine routines.

```
#include <cmath>
#include <complex>
#include <limits>
#include "nr.h"
using namespace std;

void NR::cisi(const DP x, complex<DP> &cs)
```
Computes the cosine and sine integrals $\mathrm{Ci}(x)$ and $\mathrm{Si}(x)$. The function $\mathrm{Ci}(x)$ is returned as the real part of cs, and $\mathrm{Si}(x)$ as the imaginary part. $\mathrm{Ci}(0)$ is returned as a large negative number and no error message is generated. For $x < 0$ the routine returns $\mathrm{Ci}(-x)$ and you must supply the $-i\pi$ yourself.
```
{
    const int MAXIT=100;                    Maximum number of iterations allowed.
    const DP EULER=0.577215664901533, PIBY2=1.570796326794897, TMIN=2.0;
    const DP EPS=numeric_limits<DP>::epsilon();
    const DP FPMIN=numeric_limits<DP>::min()*4.0;
    const DP BIG=numeric_limits<DP>::max()*EPS;
```
Here EULER is Euler's constant γ; PIBY2 is $\pi/2$; TMIN is the dividing line between using the series and continued fraction; EPS is the relative error, or absolute error near a zero of $\mathrm{Ci}(x)$; FPMIN is a number close to the smallest representable floating-point number; BIG is a number near the machine overflow limit.
```
    int i,k;
    bool odd;
    DP a,err,fact,sign,sum,sumc,sums,t,term;
    complex<DP> h,b,c,d,del;

    t=fabs(x);
    if (t == 0.0) {                         Special case.
```

```
        cs= -BIG;
        return;
    }
    if (t > TMIN) {                          Evaluate continued fraction by modified
        b=complex<DP>(1.0,t);                    Lentz's method (§5.2).
        c=complex<DP>(BIG,0.0);
        d=h=1.0/b;
        for (i=1;i<MAXIT;i++) {
            a= -i*i;
            b += 2.0;
            d=1.0/(a*d+b);                   Denominators cannot be zero.
            c=b+a/c;
            del=c*d;
            h *= del;
            if (fabs(real(del)-1.0)+fabs(imag(del)) <= EPS) break;
        }
        if (i >= MAXIT) nrerror("cf failed in cisi");
        h=complex<DP>(cos(t),-sin(t))*h;
        cs= -conj(h)+complex<DP>(0.0,PIBY2);
    } else {                                 Evaluate both series simultaneously.
        if (t < sqrt(FPMIN)) {              Special case: avoid failure of convergence
            sumc=0.0;                           test because of underflow.
            sums=t;
        } else {
            sum=sums=sumc=0.0;
            sign=fact=1.0;
            odd=true;
            for (k=1;k<=MAXIT;k++) {
                fact *= t/k;
                term=fact/k;
                sum += sign*term;
                err=term/fabs(sum);
                if (odd) {
                    sign = -sign;
                    sums=sum;
                    sum=sumc;
                } else {
                    sumc=sum;
                    sum=sums;
                }
                if (err < EPS) break;
                odd=!odd;
            }
            if (k > MAXIT) nrerror("maxits exceeded in cisi");
        }
        cs=complex<DP>(sumc+log(t)+EULER,sums);
    }
    if (x < 0.0) cs = conj(cs);
}
```

CITED REFERENCES AND FURTHER READING:

Stegun, I.A., and Zucker, R. 1976, *Journal of Research of the National Bureau of Standards*, vol. 80B, pp. 291–311; 1981, *op. cit.*, vol. 86, pp. 661–686.

Abramowitz, M., and Stegun, I.A. 1964, *Handbook of Mathematical Functions*, Applied Mathematics Series, Volume 55 (Washington: National Bureau of Standards; reprinted 1968 by Dover Publications, New York), Chapters 5 and 7.

6.10 Dawson's Integral

Dawson's Integral $F(x)$ is defined by

$$F(x) = e^{-x^2} \int_0^x e^{t^2}\, dt \qquad\qquad (6.10.1)$$

The function can also be related to the complex error function by

$$F(z) = \frac{i\sqrt{\pi}}{2} e^{-z^2} \left[1 - \mathrm{erfc}(-iz) \right]. \qquad\qquad (6.10.2)$$

A remarkable approximation for $F(z)$, due to Rybicki [1], is

$$F(z) = \lim_{h \to 0} \frac{1}{\sqrt{\pi}} \sum_{n\ \mathrm{odd}} \frac{e^{-(z-nh)^2}}{n} \qquad\qquad (6.10.3)$$

What makes equation (6.10.3) unusual is that its accuracy increases *exponentially* as h gets small, so that quite moderate values of h (and correspondingly quite rapid convergence of the series) give very accurate approximations.

We will discuss the theory that leads to equation (6.10.3) later, in §13.11, as an interesting application of Fourier methods. Here we simply implement a routine for real values of x based on the formula.

It is first convenient to shift the summation index to center it approximately on the maximum of the exponential term. Define n_0 to be the even integer nearest to x/h, and $x_0 \equiv n_0 h$, $x' \equiv x - x_0$, and $n' \equiv n - n_0$, so that

$$F(x) \approx \frac{1}{\sqrt{\pi}} \sum_{\substack{n'=-N \\ n'\ \mathrm{odd}}}^{N} \frac{e^{-(x'-n'h)^2}}{n' + n_0}, \qquad\qquad (6.10.4)$$

where the approximate equality is accurate when h is sufficiently small and N is sufficiently large. The computation of this formula can be greatly speeded up if we note that

$$e^{-(x'-n'h)^2} = e^{-x'^2} e^{-(n'h)^2} \left(e^{2x'h} \right)^{n'}. \qquad\qquad (6.10.5)$$

The first factor is computed once, the second is an array of constants to be stored, and the third can be computed recursively, so that only two exponentials need be evaluated. Advantage is also taken of the symmetry of the coefficients $e^{-(n'h)^2}$ by breaking the summation up into positive and negative values of n' separately.

In the following routine, the choices $h = 0.4$ and $N = 11$ are made. Because of the symmetry of the summations and the restriction to odd values of n, the limits on the for loops are 0 to 5. The accuracy of the result in this version is about 2×10^{-7}. In order to maintain relative accuracy near $x = 0$, where $F(x)$ vanishes, the program branches to the evaluation of the power series [2] for $F(x)$, for $|x| < 0.2$.

```
#include <cmath>
#include "nr.h"
using namespace std;

DP NR::dawson(const DP x)
Returns Dawson's integral F(x) = exp(-x^2) ∫₀ˣ exp(t^2)dt for any real x.
{
    const int NMAX=6;
    const DP H=0.4, A1=2.0/3.0, A2=0.4, A3=2.0/7.0;
    int i,n0;
    static bool init = true;          Flag is trueif we need to initialize, else false.
    DP d1,d2,e1,e2,sum,x2,xp,xx,ans;
    static Vec_DP c(NMAX);

    if (init) {
        init=false;
        for (i=0;i<NMAX;i++) c[i]=exp(-SQR((2.0*i+1.0)*H));
    }
    if (fabs(x) < 0.2) {               Use series expansion.
        x2=x*x;
        ans=x*(1.0-A1*x2*(1.0-A2*x2*(1.0-A3*x2)));
    } else {                          Use sampling theorem representation.
        xx=fabs(x);
        n0=2*int(0.5*xx/H+0.5);
        xp=xx-n0*H;
        e1=exp(2.0*xp*H);
        e2=e1*e1;
        d1=n0+1;
        d2=d1-2.0;
        sum=0.0;
        for (i=0;i<NMAX;i++,d1+=2.0,d2-=2.0,e1*=e2)
            sum += c[i]*(e1/d1+1.0/(d2*e1));
        ans=0.5641895835*SIGN(exp(-xp*xp),x)*sum;        Constant is 1/√π.
    }
    return ans;
}
```

Other methods for computing Dawson's integral are also known [2,3].

CITED REFERENCES AND FURTHER READING:

Rybicki, G.B. 1989, *Computers in Physics*, vol. 3, no. 2, pp. 85–87. [1]

Cody, W.J., Pociorek, K.A., and Thatcher, H.C. 1970, *Mathematics of Computation*, vol. 24, pp. 171–178. [2]

McCabe, J.H. 1974, *Mathematics of Computation*, vol. 28, pp. 811–816. [3]

6.11 Elliptic Integrals and Jacobian Elliptic Functions

Elliptic integrals occur in many applications, because any integral of the form

$$\int R(t, s)\, dt \qquad\qquad (6.11.1)$$

where R is a rational function of t and s, and s is the square root of a cubic or quartic polynomial in t, can be evaluated in terms of elliptic integrals. Standard references [1] describe how to carry out the reduction, which was originally done by Legendre. Legendre showed that only three basic elliptic integrals are required. The simplest of these is

$$I_1 = \int_y^x \frac{dt}{\sqrt{(a_1 + b_1 t)(a_2 + b_2 t)(a_3 + b_3 t)(a_4 + b_4 t)}} \tag{6.11.2}$$

where we have written the quartic s^2 in factored form. In standard integral tables [2], one of the limits of integration is always a zero of the quartic, while the other limit lies closer than the next zero, so that there is no singularity within the interval. To evaluate I_1, we simply break the interval $[y, x]$ into subintervals, each of which either begins or ends on a singularity. The tables, therefore, need only distinguish the eight cases in which each of the four zeros (ordered according to size) appears as the upper or lower limit of integration. In addition, when one of the b's in (6.11.2) tends to zero, the quartic reduces to a cubic, with the largest or smallest singularity moving to $\pm\infty$; this leads to eight more cases (actually just special cases of the first eight). The sixteen cases in total are then usually tabulated in terms of Legendre's standard elliptic integral of the 1st kind, which we will define below. By a change of the variable of integration t, the zeros of the quartic are mapped to standard locations on the real axis. Then only two dimensionless parameters are needed to tabulate Legendre's integral. However, the symmetry of the original integral (6.11.2) under permutation of the roots is concealed in Legendre's notation. We will get back to Legendre's notation below. But first, here is a better way:

Carlson [3] has given a new definition of a standard elliptic integral of the first kind,

$$R_F(x, y, z) = \frac{1}{2} \int_0^\infty \frac{dt}{\sqrt{(t + x)(t + y)(t + z)}} \tag{6.11.3}$$

where x, y, and z are nonnegative and at most one is zero. By standardizing the range of integration, he retains permutation symmetry for the zeros. (Weierstrass' canonical form also has this property.) Carlson first shows that when x or y is a zero of the quartic in (6.11.2), the integral I_1 can be written in terms of R_F in a form that is symmetric under permutation of the *remaining* three zeros. In the general case when neither x nor y is a zero, two such R_F functions can be combined into a single one by an *addition theorem*, leading to the fundamental formula

$$I_1 = 2R_F(U_{12}^2, U_{13}^2, U_{14}^2) \tag{6.11.4}$$

where

$$U_{ij} = (X_i X_j Y_k Y_m + Y_i Y_j X_k X_m)/(x - y) \tag{6.11.5}$$

$$X_i = (a_i + b_i x)^{1/2}, \qquad Y_i = (a_i + b_i y)^{1/2} \tag{6.11.6}$$

and i, j, k, m is any permutation of $1, 2, 3, 4$. A short-cut in evaluating these expressions is

$$U_{13}^2 = U_{12}^2 - (a_1 b_4 - a_4 b_1)(a_2 b_3 - a_3 b_2)$$

$$U_{14}^2 = U_{12}^2 - (a_1 b_3 - a_3 b_1)(a_2 b_4 - a_4 b_2) \tag{6.11.7}$$

The U's correspond to the three ways of pairing the four zeros, and I_1 is thus manifestly symmetric under permutation of the zeros. Equation (6.11.4) therefore reproduces all sixteen cases when one limit is a zero, and also includes the cases when neither limit is a zero.

Thus Carlson's function allows arbitrary ranges of integration and arbitrary positions of the branch points of the integrand relative to the interval of integration. To handle elliptic integrals of the second and third kind, Carlson defines the standard integral of the third kind as

$$R_J(x, y, z, p) = \frac{3}{2} \int_0^\infty \frac{dt}{(t + p)\sqrt{(t + x)(t + y)(t + z)}} \tag{6.11.8}$$

which is symmetric in x, y, and z. The degenerate case when two arguments are equal is denoted

$$R_D(x, y, z) = R_J(x, y, z, z) \tag{6.11.9}$$

and is symmetric in x and y. The function R_D replaces Legendre's integral of the second kind. The degenerate form of R_F is denoted

$$R_C(x, y) = R_F(x, y, y) \tag{6.11.10}$$

It embraces logarithmic, inverse circular, and inverse hyperbolic functions.

Carlson [4-7] gives integral tables in terms of the exponents of the linear factors of the quartic in (6.11.1). For example, the integral where the exponents are $(\frac{1}{2}, \frac{1}{2}, -\frac{1}{2}, -\frac{3}{2})$ can be expressed as a single integral in terms of R_D; it accounts for 144 separate cases in Gradshteyn and Ryzhik [2]!

Refer to Carlson's papers [3-7] for some of the practical details in reducing elliptic integrals to his standard forms, such as handling complex conjugate zeros.

Turn now to the numerical evaluation of elliptic integrals. The traditional methods [8] are Gauss or Landen transformations. *Descending* transformations decrease the modulus k of the Legendre integrals towards zero, *increasing* transformations increase it towards unity. In these limits the functions have simple analytic expressions. While these methods converge quadratically and are quite satisfactory for integrals of the first and second kinds, they generally lead to loss of significant figures in certain regimes for integrals of the third kind. Carlson's algorithms [9,10], by contrast, provide a unified method for all three kinds with no significant cancellations.

The key ingredient in these algorithms is the *duplication theorem*:

$$R_F(x, y, z) = 2R_F(x + \lambda, y + \lambda, z + \lambda)$$
$$= R_F\left(\frac{x + \lambda}{4}, \frac{y + \lambda}{4}, \frac{z + \lambda}{4}\right) \tag{6.11.11}$$

where

$$\lambda = (xy)^{1/2} + (xz)^{1/2} + (yz)^{1/2} \tag{6.11.12}$$

This theorem can be proved by a simple change of variable of integration [11]. Equation (6.11.11) is iterated until the arguments of R_F are nearly equal. For equal arguments we have

$$R_F(x, x, x) = x^{-1/2} \tag{6.11.13}$$

When the arguments are close enough, the function is evaluated from a fixed Taylor expansion about (6.11.13) through fifth-order terms. While the iterative part of the algorithm is only linearly convergent, the error ultimately decreases by a factor of $4^6 = 4096$ for each iteration. Typically only two or three iterations are required, perhaps six or seven if the initial values of the arguments have huge ratios. We list the algorithm for R_F here, and refer you to Carlson's paper [9] for the other cases.

Stage 1: For $n = 0, 1, 2, \ldots$ compute

$$\mu_n = (x_n + y_n + z_n)/3$$

$$X_n = 1 - (x_n/\mu_n), \quad Y_n = 1 - (y_n/\mu_n), \quad Z_n = 1 - (z_n/\mu_n)$$

$$\epsilon_n = \max(|X_n|, |Y_n|, |Z_n|)$$

If $\epsilon_n <$ tol go to Stage 2; else compute

$$\lambda_n = (x_n y_n)^{1/2} + (x_n z_n)^{1/2} + (y_n z_n)^{1/2}$$

$$x_{n+1} = (x_n + \lambda_n)/4, \quad y_{n+1} = (y_n + \lambda_n)/4, \quad z_{n+1} = (z_n + \lambda_n)/4$$

and repeat this stage.

Stage 2:   Compute

$$E_2 = X_n Y_n - Z_n^2, \quad E_3 = X_n Y_n Z_n$$

$$R_F = (1 - \tfrac{1}{10}E_2 + \tfrac{1}{14}E_3 + \tfrac{1}{24}E_2^2 - \tfrac{3}{44}E_2 E_3)/(\mu_n)^{1/2}$$

In some applications the argument p in R_J or the argument y in R_C is negative, and the Cauchy principal value of the integral is required. This is easily handled by using the formulas

$$R_J(x,y,z,p) =$$

$$\left[(\gamma - y)R_J(x,y,z,\gamma) - 3R_F(x,y,z) + 3R_C(xz/y, p\gamma/y)\right]/(y-p)$$
(6.11.14)

where

$$\gamma \equiv y + \frac{(z-y)(y-x)}{y-p}$$
(6.11.15)

is positive if p is negative, and

$$R_C(x,y) = \left(\frac{x}{x-y}\right)^{1/2} R_C(x-y, -y)$$
(6.11.16)

The Cauchy principal value of R_J has a zero at some value of $p < 0$, so (6.11.14) will give some loss of significant figures near the zero.

```
#include <cmath>
#include "nr.h"
using namespace std;

DP NR::rf(const DP x, const DP y, const DP z)
```
Computes Carlson's elliptic integral of the first kind, $R_F(x,y,z)$. x, y, and z must be nonnegative, and at most one can be zero. TINY must be at least 5 times the machine underflow limit, BIG at most one fifth the machine overflow limit.
```
{
    const DP ERRTOL=0.0025, TINY=1.5e-38, BIG=3.0e37, THIRD=1.0/3.0;
    const DP C1=1.0/24.0, C2=0.1, C3=3.0/44.0, C4=1.0/14.0;
    DP alamb,ave,delx,dely,delz,e2,e3,sqrtx,sqrty,sqrtz,xt,yt,zt;

    if (MIN(MIN(x,y),z) < 0.0 || MIN(MIN(x+y,x+z),y+z) < TINY ||
        MAX(MAX(x,y),z) > BIG)
            nrerror("invalid arguments in rf");
    xt=x;
    yt=y;
    zt=z;
    do {
        sqrtx=sqrt(xt);
        sqrty=sqrt(yt);
        sqrtz=sqrt(zt);
        alamb=sqrtx*(sqrty+sqrtz)+sqrty*sqrtz;
        xt=0.25*(xt+alamb);
        yt=0.25*(yt+alamb);
        zt=0.25*(zt+alamb);
        ave=THIRD*(xt+yt+zt);
        delx=(ave-xt)/ave;
```

```
        dely=(ave-yt)/ave;
        delz=(ave-zt)/ave;
    } while (MAX(MAX(fabs(delx),fabs(dely)),fabs(delz)) > ERRTOL);
    e2=delx*dely-delz*delz;
    e3=delx*dely*delz;
    return (1.0+(C1*e2-C2-C3*e3)*e2+C4*e3)/sqrt(ave);
}
```

A value of 0.0025 for the error tolerance parameter gives full double precision (16 significant digits). Since the error scales as ϵ_n^6, we see that 0.08 would be adequate for single precision (7 significant digits), but would save at most two or three more iterations. Since the coefficients of the sixth-order truncation error are different for the other elliptic functions, these values for the error tolerance should be changed to 0.04 (single precision) or 0.0012 (double precision) in the algorithm for R_C, and 0.05 or 0.0015 for R_J and R_D. As well as being an algorithm in its own right for certain combinations of elementary functions, the algorithm for R_C is used repeatedly in the computation of R_J.

The C++ implementations test the input arguments against two machine-dependent constants, TINY and BIG, to ensure that there will be no underflow or overflow during the computation. We have chosen conservative values, corresponding to a machine minimum of 3×10^{-39} and a machine maximum of 1.7×10^{38}. You can always extend the range of admissible argument values by using the homogeneity relations (6.11.22), below.

```
#include <cmath>
#include "nr.h"
using namespace std;

DP NR::rd(const DP x, const DP y, const DP z)
```
Computes Carlson's elliptic integral of the second kind, $R_D(x, y, z)$. x and y must be non-negative, and at most one can be zero. z must be positive. TINY must be at least twice the negative 2/3 power of the machine overflow limit. BIG must be at most $0.1 \times$ ERRTOL times the negative 2/3 power of the machine underflow limit.
```
{
    const DP ERRTOL=0.0015, TINY=1.0e-25, BIG=4.5e21;
    const DP C1=3.0/14.0, C2=1.0/6.0, C3=9.0/22.0;
    const DP C4=3.0/26.0, C5=0.25*C3, C6=1.5*C4;
    DP alamb,ave,delx,dely,delz,ea,eb,ec,ed,ee,fac,sqrtx,sqrty,
        sqrtz,sum,xt,yt,zt;

    if (MIN(x,y) < 0.0 || MIN(x+y,z) < TINY || MAX(MAX(x,y),z) > BIG)
        nrerror("invalid arguments in rd");
    xt=x;
    yt=y;
    zt=z;
    sum=0.0;
    fac=1.0;
    do {
        sqrtx=sqrt(xt);
        sqrty=sqrt(yt);
        sqrtz=sqrt(zt);
        alamb=sqrtx*(sqrty+sqrtz)+sqrty*sqrtz;
        sum += fac/(sqrtz*(zt+alamb));
        fac=0.25*fac;
        xt=0.25*(xt+alamb);
        yt=0.25*(yt+alamb);
        zt=0.25*(zt+alamb);
        ave=0.2*(xt+yt+3.0*zt);
        delx=(ave-xt)/ave;
        dely=(ave-yt)/ave;
        delz=(ave-zt)/ave;
    } while (MAX(MAX(fabs(delx),fabs(dely)),fabs(delz)) > ERRTOL);
    ea=delx*dely;
    eb=delz*delz;
```

```
    ec=ea-eb;
    ed=ea-6.0*eb;
    ee=ed+ec+ec;
    return 3.0*sum+fac*(1.0+ed*(-C1+C5*ed-C6*delz*ee)
        +delz*(C2*ee+delz*(-C3*ec+delz*C4*ea)))/(ave*sqrt(ave));
}




#include <cmath>
#include "nr.h"
using namespace std;
```

DP NR::rj(const DP x, const DP y, const DP z, const DP p)
Computes Carlson's elliptic integral of the third kind, $R_J(x, y, z, p)$. x, y, and z must be nonnegative, and at most one can be zero. p must be nonzero. If $p < 0$, the Cauchy principal value is returned. TINY must be at least twice the cube root of the machine underflow limit, BIG at most one fifth the cube root of the machine overflow limit.

```
{
    const DP ERRTOL=0.0015, TINY=2.5e-13, BIG=9.0e11;
    const DP C1=3.0/14.0, C2=1.0/3.0, C3=3.0/22.0, C4=3.0/26.0;
    const DP C5=0.75*C3, C6=1.5*C4, C7=0.5*C2, C8=C3+C3;
    DP a,alamb,alpha,ans,ave,b,beta,delp,delx,dely,delz,ea,eb,ec,ed,ee,
        fac,pt,rcx,rho,sqrtx,sqrty,sqrtz,sum,tau,xt,yt,zt;

    if (MIN(MIN(x,y),z) < 0.0 || MIN(MIN(x+y,x+z),MIN(y+z,fabs(p))) < TINY
        || MAX(MAX(x,y),MAX(z,fabs(p))) > BIG)
            nrerror("invalid arguments in rj");
    sum=0.0;
    fac=1.0;
    if (p > 0.0) {
        xt=x;
        yt=y;
        zt=z;
        pt=p;
    } else {
        xt=MIN(MIN(x,y),z);
        zt=MAX(MAX(x,y),z);
        yt=x+y+z-xt-zt;
        a=1.0/(yt-p);
        b=a*(zt-yt)*(yt-xt);
        pt=yt+b;
        rho=xt*zt/yt;
        tau=p*pt/yt;
        rcx=rc(rho,tau);
    }
    do {
        sqrtx=sqrt(xt);
        sqrty=sqrt(yt);
        sqrtz=sqrt(zt);
        alamb=sqrtx*(sqrty+sqrtz)+sqrty*sqrtz;
        alpha=SQR(pt*(sqrtx+sqrty+sqrtz)+sqrtx*sqrty*sqrtz);
        beta=pt*SQR(pt+alamb);
        sum += fac*rc(alpha,beta);
        fac=0.25*fac;
        xt=0.25*(xt+alamb);
        yt=0.25*(yt+alamb);
        zt=0.25*(zt+alamb);
        pt=0.25*(pt+alamb);
        ave=0.2*(xt+yt+zt+pt+pt);
        delx=(ave-xt)/ave;
        dely=(ave-yt)/ave;
        delz=(ave-zt)/ave;
        delp=(ave-pt)/ave;
```

```
    } while (MAX(MAX(fabs(delx),fabs(dely)),
        MAX(fabs(delz),fabs(delp))) > ERRTOL);
    ea=delx*(dely+delz)+dely*delz;
    eb=delx*dely*delz;
    ec=delp*delp;
    ed=ea-3.0*ec;
    ee=eb+2.0*delp*(ea-ec);
    ans=3.0*sum+fac*(1.0+ed*(-C1+C5*ed-C6*ee)+eb*(C7+delp*(-C8+delp*C4))
        +delp*ea*(C2-delp*C3)-C2*delp*ec)/(ave*sqrt(ave));
    if (p <= 0.0) ans=a*(b*ans+3.0*(rcx-rf(xt,yt,zt)));
    return ans;
}
```

```
#include <cmath>
#include "nr.h"
using namespace std;

DP NR::rc(const DP x, const DP y)
```
Computes Carlson's degenerate elliptic integral, $R_C(x,y)$. x must be nonnegative and y must be nonzero. If $y < 0$, the Cauchy principal value is returned. TINY must be at least 5 times the machine underflow limit, BIG at most one fifth the machine maximum overflow limit.
```
{
    const DP ERRTOL=0.0012, TINY=1.69e-38, SQRTNY=1.3e-19, BIG=3.0e37;
    const DP TNBG=TINY*BIG, COMP1=2.236/SQRTNY, COMP2=TNBG*TNBG/25.0;
    const DP THIRD=1.0/3.0, C1=0.3, C2=1.0/7.0, C3=0.375, C4=9.0/22.0;
    DP alamb,ave,s,w,xt,yt;

    if (x < 0.0 || y == 0.0 || (x+fabs(y)) < TINY || (x+fabs(y)) > BIG ||
        (y<-COMP1 && x > 0.0 && x < COMP2))
            nrerror("invalid arguments in rc");
    if (y > 0.0) {
        xt=x;
        yt=y;
        w=1.0;
    } else {
        xt=x-y;
        yt= -y;
        w=sqrt(x)/sqrt(xt);
    }
    do {
        alamb=2.0*sqrt(xt)*sqrt(yt)+yt;
        xt=0.25*(xt+alamb);
        yt=0.25*(yt+alamb);
        ave=THIRD*(xt+yt+yt);
        s=(yt-ave)/ave;
    } while (fabs(s) > ERRTOL);
    return w*(1.0+s*s*(C1+s*(C2+s*(C3+s*C4))))/sqrt(ave);
}
```

At times you may want to express your answer in Legendre's notation. Alternatively, you may be given results in that notation and need to compute their values with the programs given above. It is a simple matter to transform back and forth. The *Legendre elliptic integral of the 1st kind* is defined as

$$F(\phi, k) \equiv \int_0^\phi \frac{d\theta}{\sqrt{1 - k^2 \sin^2 \theta}} \qquad (6.11.17)$$

The *complete elliptic integral of the 1st kind* is given by

$$K(k) \equiv F(\pi/2, k) \qquad (6.11.18)$$

In terms of R_F,

$$F(\phi, k) = \sin \phi R_F(\cos^2 \phi, 1 - k^2 \sin^2 \phi, 1)$$
$$K(k) = R_F(0, 1 - k^2, 1)$$

(6.11.19)

The *Legendre elliptic integral of the 2nd kind* and the *complete elliptic integral of the 2nd kind* are given by

$$
\begin{aligned}
E(\phi, k) &\equiv \int_0^\phi \sqrt{1 - k^2 \sin^2 \theta} \, d\theta \\
&= \sin \phi R_F(\cos^2 \phi, 1 - k^2 \sin^2 \phi, 1) \\
&\quad - \tfrac{1}{3} k^2 \sin^3 \phi R_D(\cos^2 \phi, 1 - k^2 \sin^2 \phi, 1)
\end{aligned}
$$

$$E(k) \equiv E(\pi/2, k) = R_F(0, 1 - k^2, 1) - \tfrac{1}{3} k^2 R_D(0, 1 - k^2, 1)$$

(6.11.20)

Finally, the *Legendre elliptic integral of the 3rd kind* is

$$
\begin{aligned}
\Pi(\phi, n, k) &\equiv \int_0^\phi \frac{d\theta}{(1 + n \sin^2 \theta)\sqrt{1 - k^2 \sin^2 \theta}} \\
&= \sin \phi R_F(\cos^2 \phi, 1 - k^2 \sin^2 \phi, 1) \\
&\quad - \tfrac{1}{3} n \sin^3 \phi R_J(\cos^2 \phi, 1 - k^2 \sin^2 \phi, 1, 1 + n \sin^2 \phi)
\end{aligned}
$$

(6.11.21)

(Note that this sign convention for n is opposite that of Abramowitz and Stegun [12], and that their $\sin \alpha$ is our k.)

```
#include <cmath>
#include "nr.h"
using namespace std;

DP NR::ellf(const DP phi, const DP ak)
```
Legendre elliptic integral of the 1st kind $F(\phi, k)$, evaluated using Carlson's function R_F. The argument ranges are $0 \le \phi \le \pi/2$, $0 \le k \sin \phi \le 1$.
```
{
    DP s;

    s=sin(phi);
    return s*rf(SQR(cos(phi)),(1.0-s*ak)*(1.0+s*ak),1.0);
}
```

```
#include <cmath>
#include "nr.h"
using namespace std;

DP NR::elle(const DP phi, const DP ak)
```
Legendre elliptic integral of the 2nd kind $E(\phi, k)$, evaluated using Carlson's functions R_D and R_F. The argument ranges are $0 \le \phi \le \pi/2$, $0 \le k \sin \phi \le 1$.
```
{
    DP cc,q,s;

    s=sin(phi);
```

```
    cc=SQR(cos(phi));
    q=(1.0-s*ak)*(1.0+s*ak);
    return s*(rf(cc,q,1.0)-(SQR(s*ak))*rd(cc,q,1.0)/3.0);
}
```

```
#include <cmath>
#include "nr.h"
using namespace std;

DP NR::ellpi(const DP phi, const DP en, const DP ak)
```
Legendre elliptic integral of the 3rd kind $\Pi(\phi, n, k)$, evaluated using Carlson's functions R_J and R_F. (Note that the sign convention on n is opposite that of Abramowitz and Stegun.) The ranges of ϕ and k are $0 \le \phi \le \pi/2$, $0 \le k\sin\phi \le 1$.
```
{
    DP cc,enss,q,s;

    s=sin(phi);
    enss=en*s*s;
    cc=SQR(cos(phi));
    q=(1.0-s*ak)*(1.0+s*ak);
    return s*(rf(cc,q,1.0)-enss*rj(cc,q,1.0,1.0+enss)/3.0);
}
```

Carlson's functions are homogeneous of degree $-\frac{1}{2}$ and $-\frac{3}{2}$, so

$$R_F(\lambda x, \lambda y, \lambda z) = \lambda^{-1/2} R_F(x, y, z)$$
$$R_J(\lambda x, \lambda y, \lambda z, \lambda p) = \lambda^{-3/2} R_J(x, y, z, p)$$

(6.11.22)

Thus to express a Carlson function in Legendre's notation, permute the first three arguments into ascending order, use homogeneity to scale the third argument to be 1, and then use equations (6.11.19)–(6.11.21).

Jacobian Elliptic Functions

The Jacobian elliptic function sn is defined as follows: instead of considering the elliptic integral

$$u(y, k) \equiv u = F(\phi, k) \tag{6.11.23}$$

consider the *inverse* function

$$y = \sin\phi = \mathrm{sn}(u, k) \tag{6.11.24}$$

Equivalently,

$$u = \int_0^{\mathrm{sn}} \frac{dy}{\sqrt{(1 - y^2)(1 - k^2 y^2)}} \tag{6.11.25}$$

When $k = 0$, sn is just sin. The functions cn and dn are defined by the relations

$$\mathrm{sn}^2 + \mathrm{cn}^2 = 1, \qquad k^2\mathrm{sn}^2 + \mathrm{dn}^2 = 1 \tag{6.11.26}$$

The routine given below actually takes $m_c \equiv k_c^2 = 1 - k^2$ as an input parameter. It also computes all three functions sn, cn, and dn since computing all three is no harder than computing any one of them. For a description of the method, see [8].

```cpp
#include <cmath>
#include "nr.h"
using namespace std;

void NR::sncndn(const DP uu, const DP emmc, DP &sn, DP &cn, DP &dn)
```
Returns the Jacobian elliptic functions $\text{sn}(u,k_c)$, $\text{cn}(u,k_c)$, and $\text{dn}(u,k_c)$. Here $\mathtt{uu}=u$, while $\mathtt{emmc}=k_c^2$.
```cpp
{
    const DP CA=1.0e-8;                    The accuracy is the square of CA.
    bool bo;
    int i,ii,l;
    DP a,b,c,d,emc,u;
    Vec_DP em(13),en(13);

    emc=emmc;
    u=uu;
    if (emc != 0.0) {
        bo=(emc < 0.0);
        if (bo) {
            d=1.0-emc;
            emc /= -1.0/d;
            u *= (d=sqrt(d));
        }
        a=1.0;
        dn=1.0;
        for (i=0;i<13;i++) {
            l=i;
            em[i]=a;
            en[i]=(emc=sqrt(emc));
            c=0.5*(a+emc);
            if (fabs(a-emc) <= CA*a) break;
            emc *= a;
            a=c;
        }
        u *= c;
        sn=sin(u);
        cn=cos(u);
        if (sn != 0.0) {
            a=cn/sn;
            c *= a;
            for (ii=l;ii>=0;ii--) {
                b=em[ii];
                a *= c;
                c *= dn;
                dn=(en[ii]+a)/(b+a);
                a=c/b;
            }
            a=1.0/sqrt(c*c+1.0);
            sn=(sn >= 0.0 ? a : -a);
            cn=c*sn;
        }
        if (bo) {
            a=dn;
            dn=cn;
            cn=a;
            sn /= d;
        }
    } else {
        cn=1.0/cosh(u);
        dn=cn;
        sn=tanh(u);
    }
}
```

CITED REFERENCES AND FURTHER READING:

Erdélyi, A., Magnus, W., Oberhettinger, F., and Tricomi, F.G. 1953, *Higher Transcendental Functions*, Vol. II, (New York: McGraw-Hill). [1]

Gradshteyn, I.S., and Ryzhik, I.W. 1980, *Table of Integrals, Series, and Products* (New York: Academic Press). [2]

Carlson, B.C. 1977, *SIAM Journal on Mathematical Analysis*, vol. 8, pp. 231–242. [3]

Carlson, B.C. 1987, *Mathematics of Computation*, vol. 49, pp. 595–606 [4]; 1988, *op. cit.*, vol. 51, pp. 267–280 [5]; 1989, *op. cit.*, vol. 53, pp. 327–333 [6]; 1991, *op. cit.*, vol. 56, pp. 267–280. [7]

Bulirsch, R. 1965, *Numerische Mathematik*, vol. 7, pp. 78–90; 1965, *op. cit.*, vol. 7, pp. 353–354; 1969, *op. cit.*, vol. 13, pp. 305–315. [8]

Carlson, B.C. 1979, *Numerische Mathematik*, vol. 33, pp. 1–16. [9]

Carlson, B.C., and Notis, E.M. 1981, *ACM Transactions on Mathematical Software*, vol. 7, pp. 398–403. [10]

Carlson, B.C. 1978, *SIAM Journal on Mathematical Analysis*, vol. 9, p. 524–528. [11]

Abramowitz, M., and Stegun, I.A. 1964, *Handbook of Mathematical Functions*, Applied Mathematics Series, Volume 55 (Washington: National Bureau of Standards; reprinted 1968 by Dover Publications, New York), Chapter 17. [12]

Mathews, J., and Walker, R.L. 1970, *Mathematical Methods of Physics*, 2nd ed. (Reading, MA: W.A. Benjamin/Addison-Wesley), pp. 78–79.

6.12 Hypergeometric Functions

As was discussed in §5.14, a fast, general routine for the the the complex hypergeometric function $_2F_1(a, b, c; z)$, is difficult or impossible. The function is defined as the analytic continuation of the hypergeometric series,

$$
_2F_1(a, b, c; z) = 1 + \frac{ab}{c}\frac{z}{1!} + \frac{a(a+1)b(b+1)}{c(c+1)}\frac{z^2}{2!} + \cdots
$$
$$
+ \frac{a(a+1)\dots(a+j-1)b(b+1)\dots(b+j-1)}{c(c+1)\dots(c+j-1)}\frac{z^j}{j!} + \cdots
$$
$$(6.12.1)$$

This series converges only within the unit circle $|z| < 1$ (see [1]), but one's interest in the function is not confined to this region.

Section 5.14 discussed the method of evaluating this function by direct path integration in the complex plane. We here merely list the routines that result.

Implementation of the function hypgeo is straightforward, and is described by comments in the program. The machinery associated with Chapter 16's routine for integrating differential equations, odeint, is only minimally intrusive, and need not even be completely understood: use of odeint requires one zeroed global variable, one function call, and a prescribed format for the derivative routine hypdrv.

The function hypgeo will fail, of course, for values of z too close to the singularity at 1. (If you need to approach this singularity, or the one at ∞, use the "linear transformation formulas" in §15.3 of [1].) Away from $z = 1$, and for moderate values of a, b, c, it is often remarkable how few steps are required to integrate the equations. A half-dozen is typical.

```
#include <cmath>
#include <complex>
#include "nr.h"
using namespace std;

complex<DP> aa,bb,cc,z0,dz;                    Communicates with hypdrv.

int kmax,kount;                               Used by odeint.
DP dxsav;
Vec_DP *xp_p;
Mat_DP *yp_p;

complex<DP> NR::hypgeo(const complex<DP> &a, const complex<DP> &b,
    const complex<DP> &c, const complex<DP> &z)
```

Complex hypergeometric function $_2F_1$ for complex a, b, c, and z, by direct integration of the hypergeometric equation in the complex plane. The branch cut is taken to lie along the real axis, Re $z > 1$.

```
{
    const DP EPS=1.0e-14;                     Accuracy parameter.
    int nbad,nok;
    complex<DP> ans,y[2];
    Vec_DP yy(4);

    kmax=0;
    if (norm(z) <= 0.25) {                    Use series...
        hypser(a,b,c,z,ans,y[1]);
        return ans;
    }
```
...or pick a starting point for the path integration.
```
    else if (real(z) < 0.0) z0=complex<DP>(-0.5,0.0);
    else if (real(z) <= 1.0) z0=complex<DP>(0.5,0.0);
    else z0=complex<DP>(0.0,imag(z) >= 0.0 ? 0.5 : -0.5);
    aa=a;                                     Load the global variables to pass pa-
    bb=b;                                        rameters "over the head" of odeint
    cc=c;                                        to hypdrv.
    dz=z-z0;
    hypser(aa,bb,cc,z0,y[0],y[1]);            Get starting function and derivative.
    yy[0]=real(y[0]);
    yy[1]=imag(y[0]);
    yy[2]=real(y[1]);
    yy[3]=imag(y[1]);
    odeint(yy,0.0,1.0,EPS,0.1,0.0001,nok,nbad,hypdrv,bsstep);
```
The arguments to odeint are the vector of independent variables, the starting and ending values of the dependent variable, the accuracy parameter, an initial guess for stepsize, a minimum stepsize, the (returned) number of good and bad steps taken, and the names of the derivative routine and the (here Bulirsch-Stoer) stepping routine.
```
    y[0]=complex<DP>(yy[0],yy[1]);
    return y[0];
}
```

```
#include <complex>
#include "nr.h"
using namespace std;

void NR::hypser(const complex<DP> &a, const complex<DP> &b,
    const complex<DP> &c, const complex<DP> &z,
    complex<DP> &series, complex<DP> &deriv)
```
Returns the hypergeometric series $_2F_1$ and its derivative, iterating to machine accuracy. For $|z| \leq 1/2$ convergence is quite rapid.
```
{
    int n;
    complex<DP> aa,bb,cc,fac,temp;
```

```
      deriv=0.0;
      fac=1.0;
      temp=fac;
      aa=a;
      bb=b;
      cc=c;
      for (n=1;n<=1000;n++) {
          fac *= ((aa*bb)/cc);
          deriv += fac;
          fac *= ((1.0/n)*z);
          series=temp+fac;
          if (series == temp) return;
          temp=series;
          aa += 1.0;
          bb += 1.0;
          cc += 1.0;
      }
      nrerror("convergence failure in hypser");
}
```

```
#include <complex>
#include "nr.h"
using namespace std;

extern complex<DP> aa,bb,cc,z0,dz;          Defined in hypgeo.

void NR::hypdrv(const DP s, Vec_I_DP &yy, Vec_O_DP &dyyds)
Computes derivatives for the hypergeometric equation, see text equation (5.14.4).
{
      complex<DP> z,y[2],dyds[2];

      y[0]=complex<DP>(yy[0],yy[1]);
      y[1]=complex<DP>(yy[2],yy[3]);
      z=z0+s*dz;
      dyds[0]=y[1]*dz;
      dyds[1]=(aa*bb*y[0]-(cc-(aa+bb+1.0)*z)*y[1])*dz/(z*(1.0-z));
      dyyds[0]=real(dyds[0]);
      dyyds[1]=imag(dyds[0]);
      dyyds[2]=real(dyds[1]);
      dyyds[3]=imag(dyds[1]);
}
```

CITED REFERENCES AND FURTHER READING:

Abramowitz, M., and Stegun, I.A. 1964, *Handbook of Mathematical Functions*, Applied Mathematics Series, Volume 55 (Washington: National Bureau of Standards; reprinted 1968 by Dover Publications, New York). [1]

Chapter 7.   Random Numbers

7.0 Introduction

It may seem perverse to use a computer, that most precise and deterministic of all machines conceived by the human mind, to produce "random" numbers. More than perverse, it may seem to be a conceptual impossibility. Any program, after all, will produce output that is entirely predictable, hence not truly "random."

Nevertheless, practical computer "random number generators" are in common use. We will leave it to philosophers of the computer age to resolve the paradox in a deep way (see, e.g., Knuth [1] §3.5 for discussion and references). One sometimes hears computer-generated sequences termed *pseudo-random*, while the word *random* is reserved for the output of an intrinsically random physical process, like the elapsed time between clicks of a Geiger counter placed next to a sample of some radioactive element. We will not try to make such fine distinctions.

A working, though imprecise, definition of randomness in the context of computer-generated sequences, is to say that the deterministic program that produces a random sequence should be different from, and — in all measurable respects — statistically uncorrelated with, the computer program that *uses* its output. In other words, any two different random number generators ought to produce statistically the same results when coupled to your particular applications program. If they don't, then at least one of them is not (from your point of view) a good generator.

The above definition may seem circular, comparing, as it does, one generator to another. However, there exists a body of random number generators which mutually do satisfy the definition over a very, very broad class of applications programs. And it is also found empirically that statistically identical results are obtained from random numbers produced by physical processes. So, because such generators are known to exist, we can leave to the philosophers the problem of defining them.

A pragmatic point of view, then, is that randomness is in the eye of the beholder (or programmer). What is random enough for one application may not be random enough for another. Still, one is not entirely adrift in a sea of incommensurable applications programs: There is a certain list of statistical tests, some sensible and some merely enshrined by history, which on the whole will do a very good job of ferreting out any correlations that are likely to be detected by an applications program (in this case, yours). Good random number generators ought to pass all of these tests; or at least the user had better be aware of any that they fail, so that he or she will be able to judge whether they are relevant to the case at hand.

As for references on this subject, the one to turn to first is Knuth [1]. Then try [2]. Only a few of the standard books on numerical methods [3-4] treat topics relating to random numbers.

CITED REFERENCES AND FURTHER READING:

Knuth, D.E. 1997, *Seminumerical Algorithms*, 3rd ed., vol. 2 of *The Art of Computer Programming* (Reading, MA: Addison-Wesley), Chapter 3, especially §3.5. [1]

Bratley, P., Fox, B.L., and Schrage, E.L. 1983, *A Guide to Simulation*, 2nd ed. (New York: Springer-Verlag). [2]

Dahlquist, G., and Bjorck, A. 1974, *Numerical Methods* (Englewood Cliffs, NJ: Prentice-Hall), Chapter 11. [3]

Forsythe, G.E., Malcolm, M.A., and Moler, C.B. 1977, *Computer Methods for Mathematical Computations* (Englewood Cliffs, NJ: Prentice-Hall), Chapter 10. [4]

7.1 Uniform Deviates

Uniform deviates are just random numbers that lie within a specified range (typically 0 to 1), with any one number in the range just as likely as any other. They are, in other words, what you probably think "random numbers" are. However, we want to distinguish uniform deviates from other sorts of random numbers, for example numbers drawn from a normal (Gaussian) distribution of specified mean and standard deviation. These other sorts of deviates are almost always generated by performing appropriate operations on one or more uniform deviates, as we will see in subsequent sections. So, a reliable source of random uniform deviates, the subject of this section, is an essential building block for any sort of stochastic modeling or Monte Carlo computer work.

System-Supplied Random Number Generators

C++ has inherited from the ANSI C library a pair of routines for initializing, and then generating, "random numbers." The declarations in `cstdlib` are typically

```
#include <cstdlib>
#define RAND_MAX ...

void srand(unsigned seed);
int rand(void);
```

You initialize the random number generator by invoking `srand(seed)` with some arbitrary `seed`. Each initializing value will typically result in a different random sequence, or a least a different starting point in some one enormously long sequence. The *same* initializing value of `seed` will always return the *same* random sequence, however.

You obtain successive random numbers in the sequence by successive calls to `rand()`. That function returns an integer that is in the range 0 to `RAND_MAX` (inclusive). Typically, `RAND_MAX` is the largest representable positive value of type `int`, but sometimes it is only the largest value of type `short int`. If you want a random `double` value between 0.0 (inclusive) and 1.0 (exclusive), you get it by an expression like

```
x = rand()/(RAND_MAX+1.0);
```

Now our first, and perhaps most important, lesson in this chapter is: be *very, very* suspicious of a system-supplied `rand()` that resembles the one just described. If all scientific papers whose results are in doubt because of bad `rand()`s were to disappear from library shelves, there would be a gap on each shelf about as big as your fist. System-supplied `rand()`s are almost always *linear congruential generators*, which generate a sequence of integers $I_1, I_2, I_3, \ldots$, each between 0 and $m - 1$ (e.g., `RAND_MAX`) by the recurrence relation

$$I_{j+1} = aI_j + c \pmod{m} \tag{7.1.1}$$

Here m is called the *modulus*, and a and c are positive integers called the *multiplier* and the *increment* respectively. The recurrence (7.1.1) will eventually repeat itself, with a period that is obviously no greater than m. If m, a, and c are properly chosen, then the period will be of maximal length, i.e., of length m. In that case, all possible integers between 0 and $m - 1$ occur at some point, so any initial "seed" choice of I_0 is as good as any other: the sequence just takes off from that point.

Although this general framework is powerful enough to provide quite decent random numbers, its implementation in many, if not most, ANSI C libraries is quite flawed; quite a number of implementations are in the category "totally botched." Blame should be apportioned about equally between the ANSI C committee and the implementors. The typical problems are these: First, since the ANSI standard specifies that `rand()` return a value of type `int` — which is only a two-byte quantity on many machines — `RAND_MAX` is often *not* very large. The ANSI C standard requires only that it be at least 32767. This can be disastrous in many circumstances: for a Monte Carlo integration (§7.6 and §7.8), you might well want to evaluate 10^6 different points, but actually be evaluating the same 32767 points 30 times each, not at all the same thing! You should categorically reject any library random number routine with a two-byte returned value.

Second, the ANSI committee's published rationale includes the following mischievous passage: "The committee decided that an implementation should be allowed to provide a `rand` function which generates the best random sequence possible in that implementation, and therefore mandated no standard algorithm. It recognized the value, however, of being able to generate the same pseudo-random sequence in different implementations, and so *it has published an example*. . . . [emphasis added]" The "example" is

```
unsigned long next=1;

int rand(void) /* NOT RECOMMENDED (see text) */
{
    next = next*1103515245 + 12345;
    return (unsigned int)(next/65536) % 32768;
}

void srand(unsigned int seed)
{
    next=seed;
}
```

This corresponds to equation (7.1.1) with $a = 1103515245$, $c = 12345$, and $m = 2^{32}$ (since arithmetic done on unsigned long quantities is guaranteed to return the correct low-order bits). These are not particularly good choices for a and c, though they are not gross embarrassments by themselves. The real botches occur when implementors, taking the committee's statement above as license, try to "improve" on the published example. For example, one popular 32-bit PC-compatible compiler provides a long generator that uses the above congruence, but swaps the high-order and low-order 16 bits of the returned value. Somebody probably thought that this extra flourish added randomness; in fact it ruins the generator. While these kinds of blunders can, of course, be fixed, there remains a fundamental flaw in simple linear congruential generators, which we now discuss.

The linear congruential method has the advantage of being very fast, requiring only a few operations per call, hence its almost universal use. It has the disadvantage that it is not free of sequential correlation on successive calls. If k random numbers at a time are used to plot points in k dimensional space (with each coordinate between 0 and 1), then the points will not tend to "fill up" the k-dimensional space, but rather will lie on $(k - 1)$-dimensional "planes." There will be *at most* about $m^{1/k}$ such planes. If the constants m, a, and c are not very carefully chosen, there will be *many fewer than that.* If m is as bad as 32768, then the number of planes on which triples of points lie in three-dimensional space will be no greater than about the cube root of 32768, or 32. Even if m is close to the machine's largest representable integer, e.g., $\sim 2^{32}$, the number of planes on which triples of points lie in three-dimensional space is usually no greater than about the cube root of 2^{32}, about 1600. You might well be focusing attention on a physical process that occurs in a small fraction of the total volume, so that the discreteness of the planes can be very pronounced.

Even worse, you might be using a generator whose choices of m, a, and c have been botched. One infamous such routine, RANDU, with $a = 65539$ and $m = 2^{31}$, was widespread on IBM mainframe computers for many years, and widely copied onto other systems [1]. One of us recalls producing a "random" plot with only 11 planes, and being told by his computer center's programming consultant that he had misused the random number generator: "We guarantee that each number is random individually, but we don't guarantee that more than one of them is random." Figure that out.

Correlation in k-space is not the only weakness of linear congruential generators. Such generators often have their low-order (least significant) bits much less random than their high-order bits. If you want to generate a random integer between 1 and 10, you should always do it using high-order bits, as in

```
j=1+int(10.0*rand()/(RAND_MAX+1.0));
```

and never by anything resembling

```
j=1+(rand() % 10);
```

(which uses lower-order bits). Similarly you should never try to take apart a "rand()" number into several supposedly random pieces. Instead use separate calls for every piece.

Portable Random Number Generators

Park and Miller [1] have surveyed a large number of random number generators that have been used over the last 30 years or more. Along with a good theoretical review, they present an anecdotal sampling of a number of inadequate generators that have come into widespread use. The historical record is nothing if not appalling.

There is good evidence, both theoretical and empirical, that the simple multiplicative congruential algorithm

$$I_{j+1} = aI_j \quad (\text{mod } m) \tag{7.1.2}$$

can be as good as any of the more general linear congruential generators that have $c \neq 0$ (equation 7.1.1) — *if* the multiplier a and modulus m are chosen exquisitely carefully. Park and Miller propose a "Minimal Standard" generator based on the choices

$$a = 7^5 = 16807 \qquad m = 2^{31} - 1 = 2147483647 \tag{7.1.3}$$

First proposed by Lewis, Goodman, and Miller in 1969, this generator has in subsequent years passed all new theoretical tests, and (perhaps more importantly) has accumulated a large amount of successful use. Park and Miller do not claim that the generator is "perfect" (we will see below that it is not), but only that it is a good minimal standard against which other generators should be judged.

It is not possible to implement equations (7.1.2) and (7.1.3) directly in a high-level language, since the product of a and $m - 1$ exceeds the maximum value for a 32-bit integer. Assembly language implementation using a 64-bit product register is straightforward, but not portable from machine to machine. A trick due to Schrage [2,3] for multiplying two 32-bit integers modulo a 32-bit constant, without using any intermediates larger than 32 bits (including a sign bit) is therefore extremely interesting: It allows the Minimal Standard generator to be implemented in essentially any programming language on essentially any machine.

Schrage's algorithm is based on an *approximate factorization* of m,

$$m = aq + r, \quad \text{i.e.,} \quad q = [m/a], \ r = m \text{ mod } a \tag{7.1.4}$$

with square brackets denoting integer part. If r is small, specifically $r < q$, and $0 < z < m - 1$, it can be shown that both $a(z \text{ mod } q)$ and $r[z/q]$ lie in the range $0, \ldots, m - 1$, and that

$$az \text{ mod } m = \begin{cases} a(z \text{ mod } q) - r[z/q] & \text{if it is} \geq 0, \\ a(z \text{ mod } q) - r[z/q] + m & \text{otherwise} \end{cases} \tag{7.1.5}$$

The application of Schrage's algorithm to the constants (7.1.3) uses the values $q = 127773$ and $r = 2836$.

Here is an implementation of the Minimal Standard generator:

```
#include "nr.h"

DP NR::ran0(int &idum)
```
"Minimal" random number generator of Park and Miller. Returns a uniform random deviate between 0.0 and 1.0. Set or reset idum to any integer value (except the unlikely value MASK) to initialize the sequence; idum must not be altered between calls for successive deviates in a sequence.
```
{
    const int IA=16807,IM=2147483647,IQ=127773;
    const int IR=2836,MASK=123459876;
    const DP AM=1.0/DP(IM);
    int k;
    DP ans;

    idum ^= MASK;                     XORing with MASK allows use of zero and other
    k=idum/IQ;                          simple bit patterns for idum.
    idum=IA*(idum-k*IQ)-IR*k;         Compute idum=(IA*idum) % IM without over-
    if (idum < 0) idum += IM;           flows by Schrage's method.
    ans=AM*idum;                      Convert idum to a floating result.
    idum ^= MASK;                     Unmask before return.
    return ans;
}
```

The period of ran0 is $2^{31} - 2 \approx 2.1 \times 10^9$. A peculiarity of generators of the form (7.1.2) is that the value 0 must never be allowed as the initial seed — it perpetuates itself — and it never occurs for any nonzero initial seed. Experience has shown that users always manage to call random number generators with the seed idum=0. That is why ran0 performs its exclusive-or with an arbitrary constant both on entry and exit. If you are the first user in history to be proof against human error, you can remove the two lines with the $\wedge$ operation.

Park and Miller discuss two other multipliers a that can be used with the same $m = 2^{31} - 1$. These are $a = 48271$ (with $q = 44488$ and $r = 3399$) and $a = 69621$ (with $q = 30845$ and $r = 23902$). These can be substituted in the routine ran0 if desired; they may be slightly superior to Lewis *et al.*'s longer-tested values. No values other than these should be used.

The routine ran0 is a Minimal Standard, satisfactory for the majority of applications, but we do not recommend it as the final word on random number generators. Our reason is precisely the simplicity of the Minimal Standard. It is not hard to think of situations where successive random numbers might be used in a way that accidentally conflicts with the generation algorithm. For example, since successive numbers differ by a multiple of only 1.6×10^4 out of a modulus of more than 2×10^9, very small random numbers will tend to be followed by smaller than average values. One time in 10^6, for example, there will be a value $< 10^{-6}$ returned (as there should be), but this will *always* be followed by a value less than about 0.0168. One can easily think of applications involving rare events where this property would lead to wrong results.

There are other, more subtle, serial correlations present in ran0. For example, if successive points (I_i, I_{i+1}) are binned into a two-dimensional plane for $i = 1, 2, \ldots, N$, then the resulting distribution fails the χ^2 test when N is greater than a few $\times 10^7$, much less than the period $m - 2$. Since low-order serial correlations have historically been such a bugaboo, and since there is a very simple way to remove them, we think that it is prudent to do so.

The following routine, ran1, uses the Minimal Standard for its random value, but it shuffles the output to remove low-order serial correlations. A random deviate derived from the jth value in the sequence, I_j, is output not on the jth call, but rather on a randomized later call, $j + 32$ on average. The shuffling algorithm is due to Bays and Durham as described in Knuth [4], and is illustrated in Figure 7.1.1.

```
#include "nr.h"

DP NR::ran1(int &idum)
```
"Minimal" random number generator of Park and Miller with Bays-Durham shuffle and added safeguards. Returns a uniform random deviate between 0.0 and 1.0 (exclusive of the endpoint values). Call with idum a negative integer to initialize; thereafter, do not alter idum between successive deviates in a sequence. RNMX should approximate the largest floating value that is less than 1.
```
{
    const int IA=16807,IM=2147483647,IQ=127773,IR=2836,NTAB=32;
    const int NDIV=(1+(IM-1)/NTAB);
    const DP EPS=3.0e-16,AM=1.0/IM,RNMX=(1.0-EPS);
    static int iy=0;
    static Vec_INT iv(NTAB);
    int j,k;
    DP temp;

    if (idum <= 0 || !iy) {                 Initialize.
        if (-idum < 1) idum=1;              Be sure to prevent idum = 0.
        else idum = -idum;
        for (j=NTAB+7;j>=0;j--) {           Load the shuffle table (after 8 warm-ups).
            k=idum/IQ;
            idum=IA*(idum-k*IQ)-IR*k;
            if (idum < 0) idum += IM;
            if (j < NTAB) iv[j] = idum;
        }
        iy=iv[0];
    }
    k=idum/IQ;                              Start here when not initializing.
    idum=IA*(idum-k*IQ)-IR*k;              Compute idum=(IA*idum) % IM without over-
    if (idum < 0) idum += IM;                  flows by Schrage's method.
    j=iy/NDIV;                              Will be in the range 0..NTAB-1.
    iy=iv[j];                              Output previously stored value and refill the
    iv[j] = idum;                              shuffle table.
    if ((temp=AM*iy) > RNMX) return RNMX;  Because users don't expect endpoint values.
    else return temp;
}
```

The routine ran1 passes those statistical tests that ran0 is known to fail. In fact, we do not know of any statistical test that ran1 fails to pass, except when the number of calls starts to become on the order of the period m, say $> 10^8 \approx m/20$.

For situations when even longer random sequences are needed, L'Ecuyer [6] has given a good way of combining two different sequences with different periods so as to obtain a new sequence whose period is the least common multiple of the two periods. The basic idea is simply to add the two sequences, modulo the modulus of *either* of them (call it m). A trick to avoid an intermediate value that overflows the integer wordsize is to subtract rather than add, and then add back the constant $m - 1$ if the result is ≤ 0, so as to wrap around into the desired interval $0, \ldots, m - 1$.

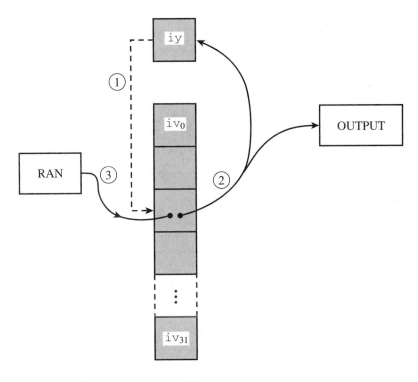

Figure 7.1.1.   Shuffling procedure used in `ran1` to break up sequential correlations in the Minimal Standard generator. Circled numbers indicate the sequence of events: On each call, the random number in `iy` is used to choose a random element in the array `iv`. That element becomes the output random number, and also is the next `iy`. Its spot in `iv` is refilled from the Minimal Standard routine.

Notice that it is not necessary that this wrapped subtraction be able to reach all values $0, \ldots, m - 1$ from *every* value of the first sequence. Consider the absurd extreme case where the value subtracted was only between 1 and 10: The resulting sequence would still be no less random than the first sequence by itself. As a practical matter it is only necessary that the second sequence have a range covering *substantially* all of the range of the first. L'Ecuyer recommends the use of the two generators $m_1 = 2147483563$ (with $a_1 = 40014$, $q_1 = 53668$, $r_1 = 12211$) and $m_2 = 2147483399$ (with $a_2 = 40692$, $q_2 = 52774$, $r_2 = 3791$). Both moduli are slightly less than 2^{31}. The periods $m_1 - 1 = 2 \times 3 \times 7 \times 631 \times 81031$ and $m_2 - 1 = 2 \times 19 \times 31 \times 1019 \times 1789$ share only the factor 2, so the period of the combined generator is $\approx 2.3 \times 10^{18}$. For present computers, period exhaustion is a practical impossibility.

Combining the two generators breaks up serial correlations to a considerable extent. We nevertheless recommend the additional shuffle that is implemented in the following routine, `ran2`. We think that, within the limits of its floating-point precision, `ran2` provides perfect random numbers; a practical definition of "perfect" is that we will pay $1000 to the first reader who convinces us otherwise (by finding a statistical test that `ran2` fails in a nontrivial way, excluding the ordinary limitations of a machine's floating-point representation).

```
#include "nr.h"

DP NR::ran2(int &idum)
```
Long period ($> 2 \times 10^{18}$) random number generator of L'Ecuyer with Bays-Durham shuffle and added safeguards. Returns a uniform random deviate between 0.0 and 1.0 (exclusive of the endpoint values). Call with idum a negative integer to initialize; thereafter, do not alter idum between successive deviates in a sequence. RNMX should approximate the largest floating value that is less than 1.

```
{
    const int IM1=2147483563,IM2=2147483399;
    const int IA1=40014,IA2=40692,IQ1=53668,IQ2=52774;
    const int IR1=12211,IR2=3791,NTAB=32,IMM1=IM1-1;
    const int NDIV=1+IMM1/NTAB;
    const DP EPS=3.0e-16,RNMX=1.0-EPS,AM=1.0/DP(IM1);
    static int idum2=123456789,iy=0;
    static Vec_INT iv(NTAB);
    int j,k;
    DP temp;

    if (idum <= 0) {                              Initialize.
        idum=(idum==0 ? 1 : -idum);               Be sure to prevent idum = 0.
        idum2=idum;
        for (j=NTAB+7;j>=0;j--) {                  Load the shuffle table (after 8 warm-ups).
            k=idum/IQ1;
            idum=IA1*(idum-k*IQ1)-k*IR1;
            if (idum < 0) idum += IM1;
            if (j < NTAB) iv[j] = idum;
        }
        iy=iv[0];
    }
    k=idum/IQ1;                                   Start here when not initializing.
    idum=IA1*(idum-k*IQ1)-k*IR1;                  Compute idum=(IA1*idum) % IM1 without
    if (idum < 0) idum += IM1;                        overflows by Schrage's method.
    k=idum2/IQ2;
    idum2=IA2*(idum2-k*IQ2)-k*IR2;               Compute idum2=(IA2*idum) % IM2 likewise.
    if (idum2 < 0) idum2 += IM2;
    j=iy/NDIV;                                     Will be in the range 0..NTAB-1.
    iy=iv[j]-idum2;                                Here idum is shuffled, idum and idum2 are
    iv[j] = idum;                                     combined to generate output.
    if (iy < 1) iy += IMM1;
    if ((temp=AM*iy) > RNMX) return RNMX;          Because users don't expect endpoint values.
    else return temp;
}
```

L'Ecuyer [6] lists additional short generators that can be combined into longer ones, including generators that can be implemented in 16-bit integer arithmetic.

Finally, we give you Knuth's suggestion [4] for a portable routine, which we have translated to the present conventions as ran3. This is not based on the linear congruential method at all, but rather on a *subtractive method* (see also [5]). One might hope that its weaknesses, if any, are therefore of a highly different character than the weaknesses, if any, of ran1 above. If you ever suspect trouble with one routine, it is a good idea to try the other in the same application. ran3 has one nice feature: if your machine is poor on integer arithmetic (i.e., is limited to 16-bit integers), you can declare mj, mk, and ma[] as double, define mbig and mseed as 4000000 and 1618033, respectively, and the routine will be rendered entirely floating-point.

```
#include <cstdlib>
#include "nr.h"
using namespace std;

DP NR::ran3(int &idum)
```
Returns a uniform random deviate between 0.0 and 1.0. Set `idum` to any negative value to initialize or reinitialize the sequence.
```
{
    static int inext,inextp;
    static int iff=0;
    const int MBIG=1000000000,MSEED=161803398,MZ=0;
    const DP FAC=(1.0/MBIG);
```
According to Knuth, any large MBIG, and any smaller (but still large) MSEED can be substituted for the above values.
```
    static Vec_INT ma(56);          The value 56 (range ma[1..55]) is special and
    int i,ii,k,mj,mk;               should not be modified; see Knuth.

    if (idum < 0 || iff == 0) {     Initialization.
        iff=1;
        mj=labs(MSEED-labs(idum));  Initialize ma[55] using the seed idum and the
        mj %= MBIG;                 large number MSEED.
        ma[55]=mj;
        mk=1;
        for (i=1;i<=54;i++) {       Now initialize the rest of the table,
            ii=(21*i) % 55;         in a slightly random order,
            ma[ii]=mk;              with numbers that are not especially random.
            mk=mj-mk;
            if (mk < int(MZ)) mk += MBIG;
            mj=ma[ii];
        }
        for (k=0;k<4;k++)           We randomize them by "warming up the gener-
            for (i=1;i<=55;i++) {   ator."
                ma[i] -= ma[1+(i+30) % 55];
                if (ma[i] < int(MZ)) ma[i] += MBIG;
            }
        inext=0;                    Prepare indices for our first generated number.
        inextp=31;                  The constant 31 is special; see Knuth.
        idum=1;
    }
```
Here is where we start, except on initialization.
```
    if (++inext == 56) inext=1;     Increment inext and inextp, wrapping around
    if (++inextp == 56) inextp=1;      56 to 1.
    mj=ma[inext]-ma[inextp];        Generate a new random number subtractively.
    if (mj < int(MZ)) mj += MBIG;   Be sure that it is in range.
    ma[inext]=mj;                   Store it,
    return mj*FAC;                  and output the derived uniform deviate.
}
```

Quick and Dirty Generators

One sometimes would like a "quick and dirty" generator to embed in a program, perhaps taking only one or two lines of code, just to *somewhat* randomize things. One might wish to process data from an experiment not always in exactly the same order, for example, so that the first output is more "typical" than might otherwise be the case.

For this kind of application, all we really need is a list of "good" choices for m, a, and c in equation (7.1.1). If we don't need a period longer than 10^4 to 10^6, say, we can keep the value of $(m-1)a + c$ small enough to avoid overflows that would otherwise mandate the extra complexity of Schrage's method (above). We can thus easily embed in our programs

Constants for Quick and Dirty Random Number Generators

overflow at	im	ia	ic
2^{20}	6075	106	1283
2^{21}	7875	211	1663
2^{22}	7875	421	1663
2^{23}	6075	1366	1283
	6655	936	1399
	11979	430	2531
2^{24}	14406	967	3041
	29282	419	6173
	53125	171	11213
2^{25}	12960	1741	2731
	14000	1541	2957
	21870	1291	4621
	31104	625	6571
	139968	205	29573
2^{26}	29282	1255	6173
	81000	421	17117
	134456	281	28411

overflow at	im	ia	ic
2^{27}	86436	1093	18257
	121500	1021	25673
	259200	421	54773
2^{28}	117128	1277	24749
	121500	2041	25673
	312500	741	66037
2^{29}	145800	3661	30809
	175000	2661	36979
	233280	1861	49297
	244944	1597	51749
2^{30}	139968	3877	29573
	214326	3613	45289
	714025	1366	150889
2^{31}	134456	8121	28411
	259200	7141	54773
2^{32}	233280	9301	49297
	714025	4096	150889

```
unsigned long jran,ia,ic,im;
DP ran;
...
jran=(jran*ia+ic) % im;
ran=DP(jran)/DP(im);
```

whenever we want a quick and dirty uniform deviate, or

```
jran=(jran*ia+ic) % im;
j=jlo+((jhi-jlo+1)*jran)/im;
```

whenever we want an integer between `jlo` and `jhi`, inclusive. (In both cases `jran` was once initialized to any seed value between 0 and `im-1`.)

Be sure to remember, however, that when `im` is small, the kth root of it, which is the number of planes in k-space, is even smaller! So a quick and dirty generator should never be used to select points in k-space with $k > 1$.

With these caveats, some "good" choices for the constants are given in the accompanying table. These constants (i) give a period of maximal length `im`, and, more important, (ii) pass Knuth's "spectral test" for dimensions 2, 3, 4, 5, and 6. The increment `ic` is a prime, close to the value $(\frac{1}{2} - \frac{1}{6}\sqrt{3})$`im`; actually almost any value of `ic` that is relatively prime to `im` will do just as well, but there is some "lore" favoring this choice (see [4], p. 84).

An Even Quicker Generator

In C++, if you multiply two unsigned long integers on a machine with a 32-bit long integer representation, the value returned is the low-order 32 bits of the true 64-bit product. If we now choose $m = 2^{32}$, the "mod" in equation (7.1.1) is free, and we have simply

$$I_{j+1} = aI_j + c \tag{7.1.6}$$

Knuth suggests $a = 1664525$ as a suitable multiplier for this value of m. H.W. Lewis has conducted extensive tests of this value of a with $c = 1013904223$, which is a prime close to $(\sqrt{5} - 2)m$. The resulting in-line generator (we will call it ranqd1) is simply

```
unsigned long idum;
...
idum = 1664525L*idum + 1013904223L;
```

This is about as good as any 32-bit linear congruential generator, entirely adequate for many uses. And, with only a single multiply and add, it is *very* fast.

To check whether your machine has the desired integer properties, see if you can generate the following sequence of 32-bit values (given here in hex): 00000000, 3C6EF35F, 47502932, D1CCF6E9, AAF95334, 6252E503, 9F2EC686, 57FE6C2D, A3D95FA8, 81FD-BEE7, 94F0AF1A, CBF633B1.

If you need floating-point values instead of 32-bit integers, and want to avoid a divide by floating-point 2^{32}, a dirty trick is to mask in an exponent that makes the value lie between 1 and 2, then subtract 1.0. The resulting in-line generator (call it ranqd2) will look something like

```
    unsigned long idum,itemp;
    float rand;
#ifdef vax
    static unsigned long jflone = 0x00004080;
    static unsigned long jflmsk = 0xffff007f;
#else
    static unsigned long jflone = 0x3f800000;
    static unsigned long jflmsk = 0x007fffff;
#endif
    ...
    idum = 1664525L*idum + 1013904223L;
    itemp = jflone | (jflmsk & idum);
    rand = (*(float *)&itemp)-1.0;
```

The hex constants 3F800000 and 007FFFFF are the appropriate ones for computers using the IEEE representation for 32-bit floating-point numbers (e.g., Pentium machines and many Linux or UNIX workstations). For DEC VAXes, the correct hex constants are, respectively, 00004080 and FFFF007F. Notice that the IEEE mask results in the floating-point number being constructed out of the 23 low-order bits of the integer, which is not ideal. (Your authors have tried very hard to make *almost all* of the material in this book machine and compiler independent — indeed, even programming language independent. This subsection is a rare aberration. Forgive us. Once in a great while the temptation to be *really dirty* is just irresistible.)

Relative Timings and Recommendations

Timings are inevitably machine dependent. Nevertheless the following table is indicative of the *relative* timings, for typical machines, of the various uniform generators discussed in this section, plus ran4 from §7.5. Smaller values in the table indicate faster generators. The generators ranqd1 and ranqd2 refer to the "quick and dirty" generators immediately above.

Generator	Relative Execution Time
ran0	$\equiv 1.0$
ran1	≈ 1.3
ran2	≈ 2.0
ran3	≈ 0.6
ranqd1	≈ 0.10
ranqd2	≈ 0.25
ran4	≈ 4.0

On balance, we recommend ran1 for general use. It is portable, based on Park and Miller's Minimal Standard generator with an additional shuffle, and has no known (to us) flaws other than period exhaustion.

If you are generating more than 100,000,000 random numbers in a single calculation (that is, more than about 5% of ran1's period), we recommend the use of ran2, with its much longer period.

Knuth's subtractive routine ran3 seems to be the timing winner among portable routines. Unfortunately the subtractive method is not so well studied, and not a standard. We like to keep ran3 in reserve for a "second opinion," substituting it when we suspect another generator of introducing unwanted correlations into a calculation.

The routine ran4 generates *extremely* good random deviates, and has some other nice properties, but it is slow. See §7.5 for discussion.

Finally, the quick and dirty in-line generators ranqd1 and ranqd2 are very fast, but they are somewhat machine dependent, and at best only as good as a 32-bit linear congruential generator ever is — in our view not good enough in many situations. We would use these only in very special cases, where speed is critical.

CITED REFERENCES AND FURTHER READING:

Park, S.K., and Miller, K.W. 1988, *Communications of the ACM*, vol. 31, pp. 1192–1201. [1]

Schrage, L. 1979, *ACM Transactions on Mathematical Software*, vol. 5, pp. 132–138. [2]

Bratley, P., Fox, B.L., and Schrage, E.L. 1983, *A Guide to Simulation*, 2nd ed. (New York: Springer-Verlag). [3]

Knuth, D.E. 1997, *Seminumerical Algorithms*, 3rd ed., vol. 2 of *The Art of Computer Programming* (Reading, MA: Addison-Wesley), §§3.2–3.3. [4]

Kahaner, D., Moler, C., and Nash, S. 1989, *Numerical Methods and Software* (Englewood Cliffs, NJ: Prentice Hall), Chapter 10. [5]

L'Ecuyer, P. 1988, *Communications of the ACM*, vol. 31, pp. 742–774. [6]

Forsythe, G.E., Malcolm, M.A., and Moler, C.B. 1977, *Computer Methods for Mathematical Computations* (Englewood Cliffs, NJ: Prentice-Hall), Chapter 10.

7.2 Transformation Method: Exponential and Normal Deviates

In the previous section, we learned how to generate random deviates with a uniform probability distribution, so that the probability of generating a number between x and $x + dx$, denoted $p(x)dx$, is given by

$$p(x)dx = \begin{cases} dx & 0 < x < 1 \\ 0 & \text{otherwise} \end{cases} \tag{7.2.1}$$

The probability distribution $p(x)$ is of course normalized, so that

$$\int_{-\infty}^{\infty} p(x)dx = 1 \tag{7.2.2}$$

Now suppose that we generate a uniform deviate x and then take some prescribed function of it, $y(x)$. The probability distribution of y, denoted $p(y)dy$, is determined by the fundamental transformation law of probabilities, which is simply

$$|p(y)dy| = |p(x)dx| \tag{7.2.3}$$

or

$$p(y) = p(x) \left| \frac{dx}{dy} \right| \tag{7.2.4}$$

Exponential Deviates

As an example, suppose that $y(x) \equiv -\ln(x)$, and that $p(x)$ is as given by equation (7.2.1) for a uniform deviate. Then

$$p(y)dy = \left| \frac{dx}{dy} \right| dy = e^{-y} dy \tag{7.2.5}$$

which is distributed exponentially. This exponential distribution occurs frequently in real problems, usually as the distribution of waiting times between independent Poisson-random events, for example the radioactive decay of nuclei. You can also easily see (from 7.2.4) that the quantity y/λ has the probability distribution $\lambda e^{-\lambda y}$.

So we have

```
#include <cmath>
#include "nr.h"
using namespace std;

DP NR::expdev(int &idum)
Returns an exponentially distributed, positive, random deviate of unit mean, using
ran1(idum) as the source of uniform deviates.
{
    DP dum;

    do
        dum=ran1(idum);
    while (dum == 0.0);
    return -log(dum);
}
```

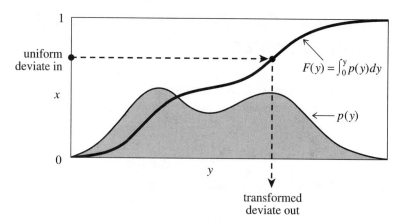

Figure 7.2.1.   Transformation method for generating a random deviate y from a known probability distribution $p(y)$. The indefinite integral of $p(y)$ must be known and invertible. A uniform deviate x is chosen between 0 and 1. Its corresponding y on the definite-integral curve is the desired deviate.

Let's see what is involved in using the above *transformation method* to generate some arbitrary desired distribution of y's, say one with $p(y) = f(y)$ for some positive function f whose integral is 1. (See Figure 7.2.1.) According to (7.2.4), we need to solve the differential equation

$$\frac{dx}{dy} = f(y) \tag{7.2.6}$$

But the solution of this is just $x = F(y)$, where $F(y)$ is the indefinite integral of $f(y)$. The desired transformation which takes a uniform deviate into one distributed as $f(y)$ is therefore

$$y(x) = F^{-1}(x) \tag{7.2.7}$$

where F^{-1} is the inverse function to F. Whether (7.2.7) is feasible to implement depends on whether the *inverse function of the integral of f(y)* is itself feasible to compute, either analytically or numerically. Sometimes it is, and sometimes it isn't.

Incidentally, (7.2.7) has an immediate geometric interpretation: Since $F(y)$ is the area under the probability curve to the left of y, (7.2.7) is just the prescription: choose a uniform random x, then find the value y that has that fraction x of probability area to its left, and return the value y.

Normal (Gaussian) Deviates

Transformation methods generalize to more than one dimension. If $x_1, x_2,$ ... are random deviates with a *joint* probability distribution $p(x_1, x_2, \ldots)$ $dx_1 dx_2 \ldots$, and if $y_1, y_2, \ldots$ are each functions of all the x's (same number of y's as x's), then the joint probability distribution of the y's is

$$p(y_1, y_2, \ldots) dy_1 dy_2 \ldots = p(x_1, x_2, \ldots) \left| \frac{\partial(x_1, x_2, \ldots)}{\partial(y_1, y_2, \ldots)} \right| dy_1 dy_2 \ldots \tag{7.2.8}$$

where $|\partial(\ )/\partial(\ )|$ is the Jacobian determinant of the x's with respect to the y's (or reciprocal of the Jacobian determinant of the y's with respect to the x's).

An important example of the use of (7.2.8) is the *Box-Muller* method for generating random deviates with a normal (Gaussian) distribution,

$$p(y)dy = \frac{1}{\sqrt{2\pi}}e^{-y^2/2}dy \tag{7.2.9}$$

Consider the transformation between two uniform deviates on $(0,1)$, x_1, x_2, and two quantities y_1, y_2,

$$
\begin{aligned}
y_1 &= \sqrt{-2\ln x_1}\cos 2\pi x_2 \\
y_2 &= \sqrt{-2\ln x_1}\sin 2\pi x_2
\end{aligned}
\tag{7.2.10}
$$

Equivalently we can write

$$
\begin{aligned}
x_1 &= \exp\left[-\frac{1}{2}(y_1^2 + y_2^2)\right] \\
x_2 &= \frac{1}{2\pi}\arctan\frac{y_2}{y_1}
\end{aligned}
\tag{7.2.11}
$$

Now the Jacobian determinant can readily be calculated (try it!):

$$\frac{\partial(x_1, x_2)}{\partial(y_1, y_2)} = \begin{vmatrix} \frac{\partial x_1}{\partial y_1} & \frac{\partial x_1}{\partial y_2} \\ \frac{\partial x_2}{\partial y_1} & \frac{\partial x_2}{\partial y_2} \end{vmatrix} = -\left[\frac{1}{\sqrt{2\pi}}e^{-y_1^2/2}\right]\left[\frac{1}{\sqrt{2\pi}}e^{-y_2^2/2}\right] \tag{7.2.12}$$

Since this is the product of a function of y_2 alone and a function of y_1 alone, we see that each y is independently distributed according to the normal distribution (7.2.9).

One further trick is useful in applying (7.2.10). Suppose that, instead of picking uniform deviates x_1 and x_2 in the unit square, we instead pick v_1 and v_2 as the ordinate and abscissa of a random point inside the unit circle around the origin. Then the sum of their squares, $R^2 \equiv v_1^2 + v_2^2$ is a uniform deviate, which can be used for x_1, while the angle that (v_1, v_2) defines with respect to the v_1 axis can serve as the random angle $2\pi x_2$. What's the advantage? It's that the cosine and sine in (7.2.10) can now be written as $v_1/\sqrt{R^2}$ and $v_2/\sqrt{R^2}$, obviating the trigonometric function calls!

We thus have

```
#include <cmath>
#include "nr.h"
using namespace std;

DP NR::gasdev(int &idum)
Returns a normally distributed deviate with zero mean and unit variance, using ran1(idum)
as the source of uniform deviates.
{
    static int iset=0;
    static DP gset;
    DP fac,rsq,v1,v2;

    if (idum < 0) iset=0;                   Reinitialize.
    if (iset == 0) {                        We don't have an extra deviate handy, so
        do {
            v1=2.0*ran1(idum)-1.0;          pick two uniform numbers in the square ex-
            v2=2.0*ran1(idum)-1.0;              tending from -1 to +1 in each direction,
```

```
        rsq=v1*v1+v2*v2;                    see if they are in the unit circle,
    } while (rsq >= 1.0 || rsq == 0.0);       and if they are not, try again.
    fac=sqrt(-2.0*log(rsq)/rsq);
    Now make the Box-Muller transformation to get two normal deviates. Return one and
    save the other for next time.
    gset=v1*fac;
    iset=1;                             Set flag.
    return v2*fac;
} else {                                We have an extra deviate handy,
    iset=0;                             so unset the flag,
    return gset;                        and return it.
}
}
```

See Devroye [1] and Bratley [2] for many additional algorithms.

CITED REFERENCES AND FURTHER READING:

Devroye, L. 1986, *Non-Uniform Random Variate Generation* (New York: Springer-Verlag), §9.1. [1]

Bratley, P., Fox, B.L., and Schrage, E.L. 1983, *A Guide to Simulation*, 2nd ed. (New York: Springer-Verlag). [2]

Knuth, D.E. 1997, *Seminumerical Algorithms*, 3rd ed., vol. 2 of *The Art of Computer Programming* (Reading, MA: Addison-Wesley), pp. 121ff.

7.3 Rejection Method: Gamma, Poisson, Binomial Deviates

The *rejection method* is a powerful, general technique for generating random deviates whose distribution function $p(x)dx$ (probability of a value occurring between x and $x + dx$) is known and computable. The rejection method does *not* require that the cumulative distribution function [indefinite integral of $p(x)$] be readily computable, much less the inverse of that function — which was required for the transformation method in the previous section.

The rejection method is based on a simple geometrical argument:

Draw a graph of the probability distribution $p(x)$ that you wish to generate, so that the area under the curve in any range of x corresponds to the desired probability of generating an x in that range. If we had some way of choosing a random point *in two dimensions*, with uniform probability in the *area* under your curve, then the x value of that random point would have the desired distribution.

Now, on the same graph, draw any other curve $f(x)$ which has finite (not infinite) area and lies everywhere *above* your original probability distribution. (This is always possible, because your original curve encloses only unit area, by definition of probability.) We will call this $f(x)$ the *comparison function*. Imagine now that you have some way of choosing a random point in two dimensions that is uniform in the area under the comparison function. Whenever that point lies outside the area under the original probability distribution, we will *reject* it and choose another random point. Whenever it lies inside the area under the original probability distribution, we will *accept* it. It should be obvious that the accepted points are uniform in the

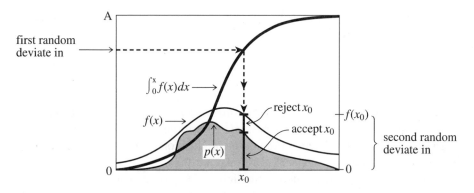

Figure 7.3.1. Rejection method for generating a random deviate x from a known probability distribution $p(x)$ that is everywhere less than some other function $f(x)$. The transformation method is first used to generate a random deviate x of the distribution f (compare Figure 7.2.1). A second uniform deviate is used to decide whether to accept or reject that x. If it is rejected, a new deviate of f is found; and so on. The ratio of accepted to rejected points is the ratio of the area under p to the area between p and f.

accepted area, so that their x values have the desired distribution. It should also be obvious that the fraction of points rejected just depends on the ratio of the area of the comparison function to the area of the probability distribution function, not on the details of shape of either function. For example, a comparison function whose area is less than 2 will reject fewer than half the points, even if it approximates the probability function very badly at some values of x, e.g., remains finite in some region where $p(x)$ is zero.

It remains only to suggest how to choose a uniform random point in two dimensions under the comparison function $f(x)$. A variant of the transformation method (§7.2) does nicely: Be sure to have chosen a comparison function whose indefinite integral is known analytically, and is also analytically invertible to give x as a function of "area under the comparison function to the left of x." Now pick a uniform deviate between 0 and A, where A is the total area under $f(x)$, and use it to get a corresponding x. Then pick a uniform deviate between 0 and $f(x)$ as the y value for the two-dimensional point. You should be able to convince yourself that the point (x, y) is uniformly distributed in the area under the comparison function $f(x)$.

An equivalent procedure is to pick the second uniform deviate between zero and one, and accept or reject according to whether it is respectively less than or greater than the ratio $p(x)/f(x)$.

So, to summarize, the rejection method for some given $p(x)$ requires that one find, once and for all, some reasonably good comparison function $f(x)$. Thereafter, each deviate generated requires two uniform random deviates, one evaluation of f (to get the coordinate y), and one evaluation of p (to decide whether to accept or reject the point x, y). Figure 7.3.1 illustrates the procedure. Then, of course, this procedure must be repeated, on the average, A times before the final deviate is obtained.

Gamma Distribution

The gamma distribution of integer order $a > 0$ is the waiting time to the ath event in a Poisson random process of unit mean. For example, when $a = 1$, it is just the exponential distribution of §7.2, the waiting time to the first event.

A gamma deviate has probability $p_a(x)dx$ of occurring with a value between x and $x + dx$, where

$$p_a(x)dx = \frac{x^{a-1}e^{-x}}{\Gamma(a)}dx \qquad x > 0 \qquad (7.3.1)$$

To generate deviates of (7.3.1) for small values of a, it is best to add up a exponentially distributed waiting times, i.e., logarithms of uniform deviates. Since the sum of logarithms is the logarithm of the product, one really has only to generate the product of a uniform deviates, then take the log.

For larger values of a, the distribution (7.3.1) has a typically "bell-shaped" form, with a peak at $x = a$ and a half-width of about $\sqrt{a}$.

We will be interested in several probability distributions with this same qualitative form. A useful comparison function in such cases is derived from the *Lorentzian distribution*

$$p(y)dy = \frac{1}{\pi}\left(\frac{1}{1 + y^2}\right)dy \qquad (7.3.2)$$

whose inverse indefinite integral is just the tangent function. It follows that the x-coordinate of an area-uniform random point under the comparison function

$$f(x) = \frac{c_0}{1 + (x - x_0)^2/a_0^2} \qquad (7.3.3)$$

for any constants a_0, c_0, and x_0, can be generated by the prescription

$$x = a_0 \tan(\pi U) + x_0 \qquad (7.3.4)$$

where U is a uniform deviate between 0 and 1. Thus, for some specific "bell-shaped" $p(x)$ probability distribution, we need only find constants a_0, c_0, x_0, with the product $a_0 c_0$ (which determines the area) as small as possible, such that (7.3.3) is everywhere greater than $p(x)$.

Ahrens has done this for the gamma distribution, yielding the following algorithm (as described in Knuth [1]):

```
#include <cmath>
#include "nr.h"
using namespace std;

DP NR::gamdev(const int ia, int &idum)
Returns a deviate distributed as a gamma distribution of integer order ia, i.e., a waiting time
to the iath event in a Poisson process of unit mean, using ran1(idum) as the source of
uniform deviates.
{
    int j;
    DP am,e,s,v1,v2,x,y;

    if (ia < 1) nrerror("Error in routine gamdev");
    if (ia < 6) {                          Use direct method, adding waiting
        x=1.0;                                          times.
        for (j=1;j<=ia;j++) x *= ran1(idum);
        x = -log(x);
    } else {                               Use rejection method.
```

```
    do {
        do {
            do {
                v1=ran1(idum);
                v2=2.0*ran1(idum)-1.0;
            } while (v1*v1+v2*v2 > 1.0);
            y=v2/v1;
            am=ia-1;
            s=sqrt(2.0*am+1.0);
            x=s*y+am;
        } while (x <= 0.0);
        e=(1.0+y*y)*exp(am*log(x/am)-s*y);
    } while (ran1(idum) > e);
    }
    return x;
}
```

These four lines generate the tangent of a random angle, i.e., they are equivalent to
y = tan(π * ran1(idum)).

We decide whether to reject x:
Reject in region of zero probability.
Ratio of prob. fn. to comparison fn.
Reject on basis of a second uniform deviate.

Poisson Deviates

The Poisson distribution is conceptually related to the gamma distribution. It gives the probability of a certain integer number m of unit rate Poisson random events occurring in a given interval of time x, while the gamma distribution was the probability of waiting time between x and $x + dx$ to the mth event. Note that m takes on only integer values ≥ 0, so that the Poisson distribution, viewed as a continuous distribution function $p_x(m)dm$, is zero everywhere except where m is an integer ≥ 0. At such places, it is infinite, such that the integrated probability over a region containing the integer is some finite number. The total probability at an integer j is

$$\text{Prob}(j) = \int_{j-\epsilon}^{j+\epsilon} p_x(m)dm = \frac{x^j e^{-x}}{j!} \qquad (7.3.5)$$

At first sight this might seem an unlikely candidate distribution for the rejection method, since no continuous comparison function can be larger than the infinitely tall, but infinitely narrow, *Dirac delta functions* in $p_x(m)$. However, there is a trick that we can do: Spread the finite area in the spike at j uniformly into the interval between j and $j + 1$. This defines a continuous distribution $q_x(m)dm$ given by

$$q_x(m)dm = \frac{x^{[m]} e^{-x}}{[m]!} dm \qquad (7.3.6)$$

where $[m]$ represents the largest integer less than m. If we now use the rejection method to generate a (noninteger) deviate from (7.3.6), and then take the integer part of that deviate, it will be as if drawn from the desired distribution (7.3.5). (See Figure 7.3.2.) This trick is general for any integer-valued probability distribution.

For x large enough, the distribution (7.3.6) is qualitatively bell-shaped (albeit with a bell made out of small, square steps), and we can use the same kind of Lorentzian comparison function as was already used above. For small x, we can generate independent exponential deviates (waiting times between events); when the sum of these first exceeds x, then the number of events that would have occurred in waiting time x becomes known and is one less than the number of terms in the sum.

These ideas produce the following routine:

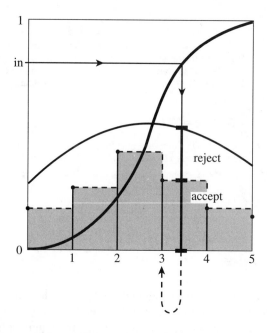

Figure 7.3.2.   Rejection method as applied to an integer-valued distribution. The method is performed on the step function shown as a dashed line, yielding a real-valued deviate. This deviate is rounded down to the next lower integer, which is output.

```
#include <cmath>
#include "nr.h"
using namespace std;

DP NR::poidev(const DP xm, int &idum)
```
Returns as a floating-point number an integer value that is a random deviate drawn from a Poisson distribution of mean `xm`, using `ran1(idum)` as a source of uniform random deviates.
```
{
    const DP PI=3.141592653589793238;
    static DP sq,alxm,g,oldm=(-1.0);      oldm is a flag for whether xm has changed since
    DP em,t,y;                            last call.

    if (xm < 12.0) {                      Use direct method.
        if (xm != oldm) {
            oldm=xm;
            g=exp(-xm);                   If xm is new, compute the exponential.
        }
        em = -1;
        t=1.0;
        do {                              Instead of adding exponential deviates it is equiv-
            ++em;                             alent to multiply uniform deviates.  We never
            t *= ran1(idum);                  actually have to take the log, merely com-
        } while (t > g);                      pare to the pre-computed exponential.
    } else {                              Use rejection method.
        if (xm != oldm) {                 If xm has changed since the last call, then pre-
            oldm=xm;                          compute some functions that occur below.
            sq=sqrt(2.0*xm);
            alxm=log(xm);
            g=xm*alxm-gammln(xm+1.0);
            The function gammln is the natural log of the gamma function, as given in §6.1.
        }
        do {
            do {                          y is a deviate from a Lorentzian comparison func-
                y=tan(PI*ran1(idum));         tion.
```

```
                em=sq*y+xm;                   em is y, shifted and scaled.
        } while (em < 0.0);                   Reject if in regime of zero probability.
            em=floor(em);                     The trick for integer-valued distributions.
            t=0.9*(1.0+y*y)*exp(em*alxm-gammln(em+1.0)-g);
            The ratio of the desired distribution to the comparison function; we accept or
            reject by comparing it to another uniform deviate. The factor 0.9 is chosen
            so that t never exceeds 1.
    } while (ran1(idum) > t);
}
return em;
}
```

Binomial Deviates

If an event occurs with probability q, and we make n trials, then the number of times m that it occurs has the binomial distribution,

$$\int_{j-\epsilon}^{j+\epsilon} p_{n,q}(m)dm = \binom{n}{j} q^j (1-q)^{n-j} \qquad (7.3.7)$$

The binomial distribution is integer valued, with m taking on possible values from 0 to n. It depends on *two* parameters, n and q, so is correspondingly a bit harder to implement than our previous examples. Nevertheless, the techniques already illustrated are sufficiently powerful to do the job:

```
#include <cmath>
#include "nr.h"
using namespace std;

DP NR::bnldev(const DP pp, const int n, int &idum)
Returns as a floating-point number an integer value that is a random deviate drawn from
a binomial distribution of n trials each of probability pp, using ran1(idum) as a source of
uniform random deviates.
{
    const DP PI=3.141592653589793238;
    int j;
    static int nold=(-1);
    DP am,em,g,angle,p,bnl,sq,t,y;
    static DP pold=(-1.0),pc,plog,pclog,en,oldg;

    p=(pp <= 0.5 ? pp : 1.0-pp);
    The binomial distribution is invariant under changing pp to 1-pp, if we also change the
    answer to n minus itself; we'll remember to do this below.
    am=n*p;                         This is the mean of the deviate to be produced.
    if (n < 25) {                   Use the direct method while n is not too large.
        bnl=0.0;                        This can require up to 25 calls to ran1.
        for (j=0;j<n;j++)
            if (ran1(idum) < p) ++bnl;
    } else if (am < 1.0) {          If fewer than one event is expected out of 25
        g=exp(-am);                     or more trials, then the distribution is quite
        t=1.0;                          accurately Poisson. Use direct Poisson method.
        for (j=0;j<=n;j++) {
            t *= ran1(idum);
            if (t < g) break;
        }
        bnl=(j <= n ? j : n);
    } else {                        Use the rejection method.
```

```
    if (n != nold) {                If n has changed, then compute useful quanti-
        en=n;                       ties.
        oldg=gammln(en+1.0);
        nold=n;
    } if (p != pold) {              If p has changed, then compute useful quanti-
        pc=1.0-p;                   ties.
        plog=log(p);
        pclog=log(pc);
        pold=p;
    }
    sq=sqrt(2.0*am*pc);             The following code should by now seem familiar:
    do {                            rejection method with a Lorentzian compar-
        do {                        ison function.
            angle=PI*ran1(idum);
            y=tan(angle);
            em=sq*y+am;
        } while (em < 0.0 || em >= (en+1.0));      Reject.
        em=floor(em);               Trick for integer-valued distribution.
        t=1.2*sq*(1.0+y*y)*exp(oldg-gammln(em+1.0)
            -gammln(en-em+1.0)+em*plog+(en-em)*pclog);
    } while (ran1(idum) > t);       Reject. This happens about 1.5 times per devi-
    bnl=em;                         ate, on average.
    }
    if (p != pp) bnl=n-bnl;         Remember to undo the symmetry transforma-
    return bnl;                     tion.
}
```

See Devroye [2] and Bratley [3] for many additional algorithms.

CITED REFERENCES AND FURTHER READING:

Knuth, D.E. 1997, *Seminumerical Algorithms*, 3rd ed., vol. 2 of *The Art of Computer Programming* (Reading, MA: Addison-Wesley), pp. 125ff. [1]

Devroye, L. 1986, *Non-Uniform Random Variate Generation* (New York: Springer-Verlag), §X.4. [2]

Bratley, P., Fox, B.L., and Schrage, E.L. 1983, *A Guide to Simulation*, 2nd ed. (New York: Springer-Verlag). [3].

7.4 Generation of Random Bits

The C++ language gives you useful access to some machine-level bitwise operations such as << (left shift). This section will show you how to put such abilities to good use.

The problem is how to generate single random bits, with 0 and 1 equally probable. Of course you can just generate uniform random deviates between zero and one and use their high-order bit (i.e., test if they are greater than or less than 0.5). However this takes a lot of arithmetic; there are special-purpose applications, such as real-time signal processing, where you want to generate bits very much faster than that.

One method for generating random bits, with two variant implementations, is based on "primitive polynomials modulo 2." The theory of these polynomials is beyond our scope (although §7.7 and §20.3 will give you small tastes of it). Here, suffice it to say that there are special polynomials among those whose coefficients are zero or one. An example is

$$x^{18} + x^5 + x^2 + x^1 + x^0 \qquad (7.4.1)$$

which we can abbreviate by just writing the nonzero powers of x, e.g.,

$$(18, 5, 2, 1, 0)$$

Every primitive polynomial modulo 2 of order n (=18 above) defines a recurrence relation for obtaining a new random bit from the n preceding ones. The recurrence relation is guaranteed to produce a sequence of maximal length, i.e., cycle through all possible sequences of n bits (except all zeros) before it repeats. Therefore one can seed the sequence with any initial bit pattern (except all zeros), and get $2^n - 1$ random bits before the sequence repeats.

Let the bits be numbered from 1 (most recently generated) through n (generated n steps ago), and denoted $a_1, a_2, \ldots, a_n$. We want to give a formula for a new bit a_0. After generating a_0 we will shift all the bits by one, so that the old a_n is finally lost, and the new a_0 becomes a_1. We then apply the formula again, and so on.

"Method I" is the easiest to implement in hardware, requiring only a single shift register n bits long and a few XOR ("exclusive or" or bit addition mod 2) gates, the operation denoted in C++ by "$\wedge$". For the primitive polynomial given above, the recurrence formula is

$$a_0 = a_{18} \wedge a_5 \wedge a_2 \wedge a_1 \qquad (7.4.2)$$

The terms that are $\wedge$'d together can be thought of as "taps" on the shift register, $\wedge$'d into the register's input. More generally, there is precisely one term for each nonzero coefficient in the primitive polynomial except the constant (zero bit) term. So the first term will always be a_n for a primitive polynomial of degree n, while the last term might or might not be a_1, depending on whether the primitive polynomial has a term in x^1.

While it is simple in hardware, Method I is somewhat cumbersome in C++, because the individual bits must be collected by a sequence of full-word masks:

```
#include "nr.h"

int NR::irbit1(unsigned long &iseed)
Returns as an integer a random bit, based on the 18 low-significance bits in iseed (which is
modified for the next call).
{
    unsigned long newbit;                    The accumulated XOR's.

    newbit = ((iseed >> 17) & 1)             Get bit 18.
        ^ ((iseed >> 4) & 1)                 XOR with bit 5.
        ^ ((iseed >> 1) & 1)                 XOR with bit 2.
        ^ (iseed & 1);                       XOR with bit 1.
    iseed=(iseed << 1) | newbit;             Leftshift the seed and put the result of the
    return int(newbit);                          XOR's in its bit 1.
}
```

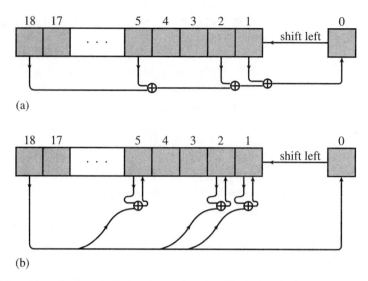

Figure 7.4.1.   Two related methods for obtaining random bits from a shift register and a primitive polynomial modulo 2. (a) The contents of selected taps are combined by exclusive-or (addition modulo 2), and the result is shifted in from the right. This method is easiest to implement in hardware. (b) Selected bits are modified by exclusive-or with the leftmost bit, which is then shifted in from the right. This method is easiest to implement in software.

"Method II" is less suited to direct hardware implementation (though still possible), but is beautifully suited to C++. It modifies more than one bit among the saved n bits as each new bit is generated (Figure 7.4.1). It generates the maximal length sequence, but not in the same order as Method I. The prescription for the primitive polynomial (7.4.1) is:

$$
\begin{aligned}
a_0 &= a_{18} \\
a_5 &= a_5 \wedge a_0 \\
a_2 &= a_2 \wedge a_0 \\
a_1 &= a_1 \wedge a_0
\end{aligned}
\qquad (7.4.3)
$$

In general there will be an exclusive-or for each nonzero term in the primitive polynomial except 0 and n. The nice feature about Method II is that all the exclusive-or's can usually be done as a single full-word exclusive-or operation:

```
#include "nr.h"

int NR::irbit2(unsigned long &iseed)
Returns as an integer a random bit, based on the 18 low-significance bits in iseed (which is
modified for the next call).
{
    const unsigned long IB1=1,IB2=2,IB5=16,IB18=131072;       Powers of 2.
    const unsigned long MASK=IB1+IB2+IB5;

    if (iseed & IB18) {             Change all masked bits, shift, and put 1 into bit 1.
        iseed=((iseed ^ MASK) << 1) | IB1;
        return 1;
    } else {                        Shift and put 0 into bit 1.
        iseed <<= 1;
```

```
      return 0;
  }
}
```

Some Primitive Polynomials Modulo 2 (after Watson)							
(1,	0)						
(2,	1,	0)					
(3,	1,	0)					
(4,	1,	0)					
(5,	2,	0)					
(6,	1,	0)					
(7,	1,	0)					
(8,	4,	3,	2,	0)			
(9,	4,	0)					
(10,	3,	0)					
(11,	2,	0)					
(12,	6,	4,	1,	0)			
(13,	4,	3,	1,	0)			
(14,	5,	3,	1,	0)			
(15,	1,	0)					
(16,	5,	3,	2,	0)			
(17,	3,	0)					
(18,	5,	2,	1,	0)			
(19,	5,	2,	1,	0)			
(20,	3,	0)					
(21,	2,	0)					
(22,	1,	0)					
(23,	5,	0)					
(24,	4,	3,	1,	0)			
(25,	3,	0)					
(26,	6,	2,	1,	0)			
(27,	5,	2,	1,	0)			
(28,	3,	0)					
(29,	2,	0)					
(30,	6,	4,	1,	0)			
(31,	3,	0)					
(32,	7,	5,	3,	2,	1,	0)	
(33,	6,	4,	1,	0)			
(34,	7,	6,	5,	2,	1,	0)	
(35,	2,	0)					
(36,	6,	5,	4,	2,	1,	0)	
(37,	5,	4,	3,	2,	1,	0)	
(38,	6,	5,	1,	0)			
(39,	4,	0)					
(40,	5,	4	3,	0)			
(41,	3,	0)					
(42,	5,	4,	3,	2,	1,	0)	
(43,	6,	4,	3,	0)			
(44,	6,	5,	2,	0)			
(45,	4,	3,	1,	0)			
(46,	8,	5,	3,	2,	1,	0)	
(47,	5,	0)					
(48,	7,	5,	4,	2,	1,	0)	
(49,	6,	5,	4,	0)			
(50,	4,	3,	2,	0)			
(51,	6,	3,	1,	0)			
(52,	3,	0)					
(53,	6,	2,	1,	0)			
(54,	6,	5,	4,	3,	2,	0)	
(55,	6,	2,	1,	0)			
(56,	7,	4,	2,	0)			
(57,	5,	3,	2,	0)			
(58,	6,	5,	1,	0)			
(59,	6,	5,	4,	3,	1,	0)	
(60,	1,	0)					
(61,	5,	2,	1,	0)			
(62,	6,	5,	3,	0)			
(63,	1,	0)					
(64,	4,	3,	1,	0)			
(65,	4,	3,	1,	0)			
(66,	8,	6,	5,	3,	2,	0)	
(67,	5,	2,	1,	0)			
(68,	7,	5,	1,	0)			
(69,	6,	5,	2,	0)			
(70,	5,	3,	1,	0)			
(71,	5,	3,	1,	0)			
(72,	6,	4,	3,	2,	1,	0)	
(73,	4,	3,	2,	0)			
(74,	7,	4,	3,	0)			
(75,	6,	3,	1,	0)			
(76,	5,	4,	2,	0)			
(77,	6,	5,	2,	0)			
(78,	7,	2,	1,	0)			
(79,	4,	3,	2,	0)			
(80,	7,	5,	3,	2,	1,	0)	
(81,	4	0)					
(82,	8,	7,	6,	4,	1,	0)	
(83,	7,	4,	2,	0)			
(84,	8,	7,	5,	3,	1,	0)	
(85,	8,	2,	1,	0)			
(86,	6,	5,	2,	0)			
(87,	7,	5,	1,	0)			
(88,	8,	5,	4,	3,	1,	0)	
(89,	6,	5,	3,	0)			
(90,	5,	3,	2,	0)			
(91,	7,	6,	5,	3,	2,	0)	
(92,	6,	5,	2,	0)			
(93,	2,	0)					
(94,	6,	5,	1,	0)			
(95,	6,	5,	4,	2,	1,	0)	
(96,	7,	6,	4,	3,	2,	0)	
(97,	6,	0)					
(98,	7,	4,	3,	2,	1,	0)	
(99,	7,	5,	4,	0)			
(100,	8,	7,	2,	0)			

A word of caution is: Don't use sequential bits from these routines as the bits of a large, supposedly random, integer, or as the bits in the mantissa of a supposedly random floating-point number. They are not very random for that purpose; see

Knuth [1]. Examples of acceptable uses of these random bits are: (i) multiplying a signal randomly by ± 1 at a rapid "chip rate," so as to spread its spectrum uniformly (but recoverably) across some desired bandpass, or (ii) Monte Carlo exploration of a binary tree, where decisions as to whether to branch left or right are to be made randomly.

Now we do not want you to go through life thinking that there is something special about the primitive polynomial of degree 18 used in the above examples. (We chose 18 because 2^{18} is small enough for you to verify our claims directly by numerical experiment.) The accompanying table [2] lists one primitive polynomial for each degree up to 100. (In fact there exist many such for each degree. For example, see §7.7 for a complete table up to degree 10.)

CITED REFERENCES AND FURTHER READING:

Knuth, D.E. 1997, *Seminumerical Algorithms*, 3rd ed., vol. 2 of *The Art of Computer Programming* (Reading, MA: Addison-Wesley), pp. 30ff. [1]

Horowitz, P., and Hill, W. 1989, *The Art of Electronics*, 2nd ed. (Cambridge: Cambridge University Press), §§9.32–9.37.

Tausworthe, R.C. 1965, *Mathematics of Computation*, vol. 19, pp. 201–209.

Watson, E.J. 1962, *Mathematics of Computation*, vol. 16, pp. 368–369. [2]

7.5 Random Sequences Based on Data Encryption

In *Numerical Recipes'* first edition, we described how to use the Data Encryption Standard (DES) [1-3] for the generation of random numbers. Unfortunately, when implemented in software in a high-level language like C++, DES is very slow, so excruciatingly slow, in fact, that our previous implementation can be viewed as more mischievous than useful. Here we give a much faster and simpler algorithm which, though it may not be secure in the cryptographic sense, generates about equally good random numbers.

DES, like its progenitor cryptographic system LUCIFER, is a so-called "block product cipher" [4]. It acts on 64 bits of input by iteratively applying (16 times, in fact) a kind of highly nonlinear bit-mixing function. Figure 7.5.1 shows the flow of information in DES during this mixing. The function g, which takes 32-bits into 32-bits, is called the "cipher function." Meyer and Matyas [4] discuss the importance of the cipher function being nonlinear, as well as other design criteria.

DES constructs its cipher function g from an intricate set of bit permutations and table lookups acting on short sequences of consecutive bits. Apparently, this function was chosen to be particularly strong cryptographically (or conceivably as some critics contend, to have an exquisitely subtle cryptographic flaw!). For our purposes, a different function g that can be rapidly computed in a high-level computer language is preferable. Such a function may weaken the algorithm cryptographically. Our purposes are not, however, cryptographic: We want to find the fastest g, and smallest number of iterations of the mixing procedure in Figure 7.5.1, such that our output random sequence passes the standard tests that are customarily applied to random number generators. The resulting algorithm will not be DES, but rather a kind of "pseudo-DES," better suited to the purpose at hand.

Following the criterion, mentioned above, that g should be nonlinear, we must give the integer multiply operation a prominent place in g. Because 64-bit registers are not generally accessible in high-level languages, we must confine ourselves to multiplying 16-bit operands into a 32-bit result. So, the general idea of g, almost forced, is to calculate the three distinct 32-bit products of the high and low 16-bit input half-words, and then to combine

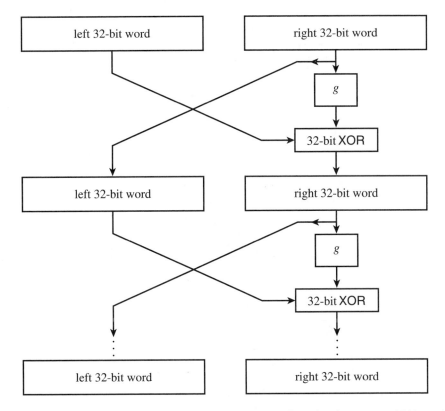

Figure 7.5.1. The Data Encryption Standard (DES) iterates a nonlinear function g on two 32-bit words, in the manner shown here (after Meyer and Matyas [4]).

these, and perhaps additional fixed constants, by fast operations (e.g., add or exclusive-or) into a single 32-bit result.

There are only a limited number of ways of effecting this general scheme, allowing systematic exploration of the alternatives. Experimentation, and tests of the randomness of the output, lead to the sequence of operations shown in Figure 7.5.2. The few new elements in the figure need explanation: The values C_1 and C_2 are fixed constants, chosen randomly with the constraint that they have exactly 16 1-bits and 16 0-bits; combining these constants via exclusive-or ensures that the overall g has no bias towards 0 or 1 bits.

The "reverse half-words" operation in Figure 7.5.2 turns out to be essential; otherwise, the very lowest and very highest bits are not properly mixed by the three multiplications. The nonobvious choices in g are therefore: where along the vertical "pipeline" to do the reverse; in what order to combine the three products and C_2; and with which operation (add or exclusive-or) should each combining be done? We tested these choices exhaustively before settling on the algorithm shown in the figure.

It remains to determine the smallest number of iterations N_{it} that we can get away with. The minimum meaningful N_{it} is evidently two, since a single iteration simply moves one 32-bit word without altering it. One can use the constants C_1 and C_2 to help determine an appropriate N_{it}: When $N_{it} = 2$ and $C_1 = C_2 = 0$ (an intentionally very poor choice), the generator fails several tests of randomness by easily measurable, though not overwhelming, amounts. When $N_{it} = 4$, on the other hand, or with $N_{it} = 2$ but with the constants C_1, C_2 nonsparse, we have been unable to find *any* statistical deviation from randomness in sequences of up to 10^9 floating numbers r_i derived from this scheme. The combined strength of $N_{it} = 4$ and nonsparse C_1, C_2 should therefore give sequences that are random to tests even far beyond those that we have actually tried. These are our recommended conservative parameter values, notwithstanding the fact that $N_{it} = 2$ (which is, of course, twice as fast) has no nonrandomness discernible (by us).

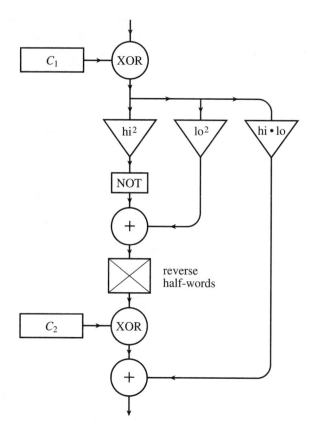

Figure 7.5.2.   The nonlinear function g used by the routine psdes.

Implementation of these ideas is straightforward. The following routine is not quite strictly portable, since it assumes that unsigned long integers are 32-bits, as is the case on most machines. However, there is no reason to believe that *longer* integers would be in any way inferior (with suitable extensions of the constants C_1, C_2). C++ does not provide a convenient, portable way to divide a long integer into half words, so we must use a combination of masking (& 0xffff) with left- and right-shifts by 16 bits (<<16 and >>16). On some machines the half-word extraction could be made faster by the use of C++'s union construction, but this would generally not be portable between "big-endian" and "little-endian" machines. (Big- and little-endian refer to the order in which the bytes are stored in a word.)

```
#include "nr.h"

void NR::psdes(unsigned long &lword, unsigned long &irword)
"Pseudo-DES" hashing of the 64-bit word (lword,irword). Both 32-bit arguments are re-
turned hashed on all bits.
{
    const int NITER=4;
    static const unsigned long c1[NITER]={
        0xbaa96887L, 0x1e17d32cL, 0x03bcdc3cL, 0x0f33d1b2L};
    static const unsigned long c2[NITER]={
        0x4b0f3b58L, 0xe874f0c3L, 0x6955c5a6L, 0x55a7ca46L};
    unsigned long i,ia,ib,iswap,itmph=0,itmpl=0;

    for (i=0;i<NITER;i++) {
```
Perform niter iterations of DES logic, using a simpler (non-cryptographic) nonlinear func-
tion instead of DES's.
```
        ia=(iswap=irword) ^ c1[i];              The bit-rich constants c1 and (below)
                                                c2 guarantee lots of nonlinear mix-
                                                ing.
```

```
                itmpl = ia & 0xffff;
                itmph = ia >> 16;
                ib=itmpl*itmpl+ ~(itmph*itmph);
                irword=lword ^ (((ia = (ib >> 16) |
                    ((ib & 0xffff) << 16)) ^ c2[i])+itmpl*itmph);
                lword=iswap;
        }
}
```

The routine `ran4`, listed below, uses `psdes` to generate uniform random deviates. We adopt the convention that a negative value of the argument `idum` sets the left 32-bit word, while a positive value i sets the right 32-bit word, returns the ith random deviate, and increments `idum` to $i + 1$. This is no more than a convenient way of defining many different sequences (negative values of `idum`), but still with random access to each sequence (positive values of `idum`). For getting a floating-point number from the 32-bit integer, we like to do it by the masking trick described at the end of §7.1, above. The hex constants 3F800000 and 007FFFFF are the appropriate ones for computers using the IEEE representation for 32-bit floating-point numbers (e.g., Pentium machines and many Linux or UNIX workstations). For DEC VAXes, the correct hex constants are, respectively, 00004080 and FFFF007F. For greater portability, you can instead construct a floating number by making the (signed) 32-bit integer nonnegative (typically, you add exactly 2^{31} if it is negative) and then multiplying it by a floating constant (typically $2.^{-31}$).

An interesting, and sometimes useful, feature of the routine `ran4`, below, is that it allows random access to the nth random value in a sequence, without the necessity of first generating values $1 \cdots n - 1$. This property is shared by any random number generator based on *hashing* (the technique of mapping data keys, which may be highly clustered in value, approximately uniformly into a storage address space) [5,6]. One might have a simulation problem in which some certain rare situation becomes recognizable by its consequences only considerably after it has occurred. One may wish to restart the simulation back at that occurrence, using identical random values but, say, varying some other control parameters. The relevant question might then be something like "what random numbers were used in cycle number 337098901?" It might already be cycle number 395100273 before the question comes up. Random generators based on recursion, rather than hashing, cannot easily answer such a question.

```
#include "nr.h"

DP NR::ran4(int &idum)
```
Returns a uniform random deviate in the range 0.0 to 1.0, generated by pseudo-DES (DES-like) hashing of the 64-bit word (`idums`,`idum`), where `idums` was set by a previous call with negative `idum`. Also increments `idum`. Routine can be used to generate a random sequence by successive calls, leaving `idum` unaltered between calls; or it can randomly access the nth deviate in a sequence by calling with `idum` = n. Different sequences are initialized by calls with differing negative values of `idum`.
```
{
```
The hexadecimal constants `jflone` and `jflmsk` below are used to produce a floating number between 1 and 2 by bitwise masking. They are machine-dependent. See text.
```
#if defined(vax) || defined(_vax_) || defined(__vax__) || defined(VAX)
    static const unsigned long jflone = 0x00004080;
    static const unsigned long jflmsk = 0xffff007f;
#else
    static const unsigned long jflone = 0x3f800000;
    static const unsigned long jflmsk = 0x007fffff;
#endif
    unsigned long irword,itemp,lword;
    static int idums = 0;

    if (idum < 0) {                          Reset idums and prepare to return the first
        idums = -idum;                             deviate in its sequence.
        idum=1;
    }
```

```
        irword=idum;
        lword=idums;
        psdes(lword,irword);                    "Pseudo-DES" encode the words.
        itemp=jflone | (jflmsk & irword);       Mask to a floating number between 1 and
        ++idum;                                  2.
        return (*(float *)&itemp)-1.0;          Subtraction moves range to 0 to 1.
    }
```

The accompanying table gives data for verifying that `ran4` and `psdes` work correctly on your machine. We do not advise the use of `ran4` unless you are able to reproduce the hex values shown. Typically, `ran4` is about 4 times slower than `ran0` (§7.1), or about 3 times slower than `ran1`.

Values for Verifying the Implementation of `psdes`					
idum	before psdes call		after psdes call (hex)		ran4(idum)
	lword	irword	lword	irword	VAX	Pentium
−1	1	1	604D1DCE	509C0C23	0.275898	0.219120
99	1	99	D97F8571	A66CB41A	0.208204	0.849246
−99	99	1	7822309D	64300984	0.034307	0.375290
99	99	99	D7F376F0	59BA89EB	0.838676	0.457334

Successive calls to `ran4` with arguments −1, 99, −99, and 99 should produce exactly the `lword` and `irword` values shown. Masking conversion to a returned floating random value is allowed to be machine dependent; values for VAX and Pentium are shown.

CITED REFERENCES AND FURTHER READING:

Data Encryption Standard, 1977 January 15, Federal Information Processing Standards Publication, number 46 (Washington: U.S. Department of Commerce, National Bureau of Standards). [1]

Guidelines for Implementing and Using the NBS Data Encryption Standard, 1981 April 1, Federal Information Processing Standards Publication, number 74 (Washington: U.S. Department of Commerce, National Bureau of Standards). [2]

Validating the Correctness of Hardware Implementations of the NBS Data Encryption Standard, 1980, NBS Special Publication 500–20 (Washington: U.S. Department of Commerce, National Bureau of Standards). [3]

Meyer, C.H. and Matyas, S.M. 1982, *Cryptography: A New Dimension in Computer Data Security* (New York: Wiley). [4]

Knuth, D.E. 1997, *Sorting and Searching*, 3rd ed., vol. 3 of *The Art of Computer Programming* (Reading, MA: Addison-Wesley), Chapter 6. [5]

Vitter, J.S., and Chen, W-C. 1987, *Design and Analysis of Coalesced Hashing* (New York: Oxford University Press). [6]

7.6 Simple Monte Carlo Integration

Inspirations for numerical methods can spring from unlikely sources. "Splines" first were flexible strips of wood used by draftsmen. "Simulated annealing" (we shall see in §10.9) is rooted in a thermodynamic analogy. And who does not feel at least a faint echo of glamor in the name "Monte Carlo method"?

Suppose that we pick N random points, uniformly distributed in a multidimensional volume V. Call them $x_0, \ldots, x_{N-1}$. Then the basic theorem of Monte Carlo integration estimates the integral of a function f over the multidimensional volume,

$$\int f \, dV \approx V \langle f \rangle \pm V \sqrt{\frac{\langle f^2 \rangle - \langle f \rangle^2}{N}} \tag{7.6.1}$$

Here the angle brackets denote taking the arithmetic mean over the N sample points,

$$\langle f \rangle \equiv \frac{1}{N} \sum_{i=0}^{N-1} f(x_i) \qquad \langle f^2 \rangle \equiv \frac{1}{N} \sum_{i=0}^{N-1} f^2(x_i) \tag{7.6.2}$$

The "plus-or-minus" term in (7.6.1) is a one standard deviation error estimate for the integral, not a rigorous bound; further, there is no guarantee that the error is distributed as a Gaussian, so the error term should be taken only as a rough indication of probable error.

Suppose that you want to integrate a function g over a region W that is not easy to sample randomly. For example, W might have a very complicated shape. No problem. Just find a region V that *includes* W and that *can* easily be sampled (Figure 7.6.1), and then define f to be equal to g for points in W and equal to zero for points outside of W (but still inside the sampled V). You want to try to make V enclose W as closely as possible, because the zero values of f will increase the error estimate term of (7.6.1). And well they should: points chosen outside of W have no information content, so the effective value of N, the number of points, is reduced. The error estimate in (7.6.1) takes this into account.

General purpose routines for Monte Carlo integration are quite complicated (see §7.8), but a worked example will show the underlying simplicity of the method. Suppose that we want to find the weight and the position of the center of mass of an object of complicated shape, namely the intersection of a torus with the edge of a large box. In particular let the object be defined by the three simultaneous conditions

$$z^2 + \left(\sqrt{x^2 + y^2} - 3 \right)^2 \leq 1 \tag{7.6.3}$$

(torus centered on the origin with major radius $= 4$, minor radius $= 2$)

$$x \geq 1 \qquad y \geq -3 \tag{7.6.4}$$

(two faces of the box, see Figure 7.6.2). Suppose for the moment that the object has a constant density ρ.

We want to estimate the following integrals over the interior of the complicated object:

$$\int \rho \, dx \, dy \, dz \qquad \int x\rho \, dx \, dy \, dz \qquad \int y\rho \, dx \, dy \, dz \qquad \int z\rho \, dx \, dy \, dz \tag{7.6.5}$$

The coordinates of the center of mass will be the ratio of the latter three integrals (linear moments) to the first one (the weight).

In the following fragment, the region V, enclosing the piece-of-torus W, is the rectangular box extending from 1 to 4 in x, -3 to 4 in y, and -1 to 1 in z.

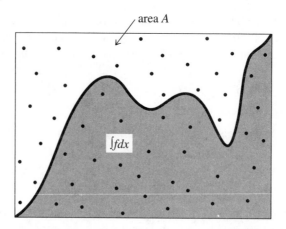

Figure 7.6.1. Monte Carlo integration. Random points are chosen within the area A. The integral of the function f is estimated as the area of A multiplied by the fraction of random points that fall below the curve f. Refinements on this procedure can improve the accuracy of the method; see text.

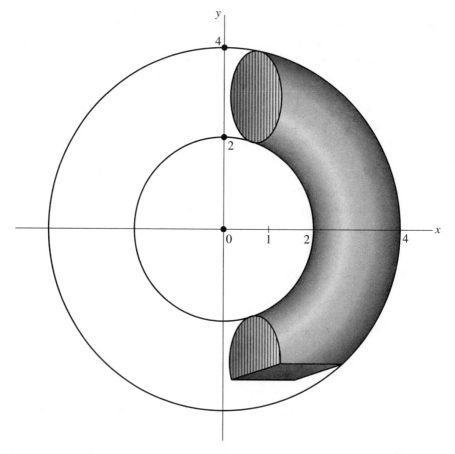

Figure 7.6.2. Example of Monte Carlo integration (see text). The region of interest is a piece of a torus, bounded by the intersection of two planes. The limits of integration of the region cannot easily be written in analytically closed form, so Monte Carlo is a useful technique.

```
#include "nr.h"
...
n=...                                    Set to the number of sample points desired.
den=...                                  Set to the constant value of the density.
sw=swx=swy=swz=0.0;                      Zero the various sums to be accumulated.
varw=varx=vary=varz=0.0;
vol=3.0*7.0*2.0;                         Volume of the sampled region.
for(j=0;j<n;j++) {
    x=1.0+3.0*ran2(idum);                Pick a point randomly in the sampled re-
    y=(-3.0)+7.0*ran2(idum);             gion.
    z=(-1.0)+2.0*ran2(idum);
    if (z*z+SQR(sqrt(x*x+y*y)-3.0) < 1.0) {      Is it in the torus?
        sw += den;                       If so, add to the various cumulants.
        swx += x*den;
        swy += y*den;
        swz += z*den;
        varw += SQR(den);
        varx += SQR(x*den);
        vary += SQR(y*den);
        varz += SQR(z*den);
    }
}
w=vol*sw/n;                              The values of the integrals (7.6.5),
x=vol*swx/n;
y=vol*swy/n;
z=vol*swz/n;
dw=vol*sqrt((varw/n-SQR(sw/n))/n);       and their corresponding error estimates.
dx=vol*sqrt((varx/n-SQR(swx/n))/n);
dy=vol*sqrt((vary/n-SQR(swy/n))/n);
dz=vol*sqrt((varz/n-SQR(swz/n))/n);
```

A change of variable can often be extremely worthwhile in Monte Carlo integration. Suppose, for example, that we want to evaluate the same integrals, but for a piece-of-torus whose density is a strong function of z, in fact varying according to

$$\rho(x, y, z) = e^{5z} \qquad (7.6.6)$$

One way to do this is to put the statement

```
den=exp(5.0*z);
```

inside the if (...) block, just before den is first used. This will work, but it is a poor way to proceed. Since (7.6.6) falls so rapidly to zero as z decreases (down to its lower limit -1), most sampled points contribute almost nothing to the sum of the weight or moments. These points are effectively wasted, almost as badly as those that fall outside of the region W. A change of variable, exactly as in the transformation methods of §7.2, solves this problem. Let

$$ds = e^{5z} dz \qquad \text{so that} \qquad s = \frac{1}{5} e^{5z}, \quad z = \frac{1}{5} \ln(5s) \qquad (7.6.7)$$

Then $\rho dz = ds$, and the limits $-1 < z < 1$ become $.00135 < s < 29.682$. The program fragment now looks like this

```
#include "nrutil.h"
...
n=...                                          Set to the number of sample points desired.
sw=swx=swy=swz=0.0;
varw=varx=vary=varz=0.0;
ss=0.2*(exp(5.0)-exp(-5.0))                    Interval of s to be random sampled.
vol=3.0*7.0*ss                                 Volume in x,y,s-space.
for(j=0;j<n;j++) {
    x=1.0+3.0*ran2(idum);
    y=(-3.0)+7.0*ran2(idum);
    s=0.00135+ss*ran2(idum);                   Pick a point in s.
    z=0.2*log(5.0*s);                          Equation (7.6.7).
    if (z*z+SQR(sqrt(x*x+y*y)-3.0) < 1.0) {
        sw += 1.0;                             Density is 1, since absorbed into definition
        swx += x;                              of s.
        swy += y;
        swz += z;
        varw += 1.0;
        varx += x*x;
        vary += y*y;
        varz += z*z;
    }
}
w=vol*sw/n;                                     The values of the integrals (7.6.5),
x=vol*swx/n;
y=vol*swy/n;
z=vol*swz/n;
dw=vol*sqrt((varw/n-SQR(sw/n))/n);             and their corresponding error estimates.
dx=vol*sqrt((varx/n-SQR(swx/n))/n);
dy=vol*sqrt((vary/n-SQR(swy/n))/n);
dz=vol*sqrt((varz/n-SQR(swz/n))/n);
```

If you think for a minute, you will realize that equation (7.6.7) was useful only because the part of the integrand that we wanted to eliminate (e^{5z}) was both integrable analytically, and had an integral that could be analytically inverted. (Compare §7.2.) In general these properties will not hold. Question: What then? Answer: Pull out of the integrand the "best" factor that *can* be integrated and inverted. The criterion for "best" is to try to reduce the remaining integrand to a function that is as close as possible to constant.

The limiting case is instructive: If you manage to make the integrand f *exactly* constant, and if the region V, of known volume, *exactly* encloses the desired region W, then the average of f that you compute will be exactly its constant value, and the error estimate in equation (7.6.1) will exactly vanish. You will, in fact, have done the integral exactly, and the Monte Carlo numerical evaluations are superfluous. So, backing off from the extreme limiting case, *to the extent* that you are able to make f approximately constant by change of variable, and *to the extent* that you can sample a region only slightly larger than W, you will increase the accuracy of the Monte Carlo integral. This technique is generically called *reduction of variance* in the literature.

The fundamental disadvantage of simple Monte Carlo integration is that its accuracy increases only as the square root of N, the number of sampled points. If your accuracy requirements are modest, or if your computer budget is large, then the technique is highly recommended as one of great generality. In the next two sections we will see that there are techniques available for "breaking the square root of N barrier" and achieving, at least in some cases, higher accuracy with fewer function evaluations.

CITED REFERENCES AND FURTHER READING:

Hammersley, J.M., and Handscomb, D.C. 1964, *Monte Carlo Methods* (London: Methuen).

Shreider, Yu. A. (ed.) 1966, *The Monte Carlo Method* (Oxford: Pergamon).

Sobol', I.M. 1974, *The Monte Carlo Method* (Chicago: University of Chicago Press).

Kalos, M.H., and Whitlock, P.A. 1986, *Monte Carlo Methods* (New York: Wiley).

7.7 Quasi- (that is, Sub-) Random Sequences

We have just seen that choosing N points uniformly randomly in an n-dimensional space leads to an error term in Monte Carlo integration that decreases as $1/\sqrt{N}$. In essence, each new point sampled adds linearly to an accumulated sum that will become the function average, and also linearly to an accumulated sum of squares that will become the variance (equation 7.6.2). The estimated error comes from the square root of this variance, hence the power $N^{-1/2}$.

Just because this square root convergence is familiar does not, however, mean that it is inevitable. A simple counterexample is to choose sample points that lie on a Cartesian grid, and to sample each grid point exactly once (in whatever order). The Monte Carlo method thus becomes a deterministic quadrature scheme — albeit a simple one — whose fractional error decreases at least as fast as N^{-1} (even faster if the function goes to zero smoothly at the boundaries of the sampled region, or is periodic in the region).

The trouble with a grid is that one has to decide *in advance* how fine it should be. One is then committed to completing all of its sample points. With a grid, it is not convenient to "sample *until*" some convergence or termination criterion is met. One might ask if there is not some intermediate scheme, some way to pick sample points "at random," yet spread out in some self-avoiding way, avoiding the chance clustering that occurs with uniformly random points.

A similar question arises for tasks other than Monte Carlo integration. We might want to search an n-dimensional space for a point where some (locally computable) condition holds. Of course, for the task to be computationally meaningful, there had better be continuity, so that the desired condition will hold in some finite n-dimensional neighborhood. We may not know *a priori* how large that neighborhood is, however. We want to "sample *until*" the desired point is found, moving smoothly to finer scales with increasing samples. Is there any way to do this that is better than uncorrelated, random samples?

The answer to the above question is "yes." Sequences of n-tuples that fill n-space more uniformly than uncorrelated random points are called *quasi-random sequences*. That term is somewhat of a misnomer, since there is nothing "random" about quasi-random sequences: They are cleverly crafted to be, in fact, *sub*-random. The sample points in a quasi-random sequence are, in a precise sense, "maximally avoiding" of each other.

A conceptually simple example is *Halton's sequence* [1]. In one dimension, the jth number H_j in the sequence is obtained by the following steps: (i) Write j as a number in base b, where b is some prime. (For example $j = 17$ in base $b = 3$ is 122.) (ii) Reverse the digits and put a radix point (i.e., a decimal point base b) in front of

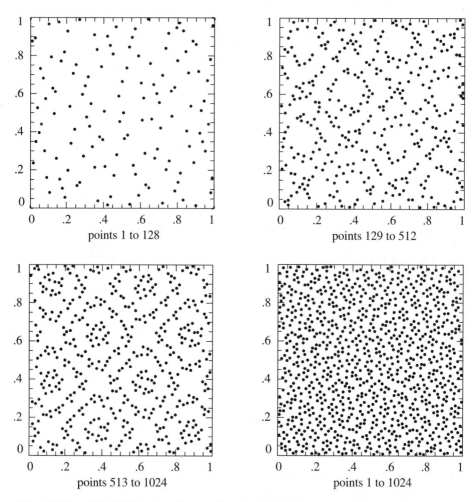

Figure 7.7.1.   First 1024 points of a two-dimensional Sobol' sequence. The sequence is generated number-theoretically, rather than randomly, so successive points at any stage "know" how to fill in the gaps in the previously generated distribution.

the sequence. (In the example, we get 0.221 base 3.) The result is H_j. To get a sequence of n-tuples in n-space, you make each component a Halton sequence with a different prime base b. Typically, the first n primes are used.

It is not hard to see how Halton's sequence works: Every time the number of digits in j increases by one place, j's digit-reversed fraction becomes a factor of b finer-meshed. Thus the process is one of filling in all the points on a sequence of finer and finer Cartesian grids — and in a kind of maximally spread-out order on each grid (since, e.g., the most rapidly changing digit in j controls the *most* significant digit of the fraction).

Other ways of generating quasi-random sequences have been suggested by Faure, Sobol', Niederreiter, and others. Bratley and Fox [2] provide a good review and references, and discuss a particularly efficient variant of the Sobol' [3] sequence suggested by Antonov and Saleev [4]. It is this Antonov-Saleev variant whose implementation we now discuss.

Degree	Primitive Polynomials Modulo 2*
1	0 (i.e., $x + 1$)
2	1 (i.e., $x^2 + x + 1$)
3	1, 2 (i.e., $x^3 + x + 1$ and $x^3 + x^2 + 1$)
4	1, 4 (i.e., $x^4 + x + 1$ and $x^4 + x^3 + 1$)
5	2, 4, 7, 11, 13, 14
6	1, 13, 16, 19, 22, 25
7	1, 4, 7, 8, 14, 19, 21, 28, 31, 32, 37, 41, 42, 50, 55, 56, 59, 62
8	14, 21, 22, 38, 47, 49, 50, 52, 56, 67, 70, 84, 97, 103, 115, 122
9	8, 13, 16, 22, 25, 44, 47, 52, 55, 59, 62, 67, 74, 81, 82, 87, 91, 94, 103, 104, 109, 122, 124, 137, 138, 143, 145, 152, 157, 167, 173, 176, 181, 182, 185, 191, 194, 199, 218, 220, 227, 229, 230, 234, 236, 241, 244, 253
10	4, 13, 19, 22, 50, 55, 64, 69, 98, 107, 115, 121, 127, 134, 140, 145, 152, 158, 161, 171, 181, 194, 199, 203, 208, 227, 242, 251, 253, 265, 266, 274, 283, 289, 295, 301, 316, 319, 324, 346, 352, 361, 367, 382, 395, 398, 400, 412, 419, 422, 426, 428, 433, 446, 454, 457, 472, 493, 505, 508

*Expressed as a decimal integer whose binary representation gives the coefficients, from the highest to lowest power of x. Only the internal terms are represented — the highest order term and the constant term always have coefficient 1.

The Sobol' sequence generates numbers between zero and one directly as binary fractions of length w bits, from a set of w special binary fractions, V_i, $i = 1, 2, \ldots, w$, called *direction numbers*. In Sobol's original method, the jth number X_j is generated by XORing (bitwise exclusive or) together the set of V_i's satisfying the criterion on i, "the ith bit of j is nonzero." As j increments, in other words, different ones of the V_i's flash in and out of X_j on different time scales. V_1 alternates between being present and absent most quickly, while V_k goes from present to absent (or vice versa) only every 2^{k-1} steps.

Antonov and Saleev's contribution was to show that instead of using the bits of the integer j to select direction numbers, one could just as well use the bits of the *Gray code* of j, $G(j)$. (For a quick review of Gray codes, look at §20.2.)

Now $G(j)$ and $G(j + 1)$ differ in exactly one bit position, namely in the position of the rightmost zero bit in the binary representation of j (adding a leading zero to j if necessary). A consequence is that the $j + 1$st Sobol'-Antonov-Saleev number can be obtained from the jth by XORing it with *a single V_i*, namely with i the position of the rightmost zero bit in j. This makes the calculation of the sequence very efficient, as we shall see.

Figure 7.7.1 plots the first 1024 points generated by a two-dimensional Sobol' sequence. One sees that successive points do "know" about the gaps left previously, and keep filling them in, hierarchically.

We have deferred to this point a discussion of how the direction numbers V_i are generated. Some nontrivial mathematics is involved in that, so we will content ourself with a cookbook summary only: Each different Sobol' sequence (or component of an n-dimensional sequence) is based on a different primitive polynomial over the integers modulo 2, that is, a polynomial whose coefficients are either 0 or 1, and which generates a maximal length shift register sequence. (Primitive polynomials modulo 2 were used in §7.4, and are further discussed in

Initializing Values Used in sobseq						
Degree	Polynomial	\multicolumn{5}{c}{Starting Values}				
1	0	1	(3)	(5)	(15)	...
2	1	1	1	(7)	(11)	...
3	1	1	3	7	(5)	...
3	2	1	3	3	(15)	...
4	1	1	1	3	13	...
4	4	1	1	5	9	...

Parenthesized values are not freely specifiable, but are forced by the required recurrence for this degree.

§20.3.) Suppose P is such a polynomial, of degree q,

$$P = x^q + a_1 x^{q-1} + a_2 x^{q-2} + \cdots + a_{q-1} x + 1 \tag{7.7.1}$$

Define a sequence of integers M_i by the q-term recurrence relation,

$$M_i = 2a_1 M_{i-1} \oplus 2^2 a_2 M_{i-2} \oplus \cdots \oplus 2^{q-1} M_{i-q+1} a_{q-1} \oplus (2^q M_{i-q} \oplus M_{i-q}) \tag{7.7.2}$$

Here bitwise XOR is denoted by $\oplus$. The starting values for this recurrence are that $M_1, \ldots, M_q$ can be arbitrary odd integers less than $2, \ldots, 2^q$, respectively. Then, the direction numbers V_i are given by

$$V_i = M_i / 2^i \qquad i = 1, \ldots, w \tag{7.7.3}$$

The accompanying table lists all primitive polynomials modulo 2 with degree $q \leq 10$. Since the coefficients are either 0 or 1, and since the coefficients of x^q and of 1 are predictably 1, it is convenient to denote a polynomial by its middle coefficients taken as the bits of a binary number (higher powers of x being more significant bits). The table uses this convention.

Turn now to the implementation of the Sobol' sequence. Successive calls to the function sobseq (after a preliminary initializing call) return successive points in an n-dimensional Sobol' sequence based on the first n primitive polynomials in the table. As given, the routine is initialized for maximum n of 6 dimensions, and for a word length w of 30 bits. These parameters can be altered by changing MAXBIT ($\equiv w$) and MAXDIM, and by adding more initializing data to the arrays ip (the primitive polynomials from the table), mdeg (their degrees), and iv (the starting values for the recurrence, equation 7.7.2). A second table, above, elucidates the initializing data in the routine.

```
#include "nr.h"

void NR::sobseq(const int n, Vec_O_DP &x)
When n is negative, internally initializes a set of MAXBIT direction numbers for each of MAXDIM
different Sobol' sequences. When n is positive (but ≤MAXDIM), returns as the vector x[0..n-1]
the next values from n of these sequences. (n must not be changed between initializations.)
{
    const int MAXBIT=30,MAXDIM=6;
    int j,k,l;
    unsigned long i,im,ipp;
    static int mdeg[MAXDIM]={1,2,3,3,4,4};
    static unsigned long in;
    static Vec_ULNG ix(MAXDIM);
    static Vec_ULNG_p iu(MAXBIT);
    static unsigned long ip[MAXDIM]={0,1,1,2,1,4};
    static unsigned long iv[MAXDIM*MAXBIT]=
```

```
    {1,1,1,1,1,1,3,1,3,3,1,1,5,7,7,3,3,5,15,11,5,15,13,9};
static DP fac;

if (n < 0) {                              Initialize, don't return a vector.
    for (k=0;k<MAXDIM;k++) ix[k]=0;
    in=0;
    if (iv[0] != 1) return;
    fac=1.0/(1 << MAXBIT);
    for (j=0,k=0;j<MAXBIT;j++,k+=MAXDIM) iu[j] = &iv[k];
    To allow both 1D and 2D addressing.
    for (k=0;k<MAXDIM;k++) {
        for (j=0;j<mdeg[k];j++) iu[j][k] <<= (MAXBIT-1-j);
        Stored values only require normalization.
        for (j=mdeg[k];j<MAXBIT;j++) {        Use the recurrence to get other val-
            ipp=ip[k];                        ues.
            i=iu[j-mdeg[k]][k];
            i ^= (i >> mdeg[k]);
            for (l=mdeg[k]-1;l>=1;l--) {
                if (ipp & 1) i ^= iu[j-l][k];
                ipp >>= 1;
            }
            iu[j][k]=i;
        }
    }
} else {                                   Calculate the next vector in the se-
    im=in++;                               quence.
    for (j=0;j<MAXBIT;j++) {               Find the rightmost zero bit.
        if (!(im & 1)) break;
        im >>= 1;
    }
    if (j >= MAXBIT) nrerror("MAXBIT too small in sobseq");
    im=j*MAXDIM;
    for (k=0;k<MIN(n,MAXDIM);k++) {        XOR the appropriate direction num-
        ix[k] ^= iv[im+k];                 ber into each component of the
        x[k]=ix[k]*fac;                    vector and convert to a floating
    }                                      number.
}
}
```

How good is a Sobol' sequence, anyway? For Monte Carlo integration of a smooth function in n dimensions, the answer is that the fractional error will decrease with N, the number of samples, as $(\ln N)^n/N$, i.e., almost as fast as $1/N$. As an example, let us integrate a function that is nonzero inside a torus (doughnut) in three-dimensional space. If the major radius of the torus is R_0, the minor radial coordinate r is defined by

$$r = \left([(x^2 + y^2)^{1/2} - R_0]^2 + z^2 \right)^{1/2} \tag{7.7.4}$$

Let us try the function

$$f(x,y,z) = \begin{cases} 1 + \cos\left(\dfrac{\pi r^2}{a^2}\right) & r < r_0 \\ 0 & r \ge r_0 \end{cases} \tag{7.7.5}$$

which can be integrated analytically in cylindrical coordinates, giving

$$\int \int \int dx\, dy\, dz\, f(x,y,z) = 2\pi^2 a^2 R_0 \tag{7.7.6}$$

With parameters $R_0 = 0.6$, $r_0 = 0.3$, we did 100 successive Monte Carlo integrations of equation (7.7.4), sampling uniformly in the region $-1 < x,y,z < 1$, for the two cases of uncorrelated random points and the Sobol' sequence generated by the routine sobseq. Figure 7.7.2 shows the results, plotting the r.m.s. average error of the 100 integrations as a function of the number of points sampled. (For any *single* integration, the error of course wanders

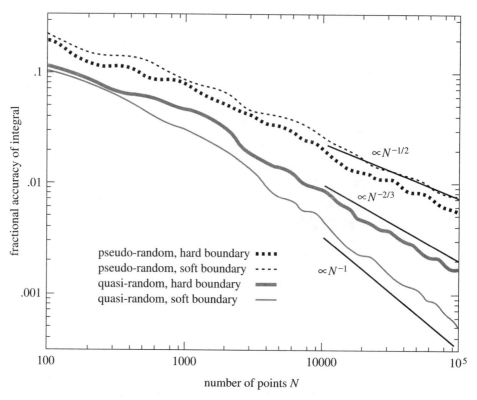

Figure 7.7.2. Fractional accuracy of Monte Carlo integrations as a function of number of points sampled, for two different integrands and two different methods of choosing random points. The quasi-random Sobol' sequence converges much more rapidly than a conventional pseudo-random sequence. Quasi-random sampling does better when the integrand is smooth ("soft boundary") than when it has step discontinuities ("hard boundary"). The curves shown are the r.m.s. average of 100 trials.

from positive to negative, or vice versa, so a logarithmic plot of fractional error is not very informative.) The thin, dashed curve corresponds to uncorrelated random points and shows the familiar $N^{-1/2}$ asymptotics. The thin, solid gray curve shows the result for the Sobol' sequence. The logarithmic term in the expected $(\ln N)^3/N$ is readily apparent as curvature in the curve, but the asymptotic N^{-1} is unmistakable.

To understand the importance of Figure 7.7.2, suppose that a Monte Carlo integration of f with 1% accuracy is desired. The Sobol' sequence achieves this accuracy in a few thousand samples, while pseudorandom sampling requires nearly 100,000 samples. The ratio would be even greater for higher desired accuracies.

A different, not quite so favorable, case occurs when the function being integrated has hard (discontinuous) boundaries inside the sampling region, for example the function that is one inside the torus, zero outside,

$$f(x,y,z) = \begin{cases} 1 & r < r_0 \\ 0 & r \geq r_0 \end{cases} \tag{7.7.7}$$

where r is defined in equation (7.7.4). Not by coincidence, this function has the same analytic integral as the function of equation (7.7.5), namely $2\pi^2 a^2 R_0$.

The carefully hierarchical Sobol' sequence is based on a set of Cartesian grids, but the boundary of the torus has no particular relation to those grids. The result is that it is essentially random whether sampled points in a thin layer at the surface of the torus, containing on the order of $N^{2/3}$ points, come out to be inside, or outside, the torus. The square root law, applied to this thin layer, gives $N^{1/3}$ fluctuations in the sum, or $N^{-2/3}$ fractional error in the Monte Carlo integral. One sees this behavior verified in Figure 7.7.2 by the thicker gray curve. The thicker dashed curve in Figure 7.7.2 is the result of integrating the function of equation (7.7.7) using independent random points. While the advantage of the Sobol' sequence is not quite so dramatic as in the case of a smooth function, it can nonetheless be a significant factor ($\sim$5) even at modest accuracies like 1%, and greater at higher accuracies.

Note that we have not provided the routine sobseq with a means of starting the sequence at a point other than the beginning, but this feature would be easy to add. Once the initialization of the direction numbers iv has been done, the jth point can be obtained directly by XORing together those direction numbers corresponding to nonzero bits in the Gray code of j, as described above.

The Latin Hypercube

We might here give passing mention the unrelated technique of *Latin square* or *Latin hypercube* sampling, which is useful when you must sample an N-dimensional space *exceedingly* sparsely, at M points. For example, you may want to test the crashworthiness of cars as a simultaneous function of 4 different design parameters, but with a budget of only three expendable cars. (The issue is not whether this is a good plan — it isn't — but rather how to make the best of the situation!)

The idea is to partition each design parameter (dimension) into M segments, so that the whole space is partitioned into M^N cells. (You can choose the segments in each dimension to be equal or unequal, according to taste.) With 4 parameters and 3 cars, for example, you end up with $3 \times 3 \times 3 \times 3 = 81$ cells.

Next, choose M cells to contain the sample points by the following algorithm: Randomly choose one of the M^N cells for the first point. Now eliminate all cells that agree with this point on *any* of its parameters (that is, cross out all cells in the same row, column, etc.), leaving $(M-1)^N$ candidates. Randomly choose one of these, eliminate new rows and columns, and continue the process until there is only one cell left, which then contains the final sample point.

The result of this construction is that *each* design parameter will have been tested in *every one* of its subranges. If the response of the system under test is dominated by *one* of the design parameters, that parameter will be found with this sampling technique. On the other hand, if there is an important interaction among different design parameters, then the Latin hypercube gives no particular advantage. Use with care.

CITED REFERENCES AND FURTHER READING:

Halton, J.H. 1960, *Numerische Mathematik*, vol. 2, pp. 84–90. [1]

Bratley P., and Fox, B.L. 1988, *ACM Transactions on Mathematical Software*, vol. 14, pp. 88–100. [2]

Lambert, J.P. 1988, in *Numerical Mathematics – Singapore 1988*, ISNM vol. 86, R.P. Agarwal, Y.M. Chow, and S.J. Wilson, eds. (Basel: Birkhaüser), pp. 273–284.

Niederreiter, H. 1988, in *Numerical Integration III*, ISNM vol. 85, H. Brass and G. Hämmerlin, eds. (Basel: Birkhaüser), pp. 157–171.

Sobol', I.M. 1967, *USSR Computational Mathematics and Mathematical Physics*, vol. 7, no. 4, pp. 86–112. [3]

Antonov, I.A., and Saleev, V.M 1979, *USSR Computational Mathematics and Mathematical Physics*, vol. 19, no. 1, pp. 252–256. [4]

Dunn, O.J., and Clark, V.A. 1974, *Applied Statistics: Analysis of Variance and Regression* (New York, Wiley) [discusses Latin Square].

7.8 Adaptive and Recursive Monte Carlo Methods

This section discusses more advanced techniques of Monte Carlo integration. As examples of the use of these techniques, we include two rather different, fairly sophisticated, multidimensional Monte Carlo codes: vegas [1,2], and miser [3]. The techniques that we discuss all fall under the general rubric of *reduction of variance* (§7.6), but are otherwise quite distinct.

Importance Sampling

The use of *importance sampling* was already implicit in equations (7.6.6) and (7.6.7). We now return to it in a slightly more formal way. Suppose that an integrand f can be written as the product of a function h that is almost constant times another, positive, function g. Then its integral over a multidimensional volume V is

$$\int f \, dV = \int (f/g) \, g dV = \int h \, g dV \qquad (7.8.1)$$

In equation (7.6.7) we interpreted equation (7.8.1) as suggesting a change of variable to G, the indefinite integral of g. That made gdV a perfect differential. We then proceeded to use the basic theorem of Monte Carlo integration, equation (7.6.1). A more general interpretation of equation (7.8.1) is that we can integrate f by instead sampling h — not, however, with uniform probability density dV, but rather with nonuniform density gdV. In this second interpretation, the first interpretation follows as the special case, where the *means* of generating the nonuniform sampling of gdV is via the transformation method, using the indefinite integral G (see §7.2).

More directly, one can go back and generalize the basic theorem (7.6.1) to the case of nonuniform sampling: Suppose that points x_i are chosen within the volume V with a probability density p satisfying

$$\int p \, dV = 1 \qquad (7.8.2)$$

The generalized fundamental theorem is that the integral of any function f is estimated, using N sample points $x_0, \ldots, x_{N-1}$, by

$$I \equiv \int f \, dV = \int \frac{f}{p} \, p dV \approx \left\langle \frac{f}{p} \right\rangle \pm \sqrt{\frac{\langle f^2/p^2 \rangle - \langle f/p \rangle^2}{N}} \qquad (7.8.3)$$

where angle brackets denote arithmetic means over the N points, exactly as in equation (7.6.2). As in equation (7.6.1), the "plus-or-minus" term is a one standard deviation error estimate. Notice that equation (7.6.1) is in fact the special case of equation (7.8.3), with $p = \text{constant} = 1/V$.

What is the best choice for the sampling density p? Intuitively, we have already seen that the idea is to make $h = f/p$ as close to constant as possible. We can be more rigorous by focusing on the numerator inside the square root in equation (7.8.3), which is the variance per sample point. Both angle brackets are themselves Monte Carlo estimators of integrals, so we can write

$$S \equiv \left\langle \frac{f^2}{p^2} \right\rangle - \left\langle \frac{f}{p} \right\rangle^2 \approx \int \frac{f^2}{p^2} \, p dV - \left[\int \frac{f}{p} \, p dV\right]^2 = \int \frac{f^2}{p} \, dV - \left[\int f \, dV\right]^2 \qquad (7.8.4)$$

We now find the optimal p subject to the constraint equation (7.8.2) by the functional variation

$$0 = \frac{\delta}{\delta p} \left(\int \frac{f^2}{p} \, dV - \left[\int f \, dV\right]^2 + \lambda \int p \, dV \right) \qquad (7.8.5)$$

with λ a Lagrange multiplier. Note that the middle term does not depend on p. The variation (which comes inside the integrals) gives $0 = -f^2/p^2 + \lambda$ or

$$p = \frac{|f|}{\sqrt{\lambda}} = \frac{|f|}{\int |f|\, dV} \tag{7.8.6}$$

where λ has been chosen to enforce the constraint (7.8.2).

If f has one sign in the region of integration, then we get the obvious result that the optimal choice of p — if one can figure out a practical way of effecting the sampling — is that it be proportional to $|f|$. Then the variance is reduced to zero. Not so obvious, but seen to be true, is the fact that $p \propto |f|$ is optimal even if f takes on both signs. In that case the variance per sample point (from equations 7.8.4 and 7.8.6) is

$$S = S_{\text{optimal}} = \left(\int |f|\, dV \right)^2 - \left(\int f\, dV \right)^2 \tag{7.8.7}$$

One curiosity is that one can add a constant to the integrand to make it all of one sign, since this changes the integral by a known amount, constant $\times V$. Then, the optimal choice of p always gives zero variance, that is, a perfectly accurate integral! The resolution of this seeming paradox (already mentioned at the end of §7.6) is that perfect knowledge of p in equation (7.8.6) requires perfect knowledge of $\int |f| dV$, which is tantamount to already knowing the integral you are trying to compute!

If your function f takes on a known constant value in most of the volume V, it is certainly a good idea to add a constant so as to make that value zero. Having done that, the accuracy attainable by importance sampling depends in practice not on how small equation (7.8.7) is, but rather on how small is equation (7.8.4) for an *implementable* p, likely only a crude approximation to the ideal.

Stratified Sampling

The idea of *stratified sampling* is quite different from importance sampling. Let us expand our notation slightly and let $\langle\langle f \rangle\rangle$ denote the true average of the function f over the volume V (namely the integral divided by V), while $\langle f \rangle$ denotes as before the simplest (uniformly sampled) Monte Carlo *estimator* of that average:

$$\langle\langle f \rangle\rangle \equiv \frac{1}{V} \int f\, dV \qquad \langle f \rangle \equiv \frac{1}{N} \sum_i f(x_i) \tag{7.8.8}$$

The variance of the estimator, $\operatorname{Var}\left(\langle f \rangle\right)$, which measures the square of the error of the Monte Carlo integration, is asymptotically related to the variance of the function, $\operatorname{Var}\left(f\right) \equiv \langle\langle f^2 \rangle\rangle - \langle\langle f \rangle\rangle^2$, by the relation

$$\operatorname{Var}\left(\langle f \rangle\right) = \frac{\operatorname{Var}\left(f\right)}{N} \tag{7.8.9}$$

(compare equation 7.6.1).

Suppose we divide the volume V into two equal, disjoint subvolumes, denoted a and b, and sample $N/2$ points in each subvolume. Then another estimator for $\langle\langle f \rangle\rangle$, different from equation (7.8.8), which we denote $\langle f \rangle'$, is

$$\langle f \rangle' \equiv \frac{1}{2} \left(\langle f \rangle_a + \langle f \rangle_b \right) \tag{7.8.10}$$

in other words, the mean of the sample averages in the two half-regions. The variance of estimator (7.8.10) is given by

$$\begin{aligned}
\operatorname{Var}\left(\langle f \rangle'\right) &= \frac{1}{4} \left[\operatorname{Var}\left(\langle f \rangle_a\right) + \operatorname{Var}\left(\langle f \rangle_b\right) \right] \\
&= \frac{1}{4} \left[\frac{\operatorname{Var}_a\left(f\right)}{N/2} + \frac{\operatorname{Var}_b\left(f\right)}{N/2} \right] \\
&= \frac{1}{2N} \left[\operatorname{Var}_a\left(f\right) + \operatorname{Var}_b\left(f\right) \right]
\end{aligned} \tag{7.8.11}$$

Here $\text{Var}_a\,(f)$ denotes the variance of f in subregion a, that is, $\langle\langle f^2\rangle\rangle_a - \langle\langle f\rangle\rangle_a^2$, and correspondingly for b.

From the definitions already given, it is not difficult to prove the relation

$$\text{Var}\,(f) = \frac{1}{2}\,[\text{Var}_a\,(f) + \text{Var}_b\,(f)] + \frac{1}{4}\,(\langle\langle f\rangle\rangle_a - \langle\langle f\rangle\rangle_b)^2 \qquad (7.8.12)$$

(In physics, this formula for combining second moments is the "parallel axis theorem.") Comparing equations (7.8.9), (7.8.11), and (7.8.12), one sees that the stratified (into two subvolumes) sampling gives a variance that is never larger than the simple Monte Carlo case — and smaller whenever the means of the stratified samples, $\langle\langle f\rangle\rangle_a$ and $\langle\langle f\rangle\rangle_b$, are different.

We have not yet exploited the possibility of sampling the two subvolumes with *different numbers* of points, say N_a in subregion a and $N_b \equiv N - N_a$ in subregion b. Let us do so now. Then the variance of the estimator is

$$\text{Var}\,(\langle f\rangle') = \frac{1}{4}\left[\frac{\text{Var}_a\,(f)}{N_a} + \frac{\text{Var}_b\,(f)}{N - N_a}\right] \qquad (7.8.13)$$

which is minimized (one can easily verify) when

$$\frac{N_a}{N} = \frac{\sigma_a}{\sigma_a + \sigma_b} \qquad (7.8.14)$$

Here we have adopted the shorthand notation $\sigma_a \equiv [\text{Var}_a\,(f)]^{1/2}$, and correspondingly for b. If N_a satisfies equation (7.8.14), then equation (7.8.13) reduces to

$$\text{Var}\,(\langle f\rangle') = \frac{(\sigma_a + \sigma_b)^2}{4N} \qquad (7.8.15)$$

Equation (7.8.15) reduces to equation (7.8.9) if $\text{Var}\,(f) = \text{Var}_a\,(f) = \text{Var}_b\,(f)$, in which case stratifying the sample makes no difference.

A standard way to generalize the above result is to consider the volume V divided into more than two equal subregions. One can readily obtain the result that the optimal allocation of sample points among the regions is to have the number of points in each region j proportional to σ_j (that is, the square root of the variance of the function f in that subregion). In spaces of high dimensionality (say $d \gtrsim 4$) this is not in practice very useful, however. Dividing a volume into K segments along each dimension implies K^d subvolumes, typically much too large a number when one contemplates estimating all the corresponding σ_j's.

Mixed Strategies

Importance sampling and stratified sampling seem, at first sight, inconsistent with each other. The former concentrates sample points where the magnitude of the integrand $|f|$ is largest, that latter where the variance of f is largest. How can both be right?

The answer is that (like so much else in life) it all depends on what you know and how well you know it. Importance sampling depends on already knowing some approximation to your integral, so that you are able to generate random points x_i with the desired probability density p. To the extent that your p is not ideal, you are left with an error that decreases only as $N^{-1/2}$. Things are particularly bad if your p is far from ideal in a region where the integrand f is changing rapidly, since then the sampled function $h = f/p$ will have a large variance. Importance sampling works by smoothing the values of the sampled function h, and is effective only to the extent that you succeed in this.

Stratified sampling, by contrast, does not necessarily require that you know anything about f. Stratified sampling works by smoothing out the fluctuations of the *number* of points in subregions, not by smoothing the values of the points. The simplest stratified strategy, dividing V into N equal subregions and choosing one point randomly in each subregion, already gives a method whose error decreases asymptotically as N^{-1}, much faster than $N^{-1/2}$. (Note that quasi-random numbers, §7.7, are another way of smoothing fluctuations in the density of points, giving nearly as good a result as the "blind" stratification strategy.)

However, "asymptotically" is an important caveat: For example, if the integrand is negligible in all but a single subregion, then the resulting one-sample integration is all but useless. Information, even very crude, allowing importance sampling to put many points in the active subregion would be much better than blind stratified sampling.

Stratified sampling really comes into its own if you have some way of estimating the variances, so that you can put unequal numbers of points in different subregions, according to (7.8.14) or its generalizations, *and* if you can find a way of dividing a region into a practical number of subregions (notably *not* K^d with large dimension d), while yet significantly reducing the variance of the function in each subregion compared to its variance in the full volume. Doing this requires a lot of knowledge about f, though different knowledge from what is required for importance sampling.

In practice, importance sampling and stratified sampling are not incompatible. In many, if not most, cases of interest, the integrand f is small everywhere in V except for a small fractional volume of "active regions." In these regions the magnitude of $|f|$ and the standard deviation $\sigma = [\text{Var}\,(f)]^{1/2}$ are comparable in size, so both techniques will give about the same concentration of points. In more sophisticated implementations, it is also possible to "nest" the two techniques, so that (e.g.) importance sampling on a crude grid is followed by stratification within each grid cell.

Adaptive Monte Carlo: VEGAS

The VEGAS algorithm, invented by Peter Lepage [1,2], is widely used for multidimensional integrals that occur in elementary particle physics. VEGAS is primarily based on importance sampling, but it also does some stratified sampling if the dimension d is small enough to avoid K^d explosion (specifically, if $(K/2)^d < N/2$, with N the number of sample points). The basic technique for importance sampling in VEGAS is to construct, adaptively, a multidimensional weight function g that is *separable*,

$$p \propto g(x, y, z, \ldots) = g_x(x)g_y(y)g_z(z) \ldots \qquad (7.8.16)$$

Such a function avoids the K^d explosion in two ways: (i) It can be stored in the computer as d separate one-dimensional functions, each defined by K tabulated values, say — so that $K \times d$ replaces K^d. (ii) It can be sampled as a probability density by consecutively sampling the d one-dimensional functions to obtain coordinate vector components $(x, y, z, \ldots)$.

The optimal separable weight function can be shown to be [1]

$$g_x(x) \propto \left[\int dy \int dz \ldots \frac{f^2(x, y, z, \ldots)}{g_y(y)g_z(z)\ldots} \right]^{1/2} \qquad (7.8.17)$$

(and correspondingly for $y, z, \ldots$). Notice that this reduces to $g \propto |f|$ (7.8.6) in one dimension. Equation (7.8.17) immediately suggests VEGAS' adaptive strategy: Given a set of g-functions (initially all constant, say), one samples the function f, accumulating not only the overall estimator of the integral, but also the Kd estimators (K subdivisions of the independent variable in each of d dimensions) of the right-hand side of equation (7.8.17). These then determine improved g functions for the next iteration.

When the integrand f is concentrated in one, or at most a few, regions in d-space, then the weight function g's quickly become large at coordinate values that are the projections of these regions onto the coordinate axes. The accuracy of the Monte Carlo integration is then enormously enhanced over what simple Monte Carlo would give.

The weakness of VEGAS is the obvious one: To the extent that the projection of the function f onto individual coordinate directions is uniform, VEGAS gives no concentration of sample points in those dimensions. The worst case for VEGAS, e.g., is an integrand that is concentrated close to a body diagonal line, e.g., one from $(0, 0, 0, \ldots)$ to $(1, 1, 1, \ldots)$. Since this geometry is completely nonseparable, VEGAS can give no advantage at all. More generally, VEGAS may not do well when the integrand is concentrated in one-dimensional (or higher) curved trajectories (or hypersurfaces), unless these happen to be oriented close to the coordinate directions.

The routine vegas that follows is essentially Lepage's standard version, minimally modified to conform to our conventions. (We thank Lepage for permission to reproduce the program here.) For consistency with other versions of the VEGAS algorithm in circulation, we have preserved original variable names. The parameter NDMX is what we have called K, the maximum number of increments along each axis; MXDIM is the maximum value of d; some other parameters are explained in the comments.

The vegas routine performs $m = $ itmx statistically independent evaluations of the desired integral, each with $N = $ ncall function evaluations. While statistically independent, these iterations do assist each other, since each one is used to refine the sampling grid for the next one. The results of all iterations are combined into a single best answer, and its estimated error, by the relations

$$I_{\text{best}} = \sum_{i=0}^{m-1} \frac{I_i}{\sigma_i^2} \bigg/ \sum_{i=0}^{m-1} \frac{1}{\sigma_i^2} \qquad \sigma_{\text{best}} = \left(\sum_{i=0}^{m-1} \frac{1}{\sigma_i^2} \right)^{-1/2} \qquad (7.8.18)$$

Also returned is the quantity

$$\chi^2/m \equiv \frac{1}{m-1} \sum_{i=0}^{m-1} \frac{(I_i - I_{\text{best}})^2}{\sigma_i^2} \qquad (7.8.19)$$

If this is significantly larger than 1, then the results of the iterations are statistically inconsistent, and the answers are suspect.

The input flag init can be used to advantage. One might have a call with init=0, ncall=1000, itmx=5 immediately followed by a call with init=1, ncall=100000, itmx=1. The effect would be to develop a sampling grid over 5 iterations of a small number of samples, then to do a single high accuracy integration on the optimized grid.

Note that the user-supplied integrand function, fxn, has an argument wgt in addition to the expected evaluation point x. In most applications you ignore wgt inside the function. Occasionally, however, you may want to integrate some additional function or functions along with the principal function f. The integral of any such function g can be estimated by

$$I_g = \sum_i w_i g(\mathbf{x}) \qquad (7.8.20)$$

where the w_i's and $\mathbf{x}$'s are the arguments wgt and x, respectively. It is straightforward to accumulate this sum inside your function fxn, and to pass the answer back to your main program via global variables. Of course, $g(\mathbf{x})$ had better resemble the principal function f to some degree, since the sampling will be optimized for f.

```
#include <iostream>
#include <iomanip>
#include <cmath>
#include "nr.h"
using namespace std;

extern int idum;                          For random number initialization in main.

void NR::vegas(Vec_I_DP &regn, DP fxn(Vec_I_DP &, const DP), const int init,
     const int ncall, const int itmx, const int nprn, DP &tgral, DP &sd,
     DP &chi2a)
```
Performs Monte Carlo integration of a user-supplied ndim-dimensional function fxn over a rectangular volume specified by regn[0..2*ndim-1], a vector consisting of ndim "lower left" coordinates of the region followed by ndim "upper right" coordinates. The integration consists of itmx iterations, each with approximately ncall calls to the function. After each iteration the grid is refined; more than 5 or 10 iterations are rarely useful. The input flag init signals whether this call is a new start, or a subsequent call for additional iterations (see comments below). The input flag nprn (normally 0) controls the amount of diagnostic output. Returned answers are tgral (the best estimate of the integral), sd (its standard deviation), and chi2a (χ^2 per degree of freedom, an indicator of whether consistent results are being obtained). See text for further details.
```
{
    const int NDMX=50, MXDIM=10;
    const DP ALPH=1.5, TINY=1.0e-30;
    static int i,it,j,k,mds,nd,ndo,ng,npg;
```

```
static DP calls,dv2g,dxg,f,f2,f2b,fb,rc,ti;
static DP tsi,wgt,xjac,xn,xnd,xo,schi,si,swgt;
static Vec_INT ia(MXDIM),kg(MXDIM);
static Vec_DP dt(MXDIM),dx(MXDIM),r(NDMX),x(MXDIM),xin(NDMX);
static Mat_DP d(NDMX,MXDIM),di(NDMX,MXDIM),xi(MXDIM,NDMX);
Best make everything static, allowing restarts.

int ndim=regn.size()/2;
if (init <= 0) {                        Normal entry. Enter here on a cold start.
    mds=ndo=1;                          Change to mds=0 to disable stratified sampling,
    for (j=0;j<ndim;j++) xi[j][0]=1.0;      i.e., use importance sampling only.
}
if (init <= 1) si=swgt=schi=0.0;
Enter here to inherit the grid from a previous call, but not its answers.
if (init <= 2) {                        Enter here to inherit the previous grid and its
    nd=NDMX;                                answers.
    ng=1;
    if (mds != 0) {                     Set up for stratification.
        ng=int(pow(ncall/2.0+0.25,1.0/ndim));
        mds=1;
        if ((2*ng-NDMX) >= 0) {
            mds = -1;
            npg=ng/NDMX+1;
            nd=ng/npg;
            ng=npg*nd;
        }
    }
    for (k=1,i=0;i<ndim;i++) k *= ng;
    npg=MAX(int(ncall/k),2);
    calls=DP(npg)*DP(k);
    dxg=1.0/ng;
    for (dv2g=1,i=0;i<ndim;i++) dv2g *= dxg;
    dv2g=SQR(calls*dv2g)/npg/npg/(npg-1.0);
    xnd=nd;
    dxg *= xnd;
    xjac=1.0/calls;
    for (j=0;j<ndim;j++) {
        dx[j]=regn[j+ndim]-regn[j];
        xjac *= dx[j];
    }
    if (nd != ndo) {                    Do binning if necessary.
        for (i=0;i<MAX(nd,ndo);i++) r[i]=1.0;
        for (j=0;j<ndim;j++)
            rebin(ndo/xnd,nd,r,xin,xi,j);
        ndo=nd;
    }
    if (nprn >= 0) {
        cout << " Input parameters for vegas";
        cout << " ndim= " << setw(4) << ndim;
        cout << " ncall= " << setw(8) << calls << endl;
        cout << setw(34) << " it=" << setw(5) << it;
        cout << " itmx=" << setw(5) << itmx << endl;
        cout << setw(34) << " nprn=" << setw(5) << nprn;
        cout << " ALPH=" << setw(9) << ALPH << endl;
        cout << setw(34) << " mds=" << setw(5) << mds;
        cout << " nd=" << setw(5) << nd << endl;
        for (j=0;j<ndim;j++) {
            cout << setw(30) << " x1[" << setw(2) << j;
            cout << "]= " << setw(11) << regn[j] << " xu[";
            cout << setw(2) << j << "]= ";
            cout << setw(11) << regn[j+ndim] << endl;
        }
    }
}
```

```
for (it=0;it<itmx;it++) {
```
Main iteration loop. Can enter here (init $\geq$ 3) to do an additional `itmx` iterations with all other parameters unchanged.
```
    ti=tsi=0.0;
    for (j=0;j<ndim;j++) {
        kg[j]=1;
        for (i=0;i<nd;i++) d[i][j]=di[i][j]=0.0;
    }
    for (;;) {
        fb=f2b=0.0;
        for (k=0;k<npg;k++) {
            wgt=xjac;
            for (j=0;j<ndim;j++) {
                xn=(kg[j]-ran2(idum))*dxg+1.0;
                ia[j]=MAX(MIN(int(xn),NDMX),1);
                if (ia[j] > 1) {
                    xo=xi[j][ia[j]-1]-xi[j][ia[j]-2];
                    rc=xi[j][ia[j]-2]+(xn-ia[j])*xo;
                } else {
                    xo=xi[j][ia[j]-1];
                    rc=(xn-ia[j])*xo;
                }
                x[j]=regn[j]+rc*dx[j];
                wgt *= xo*xnd;
            }
            f=wgt*fxn(x,wgt);
            f2=f*f;
            fb += f;
            f2b += f2;
            for (j=0;j<ndim;j++) {
                di[ia[j]-1][j] += f;
                if (mds >= 0) d[ia[j]-1][j] += f2;
            }
        }
        f2b=sqrt(f2b*npg);
        f2b=(f2b-fb)*(f2b+fb);
        if (f2b <= 0.0) f2b=TINY;
        ti += fb;
        tsi += f2b;
        if (mds < 0) {                      Use stratified sampling.
            for (j=0;j<ndim;j++) d[ia[j]-1][j] += f2b;
        }
        for (k=ndim-1;k>=0;k--) {
            kg[k] %= ng;
            if (++kg[k] != 1) break;
        }
        if (k < 0) break;
    }
    tsi *= dv2g;                        Compute final results for this iteration.
    wgt=1.0/tsi;
    si += wgt*ti;
    schi += wgt*ti*ti;
    swgt += wgt;
    tgral=si/swgt;
    chi2a=(schi-si*tgral)/(it+0.0001);
    if (chi2a < 0.0) chi2a = 0.0;
    sd=sqrt(1.0/swgt);
    tsi=sqrt(tsi);
    if (nprn >= 0) {
        cout << " iteration no. " << setw(3) << (it+1);
        cout << " : integral = " << setw(14) << ti;
        cout << " +/- " << setw(9) << tsi << endl;
        cout << " all iterations: " << " integral =";
        cout << setw(14) << tgral << "+-" << setw(9) << sd;
```

```
            cout << " chi**2/IT n =" << setw(9) << chi2a << endl;
            if (nprn != 0) {
                for (j=0;j<ndim;j++) {
                    cout << " DATA FOR axis " << setw(2) << j << endl;
                    cout << " X delta i X delta i";
                    cout << " X deltai" << endl;
                    for (i=nprn/2;i<nd;i += nprn+2) {
                        cout << setw(8) << xi[j][i] << setw(12) << di[i][j];
                        cout << setw(12) << xi[j][i+1] << setw(12) << di[i+1][j];
                        cout << setw(12) << xi[j][i+2] << setw(12) << di[i+2][j];
                        cout << endl;
                    }
                }
            }
        }
        for (j=0;j<ndim;j++) {              Refine the grid.  Consult references to understand
            xo=d[0][j];                        the subtlety of this procedure.  The refine-
            xn=d[1][j];                        ment is damped, to avoid rapid, destabiliz-
            d[0][j]=(xo+xn)/2.0;               ing changes, and also compressed in range
            dt[j]=d[0][j];                     by the exponent ALPH.
            for (i=2;i<nd;i++) {
                rc=xo+xn;
                xo=xn;
                xn=d[i][j];
                d[i-1][j] = (rc+xn)/3.0;
                dt[j] += d[i-1][j];
            }
            d[nd-1][j]=(xo+xn)/2.0;
            dt[j] += d[nd-1][j];
        }
        for (j=0;j<ndim;j++) {
            rc=0.0;
            for (i=0;i<nd;i++) {
                if (d[i][j] < TINY) d[i][j]=TINY;
                r[i]=pow((1.0-d[i][j]/dt[j])/
                    (log(dt[j])-log(d[i][j])),ALPH);
                rc += r[i];
            }
            rebin(rc/xnd,nd,r,xin,xi,j);
        }
    }
}
```

```
#include "nr.h"

void NR::rebin(const DP rc, const int nd, Vec_I_DP &r, Vec_O_DP &xin,
    Mat_IO_DP &xi, const int j)
Utility routine used by vegas, to rebin a vector of densities contained in row j of xi into new
bins defined by a vector r.
{
    int i,k=0;
    DP dr=0.0,xn=0.0,xo=0.0;

    for (i=0;i<nd-1;i++) {
        while (rc > dr)
            dr += r[(++k)-1];
        if (k > 1) xo=xi[j][k-2];
        xn=xi[j][k-1];
        dr -= rc;
        xin[i]=xn-(xn-xo)*dr/r[k-1];
    }
    for (i=0;i<nd-1;i++) xi[j][i]=xin[i];
```

```
    xi[j][nd-1]=1.0;
}
```

Recursive Stratified Sampling

The problem with stratified sampling, we have seen, is that it may not avoid the K^d explosion inherent in the obvious, Cartesian, tessellation of a d-dimensional volume. A technique called *recursive stratified sampling* [3] attempts to do this by successive bisections of a volume, not along all d dimensions, but rather along only one dimension at a time. The starting points are equations (7.8.10) and (7.8.13), applied to bisections of successively smaller subregions.

Suppose that we have a quota of N evaluations of the function f, and want to evaluate $\langle f \rangle'$ in the rectangular parallelepiped region $R = (\mathbf{x}_a, \mathbf{x}_b)$. (We denote such a region by the two coordinate vectors of its diagonally opposite corners.) First, we allocate a fraction p of N towards exploring the variance of f in R: We sample pN function values uniformly in R and accumulate the sums that will give the d different pairs of variances corresponding to the d different coordinate directions along which R can be bisected. In other words, in pN samples, we estimate $\text{Var}(f)$ in each of the regions resulting from a possible bisection of R,

$$
\begin{aligned}
R_{ai} &\equiv (\mathbf{x}_a, \mathbf{x}_b - \tfrac{1}{2}\mathbf{e}_i \cdot (\mathbf{x}_b - \mathbf{x}_a)\mathbf{e}_i) \\
R_{bi} &\equiv (\mathbf{x}_a + \tfrac{1}{2}\mathbf{e}_i \cdot (\mathbf{x}_b - \mathbf{x}_a)\mathbf{e}_i, \mathbf{x}_b)
\end{aligned}
\tag{7.8.21}
$$

Here $\mathbf{e}_i$ is the unit vector in the ith coordinate direction, $i = 1, 2, \ldots, d$.

Second, we inspect the variances to find the most favorable dimension i to bisect. By equation (7.8.15), we could, for example, choose that i for which the sum of the square roots of the variance estimators in regions R_{ai} and R_{bi} is minimized. (Actually, as we will explain, we do something slightly different.)

Third, we allocate the remaining $(1 - p)N$ function evaluations between the regions R_{ai} and R_{bi}. If we used equation (7.8.15) to choose i, we should do this allocation according to equation (7.8.14).

We now have two parallelepipeds each with its own allocation of function evaluations for estimating the mean of f. Our "RSS" algorithm now shows itself to be *recursive*: To evaluate the mean in each region, we go back to the sentence beginning "First,..." in the paragraph above equation (7.8.21). (Of course, when the allocation of points to a region falls below some number, we resort to simple Monte Carlo rather than continue with the recursion.)

Finally, we combine the means, and also estimated variances of the two subvolumes, using equation (7.8.10) and the first line of equation (7.8.11).

This completes the RSS algorithm in its simplest form. Before we describe some additional tricks under the general rubric of "implementation details," we need to return briefly to equations (7.8.13)–(7.8.15) and derive the equations that we actually use instead of these. The right-hand side of equation (7.8.13) applies the familiar scaling law of equation (7.8.9) twice, once to a and again to b. This would be correct if the estimates $\langle f \rangle_a$ and $\langle f \rangle_b$ were each made by simple Monte Carlo, with uniformly random sample points. However, the two estimates of the mean are in fact made recursively. Thus, there is no reason to expect equation (7.8.9) to hold. Rather, we might substitute for equation (7.8.13) the relation,

$$
\text{Var}\left(\langle f \rangle'\right) = \frac{1}{4}\left[\frac{\text{Var}_a(f)}{N_a^\alpha} + \frac{\text{Var}_b(f)}{(N - N_a)^\alpha}\right]
\tag{7.8.22}
$$

where α is an unknown constant ≥ 1 (the case of equality corresponding to simple Monte Carlo). In that case, a short calculation shows that $\text{Var}\left(\langle f \rangle'\right)$ is minimized when

$$
\frac{N_a}{N} = \frac{\text{Var}_a(f)^{1/(1+\alpha)}}{\text{Var}_a(f)^{1/(1+\alpha)} + \text{Var}_b(f)^{1/(1+\alpha)}}
\tag{7.8.23}
$$

and that its minimum value is

$$\text{Var}\left(\langle f \rangle'\right) \propto \left[\text{Var}_a\left(f\right)^{1/(1+\alpha)} + \text{Var}_b\left(f\right)^{1/(1+\alpha)}\right]^{1+\alpha} \qquad (7.8.24)$$

Equations (7.8.22)–(7.8.24) reduce to equations (7.8.13)–(7.8.15) when $\alpha = 1$. Numerical experiments to find a self-consistent value for α find that $\alpha \approx 2$. That is, when equation (7.8.23) with $\alpha = 2$ is used recursively to allocate sample opportunities, the observed variance of the RSS algorithm goes approximately as N^{-2}, while any other value of α in equation (7.8.23) gives a poorer fall-off. (The sensitivity to α is, however, not very great; it is not known whether $\alpha = 2$ is an analytically justifiable result, or only a useful heuristic.)

The principal difference between miser's implementation and the algorithm as described thus far lies in how the variances on the right-hand side of equation (7.8.23) are estimated. We find empirically that it is somewhat more robust to use the square of the difference of maximum and minimum sampled function values, instead of the genuine second moment of the samples. This estimator is of course increasingly biased with increasing sample size; however, equation (7.8.23) uses it only to compare two subvolumes (a and b) having approximately equal numbers of samples. The "max minus min" estimator proves its worth when the preliminary sampling yields only a single point, or small number of points, in active regions of the integrand. In many realistic cases, these are indicators of nearby regions of even greater importance, and it is useful to let them attract the greater sampling weight that "max minus min" provides.

A second modification embodied in the code is the introduction of a "dithering parameter," dith, whose nonzero value causes subvolumes to be divided not exactly down the middle, but rather into fractions $0.5\pm$dith, with the sign of the $\pm$ randomly chosen by a built-in random number routine. Normally dith can be set to zero. However, there is a large advantage in taking dith to be nonzero if some special symmetry of the integrand puts the active region exactly at the midpoint of the region, or at the center of some power-of-two submultiple of the region. One wants to avoid the extreme case of the active region being evenly divided into 2^d abutting corners of a d-dimensional space. A typical nonzero value of dith, on those occasions when it is useful, might be 0.1. Of course, when the dithering parameter is nonzero, we must take the differing sizes of the subvolumes into account; the code does this through the variable fracl.

One final feature in the code deserves mention. The RSS algorithm uses a single set of sample points to evaluate equation (7.8.23) in all d directions. At bottom levels of the recursion, the number of sample points can be quite small. Although rare, it can happen that in one direction all the samples are in one half of the volume; in that case, that direction is ignored as a candidate for bifurcation. Even more rare is the possibility that all of the samples are in one half of the volume in *all* directions. In this case, a random direction is chosen. If this happens too often in your application, then you should increase MNPT (see line if (jb == -1)... in the code).

Note that miser, as given, returns as ave an estimate of the average function value $\langle\langle f \rangle\rangle$, not the integral of f over the region. The routine vegas, adopting the other convention, returns as tgral the integral. The two conventions are of course trivially related, by equation (7.8.8), since the volume V of the rectangular region is known.

```
#include <cmath>
#include "nr.h"
using namespace std;

void NR::miser(DP func(Vec_I_DP &), Vec_I_DP &regn, const int npts,
    const DP dith, DP &ave, DP &var)
```
Monte Carlo samples a user-supplied ndim-dimensional function func in a rectangular volume specified by regn[0..2*ndim-1], a vector consisting of ndim "lower-left" coordinates of the region followed by ndim "upper-right" coordinates. The function is sampled a total of npts times, at locations determined by the method of recursive stratified sampling. The mean value of the function in the region is returned as ave; an estimate of the statistical uncertainty of ave (square of standard deviation) is returned as var. The input parameter dith should normally be set to zero, but can be set to (e.g.) 0.1 if func's active region falls on the boundary of a power-of-two subdivision of region.
```
{
```

```
const int MNPT=15, MNBS=60;
const DP PFAC=0.1, TINY=1.0e-30, BIG=1.0e30;
```
Here PFAC is the fraction of remaining function evaluations used *at each stage* to explore
the variance of func. At least MNPT function evaluations are performed in any terminal
subregion; a subregion is further bisected only if at least MNBS function evaluations are
available. We take MNBS = 4*MNPT.
```
static int iran=0;
int j,jb,n,ndim,npre,nptl,nptr;
DP avel,varl,fracl,fval,rgl,rgm,rgr,s,sigl,siglb,sigr,sigrb;
DP sum,sumb,summ,summ2;

ndim=regn.size()/2;
Vec_DP pt(ndim);
if (npts < MNBS) {                              Too few points to bisect; do straight
    summ=summ2=0.0;                                 Monte Carlo.
    for (n=0;n<npts;n++) {
        ranpt(pt,regn);
        fval=func(pt);
        summ += fval;
        summ2 += fval * fval;
    }
    ave=summ/npts;
    var=MAX(TINY,(summ2-summ*summ/npts)/(npts*npts));
} else {                                        Do the preliminary (uniform) sampling.
    Vec_DP rmid(ndim);
    npre=MAX(int(npts*PFAC),int(MNPT));
    Vec_DP fmaxl(ndim),fmaxr(ndim),fminl(ndim),fminr(ndim);
    for (j=0;j<ndim;j++) {                      Initialize the left and right bounds for
        iran=(iran*2661+36979) % 175000;           each dimension.
        s=SIGN(dith,DP(iran-87500));
        rmid[j]=(0.5+s)*regn[j]+(0.5-s)*regn[ndim+j];
        fminl[j]=fminr[j]=BIG;
        fmaxl[j]=fmaxr[j]=(-BIG);
    }
    for (n=0;n<npre;n++) {                      Loop over the points in the sample.
        ranpt(pt,regn);
        fval=func(pt);
        for (j=0;j<ndim;j++) {                  Find the left and right bounds for each
            if (pt[j]<=rmid[j]) {                   dimension.
                fminl[j]=MIN(fminl[j],fval);
                fmaxl[j]=MAX(fmaxl[j],fval);
            } else {
                fminr[j]=MIN(fminr[j],fval);
                fmaxr[j]=MAX(fmaxr[j],fval);
            }
        }
    }
    sumb=BIG;                                   Choose which dimension jb to bisect.
    jb= -1;
    siglb=sigrb=1.0;
    for (j=0;j<ndim;j++) {
        if (fmaxl[j] > fminl[j] && fmaxr[j] > fminr[j]) {
            sigl=MAX(TINY,pow(fmaxl[j]-fminl[j],2.0/3.0));
            sigr=MAX(TINY,pow(fmaxr[j]-fminr[j],2.0/3.0));
            sum=sigl+sigr;                      Equation (7.8.24), see text.
            if (sum<=sumb) {
                sumb=sum;
                jb=j;
                siglb=sigl;
                sigrb=sigr;
            }
        }
    }
    if (jb == -1) jb=(ndim*iran)/175000;   MNPT may be too small.
```

```
        rgl=regn[jb];                              Apportion the remaining points between
        rgm=rmid[jb];                              left and right.
        rgr=regn[ndim+jb];
        fracl=fabs((rgm-rgl)/(rgr-rgl));
        nptl=int(MNPT+(npts-npre-2*MNPT)*fracl*siglb
            /(fracl*siglb+(1.0-fracl)*sigrb));      Equation (7.8.23).
        nptr=npts-npre-nptl;
        Vec_DP regn_temp(2*ndim);                  Now allocate and integrate the two sub-
        for (j=0;j<ndim;j++) {                     regions.
            regn_temp[j]=regn[j];
            regn_temp[ndim+j]=regn[ndim+j];
        }
        regn_temp[ndim+jb]=rmid[jb];
        miser(func,regn_temp,nptl,dith,avel,varl);
        regn_temp[jb]=rmid[jb];                    Dispatch recursive call; will return back
        regn_temp[ndim+jb]=regn[ndim+j];           here eventually.
        miser(func,regn_temp,nptr,dith,ave,var);
        ave=fracl*avel+(1-fracl)*ave;
        var=fracl*fracl*varl+(1-fracl)*(1-fracl)*var;
        Combine left and right regions by equation (7.8.11) (1st line).
    }
}
```

The `miser` routine calls a short function `ranpt` to get a random point within a specified
d-dimensional region. The following version of `ranpt` makes consecutive calls to a uniform
random number generator and does the obvious scaling. One can easily modify `ranpt` to
generate its points via the quasi-random routine `sobseq` (§7.7). We find that `miser` with
`sobseq` can be considerably more accurate than `miser` with uniform random deviates. Since
the use of RSS and the use of quasi-random numbers are completely separable, however, we
have not made the code given here dependent on `sobseq`. A similar remark might be made
regarding importance sampling, which could in principle be combined with RSS. (One could
in principle combine `vegas` and `miser`, although the programming would be intricate.)

```
#include "nr.h"

extern int idum;

void NR::ranpt(Vec_O_DP &pt, Vec_I_DP &regn)
Returns a uniformly random point pt in an n-dimensional rectangular region. Used by miser;
calls ran1 for uniform deviates. Your main program should initialize the global variable idum
to a negative seed integer.
{
    int j;

    int n=pt.size();
    for (j=0;j<n;j++)
        pt[j]=regn[j]+(regn[n+j]-regn[j])*ran1(idum);
}
```

CITED REFERENCES AND FURTHER READING:

Hammersley, J.M. and Handscomb, D.C. 1964, *Monte Carlo Methods* (London: Methuen).

Kalos, M.H. and Whitlock, P.A. 1986, *Monte Carlo Methods* (New York: Wiley).

Bratley, P., Fox, B.L., and Schrage, E.L. 1983, *A Guide to Simulation*, 2nd ed. (New York: Springer-Verlag).

Lepage, G.P. 1978, *Journal of Computational Physics*, vol. 27, pp. 192–203. [1]

Lepage, G.P. 1980, "VEGAS: An Adaptive Multidimensional Integration Program," Publication CLNS-80/447, Cornell University. [2]

Press, W.H., and Farrar, G.R. 1990, *Computers in Physics*, vol. 4, pp. 190–195. [3]

Chapter 8.   Sorting

8.0 Introduction

This chapter almost doesn't belong in a book on *numerical* methods. However, some practical knowledge of techniques for sorting is an indispensable part of any good programmer's expertise. We would not want you to consider yourself expert in numerical techniques while remaining ignorant of so basic a subject.

In conjunction with numerical work, sorting is frequently necessary when data (either experimental or numerically generated) are being handled. One has tables or lists of numbers, representing one or more independent (or "control") variables, and one or more dependent (or "measured") variables. One may wish to arrange these data, in various circumstances, in order by one or another of these variables. Alternatively, one may simply wish to identify the "median" value, or the "upper quartile" value of one of the lists of values. This task, closely related to sorting, is called *selection*.

Here, more specifically, are the tasks that this chapter will deal with:

- Sort, i.e., rearrange, an array of numbers into numerical order.
- Rearrange an array into numerical order while performing the corresponding rearrangement of one or more additional arrays, so that the correspondence between elements in all arrays is maintained.
- Given an array, prepare an *index table* for it, i.e., a table of pointers telling which number array element comes first in numerical order, which second, and so on.
- Given an array, prepare a *rank table* for it, i.e., a table telling what is the numerical rank of the first array element, the second array element, and so on.
- Select the Mth largest element from an array.

For the basic task of sorting N elements, the best algorithms require on the order of several times $N \log_2 N$ operations. The algorithm inventor tries to reduce the constant in front of this estimate to as small a value as possible. Two of the best algorithms are *Quicksort* (§8.2), invented by the inimitable C.A.R. Hoare, and *Heapsort* (§8.3), invented by J.W.J. Williams.

For large N (say > 1000), Quicksort is faster, on most machines, by a factor of 1.5 or 2; it requires a bit of extra memory, however, and is a moderately complicated program. Heapsort is a true "sort in place," and is somewhat more compact to program and therefore a bit easier to modify for special purposes. On balance, we recommend Quicksort because of its speed, but we implement both routines.

For small N one does better to use an algorithm whose operation count goes as a higher, i.e., poorer, power of N, if the constant in front is small enough. For $N < 20$, roughly, the method of *straight insertion* (§8.1) is concise and fast enough. We include it with some trepidation: It is an N^2 algorithm, whose potential for misuse (by using it for too large an N) is great. The resultant waste of computer time is so awesome, that we were tempted not to include any N^2 routine at all. We *will* draw the line, however, at the inefficient N^2 algorithm, beloved of elementary computer science texts, called *bubble sort*. If you know what bubble sort is, wipe it from your mind; if you don't know, make a point of never finding out!

For $N < 50$, roughly, *Shell's method* (§8.1), only slightly more complicated to program than straight insertion, is competitive with the more complicated Quicksort on many machines. This method goes as $N^{3/2}$ in the worst case, but is usually faster.

See references [1,2] for further information on the subject of sorting, and for detailed references to the literature.

CITED REFERENCES AND FURTHER READING:

Knuth, D.E. 1997, *Sorting and Searching*, 3rd ed., vol. 3 of *The Art of Computer Programming* (Reading, MA: Addison-Wesley). [1]

Sedgewick, R. 1998, *Algorithms in C*, 3rd ed. (Reading, MA: Addison-Wesley), Chapters 8–13. [2]

8.1 Straight Insertion and Shell's Method

Straight insertion is an N^2 routine, and should be used only for small N, say < 20.

The technique is exactly the one used by experienced card players to sort their cards: Pick out the second card and put it in order with respect to the first; then pick out the third card and insert it into the sequence among the first two; and so on until the last card has been picked out and inserted.

```
#include "nr.h"

void NR::piksrt(Vec_IO_DP &arr)
Sorts an array arr[0..n-1] into ascending numerical order, by straight insertion. arr is
replaced on output by its sorted rearrangement.
{
    int i,j;
    DP a;

    int n=arr.size();
    for (j=1;j<n;j++) {                  Pick out each element in turn.
        a=arr[j];
        i=j;
        while (i > 0 && arr[i-1] > a) {   Look for the place to insert it.
            arr[i]=arr[i-1];
            i--;
        }
        arr[i]=a;                         Insert it.
    }
}
```

What if you also want to rearrange an array `brr` at the same time as you sort `arr`? Simply move an element of `brr` whenever you move an element of `arr`:

```
#include "nr.h"

void NR::piksr2(Vec_IO_DP &arr, Vec_IO_DP &brr)
Sorts an array arr[0..n-1] into ascending numerical order, by straight insertion, while making
the corresponding rearrangement of the array brr[0..n-1].
{
    int i,j;
    DP a,b;

    int n=arr.size();
    for (j=1;j<n;j++) {              Pick out each element in turn.
        a=arr[j];
        b=brr[j];
        i=j;
        while (i > 0 && arr[i-1] > a) {    Look for the place to insert it.
            arr[i]=arr[i-1];
            brr[i]=brr[i-1];
            i--;
        }
        arr[i]=a;                   Insert it.
        brr[i]=b;
    }
}
```

For the case of rearranging a larger number of arrays by sorting on one of them, see §8.4.

Shell's Method

This is actually a variant on straight insertion, but a very powerful variant indeed. The rough idea, e.g., for the case of sorting 16 numbers $n_0 \ldots n_{15}$, is this: First sort, by straight insertion, each of the 8 groups of 2 $(n_0, n_8), (n_1, n_9), \ldots, (n_7, n_{15})$. Next, sort each of the 4 groups of 4 $(n_0, n_4, n_8, n_{12}), \ldots, (n_3, n_7, n_{11}, n_{15})$. Next sort the 2 groups of 8 records, beginning with $(n_0, n_2, n_4, n_6, n_8, n_{10}, n_{12}, n_{14})$. Finally, sort the whole list of 16 numbers.

Of course, only the *last* sort is *necessary* for putting the numbers into order. So what is the purpose of the previous partial sorts? The answer is that the previous sorts allow numbers efficiently to filter up or down to positions close to their final resting places. Therefore, the straight insertion passes on the final sort rarely have to go past more than a "few" elements before finding the right place. (Think of sorting a hand of cards that are already almost in order.)

The spacings between the numbers sorted on each pass through the data (8,4,2,1 in the above example) are called the *increments*, and a Shell sort is sometimes called a *diminishing increment sort*. There has been a lot of research into how to choose a good set of increments, but the optimum choice is not known. The set $\ldots, 8, 4, 2, 1$ is in fact not a good choice, especially for N a power of 2. A much better choice is the sequence

$$(3^k - 1)/2, \ldots, 40, 13, 4, 1 \tag{8.1.1}$$

which can be generated by the recurrence

$$i_0 = 1, \qquad i_{k+1} = 3i_k + 1, \quad k = 0, 1, \ldots \qquad (8.1.2)$$

It can be shown (see [1]) that for this sequence of increments the number of operations required in all is of order $N^{3/2}$ for the worst possible ordering of the original data. For "randomly" ordered data, the operations count goes approximately as $N^{1.25}$, at least for $N < 60000$. For $N > 50$, however, Quicksort is generally faster. The program follows:

```
#include "nr.h"

void NR::shell(const int m, Vec_IO_DP &a)
```
Sorts an array a[0..n-1] into ascending numerical order by Shell's method (diminishing increment sort). a is replaced on output by its sorted rearrangement. Normally, the argument m should be set to the size n of array a, but if m < a.size(), then only the first m elements of a are sorted. This feature is used in selip.
```
{
    int i,j,inc;
    DP v;

    inc=1;                          Determine the starting increment.
    do {
        inc *= 3;
        inc++;
    } while (inc <= m);
    do {                            Loop over the partial sorts.
        inc /= 3;
        for (i=inc;i<m;i++) {       Outer loop of straight insertion.
            v=a[i];
            j=i;
            while (a[j-inc] > v) {  Inner loop of straight insertion.
                a[j]=a[j-inc];
                j -= inc;
                if (j < inc) break;
            }
            a[j]=v;
        }
    } while (inc > 1);
}
```

CITED REFERENCES AND FURTHER READING:

Knuth, D.E. 1997, *Sorting and Searching*, 3rd ed., vol. 3 of *The Art of Computer Programming* (Reading, MA: Addison-Wesley), §5.2.1. [1]

Sedgewick, R. 1998, *Algorithms in C*, 3rd ed. (Reading, MA: Addison-Wesley), Chapter 8.

8.2 Quicksort

Quicksort is, on most machines, on average, for large N, the fastest known sorting algorithm. It is a "partition-exchange" sorting method: A "partitioning element" a is selected from the array. Then by pairwise exchanges of elements, the original array is partitioned into two subarrays. At the end of a round of partitioning, the element a is in its final place in the array. All elements in the left subarray are $\leq$ a, while all elements in the right subarray are $\geq$ a. The process is then repeated on the left and right subarrays independently, and so on.

The partitioning process is carried out by selecting some element, say the leftmost, as the partitioning element a. Scan a pointer up the array until you find an element $>$ a, and then scan another pointer down from the end of the array until you find an element $<$ a. These two elements are clearly out of place for the final partitioned array, so exchange them. Continue this process until the pointers cross. This is the right place to insert a, and that round of partitioning is done. The question of the best strategy when an element is equal to the partitioning element is subtle; we refer you to Sedgewick [1] for a discussion. (Answer: You should stop and do an exchange.)

For speed of execution, we do not implement Quicksort using recursion. Thus the algorithm requires an auxiliary array of storage, of length $2 \log_2 N$, which it uses as a push-down stack for keeping track of the pending subarrays. When a subarray has gotten down to some size M, it becomes faster to sort it by straight insertion (§8.1), so we will do this. The optimal setting of M is machine dependent, but $M = 7$ is not too far wrong. Some people advocate leaving the short subarrays unsorted until the end, and then doing one giant insertion sort at the end. Since each element moves at most 7 places, this is just as efficient as doing the sorts immediately, and saves on the overhead. However, on modern machines with paged memory, there is increased overhead when dealing with a large array all at once. We have not found any advantage in saving the insertion sorts till the end.

As already mentioned, Quicksort's *average* running time is fast, but its *worst case* running time can be very slow: For the worst case it is, in fact, an N^2 method! And for the most straightforward implementation of Quicksort it turns out that the worst case is achieved for an input array that is already in order! This ordering of the input array might easily occur in practice. One way to avoid this is to use a little random number generator to choose a random element as the partitioning element. Another is to use instead the median of the first, middle, and last elements of the current subarray.

The great speed of Quicksort comes from the simplicity and efficiency of its inner loop. Simply adding one unnecessary test (for example, a test that your pointer has not moved off the end of the array) can almost double the running time! One avoids such unnecessary tests by placing "sentinels" at either end of the subarray being partitioned. The leftmost sentinel is $\leq$ a, the rightmost $\geq$ a. With the "median-of-three" selection of a partitioning element, we can use the two elements that were not the median to be the sentinels for that subarray.

Our implementation closely follows [1]:

```
#include "nr.h"

void NR::sort(Vec_IO_DP &arr)
```
Sorts an array `arr[0..n-1]` into ascending numerical order using the Quicksort algorithm.
`arr` is replaced on output by its sorted rearrangement.
```
{
    const int M=7,NSTACK=50;
```
Here M is the size of subarrays sorted by straight insertion and NSTACK is the required
auxiliary storage.
```
    int i,ir,j,k,jstack=-1,l=0;
    DP a;
    Vec_INT istack(NSTACK);

    int n=arr.size();
    ir=n-1;
    for (;;) {                              Insertion sort when subarray small enough.
        if (ir-l < M) {
            for (j=l+1;j<=ir;j++) {
                a=arr[j];
                for (i=j-1;i>=l;i--) {
                    if (arr[i] <= a) break;
                    arr[i+1]=arr[i];
                }
                arr[i+1]=a;
            }
            if (jstack < 0) break;
            ir=istack[jstack--];            Pop stack and begin a new round of parti-
            l=istack[jstack--];                tioning.
        } else {
            k=(l+ir) >> 1;                  Choose median of left, center, and right el-
            SWAP(arr[k],arr[l+1]);            ements as partitioning element a. Also
            if (arr[l] > arr[ir]) {           rearrange so that a[l] ≤ a[l+1] ≤ a[ir].
                SWAP(arr[l],arr[ir]);
            }
            if (arr[l+1] > arr[ir]) {
                SWAP(arr[l+1],arr[ir]);
            }
            if (arr[l] > arr[l+1]) {
                SWAP(arr[l],arr[l+1]);
            }
            i=l+1;                          Initialize pointers for partitioning.
            j=ir;
            a=arr[l+1];                      Partitioning element.
            for (;;) {                       Beginning of innermost loop.
                do i++; while (arr[i] < a);      Scan up to find element > a.
                do j--; while (arr[j] > a);      Scan down to find element < a.
                if (j < i) break;            Pointers crossed. Partitioning complete.
                SWAP(arr[i],arr[j]);         Exchange elements.
            }                               End of innermost loop.
            arr[l+1]=arr[j];                Insert partitioning element.
            arr[j]=a;
            jstack += 2;
```
Push pointers to larger subarray on stack, process smaller subarray immediately.
```
            if (jstack >= NSTACK) nrerror("NSTACK too small in sort.");
            if (ir-i+1 >= j-l) {
                istack[jstack]=ir;
                istack[jstack-1]=i;
                ir=j-1;
            } else {
                istack[jstack]=j-1;
                istack[jstack-1]=l;
                l=i;
            }
        }
    }
```

```
        }
    }
```

    As usual you can move any other arrays around at the same time as you sort
arr. At the risk of being repetitious:

```
#include "nr.h"

void NR::sort2(Vec_IO_DP &arr, Vec_IO_DP &brr)
```
Sorts an array arr[0..n-1] into ascending order using Quicksort, while making the corresponding rearrangement of the array brr[0..n-1].
```
{
    const int M=7,NSTACK=50;
    int i,ir,j,k,jstack=-1,l=0;
    DP a,b;
    Vec_INT istack(NSTACK);

    int n=arr.size();
    ir=n-1;
    for (;;) {                                  Insertion sort when subarray small enough.
        if (ir-l < M) {
            for (j=l+1;j<=ir;j++) {
                a=arr[j];
                b=brr[j];
                for (i=j-1;i>=l;i--) {
                    if (arr[i] <= a) break;
                    arr[i+1]=arr[i];
                    brr[i+1]=brr[i];
                }
                arr[i+1]=a;
                brr[i+1]=b;
            }
            if (jstack < 0) break;
            ir=istack[jstack--];                Pop stack and begin a new round of parti-
            l=istack[jstack--];                    tioning.
        } else {
            k=(l+ir) >> 1;                      Choose median of left, center and right el-
            SWAP(arr[k],arr[l+1]);                 ements as partitioning element a. Also
            SWAP(brr[k],brr[l+1]);                 rearrange so that a[l] ≤ a[l+1] ≤ a[ir].
            if (arr[l] > arr[ir]) {
                SWAP(arr[l],arr[ir]);
                SWAP(brr[l],brr[ir]);
            }
            if (arr[l+1] > arr[ir]) {
                SWAP(arr[l+1],arr[ir]);
                SWAP(brr[l+1],brr[ir]);
            }
            if (arr[l] > arr[l+1]) {
                SWAP(arr[l],arr[l+1]);
                SWAP(brr[l],brr[l+1]);
            }
            i=l+1;                              Initialize pointers for partitioning.
            j=ir;
            a=arr[l+1];                         Partitioning element.
            b=brr[l+1];
            for (;;) {                          Beginning of innermost loop.
                do i++; while (arr[i] < a);         Scan up to find element > a.
                do j--; while (arr[j] > a);         Scan down to find element < a.
                if (j < i) break;               Pointers crossed. Partitioning complete.
                SWAP(arr[i],arr[j]);            Exchange elements of both arrays.
                SWAP(brr[i],brr[j]);
            }                                   End of innermost loop.
```

```
        arr[l+1]=arr[j];              Insert partitioning element in both arrays.
        arr[j]=a;
        brr[l+1]=brr[j];
        brr[j]=b;
        jstack += 2;
        Push pointers to larger subarray on stack, process smaller subarray immediately.
        if (jstack >= NSTACK) nrerror("NSTACK too small in sort2.");
        if (ir-i+1 >= j-l) {
            istack[jstack]=ir;
            istack[jstack-1]=i;
            ir=j-1;
        } else {
            istack[jstack]=j-1;
            istack[jstack-1]=l;
            l=i;
        }
    }
}
}
```

You could, in principle, rearrange any number of additional arrays along with `brr`, but this becomes wasteful as the number of such arrays becomes large. The preferred technique is to make use of an index table, as described in §8.4.

CITED REFERENCES AND FURTHER READING:

Sedgewick, R. 1978, *Communications of the ACM*, vol. 21, pp. 847–857. [1]

8.3 Heapsort

While usually not quite as fast as Quicksort, Heapsort is one of our favorite sorting routines. It is a true "in-place" sort, requiring no auxiliary storage. It is an $N \log_2 N$ process, not only on average, but also for the worst-case order of input data. In fact, its worst case is only 20 percent or so worse than its average running time.

It is beyond our scope to give a complete exposition on the theory of Heapsort. We will mention the general principles, then let you refer to the references [1,2], or analyze the program yourself, if you want to understand the details.

A set of N numbers a_j, $j = 0, \ldots, N - 1$, is said to form a "heap" if it satisfies the relation

$$a_{(j-1)/2} \geq a_j \quad \text{for} \quad 0 \leq (j-1)/2 < j < N \qquad (8.3.1)$$

Here the division in $j/2$ means "integer divide," i.e., is an exact integer or else is rounded down to the closest integer. Definition (8.3.1) will make sense if you think of the numbers a_i as being arranged in a binary tree, with the top, "boss," node being a_0, the two "underling" nodes being a_1 and a_2, *their* four underling nodes being a_3 through a_6, etc. (See Figure 8.3.1.) In this form, a heap has every "supervisor" greater than or equal to its two "supervisees," down through the levels of the hierarchy.

If you have managed to rearrange your array into an order that forms a heap, then sorting it is very easy: You pull off the "top of the heap," which will be the

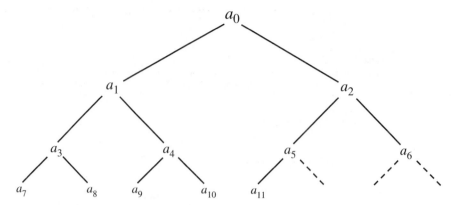

Figure 8.3.1.   Ordering implied by a "heap," here of 12 elements. Elements connected by an upward path are sorted with respect to one another, but there is not necessarily any ordering among elements related only "laterally."

largest element yet unsorted. Then you "promote" to the top of the heap its largest underling. Then you promote *its* largest underling, and so on. The process is like what happens (or is supposed to happen) in a large corporation when the chairman of the board retires. You then repeat the whole process by retiring the new chairman of the board. Evidently the whole thing is an $N \log_2 N$ process, since each retiring chairman leads to $\log_2 N$ promotions of underlings.

Well, how do you arrange the array into a heap in the first place? The answer is again a "sift-up" process like corporate promotion. Imagine that the corporation starts out with $N/2$ employees on the production line, but with no supervisors. Now a supervisor is hired to supervise two workers. If he is less capable than one of his workers, that one is promoted in his place, and he joins the production line. After supervisors are hired, then supervisors of supervisors are hired, and so on up the corporate ladder. Each employee is brought in at the top of the tree, but then immediately sifted down, with more capable workers promoted until their proper corporate level has been reached.

In the Heapsort implementation, the same "sift-up" code can be used for the initial creation of the heap and for the subsequent retirement-and-promotion phase. One execution of the Heapsort function represents the entire life-cycle of a giant corporation: $N/2$ workers are hired; $N/2$ potential supervisors are hired; there is a sifting up in the ranks, a sort of super Peter Principle: in due course, each of the original employees gets promoted to chairman of the board.

```
#include "nr.h"

namespace {
    void sift_down(Vec_IO_DP &ra, const int l, const int r)
    Carry out the sift-down on element ra(l) to maintain the heap structure. l and r determine
    the "left" and "right" range of the sift-down.
    {
        int j,jold;
        DP a;

        a=ra[l];
        jold=l;
        j=l+1;
        while (j <= r) {
```

```
            if (j < r && ra[j] < ra[j+1]) j++;     Compare to the better underling.
            if (a >= ra[j]) break;                  Found a's level. Terminate the sift-
            ra[jold]=ra[j];                            down. Otherwise, demote a and
            jold=j;                                    continue.
            j=2*j+1;
        }
        ra[jold]=a;                                 Put a into its slot.
    }
}
```

```
void NR::hpsort(Vec_IO_DP &ra)
```
Sorts an array `ra[0..n-1]` into ascending numerical order using the Heapsort algorithm. `ra`
is replaced on output by its sorted rearrangement.
```
{
    int i;

    int n=ra.size();
    for (i=n/2-1; i>=0; i--)
        The index i, which here determines the "left" range of the sift-down, i.e., the element
        to be sifted down, is decremented from n/2-1 down to 0 during the "hiring" (heap
        creation) phase.
        sift_down(ra,i,n-1);
    for (i=n-1; i>0; i--) {
        Here the "right" range of the sift-down is decremented from n-2 down to 0 during the
        "retirement-and-promotion" (heap selection) phase.
        SWAP(ra[0],ra[i]);                          Clear a space at the end of the array, and retire
        sift_down(ra,0,i-1);                            the top of the heap into it.
    }
}
```

CITED REFERENCES AND FURTHER READING:

Knuth, D.E. 1997, *Sorting and Searching*, 3rd ed., vol. 3 of *The Art of Computer Programming*
(Reading, MA: Addison-Wesley), §5.2.3. [1]

Sedgewick, R. 1998, *Algorithms in C*, 3rd ed. (Reading, MA: Addison-Wesley), Chapter 11. [2]

8.4 Indexing and Ranking

The concept of *keys* plays a prominent role in the management of data files. A
data *record* in such a file may contain several items, or *fields*. For example, a record
in a file of weather observations may have fields recording time, temperature, and
wind velocity. When we sort the records, we must decide which of these fields we
want to be brought into sorted order. The other fields in a record just come along
for the ride, and will not, in general, end up in any particular order. The field on
which the sort is performed is called the *key* field.

For a data file with many records and many fields, the actual movement of N
records into the sorted order of their keys K_i, $i = 0, \ldots, N - 1$, can be a daunting
task. Instead, one can construct an *index table* I_j, $j = 0, \ldots, N - 1$, such that the
smallest K_i has $i = I_0$, the second smallest has $i = I_1$, and so on up to the largest
K_i with $i = I_{N-1}$. In other words, the array

$$K_{I_j} \quad j = 0, 1, \ldots, N - 1 \tag{8.4.1}$$

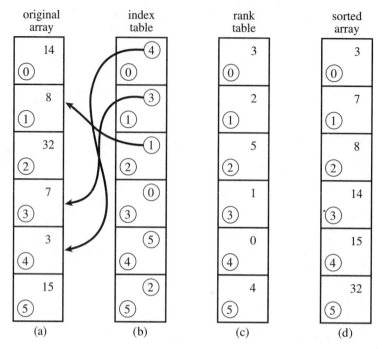

Figure 8.4.1.   (a) An unsorted array of six numbers. (b) Index table, whose entries are pointers to the elements of (a) in ascending order. (c) Rank table, whose entries are the ranks of the corresponding elements of (a). (d) Sorted array of the elements in (a).

is in sorted order when indexed by j. When an index table is available, one need not move records from their original order. Further, different index tables can be made from the same set of records, indexing them to different keys.

The algorithm for constructing an index table is straightforward: Initialize the index array with the integers from 0 to $N-1$, then perform the Quicksort algorithm, moving the elements around *as if* one were sorting the keys. The integer that initially numbered the smallest key thus ends up in the number one position, and so on.

We overload the function `indexx` with two specific implementations, one for double values, the other for integers.

```
#include "nr.h"

void NR::indexx(Vec_I_DP &arr, Vec_O_INT &indx)
```
Indexes an array `arr[0..n-1]`, i.e., outputs the array `indx[0..n-1]` such that `arr[indx[j]]` is in ascending order for $j = 0, 1, \ldots, N-1$. The input array `arr` is not changed.
```
{
    const int M=7,NSTACK=50;
    int i,indxt,ir,j,k,jstack=-1,l=0;
    DP a;
    Vec_INT istack(NSTACK);

    int n=arr.size();
    ir=n-1;
    for (j=0;j<n;j++) indx[j]=j;
    for (;;) {
        if (ir-l < M) {
            for (j=l+1;j<=ir;j++) {
                indxt=indx[j];
                a=arr[indxt];
```

```
            for (i=j-1;i>=1;i--) {
                if (arr[indx[i]] <= a) break;
                indx[i+1]=indx[i];
            }
            indx[i+1]=indxt;
        }
        if (jstack < 0) break;
        ir=istack[jstack--];
        l=istack[jstack--];
    } else {
        k=(l+ir) >> 1;
        SWAP(indx[k],indx[l+1]);
        if (arr[indx[l]] > arr[indx[ir]]) {
            SWAP(indx[l],indx[ir]);
        }
        if (arr[indx[l+1]] > arr[indx[ir]]) {
            SWAP(indx[l+1],indx[ir]);
        }
        if (arr[indx[l]] > arr[indx[l+1]]) {
            SWAP(indx[l],indx[l+1]);
        }
        i=l+1;
        j=ir;
        indxt=indx[l+1];
        a=arr[indxt];
        for (;;) {
            do i++; while (arr[indx[i]] < a);
            do j--; while (arr[indx[j]] > a);
            if (j < i) break;
            SWAP(indx[i],indx[j]);
        }
        indx[l+1]=indx[j];
        indx[j]=indxt;
        jstack += 2;
        if (jstack >= NSTACK) nrerror("NSTACK too small in indexx.");
        if (ir-i+1 >= j-l) {
            istack[jstack]=ir;
            istack[jstack-1]=i;
            ir=j-1;
        } else {
            istack[jstack]=j-1;
            istack[jstack-1]=l;
            l=i;
        }
    }
}
}
}
```

```
void NR::indexx(Vec_I_INT &arr, Vec_O_INT &indx)
```
Overloaded version for integer array.
```
{
    const int M=7,NSTACK=50;
    int i,indxt,ir,j,k,jstack=-1,l=0;
    int a;
```
The rest of the code is identical to the double version above and is omitted.

If you want to sort an array while making the corresponding rearrangement of several or many other arrays, you should first make an index table, then use it to rearrange each array in turn. This requires two arrays of working space: one to hold the index, and another into which an array is temporarily moved, and from which it is redeposited back on itself in the rearranged order. For 3 arrays, the procedure looks like this:

```
#include "nr.h"

void NR::sort3(Vec_IO_DP &ra, Vec_IO_DP &rb, Vec_IO_DP &rc)
Sorts an array ra[0..n-1] into ascending numerical order while making the corresponding
rearrangements of the arrays rb[0..n-1] and rc[0..n-1]. An index table is constructed
via the routine indexx.
{
    int j;

    int n=ra.size();
    Vec_INT iwksp(n);
    Vec_DP wksp(n);
    indexx(ra,iwksp);                           Make the index table.
    for (j=0;j<n;j++) wksp[j]=ra[j];            Save the array ra.
    for (j=0;j<n;j++) ra[j]=wksp[iwksp[j]];     Copy it back in rearranged order.
    for (j=0;j<n;j++) wksp[j]=rb[j];            Ditto rb.
    for (j=0;j<n;j++) rb[j]=wksp[iwksp[j]];
    for (j=0;j<n;j++) wksp[j]=rc[j];            Ditto rc.
    for (j=0;j<n;j++) rc[j]=wksp[iwksp[j]];
}
```

The generalization to any other number of arrays is obviously straightforward.

A *rank table* is different from an index table. A rank table's jth entry gives the rank of the jth element of the original array of keys, ranging from 0 (if that element was the smallest) to $N - 1$ (if that element was the largest). One can easily construct a rank table from an index table, however:

```
#include "nr.h"

void NR::rank(Vec_I_INT &indx, Vec_O_INT &irank)
Given indx[0..n-1] as output from the routine indexx, returns an array irank[0..n-1],
the corresponding table of ranks.
{
    int j;

    int n=indx.size();
    for (j=0;j<n;j++) irank[indx[j]]=j;
}
```

Figure 8.4.1 summarizes the concepts discussed in this section.

8.5 Selecting the Mth Largest

Selection is sorting's austere sister. (Say *that* five times quickly!) Where sorting demands the rearrangement of an entire data array, selection politely asks for a single returned value: What is the kth smallest (or, equivalently, the $m = N - 1 - k$th largest) element out of N elements? (Note that in this section $k = 0, 1, \ldots, N-1$, so for example $k = 0$ corresponds to the smallest array element.) The fastest methods for selection do, unfortunately, rearrange the array for their own computational purposes, typically putting all smaller elements to the left of the kth, all larger elements to the right, and scrambling the order within each subset. This side effect is at best innocuous, at worst downright inconvenient. When the array is very long, so that making a scratch copy of it is taxing on memory, or when the computational burden of the selection is a negligible part of a larger calculation, one turns to selection algorithms without side effects, which leave the original array undisturbed. Such *in place* selection is slower than the faster selection methods by a factor of about 10. We give routines of both types, below.

The most common use of selection is in the statistical characterization of a set of data. One often wants to know the median element in an array, or the top and bottom quartile elements. When N is odd, the median is the kth element, with $k = (N-1)/2$. When N is even, statistics books define the median as the arithmetic mean of the elements $k = N/2 - 1$ and $k = N/2$ (that is, $N/2$ from the bottom and $N/2$ from the top). If you accept such pedantry, you must perform two separate selections to find these elements. For $N > 100$ we usually define $k = N/2$ to be the median element, pedants be damned.

The fastest general method for selection, allowing rearrangement, is *partitioning*, exactly as was done in the Quicksort algorithm (§8.2). Selecting a "random" partition element, one marches through the array, forcing smaller elements to the left, larger elements to the right. As in Quicksort, it is important to optimize the inner loop, using "sentinels" (§8.2) to minimize the number of comparisons. For sorting, one would then proceed to further partition both subsets. For selection, we can ignore one subset and attend only to the one that contains our desired kth element. Selection by partitioning thus does not need a stack of pending operations, and its operations count scales as N rather than as $N \log N$ (see [1]). Comparison with sort in §8.2 should make the following routine obvious:

```
#include "nr.h"

DP NR::select(const int k, Vec_IO_DP &arr)
Given k in [0..n-1] returns an array value from arr[0..n-1] such that k array values are
less than or equal to the one returned. The input array will be rearranged to have this value in
location arr[k], with all smaller elements moved to arr[0..k-1] (in arbitrary order) and all
larger elements in arr[k+1..n-1] (also in arbitrary order).
{
    int i,ir,j,l,mid;
    DP a;

    int n=arr.size();
    l=0;
    ir=n-1;
    for (;;) {
        if (ir <= l+1) {                  Active partition contains 1 or 2 elements.
            if (ir == l+1 && arr[ir] < arr[l])        Case of 2 elements.
                SWAP(arr[l],arr[ir]);
            return arr[k];
        } else {
            mid=(l+ir) >> 1;              Choose median of left, center, and right el-
            SWAP(arr[mid],arr[l+1]);         ements as partitioning element a. Also
            if (arr[l] > arr[ir])            rearrange so that arr[l] ≤ arr[l+1],
                SWAP(arr[l],arr[ir]);        arr[ir] ≥ arr[l+1].
            if (arr[l+1] > arr[ir])
                SWAP(arr[l+1],arr[ir]);
            if (arr[l] > arr[l+1])
                SWAP(arr[l],arr[l+1]);
```

```
            i=l+1;                        Initialize pointers for partitioning.
            j=ir;
            a=arr[l+1];                   Partitioning element.
            for (;;) {                    Beginning of innermost loop.
                do i++; while (arr[i] < a);     Scan up to find element > a.
                do j--; while (arr[j] > a);     Scan down to find element < a.
                if (j < i) break;         Pointers crossed. Partitioning complete.
                SWAP(arr[i],arr[j]);
            }                             End of innermost loop.
            arr[l+1]=arr[j];              Insert partitioning element.
            arr[j]=a;
            if (j >= k) ir=j-1;           Keep active the partition that contains the
            if (j <= k) l=i;                  kth element.
        }
    }
}
```

In-place, nondestructive, selection is conceptually simple, but it requires a lot of bookkeeping, and it is correspondingly slower. The general idea is to pick some number M of elements at random, to sort them, and then to make a pass through the array *counting* how many elements fall in each of the $M + 1$ intervals defined by these elements. The kth largest will fall in one such interval — call it the "live" interval. One then does a second round, first picking M random elements in the live interval, and then determining which of the new, finer, $M + 1$ intervals all presently live elements fall into. And so on, until the kth element is finally localized within a single array of size M, at which point direct selection is possible.

How shall we pick M? The number of rounds, $\log_M N = \log_2 N / \log_2 M$, will be smaller if M is larger; but the work to locate each element among $M + 1$ subintervals will be larger, scaling as $\log_2 M$ for bisection, say. Each round requires looking at all N elements, if only to find those that are still alive, while the bisections are dominated by the N that occur in the first round. Minimizing $O(N \log_M N) + O(N \log_2 M)$ thus yields the result

$$M \sim 2^{\sqrt{\log_2 N}} \tag{8.5.1}$$

The square root of the logarithm is so slowly varying that secondary considerations of machine timing become important. We use $M = 64$ as a convenient constant value.

Two minor additional tricks in the following routine, selip, are (i) augmenting the set of M random values by an $M + 1$st, the arithmetic mean, and (ii) choosing the M random values "on the fly" in a pass through the data, by a method that makes later values no less likely to be chosen than earlier ones. (The underlying idea is to give element $m > M$ an M/m chance of being brought into the set. You can prove by induction that this yields the desired result.)

```
#include "nr.h"

DP NR::selip(const int k, Vec_I_DP &arr)
Given k in [0..n-1] returns an array value from arr[0..n-1] such that k array values are
less than or equal to the one returned. The input array is not altered.
{
    const int M=64;
    const DP BIG=1.0e30;
    int i,j,jl,jm,ju,kk,mm,nlo,nxtmm;
    DP ahi,alo,sum;
```

```
Vec_INT isel(M+2);
Vec_DP sel(M+2);

int n=arr.size();
if (k < 0 || k > n-1) nrerror("bad input to selip");
kk=k;
ahi=BIG;
alo = -BIG;
for (;;) {                              Main iteration loop, until desired ele-
    mm=nlo=0;                           ment is isolated.
    sum=0.0;
    nxtmm=M+1;
    for (i=0;i<n;i++) {                 Make a pass through the whole array.
        if (arr[i] >= alo && arr[i] <= ahi) {
            Consider only elements in the current brackets.
            mm++;
            if (arr[i] == alo) nlo++;   In case of ties for low bracket.
            Now use statistical procedure for selecting m in-range elements with equal
            probability, even without knowing in advance how many there are!
            if (mm <= M) sel[mm-1]=arr[i];
            else if (mm == nxtmm) {
                nxtmm=mm+mm/M;
                sel[(i+2+mm+kk) % M]=arr[i];    The % operation provides a some-
            }                                   what random number.
            sum += arr[i];
        }
    }
    if (kk < nlo) {                     Desired element is tied for lower bound;
        return alo;                     return it.
    }
    else if (mm < M+1) {                All in-range elements were kept. So re-
        shell(mm,sel);                  turn answer by direct method.
        ahi = sel[kk];
        return ahi;
    }
    sel[M]=sum/mm;                      Augment selected set by mean value (fixes
    shell(M+1,sel);                     degeneracies), and sort it.
    sel[M+1]=ahi;
    for (j=0;j<M+2;j++) isel[j]=0;      Zero the count array.
    for (i=0;i<n;i++) {                 Make another pass through the array.
        if (arr[i] >= alo && arr[i] <= ahi) {    For each in-range element..
            jl=0;
            ju=M+2;
            while (ju-jl > 1) {         ...find its position in the selected set by
                jm=(ju+jl)/2;           bisection...
                if (arr[i] >= sel[jm-1]) jl=jm;
                else ju=jm;
            }
            isel[ju-1]++;               ...and increment the counter.
        }
    }
    j=0;                                Now we can narrow the bounds to just
    while (kk >= isel[j]) {             one bin, that is, by a factor of order
        alo=sel[j];                     m.
        kk -= isel[j++];
    }
    ahi=sel[j];
}
}
```

Approximate timings: selip is about 10 times slower than select. Indeed,
for N in the range of $\sim 10^5$, selip is about 1.5 times slower than a full sort with

sort, while select is about 6 times faster than sort. You should weigh time against memory and convenience carefully.

Of course neither of the above routines should be used for the trivial cases of finding the largest, or smallest, element in an array. Those cases, you code by hand as simple for loops. There are also good ways to code the case where k is modest in comparison to N, so that extra memory of order k is not burdensome. An example is to use the method of Heapsort (§8.3) to make a single pass through an array of length N while saving the m *largest* elements. The advantage of the heap structure is that only $\log m$, rather than m, comparisons are required every time a new element is added to the candidate list. This becomes a real savings when $m > O(\sqrt{N})$, but it never hurts otherwise and is easy to code. The following program gives the idea.

```
#include "nr.h"

void NR::hpsel(Vec_I_DP &arr, Vec_O_DP &heap)
Returns in heap[0..m-1] the largest m elements of the array arr[0..n-1], with heap[0]
guaranteed to be the the mth largest element. The array arr is not altered. For efficiency, this
routine should be used only when m ≪ n.
{
    int i,j,k;

    int m=heap.size();
    int n=arr.size();
    if (m > n/2 || m < 1) nrerror("probable misuse of hpsel");
    for (i=0;i<m;i++) heap[i]=arr[i];
    sort(heap);                         Create initial heap by overkill! We assume m ≪ n.
    for (i=m;i<n;i++) {                  For each remaining element...
        if (arr[i] > heap[0]) {          Put it on the heap?
            heap[0]=arr[i];
            for (j=0;;) {                Sift down.
                k=(j << 1)+1;
                if (k > m-1) break;
                if (k != (m-1) && heap[k] > heap[k+1]) k++;
                if (heap[j] <= heap[k]) break;
                SWAP(heap[k],heap[j]);
                j=k;
            }
        }
    }
}
```

CITED REFERENCES AND FURTHER READING:

Sedgewick, R. 1998, *Algorithms in C*, 3rd ed. (Reading, MA: Addison-Wesley), pp. 126ff. [1]

Knuth, D.E. 1997, *Sorting and Searching*, 3rd ed., vol. 3 of *The Art of Computer Programming* (Reading, MA: Addison-Wesley).

8.6 Determination of Equivalence Classes

A number of techniques for sorting and searching relate to data structures whose details are beyond the scope of this book, for example, trees, linked lists, etc. These structures and their manipulations are the bread and butter of computer science, as distinct from numerical analysis, and there is no shortage of books on the subject.

In working with experimental data, we have found that one particular such manipulation, namely the determination of equivalence classes, arises sufficiently often to justify inclusion here.

The problem is this: There are N "elements" (or "data points" or whatever), numbered $0, \ldots, N-1$. You are given pairwise information about whether elements are in the same *equivalence class* of "sameness," by whatever criterion happens to be of interest. For example, you may have a list of facts like: "Element 3 and element 7 are in the same class; element 19 and element 4 are in the same class; element 7 and element 12 are in the same class, ..." Alternatively, you may have a procedure, given the numbers of two elements j and k, for deciding whether they are in the same class or different classes. (Recall that an equivalence relation can be anything satisfying the *RST properties*: reflexive, symmetric, transitive. This is compatible with any intuitive definition of "sameness.")

The desired output is an assignment to each of the N elements of an equivalence class number, such that two elements are in the same class if and only if they are assigned the same class number.

Efficient algorithms work like this: Let $F(j)$ be the class or "family" number of element j. Start off with each element in its own family, so that $F(j) = j$. The array $F(j)$ can be interpreted as a tree structure, where $F(j)$ denotes the parent of j. If we arrange for each family to be its own tree, disjoint from all the other "family trees," then we can label each family (equivalence class) by its most senior great-great-...grandparent. The detailed topology of the tree doesn't matter at all, as long as we graft each related element onto it *somewhere*.

Therefore, we process each elemental datum "j is equivalent to k" by (i) tracking j up to its highest ancestor, (ii) tracking k up to its highest ancestor, (iii) giving j to k as a new parent, or vice versa (it makes no difference). After processing all the relations, we go through all the elements j and reset their $F(j)$'s to their highest possible ancestors, which then label the equivalence classes.

The following routine, based on Knuth [1], assumes that there are m elemental pieces of information, stored in two arrays of length m, lista,listb, the interpretation being that lista[j] and listb[j], j=0...m-1, are the numbers of two elements which (we are thus told) are related.

```
#include "nr.h"

void NR::eclass(Vec_O_INT &nf, Vec_I_INT &lista, Vec_I_INT &listb)
Given m equivalences between pairs of n individual elements in the form of the input arrays
lista[0..m-1] and listb[0..m-1], this routine returns in nf[0..n-1] the number of
the equivalence class of each of the n elements, integers between 0 and n-1 (not all such
integers used).
{
    int l,k,j;

    int n=nf.size();
    int m=lista.size();
    for (k=0;k<n;k++) nf[k]=k;            Initialize each element its own class.
    for (l=0;l<m;l++) {                   For each piece of input information...
        j=lista[l];
        while (nf[j] != j) j=nf[j];       Track first element up to its ancestor.
        k=listb[l];
        while (nf[k] != k) k=nf[k];       Track second element up to its ancestor.
        if (j != k) nf[j]=k;              If they are not already related, make them
    }                                        so.
    for (j=0;j<n;j++)                     Final sweep up to highest ancestors.
        while (nf[j] != nf[nf[j]]) nf[j]=nf[nf[j]];
}
```

Alternatively, we may be able to construct a boolean function equiv(j,k) that returns a value true if elements j and k are related, or false if they are not. Then we want to loop over all pairs of elements to get the complete picture. D. Eardley has devised a clever way of doing this while simultaneously sweeping the tree up to high ancestors in a manner that keeps it current and obviates most of the final sweep phase:

```
#include "nr.h"

void NR::eclazz(Vec_O_INT &nf, bool equiv(const int, const int))
```
Given a user-supplied boolean function equiv which tells whether a pair of elements, each in the range $0 \ldots n-1$, are related, return in nf$[0 \ldots n-1]$ equivalence class numbers for each element.
```
{
    int kk,jj;

    int n=nf.size();
    nf[0]=0;
    for (jj=1;jj<n;jj++) {                  Loop over first element of all pairs.
        nf[jj]=jj;
        for (kk=0;kk<jj;kk++) {             Loop over second element of all pairs.
            nf[kk]=nf[nf[kk]];              Sweep it up this much.
            if (equiv(jj+1,kk+1)) nf[nf[nf[kk]]]=jj;
            Good exercise for the reader to figure out why this much ancestry is necessary!
        }
    }
    for (jj=0;jj<n;jj++) nf[jj]=nf[nf[jj]];       Only this much sweeping is needed
}                                                                    finally.
```

CITED REFERENCES AND FURTHER READING:

Knuth, D.E. 1997, *Fundamental Algorithms*, 3rd ed., vol. 1 of *The Art of Computer Programming* (Reading, MA: Addison-Wesley), §2.3.3. [1]

Sedgewick, R. 1998, *Algorithms in C*, 3rd ed. (Reading, MA: Addison-Wesley), Chapter 30.

Chapter 9.   Root Finding and Nonlinear Sets of Equations

9.0 Introduction

We now consider that most basic of tasks, solving equations numerically. While most equations are born with both a right-hand side and a left-hand side, one traditionally moves all terms to the left, leaving

$$f(x) = 0 \qquad (9.0.1)$$

whose solution or solutions are desired. When there is only one independent variable, the problem is *one-dimensional*, namely to find the root or roots of a function.

With more than one independent variable, more than one equation can be satisfied simultaneously. You likely once learned the *implicit function theorem* which (in this context) gives us the hope of satisfying N equations in N unknowns simultaneously. Note that we have only hope, not certainty. A nonlinear set of equations may have no (real) solutions at all. Contrariwise, it may have more than one solution. The implicit function theorem tells us that "generically" the solutions will be distinct, pointlike, and separated from each other. If, however, life is so unkind as to present you with a nongeneric, i.e., degenerate, case, then you can get a continuous family of solutions. In vector notation, we want to find one or more N-dimensional solution vectors $\mathbf{x}$ such that

$$\mathbf{f}(\mathbf{x}) = \mathbf{0} \qquad (9.0.2)$$

where $\mathbf{f}$ is the N-dimensional vector-valued function whose components are the individual equations to be satisfied simultaneously.

Don't be fooled by the apparent notational similarity of equations (9.0.2) and (9.0.1). Simultaneous solution of equations in N dimensions is *much* more difficult than finding roots in the one-dimensional case. The principal difference between one and many dimensions is that, in one dimension, it is possible to bracket or "trap" a root between bracketing values, and then hunt it down like a rabbit. In multidimensions, you can never be sure that the root is there at all until you have found it.

Except in linear problems, root finding invariably proceeds by iteration, and this is equally true in one or in many dimensions. Starting from some approximate trial solution, a useful algorithm will improve the solution until some predetermined convergence criterion is satisfied. For smoothly varying functions, good algorithms will always converge, *provided* that the initial guess is good enough. Indeed one can even determine in advance the rate of convergence of most algorithms.

It cannot be overemphasized, however, how crucially success depends on having a good first guess for the solution, especially for multidimensional problems. This crucial beginning usually depends on analysis rather than numerics. Carefully crafted initial estimates reward you not only with reduced computational effort, but also with understanding and increased self-esteem. Hamming's motto, "the purpose of computing is insight, not numbers," is particularly apt in the area of finding roots. You should repeat this motto aloud whenever your program converges, with ten-digit accuracy, to the wrong root of a problem, or whenever it fails to converge because there is actually *no* root, or because there is a root but your initial estimate was not sufficiently close to it.

"This talk of insight is all very well, but what do I actually do?" For one-dimensional root finding, it is possible to give some straightforward answers: You should try to get some idea of what your function looks like before trying to find its roots. If you need to mass-produce roots for many different functions, then you should at least know what some typical members of the ensemble look like. Next, you should always bracket a root, that is, know that the function changes sign in an identified interval, before trying to converge to the root's value.

Finally (this is advice with which some daring souls might disagree, but we give it nonetheless) never let your iteration method get outside of the best bracketing bounds obtained at any stage. We will see below that some pedagogically important algorithms, such as *secant method* or *Newton-Raphson*, can violate this last constraint, and are thus not recommended unless certain fixups are implemented.

Multiple roots, or very close roots, are a real problem, especially if the multiplicity is an even number. In that case, there may be no readily apparent sign change in the function, so the notion of bracketing a root — and maintaining the bracket — becomes difficult. We are hard-liners: we nevertheless insist on bracketing a root, even if it takes the minimum-searching techniques of Chapter 10 to determine whether a tantalizing dip in the function really does cross zero or not. (You can easily modify the simple golden section routine of §10.1 to return early if it detects a sign change in the function. And, if the minimum of the function is exactly zero, then you have found a *double* root.)

As usual, we want to discourage you from using routines as black boxes without understanding them. However, as a guide to beginners, here are some reasonable starting points:

- Brent's algorithm in §9.3 is the method of choice to find a bracketed root of a general one-dimensional function, when you cannot easily compute the function's derivative. Ridders' method (§9.2) is concise, and a close competitor.

- When you can compute the function's derivative, the routine `rtsafe` in §9.4, which combines the Newton-Raphson method with some bookkeeping on bounds, is recommended. Again, you must first bracket your root.

- Roots of polynomials are a special case. Laguerre's method, in §9.5, is recommended as a starting point. Beware: Some polynomials are ill-conditioned!

- Finally, for multidimensional problems, the only elementary method is Newton-Raphson (§9.6), which works *very* well if you can supply a

good first guess of the solution. Try it. Then read the more advanced
material in §9.7 for some more complicated, but globally more convergent,
alternatives.

Avoiding implementations for specific computers, this book must generally
steer clear of interactive or graphics-related routines. We make an exception right
now. The following routine, which produces a crude function plot with interactively
scaled axes, can save you a lot of grief as you enter the world of root finding.

```cpp
#include <string>
#include <iostream>
#include <iomanip>
#include "nr.h"
using namespace std;

void NR::scrsho(DP fx(const DP))
```
For interactive CRT terminal use. Produce a crude graph of the function `fx` over the prompted-
for interval `x1,x2`. Query for another plot until the user signals satisfaction.
```cpp
{
    const int ISCR=60, JSCR=21;            // Number of horizontal and vertical positions in
    const char BLANK=' ', ZERO='-', YY='l', XX='-', FF='x';     // display.
    int jz,j,i;
    DP ysml,ybig,x2,x1,x,dyj,dx;
    Vec_DP y(ISCR);
    string scr[JSCR];

    for (;;) {
        cout << endl << "Enter x1 x2 (x1=x2 to stop):" << endl;
                                           // Query for another plot, quit if x1=x2.
        cin >> x1 >> x2;
        if (x1 == x2) break;
        scr[0]=YY;                         // Fill top left corner with character 'l'.
        for (i=1;i<(ISCR-1);i++)           // Fill rest of top with character '-'.
            scr[0] += XX;
        scr[0] += YY;                      // End top with character 'l'.
        for (j=1;j<(JSCR-1);j++) {
            scr[j]=YY;                     // Fill left side with character 'l'.
            for (i=1;i<(ISCR-1);i++)       // Fill interior with blanks.
                scr[j] += BLANK;
            scr[j] += YY;                  // Fill right side with character 'l'.
        }
        scr[JSCR-1]=scr[0];                // Bottom is same as top.
        dx=(x2-x1)/(ISCR-1);
        x=x1;
        ysml=ybig=0.0;                     // Limits will include 0.
        for (i=0;i<ISCR;i++) {             // Evaluate the function at equal intervals.
            y[i]=fx(x);                    //     Find the largest and smallest val-
            if (y[i] < ysml) ysml=y[i];    // ues.
            if (y[i] > ybig) ybig=y[i];
            x += dx;
        }
        if (ybig == ysml) ybig=ysml+1.0;   // Be sure to separate top and bottom.
        dyj=(JSCR-1)/(ybig-ysml);
        jz=int(-ysml*dyj);                 // Note which row corresponds to 0.
        for (i=0;i<ISCR;i++) {             // Place an indicator at function height and
            scr[jz][i]=ZERO;               //     at 0.
            j=int((y[i]-ysml)*dyj);
            scr[j][i]=FF;
        }
        cout << fixed << setprecision(3);
        cout << setw(11) << ybig << " " << scr[JSCR-1] << endl;
        for (j=JSCR-2;j>=1;j--)            // Display.
            cout << " " << scr[j] << endl;
```

```
            cout << setw(11) << ysml << " " << scr[0] << endl;
            cout << setw(19) << x1 << setw(55) << x2;
    }
}
```

CITED REFERENCES AND FURTHER READING:

Stoer, J., and Bulirsch, R. 1993, *Introduction to Numerical Analysis*, 2nd ed. (New York: Springer-Verlag), Chapter 5.

Acton, F.S. 1970, *Numerical Methods That Work*; 1990, corrected edition (Washington: Mathematical Association of America), Chapters 2, 7, and 14.

Ralston, A., and Rabinowitz, P. 1978, *A First Course in Numerical Analysis*, 2nd ed.; reprinted 2001 (New York: Dover), Chapter 8.

Householder, A.S. 1970, *The Numerical Treatment of a Single Nonlinear Equation* (New York: McGraw-Hill).

9.1 Bracketing and Bisection

We will say that a root is *bracketed* in the interval (a, b) if $f(a)$ and $f(b)$ have opposite signs. If the function is continuous, then at least one root must lie in that interval (the *intermediate value theorem*). If the function is discontinuous, but bounded, then instead of a root there might be a step discontinuity which crosses zero (see Figure 9.1.1). For numerical purposes, that might as well be a root, since the behavior is indistinguishable from the case of a continuous function whose zero crossing occurs in between two "adjacent" floating-point numbers in a machine's finite-precision representation. Only for functions with singularities is there the possibility that a bracketed root is not really there, as for example

$$f(x) = \frac{1}{x - c} \qquad (9.1.1)$$

Some root-finding algorithms (e.g., bisection in this section) will readily converge to c in (9.1.1). Luckily there is not much possibility of your mistaking c, or any number x close to it, for a root, since mere evaluation of $|f(x)|$ will give a very large, rather than a very small, result.

If you are given a function in a black box, there is no sure way of bracketing its roots, or of even determining that it has roots. If you like pathological examples, think about the problem of locating the two real roots of equation (3.0.1), which dips below zero only in the ridiculously small interval of about $x = \pi \pm 10^{-667}$.

In the next chapter we will deal with the related problem of bracketing a function's minimum. There it is possible to give a procedure that always succeeds; in essence, "Go downhill, taking steps of increasing size, until your function starts back uphill." There is no analogous procedure for roots. The procedure "go downhill until your function changes sign," can be foiled by a function that has a simple extremum. Nevertheless, if you are prepared to deal with a "failure" outcome, this procedure is often a good first start; success is usual if your function has opposite signs in the limit $x \to \pm\infty$.

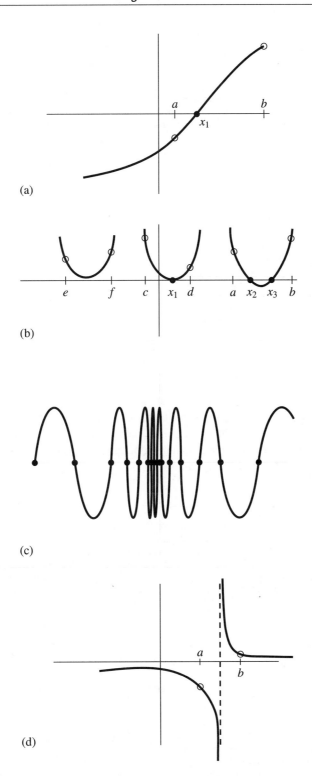

Figure 9.1.1.   Some situations encountered while root finding: (a) shows an isolated root x_1 bracketed by two points a and b at which the function has opposite signs; (b) illustrates that there is not necessarily a sign change in the function near a double root (in fact, there is not necessarily a root!); (c) is a pathological function with many roots; in (d) the function has opposite signs at points a and b, but the points bracket a singularity, not a root.

```
#include <cmath>
#include "nr.h"
using namespace std;

bool NR::zbrac(DP func(const DP), DP &x1, DP &x2)
```
Given a function `func` and an initial guessed range x1 to x2, the routine expands the range
geometrically until a root is bracketed by the returned values x1 and x2 (in which case `zbrac`
returns `true`) or until the range becomes unacceptably large (in which case `zbrac` returns
`false`).
```
{
    const int NTRY=50;
    const DP FACTOR=1.6;
    int j;
    DP f1,f2;

    if (x1 == x2) nrerror("Bad initial range in zbrac");
    f1=func(x1);
    f2=func(x2);
    for (j=0;j<NTRY;j++) {
        if (f1*f2 < 0.0) return true;
        if (fabs(f1) < fabs(f2))
            f1=func(x1 += FACTOR*(x1-x2));
        else
            f2=func(x2 += FACTOR*(x2-x1));
    }
    return false;
}
```

Alternatively, you might want to "look inward" on an initial interval, rather
than "look outward" from it, asking if there are any roots of the function $f(x)$ in
the interval from x_1 to x_2 when a search is carried out by subdivision into n equal
intervals. The following function calculates brackets for up to nb distinct intervals
which each contain one or more roots.

```
#include "nr.h"

void NR::zbrak(DP fx(const DP), const DP x1, const DP x2, const int n,
    Vec_O_DP &xb1, Vec_O_DP &xb2, int &nroot)
```
Given a function `fx` defined on the interval from x1-x2 subdivide the interval into n equally
spaced segments, and search for zero crossings of the function. The arrays xb1[0..nb-1] and
xb2[0..nb-1] will be filled sequentially with any bracketing pairs that are found, and must be
provided with a size nb that is sufficient to hold the maximum number of roots sought. `nroot`
will be set to the number of bracketing pairs actually found.
```
{
    int i;
    DP x,fp,fc,dx;

    int nb=xb1.size();
    nroot=0;
    dx=(x2-x1)/n;                        Determine the spacing appropriate to the mesh.
    fp=fx(x=x1);
    for (i=0;i<n;i++) {                  Loop over all intervals
        fc=fx(x += dx);
        if (fc*fp <= 0.0) {              If a sign change occurs then record values for the
            xb1[nroot]=x-dx;                 bounds.
            xb2[nroot++]=x;
            if(nroot == nb) return;
        }
        fp=fc;
    }
}
```

Bisection Method

Once we know that an interval contains a root, several classical procedures are available to refine it. These proceed with varying degrees of speed and sureness towards the answer. Unfortunately, the methods that are guaranteed to converge plod along most slowly, while those that rush to the solution in the best cases can also dash rapidly to infinity without warning if measures are not taken to avoid such behavior.

The *bisection method* is one that cannot fail. It is thus not to be sneered at as a method for otherwise badly behaved problems. The idea is simple. Over some interval the function is known to pass through zero because it changes sign. Evaluate the function at the interval's midpoint and examine its sign. Use the midpoint to replace whichever limit has the same sign. After each iteration the bounds containing the root decrease by a factor of two. If after n iterations the root is known to be within an interval of size ϵ_n, then after the next iteration it will be bracketed within an interval of size

$$\epsilon_{n+1} = \epsilon_n/2 \qquad\qquad (9.1.2)$$

neither more nor less. Thus, we know in advance the number of iterations required to achieve a given tolerance in the solution,

$$n = \log_2 \frac{\epsilon_0}{\epsilon} \qquad\qquad (9.1.3)$$

where ϵ_0 is the size of the initially bracketing interval, ϵ is the desired ending tolerance.

Bisection *must* succeed. If the interval happens to contain two or more roots, bisection will find one of them. If the interval contains no roots and merely straddles a singularity, it will converge on the singularity.

When a method converges as a factor (less than 1) times the previous uncertainty to the first power (as is the case for bisection), it is said to converge *linearly*. Methods that converge as a higher power,

$$\epsilon_{n+1} = \text{constant} \times (\epsilon_n)^m \qquad m > 1 \qquad\qquad (9.1.4)$$

are said to converge superlinearly. In other contexts "linear" convergence would be termed "exponential," or "geometrical." That is not too bad at all: Linear convergence means that successive significant figures are won linearly with computational effort.

It remains to discuss practical criteria for convergence. It is crucial to keep in mind that computers use a fixed number of binary digits to represent floating-point numbers. While your function might analytically pass through zero, it is possible that its computed value is never zero, for any floating-point argument. One must decide what accuracy on the root is attainable: Convergence to within 10^{-6} in absolute value is reasonable when the root lies near 1, but certainly unachievable if the root lies near 10^{26}. One might thus think to specify convergence by a relative (fractional) criterion, but this becomes unworkable for roots near zero. To be most general, the routines below will require you to specify an absolute tolerance, such that iterations continue until the interval becomes smaller than this tolerance in absolute units. Usually you may wish to take the tolerance to be $\epsilon(|x_1| + |x_2|)/2$ where ϵ is the machine precision and x_1 and x_2 are the initial brackets. When the root lies near zero you ought to consider carefully what reasonable tolerance means for your function. The following routine quits after 40 bisections in any event, with $2^{-40} \approx 10^{-12}$.

```cpp
#include <cmath>
#include "nr.h"
using namespace std;

DP NR::rtbis(DP func(const DP), const DP x1, const DP x2, const DP xacc)
```
Using bisection, find the root of a function func known to lie between x1 and x2. The root, returned as rtbis, will be refined until its accuracy is ±xacc.
```cpp
{
    const int JMAX=40;              Maximum allowed number of bisections.
    int j;
    DP dx,f,fmid,xmid,rtb;

    f=func(x1);
    fmid=func(x2);
    if (f*fmid >= 0.0) nrerror("Root must be bracketed for bisection in rtbis");
    rtb = f < 0.0 ? (dx=x2-x1,x1) : (dx=x1-x2,x2);        Orient the search so that f>0
    for (j=0;j<JMAX;j++) {                                lies at x+dx.
        fmid=func(xmid=rtb+(dx *= 0.5));                  Bisection loop.
        if (fmid <= 0.0) rtb=xmid;
        if (fabs(dx) < xacc || fmid == 0.0) return rtb;
    }
    nrerror("Too many bisections in rtbis");
    return 0.0;                                           Never get here.
}
```

9.2 Secant Method, False Position Method, and Ridders' Method

For functions that are smooth near a root, the methods known respectively as *false position* (or *regula falsi*) and *secant method* generally converge faster than bisection. In both of these methods the function is assumed to be approximately linear in the local region of interest, and the next improvement in the root is taken as the point where the approximating line crosses the axis. After each iteration one of the previous boundary points is discarded in favor of the latest estimate of the root.

The *only* difference between the methods is that secant retains the most recent of the prior estimates (Figure 9.2.1; this requires an arbitrary choice on the first iteration), while false position retains that prior estimate for which the function value has opposite sign from the function value at the current best estimate of the root, so that the two points continue to bracket the root (Figure 9.2.2). Mathematically, the secant method converges more rapidly near a root of a sufficiently continuous function. Its order of convergence can be shown to be the "golden ratio" $1.618\ldots$, so that

$$\lim_{k \to \infty} |\epsilon_{k+1}| \approx \text{const} \times |\epsilon_k|^{1.618} \tag{9.2.1}$$

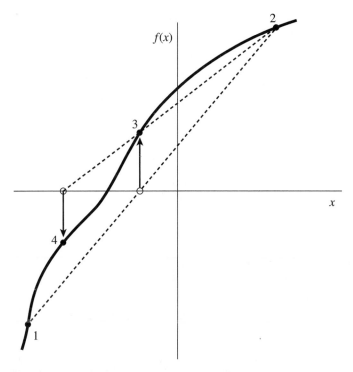

Figure 9.2.1.   Secant method. Extrapolation or interpolation lines (dashed) are drawn through the two most recently evaluated points, whether or not they bracket the function. The points are numbered in the order that they are used.

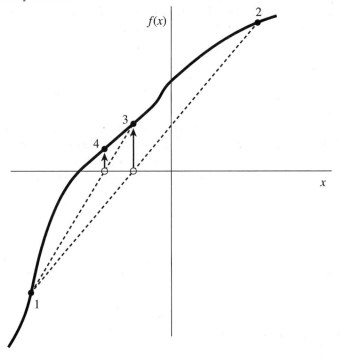

Figure 9.2.2.   False position method. Interpolation lines (dashed) are drawn through the most recent points *that bracket the root*. In this example, point 1 thus remains "active" for many steps. False position converges less rapidly than the secant method, but it is more certain.

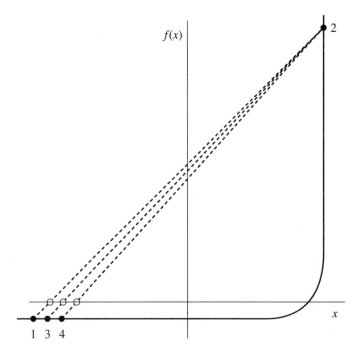

Figure 9.2.3.   Example where both the secant and false position methods will take many iterations to arrive at the true root. This function would be difficult for many other root-finding methods.

The secant method has, however, the disadvantage that the root does not necessarily remain bracketed. For functions that are *not* sufficiently continuous, the algorithm can therefore not be guaranteed to converge: Local behavior might send it off towards infinity.

False position, since it sometimes keeps an older rather than newer function evaluation, has a lower order of convergence. Since the newer function value will *sometimes* be kept, the method is often superlinear, but estimation of its exact order is not so easy.

Here are sample implementations of these two related methods. While these methods are standard textbook fare, *Ridders' method*, described below, or *Brent's method*, in the next section, are almost always better choices. Figure 9.2.3 shows the behavior of secant and false-position methods in a difficult situation.

```
#include <cmath>
#include "nr.h"
using namespace std;

DP NR::rtflsp(DP func(const DP), const DP x1, const DP x2, const DP xacc)
Using the false position method, find the root of a function func known to lie between x1 and
x2. The root, returned as rtflsp, is refined until its accuracy is ±xacc.
{
    const int MAXIT=30;              Set to the maximum allowed number of iterations.
    int j;
    DP fl,fh,xl,xh,dx,del,f,rtf;

    fl=func(x1);
    fh=func(x2);                     Be sure the interval brackets a root.
    if (fl*fh > 0.0) nrerror("Root must be bracketed in rtflsp");
    if (fl < 0.0) {                  Identify the limits so that x1 corresponds to the low
                                       side.
```

```
        xl=x1;
        xh=x2;
    } else {
        xl=x2;
        xh=x1;
        SWAP(fl,fh);
    }
    dx=xh-xl;
    for (j=0;j<MAXIT;j++) {          False position loop.
        rtf=xl+dx*fl/(fl-fh);        Increment with respect to latest value.
        f=func(rtf);
        if (f < 0.0) {               Replace appropriate limit.
            del=xl-rtf;
            xl=rtf;
            fl=f;
        } else {
            del=xh-rtf;
            xh=rtf;
            fh=f;
        }
        dx=xh-xl;
        if (fabs(del) < xacc || f == 0.0) return rtf;        Convergence.
    }
    nrerror("Maximum number of iterations exceeded in rtflsp");
    return 0.0;                      Never get here.
}




#include <cmath>
#include "nr.h"
using namespace std;

DP NR::rtsec(DP func(const DP), const DP x1, const DP x2, const DP xacc)
Using the secant method, find the root of a function func thought to lie between x1 and x2.
The root, returned as rtsec, is refined until its accuracy is ±xacc.
{
    const int MAXIT=30;              Maximum allowed number of iterations.
    int j;
    DP fl,f,dx,xl,rts;

    fl=func(x1);
    f=func(x2);
    if (fabs(fl) < fabs(f)) {        Pick the bound with the smaller function value as
        rts=x1;                          the most recent guess.
        xl=x2;
        SWAP(fl,f);
    } else {
        xl=x1;
        rts=x2;
    }
    for (j=0;j<MAXIT;j++) {          Secant loop.
        dx=(xl-rts)*f/(f-fl);        Increment with respect to latest value.
        xl=rts;
        fl=f;
        rts += dx;
        f=func(rts);
        if (fabs(dx) < xacc || f == 0.0) return rts;      Convergence.
    }
    nrerror("Maximum number of iterations exceeded in rtsec");
    return 0.0;                      Never get here.
}
```

Ridders' Method

A powerful variant on false position is due to Ridders [1]. When a root is bracketed between x_1 and x_2, Ridders' method first evaluates the function at the midpoint $x_3 = (x_1 + x_2)/2$. It then factors out that unique exponential function which turns the residual function into a straight line. Specifically, it solves for a factor e^Q that gives

$$f(x_1) - 2f(x_3)e^Q + f(x_2)e^{2Q} = 0 \qquad (9.2.2)$$

This is a quadratic equation in e^Q, which can be solved to give

$$e^Q = \frac{f(x_3) + \text{sign}[f(x_2)]\sqrt{f(x_3)^2 - f(x_1)f(x_2)}}{f(x_2)} \qquad (9.2.3)$$

Now the false position method is applied, not to the values $f(x_1), f(x_3), f(x_2)$, but to the values $f(x_1), f(x_3)e^Q, f(x_2)e^{2Q}$, yielding a new guess for the root, x_4. The overall updating formula (incorporating the solution 9.2.3) is

$$x_4 = x_3 + (x_3 - x_1)\frac{\text{sign}[f(x_1) - f(x_2)]f(x_3)}{\sqrt{f(x_3)^2 - f(x_1)f(x_2)}} \qquad (9.2.4)$$

Equation (9.2.4) has some very nice properties. First, x_4 is guaranteed to lie in the interval (x_1, x_2), so the method never jumps out of its brackets. Second, the convergence of successive applications of equation (9.2.4) is *quadratic*, that is, $m = 2$ in equation (9.1.4). Since each application of (9.2.4) requires two function evaluations, the actual order of the method is $\sqrt{2}$, not 2; but this is still quite respectably superlinear: the number of significant digits in the answer approximately *doubles* with each two function evaluations. Third, taking out the function's "bend" via exponential (that is, ratio) factors, rather than via a polynomial technique (e.g., fitting a parabola), turns out to give an extraordinarily robust algorithm. In both reliability and speed, Ridders' method is generally competitive with the more highly developed and better established (but more complicated) method of Van Wijngaarden, Dekker, and Brent, which we next discuss.

```
#include <cmath>
#include "nr.h"
using namespace std;

DP NR::zriddr(DP func(const DP), const DP x1, const DP x2, const DP xacc)
Using Ridders' method, return the root of a function func known to lie between x1 and x2.
The root, returned as zriddr, will be refined to an approximate accuracy xacc.
{
    const int MAXIT=60;
    const DP UNUSED=-1.11e30;
    int j;
    DP ans,fh,fl,fm,fnew,s,xh,xl,xm,xnew;

    fl=func(x1);
    fh=func(x2);
    if ((fl > 0.0 && fh < 0.0) || (fl < 0.0 && fh > 0.0)) {
        xl=x1;
        xh=x2;
```

```
    ans=UNUSED;                                    Any highly unlikely value, to simplify logic
    for (j=0;j<MAXIT;j++) {                            below.
        xm=0.5*(xl+xh);
        fm=func(xm);                               First of two function evaluations per it-
        s=sqrt(fm*fm-fl*fh);                            eration.
        if (s == 0.0) return ans;
        xnew=xm+(xm-xl)*((fl >= fh ? 1.0 : -1.0)*fm/s);      Updating formula.
        if (fabs(xnew-ans) <= xacc) return ans;
        ans=xnew;
        fnew=func(ans);                            Second of two function evaluations per
        if (fnew == 0.0) return ans;                    iteration.
        if (SIGN(fm,fnew) != fm) {                 Bookkeeping to keep the root bracketed
            xl=xm;                                      on next iteration.
            fl=fm;
            xh=ans;
            fh=fnew;
        } else if (SIGN(fl,fnew) != fl) {
            xh=ans;
            fh=fnew;
        } else if (SIGN(fh,fnew) != fh) {
            xl=ans;
            fl=fnew;
        } else nrerror("never get here.");
        if (fabs(xh-xl) <= xacc) return ans;
    }
    nrerror("zriddr exceed maximum iterations");
}
else {
    if (fl == 0.0) return x1;
    if (fh == 0.0) return x2;
    nrerror("root must be bracketed in zriddr.");
}
return 0.0;                                         Never get here.
}
```

CITED REFERENCES AND FURTHER READING:

Ralston, A., and Rabinowitz, P. 1978, *A First Course in Numerical Analysis*, 2nd ed.; reprinted 2001 (New York: Dover), §8.3.

Ostrowski, A.M. 1966, *Solutions of Equations and Systems of Equations*, 2nd ed. (New York: Academic Press), Chapter 12.

Ridders, C.J.F. 1979, *IEEE Transactions on Circuits and Systems*, vol. CAS-26, pp. 979–980. [1]

9.3 Van Wijngaarden–Dekker–Brent Method

While secant and false position formally converge faster than bisection, one finds in practice pathological functions for which bisection converges more rapidly. These can be choppy, discontinuous functions, or even smooth functions if the second derivative changes sharply near the root. Bisection always halves the interval, while secant and false position can sometimes spend many cycles slowly pulling distant bounds closer to a root. Ridders' method does a much better job, but it too can sometimes be fooled. Is there a way to combine superlinear convergence with the sureness of bisection?

Yes. We can keep track of whether a supposedly superlinear method is actually converging the way it is supposed to, and, if it is not, we can intersperse bisection steps so as to guarantee *at least* linear convergence. This kind of super-strategy requires attention to bookkeeping detail, and also careful consideration of how roundoff errors can affect the guiding strategy. Also, we must be able to determine reliably when convergence has been achieved.

An excellent algorithm that pays close attention to these matters was developed in the 1960s by van Wijngaarden, Dekker, and others at the Mathematical Center in Amsterdam, and later improved by Brent [1]. For brevity, we refer to the final form of the algorithm as *Brent's method*. The method is *guaranteed* (by Brent) to converge, so long as the function can be evaluated within the initial interval known to contain a root.

Brent's method combines root bracketing, bisection, and *inverse quadratic interpolation* to converge from the neighborhood of a zero crossing. While the false position and secant methods assume approximately linear behavior between two prior root estimates, inverse quadratic interpolation uses three prior points to fit an inverse quadratic function (x as a quadratic function of y) whose value at $y = 0$ is taken as the next estimate of the root x. Of course one must have contingency plans for what to do if the root falls outside of the brackets. Brent's method takes care of all that. If the three point pairs are $[a, f(a)], [b, f(b)], [c, f(c)]$ then the interpolation formula (cf. equation 3.1.1) is

$$x = \frac{[y - f(a)][y - f(b)]c}{[f(c) - f(a)][f(c) - f(b)]} + \frac{[y - f(b)][y - f(c)]a}{[f(a) - f(b)][f(a) - f(c)]}$$
$$+ \frac{[y - f(c)][y - f(a)]b}{[f(b) - f(c)][f(b) - f(a)]} \tag{9.3.1}$$

Setting y to zero gives a result for the next root estimate, which can be written as

$$x = b + P/Q \tag{9.3.2}$$

where, in terms of

$$R \equiv f(b)/f(c), \qquad S \equiv f(b)/f(a), \qquad T \equiv f(a)/f(c) \tag{9.3.3}$$

we have

$$P = S\left[T(R - T)(c - b) - (1 - R)(b - a)\right] \tag{9.3.4}$$
$$Q = (T - 1)(R - 1)(S - 1) \tag{9.3.5}$$

In practice b is the current best estimate of the root and P/Q ought to be a "small" correction. Quadratic methods work well only when the function behaves smoothly; they run the serious risk of giving very bad estimates of the next root or causing machine failure by an inappropriate division by a very small number ($Q \approx 0$). Brent's method guards against this problem by maintaining brackets on the root and checking where the interpolation would land before carrying out the division. When the correction P/Q would not land within the bounds, or when the bounds are not collapsing rapidly enough, the algorithm takes a bisection step. Thus,

Brent's method combines the sureness of bisection with the speed of a higher-order method when appropriate. We recommend it as the method of choice for general one-dimensional root finding where a function's values only (and not its derivative or functional form) are available.

```cpp
#include <cmath>
#include <limits>
#include "nr.h"
using namespace std;

DP NR::zbrent(DP func(const DP), const DP x1, const DP x2, const DP tol)
```
Using Brent's method, find the root of a function func known to lie between x1 and x2. The root, returned as zbrent, will be refined until its accuracy is tol.
```cpp
{
    const int ITMAX=100;                        Maximum allowed number of iterations.
    const DP EPS=numeric_limits<DP>::epsilon();
    Machine floating-point precision.
    int iter;
    DP a=x1,b=x2,c=x2,d,e,min1,min2;
    DP fa=func(a),fb=func(b),fc,p,q,r,s,tol1,xm;

    if ((fa > 0.0 && fb > 0.0) || (fa < 0.0 && fb < 0.0))
        nrerror("Root must be bracketed in zbrent");
    fc=fb;
    for (iter=0;iter<ITMAX;iter++) {
        if ((fb > 0.0 && fc > 0.0) || (fb < 0.0 && fc < 0.0)) {
            c=a;                                Rename a, b, c and adjust bounding interval
            fc=fa;                              d.
            e=d=b-a;
        }
        if (fabs(fc) < fabs(fb)) {
            a=b;
            b=c;
            c=a;
            fa=fb;
            fb=fc;
            fc=fa;
        }
        tol1=2.0*EPS*fabs(b)+0.5*tol;           Convergence check.
        xm=0.5*(c-b);
        if (fabs(xm) <= tol1 || fb == 0.0) return b;
        if (fabs(e) >= tol1 && fabs(fa) > fabs(fb)) {
            s=fb/fa;                            Attempt inverse quadratic interpolation.
            if (a == c) {
                p=2.0*xm*s;
                q=1.0-s;
            } else {
                q=fa/fc;
                r=fb/fc;
                p=s*(2.0*xm*q*(q-r)-(b-a)*(r-1.0));
                q=(q-1.0)*(r-1.0)*(s-1.0);
            }
            if (p > 0.0) q = -q;                Check whether in bounds.
            p=fabs(p);
            min1=3.0*xm*q-fabs(tol1*q);
            min2=fabs(e*q);
            if (2.0*p < (min1 < min2 ? min1 : min2)) {
                e=d;                            Accept interpolation.
                d=p/q;
            } else {
                d=xm;                           Interpolation failed, use bisection.
                e=d;
            }
```

```
      } else {                          Bounds decreasing too slowly, use bisection.
          d=xm;
          e=d;
      }
      a=b;                              Move last best guess to a.
      fa=fb;
      if (fabs(d) > tol1)               Evaluate new trial root.
          b += d;
      else
          b += SIGN(tol1,xm);
          fb=func(b);
  }
  nrerror("Maximum number of iterations exceeded in zbrent");
  return 0.0;                           Never get here.
}
```

CITED REFERENCES AND FURTHER READING:

Brent, R.P. 1973, *Algorithms for Minimization without Derivatives* (Englewood Cliffs, NJ: Prentice-Hall), Chapters 3, 4. [1]

Forsythe, G.E., Malcolm, M.A., and Moler, C.B. 1977, *Computer Methods for Mathematical Computations* (Englewood Cliffs, NJ: Prentice-Hall), §7.2.

9.4 Newton-Raphson Method Using Derivative

Perhaps the most celebrated of all one-dimensional root-finding routines is *Newton's method*, also called the *Newton-Raphson method*. This method is distinguished from the methods of previous sections by the fact that it requires the evaluation of both the function $f(x)$, *and* the derivative $f'(x)$, at arbitrary points x. The Newton-Raphson formula consists geometrically of extending the tangent line at a current point x_i until it crosses zero, then setting the next guess x_{i+1} to the abscissa of that zero-crossing (see Figure 9.4.1). Algebraically, the method derives from the familiar Taylor series expansion of a function in the neighborhood of a point,

$$f(x + \delta) \approx f(x) + f'(x)\delta + \frac{f''(x)}{2}\delta^2 + \dots . \tag{9.4.1}$$

For small enough values of δ, and for well-behaved functions, the terms beyond linear are unimportant, hence $f(x + \delta) = 0$ implies

$$\delta = -\frac{f(x)}{f'(x)}. \tag{9.4.2}$$

Newton-Raphson is not restricted to one dimension. The method readily generalizes to multiple dimensions, as we shall see in §9.6 and §9.7, below.

Far from a root, where the higher-order terms in the series *are* important, the Newton-Raphson formula can give grossly inaccurate, meaningless corrections. For instance, the initial guess for the root might be so far from the true root as to let the search interval include a local maximum or minimum of the function. This can be

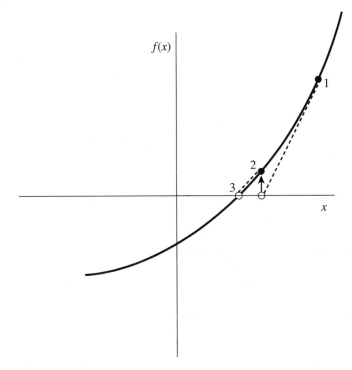

Figure 9.4.1.   Newton's method extrapolates the local derivative to find the next estimate of the root. In this example it works well and converges quadratically.

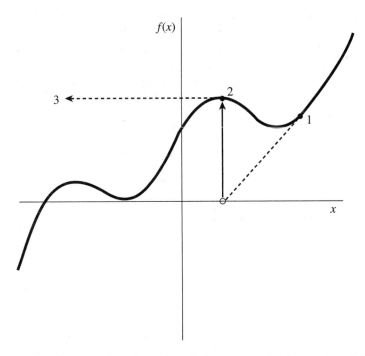

Figure 9.4.2.   Unfortunate case where Newton's method encounters a local extremum and shoots off to outer space. Here bracketing bounds, as in rtsafe, would save the day.

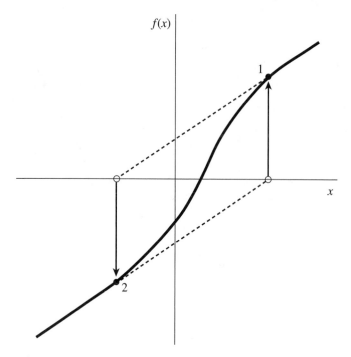

Figure 9.4.3.   Unfortunate case where Newton's method enters a nonconvergent cycle. This behavior is often encountered when the function f is obtained, in whole or in part, by table interpolation. With a better initial guess, the method would have succeeded.

death to the method (see Figure 9.4.2). If an iteration places a trial guess near such a local extremum, so that the first derivative nearly vanishes, then Newton-Raphson sends its solution off to limbo, with vanishingly small hope of recovery. Like most powerful tools, Newton-Raphson can be destructive when used in inappropriate circumstances. Figure 9.4.3 demonstrates another possible pathology.

Why do we call Newton-Raphson powerful? The answer lies in its rate of convergence: Within a small distance ϵ of x the function and its derivative are approximately:

$$
\begin{aligned}
f(x + \epsilon) &= f(x) + \epsilon f'(x) + \epsilon^2 \frac{f''(x)}{2} + \cdots , \\
f'(x + \epsilon) &= f'(x) + \epsilon f''(x) + \cdots
\end{aligned}
\tag{9.4.3}
$$

By the Newton-Raphson formula,

$$
x_{i+1} = x_i - \frac{f(x_i)}{f'(x_i)},
\tag{9.4.4}
$$

so that

$$
\epsilon_{i+1} = \epsilon_i - \frac{f(x_i)}{f'(x_i)}.
\tag{9.4.5}
$$

When a trial solution x_i differs from the true root by ϵ_i, we can use (9.4.3) to express $f(x_i), f'(x_i)$ in (9.4.4) in terms of ϵ_i and derivatives at the root itself. The result is a recurrence relation for the deviations of the trial solutions

$$\epsilon_{i+1} = -\epsilon_i^2 \frac{f''(x)}{2f'(x)}. \tag{9.4.6}$$

Equation (9.4.6) says that Newton-Raphson converges *quadratically* (cf. equation 9.2.3). Near a root, the number of significant digits approximately *doubles* with each step. This very strong convergence property makes Newton-Raphson the method of choice for any function whose derivative can be evaluated efficiently, and whose derivative is continuous and nonzero in the neighborhood of a root.

Even where Newton-Raphson is rejected for the early stages of convergence (because of its poor global convergence properties), it is very common to "polish up" a root with one or two steps of Newton-Raphson, which can multiply by two or four its number of significant figures!

For an efficient realization of Newton-Raphson the user provides a routine that evaluates both $f(x)$ and its first derivative $f'(x)$ at the point x. The Newton-Raphson formula can also be applied using a numerical difference to approximate the true local derivative,

$$f'(x) \approx \frac{f(x + dx) - f(x)}{dx}. \tag{9.4.7}$$

This is not, however, a recommended procedure for the following reasons: (i) You are doing two function evaluations per step, so *at best* the superlinear order of convergence will be only $\sqrt{2}$. (ii) If you take dx too small you will be wiped out by roundoff, while if you take it too large your order of convergence will be only linear, no better than using the *initial* evaluation $f'(x_0)$ for all subsequent steps. Therefore, Newton-Raphson with numerical derivatives is (in one dimension) always dominated by the secant method of §9.2. (In multidimensions, where there is a paucity of available methods, Newton-Raphson with numerical derivatives must be taken more seriously. See §§9.6–9.7.)

The following function calls a user supplied function `funcd(x,fn,df)` which supplies the function value as `fn` and the derivative as `df`. We have included input bounds on the root simply to be consistent with previous root-finding routines: Newton does not adjust bounds, and works only on local information at the point `x`. The bounds are used only to pick the midpoint as the first guess, and to reject the solution if it wanders outside of the bounds.

```
#include <cmath>
#include "nr.h"
using namespace std;

DP NR::rtnewt(void funcd(const DP, DP &, DP &), const DP x1, const DP x2,
    const DP xacc)
```
Using the Newton-Raphson method, find the root of a function known to lie in the interval [x1, x2]. The root `rtnewt` will be refined until its accuracy is known within ±xacc. `funcd` is a user-supplied routine that returns both the function value and the first derivative of the function at the point x.
```
{
    const int JMAX=20;                    Set to maximum number of iterations.
    int j;
    DP df,dx,f,rtn;
```

```
    rtn=0.5*(x1+x2);                              Initial guess.
    for (j=0;j<JMAX;j++) {
        funcd(rtn,f,df);
        dx=f/df;
        rtn -= dx;
        if ((x1-rtn)*(rtn-x2) < 0.0)
            nrerror("Jumped out of brackets in rtnewt");
        if (fabs(dx) < xacc) return rtn;        Convergence.
    }
    nrerror("Maximum number of iterations exceeded in rtnewt");
    return 0.0;                                    Never get here.
}
```

While Newton-Raphson's global convergence properties are poor, it is fairly easy to design a fail-safe routine that utilizes a combination of bisection and Newton-Raphson. The hybrid algorithm takes a bisection step whenever Newton-Raphson would take the solution out of bounds, or whenever Newton-Raphson is not reducing the size of the brackets rapidly enough.

```
#include <cmath>
#include "nr.h"
using namespace std;

DP NR::rtsafe(void funcd(const DP, DP &, DP &), const DP x1, const DP x2,
    const DP xacc)
```

Using a combination of Newton-Raphson and bisection, find the root of a function bracketed between x1 and x2. The root, returned as the function value rtsafe, will be refined until its accuracy is known within ±xacc. funcd is a user-supplied routine that returns both the function value and the first derivative of the function.

```
{
    const int MAXIT=100;                         Maximum allowed number of iterations.
    int j;
    DP df,dx,dxold,f,fh,fl,temp,xh,xl,rts;

    funcd(x1,fl,df);
    funcd(x2,fh,df);
    if ((fl > 0.0 && fh > 0.0) || (fl < 0.0 && fh < 0.0))
        nrerror("Root must be bracketed in rtsafe");
    if (fl == 0.0) return x1;
    if (fh == 0.0) return x2;
    if (fl < 0.0) {                              Orient the search so that f(xl) < 0.
        xl=x1;
        xh=x2;
    } else {
        xh=x1;
        xl=x2;
    }
    rts=0.5*(x1+x2);                             Initialize the guess for root,
    dxold=fabs(x2-x1);                           the "stepsize before last,"
    dx=dxold;                                    and the last step.
    funcd(rts,f,df);
    for (j=0;j<MAXIT;j++) {                      Loop over allowed iterations.
        if ((((rts-xh)*df-f)*((rts-xl)*df-f) > 0.0)      Bisect if Newton out of range,
            || (fabs(2.0*f) > fabs(dxold*df))) {         or not decreasing fast enough.
            dxold=dx;
            dx=0.5*(xh-xl);
            rts=xl+dx;
            if (xl == rts) return rts;           Change in root is negligible.
        } else {                                 Newton step acceptable. Take it.
            dxold=dx;
            dx=f/df;
```

```
        temp=rts;
        rts -= dx;
        if (temp == rts) return rts;
    }
    if (fabs(dx) < xacc) return rts;          Convergence criterion.
    funcd(rts,f,df);
    The one new function evaluation per iteration.
    if (f < 0.0)                              Maintain the bracket on the root.
        xl=rts;
    else
        xh=rts;
}
nrerror("Maximum number of iterations exceeded in rtsafe");
return 0.0;                                   Never get here.
}
```

For many functions the derivative $f'(x)$ often converges to machine accuracy before the function $f(x)$ itself does. When that is the case one need not subsequently update $f'(x)$. This shortcut is recommended only when you confidently understand the generic behavior of your function, but it speeds computations when the derivative calculation is laborious. (Formally this makes the convergence only linear, but if the derivative isn't changing anyway, you can do no better.)

Newton-Raphson and Fractals

An interesting sidelight to our repeated warnings about Newton-Raphson's unpredictable global convergence properties — its very rapid local convergence notwithstanding — is to investigate, for some particular equation, the set of starting values from which the method does, or doesn't converge to a root.

Consider the simple equation

$$z^3 - 1 = 0 \qquad (9.4.8)$$

whose single real root is $z = 1$, but which also has complex roots at the other two cube roots of unity, $\exp(\pm 2\pi i/3)$. Newton's method gives the iteration

$$z_{j+1} = z_j - \frac{z_j^3 - 1}{3z_j^2} \qquad (9.4.9)$$

Up to now, we have applied an iteration like equation (9.4.9) only for real starting values z_0, but in fact all of the equations in this section also apply in the complex plane. We can therefore map out the complex plane into regions from which a starting value z_0, iterated in equation (9.4.9), will, or won't, converge to $z = 1$. Naively, we might expect to find a "basin of convergence" somehow surrounding the root $z = 1$. We surely do not expect the basin of convergence to fill the whole plane, because the plane must also contain regions that converge to each of the two complex roots. In fact, by symmetry, the three regions must have identical shapes. Perhaps they will be three symmetric $120°$ wedges, with one root centered in each?

Now take a look at Figure 9.4.4, which shows the result of a numerical exploration. The basin of convergence does indeed cover $1/3$ the area of the complex plane, but its boundary is highly irregular — in fact, *fractal*. (A fractal, so called, has self-similar structure that repeats on all scales of magnification.) How does this

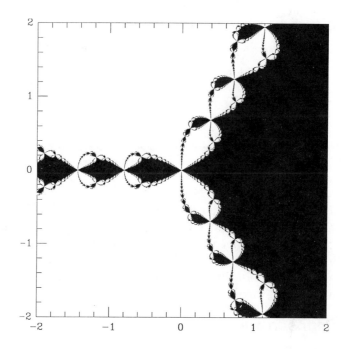

Figure 9.4.4.   The complex z plane with real and imaginary components in the range $(-2, 2)$. The black region is the set of points from which Newton's method converges to the root $z = 1$ of the equation $z^3 - 1 = 0$.   Its shape is fractal.

fractal emerge from something as simple as Newton's method, and an equation as simple as (9.4.8)? The answer is already implicit in Figure 9.4.2, which showed how, on the real line, a local extremum causes Newton's method to shoot off to infinity. Suppose one is *slightly* removed from such a point. Then one might be shot off not to infinity, but — by luck — right into the basin of convergence of the desired root. But that means that in the neighborhood of an extremum there must be a tiny, perhaps distorted, copy of the basin of convergence — a kind of "one-bounce away" copy. Similar logic shows that there can be "two-bounce" copies, "three-bounce" copies, and so on. A fractal thus emerges.

Notice that, for equation (9.4.8), almost the whole real axis is in the domain of convergence for the root $z = 1$. We say "almost" because of the peculiar discrete points on the negative real axis whose convergence is indeterminate (see figure). What happens if you start Newton's method from one of these points? (Try it.)

CITED REFERENCES AND FURTHER READING:

Acton, F.S. 1970, *Numerical Methods That Work*; 1990, corrected edition (Washington: Mathematical Association of America), Chapter 2.

Ralston, A., and Rabinowitz, P. 1978, *A First Course in Numerical Analysis*, 2nd ed.; reprinted 2001 (New York: Dover), §8.4.

Ortega, J., and Rheinboldt, W. 1970, *Iterative Solution of Nonlinear Equations in Several Variables* (New York: Academic Press).

Mandelbrot, B.B. 1983, *The Fractal Geometry of Nature* (San Francisco: W.H. Freeman).

Peitgen, H.-O., and Saupe, D. (eds.) 1988, *The Science of Fractal Images* (New York: Springer-Verlag).

9.5 Roots of Polynomials

Here we present a few methods for finding roots of polynomials. These will serve for most practical problems involving polynomials of low-to-moderate degree or for well-conditioned polynomials of higher degree. Not as well appreciated as it ought to be is the fact that some polynomials are exceedingly ill-conditioned. The tiniest changes in a polynomial's coefficients can, in the worst case, send its roots sprawling all over the complex plane. (An infamous example due to Wilkinson is detailed by Acton [1].)

Recall that a polynomial of degree n will have n roots. The roots can be real or complex, and they might not be distinct. If the coefficients of the polynomial are real, then complex roots will occur in pairs that are conjugate, i.e., if $x_1 = a + bi$ is a root then $x_2 = a - bi$ will also be a root. When the coefficients are complex, the complex roots need not be related.

Multiple roots, or closely spaced roots, produce the most difficulty for numerical algorithms (see Figure 9.5.1). For example, $P(x) = (x - a)^2$ has a double real root at $x = a$. However, we cannot bracket the root by the usual technique of identifying neighborhoods where the function changes sign, nor will slope-following methods such as Newton-Raphson work well, because both the function and its derivative vanish at a multiple root. Newton-Raphson *may* work, but slowly, since large roundoff errors can occur. When a root is known in advance to be multiple, then special methods of attack are readily devised. Problems arise when (as is generally the case) we do not know in advance what pathology a root will display.

Deflation of Polynomials

When seeking several or all roots of a polynomial, the total effort can be significantly reduced by the use of *deflation*. As each root r is found, the polynomial is factored into a product involving the root and a reduced polynomial of degree one less than the original, i.e., $P(x) = (x - r)Q(x)$. Since the roots of Q are exactly the remaining roots of P, the effort of finding additional roots decreases, because we work with polynomials of lower and lower degree as we find successive roots. Even more important, with deflation we can avoid the blunder of having our iterative method converge twice to the same (nonmultiple) root instead of separately to two different roots.

Deflation, which amounts to synthetic division, is a simple operation that acts on the array of polynomial coefficients. The concise code for synthetic division by a monomial factor was given in §5.3 above. You can deflate complex roots either by converting that code to complex data type, or else — in the case of a polynomial with real coefficients but possibly complex roots — by deflating by a quadratic factor,

$$[x - (a + ib)]\,[x - (a - ib)] = x^2 - 2ax + (a^2 + b^2) \qquad (9.5.1)$$

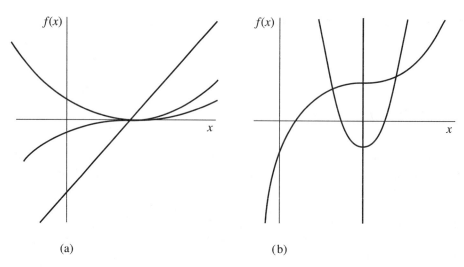

Figure 9.5.1.   (a) Linear, quadratic, and cubic behavior at the roots of polynomials. Only under high magnification (b) does it become apparent that the cubic has one, not three, roots, and that the quadratic has two roots rather than none.

The routine poldiv in §5.3 can be used to divide the polynomial by this factor.

Deflation must, however, be utilized with care. Because each new root is known with only finite accuracy, errors creep into the determination of the coefficients of the successively deflated polynomial. Consequently, the roots can become more and more inaccurate. It matters a lot whether the inaccuracy creeps in stably (plus or minus a few multiples of the machine precision at each stage) or unstably (erosion of successive significant figures until the results become meaningless). Which behavior occurs depends on just how the root is divided out. *Forward deflation*, where the new polynomial coefficients are computed in the order from the highest power of x down to the constant term, was illustrated in §5.3. This turns out to be stable if the root of smallest absolute value is divided out at each stage. Alternatively, one can do *backward deflation*, where new coefficients are computed in order from the constant term up to the coefficient of the highest power of x. This is stable if the remaining root of *largest* absolute value is divided out at each stage.

A polynomial whose coefficients are interchanged "end-to-end," so that the constant becomes the highest coefficient, etc., has its roots mapped into their reciprocals. (Proof: Divide the whole polynomial by its highest power x^n and rewrite it as a polynomial in $1/x$.) The algorithm for backward deflation is therefore virtually identical to that of forward deflation, except that the original coefficients are taken in reverse order and the reciprocal of the deflating root is used. Since we will use forward deflation below, we leave to you the exercise of writing a concise coding for backward deflation (as in §5.3). For more on the stability of deflation, consult [2].

To minimize the impact of increasing errors (even stable ones) when using deflation, it is advisable to treat roots of the successively deflated polynomials as only *tentative* roots of the original polynomial. One then *polishes* these tentative roots by taking them as initial guesses that are to be re-solved for, using the *nondeflated* original polynomial P. Again you must beware lest two deflated roots are inaccurate enough that, under polishing, they both converge to the same undeflated root; in that case you gain a spurious root-multiplicity and lose a distinct root. This is detectable, since you can compare each polished root for equality to previous ones from distinct tentative roots. When it happens, you are advised to deflate the polynomial just once (and for this root only), then again polish the tentative root, or to use Maehly's procedure (see equation 9.5.29 below).

Below we say more about techniques for polishing real and complex-conjugate tentative roots. First, let's get back to overall strategy.

There are two schools of thought about how to proceed when faced with a polynomial of real coefficients. One school says to go after the easiest quarry, the real, distinct roots, by the same kinds of methods that we have discussed in previous sections for general functions, i.e., trial-and-error bracketing followed by a safe Newton-Raphson as in `rtsafe`. Sometimes you are *only* interested in real roots, in which case the strategy is complete. Otherwise, you then go after quadratic factors of the form (9.5.1) by any of a variety of methods. One such is Bairstow's method, which we will discuss below in the context of root polishing. Another is Muller's method, which we here briefly discuss.

Muller's Method

Muller's method generalizes the secant method, but uses quadratic interpolation among three points instead of linear interpolation between two. Solving for the zeros of the quadratic allows the method to find complex pairs of roots. Given *three* previous guesses for the root x_{i-2}, x_{i-1}, x_i, and the values of the polynomial $P(x)$ at those points, the next approximation x_{i+1} is produced by the following formulas,

$$
\begin{aligned}
q &\equiv \frac{x_i - x_{i-1}}{x_{i-1} - x_{i-2}} \\
A &\equiv qP(x_i) - q(1+q)P(x_{i-1}) + q^2 P(x_{i-2}) \\
B &\equiv (2q+1)P(x_i) - (1+q)^2 P(x_{i-1}) + q^2 P(x_{i-2}) \\
C &\equiv (1+q)P(x_i)
\end{aligned}
\tag{9.5.2}
$$

followed by

$$
x_{i+1} = x_i - (x_i - x_{i-1}) \left[\frac{2C}{B \pm \sqrt{B^2 - 4AC}} \right]
\tag{9.5.3}
$$

where the sign in the denominator is chosen to make its absolute value or modulus as large as possible. You can start the iterations with any three values of x that you like, e.g., three equally spaced values on the real axis. Note that you must allow for the possibility of a complex denominator, and subsequent complex arithmetic, in implementing the method.

Muller's method is sometimes also used for finding complex zeros of analytic functions (not just polynomials) in the complex plane, for example in the IMSL routine ZANLY [3].

Laguerre's Method

The second school regarding overall strategy happens to be the one to which we belong. That school advises you to use one of a very small number of methods that will converge (though with greater or lesser efficiency) to all types of roots: real, complex, single, or multiple. Use such a method to get tentative values for all n roots of your nth degree polynomial. Then go back and polish them as you desire.

Laguerre's method is by far the most straightforward of these general, complex methods. It does require complex arithmetic, even while converging to real roots; however, for polynomials with all real roots, it is guaranteed to converge to a root from any starting point. For polynomials with some complex roots, little is theoretically proved about the method's convergence. Much empirical experience, however, suggests that nonconvergence is extremely unusual, and, further, can almost always be fixed by a simple scheme to break a nonconverging limit cycle. (This is implemented in our routine, below.) An example of a polynomial that requires this cycle-breaking scheme is one of high degree ($\gtrsim 20$), with all its roots just outside of the complex unit circle, approximately equally spaced around it. When the method converges on a simple complex zero, it is known that its convergence is third order.

In some instances the complex arithmetic in the Laguerre method is no disadvantage, since the polynomial itself may have complex coefficients.

To motivate (although not rigorously derive) the Laguerre formulas we can note the following relations between the polynomial and its roots and derivatives

$$P_n(x) = (x - x_0)(x - x_1)\ldots(x - x_{n-1}) \tag{9.5.4}$$

$$\ln|P_n(x)| = \ln|x - x_0| + \ln|x - x_1| + \ldots + \ln|x - x_{n-1}| \tag{9.5.5}$$

$$\frac{d\ln|P_n(x)|}{dx} = +\frac{1}{x - x_0} + \frac{1}{x - x_1} + \ldots + \frac{1}{x - x_{n-1}} = \frac{P_n'}{P_n} \equiv G \tag{9.5.6}$$

$$-\frac{d^2\ln|P_n(x)|}{dx^2} = +\frac{1}{(x - x_0)^2} + \frac{1}{(x - x_1)^2} + \ldots + \frac{1}{(x - x_{n-1})^2}$$

$$= \left[\frac{P_n'}{P_n}\right]^2 - \frac{P_n''}{P_n} \equiv H \tag{9.5.7}$$

Starting from these relations, the Laguerre formulas make what Acton [1] nicely calls "a rather drastic set of assumptions": The root x_0 that we seek is assumed to be located some distance a from our current guess x, while *all other roots* are assumed to be located at a distance b

$$x - x_0 = a \quad ; \quad x - x_i = b \quad i = 1, 2, \ldots, n - 1 \tag{9.5.8}$$

Then we can express (9.5.6), (9.5.7) as

$$\frac{1}{a} + \frac{n - 1}{b} = G \tag{9.5.9}$$

$$\frac{1}{a^2} + \frac{n - 1}{b^2} = H \tag{9.5.10}$$

which yields as the solution for a

$$a = \frac{n}{G \pm \sqrt{(n - 1)(nH - G^2)}} \tag{9.5.11}$$

where the sign should be taken to yield the largest magnitude for the denominator. Since the factor inside the square root can be negative, a can be complex. (A more rigorous justification of equation 9.5.11 is in [4].)

The method operates iteratively: For a trial value x, a is calculated by equation (9.5.11). Then $x - a$ becomes the next trial value. This continues until a is sufficiently small.

The following routine implements the Laguerre method to find one root of a given polynomial of degree m, whose coefficients can be complex. As usual, the first coefficient a[0] is the constant term, while a[m] is the coefficient of the highest power of x. The routine implements a simplified version of an elegant stopping criterion due to Adams [5], which neatly balances the desire to achieve full machine accuracy, on the one hand, with the danger of iterating forever in the presence of roundoff error, on the other.

```cpp
#include <cmath>
#include <complex>
#include <limits>
#include "nr.h"
using namespace std;
```

```cpp
void NR::laguer(Vec_I_CPLX_DP &a, complex<DP> &x, int &its)
```
Given the m+1 complex coefficients a[0..m] of the polynomial $\sum_{i=0}^{m} a[i]x^i$, and given a complex value x, this routine improves x by Laguerre's method until it converges, within the achievable roundoff limit, to a root of the given polynomial. The number of iterations taken is returned as its.
```cpp
{
    const int MR=8,MT=10,MAXIT=MT*MR;
    const DP EPS=numeric_limits<DP>::epsilon();
```
Here EPS is the estimated fractional roundoff error. We try to break (rare) limit cycles with MR different fractional values, once every MT steps, for MAXIT total allowed iterations.
```cpp
    static const DP frac[MR+1]=
        {0.0,0.5,0.25,0.75,0.13,0.38,0.62,0.88,1.0};
```
Fractions used to break a limit cycle.
```cpp
    int iter,j;
    DP abx,abp,abm,err;
    complex<DP> dx,x1,b,d,f,g,h,sq,gp,gm,g2;

    int m=a.size()-1;
    for (iter=1;iter<=MAXIT;iter++) {          Loop over iterations up to allowed maximum.
        its=iter;
        b=a[m];
        err=abs(b);
        d=f=0.0;
        abx=abs(x);
        for (j=m-1;j>=0;j--) {                 Efficient computation of the polynomial and
            f=x*f+d;                               its first two derivatives. f stores P''/2.
            d=x*d+b;
            b=x*b+a[j];
            err=abs(b)+abx*err;
        }
        err *= EPS;
```
Estimate of roundoff error in evaluating polynomial.
```cpp
        if (abs(b) <= err) return;             We are on the root.
        g=d/b;                                 The generic case: use Laguerre's formula.
        g2=g*g;
        h=g2-2.0*f/b;
        sq=sqrt(DP(m-1)*(DP(m)*h-g2));
        gp=g+sq;
        gm=g-sq;
```

```
        abp=abs(gp);
        abm=abs(gm);
        if (abp < abm) gp=gm;
        dx=MAX(abp,abm) > 0.0 ? DP(m)/gp : polar(1+abx,DP(iter));
        x1=x-dx;
        if (x == x1) return;          Converged.
        if (iter % MT != 0) x=x1;
        else x -= frac[iter/MT]*dx;
```
        Every so often we take a fractional step, to break any limit cycle (itself a rare occur-
        rence.
```
    }
    nrerror("too many iterations in laguer");
```
    Very unusual — can occur only for complex roots. Try a different starting guess for the
    root.
```
    return;
}
```

Here is a driver routine that calls `laguer` in succession for each root, performs the deflation, optionally polishes the roots by the same Laguerre method — if you are not going to polish in some other way — and finally sorts the roots by their real parts. (We will use this routine in Chapter 13.)

```
#include <cmath>
#include <complex>
#include "nr.h"
using namespace std;

void NR::zroots(Vec_I_CPLX_DP &a, Vec_O_CPLX_DP &roots, const bool &polish)
```
Given the `m+1` complex coefficients `a[0..m]` of the polynomial $\sum_{i=0}^{m} a(i)x^i$, this routine suc-
cessively calls `laguer` and finds all m complex roots in `roots[0..m-1]`. The boolean variable
`polish` should be input as `true` if polishing (also by Laguerre's method) is desired, `false` if
the roots will be subsequently polished by other means.
```
{
    const DP EPS=1.0e-14;              A small number.
    int i,its,j,jj;
    complex<DP> x,b,c;

    int m=a.size()-1;
    Vec_CPLX_DP ad(m+1);
    for (j=0;j<=m;j++) ad[j]=a[j];        Copy of coefficients for successive deflation.
    for (j=m-1;j>=0;j--) {                Loop over each root to be found.
        x=0.0;                           Start at zero to favor convergence to small-
        Vec_CPLX_DP ad_v(j+2);               est remaining root, and find the root.
        for (jj=0;jj<j+2;jj++) ad_v[jj]=ad[jj];
        laguer(ad_v,x,its);
        if (fabs(imag(x)) <= 2.0*EPS*fabs(real(x)))
            x=complex<DP>(real(x),0.0);
        roots[j]=x;
        b=ad[j+1];                        Forward deflation.
        for (jj=j;jj>=0;jj--) {
            c=ad[jj];
            ad[jj]=b;
            b=x*b+c;
        }
    }
    if (polish)
        for (j=0;j<m;j++)                 Polish the roots using the undeflated coeffi-
            laguer(a,roots[j],its);            cients.
    for (j=1;j<m;j++) {                   Sort roots by their real parts by straight in-
        x=roots[j];                          sertion.
        for (i=j-1;i>=0;i--) {
```

```
              if (real(roots[i]) <= real(x)) break;
              roots[i+1]=roots[i];
          }
          roots[i+1]=x;
      }
}
```

Eigenvalue Methods

The eigenvalues of a matrix $\mathbf{A}$ are the roots of the "characteristic polynomial" $P(x) = \det[\mathbf{A} - x\mathbf{I}]$. However, as we will see in Chapter 11, root-finding is not generally an efficient way to find eigenvalues. Turning matters around, we can use the more efficient eigenvalue methods that are discussed in Chapter 11 to find the roots of arbitrary polynomials. You can easily verify (see, e.g., [6]) that the characteristic polynomial of the special $m \times m$ *companion matrix*

$$\mathbf{A} = \begin{pmatrix} -\frac{a_{m-1}}{a_m} & -\frac{a_{m-2}}{a_m} & \cdots & -\frac{a_1}{a_m} & -\frac{a_0}{a_m} \\ 1 & 0 & \cdots & 0 & 0 \\ 0 & 1 & \cdots & 0 & 0 \\ \vdots & & & & \vdots \\ 0 & 0 & \cdots & 1 & 0 \end{pmatrix} \tag{9.5.12}$$

is equivalent to the general polynomial

$$P(x) = \sum_{i=0}^{m} a_i x^i \tag{9.5.13}$$

If the coefficients a_i are real, rather than complex, then the eigenvalues of $\mathbf{A}$ can be found using the routines balanc and hqr in §§11.5–11.6 (see discussion there). This method, implemented in the routine zrhqr following, is typically about a factor 2 slower than zroots (above). However, for some classes of polynomials, it is a more robust technique, largely because of the fairly sophisticated convergence methods embodied in hqr. If your polynomial has real coefficients, and you are having trouble with zroots, then zrhqr is a recommended alternative.

```
#include <complex>
#include "nr.h"
using namespace std;

void NR::zrhqr(Vec_I_DP &a, Vec_O_CPLX_DP &rt)
```
Find all the roots of a polynomial with real coefficients, $\sum_{i=0}^{m} a(i)x^i$, given the coefficients a[0..m]. The method is to construct an upper Hessenberg matrix whose eigenvalues are the desired roots, and then use the routines balanc and hqr. The real and imaginary parts of the roots are returned in rtr[0..m-1] and rti[0..m-1], respectively.
```
{
    int j,k;
    complex<DP> x;

    int m=a.size()-1;
    Mat_DP hess(m,m);
    for (k=0;k<m;k++) {          Construct the matrix.
        hess[0][k] = -a[m-k-1]/a[m];
```

```
        for (j=1;j<m;j++) hess[j][k]=0.0;
        if (k != m-1) hess[k+1][k]=1.0;
    }
    balanc(hess);                    Find its eigenvalues.
    hqr(hess,rt);
    for (j=1;j<m;j++) {              Sort roots by their real parts by straight insertion.
        x=rt[j];
        for (k=j-1;k>=0;k--) {
            if (real(rt[k]) <= real(x)) break;
            rt[k+1]=rt[k];
        }
        rt[k+1]=x;
    }
}
```

Other Sure-Fire Techniques

The *Jenkins-Traub method* has become practically a standard in black-box polynomial root-finders, e.g., in the IMSL library [3]. The method is too complicated to discuss here, but is detailed, with references to the primary literature, in [4].

The *Lehmer-Schur algorithm* is one of a class of methods that isolate roots in the complex plane by generalizing the notion of one-dimensional bracketing. It is possible to determine efficiently whether there are any polynomial roots within a circle of given center and radius. From then on it is a matter of bookkeeping to hunt down all the roots by a series of decisions regarding where to place new trial circles. Consult [1] for an introduction.

Techniques for Root-Polishing

Newton-Raphson works very well for real roots once the neighborhood of a root has been identified. The polynomial and its derivative can be efficiently simultaneously evaluated as in §5.3. For a polynomial of degree n with coefficients c[0]...c[n], the following segment of code embodies one cycle of Newton-Raphson:

```
p=c[n]*x+c[n-1];
p1=c[n];
for(i=n-2;i>=0;i--) {
    p1=p+p1*x;
    p=c[i]+p*x;
}
if (p1 == 0.0) NR::nrerror("derivative should not vanish");
x -= p/p1;
```

Once all real roots of a polynomial have been polished, one must polish the complex roots, either directly, or by looking for quadratic factors.

Direct polishing by Newton-Raphson is straightforward for complex roots if the above code is converted to complex data types. With real polynomial coefficients, note that your starting guess (tentative root) *must* be off the real axis, otherwise you will never get off that axis — and may get shot off to infinity by a minimum or maximum of the polynomial.

For real polynomials, the alternative means of polishing complex roots (or, for that matter, double real roots) is *Bairstow's method*, which seeks quadratic factors. The advantage of going after quadratic factors is that it avoids all complex arithmetic. Bairstow's method seeks a quadratic factor that embodies the two roots $x = a \pm ib$, namely

$$x^2 - 2ax + (a^2 + b^2) \equiv x^2 + Bx + C \tag{9.5.14}$$

In general if we divide a polynomial by a quadratic factor, there will be a linear remainder

$$P(x) = (x^2 + Bx + C)Q(x) + Rx + S. \tag{9.5.15}$$

Given B and C, R and S can be readily found, by polynomial division (§5.3). We can consider R and S to be adjustable functions of B and C, and they will be zero if the quadratic factor is a divisor of $P(x)$.

In the neighborhood of a root a first-order Taylor series expansion approximates the variation of R, S with respect to small changes in B, C

$$R(B + \delta B, C + \delta C) \approx R(B, C) + \frac{\partial R}{\partial B}\delta B + \frac{\partial R}{\partial C}\delta C \tag{9.5.16}$$

$$S(B + \delta B, C + \delta C) \approx S(B, C) + \frac{\partial S}{\partial B}\delta B + \frac{\partial S}{\partial C}\delta C \tag{9.5.17}$$

To evaluate the partial derivatives, consider the derivative of (9.5.15) with respect to C. Since $P(x)$ is a fixed polynomial, it is independent of C, hence

$$0 = (x^2 + Bx + C)\frac{\partial Q}{\partial C} + Q(x) + \frac{\partial R}{\partial C}x + \frac{\partial S}{\partial C} \tag{9.5.18}$$

which can be rewritten as

$$-Q(x) = (x^2 + Bx + C)\frac{\partial Q}{\partial C} + \frac{\partial R}{\partial C}x + \frac{\partial S}{\partial C} \tag{9.5.19}$$

Similarly, $P(x)$ is independent of B, so differentiating (9.5.15) with respect to B gives

$$-xQ(x) = (x^2 + Bx + C)\frac{\partial Q}{\partial B} + \frac{\partial R}{\partial B}x + \frac{\partial S}{\partial B} \tag{9.5.20}$$

Now note that equation (9.5.19) matches equation (9.5.15) in form. Thus if we perform a second synthetic division of $P(x)$, i.e., a division of $Q(x)$ by the same quadratic factor, yielding a remainder $R_1 x + S_1$, then

$$\frac{\partial R}{\partial C} = -R_1 \qquad \frac{\partial S}{\partial C} = -S_1 \tag{9.5.21}$$

To get the remaining partial derivatives, evaluate equation (9.5.20) at the two roots of the quadratic, x_+ and x_-. Since

$$Q(x_\pm) = R_1 x_\pm + S_1 \tag{9.5.22}$$

we get

$$\frac{\partial R}{\partial B}x_+ + \frac{\partial S}{\partial B} = -x_+(R_1 x_+ + S_1) \tag{9.5.23}$$

$$\frac{\partial R}{\partial B}x_- + \frac{\partial S}{\partial B} = -x_-(R_1 x_- + S_1) \tag{9.5.24}$$

Solve these two equations for the partial derivatives, using

$$x_+ + x_- = -B \qquad x_+ x_- = C \tag{9.5.25}$$

and find

$$\frac{\partial R}{\partial B} = BR_1 - S_1 \qquad \frac{\partial S}{\partial B} = CR_1 \tag{9.5.26}$$

Bairstow's method now consists of using Newton-Raphson in two dimensions (which is actually the subject of the *next* section) to find a simultaneous zero of R and S. Synthetic division is used twice per cycle to evaluate R, S and their partial derivatives with respect to B, C. Like one-dimensional Newton-Raphson, the method works well in the vicinity of a root pair (real or complex), but it can fail miserably when started at a random point. We therefore recommend it only in the context of polishing tentative complex roots.

```
#include <cmath>
#include "nr.h"
using namespace std;

void NR::qroot(Vec_I_DP &p, DP &b, DP &c, const DP eps)
Given n+1 coefficients p[0..n] of a polynomial of degree n, and trial values for the coefficients
of a quadratic factor x*x+b*x+c, improve the solution until the coefficients b,c change by less
than eps. The routine poldiv §5.3 is used.
{
    const int ITMAX=20;                         At most ITMAX iterations.
    const DP TINY=1.0e-14;
    int iter;
    DP sc,sb,s,rc,rb,r,dv,delc,delb;
    Vec_DP d(3);

    int n=p.size()-1;
    Vec_DP q(n+1),qq(n+1),rem(n+1);
    d[2]=1.0;
    for (iter=0;iter<ITMAX;iter++) {
        d[1]=b;
        d[0]=c;
        poldiv(p,d,q,rem);
        s=rem[0];                               First division r,s.
        r=rem[1];
        poldiv(q,d,qq,rem);
        sb = -c*(rc = -rem[1]);                 Second division partial r,s with respect to
        rb = -b*rc+(sc = -rem[0]);                c.
        dv=1.0/(sb*rc-sc*rb);                   Solve 2x2 equation.
        delb=(r*sc-s*rc)*dv;
        delc=(-r*sb+s*rb)*dv;
        b += (delb=(r*sc-s*rc)*dv);
        c += (delc=(-r*sb+s*rb)*dv);
        if ((fabs(delb) <= eps*fabs(b) || fabs(b) < TINY)
            && (fabs(delc) <= eps*fabs(c) || fabs(c) < TINY)) {
            return;                             Coefficients converged.
        }
    }
    nrerror("Too many iterations in routine qroot");
}
```

We have already remarked on the annoyance of having two tentative roots collapse to one value under polishing. You are left not knowing whether your polishing procedure has lost a root, or whether there *is* actually a double root, which was split only by roundoff errors in your previous deflation. One solution is deflate-and-repolish; but deflation is what we are trying to avoid at the polishing stage. An alternative is *Maehly's procedure.* Maehly pointed out that the derivative of the reduced polynomial

$$P_j(x) \equiv \frac{P(x)}{(x - x_0) \cdots (x - x_{j-1})} \tag{9.5.27}$$

can be written as

$$P'_j(x) = \frac{P'(x)}{(x - x_0) \cdots (x - x_{j-1})} - \frac{P(x)}{(x - x_0) \cdots (x - x_{j-1})} \sum_{i=0}^{j-1} (x - x_i)^{-1} \tag{9.5.28}$$

Hence one step of Newton-Raphson, taking a guess x_k into a new guess x_{k+1}, can be written as

$$x_{k+1} = x_k - \frac{P(x_k)}{P'(x_k) - P(x_k)\sum_{i=0}^{j-1}(x_k - x_i)^{-1}} \tag{9.5.29}$$

This equation, if used with i ranging over the roots already polished, will prevent a tentative root from spuriously hopping to another one's true root. It is an example of so-called *zero suppression* as an alternative to true deflation.

Muller's method, which was described above, can also be useful at the polishing stage.

CITED REFERENCES AND FURTHER READING:

Acton, F.S. 1970, *Numerical Methods That Work*; 1990, corrected edition (Washington: Mathematical Association of America), Chapter 7. [1]

Peters G., and Wilkinson, J.H. 1971, *Journal of the Institute of Mathematics and its Applications*, vol. 8, pp. 16–35. [2]

IMSL Math/Library Users Manual (IMSL Inc., 2500 CityWest Boulevard, Houston TX 77042). [3]

Ralston, A., and Rabinowitz, P. 1978, *A First Course in Numerical Analysis*, 2nd ed.; reprinted 2001 (New York: Dover), §8.9–8.13. [4]

Adams, D.A. 1967, *Communications of the ACM*, vol. 10, pp. 655–658. [5]

Johnson, L.W., and Riess, R.D. 1982, *Numerical Analysis*, 2nd ed. (Reading, MA: Addison-Wesley), §4.4.3. [6]

Henrici, P. 1974, *Applied and Computational Complex Analysis*, vol. 1 (New York: Wiley).

Stoer, J., and Bulirsch, R. 1993, *Introduction to Numerical Analysis*, 2nd ed. (New York: Springer-Verlag), §§5.5–5.9.

9.6 Newton-Raphson Method for Nonlinear Systems of Equations

We make an extreme, but wholly defensible, statement: There are *no* good, general methods for solving systems of more than one nonlinear equation. Furthermore, it is not hard to see why (very likely) there *never will be* any good, general methods: Consider the case of two dimensions, where we want to solve simultaneously

$$f(x, y) = 0$$
$$g(x, y) = 0 \tag{9.6.1}$$

The functions f and g are two arbitrary functions, each of which has zero contour lines that divide the (x, y) plane into regions where their respective function is positive or negative. These zero contour boundaries are of interest to us. The solutions that we seek are those points (if any) that are common to the zero contours of f and g (see Figure 9.6.1). Unfortunately, the functions f and g have, in general, no relation to each other at all! There is nothing special about a common point from either f's point of view, or from g's. In order to find all common points, which are

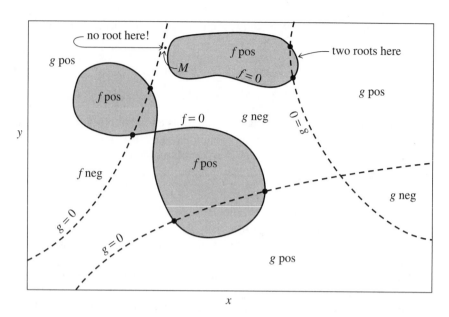

Figure 9.6.1.   Solution of two nonlinear equations in two unknowns. Solid curves refer to $f(x, y)$, dashed curves to $g(x, y)$. Each equation divides the (x, y) plane into positive and negative regions, bounded by zero curves. The desired solutions are the intersections of these unrelated zero curves. The number of solutions is *a priori* unknown.

the solutions of our nonlinear equations, we will (in general) have to do neither more nor less than map out the full zero contours of both functions. Note further that the zero contours will (in general) consist of an unknown number of disjoint closed curves. How can we ever hope to know when we have found all such disjoint pieces?

For problems in more than two dimensions, we need to find points mutually common to N unrelated zero-contour hypersurfaces, each of dimension $N - 1$. You see that root finding becomes virtually impossible without insight! You will almost always have to use additional information, specific to your particular problem, to answer such basic questions as, "Do I expect a unique solution?" and "Approximately where?" Acton [1] has a good discussion of some of the particular strategies that can be tried.

In this section we will discuss the simplest multidimensional root finding method, Newton-Raphson. This method gives you a very efficient means of converging to a root, if you have a sufficiently good initial guess. It can also spectacularly fail to converge, indicating (though not proving) that your putative root does not exist nearby. In §9.7 we discuss more sophisticated implementations of the Newton-Raphson method, which try to improve on Newton-Raphson's poor global convergence. A multidimensional generalization of the secant method, called Broyden's method, is also discussed in §9.7.

A typical problem gives N functional relations to be zeroed, involving variables $x_i, i = 0, 1, \ldots, N - 1$:

$$F_i(x_0, x_1, \ldots, x_{N-1}) = 0 \qquad i = 0, 1, \ldots, N - 1. \tag{9.6.2}$$

We let **x** denote the entire vector of values x_i and **F** denote the entire vector of functions F_i. In the neighborhood of **x**, each of the functions F_i can be expanded in Taylor series

$$F_i(\mathbf{x} + \delta\mathbf{x}) = F_i(\mathbf{x}) + \sum_{j=0}^{N-1} \frac{\partial F_i}{\partial x_j} \delta x_j + O(\delta\mathbf{x}^2). \tag{9.6.3}$$

The matrix of partial derivatives appearing in equation (9.6.3) is the *Jacobian* matrix $\mathbf{J}$:

$$J_{ij} \equiv \frac{\partial F_i}{\partial x_j}. \tag{9.6.4}$$

In matrix notation equation (9.6.3) is

$$\mathbf{F}(\mathbf{x} + \delta\mathbf{x}) = \mathbf{F}(\mathbf{x}) + \mathbf{J} \cdot \delta\mathbf{x} + O(\delta\mathbf{x}^2). \tag{9.6.5}$$

By neglecting terms of order $\delta\mathbf{x}^2$ and higher and by setting $\mathbf{F}(\mathbf{x} + \delta\mathbf{x}) = 0$, we obtain a set of linear equations for the corrections $\delta\mathbf{x}$ that move each function closer to zero simultaneously, namely

$$\mathbf{J} \cdot \delta\mathbf{x} = -\mathbf{F}. \tag{9.6.6}$$

Matrix equation (9.6.6) can be solved by *LU* decomposition as described in §2.3. The corrections are then added to the solution vector,

$$\mathbf{x}_{\text{new}} = \mathbf{x}_{\text{old}} + \delta\mathbf{x} \tag{9.6.7}$$

and the process is iterated to convergence. In general it is a good idea to check the degree to which both functions and variables have converged. Once either reaches machine accuracy, the other won't change.

The following routine mnewt performs ntrial iterations starting from an initial guess at the solution vector x[0..n-1]. Iteration stops if either the sum of the magnitudes of the functions F_i is less than some tolerance tolf, or the sum of the absolute values of the corrections to δx_i is less than some tolerance tolx. mnewt calls a user supplied function usrfun which must provide the function values $\mathbf{F}$ and the Jacobian matrix $\mathbf{J}$. If $\mathbf{J}$ is difficult to compute analytically, you can try having usrfun call the routine fdjac of §9.7 to compute the partial derivatives by finite differences. You should not make ntrial too big; rather inspect to see what is happening before continuing for some further iterations.

```cpp
#include <cmath>
#include "nr.h"
using namespace std;

void usrfun(Vec_I_DP &x, Vec_O_DP &fvec, Mat_O_DP &fjac);

void NR::mnewt(const int ntrial, Vec_IO_DP &x, const DP tolx, const DP tolf)
Given an initial guess x[0..n-1] for a root in n dimensions, take ntrial Newton-Raphson
steps to improve the root. Stop if the root converges in either summed absolute variable incre-
ments tolx or summed absolute function values tolf.
{
    int k,i;
    DP errx,errf,d;
```

```
int n=x.size();
Vec_INT indx(n);
Vec_DP p(n),fvec(n);
Mat_DP fjac(n,n);
for (k=0;k<ntrial;k++) {
    usrfun(x,fvec,fjac);                    User function supplies function values at x in
    errf=0.0;                                   fvec and Jacobian matrix in fjac.
    for (i=0;i<n;i++) errf += fabs(fvec[i]);     Check function convergence.
    if (errf <= tolf) return;
    for (i=0;i<n;i++) p[i] = -fvec[i];      Right-hand side of linear equations.
    ludcmp(fjac,indx,d);                    Solve linear equations using LU decomposition.
    lubksb(fjac,indx,p);
    errx=0.0;                               Check root convergence.
    for (i=0;i<n;i++) {                      Update solution.
        errx += fabs(p[i]);
        x[i] += p[i];
    }
    if (errx <= tolx) return;
}
return;
}
```

Newton's Method versus Minimization

In the next chapter, we will find that there *are* efficient general techniques for finding a minimum of a function of many variables. Why is that task (relatively) easy, while multidimensional root finding is often quite hard? Isn't minimization equivalent to finding a zero of an N-dimensional gradient vector, not so different from zeroing an N-dimensional function? No! The components of a gradient vector are not independent, arbitrary functions. Rather, they obey so-called integrability conditions that are highly restrictive. Put crudely, you can always find a minimum by sliding downhill on a single surface. The test of "downhillness" is thus one-dimensional. There is no analogous conceptual procedure for finding a multidimensional root, where "downhill" must mean simultaneously downhill in N separate function spaces, thus allowing a multitude of trade-offs, as to how much progress in one dimension is worth compared with progress in another.

It might occur to you to carry out multidimensional root finding by collapsing all these dimensions into one: Add up the sums of squares of the individual functions F_i to get a master function F which (i) is positive definite, and (ii) has a global minimum of zero exactly at all solutions of the original set of nonlinear equations. Unfortunately, as you will see in the next chapter, the efficient algorithms for finding minima come to rest on global and local minima indiscriminately. You will often find, to your great dissatisfaction, that your function F has a great number of local minima. In Figure 9.6.1, for example, there is likely to be a local minimum wherever the zero contours of f and g make a close approach to each other. The point labeled M is such a point, and one sees that there are no nearby roots.

However, we will now see that sophisticated strategies for multidimensional root finding can in fact make use of the idea of minimizing a master function F, by *combining* it with Newton's method applied to the full set of functions F_i. While such methods can still occasionally fail by coming to rest on a local minimum of F, they often succeed where a direct attack via Newton's method alone fails. The next section deals with these methods.

CITED REFERENCES AND FURTHER READING:

Acton, F.S. 1970, *Numerical Methods That Work*; 1990, corrected edition (Washington: Mathematical Association of America), Chapter 14. [1]

Ostrowski, A.M. 1966, *Solutions of Equations and Systems of Equations*, 2nd ed. (New York: Academic Press).

Ortega, J., and Rheinboldt, W. 1970, *Iterative Solution of Nonlinear Equations in Several Variables* (New York: Academic Press).

9.7 Globally Convergent Methods for Nonlinear Systems of Equations

We have seen that Newton's method for solving nonlinear equations has an unfortunate tendency to wander off into the wild blue yonder if the initial guess is not sufficiently close to the root. A *global* method is one that converges to a solution from almost any starting point. In this section we will develop an algorithm that combines the rapid local convergence of Newton's method with a globally convergent strategy that will guarantee some progress towards the solution at each iteration. The algorithm is closely related to the quasi-Newton method of minimization which we will describe in §10.7.

Recall our discussion of §9.6: the Newton step for the set of equations

$$\mathbf{F}(\mathbf{x}) = 0 \tag{9.7.1}$$

is

$$\mathbf{x}_{\text{new}} = \mathbf{x}_{\text{old}} + \delta\mathbf{x} \tag{9.7.2}$$

where

$$\delta\mathbf{x} = -\mathbf{J}^{-1} \cdot \mathbf{F} \tag{9.7.3}$$

Here $\mathbf{J}$ is the Jacobian matrix. How do we decide whether to accept the Newton step $\delta\mathbf{x}$? A reasonable strategy is to require that the step decrease $|\mathbf{F}|^2 = \mathbf{F} \cdot \mathbf{F}$. This is the same requirement we would impose if we were trying to minimize

$$f = \frac{1}{2}\mathbf{F} \cdot \mathbf{F} \tag{9.7.4}$$

(The $\frac{1}{2}$ is for later convenience.) Every solution to (9.7.1) minimizes (9.7.4), but there may be local minima of (9.7.4) that are not solutions to (9.7.1). Thus, as already mentioned, simply applying one of our minimum finding algorithms from Chapter 10 to (9.7.4) is *not* a good idea.

To develop a better strategy, note that the Newton step (9.7.3) is a *descent direction* for f:

$$\nabla f \cdot \delta\mathbf{x} = (\mathbf{F} \cdot \mathbf{J}) \cdot (-\mathbf{J}^{-1} \cdot \mathbf{F}) = -\mathbf{F} \cdot \mathbf{F} < 0 \tag{9.7.5}$$

Thus our strategy is quite simple: We always first try the full Newton step, because once we are close enough to the solution we will get quadratic convergence. However, we check at each iteration that the proposed step reduces f. If not, we *backtrack* along the Newton direction until we have an acceptable step. Because the Newton step is a descent direction for f, we are guaranteed to find an acceptable step by backtracking. We will discuss the backtracking algorithm in more detail below.

Note that this method essentially minimizes f by taking Newton steps designed to bring $\mathbf{F}$ to zero. This is *not* equivalent to minimizing f directly by taking Newton steps designed to bring ∇f to zero. While the method can still occasionally fail by landing on a local minimum of f, this is quite rare in practice. The routine newt below will warn you if this happens. The remedy is to try a new starting point.

Line Searches and Backtracking

When we are not close enough to the minimum of f, taking the full Newton step $\mathbf{p} = \delta\mathbf{x}$ need not decrease the function; we may move too far for the quadratic approximation to be valid. All we are guaranteed is that *initially* f decreases as we move in the Newton direction. So the goal is to move to a new point $\mathbf{x}_{\text{new}}$ along the *direction* of the Newton step $\mathbf{p}$, but not necessarily all the way:

$$\mathbf{x}_{\text{new}} = \mathbf{x}_{\text{old}} + \lambda\mathbf{p}, \qquad 0 < \lambda \leq 1 \tag{9.7.6}$$

The aim is to find λ so that $f(\mathbf{x}_{\text{old}} + \lambda\mathbf{p})$ has decreased sufficiently. Until the early 1970s, standard practice was to choose λ so that $\mathbf{x}_{\text{new}}$ exactly minimizes f in the direction $\mathbf{p}$. However, we now know that it is extremely wasteful of function evaluations to do so. A better strategy is as follows: Since $\mathbf{p}$ is always the Newton direction in our algorithms, we first try $\lambda = 1$, the full Newton step. This will lead to quadratic convergence when $\mathbf{x}$ is sufficiently close to the solution. However, if $f(\mathbf{x}_{\text{new}})$ does not meet our acceptance criteria, we *backtrack* along the Newton direction, trying a smaller value of λ, until we find a suitable point. Since the Newton direction is a descent direction, we are guaranteed to decrease f for sufficiently small λ.

What should the criterion for accepting a step be? It is *not* sufficient to require merely that $f(\mathbf{x}_{\text{new}}) < f(\mathbf{x}_{\text{old}})$. This criterion can fail to converge to a minimum of f in one of two ways. First, it is possible to construct a sequence of steps satisfying this criterion with f decreasing too slowly relative to the step lengths. Second, one can have a sequence where the step lengths are too small relative to the initial rate of decrease of f. (For examples of such sequences, see [1], p. 117.)

A simple way to fix the first problem is to require the *average* rate of decrease of f to be at least some fraction α of the *initial* rate of decrease $\nabla f \cdot \mathbf{p}$:

$$f(\mathbf{x}_{\text{new}}) \leq f(\mathbf{x}_{\text{old}}) + \alpha\nabla f \cdot (\mathbf{x}_{\text{new}} - \mathbf{x}_{\text{old}}) \tag{9.7.7}$$

Here the parameter α satisfies $0 < \alpha < 1$. We can get away with quite small values of α; $\alpha = 10^{-4}$ is a good choice.

The second problem can be fixed by requiring the rate of decrease of f at $\mathbf{x}_{\text{new}}$ to be greater than some fraction β of the rate of decrease of f at $\mathbf{x}_{\text{old}}$. In practice, we will not need to impose this second constraint because our backtracking algorithm will have a built-in cutoff to avoid taking steps that are too small.

Here is the strategy for a practical backtracking routine: Define

$$g(\lambda) \equiv f(\mathbf{x}_{\text{old}} + \lambda\mathbf{p}) \tag{9.7.8}$$

so that

$$g'(\lambda) = \nabla f \cdot \mathbf{p} \tag{9.7.9}$$

If we need to backtrack, then we model g with the most current information we have and choose λ to minimize the model. We start with $g(0)$ and $g'(0)$ available. The first step is always the Newton step, $\lambda = 1$. If this step is not acceptable, we have available $g(1)$ as well. We can therefore model $g(\lambda)$ as a quadratic:

$$g(\lambda) \approx [g(1) - g(0) - g'(0)]\lambda^2 + g'(0)\lambda + g(0) \tag{9.7.10}$$

Taking the derivative of this quadratic, we find that it is a minimum when

$$\lambda = -\frac{g'(0)}{2[g(1) - g(0) - g'(0)]} \tag{9.7.11}$$

Since the Newton step failed, we can show that $\lambda \lesssim \frac{1}{2}$ for small α. We need to guard against too small a value of λ, however. We set $\lambda_{\min} = 0.1$.

On second and subsequent backtracks, we model g as a cubic in λ, using the previous value $g(\lambda_1)$ and the second most recent value $g(\lambda_2)$:

$$g(\lambda) = a\lambda^3 + b\lambda^2 + g'(0)\lambda + g(0) \tag{9.7.12}$$

Requiring this expression to give the correct values of g at λ_1 and λ_2 gives two equations that can be solved for the coefficients a and b:

$$\begin{bmatrix} a \\ b \end{bmatrix} = \frac{1}{\lambda_1 - \lambda_2} \begin{bmatrix} 1/\lambda_1^2 & -1/\lambda_2^2 \\ -\lambda_2/\lambda_1^2 & \lambda_1/\lambda_2^2 \end{bmatrix} \cdot \begin{bmatrix} g(\lambda_1) - g'(0)\lambda_1 - g(0) \\ g(\lambda_2) - g'(0)\lambda_2 - g(0) \end{bmatrix} \tag{9.7.13}$$

The minimum of the cubic (9.7.12) is at

$$\lambda = \frac{-b + \sqrt{b^2 - 3ag'(0)}}{3a} \tag{9.7.14}$$

We enforce that λ lie between $\lambda_{\max} = 0.5\lambda_1$ and $\lambda_{\min} = 0.1\lambda_1$.

The routine has two additional features, a minimum step length `alamin` and a maximum step length `stpmax`. `lnsrch` will also be used in the quasi-Newton minimization routine `dfpmin` in the next section.

```
#include <cmath>
#include <limits>
#include "nr.h"
using namespace std;

void NR::lnsrch(Vec_I_DP &xold, const DP fold, Vec_I_DP &g, Vec_IO_DP &p,
    Vec_O_DP &x, DP &f, const DP stpmax, bool &check, DP func(Vec_I_DP &))
```
Given an n-dimensional point xold[0..n-1], the value of the function and gradient there, fold and g[0..n-1], and a direction p[0..n-1], finds a new point x[0..n-1] along the direction p from xold where the function func has decreased "sufficiently." The new function value is returned in f. stpmax is an input quantity that limits the length of the steps so that you do not try to evaluate the function in regions where it is undefined or subject to overflow. p is usually the Newton direction. The output quantity check is false on a normal exit. It is true when x is too close to xold. In a minimization algorithm, this usually signals convergence and can be ignored. However, in a zero-finding algorithm the calling program should check whether the convergence is spurious.
```
{
    const DP ALF=1.0e-4, TOLX=numeric_limits<DP>::epsilon();
```
    ALF ensures sufficient decrease in function value; TOLX is the convergence criterion on Δx.
```
    int i;
    DP a,alam,alam2=0.0,alamin,b,disc,f2=0.0;
    DP rhs1,rhs2,slope,sum,temp,test,tmplam;

    int n=xold.size();
    check=false;
    sum=0.0;
    for (i=0;i<n;i++) sum += p[i]*p[i];
    sum=sqrt(sum);
    if (sum > stpmax)
        for (i=0;i<n;i++) p[i] *= stpmax/sum;    Scale if attempted step is too big.
    slope=0.0;
    for (i=0;i<n;i++)
        slope += g[i]*p[i];
    if (slope >= 0.0) nrerror("Roundoff problem in lnsrch.");
    test=0.0;                                    Compute λmin.
    for (i=0;i<n;i++) {
```

```
            temp=fabs(p[i])/MAX(fabs(xold[i]),1.0);
            if (temp > test) test=temp;
        }
        alamin=TOLX/test;
        alam=1.0;                                    Always try full Newton step first.
        for (;;) {                                   Start of iteration loop.
            for (i=0;i<n;i++) x[i]=xold[i]+alam*p[i];
            f=func(x);
            if (alam < alamin) {                     Convergence on Δx. For zero find-
                for (i=0;i<n;i++) x[i]=xold[i];         ing, the calling program should
                check=true;                             verify the convergence.
                return;
            } else if (f <= fold+ALF*alam*slope) return;   Sufficient function decrease.
            else {                                   Backtrack.
                if (alam == 1.0)
                    tmplam = -slope/(2.0*(f-fold-slope));   First time.
                else {                               Subsequent backtracks.
                    rhs1=f-fold-alam*slope;
                    rhs2=f2-fold-alam2*slope;
                    a=(rhs1/(alam*alam)-rhs2/(alam2*alam2))/(alam-alam2);
                    b=(-alam2*rhs1/(alam*alam)+alam*rhs2/(alam2*alam2))/(alam-alam2);
                    if (a == 0.0) tmplam = -slope/(2.0*b);
                    else {
                        disc=b*b-3.0*a*slope;
                        if (disc < 0.0) tmplam=0.5*alam;
                        else if (b <= 0.0) tmplam=(-b+sqrt(disc))/(3.0*a);
                        else tmplam=-slope/(b+sqrt(disc));
                    }
                    if (tmplam>0.5*alam)
                        tmplam=0.5*alam;             λ ≤ 0.5λ₁.
                }
            }
            alam2=alam;
            f2 = f;
            alam=MAX(tmplam,0.1*alam);               λ ≥ 0.1λ₁.
        }                                            Try again.
    }
```

Here now is the globally convergent Newton routine `newt` that uses `lnsrch`. A feature of `newt` is that you need not supply the Jacobian matrix analytically; the routine will attempt to compute the necessary partial derivatives of **F** by finite differences in the routine `fdjac`. This routine uses some of the techniques described in §5.7 for computing numerical derivatives. Of course, you can always replace `fdjac` with a routine that calculates the Jacobian analytically if this is easy for you to do.

```
#include <cmath>
#include <limits>
#include "nr.h"
using namespace std;

Vec_DP *fvec_p;                              Global variables to communicate with fmin.
void (*nrfuncv)(Vec_I_DP &v, Vec_O_DP &f);

void NR::newt(Vec_IO_DP &x, bool &check, void vecfunc(Vec_I_DP &, Vec_O_DP &))
```
Given an initial guess `x[0..n-1]` for a root in `n` dimensions, find the root by a globally convergent Newton's method. The vector of functions to be zeroed, called `fvec[0..n-1]` in the routine below, is returned by the user-supplied routine `vecfunc(x,fvec)`. The output quantity `check` is `false` on a normal return and `true` if the routine has converged to a local minimum of the function `fmin` defined below. In this case try restarting from a different initial guess.
```
{
    const int MAXITS=200;
```

```
const DP TOLF=1.0e-8,TOLMIN=1.0e-12,STPMX=100.0;
const DP TOLX=numeric_limits<DP>::epsilon();
```
Here MAXITS is the maximum number of iterations; TOLF sets the convergence criterion on function values; TOLMIN sets the criterion for deciding whether spurious convergence to a minimum of fmin has occurred; STPMX is the scaled maximum step length allowed in line searches; TOLX is the convergence criterion on $\delta\mathbf{x}$.
```
int i,j,its;
DP d,den,f,fold,stpmax,sum,temp,test;

int n=x.size();
Vec_INT indx(n);
Vec_DP g(n),p(n),xold(n);
Mat_DP fjac(n,n);
fvec_p=new Vec_DP(n);                    Define global variables.
nrfuncv=vecfunc;
Vec_DP &fvec=*fvec_p;                     Make an alias to simplify coding.
f=fmin(x);                               fvec is also computed by this call.
test=0.0;                                Test for initial guess being a root. Use
for (i=0;i<n;i++)                            more stringent test than simply TOLF.
    if (fabs(fvec[i]) > test) test=fabs(fvec[i]);
if (test < 0.01*TOLF) {
    check=false;
    delete fvec_p;
    return;
}
sum=0.0;
for (i=0;i<n;i++) sum += SQR(x[i]);      Calculate stpmax for line searches.
stpmax=STPMX*MAX(sqrt(sum),DP(n));
for (its=0;its<MAXITS;its++) {           Start of iteration loop.
    fdjac(x,fvec,fjac,vecfunc);
    If analytic Jacobian is available, you can replace the routine fdjac below with your
    own routine.
    for (i=0;i<n;i++) {                   Compute ∇f for the line search.
        sum=0.0;
        for (j=0;j<n;j++) sum += fjac[j][i]*fvec[j];
        g[i]=sum;
    }
    for (i=0;i<n;i++) xold[i]=x[i];       Store x,
    fold=f;                               and f.
    for (i=0;i<n;i++) p[i] = -fvec[i];    Right-hand side for linear equations.
    ludcmp(fjac,indx,d);                  Solve linear equations by LU decompo-
    lubksb(fjac,indx,p);                      sition.
    lnsrch(xold,fold,g,p,x,f,stpmax,check,fmin);
    lnsrch returns new x and f. It also calculates fvec at the new x when it calls fmin.
    test=0.0;                             Test for convergence on function val-
    for (i=0;i<n;i++)                         ues.
        if (fabs(fvec[i]) > test) test=fabs(fvec[i]);
    if (test < TOLF) {
        check=false;
        delete fvec_p;
        return;
    }
    if (check) {                          Check for gradient of f zero, i.e., spuri-
        test=0.0;                             ous convergence.
        den=MAX(f,0.5*n);
        for (i=0;i<n;i++) {
            temp=fabs(g[i])*MAX(fabs(x[i]),1.0)/den;
            if (temp > test) test=temp;
        }
        check=(test < TOLMIN);
        delete fvec_p;
        return;
    }
    test=0.0;                             Test for convergence on δx.
```

```
    for (i=0;i<n;i++) {
        temp=(fabs(x[i]-xold[i]))/MAX(fabs(x[i]),1.0);
        if (temp > test) test=temp;
    }
    if (test < TOLX) {
        delete fvec_p;
        return;
    }
}
nrerror("MAXITS exceeded in newt");
}
```

```
#include <cmath>
#include "nr.h"
using namespace std;

void NR::fdjac(Vec_IO_DP &x, Vec_I_DP &fvec, Mat_O_DP &df,
    void vecfunc(Vec_I_DP &, Vec_O_DP &))
```
Computes forward-difference approximation to Jacobian. On input, $x[0..n-1]$ is the point at which the Jacobian is to be evaluated, $fvec[0..n-1]$ is the vector of function values at the point, and $vecfunc(x,f)$ is a user-supplied routine that returns the vector of functions at x. On output, $df[0..n-1][0..n-1]$ is the Jacobian array.
```
{
    const DP EPS=1.0e-8;              Approximate square root of the machine precision.
    int i,j;
    DP h,temp;

    int n=x.size();
    Vec_DP f(n);
    for (j=0;j<n;j++) {
        temp=x[j];
        h=EPS*fabs(temp);
        if (h == 0.0) h=EPS;
        x[j]=temp+h;                  Trick to reduce finite precision error.
        h=x[j]-temp;
        vecfunc(x,f);
        x[j]=temp;
        for (i=0;i<n;i++)             Forward difference formula.
            df[i][j]=(f[i]-fvec[i])/h;
    }
}
```

```
#include "nr.h"

extern Vec_DP *fvec_p;
extern void (*nrfuncv)(Vec_I_DP &v, Vec_O_DP &f);

DP NR::fmin(Vec_I_DP &x)
```
Returns $f = \frac{1}{2}\mathbf{F} \cdot \mathbf{F}$ at x. The global pointer $*nrfuncv$ points to a routine that returns the vector of functions at x. It is set to point to a user-supplied routine in the calling program. Global variables also communicate the function values back to the calling program.
```
{
    int i;
    DP sum;

    Vec_DP &fvec=*fvec_p;            Make an alias to simplify coding.
    nrfuncv(x,fvec);
    int n=x.size();
    for (sum=0.0,i=0;i<n;i++) sum += SQR(fvec[i]);
    return 0.5*sum;
}
```

The routine `newt` assumes that typical values of all components of **x** and of **F** are of order unity, and it can fail if this assumption is badly violated. You should rescale the variables by their typical values before invoking `newt` if this problem occurs.

Multidimensional Secant Methods: Broyden's Method

Newton's method as implemented above is quite powerful, but it still has several disadvantages. One drawback is that the Jacobian matrix is needed. In many problems analytic derivatives are unavailable. If function evaluation is expensive, then the cost of finite-difference determination of the Jacobian can be prohibitive.

Just as the quasi-Newton methods to be discussed in §10.7 provide cheap approximations for the Hessian matrix in minimization algorithms, there are quasi-Newton methods that provide cheap approximations to the Jacobian for zero finding. These methods are often called *secant methods*, since they reduce to the secant method (§9.2) in one dimension (see, e.g., [1]). The best of these methods still seems to be the first one introduced, *Broyden's method* [2].

Let us denote the approximate Jacobian by **B**. Then the ith quasi-Newton step $\delta\mathbf{x}_i$ is the solution of

$$\mathbf{B}_i \cdot \delta\mathbf{x}_i = -\mathbf{F}_i \tag{9.7.15}$$

where $\delta\mathbf{x}_i = \mathbf{x}_{i+1} - \mathbf{x}_i$ (cf. equation 9.7.3). The quasi-Newton or secant condition is that $\mathbf{B}_{i+1}$ satisfy

$$\mathbf{B}_{i+1} \cdot \delta\mathbf{x}_i = \delta\mathbf{F}_i \tag{9.7.16}$$

where $\delta\mathbf{F}_i = \mathbf{F}_{i+1} - \mathbf{F}_i$. This is the generalization of the one-dimensional secant approximation to the derivative, $\delta F/\delta x$. However, equation (9.7.16) does not determine $\mathbf{B}_{i+1}$ uniquely in more than one dimension.

Many different auxiliary conditions to pin down $\mathbf{B}_{i+1}$ have been explored, but the best-performing algorithm in practice results from Broyden's formula. This formula is based on the idea of getting $\mathbf{B}_{i+1}$ by making the least change to $\mathbf{B}_i$ consistent with the secant equation (9.7.16). Broyden showed that the resulting formula is

$$\mathbf{B}_{i+1} = \mathbf{B}_i + \frac{(\delta\mathbf{F}_i - \mathbf{B}_i \cdot \delta\mathbf{x}_i) \otimes \delta\mathbf{x}_i}{\delta\mathbf{x}_i \cdot \delta\mathbf{x}_i} \tag{9.7.17}$$

You can easily check that $\mathbf{B}_{i+1}$ satisfies (9.7.16).

Early implementations of Broyden's method used the Sherman-Morrison formula, equation (2.7.2), to invert equation (9.7.17) analytically,

$$\mathbf{B}_{i+1}^{-1} = \mathbf{B}_i^{-1} + \frac{(\delta\mathbf{x}_i - \mathbf{B}_i^{-1} \cdot \delta\mathbf{F}_i) \otimes \delta\mathbf{x}_i \cdot \mathbf{B}_i^{-1}}{\delta\mathbf{x}_i \cdot \mathbf{B}_i^{-1} \cdot \delta\mathbf{F}_i} \tag{9.7.18}$$

Then instead of solving equation (9.7.3) by e.g., LU decomposition, one determined

$$\delta\mathbf{x}_i = -\mathbf{B}_i^{-1} \cdot \mathbf{F}_i \tag{9.7.19}$$

by matrix multiplication in $O(N^2)$ operations. The disadvantage of this method is that it cannot easily be embedded in a globally convergent strategy, for which the gradient of equation (9.7.4) requires **B**, not $\mathbf{B}^{-1}$,

$$\nabla(\tfrac{1}{2}\mathbf{F} \cdot \mathbf{F}) \simeq \mathbf{B}^T \cdot \mathbf{F} \tag{9.7.20}$$

Accordingly, we implement the update formula in the form (9.7.17).

However, we can still preserve the $O(N^2)$ solution of (9.7.3) by using QR decomposition (§2.10) instead of LU decomposition. The reason is that because of the special form of equation (9.7.17), the QR decomposition of $\mathbf{B}_i$ can be updated into the QR decomposition of $\mathbf{B}_{i+1}$ in $O(N^2)$ operations (§2.10). All we need is an initial approximation $\mathbf{B}_0$ to start the ball rolling. It is often acceptable to start simply with the identity matrix, and then allow $O(N)$ updates to produce a reasonable approximation to the Jacobian. We prefer to spend the first N function evaluations on a finite-difference approximation to initialize **B** via a call to `fdjac`.

Since **B** is not the exact Jacobian, we are not guaranteed that $\delta\mathbf{x}$ is a descent direction for $f = \frac{1}{2}\mathbf{F}\cdot\mathbf{F}$ (cf. equation 9.7.5). Thus the line search algorithm can fail to return a suitable step if **B** wanders far from the true Jacobian. In this case, we reinitialize **B** by another call to fdjac.

Like the secant method in one dimension, Broyden's method converges superlinearly once you get close enough to the root. Embedded in a global strategy, it is almost as robust as Newton's method, and often needs far fewer function evaluations to determine a zero. Note that the final value of **B** is *not* always close to the true Jacobian at the root, even when the method converges.

The routine broydn given below is very similar to newt in organization. The principal differences are the use of QR decomposition instead of LU, and the updating formula instead of directly determining the Jacobian. The remarks at the end of newt about scaling the variables apply equally to broydn.

```
#include <cmath>
#include <limits>
#include "nr.h"
using namespace std;

Vec_DP *fvec_p;                        Global variables to communicate with fmin.
void (*nrfuncv)(Vec_I_DP &v, Vec_O_DP &f);

void NR::broydn(Vec_IO_DP &x, bool &check, void vecfunc(Vec_I_DP &, Vec_O_DP &))
```
Given an initial guess x[0..n-1] for a root in n dimensions, find the root by Broyden's method embedded in a globally convergent strategy. The vector of functions to be zeroed, called fvec[0..n-1] in the routine below, is returned by the user-supplied routine vecfunc(x,fvec). The routine fdjac and the function fmin from newt are used. The output quantity check is false on a normal return and true if the routine has converged to a local minimum of the function fmin or if Broyden's method can make no further progress. In this case try restarting from a different initial guess.
```
{
    const int MAXITS=200;
    const DP EPS=numeric_limits<DP>::epsilon();
    const DP TOLF=1.0e-8, TOLX=EPS, STPMX=100.0, TOLMIN=1.0e-12;
```
Here MAXITS is the maximum number of iterations; EPS is the machine precision; TOLF is the convergence criterion on function values; TOLX is the convergence criterion on $\delta\mathbf{x}$; STPMX is the scaled maximum step length allowed in line searches; TOLMIN is used to decide whether spurious convergence to a minimum of fmin has occurred.
```
    bool restrt,sing,skip;
    int i,its,j,k;
    DP den,f,fold,stpmax,sum,temp,test;

    int n=x.size();
    Mat_DP qt(n,n),r(n,n);
    Vec_DP c(n),d(n),fvcold(n),g(n),p(n),s(n),t(n),w(n),xold(n);
    fvec_p=new Vec_DP(n);              Define global variables.
    nrfuncv=vecfunc;
    Vec_DP &fvec=*fvec_p;              Make an alias to simplify coding.
    f=fmin(x);                         The vector fvec is also computed by this
    test=0.0;                          call.
    for (i=0;i<n;i++)                  Test for initial guess being a root. Use more
        if (fabs(fvec[i]) > test) test=fabs(fvec[i]);    stringent test than sim-
    if (test < 0.01*TOLF) {            ply TOLF.
        check=false;
        delete fvec_p;
        return;
    }
    for (sum=0.0,i=0;i<n;i++) sum += SQR(x[i]);    Calculate stpmax for line searches.
    stpmax=STPMX*MAX(sqrt(sum),DP(n));
    restrt=true;                       Ensure initial Jacobian gets computed.
    for (its=1;its<=MAXITS;its++) {    Start of iteration loop.
        if (restrt) {
            fdjac(x,fvec,r,vecfunc);   Initialize or reinitialize Jacobian in r.
```

```
    qrdcmp(r,c,d,sing);                QR decomposition of Jacobian.
    if (sing) nrerror("singular Jacobian in broydn");
    for (i=0;i<n;i++) {                 Form Q^T explicitly.
        for (j=0;j<n;j++) qt[i][j]=0.0;
        qt[i][i]=1.0;
    }
    for (k=0;k<n-1;k++) {
        if (c[k] != 0.0) {
            for (j=0;j<n;j++) {
                sum=0.0;
                for (i=k;i<n;i++)
                    sum += r[i][k]*qt[i][j];
                sum /= c[k];
                for (i=k;i<n;i++)
                    qt[i][j] -= sum*r[i][k];
            }
        }
    }
    for (i=0;i<n;i++) {                 Form R explicitly.
        r[i][i]=d[i];
        for (j=0;j<i;j++) r[i][j]=0.0;
    }
} else {                               Carry out Broyden update.
    for (i=0;i<n;i++) s[i]=x[i]-xold[i];    s = δx.
    for (i=0;i<n;i++) {                 t = R · s.
        for (sum=0.0,j=i;j<n;j++) sum += r[i][j]*s[j];
        t[i]=sum;
    }
    skip=true;
    for (i=0;i<n;i++) {                 w = δF - B · s.
        for (sum=0.0,j=0;j<n;j++) sum += qt[j][i]*t[j];
        w[i]=fvec[i]-fvcold[i]-sum;
        if (fabs(w[i]) >= EPS*(fabs(fvec[i])+fabs(fvcold[i]))) skip=false;
        Don't update with noisy components of w.
        else w[i]=0.0;
    }
    if (!skip) {
        for (i=0;i<n;i++) {            t = Q^T · w.
            for (sum=0.0,j=0;j<n;j++) sum += qt[i][j]*w[j];
            t[i]=sum;
        }
        for (den=0.0,i=0;i<n;i++) den += SQR(s[i]);
        for (i=0;i<n;i++) s[i] /= den;    Store s/(s · s) in s.
        qrupdt(r,qt,t,s);              Update R and Q^T.
        for (i=0;i<n;i++) {
            if (r[i][i] == 0.0) nrerror("r singular in broydn");
            d[i]=r[i][i];             Diagonal of R stored in d.
        }
    }
}
for (i=0;i<n;i++) {                    Right-hand side for linear equations is -Q^T · F.
    for (sum=0.0,j=0;j<n;j++) sum += qt[i][j]*fvec[j];
    p[i] = -sum;
}
for (i=n-1;i>=0;i--) {                 Compute ∇f ≈ (Q · R)^T · F for the line search.
    for (sum=0.0,j=0;j<=i;j++) sum -= r[j][i]*p[j];
    g[i]=sum;
}
for (i=0;i<n;i++) {                    Store x and F.
    xold[i]=x[i];
    fvcold[i]=fvec[i];
}
fold=f;                               Store f.
rsolv(r,d,p);                         Solve linear equations.
```

```
lnsrch(xold,fold,g,p,x,f,stpmax,check,fmin);
lnsrch returns new x and f. It also calculates fvec at the new x when it calls fmin.
test=0.0;                          Test for convergence on function values.
for (i=0;i<n;i++)
    if (fabs(fvec[i]) > test) test=fabs(fvec[i]);
if (test < TOLF) {
    check=false;
    delete fvec_p;
    return;
}
if (check) {                       True if line search failed to find a new x.
    if (restrt) {                  Failure; already tried reinitializing the Jacobian.
        delete fvec_p;
        return;
    } else {
        test=0.0;                  Check for gradient of f zero, i.e., spurious con-
        den=MAX(f,0.5*n);             vergence.
        for (i=0;i<n;i++) {
            temp=fabs(g[i])*MAX(fabs(x[i]),1.0)/den;
            if (temp > test) test=temp;
        }
        if (test < TOLMIN) {
            delete fvec_p;
            return;
        }
        else restrt=true;          Try reinitializing the Jacobian.
    }
} else {                           Successful step; will use Broyden update for next
    restrt=false;                     step.
    test=0.0;                      Test for convergence on δx.
    for (i=0;i<n;i++) {
        temp=(fabs(x[i]-xold[i]))/MAX(fabs(x[i]),1.0);
        if (temp > test) test=temp;
    }
    if (test < TOLX) {
        delete fvec_p;
        return;
    }
}
}
nrerror("MAXITS exceeded in broydn");
return;
}
```

More Advanced Implementations

One of the principal ways that the methods described so far can fail is if $\mathbf{J}$ (in Newton's method) or $\mathbf{B}$ in (Broyden's method) becomes singular or nearly singular, so that $\delta\mathbf{x}$ cannot be determined. If you are lucky, this situation will not occur very often in practice. Methods developed so far to deal with this problem involve monitoring the condition number of $\mathbf{J}$ and perturbing $\mathbf{J}$ if singularity or near singularity is detected. This is most easily implemented if the QR decomposition is used instead of LU in Newton's method (see [1] for details). Our personal experience is that, while such an algorithm can solve problems where $\mathbf{J}$ is exactly singular and the standard Newton's method fails, it is occasionally less robust on other problems where LU decomposition succeeds. Clearly implementation details involving roundoff, underflow, etc., are important here and the last word is yet to be written.

Our global strategies both for minimization and zero finding have been based on line searches. Other global algorithms, such as the *hook step* and *dogleg step* methods, are based instead on the *model-trust region* approach, which is related to the Levenberg-Marquardt algorithm for nonlinear least-squares (§15.5). While somewhat more complicated than line searches, these methods have a reputation for robustness even when starting far from the desired zero or minimum [1].

CITED REFERENCES AND FURTHER READING:

Dennis, J.E., and Schnabel, R.B. 1983, *Numerical Methods for Unconstrained Optimization and Nonlinear Equations*; reprinted 1996 (Philadelphia: S.I.A.M.). [1]

Broyden, C.G. 1965, *Mathematics of Computation*, vol. 19, pp. 577–593. [2]

Chapter 10. Minimization or Maximization of Functions

10.0 Introduction

In a nutshell: You are given a single function f that depends on one or more independent variables. You want to find the value of those variables where f takes on a maximum or a minimum value. You can then calculate what value of f is achieved at the maximum or minimum. The tasks of maximization and minimization are trivially related to each other, since one person's function f could just as well be another's $-f$. The computational desiderata are the usual ones: Do it quickly, cheaply, and in small memory. Often the computational effort is dominated by the cost of evaluating f (and also perhaps its partial derivatives with respect to all variables, if the chosen algorithm requires them). In such cases the desiderata are sometimes replaced by the simple surrogate: Evaluate f as few times as possible.

An extremum (maximum or minimum point) can be either *global* (truly the highest or lowest function value) or *local* (the highest or lowest in a finite neighborhood and not on the boundary of that neighborhood). (See Figure 10.0.1.) Finding a global extremum is, in general, a very difficult problem. Two standard heuristics are widely used: (i) find local extrema starting from widely varying starting values of the independent variables (perhaps chosen quasi-randomly, as in §7.7), and then pick the most extreme of these (if they are not all the same); or (ii) perturb a local extremum by taking a finite amplitude step away from it, and then see if your routine returns you to a better point, or "always" to the same one. Relatively recently, so-called "simulated annealing methods" (§10.9) have demonstrated important successes on a variety of global extremization problems.

Our chapter title could just as well be *optimization*, which is the usual name for this very large field of numerical research. The importance ascribed to the various tasks in this field depends strongly on the particular interests of whom you talk to. Economists, and some engineers, are particularly concerned with *constrained optimization*, where there are *a priori* limitations on the allowed values of independent variables. For example, the production of wheat in the U.S. must be a nonnegative number. One particularly well-developed area of constrained optimization is *linear programming*, where both the function to be optimized and the constraints happen to be linear functions of the independent variables. Section 10.8, which is otherwise somewhat disconnected from the rest of the material that we have chosen to include in this chapter, implements the so-called "simplex algorithm" for linear programming problems.

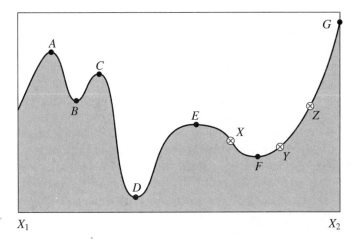

Figure 10.0.1. Extrema of a function in an interval. Points A, C, and E are local, but not global maxima. Points B and F are local, but not global minima. The global maximum occurs at G, which is on the boundary of the interval so that the derivative of the function need not vanish there. The global minimum is at D. At point E, derivatives higher than the first vanish, a situation which can cause difficulty for some algorithms. The points X, Y, and Z are said to "bracket" the minimum F, since Y is less than both X and Z.

One other section, §10.9, also lies outside of our main thrust, but for a different reason: so-called "annealing methods" are relatively new, so we do not yet know where they will ultimately fit into the scheme of things. However, these methods have solved some problems previously thought to be practically insoluble; they address directly the problem of finding global extrema in the presence of large numbers of undesired local extrema.

The other sections in this chapter constitute a selection of the best established algorithms in unconstrained minimization. (For definiteness, we will henceforth regard the optimization problem as that of minimization.) These sections are connected, with later ones depending on earlier ones. If you are just looking for the one "perfect" algorithm to solve your particular application, you may feel that we are telling you more than you want to know. Unfortunately, there is *no* perfect optimization algorithm. This is a case where we strongly urge you to try more than one method in comparative fashion. Your initial choice of method can be based on the following considerations:

- You must choose between methods that need only evaluations of the function to be minimized and methods that also require evaluations of the derivative of that function. In the multidimensional case, this derivative is the gradient, a vector quantity. Algorithms using the derivative are somewhat more powerful than those using only the function, but not always enough so as to compensate for the additional calculations of derivatives. We can easily construct examples favoring one approach or favoring the other. However, if you *can* compute derivatives, be prepared to try using them.
- For one-dimensional minimization (minimize a function of one variable) *without* calculation of the derivative, bracket the minimum as described in §10.1, and then use *Brent's method* as described in §10.2. If your function has a discontinuous second (or lower) derivative, then the parabolic

interpolations of Brent's method are of no advantage, and you might wish to use the simplest form of *golden section search*, as described in §10.1.

- For one-dimensional minimization *with* calculation of the derivative, §10.3 supplies a variant of Brent's method which makes limited use of the first derivative information. We shy away from the alternative of using derivative information to construct high-order interpolating polynomials. In our experience the improvement in convergence very near a smooth, analytic minimum does not make up for the tendency of polynomials sometimes to give wildly wrong interpolations at early stages, especially for functions that may have sharp, "exponential" features.

We now turn to the multidimensional case, both with and without computation of first derivatives.

- You must choose between methods that require storage of order N^2 and those that require only of order N, where N is the number of dimensions. For moderate values of N and reasonable memory sizes this is not a serious constraint. There will be, however, the occasional application where storage may be critical.
- We give in §10.4 a sometimes overlooked *downhill simplex method* due to Nelder and Mead. (This use of the word "simplex" is not to be confused with the simplex method of linear programming.) This method just crawls downhill in a straightforward fashion that makes almost no special assumptions about your function. This can be extremely slow, but it can also, in some cases, be extremely robust. Not to be overlooked is the fact that the code is concise and completely self-contained: a general N-dimensional minimization program in under 100 program lines! This method is most useful when the minimization calculation is only an incidental part of your overall problem. The storage requirement is of order N^2, and derivative calculations are not required.
- Section 10.5 deals with *direction-set methods*, of which *Powell's method* is the prototype. These are the methods of choice when you cannot easily calculate derivatives, and are not necessarily to be sneered at even if you can. Although derivatives are not needed, the method does require a one-dimensional minimization sub-algorithm such as Brent's method (see above). Storage is of order N^2.

There are two major families of algorithms for multidimensional minimization *with* calculation of first derivatives. Both families require a one-dimensional minimization sub-algorithm, which can itself either use, or not use, the derivative information, as you see fit (depending on the relative effort of computing the function and of its gradient vector). We do not think that either family dominates the other in all applications; you should think of them as available alternatives:

- The first family goes under the name *conjugate gradient methods*, as typi-fied by the *Fletcher-Reeves algorithm* and the closely related and probably superior *Polak-Ribiere algorithm*. Conjugate gradient methods require only of order a few times N storage, require derivative calculations and one-dimensional sub-minimization. Turn to §10.6 for detailed discussion and implementation.

- The second family goes under the names *quasi-Newton* or *variable metric* methods, as typified by the *Davidon-Fletcher-Powell (DFP)* algorithm (sometimes referred to just as *Fletcher-Powell*) or the closely related *Broyden-Fletcher-Goldfarb-Shanno (BFGS)* algorithm. These methods require of order N^2 storage, require derivative calculations and one-dimensional sub-minimization. Details are in §10.7.

You are now ready to proceed with scaling the peaks (and/or plumbing the depths) of practical optimization.

CITED REFERENCES AND FURTHER READING:

Dennis, J.E., and Schnabel, R.B. 1983, *Numerical Methods for Unconstrained Optimization and Nonlinear Equations*; reprinted 1996 (Philadelphia: S.I.A.M.).

Polak, E. 1971, *Computational Methods in Optimization* (New York: Academic Press).

Gill, P.E., Murray, W., and Wright, M.H. 1981, *Practical Optimization* (New York: Academic Press).

Acton, F.S. 1970, *Numerical Methods That Work*; 1990, corrected edition (Washington: Mathematical Association of America), Chapter 17.

Jacobs, D.A.H. (ed.) 1977, *The State of the Art in Numerical Analysis* (London: Academic Press), Chapter III.1.

Brent, R.P. 1973, *Algorithms for Minimization without Derivatives* (Englewood Cliffs, NJ: Prentice-Hall).

Dahlquist, G., and Bjorck, A. 1974, *Numerical Methods* (Englewood Cliffs, NJ: Prentice-Hall), Chapter 10.

10.1 Golden Section Search in One Dimension

Recall how the bisection method finds roots of functions in one dimension (§9.1): The root is supposed to have been bracketed in an interval (a, b). One then evaluates the function at an intermediate point x and obtains a new, smaller bracketing interval, either (a, x) or (x, b). The process continues until the bracketing interval is acceptably small. It is optimal to choose x to be the midpoint of (a, b) so that the decrease in the interval length is maximized when the function is as uncooperative as it can be, i.e., when the luck of the draw forces you to take the bigger bisected segment.

There is a precise, though slightly subtle, translation of these considerations to the minimization problem: What does it mean to *bracket* a minimum? A root of a function is known to be bracketed by a pair of points, a and b, when the function has opposite sign at those two points. A minimum, by contrast, is known to be bracketed only when there is a *triplet* of points, $a < b < c$ (or $c < b < a$), such that $f(b)$ is less than both $f(a)$ and $f(c)$. In this case we know that the function (if it is nonsingular) has a minimum in the interval (a, c).

The analog of bisection is to choose a new point x, either between a and b or between b and c. Suppose, to be specific, that we make the latter choice. Then we evaluate $f(x)$. If $f(b) < f(x)$, then the new bracketing triplet of points is (a, b, x);

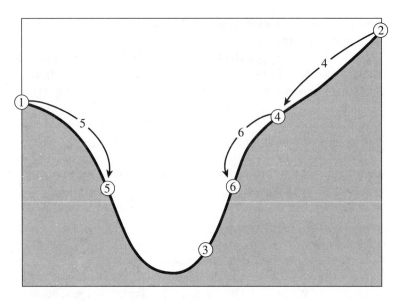

Figure 10.1.1.   Successive bracketing of a minimum. The minimum is originally bracketed by points 1,3,2. The function is evaluated at 4, which replaces 2; then at 5, which replaces 1; then at 6, which replaces 4. The rule at each stage is to keep a center point that is lower than the two outside points. After the steps shown, the minimum is bracketed by points 5,3,6.

contrariwise, if $f(b) > f(x)$, then the new bracketing triplet is (b, x, c). In all cases the middle point of the new triplet is the abscissa whose ordinate is the best minimum achieved so far; see Figure 10.1.1. We continue the process of bracketing until the distance between the two outer points of the triplet is tolerably small.

How small is "tolerably" small? For a minimum located at a value b, you might naively think that you will be able to bracket it in as small a range as $(1 - \epsilon)b < b < (1 + \epsilon)b$, where ϵ is your computer's floating-point precision, a number like 3×10^{-8} (for float) or 10^{-15} (for double). Not so! In general, the shape of your function $f(x)$ near b will be given by Taylor's theorem

$$f(x) \approx f(b) + \frac{1}{2}f''(b)(x - b)^2 \qquad (10.1.1)$$

The second term will be negligible compared to the first (that is, will be a factor ϵ smaller and will act just like zero when added to it) whenever

$$|x - b| < \sqrt{\epsilon}|b| \sqrt{\frac{2|f(b)|}{b^2 f''(b)}} \qquad (10.1.2)$$

The reason for writing the right-hand side in this way is that, for most functions, the final square root is a number of order unity. Therefore, as a rule of thumb, it is hopeless to ask for a bracketing interval of width less than $\sqrt{\epsilon}$ times its central value, a fractional width of only about 10^{-4} (single precision) or 3×10^{-8} (double precision). Knowing this inescapable fact will save you a lot of useless bisections!

The minimum-finding routines of this chapter will often call for a user-supplied argument tol, and return with an abscissa whose fractional precision is about $\pm$tol (bracketing interval of fractional size about 2$\times$tol). Unless you have a better estimate for the right-hand side of equation (10.1.2), you should set `tol` equal to (not much less than) the square root of your machine's floating-point precision, since smaller values will gain you nothing.

It remains to decide on a strategy for choosing the new point x, given (a, b, c). Suppose that b is a fraction w of the way between a and c, i.e.

$$\frac{b-a}{c-a} = w \qquad \frac{c-b}{c-a} = 1-w \qquad (10.1.3)$$

Also suppose that our next trial point x is an additional fraction z beyond b,

$$\frac{x-b}{c-a} = z \qquad (10.1.4)$$

Then the next bracketing segment will either be of length $w + z$ relative to the current one, or else of length $1 - w$. If we want to minimize the worst case possibility, then we will choose z to make these equal, namely

$$z = 1 - 2w \qquad (10.1.5)$$

We see at once that the new point is the symmetric point to b in the original interval, namely with $|b - a|$ equal to $|x - c|$. This implies that the point x lies in the larger of the two segments (z is positive only if $w < 1/2$).

But where in the larger segment? Where did the value of w itself come from? Presumably from the previous stage of applying our same strategy. Therefore, if z is chosen to be optimal, then so was w before it. This *scale similarity* implies that x should be the same fraction of the way from b to c (if that is the bigger segment) as was b from a to c, in other words,

$$\frac{z}{1-w} = w \qquad (10.1.6)$$

Equations (10.1.5) and (10.1.6) give the quadratic equation

$$w^2 - 3w + 1 = 0 \qquad \text{yielding} \qquad w = \frac{3 - \sqrt{5}}{2} \approx 0.38197 \qquad (10.1.7)$$

In other words, the optimal bracketing interval (a, b, c) has its middle point b a fractional distance 0.38197 from one end (say, a), and 0.61803 from the other end (say, b). These fractions are those of the so-called *golden mean* or *golden section*, whose supposedly aesthetic properties hark back to the ancient Pythagoreans. This optimal method of function minimization, the analog of the bisection method for finding zeros, is thus called the *golden section search*, summarized as follows:

Given, at each stage, a bracketing triplet of points, the next point to be tried is that which is a fraction 0.38197 into the larger of the two intervals (measuring from the central point of the triplet). If you start out with a bracketing triplet whose segments are not in the golden ratios, the procedure of choosing successive points at the golden mean point of the larger segment will quickly converge you to the proper, self-replicating ratios.

The golden section search guarantees that each new function evaluation will (after self-replicating ratios have been achieved) bracket the minimum to an interval just 0.61803 times the size of the preceding interval. This is comparable to, but not quite as good as, the 0.50000 that holds when finding roots by bisection. Note that the convergence is *linear* (in the language of Chapter 9), meaning that successive significant figures are won linearly with additional function evaluations. In the next section we will give a superlinear method, where the rate at which successive significant figures are liberated increases with each successive function evaluation.

Routine for Initially Bracketing a Minimum

The preceding discussion has assumed that you are able to bracket the minimum in the first place. We consider this initial bracketing to be an essential part of any one-dimensional minimization. There are some one-dimensional algorithms that do not require a rigorous initial bracketing. However, we would *never* trade the secure feeling of *knowing* that a minimum is "in there somewhere" for the dubious reduction of function evaluations that these nonbracketing routines may promise. Please bracket your minima (or, for that matter, your zeros) before isolating them!

There is not much theory as to how to do this bracketing. Obviously you want to step downhill. But how far? We like to take larger and larger steps, starting with some (wild?) initial guess and then increasing the stepsize at each step either by a constant factor, or else by the result of a parabolic extrapolation of the preceding points that is designed to take us to the extrapolated turning point. It doesn't much matter if the steps get big. After all, we are stepping downhill, so we already have the left and middle points of the bracketing triplet. We just need to take a big enough step to stop the downhill trend and get a high third point.

Our standard routine is this:

```
#include <cmath>
#include "nr.h"
using namespace std;

namespace {
    inline void shft3(DP &a, DP &b, DP &c, const DP d)
    {
        a=b;
        b=c;
        c=d;
    }
}
```

```
void NR::mnbrak(DP &ax, DP &bx, DP &cx, DP &fa, DP &fb, DP &fc,
    DP func(const DP))
```
Given a function func, and given distinct initial points ax and bx, this routine searches in the downhill direction (defined by the function as evaluated at the initial points) and returns new points ax, bx, cx that bracket a minimum of the function. Also returned are the function values at the three points, fa, fb, and fc.
```
{
    const DP GOLD=1.618034,GLIMIT=100.0,TINY=1.0e-20;
```
Here GOLD is the default ratio by which successive intervals are magnified; GLIMIT is the maximum magnification allowed for a parabolic-fit step.
```
    DP ulim,u,r,q,fu;

    fa=func(ax);
    fb=func(bx);
    if (fb > fa) {                       Switch roles of a and b so that we can go
        SWAP(ax,bx);                     downhill in the direction from a to b.
        SWAP(fb,fa);
```

```
}
cx=bx+GOLD*(bx-ax);                    First guess for c.
fc=func(cx);
while (fb > fc) {                      Keep returning here until we bracket.
    r=(bx-ax)*(fb-fc);                 Compute u by parabolic extrapolation from
    q=(bx-cx)*(fb-fa);                     a, b, c. TINY is used to prevent any pos-
    u=bx-((bx-cx)*q-(bx-ax)*r)/             sible division by zero.
        (2.0*SIGN(MAX(fabs(q-r),TINY),q-r));
    ulim=bx+GLIMIT*(cx-bx);
    We won't go farther than this. Test various possibilities:
    if ((bx-u)*(u-cx) > 0.0) {         Parabolic u is between b and c: try it.
        fu=func(u);
        if (fu < fc) {                 Got a minimum between b and c.
            ax=bx;
            bx=u;
            fa=fb;
            fb=fu;
            return;
        } else if (fu > fb) {          Got a minimum between between a and u.
            cx=u;
            fc=fu;
            return;
        }
        u=cx+GOLD*(cx-bx);             Parabolic fit was no use. Use default mag-
        fu=func(u);                        nification.
    } else if ((cx-u)*(u-ulim) > 0.0) {     Parabolic fit is between c and its
        fu=func(u);                             allowed limit.
        if (fu < fc) {
            shft3(bx,cx,u,cx+GOLD*(cx-bx));
            shft3(fb,fc,fu,func(u));
        }
    } else if ((u-ulim)*(ulim-cx) >= 0.0) {  Limit parabolic u to maximum
        u=ulim;                                  allowed value.
        fu=func(u);
    } else {                           Reject parabolic u, use default magnifica-
        u=cx+GOLD*(cx-bx);                 tion.
        fu=func(u);
    }
    shft3(ax,bx,cx,u);                 Eliminate oldest point and continue.
    shft3(fa,fb,fc,fu);
}
}
```

(Because of the housekeeping involved in moving around three or four points and their function values, the above program ends up looking deceptively formidable. That is true of several other programs in this chapter as well. The underlying ideas, however, are quite simple.)

Routine for Golden Section Search

```
#include <cmath>
#include "nr.h"
using namespace std;

namespace {
    inline void shft2(DP &a, DP &b, const DP c)
    {
        a=b;
        b=c;
```

```
    }

    inline void shft3(DP &a, DP &b, DP &c, const DP d)
    {
        a=b;
        b=c;
        c=d;
    }
}
```

```
DP NR::golden(const DP ax, const DP bx, const DP cx, DP f(const DP),
    const DP tol, DP &xmin)
```
Given a function f, and given a bracketing triplet of abscissas ax, bx, cx (such that bx is between ax and cx, and f(bx) is less than both f(ax) and f(cx)), this routine performs a golden section search for the minimum, isolating it to a fractional precision of about tol. The abscissa of the minimum is returned as xmin, and the minimum function value is returned as golden, the returned function value.
```
{
    const DP R=0.61803399,C=1.0-R;       The golden ratios.
    DP f1,f2,x0,x1,x2,x3;

    x0=ax;                               At any given time we will keep track of four
    x3=cx;                                   points, x0,x1,x2,x3.
    if (fabs(cx-bx) > fabs(bx-ax)) {     Make x0 to x1 the smaller segment,
        x1=bx;
        x2=bx+C*(cx-bx);                 and fill in the new point to be tried.
    } else {
        x2=bx;
        x1=bx-C*(bx-ax);
    }
    f1=f(x1);                            The initial function evaluations. Note that
    f2=f(x2);                                we never need to evaluate the function
    while (fabs(x3-x0) > tol*(fabs(x1)+fabs(x2))) {      at the original endpoints.
        if (f2 < f1) {                   One possible outcome,
            shft3(x0,x1,x2,R*x2+C*x3);   its housekeeping,
            shft2(f1,f2,f(x2));          and a new function evaluation.
        } else {                         The other outcome,
            shft3(x3,x2,x1,R*x1+C*x0);
            shft2(f2,f1,f(x1));          and its new function evaluation.
        }
    }                                    Back to see if we are done.
    if (f1 < f2) {                       We are done. Output the best of the two
        xmin=x1;                             current values.
        return f1;
    } else {
        xmin=x2;
        return f2;
    }
}
```

10.2 Parabolic Interpolation and Brent's Method in One Dimension

We already tipped our hand about the desirability of parabolic interpolation in the previous section's mnbrak routine, but it is now time to be more explicit. A golden section search is designed to handle, in effect, the worst possible case of

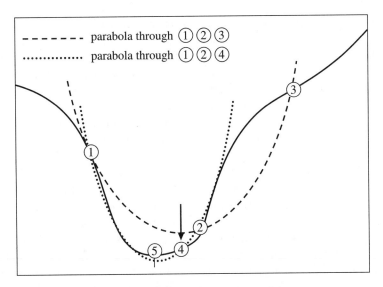

Figure 10.2.1. Convergence to a minimum by inverse parabolic interpolation. A parabola (dashed line) is drawn through the three original points 1,2,3 on the given function (solid line). The function is evaluated at the parabola's minimum, 4, which replaces point 3. A new parabola (dotted line) is drawn through points 1,4,2. The minimum of this parabola is at 5, which is close to the minimum of the function.

function minimization, with the uncooperative minimum hunted down and cornered like a scared rabbit. But why assume the worst? If the function is nicely parabolic near to the minimum — surely the generic case for sufficiently smooth functions — then the parabola fitted through any three points ought to take us in a single leap to the minimum, or at least very near to it (see Figure 10.2.1). Since we want to find an abscissa rather than an ordinate, the procedure is technically called *inverse parabolic interpolation*.

The formula for the abscissa x that is the minimum of a parabola through three points $f(a)$, $f(b)$, and $f(c)$ is

$$x = b - \frac{1}{2} \frac{(b-a)^2[f(b)-f(c)] - (b-c)^2[f(b)-f(a)]}{(b-a)[f(b)-f(c)] - (b-c)[f(b)-f(a)]} \qquad (10.2.1)$$

as you can easily derive. This formula fails only if the three points are collinear, in which case the denominator is zero (minimum of the parabola is infinitely far away). Note, however, that (10.2.1) is as happy jumping to a parabolic maximum as to a minimum. No minimization scheme that depends solely on (10.2.1) is likely to succeed in practice.

The exacting task is to invent a scheme that relies on a sure-but-slow technique, like golden section search, when the function is not cooperative, but that switches over to (10.2.1) when the function allows. The task is nontrivial for several reasons, including these: (i) The housekeeping needed to avoid unnecessary function evaluations in switching between the two methods can be complicated. (ii) Careful attention must be given to the "endgame," where the function is being evaluated very near to the roundoff limit of equation (10.1.2). (iii) The scheme for detecting a cooperative versus noncooperative function must be very robust.

Brent's method [1] is up to the task in all particulars. At any particular stage, it is keeping track of six function points (not necessarily all distinct), a, b, u, v,

w and x, defined as follows: the minimum is bracketed between a and b; x is the point with the very least function value found so far (or the most recent one in case of a tie); w is the point with the second least function value; v is the previous value of w; u is the point at which the function was evaluated most recently. Also appearing in the algorithm is the point x_m, the midpoint between a and b; however, the function is not evaluated there.

You can read the code below to understand the method's logical organization. Mention of a few general principles here may, however, be helpful: Parabolic interpolation is attempted, fitting through the points x, v, and w. To be acceptable, the parabolic step must (i) fall within the bounding interval (a, b), and (ii) imply a movement from the best current value x that is *less* than half the movement of the *step before last*. This second criterion insures that the parabolic steps are actually converging to something, rather than, say, bouncing around in some nonconvergent limit cycle. In the worst possible case, where the parabolic steps are acceptable but useless, the method will approximately alternate between parabolic steps and golden sections, converging in due course by virtue of the latter. The reason for comparing to the step *before* last seems essentially heuristic: Experience shows that it is better not to "punish" the algorithm for a single bad step if it can make it up on the next one.

Another principle exemplified in the code is never to evaluate the function less than a distance `tol` from a point already evaluated (or from a known bracketing point). The reason is that, as we saw in equation (10.1.2), there is simply no information content in doing so: the function will differ from the value already evaluated only by an amount of order the roundoff error. Therefore in the code below you will find several tests and modifications of a potential new point, imposing this restriction. This restriction also interacts subtly with the test for "doneness," which the method takes into account.

A typical ending configuration for Brent's method is that a and b are $2 \times x \times$ `tol` apart, with x (the best abscissa) at the midpoint of a and b, and therefore fractionally accurate to $\pm$`tol`.

Indulge us a final reminder that `tol` should generally be no smaller than the square root of your machine's floating-point precision.

```
#include <cmath>
#include <limits>
#include "nr.h"
using namespace std;

namespace {
    inline void shft3(DP &a, DP &b, DP &c, const DP d)
    {
        a=b;
        b=c;
        c=d;
    }
}
```

```
DP NR::brent(const DP ax, const DP bx, const DP cx, DP f(const DP),
    const DP tol, DP &xmin)
```
Given a function f, and given a bracketing triplet of abscissas ax, bx, cx (such that bx is between ax and cx, and f(bx) is less than both f(ax) and f(cx)), this routine isolates the minimum to a fractional precision of about tol using Brent's method. The abscissa of the minimum is returned as xmin, and the minimum function value is returned as brent, the returned function value.
```
{
```

```
const int ITMAX=100;
const DP CGOLD=0.3819660;
const DP ZEPS=numeric_limits<DP>::epsilon()*1.0e-3;
```
Here ITMAX is the maximum allowed number of iterations; CGOLD is the golden ratio; ZEPS is a small number that protects against trying to achieve fractional accuracy for a minimum that happens to be exactly zero.
```
int iter;
DP a,b,d=0.0,etemp,fu,fv,fw,fx;
DP p,q,r,tol1,tol2,u,v,w,x,xm;
DP e=0.0;                               This will be the distance moved on
                                          the step before last.
a=(ax < cx ? ax : cx);                  a and b must be in ascending order,
b=(ax > cx ? ax : cx);                    but input abscissas need not be.
x=w=v=bx;                               Initializations...
fw=fv=fx=f(x);
for (iter=0;iter<ITMAX;iter++) {        Main program loop.
    xm=0.5*(a+b);
    tol2=2.0*(tol1=tol*fabs(x)+ZEPS);
    if (fabs(x-xm) <= (tol2-0.5*(b-a))) {   Test for done here.
        xmin=x;
        return fx;
    }
    if (fabs(e) > tol1) {               Construct a trial parabolic fit.
        r=(x-w)*(fx-fv);
        q=(x-v)*(fx-fw);
        p=(x-v)*q-(x-w)*r;
        q=2.0*(q-r);
        if (q > 0.0) p = -p;
        q=fabs(q);
        etemp=e;
        e=d;
        if (fabs(p) >= fabs(0.5*q*etemp) || p <= q*(a-x) || p >= q*(b-x))
            d=CGOLD*(e=(x >= xm ? a-x : b-x));
```
The above conditions determine the acceptability of the parabolic fit. Here we take the golden section step into the larger of the two segments.
```
        else {
            d=p/q;                      Take the parabolic step.
            u=x+d;
            if (u-a < tol2 || b-u < tol2)
                d=SIGN(tol1,xm-x);
        }
    } else {
        d=CGOLD*(e=(x >= xm ? a-x : b-x));
    }
    u=(fabs(d) >= tol1 ? x+d : x+SIGN(tol1,d));
    fu=f(u);
```
This is the one function evaluation per iteration.
```
    if (fu <= fx) {                     Now decide what to do with our func-
        if (u >= x) a=x; else b=x;        tion evaluation.
        shft3(v,w,x,u);                 Housekeeping follows:
        shft3(fv,fw,fx,fu);
    } else {
        if (u < x) a=u; else b=u;
        if (fu <= fw || w == x) {
            v=w;
            w=u;
            fv=fw;
            fw=fu;
        } else if (fu <= fv || v == x || v == w) {
            v=u;
            fv=fu;
        }
    }                                   Done with housekeeping. Back for
}                                         another iteration.
```

```
nrerror("Too many iterations in brent");
xmin=x;                                    Never get here.
return fx;
}
```

CITED REFERENCES AND FURTHER READING:

Brent, R.P. 1973, *Algorithms for Minimization without Derivatives* (Englewood Cliffs, NJ: Prentice-Hall), Chapter 5. [1]

Forsythe, G.E., Malcolm, M.A., and Moler, C.B. 1977, *Computer Methods for Mathematical Computations* (Englewood Cliffs, NJ: Prentice-Hall), §8.2.

10.3 One-Dimensional Search with First Derivatives

Here we want to accomplish precisely the same goal as in the previous section, namely to isolate a functional minimum that is bracketed by the triplet of abscissas (a, b, c), but utilizing an additional capability to compute the function's first derivative as well as its value.

In principle, we might simply search for a zero of the derivative, ignoring the function value information, using a root finder like `rtflsp` or `zbrent` (§§9.2–9.3). It doesn't take long to reject *that* idea: How do we distinguish maxima from minima? Where do we go from initial conditions where the derivatives on one or both of the outer bracketing points indicate that "downhill" is in the direction *out* of the bracketed interval?

We don't want to give up our strategy of maintaining a rigorous bracket on the minimum at all times. The only way to keep such a bracket is to update it using function (not derivative) information, with the central point in the bracketing triplet always that with the lowest function value. Therefore the role of the derivatives can only be to help us choose new trial points within the bracket.

One school of thought is to "use everything you've got": Compute a polynomial of relatively high order (cubic or above) that agrees with some number of previous function and derivative evaluations. For example, there is a unique cubic that agrees with function and derivative at two points, and one can jump to the interpolated minimum of that cubic (if there is a minimum within the bracket). Suggested by Davidon and others, formulas for this tactic are given in [1].

We like to be more conservative than this. Once superlinear convergence sets in, it hardly matters whether its order is moderately lower or higher. In practical problems that we have met, most function evaluations are spent in getting globally close enough to the minimum for superlinear convergence to commence. So we are more worried about all the funny "stiff" things that high-order polynomials can do (cf. Figure 3.0.1b), and about their sensitivities to roundoff error.

This leads us to use derivative information only as follows: The sign of the derivative at the central point of the bracketing triplet (a, b, c) indicates uniquely whether the next test point should be taken in the interval (a, b) or in the interval (b, c). The value of this derivative and of the derivative at the second-best-so-far point are extrapolated to zero by the secant method (inverse linear interpolation), which by itself is superlinear of order 1.618. (The golden mean again: see [1], p. 57.) We impose the same sort of restrictions on this new trial point as in Brent's method. If the trial point must be rejected, we *bisect* the interval under scrutiny.

Yes, we are fuddy-duddies when it comes to making flamboyant use of derivative information in one-dimensional minimization. But we have met too many functions whose computed "derivatives" *don't* integrate up to the function value and *don't* accurately point the way to the minimum, usually because of roundoff errors, sometimes because of truncation error in the method of derivative evaluation.

You will see that the following routine is closely modeled on `brent` in the previous section.

```cpp
#include <cmath>
#include <limits>
#include "nr.h"
using namespace std;

namespace {
    inline void mov3(DP &a, DP &b, DP &c, const DP d, const DP e,
        const DP f)
    {
        a=d; b=e; c=f;
    }
}
```

```cpp
DP NR::dbrent(const DP ax, const DP bx, const DP cx, DP f(const DP),
    DP df(const DP), const DP tol, DP &xmin)
```
Given a function f and its derivative function df, and given a bracketing triplet of abscissas ax, bx, cx [such that bx is between ax and cx, and f(bx) is less than both f(ax) and f(cx)], this routine isolates the minimum to a fractional precision of about tol using a modification of Brent's method that uses derivatives. The abscissa of the minimum is returned as xmin, and the minimum function value is returned as dbrent, the returned function value.
```cpp
{
    const int ITMAX=100;
    const DP ZEPS=numeric_limits<DP>::epsilon()*1.0e-3;
    bool ok1,ok2;                           Will be used as flags for whether pro-
    int iter;                               posed steps are acceptable or not.
    DP a,b,d=0.0,d1,d2,du,dv,dw,dx,e=0.0;
    DP fu,fv,fw,fx,olde,tol1,tol2,u,u1,u2,v,w,x,xm;

    Comments following will point out only differences from the routine brent. Read that
    routine first.
    a=(ax < cx ? ax : cx);
    b=(ax > cx ? ax : cx);
    x=w=v=bx;
    fw=fv=fx=f(x);
    dw=dv=dx=df(x);                         All our housekeeping chores are dou-
    for (iter=0;iter<ITMAX;iter++) {            bled by the necessity of moving
        xm=0.5*(a+b);                           derivative values around as well
        tol1=tol*fabs(x)+ZEPS;                  as function values.
        tol2=2.0*tol1;
        if (fabs(x-xm) <= (tol2-0.5*(b-a))) {
            xmin=x;
            return fx;
        }
        if (fabs(e) > tol1) {
            d1=2.0*(b-a);                   Initialize these d's to an out-of-bracket
            d2=d1;                              value.
            if (dw != dx) d1=(w-x)*dx/(dx-dw);  Secant method with one point.
            if (dv != dx) d2=(v-x)*dx/(dx-dv);  And the other.
```

Which of these two estimates of d shall we take? We will insist that they be within the bracket, and on the side pointed to by the derivative at x:

```
u1=x+d1;
u2=x+d2;
ok1 = (a-u1)*(u1-b) > 0.0 && dx*d1 <= 0.0;
ok2 = (a-u2)*(u2-b) > 0.0 && dx*d2 <= 0.0;
olde=e;                                      Movement on the step before last.
e=d;
if (ok1 || ok2) {                            Take only an acceptable d, and if
    if (ok1 && ok2)                              both are acceptable, then take
        d=(fabs(d1) < fabs(d2) ? d1 : d2);   the smallest one.
    else if (ok1)
        d=d1;
    else
        d=d2;
    if (fabs(d) <= fabs(0.5*olde)) {
        u=x+d;
        if (u-a < tol2 || b-u < tol2)
            d=SIGN(tol1,xm-x);
    } else {                                 Bisect, not golden section.
        d=0.5*(e=(dx >= 0.0 ? a-x : b-x));
        Decide which segment by the sign of the derivative.
    }
} else {
    d=0.5*(e=(dx >= 0.0 ? a-x : b-x));
}
if (fabs(d) >= tol1) {
    u=x+d;
    fu=f(u);
} else {
    u=x+SIGN(tol1,d);
    fu=f(u);
    if (fu > fx) {                           If the minimum step in the downhill
        xmin=x;                                 direction takes us uphill, then
        return fx;                              we are done.
    }
}
du=df(u);                                    Now all the housekeeping, sigh.
if (fu <= fx) {
    if (u >= x) a=x; else b=x;
    mov3(v,fv,dv,w,fw,dw);
    mov3(w,fw,dw,x,fx,dx);
    mov3(x,fx,dx,u,fu,du);
} else {
    if (u < x) a=u; else b=u;
    if (fu <= fw || w == x) {
        mov3(v,fv,dv,w,fw,dw);
        mov3(w,fw,dw,u,fu,du);
    } else if (fu < fv || v == x || v == w) {
        mov3(v,fv,dv,u,fu,du);
    }
}
}
nrerror("Too many iterations in routine dbrent");
return 0.0;                                  Never get here.
}
```

CITED REFERENCES AND FURTHER READING:

Acton, F.S. 1970, *Numerical Methods That Work*; 1990, corrected edition (Washington: Mathematical Association of America), pp. 55; 454–458. [1]

Brent, R.P. 1973, *Algorithms for Minimization without Derivatives* (Englewood Cliffs, NJ: Prentice-Hall), p. 78.

10.4 Downhill Simplex Method in Multidimensions

With this section we begin consideration of multidimensional minimization, that is, finding the minimum of a function of more than one independent variable. This section stands apart from those which follow, however: All of the algorithms after this section will make explicit use of a one-dimensional minimization algorithm as a part of their computational strategy. This section implements an entirely self-contained strategy, in which one-dimensional minimization does not figure.

The *downhill simplex method* is due to Nelder and Mead [1]. The method requires only function evaluations, not derivatives. It is not very efficient in terms of the number of function evaluations that it requires. Powell's method (§10.5) is almost surely faster in all likely applications. However, the downhill simplex method may frequently be the *best* method to use if the figure of merit is "get something working quickly" for a problem whose computational burden is small.

The method has a geometrical naturalness about it which makes it delightful to describe or work through:

A *simplex* is the geometrical figure consisting, in N dimensions, of $N + 1$ points (or vertices) and all their interconnecting line segments, polygonal faces, etc. In two dimensions, a simplex is a triangle. In three dimensions it is a tetrahedron, not necessarily the regular tetrahedron. (The *simplex method* of linear programming, described in §10.8, also makes use of the geometrical concept of a simplex. Otherwise it is completely unrelated to the algorithm that we are describing in this section.) In general we are only interested in simplexes that are nondegenerate, i.e., that enclose a finite inner N-dimensional volume. If any point of a nondegenerate simplex is taken as the origin, then the N other points define vector directions that span the N-dimensional vector space.

In one-dimensional minimization, it was possible to bracket a minimum, so that the success of a subsequent isolation was guaranteed. Alas! There is no analogous procedure in multidimensional space. For multidimensional minimization, the best we can do is give our algorithm a starting guess, that is, an N-vector of independent variables as the first point to try. The algorithm is then supposed to make its own way downhill through the unimaginable complexity of an N-dimensional topography, until it encounters a (local, at least) minimum.

The downhill simplex method must be started not just with a single point, but with $N + 1$ points, defining an initial simplex. If you think of one of these points (it matters not which) as being your initial starting point $\mathbf{P}_0$, then you can take the other N points to be

$$\mathbf{P}_i = \mathbf{P}_0 + \lambda \mathbf{e}_i \qquad (10.4.1)$$

where the $\mathbf{e}_i$'s are N unit vectors, and where λ is a constant which is your guess of the problem's characteristic length scale. (Or, you could have different λ_i's for each vector direction.)

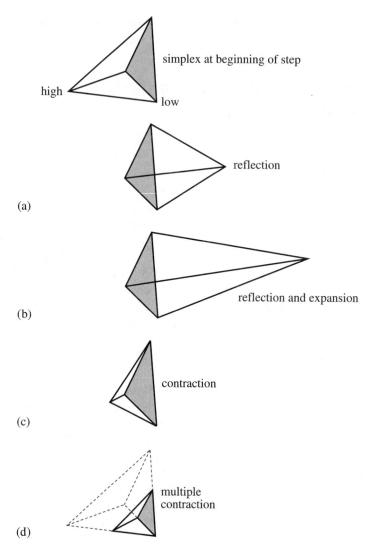

simplex at beginning of step

high

low

reflection

(a)

reflection and expansion

(b)

contraction

(c)

multiple
contraction

(d)

Figure 10.4.1.   Possible outcomes for a step in the downhill simplex method. The simplex at the
beginning of the step, here a tetrahedron, is shown, top. The simplex at the end of the step can be any one
of (a) a reflection away from the high point, (b) a reflection and expansion away from the high point, (c)
a contraction along one dimension from the high point, or (d) a contraction along all dimensions towards
the low point. An appropriate sequence of such steps will always converge to a minimum of the function.

The downhill simplex method now takes a series of steps, most steps just moving
the point of the simplex where the function is largest ("highest point") through the
opposite face of the simplex to a lower point. These steps are called reflections,
and they are constructed to conserve the volume of the simplex (hence maintain
its nondegeneracy). When it can do so, the method expands the simplex in one or
another direction to take larger steps. When it reaches a "valley floor," the method
contracts itself in the transverse direction and tries to ooze down the valley. If there
is a situation where the simplex is trying to "pass through the eye of a needle," it
contracts itself in all directions, pulling itself in around its lowest (best) point. The
routine name amoeba is intended to be descriptive of this kind of behavior; the basic
moves are summarized in Figure 10.4.1.

Termination criteria can be delicate in any multidimensional minimization routine. Without bracketing, and with more than one independent variable, we no longer have the option of requiring a certain tolerance for a single independent variable. We typically can identify one "cycle" or "step" of our multidimensional algorithm. It is then possible to terminate when the vector distance moved in that step is fractionally smaller in magnitude than some tolerance `tol`. Alternatively, we could require that the decrease in the function value in the terminating step be fractionally smaller than some tolerance `ftol`. Note that while `tol` should not usually be smaller than the square root of the machine precision, it is perfectly appropriate to let `ftol` be of order the machine precision (or perhaps slightly larger so as not to be diddled by roundoff).

Note well that either of the above criteria might be fooled by a single anomalous step that, for one reason or another, failed to get anywhere. Therefore, it is frequently a good idea to *restart* a multidimensional minimization routine at a point where it claims to have found a minimum. For this restart, you should reinitialize any ancillary input quantities. In the downhill simplex method, for example, you should reinitialize N of the $N + 1$ vertices of the simplex again by equation (10.4.1), with $\mathbf{P}_0$ being one of the vertices of the claimed minimum.

Restarts should never be very expensive; your algorithm did, after all, converge to the restart point once, and now you are starting the algorithm already there.

Consider, then, our N-dimensional amoeba:

```
#include <cmath>
#include "nr.h"
using namespace std;

namespace {
    inline void get_psum(Mat_I_DP &p, Vec_O_DP &psum)
    {
        int i,j;
        DP sum;

        int mpts=p.nrows();
        int ndim=p.ncols();
        for (j=0;j<ndim;j++) {
            for (sum=0.0,i=0;i<mpts;i++)
                sum += p[i][j];
            psum[j]=sum;
        }
    }
}

void NR::amoeba(Mat_IO_DP &p, Vec_IO_DP &y, const DP ftol, DP funk(Vec_I_DP &),
    int &nfunk)
```
Multidimensional minimization of the function `funk(x)` where `x[0..ndim-1]` is a vector in `ndim` dimensions, by the downhill simplex method of Nelder and Mead. The matrix `p[0..ndim]` `[0..ndim-1]` is input. Its `ndim+1` rows are `ndim`-dimensional vectors that are the vertices of the starting simplex. Also input is the vector `y[0..ndim]`, whose components must be pre-initialized to the values of `funk` evaluated at the `ndim+1` vertices (rows) of p; and `ftol` the fractional convergence tolerance to be achieved in the function value (n.b.!). On output, p and y will have been reset to `ndim+1` new points all within `ftol` of a minimum function value, and `nfunk` gives the number of function evaluations taken.
```
{
    const int NMAX=5000;            Maximum allowed number of function evalua-
    const DP TINY=1.0e-10;          tions.
    int i,ihi,ilo,inhi,j;
    DP rtol,ysave,ytry;
```

```
    int mpts=p.nrows();
    int ndim=p.ncols();
    Vec_DP psum(ndim);
    nfunk=0;
    get_psum(p,psum);
    for (;;) {
        ilo=0;
```
First we must determine which point is the highest (worst), next-highest, and lowest
(best), by looping over the points in the simplex.
```
        ihi = y[0]>y[1] ? (inhi=1,0) : (inhi=0,1);
        for (i=0;i<mpts;i++) {
            if (y[i] <= y[ilo]) ilo=i;
            if (y[i] > y[ihi]) {
                inhi=ihi;
                ihi=i;
            } else if (y[i] > y[inhi] && i != ihi) inhi=i;
        }
        rtol=2.0*fabs(y[ihi]-y[ilo])/(fabs(y[ihi])+fabs(y[ilo])+TINY);
```
Compute the fractional range from highest to lowest and return if satisfactory.
```
        if (rtol < ftol) {                If returning, put best point and value in slot 1.
            SWAP(y[0],y[ilo]);
            for (i=0;i<ndim;i++) SWAP(p[0][i],p[ilo][i]);
            break;
        }
        if (nfunk >= NMAX) nrerror("NMAX exceeded");
        nfunk += 2;
```
Begin a new iteration. First extrapolate by a factor -1 through the face of the simplex
across from the high point, i.e., reflect the simplex from the high point.
```
        ytry=amotry(p,y,psum,funk,ihi,-1.0);
        if (ytry <= y[ilo])
```
Gives a result better than the best point, so try an additional extrapolation by a
factor 2.
```
            ytry=amotry(p,y,psum,funk,ihi,2.0);
        else if (ytry >= y[inhi]) {
```
The reflected point is worse than the second-highest, so look for an intermediate
lower point, i.e., do a one-dimensional contraction.
```
            ysave=y[ihi];
            ytry=amotry(p,y,psum,funk,ihi,0.5);
            if (ytry >= ysave) {          Can't seem to get rid of that high point. Better
                for (i=0;i<mpts;i++) {        contract around the lowest (best) point.
                    if (i != ilo) {
                        for (j=0;j<ndim;j++)
                            p[i][j]=psum[j]=0.5*(p[i][j]+p[ilo][j]);
                        y[i]=funk(psum);
                    }
                }
                nfunk += ndim;            Keep track of function evaluations.
                get_psum(p,psum);         Recompute psum.
            }
        } else --nfunk;                   Correct the evaluation count.
    }                                     Go back for the test of doneness and the next
}                                             iteration.
```

```
#include "nr.h"

DP NR::amotry(Mat_IO_DP &p, Vec_O_DP &y, Vec_IO_DP &psum, DP funk(Vec_I_DP &),
    const int ihi, const DP fac)
```
Extrapolates by a factor fac through the face of the simplex across from the high point, tries
it, and replaces the high point if the new point is better.
```
{
    int j;
```

```
DP fac1,fac2,ytry;

int ndim=p.ncols();
Vec_DP ptry(ndim);
fac1=(1.0-fac)/ndim;
fac2=fac1-fac;
for (j=0;j<ndim;j++)
    ptry[j]=psum[j]*fac1-p[ihi][j]*fac2;
ytry=funk(ptry);              Evaluate the function at the trial point.
if (ytry < y[ihi]) {          If it's better than the highest, then replace the highest.
    y[ihi]=ytry;
    for (j=0;j<ndim;j++) {
        psum[j] += ptry[j]-p[ihi][j];
        p[ihi][j]=ptry[j];
    }
}
return ytry;
}
```

CITED REFERENCES AND FURTHER READING:

Nelder, J.A., and Mead, R. 1965, *Computer Journal*, vol. 7, pp. 308–313. [1]

Yarbro, L.A., and Deming, S.N. 1974, *Analytica Chimica Acta*, vol. 73, pp. 391–398.

Jacoby, S.L.S, Kowalik, J.S., and Pizzo, J.T. 1972, *Iterative Methods for Nonlinear Optimization Problems* (Englewood Cliffs, NJ: Prentice-Hall).

10.5 Direction Set (Powell's) Methods in Multidimensions

We know (§10.1–§10.3) how to minimize a function of one variable. If we start at a point $\mathbf{P}$ in N-dimensional space, and proceed from there in some vector direction $\mathbf{n}$, then any function of N variables $f(\mathbf{P})$ can be minimized along the line $\mathbf{n}$ by our one-dimensional methods. One can dream up various multidimensional minimization methods that consist of sequences of such line minimizations. Different methods will differ only by how, at each stage, they choose the next direction $\mathbf{n}$ to try. All such methods presume the existence of a "black-box" sub-algorithm, which we might call linmin (given as an explicit routine at the end of this section), whose definition can be taken for now as

> linmin: Given as input the vectors $\mathbf{P}$ and $\mathbf{n}$, and the function f, find the scalar λ that minimizes $f(\mathbf{P} + \lambda\mathbf{n})$. Replace $\mathbf{P}$ by $\mathbf{P} + \lambda\mathbf{n}$. Replace $\mathbf{n}$ by $\lambda\mathbf{n}$. Done.

All the minimization methods in this section and in the two sections following fall under this general schema of successive line minimizations. (The algorithm in §10.7 does not need very accurate line minimizations. Accordingly, it has its own approximate line minimization routine, lnsrch.) In this section we consider a class of methods whose choice of successive directions does not involve explicit computation of the function's gradient; the next two sections do require such gradient calculations. You will note that we need not specify whether `linmin` uses gradient information or not. That choice is up to you, and its optimization depends on your particular function. You would be crazy, however, to use gradients in `linmin` and *not* use them in the choice of directions, since in this latter role they can drastically reduce the total computational burden.

But what if, in your application, calculation of the gradient is out of the question. You might first think of this simple method: Take the unit vectors $\mathbf{e}_0, \mathbf{e}_1, \ldots \mathbf{e}_{N-1}$ as a *set of directions*. Using `linmin`, move along the first direction to its minimum, then *from there* along the second direction to *its* minimum, and so on, cycling through the whole set of directions as many times as necessary, until the function stops decreasing.

This simple method is actually not too bad for many functions. Even more interesting is why it *is* bad, i.e. very inefficient, for some other functions. Consider a function of two dimensions whose contour map (level lines) happens to define a long, narrow valley at some angle to the coordinate basis vectors (see Figure 10.5.1). Then the only way "down the length of the valley" going along the basis vectors at each stage is by a series of many tiny steps. More generally, in N dimensions, if the function's second derivatives are much larger in magnitude in some directions than in others, then many cycles through all N basis vectors will be required in order to get anywhere. This condition is not all that unusual; according to Murphy's Law, you should count on it.

Obviously what we need is a better set of directions than the $\mathbf{e}_i$'s. All *direction set methods* consist of prescriptions for updating the set of directions as the method proceeds, attempting to come up with a set which either (i) includes some very good directions that will take us far along narrow valleys, or else (more subtly) (ii) includes some number of "non-interfering" directions with the special property that minimization along one is not "spoiled" by subsequent minimization along another, so that interminable cycling through the set of directions can be avoided.

Conjugate Directions

This concept of "non-interfering" directions, more conventionally called *conjugate directions*, is worth making mathematically explicit.

First, note that if we minimize a function along some direction $\mathbf{u}$, then the gradient of the function must be perpendicular to $\mathbf{u}$ at the line minimum; if not, then there would still be a nonzero directional derivative along $\mathbf{u}$.

Next take some particular point $\mathbf{P}$ as the origin of the coordinate system with coordinates $\mathbf{x}$. Then any function f can be approximated by its Taylor series

$$f(\mathbf{x}) = f(\mathbf{P}) + \sum_i \frac{\partial f}{\partial x_i} x_i + \frac{1}{2} \sum_{i,j} \frac{\partial^2 f}{\partial x_i \partial x_j} x_i x_j + \cdots$$

$$\approx c - \mathbf{b} \cdot \mathbf{x} + \frac{1}{2} \mathbf{x} \cdot \mathbf{A} \cdot \mathbf{x}$$

(10.5.1)

where

$$c \equiv f(\mathbf{P}) \qquad \mathbf{b} \equiv -\nabla f|_{\mathbf{P}} \qquad [\mathbf{A}]_{ij} \equiv \left. \frac{\partial^2 f}{\partial x_i \partial x_j} \right|_{\mathbf{P}}$$

(10.5.2)

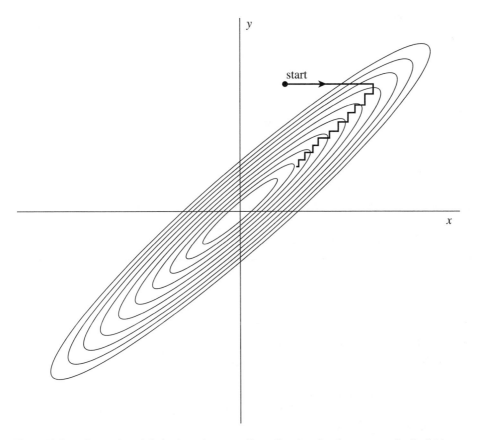

Figure 10.5.1. Successive minimizations along coordinate directions in a long, narrow "valley" (shown as contour lines). Unless the valley is optimally oriented, this method is extremely inefficient, taking many tiny steps to get to the minimum, crossing and re-crossing the principal axis.

The matrix **A** whose components are the second partial derivative matrix of the function is called the *Hessian matrix* of the function at **P**.

In the approximation of (10.5.1), the gradient of f is easily calculated as

$$\nabla f = \mathbf{A} \cdot \mathbf{x} - \mathbf{b} \qquad (10.5.3)$$

(This implies that the gradient will vanish — the function will be at an extremum — at a value of **x** obtained by solving $\mathbf{A} \cdot \mathbf{x} = \mathbf{b}$. This idea we will return to in §10.7!)

How does the gradient ∇f *change* as we move along some direction? Evidently

$$\delta(\nabla f) = \mathbf{A} \cdot (\delta \mathbf{x}) \qquad (10.5.4)$$

Suppose that we have moved along some direction **u** to a minimum and now propose to move along some new direction **v**. The condition that motion along **v** not *spoil* our minimization along **u** is just that the gradient stay perpendicular to **u**, i.e., that the change in the gradient be perpendicular to **u**. By equation (10.5.4) this is just

$$0 = \mathbf{u} \cdot \delta(\nabla f) = \mathbf{u} \cdot \mathbf{A} \cdot \mathbf{v} \qquad (10.5.5)$$

When (10.5.5) holds for two vectors **u** and **v**, they are said to be *conjugate*. When the relation holds pairwise for all members of a set of vectors, they are said to be a conjugate set. If you do successive line minimization of a function along a conjugate set of directions, then you don't need to redo any of those directions (unless, of course, you spoil things by minimizing along a direction that they are *not* conjugate to).

A triumph for a direction set method is to come up with a set of N linearly independent, mutually conjugate directions. Then, one pass of N line minimizations will put it exactly at the minimum of a quadratic form like (10.5.1). For functions f that are not exactly quadratic forms, it won't be exactly at the minimum; but repeated cycles of N line minimizations will in due course converge *quadratically* to the minimum.

Powell's Quadratically Convergent Method

Powell first discovered a direction set method that does produce N mutually conjugate directions. Here is how it goes: Initialize the set of directions $\mathbf{u}_i$ to the basis vectors,

$$\mathbf{u}_i = \mathbf{e}_i \qquad i = 0, \ldots, N - 1 \tag{10.5.6}$$

Now repeat the following sequence of steps ("basic procedure") until your function stops decreasing:

- Save your starting position as $\mathbf{P}_0$.
- For $i = 0, \ldots, N - 1$, move $\mathbf{P}_i$ to the minimum along direction $\mathbf{u}_i$ and call this point $\mathbf{P}_{i+1}$.
- For $i = 0, \ldots, N - 2$, set $\mathbf{u}_i \leftarrow \mathbf{u}_{i+1}$.
- Set $\mathbf{u}_{N-1} \leftarrow \mathbf{P}_N - \mathbf{P}_0$.
- Move $\mathbf{P}_N$ to the minimum along direction $\mathbf{u}_{N-1}$ and call this point $\mathbf{P}_0$.

Powell, in 1964, showed that, for a quadratic form like (10.5.1), k iterations of the above basic procedure produce a set of directions $\mathbf{u}_i$ whose last k members are mutually conjugate. Therefore, N iterations of the basic procedure, amounting to $N(N + 1)$ line minimizations in all, will exactly minimize a quadratic form. Brent [1] gives proofs of these statements in accessible form.

Unfortunately, there is a problem with Powell's quadratically convergent algorithm. The procedure of throwing away, at each stage, $\mathbf{u}_0$ in favor of $\mathbf{P}_N - \mathbf{P}_0$ tends to produce sets of directions that "fold up on each other" and become linearly dependent. Once this happens, then the procedure finds the minimum of the function f only over a subspace of the full N-dimensional case; in other words, it gives the wrong answer. Therefore, the algorithm must not be used in the form given above.

There are a number of ways to fix up the problem of linear dependence in Powell's algorithm, among them:

1. You can reinitialize the set of directions $\mathbf{u}_i$ to the basis vectors $\mathbf{e}_i$ after every N or $N + 1$ iterations of the basic procedure. This produces a serviceable method, which we commend to you if quadratic convergence is important for your application (i.e., if your functions are close to quadratic forms and if you desire high accuracy).

2. Brent points out that the set of directions can equally well be reset to the columns of any orthogonal matrix. Rather than throw away the information

on conjugate directions already built up, he resets the direction set to calculated principal directions of the matrix $\mathbf{A}$ (which he gives a procedure for determining). The calculation is essentially a singular value decomposition algorithm (see §2.6). Brent has a number of other cute tricks up his sleeve, and his modification of Powell's method is probably the best presently known. Consult [1] for a detailed description and listing of the program. Unfortunately it is rather too elaborate for us to include here.

3. You can give up the property of quadratic convergence in favor of a more heuristic scheme (due to Powell) which tries to find a few good directions along narrow valleys instead of N necessarily conjugate directions. This is the method that we now implement. (It is also the version of Powell's method given in Acton [2], from which parts of the following discussion are drawn.)

Discarding the Direction of Largest Decrease

The fox and the grapes: Now that we are going to give up the property of quadratic convergence, was it so important after all? That depends on the function that you are minimizing. Some applications produce functions with long, twisty valleys. Quadratic convergence is of no particular advantage to a program which must slalom down the length of a valley floor that twists one way and another (and another, and another, ... – there are N dimensions!). Along the long direction, a quadratically convergent method is trying to extrapolate to the minimum of a parabola which just isn't (yet) there; while the conjugacy of the $N - 1$ transverse directions keeps getting spoiled by the twists.

Sooner or later, however, we do arrive at an approximately ellipsoidal minimum (cf. equation 10.5.1 when $\mathbf{b}$, the gradient, is zero). Then, depending on how much accuracy we require, a method with quadratic convergence can save us several times N^2 extra line minimizations, since quadratic convergence *doubles* the number of significant figures at each iteration.

The basic idea of our now-modified Powell's method is still to take $\mathbf{P}_N - \mathbf{P}_0$ as a new direction; it is, after all, the average direction moved after trying all N possible directions. For a valley whose long direction is twisting slowly, this direction is likely to give us a good run along the new long direction. The change is to discard the old direction along which the function f made its *largest decrease*. This seems paradoxical, since that direction was the *best* of the previous iteration. However, it is also likely to be a major component of the new direction that we are adding, so dropping it gives us the best chance of avoiding a buildup of linear dependence.

There are a couple of exceptions to this basic idea. Sometimes it is better *not* to add a new direction at all. Define

$$f_0 \equiv f(\mathbf{P}_0) \qquad f_N \equiv f(\mathbf{P}_N) \qquad f_E \equiv f(2\mathbf{P}_N - \mathbf{P}_0) \qquad (10.5.7)$$

Here f_E is the function value at an "extrapolated" point somewhat further along the proposed new direction. Also define Δf to be the magnitude of the largest decrease along one particular direction of the present basic procedure iteration. (Δf is a positive number.) Then:

1. If $f_E \geq f_0$, then keep the old set of directions for the next basic procedure, because the average direction $\mathbf{P}_N - \mathbf{P}_0$ is all played out.

2. If $2\,(f_0 - 2f_N + f_E)\,[(f_0 - f_N) - \Delta f]^2 \geq (f_0 - f_E)^2 \Delta f$, then keep the old set of directions for the next basic procedure, because either (i) the decrease along the average direction was not primarily due to any single direction's decrease, or (ii) there is a substantial second derivative along the average direction and we seem to be near to the bottom of its minimum.

The following routine implements Powell's method in the version just described. In the routine, xi is the matrix whose columns are the set of directions $\mathbf{n}_i$; otherwise the correspondence of notation should be self-evident.

```
#include <cmath>
#include "nr.h"
using namespace std;

void NR::powell(Vec_IO_DP &p, Mat_IO_DP &xi, const DP ftol, int &iter,
    DP &fret, DP func(Vec_I_DP &))
```
Minimization of a function func of n variables. Input consists of an initial starting point p[0..n-1]; an initial matrix xi[0..n-1][0..n-1], whose columns contain the initial set of directions (usually the n unit vectors); and ftol, the fractional tolerance in the function value such that failure to decrease by more than this amount on one iteration signals doneness. On output, p is set to the best point found, xi is the then-current direction set, fret is the returned function value at p, and iter is the number of iterations taken. The routine linmin is used.
```
{
    const int ITMAX=200;              Maximum allowed iterations.
    const DP TINY=1.0e-25;            A small number.
    int i,j,ibig;
    DP del,fp,fptt,t;

    int n=p.size();
    Vec_DP pt(n),ptt(n),xit(n);
    fret=func(p);
    for (j=0;j<n;j++) pt[j]=p[j];     Save the initial point.
    for (iter=0;;++iter) {
        fp=fret;
        ibig=0;
        del=0.0;                      Will be the biggest function decrease.
        for (i=0;i<n;i++) {           In each iteration, loop over all directions in the set.
            for (j=0;j<n;j++) xit[j]=xi[j][i];   Copy the direction,
            fptt=fret;
            linmin(p,xit,fret,func);          minimize along it,
            if (fptt-fret > del) {            and record it if it is the largest decrease
                del=fptt-fret;               so far.
                ibig=i+1;
            }
        }                                     Here comes the termination criterion:
        if (2.0*(fp-fret) <= ftol*(fabs(fp)+fabs(fret))+TINY) {
            return;
        }
        if (iter == ITMAX) nrerror("powell exceeding maximum iterations.");
        for (j=0;j<n;j++) {                Construct the extrapolated point and the
            ptt[j]=2.0*p[j]-pt[j];        average direction moved. Save the
            xit[j]=p[j]-pt[j];            old starting point.
            pt[j]=p[j];
        }
        fptt=func(ptt);                   Function value at extrapolated point.
        if (fptt < fp) {
            t=2.0*(fp-2.0*fret+fptt)*SQR(fp-fret-del)-del*SQR(fp-fptt);
            if (t < 0.0) {
                linmin(p,xit,fret,func);  Move to the minimum of the new direc-
                for (j=0;j<n;j++) {       tion, and save the new direction.
                    xi[j][ibig-1]=xi[j][n-1];
                    xi[j][n-1]=xit[j];
```

```
                    }
                 }
              }
           }
     }
```

Implementation of Line Minimization

Make no mistake, there is a *right* way to implement `linmin`: It is to use the *methods* of one-dimensional minimization described in §10.1–§10.3, but to rewrite the programs of those sections so that their bookkeeping is done on vector-valued points **P** (all lying along a given direction **n**) rather than scalar-valued abscissas x. That straightforward task produces long routines densely populated with "`for(k=0;k<n;k++)`" loops.

We do not have space to include such routines in this book. Our `linmin`, which works just fine, is instead a kind of bookkeeping swindle. It constructs an "artificial" function of one variable called `f1dim`, which is the value of your function, say, `func`, along the line going through the point p in the direction xi. `linmin` calls our familiar one-dimensional routines `mnbrak` (§10.1) and `brent` (§10.3) and instructs them to minimize `f1dim`. `linmin` communicates with `f1dim` "over the head" of `mnbrak` and `brent`, through global (external) variables. That is also how it passes to `f1dim` a pointer to your user-supplied function.

The only thing inefficient about `linmin` is this: Its use as an interface between a multidimensional minimization strategy and a one-dimensional minimization routine results in some unnecessary copying of vectors hither and yon. That should not normally be a significant addition to the overall computational burden, but we cannot disguise its inelegance.

```
#include "nr.h"

int ncom;                                Global variables communicate with f1dim.
DP (*nrfunc)(Vec_I_DP &);
Vec_DP *pcom_p,*xicom_p;
```

```
void NR::linmin(Vec_IO_DP &p, Vec_IO_DP &xi, DP &fret, DP func(Vec_I_DP &))
```
Given an n-dimensional point p[0..n-1] and an n-dimensional direction xi[0..n-1], moves and resets p to where the function func(p) takes on a minimum along the direction xi from p, and replaces xi by the actual vector displacement that p was moved. Also returns as fret the value of func at the returned location p. This is actually all accomplished by calling the routines mnbrak and brent.
```
{
    int j;
    const DP TOL=1.0e-8;                 Tolerance passed to brent.
    DP xx,xmin,fx,fb,fa,bx,ax;

    int n=p.size();
    ncom=n;                              Define the global variables.
    pcom_p=new Vec_DP(n);
    xicom_p=new Vec_DP(n);
    nrfunc=func;
    Vec_DP &pcom=*pcom_p,&xicom=*xicom_p;   Make aliases to simplify coding.
    for (j=0;j<n;j++) {
        pcom[j]=p[j];
        xicom[j]=xi[j];
    }
```

```
    ax=0.0;                                         Initial guess for brackets.
    xx=1.0;
    mnbrak(ax,xx,bx,fa,fx,fb,f1dim);
    fret=brent(ax,xx,bx,f1dim,TOL,xmin);
    for (j=0;j<n;j++) {                             Construct the vector results to return.
        xi[j] *= xmin;
        p[j] += xi[j];
    }
    delete xicom_p;
    delete pcom_p;
}


#include "nr.h"

extern int ncom;                                    Defined in linmin.
extern DP (*nrfunc)(Vec_I_DP &);
extern Vec_DP *pcom_p,*xicom_p;

DP NR::f1dim(const DP x)
Must accompany linmin.
{
    int j;

    Vec_DP xt(ncom);
    Vec_DP &pcom=*pcom_p,&xicom=*xicom_p;           Make aliases to simplify coding.
    for (j=0;j<ncom;j++)
        xt[j]=pcom[j]+x*xicom[j];
    return nrfunc(xt);
}
```

CITED REFERENCES AND FURTHER READING:

Brent, R.P. 1973, *Algorithms for Minimization without Derivatives* (Englewood Cliffs, NJ: Prentice-Hall), Chapter 7. [1]

Acton, F.S. 1970, *Numerical Methods That Work*; 1990, corrected edition (Washington: Mathematical Association of America), pp. 464–467. [2]

Jacobs, D.A.H. (ed.) 1977, *The State of the Art in Numerical Analysis* (London: Academic Press), pp. 259–262.

10.6 Conjugate Gradient Methods in Multidimensions

We consider now the case where you are able to calculate, at a given N-dimensional point $\mathbf{P}$, not just the value of a function $f(\mathbf{P})$ but also the gradient (vector of first partial derivatives) $\nabla f(\mathbf{P})$.

A rough counting argument will show how advantageous it is to use the gradient information: Suppose that the function f is roughly approximated as a quadratic form, as above in equation (10.5.1),

$$f(\mathbf{x}) \approx c - \mathbf{b} \cdot \mathbf{x} + \frac{1}{2}\, \mathbf{x} \cdot \mathbf{A} \cdot \mathbf{x} \qquad (10.6.1)$$

Then the number of unknown parameters in f is equal to the number of free parameters in $\mathbf{A}$ and $\mathbf{b}$, which is $\frac{1}{2}N(N+1)$, which we see to be of order N^2. Changing any one of these parameters can move the location of the minimum. Therefore, we should not expect to be able to *find* the minimum until we have collected an equivalent information content, of order N^2 numbers.

In the direction set methods of §10.5, we collected the necessary information by making on the order of N^2 separate line minimizations, each requiring "a few" (but sometimes a *big* few!) function evaluations. Now, each evaluation of the gradient will bring us N new components of information. If we use them wisely, we should need to make only of order N separate line minimizations. That is in fact the case for the algorithms in this section and the next.

A factor of N improvement in computational speed is not necessarily implied. As a rough estimate, we might imagine that the calculation of *each component* of the gradient takes about as long as evaluating the function itself. In that case there will be of order N^2 equivalent function evaluations both with and without gradient information. Even if the advantage is not of order N, however, it is nevertheless quite substantial: (i) Each calculated component of the gradient will typically save not just one function evaluation, but a number of them, equivalent to, say, a whole line minimization. (ii) There is often a high degree of redundancy in the formulas for the various components of a function's gradient; when this is so, especially when there is also redundancy with the calculation of the function, then the calculation of the gradient may cost significantly less than N function evaluations.

A common beginner's error is to assume that any reasonable way of incorporating gradient information should be about as good as any other. This line of thought leads to the following *not very good* algorithm, the *steepest descent method*:

> Steepest Descent: Start at a point $\mathbf{P}_0$. As many times as needed, move from point $\mathbf{P}_i$ to the point $\mathbf{P}_{i+1}$ by minimizing along the line from $\mathbf{P}_i$ in the direction of the local downhill gradient $-\nabla f(\mathbf{P}_i)$.

The problem with the steepest descent method (which, incidentally, goes back to Cauchy), is similar to the problem that was shown in Figure 10.5.1. The method will perform many small steps in going down a long, narrow valley, even if the valley is a perfect quadratic form. You might have hoped that, say in two dimensions, your first step would take you to the valley floor, the second step directly down the long axis; but remember that the new gradient at the minimum point of any line minimization is perpendicular to the direction just traversed. Therefore, with the steepest descent method, you *must* make a right angle turn, which does *not*, in general, take you to the minimum. (See Figure 10.6.1.)

Just as in the discussion that led up to equation (10.5.5), we really want a way of proceeding not down the new gradient, but rather in a direction that is somehow constructed to be *conjugate* to the old gradient, and, insofar as possible, to all previous directions traversed. Methods that accomplish this construction are called *conjugate gradient* methods.

In §2.7 we discussed the conjugate gradient method as a technique for solving linear algebraic equations by minimizing a quadratic form. That formalism can also be applied to the problem of minimizing a function *approximated* by the quadratic

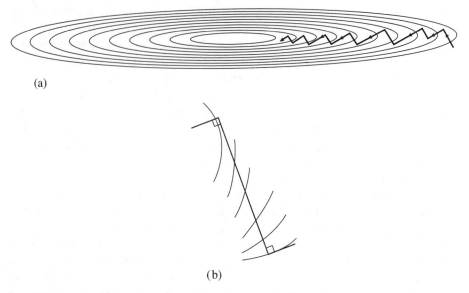

(a)

(b)

Figure 10.6.1.   (a) Steepest descent method in a long, narrow "valley." While more efficient than the strategy of Figure 10.5.1, steepest descent is nonetheless an inefficient strategy, taking many steps to reach the valley floor. (b) Magnified view of one step: A step starts off in the local gradient direction, perpendicular to the contour lines, and traverses a straight line until a local minimum is reached, where the traverse is parallel to the local contour lines.

form (10.6.1). Recall that, starting with an arbitrary initial vector $\mathbf{g}_0$ and letting $\mathbf{h}_0 = \mathbf{g}_0$, the conjugate gradient method constructs two sequences of vectors from the recurrence

$$\mathbf{g}_{i+1} = \mathbf{g}_i - \lambda_i \mathbf{A} \cdot \mathbf{h}_i \qquad \mathbf{h}_{i+1} = \mathbf{g}_{i+1} + \gamma_i \mathbf{h}_i \qquad i = 0, 1, 2, \ldots \qquad (10.6.2)$$

The vectors satisfy the orthogonality and conjugacy conditions

$$\mathbf{g}_i \cdot \mathbf{g}_j = 0 \qquad \mathbf{h}_i \cdot \mathbf{A} \cdot \mathbf{h}_j = 0 \qquad \mathbf{g}_i \cdot \mathbf{h}_j = 0 \qquad j < i \qquad (10.6.3)$$

The scalars λ_i and γ_i are given by

$$\lambda_i = \frac{\mathbf{g}_i \cdot \mathbf{g}_i}{\mathbf{h}_i \cdot \mathbf{A} \cdot \mathbf{h}_i} = \frac{\mathbf{g}_i \cdot \mathbf{h}_i}{\mathbf{h}_i \cdot \mathbf{A} \cdot \mathbf{h}_i} \qquad (10.6.4)$$

$$\gamma_i = \frac{\mathbf{g}_{i+1} \cdot \mathbf{g}_{i+1}}{\mathbf{g}_i \cdot \mathbf{g}_i} \qquad (10.6.5)$$

Equations (10.6.2)–(10.6.5) are simply equations (2.7.32)–(2.7.35) for a symmetric $\mathbf{A}$ in a new notation. (A self-contained derivation of these results in the context of function minimization is given by Polak [1].)

Now suppose that we knew the Hessian matrix $\mathbf{A}$ in equation (10.6.1). Then we could use the construction (10.6.2) to find successively conjugate directions $\mathbf{h}_i$ along which to line-minimize. After N such, we would efficiently have arrived at the minimum of the quadratic form. But we don't know $\mathbf{A}$.

Here is a remarkable theorem to save the day: Suppose we happen to have $\mathbf{g}_i = -\nabla f(\mathbf{P}_i)$, for some point $\mathbf{P}_i$, where f is of the form (10.6.1). Suppose also that we proceed from $\mathbf{P}_i$ along the direction $\mathbf{h}_i$ to the local minimum of f located at some point $\mathbf{P}_{i+1}$ and then set $\mathbf{g}_{i+1} = -\nabla f(\mathbf{P}_{i+1})$. Then, this $\mathbf{g}_{i+1}$ is the same vector as would have been constructed by equation (10.6.2). (And we have constructed it without knowledge of $\mathbf{A}$!)

Proof: By equation (10.5.3), $\mathbf{g}_i = -\mathbf{A} \cdot \mathbf{P}_i + \mathbf{b}$, and

$$\mathbf{g}_{i+1} = -\mathbf{A} \cdot (\mathbf{P}_i + \lambda \mathbf{h}_i) + \mathbf{b} = \mathbf{g}_i - \lambda \mathbf{A} \cdot \mathbf{h}_i \qquad (10.6.6)$$

with λ chosen to take us to the line minimum. But at the line minimum $\mathbf{h}_i \cdot \nabla f = -\mathbf{h}_i \cdot \mathbf{g}_{i+1} = 0$. This latter condition is easily combined with (10.6.6) to solve for λ. The result is exactly the expression (10.6.4). But with this value of λ, (10.6.6) is the same as (10.6.2), q.e.d.

We have, then, the basis of an algorithm that requires neither knowledge of the Hessian matrix $\mathbf{A}$, nor even the storage necessary to store such a matrix. A sequence of directions $\mathbf{h}_i$ is constructed, using only line minimizations, evaluations of the gradient vector, and an auxiliary vector to store the latest in the sequence of $\mathbf{g}$'s.

The algorithm described so far is the original Fletcher-Reeves version of the conjugate gradient algorithm. Later, Polak and Ribiere introduced one tiny, but sometimes significant, change. They proposed using the form

$$\gamma_i = \frac{(\mathbf{g}_{i+1} - \mathbf{g}_i) \cdot \mathbf{g}_{i+1}}{\mathbf{g}_i \cdot \mathbf{g}_i} \qquad (10.6.7)$$

instead of equation (10.6.5). "Wait," you say, "aren't they equal by the orthogonality conditions (10.6.3)?" They are equal for exact quadratic forms. In the real world, however, your function is not exactly a quadratic form. Arriving at the supposed minimum of the quadratic form, you may still need to proceed for another set of iterations. There is some evidence [2] that the Polak-Ribiere formula accomplishes the transition to further iterations more gracefully: When it runs out of steam, it tends to reset $\mathbf{h}$ to be down the local gradient, which is equivalent to beginning the conjugate-gradient procedure anew.

The following routine implements the Polak-Ribiere variant, which we recommend; but changing one program line, as shown, will give you Fletcher-Reeves. The routine presumes the existence of a function func(p), where p[0..n-1] is a vector of length n, and also presumes the existence of a function dfunc(p,df) that sets the vector gradient df[0..n-1] evaluated at the input point p.

The routine calls linmin to do the line minimizations. As already discussed, you may wish to use a modified version of linmin that uses dbrent instead of brent, i.e., that uses the gradient in doing the line minimizations. See note below.

```
#include <cmath>
#include "nr.h"
using namespace std;

void NR::frprmn(Vec_IO_DP &p, const DP ftol, int &iter, DP &fret,
    DP func(Vec_I_DP &), void dfunc(Vec_I_DP &, Vec_O_DP &))
```
Given a starting point p[0..n-1], Fletcher-Reeves-Polak-Ribiere minimization is performed on a function func, using its gradient as calculated by a routine dfunc. The convergence tolerance on the function value is input as ftol. Returned quantities are p (the location of the minimum), iter (the number of iterations that were performed), and fret (the minimum value of the function). The routine linmin is called to perform line minimizations.
```
{
```

```
const int ITMAX=200;
const DP EPS=1.0e-18;
```
Here ITMAX is the maximum allowed number of iterations, while EPS is a small number to
rectify the special case of converging to exactly zero function value.
```
int j,its;
DP gg,gam,fp,dgg;

int n=p.size();
Vec_DP g(n),h(n),xi(n);
fp=func(p);                                  Initializations.
dfunc(p,xi);
for (j=0;j<n;j++) {
    g[j] = -xi[j];
    xi[j]=h[j]=g[j];
}
for (its=0;its<ITMAX;its++) {                Loop over iterations.
    iter=its;
    linmin(p,xi,fret,func);                  Next statement is the normal return:
    if (2.0*fabs(fret-fp) <= ftol*(fabs(fret)+fabs(fp)+EPS))
        return;
    fp=fret;
    dfunc(p,xi);
    dgg=gg=0.0;
    for (j=0;j<n;j++) {
        gg += g[j]*g[j];
//      dgg += xi[j]*xi[j];                  This statement for Fletcher-Reeves.
        dgg += (xi[j]+g[j])*xi[j];           This statement for Polak-Ribiere.
    }
    if (gg == 0.0)                           Unlikely. If gradient is exactly zero then
        return;                                  we are already done.
    gam=dgg/gg;
    for (j=0;j<n;j++) {
        g[j] = -xi[j];
        xi[j]=h[j]=g[j]+gam*h[j];
    }
}
nrerror("Too many iterations in frprmn");
}
```

Note on Line Minimization Using Derivatives

Kindly reread the last part of §10.5. We here want to do the same thing, but
using derivative information in performing the line minimization.

The modified version of linmin, called dlinmin, and its required companion
routine df1dim follow:

```
#include "nr.h"

int ncom;                                    Global variables communicate with df1dim.
DP (*nrfunc)(Vec_I_DP &);
void (*nrdfun)(Vec_I_DP &, Vec_O_DP &);
Vec_DP *pcom_p,*xicom_p;

void NR::dlinmin(Vec_IO_DP &p, Vec_IO_DP &xi, DP &fret, DP func(Vec_I_DP &),
    void dfunc(Vec_I_DP &, Vec_O_DP &))
```
Given an n-dimensional point p[0..n-1] and an n-dimensional direction xi[0..n-1], moves
and resets p to where the function func(p) takes on a minimum along the direction xi from
p, and replaces xi by the actual vector displacement that p was moved. Also returns as fret
the value of func at the returned location p. This is actually all accomplished by calling the
routines mnbrak and dbrent.

```
{
    const DP TOL=2.0e-8;                          Tolerance passed to dbrent.
    int j;
    DP xx,xmin,fx,fb,fa,bx,ax;

    int n=p.size();
    ncom=n;                                       Define the global variables.
    pcom_p=new Vec_DP(n);
    xicom_p=new Vec_DP(n);
    nrfunc=func;
    nrdfun=dfunc;
    Vec_DP &pcom=*pcom_p,&xicom=*xicom_p;         Make aliases to simplify coding.
    for (j=0;j<n;j++) {
        pcom[j]=p[j];
        xicom[j]=xi[j];
    }
    ax=0.0;                                       Initial guess for brackets.
    xx=1.0;
    mnbrak(ax,xx,bx,fa,fx,fb,f1dim);
    fret=dbrent(ax,xx,bx,f1dim,df1dim,TOL,xmin);
    for (j=0;j<n;j++) {                           Construct the vector results to return.
        xi[j] *= xmin;
        p[j] += xi[j];
    }
    delete xicom_p;
    delete pcom_p;
}



#include "nr.h"

extern int ncom;                                  Defined in dlinmin.
extern DP (*nrfunc)(Vec_I_DP &);
extern void (*nrdfun)(Vec_I_DP &, Vec_O_DP &);
extern Vec_DP *pcom_p,*xicom_p;

DP NR::df1dim(const DP x)
{
    int j;
    DP df1=0.0;
    Vec_DP xt(ncom),df(ncom);

    Vec_DP &pcom=*pcom_p,&xicom=*xicom_p;         Make aliases to simplify coding.
    for (j=0;j<ncom;j++) xt[j]=pcom[j]+x*xicom[j];
    nrdfun(xt,df);
    for (j=0;j<ncom;j++) df1 += df[j]*xicom[j];
    return df1;
}
```

CITED REFERENCES AND FURTHER READING:

Polak, E. 1971, *Computational Methods in Optimization* (New York: Academic Press), §2.3. [1]

Jacobs, D.A.H. (ed.) 1977, *The State of the Art in Numerical Analysis* (London: Academic Press), Chapter III.1.7 (by K.W. Brodlie). [2]

Stoer, J., and Bulirsch, R. 1993, *Introduction to Numerical Analysis*, 2nd ed. (New York: Springer-Verlag), §8.7.

10.7 Variable Metric Methods in Multidimensions

The goal of *variable metric* methods, which are sometimes called *quasi-Newton* methods, is not different from the goal of conjugate gradient methods: to accumulate information from successive line minimizations so that N such line minimizations lead to the exact minimum of a quadratic form in N dimensions. In that case, the method will also be quadratically convergent for more general smooth functions.

Both variable metric and conjugate gradient methods require that you are able to compute your function's gradient, or first partial derivatives, at arbitrary points. The variable metric approach differs from the conjugate gradient in the way that it stores and updates the information that is accumulated. Instead of requiring intermediate storage on the order of N, the number of dimensions, it requires a matrix of size $N \times N$. Generally, for any moderate N, this is an entirely trivial disadvantage.

On the other hand, there is not, as far as we know, any overwhelming advantage that the variable metric methods hold over the conjugate gradient techniques, except perhaps a historical one. Developed somewhat earlier, and more widely propagated, the variable metric methods have by now developed a wider constituency of satisfied users. Likewise, some fancier implementations of variable metric methods (going beyond the scope of this book, see below) have been developed to a greater level of sophistication on issues like the minimization of roundoff error, handling of special conditions, and so on. *We* tend to use variable metric rather than conjugate gradient, but we have no reason to urge this habit on you.

Variable metric methods come in two main flavors. One is the *Davidon-Fletcher-Powell (DFP)* algorithm (sometimes referred to as simply *Fletcher-Powell*). The other goes by the name *Broyden-Fletcher-Goldfarb-Shanno (BFGS)*. The BFGS and DFP schemes differ only in details of their roundoff error, convergence tolerances, and similar "dirty" issues which are outside of our scope [1,2]. However, it has become generally recognized that, empirically, the BFGS scheme is superior in these details. We will implement BFGS in this section.

As before, we imagine that our arbitrary function $f(\mathbf{x})$ can be locally approximated by the quadratic form of equation (10.6.1). We don't, however, have any information about the values of the quadratic form's parameters $\mathbf{A}$ and $\mathbf{b}$, except insofar as we can glean such information from our function evaluations and line minimizations.

The basic idea of the variable metric method is to build up, iteratively, a good approximation to the inverse Hessian matrix $\mathbf{A}^{-1}$, that is, to construct a sequence of matrices $\mathbf{H}_i$ with the property,

$$\lim_{i \to \infty} \mathbf{H}_i = \mathbf{A}^{-1} \tag{10.7.1}$$

Even better if the limit is achieved after N iterations instead of ∞.

The reason that variable metric methods are sometimes called quasi-Newton methods can now be explained. Consider finding a minimum by using Newton's method to search for a zero of the gradient of the function. Near the current point $\mathbf{x}_i$, we have to second order

$$f(\mathbf{x}) = f(\mathbf{x}_i) + (\mathbf{x} - \mathbf{x}_i) \cdot \nabla f(\mathbf{x}_i) + \tfrac{1}{2}(\mathbf{x} - \mathbf{x}_i) \cdot \mathbf{A} \cdot (\mathbf{x} - \mathbf{x}_i) \tag{10.7.2}$$

so

$$\nabla f(\mathbf{x}) = \nabla f(\mathbf{x}_i) + \mathbf{A} \cdot (\mathbf{x} - \mathbf{x}_i) \tag{10.7.3}$$

In Newton's method we set $\nabla f(\mathbf{x}) = 0$ to determine the next iteration point:

$$\mathbf{x} - \mathbf{x}_i = -\mathbf{A}^{-1} \cdot \nabla f(\mathbf{x}_i) \tag{10.7.4}$$

The left-hand side is the finite step we need take to get to the exact minimum; the right-hand side is known once we have accumulated an accurate $\mathbf{H} \approx \mathbf{A}^{-1}$.

The "quasi" in quasi-Newton is because we don't use the actual Hessian matrix of f, but instead use our current approximation of it. This is often *better* than using the true Hessian. We can understand this paradoxical result by considering the *descent directions* of f at $\mathbf{x}_i$. These are the directions $\mathbf{p}$ along which f decreases: $\nabla f \cdot \mathbf{p} < 0$. For the Newton direction (10.7.4) to be a descent direction, we must have

$$\nabla f(\mathbf{x}_i) \cdot (\mathbf{x} - \mathbf{x}_i) = -(\mathbf{x} - \mathbf{x}_i) \cdot \mathbf{A} \cdot (\mathbf{x} - \mathbf{x}_i) < 0 \tag{10.7.5}$$

which is true if $\mathbf{A}$ is positive definite. In general, far from a minimum, we have no guarantee that the Hessian is positive definite. Taking the actual Newton step with the real Hessian can move us to points where the function is *increasing* in value. The idea behind quasi-Newton methods is to start with a positive definite, symmetric approximation to $\mathbf{A}$ (usually the unit matrix) and build up the approximating $\mathbf{H}_i$'s in such a way that the matrix $\mathbf{H}_i$ remains positive definite and symmetric. Far from the minimum, this guarantees that we always move in a downhill direction. Close to the minimum, the updating formula approaches the true Hessian and we enjoy the quadratic convergence of Newton's method.

When we are not close enough to the minimum, taking the full Newton step $\mathbf{p}$ even with a positive definite $\mathbf{A}$ need not decrease the function; we may move too far for the quadratic approximation to be valid. All we are guaranteed is that *initially* f decreases as we move in the Newton direction. Once again we can use the backtracking strategy described in §9.7 to choose a step along the *direction* of the Newton step $\mathbf{p}$, but not necessarily all the way.

We won't rigorously derive the DFP algorithm for taking $\mathbf{H}_i$ into $\mathbf{H}_{i+1}$; you can consult [3] for clear derivations. Following Brodlie (in [2]), we will give the following heuristic motivation of the procedure.

Subtracting equation (10.7.4) at $\mathbf{x}_{i+1}$ from that same equation at $\mathbf{x}_i$ gives

$$\mathbf{x}_{i+1} - \mathbf{x}_i = \mathbf{A}^{-1} \cdot (\nabla f_{i+1} - \nabla f_i) \tag{10.7.6}$$

where $\nabla f_j \equiv \nabla f(\mathbf{x}_j)$. Having made the step from $\mathbf{x}_i$ to $\mathbf{x}_{i+1}$, we might reasonably want to require that the new approximation $\mathbf{H}_{i+1}$ satisfy (10.7.6) as if it were actually $\mathbf{A}^{-1}$, that is,

$$\mathbf{x}_{i+1} - \mathbf{x}_i = \mathbf{H}_{i+1} \cdot (\nabla f_{i+1} - \nabla f_i) \tag{10.7.7}$$

We might also imagine that the updating formula should be of the form $\mathbf{H}_{i+1} = \mathbf{H}_i +$ correction.

What "objects" are around out of which to construct a correction term? Most notable are the two vectors $\mathbf{x}_{i+1} - \mathbf{x}_i$ and $\nabla f_{i+1} - \nabla f_i$; and there is also $\mathbf{H}_i$.

There are not infinitely many natural ways of making a matrix out of these objects, especially if (10.7.7) must hold! One such way, the *DFP updating formula*, is

$$
\mathbf{H}_{i+1} = \mathbf{H}_i + \frac{(\mathbf{x}_{i+1} - \mathbf{x}_i) \otimes (\mathbf{x}_{i+1} - \mathbf{x}_i)}{(\mathbf{x}_{i+1} - \mathbf{x}_i) \cdot (\nabla f_{i+1} - \nabla f_i)}
$$
$$
- \frac{[\mathbf{H}_i \cdot (\nabla f_{i+1} - \nabla f_i)] \otimes [\mathbf{H}_i \cdot (\nabla f_{i+1} - \nabla f_i)]}{(\nabla f_{i+1} - \nabla f_i) \cdot \mathbf{H}_i \cdot (\nabla f_{i+1} - \nabla f_i)}
\tag{10.7.8}
$$

where $\otimes$ denotes the "outer" or "direct" product of two vectors, a matrix: The ij component of $\mathbf{u} \otimes \mathbf{v}$ is $u_i v_j$. (You might want to verify that 10.7.8 does satisfy 10.7.7.)

The *BFGS updating formula* is exactly the same, but with one additional term,

$$
\cdots + [(\nabla f_{i+1} - \nabla f_i) \cdot \mathbf{H}_i \cdot (\nabla f_{i+1} - \nabla f_i)] \, \mathbf{u} \otimes \mathbf{u}
\tag{10.7.9}
$$

where $\mathbf{u}$ is defined as the vector

$$
\mathbf{u} \equiv \frac{(\mathbf{x}_{i+1} - \mathbf{x}_i)}{(\mathbf{x}_{i+1} - \mathbf{x}_i) \cdot (\nabla f_{i+1} - \nabla f_i)}
$$
$$
- \frac{\mathbf{H}_i \cdot (\nabla f_{i+1} - \nabla f_i)}{(\nabla f_{i+1} - \nabla f_i) \cdot \mathbf{H}_i \cdot (\nabla f_{i+1} - \nabla f_i)}
\tag{10.7.10}
$$

(You might also verify that this satisfies 10.7.7.)

You will have to take on faith — or else consult [3] for details of — the "deep" result that equation (10.7.8), with or without (10.7.9), does in fact converge to $\mathbf{A}^{-1}$ in N steps, if f is a quadratic form.

Here now is the routine dfpmin that implements the quasi-Newton method, and uses lnsrch from §9.7. As mentioned at the end of newt in §9.7, this algorithm can fail if your variables are badly scaled.

```
#include <cmath>
#include <limits>
#include "nr.h"
using namespace std;

void NR::dfpmin(Vec_IO_DP &p, const DP gtol, int &iter, DP &fret,
    DP func(Vec_I_DP &), void dfunc(Vec_I_DP &, Vec_O_DP &))
```
Given a starting point p[0..n-1] that is a vector of length n, the Broyden-Fletcher-Goldfarb-Shanno variant of Davidon-Fletcher-Powell minimization is performed on a function func, using its gradient as calculated by a routine dfunc. The convergence requirement on zeroing the gradient is input as gtol. Returned quantities are p[0..n-1] (the location of the minimum), iter (the number of iterations that were performed), and fret (the minimum value of the function). The routine lnsrch is called to perform approximate line minimizations.
```
{
    const int ITMAX=200;
    const DP EPS=numeric_limits<DP>::epsilon();
    const DP TOLX=4*EPS,STPMX=100.0;
```
Here ITMAX is the maximum allowed number of iterations; EPS is the machine precision; TOLX is the convergence criterion on x values; and STPMX is the scaled maximum step length allowed in line searches.
```
    bool check;
    int i,its,j;
    DP den,fac,fad,fae,fp,stpmax,sum=0.0,sumdg,sumxi,temp,test;

    int n=p.size();
    Vec_DP dg(n),g(n),hdg(n),pnew(n),xi(n);
```

```
Mat_DP hessin(n,n);
fp=func(p);                                 Calculate starting function value and gra-
dfunc(p,g);                                     dient,
for (i=0;i<n;i++) {                          and initialize the inverse Hessian to the
    for (j=0;j<n;j++) hessin[i][j]=0.0;          unit matrix.
    hessin[i][i]=1.0;
    xi[i] = -g[i];                          Initial line direction.
    sum += p[i]*p[i];
}
stpmax=STPMX*MAX(sqrt(sum),DP(n));
for (its=0;its<ITMAX;its++) {               Main loop over the iterations.
    iter=its;
    lnsrch(p,fp,g,xi,pnew,fret,stpmax,check,func);
    The new function evaluation occurs in lnsrch; save the function value in fp for the
    next line search. It is usually safe to ignore the value of check.
    fp=fret;
    for (i=0;i<n;i++) {
        xi[i]=pnew[i]-p[i];                 Update the line direction,
        p[i]=pnew[i];                       and the current point.
    }
    test=0.0;                               Test for convergence on Δx.
    for (i=0;i<n;i++) {
        temp=fabs(xi[i])/MAX(fabs(p[i]),1.0);
        if (temp > test) test=temp;
    }
    if (test < TOLX)
        return;
    for (i=0;i<n;i++) dg[i]=g[i];           Save the old gradient,
    dfunc(p,g);                             and get the new gradient.
    test=0.0;                               Test for convergence on zero gradient.
    den=MAX(fret,1.0);
    for (i=0;i<n;i++) {
        temp=fabs(g[i])*MAX(fabs(p[i]),1.0)/den;
        if (temp > test) test=temp;
    }
    if (test < gtol)
        return;
    for (i=0;i<n;i++) dg[i]=g[i]-dg[i];     Compute difference of gradients,
    for (i=0;i<n;i++) {                      and difference times current matrix.
        hdg[i]=0.0;
        for (j=0;j<n;j++) hdg[i] += hessin[i][j]*dg[j];
    }
    fac=fae=sumdg=sumxi=0.0;                Calculate dot products for the denomi-
    for (i=0;i<n;i++) {                          nators.
        fac += dg[i]*xi[i];
        fae += dg[i]*hdg[i];
        sumdg += SQR(dg[i]);
        sumxi += SQR(xi[i]);
    }
    if (fac > sqrt(EPS*sumdg*sumxi)) {       Skip update if fac not sufficiently posi-
        fac=1.0/fac;                             tive.
        fad=1.0/fae;
        The vector that makes BFGS different from DFP:
        for (i=0;i<n;i++) dg[i]=fac*xi[i]-fad*hdg[i];
        for (i=0;i<n;i++) {                  The BFGS updating formula:
            for (j=i;j<n;j++) {
                hessin[i][j] += fac*xi[i]*xi[j]
                    -fad*hdg[i]*hdg[j]+fae*dg[i]*dg[j];
                hessin[j][i]=hessin[i][j];
            }
        }
    }
    for (i=0;i<n;i++) {                      Now calculate the next direction to go,
        xi[i]=0.0;
```

```
            for (j=0;j<n;j++) xi[i] -= hessin[i][j]*g[j];
    }
}                                              and go back for another iteration.
nrerror("too many iterations in dfpmin");
}
```

Quasi-Newton methods like dfpmin work well with the approximate line minimization done by lnsrch. The routines powell (§10.5) and frprmn (§10.6), however, need more accurate line minimization, which is carried out by the routine linmin.

Advanced Implementations of Variable Metric Methods

Although rare, it can conceivably happen that roundoff errors cause the matrix $\mathbf{H}_i$ to become nearly singular or non-positive-definite. This can be serious, because the supposed search directions might then not lead downhill, and because nearly singular $\mathbf{H}_i$'s tend to give subsequent $\mathbf{H}_i$'s that are also nearly singular.

There is a simple fix for this rare problem, the same as was mentioned in §10.4: In case of any doubt, you should *restart* the algorithm at the claimed minimum point, and see if it goes anywhere. Simple, but not very elegant. Modern implementations of variable metric methods deal with the problem in a more sophisticated way.

Instead of building up an approximation to $\mathbf{A}^{-1}$, it is possible to build up an approximation of $\mathbf{A}$ itself. Then, instead of calculating the left-hand side of (10.7.4) directly, one solves the set of linear equations

$$\mathbf{A} \cdot (\mathbf{x} - \mathbf{x}_i) = -\nabla f(\mathbf{x}_i) \tag{10.7.11}$$

At first glance this seems like a bad idea, since solving (10.7.11) is a process of order N^3 — and anyway, how does this help the roundoff problem? The trick is not to store $\mathbf{A}$ but rather a triangular decomposition of $\mathbf{A}$, its *Cholesky decomposition* (cf. §2.9). The updating formula used for the Cholesky decomposition of $\mathbf{A}$ is of order N^2 and can be arranged to guarantee that the matrix remains positive definite and nonsingular, even in the presence of finite roundoff. This method is due to Gill and Murray [1,2].

CITED REFERENCES AND FURTHER READING:

Dennis, J.E., and Schnabel, R.B. 1983, *Numerical Methods for Unconstrained Optimization and Nonlinear Equations*; reprinted 1996 (Philadelphia: S.I.A.M.). [1]

Jacobs, D.A.H. (ed.) 1977, *The State of the Art in Numerical Analysis* (London: Academic Press), Chapter III.1, §§3–6 (by K. W. Brodlie). [2]

Polak, E. 1971, *Computational Methods in Optimization* (New York: Academic Press), pp. 56ff. [3]

Acton, F.S. 1970, *Numerical Methods That Work*; 1990, corrected edition (Washington: Mathematical Association of America), pp. 467–468.

10.8 Linear Programming and the Simplex Method

The subject of *linear programming*, sometimes called *linear optimization*, concerns itself with the following problem: For N independent variables $x_0, \ldots, x_{N-1}$, *maximize* the function

$$z = a_{00}x_0 + a_{01}x_1 + \cdots + a_{0,N-1}x_{N-1} \tag{10.8.1}$$

subject to the primary constraints

$$x_0 \geq 0, \quad x_1 \geq 0, \quad \ldots \quad x_{N-1} \geq 0 \qquad (10.8.2)$$

and simultaneously subject to $M = m_1 + m_2 + m_3$ additional constraints, m_1 of them of the form

$$a_{i0}x_0 + a_{i1}x_1 + \cdots + a_{i,N-1}x_{N-1} \leq b_i \qquad (b_i \geq 0) \qquad i = 1, \ldots, m_1 \quad (10.8.3)$$

m_2 of them of the form

$$a_{j0}x_0 + a_{j1}x_1 + \cdots + a_{j,N-1}x_{N-1} \geq b_j \geq 0 \qquad j = m_1 + 1, \ldots, m_1 + m_2$$
$$(10.8.4)$$

and m_3 of them of the form

$$a_{k0}x_0 + a_{k1}x_1 + \cdots + a_{k,N-1}x_{N-1} = b_k \geq 0$$
$$k = m_1 + m_2 + 1, \ldots, m_1 + m_2 + m_3 \qquad (10.8.5)$$

The various a_{ij}'s can have either sign, or be zero. The fact that the b's must all be nonnegative (as indicated by the final inequality in the above three equations) is a matter of convention only, since you can multiply any contrary inequality by -1. There is no particular significance in the number of constraints M being less than, equal to, or greater than the number of unknowns N.

A set of values $x_0 \ldots x_{N-1}$ that satisfies the constraints (10.8.2)–(10.8.5) is called a *feasible vector*. The function that we are trying to maximize is called the *objective function*. The feasible vector that maximizes the objective function is called the *optimal feasible vector*. An optimal feasible vector can fail to exist for two distinct reasons: (i) there are *no* feasible vectors, i.e., the given constraints are incompatible, or (ii) there is no maximum, i.e., there is a direction in N space where one or more of the variables can be taken to infinity while still satisfying the constraints, giving an unbounded value for the objective function.

As you see, the subject of linear programming is surrounded by notational and terminological thickets. Both of these thorny defenses are lovingly cultivated by a coterie of stern acolytes who have devoted themselves to the field. Actually, the basic ideas of linear programming are quite simple. Avoiding the shrubbery, we want to teach you the basics by means of a couple of specific examples; it should then be quite obvious how to generalize.

Why is linear programming so important? (i) Because "nonnegativity" is the usual constraint on any variable x_i that represents the tangible amount of some physical commodity, like guns, butter, dollars, units of vitamin E, food calories, kilowatt hours, mass, etc. Hence equation (10.8.2). (ii) Because one is often interested in additive (linear) limitations or bounds imposed by man or nature: minimum nutritional requirement, maximum affordable cost, maximum on available labor or capital, minimum tolerable level of voter approval, etc. Hence equations (10.8.3)–(10.8.5). (iii) Because the function that one wants to optimize may be linear, or else may at least be approximated by a linear function — since that is the problem that linear programming *can* solve. Hence equation (10.8.1). For a short, semipopular survey of linear programming applications, see Bland [1].

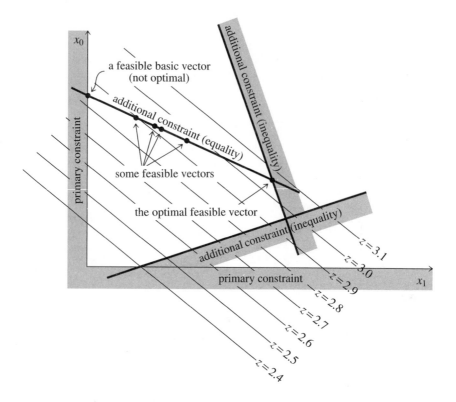

Figure 10.8.1.   Basic concepts of linear programming. The case of only two independent variables, x_0, x_1, is shown. The linear function z, to be maximized, is represented by its contour lines. Primary constraints require x_0 and x_1 to be positive. Additional constraints may restrict the solution to regions (inequality constraints) or to surfaces of lower dimensionality (equality constraints). Feasible vectors satisfy all constraints. Feasible basic vectors also lie on the boundary of the allowed region. The simplex method steps among feasible basic vectors until the optimal feasible vector is found.

Here is a specific example of a problem in linear programming, which has $N = 4$, $m_1 = 2$, $m_2 = m_3 = 1$, hence $M = 4$:

$$\text{Maximize}\quad z = x_0 + x_1 + 3x_2 - \tfrac{1}{2}x_3 \qquad (10.8.6)$$

with all the x's nonnegative and also with

$$x_0 + 2x_2 \leq 740$$

$$2x_1 - 7x_3 \leq 0$$

$$x_1 - x_2 + 2x_3 \geq \tfrac{1}{2} \qquad (10.8.7)$$

$$x_0 + x_1 + x_2 + x_3 = 9$$

The answer turns out to be (to 2 decimals) $x_0 = 0$, $x_1 = 3.33$, $x_2 = 4.73$, $x_3 = 0.95$. In the rest of this section we will learn how this answer is obtained. Figure 10.8.1 summarizes some of the terminology thus far.

Fundamental Theorem of Linear Optimization

Imagine that we start with a full N-dimensional space of candidate vectors. Then (in mind's eye, at least) we carve away the regions that are eliminated in turn by each imposed constraint. Since the constraints are linear, every boundary introduced by this process is a plane, or rather hyperplane. Equality constraints of the form (10.8.5) force the feasible region onto hyperplanes of smaller dimension, while inequalities simply divide the then-feasible region into allowed and nonallowed pieces.

When all the constraints are imposed, either we are left with some feasible region or else there are no feasible vectors. Since the feasible region is bounded by hyperplanes, it is geometrically a kind of convex polyhedron or simplex (cf. §10.4). If there is a feasible region, can the optimal feasible vector be somewhere in its interior, away from the boundaries? No, because the objective function is linear. This means that it always has a nonzero vector gradient. This, in turn, means that we could always increase the objective function by running up the gradient until we hit a boundary wall.

The boundary of any geometrical region has one less dimension than its interior. Therefore, we can now run up the gradient projected into the boundary wall until we reach an edge of that wall. We can then run up that edge, and so on, down through whatever number of dimensions, until we finally arrive at a point, a *vertex* of the original simplex. Since this point has all N of its coordinates defined, it must be the solution of N simultaneous *equalities* drawn from the original set of equalities and inequalities (10.8.2)–(10.8.5).

Points that are feasible vectors and that satisfy N of the original constraints as equalities, are termed *feasible basic vectors*. If $N > M$, then a feasible basic vector has *at least* $N - M$ of its components equal to zero, since at least that many of the constraints (10.8.2) will be needed to make up the total of N. Put the other way, *at most* M components of a feasible basic vector are nonzero. In the example (10.8.6)–(10.8.7), you can check that the solution as given satisfies as equalities the last three constraints of (10.8.7) and the constraint $x_0 \geq 0$, for the required total of 4.

Put together the two preceding paragraphs and you have the *Fundamental Theorem of Linear Optimization*: If an optimal feasible vector exists, then there is a feasible basic vector that is optimal. (Didn't we warn you about the terminological thicket?)

The importance of the fundamental theorem is that it reduces the optimization problem to a "combinatorial" problem, that of determining which N constraints (out of the $M + N$ constraints in 10.8.2–10.8.5) should be satisfied by the optimal feasible vector. We have only to keep trying different combinations, and computing the objective function for each trial, until we find the best.

Doing this blindly would take halfway to forever. The *simplex method*, first published by Dantzig in 1948 (see [2]), is a way of organizing the procedure so that (i) a series of combinations is tried for which the objective function increases at each step, and (ii) the optimal feasible vector is reached after a number of iterations that is almost always no larger than of order M or N, whichever is larger. An interesting mathematical sidelight is that this second property, although known empirically ever since the simplex method was devised, was not proved to be true until the 1982 work of Stephen Smale. (For a contemporary account, see [3].)

Simplex Method for a Restricted Normal Form

A linear programming problem is said to be in *normal form* if it has no constraints in the form (10.8.3) or (10.8.4), but rather only equality constraints of the form (10.8.5) and nonnegativity constraints of the form (10.8.2).

For our purposes it will be useful to consider an even more restricted set of cases, with this additional property: Each equality constraint of the form (10.8.5) must have at least one variable that has a positive coefficient and *that appears uniquely in that one constraint only*. We can then choose one such variable in each constraint equation, and solve that constraint equation for it. The variables thus chosen are called *left-hand variables* or *basic variables*, and there are exactly M ($= m_3$) of them. The remaining $N - M$ variables are called *right-hand variables* or *nonbasic variables*. Obviously this *restricted normal form* can be achieved only in the case $M \leq N$, so that is the case that we will consider.

You may be thinking that our restricted normal form is so specialized that it is unlikely to include the linear programming problem that you wish to solve. Not at all! We will presently show how *any* linear programming problem can be transformed into restricted normal form. Therefore bear with us and learn how to apply the simplex method to a restricted normal form.

Here is an example of a problem in restricted normal form:

$$\text{Maximize} \quad z = 2x_1 - 4x_2 \tag{10.8.8}$$

with x_0, x_1, x_2, and x_3 all nonnegative and also with

$$
\begin{aligned}
x_0 &= 2 - 6x_1 + x_2 \\
x_3 &= 8 + 3x_1 - 4x_2
\end{aligned}
\tag{10.8.9}
$$

This example has $N = 4$, $M = 2$; the left-hand variables are x_0 and x_3; the right-hand variables are x_1 and x_2. The objective function (10.8.8) is written so as to depend only on right-hand variables; note, however, that this is not an actual restriction on objective functions in restricted normal form, since any left-hand variables appearing in the objective function could be eliminated algebraically by use of (10.8.9) or its analogs.

For any problem in restricted normal form, we can instantly read off a feasible basic vector (although not necessarily the *optimal* feasible basic vector). Simply set all right-hand variables equal to zero, and equation (10.8.9) then gives the values of the left-hand variables for which the constraints are satisfied. The idea of the simplex method is to proceed by a series of exchanges. In each exchange, a right-hand variable and a left-hand variable change places. At each stage we maintain a problem in restricted normal form that is equivalent to the original problem.

It is notationally convenient to record the information content of equations (10.8.8) and (10.8.9) in a so-called *tableau*, as follows:

		x_1	x_2
z	0	2	-4
x_0	2	-6	1
x_3	8	3	-4

$$\tag{10.8.10}$$

You should study (10.8.10) to be sure that you understand where each entry comes from, and how to translate back and forth between the tableau and equation formats of a problem in restricted normal form.

The first step in the simplex method is to examine the top row of the tableau, which we will call the "z-row." Look at the entries in columns labeled by right-hand variables (we will call these "right-columns"). We want to imagine in turn the effect of increasing each right-hand variable from its present value of zero, while leaving all the other right-hand variables at zero. Will the objective function increase or decrease? The answer is given by the sign of the entry in the z-row. Since we want to increase the objective function, only right columns having positive z-row entries are of interest. In (10.8.10) there is only one such column, whose z-row entry is 2.

The second step is to examine the column entries below each z-row entry that was selected by step one. We want to ask how much we can increase the right-hand variable before one of the left-hand variables is driven negative, which is not allowed. If the tableau element at the intersection of the right-hand column and the left-hand variable's row is positive, then it poses no restriction: the corresponding left-hand variable will just be driven more and more positive. If *all* the entries in any right-hand column are positive, then there is no bound on the objective function and (having said so) we are done with the problem.

If one or more entries below a positive z-row entry are negative, then we have to figure out which such entry first limits the increase of that column's right-hand variable. Evidently the limiting increase is given by dividing the element in the right-hand column (which is called the *pivot element*) into the element in the "constant column" (leftmost column) of the pivot element's row. A value that is small in magnitude is most restrictive. The increase in the objective function for this choice of pivot element is then that value multiplied by the z-row entry of that column. We repeat this procedure on all possible right-hand columns to find the pivot element with the largest such increase. That completes our "choice of a pivot element."

In the above example, the only positive z-row entry is 2. There is only one negative entry below it, namely -6, so this is the pivot element. Its constant-column entry is 2. This pivot will therefore allow x_1 to be increased by $2 \div |6|$, which results in an increase of the objective function by an amount $(2 \times 2) \div |6|$.

The third step is to *do* the increase of the selected right-hand variable, thus making it a left-hand variable; and simultaneously to modify the left-hand variables, reducing the pivot-row element to zero and thus making it a right-hand variable. For our above example let's do this first by hand: We begin by solving the pivot-row equation for the new left-hand variable x_1 in favor of the old one x_0, namely

$$x_0 = 2 - 6x_1 + x_2 \qquad \rightarrow \qquad x_1 = \tfrac{1}{3} - \tfrac{1}{6}x_0 + \tfrac{1}{6}x_2 \qquad (10.8.11)$$

We then substitute this into the old z-row,

$$z = 2x_1 - 4x_2 = 2\left[\tfrac{1}{3} - \tfrac{1}{6}x_0 + \tfrac{1}{6}x_2\right] - 4x_2 = \tfrac{2}{3} - \tfrac{1}{3}x_0 - \tfrac{11}{3}x_2 \qquad (10.8.12)$$

and into all other left-variable rows, in this case only x_3,

$$x_3 = 8 + 3\left[\tfrac{1}{3} - \tfrac{1}{6}x_0 + \tfrac{1}{6}x_2\right] - 4x_2 = 9 - \tfrac{1}{2}x_0 - \tfrac{7}{2}x_2 \qquad (10.8.13)$$

Equations (10.8.11)–(10.8.13) form the new tableau

		x_0	x_2
z	$\frac{2}{3}$	$-\frac{1}{3}$	$-\frac{11}{3}$
x_1	$\frac{1}{3}$	$-\frac{1}{6}$	$\frac{1}{6}$
x_3	9	$-\frac{1}{2}$	$-\frac{7}{2}$

$$(10.8.14)$$

The fourth step is to go back and repeat the first step, looking for another possible increase of the objective function. We do this as many times as possible, that is, until all the right-hand entries in the z-row are negative, signaling that no further increase is possible. In the present example, this already occurs in (10.8.14), so we are done.

The answer can now be read from the constant column of the final tableau. In (10.8.14) we see that the objective function is maximized to a value of $2/3$ for the solution vector $x_1 = 1/3$, $x_3 = 9$, $x_0 = x_2 = 0$.

Now look back over the procedure that led from (10.8.10) to (10.8.14). You will find that it could be summarized entirely in tableau format as a series of prescribed elementary matrix operations:

- Locate the pivot element and save it.
- Save the whole pivot column.
- Replace each row, except the pivot row, by that linear combination of itself and the pivot row which makes its pivot-column entry zero.
- Divide the pivot row by the negative of the pivot.
- Replace the pivot element by the reciprocal of its saved value.
- Replace the rest of the pivot column by its saved values divided by the saved pivot element.

This is the sequence of operations actually performed by a linear programming routine, such as the one that we will presently give.

You should now be able to solve almost any linear programming problem that starts in restricted normal form. The only special case that might stump you is if an entry in the constant column turns out to be zero at some stage, so that a left-hand variable is zero at the same time as all the right-hand variables are zero. This is called a *degenerate feasible vector*. To proceed, you may need to exchange the degenerate left-hand variable for one of the right-hand variables, perhaps even making several such exchanges.

Writing the General Problem in Restricted Normal Form

Here is a pleasant surprise. There exist a couple of clever tricks that render trivial the task of translating a general linear programming problem into restricted normal form!

First, we need to get rid of the inequalities of the form (10.8.3) or (10.8.4), for example, the first three constraints in (10.8.7). We do this by adding to the problem so-called *slack variables* which, when their nonnegativity is required, convert the inequalities to equalities. We will denote slack variables as y_i. There will be $m_1 + m_2$ of them. Once they are introduced, you treat them on an equal footing with the original variables x_i; then, at the very end, you simply ignore them.

For example, introducing slack variables leaves (10.8.6) unchanged but turns (10.8.7) into

$$x_0 + 2x_2 + y_0 = 740$$

$$2x_1 - 7x_3 + y_1 = 0$$

$$x_1 - x_2 + 2x_3 - y_2 = \tfrac{1}{2} \qquad (10.8.15)$$

$$x_0 + x_1 + x_2 + x_3 = 9$$

(Notice how the sign of the coefficient of the slack variable is determined by which sense of inequality it is replacing.)

Second, we need to insure that there is a set of M left-hand vectors, so that we can set up a starting tableau in restricted normal form. (In other words, we need to find a "feasible basic starting vector.") The trick is again to invent new variables! There are M of these, and they are called *artificial variables*; we denote them by z_i. You put exactly one artificial variable into each constraint equation on the following model for the example (10.8.15):

$$z_0 = 740 - x_0 - 2x_2 - y_0$$

$$z_1 = -2x_1 + 7x_3 - y_1$$

$$z_2 = \tfrac{1}{2} - x_1 + x_2 - 2x_3 + y_2 \qquad (10.8.16)$$

$$z_3 = 9 - x_0 - x_1 - x_2 - x_3$$

Our example is now in restricted normal form.

Now you may object that (10.8.16) is not the same problem as (10.8.15) or (10.8.7) *unless all the z_i's are zero.* Right you are! There is some subtlety here! We must proceed to solve our problem in two phases. First phase: We replace our objective function (10.8.6) by a so-called *auxiliary objective function*

$$z' \equiv -z_0 - z_1 - z_2 - z_3 = -(749\tfrac{1}{2} - 2x_0 - 4x_1 - 2x_2 + 4x_3 - y_0 - y_1 + y_2) \qquad (10.8.17)$$

(where the last equality follows from using 10.8.16). We now perform the simplex method on the auxiliary objective function (10.8.17) with the constraints (10.8.16). Obviously the auxiliary objective function will be maximized for nonnegative z_i's if all the z_i's are zero. We therefore expect the simplex method in this first phase to produce a set of left-hand variables drawn from the x_i's and y_i's only, with all the z_i's being right-hand variables. Aha! We then cross out the z_i's, leaving a problem involving only x_i's and y_i's in restricted normal form. In other words, the first phase produces an initial feasible basic vector. Second phase: Solve the problem produced by the first phase, using the original objective function, not the auxiliary.

And what if the first phase *doesn't* produce zero values for all the z_i's? That signals that there is *no* initial feasible basic vector, i.e., that the constraints given to us are inconsistent among themselves. Report that fact, and you are done.

Here is how to translate into tableau format the information needed for both the first and second phases of the overall method. As before, the underlying problem to be solved is as posed in equations (10.8.6)–(10.8.7).

		x_0	x_1	x_2	x_3	y_0	y_1	y_2
z	0	1	1	3	$-\frac{1}{2}$	0	0	0
z_0	740	-1	0	-2	0	-1	0	0
z_1	0	0	-2	0	7	0	-1	0
z_2	$\frac{1}{2}$	0	-1	1	-2	0	0	1
z_3	9	-1	-1	-1	-1	0	0	0
z'	$-749\frac{1}{2}$	2	4	2	-4	1	1	-1

$$(10.8.18)$$

This is not as daunting as it may, at first sight, appear. The table entries inside the box of double lines are no more than the coefficients of the original problem (10.8.6)–(10.8.7) organized into a tabular form. In fact, these entries, along with the values of N, M, m_1, m_2, and m_3, are the only input that is needed by the simplex method routine below. The columns under the slack variables y_i simply record whether each of the M constraints is of the form $\leq$, $\geq$, or $=$; this is redundant information with the values m_1, m_2, m_3, as long as we are sure to enter the rows of the tableau in the correct respective order. The coefficients of the auxiliary objective function (bottom row) are just the negatives of the column sums of the rows above (excluding the z-row), so these are easily calculated automatically.

The output from a simplex routine will be (i) a flag telling whether a finite solution, no solution, or an unbounded solution was found, and (ii) an updated tableau. The output tableau that derives from (10.8.18), given to two significant figures, is

		x_0	y_1	y_2	$\cdots$
z	17.03	$-.95$	$-.05$	-1.05	$\cdots$
x_1	3.33	$-.35$	$-.15$	$.35$	$\cdots$
x_2	4.73	$-.55$	$.05$	$-.45$	$\cdots$
x_3	.95	$-.10$	$.10$	$.10$	$\cdots$
y_0	730.55	$.10$	$-.10$	$.90$	$\cdots$

$$(10.8.19)$$

A little counting of the x_i's and y_i's will convince you that there are $M + 1$ rows (including the z-row) in both the input and the output tableaux, but that only $N + 1 - m_3$ columns of the output tableau (including the constant column) contain any useful information, the other columns belonging to now-discarded artificial variables. In the output, the first numerical column contains the solution vector, along with the maximum value of the objective function. Where a slack variable (y_i) appears on the left, the corresponding value is the amount by which its inequality is safely satisfied. Variables that are not left-hand variables in the output tableau have zero values. Slack variables with zero values represent constraints that are satisfied as equalities.

Routine Implementing the Simplex Method

The following routine is based algorithmically on the implementation of Kuenzi, Tzschach, and Zehnder [4]. Aside from input values of M, N, m_1, m_2, m_3, the principal input to the routine is a two-dimensional array a containing the portion of the tableau (10.8.18) that is contained between the double lines. This input occupies the $M + 1$ rows and $N + 1$ columns of the matrix a. Note, however, that reference is made internally to an additional row of a (used for the auxiliary objective function, just as in 10.8.18). Therefore the variable declared as Mat_IO_DP &a must allow references in the range

$$a[i][k], \quad i = 0 \ldots \text{m+1}, \, k = 0 \ldots \text{n} \qquad (10.8.20)$$

You will suffer endless agonies if you fail to understand this simple point. Also do not neglect to order the rows of a in the same order as equations (10.8.1), (10.8.3), (10.8.4), and (10.8.5), that is, objective function, $\leq$-constraints, $\geq$-constraints, $=$-constraints.

On output, the tableau a is indexed by two returned arrays of integers. iposv[j] contains, for j$= 0 \ldots M - 1$, the number i whose original variable x_i is now represented by row j+1 of a. These are thus the left-hand variables in the solution. (The first row of a is of course the z-row.) A value $i > N - 1$ indicates that the variable is a y_i rather than an x_i, $x_{N+j} \equiv y_j$. Likewise, izrov[j] contains, for j$= 0 \ldots N - 1$, the number i whose original variable x_i is now a right-hand variable, represented by column j+1 of a. These variables are all zero in the solution. The meaning of $i > N - 1$ is the same as above, except that $i \geq N + m_1 + m_2$ denotes an artificial or slack variable which was used only internally and should now be entirely ignored.

The flag icase is set to zero if a finite solution is found, $+1$ if the objective function is unbounded, -1 if no solution satisfies the given constraints.

The routine treats the case of degenerate feasible vectors, so don't worry about them. You may also wish to admire the fact that the routine does not require storage for the columns of the tableau (10.8.18) that are to the right of the double line; it keeps track of slack variables by more efficient bookkeeping.

Please note that, as given, the routine is only "semi-sophisticated" in its tests for convergence. While the routine properly implements tests for inequality with zero as tests against some small parameter EPS, it does not adjust this parameter to reflect the scale of the input data. This is adequate for many problems, where the input data do not differ from unity by too many orders of magnitude. If, however, you encounter endless cycling, then you should modify EPS in the routines simplx and simp2. Permuting your variables can also help. Finally, consult [5].

```
#include "nr.h"

void NR::simplx(Mat_IO_DP &a, const int m1, const int m2, const int m3,
    int &icase, Vec_O_INT &izrov, Vec_O_INT &iposv)
Simplex method for linear programming. Input parameters a, m1, m2, and m3, and output
parameters a, icase, izrov, and iposv are described above.
{
    const DP EPS=1.0e-14;              Here EPS is the absolute precision,
    int i,k,ip,is,kh,kp,nl1;               which should be adjusted to the
    DP q1,bmax;                            scale of your variables.
```

```
int m=a.nrows()-2;
int n=a.ncols()-1;
if (m != (m1+m2+m3)) nrerror("Bad input constraint counts in simplx");
Vec_INT l1(n+1),l3(m);
nl1=n;
for (k=0;k<n;k++) {                              Initialize index list of columns ad-
    l1[k]=k+1;                                   missible for exchange, and make
    izrov[k]=k;                                  all variables initially right-hand.
}
for (i=1;i<=m;i++) {
    if (a[i][0] < 0.0) nrerror("Bad input tableau in simplx");
    Constants bi must be nonnegative.
    iposv[i-1]=n+i-1;
    Initial left-hand variables. m1 type constraints are represented by having their slack
    variable initially left-hand, with no artificial variable. m2 type constraints have their
    slack variable initially left-hand, with a minus sign, and their artificial variable handled
    implicitly during their first exchange. m3 type constraints have their artificial variable
    initially left-hand.
}
if (m2+m3 != 0) {                               Origin is not a feasible starting so-
    for (i=0;i<m2;i++) l3[i]=1;                 lution: we must do phase one.
    Initialize list of m2 constraints whose slack variables have never been exchanged out
    of the initial basis.
    for (k=0;k<(n+1);k++) {                     Compute the auxiliary objective func-
        q1=0.0;                                 tion.
        for (i=m1+1;i<m+1;i++) q1 += a[i][k];
        a[m+1][k] = -q1;
    }
    for (;;) {
        simp1(a,m+1,l1,nl1,0,kp,bmax);          Find max. coeff. of auxiliary objec-
        if (bmax <= EPS && a[m+1][0] < -EPS) {  tive fn.
            icase = -1;
            Auxiliary objective function is still negative and can't be improved, hence no
            feasible solution exists.
            return;
        } else if (bmax <= EPS && a[m+1][0] <= EPS) {
        Auxiliary objective function is zero and can't be improved; we have a feasible
        starting vector. Clean out the artificial variables corresponding to any remaining
        equality constraints by goto one and then move on to phase two.
            for (ip=m1+m2+1;ip<m+1;ip++) {
                if (iposv[ip-1] == (ip+n-1)) {  Found an artificial variable for
                    simp1(a,ip,l1,nl1,1,kp,bmax);  an equality constraint.
                    if (bmax > EPS)             Exchange with column correspond-
                        goto one;               ing to maximum pivot element
                }                               in row.
            }
            for (i=m1+1;i<=m1+m2;i++)           Change sign of row for any m2 con-
                if (l3[i-m1-1] == 1)            straints still present from the ini-
                    for (k=0;k<n+1;k++)         tial basis.
                        a[i][k]= -a[i][k];
            break;                              Go to phase two.
        }
        simp2(a,m,n,ip,kp);                     Locate a pivot element (phase one).
        if (ip == 0) {                          Maximum of auxiliary objective func-
            icase = -1;                         tion is unbounded, so no feasi-
            return;                             ble solution exists.
        }
one:    simp3(a,m+1,n,ip,kp);
        Exchange a left- and a right-hand variable (phase one), then update lists.
        if (iposv[ip-1] >= (n+m1+m2)) {         Exchanged out an artificial variable
            for (k=0;k<nl1;k++)                 for an equality constraint. Make
                if (l1[k] == kp) break;         sure it stays out by removing it
            --nl1;                              from the l1 list.
```

```
                for (is=k;is<nl1;is++) ll[is]=ll[is+1];
        } else {
            kh=iposv[ip-1]-m1-n+1;
            if (kh >= 1 && l3[kh-1]) {          Exchanged out an m2 type constraint
                l3[kh-1]=0;                      for the first time. Correct the
                ++a[m+1][kp];                    pivot column for the minus sign
                for (i=0;i<m+2;i++)              and the implicit artificial vari-
                    a[i][kp]= -a[i][kp];         able.
            }
        }
        SWAP(izrov[kp-1],iposv[ip-1]);
        Update lists of left- and right-hand variables.
    }                                          Still in phase one, go back to the
}                                                  for(;;).
```
End of phase one code for finding an initial feasible solution. Now, in phase two, optimize
it.
```
for (;;) {
    simp1(a,0,ll,nl1,0,kp,bmax);              Test the z-row for doneness.
    if (bmax <= EPS) {                        Done. Solution found. Return with
        icase=0;                                  the good news.
        return;
    }
    simp2(a,m,n,ip,kp);                       Locate a pivot element (phase two).
    if (ip == 0) {                            Objective function is unbounded. Re-
        icase=1;                                  port and return.
        return;
    }
    simp3(a,m,n,ip,kp);                       Exchange a left- and a right-hand
    SWAP(izrov[kp-1],iposv[ip-1]);                variable (phase two),
}                                             and return for another iteration.
}
```

The preceding routine makes use of the following utility functions.

```
#include <cmath>
#include "nr.h"
using namespace std;

void NR::simp1(Mat_I_DP &a, const int mm, Vec_I_INT &ll, const int nll,
    const int iabf, int &kp, DP &bmax)
```
Determines the maximum of those elements whose index is contained in the supplied list ll,
either with or without taking the absolute value, as flagged by iabf.
```
{
    int k;
    DP test;

    if (nll <= 0)                    No eligible columns.
        bmax=0.0;
    else {
        kp=ll[0];
        bmax=a[mm][kp];
        for (k=1;k<nll;k++) {
            if (iabf == 0)
                test=a[mm][ll[k]]-bmax;
            else
                test=fabs(a[mm][ll[k]])-fabs(bmax);
            if (test > 0.0) {
                bmax=a[mm][ll[k]];
                kp=ll[k];
            }
        }
    }
}
```

```
#include "nr.h"

void NR::simp2(Mat_I_DP &a, const int m, const int n, int &ip, const int kp)
Locate a pivot element, taking degeneracy into account.
{
    const DP EPS=1.0e-14;
    int k,i;
    DP qp,q0,q,q1;

    ip=0;
    for (i=0;i<m;i++)
        if (a[i+1][kp] < -EPS) break;            Any possible pivots?
    if (i+1>m) return;
    q1 = -a[i+1][0]/a[i+1][kp];
    ip=i+1;
    for (i=ip;i<m;i++) {
        if (a[i+1][kp] < -EPS) {
            q = -a[i+1][0]/a[i+1][kp];
            if (q < q1) {
                ip=i+1;
                q1=q;
            } else if (q == q1) {                We have a degeneracy.
                for (k=0;k<n;k++) {
                    qp = -a[ip][k+1]/a[ip][kp];
                    q0 = -a[i][k+1]/a[i][kp];
                    if (q0 != qp) break;
                }
                if (q0 < qp) ip=i+1;
            }
        }
    }
}
```

```
#include "nr.h"

void NR::simp3(Mat_IO_DP &a, const int i1, const int k1, const int ip,
    const int kp)
Matrix operations to exchange a left-hand and right-hand variable (see text).
{
    int ii,kk;
    DP piv;

    piv=1.0/a[ip][kp];
    for (ii=0;ii<i1+1;ii++)
        if (ii != ip) {
            a[ii][kp] *= piv;
            for (kk=0;kk<k1+1;kk++)
                if (kk != kp)
                    a[ii][kk] -= a[ip][kk]*a[ii][kp];
        }
    for (kk=0;kk<k1+1;kk++)
        if (kk != kp) a[ip][kk] *= -piv;
    a[ip][kp]=piv;
}
```

Other Topics Briefly Mentioned

Every linear programming problem in normal form with N variables and M constraints has a corresponding *dual* problem with M variables and N constraints.

The tableau of the dual problem is, in essence, the transpose of the tableau of the original (sometimes called *primal*) problem. It is possible to go from a solution of the dual to a solution of the primal. This can occasionally be computationally useful, but generally it is no big deal.

The *revised simplex method* is exactly equivalent to the simplex method in its choice of which left-hand and right-hand variables are exchanged. Its computational effort is not significantly less than that of the simplex method. It does differ in the organization of its storage, requiring only a matrix of size $M \times M$, rather than $M \times N$, in its intermediate stages. If you have a lot of constraints, and memory size is one of them, then you should look into it.

The *primal-dual algorithm* and the *composite simplex algorithm* are two different methods for avoiding the two phases of the usual simplex method: Progress is made simultaneously towards finding a feasible solution and finding an optimal solution. There seems to be no clearcut evidence that these methods are superior to the usual method by any factor substantially larger than the "tender-loving-care factor" (which reflects the programming effort of the proponents).

Problems where the objective function and/or one or more of the constraints are replaced by expressions nonlinear in the variables are called *nonlinear programming problems*. The literature on such problems is vast, but outside our scope. The special case of quadratic expressions is called *quadratic programming*. Optimization problems where the variables take on only integer values are called *integer programming problems*, a special case of *discrete optimization* generally. The next section looks at a particular kind of discrete optimization problem.

CITED REFERENCES AND FURTHER READING:

Bland, R.G. 1981, *Scientific American*, vol. 244 (June), pp. 126–144. [1]

Dantzig, G.B. 1963, *Linear Programming and Extensions* (Princeton, NJ: Princeton University Press). [2]

Kolata, G. 1982, *Science*, vol. 217, p. 39. [3]

Gill, P.E., Murray, W., and Wright, M.H. 1991, *Numerical Linear Algebra and Optimization*, vol. 1 (Redwood City, CA: Addison-Wesley), Chapters 7–8.

Cooper, L., and Steinberg, D. 1970, *Introduction to Methods of Optimization* (Philadelphia: Saunders).

Gass, S.I. 1985, *Linear Programming*, 5th ed. (New York: McGraw-Hill).

Murty, K.G. 1976, *Linear and Combinatorial Programming* (New York: Wiley).

Land, A.H., and Powell, S. 1973, *Fortran Codes for Mathematical Programming* (London: Wiley-Interscience).

Kuenzi, H.P., Tzschach, H.G., and Zehnder, C.A. 1971, *Numerical Methods of Mathematical Optimization* (New York: Academic Press). [4]

Stoer, J., and Bulirsch, R. 1993, *Introduction to Numerical Analysis*, 2nd ed. (New York: Springer-Verlag), §4.10.

Wilkinson, J.H., and Reinsch, C. 1971, *Linear Algebra*, vol. II of *Handbook for Automatic Computation* (New York: Springer-Verlag). [5]

10.9 Simulated Annealing Methods

The *method of simulated annealing* [1,2] is a technique that has attracted significant attention as suitable for optimization problems of large scale, especially ones where a desired global extremum is hidden among many, poorer, local extrema. For practical purposes, simulated annealing has effectively "solved" the famous *traveling salesman problem* of finding the shortest cyclical itinerary for a traveling salesman who must visit each of N cities in turn. (Other practical methods have also been found.) The method has also been used successfully for designing complex integrated circuits: The arrangement of several hundred thousand circuit elements on a tiny silicon substrate is optimized so as to minimize interference among their connecting wires [3,4]. Surprisingly, the implementation of the algorithm is relatively simple.

Notice that the two applications cited are both examples of *combinatorial minimization*. There is an objective function to be minimized, as usual; but the space over which that function is defined is not simply the N-dimensional space of N continuously variable parameters. Rather, it is a discrete, but very large, configuration space, like the set of possible orders of cities, or the set of possible allocations of silicon "real estate" blocks to circuit elements. The number of elements in the configuration space is factorially large, so that they cannot be explored exhaustively. Furthermore, since the set is discrete, we are deprived of any notion of "continuing downhill in a favorable direction." The concept of "direction" may not have any meaning in the configuration space.

Below, we will also discuss how to use simulated annealing methods for spaces with continuous control parameters, like those of §§10.4–10.7. This application is actually more complicated than the combinatorial one, since the familiar problem of "long, narrow valleys" again asserts itself. Simulated annealing, as we will see, tries "random" steps; but in a long, narrow valley, almost all random steps are uphill! Some additional finesse is therefore required.

At the heart of the method of simulated annealing is an analogy with thermodynamics, specifically with the way that liquids freeze and crystallize, or metals cool and anneal. At high temperatures, the molecules of a liquid move freely with respect to one another. If the liquid is cooled slowly, thermal mobility is lost. The atoms are often able to line themselves up and form a pure crystal that is completely ordered over a distance up to billions of times the size of an individual atom in all directions. This crystal is the state of minimum energy for this system. The amazing fact is that, for slowly cooled systems, nature is able to find this minimum energy state. In fact, if a liquid metal is cooled quickly or "quenched," it does not reach this state but rather ends up in a polycrystalline or amorphous state having somewhat higher energy.

So the essence of the process is *slow* cooling, allowing ample time for redistribution of the atoms as they lose mobility. This is the technical definition of *annealing*, and it is essential for ensuring that a low energy state will be achieved.

Although the analogy is not perfect, there is a sense in which all of the minimization algorithms thus far in this chapter correspond to rapid cooling or quenching. In all cases, we have gone greedily for the quick, nearby solution: From the starting point, go immediately downhill as far as you can go. This, as often remarked above, leads to a local, but not necessarily a global, minimum. Nature's own minimization algorithm is based on quite a different procedure. The so-called

Boltzmann probability distribution,

$$\text{Prob}\,(E) \sim \exp(-E/kT) \tag{10.9.1}$$

expresses the idea that a system in thermal equilibrium at temperature T has its energy probabilistically distributed among all different energy states E. Even at low temperature, there is a chance, albeit very small, of a system being in a high energy state. Therefore, there is a corresponding chance for the system to get out of a local energy minimum in favor of finding a better, more global, one. The quantity k (Boltzmann's constant) is a constant of nature that relates temperature to energy. In other words, the system sometimes goes *uphill* as well as downhill; but the lower the temperature, the less likely is any significant uphill excursion.

In 1953, Metropolis and coworkers [5] first incorporated these kinds of principles into numerical calculations. Offered a succession of options, a simulated thermodynamic system was assumed to change its configuration from energy E_1 to energy E_2 with probability $p = \exp[-(E_2 - E_1)/kT]$. Notice that if $E_2 < E_1$, this probability is greater than unity; in such cases the change is arbitrarily assigned a probability $p = 1$, i.e., the system *always* took such an option. This general scheme, of always taking a downhill step while *sometimes* taking an uphill step, has come to be known as the Metropolis algorithm.

To make use of the Metropolis algorithm for other than thermodynamic systems, one must provide the following elements:

1. A description of possible system configurations.

2. A generator of random changes in the configuration; these changes are the "options" presented to the system.

3. An objective function E (analog of energy) whose minimization is the goal of the procedure.

4. A control parameter T (analog of temperature) and an *annealing schedule* which tells how it is lowered from high to low values, e.g., after how many random changes in configuration is each downward step in T taken, and how large is that step. The meaning of "high" and "low" in this context, and the assignment of a schedule, may require physical insight and/or trial-and-error experiments.

Combinatorial Minimization: The Traveling Salesman

A concrete illustration is provided by the traveling salesman problem. The proverbial seller visits N cities with given positions (x_i, y_i), returning finally to his or her city of origin. Each city is to be visited only once, and the route is to be made as short as possible. This problem belongs to a class known as *NP-complete* problems, whose computation time for an *exact* solution increases with N as $\exp(\text{const.} \times N)$, becoming rapidly prohibitive in cost as N increases. The traveling salesman problem also belongs to a class of minimization problems for which the objective function E has many local minima. In practical cases, it is often enough to be able to choose from these a minimum which, even if not absolute, cannot be significantly improved upon. The annealing method manages to achieve this, while limiting its calculations to scale as a small power of N.

As a problem in simulated annealing, the traveling salesman problem is handled as follows:

1. *Configuration.* The cities are numbered $i = 0 \ldots N - 1$ and each has coordinates (x_i, y_i). A configuration is a permutation of the number $0 \ldots N - 1$, interpreted as the order in which the cities are visited.

2. *Rearrangements.* An efficient set of moves has been suggested by Lin [6]. The moves consist of two types: (a) A section of path is removed and then replaced with the same cities running in the opposite order; or (b) a section of path is removed and then replaced in between two cities on another, randomly chosen, part of the path.

3. *Objective Function.* In the simplest form of the problem, E is taken just as the total length of journey,

$$E = L \equiv \sum_{i=0}^{N-1} \sqrt{(x_i - x_{i+1})^2 + (y_i - y_{i+1})^2} \qquad (10.9.2)$$

with the convention that point N is identified with point 0. To illustrate the flexibility of the method, however, we can add the following additional wrinkle: Suppose that the salesman has an irrational fear of flying over the Mississippi River. In that case, we would assign each city a parameter μ_i, equal to $+1$ if it is east of the Mississippi, -1 if it is west, and take the objective function to be

$$E = \sum_{i=0}^{N-1} \left[\sqrt{(x_i - x_{i+1})^2 + (y_i - y_{i+1})^2} + \lambda(\mu_i - \mu_{i+1})^2 \right] \qquad (10.9.3)$$

A penalty 4λ is thereby assigned to any river crossing. The algorithm now finds the shortest path that avoids crossings. The relative importance that it assigns to length of path versus river crossings is determined by our choice of λ. Figure 10.9.1 shows the results obtained. Clearly, this technique can be generalized to include many conflicting goals in the minimization.

4. *Annealing schedule.* This requires experimentation. We first generate some random rearrangements, and use them to determine the range of values of ΔE that will be encountered from move to move. Choosing a starting value for the parameter T which is considerably larger than the largest ΔE normally encountered, we proceed downward in multiplicative steps each amounting to a 10 percent decrease in T. We hold each new value of T constant for, say, $100N$ reconfigurations, or for $10N$ successful reconfigurations, whichever comes first. When efforts to reduce E further become sufficiently discouraging, we stop.

The following traveling salesman program, using the Metropolis algorithm, illustrates the main aspects of the simulated annealing technique for combinatorial problems.

```
#include <iostream>
#include <iomanip>
#include <cmath>
#include <cstdlib>
#include "nr.h"
using namespace std;

namespace {
    inline DP alen(const DP a, const DP b, const DP c, const DP d)
    {
        return sqrt((b-a)*(b-a)+(d-c)*(d-c));
```

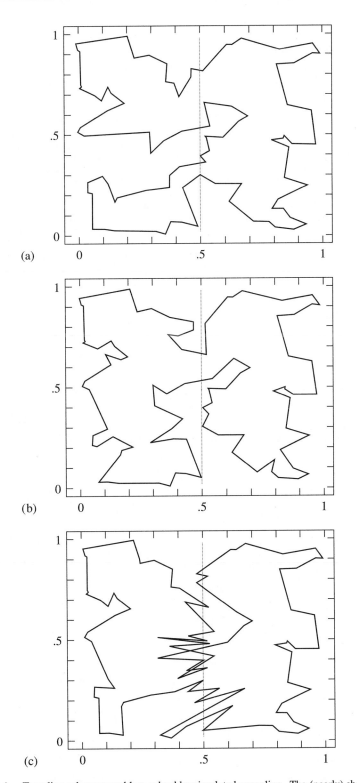

Figure 10.9.1.   Traveling salesman problem solved by simulated annealing. The (nearly) shortest path among 100 randomly positioned cities is shown in (a). The dotted line is a river, but there is no penalty in crossing. In (b) the river-crossing penalty is made large, and the solution restricts itself to the minimum number of crossings, two. In (c) the penalty has been made negative: the salesman is actually a smuggler who crosses the river on the flimsiest excuse!

```
          }
     }

     void NR::anneal(Vec_I_DP &x, Vec_I_DP &y, Vec_IO_INT &iorder)
```
This algorithm finds the shortest round-trip path to ncity cities whose coordinates are in the arrays x[0..ncity-1],y[0..ncity-1]. The array iorder[0..ncity-1] specifies the order in which the cities are visited. On input, the elements of iorder may be set to any permutation of the numbers 0 to ncity-1. This routine will return the best alternative path it can find.
```
     {
          const DP TFACTR=0.9;          Annealing schedule: reduce t by this factor on each step.
          bool ans;
          int i,i1,i2,idec,idum,j,k,nn,nover,nlimit,nsucc;
          static Vec_INT n(6);
          unsigned long iseed;
          DP path,de,t;

          int ncity=x.size();
          nover=100*ncity;              Maximum number of paths tried at any temperature.
          nlimit=10*ncity;              Maximum number of successful path changes before con-
          path=0.0;                          tinuing.
          t=0.5;
          for (i=0;i<ncity-1;i++) {    Calculate initial path length.
               i1=iorder[i];
               i2=iorder[i+1];
               path += alen(x[i1],x[i2],y[i1],y[i2]);
          }
          i1=iorder[ncity-1];
          i2=iorder[0];
          path += alen(x[i1],x[i2],y[i1],y[i2]);
          idum = -1;
          iseed=111;
          cout << fixed << setprecision(6);
          for (j=0;j<100;j++) {        Try up to 100 temperature steps.
               nsucc=0;
               for (k=0;k<nover;k++) {
                    do {
                         n[0]=int(ncity*ran3(idum));     Choose beginning of segment ..
                         n[1]=int((ncity-1)*ran3(idum));      ..and end of segment.
                         if (n[1] >= n[0]) ++n[1];
                         nn=(n[0]-n[1]+ncity-1) % ncity;      nn is the number of cities
                    } while (nn<2);                           not on the segment.
                    idec=irbit1(iseed);
                    Decide whether to do a segment reversal or transport.
                    if (idec == 0) {                    Do a transport.
                         n[2]=n[1]+int(abs(nn-1)*ran3(idum))+1;
                         n[2] %= ncity;                 Transport to location not on the path.
                         de=trncst(x,y,iorder,n);       Calculate cost.
                         ans=metrop(de,t);              Consult the oracle.
                         if (ans) {
                              ++nsucc;
                              path += de;
                              trnspt(iorder,n);          Carry out the transport.
                         }
                    } else {                            Do a path reversal.
                         de=revcst(x,y,iorder,n);       Calculate cost.
                         ans=metrop(de,t);              Consult the oracle.
                         if (ans) {
                              ++nsucc;
                              path += de;
                              reverse(iorder,n);         Carry out the reversal.
                         }
                    }
                    if (nsucc >= nlimit) break;          Finish early if we have enough suc-
                                                         cessful changes.
```

```
        }
        cout << endl << "T = " << setw(12) << t;
        cout << "        Path Length = " << setw(12) << path << endl;
        cout << "Successful Moves: " << nsucc << endl;
        t *= TFACTR;                          Annealing schedule.
        if (nsucc == 0) return;               If no success, we are done.
    }
}
```

```
#include <cmath>
#include "nr.h"
using namespace std;

namespace {
    inline DP alen(const DP a, const DP b, const DP c, const DP d)
    {
        return sqrt((b-a)*(b-a)+(d-c)*(d-c));
    }
}
```

`DP NR::revcst(Vec_I_DP &x, Vec_I_DP &y, Vec_I_INT &iorder, Vec_IO_INT &n)`
This function returns the value of the cost function for a proposed path reversal. `ncity` is the number of cities, and arrays `x[0..ncity-1]`,`y[0..ncity-1]` give the coordinates of these cities. `iorder[0..ncity-1]` holds the present itinerary. On input, the first two values `n[0]` and `n[1]` of array `n` should be set to the starting and ending cities along the path segment to be reversed. On output, the function value returns the cost of making the reversal. The actual reversal is not performed by this routine.

```
{
    int j,ii;
    DP de;
    Vec_DP xx(4),yy(4);

    int ncity=x.size();
    n[2]=(n[0]+ncity-1) % ncity;            Find the city before n[0] ..
    n[3]=(n[1]+1) % ncity;                  .. and the city after n[1].
    for (j=0;j<4;j++) {
        ii=iorder[n[j]];                    Find coordinates for the four cities in-
        xx[j]=x[ii];                        volved.
        yy[j]=y[ii];
    }
    de = -alen(xx[0],xx[2],yy[0],yy[2]);    Calculate cost of disconnecting the seg-
    de -= alen(xx[1],xx[3],yy[1],yy[3]);    ment at both ends and reconnecting
    de += alen(xx[0],xx[3],yy[0],yy[3]);    in the opposite order.
    de += alen(xx[1],xx[2],yy[1],yy[2]);
    return de;
}
```

```
#include "nr.h"
```

`void NR::reverse(Vec_IO_INT &iorder, Vec_I_INT &n)`
This routine performs a path segment reversal. `iorder[0..ncity-1]` is an input array giving the present itinerary. The vector `n` has as its first four elements the first and last cities `n[0]`,`n[1]` of the path segment to be reversed, and the two cities `n[2]` and `n[3]` that immediately precede and follow this segment. `n[2]` and `n[3]` are found by function `revcst`. On output, `iorder[0..ncity-1]` contains the segment from `n[0]` to `n[1]` in reversed order.

```
{
    int nn,j,k,l,itmp;

    int ncity=iorder.size();
    nn=(1+((n[1]-n[0]+ncity) % ncity))/2;   This many cities must be swapped to
                                            effect the reversal.
```

```
    for (j=0;j<nn;j++) {
        k=(n[0]+j) % ncity;
        l=(n[1]-j+ncity) % ncity;
        itmp=iorder[k];
        iorder[k]=iorder[l];
        iorder[l]=itmp;
    }
}
```
Start at the ends of the segment and swap pairs of cities, moving toward the center.

```
#include <cmath>
#include "nr.h"
using namespace std;

namespace {
    inline DP alen(const DP a, const DP b, const DP c, const DP d)
    {
        return sqrt((b-a)*(b-a)+(d-c)*(d-c));
    }
}
```

DP NR::trncst(Vec_I_DP &x, Vec_I_DP &y, Vec_I_INT &iorder, Vec_IO_INT &n)
This routine returns the value of the cost function for a proposed path segment transport. ncity is the number of cities, and arrays x[0..ncity-1] and y[0..ncity-1] give the city coordinates. iorder[0..ncity-1] is an array giving the present itinerary. On input, the first three elements of array n should be set to the starting and ending cities of the path to be transported, and the point among the remaining cities after which it is to be inserted. On output, the function value returns the cost of the change. The actual transport is not performed by this routine.

```
{
    int j,ii;
    DP de;
    Vec_DP xx(6),yy(6);

    int ncity=x.size();
    n[3]=(n[2]+1) % ncity;
    n[4]=(n[0]+ncity-1) % ncity;
    n[5]=(n[1]+1) % ncity;
    for (j=0;j<6;j++) {
        ii=iorder[n[j]];
        xx[j]=x[ii];
        yy[j]=y[ii];
    }
    de = -alen(xx[1],xx[5],yy[1],yy[5]);
    de -= alen(xx[0],xx[4],yy[0],yy[4]);
    de -= alen(xx[2],xx[3],yy[2],yy[3]);
    de += alen(xx[0],xx[2],yy[0],yy[2]);
    de += alen(xx[1],xx[3],yy[1],yy[3]);
    de += alen(xx[4],xx[5],yy[4],yy[5]);
    return de;
}
```
Find the city following n[2]..
..and the one preceding n[0]..
..and the one following n[1].

Determine coordinates for the six cities involved.

Calculate the cost of disconnecting the path segment from n[0] to n[1], opening a space between n[2] and n[3], connecting the segment in the space, and connecting n[4] to n[5].

```
#include "nr.h"
```

void NR::trnspt(Vec_IO_INT &iorder, Vec_I_INT &n)
This routine does the actual path transport, once metrop has approved. iorder[0..ncity-1] is an input array giving the present itinerary. The array n has as its six elements the beginning n[0] and end n[1] of the path to be transported, the adjacent cities n[2] and n[3] between which the path is to be placed, and the cities n[4] and n[5] that precede and follow the path. n[3], n[4], and n[5] are calculated by function trncst. On output, iorder is modified to reflect the movement of the path segment.

```
{
    int m1,m2,m3,nn,j,jj;

    int ncity=iorder.size();
    Vec_INT jorder(ncity);
    m1=(n[1]-n[0]+ncity) % ncity;          Find number of cities from n[0] to n[1]
    m2=(n[4]-n[3]+ncity) % ncity;          ...and the number from n[3] to n[4]
    m3=(n[2]-n[5]+ncity) % ncity;          ...and the number from n[5] to n[2].
    nn=0;
    for (j=0;j<=m1;j++) {
        jj=(j+n[0]) % ncity;               Copy the chosen segment.
        jorder[nn++]=iorder[jj];
    }
    for (j=0;j<=m2;j++) {                   Then copy the segment from n[3] to
        jj=(j+n[3]) % ncity;                   n[4].
        jorder[nn++]=iorder[jj];
    }
    for (j=0;j<=m3;j++) {                   Finally, the segment from n[5] to n[2].
        jj=(j+n[5]) % ncity;
        jorder[nn++]=iorder[jj];
    }
    for (j=0;j<ncity;j++)                   Copy jorder back into iorder.
        iorder[j]=jorder[j];
}
```

```
#include <cmath>
#include "nr.h"
using namespace std;

bool NR::metrop(const DP de, const DP t)
```
Metropolis algorithm. `metrop` returns a boolean variable that issues a verdict on whether to accept a reconfiguration that leads to a change de in the objective function e. If de<0, metrop = true, while if de>0, metrop is only true with probability exp(-de/t), where t is a temperature determined by the annealing schedule.
```
{
    static int gljdum=1;

    return de < 0.0 || ran3(gljdum) < exp(-de/t);
}
```

Continuous Minimization by Simulated Annealing

The basic ideas of simulated annealing are also applicable to optimization problems with continuous N-dimensional control spaces, e.g., finding the (ideally, global) minimum of some function $f(\mathbf{x})$, in the presence of many local minima, where $\mathbf{x}$ is an N-dimensional vector. The four elements required by the Metropolis procedure are now as follows: The value of f is the objective function. The system state is the point $\mathbf{x}$. The control parameter T is, as before, something like a temperature, with an annealing schedule by which it is gradually reduced. And there must be a generator of random changes in the configuration, that is, a procedure for taking a random step from $\mathbf{x}$ to $\mathbf{x} + \Delta\mathbf{x}$.

The last of these elements is the most problematical. The literature to date [7-10] describes several different schemes for choosing $\Delta\mathbf{x}$, none of which, in our view, inspire complete confidence. The problem is one of efficiency: A generator of random changes is inefficient if, *when local downhill moves exist*, it nevertheless almost always proposes an uphill move. A good generator, we think, should not become inefficient in narrow valleys; nor should it become more and more inefficient as convergence to a minimum is approached. Except possibly for [7], all of the schemes that we have seen are inefficient in one or both of these situations.

Our own way of doing simulated annealing minimization on continuous control spaces is to use a modification of the downhill simplex method (§10.4). This amounts to replacing the single point $\mathbf{x}$ as a description of the system state by a simplex of $N + 1$ points. The "moves" are the same as described in §10.4, namely reflections, expansions, and contractions of the simplex. The implementation of the Metropolis procedure is slightly subtle: We *add* a positive, logarithmically distributed random variable, proportional to the temperature T, to the stored function value associated with every vertex of the simplex, and we *subtract* a similar random variable from the function value of every new point that is tried as a replacement point. Like the ordinary Metropolis procedure, this method always accepts a true downhill step, but sometimes accepts an uphill one. In the limit $T \rightarrow 0$, this algorithm reduces exactly to the downhill simplex method and converges to a local minimum.

At a finite value of T, the simplex expands to a scale that approximates the size of the region that can be reached at this temperature, and then executes a stochastic, tumbling Brownian motion within that region, sampling new, approximately random, points as it does so. The efficiency with which a region is explored is independent of its narrowness (for an ellipsoidal valley, the ratio of its principal axes) and orientation. If the temperature is reduced sufficiently slowly, it becomes highly likely that the simplex will shrink into that region containing the lowest relative minimum encountered.

As in all applications of simulated annealing, there can be quite a lot of problem-dependent subtlety in the phrase "sufficiently slowly"; success or failure is quite often determined by the choice of annealing schedule. Here are some possibilities worth trying:

- Reduce T to $(1 - \epsilon)T$ after every m moves, where ϵ/m is determined by experiment.
- Budget a total of K moves, and reduce T after every m moves to a value $T = T_0(1 - k/K)^\alpha$, where k is the cumulative number of moves thus far, and α is a constant, say 1, 2, or 4. The optimal value for α depends on the statistical distribution of relative minima of various depths. Larger values of α spend more iterations at lower temperature.
- After every m moves, set T to β times $f_1 - f_b$, where β is an experimentally determined constant of order 1, f_1 is the smallest function value currently represented in the simplex, and f_b is the best function ever encountered. However, never reduce T by more than some fraction γ at a time.

Another strategic question is whether to do an occasional *restart*, where a vertex of the simplex is discarded in favor of the "best-ever" point. (You must be sure that the best-ever point is not one of the other vertices of the simplex when you do this!) We have found problems for which restarts — every time the temperature has decreased by a factor of 3, say — are highly beneficial; we have found other problems for which restarts have no positive, or a somewhat negative, effect.

You should compare the following routine, amebsa, with its counterpart amoeba in §10.4. Note that the argument iter is used in a somewhat different manner.

```
#include <cmath>
#include "nr.h"
using namespace std;

namespace {
    inline void get_psum(Mat_I_DP &p, Vec_O_DP &psum)
    {
        int n,m;
        DP sum;

        int mpts=p.nrows();
        int ndim=p.ncols();
        for (n=0;n<ndim;n++) {
            for (sum=0.0,m=0;m<mpts;m++) sum += p[m][n];
            psum[n]=sum;
        }
    }
}
```

```
extern int idum;                          Should be defined and initialized in main.
DP tt;                                     Communicates with amotsa.
```

```
void NR::amebsa(Mat_IO_DP &p, Vec_IO_DP &y, Vec_O_DP &pb, DP &yb, const DP ftol,
    DP funk(Vec_I_DP &), int &iter, const DP temptr)
```
Multidimensional minimization of the function `funk(x)` where `x[0..ndim-1]` is a vector in `ndim` dimensions, by simulated annealing combined with the downhill simplex method of Nelder and Mead. The input matrix `p[0..ndim][0..ndim-1]` has `ndim+1` rows, each an `ndim`-dimensional vector which is a vertex of the starting simplex. Also input are the following: the vector `y[0..ndim]`, whose components must be pre-initialized to the values of `funk` evaluated at the `ndim+1` vertices (rows) of `p`; `ftol`, the fractional convergence tolerance to be achieved in the function value for an early return; `iter`, and `temptr`. The routine makes `iter` function evaluations at an annealing temperature `temptr`, then returns. You should then decrease `temptr` according to your annealing schedule, reset `iter`, and call the routine again (leaving other arguments unaltered between calls). If `iter` is returned with a positive value, then early convergence and return occurred. If you initialize `yb` to a very large value on the first call, then `yb` and `pb[0..ndim-1]` will subsequently return the best function value and point ever encountered (even if it is no longer a point in the simplex).

```
{
    int i,ihi,ilo,j,n;
    DP rtol,yhi,ylo,ynhi,ysave,yt,ytry;

    int mpts=p.nrows();
    int ndim=p.ncols();
    Vec_DP psum(ndim);
    tt = -temptr;
    get_psum(p,psum);
    for (;;) {
        ilo=0;                            Determine which point is the highest (worst),
        ihi=1;                                next-highest, and lowest (best).
        ynhi=ylo=y[0]+tt*log(ran1(idum));  Whenever we "look at" a vertex, it gets
        yhi=y[1]+tt*log(ran1(idum));          a random thermal fluctuation.
        if (ylo > yhi) {
            ihi=0;
            ilo=1;
            ynhi=yhi;
            yhi=ylo;
            ylo=ynhi;
        }
        for (i=3;i<=mpts;i++) {           Loop over the points in the simplex.
            yt=y[i-1]+tt*log(ran1(idum)); More thermal fluctuations.
            if (yt <= ylo) {
                ilo=i-1;
                ylo=yt;
            }
```

```
                if (yt > yhi) {
                    ynhi=yhi;
                    ihi=i-1;
                    yhi=yt;
                } else if (yt > ynhi) {
                    ynhi=yt;
                }
            }
            rtol=2.0*fabs(yhi-ylo)/(fabs(yhi)+fabs(ylo));
```
Compute the fractional range from highest to lowest and return if satisfactory.
```
            if (rtol < ftol || iter < 0) {          If returning, put best point and value in
                SWAP(y[0],y[ilo]);                       slot 0.
                for (n=0;n<ndim;n++)
                    SWAP(p[0][n],p[ilo][n]);
                break;
            }
            iter -= 2;
```
Begin a new iteration. First extrapolate by a factor −1 through the face of the simplex across from the high point, i.e., reflect the simplex from the high point.
```
            ytry=amotsa(p,y,psum,pb,yb,funk,ihi,yhi,-1.0);
            if (ytry <= ylo) {
```
Gives a result better than the best point, so try an additional extrapolation by a factor of 2.
```
                ytry=amotsa(p,y,psum,pb,yb,funk,ihi,yhi,2.0);
            } else if (ytry >= ynhi) {
```
The reflected point is worse than the second-highest, so look for an intermediate lower point, i.e., do a one-dimensional contraction.
```
                ysave=yhi;
                ytry=amotsa(p,y,psum,pb,yb,funk,ihi,yhi,0.5);
                if (ytry >= ysave) {                 Can't seem to get rid of that high point.
                    for (i=0;i<mpts;i++) {              Better contract around the lowest
                        if (i != ilo) {                    (best) point.
                            for (j=0;j<ndim;j++) {
                                psum[j]=0.5*(p[i][j]+p[ilo][j]);
                                p[i][j]=psum[j];
                            }
                            y[i]=funk(psum);
                        }
                    }
                    iter -= ndim;
                    get_psum(p,psum);                Recompute psum.
                }
            } else ++iter;                           Correct the evaluation count.
        }
    }
```

```
#include <cmath>
#include "nr.h"
using namespace std;

extern int idum;                        Defined and initialized in main.
extern DP tt;                           Defined in amebsa.

DP NR::amotsa(Mat_IO_DP &p, Vec_O_DP &y, Vec_IO_DP &psum, Vec_O_DP &pb, DP &yb,
    DP funk(Vec_I_DP &), const int ihi, DP &yhi, const DP fac)
```
Extrapolates by a factor fac through the face of the simplex across from the high point, tries it, and replaces the high point if the new point is better.
```
{
    int j;
    DP fac1,fac2,yflu,ytry;

    int ndim=p.ncols();
```

```
Vec_DP ptry(ndim);
fac1=(1.0-fac)/ndim;
fac2=fac1-fac;
for (j=0;j<ndim;j++)
    ptry[j]=psum[j]*fac1-p[ihi][j]*fac2;
ytry=funk(ptry);
if (ytry <= yb) {                         Save the best-ever.
    for (j=0;j<ndim;j++) pb[j]=ptry[j];
    yb=ytry;
}
yflu=ytry-tt*log(ran1(idum));             We added a thermal fluctuation to all the current
if (yflu < yhi) {                         vertices, but we subtract it here, so as to give
    y[ihi]=ytry;                          the simplex a thermal Brownian motion: It
    yhi=yflu;                             likes to accept any suggested change.
    for (j=0;j<ndim;j++) {
        psum[j] += ptry[j]-p[ihi][j];
        p[ihi][j]=ptry[j];
    }
}
    return yflu;
}
```

There is not yet enough practical experience with the method of simulated annealing to say definitively what its future place among optimization methods will be. The method has several extremely attractive features, rather unique when compared with other optimization techniques.

First, it is not "greedy," in the sense that it is not easily fooled by the quick payoff achieved by falling into unfavorable local minima. Provided that sufficiently general reconfigurations are given, it wanders freely among local minima of depth less than about T. As T is lowered, the number of such minima qualifying for frequent visits is gradually reduced.

Second, configuration decisions tend to proceed in a logical order. Changes that cause the greatest energy differences are sifted over when the control parameter T is large. These decisions become more permanent as T is lowered, and attention then shifts more to smaller refinements in the solution. For example, in the traveling salesman problem with the Mississippi River twist, if λ is large, a decision to cross the Mississippi only twice is made at high T, while the specific routes on each side of the river are determined only at later stages.

The analogies to thermodynamics may be pursued to a greater extent than we have done here. Quantities analogous to specific heat and entropy may be defined, and these can be useful in monitoring the progress of the algorithm towards an acceptable solution. Information on this subject is found in [1].

CITED REFERENCES AND FURTHER READING:

Kirkpatrick, S., Gelatt, C.D., and Vecchi, M.P. 1983, *Science*, vol. 220, pp. 671–680. [1]

Kirkpatrick, S. 1984, *Journal of Statistical Physics*, vol. 34, pp. 975–986. [2]

Vecchi, M.P. and Kirkpatrick, S. 1983, *IEEE Transactions on Computer Aided Design*, vol. CAD-2, pp. 215–222. [3]

Otten, R.H.J.M., and van Ginneken, L.P.P.P. 1989, *The Annealing Algorithm* (Boston: Kluwer) [contains many references to the literature]. [4]

Metropolis, N., Rosenbluth, A., Rosenbluth, M., Teller A., and Teller, E. 1953, *Journal of Chemical Physics*, vol. 21, pp. 1087–1092. [5]

Lin, S. 1965, *Bell System Technical Journal*, vol. 44, pp. 2245–2269. [6]

Vanderbilt, D., and Louie, S.G. 1984, *Journal of Computational Physics*, vol. 56, pp. 259–271. [7]

Bohachevsky, I.O., Johnson, M.E., and Stein, M.L. 1986, *Technometrics*, vol. 28, pp. 209–217. [8]

Corana, A., Marchesi, M., Martini, C., and Ridella, S. 1987, *ACM Transactions on Mathematical Software*, vol. 13, pp. 262–280. [9]

Bélisle, C.J.P., Romeijn, H.E., and Smith, R.L. 1990, Technical Report 90–25, Department of Industrial and Operations Engineering, University of Michigan, submitted to *Mathematical Programming*. [10]

Christofides, N., Mingozzi, A., Toth, P., and Sandi, C. (eds.) 1979, *Combinatorial Optimization* (London and New York: Wiley-Interscience) [not simulated annealing, but other topics and algorithms].

Chapter 11.   Eigensystems

11.0 Introduction

An $N \times N$ matrix $\mathbf{A}$ is said to have an *eigenvector* $\mathbf{x}$ and corresponding *eigenvalue* λ if

$$\mathbf{A} \cdot \mathbf{x} = \lambda \mathbf{x} \tag{11.0.1}$$

Obviously any multiple of an eigenvector $\mathbf{x}$ will also be an eigenvector, but we won't consider such multiples as being distinct eigenvectors. (The zero vector is not considered to be an eigenvector at all.) Evidently (11.0.1) can hold only if

$$\det |\mathbf{A} - \lambda \mathbf{1}| = 0 \tag{11.0.2}$$

which, if expanded out, is an Nth degree polynomial in λ whose roots are the eigenvalues. This proves that there are always N (not necessarily distinct) eigenvalues. Equal eigenvalues coming from multiple roots are called *degenerate*. Root-searching in the characteristic equation (11.0.2) is usually a very poor computational method for finding eigenvalues. We will learn much better ways in this chapter, as well as efficient ways for finding corresponding eigenvectors.

The above two equations also prove that every one of the N eigenvalues has a (not necessarily distinct) corresponding eigenvector: If λ is set to an eigenvalue, then the matrix $\mathbf{A} - \lambda \mathbf{1}$ is singular, and we know that every singular matrix has at least one nonzero vector in its nullspace (see §2.6 on singular value decomposition).

If you add $\tau \mathbf{x}$ to both sides of (11.0.1), you will easily see that the eigenvalues of any matrix can be changed or *shifted* by an additive constant τ by adding to the matrix that constant times the identity matrix. The eigenvectors are unchanged by this shift. Shifting, as we will see, is an important part of many algorithms for computing eigenvalues. We see also that there is no special significance to a zero eigenvalue. Any eigenvalue can be shifted to zero, or any zero eigenvalue can be shifted away from zero.

Definitions and Basic Facts

A matrix is called *symmetric* if it is equal to its transpose,

$$\mathbf{A} = \mathbf{A}^T \qquad \text{or} \qquad a_{ij} = a_{ji} \qquad (11.0.3)$$

It is called *Hermitian* or *self-adjoint* if it equals the complex-conjugate of its transpose (its *Hermitian conjugate*, denoted by "†")

$$\mathbf{A} = \mathbf{A}^\dagger \qquad \text{or} \qquad a_{ij} = a_{ji}{}^* \qquad (11.0.4)$$

It is termed *orthogonal* if its transpose equals its inverse,

$$\mathbf{A}^T \cdot \mathbf{A} = \mathbf{A} \cdot \mathbf{A}^T = \mathbf{1} \qquad (11.0.5)$$

and *unitary* if its Hermitian conjugate equals its inverse. Finally, a matrix is called *normal* if it *commutes* with its Hermitian conjugate,

$$\mathbf{A} \cdot \mathbf{A}^\dagger = \mathbf{A}^\dagger \cdot \mathbf{A} \qquad (11.0.6)$$

For real matrices, Hermitian means the same as symmetric, unitary means the same as orthogonal, and *both* of these distinct classes are normal.

The reason that "Hermitian" is an important concept has to do with eigenvalues. The eigenvalues of a Hermitian matrix are all real. In particular, the eigenvalues of a real symmetric matrix are all real. Contrariwise, the eigenvalues of a real nonsymmetric matrix may include real values, but may also include pairs of complex conjugate values; and the eigenvalues of a complex matrix that is not Hermitian will in general be complex.

The reason that "normal" is an important concept has to do with the eigenvectors. The eigenvectors of a normal matrix with nondegenerate (i.e., distinct) eigenvalues are complete and orthogonal, spanning the N-dimensional vector space. For a normal matrix with degenerate eigenvalues, we have the additional freedom of replacing the eigenvectors corresponding to a degenerate eigenvalue by linear combinations of themselves. Using this freedom, we can always perform Gram-Schmidt orthogonalization (consult any linear algebra text) and *find* a set of eigenvectors that are complete and orthogonal, just as in the nondegenerate case. The matrix whose columns are an orthonormal set of eigenvectors is evidently unitary. A special case is that the matrix of eigenvectors of a real, symmetric matrix is orthogonal, since the eigenvectors of that matrix are all real.

When a matrix is not normal, as typified by any random, nonsymmetric, real matrix, then in general we cannot find *any* orthonormal set of eigenvectors, nor even any pairs of eigenvectors that are orthogonal (except perhaps by rare chance). While the N non-orthonormal eigenvectors will "usually" span the N-dimensional vector space, they do not always do so; that is, the eigenvectors are not always complete. Such a matrix is said to be *defective*.

Left and Right Eigenvectors

While the eigenvectors of a non-normal matrix are not particularly orthogonal among themselves, they *do* have an orthogonality relation with a different set of vectors, which we must now define. Up to now our eigenvectors have been column vectors that are multiplied to the right of a matrix $\mathbf{A}$, as in (11.0.1). These, more explicitly, are termed *right eigenvectors*. We could also, however, try to find row vectors, which multiply $\mathbf{A}$ to the left and satisfy

$$\mathbf{x} \cdot \mathbf{A} = \lambda \mathbf{x} \qquad (11.0.7)$$

These are called *left eigenvectors*. By taking the transpose of equation (11.0.7), we see that every left eigenvector is the transpose of a right eigenvector *of the transpose of* $\mathbf{A}$. Now by comparing to (11.0.2), and using the fact that the determinant of a matrix equals the determinant of its transpose, we also see that the left and right eigen*values* of $\mathbf{A}$ are identical.

If the matrix $\mathbf{A}$ is symmetric, then the left and right eigenvectors are just transposes of each other, that is, have the same numerical components. Likewise, if the matrix is self-adjoint, the left and right eigenvectors are Hermitian conjugates of each other. For the general nonnormal case, however, we have the following calculation: Let $\mathbf{X}_R$ be the matrix formed by columns from the right eigenvectors, and $\mathbf{X}_L$ be the matrix formed by rows from the left eigenvectors. Then (11.0.1) and (11.0.7) can be rewritten as

$$\mathbf{A} \cdot \mathbf{X}_R = \mathbf{X}_R \cdot \mathrm{diag}(\lambda_0 \ldots \lambda_{N-1}) \qquad \mathbf{X}_L \cdot \mathbf{A} = \mathrm{diag}(\lambda_0 \ldots \lambda_{N-1}) \cdot \mathbf{X}_L \quad (11.0.8)$$

Multiplying the first of these equations on the left by $\mathbf{X}_L$, the second on the right by $\mathbf{X}_R$, and subtracting the two, gives

$$(\mathbf{X}_L \cdot \mathbf{X}_R) \cdot \mathrm{diag}(\lambda_0 \ldots \lambda_{N-1}) = \mathrm{diag}(\lambda_0 \ldots \lambda_{N-1}) \cdot (\mathbf{X}_L \cdot \mathbf{X}_R) \qquad (11.0.9)$$

This says that the matrix of dot products of the left and right eigenvectors commutes with the diagonal matrix of eigenvalues. But the only matrices that commute with a diagonal matrix *of distinct elements* are themselves diagonal. Thus, if the eigenvalues are nondegenerate, each left eigenvector is orthogonal to all right eigenvectors except its corresponding one, and vice versa. By choice of normalization, the dot products of corresponding left and right eigenvectors can always be made unity for any matrix with nondegenerate eigenvalues.

If some eigenvalues are degenerate, then either the left or the right eigenvectors corresponding to a degenerate eigenvalue must be linearly combined among themselves to achieve orthogonality with the right or left ones, respectively. This can always be done by a procedure akin to Gram-Schmidt orthogonalization. The normalization can then be adjusted to give unity for the nonzero dot products between corresponding left and right eigenvectors. If the dot product of corresponding left and right eigenvectors is zero at this stage, then you have a case where the eigenvectors are incomplete! Note that incomplete eigenvectors can occur only where there are degenerate eigenvalues, but do not always occur in such cases (in fact, never occur for the class of "normal" matrices). See [1] for a clear discussion.

In both the degenerate and nondegenerate cases, the final normalization to unity of all nonzero dot products produces the result: The matrix whose rows are left eigenvectors is the inverse matrix of the matrix whose columns are right eigenvectors, *if the inverse exists.*

Diagonalization of a Matrix

Multiplying the first equation in (11.0.8) by $\mathbf{X}_L$, and using the fact that $\mathbf{X}_L$ and $\mathbf{X}_R$ are matrix inverses, we get

$$\mathbf{X}_R^{-1} \cdot \mathbf{A} \cdot \mathbf{X}_R = \mathrm{diag}(\lambda_0 \dots \lambda_{N-1}) \qquad (11.0.10)$$

This is a particular case of a *similarity transform* of the matrix $\mathbf{A}$,

$$\mathbf{A} \quad \rightarrow \quad \mathbf{Z}^{-1} \cdot \mathbf{A} \cdot \mathbf{Z} \qquad (11.0.11)$$

for some transformation matrix $\mathbf{Z}$. Similarity transformations play a crucial role in the computation of eigenvalues, because they leave the eigenvalues of a matrix unchanged. This is easily seen from

$$\det \left| \mathbf{Z}^{-1} \cdot \mathbf{A} \cdot \mathbf{Z} - \lambda \mathbf{1} \right| = \det \left| \mathbf{Z}^{-1} \cdot (\mathbf{A} - \lambda \mathbf{1}) \cdot \mathbf{Z} \right|$$

$$= \det |\mathbf{Z}| \; \det |\mathbf{A} - \lambda \mathbf{1}| \; \det \left| \mathbf{Z}^{-1} \right| \qquad (11.0.12)$$

$$= \det |\mathbf{A} - \lambda \mathbf{1}|$$

Equation (11.0.10) shows that any matrix with complete eigenvectors (which includes all normal matrices and "most" random nonnormal ones) can be diagonalized by a similarity transformation, that the columns of the transformation matrix that effects the diagonalization are the right eigenvectors, and that the rows of its inverse are the left eigenvectors.

For real, symmetric matrices, the eigenvectors are real and orthonormal, so the transformation matrix is orthogonal. The similarity transformation is then also an *orthogonal transformation* of the form

$$\mathbf{A} \quad \rightarrow \quad \mathbf{Z}^T \cdot \mathbf{A} \cdot \mathbf{Z} \qquad (11.0.13)$$

While real nonsymmetric matrices can be diagonalized in their usual case of complete eigenvectors, the transformation matrix is not necessarily real. It turns out, however, that a real similarity transformation can "almost" do the job. It can reduce the matrix down to a form with little two-by-two blocks along the diagonal, all other elements zero. Each two-by-two block corresponds to a complex-conjugate pair of complex eigenvalues. We will see this idea exploited in some routines given later in the chapter.

The "grand strategy" of virtually all modern eigensystem routines is to nudge the matrix $\mathbf{A}$ towards diagonal form by a sequence of similarity transformations,

$$\mathbf{A} \quad \rightarrow \quad \mathbf{P}_1^{-1} \cdot \mathbf{A} \cdot \mathbf{P}_1 \quad \rightarrow \quad \mathbf{P}_2^{-1} \cdot \mathbf{P}_1^{-1} \cdot \mathbf{A} \cdot \mathbf{P}_1 \cdot \mathbf{P}_2$$

$$\rightarrow \quad \mathbf{P}_3^{-1} \cdot \mathbf{P}_2^{-1} \cdot \mathbf{P}_1^{-1} \cdot \mathbf{A} \cdot \mathbf{P}_1 \cdot \mathbf{P}_2 \cdot \mathbf{P}_3 \quad \rightarrow \quad \text{etc.} \qquad (11.0.14)$$

If we get all the way to diagonal form, then the eigenvectors are the columns of the accumulated transformation

$$\mathbf{X}_R = \mathbf{P}_1 \cdot \mathbf{P}_2 \cdot \mathbf{P}_3 \cdots \qquad (11.0.15)$$

Sometimes we do not want to go all the way to diagonal form. For example, if we are interested only in eigenvalues, not eigenvectors, it is enough to transform the matrix $\mathbf{A}$ to be triangular, with all elements below (or above) the diagonal zero. In this case the diagonal elements are already the eigenvalues, as you can see by mentally evaluating (11.0.2) using expansion by minors.

There are two rather different sets of techniques for implementing the grand strategy (11.0.14). It turns out that they work rather well in combination, so most modern eigensystem routines use both. The first set of techniques constructs individual $\mathbf{P}_i$'s as explicit "atomic" transformations designed to perform specific tasks, for example zeroing a particular off-diagonal element (Jacobi transformation, §11.1), or a whole particular row or column (Householder transformation, §11.2; elimination method, §11.5). In general, a finite sequence of these simple transformations cannot completely diagonalize a matrix. There are then two choices: either use the finite sequence of transformations to go most of the way (e.g., to some special form like *tridiagonal* or *Hessenberg*, see §11.2 and §11.5 below) and follow up with the second set of techniques about to be mentioned; or else iterate the finite sequence of simple transformations over and over until the deviation of the matrix from diagonal is negligibly small. This latter approach is conceptually simplest, so we will discuss it in the next section; however, for N greater than ~ 10, it is computationally inefficient by a roughly constant factor ~ 5.

The second set of techniques, called *factorization methods*, is more subtle. Suppose that the matrix $\mathbf{A}$ can be factored into a left factor $\mathbf{F}_L$ and a right factor $\mathbf{F}_R$. Then

$$\mathbf{A} = \mathbf{F}_L \cdot \mathbf{F}_R \qquad \text{or equivalently} \qquad \mathbf{F}_L^{-1} \cdot \mathbf{A} = \mathbf{F}_R \qquad (11.0.16)$$

If we now multiply back together the factors in the reverse order, and use the second equation in (11.0.16) we get

$$\mathbf{F}_R \cdot \mathbf{F}_L = \mathbf{F}_L^{-1} \cdot \mathbf{A} \cdot \mathbf{F}_L \qquad (11.0.17)$$

which we recognize as having effected a similarity transformation on $\mathbf{A}$ with the transformation matrix being $\mathbf{F}_L$! In §11.3 and §11.6 we will discuss the *QR method* which exploits this idea.

Factorization methods also do not converge exactly in a finite number of transformations. But the better ones do converge rapidly and reliably, and, when following an appropriate initial reduction by simple similarity transformations, they are the methods of choice.

"Eigenpackages of Canned Eigenroutines"

You have probably gathered by now that the solution of eigensystems is a fairly complicated business. It is. It is one of the few subjects covered in this book for which we do *not* recommend that you avoid canned routines. On the contrary, the purpose of this chapter is precisely to give you some appreciation of what is going on inside such canned routines, so that you can make intelligent choices about using them, and intelligent diagnoses when something goes wrong.

You will find that almost all canned routines in use nowadays trace their ancestry back to routines published in Wilkinson and Reinsch's *Handbook for Automatic Computation, Vol. II, Linear Algebra* [2]. This excellent reference, containing papers by a number of authors, is the Bible of the field. A public-domain implementation of the *Handbook* routines in Fortran is the EISPACK set of programs [3]. The routines in this chapter are translations of either the *Handbook* or EISPACK routines, so understanding these will take you a lot of the way towards understanding those canonical packages.

IMSL [4] and NAG [5] each provide proprietary implementations, in Fortran and C, of what are essentially the Handbook routines.

A good "eigenpackage" will provide separate routines, or separate paths through sequences of routines, for the following desired calculations:
- all eigenvalues and no eigenvectors
- all eigenvalues and some corresponding eigenvectors
- all eigenvalues and all corresponding eigenvectors

The purpose of these distinctions is to save compute time and storage; it is wasteful to calculate eigenvectors that you don't need. Often one is interested only in the eigenvectors corresponding to the largest few eigenvalues, or largest few in magnitude, or few that are negative. The method usually used to calculate "some" eigenvectors is typically more efficient than calculating all eigenvectors if you desire fewer than about a quarter of the eigenvectors.

A good eigenpackage also provides separate paths for each of the above calculations for each of the following special forms of the matrix:
- real, symmetric, tridiagonal
- real, symmetric, banded (only a small number of sub- and superdiagonals are nonzero)
- real, symmetric
- real, nonsymmetric
- complex, Hermitian
- complex, non-Hermitian

Again, the purpose of these distinctions is to save time and storage by using the *least* general routine that will serve in any particular application.

In this chapter, as a bare introduction, we give good routines for the following paths:
- all eigenvalues and eigenvectors of a real, symmetric, tridiagonal matrix (§11.3)
- all eigenvalues and eigenvectors of a real, symmetric, matrix (§11.1–§11.3)
- all eigenvalues and eigenvectors of a complex, Hermitian matrix (§11.4)
- all eigenvalues and no eigenvectors of a real, nonsymmetric matrix

(§11.5–§11.6)

We also discuss, in §11.7, how to obtain some eigenvectors of nonsymmetric matrices by the method of inverse iteration.

Generalized and Nonlinear Eigenvalue Problems

Many eigenpackages also deal with the so-called *generalized eigenproblem*, [6]

$$\mathbf{A} \cdot \mathbf{x} = \lambda \mathbf{B} \cdot \mathbf{x} \tag{11.0.18}$$

where $\mathbf{A}$ and $\mathbf{B}$ are both matrices. Most such problems, where $\mathbf{B}$ is nonsingular, can be handled by the equivalent

$$(\mathbf{B}^{-1} \cdot \mathbf{A}) \cdot \mathbf{x} = \lambda \mathbf{x} \tag{11.0.19}$$

Often $\mathbf{A}$ and $\mathbf{B}$ are symmetric and $\mathbf{B}$ is positive definite. The matrix $\mathbf{B}^{-1} \cdot \mathbf{A}$ in (11.0.19) is not symmetric, but we can recover a symmetric eigenvalue problem by using the Cholesky decomposition $\mathbf{B} = \mathbf{L} \cdot \mathbf{L}^T$ of §2.9. Multiplying equation (11.0.18) by $\mathbf{L}^{-1}$, we get

$$\mathbf{C} \cdot (\mathbf{L}^T \cdot \mathbf{x}) = \lambda (\mathbf{L}^T \cdot \mathbf{x}) \tag{11.0.20}$$

where

$$\mathbf{C} = \mathbf{L}^{-1} \cdot \mathbf{A} \cdot (\mathbf{L}^{-1})^T \tag{11.0.21}$$

The matrix $\mathbf{C}$ is symmetric and its eigenvalues are the same as those of the original problem (11.0.18); its eigenfunctions are $\mathbf{L}^T \cdot \mathbf{x}$. The efficient way to form $\mathbf{C}$ is first to solve the equation

$$\mathbf{Y} \cdot \mathbf{L}^T = \mathbf{A} \tag{11.0.22}$$

for the lower triangle of the matrix $\mathbf{Y}$. Then solve

$$\mathbf{L} \cdot \mathbf{C} = \mathbf{Y} \tag{11.0.23}$$

for the lower triangle of the symmetric matrix $\mathbf{C}$.

Another generalization of the standard eigenvalue problem is to problems nonlinear in the eigenvalue λ, for example,

$$(\mathbf{A}\lambda^2 + \mathbf{B}\lambda + \mathbf{C}) \cdot \mathbf{x} = 0 \tag{11.0.24}$$

This can be turned into a linear problem by introducing an additional unknown eigenvector $\mathbf{y}$ and solving the $2N \times 2N$ eigensystem,

$$\begin{pmatrix} 0 & 1 \\ -\mathbf{A}^{-1} \cdot \mathbf{C} & -\mathbf{A}^{-1} \cdot \mathbf{B} \end{pmatrix} \cdot \begin{pmatrix} \mathbf{x} \\ \mathbf{y} \end{pmatrix} = \lambda \begin{pmatrix} \mathbf{x} \\ \mathbf{y} \end{pmatrix} \tag{11.0.25}$$

This technique generalizes to higher-order polynomials in λ. A polynomial of degree M produces a linear $MN \times MN$ eigensystem (see [7]).

CITED REFERENCES AND FURTHER READING:

Stoer, J., and Bulirsch, R. 1993, *Introduction to Numerical Analysis*, 2nd ed. (New York: Springer-Verlag), Chapter 6. [1]

Wilkinson, J.H., and Reinsch, C. 1971, *Linear Algebra*, vol. II of *Handbook for Automatic Computation* (New York: Springer-Verlag). [2]

Smith, B.T., et al. 1976, *Matrix Eigensystem Routines — EISPACK Guide*, 2nd ed., vol. 6 of Lecture Notes in Computer Science (New York: Springer-Verlag). [3]

IMSL Math/Library Users Manual (IMSL Inc., 2500 CityWest Boulevard, Houston TX 77042). [4]

NAG Fortran Library (Numerical Algorithms Group, 256 Banbury Road, Oxford OX27DE, U.K.), Chapter F02. [5]

Golub, G.H., and Van Loan, C.F. 1996, *Matrix Computations*, 3rd ed. (Baltimore: Johns Hopkins University Press), §7.7. [6]

Wilkinson, J.H. 1965, *The Algebraic Eigenvalue Problem* (New York: Oxford University Press). [7]

Acton, F.S. 1970, *Numerical Methods That Work*; 1990, corrected edition (Washington: Mathematical Association of America), Chapter 13.

Horn, R.A., and Johnson, C.R. 1985, *Matrix Analysis* (Cambridge: Cambridge University Press).

11.1 Jacobi Transformations of a Symmetric Matrix

The Jacobi method consists of a sequence of orthogonal similarity transformations of the form of equation (11.0.14). Each transformation (a *Jacobi rotation*) is just a plane rotation designed to annihilate one of the off-diagonal matrix elements. Successive transformations undo previously set zeros, but the off-diagonal elements nevertheless get smaller and smaller, until the matrix is diagonal to machine precision. Accumulating the product of the transformations as you go gives the matrix of eigenvectors, equation (11.0.15), while the elements of the final diagonal matrix are the eigenvalues.

The Jacobi method is absolutely foolproof for all real symmetric matrices. For matrices of order greater than about 10, say, the algorithm is slower, by a significant constant factor, than the QR method we shall give in §11.3. However, the Jacobi algorithm is much simpler than the more efficient methods. We thus recommend it for matrices of moderate order, where expense is not a major consideration.

The basic Jacobi rotation $\mathbf{P}_{pq}$ is a matrix of the form

$$
\mathbf{P}_{pq} =
\begin{bmatrix}
1 & & & & & & \\
 & \cdots & & & & & \\
 & & c & \cdots & s & & \\
 & & \vdots & 1 & \vdots & & \\
 & & -s & \cdots & c & & \\
 & & & & & \cdots & \\
 & & & & & & 1
\end{bmatrix}
\tag{11.1.1}
$$

Here all the diagonal elements are unity except for the two elements c in rows (and columns) p and q. All off-diagonal elements are zero except the two elements s and $-s$. The numbers c and s are the cosine and sine of a rotation angle ϕ, so $c^2 + s^2 = 1$.

A plane rotation such as (11.1.1) is used to transform the matrix $\mathbf{A}$ according to

$$\mathbf{A}' = \mathbf{P}_{pq}^T \cdot \mathbf{A} \cdot \mathbf{P}_{pq} \qquad (11.1.2)$$

Now, $\mathbf{P}_{pq}^T \cdot \mathbf{A}$ changes only rows p and q of $\mathbf{A}$, while $\mathbf{A} \cdot \mathbf{P}_{pq}$ changes only columns p and q. Notice that the subscripts p and q do not denote components of $\mathbf{P}_{pq}$, but rather label which kind of rotation the matrix is, i.e., which rows and columns it affects. Thus the changed elements of $\mathbf{A}$ in (11.1.2) are only in rows p and q, and columns p and q, as indicated below:

$$\mathbf{A}' = \begin{bmatrix} & & a'_{0p} & & a'_{0q} & & \\ & & \vdots & & \vdots & & \\ a'_{p0} & \cdots & a'_{pp} & \cdots & a'_{pq} & \cdots & a'_{p,n-1} \\ & & \vdots & & \vdots & & \\ a'_{q0} & \cdots & a'_{qp} & \cdots & a'_{qq} & \cdots & a'_{q,n-1} \\ & & \vdots & & \vdots & & \\ & & a'_{n-1,p} & & a'_{n-1,q} & & \end{bmatrix} \qquad (11.1.3)$$

Multiplying out equation (11.1.2) and using the symmetry of $\mathbf{A}$, we get the explicit formulas

$$\begin{aligned} a'_{rp} &= ca_{rp} - sa_{rq} \\ a'_{rq} &= ca_{rq} + sa_{rp} \end{aligned} \qquad r \neq p,\ r \neq q \qquad (11.1.4)$$

$$a'_{pp} = c^2 a_{pp} + s^2 a_{qq} - 2sc\, a_{pq} \qquad (11.1.5)$$

$$a'_{qq} = s^2 a_{pp} + c^2 a_{qq} + 2sc\, a_{pq} \qquad (11.1.6)$$

$$a'_{pq} = (c^2 - s^2) a_{pq} + sc(a_{pp} - a_{qq}) \qquad (11.1.7)$$

The idea of the Jacobi method is to try to zero the off-diagonal elements by a series of plane rotations. Accordingly, to set $a'_{pq} = 0$, equation (11.1.7) gives the following expression for the rotation angle ϕ

$$\theta \equiv \cot 2\phi \equiv \frac{c^2 - s^2}{2sc} = \frac{a_{qq} - a_{pp}}{2a_{pq}} \qquad (11.1.8)$$

If we let $t \equiv s/c$, the definition of θ can be rewritten

$$t^2 + 2t\theta - 1 = 0 \qquad (11.1.9)$$

The smaller root of this equation corresponds to a rotation angle less than $\pi/4$ in magnitude; this choice at each stage gives the most stable reduction. Using the form of the quadratic formula with the discriminant in the denominator, we can write this smaller root as

$$t = \frac{\operatorname{sgn}(\theta)}{|\theta| + \sqrt{\theta^2 + 1}} \qquad (11.1.10)$$

If θ is so large that θ^2 would overflow on the computer, we set $t = 1/(2\theta)$. It now follows that

$$c = \frac{1}{\sqrt{t^2 + 1}} \tag{11.1.11}$$

$$s = tc \tag{11.1.12}$$

When we actually use equations (11.1.4)–(11.1.7) numerically, we rewrite them to minimize roundoff error. Equation (11.1.7) is replaced by

$$a'_{pq} = 0 \tag{11.1.13}$$

The idea in the remaining equations is to set the new quantity equal to the old quantity plus a small correction. Thus we can use (11.1.7) and (11.1.13) to eliminate a_{qq} from (11.1.5), giving

$$a'_{pp} = a_{pp} - t a_{pq} \tag{11.1.14}$$

Similarly,

$$a'_{qq} = a_{qq} + t a_{pq} \tag{11.1.15}$$

$$a'_{rp} = a_{rp} - s(a_{rq} + \tau a_{rp}) \tag{11.1.16}$$

$$a'_{rq} = a_{rq} + s(a_{rp} - \tau a_{rq}) \tag{11.1.17}$$

where τ $(= \tan \phi/2)$ is defined by

$$\tau \equiv \frac{s}{1 + c} \tag{11.1.18}$$

One can see the convergence of the Jacobi method by considering the sum of the squares of the off-diagonal elements

$$S = \sum_{r \neq s} |a_{rs}|^2 \tag{11.1.19}$$

Equations (11.1.4)–(11.1.7) imply that

$$S' = S - 2|a_{pq}|^2 \tag{11.1.20}$$

(Since the transformation is orthogonal, the sum of the squares of the diagonal elements increases correspondingly by $2|a_{pq}|^2$.) The sequence of S's thus decreases monotonically. Since the sequence is bounded below by zero, and since we can choose a_{pq} to be whatever element we want, the sequence can be made to converge to zero.

Eventually one obtains a matrix $\mathbf{D}$ that is diagonal to machine precision. The diagonal elements give the eigenvalues of the original matrix $\mathbf{A}$, since

$$\mathbf{D} = \mathbf{V}^T \cdot \mathbf{A} \cdot \mathbf{V} \tag{11.1.21}$$

where
$$\mathbf{V} = \mathbf{P}_1 \cdot \mathbf{P}_2 \cdot \mathbf{P}_3 \cdots \tag{11.1.22}$$

the $\mathbf{P}_i$'s being the successive Jacobi rotation matrices. The columns of $\mathbf{V}$ are the eigenvectors (since $\mathbf{A} \cdot \mathbf{V} = \mathbf{V} \cdot \mathbf{D}$). They can be computed by applying

$$\mathbf{V}' = \mathbf{V} \cdot \mathbf{P}_i \tag{11.1.23}$$

at each stage of calculation, where initially $\mathbf{V}$ is the identity matrix. In detail, equation (11.1.23) is

$$
\begin{aligned}
v'_{rs} &= v_{rs} \qquad (s \neq p,\ s \neq q) \\
v'_{rp} &= cv_{rp} - sv_{rq} \\
v'_{rq} &= sv_{rp} + cv_{rq}
\end{aligned}
\tag{11.1.24}
$$

We rewrite these equations in terms of τ as in equations (11.1.16) and (11.1.17) to minimize roundoff.

The only remaining question is the strategy one should adopt for the order in which the elements are to be annihilated. Jacobi's original algorithm of 1846 searched the whole upper triangle at each stage and set the largest off-diagonal element to zero. This is a reasonable strategy for hand calculation, but it is prohibitive on a computer since the search alone makes each Jacobi rotation a process of order N^2 instead of N.

A better strategy for our purposes is the *cyclic Jacobi method*, where one annihilates elements in strict order. For example, one can simply proceed down the rows: $\mathbf{P}_{01}, \mathbf{P}_{02}, ..., \mathbf{P}_{0,n-1}$; then $\mathbf{P}_{12}, \mathbf{P}_{13}$, etc. One can show that convergence is generally quadratic for both the original or the cyclic Jacobi methods, for nondegenerate eigenvalues. One such set of $n(n-1)/2$ Jacobi rotations is called a *sweep*.

The program below, based on the implementations in [1,2], uses two further refinements:

- In the first three sweeps, we carry out the pq rotation only if $|a_{pq}| > \epsilon$ for some threshold value

$$\epsilon = \frac{1}{5}\frac{S_0}{n^2} \tag{11.1.25}$$

where S_0 is the sum of the off-diagonal moduli,

$$S_0 = \sum_{r<s} |a_{rs}| \tag{11.1.26}$$

- After four sweeps, if $|a_{pq}| \ll |a_{pp}|$ and $|a_{pq}| \ll |a_{qq}|$, we set $|a_{pq}| = 0$ and skip the rotation. The criterion used in the comparison is $|a_{pq}| < 10^{-(D+2)}|a_{pp}|$, where D is the number of significant decimal digits on the machine, and similarly for $|a_{qq}|$.

In the following routine the n×n symmetric matrix a is stored as a[0..n-1] [0..n-1]. On output, the superdiagonal elements of a are destroyed, but the diagonal and subdiagonal are unchanged and give full information on the original symmetric matrix a. The vector d[0..n-1] returns the eigenvalues of a. During the computation, it contains the current diagonal of a. The matrix v[0..n-1][0..n-1] outputs the normalized eigenvector belonging to d[k] in column k. The parameter nrot is the number of Jacobi rotations that were needed to achieve convergence.

Typical matrices require 6 to 10 sweeps to achieve convergence, or $3n^2$ to $5n^2$ Jacobi rotations. Each rotation requires of order $4n$ operations, each consisting of a multiply and an add, so the total labor is of order $12n^3$ to $20n^3$ operations. Calculation of the eigenvectors as well as the eigenvalues changes the operation count from $4n$ to $6n$ per rotation, which is only a 50 percent overhead.

```
#include <cmath>
#include "nr.h"
using namespace std;

namespace {
    inline void rot(Mat_IO_DP &a, const DP s, const DP tau, const int i,
        const int j, const int k, const int l)
    {
        DP g,h;

        g=a[i][j];
        h=a[k][l];
        a[i][j]=g-s*(h+g*tau);
        a[k][l]=h+s*(g-h*tau);
    }
}

void NR::jacobi(Mat_IO_DP &a, Vec_O_DP &d, Mat_O_DP &v, int &nrot)
```
Computes all eigenvalues and eigenvectors of a real symmetric matrix a[0..n-1][0..n-1]. On output, elements of a above the diagonal are destroyed. d[0..n-1] returns the eigenvalues of a. v[0..n-1][0..n-1] is a matrix whose columns contain, on output, the normalized eigenvectors of a. nrot returns the number of Jacobi rotations that were required.
```
{
    int i,j,ip,iq;
    DP tresh,theta,tau,t,sm,s,h,g,c;

    int n=d.size();
    Vec_DP b(n),z(n);
    for (ip=0;ip<n;ip++) {                        Initialize to the identity matrix.
        for (iq=0;iq<n;iq++) v[ip][iq]=0.0;
        v[ip][ip]=1.0;
    }
    for (ip=0;ip<n;ip++) {                        Initialize b and d to the diagonal
        b[ip]=d[ip]=a[ip][ip];                        of a.
        z[ip]=0.0;                                This vector will accumulate terms
    }                                                 of the form $ta_{pq}$ as in equa-
    nrot=0;                                           tion (11.1.14).
    for (i=1;i<=50;i++) {
        sm=0.0;
        for (ip=0;ip<n-1;ip++) {                  Sum magnitude of off-diagonal
            for (iq=ip+1;iq<n;iq++)                   elements.
                sm += fabs(a[ip][iq]);
        }
```

```
    if (sm == 0.0)                          The normal return, which relies
        return;                             on quadratic convergence to
    if (i < 4)                              machine underflow.
        tresh=0.2*sm/(n*n);                 ...on the first three sweeps.
    else
        tresh=0.0;                          ...thereafter.
    for (ip=0;ip<n-1;ip++) {
        for (iq=ip+1;iq<n;iq++) {
            g=100.0*fabs(a[ip][iq]);
            After four sweeps, skip the rotation if the off-diagonal element is small.
            if (i > 4 && (fabs(d[ip])+g) == fabs(d[ip])
                && (fabs(d[iq])+g) == fabs(d[iq]))
                    a[ip][iq]=0.0;
            else if (fabs(a[ip][iq]) > tresh) {
                h=d[iq]-d[ip];
                if ((fabs(h)+g) == fabs(h))
                    t=(a[ip][iq])/h;                    t = 1/(2θ)
                else {
                    theta=0.5*h/(a[ip][iq]);     Equation (11.1.10).
                    t=1.0/(fabs(theta)+sqrt(1.0+theta*theta));
                    if (theta < 0.0) t = -t;
                }
                c=1.0/sqrt(1+t*t);
                s=t*c;
                tau=s/(1.0+c);
                h=t*a[ip][iq];
                z[ip] -= h;
                z[iq] += h;
                d[ip] -= h;
                d[iq] += h;
                a[ip][iq]=0.0;
                for (j=0;j<ip;j++)                    Case of rotations 0 ≤ j < p.
                    rot(a,s,tau,j,ip,j,iq);
                for (j=ip+1;j<iq;j++)                 Case of rotations p < j < q.
                    rot(a,s,tau,ip,j,j,iq);
                for (j=iq+1;j<n;j++)                  Case of rotations q < j < n.
                    rot(a,s,tau,ip,j,iq,j);
                for (j=0;j<n;j++)
                    rot(v,s,tau,j,ip,j,iq);
                ++nrot;
            }
        }
    }
    for (ip=0;ip<n;ip++) {
        b[ip] += z[ip];
        d[ip]=b[ip];                         Update d with the sum of ta_pq,
        z[ip]=0.0;                           and reinitialize z.
    }
}
nrerror("Too many iterations in routine jacobi");
}
```

Note that the above routine assumes that underflows are set to zero. On machines where this is not true, the program must be modified.

The eigenvalues are not ordered on output. If sorting is desired, the following routine can be invoked to reorder the output of jacobi or of later routines in this chapter. (The method, straight insertion, is N^2 rather than $N \log N$; but since you have just done an N^3 procedure to get the eigenvalues, you can afford yourself this little indulgence.)

```
#include "nr.h"

void NR::eigsrt(Vec_IO_DP &d, Mat_IO_DP &v)
```
Given the eigenvalues d[0..n-1] and eigenvectors v[0..n-1][0..n-1] as output from
jacobi (§11.1) or tqli (§11.3), this routine sorts the eigenvalues into descending order, and
rearranges the columns of v correspondingly. The method is straight insertion.
```
{
    int i,j,k;
    DP p;

    int n=d.size();
    for (i=0;i<n-1;i++) {
        p=d[k=i];
        for (j=i;j<n;j++)
            if (d[j] >= p) p=d[k=j];
        if (k != i) {
            d[k]=d[i];
            d[i]=p;
            for (j=0;j<n;j++) {
                p=v[j][i];
                v[j][i]=v[j][k];
                v[j][k]=p;
            }
        }
    }
}
```

CITED REFERENCES AND FURTHER READING:

Golub, G.H., and Van Loan, C.F. 1996, *Matrix Computations*, 3rd ed. (Baltimore: Johns Hopkins
    University Press), §8.4.

Smith, B.T., et al. 1976, *Matrix Eigensystem Routines — EISPACK Guide*, 2nd ed., vol. 6 of
    Lecture Notes in Computer Science (New York: Springer-Verlag). [1]

Wilkinson, J.H., and Reinsch, C. 1971, *Linear Algebra*, vol. II of *Handbook for Automatic Com-
    putation* (New York: Springer-Verlag). [2]

11.2 Reduction of a Symmetric Matrix to Tridiagonal Form: Givens and Householder Reductions

As already mentioned, the optimum strategy for finding eigenvalues and
eigenvectors is, first, to reduce the matrix to a simple form, only then beginning an
iterative procedure. For symmetric matrices, the preferred simple form is tridiagonal.
The *Givens reduction* is a modification of the Jacobi method. Instead of trying to
reduce the matrix all the way to diagonal form, we are content to stop when the
matrix is tridiagonal. This allows the procedure to be carried out *in a finite number
of steps*, unlike the Jacobi method, which requires iteration to convergence.

Givens Method

For the Givens method, we choose the rotation angle in equation (11.1.1) so as to zero an element that is *not* at one of the four "corners," i.e., not a_{pp}, a_{pq}, or a_{qq} in equation (11.1.3). Specifically, we first choose $\mathbf{P}_{12}$ to annihilate a_{20} (and, by symmetry, a_{02}). Then we choose $\mathbf{P}_{13}$ to annihilate a_{30}. In general, we choose the sequence

$$\mathbf{P}_{12}, \mathbf{P}_{13}, \ldots, \mathbf{P}_{1,n-1}; \mathbf{P}_{23}, \ldots, \mathbf{P}_{2,n-1}; \cdots ; \mathbf{P}_{n-2,n-1}$$

where $\mathbf{P}_{jk}$ annihilates $a_{k,j-1}$. The method works because elements such as a'_{rp} and a'_{rq}, with $r \neq p$ $r \neq q$, are linear combinations of the old quantities a_{rp} and a_{rq}, by equation (11.1.4). Thus, if a_{rp} and a_{rq} have already been set to zero, they remain zero as the reduction proceeds. Evidently, of order $n^2/2$ rotations are required, and the number of multiplications in a straightforward implementation is of order $4n^3/3$, not counting those for keeping track of the product of the transformation matrices, required for the eigenvectors.

The Householder method, to be discussed next, is just as stable as the Givens reduction and it is a factor of 2 more efficient, so the Givens method is not generally used. Recent work (see [1]) has shown that the Givens reduction can be reformulated to reduce the number of operations by a factor of 2, and also avoid the necessity of taking square roots. This appears to make the algorithm competitive with the Householder reduction. However, this "fast Givens" reduction has to be monitored to avoid overflows, and the variables have to be periodically rescaled. There does not seem to be any compelling reason to prefer the Givens reduction over the Householder method.

Householder Method

The Householder algorithm reduces an $n \times n$ symmetric matrix $\mathbf{A}$ to tridiagonal form by $n - 2$ orthogonal transformations. Each transformation annihilates the required part of a whole column and whole corresponding row. The basic ingredient is a Householder matrix $\mathbf{P}$, which has the form

$$\mathbf{P} = \mathbf{1} - 2\mathbf{w} \cdot \mathbf{w}^T \tag{11.2.1}$$

where $\mathbf{w}$ is a real vector with $|\mathbf{w}|^2 = 1$. (In the present notation, the *outer* or matrix product of two vectors, $\mathbf{a}$ and $\mathbf{b}$ is written $\mathbf{a} \cdot \mathbf{b}^T$, while the *inner* or scalar product of the vectors is written as $\mathbf{a}^T \cdot \mathbf{b}$.) The matrix $\mathbf{P}$ is orthogonal, because

$$\begin{aligned}
\mathbf{P}^2 &= (\mathbf{1} - 2\mathbf{w} \cdot \mathbf{w}^T) \cdot (\mathbf{1} - 2\mathbf{w} \cdot \mathbf{w}^T) \\
&= \mathbf{1} - 4\mathbf{w} \cdot \mathbf{w}^T + 4\mathbf{w} \cdot (\mathbf{w}^T \cdot \mathbf{w}) \cdot \mathbf{w}^T \\
&= \mathbf{1}
\end{aligned} \tag{11.2.2}$$

Therefore $\mathbf{P} = \mathbf{P}^{-1}$. But $\mathbf{P}^T = \mathbf{P}$, and so $\mathbf{P}^T = \mathbf{P}^{-1}$, proving orthogonality.

Rewrite $\mathbf{P}$ as

$$\mathbf{P} = \mathbf{1} - \frac{\mathbf{u} \cdot \mathbf{u}^T}{H} \tag{11.2.3}$$

where the scalar H is

$$H \equiv \frac{1}{2}|\mathbf{u}|^2 \tag{11.2.4}$$

and $\mathbf{u}$ can now be any vector. Suppose $\mathbf{x}$ is the vector composed of the first column of $\mathbf{A}$. Choose

$$\mathbf{u} = \mathbf{x} \mp |\mathbf{x}|\mathbf{e}_0 \tag{11.2.5}$$

where $\mathbf{e}_0$ is the unit vector $[1, 0, \ldots, 0]^T$, and the choice of signs will be made later. Then

$$
\begin{aligned}
\mathbf{P} \cdot \mathbf{x} &= \mathbf{x} - \frac{\mathbf{u}}{H} \cdot (\mathbf{x} \mp |\mathbf{x}|\mathbf{e}_0)^T \cdot \mathbf{x} \\
&= \mathbf{x} - \frac{2\mathbf{u} \cdot (|\mathbf{x}|^2 \mp |\mathbf{x}|x_0)}{2|\mathbf{x}|^2 \mp 2|\mathbf{x}|x_0} \\
&= \mathbf{x} - \mathbf{u} \\
&= \pm|\mathbf{x}|\mathbf{e}_0
\end{aligned}
\tag{11.2.6}
$$

This shows that the Householder matrix $\mathbf{P}$ acts on a given vector $\mathbf{x}$ to zero all its elements except the first one.

To reduce a symmetric matrix $\mathbf{A}$ to tridiagonal form, we choose the vector $\mathbf{x}$ for the first Householder matrix to be the lower $n-1$ elements of column 0. Then the lower $n-2$ elements will be zeroed:

$$
\mathbf{P}_1 \cdot \mathbf{A} =
\begin{bmatrix}
1 & 0 & 0 & \cdots & 0 \\
0 & & & & \\
0 & & & & \\
\vdots & & {}^{(n-1)}\mathbf{P}_1 & & \\
0 & & & &
\end{bmatrix}
\cdot
\begin{bmatrix}
a_{00} & a_{01} & a_{02} & \cdots & a_{0,n-1} \\
a_{10} & & & & \\
a_{20} & & & & \\
\vdots & & & \text{irrelevant} & \\
a_{n-1,0} & & & &
\end{bmatrix}
$$

$$
=
\begin{bmatrix}
a_{00} & a_{01} & a_{02} & \cdots & a_{0,n-1} \\
k & & & & \\
0 & & & & \\
\vdots & & & \text{irrelevant} & \\
0 & & & &
\end{bmatrix}
\tag{11.2.7}
$$

Here we have written the matrices in partitioned form, with ${}^{(n-1)}\mathbf{P}$ denoting a Householder matrix with dimensions $(n-1) \times (n-1)$. The quantity k is simply plus or minus the magnitude of the vector $[a_{10}, \ldots, a_{n-1,0}]^T$.

The complete orthogonal transformation is now

$$
\mathbf{A}' = \mathbf{P} \cdot \mathbf{A} \cdot \mathbf{P} =
\begin{bmatrix}
a_{00} & k & 0 & \cdots & 0 \\
k & & & & \\
0 & & & & \\
\vdots & & & \text{irrelevant} & \\
0 & & & &
\end{bmatrix}
\tag{11.2.8}
$$

We have used the fact that $\mathbf{P}^T = \mathbf{P}$.

Now choose the vector $\mathbf{x}$ for the second Householder matrix to be the bottom $n-2$ elements of column 1, and from it construct

$$
\mathbf{P}_2 \equiv
\begin{bmatrix}
1 & 0 & 0 & \cdots & & 0 \\
0 & 1 & 0 & \cdots & & 0 \\
0 & 0 & & & & \\
\vdots & \vdots & & {}^{(n-2)}\mathbf{P}_2 & & \\
0 & 0 & & & &
\end{bmatrix}
\tag{11.2.9}
$$

The identity block in the upper left corner insures that the tridiagonalization achieved in the first step will not be spoiled by this one, while the $(n-2)$-dimensional Householder matrix ${}^{(n-2)}\mathbf{P}_2$ creates one additional row and column of the tridiagonal output. Clearly, a sequence of $n-2$ such transformations will reduce the matrix $\mathbf{A}$ to tridiagonal form.

Instead of actually carrying out the matrix multiplications in $\mathbf{P} \cdot \mathbf{A} \cdot \mathbf{P}$, we compute a vector

$$
\mathbf{p} \equiv \frac{\mathbf{A} \cdot \mathbf{u}}{H}
\tag{11.2.10}
$$

Then

$$
\mathbf{A} \cdot \mathbf{P} = \mathbf{A} \cdot (1 - \frac{\mathbf{u} \cdot \mathbf{u}^T}{H}) = \mathbf{A} - \mathbf{p} \cdot \mathbf{u}^T
$$
$$
\mathbf{A}' = \mathbf{P} \cdot \mathbf{A} \cdot \mathbf{P} = \mathbf{A} - \mathbf{p} \cdot \mathbf{u}^T - \mathbf{u} \cdot \mathbf{p}^T + 2K\mathbf{u} \cdot \mathbf{u}^T
$$

where the scalar K is defined by

$$
K = \frac{\mathbf{u}^T \cdot \mathbf{p}}{2H}
\tag{11.2.11}
$$

If we write

$$
\mathbf{q} \equiv \mathbf{p} - K\mathbf{u}
\tag{11.2.12}
$$

then we have

$$
\mathbf{A}' = \mathbf{A} - \mathbf{q} \cdot \mathbf{u}^T - \mathbf{u} \cdot \mathbf{q}^T
\tag{11.2.13}
$$

This is the computationally useful formula.

Following [2], the routine for Householder reduction given below actually starts in the column $n-1$ of $\mathbf{A}$, not column 0 as in the explanation above. In detail, the equations are as follows: At stage m $(m = 1, 2, \ldots, n-2)$ the vector $\mathbf{u}$ has the form

$$
\mathbf{u}^T = [a_{i0}, a_{i1}, \ldots, a_{i,i-2}, \ a_{i,i-1} \pm \sqrt{\sigma}, 0, \ldots, 0]
\tag{11.2.14}
$$

Here

$$i \equiv n - m = n - 1, n - 2, \ldots, 2 \qquad (11.2.15)$$

and the quantity σ ($|x|^2$ in our earlier notation) is

$$\sigma = (a_{i0})^2 + \cdots + (a_{i,i-1})^2 \qquad (11.2.16)$$

We choose the sign of σ in (11.2.14) to be the same as the sign of $a_{i,i-1}$ to lessen roundoff error.

Variables are thus computed in the following order: $\sigma, \mathbf{u}, H, \mathbf{p}, K, \mathbf{q}, \mathbf{A}'$. At any stage m, $\mathbf{A}$ is tridiagonal in its last $m - 1$ rows and columns.

If the eigenvectors of the final tridiagonal matrix are found (for example, by the routine in the next section), then the eigenvectors of $\mathbf{A}$ can be obtained by applying the accumulated transformation

$$\mathbf{Q} = \mathbf{P}_1 \cdot \mathbf{P}_2 \cdots \mathbf{P}_{n-2} \qquad (11.2.17)$$

to those eigenvectors. We therefore form $\mathbf{Q}$ by recursion after all the $\mathbf{P}$'s have been determined:

$$\mathbf{Q}_{n-2} = \mathbf{P}_{n-2}$$
$$\mathbf{Q}_j = \mathbf{P}_j \cdot \mathbf{Q}_{j+1}, \qquad j = n - 3, \ldots, 1 \qquad (11.2.18)$$
$$\mathbf{Q} = \mathbf{Q}_1$$

Input for the routine below is the real, symmetric matrix a[0..n-1][0..n-1]. On output, a contains the elements of the orthogonal matrix q. The vector d[0..n-1] is set to the diagonal elements of the tridiagonal matrix $\mathbf{A}'$, while the vector e[0..n-1] is set to the off-diagonal elements in its components 1 through n-1, with e[0]=0. Note that since a is overwritten, you should copy it before calling the routine, if it is required for subsequent computations.

No extra storage arrays are needed for the intermediate results. At stage m, the vectors $\mathbf{p}$ and $\mathbf{q}$ are nonzero only in elements $0, \ldots, i$ (recall that $i = n - m$), while $\mathbf{u}$ is nonzero only in elements $0, \ldots, i - 1$. The elements of the vector e are being determined in the order $n - 1, n - 2, \ldots$, so we can store $\mathbf{p}$ in the elements of e not already determined. The vector $\mathbf{q}$ can overwrite $\mathbf{p}$ once $\mathbf{p}$ is no longer needed. We store $\mathbf{u}$ in row i of a and $\mathbf{u}/H$ in column i of a. Once the reduction is complete, we compute the matrices $\mathbf{Q}_j$ using the quantities $\mathbf{u}$ and $\mathbf{u}/H$ that have been stored in a. Since $\mathbf{Q}_j$ is an identity matrix from row and column $n - j$ on, we only need compute its elements up to row and column $n - j - 1$. These can overwrite the $\mathbf{u}$'s and $\mathbf{u}/H$'s in the corresponding rows and columns of a, which are no longer required for subsequent $\mathbf{Q}$'s.

The routine tred2, given below, includes one further refinement. If the quantity σ is zero or "small" at any stage, one can skip the corresponding transformation. A simple criterion, such as

$$\sigma < \frac{\text{smallest positive number representable on machine}}{\text{machine precision}}$$

would be fine most of the time. A more careful criterion is actually used. At stage i, define the quantity

$$\epsilon = \sum_{k=0}^{i-1} |a_{ik}| \qquad (11.2.19)$$

If $\epsilon = 0$ to machine precision, we skip the transformation. Otherwise we redefine

$$a_{ik} \quad \text{becomes} \quad a_{ik}/\epsilon \qquad (11.2.20)$$

and use the scaled variables for the transformation. (A Householder transformation depends only on the ratios of the elements.)

Note that when dealing with a matrix whose elements vary over many orders of magnitude, it is important that the matrix be permuted, insofar as possible, so that the smaller elements are in the top left-hand corner. This is because the reduction is performed starting from the bottom right-hand corner, and a mixture of small and large elements there can lead to considerable rounding errors.

The routine tred2 is designed for use with the routine tqli of the next section. tqli finds the eigenvalues and eigenvectors of a symmetric, tridiagonal matrix. The combination of tred2 and tqli is the most efficient known technique for finding all the eigenvalues and eigenvectors (or just all the eigenvalues) of a real, symmetric matrix.

In the listing below, the statements indicated by comments are required only for subsequent computation of eigenvectors. If only eigenvalues are required, omission of the commented statements speeds up the execution time of tred2 by a factor of 2 for large n. In the limit of large n, the operation count of the Householder reduction is $2n^3/3$ for eigenvalues only, and $4n^3/3$ for both eigenvalues and eigenvectors.

```cpp
#include <cmath>
#include "nr.h"
using namespace std;

void NR::tred2(Mat_IO_DP &a, Vec_O_DP &d, Vec_O_DP &e)
```
Householder reduction of a real, symmetric matrix a[0..n-1][0..n-1]. On output, a is replaced by the orthogonal matrix **Q** effecting the transformation. d[0..n-1] returns the diagonal elements of the tridiagonal matrix, and e[0..n-1] the off-diagonal elements, with e[0]=0.
Several statements, as noted in comments, can be omitted if only eigenvalues are to be found, in which case a contains no useful information on output. Otherwise they are to be included.
```cpp
{
    int l,k,j,i;
    DP scale,hh,h,g,f;

    int n=d.size();
    for (i=n-1;i>0;i--) {
        l=i-1;
        h=scale=0.0;
        if (l > 0) {
            for (k=0;k<l+1;k++)
                scale += fabs(a[i][k]);
            if (scale == 0.0)               Skip transformation.
                e[i]=a[i][l];
            else {
                for (k=0;k<l+1;k++) {
                    a[i][k] /= scale;       Use scaled a's for transformation.
                    h += a[i][k]*a[i][k];   Form σ in h.
```

```
                    }
                    f=a[i][l];
                    g=(f >= 0.0 ? -sqrt(h) : sqrt(h));
                    e[i]=scale*g;
                    h -= f*g;                          Now h is equation (11.2.4).
                    a[i][l]=f-g;                       Store u in row i of a.
                    f=0.0;
                    for (j=0;j<l+1;j++) {
                    // Next statement can be omitted if eigenvectors not wanted
                        a[j][i]=a[i][j]/h;             Store u/H in column i of a.
                        g=0.0;                         Form an element of A · u in g.
                        for (k=0;k<j+1;k++)
                            g += a[j][k]*a[i][k];
                        for (k=j+1;k<l+1;k++)
                            g += a[k][j]*a[i][k];
                        e[j]=g/h;                      Form element of p in temporarily unused
                        f += e[j]*a[i][j];                 element of e.
                    }
                    hh=f/(h+h);                        Form K, equation (11.2.11).
                    for (j=0;j<l+1;j++) {              Form q and store in e overwriting p.
                        f=a[i][j];
                        e[j]=g=e[j]-hh*f;
                        for (k=0;k<j+1;k++)            Reduce a, equation (11.2.13).
                            a[j][k] -= (f*e[k]+g*a[i][k]);
                    }
                }
            } else
                e[i]=a[i][l];
            d[i]=h;
    }
    // Next statement can be omitted if eigenvectors not wanted
    d[0]=0.0;
    e[0]=0.0;
    // Contents of this loop can be omitted if eigenvectors not
    //    wanted except for statement d[i]=a[i][i];
    for (i=0;i<n;i++) {                                Begin accumulation of transformation ma-
        l=i;                                            trices.
        if (d[i] != 0.0) {                            This block skipped when i=0.
            for (j=0;j<l;j++) {
                g=0.0;
                for (k=0;k<l;k++)                     Use u and u/H stored in a to form P·Q.
                    g += a[i][k]*a[k][j];
                for (k=0;k<l;k++)
                    a[k][j] -= g*a[k][i];
            }
        }
        d[i]=a[i][i];                                 This statement remains.
        a[i][i]=1.0;                                  Reset row and column of a to identity
        for (j=0;j<l;j++) a[j][i]=a[i][j]=0.0;          matrix for next iteration.
    }
}
```

CITED REFERENCES AND FURTHER READING:

Golub, G.H., and Van Loan, C.F. 1996, *Matrix Computations*, 3rd ed. (Baltimore: Johns Hopkins University Press), §5.1. [1]

Smith, B.T., et al. 1976, *Matrix Eigensystem Routines — EISPACK Guide*, 2nd ed., vol. 6 of Lecture Notes in Computer Science (New York: Springer-Verlag).

Wilkinson, J.H., and Reinsch, C. 1971, *Linear Algebra*, vol. II of *Handbook for Automatic Computation* (New York: Springer-Verlag). [2]

11.3 Eigenvalues and Eigenvectors of a Tridiagonal Matrix

Evaluation of the Characteristic Polynomial

Once our original, real, symmetric matrix has been reduced to tridiagonal form, one possible way to determine its eigenvalues is to find the roots of the characteristic polynomial $p_n(\lambda)$ directly. The characteristic polynomial of a tridiagonal matrix can be evaluated for any trial value of λ by an efficient recursion relation (see [1], for example). The polynomials of lower degree produced during the recurrence form a Sturmian sequence that can be used to localize the eigenvalues to intervals on the real axis. A root-finding method such as bisection or Newton's method can then be employed to refine the intervals. The corresponding eigenvectors can then be found by inverse iteration (see §11.7).

Procedures based on these ideas can be found in [2,3]. If, however, more than a small fraction of all the eigenvalues and eigenvectors are required, then the factorization method next considered is much more efficient.

The QR and QL Algorithms

The basic idea behind the QR algorithm is that any real matrix can be decomposed in the form

$$\mathbf{A} = \mathbf{Q} \cdot \mathbf{R} \qquad (11.3.1)$$

where $\mathbf{Q}$ is orthogonal and $\mathbf{R}$ is upper triangular. For a general matrix, the decomposition is constructed by applying Householder transformations to annihilate successive columns of $\mathbf{A}$ below the diagonal (see §2.10).

Now consider the matrix formed by writing the factors in (11.3.1) in the opposite order:

$$\mathbf{A}' = \mathbf{R} \cdot \mathbf{Q} \qquad (11.3.2)$$

Since $\mathbf{Q}$ is orthogonal, equation (11.3.1) gives $\mathbf{R} = \mathbf{Q}^T \cdot \mathbf{A}$. Thus equation (11.3.2) becomes

$$\mathbf{A}' = \mathbf{Q}^T \cdot \mathbf{A} \cdot \mathbf{Q} \qquad (11.3.3)$$

We see that $\mathbf{A}'$ is an orthogonal transformation of $\mathbf{A}$.

You can verify that a QR transformation preserves the following properties of a matrix: symmetry, tridiagonal form, and Hessenberg form (to be defined in §11.5).

There is nothing special about choosing one of the factors of $\mathbf{A}$ to be upper triangular; one could equally well make it lower triangular. This is called the QL algorithm, since

$$\mathbf{A} = \mathbf{Q} \cdot \mathbf{L} \qquad (11.3.4)$$

where $\mathbf{L}$ is lower triangular. (The standard, but confusing, nomenclature R and L stands for whether the *right* or *left* of the matrix is nonzero.)

Recall that in the Householder reduction to tridiagonal form in §11.2, we started in column $n-1$ of the original matrix. To minimize roundoff, we then exhorted you to put the biggest elements of the matrix in the lower right-hand corner, if you can. If we now wish to diagonalize the resulting tridiagonal matrix, the QL algorithm will have smaller roundoff than the QR algorithm, so we shall use QL henceforth.

The QL algorithm consists of a *sequence* of orthogonal transformations:

$$\mathbf{A}_s = \mathbf{Q}_s \cdot \mathbf{L}_s$$
$$\mathbf{A}_{s+1} = \mathbf{L}_s \cdot \mathbf{Q}_s \qquad (= \mathbf{Q}_s^T \cdot \mathbf{A}_s \cdot \mathbf{Q}_s) \tag{11.3.5}$$

The following (nonobvious!) theorem is the basis of the algorithm for a general matrix $\mathbf{A}$: (i) If $\mathbf{A}$ has eigenvalues of different absolute value $|\lambda_i|$, then $\mathbf{A}_s \rightarrow$ [lower triangular form] as $s \rightarrow \infty$. The eigenvalues appear on the diagonal in increasing order of absolute magnitude. (ii) If $\mathbf{A}$ has an eigenvalue $|\lambda_i|$ of multiplicity p, $\mathbf{A}_s \rightarrow$ [lower triangular form] as $s \rightarrow \infty$, except for a diagonal block matrix of order p, whose eigenvalues $\rightarrow \lambda_i$. The proof of this theorem is fairly lengthy; see, for example, [4].

The workload in the QL algorithm is $O(n^3)$ per iteration for a general matrix, which is prohibitive. However, the workload is only $O(n)$ per iteration for a tridiagonal matrix and $O(n^2)$ for a Hessenberg matrix, which makes it highly efficient on these forms.

In this section we are concerned only with the case where $\mathbf{A}$ is a real, symmetric, tridiagonal matrix. All the eigenvalues λ_i are thus real. According to the theorem, if any λ_i has a multiplicity p, then there must be at least $p-1$ zeros on the sub- and superdiagonal. Thus the matrix can be split into submatrices that can be diagonalized separately, and the complication of diagonal blocks that can arise in the general case is irrelevant.

In the proof of the theorem quoted above, one finds that in general a super-diagonal element converges to zero like

$$a_{ij}^{(s)} \sim \left(\frac{\lambda_i}{\lambda_j}\right)^s \tag{11.3.6}$$

Although $\lambda_i < \lambda_j$, convergence can be slow if λ_i is close to λ_j. Convergence can be accelerated by the technique of *shifting*: If k is any constant, then $\mathbf{A} - k\mathbf{1}$ has eigenvalues $\lambda_i - k$. If we decompose

$$\mathbf{A}_s - k_s\mathbf{1} = \mathbf{Q}_s \cdot \mathbf{L}_s \tag{11.3.7}$$

so that

$$\mathbf{A}_{s+1} = \mathbf{L}_s \cdot \mathbf{Q}_s + k_s\mathbf{1}$$
$$= \mathbf{Q}_s^T \cdot \mathbf{A}_s \cdot \mathbf{Q}_s \tag{11.3.8}$$

then the convergence is determined by the ratio

$$\frac{\lambda_i - k_s}{\lambda_j - k_s} \tag{11.3.9}$$

The idea is to choose the shift k_s at each stage to maximize the rate of convergence. A good choice for the shift initially would be k_s close to λ_0, the smallest eigenvalue. Then the first row of off-diagonal elements would tend rapidly to zero. However, λ_0 is not usually known *a priori*. A very effective strategy in practice (although there is no proof that it is optimal) is to compute the eigenvalues of the leading 2×2 diagonal submatrix of $\mathbf{A}$. Then set k_s equal to the eigenvalue closer to a_{00}.

More generally, suppose you have already found r eigenvalues of $\mathbf{A}$. Then you can *deflate* the matrix by crossing out the first r rows and columns, leaving

$$
\mathbf{A} = \begin{bmatrix}
0 & & \cdots & \cdots & & & 0 \\
& \cdots & & & & & \\
& & 0 & & & & \\
\vdots & & & d_r & e_r & & \vdots \\
\vdots & & & e_r & d_{r+1} & & \\
& & & & & \cdots & 0 \\
& & & & & d_{n-2} & e_{n-2} \\
0 & & \cdots & & 0 & e_{n-2} & d_{n-1}
\end{bmatrix}
\tag{11.3.10}
$$

Choose k_s equal to the eigenvalue of the leading 2×2 submatrix that is closer to d_r. One can show that the convergence of the algorithm with this strategy is generally cubic (and at worst quadratic for degenerate eigenvalues). This rapid convergence is what makes the algorithm so attractive.

Note that with shifting, the eigenvalues no longer necessarily appear on the diagonal in order of increasing absolute magnitude. The routine eigsrt (§11.1) can be used if required.

As we mentioned earlier, the QL decomposition of a general matrix is effected by a sequence of Householder transformations. For a tridiagonal matrix, however, it is more efficient to use plane rotations $\mathbf{P}_{pq}$. One uses the sequence $\mathbf{P}_{01}, \mathbf{P}_{12}, \ldots, \mathbf{P}_{n-2,n-1}$ to annihilate the elements $a_{01}, a_{12}, \ldots, a_{n-2,n-1}$. By symmetry, the subdiagonal elements $a_{10}, a_{21}, \ldots, a_{n-1,n-2}$ will be annihilated too. Thus each $\mathbf{Q}_s$ is a product of plane rotations:

$$
\mathbf{Q}_s^T = \mathbf{P}_1^{(s)} \cdot \mathbf{P}_2^{(s)} \cdots \mathbf{P}_{n-1}^{(s)}
\tag{11.3.11}
$$

where $\mathbf{P}_i$ annihilates $a_{i-1,i}$. Note that it is $\mathbf{Q}^T$ in equation (11.3.11), not $\mathbf{Q}$, because we defined $\mathbf{L} = \mathbf{Q}^T \cdot \mathbf{A}$.

QL Algorithm with Implicit Shifts

The algorithm as described so far can be very successful. However, when the elements of $\mathbf{A}$ differ widely in order of magnitude, subtracting a large k_s from the diagonal elements can lead to loss of accuracy for the small eigenvalues. This difficulty is avoided by the QL algorithm with *implicit shifts*. The implicit QL algorithm is mathematically equivalent to the original QL algorithm, but the computation does not require $k_s \mathbf{1}$ to be actually subtracted from $\mathbf{A}$.

The algorithm is based on the following lemma: If $\mathbf{A}$ is a symmetric nonsingular matrix and $\mathbf{B} = \mathbf{Q}^T \cdot \mathbf{A} \cdot \mathbf{Q}$, where $\mathbf{Q}$ is orthogonal and $\mathbf{B}$ is tridiagonal with positive off-diagonal elements, then $\mathbf{Q}$ and $\mathbf{B}$ are fully determined when the last row of $\mathbf{Q}^T$ is specified. Proof: Let $\mathbf{q}_i^T$ denote the row vector i of the matrix $\mathbf{Q}^T$. Then $\mathbf{q}_i$ is the column vector i of the matrix $\mathbf{Q}$. The relation $\mathbf{B} \cdot \mathbf{Q}^T = \mathbf{Q}^T \cdot \mathbf{A}$ can be written

$$
\begin{bmatrix}
\beta_0 & \gamma_0 \\
\alpha_1 & \beta_1 & \gamma_1 \\
& & \vdots \\
& & \alpha_{n-2} & \beta_{n-2} & \gamma_{n-2} \\
& & & \alpha_{n-1} & \beta_{n-1}
\end{bmatrix}
\cdot
\begin{bmatrix}
\mathbf{q}_0^T \\
\mathbf{q}_1^T \\
\vdots \\
\mathbf{q}_{n-2}^T \\
\mathbf{q}_{n-1}^T
\end{bmatrix}
=
\begin{bmatrix}
\mathbf{q}_0^T \\
\mathbf{q}_1^T \\
\vdots \\
\mathbf{q}_{n-2}^T \\
\mathbf{q}_{n-1}^T
\end{bmatrix}
\cdot \mathbf{A}
\qquad (11.3.12)
$$

Row $n-1$ of this matrix equation is

$$
\alpha_{n-1}\mathbf{q}_{n-2}^T + \beta_{n-1}\mathbf{q}_{n-1}^T = \mathbf{q}_{n-1}^T \cdot \mathbf{A}
\qquad (11.3.13)
$$

Since $\mathbf{Q}$ is orthogonal,

$$
\mathbf{q}_{n-1}^T \cdot \mathbf{q}_m = \delta_{n-1,m}
\qquad (11.3.14)
$$

Thus if we postmultiply equation (11.3.13) by $\mathbf{q}_{n-1}$, we find

$$
\beta_{n-1} = \mathbf{q}_{n-1}^T \cdot \mathbf{A} \cdot \mathbf{q}_{n-1}
\qquad (11.3.15)
$$

which is known since $\mathbf{q}_{n-1}$ is known. Then equation (11.3.13) gives

$$
\alpha_{n-1}\mathbf{q}_{n-2}^T = \mathbf{z}_{n-2}^T
\qquad (11.3.16)
$$

where

$$
\mathbf{z}_{n-2}^T \equiv \mathbf{q}_{n-1}^T \cdot \mathbf{A} - \beta_{n-1}\mathbf{q}_{n-1}^T
\qquad (11.3.17)
$$

is known. Therefore

$$
\alpha_{n-1}^2 = \mathbf{z}_{n-2}^T \mathbf{z}_{n-2},
\qquad (11.3.18)
$$

or

$$
\alpha_{n-1} = |\mathbf{z}_{n-2}|
\qquad (11.3.19)
$$

and

$$
\mathbf{q}_{n-2}^T = \mathbf{z}_{n-2}^T / \alpha_{n-1}
\qquad (11.3.20)
$$

(where α_{n-1} is nonzero by hypothesis). Similarly, one can show by induction that if we know $\mathbf{q}_{n-1}, \mathbf{q}_{n-2}, \ldots, \mathbf{q}_{n-j}$ and the α's, β's, and γ's up to level $n-j$, one can determine the quantities at level $n-(j+1)$.

To apply the lemma in practice, suppose one can somehow find a tridiagonal matrix $\overline{\mathbf{A}}_{s+1}$ such that

$$
\overline{\mathbf{A}}_{s+1} = \overline{\mathbf{Q}}_s^T \cdot \overline{\mathbf{A}}_s \cdot \overline{\mathbf{Q}}_s
\qquad (11.3.21)
$$

where $\overline{\mathbf{Q}}_s^T$ is orthogonal and has the same last row as $\mathbf{Q}_s^T$ in the original QL algorithm. Then $\overline{\mathbf{Q}}_s = \mathbf{Q}_s$ and $\overline{\mathbf{A}}_{s+1} = \mathbf{A}_{s+1}$.

Now, in the original algorithm, from equation (11.3.11) we see that the last row of $\mathbf{Q}_s^T$ is the same as the last row of $\mathbf{P}_{n-1}^{(s)}$. But recall that $\mathbf{P}_{n-1}^{(s)}$ is a plane rotation designed to annihilate the $(n-2, n-1)$ element of $\mathbf{A}_s - k_s \mathbf{1}$. A simple calculation using the expression (11.1.1) shows that it has parameters

$$
c = \frac{d_{n-1} - k_s}{\sqrt{e_{n-1}^2 + (d_{n-1} - k_s)^2}} \quad , \quad s = \frac{-e_{n-2}}{\sqrt{e_{n-1}^2 + (d_{n-1} - k_s)^2}}
\qquad (11.3.22)
$$

The matrix $\mathbf{P}_{n-1}^{(s)} \cdot \mathbf{A}_s \cdot \mathbf{P}_{n-1}^{(s)T}$ is tridiagonal with 2 extra elements:

$$
\begin{bmatrix}
\cdots \\
& \times & \times & \times \\
& \times & \times & \times & \mathbf{x} \\
& & \times & \times & \times \\
& & \mathbf{x} & \times & \times
\end{bmatrix}
\qquad (11.3.23)
$$

We must now reduce this to tridiagonal form with an orthogonal matrix whose last row is $[0, 0, \ldots, 0, 1]$ so that the last row of $\overline{\mathbf{Q}}_s^T$ will stay equal to $\mathbf{P}_{n-1}^{(s)}$. This can be done by a sequence of Householder or Givens transformations. For the special form of the matrix (11.3.23), Givens is better. We rotate in the plane $(n-3, n-2)$ to annihilate the $(n-3, n-1)$ element. [By symmetry, the $(n-1, n-3)$ element will also be zeroed.] This leaves us with tridiagonal form except for extra elements $(n-4, n-2)$ and $(n-2, n-4)$. We annihilate these with a rotation in the $(n-4, n-3)$ plane, and so on. Thus a sequence of $n-2$ Givens rotations is required. The result is that

$$\mathbf{Q}_s^T = \overline{\mathbf{Q}}_s^T = \overline{\mathbf{P}}_1^{(s)} \cdot \overline{\mathbf{P}}_2^{(s)} \cdots \overline{\mathbf{P}}_{n-2}^{(s)} \cdot \mathbf{P}_{n-1}^{(s)} \qquad (11.3.24)$$

where the $\overline{\mathbf{P}}$'s are the Givens rotations and $\mathbf{P}_{n-1}$ is the same plane rotation as in the original algorithm. Then equation (11.3.21) gives the next iterate of $\mathbf{A}$. Note that the shift k_s enters implicitly through the parameters (11.3.22).

The following routine `tqli` ("Tridiagonal QL Implicit"), based algorithmically on the implementations in [2,3], works extremely well in practice. The number of iterations for the first few eigenvalues might be 4 or 5, say, but meanwhile the off-diagonal elements in the lower right-hand corner have been reduced too. The later eigenvalues are liberated with very little work. The average number of iterations per eigenvalue is typically $1.3 - 1.6$. The operation count per iteration is $O(n)$, with a fairly large effective coefficient, say, $\sim 20n$. The total operation count for the diagonalization is then $\sim 20n \times (1.3 - 1.6)n \sim 30n^2$. If the eigenvectors are required, the statements indicated by comments are included and there is an additional, much larger, workload of about $3n^3$ operations.

```
#include <cmath>
#include "nr.h"
using namespace std;

void NR::tqli(Vec_IO_DP &d, Vec_IO_DP &e, Mat_IO_DP &z)
```
QL algorithm with implicit shifts, to determine the eigenvalues and eigenvectors of a real, symmetric, tridiagonal matrix, or of a real, symmetric matrix previously reduced by `tred2` §11.2. On input, `d[0..n-1]` contains the diagonal elements of the tridiagonal matrix. On output, it returns the eigenvalues. The vector `e[0..n-1]` inputs the subdiagonal elements of the tridiagonal matrix, with `e[0]` arbitrary. On output `e` is destroyed. When finding only the eigenvalues, several lines may be omitted, as noted in the comments. If the eigenvectors of a tridiagonal matrix are desired, the matrix `z[0..n-1][0..n-1]` is input as the identity matrix. If the eigenvectors of a matrix that has been reduced by `tred2` are required, then `z` is input as the matrix output by `tred2`. In either case, column `k` of `z` returns the normalized eigenvector corresponding to `d[k]`.
```
{
    int m,l,iter,i,k;
    DP s,r,p,g,f,dd,c,b;

    int n=d.size();
    for (i=1;i<n;i++) e[i-1]=e[i];        Convenient to renumber the el-
    e[n-1]=0.0;                           ements of e.
    for (l=0;l<n;l++) {
        iter=0;
        do {
            for (m=l;m<n-1;m++) {         Look for a single small subdi-
                dd=fabs(d[m])+fabs(d[m+1]);   agonal element to split the
                if (fabs(e[m])+dd == dd) break;   matrix.
            }
            if (m != l) {
                if (iter++ == 30) nrerror("Too many iterations in tqli");
                g=(d[l+1]-d[l])/(2.0*e[l]);        Form shift.
                r=pythag(g,1.0);
```

```
g=d[m]-d[l]+e[l]/(g+SIGN(r,g));        This is d_m − k_s.
s=c=1.0;
p=0.0;
for (i=m-1;i>=1;i--) {                  A plane rotation as in the origi-
    f=s*e[i];                               nal QL, followed by Givens
    b=c*e[i];                               rotations to restore tridiag-
    e[i+1]=(r=pythag(f,g));                 onal form.
    if (r == 0.0) {                     Recover from underflow.
        d[i+1] -= p;
        e[m]=0.0;
        break;
    }
    s=f/r;
    c=g/r;
    g=d[i+1]-p;
    r=(d[i]-g)*s+2.0*c*b;
    d[i+1]=g+(p=s*r);
    g=c*r-b;
    // Next loop can be omitted if eigenvectors not wanted
    for (k=0;k<n;k++) {                  Form eigenvectors.
        f=z[k][i+1];
        z[k][i+1]=s*z[k][i]+c*f;
        z[k][i]=c*z[k][i]-s*f;
    }
}
if (r == 0.0 && i >= 1) continue;
d[l] -= p;
e[l]=g;
e[m]=0.0;
        }
    } while (m != 1);
}
}
```

CITED REFERENCES AND FURTHER READING:

Acton, F.S. 1970, *Numerical Methods That Work*; 1990, corrected edition (Washington: Mathematical Association of America), pp. 331–335. [1]

Wilkinson, J.H., and Reinsch, C. 1971, *Linear Algebra*, vol. II of *Handbook for Automatic Computation* (New York: Springer-Verlag). [2]

Smith, B.T., et al. 1976, *Matrix Eigensystem Routines — EISPACK Guide*, 2nd ed., vol. 6 of Lecture Notes in Computer Science (New York: Springer-Verlag). [3]

Stoer, J., and Bulirsch, R. 1993, *Introduction to Numerical Analysis*, 2nd ed. (New York: Springer-Verlag), §6.6.4. [4]

11.4 Hermitian Matrices

The complex analog of a real, symmetric matrix is a Hermitian matrix, satisfying equation (11.0.4). Jacobi transformations can be used to find eigenvalues and eigenvectors, as can Householder reduction to tridiagonal form followed by QL iteration. Complex versions of the previous routines jacobi, tred2, and tqli are quite analogous to their real counterparts. For working routines, consult [1,2].

An alternative, using the routines in this book, is to convert the Hermitian problem to a real, symmetric one: If $\mathbf{C} = \mathbf{A} + i\mathbf{B}$ is a Hermitian matrix, then the $n \times n$ complex eigenvalue problem

$$(\mathbf{A} + i\mathbf{B}) \cdot (\mathbf{u} + i\mathbf{v}) = \lambda(\mathbf{u} + i\mathbf{v}) \qquad (11.4.1)$$

is equivalent to the $2n \times 2n$ real problem

$$\begin{bmatrix} \mathbf{A} & -\mathbf{B} \\ \mathbf{B} & \mathbf{A} \end{bmatrix} \cdot \begin{bmatrix} \mathbf{u} \\ \mathbf{v} \end{bmatrix} = \lambda \begin{bmatrix} \mathbf{u} \\ \mathbf{v} \end{bmatrix} \qquad (11.4.2)$$

Note that the $2n \times 2n$ matrix in (11.4.2) is symmetric: $\mathbf{A}^T = \mathbf{A}$ and $\mathbf{B}^T = -\mathbf{B}$ if $\mathbf{C}$ is Hermitian.

Corresponding to a given eigenvalue λ, the vector

$$\begin{bmatrix} -\mathbf{v} \\ \mathbf{u} \end{bmatrix} \qquad (11.4.3)$$

is also an eigenvector, as you can verify by writing out the two matrix equations implied by (11.4.2). Thus if $\lambda_0, \lambda_1, \ldots, \lambda_{n-1}$ are the eigenvalues of $\mathbf{C}$, then the $2n$ eigenvalues of the augmented problem (11.4.2) are $\lambda_0, \lambda_0, \lambda_1, \lambda_1, \ldots,$ $\lambda_{n-1}, \lambda_{n-1}$; each, in other words, is repeated twice. The eigenvectors are pairs of the form $\mathbf{u} + i\mathbf{v}$ and $i(\mathbf{u} + i\mathbf{v})$; that is, they are the same up to an inessential phase. Thus we solve the augmented problem (11.4.2), and choose one eigenvalue and eigenvector from each pair. These give the eigenvalues and eigenvectors of the original matrix $\mathbf{C}$.

Working with the augmented matrix requires a factor of 2 more storage than the original complex matrix. In principle, a complex algorithm is also a factor of 2 more efficient in computer time than is the solution of the augmented problem. In practice, most complex implementations do not achieve this factor unless they are written entirely in real arithmetic. (Good library routines always do this.)

CITED REFERENCES AND FURTHER READING:

Wilkinson, J.H., and Reinsch, C. 1971, *Linear Algebra*, vol. II of *Handbook for Automatic Computation* (New York: Springer-Verlag). [1]

Smith, B.T., et al. 1976, *Matrix Eigensystem Routines — EISPACK Guide*, 2nd ed., vol. 6 of Lecture Notes in Computer Science (New York: Springer-Verlag). [2]

11.5 Reduction of a General Matrix to Hessenberg Form

The algorithms for symmetric matrices, given in the preceding sections, are highly satisfactory in practice. By contrast, it is impossible to design equally satisfactory algorithms for the nonsymmetric case. There are two reasons for this. First, the eigenvalues of a nonsymmetric matrix can be very sensitive to small changes in the matrix elements. Second, the matrix itself can be defective, so that there is no complete set of eigenvectors. We emphasize that these difficulties are intrinsic properties of certain nonsymmetric matrices, and no numerical procedure can "cure" them. The best we can hope for are procedures that don't exacerbate such problems.

The presence of rounding error can only make the situation worse. With finite-precision arithmetic, one cannot even design a foolproof algorithm to determine whether a given matrix is defective or not. Thus current algorithms generally *try* to find a *complete* set of eigenvectors, and rely on the user to inspect the results. If any eigenvectors are almost parallel, the matrix is probably defective.

Apart from referring you to the literature, and to the collected routines in [1,2], we are going to sidestep the problem of eigenvectors, giving algorithms for eigenvalues only. If you require just a few eigenvectors, you can read §11.7 and consider finding them by inverse iteration. We consider the problem of finding *all* eigenvectors of a nonsymmetric matrix as lying beyond the scope of this book.

Balancing

The sensitivity of eigenvalues to rounding errors during the execution of some algorithms can be reduced by the procedure of *balancing*. The errors in the eigensystem found by a numerical procedure are generally proportional to the Euclidean norm of the matrix, that is, to the square root of the sum of the squares of the elements. The idea of balancing is to use similarity transformations to make corresponding rows and columns of the matrix have comparable norms, thus reducing the overall norm of the matrix while leaving the eigenvalues unchanged. A symmetric matrix is already balanced.

Balancing is a procedure with of order N^2 operations. Thus, the time taken by the procedure `balanc`, given below, should never be more than a few percent of the total time required to find the eigenvalues. It is therefore recommended that you *always* balance nonsymmetric matrices. It never hurts, and it can substantially improve the accuracy of the eigenvalues computed for a badly balanced matrix.

The actual algorithm used is due to Osborne, as discussed in [1]. It consists of a sequence of similarity transformations by diagonal matrices **D**. To avoid introducing rounding errors during the balancing process, the elements of **D** are restricted to be exact powers of the radix base employed for floating-point arithmetic (i.e., 2 for most machines, but 16 for IBM mainframe architectures). The output is a matrix that is balanced in the norm given by summing the absolute magnitudes of the matrix elements. This is more efficient than using the Euclidean norm, and equally effective: A large reduction in one norm implies a large reduction in the other.

Note that if the off-diagonal elements of any row or column of a matrix are all zero, then the diagonal element is an eigenvalue. If the eigenvalue happens to be ill-conditioned (sensitive to small changes in the matrix elements), it will have relatively large errors when determined by the routine `hqr` (§11.6). Had we merely inspected the matrix beforehand, we could have determined the isolated eigenvalue exactly and then deleted the corresponding row and column from the matrix. You should consider whether such a pre-inspection might be useful in your application. (For symmetric matrices, the routines we gave will determine isolated eigenvalues accurately in all cases.)

The routine `balanc` does not keep track of the accumulated similarity transformation of the original matrix, since we will only be concerned with finding eigenvalues of nonsymmetric matrices, not eigenvectors. Consult [1-3] if you want to keep track of the transformation.

```cpp
#include <cmath>
#include <limits>
#include "nr.h"
using namespace std;

void NR::balanc(Mat_IO_DP &a)
```
Given a matrix a[0..n-1][0..n-1], this routine replaces it by a balanced matrix with identical eigenvalues. A symmetric matrix is already balanced and is unaffected by this procedure. The constant RADIX should be the machine's floating-point radix.
```cpp
{
    const DP RADIX = numeric_limits<DP>::radix;
    int i,j,last=0;
    DP s,r,g,f,c,sqrdx;

    int n=a.nrows();
    sqrdx=RADIX*RADIX;
    while (last == 0) {
        last=1;
        for (i=0;i<n;i++) {                      Calculate row and column norms.
            r=c=0.0;
            for (j=0;j<n;j++)
                if (j != i) {
                    c += fabs(a[j][i]);
                    r += fabs(a[i][j]);
                }
            if (c != 0.0 && r != 0.0) {          If both are nonzero,
                g=r/RADIX;
                f=1.0;
                s=c+r;
                while (c<g) {                    find the integer power of the machine
                    f *= RADIX;                  radix that comes closest to balanc-
                    c *= sqrdx;                  ing the matrix.
                }
                g=r*RADIX;
                while (c>g) {
                    f /= RADIX;
                    c /= sqrdx;
                }
                if ((c+r)/f < 0.95*s) {
                    last=0;
                    g=1.0/f;
                    for (j=0;j<n;j++) a[i][j] *= g;     Apply similarity transforma-
                    for (j=0;j<n;j++) a[j][i] *= f;     tion.
                }
            }
        }
    }
}
```

Reduction to Hessenberg Form

The strategy for finding the eigensystem of a general matrix parallels that of the symmetric case. First we reduce the matrix to a simpler form, and then we perform an iterative procedure on the simplified matrix. The simpler structure we use here is

called *Hessenberg* form. An *upper Hessenberg* matrix has zeros everywhere below
the diagonal except for the first subdiagonal row. For example, in the 6×6 case,
the nonzero elements are:

$$
\begin{bmatrix}
\times & \times & \times & \times & \times & \times \\
\times & \times & \times & \times & \times & \times \\
 & \times & \times & \times & \times & \times \\
 & & \times & \times & \times & \times \\
 & & & \times & \times & \times \\
 & & & & \times & \times
\end{bmatrix}
$$

By now you should be able to tell at a glance that such a structure can
be achieved by a sequence of Householder transformations, each one zeroing the
required elements in a column of the matrix. Householder reduction to Hessenberg
form is in fact an accepted technique. An alternative, however, is a procedure
analogous to Gaussian elimination with pivoting. We will use this elimination
procedure since it is about a factor of 2 more efficient than the Householder method,
and also since we want to teach you the method. It is possible to construct matrices
for which the Householder reduction, being orthogonal, is stable and elimination is
not, but such matrices are extremely rare in practice.

Straight Gaussian elimination is not a similarity transformation of the matrix.
Accordingly, the actual elimination procedure used is slightly different. We proceed
in a series of stages $r = 1, 2, \ldots, N - 2$. Before the rth stage, the original matrix
$\mathbf{A} \equiv \mathbf{A}_1$ has become $\mathbf{A}_r$, which is upper Hessenberg up to, but not including, row and
column $r - 1$. The rth stage then consists of the following sequence of operations:

- Find the element of maximum magnitude in column $r - 1$ below the
  diagonal. If it is zero, skip the next two "bullets" and the stage is done.
  Otherwise, suppose the maximum element was in row r'.
- Interchange rows r' and r. This is the pivoting procedure. To make the
  permutation a similarity transformation, also interchange columns r' and r.
- For $i = r + 1, r + 2, \ldots, N - 1$, compute the multiplier

$$
n_{ir} \equiv \frac{a_{i,r-1}}{a_{r,r-1}}
$$

  Subtract n_{ir} times row r from row i. To make the elimination a similarity
  transformation, also *add* n_{ir} times column i to column r.

A total of $N - 2$ such stages are required.

When the magnitudes of the matrix elements vary over many orders, you should
try to rearrange the matrix so that the largest elements are in the top left-hand corner.
This reduces the roundoff error, since the reduction proceeds from left to right.

Since we are concerned only with eigenvalues, the routine `elmhes` does not
keep track of the accumulated similarity transformation. The operation count is
about $5N^3/6$ for large N.

```
#include <cmath>
#include "nr.h"
using namespace std;
```

```
void NR::elmhes(Mat_IO_DP &a)
```
Reduction to Hessenberg form by the elimination method. The real, nonsymmetric matrix
a[0..n-1][0..n-1] is replaced by an upper Hessenberg matrix with identical eigenvalues.

Recommended, but not required, is that this routine be preceded by `balanc`. On output, the Hessenberg matrix is in elements `a[i][j]` with i ≤ j+1. Elements with i > j+1 are to be thought of as zero, but are returned with random values.

```
{
    int i,j,m;
    DP y,x;

    int n=a.nrows();
    for (m=1;m<n-1;m++) {                m is called r in the text.
        x=0.0;
        i=m;
        for (j=m;j<n;j++) {              Find the pivot.
            if (fabs(a[j][m-1]) > fabs(x)) {
                x=a[j][m-1];
                i=j;
            }
        }
        if (i != m) {                    Interchange rows and columns.
            for (j=m-1;j<n;j++) SWAP(a[i][j],a[m][j]);
            for (j=0;j<n;j++) SWAP(a[j][i],a[j][m]);
        }
        if (x != 0.0) {                  Carry out the elimination.
            for (i=m+1;i<n;i++) {
                if ((y=a[i][m-1]) != 0.0) {
                    y /= x;
                    a[i][m-1]=y;
                    for (j=m;j<n;j++) a[i][j] -= y*a[m][j];
                    for (j=0;j<n;j++) a[j][m] += y*a[j][i];
                }
            }
        }
    }
}
```

CITED REFERENCES AND FURTHER READING:

Wilkinson, J.H., and Reinsch, C. 1971, *Linear Algebra*, vol. II of *Handbook for Automatic Computation* (New York: Springer-Verlag). [1]

Smith, B.T., et al. 1976, *Matrix Eigensystem Routines — EISPACK Guide*, 2nd ed., vol. 6 of Lecture Notes in Computer Science (New York: Springer-Verlag). [2]

Stoer, J., and Bulirsch, R. 1993, *Introduction to Numerical Analysis*, 2nd ed. (New York: Springer-Verlag), §6.5.4. [3]

11.6 The QR Algorithm for Real Hessenberg Matrices

Recall the following relations for the QR algorithm with shifts:

$$\mathbf{Q}_s \cdot (\mathbf{A}_s - k_s \mathbf{1}) = \mathbf{R}_s \qquad (11.6.1)$$

where $\mathbf{Q}$ is orthogonal and $\mathbf{R}$ is upper triangular, and

$$\mathbf{A}_{s+1} = \mathbf{R}_s \cdot \mathbf{Q}_s^T + k_s \mathbf{1}$$
$$= \mathbf{Q}_s \cdot \mathbf{A}_s \cdot \mathbf{Q}_s^T \qquad (11.6.2)$$

The QR transformation preserves the upper Hessenberg form of the original matrix $\mathbf{A} \equiv \mathbf{A}_1$, and the workload on such a matrix is $O(n^2)$ per iteration as opposed to $O(n^3)$ on a general matrix. As $s \to \infty$, $\mathbf{A}_s$ converges to a form where the eigenvalues are either isolated on the diagonal or are eigenvalues of a 2×2 submatrix on the diagonal.

As we pointed out in §11.3, shifting is essential for rapid convergence. A key difference here is that a nonsymmetric real matrix can have complex eigenvalues. This means that good choices for the shifts k_s may be complex, apparently necessitating complex arithmetic.

Complex arithmetic can be avoided, however, by a clever trick. The trick depends on a result analogous to the lemma we used for implicit shifts in §11.3. The lemma we need here states that if $\mathbf{B}$ is a nonsingular matrix such that

$$\mathbf{B} \cdot \mathbf{Q} = \mathbf{Q} \cdot \mathbf{H} \tag{11.6.3}$$

where $\mathbf{Q}$ is orthogonal and $\mathbf{H}$ is upper Hessenberg, then $\mathbf{Q}$ and $\mathbf{H}$ are fully determined by column 0 of $\mathbf{Q}$. (The determination is unique if $\mathbf{H}$ has positive subdiagonal elements.) The lemma can be proved by induction analogously to the proof given for tridiagonal matrices in §11.3.

The lemma is used in practice by taking two steps of the QR algorithm, either with two real shifts k_s and k_{s+1}, or with complex conjugate values k_s and $k_{s+1} = k_s{}^*$. This gives a real matrix $\mathbf{A}_{s+2}$, where

$$\mathbf{A}_{s+2} = \mathbf{Q}_{s+1} \cdot \mathbf{Q}_s \cdot \mathbf{A}_s \cdot \mathbf{Q}_s^T \cdot \mathbf{Q}_{s+1}^T. \tag{11.6.4}$$

The $\mathbf{Q}$'s are determined by

$$\mathbf{A}_s - k_s \mathbf{1} = \mathbf{Q}_s^T \cdot \mathbf{R}_s \tag{11.6.5}$$

$$\mathbf{A}_{s+1} = \mathbf{Q}_s \cdot \mathbf{A}_s \cdot \mathbf{Q}_s^T \tag{11.6.6}$$

$$\mathbf{A}_{s+1} - k_{s+1} \mathbf{1} = \mathbf{Q}_{s+1}^T \cdot \mathbf{R}_{s+1} \tag{11.6.7}$$

Using (11.6.6), equation (11.6.7) can be rewritten

$$\mathbf{A}_s - k_{s+1} \mathbf{1} = \mathbf{Q}_s^T \cdot \mathbf{Q}_{s+1}^T \cdot \mathbf{R}_{s+1} \cdot \mathbf{Q}_s \tag{11.6.8}$$

Hence, if we define

$$\mathbf{M} = (\mathbf{A}_s - k_{s+1} \mathbf{1}) \cdot (\mathbf{A}_s - k_s \mathbf{1}) \tag{11.6.9}$$

equations (11.6.5) and (11.6.8) give

$$\mathbf{R} = \mathbf{Q} \cdot \mathbf{M} \tag{11.6.10}$$

where

$$\mathbf{Q} = \mathbf{Q}_{s+1} \cdot \mathbf{Q}_s \tag{11.6.11}$$

$$\mathbf{R} = \mathbf{R}_{s+1} \cdot \mathbf{R}_s \tag{11.6.12}$$

Equation (11.6.4) can be rewritten

$$\mathbf{A}_s \cdot \mathbf{Q}^T = \mathbf{Q}^T \cdot \mathbf{A}_{s+2} \tag{11.6.13}$$

Thus suppose we can somehow find an upper Hessenberg matrix $\mathbf{H}$ such that

$$\mathbf{A}_s \cdot \overline{\mathbf{Q}}^T = \overline{\mathbf{Q}}^T \cdot \mathbf{H} \qquad (11.6.14)$$

where $\overline{\mathbf{Q}}$ is orthogonal. If $\overline{\mathbf{Q}}^T$ has the same column 0 as $\mathbf{Q}^T$ (i.e., $\overline{\mathbf{Q}}$ has the same row 0 as $\mathbf{Q}$), then $\overline{\mathbf{Q}} = \mathbf{Q}$ and $\mathbf{A}_{s+2} = \mathbf{H}$.

Row 0 of $\mathbf{Q}$ is found as follows. Equation (11.6.10) shows that $\mathbf{Q}$ is the orthogonal matrix that triangularizes the real matrix $\mathbf{M}$. Any real matrix can be triangularized by premultiplying it by a sequence of Householder matrices $\mathbf{P}_1$ (acting on column 0), $\mathbf{P}_2$ (acting on column 1), ..., $\mathbf{P}_{n-1}$. Thus $\mathbf{Q} = \mathbf{P}_{n-1} \cdots \mathbf{P}_2 \cdot \mathbf{P}_1$, and row 0 of $\mathbf{Q}$ is row 0 of $\mathbf{P}_1$ since $\mathbf{P}_i$ is an $(i-1) \times (i-1)$ identity matrix in the top left-hand corner. We now must find $\overline{\mathbf{Q}}$ satisfying (11.6.14) whose row 0 is that of $\mathbf{P}_1$.

The Householder matrix $\mathbf{P}_1$ is determined by column 0 of $\mathbf{M}$. Since $\mathbf{A}_s$ is upper Hessenberg, equation (11.6.9) shows that column 0 of $\mathbf{M}$ has the form $[p_1, q_1, r_1, 0, ..., 0]^T$, where

$$p_1 = a_{00}^2 - a_{00}(k_s + k_{s+1}) + k_s k_{s+1} + a_{01} a_{10}$$

$$q_1 = a_{10}(a_{00} + a_{11} - k_s - k_{s+1}) \qquad (11.6.15)$$

$$r_1 = a_{10} a_{21}$$

Hence

$$\mathbf{P}_1 = 1 - 2\mathbf{w}_1 \cdot \mathbf{w}_1^T \qquad (11.6.16)$$

where $\mathbf{w}_1$ has only its first 3 elements nonzero (cf. equation 11.2.5). The matrix $\mathbf{P}_1 \cdot \mathbf{A}_s \cdot \mathbf{P}_1^T$ is therefore upper Hessenberg with 3 extra elements:

$$\mathbf{P}_1 \cdot \mathbf{A}_1 \cdot \mathbf{P}_1^T = \begin{bmatrix} \times & \times & \times & \times & \times & \times & \times \\ \times & \times & \times & \times & \times & \times & \times \\ \mathbf{x} & \times & \times & \times & \times & \times & \times \\ \mathbf{x} & \mathbf{x} & \times & \times & \times & \times & \times \\ & & & \times & \times & \times & \times \\ & & & & \times & \times & \times \\ & & & & & \times & \times \end{bmatrix} \qquad (11.6.17)$$

This matrix can be restored to upper Hessenberg form without affecting the first row by a sequence of Householder similarity transformations. The first such Householder matrix, $\mathbf{P}_2$, acts on elements 1, 2, and 3 in column 0, annihilating elements 2 and 3. This produces a matrix of the same form as (11.6.17), with the 3 extra elements appearing one column over:

$$\begin{bmatrix} \times & \times & \times & \times & \times & \times & \times \\ \times & \times & \times & \times & \times & \times & \times \\ & \times & \times & \times & \times & \times & \times \\ & \mathbf{x} & \times & \times & \times & \times & \times \\ & \mathbf{x} & \mathbf{x} & \times & \times & \times & \times \\ & & & \times & \times & \times & \times \\ & & & & \times & \times & \times \end{bmatrix} \qquad (11.6.18)$$

Proceeding in this way up to $\mathbf{P}_{n-1}$, we see that at each stage the Householder matrix $\mathbf{P}_r$ has a vector $\mathbf{w}_r$ that is nonzero only in elements $r-1$, r, and $r+1$. These elements are determined by the elements $r-1$, r, and $r+1$ in column $r-2$ of the current matrix. Note that the preliminary matrix $\mathbf{P}_1$ has the same structure as $\mathbf{P}_2, \ldots, \mathbf{P}_{n-1}$.

The result is that

$$\mathbf{P}_{n-1} \cdots \mathbf{P}_2 \cdot \mathbf{P}_1 \cdot \mathbf{A}_s \cdot \mathbf{P}_1^T \cdot \mathbf{P}_2^T \cdots \mathbf{P}_{n-1}^T = \mathbf{H} \tag{11.6.19}$$

where $\mathbf{H}$ is upper Hessenberg. Thus

$$\overline{\mathbf{Q}} = \mathbf{Q} = \mathbf{P}_{n-1} \cdots \mathbf{P}_2 \cdot \mathbf{P}_1 \tag{11.6.20}$$

and

$$\mathbf{A}_{s+2} = \mathbf{H} \tag{11.6.21}$$

The shifts of origin at each stage are taken to be the eigenvalues of the 2×2 matrix in the bottom right-hand corner of the current $\mathbf{A}_s$. This gives

$$
\begin{aligned}
k_s + k_{s+2} &= a_{n-2,n-2} + a_{n-1,n-1} \\
k_s k_{s+1} &= a_{n-2,n-2} a_{n-1,n-1} - a_{n-2,n-1} a_{n-1,n-2}
\end{aligned}
\tag{11.6.22}
$$

Substituting (11.6.22) in (11.6.15), we get

$$
\begin{aligned}
p_1 &= a_{10} \left\{ [(a_{n-1,n-1} - a_{00})(a_{n-2,n-2} - a_{00}) - a_{n-2,n-1} a_{n-1,n-2}]/a_{10} + a_{01} \right\} \\
q_1 &= a_{10}[a_{11} - a_{00} - (a_{n-1,n-1} - a_{00}) - (a_{n-2,n-2} - a_{00})] \\
r_1 &= a_{10} a_{21}
\end{aligned}
\tag{11.6.23}
$$

We have judiciously grouped terms to reduce possible roundoff when there are small off-diagonal elements. Since only the ratios of elements are relevant for a Householder transformation, we can omit the factor a_{10} from (11.6.23).

In summary, to carry out a double QR step we construct the Householder matrices $\mathbf{P}_r$, $r = 1, \ldots, n-1$. For $\mathbf{P}_1$ we use p_1, q_1, and r_1 given by (11.6.23). For the remaining matrices, p_r, q_r, and r_r are determined by the $(r-1, r-2)$, $(r, r-2)$, and $(r+1, r-2)$ elements of the current matrix. The number of arithmetic operations can be reduced by writing the nonzero elements of the $2\mathbf{w} \cdot \mathbf{w}^T$ part of the Householder matrix in the form

$$2\mathbf{w} \cdot \mathbf{w}^T = \begin{bmatrix} (p \pm s)/(\pm s) \\ q/(\pm s) \\ r/(\pm s) \end{bmatrix} \cdot [1 \quad q/(p \pm s) \quad r/(p \pm s)] \tag{11.6.24}$$

where

$$s^2 = p^2 + q^2 + r^2 \tag{11.6.25}$$

(We have simply divided each element by a piece of the normalizing factor; cf. the equations in §11.2.)

If we proceed in this way, convergence is usually very fast. There are two possible ways of terminating the iteration for an eigenvalue. First, if $a_{n-1,n-2}$ becomes "negligible," then $a_{n-1,n-1}$ is an eigenvalue. We can then delete row and column $n-1$ of the matrix and look for the next eigenvalue. Alternatively, $a_{n-2,n-3}$ may become negligible. In this case the eigenvalues of the 2×2 matrix in the lower right-hand corner may be taken to be eigenvalues. We delete rows and columns $n-1$ and $n-2$ of the matrix and continue.

The test for convergence to an eigenvalue is combined with a test for negligible subdiagonal elements that allows splitting of the matrix into submatrices. We find the largest i such that $a_{i,i-1}$ is negligible. If $i = n-1$, we have found a single eigenvalue. If $i = n-2$, we have found two eigenvalues. Otherwise we continue the iteration on the submatrix in rows i to $n-1$ (i being set to zero if there is no small subdiagonal element).

After determining i, the submatrix in rows i to $n-1$ is examined to see if the *product* of any two consecutive subdiagonal elements is small enough that we can work with an even smaller submatrix, starting say in row m. We start with $m = n-3$ and decrement it down to $i+1$, computing p, q, and r according to equations (11.6.23) with 0 replaced by m and 1 by $m+1$. If these were indeed the elements of the special "first" Householder matrix in a double QR step, then applying the Householder matrix would lead to nonzero elements in positions $(m+1, m-1)$, $(m+2, m-1)$, and $(m+2, m)$. We require that the first two of these elements be small compared with the local diagonal elements $a_{m-1,m-1}$, a_{mm} and $a_{m+1,m+1}$. A satisfactory approximate criterion is

$$|a_{m,m-1}|(|q| + |r|) \ll |p|(|a_{m+1,m+1}| + |a_{mm}| + |a_{m-1,m-1}|) \qquad (11.6.26)$$

Very rarely, the procedure described so far will fail to converge. On such matrices, experience shows that if one double step is performed with any shifts that are of order the norm of the matrix, convergence is subsequently very rapid. Accordingly, if ten iterations occur without determining an eigenvalue, the usual shifts are replaced for the next iteration by shifts defined by

$$k_s + k_{s+1} = 1.5 \times (|a_{n-1,n-2}| + |a_{n-2,n-3}|)$$
$$k_s k_{s+1} = (|a_{n-1,n-2}| + |a_{n-2,n-3}|)^2 \qquad (11.6.27)$$

The factor 1.5 was arbitrarily chosen to lessen the likelihood of an "unfortunate" choice of shifts. This strategy is repeated after 20 unsuccessful iterations. After 30 unsuccessful iterations, the routine reports failure.

The operation count for the QR algorithm described here is $\sim 5k^2$ per iteration, where k is the current size of the matrix. The typical average number of iterations per eigenvalue is ~ 1.8, so the total operation count for all the eigenvalues is $\sim 3n^3$. This estimate neglects any possible efficiency due to splitting or sparseness of the matrix.

The following routine hqr is based algorithmically on the above description, in turn following the implementations in [1,2].

```
#include <cmath>
#include <complex>
#include "nr.h"
using namespace std;
```

```
void NR::hqr(Mat_IO_DP &a, Vec_O_CPLX_DP &wri)
```
Finds all eigenvalues of an upper Hessenberg matrix a[0..n-1][0..n-1]. On input a can
be exactly as output from elmhes §11.5; on output it is destroyed. The complex eigenvalues
are returned in wri[0..n-1].

```
{
    int nn,m,l,k,j,its,i,mmin;
    DP z,y,x,w,v,u,t,s,r,q,p,anorm;

    int n=a.nrows();
    anorm=0.0;                              Compute matrix norm for possible use in lo-
    for (i=0;i<n;i++)                          cating single small subdiagonal element.
        for (j=MAX(i-1,0);j<n;j++)
            anorm += fabs(a[i][j]);
    nn=n-1;
    t=0.0;                                  Gets changed only by an exceptional shift.
    while (nn >= 0) {                       Begin search for next eigenvalue.
        its=0;
        do {
            for (l=nn;l>0;l--) {            Begin iteration: look for single small subdi-
                s=fabs(a[l-1][l-1])+fabs(a[l][l]);    agonal element.
                if (s == 0.0) s=anorm;
                if (fabs(a[l][l-1]) + s == s) {
                    a[l][l-1] = 0.0;
                    break;
                }
            }
            x=a[nn][nn];
            if (l == nn) {                  One root found.
                wri[nn--]=x+t;
            } else {
                y=a[nn-1][nn-1];
                w=a[nn][nn-1]*a[nn-1][nn];
                if (l == nn-1) {            Two roots found...
                    p=0.5*(y-x);
                    q=p*p+w;
                    z=sqrt(fabs(q));
                    x += t;
                    if (q >= 0.0) {         ...a real pair.
                        z=p+SIGN(z,p);
                        wri[nn-1]=wri[nn]=x+z;
                        if (z != 0.0) wri[nn]=x-w/z;
                    } else {                ...a complex pair.
                        wri[nn]=complex<DP>(x+p,z);
                        wri[nn-1]=conj(wri[nn]);
                    }
                    nn -= 2;
                } else {                    No roots found. Continue iteration.
                    if (its == 30) nrerror("Too many iterations in hqr");
                    if (its == 10 || its == 20) {    Form exceptional shift.
                        t += x;
                        for (i=0;i<nn+1;i++) a[i][i] -= x;
                        s=fabs(a[nn][nn-1])+fabs(a[nn-1][nn-2]);
                        y=x=0.75*s;
                        w = -0.4375*s*s;
                    }
                    ++its;
                    for (m=nn-2;m>=l;m--) {     Form shift and then look for
                        z=a[m][m];                 2 consecutive small sub-
                        r=x-z;                     diagonal elements.
                        s=y-z;
                        p=(r*s-w)/a[m+1][m]+a[m][m+1];    Equation (11.6.23).
                        q=a[m+1][m+1]-z-r-s;
                        r=a[m+2][m+1];
                        s=fabs(p)+fabs(q)+fabs(r);     Scale to prevent overflow or
                                                          underflow.
```

```
            p /= s;
            q /= s;
            r /= s;
            if (m == 1) break;
            u=fabs(a[m][m-1])*(fabs(q)+fabs(r));
            v=fabs(p)*(fabs(a[m-1][m-1])+fabs(z)+fabs(a[m+1][m+1]));
            if (u+v == v) break;                 Equation (11.6.26).
        }
        for (i=m;i<nn-1;i++) {
            a[i+2][i]=0.0;
            if (i != m) a[i+2][i-1]=0.0;
        }
        for (k=m;k<nn;k++) {
        Double QR step on rows l to nn and columns m to nn.
            if (k != m) {
                p=a[k][k-1];                      Begin setup of Householder
                q=a[k+1][k-1];                        vector.
                r=0.0;
                if (k+1 != nn) r=a[k+2][k-1];
                if ((x=fabs(p)+fabs(q)+fabs(r)) != 0.0) {
                    p /= x;                        Scale to prevent overflow or
                    q /= x;                            underflow.
                    r /= x;
                }
            }
            if ((s=SIGN(sqrt(p*p+q*q+r*r),p)) != 0.0) {
                if (k == m) {
                    if (l != m)
                    a[k][k-1] = -a[k][k-1];
                } else
                    a[k][k-1] = -s*x;
                p += s;                            Equations (11.6.24).
                x=p/s;
                y=q/s;
                z=r/s;
                q /= p;
                r /= p;
                for (j=k;j<nn+1;j++) {            Row modification.
                    p=a[k][j]+q*a[k+1][j];
                    if (k+1 != nn) {
                        p += r*a[k+2][j];
                        a[k+2][j] -= p*z;
                    }
                    a[k+1][j] -= p*y;
                    a[k][j] -= p*x;
                }
                mmin = nn < k+3 ? nn : k+3;
                for (i=1;i<mmin+1;i++) {          Column modification.
                    p=x*a[i][k]+y*a[i][k+1];
                    if (k != (nn)) {
                        p += z*a[i][k+2];
                        a[i][k+2] -= p*r;
                    }
                    a[i][k+1] -= p*q;
                    a[i][k] -= p;
                }
            }
        }
    }
    } while (l+1 < nn);
}
}
```

CITED REFERENCES AND FURTHER READING:

Wilkinson, J.H., and Reinsch, C. 1971, *Linear Algebra*, vol. II of *Handbook for Automatic Computation* (New York: Springer-Verlag). [1]

Golub, G.H., and Van Loan, C.F. 1996, *Matrix Computations*, 3rd ed. (Baltimore: Johns Hopkins University Press), §7.5.

Smith, B.T., et al. 1976, *Matrix Eigensystem Routines — EISPACK Guide*, 2nd ed., vol. 6 of Lecture Notes in Computer Science (New York: Springer-Verlag). [2]

11.7 Improving Eigenvalues and/or Finding Eigenvectors by Inverse Iteration

The basic idea behind inverse iteration is quite simple. Let $\mathbf{y}$ be the solution of the linear system

$$(\mathbf{A} - \tau\mathbf{1}) \cdot \mathbf{y} = \mathbf{b} \tag{11.7.1}$$

where $\mathbf{b}$ is a random vector and τ is close to some eigenvalue λ of $\mathbf{A}$. Then the solution $\mathbf{y}$ will be close to the eigenvector corresponding to λ. The procedure can be iterated: Replace $\mathbf{b}$ by $\mathbf{y}$ and solve for a new $\mathbf{y}$, which will be even closer to the true eigenvector.

We can see why this works by expanding both $\mathbf{y}$ and $\mathbf{b}$ as linear combinations of the eigenvectors $\mathbf{x}_j$ of $\mathbf{A}$:

$$\mathbf{y} = \sum_j \alpha_j \mathbf{x}_j \qquad \mathbf{b} = \sum_j \beta_j \mathbf{x}_j \tag{11.7.2}$$

Then (11.7.1) gives

$$\sum_j \alpha_j (\lambda_j - \tau)\mathbf{x}_j = \sum_j \beta_j \mathbf{x}_j \tag{11.7.3}$$

so that

$$\alpha_j = \frac{\beta_j}{\lambda_j - \tau} \tag{11.7.4}$$

and

$$\mathbf{y} = \sum_j \frac{\beta_j \mathbf{x}_j}{\lambda_j - \tau} \tag{11.7.5}$$

If τ is close to λ_n, say, then provided β_n is not accidentally too small, $\mathbf{y}$ will be approximately $\mathbf{x}_n$, up to a normalization. Moreover, each iteration of this procedure gives another power of $\lambda_j - \tau$ in the denominator of (11.7.5). Thus the convergence is rapid for well-separated eigenvalues.

Suppose at the kth stage of iteration we are solving the equation

$$(\mathbf{A} - \tau_k\mathbf{1}) \cdot \mathbf{y} = \mathbf{b}_k \tag{11.7.6}$$

where $\mathbf{b}_k$ and τ_k are our current guesses for some eigenvector and eigenvalue of interest (let's say, $\mathbf{x}_n$ and λ_n). Normalize $\mathbf{b}_k$ so that $\mathbf{b}_k \cdot \mathbf{b}_k = 1$. The exact eigenvector and eigenvalue satisfy

$$\mathbf{A} \cdot \mathbf{x}_n = \lambda_n \mathbf{x}_n \tag{11.7.7}$$

so

$$(\mathbf{A} - \tau_k \mathbf{1}) \cdot \mathbf{x}_n = (\lambda_n - \tau_k)\mathbf{x}_n \tag{11.7.8}$$

Since $\mathbf{y}$ of (11.7.6) is an improved approximation to $\mathbf{x}_n$, we normalize it and set

$$\mathbf{b}_{k+1} = \frac{\mathbf{y}}{|\mathbf{y}|} \tag{11.7.9}$$

We get an improved estimate of the eigenvalue by substituting our improved guess $\mathbf{y}$ for $\mathbf{x}_n$ in (11.7.8). By (11.7.6), the left-hand side is $\mathbf{b}_k$, so calling λ_n our new value τ_{k+1}, we find

$$\tau_{k+1} = \tau_k + \frac{1}{\mathbf{b}_k \cdot \mathbf{y}} \tag{11.7.10}$$

While the above formulas look simple enough, in practice the implementation can be quite tricky. The first question to be resolved is *when* to use inverse iteration. Most of the computational load occurs in solving the linear system (11.7.6). Thus a possible strategy is first to reduce the matrix $\mathbf{A}$ to a special form that allows easy solution of (11.7.6). Tridiagonal form for symmetric matrices or Hessenberg for nonsymmetric are the obvious choices. Then apply inverse iteration to generate all the eigenvectors. While this is an $O(N^3)$ method for symmetric matrices, it is many times less efficient than the QL method given earlier. In fact, even the best inverse iteration packages are less efficient than the QL method as soon as more than about 25 percent of the eigenvectors are required. Accordingly, inverse iteration is generally used when one already has good eigenvalues and wants only a few selected eigenvectors.

You can write a simple inverse iteration routine yourself using LU decomposition to solve (11.7.6). You can decide whether to use the general LU algorithm we gave in Chapter 2 or whether to take advantage of tridiagonal or Hessenberg form. Note that, since the linear system (11.7.6) is nearly singular, you must be careful to use a version of LU decomposition like that in §2.3 which replaces a zero pivot with a very small number.

We have chosen not to give a general inverse iteration routine in this book, because it is quite cumbersome to take account of all the cases that can arise. Routines are given, for example, in [1,2]. If you use these, or write your own routine, you may appreciate the following pointers.

One starts by supplying an initial value τ_0 for the eigenvalue λ_n of interest. Choose a random normalized vector $\mathbf{b}_0$ as the initial guess for the eigenvector $\mathbf{x}_n$, and solve (11.7.6). The new vector $\mathbf{y}$ is bigger than $\mathbf{b}_0$ by a "growth factor" $|\mathbf{y}|$, which ideally should be large. Equivalently, the change in the eigenvalue, which by (11.7.10) is essentially $1/|\mathbf{y}|$, should be small. The following cases can arise:

- If the growth factor is too small initially, then we assume we have made a "bad" choice of random vector. This can happen not just because of a small β_n in (11.7.5), but also in the case of a defective matrix, when (11.7.5) does not even apply (see, e.g., [1] or [3] for details). We go back to the beginning and choose a new initial vector.

- The change $|\mathbf{b}_1 - \mathbf{b}_0|$ might be less than some tolerance ϵ. We can use this as a criterion for stopping, iterating until it is satisfied, with a maximum of 5 – 10 iterations, say.

- After a few iterations, if $|\mathbf{b}_{k+1} - \mathbf{b}_k|$ is not decreasing rapidly enough, we can try updating the eigenvalue according to (11.7.10). If $\tau_{k+1} = \tau_k$ to machine accuracy, we are not going to improve the eigenvector much more and can quit. Otherwise start another cycle of iterations with the new eigenvalue.

The reason we do not update the eigenvalue at every step is that when we solve the linear system (11.7.6) by LU decomposition, we can save the decomposition if τ_k is fixed. We only need do the backsubstitution step each time we update $\mathbf{b}_k$. The number of iterations we decide to do with a fixed τ_k is a trade-off between the quadratic convergence but $O(N^3)$ workload for updating τ_k at each step and the linear convergence but $O(N^2)$ load for keeping τ_k fixed. If you have determined the eigenvalue by one of the routines given earlier in the chapter, it is probably correct to machine accuracy anyway, and you can omit updating it.

There are two different pathologies that can arise during inverse iteration. The first is multiple or closely spaced roots. This is more often a problem with symmetric matrices. Inverse iteration will find only one eigenvector for a given initial guess τ_0. A good strategy is to perturb the last few significant digits in τ_0 and then repeat the iteration. Usually this provides an independent eigenvector. Special steps generally have to be taken to ensure orthogonality of the linearly independent eigenvectors, whereas the Jacobi and QL algorithms automatically yield orthogonal eigenvectors even in the case of multiple eigenvalues.

The second problem, peculiar to nonsymmetric matrices, is the defective case. Unless one makes a "good" initial guess, the growth factor is small. Moreover, iteration does not improve matters. In this case, the remedy is to choose random initial vectors, solve (11.7.6) once, and quit as soon as *any* vector gives an acceptably large growth factor. Typically only a few trials are necessary.

One further complication in the nonsymmetric case is that a real matrix can have complex-conjugate pairs of eigenvalues. You will then have to use complex arithmetic to solve (11.7.6) for the complex eigenvectors. For any moderate-sized (or larger) nonsymmetric matrix, our recommendation is to avoid inverse iteration in favor of a QR method that includes the eigenvector computation in complex arithmetic. You will find routines for this in [1,2] and other places.

CITED REFERENCES AND FURTHER READING:

Acton, F.S. 1970, *Numerical Methods That Work*; 1990, corrected edition (Washington: Mathematical Association of America).

Wilkinson, J.H., and Reinsch, C. 1971, *Linear Algebra*, vol. II of *Handbook for Automatic Computation* (New York: Springer-Verlag), p. 418. [1]

Smith, B.T., et al. 1976, *Matrix Eigensystem Routines — EISPACK Guide*, 2nd ed., vol. 6 of Lecture Notes in Computer Science (New York: Springer-Verlag). [2]

Stoer, J., and Bulirsch, R. 1993, *Introduction to Numerical Analysis*, 2nd ed. (New York: Springer-Verlag), p. 375. [3]

Chapter 12.   Fast Fourier Transform

12.0 Introduction

A very large class of important computational problems falls under the general rubric of "Fourier transform methods" or "spectral methods." For some of these problems, the Fourier transform is simply an efficient computational tool for accomplishing certain common manipulations of data. In other cases, we have problems for which the Fourier transform (or the related "power spectrum") is itself of intrinsic interest. These two kinds of problems share a common methodology.

Largely for historical reasons the literature on Fourier and spectral methods has been disjoint from the literature on "classical" numerical analysis. Nowadays there is no justification for such a split. Fourier methods are commonplace in research and we shall not treat them as specialized or arcane. At the same time, we realize that many computer users have had relatively less experience with this field than with, say, differential equations or numerical integration. Therefore our summary of analytical results will be more complete. Numerical algorithms, per se, begin in §12.2. Various applications of Fourier transform methods are discussed in Chapter 13.

A physical process can be described either in the *time domain*, by the values of some quantity h as a function of time t, e.g., $h(t)$, or else in the *frequency domain*, where the process is specified by giving its amplitude H (generally a complex number indicating phase also) as a function of frequency f, that is $H(f)$, with $-\infty < f < \infty$. For many purposes it is useful to think of $h(t)$ and $H(f)$ as being two different *representations* of the *same* function. One goes back and forth between these two representations by means of the *Fourier transform* equations,

$$
\begin{aligned}
H(f) &= \int_{-\infty}^{\infty} h(t)e^{2\pi i f t}dt \\
h(t) &= \int_{-\infty}^{\infty} H(f)e^{-2\pi i f t}df
\end{aligned}
\tag{12.0.1}
$$

If t is measured in seconds, then f in equation (12.0.1) is in cycles per second, or Hertz (the unit of frequency). However, the equations work with other units too. If h is a function of position x (in meters), H will be a function of inverse wavelength (cycles per meter), and so on. If you are trained as a physicist or mathematician, you are probably more used to using *angular frequency* ω, which is given in *radians* per sec. The relation between ω and f, $H(\omega)$ and $H(f)$ is

$$
\omega \equiv 2\pi f \qquad H(\omega) \equiv [H(f)]_{f=\omega/2\pi}
\tag{12.0.2}
$$

and equation (12.0.1) looks like this

$$H(\omega) = \int_{-\infty}^{\infty} h(t)e^{i\omega t}dt$$

$$h(t) = \frac{1}{2\pi} \int_{-\infty}^{\infty} H(\omega)e^{-i\omega t}d\omega$$

(12.0.3)

We were raised on the ω-convention, but we changed! There are fewer factors of 2π to remember if you use the f-convention, especially when we get to discretely sampled data in §12.1.

From equation (12.0.1) it is evident at once that Fourier transformation is a *linear* operation. The transform of the sum of two functions is equal to the sum of the transforms. The transform of a constant times a function is that same constant times the transform of the function.

In the time domain, function $h(t)$ may happen to have one or more special symmetries It might be *purely real* or *purely imaginary* or it might be *even*, $h(t) = h(-t)$, or *odd*, $h(t) = -h(-t)$. In the frequency domain, these symmetries lead to relationships between $H(f)$ and $H(-f)$. The following table gives the correspondence between symmetries in the two domains:

If ...	then ...
$h(t)$ is real	$H(-f) = [H(f)]^*$
$h(t)$ is imaginary	$H(-f) = -[H(f)]^*$
$h(t)$ is even	$H(-f) = H(f)$ [i.e., $H(f)$ is even]
$h(t)$ is odd	$H(-f) = -H(f)$ [i.e., $H(f)$ is odd]
$h(t)$ is real and even	$H(f)$ is real and even
$h(t)$ is real and odd	$H(f)$ is imaginary and odd
$h(t)$ is imaginary and even	$H(f)$ is imaginary and even
$h(t)$ is imaginary and odd	$H(f)$ is real and odd

In subsequent sections we shall see how to use these symmetries to increase computational efficiency.

Here are some other elementary properties of the Fourier transform. (We'll use the "$\Longleftrightarrow$" symbol to indicate transform pairs.) If

$$h(t) \Longleftrightarrow H(f)$$

is such a pair, then other transform pairs are

$$h(at) \Longleftrightarrow \frac{1}{|a|}H(\frac{f}{a}) \qquad \text{"time scaling"} \qquad (12.0.4)$$

$$\frac{1}{|b|}h(\frac{t}{b}) \Longleftrightarrow H(bf) \qquad \text{"frequency scaling"} \qquad (12.0.5)$$

$$h(t - t_0) \Longleftrightarrow H(f)\, e^{2\pi i f t_0} \qquad \text{"time shifting"} \qquad (12.0.6)$$

$$h(t)\, e^{-2\pi i f_0 t} \Longleftrightarrow H(f - f_0) \qquad \text{"frequency shifting"} \qquad (12.0.7)$$

With two functions $h(t)$ and $g(t)$, and their corresponding Fourier transforms $H(f)$ and $G(f)$, we can form two combinations of special interest. The *convolution* of the two functions, denoted $g * h$, is defined by

$$g * h \equiv \int_{-\infty}^{\infty} g(\tau)h(t - \tau)\, d\tau \qquad (12.0.8)$$

Note that $g * h$ is a function in the time domain and that $g * h = h * g$. It turns out that the function $g * h$ is one member of a simple transform pair

$$g * h \Longleftrightarrow G(f)H(f) \qquad \text{"Convolution Theorem"} \qquad (12.0.9)$$

In other words, the Fourier transform of the convolution is just the product of the individual Fourier transforms.

The *correlation* of two functions, denoted $\text{Corr}(g, h)$, is defined by

$$\text{Corr}(g, h) \equiv \int_{-\infty}^{\infty} g(\tau + t)h(\tau)\, d\tau \qquad (12.0.10)$$

The correlation is a function of t, which is called the *lag*. It therefore lies in the time domain, and it turns out to be one member of the transform pair:

$$\text{Corr}(g, h) \Longleftrightarrow G(f)H^*(f) \qquad \text{"Correlation Theorem"} \qquad (12.0.11)$$

[More generally, the second member of the pair is $G(f)H(-f)$, but we are restricting ourselves to the usual case in which g and h are real functions, so we take the liberty of setting $H(-f) = H^*(f)$.] This result shows that multiplying the Fourier transform of one function by the complex conjugate of the Fourier transform of the other gives the Fourier transform of their correlation. The correlation of a function with itself is called its *autocorrelation*. In this case (12.0.11) becomes the transform pair

$$\text{Corr}(g, g) \Longleftrightarrow |G(f)|^2 \qquad \text{"Wiener-Khinchin Theorem"} \qquad (12.0.12)$$

The *total power* in a signal is the same whether we compute it in the time domain or in the frequency domain. This result is known as *Parseval's theorem*:

$$\text{Total Power} \equiv \int_{-\infty}^{\infty} |h(t)|^2\, dt = \int_{-\infty}^{\infty} |H(f)|^2\, df \qquad (12.0.13)$$

Frequently one wants to know "how much power" is contained in the frequency interval between f and $f + df$. In such circumstances one does not usually distinguish between positive and negative f, but rather regards f as varying from 0 ("zero frequency" or D.C.) to $+\infty$. In such cases, one defines the *one-sided power spectral density (PSD)* of the function h as

$$P_h(f) \equiv |H(f)|^2 + |H(-f)|^2 \qquad 0 \le f < \infty \qquad (12.0.14)$$

so that the total power is just the integral of $P_h(f)$ from $f = 0$ to $f = \infty$. When the function $h(t)$ is real, then the two terms in (12.0.14) are equal, so $P_h(f) = 2\,|H(f)|^2$.

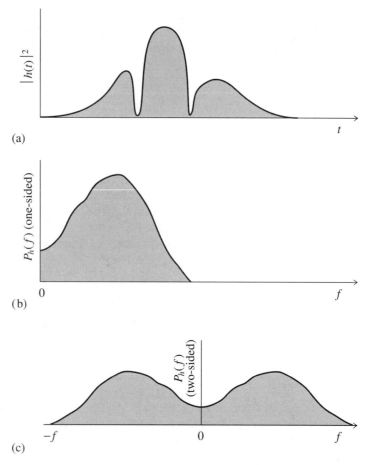

Figure 12.0.1.   Normalizations of one- and two-sided power spectra. The area under the square of the function, (a), equals the area under its one-sided power spectrum at positive frequencies, (b), and also equals the area under its two-sided power spectrum at positive and negative frequencies, (c).

Be warned that one occasionally sees PSDs defined without this factor two. These, strictly speaking, are called *two-sided power spectral densities*, but some books are not careful about stating whether one- or two-sided is to be assumed. We will always use the one-sided density given by equation (12.0.14). Figure 12.0.1 contrasts the two conventions.

   If the function $h(t)$ goes endlessly from $-\infty < t < \infty$, then its total power and power spectral density will, in general, be infinite. Of interest then is the *(one- or two-sided) power spectral density per unit time*. This is computed by taking a long, but finite, stretch of the function $h(t)$, computing its PSD [that is, the PSD of a function that equals $h(t)$ in the finite stretch but is zero everywhere else], and then dividing the resulting PSD by the length of the stretch used. Parseval's theorem in this case states that the integral of the one-sided PSD-per-unit-time over positive frequency is equal to the mean square amplitude of the signal $h(t)$.

   You might well worry about how the PSD-per-unit-time, which is a function of frequency f, converges as one evaluates it using longer and longer stretches of data. This interesting question is the content of the subject of "power spectrum estimation," and will be considered below in §13.4–§13.7. A crude answer for now is: The

PSD-per-unit-time converges to finite values at all frequencies *except* those where $h(t)$ has a discrete sine-wave (or cosine-wave) component of finite amplitude. At those frequencies, it becomes a delta-function, i.e., a sharp spike, whose width gets narrower and narrower, but whose area converges to be the mean square amplitude of the discrete sine or cosine component at that frequency.

We have by now stated all of the analytical formalism that we will need in this chapter with one exception: In computational work, especially with experimental data, we are almost never given a continuous function $h(t)$ to work with, but are given, rather, a list of measurements of $h(t_i)$ for a discrete set of t_i's. The profound implications of this seemingly unimportant fact are the subject of the next section.

CITED REFERENCES AND FURTHER READING:

Champeney, D.C. 1973, *Fourier Transforms and Their Physical Applications* (New York: Academic Press).

Elliott, D.F., and Rao, K.R. 1982, *Fast Transforms: Algorithms, Analyses, Applications* (New York: Academic Press).

12.1 Fourier Transform of Discretely Sampled Data

In the most common situations, function $h(t)$ is sampled (i.e., its value is recorded) at evenly spaced intervals in time. Let Δ denote the time interval between consecutive samples, so that the sequence of sampled values is

$$h_n = h(n\Delta) \qquad n = \ldots, -3, -2, -1, 0, 1, 2, 3, \ldots \qquad (12.1.1)$$

The reciprocal of the time interval Δ is called the *sampling rate*; if Δ is measured in seconds, for example, then the sampling rate is the number of samples recorded per second.

Sampling Theorem and Aliasing

For any sampling interval Δ, there is also a special frequency f_c, called the *Nyquist critical frequency*, given by

$$f_c \equiv \frac{1}{2\Delta} \qquad (12.1.2)$$

If a sine wave of the Nyquist critical frequency is sampled at its positive peak value, then the next sample will be at its negative trough value, the sample after that at the positive peak again, and so on. Expressed otherwise: *Critical sampling of a sine wave is two sample points per cycle.* One frequently chooses to measure time in units of the sampling interval Δ. In this case the Nyquist critical frequency is just the constant 1/2.

The Nyquist critical frequency is important for two related, but distinct, reasons. One is good news, and the other bad news. First the good news. It is the remarkable fact known as the *sampling theorem*: If a continuous function $h(t)$, sampled at an interval Δ, happens to be *bandwidth limited* to frequencies smaller in magnitude than f_c, i.e., if $H(f) = 0$ for all $|f| \geq f_c$, then the function $h(t)$ is *completely determined* by its samples h_n. In fact, $h(t)$ is given explicitly by the formula

$$h(t) = \Delta \sum_{n=-\infty}^{+\infty} h_n \frac{\sin[2\pi f_c(t - n\Delta)]}{\pi(t - n\Delta)} \qquad (12.1.3)$$

This is a remarkable theorem for many reasons, among them that it shows that the "information content" of a bandwidth limited function is, in some sense, infinitely smaller than that of a general continuous function. Fairly often, one is dealing with a signal that is known on physical grounds to be bandwidth limited (or at least approximately bandwidth limited). For example, the signal may have passed through an amplifier with a known, finite frequency response. In this case, the sampling theorem tells us that the entire information content of the signal can be recorded by sampling it at a rate Δ^{-1} equal to twice the maximum frequency passed by the amplifier (cf. 12.1.2).

Now the bad news. The bad news concerns the effect of sampling a continuous function that is *not* bandwidth limited to less than the Nyquist critical frequency. In that case, it turns out that all of the power spectral density that lies outside of the frequency range $-f_c < f < f_c$ is spuriously moved into that range. This phenomenon is called *aliasing*. Any frequency component outside of the frequency range $(-f_c, f_c)$ is *aliased* (falsely translated) into that range by the very act of discrete sampling. You can readily convince yourself that two waves $\exp(2\pi i f_1 t)$ and $\exp(2\pi i f_2 t)$ give the same samples at an interval Δ if and only if f_1 and f_2 differ by a multiple of $1/\Delta$, which is just the width in frequency of the range $(-f_c, f_c)$. There is little that you can do to remove aliased power once you have discretely sampled a signal. The way to overcome aliasing is to (i) know the natural bandwidth limit of the signal — or else enforce a known limit by analog filtering of the continuous signal, and then (ii) sample at a rate sufficiently rapid to give at least two points per cycle of the highest frequency present. Figure 12.1.1 illustrates these considerations.

To put the best face on this, we can take the alternative point of view: If a continuous function has been competently sampled, then, when we come to estimate its Fourier transform from the discrete samples, we can *assume* (or rather we *might as well* assume) that its Fourier transform is equal to zero outside of the frequency range in between $-f_c$ and f_c. Then we look to the Fourier transform to tell whether the continuous function *has* been competently sampled (aliasing effects minimized). We do this by looking to see whether the Fourier transform is already approaching zero as the frequency approaches f_c from below, or $-f_c$ from above. If, on the contrary, the transform is going towards some finite value, then chances are that components outside of the range have been folded back over onto the critical range.

Discrete Fourier Transform

We now estimate the Fourier transform of a function from a finite number of its sampled points. Suppose that we have N consecutive sampled values

$$h_k \equiv h(t_k), \qquad t_k \equiv k\Delta, \qquad k = 0, 1, 2, \ldots, N-1 \qquad (12.1.4)$$

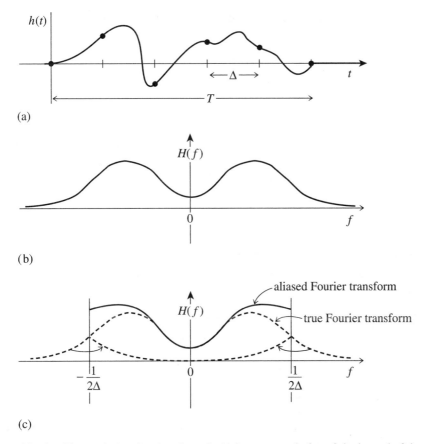

Figure 12.1.1.   The continuous function shown in (a) is nonzero only for a finite interval of time T. It follows that its Fourier transform, whose modulus is shown schematically in (b), is not bandwidth limited but has finite amplitude for all frequencies. If the original function is sampled with a sampling interval Δ, as in (a), then the Fourier transform (c) is defined only between plus and minus the Nyquist critical frequency. Power outside that range is folded over or "aliased" into the range. The effect can be eliminated only by low-pass filtering the original function *before sampling*.

so that the sampling interval is Δ. To make things simpler, let us also suppose that N is even. If the function $h(t)$ is nonzero only in a finite interval of time, then that whole interval of time is supposed to be contained in the range of the N points given. Alternatively, if the function $h(t)$ goes on forever, then the sampled points are supposed to be at least "typical" of what $h(t)$ looks like at all other times.

With N numbers of input, we will evidently be able to produce no more than N independent numbers of output. So, instead of trying to estimate the Fourier transform $H(f)$ at all values of f in the range $-f_c$ to f_c, let us seek estimates only at the discrete values

$$f_n \equiv \frac{n}{N\Delta}, \qquad n = -\frac{N}{2}, \dots, \frac{N}{2} \qquad (12.1.5)$$

The extreme values of n in (12.1.5) correspond exactly to the lower and upper limits of the Nyquist critical frequency range. If you are really on the ball, you will have noticed that there are $N + 1$, not N, values of n in (12.1.5); it will turn out that the two extreme values of n are not independent (in fact they are equal), but all the others are. This reduces the count to N.

The remaining step is to approximate the integral in (12.0.1) by a discrete sum:

$$H(f_n) = \int_{-\infty}^{\infty} h(t)e^{2\pi i f_n t} dt \approx \sum_{k=0}^{N-1} h_k \, e^{2\pi i f_n t_k} \Delta = \Delta \sum_{k=0}^{N-1} h_k \, e^{2\pi i kn/N}$$

$$(12.1.6)$$

Here equations (12.1.4) and (12.1.5) have been used in the final equality. The final summation in equation (12.1.6) is called the *discrete Fourier transform* of the N points h_k. Let us denote it by H_n,

$$H_n \equiv \sum_{k=0}^{N-1} h_k \, e^{2\pi i kn/N} \qquad\qquad (12.1.7)$$

The discrete Fourier transform maps N complex numbers (the h_k's) into N complex numbers (the H_n's). It does not depend on any dimensional parameter, such as the time scale Δ. The relation (12.1.6) between the discrete Fourier transform of a set of numbers and their continuous Fourier transform when they are viewed as samples of a continuous function sampled at an interval Δ can be rewritten as

$$H(f_n) \approx \Delta H_n \qquad\qquad (12.1.8)$$

where f_n is given by (12.1.5).

Up to now we have taken the view that the index n in (12.1.7) varies from $-N/2$ to $N/2$ (cf. 12.1.5). You can easily see, however, that (12.1.7) is periodic in n, with period N. Therefore, $H_{-n} = H_{N-n}$ $n = 1, 2, \ldots$. With this conversion in mind, one generally lets the n in H_n vary from 0 to $N-1$ (one complete period). Then n and k (in h_k) vary exactly over the same range, so the mapping of N numbers into N numbers is manifest. When this convention is followed, you must remember that zero frequency corresponds to $n = 0$, positive frequencies $0 < f < f_c$ correspond to values $1 \le n \le N/2 - 1$, while negative frequencies $-f_c < f < 0$ correspond to $N/2 + 1 \le n \le N-1$. The value $n = N/2$ corresponds to *both* $f = f_c$ and $f = -f_c$.

The discrete Fourier transform has symmetry properties almost exactly the same as the continuous Fourier transform. For example, all the symmetries in the table following equation (12.0.3) hold if we read h_k for $h(t)$, H_n for $H(f)$, and H_{N-n} for $H(-f)$. (Likewise, "even" and "odd" in time refer to whether the values h_k at k and $N-k$ are identical or the negative of each other.)

The formula for the discrete *inverse* Fourier transform, which recovers the set of h_k's exactly from the H_n's is:

$$h_k = \frac{1}{N} \sum_{n=0}^{N-1} H_n \, e^{-2\pi i kn/N} \qquad\qquad (12.1.9)$$

Notice that the only differences between (12.1.9) and (12.1.7) are (i) changing the sign in the exponential, and (ii) dividing the answer by N. This means that a routine for calculating discrete Fourier transforms can also, with slight modification, calculate the inverse transforms.

The discrete form of Parseval's theorem is

$$\sum_{k=0}^{N-1} |h_k|^2 = \frac{1}{N} \sum_{n=0}^{N-1} |H_n|^2 \qquad (12.1.10)$$

There are also discrete analogs to the convolution and correlation theorems (equations 12.0.9 and 12.0.11), but we shall defer them to §13.1 and §13.2, respectively.

CITED REFERENCES AND FURTHER READING:

Brigham, E.O. 1974, *The Fast Fourier Transform* (Englewood Cliffs, NJ: Prentice-Hall).

Elliott, D.F., and Rao, K.R. 1982, *Fast Transforms: Algorithms, Analyses, Applications* (New York: Academic Press).

12.2 Fast Fourier Transform (FFT)

How much computation is involved in computing the discrete Fourier transform (12.1.7) of N points? For many years, until the mid-1960s, the standard answer was this: Define W as the complex number

$$W \equiv e^{2\pi i/N} \qquad (12.2.1)$$

Then (12.1.7) can be written as

$$H_n = \sum_{k=0}^{N-1} W^{nk} h_k \qquad (12.2.2)$$

In other words, the vector of h_k's is multiplied by a matrix whose (n,k)th element is the constant W to the power $n \times k$. The matrix multiplication produces a vector result whose components are the H_n's. This matrix multiplication evidently requires N^2 complex multiplications, plus a smaller number of operations to generate the required powers of W. So, the discrete Fourier transform appears to be an $O(N^2)$ process. These appearances are deceiving! The discrete Fourier transform can, in fact, be computed in $O(N \log_2 N)$ operations with an algorithm called the *fast Fourier transform*, or *FFT*. The difference between $N \log_2 N$ and N^2 is immense. With $N = 10^6$, for example, it is the difference between, roughly, 30 seconds of CPU time and 2 weeks of CPU time on a microsecond cycle time computer. The existence of an FFT algorithm became generally known only in the mid-1960s, from the work of J.W. Cooley and J.W. Tukey. Retrospectively, we now know (see [1]) that efficient methods for computing the DFT had been independently discovered, and in some cases implemented, by as many as a dozen individuals, starting with Gauss in 1805!

One "rediscovery" of the FFT, that of Danielson and Lanczos in 1942, provides one of the clearest derivations of the algorithm. Danielson and Lanczos showed that a discrete Fourier transform of length N can be rewritten as the sum of two discrete Fourier transforms, each of length $N/2$. One of the two is formed from the even-numbered points of the original N, the other from the odd-numbered points. The proof is simply this:

$$
\begin{aligned}
F_k &= \sum_{j=0}^{N-1} e^{2\pi ijk/N} f_j \\
&= \sum_{j=0}^{N/2-1} e^{2\pi ik(2j)/N} f_{2j} + \sum_{j=0}^{N/2-1} e^{2\pi ik(2j+1)/N} f_{2j+1} \\
&= \sum_{j=0}^{N/2-1} e^{2\pi ikj/(N/2)} f_{2j} + W^k \sum_{j=0}^{N/2-1} e^{2\pi ikj/(N/2)} f_{2j+1} \\
&= F_k^e + W^k F_k^o
\end{aligned}
\tag{12.2.3}
$$

In the last line, W is the same complex constant as in (12.2.1), F_k^e denotes the kth component of the Fourier transform of length $N/2$ formed from the even components of the original f_j's, while F_k^o is the corresponding transform of length $N/2$ formed from the odd components. Notice also that k in the last line of (12.2.3) varies from 0 to N, not just to $N/2$. Nevertheless, the transforms F_k^e and F_k^o are periodic in k with length $N/2$. So each is repeated through two cycles to obtain F_k.

The wonderful thing about the *Danielson-Lanczos Lemma* is that it can be used recursively. Having reduced the problem of computing F_k to that of computing F_k^e and F_k^o, we can do the same reduction of F_k^e to the problem of computing the transform of *its* $N/4$ even-numbered input data and $N/4$ odd-numbered data. In other words, we can define F_k^{ee} and F_k^{eo} to be the discrete Fourier transforms of the points which are respectively even-even and even-odd on the successive subdivisions of the data.

Although there are ways of treating other cases, by far the easiest case is the one in which the original N is an integer power of 2. In fact, we categorically recommend that you *only* use FFTs with N a power of two. If the length of your data set is not a power of two, pad it with zeros up to the next power of two. (We will give more sophisticated suggestions in subsequent sections below.) With this restriction on N, it is evident that we can continue applying the Danielson-Lanczos Lemma until we have subdivided the data all the way down to transforms of length 1. What is the Fourier transform of length one? It is just the identity operation that copies its one input number into its one output slot! In other words, for every pattern of $\log_2 N$ e's and o's, there is a one-point transform that is just one of the input numbers f_n

$$
F_k^{eoeeoeo\cdots oee} = f_n \qquad \text{for some } n
\tag{12.2.4}
$$

(Of course this one-point transform actually does not depend on k, since it is periodic in k with period 1.)

The next trick is to figure out which value of n corresponds to which pattern of e's and o's in equation (12.2.4). The answer is: Reverse the pattern of e's and o's, then let $e = 0$ and $o = 1$, and you will have, *in binary* the value of n. Do you see why it works? It is because the successive subdivisions of the data into even and odd are tests of successive low-order (least significant) bits of n. This idea of *bit reversal* can be exploited in a very clever way which, along with the Danielson-Lanczos

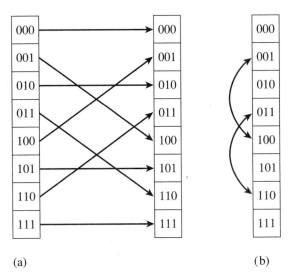

(a)　　　　　　　　　　　　　　　　　(b)

Figure 12.2.1.   Reordering an array (here of length 8) by bit reversal, (a) between two arrays, versus (b) in place. Bit reversal reordering is a necessary part of the fast Fourier transform (FFT) algorithm.

Lemma, makes FFTs practical: Suppose we take the original vector of data f_j and rearrange it into bit-reversed order (see Figure 12.2.1), so that the individual numbers are in the order not of j, but of the number obtained by bit-reversing j. Then the bookkeeping on the recursive application of the Danielson-Lanczos Lemma becomes extraordinarily simple. The points as given are the one-point transforms. We combine adjacent pairs to get two-point transforms, then combine adjacent pairs of pairs to get 4-point transforms, and so on, until the first and second halves of the whole data set are combined into the final transform. Each combination takes of order N operations, and there are evidently $\log_2 N$ combinations, so the whole algorithm is of order $N \log_2 N$ (assuming, as is the case, that the process of sorting into bit-reversed order is no greater in order than $N \log_2 N$).

This, then, is the structure of an FFT algorithm: It has two sections. The first section sorts the data into bit-reversed order. Luckily this takes no additional storage, since it involves only swapping pairs of elements. (If k_1 is the bit reverse of k_2, then k_2 is the bit reverse of k_1.) The second section has an outer loop that is executed $\log_2 N$ times and calculates, in turn, transforms of length $2, 4, 8, \ldots, N$. For each stage of this process, two nested inner loops range over the subtransforms already computed and the elements of each transform, implementing the Danielson-Lanczos Lemma. The operation is made more efficient by restricting external calls for trigonometric sines and cosines to the outer loop, where they are made only $\log_2 N$ times. Computation of the sines and cosines of multiple angles is through simple recurrence relations in the inner loops (cf. 5.5.6).

The FFT routine given below is based on one originally written by N. M. Brenner. The input quantities are the number of complex data points (nn), the data array (data[0..2*nn-1]), and isign, which should be set to either ±1 and is the sign of i in the exponential of equation (12.1.7). When isign is set to −1, the routine thus calculates the inverse transform (12.1.9) — except that it does not multiply by the normalizing factor $1/N$ that appears in that equation. You can do that yourself.

Notice that the argument nn is the number of *complex* data points, although

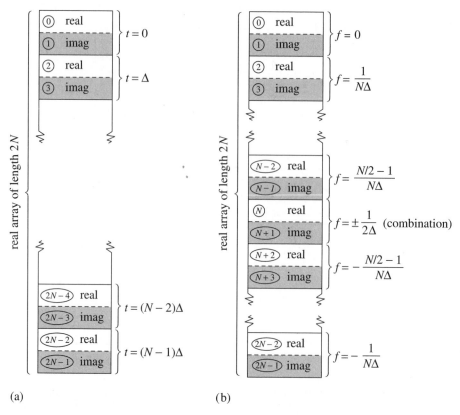

Figure 12.2.2. Input and output arrays for FFT. (a) The input array contains N (a power of 2) complex time samples in a real array of length $2N$, with real and imaginary parts alternating. (b) The output array contains the complex Fourier spectrum at N values of frequency. Real and imaginary parts again alternate. The array starts with zero frequency, works up to the most positive frequency (which is aliased with the most negative frequency). Negative frequencies follow, from the second-most negative up to the frequency just below zero.

we avoid the use of complex arithmetic because of the inefficient implementations found on many computers. The actual length of the real array (data[0..2*nn-1]) is 2 times nn, with each complex value occupying two consecutive locations. In other words, data[0] is the real part of f_0, data[1] is the imaginary part of f_0, and so on up to data[2*nn-2], which is the real part of f_{N-1}, and data[2*nn-1], which is the imaginary part of f_{N-1}. The FFT routine gives back the F_n's packed in exactly the same fashion, as nn complex numbers.

The real and imaginary parts of the zero frequency component F_0 are in data[0] and data[1]; the smallest nonzero positive frequency has real and imaginary parts in data[2] and data[3]; the smallest (in magnitude) nonzero negative frequency has real and imaginary parts in data[2*nn-2] and data[2*nn-1]. Positive frequencies increasing in magnitude are stored in the real-imaginary pairs data[4], data[5] up to data[nn-2], data[nn-1]. Negative frequencies of increasing magnitude are stored in data[2*nn-4], data[2*nn-3] down to data[nn+2], data[nn+3]. Finally, the pair data[nn], data[nn+1] contain the real and imaginary parts of the one aliased point that contains the most positive and the most negative frequency. You should try to develop a familiarity with this storage arrangement of complex spectra, also shown in Figure 12.2.2, since it is the practical standard.

```
#include <cmath>
#include "nr.h"
using namespace std;

void NR::four1(Vec_IO_DP &data, const int isign)
```
Replaces data[0..2*nn-1] by its discrete Fourier transform, if isign is input as 1; or replaces data[0..2*nn-1] by nn times its inverse discrete Fourier transform, if isign is input as -1. data is a complex array of length nn stored as a real array of length 2*nn. nn MUST be an integer power of 2 (this is not checked for!).
```
{
    int n,mmax,m,j,istep,i;
    DP wtemp,wr,wpr,wpi,wi,theta,tempr,tempi;

    int nn=data.size()/2;
    n=nn << 1;
    j=1;
    for (i=1;i<n;i+=2) {                       This is the bit-reversal section of the
        if (j > i) {                           routine.
            SWAP(data[j-1],data[i-1]);         Exchange the two complex numbers.
            SWAP(data[j],data[i]);
        }
        m=nn;
        while (m >= 2 && j > m) {
            j -= m;
            m >>= 1;
        }
        j += m;
    }
```
Here begins the Danielson-Lanczos section of the routine.
```
    mmax=2;
    while (n > mmax) {                         Outer loop executed log₂ nn times.
        istep=mmax << 1;
        theta=isign*(6.28318530717959/mmax);   Initialize the trigonometric recurrence.
        wtemp=sin(0.5*theta);
        wpr = -2.0*wtemp*wtemp;
        wpi=sin(theta);
        wr=1.0;
        wi=0.0;
        for (m=1;m<mmax;m+=2) {                 Here are the two nested inner loops.
            for (i=m;i<=n;i+=istep) {
                j=i+mmax;                        This is the Danielson-Lanczos for-
                tempr=wr*data[j-1]-wi*data[j];   mula:
                tempi=wr*data[j]+wi*data[j-1];
                data[j-1]=data[i-1]-tempr;
                data[j]=data[i]-tempi;
                data[i-1] += tempr;
                data[i] += tempi;
            }
            wr=(wtemp=wr)*wpr-wi*wpi+wr;         Trigonometric recurrence.
            wi=wi*wpr+wtemp*wpi+wi;
        }
        mmax=istep;
    }
}
```

Other FFT Algorithms

We should mention that there are a number of variants on the basic FFT algorithm given above. As we have seen, that algorithm first rearranges the input elements into bit-reverse order, then builds up the output transform in $\log_2 N$ iterations. In the literature, this sequence is called a *decimation-in-time* or *Cooley-Tukey* FFT algorithm. It is also possible to derive FFT algorithms that first go through a set of $\log_2 N$ iterations on the input data, and rearrange the *output* values into bit-reverse order. These are called *decimation-in-frequency* or *Sande-Tukey* FFT algorithms. For some applications, such as convolution (§13.1), one takes a data set into the Fourier domain and then, after some manipulation, back out again. In these cases it is possible to avoid all bit reversing. You use a decimation-in-frequency algorithm (without its bit reversing) to get into the "scrambled" Fourier domain, do your operations there, and then use an inverse algorithm (without *its* bit reversing) to get back to the time domain. While elegant in principle, this procedure does not in practice save much computation time, since the bit reversals represent only a small fraction of an FFT's operations count, and since most useful operations in the frequency domain require a knowledge of which points correspond to which frequencies.

Another class of FFTs subdivides the initial data set of length N not all the way down to the trivial transform of length 1, but rather only down to some other small power of 2, for example $N = 4$, *base-4 FFTs*, or $N = 8$, *base-8 FFTs*. These small transforms are then done by small sections of highly optimized coding which take advantage of special symmetries of that particular small N. For example, for $N = 4$, the trigonometric sines and cosines that enter are all ± 1 or 0, so many multiplications are eliminated, leaving largely additions and subtractions. These can be faster than simpler FFTs by some significant, but not overwhelming, factor, e.g., 20 or 30 percent.

There are also FFT algorithms for data sets of length N not a power of two. They work by using relations analogous to the Danielson-Lanczos Lemma to subdivide the initial problem into successively smaller problems, not by factors of 2, but by whatever small prime factors happen to divide N. The larger that the largest prime factor of N is, the worse this method works. If N is prime, then no subdivision is possible, and the user (whether he knows it or not) is taking a *slow* Fourier transform, of order N^2 instead of order $N \log_2 N$. Our advice is to stay clear of such FFT implementations, with perhaps one class of exceptions, the *Winograd Fourier transform algorithms*. Winograd algorithms are in some ways analogous to the base-4 and base-8 FFTs. Winograd has derived highly optimized codings for taking small-N discrete Fourier transforms, e.g., for $N = 2, 3, 4, 5, 7, 8, 11, 13, 16$. The algorithms also use a new and clever way of combining the subfactors. The method involves a reordering of the data both before the hierarchical processing and after it, but it allows a significant reduction in the number of multiplications in the algorithm. For some especially favorable values of N, the Winograd algorithms can be significantly (e.g., up to a factor of 2) faster than the simpler FFT algorithms of the nearest integer power of 2. This advantage in speed, however, must be weighed against the considerably more complicated data indexing involved in these transforms, and the fact that the Winograd transform cannot be done "in place."

Finally, an interesting class of transforms for doing convolutions quickly are number theoretic transforms. These schemes replace floating-point arithmetic with integer arithmetic modulo some large prime $N+1$, and the Nth root of 1 by the modulo arithmetic equivalent. Strictly speaking, these are not *Fourier* transforms at all, but the properties are quite similar and computational speed can be far superior. On the other hand, their use is somewhat restricted to quantities like correlations and convolutions since the transform itself is not easily interpretable as a "frequency" spectrum.

CITED REFERENCES AND FURTHER READING:

Nussbaumer, H.J. 1982, *Fast Fourier Transform and Convolution Algorithms* (New York: Springer-Verlag).

Elliott, D.F., and Rao, K.R. 1982, *Fast Transforms: Algorithms, Analyses, Applications* (New York: Academic Press).

Brigham, E.O. 1974, *The Fast Fourier Transform* (Englewood Cliffs, NJ: Prentice-Hall). [1]

Bloomfield, P. 1976, *Fourier Analysis of Time Series – An Introduction* (New York: Wiley).

Van Loan, C. 1992, *Computational Frameworks for the Fast Fourier Transform* (Philadelphia: S.I.A.M.).

Beauchamp, K.G. 1984, *Applications of Walsh Functions and Related Functions* (New York: Academic Press) [non-Fourier transforms].

Heideman, M.T., Johnson, D.H., and Burris, C.S. 1984, *IEEE ASSP Magazine*, pp. 14–21 (October).

12.3 FFT of Real Functions, Sine and Cosine Transforms

It happens frequently that the data whose FFT is desired consist of real-valued samples f_j, $j = 0 \ldots N - 1$. To use four1, we put these into a complex array with all imaginary parts set to zero. The resulting transform F_n, $n = 0 \ldots N - 1$ satisfies $F_{N-n}{}^* = F_n$. Since this complex-valued array has real values for F_0 and $F_{N/2}$, and $(N/2) - 1$ other independent values $F_1 \ldots F_{N/2-1}$, it has the same $2(N/2 - 1) + 2 = N$ "degrees of freedom" as the original, real data set. However, the use of the full complex FFT algorithm for real data is inefficient, both in execution time and in storage required. You would think that there is a better way.

There are *two* better ways. The first is "mass production": Pack two separate real functions into the input array in such a way that their individual transforms can be separated from the result. This is implemented in the program twofft below. This may remind you of a one-cent sale, at which you are coerced to purchase two of an item when you only need one. However, remember that for correlations and convolutions the Fourier transforms of two functions are involved, and this is a handy way to do them both at once. The second method is to pack the real input array cleverly, without extra zeros, into a complex array of half its length. One then performs a complex FFT on this shorter length; the trick is then to get the required answer out of the result. This is done in the program realft below.

Transform of Two Real Functions Simultaneously

First we show how to exploit the symmetry of the transform F_n to handle two real functions at once: Since the input data f_j are real. the components of the discrete Fourier transform satisfy

$$F_{N-n} = (F_n)^*$$ (12.3.1)

where the asterisk denotes complex conjugation. By the same token, the discrete
Fourier transform of a purely imaginary set of g_j's has the opposite symmetry.

$$G_{N-n} = -(G_n)^* \qquad (12.3.2)$$

Therefore we can take the discrete Fourier transform of two real functions each of
length N simultaneously by packing the two data arrays as the real and imaginary
parts, respectively, of the complex input array of four1. Then the resulting transform
array can be unpacked into two complex arrays with the aid of the two symmetries.
Routine twofft works out these ideas.

```
#include "nr.h"

void NR::twofft(Vec_I_DP &data1, Vec_I_DP &data2, Vec_O_DP &fft1,
    Vec_O_DP &fft2)
```
Given two real input arrays data1[0..n-1] and data2[0..n-1], this routine calls four1 and
returns two complex output arrays, fft1[0..2n-1] and fft2[0..2n-1], each of complex
length n (i.e., real length 2*n), which contain the discrete Fourier transforms of the respective
data arrays. n MUST be an integer power of 2.
```
{
    int nn3,nn2,jj,j;
    DP rep,rem,aip,aim;

    int n=data1.size();
    nn3=1+(nn2=n+n);
    for (j=0,jj=0;j<n;j++,jj+=2) {          Pack the two real arrays into one com-
        fft1[jj]=data1[j];                      plex array.
        fft1[jj+1]=data2[j];
    }
    four1(fft1,1);                          Transform the complex array.
    fft2[0]=fft1[1];
    fft1[1]=fft2[1]=0.0;
    for (j=2;j<n+1;j+=2) {
        rep=0.5*(fft1[j]+fft1[nn2-j]);      Use symmetries to separate the two trans-
        rem=0.5*(fft1[j]-fft1[nn2-j]);          forms.
        aip=0.5*(fft1[j+1]+fft1[nn3-j]);
        aim=0.5*(fft1[j+1]-fft1[nn3-j]);
        fft1[j]=rep;                        Ship them out in two complex arrays.
        fft1[j+1]=aim;
        fft1[nn2-j]=rep;
        fft1[nn3-j]= -aim;
        fft2[j]=aip;
        fft2[j+1]= -rem;
        fft2[nn2-j]=aip;
        fft2[nn3-j]=rem;
    }
}
```

What about the reverse process? Suppose you have two complex transform
arrays, each of which has the symmetry (12.3.1), so that you know that the inverses
of both transforms are real functions. Can you invert both in a single FFT? This is
even easier than the other direction. Use the fact that the FFT is linear and form
the sum of the first transform plus i times the second. Invert using four1 with
$\text{isign} = -1$. The real and imaginary parts of the resulting complex array are the
two desired real functions.

FFT of Single Real Function

To implement the second method, which allows us to perform the FFT of a *single* real function without redundancy, we split the data set in half, thereby forming two real arrays of half the size. We can apply the program above to these two, but of course the result will not be the transform of the original data. It will be a schizophrenic combination of two transforms, each of which has half of the information we need. Fortunately, this schizophrenia is treatable. It works like this:

The right way to split the original data is to take the even-numbered f_j as one data set, and the odd-numbered f_j as the other. The beauty of this is that we can take the original real array and treat it as a complex array h_j of half the length. The first data set is the real part of this array, and the second is the imaginary part, as prescribed for twofft. No repacking is required. In other words $h_j = f_{2j} + i f_{2j+1}$, $j = 0, \ldots, N/2 - 1$. We submit this to four1, and it will give back a complex array $H_n = F_n^e + i F_n^o$, $n = 0, \ldots, N/2 - 1$ with

$$
\begin{aligned}
F_n^e &= \sum_{k=0}^{N/2-1} f_{2k}\, e^{2\pi i k n/(N/2)} \\
F_n^o &= \sum_{k=0}^{N/2-1} f_{2k+1}\, e^{2\pi i k n/(N/2)}
\end{aligned}
\tag{12.3.3}
$$

The discussion of program twofft tells you how to separate the two transforms F_n^e and F_n^o out of H_n. How do you work them into the transform F_n of the original data set f_j? Simply glance back at equation (12.2.3):

$$
F_n = F_n^e + e^{2\pi i n/N} F_n^o \qquad n = 0, \ldots, N - 1 \tag{12.3.4}
$$

Expressed directly in terms of the transform H_n of our real (masquerading as complex) data set, the result is

$$
F_n = \frac{1}{2}(H_n + H_{N/2-n}^*) - \frac{i}{2}(H_n - H_{N/2-n}^*)e^{2\pi i n/N} \qquad n = 0, \ldots, N - 1 \tag{12.3.5}
$$

A few remarks:
- Since $F_{N-n}^* = F_n$ there is no point in saving the entire spectrum. The positive frequency half is sufficient and can be stored in the same array as the original data. The operation can, in fact, be done in place.
- Even so, we need values H_n, $n = 0, \ldots, N/2$ whereas four1 gives only the values $n = 0, \ldots, N/2 - 1$. Symmetry to the rescue, $H_{N/2} = H_0$.
- The values F_0 and $F_{N/2}$ are real and independent. In order to actually get the entire F_n in the original array space, it is convenient to put $F_{N/2}$ into the imaginary part of F_0.
- Despite its complicated form, the process above is invertible. First peel $F_{N/2}$ out of F_0. Then construct

$$
\begin{aligned}
F_n^e &= \frac{1}{2}(F_n + F_{N/2-n}^*) \\
F_n^o &= \frac{1}{2}e^{-2\pi i n/N}(F_n - F_{N/2-n}^*)
\end{aligned}
\qquad n = 0, \ldots, N/2 - 1 \tag{12.3.6}
$$

and use `four1` to find the inverse transform of $H_n = F_n^{(1)} + iF_n^{(2)}$. Surprisingly, the actual algebraic steps are virtually identical to those of the forward transform.

Here is a representation of what we have said:

```
#include <cmath>
#include "nr.h"
using namespace std;

void NR::realft(Vec_IO_DP &data, const int isign)
```
Calculates the Fourier transform of a set of n real-valued data points. Replaces this data (which is stored in array `data[0..n-1]`) by the positive frequency half of its complex Fourier transform. The real-valued first and last components of the complex transform are returned as elements `data[0]` and `data[1]`, respectively. n must be a power of 2. This routine also calculates the inverse transform of a complex data array if it is the transform of real data. (Result in this case must be multiplied by 2/n.)
```
{
    int i,i1,i2,i3,i4;
    DP c1=0.5,c2,h1r,h1i,h2r,h2i,wr,wi,wpr,wpi,wtemp,theta;

    int n=data.size();
    theta=3.141592653589793238/DP(n>>1);         Initialize the recurrence.
    if (isign == 1) {
        c2 = -0.5;
        four1(data,1);                           The forward transform is here.
    } else {
        c2=0.5;                                  Otherwise set up for an inverse trans-
        theta = -theta;                            form.
    }
    wtemp=sin(0.5*theta);
    wpr = -2.0*wtemp*wtemp;
    wpi=sin(theta);
    wr=1.0+wpr;
    wi=wpi;
    for (i=1;i<(n>>2);i++) {                      Case i=0 done separately below.
        i2=1+(i1=i+i);
        i4=1+(i3=n-i1);
        h1r=c1*(data[i1]+data[i3]);              The two separate transforms are sep-
        h1i=c1*(data[i2]-data[i4]);                arated out of data.
        h2r= -c2*(data[i2]+data[i4]);
        h2i=c2*(data[i1]-data[i3]);
        data[i1]=h1r+wr*h2r-wi*h2i;              Here they are recombined to form
        data[i2]=h1i+wr*h2i+wi*h2r;                the true transform of the origi-
        data[i3]=h1r-wr*h2r+wi*h2i;                nal real data.
        data[i4]= -h1i+wr*h2i+wi*h2r;
        wr=(wtemp=wr)*wpr-wi*wpi+wr;             The recurrence.
        wi=wi*wpr+wtemp*wpi+wi;
    }
    if (isign == 1) {
        data[0] = (h1r=data[0])+data[1];         Squeeze the first and last data to-
        data[1] = h1r-data[1];                     gether to get them all within the
    } else {                                       original array.
        data[0]=c1*((h1r=data[0])+data[1]);
        data[1]=c1*(h1r-data[1]);
        four1(data,-1);                          This is the inverse transform for the
    }                                              case isign=-1.
}
```

Fast Sine and Cosine Transforms

Among their other uses, the Fourier transforms of functions can be used to solve differential equations (see §19.4). The most common boundary conditions for the solutions are 1) they have the value zero at the boundaries, or 2) their derivatives are zero at the boundaries. In the first instance, the natural transform to use is the *sine* transform, given by

$$F_k = \sum_{j=1}^{N-1} f_j \sin(\pi jk/N) \qquad \text{sine transform} \qquad (12.3.7)$$

where f_j, $j = 0, \ldots, N-1$ is the data array, and $f_0 \equiv 0$.

At first blush this appears to be simply the imaginary part of the discrete Fourier transform. However, the argument of the sine differs by a factor of two from the value that would make this so. The sine transform uses *sines only* as a complete set of functions in the interval from 0 to 2π, and, as we shall see, the cosine transform uses *cosines only*. By contrast, the normal FFT uses both sines and cosines, but only half as many of each. (See Figure 12.3.1.)

The expression (12.3.7) can be "force-fit" into a form that allows its calculation via the FFT. The idea is to extend the given function rightward past its last tabulated value. We extend the data to twice their length in such a way as to make them an *odd* function about $j = N$, with $f_N = 0$,

$$f_{2N-j} \equiv -f_j \qquad j = 0, \ldots, N-1 \qquad (12.3.8)$$

Consider the FFT of this extended function:

$$F_k = \sum_{j=0}^{2N-1} f_j e^{2\pi ijk/(2N)} \qquad (12.3.9)$$

The half of this sum from $j = N$ to $j = 2N - 1$ can be rewritten with the substitution $j' = 2N - j$

$$
\begin{aligned}
\sum_{j=N}^{2N-1} f_j e^{2\pi ijk/(2N)} &= \sum_{j'=1}^{N} f_{2N-j'} e^{2\pi i(2N-j')k/(2N)} \\
&= -\sum_{j'=0}^{N-1} f_{j'} e^{-2\pi ij'k/(2N)}
\end{aligned}
\qquad (12.3.10)
$$

so that

$$
\begin{aligned}
F_k &= \sum_{j=0}^{N-1} f_j \left[e^{2\pi ijk/(2N)} - e^{-2\pi ijk/(2N)} \right] \\
&= 2i \sum_{j=0}^{N-1} f_j \sin(\pi jk/N)
\end{aligned}
\qquad (12.3.11)
$$

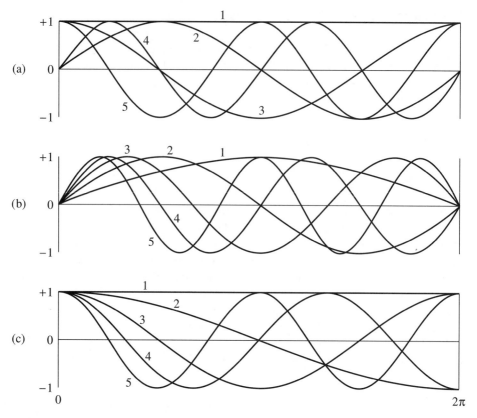

Figure 12.3.1. Basis functions used by the Fourier transform (a), sine transform (b), and cosine transform (c) are plotted. The first five basis functions are shown in each case. (For the Fourier transform, the real and imaginary parts of the basis functions are both shown.) While some basis functions occur in more than one transform, the basis sets are distinct. For example, the sine transform functions labeled (1), (3), (5) are not present in the Fourier basis. Any of the three sets can expand any function in the interval shown; however, the sine or cosine transform best expands functions matching the boundary conditions of the respective basis functions, namely zero function values for sine, zero derivatives for cosine.

Thus, up to a factor $2i$ we get the sine transform from the FFT of the extended function.

  This method introduces a factor of two inefficiency into the computation by extending the data. This inefficiency shows up in the FFT output, which has zeros for the real part of every element of the transform. For a one-dimensional problem, the factor of two may be bearable, especially in view of the simplicity of the method. When we work with partial differential equations in two or three dimensions, though, the factor becomes four or eight, so efforts to eliminate the inefficiency are well rewarded.

  From the original real data array f_j we will construct an auxiliary array y_j and apply to it the routine `realft`. The output will then be used to construct the desired transform. For the sine transform of data f_j, $j = 1, \ldots, N-1$, the auxiliary array is

$$y_0 = 0$$

$$y_j = \sin(j\pi/N)(f_j + f_{N-j}) + \frac{1}{2}(f_j - f_{N-j}) \qquad j = 1, \ldots, N-1 \tag{12.3.12}$$

This array is of the same dimension as the original. Notice that the first term is symmetric about $j = N/2$ and the second is antisymmetric. Consequently, when `realft` is applied to y_j, the result has real parts R_k and imaginary parts I_k given by

$$R_k = \sum_{j=0}^{N-1} y_j \cos(2\pi jk/N)$$

$$= \sum_{j=1}^{N-1} (f_j + f_{N-j}) \sin(j\pi/N) \cos(2\pi jk/N)$$

$$= \sum_{j=0}^{N-1} 2f_j \sin(j\pi/N) \cos(2\pi jk/N)$$

$$= \sum_{j=0}^{N-1} f_j \left[\sin\frac{(2k+1)j\pi}{N} - \sin\frac{(2k-1)j\pi}{N} \right]$$

$$= F_{2k+1} - F_{2k-1} \tag{12.3.13}$$

$$I_k = \sum_{j=0}^{N-1} y_j \sin(2\pi jk/N)$$

$$= \sum_{j=1}^{N-1} (f_j - f_{N-j})\frac{1}{2} \sin(2\pi jk/N)$$

$$= \sum_{j=0}^{N-1} f_j \sin(2\pi jk/N)$$

$$= F_{2k} \tag{12.3.14}$$

Therefore F_k can be determined as follows:

$$F_{2k} = I_k \qquad F_{2k+1} = F_{2k-1} + R_k \qquad k = 0, \ldots, (N/2 - 1) \tag{12.3.15}$$

The even terms of F_k are thus determined very directly. The odd terms require a recursion, the starting point of which follows from setting $k = 0$ in equation (12.3.15) and using $F_1 = -F_{-1}$:

$$F_1 = \frac{1}{2}R_0 \tag{12.3.16}$$

The implementing program is

```
#include <cmath>
#include "nr.h"
using namespace std;

void NR::sinft(Vec_IO_DP &y)
```
Calculates the sine transform of a set of n real-valued data points stored in array y[0..n-1].
The number n must be a power of 2. On exit y is replaced by its transform. This program,
without changes, also calculates the inverse sine transform, but in this case the output array
should be multiplied by 2/n.
```
{
    int j;
    DP sum,y1,y2,theta,wi=0.0,wr=1.0,wpi,wpr,wtemp;
```

```
int n=y.size();
theta=3.141592653589793238/DP(n);          Initialize the recurrence.
wtemp=sin(0.5*theta);
wpr= -2.0*wtemp*wtemp;
wpi=sin(theta);
y[0]=0.0;
for (j=1;j<(n>>1)+1;j++) {
    wr=(wtemp=wr)*wpr-wi*wpi+wr;           Calculate the sine for the auxiliary array.
    wi=wi*wpr+wtemp*wpi+wi;                The cosine is needed to continue the recurrence.
    y1=wi*(y[j]+y[n-j]);                   Construct the auxiliary array.
    y2=0.5*(y[j]-y[n-j]);
    y[j]=y1+y2;                            Terms j and N − j are related.
    y[n-j]=y1-y2;
}
realft(y,1);                              Transform the auxiliary array.
y[0]*=0.5;                                Initialize the sum used for odd terms below.
sum=y[1]=0.0;
for (j=0;j<n-1;j+=2) {
    sum += y[j];
    y[j]=y[j+1];                          Even terms determined directly.
    y[j+1]=sum;                           Odd terms determined by this running sum.
}
}
```

The sine transform, curiously, is its own inverse. If you apply it twice, you get the original data, but multiplied by a factor of $N/2$.

The other common boundary condition for differential equations is that the derivative of the function is zero at the boundary. In this case the natural transform is the *cosine* transform. There are several possible ways of defining the transform. Each can be thought of as resulting from a different way of extending a given array to create an even array of double the length, and/or from whether the extended array contains $2N - 1$, $2N$, or some other number of points. In practice, only two of the numerous possibilities are useful so we will restrict ourselves to just these two.

The first form of the cosine transform uses $N + 1$ data points:

$$F_k = \frac{1}{2}[f_0 + (-1)^k f_N] + \sum_{j=1}^{N-1} f_j \cos(\pi jk/N) \qquad (12.3.17)$$

It results from extending the given array to an even array about $j = N$, with

$$f_{2N-j} = f_j, \qquad j = 0, \ldots, N - 1 \qquad (12.3.18)$$

If you substitute this extended array into equation (12.3.9), and follow steps analogous to those leading up to equation (12.3.11), you will find that the Fourier transform is just twice the cosine transform (12.3.17). Another way of thinking about the formula (12.3.17) is to notice that it is the Chebyshev Gauss-Lobatto quadrature formula (see §4.5), often used in Clenshaw-Curtis adaptive quadrature (see §5.9, equation 5.9.4).

Once again the transform can be computed without the factor of two inefficiency. In this case the auxiliary function is

$$y_j = \frac{1}{2}(f_j + f_{N-j}) - \sin(j\pi/N)(f_j - f_{N-j}) \qquad j = 0, \ldots, N - 1 \quad (12.3.19)$$

Instead of equation (12.3.15), `realft` now gives

$$F_{2k} = R_k \qquad F_{2k+1} = F_{2k-1} + I_k \qquad k = 0, \ldots, (N/2 - 1) \qquad (12.3.20)$$

The starting value for the recursion for odd k in this case is

$$F_1 = \frac{1}{2}(f_0 - f_N) + \sum_{j=1}^{N-1} f_j \cos(j\pi/N) \qquad (12.3.21)$$

This sum does not appear naturally among the R_k and I_k, and so we accumulate it during the generation of the array y_j.

Once again this transform is its own inverse, and so the following routine works for both directions of the transformation. Note that although this form of the cosine transform has $N + 1$ input and output values, it passes an array only of length N to `realft`.

```
#include <cmath>
#include "nr.h"
using namespace std;

void NR::cosft1(Vec_IO_DP &y)
```
Calculates the cosine transform of a set `y[0..n]` of real-valued data points. The transformed data replace the original data in array `y`. `n` must be a power of 2. This program, without changes, also calculates the inverse cosine transform, but in this case the output array should be multiplied by `2/n`.
```
{
    const DP PI=3.141592653589793238;
    int j;
    DP sum,y1,y2,theta,wi=0.0,wpi,wpr,wr=1.0,wtemp;

    int n=y.size()-1;
    Vec_DP yy(n);                          Need array of length n, not n+1, for realft.
    theta=PI/n;                            Initialize the recurrence.
    wtemp=sin(0.5*theta);
    wpr = -2.0*wtemp*wtemp;
    wpi=sin(theta);
    sum=0.5*(y[0]-y[n]);
    yy[0]=0.5*(y[0]+y[n]);
    for (j=1;j<n/2;j++) {
        wr=(wtemp=wr)*wpr-wi*wpi+wr;       Carry out the recurrence.
        wi=wi*wpr+wtemp*wpi+wi;
        y1=0.5*(y[j]+y[n-j]);              Calculate the auxiliary function.
        y2=(y[j]-y[n-j]);
        yy[j]=y1-wi*y2;                    The values for j and N − j are related.
        yy[n-j]=y1+wi*y2;
        sum += wr*y2;                      Carry along this sum for later use in unfold-
    }                                        ing the transform.
    yy[n/2]=y[n/2];                        y[n/2] unchanged.
    realft(yy,1);                          Calculate the transform of the auxiliary func-
    for (j=0;j<n;j++) y[j]=yy[j];            tion.
    y[n]=y[1];
    y[1]=sum;                              sum is the value of F₁ in equation (12.3.21).
    for (j=3;j<n;j+=2) {
        sum += y[j];                       Equation (12.3.20).
        y[j]=sum;
    }
}
```

The second important form of the cosine transform is defined by

$$F_k = \sum_{j=0}^{N-1} f_j \cos \frac{\pi k(j + \frac{1}{2})}{N} \tag{12.3.22}$$

with inverse

$$f_j = \frac{2}{N} \sum_{k=0}^{N-1}{}' F_k \cos \frac{\pi k(j + \frac{1}{2})}{N} \tag{12.3.23}$$

Here the prime on the summation symbol means that the term for $k = 0$ has a coefficient of $\frac{1}{2}$ in front. This form arises by extending the given data, defined for $j = 0, \ldots, N-1$, to $j = N, \ldots, 2N-1$ in such a way that it is even about the point $N - \frac{1}{2}$ and periodic. (It is therefore also even about $j = -\frac{1}{2}$.) The form (12.3.23) is related to Gauss-Chebyshev quadrature (see equation 4.5.19), to Chebyshev approximation (§5.8, equation 5.8.7), and Clenshaw-Curtis quadrature (§5.9).

This form of the cosine transform is useful when solving differential equations on "staggered" grids, where the variables are centered midway between mesh points. It is also the standard form in the field of data compression and image processing.

The auxiliary function used in this case is similar to equation (12.3.19):

$$y_j = \frac{1}{2}(f_j + f_{N-j-1}) + \sin \frac{\pi(j + \frac{1}{2})}{N}(f_j - f_{N-j-1}) \qquad j = 0, \ldots, N-1 \tag{12.3.24}$$

Carrying out the steps similar to those used to get from (12.3.12) to (12.3.15), we find

$$F_{2k} = \cos \frac{\pi k}{N} R_k - \sin \frac{\pi k}{N} I_k \tag{12.3.25}$$

$$F_{2k-1} = \sin \frac{\pi k}{N} R_k + \cos \frac{\pi k}{N} I_k + F_{2k+1} \tag{12.3.26}$$

Note that equation (12.3.26) gives

$$F_{N-1} = \frac{1}{2} R_{N/2} \tag{12.3.27}$$

Thus the even components are found directly from (12.3.25), while the odd components are found by recursing (12.3.26) down from $k = N/2 - 1$, using (12.3.27) to start.

Since the transform is not self-inverting, we have to reverse the above steps to find the inverse. Here is the routine:

```
#include <cmath>
#include "nr.h"
using namespace std;

void NR::cosft2(Vec_IO_DP &y, const int isign)
```
Calculates the "staggered" cosine transform of a set y[0..n-1] of real-valued data points. The transformed data replace the original data in array y. n must be a power of 2. Set isign

to $+1$ for a transform, and to -1 for an inverse transform. For an inverse transform, the output array should be multiplied by $2/n$.

```
{
    const DP PI=3.141592653589793238;
    int i;
    DP sum,sum1,y1,y2,ytemp,theta,wi=0.0,wi1,wpi,wpr,wr=1.0,wr1,wtemp;

    int n=y.size();
    theta=0.5*PI/n;                          Initialize the recurrences.
    wr1=cos(theta);
    wi1=sin(theta);
    wpr = -2.0*wi1*wi1;
    wpi=sin(2.0*theta);
    if (isign == 1) {                        Forward transform.
        for (i=0;i<n/2;i++) {
            y1=0.5*(y[i]+y[n-1-i]);          Calculate the auxiliary function.
            y2=wi1*(y[i]-y[n-1-i]);
            y[i]=y1+y2;
            y[n-1-i]=y1-y2;
            wr1=(wtemp=wr1)*wpr-wi1*wpi+wr1;  Carry out the recurrence.
            wi1=wi1*wpr+wtemp*wpi+wi1;
        }
        realft(y,1);                         Transform the auxiliary function.
        for (i=2;i<n;i+=2) {                 Even terms.
            wr=(wtemp=wr)*wpr-wi*wpi+wr;
            wi=wi*wpr+wtemp*wpi+wi;
            y1=y[i]*wr-y[i+1]*wi;
            y2=y[i+1]*wr+y[i]*wi;
            y[i]=y1;
            y[i+1]=y2;
        }
        sum=0.5*y[1];                        Initialize recurrence for odd terms
        for (i=n-1;i>0;i-=2) {                   with $\frac{1}{2}R_{N/2}$.
            sum1=sum;                        Carry out recurrence for odd terms.
            sum += y[i];
            y[i]=sum1;
        }
    } else if (isign == -1) {                Inverse transform.
        ytemp=y[n-1];
        for (i=n-1;i>2;i-=2)                 Form difference of odd terms.
            y[i]=y[i-2]-y[i];
        y[1]=2.0*ytemp;
        for (i=2;i<n;i+=2) {                 Calculate $R_k$ and $I_k$.
            wr=(wtemp=wr)*wpr-wi*wpi+wr;
            wi=wi*wpr+wtemp*wpi+wi;
            y1=y[i]*wr+y[i+1]*wi;
            y2=y[i+1]*wr-y[i]*wi;
            y[i]=y1;
            y[i+1]=y2;
        }
        realft(y,-1);
        for (i=0;i<n/2;i++) {                Invert auxiliary array.
            y1=y[i]+y[n-1-i];
            y2=(0.5/wi1)*(y[i]-y[n-1-i]);
            y[i]=0.5*(y1+y2);
            y[n-1-i]=0.5*(y1-y2);
            wr1=(wtemp=wr1)*wpr-wi1*wpi+wr1;
            wi1=wi1*wpr+wtemp*wpi+wi1;
        }
    }
}
```

An alternative way of implementing this algorithm is to form an auxiliary function by copying the even elements of f_j into the first $N/2$ locations, and the odd elements into the next $N/2$ elements in reverse order. However, it is not easy to implement the alternative algorithm without a temporary storage array and we prefer the above in-place algorithm.

Finally, we mention that there exist fast cosine transforms for small N that do not rely on an auxiliary function or use an FFT routine. Instead, they carry out the transform directly, often coded in hardware for fixed N of small dimension [1].

CITED REFERENCES AND FURTHER READING:

Brigham, E.O. 1974, *The Fast Fourier Transform* (Englewood Cliffs, NJ: Prentice-Hall), §10–10.

Sorensen, H.V., Jones, D.L., Heideman, M.T., and Burris, C.S. 1987, *IEEE Transactions on Acoustics, Speech, and Signal Processing*, vol. ASSP-35, pp. 849–863.

Hou, H.S. 1987, *IEEE Transactions on Acoustics, Speech, and Signal Processing*, vol. ASSP-35, pp. 1455–1461 [see for additional references].

Hockney, R.W. 1971, in *Methods in Computational Physics*, vol. 9 (New York: Academic Press).

Temperton, C. 1980, *Journal of Computational Physics*, vol. 34, pp. 314–329.

Clarke, R.J. 1985, *Transform Coding of Images*, (Reading, MA: Addison-Wesley).

Gonzalez, R.C., and Wintz, P. 1987, *Digital Image Processing*, (Reading, MA: Addison-Wesley).

Chen, W., Smith, C.H., and Fralick, S.C. 1977, *IEEE Transactions on Communications*, vol. COM-25, pp. 1004–1009. [1]

12.4 FFT in Two or More Dimensions

Given a complex function $h(k_1, k_2)$ defined over the two-dimensional grid $0 \le k_1 \le N_1 - 1$, $0 \le k_2 \le N_2 - 1$, we can define its two-dimensional discrete Fourier transform as a complex function $H(n_1, n_2)$, defined over the same grid,

$$H(n_1, n_2) \equiv \sum_{k_2=0}^{N_2-1} \sum_{k_1=0}^{N_1-1} \exp(2\pi i k_2 n_2 / N_2) \, \exp(2\pi i k_1 n_1 / N_1) \, h(k_1, k_2)$$

$$(12.4.1)$$

By pulling the "subscripts 2" exponential outside of the sum over k_1, or by reversing the order of summation and pulling the "subscripts 1" outside of the sum over k_2, we can see instantly that the two-dimensional FFT can be computed by taking one-dimensional FFTs sequentially on each index of the original function. Symbolically,

$$H(n_1, n_2) = \text{FFT-on-index-1 (FFT-on-index-2 } [h(k_1, k_2)])$$
$$= \text{FFT-on-index-2 (FFT-on-index-1 } [h(k_1, k_2)]) \qquad (12.4.2)$$

For this to be practical, of course, both N_1 and N_2 should be some efficient length for an FFT, usually a power of 2. Programming a two-dimensional FFT, using (12.4.2) with a one-dimensional FFT routine, is a bit clumsier than it seems at first. Because the one-dimensional routine requires that its input be in consecutive order as a one-dimensional complex array, you find that you are endlessly copying things out of the multidimensional input array and then copying things back into it. This is not recommended technique. Rather, you should use a multidimensional FFT routine, such as the one we give below.

The generalization of (12.4.1) to more than two dimensions, say to L-dimensions, is evidently

$$
H(n_1, \ldots, n_L) \equiv \sum_{k_L=0}^{N_L-1} \cdots \sum_{k_1=0}^{N_1-1} \exp(2\pi i k_L n_L / N_L) \times \cdots
$$
$$
\times \exp(2\pi i k_1 n_1 / N_1) \, h(k_1, \ldots, k_L) \tag{12.4.3}
$$

where n_1 and k_1 range from 0 to $N_1 - 1, \ldots$, n_L and k_L range from 0 to $N_L - 1$. How many calls to a one-dimensional FFT are in (12.4.3)? Quite a few! For each value of $k_1, k_2, \ldots, k_{L-1}$ you FFT to transform the L index. Then for each value of $k_1, k_2, \ldots, k_{L-2}$ and n_L you FFT to transform the $L - 1$ index. And so on. It is best to rely on someone else having done the bookkeeping for once and for all.

The inverse transforms of (12.4.1) or (12.4.3) are just what you would expect them to be: Change the i's in the exponentials to $-i$'s, and put an overall factor of $1/(N_1 \times \cdots \times N_L)$ in front of the whole thing. Most other features of multidimensional FFTs are also analogous to features already discussed in the one-dimensional case:

- Frequencies are arranged in wrap-around order in the transform, but now for each separate dimension.
- The input data are also treated as if they were wrapped around. If they are discontinuous across this periodic identification (in any dimension) then the spectrum will have some excess power at high frequencies because of the discontinuity. The fix, if you care, is to remove multidimensional linear trends.
- If you are doing spatial filtering and are worried about wrap-around effects, then you need to zero-pad all around the border of the multidimensional array. However, be sure to notice how costly zero-padding is in multidimensional transforms. If you use too thick a zero-pad, you are going to waste a *lot* of storage, especially in 3 or more dimensions!
- Aliasing occurs as always if sufficient bandwidth limiting does not exist along one or more of the dimensions of the transform.

The routine fourn that we furnish herewith is a descendant of one written by N. M. Brenner. It requires as input (i) a vector, telling the length of the array in each dimension, e.g., (32,64). Note that these lengths *must all* be powers of 2, and are the numbers of *complex* values in each direction; (ii) the usual scalar equal to ± 1 indicating whether you want the transform or its inverse; and, finally (iii) the array of data. The number of dimensions is determined from the length of the vector in (i).

A few words about the data array: fourn accesses it as a one-dimensional array of real numbers, that is, data[0..$(2N_1 N_2 \ldots N_L)$-1], of length equal to twice the product of the lengths of the L dimensions. It assumes that the array represents an L-dimensional complex array, with individual components ordered as follows: (i) each complex value occupies two sequential locations, real part followed by imaginary; (ii) the first subscript changes least rapidly as one goes through the

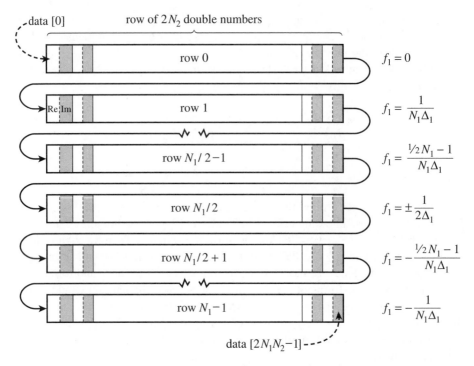

Figure 12.4.1. Storage arrangement of frequencies in the output $H(f_1, f_2)$ of a two-dimensional FFT. The input data is a two-dimensional $N_1 \times N_2$ array $h(t_1, t_2)$ (stored by rows of complex numbers). The output is also stored by complex rows. Each row corresponds to a particular value of f_1, as shown in the figure. Within each row, the arrangement of frequencies f_2 is exactly as shown in Figure 12.2.2. Δ_1 and Δ_2 are the sampling intervals in the 1 and 2 directions, respectively. The total number of (real) array elements is $2N_1 N_2$. The program fourn can also do more than two dimensions, and the storage arrangement generalizes in the obvious way.

array; the last subscript changes most rapidly (that is, "store by rows," the C++ norm). Almost all failures to get fourn to work result from improper understanding of the above ordering of the data array, so take care! (Figure 12.4.1 illustrates the format of the output array.)

```
#include <cmath>
#include "nr.h"
using namespace std;

void NR::fourn(Vec_IO_DP &data, Vec_I_INT &nn, const int isign)
```
Replaces data by its ndim-dimensional discrete Fourier transform, if isign is input as 1. nn[0..ndim-1] is an integer array containing the lengths of each dimension (number of complex values), which MUST all be powers of 2. data is a real array of length twice the product of these lengths, in which the data are stored as in a multidimensional complex array: real and imaginary parts of each element are in consecutive locations, and the rightmost index of the array increases most rapidly as one proceeds along data. For a two-dimensional array, this is equivalent to storing the array by rows. If isign is input as -1, data is replaced by its inverse transform times the product of the lengths of all dimensions.
```
{
    int idim,i1,i2,i3,i2rev,i3rev,ip1,ip2,ip3,ifp1,ifp2;
    int ibit,k1,k2,n,nprev,nrem,ntot;
    DP tempi,tempr,theta,wi,wpi,wpr,wr,wtemp;

    int ndim=nn.size();
    ntot=data.size()/2;                        Total number of complex values.
```

```
nprev=1;
for (idim=ndim-1;idim>=0;idim--) {              Main loop over the dimensions.
    n=nn[idim];
    nrem=ntot/(n*nprev);
    ip1=nprev << 1;
    ip2=ip1*n;
    ip3=ip2*nrem;
    i2rev=0;
    for (i2=0;i2<ip2;i2+=ip1) {                 This is the bit-reversal section of the
        if (i2 < i2rev) {                       routine.
            for (i1=i2;i1<i2+ip1-1;i1+=2) {
                for (i3=i1;i3<ip3;i3+=ip2) {
                    i3rev=i2rev+i3-i2;
                    SWAP(data[i3],data[i3rev]);
                    SWAP(data[i3+1],data[i3rev+1]);
                }
            }
        }
        ibit=ip2 >> 1;
        while (ibit >= ip1 && i2rev+1 > ibit) {
            i2rev -= ibit;
            ibit >>= 1;
        }
        i2rev += ibit;
    }
    ifp1=ip1;                                    Here begins the Danielson-Lanczos sec-
    while (ifp1 < ip2) {                         tion of the routine.
        ifp2=ifp1 << 1;
        theta=isign*6.28318530717959/(ifp2/ip1);    Initialize for the trig. recur-
        wtemp=sin(0.5*theta);                       rence.
        wpr= -2.0*wtemp*wtemp;
        wpi=sin(theta);
        wr=1.0;
        wi=0.0;
        for (i3=0;i3<ifp1;i3+=ip1) {
            for (i1=i3;i1<i3+ip1-1;i1+=2) {
                for (i2=i1;i2<ip3;i2+=ifp2) {
                    k1=i2;                          Danielson-Lanczos formula:
                    k2=k1+ifp1;
                    tempr=wr*data[k2]-wi*data[k2+1];
                    tempi=wr*data[k2+1]+wi*data[k2];
                    data[k2]=data[k1]-tempr;
                    data[k2+1]=data[k1+1]-tempi;
                    data[k1] += tempr;
                    data[k1+1] += tempi;
                }
            }
            wr=(wtemp=wr)*wpr-wi*wpi+wr;    Trigonometric recurrence.
            wi=wi*wpr+wtemp*wpi+wi;
        }
        ifp1=ifp2;
    }
    nprev *= n;
}
}
```

CITED REFERENCES AND FURTHER READING:

Nussbaumer, H.J. 1982, *Fast Fourier Transform and Convolution Algorithms* (New York: Springer-
    Verlag).

12.5 Fourier Transforms of Real Data in Two and Three Dimensions

Two-dimensional FFTs are particularly important in the field of image processing. An image is usually represented as a two-dimensional array of pixel intensities, real (and usually positive) numbers. One commonly desires to filter high, or low, frequency spatial components from an image; or to convolve or deconvolve the image with some instrumental point spread function. Use of the FFT is by far the most efficient technique.

In three dimensions, a common use of the FFT is to solve Poisson's equation for a potential (e.g., electromagnetic or gravitational) on a three-dimensional lattice that represents the discretization of three-dimensional space. Here the source terms (mass or charge distribution) and the desired potentials are also real. In two and three dimensions, with large arrays, memory is often at a premium. It is therefore important to perform the FFTs, insofar as possible, on the data "in place." We want a routine with functionality similar to the multidimensional FFT routine fourn (§12.4), but which operates on real, not complex, input data. We give such a routine in this section. The development is analogous to that of §12.3 leading to the one-dimensional routine realft. (You might wish to review that material at this point, particularly equation 12.3.5.)

It is convenient to think of the independent variables $n_1, \ldots, n_L$ in equation (12.4.3) as representing an L-dimensional vector $\vec{n}$ in wave-number space, with values on the lattice of integers. The transform $H(n_1, \ldots, n_L)$ is then denoted $H(\vec{n})$.

It is easy to see that the transform $H(\vec{n})$ is periodic in each of its L dimensions. Specifically, if $\vec{P}_1, \vec{P}_2, \vec{P}_3, \ldots$ denote the vectors $(N_1, 0, 0, \ldots)$, $(0, N_2, 0, \ldots)$, $(0, 0, N_3, \ldots)$, and so forth, then

$$H(\vec{n} \pm \vec{P}_j) = H(\vec{n}) \qquad j = 1, \ldots, L \qquad (12.5.1)$$

Equation (12.5.1) holds for any input data, real or complex. When the data is real, we have the additional symmetry

$$H(-\vec{n}) = H(\vec{n})^* \qquad (12.5.2)$$

Equations (12.5.1) and (12.5.2) imply that the full transform can be trivially obtained from the subset of lattice values $\vec{n}$ that have

$$0 \leq n_1 \leq N_1 - 1$$
$$0 \leq n_2 \leq N_2 - 1$$
$$\cdots \qquad (12.5.3)$$
$$0 \leq n_L \leq \frac{N_L}{2}$$

In fact, this set of values is overcomplete, because there are additional symmetry relations among the transform values that have $n_L = 0$ and $n_L = N_L/2$. However these symmetries are complicated and their use becomes extremely confusing. Therefore, we will compute our FFT on the lattice subset of equation (12.5.3), even though this requires a small amount of extra storage for the answer, i.e., the transform is not *quite* "in place." (Although an in-place transform is in fact possible, we have found it virtually impossible to explain to any user how to unscramble its output, i.e., where to find the real and imaginary components of the transform at some particular frequency!)

We will implement the multidimensional real Fourier transform for the three dimensional case $L = 3$, with the input data stored as a real, three-dimensional array data[0..nn1-1][0..nn2-1][0..nn3-1]. This scheme will allow two-dimensional data to be processed with effectively no loss of efficiency simply by choosing nn1 = 1. (Note that it must be the *first* dimension that is set to 1.) The output spectrum comes back packaged, logically at least, as a *complex*, three-dimensional array that we can call SPEC[0..nn1-1][0..nn2-1][0..nn3/2] (cf. equation 12.5.3). In the first two of its three dimensions, the respective frequency values f_1 or f_2 are stored in wrap-around order, that is with zero frequency in the first index value, the smallest positive frequency in the second index value, the smallest *negative* frequency in the *last* index value, and so on (cf. the discussion leading up to routines four1 and fourn). The third of the three dimensions returns only the positive half of the frequency spectrum. Figure 12.5.1 shows the logical storage scheme. The returned portion of the complex output spectrum is shown as the unshaded part of the lower figure.

The physical, as opposed to logical, packaging of the output spectrum is necessarily a bit different from the logical packaging, because C++ does not have a convenient, portable mechanism for equivalencing real and complex arrays. The subscript range SPEC[0..nn1-1][0..nn2-1][0..nn3/2-1] is returned in the input array data[0..nn1-1][0..nn2-1][0..nn3-1], with the correspondence

$$\text{Re}(\text{SPEC[i1][i2][i3]}) = \text{data[i1][i2][2*i3]}$$
$$\text{Im}(\text{SPEC[i1][i2][i3]}) = \text{data[i1][i2][2*i3+1]} \tag{12.5.4}$$

The remaining "plane" of values, SPEC[0..nn1-1][0..nn2-1][nn3/2], is returned in the two-dimensional double array speq[0..nn1-1][0..2*nn2-1], with the correspondence

$$\text{Re}(\text{SPEC[i1][i2][nn3/2]}) = \text{speq[i1][2*i2]}$$
$$\text{Im}(\text{SPEC[i1][i2][nn3/2]}) = \text{speq[i1][2*i2+1]} \tag{12.5.5}$$

Note that speq contains frequency components whose third component f_3 is at the Nyquist critical frequency $\pm f_c$. In some applications these values will in fact be ignored or set to zero, since they are intrinsically aliased between positive and negative frequencies.

With this much introduction, the implementing procedure, called rlft3, is something of an anticlimax. Look in the innermost loop in the procedure, and you will see equation (12.3.5) implemented on the *last* transform index. The case of i3=0 is coded separately, to account for the fact that speq is to be filled instead of overwriting the input array of data. The three enclosing for loops (indices i2, i3,

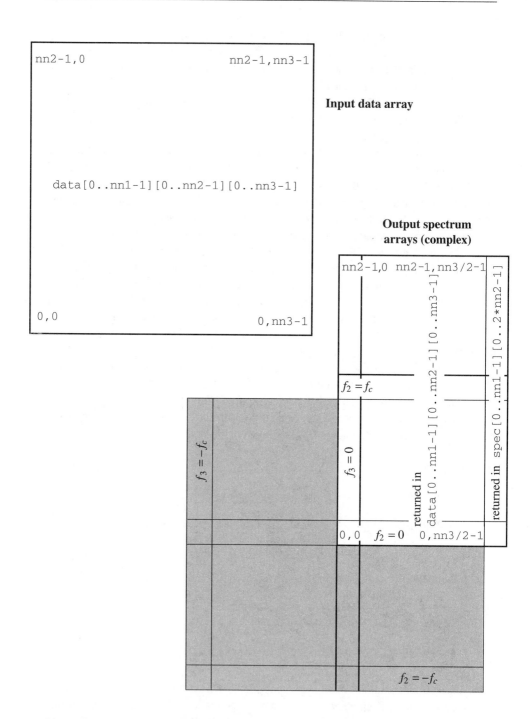

Figure 12.5.1.   Input and output data arrangement for rlft3. All arrays shown are presumed to have a first (leftmost) dimension of range [0..nn1-1], coming out of the page. The input data array is a real, three-dimensional array data[0..nn1-1][0..nn2-1][0..nn3-1]. (For two-dimensional data, one sets nn1 = 1.) The output data can be viewed as a single complex array with dimensions [0..nn1-1][0..nn2-1][0..nn3/2] (cf. equation 12.5.3), corresponding to the frequency components f_1 and f_2 being stored in wrap-around order, but only positive f_3 values being stored (others being obtainable by symmetry). The output data is actually returned mostly in the input array data, but partly stored in the real array speq[0..nn1-1][0..2*nn2-1]. See text for details.

and i1, from inside to outside) could in fact be done in any order — their actions all commute. We chose the order shown because of the following considerations: (i) i3 should not be the inner loop, because if it is, then the recurrence relations on wr and wi become burdensome. (ii) On virtual-memory machines, i1 should be the outer loop, because (with C++ order of array storage) this results in the array data, which might be very large, being accessed in block sequential order.

Keep in mind that all the computing in rlft3 is negligible, by a logarithmic factor, compared with the actual work of computing the associated complex FFT, done in the routine fourn. Since C++ complex operations are often implemented inefficiently, the operations are carried out explicitly below in terms of real and imaginary parts. The routine rlft3 is based on an earlier routine by G.B. Rybicki.

```
#include <cmath>
#include "nr.h"
using namespace std;

void NR::rlft3(Mat3D_IO_DP &data, Mat_IO_DP &speq, const int isign)
```
Given a three-dimensional real array data[0..nn1-1][0..nn2-1][0..nn3-1] (where nn1 = 1 for the case of a logically two-dimensional array), this routine returns (for isign=1) the complex fast Fourier transform as two complex arrays: On output, data contains the zero and positive frequency values of the third frequency component, while speq[0..nn1-1][0..2*nn2-1] contains the Nyquist critical frequency values of the third frequency component. First (and second) frequency components are stored for zero, positive, and negative frequencies, in standard wrap-around order. See text for description of how complex values are arranged. For isign=-1, the inverse transform (times nn1*nn2*nn3/2 as a constant multiplicative factor) is performed, with output data (viewed as a real array) deriving from input data (viewed as complex) and speq. For inverse transforms on data not generated first by a forward transform, make sure the complex input data array satisfies property (12.5.2). The dimensions nn1, nn2, nn3 must always be integer powers of 2.
```
{
    int i1,i2,i3,j1,j2,j3,ii3,k1,k2,k3,k4;
    DP theta,wi,wpi,wpr,wr,wtemp;
    DP c1,c2,h1r,h1i,h2r,h2i;
    Vec_INT nn(3);

    int nn1=data.dim1();
    int nn2=data.dim2();
    int nn3=data.dim3();
    c1=0.5;
    c2= -0.5*isign;
    theta=isign*(6.28318530717959/nn3);
    wtemp=sin(0.5*theta);
    wpr= -2.0*wtemp*wtemp;
    wpi=sin(theta);
    nn[0]=nn1;
    nn[1]=nn2;
    nn[2]=nn3 >> 1;
    Vec_DP data_v(&data[0][0][0],nn1*nn2*nn3);
    if (isign == 1) {                              Case of forward transform.
        fourn(data_v,nn,isign);                    Here is where most all of the com-
        k1=0;                                         pute time is spent.
        for (i1=0;i1<nn1;i1++)                      Extend data periodically into speq.
            for (i2=0,j2=0;i2<nn2;i2++,k1+=nn3) {
                speq[i1][j2++]=data_v[k1];
                speq[i1][j2++]=data_v[k1+1];
            }
    }
    for (i1=0;i1<nn1;i1++) {
        j1=(i1 != 0 ? nn1-i1 : 0);
```

Zero frequency is its own reflection, otherwise locate corresponding negative frequency in wrap-around order.

```
wr=1.0;                                        Initialize trigonometric recurrence.
wi=0.0;
for (ii3=0;ii3<=(nn3>>1);ii3+=2) {
    k1=i1*nn2*nn3;
    k3=j1*nn2*nn3;
    for (i2=0;i2<nn2;i2++,k1+=nn3) {
        if (ii3 == 0) {                        Equation (12.3.5).
            j2=(i2 != 0 ? ((nn2-i2)<<1) : 0);
            h1r=c1*(data_v[k1]+speq[j1][j2]);
            h1i=c1*(data_v[k1+1]-speq[j1][j2+1]);
            h2i=c2*(data_v[k1]-speq[j1][j2]);
            h2r= -c2*(data_v[k1+1]+speq[j1][j2+1]);
            data_v[k1]=h1r+h2r;
            data_v[k1+1]=h1i+h2i;
            speq[j1][j2]=h1r-h2r;
            speq[j1][j2+1]=h2i-h1i;
        } else {
            j2=(i2 != 0 ? nn2-i2 : 0);
            j3=nn3-ii3;
            k2=k1+ii3;
            k4=k3+j2*nn3+j3;
            h1r=c1*(data_v[k2]+data_v[k4]);
            h1i=c1*(data_v[k2+1]-data_v[k4+1]);
            h2i=c2*(data_v[k2]-data_v[k4]);
            h2r= -c2*(data_v[k2+1]+data_v[k4+1]);
            data_v[k2]=h1r+wr*h2r-wi*h2i;
            data_v[k2+1]=h1i+wr*h2i+wi*h2r;
            data_v[k4]=h1r-wr*h2r+wi*h2i;
            data_v[k4+1]= -h1i+wr*h2i+wi*h2r;
        }
    }
    wr=(wtemp=wr)*wpr-wi*wpi+wr;               Do the recurrence.
    wi=wi*wpr+wtemp*wpi+wi;
}
}
if (isign == -1) fourn(data_v,nn,isign);       Case of reverse transform.
k1=0;
for (i1=0;i1<nn1;i1++)
    for (i2=0;i2<nn2;i2++)
        for (i3=0;i3<nn3;i3++) data[i1][i2][i3]=data_v[k1++];
}
```

We now give some fragments from notional calling programs, to clarify the use of rlft3 for two- and three-dimensional data. Note again that the routine does not actually distinguish between two and three dimensions; two is treated like three, but with the first dimension having length 1. Since the first dimension is the outer loop, virtually no inefficiency is introduced.

The first program fragment FFTs a two-dimensional data array, allows for some processing on it, e.g., filtering, and then takes the inverse transform. Figure 12.5.2 shows an example of the use of this kind of code: A sharp image becomes blurry when its high-frequency spatial components are suppressed by the factor (here) $\max(1 - 6f^2/f_c^2, 0)$. The second program example illustrates a three-dimensional transform, where the three dimensions have different lengths. The third program example is an example of convolution, as it might occur in a program to compute the potential generated by a three-dimensional distribution of sources.

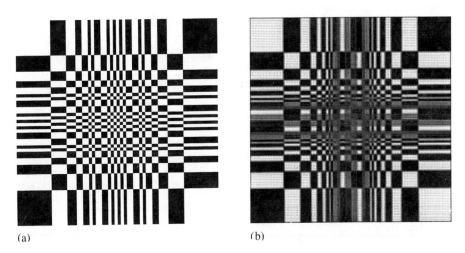

(a)   (b)

Figure 12.5.2.   (a) A two-dimensional image with intensities either purely black or purely white. (b) The same image, after it has been low-pass filtered using rlft3. Regions with fine-scale features become gray.

```
#include "nr.h"

int main(void)          // example1
This fragment shows how one might filter a 256 by 256 digital image.
{
    const int N2=256,N3=256;
    NRMat3d<DP> data(1,N2,N3);            Note that the first component must be set to 1.
    Mat_DP speq(1,2*N2);

//    ...                                 Here the image would be loaded into data.
    NR::rlft3(data,speq,1);
//    ...                                 Here the arrays data and speq would be multiplied
    NR::rlft3(data,speq,-1);                by a suitable filter function (of frequency).
//    ...                                 Here the filtered image would be unloaded from data.
    return 0;
}


int main(void)          // example2
This fragment shows how one might FFT a real three-dimensional array of size 32 by 64 by 16.
{
    const int N1=32,N2=64,N3=16;
    NRMat3d<DP> data(N1,N2,N3);
    Mat_DP speq(N1,2*N2);
//    ...                                 Here load data.
    NR::rlft3(data,speq,1);
//    ...                                 Here unload data and speq.
    return 0;
}


int main(void)          // example3
This fragment shows how one might convolve two real, three-dimensional arrays of size 32 by
32 by 32, replacing the first array by the result.
{
    int j;
    DP fac,r,i,*sp1,*sp2;
    const int N=32;
    NRMat3d<DP> data1(N,N,N),data2(N,N,N);
    Mat_DP speq1(N,2*N),speq2(N,2*N);
//    ...
```

```
NR::rlft3(data1,speq1,1);                FFT both input arrays.
NR::rlft3(data2,speq2,1);
fac=2.0/(N*N*N);                         Factor needed to get normalized inverse.
sp1 = &data1[0][0][0];
sp2 = &data1[0][0][0];
for (j=0;j<N*N*N/2;j++) {                Note how this can be made a single for-loop
    r = sp1[0]*sp2[0] - sp1[1]*sp2[1];    instead of three nested ones by using
    i = sp1[0]*sp2[1] + sp1[1]*sp2[0];    the pointers sp1 and sp2.
    sp1[0] = fac*r;
    sp1[1] = fac*i;
    sp1 += 2;
    sp2 += 2;
}
sp1 = &speq1[0][0];                      If your implementation of NRMat doesn't store el-
sp2 = &speq2[0][0];                      ements contiguously, you'll have to
for (j=0;j<N*N;j++) {                    rewrite this loop as two explicit loops
    r = sp1[0]*sp2[0] - sp1[1]*sp2[1];    over the indices of speq1 and spec2,
    i = sp1[0]*sp2[1] + sp1[1]*sp2[0];    without any pointers.
    sp1[0] = fac*r;
    sp1[1] = fac*i;
    sp1 += 2;
    sp2 += 2;
}
NR::rlft3(data1,speq1,-1);              Inverse FFT the product of the two FFTs.
//   ...
return 0;
}
```

To extend rlft3 to four dimensions, you simply add an additional (outer) nested for loop in i0, analogous to the present i1. (Modifying the routine to do an *arbitrary* number of dimensions, as in fourn, is a good programming exercise for the reader.)

CITED REFERENCES AND FURTHER READING:

Brigham, E.O. 1974, *The Fast Fourier Transform* (Englewood Cliffs, NJ: Prentice-Hall).

Swartztrauber, P. N. 1986, *Mathematics of Computation*, vol. 47, pp. 323–346.

12.6 External Storage or Memory-Local FFTs

Sometime in your life, you might have to compute the Fourier transform of a *really large* data set, larger than the size of your computer's physical memory. In such a case, the data will be stored on some external medium, such as magnetic or optical tape or disk. Needed is an algorithm that makes some manageable number of sequential passes through the external data, processing it on the fly and outputting intermediate results to other external media, which can be read on subsequent passes.

In fact, an algorithm of just this description was developed by Singleton [1] very soon after the discovery of the FFT. The algorithm requires four sequential storage devices, each capable of holding half of the input data. The first half of the input data is initially on one device, the second half on another.

Singleton's algorithm is based on the observation that it is possible to bit-reverse 2^M values by the following sequence of operations: On the first pass, values are read alternately from the two input devices, and written to a single output device (until it holds half the data), and then to the other output device. On the second pass, the output devices become input devices, and vice versa. Now, we copy *two* values from the first device, then *two* values from the second, writing them (as before) first to fill one output device, then to fill a second.

Subsequent passes read 4, 8, etc., input values at a time. After completion of pass $M - 1$, the data are in bit-reverse order.

Singleton's next observation is that it is possible to alternate the passes of essentially this bit-reversal technique with passes that implement one stage of the Danielson-Lanczos combination formula (12.2.3). The scheme, roughly, is this: One starts as before with half the input data on one device, half on another. In the first pass, one complex value is read from each input device. Two combinations are formed, and one is written to each of two output devices. After this "computing" pass, the devices are rewound, and a "permutation" pass is performed, where groups of values are read from the first input device and alternately written to the first and second output devices; when the first input device is exhausted, the second is similarly processed. This sequence of computing and permutation passes is repeated $M - K - 1$ times, where 2^K is the size of internal buffer available to the program. The second phase of the computation consists of a final K computation passes. What distinguishes the second phase from the first is that, now, the permutations are local enough to do in place during the computation. There are thus no separate permutation passes in the second phase. In all, there are $2M - K - 2$ passes through the data.

Here is an implementation of Singleton's algorithm, based on [1]:

```cpp
#include <iostream>
#include <fstream>
#include <cmath>
#include "nr.h"
using namespace std;
```

```cpp
void NR::fourfs(Vec_FSTREAM_p &file, Vec_I_INT &nn, const int isign)
```
One- or multi-dimensional Fourier transform of a large data set stored on external media. On input, `nn[0..ndim-1]` contains the lengths of each dimension (number of real and imaginary value pairs), which must be powers of two. Here `ndim` is the number of dimensions. `file[0..3]` contains the stream pointers to 4 temporary files, each large enough to hold half of the data. The four streams must be opened in the system's "binary" (as opposed to "text") mode. The input data must be in normal C++ array order (rightmost index changing most quickly), with its first half stored in file `file[0]`, its second half in `file[1]`, in native floating point form. KBF real numbers are processed per buffered read or write. `isign` should be set to 1 for the Fourier transform, to -1 for its inverse. On output, values in the array `file` may have been permuted; the first half of the result is stored in `file[2]`, the second half in `file[3]`. N.B.: For `ndim` > 1, the output is *not* stored in normal C++ array order. Instead it is the leftmost array index that cycles most quickly, then the second leftmost, and so on. If `ndim` = 2, this means that the output is stored by columns rather than rows; it is the transpose of the output that would have been produced by `fourn`.

```cpp
{
    const int KBF=128;
    static int mate[4]={1,0,3,2};
    int cc,cc0,j,j12,jk,k,kk,n=1,mm,kc=0,kd,ks,kr,na,nb,nc,nd,nr,ns,nv;
    DP tempr,tempi,wr,wi,wpr,wpi,wtemp,theta;
    Vec_DP afa(KBF),afb(KBF),afc(KBF);

    int ndim=nn.size();
    for (j=0;j<ndim;j++) {
        n *= nn[j];
        if (nn[j] <= 1) nrerror("invalid DP or wrong ndim in fourfs");
    }
    nv=0;
    jk=nn[nv];
    mm=n;
    ns=n/KBF;
    nr=ns >> 1;
    kd=KBF >> 1;
    ks=n;
    fourew(file,na,nb,nc,nd);
```
The first phase of the transform starts here.
```cpp
    for (;;) {                                          Start of the computing pass.
        theta=isign*3.141592653589793/(n/mm);
```

```
wtemp=sin(0.5*theta);
wpr = -2.0*wtemp*wtemp;
wpi=sin(theta);
wr=1.0;
wi=0.0;
mm >>= 1;
for (j12=0;j12<2;j12++) {
    kr=0;
    do {
        cc0=(*file[na]).tellg()/sizeof(DP);
        (*file[na]).read((char *) &afa[0],KBF*sizeof(DP));
        cc=(*file[na]).tellg()/sizeof(DP);
        if ((cc-cc0) != KBF) nrerror("read error 1 in fourfs");
        cc0=(*file[nb]).tellg()/sizeof(DP);
        (*file[nb]).read((char *) &afb[0],KBF*sizeof(DP));
        cc=(*file[nb]).tellg()/sizeof(DP);
        if ((cc-cc0) != KBF) nrerror("read error 2 in fourfs");
        for (j=0;j<KBF;j+=2) {
            tempr=wr*afb[j]-wi*afb[j+1];
            tempi=wi*afb[j]+wr*afb[j+1];
            afb[j]=afa[j]-tempr;
            afa[j] += tempr;
            afb[j+1]=afa[j+1]-tempi;
            afa[j+1] += tempi;
        }
        kc += kd;
        if (kc == mm) {
            kc=0;
            wr=(wtemp=wr)*wpr-wi*wpi+wr;
            wi=wi*wpr+wtemp*wpi+wi;
        }
        cc0=(*file[nc]).tellp()/sizeof(DP);
        (*file[nc]).write((char *) &afa[0],KBF*sizeof(DP));
        cc=(*file[nc]).tellp()/sizeof(DP);
        if ((cc-cc0) != KBF) nrerror("write error 1 in fourfs");
        cc0=(*file[nd]).tellp()/sizeof(DP);
        (*file[nd]).write((char *) &afb[0],KBF*sizeof(DP));
        cc=(*file[nd]).tellp()/sizeof(DP);
        if ((cc-cc0) != KBF) nrerror("write error 2 in fourfs");
    } while (++kr < nr);
    if (j12 == 0 && ks != n && ks == KBF) {
        na=mate[na];
        nb=na;
    }
    if (nr == 0) break;
}
fourew(file,na,nb,nc,nd);               Start of the permutation pass.
jk >>= 1;
while (jk == 1) {
    mm=n;
    jk=nn[++nv];
}
ks >>= 1;
if (ks > KBF) {
    for (j12=0;j12<2;j12++) {
        for (kr=0;kr<ns;kr+=ks/KBF) {
            for (k=0;k<ks;k+=KBF) {
                cc0=(*file[na]).tellg()/sizeof(DP);
                (*file[na]).read((char *) &afa[0],KBF*sizeof(DP));
                cc=(*file[na]).tellg()/sizeof(DP);
                if ((cc-cc0) != KBF) nrerror("read error 3 in fourfs");
                cc0=(*file[nc]).tellp()/sizeof(DP);
                (*file[nc]).write((char *) &afa[0],KBF*sizeof(DP));
                cc=(*file[nc]).tellp()/sizeof(DP);
```

```
                    if ((cc-cc0) != KBF) nrerror("write error 3 in fourfs");
                }
                nc=mate[nc];
            }
            na=mate[na];
        }
        fourew(file,na,nb,nc,nd);
    } else if (ks == KBF) nb=na;
    else break;
}
j=0;
```

The second phase of the transform starts here. Now, the remaining permutations are sufficiently local to be done in place.

```
for (;;) {
    theta=isign*3.141592653589793/(n/mm);
    wtemp=sin(0.5*theta);
    wpr = -2.0*wtemp*wtemp;
    wpi=sin(theta);
    wr=1.0;
    wi=0.0;
    mm >>= 1;
    ks=kd;
    kd >>= 1;
    for (j12=0;j12<2;j12++) {
        for (kr=0;kr<ns;kr++) {
            cc0=(*file[na]).tellg()/sizeof(DP);
            (*file[na]).read((char *) &afc[0],KBF*sizeof(DP));
            cc=(*file[na]).tellg()/sizeof(DP);
            if ((cc-cc0) != KBF) nrerror("read error 4 in fourfs");
            kk=0;
            k=ks;
            for (;;) {
                tempr=wr*afc[kk+ks]-wi*afc[kk+ks+1];
                tempi=wi*afc[kk+ks]+wr*afc[kk+ks+1];
                afa[j]=afc[kk]+tempr;
                afb[j]=afc[kk]-tempr;
                afa[++j]=afc[++kk]+tempi;
                afb[j++]=afc[kk++]-tempi;
                if (kk < k) continue;
                kc += kd;
                if (kc == mm) {
                    kc=0;
                    wr=(wtemp=wr)*wpr-wi*wpi+wr;
                    wi=wi*wpr+wtemp*wpi+wi;
                }
                kk += ks;
                if (kk > KBF-1) break;
                else k=kk+ks;
            }
            if (j > KBF-1) {
                cc0=(*file[nc]).tellp()/sizeof(DP);
                (*file[nc]).write((char *) &afa[0],KBF*sizeof(DP));
                cc=(*file[nc]).tellp()/sizeof(DP);
                if ((cc-cc0) != KBF) nrerror("write error 4 in fourfs");
                cc0=(*file[nd]).tellp()/sizeof(DP);
                (*file[nd]).write((char *) &afb[0],KBF*sizeof(DP));
                cc=(*file[nd]).tellp()/sizeof(DP);
                if ((cc-cc0) != KBF) nrerror("write error 5 in fourfs");
                j=0;
            }
        }
        na=mate[na];
    }
    fourew(file,na,nb,nc,nd);
```

```
            jk >>= 1;
            if (jk > 1) continue;
            mm=n;
            do {
                if (nv < ndim-1) jk=nn[++nv];
                else return;
            } while (jk == 1);
        }
}
```

```
#include <fstream>
#include "nr.h"
using namespace std;

void NR::fourew(Vec_FSTREAM_p &file, int &na, int &nb, int &nc, int &nd)
Utility used by fourfs. Rewinds and renumbers the four files.
{
    int i;

    for (i=0;i<4;i++) (*file[i]).seekp(0);
    for (i=0;i<4;i++) (*file[i]).seekg(0);
    SWAP(file[1],file[3]);
    SWAP(file[0],file[2]);
    na=2;
    nb=3;
    nc=0;
    nd=1;
}
```

For one-dimensional data, Singleton's algorithm produces output in exactly the same order as a standard FFT (e.g., four1). For multidimensional data, the output is in *transpose* order rather than in the conventional C++ array order output by fourn. That is, in scanning through the data, it is the leftmost array index that cycles most quickly, then the second leftmost, and so on. This peculiarity, which is intrinsic to the method, is generally only a minor inconvenience. For convolutions, one simply computes the component-by-component product of two transforms in their nonstandard arrangement, and then does an inverse transform on the result. Note that, if the lengths of the different dimensions are not all the same, then you must reverse the order of the values in nn[0..ndim-1] (thus giving the dimensions of the transpose-order output array) before performing the inverse transform. Note also that, just like fourn, performing a transform and then an inverse results in multiplying the original data by the product of the lengths of all dimensions.

We leave it as an exercise for the reader to figure out how to reorder fourfs's output into normal order, taking additional passes through the externally stored data. We doubt that such reordering is ever really needed.

You will likely want to modify fourfs to fit your particular application. For example, as written, KBF $\equiv 2^K$ plays the dual role of being the size of the internal buffers, and the record size of the unformatted reads and writes. The latter role limits its size to that allowed by your machine's I/O facility. It is a simple matter to perform multiple reads for a much larger KBF, thus reducing the number of passes by a few.

Another modification of fourfs would be for the case where your virtual memory machine has sufficient address space, but not sufficient physical memory, to do an efficient FFT by the conventional algorithm (whose memory references are extremely nonlocal). In that case, you will need to replace the reads, writes, and rewinds by mappings of the arrays afa, afb, and afc into your address space. In other words, these arrays are replaced by references to a single data array, with offsets that get modified wherever fourfs performs an I/O operation. The resulting algorithm will have its memory references local within blocks of size KBF. Execution speed is thereby sometimes increased enormously, albeit at the cost of requiring twice as much virtual memory as an in-place FFT.

CITED REFERENCES AND FURTHER READING:

Singleton, R.C. 1967, *IEEE Transactions on Audio and Electroacoustics*, vol. AU-15, pp. 91–97. [1]

Oppenheim, A.V., and Schafer, R.W. 1989, *Discrete-Time Signal Processing* (Englewood Cliffs, NJ: Prentice-Hall), Chapter 9.

Chapter 13.   Fourier and Spectral Applications

13.0 Introduction

Fourier methods have revolutionized fields of science and engineering, from radio astronomy to medical imaging, from seismology to spectroscopy. In this chapter, we present some of the basic applications of Fourier and spectral methods that have made these revolutions possible.

Say the word "Fourier" to a numericist, and the response, as if by Pavlovian conditioning, will likely be "FFT." Indeed, the wide application of Fourier methods must be credited principally to the existence of the fast Fourier transform. Better mousetraps stand aside: If you speed up *any* nontrivial algorithm by a factor of a million or so, the world will beat a path towards finding useful applications for it. The most direct applications of the FFT are to the convolution or deconvolution of data (§13.1), correlation and autocorrelation (§13.2), optimal filtering (§13.3), power spectrum estimation (§13.4), and the computation of Fourier integrals (§13.9).

As important as they are, however, FFT methods are not the be-all and end-all of spectral analysis. Section 13.5 is a brief introduction to the field of time-domain digital filters. In the spectral domain, one limitation of the FFT is that it always represents a function's Fourier transform as a polynomial in $z = \exp(2\pi i f \Delta)$ (cf. equation 12.1.7). Sometimes, processes have spectra whose shapes are not well represented by this form. An alternative form, which allows the spectrum to have poles in z, is used in the techniques of linear prediction (§13.6) and maximum entropy spectral estimation (§13.7).

Another significant limitation of all FFT methods is that they require the input data to be sampled at evenly spaced intervals. For irregularly or incompletely sampled data, other (albeit slower) methods are available, as discussed in §13.8.

So-called wavelet methods inhabit a representation of function space that is neither in the temporal, nor in the spectral, domain, but rather something in-between. Section 13.10 is an introduction to this subject. Finally §13.11 is an excursion into numerical use of the Fourier sampling theorem.

13.1 Convolution and Deconvolution Using the FFT

We have defined the *convolution* of two functions for the continuous case in equation (12.0.8), and have given the *convolution theorem* as equation (12.0.9). The theorem says that the Fourier transform of the convolution of two functions is equal to the product of their individual Fourier transforms. Now, we want to deal with the discrete case. We will mention first the context in which convolution is a useful procedure, and then discuss how to compute it efficiently using the FFT.

The convolution of two functions $r(t)$ and $s(t)$, denoted $r * s$, is mathematically equal to their convolution in the opposite order, $s * r$. Nevertheless, in most applications the two functions have quite different meanings and characters. One of the functions, say s, is typically a signal or data stream, which goes on indefinitely in time (or in whatever the appropriate independent variable may be). The other function r is a "response function," typically a peaked function that falls to zero in both directions from its maximum. The effect of convolution is to smear the signal $s(t)$ in time according to the recipe provided by the response function $r(t)$, as shown in Figure 13.1.1. In particular, a spike or delta-function of unit area in s which occurs at some time t_0 is supposed to be smeared into the shape of the response function itself, but translated from time 0 to time t_0 as $r(t - t_0)$.

In the discrete case, the signal $s(t)$ is represented by its sampled values at equal time intervals s_j. The response function is also a discrete set of numbers r_k, with the following interpretation: r_0 tells what multiple of the input signal in one channel (one particular value of j) is copied into the identical output channel (same value of j); r_1 tells what multiple of input signal in channel j is additionally copied into output channel $j + 1$; r_{-1} tells the multiple that is copied into channel $j - 1$; and so on for both positive and negative values of k in r_k. Figure 13.1.2 illustrates the situation.

Example: a response function with $r_0 = 1$ and all other r_k's equal to zero is just the identity filter: convolution of a signal with this response function gives identically the signal. Another example is the response function with $r_{14} = 1.5$ and all other r_k's equal to zero. This produces convolved output that is the input signal multiplied by 1.5 and delayed by 14 sample intervals.

Evidently, we have just described in words the following definition of discrete convolution with a response function of finite duration M:

$$(r * s)_j \equiv \sum_{k=-M/2+1}^{M/2} s_{j-k}\, r_k \tag{13.1.1}$$

If a discrete response function is nonzero only in some range $-M/2 < k \le M/2$, where M is a sufficiently large even integer, then the response function is called a *finite impulse response (FIR)*, and its *duration* is M. (Notice that we are defining M as the number of nonzero *values* of r_k; these values span a time interval of $M - 1$ sampling times.) In most practical circumstances the case of finite M is the case of interest, either because the response really has a finite duration, or because we choose to truncate it at some point and approximate it by a finite-duration response function.

The *discrete convolution theorem* is this: If a signal s_j is *periodic* with period N, so that it is completely determined by the N values $s_0, \ldots, s_{N-1}$, then its

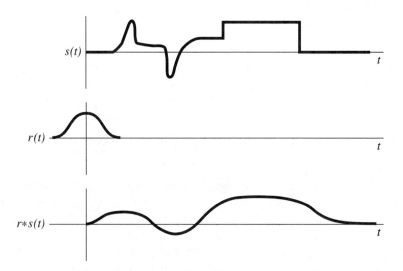

Figure 13.1.1.   Example of the convolution of two functions. A signal $s(t)$ is convolved with a response function $r(t)$. Since the response function is broader than some features in the original signal, these are "washed out" in the convolution. In the absence of any additional noise, the process can be reversed by deconvolution.

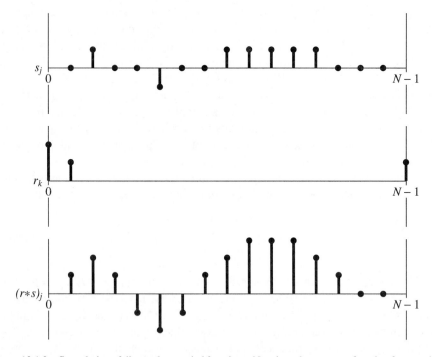

Figure 13.1.2. Convolution of discretely sampled functions. Note how the response function for negative times is wrapped around and stored at the extreme right end of the array r_k.

discrete convolution with a response function *of finite duration* N is a member of the discrete Fourier transform pair,

$$\sum_{k=-N/2+1}^{N/2} s_{j-k} \, r_k \quad \Longleftrightarrow \quad S_n R_n \qquad (13.1.2)$$

Here S_n, $(n = 0, \ldots, N - 1)$ is the discrete Fourier transform of the values s_j, $(j = 0, \ldots, N - 1)$, while R_n, $(n = 0, \ldots, N - 1)$ is the discrete Fourier transform of the values r_k, $(k = 0, \ldots, N - 1)$. These values of r_k are the same ones as for the range $k = -N/2 + 1, \ldots, N/2$, but in wrap-around order, exactly as was described at the end of §12.2.

Treatment of End Effects by Zero Padding

The discrete convolution theorem presumes a set of two circumstances that are not universal. First, it assumes that the input signal is periodic, whereas real data often either go forever without repetition or else consist of one nonperiodic stretch of finite length. Second, the convolution theorem takes the duration of the response to be the same as the period of the data; they are both N. We need to work around these two constraints.

The second is very straightforward. Almost always, one is interested in a response function whose duration M is much shorter than the length of the data set N. In this case, you simply extend the response function to length N by padding it with zeros, i.e., define $r_k = 0$ for $M/2 \le k \le N/2$ and also for $-N/2 + 1 \le k \le -M/2 + 1$. Dealing with the first constraint is more challenging. Since the convolution theorem rashly assumes that the data are periodic, it will falsely "pollute" the first output channel $(r * s)_0$ with some wrapped-around data from the far end of the data stream s_{N-1}, s_{N-2}, etc. (See Figure 13.1.3.) So, we need to set up a buffer zone of zero-padded values at the end of the s_j vector, in order to make this pollution zero. How many zero values do we need in this buffer? Exactly as many as the most negative index for which the response function is nonzero. For example, if r_{-3} is nonzero, while $r_{-4}, r_{-5}, \ldots$ are all zero, then we need three zero pads at the end of the data: $s_{N-3} = s_{N-2} = s_{N-1} = 0$. These zeros will protect the first output channel $(r * s)_0$ from wrap-around pollution. It should be obvious that the second output channel $(r * s)_1$ and subsequent ones will also be protected by these same zeros. Let K denote the number of padding zeros, so that the last actual input data point is s_{N-K-1}.

What now about pollution of the very *last* output channel? Since the data now end with s_{N-K-1}, the last output channel of interest is $(r * s)_{N-K-1}$. This channel can be polluted by wrap-around from input channel s_0 unless the number K is also large enough to take care of the most positive index k for which the response function r_k is nonzero. For example, if r_0 through r_6 are nonzero, while $r_7, r_8 \ldots$ are all zero, then we need at least $K = 6$ padding zeros at the end of the data: $s_{N-6} = \ldots = s_{N-1} = 0$.

To summarize — we need to pad the data with a number of zeros *on one end* equal to the maximum positive duration *or* maximum negative duration of the response function, *whichever is larger*. (For a symmetric response function of duration M, you will need only $M/2$ zero pads.) Combining this operation with the

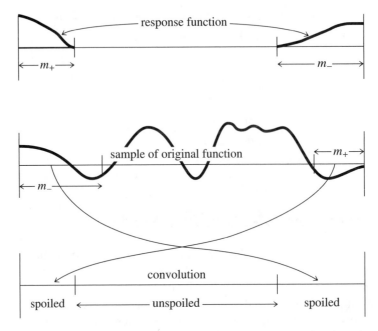

Figure 13.1.3.   The wrap-around problem in convolving finite segments of a function. Not only must the response function wrap be viewed as cyclic, but so must the sampled original function. Therefore a portion at each end of the original function is erroneously wrapped around by convolution with the response function.

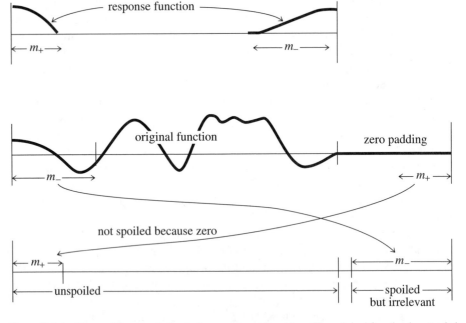

Figure 13.1.4.   Zero padding as solution to the wrap-around problem. The original function is extended by zeros, serving a dual purpose: When the zeros wrap around, they do not disturb the true convolution; and while the original function wraps around onto the zero region, that region can be discarded.

padding of the response r_k described above, we effectively insulate the data from artifacts of undesired periodicity. Figure 13.1.4 illustrates matters.

Use of FFT for Convolution

The data, complete with zero padding, are now a set of real numbers s_j, $j = 0, \ldots, N - 1$, and the response function is zero padded out to duration N and arranged in wrap-around order. (Generally this means that a large contiguous section of the r_k's, in the middle of that array, is zero, with nonzero values clustered at the two extreme ends of the array.) You now compute the discrete convolution as follows: Use the FFT algorithm to compute the discrete Fourier transform of s and of r. Multiply the two transforms together component by component, remembering that the transforms consist of complex numbers. Then use the FFT algorithm to take the inverse discrete Fourier transform of the products. The answer is the convolution $r*s$.

What about *deconvolution*? Deconvolution is the process of *undoing* the smearing in a data set that has occurred under the influence of a known response function, for example, because of the known effect of a less-than-perfect measuring apparatus. The defining equation of deconvolution is the same as that for convolution, namely (13.1.1), except now the left-hand side is taken to be known, and (13.1.1) is to be considered as a set of N linear equations for the unknown quantities s_j. Solving these simultaneous linear equations in the time domain of (13.1.1) is unrealistic in most cases, but the FFT renders the problem almost trivial. Instead of multiplying the transform of the signal and response to get the transform of the convolution, we just divide the transform of the (known) convolution by the transform of the response to get the transform of the deconvolved signal.

This procedure can go wrong *mathematically* if the transform of the response function is exactly zero for some value R_n, so that we can't divide by it. This indicates that the original convolution has truly lost all information at that one frequency, so that a reconstruction of that frequency component is not possible. You should be aware, however, that apart from mathematical problems, the process of deconvolution has other practical shortcomings. The process is generally quite sensitive to noise in the input data, and to the accuracy to which the response function r_k is known. Perfectly reasonable attempts at deconvolution can sometimes produce nonsense for these reasons. In such cases you may want to make use of the additional process of *optimal filtering*, which is discussed in §13.3.

Here is our routine for convolution and deconvolution, using the FFT as implemented in `four1` of §12.2. Since the data and response functions are real, not complex, both of their transforms could be taken simultaneously using `twofft`. However, since `data` and `respns` often have very different magnitudes, we instead make separate calls to `realft` to minimize roundoff. The data are assumed to be stored in a double array `data[0..n-1]`, with n an integer power of two. The response function is assumed to be stored in wrap-around order in a double array `respns[0..m-1]`. The value of m can be any *odd* integer less than or equal to n, since the first thing the program does is to recopy the response function into the appropriate wrap-around order in an array of length n. The answer is provided in `ans`, which is also used as working space.

```
#include "nr.h"

void NR::convlv(Vec_I_DP &data, Vec_I_DP &respns, const int isign,
    Vec_O_DP &ans)
```
Convolves or deconvolves a real data set data[0..n-1] (including any user-supplied zero padding) with a response function respns[0..m-1], where m is an odd integer $\leq$ n. The response function must be stored in wrap-around order: the first half of the array respns contains the impulse response function at positive times, while the second half of the array contains the impulse response function at negative times, counting down from the highest element respns[m-1]. On input isign is +1 for convolution, −1 for deconvolution. The answer is returned in ans[0..n-1]. n MUST be an integer power of two.

```
{
    int i,no2;
    DP mag2,tmp;

    int n=data.size();
    int m=respns.size();
    Vec_DP temp(n);
    temp[0]=respns[0];
    for (i=1;i<(m+1)/2;i++) {             Put respns in array of length n.
        temp[i]=respns[i];
        temp[n-i]=respns[m-i];
    }
    for (i=(m+1)/2;i<n-(m-1)/2;i++)       Pad with zeros.
        temp[i]=0.0;
    for (i=0;i<n;i++)
        ans[i]=data[i];
    realft(ans,1);                        FFT both arrays.
    realft(temp,1);
    no2=n>>1;
    if (isign == 1) {
        for (i=2;i<n;i+=2) {              Multiply FFTs to convolve.
            tmp=ans[i];
            ans[i]=(ans[i]*temp[i]-ans[i+1]*temp[i+1])/no2;
            ans[i+1]=(ans[i+1]*temp[i]+tmp*temp[i+1])/no2;
        }
        ans[0]=ans[0]*temp[0]/no2;
        ans[1]=ans[1]*temp[1]/no2;
    } else if (isign == -1) {
        for (i=2;i<n;i+=2) {             Divide FFTs to deconvolve.
            if ((mag2=SQR(temp[i])+SQR(temp[i+1])) == 0.0)
                nrerror("Deconvolving at response zero in convlv");
            tmp=ans[i];
            ans[i]=(ans[i]*temp[i]+ans[i+1]*temp[i+1])/mag2/no2;
            ans[i+1]=(ans[i+1]*temp[i]-tmp*temp[i+1])/mag2/no2;
        }
        if (temp[0] == 0.0 || temp[1] == 0.0)
            nrerror("Deconvolving at response zero in convlv");
        ans[0]=ans[0]/temp[0]/no2;
        ans[1]=ans[1]/temp[1]/no2;
    } else nrerror("No meaning for isign in convlv");
    realft(ans,-1);                       Inverse transform back to time domain.
}
```

Convolving or Deconvolving Very Large Data Sets

If your data set is so long that you do not want to fit it into memory all at once, then you must break it up into sections and convolve each section separately. Now, however, the treatment of end effects is a bit different. You have to worry not only about spurious wrap-around effects, but also about the fact that the ends of

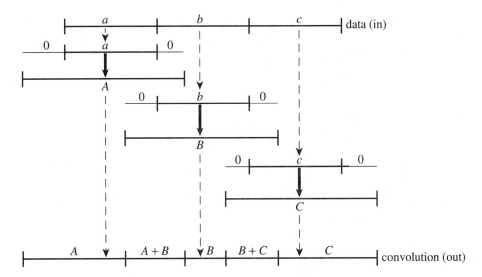

Figure 13.1.5. The overlap-add method for convolving a response with a very long signal. The signal data is broken up into smaller pieces. Each is zero padded at both ends and convolved (denoted by bold arrows in the figure). Finally the pieces are added back together, including the overlapping regions formed by the zero pads.

each section of data *should* have been influenced by data at the nearby ends of the immediately preceding and following sections of data, but were not so influenced since only one section of data is in the machine at a time.

There are two, related, standard solutions to this problem. Both are fairly obvious, so with a few words of description here, you ought to be able to implement them for yourself. The first solution is called the *overlap-save method*. In this technique you pad only the very beginning of the data with enough zeros to avoid wrap-around pollution. After this initial padding, you forget about zero padding altogether. Bring in a section of data and convolve or deconvolve it. Then throw out the points at each end that are polluted by wrap-around end effects. Output only the remaining good points in the middle. Now bring in the next section of data, but not all new data. The first points in each new section overlap the last points from the preceding section of data. The sections must be overlapped sufficiently so that the polluted output points at the end of one section are recomputed as the first of the unpolluted output points from the subsequent section. With a bit of thought you can easily determine how many points to overlap and save.

The second solution, called the *overlap-add method*, is illustrated in Figure 13.1.5. Here you *don't* overlap the input data. Each section of data is disjoint from the others and is used exactly once. However, you carefully zero-pad it at both ends so that there is no wrap-around ambiguity in the output convolution or deconvolution. Now you overlap *and add* these sections of output. Thus, an output point near the end of one section will have the response due to the input points at the beginning of the next section of data properly added in to it, and likewise for an output point near the beginning of a section, *mutatis mutandis*.

Even when computer memory is available, there is some slight gain in computing speed in segmenting a long data set, since the FFTs' $N \log_2 N$ is slightly slower than linear in N. However, the log term is so slowly varying that you will often be much happier to avoid the bookkeeping complexities of the overlap-add or overlap-save methods: If it is practical to do so, just cram the whole data set into memory and FFT away. Then you will have more time for the finer things in life, some of which are described in succeeding sections of this chapter.

CITED REFERENCES AND FURTHER READING:

Nussbaumer, H.J. 1982, *Fast Fourier Transform and Convolution Algorithms* (New York: Springer-Verlag).

Elliott, D.F., and Rao, K.R. 1982, *Fast Transforms: Algorithms, Analyses, Applications* (New York: Academic Press).

Brigham, E.O. 1974, *The Fast Fourier Transform* (Englewood Cliffs, NJ: Prentice-Hall), Chapter 13.

13.2 Correlation and Autocorrelation Using the FFT

Correlation is the close mathematical cousin of convolution. It is in some ways simpler, however, because the two functions that go into a correlation are not as conceptually distinct as were the data and response functions that entered into convolution. Rather, in correlation, the functions are represented by different, but generally similar, data sets. We investigate their "correlation," by comparing them both directly superposed, and with one of them shifted left or right.

We have already defined in equation (12.0.10) the correlation between two continuous functions $g(t)$ and $h(t)$, which is denoted $\text{Corr}(g, h)$, and is a function of *lag t*. We will occasionally show this time dependence explicitly, with the rather awkward notation $\text{Corr}(g, h)(t)$. The correlation will be large at some value of t if the first function (g) is a close copy of the second (h) but lags it in time by t, i.e., if the first function is shifted to the right of the second. Likewise, the correlation will be large for some negative value of t if the first function *leads* the second, i.e., is shifted to the left of the second. The relation that holds when the two functions are interchanged is

$$\text{Corr}(g, h)(t) = \text{Corr}(h, g)(-t) \qquad (13.2.1)$$

The discrete correlation of two sampled functions g_k and h_k, each periodic with period N, is defined by

$$\text{Corr}(g, h)_j \equiv \sum_{k=0}^{N-1} g_{j+k} h_k \qquad (13.2.2)$$

The *discrete correlation theorem* says that this discrete correlation of two real functions g and h is one member of the discrete Fourier transform pair

$$\text{Corr}(g, h)_j \Longleftrightarrow G_k H_k{}^* \qquad (13.2.3)$$

where G_k and H_k are the discrete Fourier transforms of g_j and h_j, and the asterisk denotes complex conjugation. This theorem makes the same presumptions about the functions as those encountered for the discrete convolution theorem.

We can compute correlations using the FFT as follows: FFT the two data sets, multiply one resulting transform by the complex conjugate of the other, and inverse transform the product. The result (call it r_k) will formally be a complex vector of length N. However, it will turn out to have all its imaginary parts zero since the original data sets were both real. The components of r_k are the values of the correlation at different lags, with positive and negative lags stored in the by now familiar wrap-around order: The correlation at zero lag is in r_0, the first component; the correlation at lag 1 is in r_1, the second component; the correlation at lag -1 is in r_{N-1}, the last component; etc.

Just as in the case of convolution we have to consider end effects, since our data will not, in general, be periodic as intended by the correlation theorem. Here again, we can use zero padding. If you are interested in the correlation for lags as large as $\pm K$, then you must append a buffer zone of K zeros at the end of both input data sets. If you want all possible lags from N data points (not a usual thing), then you will need to pad the data with an equal number of zeros; this is the extreme case. So here is the program:

```
#include "nr.h"

void NR::correl(Vec_I_DP &data1, Vec_I_DP &data2, Vec_O_DP &ans)
```
Computes the correlation of two real data sets data1[0..n-1] and data2[0..n-1] (including any user-supplied zero padding). n MUST be an integer power of two. The answer is returned in ans[0..n-1] stored in wrap-around order, i.e., correlations at increasingly negative lags are in ans[n-1] on down to ans[n/2], while correlations at increasingly positive lags are in ans[0] (zero lag) on up to ans[n/2-1]. Sign convention of this routine: if data1 lags data2, i.e., is shifted to the right of it, then ans will show a peak at positive lags.
```
{
    int no2,i;
    DP tmp;

    int n=data1.size();
    Vec_DP temp(n);
    for (i=0;i<n;i++) {
        ans[i]=data1[i];
        temp[i]=data2[i];
    }
    realft(ans,1);                    Transform both data vectors.
    realft(temp,1);
    no2=n>>1;                         Normalization for inverse FFT.
    for (i=2;i<n;i+=2) {              Multiply to find FFT of their correlation.
        tmp=ans[i];
        ans[i]=(ans[i]*temp[i]+ans[i+1]*temp[i+1])/no2;
        ans[i+1]=(ans[i+1]*temp[i]-tmp*temp[i+1])/no2;
    }
    ans[0]=ans[0]*temp[0]/no2;
    ans[1]=ans[1]*temp[1]/no2;
    realft(ans,-1);                   Inverse transform gives correlation.
}
```

As in convlv, we make two calls to realft instead of one call to twofft, to minimize roundoff error if data1 and data2 have very different magnitudes.

The *discrete autocorrelation* of a sampled function g_j is just the discrete correlation of the function with itself. Obviously this is always symmetric with respect to positive and negative lags. Feel free to use the above routine `correl` to obtain autocorrelations, simply calling it with the same `data` vector in both arguments. If the inefficiency bothers you, you can edit the program so that only one call is made to `realft` for the forward transform.

CITED REFERENCES AND FURTHER READING:

Brigham, E.O. 1974, *The Fast Fourier Transform* (Englewood Cliffs, NJ: Prentice-Hall), §13–2.

13.3 Optimal (Wiener) Filtering with the FFT

There are a number of other tasks in numerical processing that are routinely handled with Fourier techniques. One of these is filtering for the removal of noise from a "corrupted" signal. The particular situation we consider is this: There is some underlying, uncorrupted signal $u(t)$ that we want to measure. The measurement process is imperfect, however, and what comes out of our measurement device is a corrupted signal $c(t)$. The signal $c(t)$ may be less than perfect in either or both of two respects. First, the apparatus may not have a perfect "delta-function" response, so that the true signal $u(t)$ is convolved with (smeared out by) some known response function $r(t)$ to give a smeared signal $s(t)$,

$$ s(t) = \int_{-\infty}^{\infty} r(t - \tau)u(\tau)\, d\tau \quad \text{or} \quad S(f) = R(f)U(f) \qquad (13.3.1) $$

where S, R, U are the Fourier transforms of s, r, u, respectively. Second, the measured signal $c(t)$ may contain an additional component of noise $n(t)$,

$$ c(t) = s(t) + n(t) \qquad (13.3.2) $$

We already know how to deconvolve the effects of the response function r in the absence of any noise (§13.1); we just divide $C(f)$ by $R(f)$ to get a deconvolved signal. We now want to treat the analogous problem when noise is present. Our task is to find the *optimal filter*, $\phi(t)$ or $\Phi(f)$, which, when applied to the measured signal $c(t)$ or $C(f)$, and then deconvolved by $r(t)$ or $R(f)$, produces a signal $\tilde{u}(t)$ or $\tilde{U}(f)$ that is as close as possible to the uncorrupted signal $u(t)$ or $U(f)$. In other words we will estimate the true signal U by

$$ \tilde{U}(f) = \frac{C(f)\Phi(f)}{R(f)} \qquad (13.3.3) $$

In what sense is $\tilde{U}$ to be close to U? We ask that they be *close in the least-square sense*

$$ \int_{-\infty}^{\infty} |\tilde{u}(t) - u(t)|^2\, dt = \int_{-\infty}^{\infty} \left|\tilde{U}(f) - U(f)\right|^2\, df \quad \text{is minimized.} \quad (13.3.4) $$

Substituting equations (13.3.3) and (13.3.2), the right-hand side of (13.3.4) becomes

$$\int_{-\infty}^{\infty} \left| \frac{[S(f) + N(f)]\Phi(f)}{R(f)} - \frac{S(f)}{R(f)} \right|^2 \, df$$

$$= \int_{-\infty}^{\infty} |R(f)|^{-2} \left\{ |S(f)|^2 \, |1 - \Phi(f)|^2 + |N(f)|^2 \, |\Phi(f)|^2 \right\} \, df$$

$$(13.3.5)$$

The signal S and the noise N are *uncorrelated*, so their cross product, when integrated over frequency f, gave zero. (This is practically the *definition* of what we mean by noise!) Obviously (13.3.5) will be a minimum if and only if the integrand is minimized with respect to $\Phi(f)$ at every value of f. Let us search for such a solution where $\Phi(f)$ is a real function. Differentiating with respect to Φ, and setting the result equal to zero gives

$$\Phi(f) = \frac{|S(f)|^2}{|S(f)|^2 + |N(f)|^2} \qquad (13.3.6)$$

This is the formula for the optimal filter $\Phi(f)$.

Notice that equation (13.3.6) involves S, the smeared signal, and N, the noise. The two of these add up to be C, the measured signal. Equation (13.3.6) does not contain U, the "true" signal. This makes for an important simplification: The optimal filter can be determined independently of the determination of the deconvolution function that relates S and U.

To determine the optimal filter from equation (13.3.6) we need some way of separately estimating $|S|^2$ and $|N|^2$. There is no way to do this from the measured signal C alone without some other information, or some assumption or guess. Luckily, the extra information is often easy to obtain. For example, we can sample a long stretch of data $c(t)$ and plot its power spectral density using equations (12.0.14), (12.1.8), and (12.1.5). This quantity is proportional to the sum $|S|^2 + |N|^2$, so we have

$$|S(f)|^2 + |N(f)|^2 \approx P_c(f) = |C(f)|^2 \qquad 0 \le f < f_c \qquad (13.3.7)$$

(More sophisticated methods of estimating the power spectral density will be discussed in §13.4 and §13.7, but the estimation above is almost always good enough for the optimal filter problem.) The resulting plot (see Figure 13.3.1) will often immediately show the spectral signature of a signal sticking up above a continuous noise spectrum. The noise spectrum may be flat, or tilted, or smoothly varying; it doesn't matter, as long as we can guess a reasonable hypothesis as to what it is. Draw a smooth curve through the noise spectrum, extrapolating it into the region dominated by the signal as well. Now draw a smooth curve through the signal plus noise power. The difference between these two curves is your smooth "model" of the signal power. The quotient of your model of signal power to your model of signal plus noise power is the optimal filter $\Phi(f)$. [Extend it to negative values of f by the formula $\Phi(-f) = \Phi(f)$.] Notice that $\Phi(f)$ will be close to unity where the noise is negligible, and close to zero where the noise is dominant. That is how it does its job! The intermediate dependence given by equation (13.3.6) just turns out to be the optimal way of going in between these two extremes.

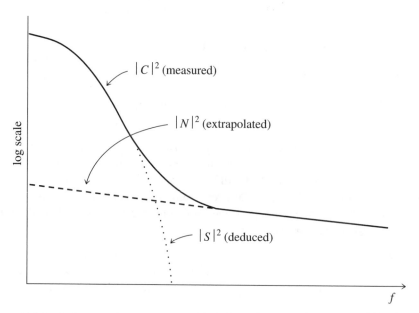

Figure 13.3.1. Optimal (Wiener) filtering. The power spectrum of signal plus noise shows a signal peak added to a noise tail. The tail is extrapolated back into the signal region as a "noise model." Subtracting gives the "signal model." The models need not be accurate for the method to be useful. A simple algebraic combination of the models gives the optimal filter (see text).

Because the optimal filter results from a minimization problem, the quality of the results obtained by optimal filtering differs from the true optimum by an amount that is *second order* in the precision to which the optimal filter is determined. In other words, even a fairly crudely determined optimal filter (sloppy, say, at the 10 percent level) can give excellent results when it is applied to data. That is why the separation of the measured signal C into signal and noise components S and N can usefully be done "by eye" from a crude plot of power spectral density. All of this may give you thoughts about iterating the procedure we have just described. For example, after designing a filter with response $\Phi(f)$ and using it to make a respectable guess at the signal $\widetilde{U}(f) = \Phi(f)C(f)/R(f)$, you might turn about and regard $\widetilde{U}(f)$ as a fresh new signal which you could improve even further with the same filtering technique. Don't waste your time on this line of thought. The scheme converges to a signal of $S(f) = 0$. Converging iterative methods do exist; this just isn't one of them.

You can use the routine four1 (§12.2) or realft (§12.3) to FFT your data when you are constructing an optimal filter. To apply the filter to your data, you can use the methods described in §13.1. The specific routine convlv is not needed for optimal filtering, since your filter is constructed in the frequency domain to begin with. If you are also deconvolving your data with a known response function, however, you can modify convlv to multiply by your optimal filter just before it takes the inverse Fourier transform.

CITED REFERENCES AND FURTHER READING:

Rabiner, L.R., and Gold, B. 1975, *Theory and Application of Digital Signal Processing* (Englewood Cliffs, NJ: Prentice-Hall).

Nussbaumer, H.J. 1982, *Fast Fourier Transform and Convolution Algorithms* (New York: Springer-Verlag).

Elliott, D.F., and Rao, K.R. 1982, *Fast Transforms: Algorithms, Analyses, Applications* (New York: Academic Press).

13.4 Power Spectrum Estimation Using the FFT

In the previous section we "informally" estimated the power spectral density of a function $c(t)$ by taking the modulus-squared of the discrete Fourier transform of some finite, sampled stretch of it. In this section we'll do roughly the same thing, but with considerably greater attention to details. Our attention will uncover some surprises.

The first detail is power spectrum (also called a power spectral density or PSD) normalization. In general there is *some* relation of proportionality between a measure of the squared amplitude of the function and a measure of the amplitude of the PSD. Unfortunately there are several different conventions for describing the normalization in each domain, and many opportunities for getting wrong the relationship between the two domains. Suppose that our function $c(t)$ is sampled at N points to produce values $c_0 \ldots c_{N-1}$, and that these points span a range of time T, that is $T = (N-1)\Delta$, where Δ is the sampling interval. Then here are several different descriptions of the total power:

$$\sum_{j=0}^{N-1} |c_j|^2 \equiv \text{"sum squared amplitude"} \qquad (13.4.1)$$

$$\frac{1}{T} \int_0^T |c(t)|^2 \, dt \approx \frac{1}{N} \sum_{j=0}^{N-1} |c_j|^2 \equiv \text{"mean squared amplitude"} \qquad (13.4.2)$$

$$\int_0^T |c(t)|^2 \, dt \approx \Delta \sum_{j=0}^{N-1} |c_j|^2 \equiv \text{"time-integral squared amplitude"} \qquad (13.4.3)$$

PSD estimators, as we shall see, have an even greater variety. In this section, we consider a class of them that give estimates at discrete values of frequency f_i, where i will range over integer values. In the next section, we will learn about a different class of estimators that produce estimates that are continuous functions of frequency f. Even if it is agreed always to relate the PSD normalization to a particular description of the function normalization (e.g., 13.4.2), there are at least the following possibilities: The PSD is

- defined for discrete positive, zero, and negative frequencies, and its sum over these is the function mean squared amplitude
- defined for zero and discrete positive frequencies only, and its sum over these is the function mean squared amplitude
- defined in the Nyquist interval from $-f_c$ to f_c, and its integral over this range is the function mean squared amplitude

- defined from 0 to f_c, and its integral over this range is the function mean squared amplitude

It *never* makes sense to integrate the PSD of a sampled function outside of the Nyquist interval $-f_c$ and f_c since, according to the sampling theorem, power there will have been aliased into the Nyquist interval.

It is hopeless to define enough notation to distinguish all possible combinations of normalizations. In what follows, we use the notation $P(f)$ to mean *any* of the above PSDs, stating in each instance how the particular $P(f)$ is normalized. Beware the inconsistent notation in the literature.

The method of power spectrum estimation used in the previous section is a simple version of an estimator called, historically, the *periodogram*. If we take an N-point sample of the function $c(t)$ at equal intervals and use the FFT to compute its discrete Fourier transform

$$C_k = \sum_{j=0}^{N-1} c_j \, e^{2\pi i j k/N} \qquad k = 0, \ldots, N - 1 \qquad (13.4.4)$$

then the periodogram estimate of the power spectrum is defined at $N/2 + 1$ frequencies as

$$P(0) = P(f_0) = \frac{1}{N^2} \left|C_0\right|^2$$

$$P(f_k) = \frac{1}{N^2} \left[\left|C_k\right|^2 + \left|C_{N-k}\right|^2 \right] \qquad k = 1, 2, \ldots, \left(\frac{N}{2} - 1\right) \qquad (13.4.5)$$

$$P(f_c) = P(f_{N/2}) = \frac{1}{N^2} \left|C_{N/2}\right|^2$$

where f_k is defined only for the zero and positive frequencies

$$f_k \equiv \frac{k}{N\Delta} = 2f_c \frac{k}{N} \qquad k = 0, 1, \ldots, \frac{N}{2} \qquad (13.4.6)$$

By Parseval's theorem, equation (12.1.10), we see immediately that equation (13.4.5) is normalized so that the sum of the $N/2 + 1$ values of P is equal to the mean squared amplitude of the function c_j.

We must now ask this question. In what sense is the periodogram estimate (13.4.5) a "true" estimator of the power spectrum of the underlying function $c(t)$? You can find the answer treated in considerable detail in the literature cited (see, e.g., [1] for an introduction). Here is a summary.

First, is the *expectation value* of the periodogram estimate equal to the power spectrum, i.e., is the estimator correct on average? Well, yes and no. We wouldn't really expect one of the $P(f_k)$'s to equal the continuous $P(f)$ at *exactly* f_k, since f_k is supposed to be representative of a whole frequency "bin" extending from halfway from the preceding discrete frequency to halfway to the next one. We *should* be expecting the $P(f_k)$ to be some kind of average of $P(f)$ over a narrow window function centered on its f_k. For the periodogram estimate (13.4.6) that window function, as a function of s the frequency offset *in bins*, is

$$W(s) = \frac{1}{N^2} \left[\frac{\sin(\pi s)}{\sin(\pi s/N)} \right]^2 \qquad (13.4.7)$$

Notice that $W(s)$ has oscillatory lobes but, apart from these, falls off only about as $W(s) \approx (\pi s)^{-2}$. This is not a very rapid fall-off, and it results in significant *leakage* (that is the technical term) from one frequency to another in the periodogram estimate. Notice also that $W(s)$ happens to be zero for s equal to a nonzero integer. This means that if the function $c(t)$ is a pure sine wave of frequency exactly equal to one of the f_k's, then there will be *no* leakage to adjacent f_k's. But this is not the characteristic case! If the frequency is, say, one-third of the way between two adjacent f_k's, then the leakage will extend *well* beyond those two adjacent bins. The solution to the problem of leakage is called *data windowing*, and we will discuss it below.

Turn now to another question about the periodogram estimate. What is the variance of that estimate as N goes to infinity? In other words, as we take more sampled points from the original function (either sampling a longer stretch of data at the same sampling rate, or else by resampling the same stretch of data with a faster sampling rate), then how much more accurate do the estimates P_k become? The unpleasant answer is that the periodogram estimates *do not become more accurate at all!* In fact, the variance of the periodogram estimate at a frequency f_k is always equal to the square of its expectation value at that frequency. In other words, the standard deviation is always 100 percent of the value, independent of N! How can this be? Where did all the information go as we added points? It all went into producing estimates at a greater number of discrete frequencies f_k. If we sample a longer run of data using the same sampling rate, then the Nyquist critical frequency f_c is unchanged, but we now have finer frequency resolution (more f_k's) within the Nyquist frequency interval; alternatively, if we sample the same length of data with a finer sampling interval, then our frequency resolution is unchanged, but the Nyquist range now extends up to a higher frequency. In neither case do the additional samples reduce the variance of any one particular frequency's estimated PSD.

You don't have to live with PSD estimates with 100 percent standard deviations, however. You simply have to know some techniques for reducing the variance of the estimates. Here are two techniques that are very nearly identical mathematically, though different in implementation. The first is to compute a periodogram estimate with finer discrete frequency spacing than you really need, and then to sum the periodogram estimates at K consecutive discrete frequencies to get one "smoother" estimate at the mid frequency of those K. The variance of that summed estimate will be smaller than the estimate itself by a factor of exactly $1/K$, i.e., the standard deviation will be smaller than 100 percent by a factor $1/\sqrt{K}$. Thus, to estimate the power spectrum at $M + 1$ discrete frequencies between 0 and f_c inclusive, you begin by taking the FFT of $2MK$ points (which number had better be an integer power of two!). You then take the modulus square of the resulting coefficients, add positive and negative frequency pairs, and divide by $(2MK)^2$, all according to equation (13.4.5) with $N = 2MK$. Finally, you "bin" the results into summed (not averaged) groups of K. This procedure is very easy to program, so we will not bother to give a routine for it. The reason that you sum, rather than average, K consecutive points is so that your final PSD estimate will preserve the normalization property that the sum of its $M + 1$ values equals the mean square value of the function.

A second technique for estimating the PSD at $M + 1$ discrete frequencies in the range 0 to f_c is to partition the original sampled data into K segments each of $2M$ consecutive sampled points. Each segment is separately FFT'd to produce a periodogram estimate (equation 13.4.5 with $N \equiv 2M$). Finally, the K periodogram estimates are averaged at each frequency. It is this final averaging that reduces the variance of the estimate by a factor K (standard deviation by $\sqrt{K}$). This second technique is computationally more efficient than the first technique above by a modest factor, since it is logarithmically more efficient to take many shorter FFTs than one longer one. The principal advantage of the second technique, however, is that only $2M$ data points are manipulated at a single time, not $2KM$ as in the first technique. This means that the second technique is the natural choice for processing long runs of data, as from a magnetic tape or other data record. We will give a routine later for implementing this second technique, but we need first to return to the matters of leakage and data windowing which were brought up after equation (13.4.7) above.

Data Windowing

The purpose of data windowing is to modify equation (13.4.7), which expresses the relation between the spectral estimate P_k at a discrete frequency and the actual underlying continuous spectrum $P(f)$ at nearby frequencies. In general, the spectral power in one "bin" k contains leakage from frequency components that are actually s bins away, where s is the independent variable in equation (13.4.7). There is, as we pointed out, quite substantial leakage even from moderately large values of s.

When we select a run of N sampled points for periodogram spectral estimation, we are in effect multiplying an infinite run of sampled data c_j by a window function in time, one that is zero except during the total sampling time $N\Delta$, and is unity during that time. In other words, the data are windowed by a square window function. By the convolution theorem (12.0.9; but interchanging the roles of f and t), the Fourier transform of the product of the data with this square window function is equal to the convolution of the data's Fourier transform with the window's Fourier transform. In fact, we determined equation (13.4.7) as nothing more than the square of the discrete Fourier transform of the unity window function.

$$W(s) = \frac{1}{N^2} \left[\frac{\sin(\pi s)}{\sin(\pi s/N)} \right]^2 = \frac{1}{N^2} \left| \sum_{k=0}^{N-1} e^{2\pi i s k/N} \right|^2 \tag{13.4.8}$$

The reason for the leakage at large values of s, is that the square window function turns on and off so rapidly. Its Fourier transform has substantial components at high frequencies. To remedy this situation, we can multiply the input data c_j, $j = 0, \ldots, N-1$ by a window function w_j that changes more gradually from zero to a maximum and then back to zero as j ranges from 0 to N. In this case, the equations for the periodogram estimator (13.4.4–13.4.5) become

$$D_k \equiv \sum_{j=0}^{N-1} c_j w_j \, e^{2\pi i j k/N} \qquad k = 0, \ldots, N-1 \tag{13.4.9}$$

$$P(0) = P(f_0) = \frac{1}{W_{ss}} |D_0|^2$$

$$P(f_k) = \frac{1}{W_{ss}} \left[|D_k|^2 + |D_{N-k}|^2 \right] \qquad k = 1, 2, \ldots, \left(\frac{N}{2} - 1 \right)$$

$$P(f_c) = P(f_{N/2}) = \frac{1}{W_{ss}} |D_{N/2}|^2 \tag{13.4.10}$$

where W_{ss} stands for "window squared and summed,"

$$W_{ss} \equiv N \sum_{j=0}^{N-1} w_j^2 \qquad (13.4.11)$$

and f_k is given by (13.4.6). The more general form of (13.4.7) can now be written in terms of the window function w_j as

$$W(s) = \frac{1}{W_{ss}} \left| \sum_{k=0}^{N-1} e^{2\pi i s k/N} w_k \right|^2$$

$$\approx \frac{1}{W_{ss}} \left| \int_{-N/2}^{N/2} \cos(2\pi s k/N) w(k - N/2) \, dk \right|^2 \qquad (13.4.12)$$

Here the approximate equality is useful for practical estimates, and holds for any window that is left-right symmetric (the usual case), and for $s \ll N$ (the case of interest for estimating leakage into nearby bins). The continuous function $w(k-N/2)$ in the integral is meant to be some smooth function that passes through the points w_k.

There is a lot of perhaps unnecessary lore about choice of a window function, and practically every function that rises from zero to a peak and then falls again has been named after someone. A few of the more common (also shown in Figure 13.4.1) are:

$$w_j = 1 - \left| \frac{j - \frac{1}{2} N}{\frac{1}{2} N} \right| \equiv \text{"Bartlett window"} \qquad (13.4.13)$$

(The "Parzen window" is very similar to this.)

$$w_j = \frac{1}{2} \left[1 - \cos\left(\frac{2\pi j}{N} \right) \right] \equiv \text{"Hann window"} \qquad (13.4.14)$$

(The "Hamming window" is similar but does not go exactly to zero at the ends.)

$$w_j = 1 - \left(\frac{j - \frac{1}{2} N}{\frac{1}{2} N} \right)^2 \equiv \text{"Welch window"} \qquad (13.4.15)$$

We are inclined to follow Welch in recommending that you use either (13.4.13) or (13.4.15) in practical work. However, at the level of this book, there is effectively *no difference* between any of these (or similar) window functions. Their difference lies in subtle trade-offs among the various figures of merit that can be used to describe the narrowness or peakedness of the spectral leakage functions computed by (13.4.12). These figures of merit have such names as: *highest sidelobe level (dB), sidelobe fall-off (dB per octave), equivalent noise bandwidth (bins), 3-dB bandwidth (bins), scallop loss (dB), worst case process loss (dB)*. Roughly speaking, the principal trade-off is between making the central peak as narrow as possible versus making the tails of the distribution fall off as rapidly as possible. For details, see (e.g.) [2]. Figure 13.4.2 plots the leakage amplitudes for several windows already discussed.

There is particularly a lore about window functions that rise smoothly from zero to unity in the first small fraction (say 10 percent) of the data, then stay at

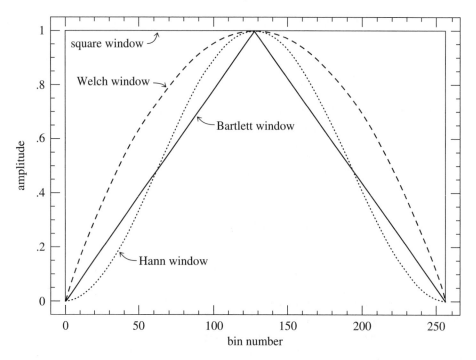

Figure 13.4.1. Window functions commonly used in FFT power spectral estimation. The data segment, here of length 256, is multiplied (bin by bin) by the window function before the FFT is computed. The square window, which is equivalent to no windowing, is least recommended. The Welch and Bartlett windows are good choices.

unity until the last small fraction (again say 10 percent) of the data, during which the window function falls smoothly back to zero. These windows will squeeze a little bit of extra narrowness out of the main lobe of the leakage function (never as much as a factor of two, however), but trade this off by widening the leakage tail by a significant factor (e.g., the reciprocal of 10 percent, a factor of ten). If we distinguish between the *width* of a window (number of samples for which it is at its maximum value) and its *rise/fall time* (number of samples during which it rises and falls); and if we distinguish between the *FWHM* (full width to half maximum value) of the leakage function's main lobe and the *leakage width* (full width that contains half of the spectral power that is not contained in the main lobe); then these quantities are related roughly by

$$(\text{FWHM in bins}) \approx \frac{N}{(\text{window width})} \qquad (13.4.16)$$

$$(\text{leakage width in bins}) \approx \frac{N}{(\text{window rise/fall time})} \qquad (13.4.17)$$

For the windows given above in (13.4.13)–(13.4.15), the effective window widths and the effective window rise/fall times are both of order $\frac{1}{2}N$. Generally speaking, we feel that the advantages of windows whose rise and fall times are only small fractions of the data length are minor or nonexistent, and we avoid using them. One sometimes hears it said that flat-topped windows "throw away less of the data," but we will now show you a better way of dealing with that problem by use of overlapping data segments.

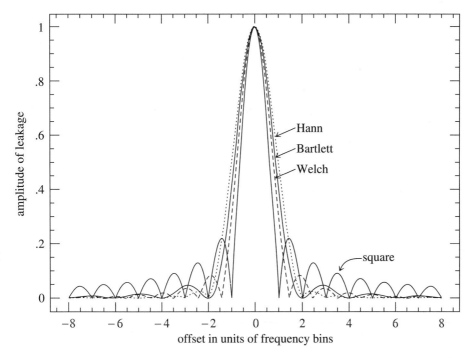

Figure 13.4.2. Leakage functions for the window functions of Figure 13.4.1. A signal whose frequency is actually located at zero offset "leaks" into neighboring bins with the amplitude shown. The purpose of windowing is to reduce the leakage at large offsets, where square (no) windowing has large sidelobes. Offset can have a fractional value, since the actual signal frequency can be located between two frequency bins of the FFT.

Let us now suppose that we have chosen a window function, and that we are ready to segment the data into K segments of $N = 2M$ points. Each segment will be FFT'd, and the resulting K periodograms will be averaged together to obtain a PSD estimate at $M + 1$ frequency values from 0 to f_c. We must now distinguish between two possible situations. We might want to obtain the smallest variance from a fixed amount of computation, without regard to the number of data points used. This will generally be the goal when the data are being gathered in real time, with the data-reduction being computer-limited. Alternatively, we might want to obtain the smallest variance from a fixed number of available sampled data points. This will generally be the goal in cases where the data are already recorded and we are analyzing it after the fact.

In the first situation (smallest spectral variance per computer operation), it is best to segment the data without any overlapping. The first $2M$ data points constitute segment number 1; the next $2M$ data points constitute segment number 2; and so on, up to segment number K, for a total of $2KM$ sampled points. The variance in this case, relative to a single segment, is reduced by a factor K.

In the second situation (smallest spectral variance per data point), it turns out to be optimal, or very nearly optimal, to overlap the segments by one half of their length. The first and second sets of M points are segment number 1; the second and third sets of M points are segment number 2; and so on, up to segment number K, which is made of the Kth and $K + 1$st sets of M points. The total number of sampled points is therefore $(K+1)M$, just over half as many as with nonoverlapping

segments. The reduction in the variance is not a full factor of K, since the segments
are not statistically independent. It can be shown that the variance is instead reduced
by a factor of about $9K/11$ (see the paper by Welch in [3]). This is, however,
significantly better than the reduction of about $K/2$ that would have resulted if the
same *number* of data points were segmented without overlapping.

    We can now codify these ideas into a routine for spectral estimation. While
we generally avoid input/output coding, we make an exception here to show how
data are read sequentially in one pass through a data file (referenced through the
parameter ifstream &fp). Only a small fraction of the data is in memory at any
one time. Note that spctrm returns the power at M, not $M+1$, frequencies, omitting
the component $P(f_c)$ at the Nyquist frequency. It would also be straightforward
to include that component.

```
#include <fstream>
#include <cmath>
#include "nr.h"
using namespace std;

namespace {
    inline DP window(const int j, const DP a, const DP b)
    {
        return 1.0-fabs((j-a)*b); // Bartlett
        // return 1.0; // Square
        // return 1.0-SQR((j-a)*b); // Welch
    }
}
```

```
void NR::spctrm(ifstream &fp, Vec_O_DP &p, const int k, const bool ovrlap)
```
Reads data from input stream specified by file pointer fp and returns as p[j] the data's power
(mean square amplitude) at frequency j/(2*m) cycles per gridpoint, for j=0,1,...,m-1,
based on (2*k+1)*m data points if ovrlap is set true, or 4*k*m data points if ovrlap is set
false. The number of segments of the data is 2*k in both cases: The routine calls four1 k
times, each call with 2 partitions each of 2*m real data points.
```
{
    int mm,m4,kk,joffn,joff,j2,j;
    DP w,facp,facm,sumw=0.0,den=0.0;

    int m=p.size();
    mm=m << 1;                                    Useful factors.
    m4=mm << 1;
    Vec_DP w1(m4),w2(m);
    facm=m;                                       Factors used by the window function.
    facp=1.0/m;
    for (j=0;j<mm;j++)                            Accumulate the squared sum of the weights.
        sumw += SQR(window(j,facm,facp));
    for (j=0;j<m;j++) p[j]=0.0;                   Initialize the spectrum to zero.
    if (ovrlap)                                   Initialize the "save" half-buffer.
        for (j=0;j<m;j++)
            fp >> w2[j];
    for (kk=0;kk<k;kk++) {
    Loop over data set segments in groups of two.
        for (joff=0;joff<2;joff++) {              Get two complete segments into workspace.
            if (ovrlap) {
                for (j=0;j<m;j++) w1[joff+j+j]=w2[j];
                for (j=0;j<m;j++)
                    fp >> w2[j];
                joffn=joff+mm-1;
                for (j=0;j<m;j++) w1[joffn+j+j+1]=w2[j];
            } else {
                for (j=joff;j<m4;j+=2)
```

```
            fp >> w1[j];
    }
}
for (j=0;j<mm;j++) {                    Apply the window to the data.
    j2=j+j;
    w=window(j,facm,facp);
    w1[j2+1] *= w;
    w1[j2] *= w;
}
four1(w1,1);                            Fourier transform the windowed data.
p[0] += (SQR(w1[0])+SQR(w1[1]));        Sum results into previous segments.
for (j=1;j<m;j++) {
    j2=j+j;
    p[j] += (SQR(w1[j2+1])+SQR(w1[j2])
        +SQR(w1[m4-j2+1])+SQR(w1[m4-j2]));
}
den += sumw;
}
den *= m4;                              Correct normalization.
for (j=0;j<m;j++) p[j] /= den;          Normalize the output.
}
```

CITED REFERENCES AND FURTHER READING:

Oppenheim, A.V., and Schafer, R.W. 1989, *Discrete-Time Signal Processing* (Englewood Cliffs, NJ: Prentice-Hall). [1]

Harris, F.J. 1978, *Proceedings of the IEEE*, vol. 66, pp. 51–83. [2]

Childers, D.G. (ed.) 1978, *Modern Spectrum Analysis* (New York: IEEE Press), paper by P.D. Welch. [3]

Champeney, D.C. 1973, *Fourier Transforms and Their Physical Applications* (New York: Academic Press).

Elliott, D.F., and Rao, K.R. 1982, *Fast Transforms: Algorithms, Analyses, Applications* (New York: Academic Press).

Bloomfield, P. 1976, *Fourier Analysis of Time Series – An Introduction* (New York: Wiley).

Rabiner, L.R., and Gold, B. 1975, *Theory and Application of Digital Signal Processing* (Englewood Cliffs, NJ: Prentice-Hall).

13.5 Digital Filtering in the Time Domain

Suppose that you have a signal that you want to filter digitally. For example, perhaps you want to apply *high-pass* or *low-pass* filtering, to eliminate noise at low or high frequencies respectively; or perhaps the interesting part of your signal lies only in a certain frequency band, so that you need a *bandpass* filter. Or, if your measurements are contaminated by 60 Hz power-line interference, you may need a *notch filter* to remove only a narrow band around that frequency. This section speaks particularly about the case in which you have chosen to do such filtering in the time domain.

Before continuing, we hope you will reconsider this choice. Remember how convenient it is to filter in the Fourier domain. You just take your whole data record, FFT it, multiply the FFT output by a filter function $\mathcal{H}(f)$, and then do an inverse FFT to get back a filtered data set in time domain. Here is some additional background on the Fourier technique that you will want to take into account.

- Remember that you must define your filter function $\mathcal{H}(f)$ for both positive and negative frequencies, and that the magnitude of the frequency extremes is always the Nyquist frequency $1/(2\Delta)$, where Δ is the sampling interval. The magnitude of the smallest nonzero frequencies in the FFT is $\pm 1/(N\Delta)$, where N is the number of (complex) points in the FFT. The positive and negative frequencies to which this filter are applied are arranged in wrap-around order.

- If the measured data are real, and you want the filtered output also to be real, then your arbitrary filter function should obey $\mathcal{H}(-f) = \mathcal{H}(f)^*$. You can arrange this most easily by picking an $\mathcal{H}$ that is real and even in f.

- If your chosen $\mathcal{H}(f)$ has sharp vertical edges in it, then the *impulse response* of your filter (the output arising from a short impulse as input) will have damped "ringing" at frequencies corresponding to these edges. There is nothing wrong with this, but if you don't like it, then pick a smoother $\mathcal{H}(f)$. To get a first-hand look at the impulse response of your filter, just take the inverse FFT of your $\mathcal{H}(f)$. If you smooth all edges of the filter function over some number k of points, then the impulse response function of your filter will have a span on the order of a fraction $1/k$ of the whole data record.

- If your data set is too long to FFT all at once, then break it up into segments of any convenient size, as long as they are much longer than the impulse response function of the filter. Use zero-padding, if necessary.

- You should probably remove any trend from the data, by subtracting from it a straight line through the first and last points (i.e., make the first and last points equal to zero). If you are segmenting the data, then you can pick overlapping segments and use only the middle section of each, comfortably distant from edge effects.

- A digital filter is said to be *causal* or *physically realizable* if its output for a particular time-step depends only on inputs at that particular time-step or earlier. It is said to be *acausal* if its output can depend on both earlier and later inputs. Filtering in the Fourier domain is, in general, acausal, since the data are processed "in a batch," without regard to time ordering. Don't let this bother you! Acausal filters can generally give superior performance (e.g., less dispersion of phases, sharper edges, less asymmetric impulse response functions). People use causal filters not because they are better, but because some situations just don't allow access to out-of-time-order data. Time domain filters can, in principle, be either causal or acausal, but they are most often used in applications where physical realizability is a constraint. For this reason we will restrict ourselves to the causal case in what follows.

If you are still favoring time-domain filtering after all we have said, it is probably because you have a real-time application, for which you must process a continuous data stream and wish to output filtered values at the same rate as you receive raw data. Otherwise, it may be that the quantity of data to be processed is so large that you can afford only a very small number of floating operations on each data point and cannot afford even a modest-sized FFT (with a number of floating operations per data point several times the logarithm of the number of points in the data set or segment).

Linear Filters

The most general linear filter takes a sequence x_k of input points and produces a sequence y_n of output points by the formula

$$y_n = \sum_{k=0}^{M} c_k \, x_{n-k} + \sum_{j=0}^{N-1} d_j \, y_{n-j-1} \qquad (13.5.1)$$

Here the $M + 1$ coefficients c_k and the N coefficients d_j are fixed and define the filter response. The filter (13.5.1) produces each new output value from the current and M previous input values, and from its own N previous output values. If $N = 0$, so that there is no second sum in (13.5.1), then the filter is called *nonrecursive* or *finite impulse response (FIR)*. If

$N \neq 0$, then it is called *recursive* or *infinite impulse response (IIR)*. (The term "IIR" connotes only that such filters are *capable* of having infinitely long impulse responses, not that their impulse response is necessarily long in a particular application. Typically the response of an IIR filter will drop off exponentially at late times, rapidly becoming negligible.)

The relation between the c_k's and d_j's and the filter response function $\mathcal{H}(f)$ is

$$\mathcal{H}(f) = \frac{\sum\limits_{k=0}^{M} c_k e^{-2\pi i k (f\Delta)}}{1 - \sum\limits_{j=0}^{N-1} d_j e^{-2\pi i (j+1)(f\Delta)}} \tag{13.5.2}$$

where Δ is, as usual, the sampling interval. The Nyquist interval corresponds to $f\Delta$ between $-1/2$ and $1/2$. For FIR filters the denominator of (13.5.2) is just unity.

Equation (13.5.2) tells how to determine $\mathcal{H}(f)$ from the c's and d's. To design a filter, though, we need a way of doing the inverse, getting a suitable set of c's and d's — as small a set as possible, to minimize the computational burden — from a desired $\mathcal{H}(f)$. Entire books are devoted to this issue. Like many other "inverse problems," it has no all-purpose solution. One clearly has to make compromises, since $\mathcal{H}(f)$ is a full continuous function, while the short list of c's and d's represents only a few adjustable parameters. The subject of digital filter design concerns itself with the various ways of making these compromises. We cannot hope to give any sort of complete treatment of the subject. We can, however, sketch a couple of basic techniques to get you started. For further details, you will have to consult some specialized books (see references).

FIR (Nonrecursive) Filters

When the denominator in (13.5.2) is unity, the right-hand side is just a discrete Fourier transform. The transform is easily invertible, giving the desired small number of c_k coefficients in terms of the same small number of values of $\mathcal{H}(f_i)$ at some discrete frequencies f_i. This fact, however, is not very useful. The reason is that, for values of c_k computed in this way, $\mathcal{H}(f)$ will tend to oscillate wildly in between the discrete frequencies where it is pinned down to specific values.

A better strategy, and one which is the basis of several formal methods in the literature, is this: Start by pretending that you are willing to have a relatively large number of filter coefficients, that is, a relatively large value of M. Then $\mathcal{H}(f)$ can be fixed to desired values on a relatively fine mesh, and the M coefficients c_k, $k = 0, \ldots, M - 1$ can be found by an FFT. Next, truncate (set to zero) most of the c_k's, leaving nonzero only the first, say, K, $(c_0, c_1, \ldots, c_{K-1})$ and last $K - 1$, $(c_{M-K+1}, \ldots, c_{M-1})$. The last few c_k's are filter coefficients at *negative lag*, because of the wrap-around property of the FFT. But we don't want coefficients at negative lag. Therefore we cyclically shift the array of c_k's, to bring everything to positive lag. (This corresponds to introducing a time-delay into the filter.) Do this by copying the c_k's into a new array of length M in the following order:

$$(c_{M-K+1}, \ldots, c_{M-1}, \ c_0, \ c_1, \ldots, c_{K-1}, \ 0, \ 0, \ldots, 0) \tag{13.5.3}$$

To see if your truncation is acceptable, take the FFT of the array (13.5.3), giving an approximation to your original $\mathcal{H}(f)$. You will generally want to compare the *modulus* $|\mathcal{H}(f)|$ to your original function, since the time-delay will have introduced complex phases into the filter response.

If the new filter function is acceptable, then you are done and have a set of $2K - 1$ filter coefficients. If it is not acceptable, then you can either (i) increase K and try again, or (ii) do something fancier to improve the acceptability for the same K. An example of something fancier is to modify the magnitudes (but not the phases) of the unacceptable $\mathcal{H}(f)$ to bring it more in line with your ideal, and then to FFT to get new c_k's. Once again set to zero all but the first $2K - 1$ values of these (no need to cyclically shift since you have preserved the time-delaying phases), then inverse transform to get a new $\mathcal{H}(f)$, which will often be more acceptable. You can iterate this procedure. Note, however, that the procedure will not converge if your requirements for acceptability are more stringent than your $2K - 1$ coefficients can handle.

The key idea, in other words, is to iterate between the space of coefficients and the space of functions $\mathcal{H}(f)$, until a Fourier conjugate pair that satisfies the imposed constraints *in both spaces* is found. A more formal technique for this kind of iteration is the *Remes Exchange Algorithm* which produces the best Chebyshev approximation to a given desired frequency response with a fixed number of filter coefficients (cf. §5.13).

IIR (Recursive) Filters

Recursive filters, whose output at a given time depends both on the current and previous inputs and on previous outputs, can generally have performance that is superior to nonrecursive filters with the same total number of coefficients (or same number of floating operations per input point). The reason is fairly clear by inspection of (13.5.2): A nonrecursive filter has a frequency response that is a polynomial in the variable $1/z$, where

$$z \equiv e^{2\pi i(f\Delta)} \tag{13.5.4}$$

By contrast, a recursive filter's frequency response is a *rational function* in $1/z$. The class of rational functions is especially good at fitting functions with sharp edges or narrow features, and most desired filter functions are in this category.

Nonrecursive filters are always stable. If you turn off the sequence of incoming x_i's, then after no more than M steps the sequence of y_j's produced by (13.5.1) will also turn off. Recursive filters, feeding as they do on their own output, are not necessarily stable. If the coefficients d_j are badly chosen, a recursive filter can have exponentially growing, so-called *homogeneous*, modes, which become huge even after the input sequence has been turned off. This is not good. The problem of designing recursive filters, therefore, is not just an inverse problem; it is an inverse problem with an additional stability constraint.

How do you tell if the filter (13.5.1) is stable for a given set of c_k and d_j coefficients? Stability depends only on the d_j's. The filter is stable if and only if all N complex roots of the *characteristic polynomial* equation

$$z^N - \sum_{j=0}^{N-1} d_j z^{(N-1)-j} = 0 \tag{13.5.5}$$

are inside the unit circle, i.e., satisfy

$$|z| \leq 1 \tag{13.5.6}$$

The various methods for constructing stable recursive filters again form a subject area for which you will need more specialized books. One very useful technique, however, is the *bilinear transformation method*. For this topic we define a new variable w that reparametrizes the frequency f,

$$w \equiv \tan[\pi(f\Delta)] = i\left(\frac{1 - e^{2\pi i(f\Delta)}}{1 + e^{2\pi i(f\Delta)}}\right) = i\left(\frac{1-z}{1+z}\right) \tag{13.5.7}$$

Don't be fooled by the i's in (13.5.7). This equation maps real frequencies f into real values of w. In fact, it maps the Nyquist interval $-\frac{1}{2} < f\Delta < \frac{1}{2}$ onto the real w axis $-\infty < w < +\infty$. The inverse equation to (13.5.7) is

$$z = e^{2\pi i(f\Delta)} = \frac{1 + iw}{1 - iw} \tag{13.5.8}$$

In reparametrizing f, w also reparametrizes z, of course. Therefore, the condition for stability (13.5.5)–(13.5.6) can be rephrased in terms of w: If the filter response $\mathcal{H}(f)$ is written as a function of w, then the filter is stable if and only if the poles of the filter function (zeros of its denominator) are all in the upper half complex plane,

$$\text{Im}(w) \geq 0 \tag{13.5.9}$$

The idea of the bilinear transformation method is that instead of specifying your desired $\mathcal{H}(f)$, you specify only its desired modulus square, $|\mathcal{H}(f)|^2 = \mathcal{H}(f)\mathcal{H}(f)^* = \mathcal{H}(f)\mathcal{H}(-f)$. Pick this to be approximated by some rational function in w^2. Then find all the poles of this function in the w complex plane. Every pole in the lower half-plane will have a corresponding pole in the upper half-plane, by symmetry. The idea is to form a product only of the factors with good poles, ones in the upper half-plane. This product is your *stably realizable* $\mathcal{H}(f)$. Now substitute equation (13.5.7) to write the function as a rational function in z, and compare with equation (13.5.2) to read off the c's and d's.

The procedure becomes clearer when we go through an example. Suppose we want to design a simple bandpass filter, whose lower cutoff frequency corresponds to a value $w = a$, and whose upper cutoff frequency corresponds to a value $w = b$, with a and b both positive numbers. A simple rational function that accomplishes this is

$$|\mathcal{H}(f)|^2 = \left(\frac{w^2}{w^2 + a^2} \right) \left(\frac{b^2}{w^2 + b^2} \right) \tag{13.5.10}$$

This function does not have a very sharp cutoff, but it is illustrative of the more general case. To obtain sharper edges, one could take the function (13.5.10) to some positive integer power, or, equivalently, run the data sequentially through some number of copies of the filter that we will obtain from (13.5.10).

The poles of (13.5.10) are evidently at $w = \pm ia$ and $w = \pm ib$. Therefore the stably realizable $\mathcal{H}(f)$ is

$$\mathcal{H}(f) = \left(\frac{w}{w - ia} \right) \left(\frac{ib}{w - ib} \right) = \frac{\left(\frac{1-z}{1+z} \right) b}{\left[\left(\frac{1-z}{1+z} \right) - a \right] \left[\left(\frac{1-z}{1+z} \right) - b \right]} \tag{13.5.11}$$

We put the i in the numerator of the second factor in order to end up with real-valued coefficients. If we multiply out all the denominators, (13.5.11) can be rewritten in the form

$$\mathcal{H}(f) = \frac{-\frac{b}{(1+a)(1+b)} + \frac{b}{(1+a)(1+b)} z^{-2}}{1 - \frac{(1+a)(1-b)+(1-a)(1+b)}{(1+a)(1+b)} z^{-1} + \frac{(1-a)(1-b)}{(1+a)(1+b)} z^{-2}} \tag{13.5.12}$$

from which one reads off the filter coefficients for equation (13.5.1),

$$c_0 = -\frac{b}{(1+a)(1+b)}$$

$$c_1 = 0$$

$$c_2 = \frac{b}{(1+a)(1+b)}$$

$$d_0 = \frac{(1+a)(1-b)+(1-a)(1+b)}{(1+a)(1+b)}$$

$$d_1 = -\frac{(1-a)(1-b)}{(1+a)(1+b)} \tag{13.5.13}$$

This completes the design of the bandpass filter.

Sometimes you can figure out how to construct directly a rational function in w for $\mathcal{H}(f)$, rather than having to start with its modulus square. The function that you construct has to have its poles only in the upper half-plane, for stability. It should also have the property of going into its own complex conjugate if you substitute $-w$ for w, so that the filter coefficients will be real.

For example, here is a function for a notch filter, designed to remove only a narrow frequency band around some fiducial frequency $w = w_0$, where w_0 is a positive number,

$$\mathcal{H}(f) = \left(\frac{w - w_0}{w - w_0 - i\epsilon w_0} \right) \left(\frac{w + w_0}{w + w_0 - i\epsilon w_0} \right)$$

$$= \frac{w^2 - w_0^2}{(w - i\epsilon w_0)^2 - w_0^2} \tag{13.5.14}$$

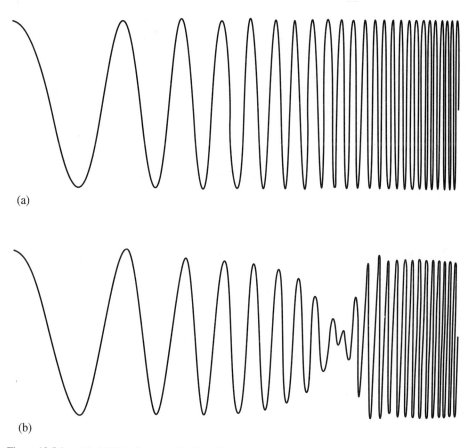

(a)

(b)

Figure 13.5.1.   (a) A "chirp," or signal whose frequency increases continuously with time. (b) Same signal after it has passed through the notch filter (13.5.15). The parameter ϵ is here 0.2.

In (13.5.14) the parameter ϵ is a small positive number that is the desired width of the notch, as a fraction of w_0. Going through the arithmetic of substituting z for w gives the filter coefficients

$$c_0 = \frac{1 + w_0^2}{(1 + \epsilon w_0)^2 + w_0^2}$$

$$c_1 = -2\frac{1 - w_0^2}{(1 + \epsilon w_0)^2 + w_0^2}$$

$$c_2 = \frac{1 + w_0^2}{(1 + \epsilon w_0)^2 + w_0^2} \qquad (13.5.15)$$

$$d_0 = 2\frac{1 - \epsilon^2 w_0^2 - w_0^2}{(1 + \epsilon w_0)^2 + w_0^2}$$

$$d_1 = -\frac{(1 - \epsilon w_0)^2 + w_0^2}{(1 + \epsilon w_0)^2 + w_0^2}$$

Figure 13.5.1 shows the results of using a filter of the form (13.5.15) on a "chirp" input signal, one that glides upwards in frequency, crossing the notch frequency along the way.

While the bilinear transformation may seem very general, its applications are limited by some features of the resulting filters. The method is good at getting the general shape of the desired filter, and good where "flatness" is a desired goal. However, the nonlinear mapping between w and f makes it difficult to design to a desired shape for a cutoff, and may move cutoff frequencies (defined by a certain number of dB) from their desired places. Consequently, practitioners of the art of digital filter design reserve the bilinear transformation for specific situations, and arm themselves with a variety of other tricks. We suggest that you do likewise, as your projects demand.

CITED REFERENCES AND FURTHER READING:

Hamming, R.W. 1983, *Digital Filters*, 2nd ed. (Englewood Cliffs, NJ: Prentice-Hall).

Antoniou, A. 1979, *Digital Filters: Analysis and Design* (New York: McGraw-Hill).

Parks, T.W., and Burrus, C.S. 1987, *Digital Filter Design* (New York: Wiley).

Oppenheim, A.V., and Schafer, R.W. 1989, *Discrete-Time Signal Processing* (Englewood Cliffs, NJ: Prentice-Hall).

Rice, J.R. 1964, *The Approximation of Functions* (Reading, MA: Addison-Wesley); also 1969, *op. cit.*, Vol. 2.

Rabiner, L.R., and Gold, B. 1975, *Theory and Application of Digital Signal Processing* (Englewood Cliffs, NJ: Prentice-Hall).

13.6 Linear Prediction and Linear Predictive Coding

We begin with a very general formulation that will allow us to make connections to various special cases. Let $\{y'_\alpha\}$ be a set of measured values for some underlying set of true values of a quantity y, denoted $\{y_\alpha\}$, related to these true values by the addition of random noise,

$$y'_\alpha = y_\alpha + n_\alpha \qquad (13.6.1)$$

(compare equation 13.3.2, with a somewhat different notation). Our use of a Greek subscript to index the members of the set is meant to indicate that the data points are not necessarily equally spaced along a line, or even ordered: they might be "random" points in three-dimensional space, for example. Now, suppose we want to construct the "best" estimate of the true value of some particular point $y_\star$ as a linear combination of the known, noisy, values. Writing

$$y_\star = \sum_\alpha d_{\star\alpha} y'_\alpha + x_\star \qquad (13.6.2)$$

we want to find coefficients $d_{\star\alpha}$ that minimize, in some way, the *discrepancy* $x_\star$. The coefficients $d_{\star\alpha}$ have a "star" subscript to indicate that they depend on the choice of point $y_\star$. Later, we might want to let $y_\star$ be one of the existing y_α's. In that case, our problem becomes one of optimal filtering or estimation, closely related to the discussion in §13.3. On the other hand, we might want $y_\star$ to be a completely new point. In that case, our problem will be one of *linear prediction*.

A natural way to minimize the discrepancy $x_\star$ is in the statistical mean square sense. If angle brackets denote statistical averages, then we seek $d_{\star\alpha}$'s that minimize

$$
\begin{aligned}
\langle x_\star^2 \rangle &= \left\langle \left[\sum_\alpha d_{\star\alpha}(y_\alpha + n_\alpha) - y_\star \right]^2 \right\rangle \\
&= \sum_{\alpha\beta} (\langle y_\alpha y_\beta \rangle + \langle n_\alpha n_\beta \rangle) d_{\star\alpha} d_{\star\beta} - 2 \sum_\alpha \langle y_\star y_\alpha \rangle d_{\star\alpha} + \langle y_\star^2 \rangle
\end{aligned}
\qquad (13.6.3)
$$

Here we have used the fact that noise is uncorrelated with signal, e.g., $\langle n_\alpha y_\beta \rangle = 0$. The quantities $\langle y_\alpha y_\beta \rangle$ and $\langle y_\star y_\alpha \rangle$ describe the autocorrelation structure of the underlying data. We have already seen an analogous expression, (13.2.2), for the case of equally spaced data points on a line; we will meet correlation several times again in its statistical sense in Chapters 14 and 15. The quantities $\langle n_\alpha n_\beta \rangle$ describe the autocorrelation properties of the noise. Often, for point-to-point uncorrelated noise, we have $\langle n_\alpha n_\beta \rangle = \langle n_\alpha^2 \rangle \delta_{\alpha\beta}$. It is convenient to think of the various correlation quantities as comprising matrices and vectors,

$$\phi_{\alpha\beta} \equiv \langle y_\alpha y_\beta \rangle \qquad \phi_{\star\alpha} \equiv \langle y_\star y_\alpha \rangle \qquad \eta_{\alpha\beta} \equiv \langle n_\alpha n_\beta \rangle \ \text{ or } \ \langle n_\alpha^2 \rangle \delta_{\alpha\beta} \quad (13.6.4)$$

Setting the derivative of equation (13.6.3) with respect to the $d_{\star\alpha}$'s equal to zero, one readily obtains the set of linear equations,

$$\sum_\beta [\phi_{\alpha\beta} + \eta_{\alpha\beta}]\, d_{\star\beta} = \phi_{\star\alpha} \qquad (13.6.5)$$

If we write the solution as a matrix inverse, then the estimation equation (13.6.2) becomes, omitting the minimized discrepancy $x_\star$,

$$y_\star \approx \sum_{\alpha\beta} \phi_{\star\alpha} [\phi_{\mu\nu} + \eta_{\mu\nu}]^{-1}_{\alpha\beta}\, y'_\beta \qquad (13.6.6)$$

From equations (13.6.3) and (13.6.5) one can also calculate the expected mean square value of the discrepancy at its minimum, denoted $\langle x_\star^2 \rangle_0$,

$$\langle x_\star^2 \rangle_0 = \langle y_\star^2 \rangle - \sum_\beta d_{\star\beta}\phi_{\star\beta} = \langle y_\star^2 \rangle - \sum_{\alpha\beta} \phi_{\star\alpha} [\phi_{\mu\nu} + \eta_{\mu\nu}]^{-1}_{\alpha\beta} \phi_{\star\beta} \qquad (13.6.7)$$

A final general result tells how much the mean square discrepancy $\langle x_\star^2 \rangle$ is increased if we use the estimation equation (13.6.2) not with the best values $d_{\star\beta}$, but with some other values $\widehat{d}_{\star\beta}$. The above equations then imply

$$\langle x_\star^2 \rangle = \langle x_\star^2 \rangle_0 + \sum_{\alpha\beta} (\widehat{d}_{\star\alpha} - d_{\star\alpha}) [\phi_{\alpha\beta} + \eta_{\alpha\beta}] (\widehat{d}_{\star\beta} - d_{\star\beta}) \qquad (13.6.8)$$

Since the second term is a pure quadratic form, we see that the increase in the discrepancy is only second order in any error made in estimating the $d_{\star\beta}$'s.

Connection to Optimal Filtering

If we change "star" to a Greek index, say γ, then the above formulas describe optimal filtering, generalizing the discussion of §13.3. One sees, for example, that if the noise amplitudes n_α go to zero, so likewise do the noise autocorrelations $\eta_{\alpha\beta}$, and, canceling a matrix times its inverse, equation (13.6.6) simply becomes $y_\gamma = y'_\gamma$. Another special case occurs if the matrices $\phi_{\alpha\beta}$ and $\eta_{\alpha\beta}$ are diagonal. In that case, equation (13.6.6) becomes

$$y_\gamma = \frac{\phi_{\gamma\gamma}}{\phi_{\gamma\gamma} + \eta_{\gamma\gamma}} y'_\gamma \qquad (13.6.9)$$

which is readily recognizable as equation (13.3.6) with $S^2 \to \phi_{\gamma\gamma}$, $N^2 \to \eta_{\gamma\gamma}$. What is going on is this: For the case of equally spaced data points, and in the Fourier domain, autocorrelations become simply squares of Fourier amplitudes (Wiener-Khinchin theorem, equation 12.0.12), and the optimal filter can be constructed algebraically, as equation (13.6.9), without inverting any matrix.

More generally, in the time domain, or any other domain, an optimal filter (one that minimizes the square of the discrepancy from the underlying true value in the presence of measurement noise) can be constructed by estimating the autocorrelation matrices $\phi_{\alpha\beta}$ and $\eta_{\alpha\beta}$, and applying equation (13.6.6) with $\star \to \gamma$. (Equation 13.6.8 is in fact the basis for the §13.3's statement that even crude optimal filtering can be quite effective.)

Linear Prediction

Classical *linear prediction* specializes to the case where the data points y_β are equally spaced along a line, y_i, $i = 0, 1, \ldots, N - 1$, and we want to use M consecutive values of y_i to predict an $M + 1$st. Stationarity is assumed. That is, the autocorrelation $\langle y_j y_k \rangle$ is assumed to depend only on the difference $|j - k|$, and not on j or k individually, so that the autocorrelation ϕ has only a single index,

$$\phi_j \equiv \langle y_i y_{i+j} \rangle \approx \frac{1}{N - j} \sum_{i=0}^{N-j-1} y_i y_{i+j} \qquad (13.6.10)$$

Here, the approximate equality shows one way to use the actual data set values to estimate the autocorrelation components. (In fact, there is a better way to make these estimates; see below.) In the situation described, the estimation equation (13.6.2) is

$$y_n = \sum_{j=0}^{M-1} d_j y_{n-j-1} + x_n \qquad (13.6.11)$$

(compare equation 13.5.1) and equation (13.6.5) becomes the set of M equations for the M unknown d_j's, now called the *linear prediction (LP) coefficients*,

$$\sum_{j=0}^{M-1} \phi_{|j-k-1|} \, d_j = \phi_k \qquad (k = 1, \ldots, M) \qquad (13.6.12)$$

Notice that while noise is not explicitly included in the equations, it is properly accounted for, *if* it is point-to-point uncorrelated: ϕ_0, as estimated by equation (13.6.10) using *measured* values y_i', actually estimates the diagonal part of $\phi_{\alpha\alpha} + \eta_{\alpha\alpha}$, above. The mean square discrepancy $\langle x_n^2 \rangle$ is estimated by equation (13.6.7) as

$$\langle x_n^2 \rangle = \phi_0 - \phi_1 d_0 - \phi_2 d_1 - \cdots - \phi_M d_{M-1} \qquad (13.6.13)$$

To use linear prediction, we first compute the d_j's, using equations (13.6.10) and (13.6.12). We then calculate equation (13.6.13) or, more concretely, apply (13.6.11) to the known record to get an idea of how large are the discrepancies x_i. If the discrepancies are small, then we can continue applying (13.6.11) right on into the future, imagining the unknown "future" discrepancies x_i to be zero. In this application, (13.6.11) is a kind of extrapolation formula. In many situations, this extrapolation turns out to be vastly more powerful than any kind of simple polynomial extrapolation. (By the way, you should not confuse the terms "linear prediction" and "linear extrapolation"; the general functional form used by linear prediction is *much* more complex than a straight line, or even a low-order polynomial!)

However, to achieve its full usefulness, linear prediction must be constrained in one additional respect: One must take additional measures to guarantee its *stability*. Equation (13.6.11) is a special case of the general linear filter (13.5.1). The condition that (13.6.11) be stable as a linear predictor is precisely that given in equations (13.5.5) and (13.5.6), namely that the characteristic polynomial

$$z^N - \sum_{j=0}^{N-1} d_j z^{(N-1)-j} = 0 \tag{13.6.14}$$

have all N of its roots inside the unit circle,

$$|z| \le 1 \tag{13.6.15}$$

There is no guarantee that the coefficients produced by equation (13.6.12) will have this property. If the data contain many oscillations without any particular trend towards increasing or decreasing amplitude, then the complex roots of (13.6.14) will generally all be rather close to the unit circle. The finite length of the data set will cause some of these roots to be inside the unit circle, others outside. In some applications, where the resulting instabilities are slowly growing and the linear prediction is not pushed too far, it is best to use the "unmassaged" LP coefficients that come directly out of (13.6.12). For example, one might be extrapolating to fill a short gap in a data set; then one might extrapolate both forwards across the gap and backwards from the data beyond the gap. If the two extrapolations agree tolerably well, then instability is not a problem.

When instability *is* a problem, you have to "massage" the LP coefficients. You do this by (i) solving (numerically) equation (13.6.14) for its N complex roots; (ii) moving the roots to where you think they ought to be inside or on the unit circle; (iii) reconstituting the now-modified LP coefficients. You may think that step (ii) sounds a little vague. It is. There is no "best" procedure. If you think that your signal is truly a sum of undamped sine and cosine waves (perhaps with incommensurate periods), then you will want simply to move each root z_i onto the unit circle,

$$z_i \; \rightarrow \; z_i / |z_i| \tag{13.6.16}$$

In other circumstances it may seem appropriate to reflect a bad root across the unit circle

$$z_i \; \rightarrow \; 1/z_i{}^* \tag{13.6.17}$$

This alternative has the property that it preserves the amplitude of the output of (13.6.11) when it is driven by a sinusoidal set of x_i's. It assumes that (13.6.12) has correctly identified the spectral width of a resonance, but only slipped up on identifying its time sense so that signals that should be damped as time proceeds end up growing in amplitude. The choice between (13.6.16) and (13.6.17) sometimes might as well be based on voodoo. We prefer (13.6.17).

Also magical is the choice of M, the number of LP coefficients to use. You should choose M to be as small as works for you, that is, you should choose it by experimenting with your data. Try $M = 5, 10, 20, 40$. If you need larger M's than this, be aware that the procedure of "massaging" all those complex roots is quite sensitive to roundoff error. Double precision is crucial.

Linear prediction is especially successful at extrapolating signals that are smooth and oscillatory, though not necessarily periodic. In such cases, linear prediction often extrapolates accurately through *many cycles* of the signal. By contrast, polynomial extrapolation in general becomes seriously inaccurate after at most a cycle or two. A prototypical example of a signal that can successfully be linearly predicted is the height of ocean tides, for which the fundamental 12-hour period is modulated in phase and amplitude over the course of the month and year, and for which local hydrodynamic effects may make even one cycle of the curve look rather different in shape from a sine wave.

We already remarked that equation (13.6.10) is not necessarily the best way to estimate the covariances ϕ_k from the data set. In fact, results obtained from linear prediction are remarkably sensitive to exactly how the ϕ_k's are estimated. One particularly good method is due to Burg [1], and involves a recursive procedure for increasing the order M by one unit at a time, at each stage re-estimating the coefficients d_j, $j = 0, \ldots, M - 1$ so as to minimize the residual in equation (13.6.13). Although further discussion of the Burg method is beyond our scope here, the method is implemented in the following routine [1,2] for estimating the LP coefficients d_j of a data set.

```
#include <cmath>
#include "nr.h"
using namespace std;

void NR::memcof(Vec_I_DP &data, DP &xms, Vec_O_DP &d)
Given a real vector of data[0..n-1], this routine returns m linear prediction coefficients as
d[0..m-1], and returns the mean square discrepancy as xms.
{
    int k,j,i;
    DP p=0.0;

    int n=data.size();
    int m=d.size();
    Vec_DP wk1(n),wk2(n),wkm(m);
    for (j=0;j<n;j++) p += SQR(data[j]);
    xms=p/n;
    wk1[0]=data[0];
    wk2[n-2]=data[n-1];
    for (j=1;j<n-1;j++) {
        wk1[j]=data[j];
        wk2[j-1]=data[j];
    }
    for (k=0;k<m;k++) {
        DP num=0.0,denom=0.0;
        for (j=0;j<(n-k-1);j++) {
            num += (wk1[j]*wk2[j]);
            denom += (SQR(wk1[j])+SQR(wk2[j]));
        }
```

```
          d[k]=2.0*num/denom;
          xms *= (1.0-SQR(d[k]));
          for (i=0;i<k;i++)
              d[i]=wkm[i]-d[k]*wkm[k-1-i];
```
The algorithm is recursive, building up the answer for larger and larger values of m until the desired value is reached. At this point in the algorithm, one could return the vector d and scalar xms for a set of LP coefficients with k (rather than m) terms.
```
          if (k == m-1)
              return;
          for (i=0;i<=k;i++) wkm[i]=d[i];
          for (j=0;j<(n-k-2);j++) {
              wk1[j] -= (wkm[k]*wk2[j]);
              wk2[j]=wk2[j+1]-wkm[k]*wk1[j+1];
          }
      }
      nrerror("never get here in memcof.");
}
```

Here are procedures for rendering the LP coefficients stable (if you choose to do so), and for extrapolating a data set by linear prediction, using the original or massaged LP coefficients. The routine zroots (§9.5) is used to find all complex roots of a polynomial.

```
#include <cmath>
#include <complex>
#include "nr.h"
using namespace std;

void NR::fixrts(Vec_IO_DP &d)
```
Given the LP coefficients d[0..m-1], this routine finds all roots of the characteristic polynomial (13.6.14), reflects any roots that are outside the unit circle back inside, and then returns a modified set of coefficients d[0..m-1].
```
{
    bool polish=true;
    int i,j;

    int m=d.size();
    Vec_CPLX_DP a(m+1),roots(m);
    a[m]=1.0;
    for (j=0;j<m;j++)                       Set up complex coefficients for polynomial root
        a[j]= -d[m-1-j];                        finder.
    zroots(a,roots,polish);                Find all the roots.
    for (j=0;j<m;j++)                       Look for a root outside the unit circle, and reflect
        if (abs(roots[j]) > 1.0)                it back inside.
            roots[j]=1.0/conj(roots[j]);
    a[0]= -roots[0];                       Now reconstruct the polynomial coefficients,
    a[1]=1.0;
    for (j=1;j<m;j++) {                     by looping over the roots
        a[j+1]=1.0;
        for (i=j;i>=1;i--)                  and synthetically multiplying.
            a[i]=a[i-1]-roots[j]*a[i];
        a[0]= -roots[j]*a[0];
    }
    for (j=0;j<m;j++)                       The polynomial coefficients are guaranteed to be
        d[m-1-j] = -real(a[j]);                 real, so we need only return the real part as
}                                               new LP coefficients.
```

```
#include "nr.h"

void NR::predic(Vec_I_DP &data, Vec_I_DP &d, Vec_O_DP &future)
```
Given data[0..ndata-1], and given the data's LP coefficients d[0..m-1], this routine applies equation (13.6.11) to predict the next nfut data points, which it returns in the array future[0..nfut-1]. Note that the routine references only the last m values of data, as initial values for the prediction.
```
{
    int k,j;
    DP sum,discrp;

    int ndata=data.size();
    int m=d.size();
    int nfut=future.size();
    Vec_DP reg(m);
    for (j=0;j<m;j++) reg[j]=data[ndata-1-j];
    for (j=0;j<nfut;j++) {
        discrp=0.0;
```
This is where you would put in a known discrepancy if you were reconstructing a function by linear predictive coding rather than extrapolating a function by linear prediction. See text.
```
        sum=discrp;
        for (k=0;k<m;k++) sum += d[k]*reg[k];
        for (k=m-1;k>=1;k--) reg[k]=reg[k-1];          [If you want to implement circular
        future[j]=reg[0]=sum;                           arrays, you can avoid this shift-
    }                                                   ing of coefficients.]
}
```

Removing the Bias in Linear Prediction

You might expect that the sum of the d_j's in equation (13.6.11) (or, more generally, in equation 13.6.2) should be 1, so that (e.g.) adding a constant to all the data points y_i yields a prediction that is increased by the same constant. However, the d_j's do not sum to 1 but, in general, to a value slightly less than one. This fact reveals a subtle point, that the estimator of classical linear prediction is not *unbiased*, even though it does minimize the mean square discrepancy. At any place where the measured autocorrelation does not imply a better estimate, the equations of linear prediction tend to predict a value that tends towards zero.

Sometimes, that is just what you want. If the process that generates the y_i's in fact has zero mean, then zero is the best guess absent other information. At other times, however, this behavior is unwarranted. If you have data that show only small variations around a positive value, you don't want linear predictions that droop towards zero.

Often it is a workable approximation to subtract the mean off your data set, perform the linear prediction, and then add the mean back. This procedure contains the germ of the correct solution; but the simple arithmetic mean is not quite the correct constant to subtract. In fact, an unbiased estimator is obtained by subtracting from every data point an autocorrelation-weighted mean defined by [3,4]

$$\bar{y} \equiv \sum_{\beta} [\phi_{\mu\nu} + \eta_{\mu\nu}]^{-1}_{\alpha\beta} y_\beta \bigg/ \sum_{\alpha\beta} [\phi_{\mu\nu} + \eta_{\mu\nu}]^{-1}_{\alpha\beta} \qquad (13.6.18)$$

With this subtraction, the sum of the LP coefficients should be unity, up to roundoff and differences in how the ϕ_k's are estimated.

Linear Predictive Coding (LPC)

A different, though related, method to which the formalism above can be applied is the "compression" of a sampled signal so that it can be stored more compactly. The original form should be *exactly* recoverable from the compressed version. Obviously, compression can be accomplished only if there is redundancy in the signal. Equation (13.6.11) describes one kind of redundancy: It says that the signal, except for a small discrepancy, is predictable from its previous values and from a small number of LP coefficients. Compression of a signal by the use of (13.6.11) is thus called *linear predictive coding*, or *LPC*.

The basic idea of LPC (in its simplest form) is to record as a compressed file (i) the number of LP coefficients M, (ii) their M values, e.g., as obtained by `memcof`, (iii) the first M data points, and then (iv) for each subsequent data point only its residual discrepancy x_i (equation 13.6.1). When you are creating the compressed file, you find the residual by applying (13.6.1) to the previous M points, subtracting the sum from the actual value of the current point. When you are reconstructing the original file, you add the residual back in, at the point indicated in the routine `predic`.

It may not be obvious why there is any compression at all in this scheme. After all, we are storing one value of residual per data point! Why not just store the original data point? The answer depends on the relative sizes of the numbers involved. The residual is obtained by subtracting two very nearly equal numbers (the data and the linear prediction). Therefore, the discrepancy typically has only a very small number of nonzero bits. These can be stored in a compressed file. How do you do it in a high-level language? Here is one way: Scale your data to have integer values, say between $+1000000$ and -1000000 (supposing that you need six significant figures). Modify equation (13.6.1) by enclosing the sum term in an "integer part of" operator. The discrepancy will now, by definition, be an integer. Experiment with different values of M, to find LP coefficients that make the range of the discrepancy as small as you can. If you can get to within a range of ±127 (and in our experience this is not at all difficult) then you can write it to a file as a single byte. This is a compression factor of 4, compared to 4-byte integer or floating formats.

Notice that the LP coefficients are computed using the *quantized* data, and that the discrepancy is also quantized, i.e., quantization is done both outside and inside the LPC loop. If you are careful in following this prescription, then, apart from the initial quantization of the data, you will not introduce even a single bit of roundoff error into the compression-reconstruction process: While the evaluation of the sum in (13.6.11) may have roundoff errors, the residual that you store is the value which, when added back to the sum, gives *exactly* the original (quantized) data value. Notice also that you do not need to massage the LP coefficients for stability; by adding the residual back in to each point, you never depart from the original data, so instabilities cannot grow. There is therefore no need for `fixrts`, above.

Look at §20.4 to learn about *Huffman coding*, which will further compress the residuals by taking advantage of the fact that smaller values of discrepancy will occur more often than larger values. A very primitive version of Huffman coding would be this: If most of the discrepancies are in the range ±127, but an occasional one is outside, then reserve the value 127 to mean "out of range," and then record on the file (immediately following the 127) a full-word value of the out-of-range discrepancy. §20.4 explains how to do much better.

There are many variant procedures that all fall under the rubric of LPC.

- If the spectral character of the data is time-variable, then it is best not to use a single set of LP coefficients for the whole data set, but rather to partition the data into segments, computing and storing different LP coefficients for each segment.

- If the data are really well characterized by their LP coefficients, and you can tolerate some small amount of error, then don't bother storing all of the residuals. Just do linear prediction until you are outside of tolerances, then reinitialize (using M sequential stored residuals) and continue predicting.

- In some applications, most notably speech synthesis, one cares only about the spectral content of the reconstructed signal, not the relative phases. In this case, one need not store any starting values at all, but only the LP coefficients for each segment of the data. The output is reconstructed by driving these coefficients with initial conditions consisting of all zeros except for one nonzero spike. A speech synthesizer chip may have of order 10 LP coefficients, which change perhaps 20 to 50 times per second.

- Some people believe that it is interesting to analyze a signal by LPC, even when the residuals x_i are *not* small. The x_i's are then interpreted as the underlying "input signal" which, when filtered through the all-poles filter defined by the LP coefficients (see §13.7), produces the observed "output signal." LPC reveals simultaneously, it is said, the nature of the filter *and* the particular input that is driving it. We are skeptical of these applications; the literature, however, is full of extravagant claims.

CITED REFERENCES AND FURTHER READING:

Childers, D.G. (ed.) 1978, *Modern Spectrum Analysis* (New York: IEEE Press), especially the paper by J. Makhoul (reprinted from *Proceedings of the IEEE*, vol. 63, p. 561, 1975).

Burg, J.P. 1968, reprinted in Childers, 1978. [1]

Anderson, N. 1974, reprinted in Childers, 1978. [2]

Cressie, N. 1991, in *Spatial Statistics and Digital Image Analysis* (Washington: National Academy Press). [3]

Press, W.H., and Rybicki, G.B. 1992, *Astrophysical Journal*, vol. 398, pp. 169–176. [4]

13.7 Power Spectrum Estimation by the Maximum Entropy (All Poles) Method

The FFT is not the only way to estimate the power spectrum of a process, nor is it necessarily the best way for all purposes. To see how one might devise another method, let us enlarge our view for a moment, so that it includes not only real frequencies in the Nyquist interval $-f_c < f < f_c$, but also the entire complex frequency plane. From that vantage point, let us transform the complex f-plane to a new plane, called the *z-transform plane* or *z-plane*, by the relation

$$z \equiv e^{2\pi i f \Delta} \tag{13.7.1}$$

where Δ is, as usual, the sampling interval in the time domain. Notice that the Nyquist interval on the real axis of the f-plane maps one-to-one onto the unit circle in the complex z-plane.

If we now compare (13.7.1) to equations (13.4.4) and (13.4.6), we see that the FFT power spectrum estimate (13.4.5) for any real sampled function $c_k \equiv c(t_k)$ can be written, except for normalization convention, as

$$P(f) = \left| \sum_{k=-N/2}^{N/2-1} c_k z^k \right|^2 \tag{13.7.2}$$

Of course, (13.7.2) is not the *true* power spectrum of the underlying function $c(t)$, but only an estimate. We can see in two related ways why the estimate is not likely to be exact. First, in the time domain, the estimate is based on only a finite range of the function $c(t)$ which may, for all we know, have continued from $t = -\infty$ to ∞. Second, in the z-plane of equation (13.7.2), the finite Laurent series offers, in general, only an approximation to a general analytic function of z. In fact, a formal expression for representing "true" power spectra (up to normalization) is

$$P(f) = \left| \sum_{k=-\infty}^{\infty} c_k z^k \right|^2 \tag{13.7.3}$$

This is an infinite Laurent series which depends on an infinite number of values c_k. Equation (13.7.2) is just one kind of analytic approximation to the analytic function of z represented by (13.7.3); the kind, in fact, that is implicit in the use of FFTs to estimate power spectra by periodogram methods. It goes under several names, including *direct method, all-zero model*, and *moving average (MA) model*. The term "all-zero" in particular refers to the fact that the model spectrum can have zeros in the z-plane, but not poles.

If we look at the problem of approximating (13.7.3) more generally it seems clear that we could do a better job with a rational function, one with a series of type (13.7.2) in both the numerator and the denominator. Less obviously, it turns out that there are some advantages in an approximation whose free parameters all lie in the *denominator*, namely,

$$P(f) \approx \frac{1}{\left| \sum\limits_{k=-M/2}^{M/2} b_k z^k \right|^2} = \frac{a_0}{\left| 1 + \sum\limits_{k=1}^{M} a_k z^k \right|^2} \tag{13.7.4}$$

Here the second equality brings in a new set of coefficients a_k's, which can be determined from the b_k's using the fact that z lies on the unit circle. The b_k's can be thought of as being determined by the condition that power series expansion of (13.7.4) agree with the first $M + 1$ terms of (13.7.3). In practice, as we shall see, one determines the b_k's or a_k's by another method.

The differences between the approximations (13.7.2) and (13.7.4) are not just cosmetic. They are approximations with very different character. Most notable is the fact that (13.7.4) can have *poles*, corresponding to infinite power spectral density, on the unit z-circle, i.e., at real frequencies in the Nyquist interval. Such poles can provide an accurate representation for underlying power spectra that have sharp, discrete "lines" or delta-functions. By contrast, (13.7.2) can have only zeros, not poles, at real frequencies in the Nyquist interval, and must thus attempt to fit sharp spectral features with, essentially, a polynomial. The approximation (13.7.4) goes under several names: *all-poles model, maximum entropy method (MEM), autoregressive model (AR)*. We need only find out how to compute the coefficients a_0 and the a_k's from a data set, so that we can actually use (13.7.4) to obtain spectral estimates.

A pleasant surprise is that we already know how! Look at equation (13.6.11) for linear prediction. Compare it with linear filter equations (13.5.1) and (13.5.2), and you will see that, viewed as a filter that takes input x's into output y's, linear prediction has a filter function

$$\mathcal{H}(f) = \frac{1}{1 - \sum\limits_{j=0}^{N-1} d_j z^{-(j+1)}} \tag{13.7.5}$$

Thus, the power spectrum of the y's should be equal to the power spectrum of the x's multiplied by $|\mathcal{H}(f)|^2$. Now let us think about what the spectrum of the input x's is, when they are residual discrepancies from linear prediction. Although we will not prove it formally, it is intuitively believable that the x's are independently random and therefore have a flat (white noise) spectrum. (Roughly speaking, any residual correlations left in the x's would have allowed a more accurate linear prediction, and would have been removed.) The overall normalization of this flat spectrum is just the mean square amplitude of the x's. But this is exactly the quantity computed in equation (13.6.13) and returned by the routine `memcof` as `xms`. Thus, the coefficients a_0 and a_k in equation (13.7.4) are related to the LP coefficients returned by `memcof` simply by

$$a_0 = \texttt{xms} \qquad a_k = -\texttt{d}(k-1), \quad k = 1, \dots, M \qquad (13.7.6)$$

There is also another way to describe the relation between the a_k's and the autocorrelation components ϕ_k. The Wiener-Khinchin theorem (12.0.12) says that the Fourier transform of the autocorrelation is equal to the power spectrum. In z-transform language, this Fourier transform is just a Laurent series in z. The equation that is to be satisfied by the coefficients in equation (13.7.4) is thus

$$\frac{a_0}{\left|1 + \sum\limits_{k=1}^{M} a_k z^k\right|^2} \approx \sum_{j=-M}^{M} \phi_j z^j \qquad (13.7.7)$$

The approximately equal sign in (13.7.7) has a somewhat special interpretation. It means that the series expansion of the left-hand side is supposed to agree with the right-hand side term by term from z^{-M} to z^M. Outside this range of terms, the right-hand side is obviously zero, while the left-hand side will still have nonzero terms. Notice that M, the number of coefficients in the approximation on the left-hand side, can be any integer up to N, the total number of autocorrelations available. (In practice, one often chooses M much smaller than N.) M is called the *order* or *number of poles* of the approximation.

Whatever the chosen value of M, the series expansion of the left-hand side of (13.7.7) defines a certain sort of *extrapolation* of the autocorrelation function to lags larger than M, in fact even to lags larger than N, i.e., *larger than the run of data can actually measure*. It turns out that this particular extrapolation can be shown to have, among all possible extrapolations, the maximum *entropy* in a definable information-theoretic sense. Hence the name *maximum entropy method*, or MEM. The maximum entropy property has caused MEM to acquire a certain "cult" popularity; one sometimes hears that it gives an intrinsically "better" estimate than is given by other methods. Don't believe it. MEM has the very cute property of being able to fit sharp spectral features, but there is nothing else magical about its power spectrum estimates.

The operations count in `memcof` scales as the product of N (the number of data points) and M (the desired order of the MEM approximation). If M were chosen to be as large as N, then the method would be much slower than the $N \log N$ FFT methods of the previous section. In practice, however, one usually wants to limit the order (or number of poles) of the MEM approximation to a few times the number of sharp spectral features that one desires it to fit. With this restricted number of poles, the method will smooth the spectrum somewhat, but this is often a desirable property. While exact values depend on the application, one might take $M = 10$ or 20 or 50 for $N = 1000$ or 10000. In that case MEM estimation is not much slower than FFT estimation.

We feel obliged to warn you that `memcof` can be a bit quirky at times. If the number of poles or number of data points is too large, roundoff error can be a problem, even in double precision. With "peaky" data (i.e., data with extremely sharp spectral features), the algorithm may suggest split peaks even at modest orders, and the peaks may shift with the phase of the sine wave. Also, with noisy input functions, if you choose too high an order, you will find spurious peaks galore! Some experts recommend the use of this algorithm in conjunction with more conservative methods, like periodograms, to help choose the correct model order, and to avoid getting too fooled by spurious spectral features. MEM can be finicky, but it can also do remarkable things. We recommend that you try it out, cautiously, on your own problems. We now turn to the evaluation of the MEM spectral estimate from its coefficients.

The MEM estimation (13.7.4) is a function of continuously varying frequency f. There is no special significance to specific equally spaced frequencies as there was in the FFT case.

In fact, since the MEM estimate may have very sharp spectral features, one wants to be able to evaluate it on a very fine mesh near to those features, but perhaps only more coarsely farther away from them. Here is a function which, given the coefficients already computed, evaluates (13.7.4) and returns the estimated power spectrum as a function of $f\Delta$ (the frequency times the sampling interval). Of course, $f\Delta$ should lie in the Nyquist range between $-1/2$ and $1/2$.

```
#include <cmath>
#include "nr.h"
using namespace std;

DP NR::evlmem(const DP fdt, Vec_I_DP &d, const DP xms)
Given d[0..m-1] and xms as returned by memcof, this function returns the power spectrum
estimate P(f) as a function of fdt = fΔ.
{
    int i;
    DP sumr=1.0,sumi=0.0,wr=1.0,wi=0.0,wpr,wpi,wtemp,theta;

    int m=d.size();
    theta=6.28318530717959*fdt;
    wpr=cos(theta);                        Set up for recurrence relations.
    wpi=sin(theta);
    for (i=0;i<m;i++) {                    Loop over the terms in the sum.
        wr=(wtemp=wr)*wpr-wi*wpi;
        wi=wi*wpr+wtemp*wpi;
        sumr -= d[i]*wr;                   These accumulate the denominator of (13.7.4).
        sumi -= d[i]*wi;
    }
    return xms/(sumr*sumr+sumi*sumi);
}
```

Be sure to evaluate $P(f)$ on a fine enough grid to *find* any narrow features that may be there! Such narrow features, if present, can contain virtually all of the power in the data. You might also wish to know how the $P(f)$ produced by the routines memcof and evlmem is normalized with respect to the mean square value of the input data vector. The answer is

$$\int_{-1/2}^{1/2} P(f\Delta)d(f\Delta) = 2\int_{0}^{1/2} P(f\Delta)d(f\Delta) = \text{mean square value of data} \qquad (13.7.8)$$

Sample spectra produced by the routines memcof and evlmem are shown in Figure 13.7.1.

CITED REFERENCES AND FURTHER READING:

Childers, D.G. (ed.) 1978, *Modern Spectrum Analysis* (New York: IEEE Press), Chapter II.

Kay, S.M., and Marple, S.L. 1981, *Proceedings of the IEEE*, vol. 69, pp. 1380–1419.

13.8 Spectral Analysis of Unevenly Sampled Data

Thus far, we have been dealing exclusively with evenly sampled data,

$$h_n = h(n\Delta) \qquad n = \ldots, -3, -2, -1, 0, 1, 2, 3, \ldots \qquad (13.8.1)$$

where Δ is the sampling interval, whose reciprocal is the sampling rate. Recall also (§12.1) the significance of the Nyquist critical frequency

$$f_c \equiv \frac{1}{2\Delta} \qquad (13.8.2)$$

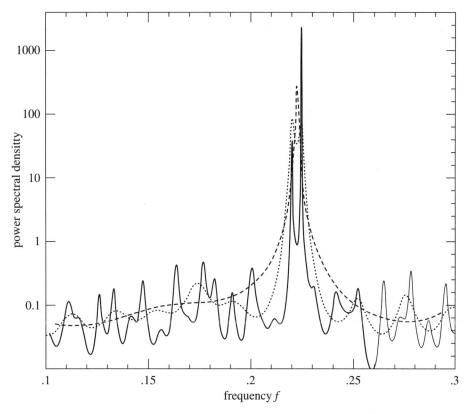

Figure 13.7.1.   Sample output of maximum entropy spectral estimation. The input signal consists of 512 samples of the sum of two sinusoids of very nearly the same frequency, plus white noise with about equal power. Shown is an expanded portion of the full Nyquist frequency interval (which would extend from zero to 0.5). The dashed spectral estimate uses 20 poles; the dotted, 40; the solid, 150. With the larger number of poles, the method can resolve the distinct sinusoids; but the flat noise background is beginning to show spurious peaks. (Note logarithmic scale.)

as codified by the sampling theorem: A sampled data set like equation (13.8.1) contains *complete* information about all spectral components for a signal $h(t)$ containing only frequencies below the Nyquist frequency, and scrambled or *aliased* information about any signal components containing frequencies larger than the Nyquist frequency. The sampling theorem thus defines both the attractiveness, and the limitation, of any analysis of an evenly spaced data set.

There are situations, however, where evenly spaced data cannot be obtained. A common case is where instrumental drop-outs occur, so that data is obtained only on a (not consecutive integer) subset of equation (13.8.1), the so-called *missing data* problem. Another case, common in observational sciences like astronomy, is that the observer cannot completely control the time of the observations, but must simply accept a certain dictated set of t_i's.

There are some obvious ways to get from unevenly spaced t_i's to evenly spaced ones, as in equation (13.8.1). Interpolation is one way: lay down a grid of evenly spaced times on your data and interpolate values onto that grid; then use FFT methods. In the missing data problem, you only have to interpolate on missing data points. If a lot of consecutive points are missing, you might as well just set them to zero, or perhaps "clamp" the value at the last measured point. However, the experience of practitioners of such interpolation techniques *is not reassuring*. Generally speaking, such techniques perform poorly. Long gaps in the data, for example, often produce a spurious bulge of power at low frequencies (wavelengths comparable to gaps).

A completely different method of spectral analysis for unevenly sampled data, one that mitigates these difficulties and has some other very desirable properties, was developed by Lomb [1], based in part on earlier work by Barning [2] and Vaníček [3], and additionally elaborated by Scargle [4]. The Lomb method (as we will call it) evaluates data, and sines and cosines, only at times t_i that are actually measured. Suppose that there are N data points $h_i \equiv h(t_i)$, $i = 0, \ldots, N - 1$. Then first find the mean and variance of the data by the usual formulas,

$$\overline{h} \equiv \frac{1}{N} \sum_{i=0}^{N-1} h_i \qquad \sigma^2 \equiv \frac{1}{N-1} \sum_{i=0}^{N-1} (h_i - \overline{h})^2 \tag{13.8.3}$$

Now, the Lomb *normalized periodogram* (spectral power as a function of angular frequency $\omega \equiv 2\pi f > 0$) is defined by

$$P_N(\omega) \equiv \frac{1}{2\sigma^2} \left\{ \frac{\left[\sum_j (h_j - \overline{h}) \cos \omega(t_j - \tau) \right]^2}{\sum_j \cos^2 \omega(t_j - \tau)} + \frac{\left[\sum_j (h_j - \overline{h}) \sin \omega(t_j - \tau) \right]^2}{\sum_j \sin^2 \omega(t_j - \tau)} \right\} \tag{13.8.4}$$

Here τ is defined by the relation

$$\tan(2\omega\tau) = \frac{\sum_j \sin 2\omega t_j}{\sum_j \cos 2\omega t_j} \tag{13.8.5}$$

The constant τ is a kind of offset that makes $P_N(\omega)$ completely independent of shifting all the t_i's by any constant. Lomb shows that this particular choice of offset has another, deeper, effect: It makes equation (13.8.4) identical to the equation that one would obtain if one estimated the harmonic content of a data set, at a given frequency ω, by linear least-squares fitting to the model

$$h(t) = A \cos \omega t + B \sin \omega t \tag{13.8.6}$$

This fact gives some insight into why the method can give results superior to FFT methods: It weights the data on a "per point" basis instead of on a "per time interval" basis, when uneven sampling can render the latter seriously in error.

A very common occurrence is that the measured data points h_i are the sum of a periodic signal and independent (white) Gaussian noise. If we are trying to determine the presence or absence of such a periodic signal, we want to be able to give a quantitative answer to the question, "How significant is a peak in the spectrum $P_N(\omega)$?" In this question, the null hypothesis is that the data values are independent Gaussian random values. A very nice property of the Lomb normalized periodogram is that the viability of the null hypothesis can be tested fairly rigorously, as we now discuss.

The word "normalized" refers to the factor σ^2 in the denominator of equation (13.8.4). Scargle [4] shows that with this normalization, at any particular ω and *in the case of the null hypothesis*, $P_N(\omega)$ has an exponential probability distribution with unit mean. In other words, the probability that $P_N(\omega)$ will be between some positive z and $z + dz$ is $\exp(-z)dz$. It readily follows that, if we scan some M *independent* frequencies, the probability that none give values larger than z is $(1 - e^{-z})^M$. So

$$P(> z) \equiv 1 - (1 - e^{-z})^M \tag{13.8.7}$$

is the false-alarm probability of the null hypothesis, that is, the *significance level* of any peak in $P_N(\omega)$ that we do see. A small value for the false-alarm probability indicates a highly significant periodic signal.

To evaluate this significance, we need to know M. After all, the more frequencies we look at, the less significant is some one modest bump in the spectrum. (Look long enough, find anything!) A typical procedure will be to plot $P_N(\omega)$ as a function of many closely spaced frequencies in some large frequency range. How many of these are independent?

Before answering, let us first see how accurately we need to know M. The interesting region is where the significance is a small (significant) number, $\ll 1$. There, equation (13.8.7) can be series expanded to give

$$P(> z) \approx M e^{-z} \tag{13.8.8}$$

We see that the significance scales linearly with M. Practical significance levels are numbers like 0.05, 0.01, 0.001, etc. An error of even $\pm 50\%$ in the estimated significance is often

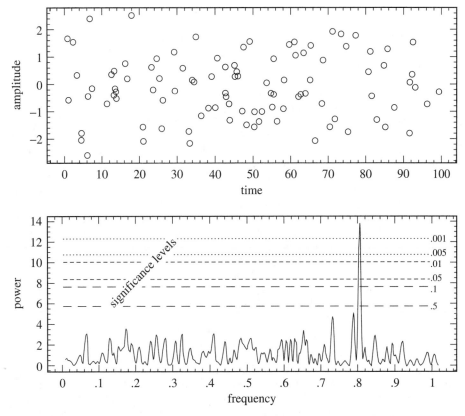

Figure 13.8.1.   Example of the Lomb algorithm in action. The 100 data points (upper figure) are at random times between 0 and 100. Their sinusoidal component is readily uncovered (lower figure) by the algorithm, at a significance level better than 0.001. If the 100 data points had been evenly spaced at unit interval, the Nyquist critical frequency would have been 0.5. Note that, for these unevenly spaced points, there is no visible aliasing into the Nyquist range.

tolerable, since quoted significance levels are typically spaced apart by factors of 5 or 10. So our estimate of M need not be very accurate.

Horne and Baliunas [5] give results from extensive Monte Carlo experiments for determining M in various cases. In general M depends on the number of frequencies sampled, the number of data points N, and their detailed spacing. It turns out that M is very nearly equal to N when the data points are approximately equally spaced, and when the sampled frequencies "fill" (oversample) the frequency range from 0 to the Nyquist frequency f_c (equation 13.8.2). Further, the value of M is not importantly different for random spacing of the data points than for equal spacing. When a larger frequency range than the Nyquist range is sampled, M increases proportionally. About the only case where M differs significantly from the case of evenly spaced points is when the points are closely clumped, say into groups of 3; then (as one would expect) the number of independent frequencies is reduced by a factor of about 3.

The program period, below, calculates an effective value for M based on the above rough-and-ready rules and assumes that there is no important clumping. This will be adequate for most purposes. In any particular case, if it really matters, it is not too difficult to compute a better value of M by simple Monte Carlo: Holding fixed the number of data points and their locations t_i, generate synthetic data sets of Gaussian (normal) deviates, find the largest values of $P_N(\omega)$ for each such data set (using the accompanying program), and fit the resulting distribution for M in equation (13.8.7).

Figure 13.8.1 shows the results of applying the method as discussed so far. In the upper figure, the data points are plotted against time. Their number is $N = 100$, and their distribution in t is Poisson random. There is certainly no sinusoidal signal evident to the eye.

The lower figure plots $P_N(\omega)$ against frequency $f = \omega/2\pi$. The Nyquist critical frequency that would obtain if the points were evenly spaced is at $f = f_c = 0.5$. Since we have searched up to about twice that frequency, and oversampled the f's to the point where successive values of $P_N(\omega)$ vary smoothly, we take $M = 2N$. The horizontal dashed and dotted lines are (respectively from bottom to top) significance levels 0.5, 0.1, 0.05, 0.01, 0.005, and 0.001. One sees a highly significant peak at a frequency of 0.81. That is in fact the frequency of the sine wave that is present in the data. (You will have to take our word for this!)

Note that two other peaks approach, but do not exceed the 50% significance level; that is about what one might expect by chance. It is also worth commenting on the fact that the significant peak was found (correctly) *above the Nyquist frequency* and without any significant aliasing down into the Nyquist interval! That would not be possible for evenly spaced data. It is possible here because the randomly spaced data has *some* points spaced much closer than the "average" sampling rate, and these remove ambiguity from any aliasing.

Implementation of the normalized periodogram in code is straightforward, with, however, a few points to be kept in mind. We are dealing with a *slow* algorithm. Typically, for N data points, we may wish to examine on the order of $2N$ or $4N$ frequencies. Each combination of frequency and data point has, in equations (13.8.4) and (13.8.5), not just a few adds or multiplies, but four calls to trigonometric functions; the operations count can easily reach several hundred times N^2. It is highly desirable — in fact results in a factor 4 speedup — to replace these trigonometric calls by recurrences. That is possible only if the sequence of frequencies examined is a linear sequence. Since such a sequence is probably what most users would want anyway, we have built this into the implementation.

At the end of this section we describe a way to evaluate equations (13.8.4) and (13.8.5) — approximately, but to any desired degree of approximation — by a fast method [6] whose operation count goes only as $N \log N$. This faster method should be used for long data sets.

The lowest independent frequency f to be examined is the inverse of the span of the input data, $\max_i(t_i) - \min_i(t_i) \equiv T$. This is the frequency such that the data can include one complete cycle. In subtracting off the data's mean, equation (13.8.4) already assumed that you are not interested in the data's zero-frequency piece — which is just that mean value. In an FFT method, higher independent frequencies would be integer multiples of $1/T$. Because we are interested in the statistical significance of any peak that may occur, however, we had better (over-) sample more finely than at interval $1/T$, so that sample points lie close to the top of any peak. Thus, the accompanying program includes an oversampling parameter, called ofac; a value ofac $\gtrsim 4$ might be typical in use. We also want to specify how high in frequency to go, say f_{hi}. One guide to choosing f_{hi} is to compare it with the Nyquist frequency f_c which would obtain if the N data points were evenly spaced over the same span T, that is $f_c = N/(2T)$. The accompanying program includes an input parameter hifac, defined as f_{hi}/f_c. The number of different frequencies N_P returned by the program is then given by

$$N_P = \frac{\text{ofac} \times \text{hifac}}{2} N \tag{13.8.9}$$

(You have to remember to dimension the output arrays to at least this size.)

The trigonometric recurrences should be done in double precision even if you convert the rest of the routine to single precision. The code embodies a few tricks with trigonometric identities, to decrease roundoff errors. If you are an aficionado of such things you can puzzle it out. A final detail is that equation (13.8.7) will fail because of roundoff error if z is too large; but equation (13.8.8) is fine in this regime.

```
#include <cmath>
#include "nr.h"
using namespace std;

void NR::period(Vec_I_DP &x, Vec_I_DP &y, const DP ofac, const DP hifac,
    Vec_O_DP &px, Vec_O_DP &py, int &nout, int &jmax, DP &prob)
```
Given n data points with abscissas x[0..n-1] (which need not be equally spaced) and ordinates y[0..n-1], and given a desired oversampling factor ofac (a typical value being 4 or larger), this routine fills array px[0..np-1] with an increasing sequence of frequencies (not angular frequencies) up to hifac times the "average" Nyquist frequency, and fills array py[0..np-1] with the values of the Lomb normalized periodogram at those frequencies. The arrays x and y

are not altered. np, the dimension of px and py, must be at least as large as nout, the number of frequencies returned (eq. 13.8.9). The routine also returns jmax such that py[jmax] is the maximum element in py, and prob, an estimate of the significance of that maximum against the hypothesis of random noise. A small value of prob indicates that a significant periodic signal is present.

```
{
    const DP TWOPI=6.283185307179586476;
    int i,j;
    DP ave,c,cc,cwtau,effm,expy,pnow,pymax,s,ss,sumc,sumcy,sums,sumsh,
        sumsy,swtau,var,wtau,xave,xdif,xmax,xmin,yy,arg,wtemp;

    int n=x.size();
    int np=px.size();
    Vec_DP wi(n),wpi(n),wpr(n),wr(n);
    nout=0.5*ofac*hifac*n;
    if (nout > np) nrerror("output arrays too short in period");
    avevar(y,ave,var);                      Get mean and variance of the input data.
    if (var == 0.0) nrerror("zero variance in period");
    xmax=xmin=x[0];                         Go through data to get the range of abscis-
    for (j=0;j<n;j++) {                        sas.
        if (x[j] > xmax) xmax=x[j];
        if (x[j] < xmin) xmin=x[j];
    }
    xdif=xmax-xmin;
    xave=0.5*(xmax+xmin);
    pymax=0.0;
    pnow=1.0/(xdif*ofac);                   Starting frequency.
    for (j=0;j<n;j++) {                      Initialize values for the trigonometric recur-
        arg=TWOPI*((x[j]-xave)*pnow);          rences at each data point.
        wpr[j]= -2.0*SQR(sin(0.5*arg));
        wpi[j]=sin(arg);
        wr[j]=cos(arg);
        wi[j]=wpi[j];
    }
    for (i=0;i<nout;i++) {                   Main loop over the frequencies to be evalu-
        px[i]=pnow;                            ated.
        sumsh=sumc=0.0;                      First, loop over the data to get $\tau$ and related
        for (j=0;j<n;j++) {                    quantities.
            c=wr[j];
            s=wi[j];
            sumsh += s*c;
            sumc += (c-s)*(c+s);
        }
        wtau=0.5*atan2(2.0*sumsh,sumc);
        swtau=sin(wtau);
        cwtau=cos(wtau);
        sums=sumc=sumsy=sumcy=0.0;           Then, loop over the data again to get the
        for (j=0;j<n;j++) {                    periodogram value.
            s=wi[j];
            c=wr[j];
            ss=s*cwtau-c*swtau;
            cc=c*cwtau+s*swtau;
            sums += ss*ss;
            sumc += cc*cc;
            yy=y[j]-ave;
            sumsy += yy*ss;
            sumcy += yy*cc;
            wr[j]=((wtemp=wr[j])*wpr[j]-wi[j]*wpi[j])+wr[j];    Update the trigono-
            wi[j]=(wi[j]*wpr[j]+wtemp*wpi[j])+wi[j];           metric recurrences.
        }
        py[i]=0.5*(sumcy*sumcy/sumc+sumsy*sumsy/sums)/var;
        if (py[i] >= pymax) pymax=py[jmax=i];
        pnow += 1.0/(ofac*xdif);            The next frequency.
    }
```

```
    expy=exp(-pymax);                    Evaluate statistical significance of the max-
    effm=2.0*nout/ofac;                  imum.
    prob=effm*expy;
    if (prob > 0.01) prob=1.0-pow(1.0-expy,effm);
}
```

Fast Computation of the Lomb Periodogram

We here show how equations (13.8.4) and (13.8.5) can be calculated — approximately, but to any desired precision — with an operation count only of order $N_P \log N_P$. The method uses the FFT, but it is in no sense an FFT periodogram of the data. It is an actual evaluation of equations (13.8.4) and (13.8.5), the Lomb normalized periodogram, with exactly that method's strengths and weaknesses. This fast algorithm, due to Press and Rybicki [6], makes feasible the application of the Lomb method to data sets at least as large as 10^6 points; it is already faster than straightforward evaluation of equations (13.8.4) and (13.8.5) for data sets as small as 60 or 100 points.

Notice that the trigonometric sums that occur in equations (13.8.5) and (13.8.4) can be reduced to four simpler sums. If we define

$$S_h \equiv \sum_{j=0}^{N-1} (h_j - \bar{h}) \sin(\omega t_j) \qquad C_h \equiv \sum_{j=0}^{N-1} (h_j - \bar{h}) \cos(\omega t_j) \qquad (13.8.10)$$

and

$$S_2 \equiv \sum_{j=0}^{N-1} \sin(2\omega t_j) \qquad C_2 \equiv \sum_{j=0}^{N-1} \cos(2\omega t_j) \qquad (13.8.11)$$

then

$$\sum_{j=0}^{N-1} (h_j - \bar{h}) \cos \omega(t_j - \tau) = C_h \cos \omega\tau + S_h \sin \omega\tau$$

$$\sum_{j=0}^{N-1} (h_j - \bar{h}) \sin \omega(t_j - \tau) = S_h \cos \omega\tau - C_h \sin \omega\tau$$

$$\qquad (13.8.12)$$

$$\sum_{j=0}^{N-1} \cos^2 \omega(t_j - \tau) = \frac{N}{2} + \frac{1}{2}C_2 \cos(2\omega\tau) + \frac{1}{2}S_2 \sin(2\omega\tau)$$

$$\sum_{j=0}^{N-1} \sin^2 \omega(t_j - \tau) = \frac{N}{2} - \frac{1}{2}C_2 \cos(2\omega\tau) - \frac{1}{2}S_2 \sin(2\omega\tau)$$

Now notice that *if* the t_js *were* evenly spaced, then the four quantities S_h, C_h, S_2, and C_2 could be evaluated by two complex FFTs, and the results could then be substituted back through equation (13.8.12) to evaluate equations (13.8.5) and (13.8.4). The problem is therefore only to evaluate equations (13.8.10) and (13.8.11) for unevenly spaced data.

Interpolation, or rather reverse interpolation — we will here call it *extirpolation* — provides the key. Interpolation, as classically understood, uses several function values on a regular mesh to construct an accurate approximation at an arbitrary point. Extirpolation, just the opposite, *replaces* a function value at an arbitrary point by several function values on a regular mesh, doing this in such a way that sums over the mesh are an accurate approximation to sums over the original arbitrary point.

It is not hard to see that the weight functions for extirpolation are identical to those for interpolation. Suppose that the function $h(t)$ to be extirpolated is known only at the discrete (unevenly spaced) points $h(t_i) \equiv h_i$, and that the function $g(t)$ (which will be, e.g., $\cos \omega t$) can be evaluated anywhere. Let $\hat{t}_k$ be a sequence of evenly spaced points on a regular mesh. Then Lagrange interpolation (§3.1) gives an approximation of the form

$$g(t) \approx \sum_k w_k(t)g(\hat{t}_k) \qquad (13.8.13)$$

where $w_k(t)$ are interpolation weights. Now let us evaluate a sum of interest by the following scheme:

$$\sum_{j=0}^{N-1} h_j g(t_j) \approx \sum_{j=0}^{N-1} h_j \left[\sum_k w_k(t_j) g(\hat{t}_k) \right] = \sum_k \left[\sum_{j=0}^{N-1} h_j w_k(t_j) \right] g(\hat{t}_k) \equiv \sum_k \hat{h}_k \, g(\hat{t}_k)$$

$$(13.8.14)$$

Here $\hat{h}_k \equiv \sum_j h_j w_k(t_j)$. Notice that equation (13.8.14) replaces the original sum by one on the regular mesh. Notice also that the accuracy of equation (13.8.13) depends only on the fineness of the mesh with respect to the function g and has nothing to do with the spacing of the points t_j or the function h; therefore the accuracy of equation (13.8.14) also has this property.

The general outline of the fast evaluation method is therefore this: (i) Choose a mesh size large enough to accommodate some desired oversampling factor, and large enough to have several extirpation points per half-wavelength of the highest frequency of interest. (ii) Extirpolate the values h_i onto the mesh and take the FFT; this gives S_h and C_h in equation (13.8.10). (iii) Extirpolate the constant values 1 onto another mesh, and take its FFT; this, with some manipulation, gives S_2 and C_2 in equation (13.8.11). (iv) Evaluate equations (13.8.12), (13.8.5), and (13.8.4), in that order.

There are several other tricks involved in implementing this algorithm efficiently. You can figure most out from the code, but we will mention the following points: (a) A nice way to get transform values at frequencies 2ω instead of ω is to stretch the time-domain data by a factor 2, and then wrap it to double-cover the original length. (This trick goes back to Tukey.) In the program, this appears as a modulo function. (b) Trigonometric identities are used to get from the left-hand side of equation (13.8.5) to the various needed trigonometric functions of $\omega\tau$. C++ identifiers like (e.g.) `cwt` and `hs2wt` represent quantities like (e.g.) $\cos\omega\tau$ and $\frac{1}{2}\sin(2\omega\tau)$. (c) The function `spread` does extirpolation onto the M most nearly centered mesh points around an arbitrary point; its turgid code evaluates coefficients of the Lagrange interpolating polynomials, in an efficient manner.

```
#include <cmath>
#include "nr.h"
using namespace std;

void NR::fasper(Vec_I_DP &x, Vec_I_DP &y, const DP ofac, const DP hifac,
    Vec_O_DP &wk1, Vec_O_DP &wk2, int &nout, int &jmax, DP &prob)
```
Given n data points with abscissas x[0..n-1] (which need not be equally spaced) and ordinates y[0..n-1], and given a desired oversampling factor ofac (a typical value being 4 or larger), this routine fills array wk1[0..nwk-1] with a sequence of nout increasing frequencies (not angular frequencies) up to hifac times the "average" Nyquist frequency, and fills array wk2[0..nwk-1] with the values of the Lomb normalized periodogram at those frequencies. The arrays x and y are not altered. nwk, the dimension of wk1 and wk2, must be large enough for intermediate work space, or an error results. The routine also returns jmax such that wk2[jmax] is the maximum element in wk2, and prob, an estimate of the significance of that maximum against the hypothesis of random noise. A small value of prob indicates that a significant periodic signal is present.
```
{
    const int MACC=4;
    int j,k,ndim,nfreq,nfreqt;
    DP ave,ck,ckk,cterm,cwt,den,df,effm,expy,fac,fndim,hc2wt,hs2wt,
        hypo,pmax,sterm,swt,var,xdif,xmax,xmin;

    int n=x.size();
    int nwk=wk1.size();
    nout=0.5*ofac*hifac*n;
    nfreqt=ofac*hifac*n*MACC;               Size the FFT as next power of 2 above
    nfreq=64;                                 nfreqt.
    while (nfreq < nfreqt) nfreq <<= 1;
    ndim=nfreq << 1;
    if (ndim > nwk) nrerror("workspaces too small in fasper");
    avevar(y,ave,var);                      Compute the mean, variance, and range
    if (var == 0.0) nrerror("zero variance in fasper");   of the data.
```

```
xmin=x[0];
xmax=xmin;
for (j=1;j<n;j++) {
    if (x[j] < xmin) xmin=x[j];
    if (x[j] > xmax) xmax=x[j];
}
xdif=xmax-xmin;
Vec_DP wk1_t(0.0,ndim);                          Zero the workspaces.
Vec_DP wk2_t(0.0,ndim);
fac=ndim/(xdif*ofac);
fndim=ndim;
for (j=0;j<n;j++) {                              Extirpolate the data into the workspaces.
    ck=fmod((x[j]-xmin)*fac,fndim);
    ckk=2.0*(ck++);
    ckk=fmod(ckk,fndim);
    ++ckk;
    spread(y[j]-ave,wk1_t,ck,MACC);
    spread(1.0,wk2_t,ckk,MACC);
}
realft(wk1_t,1);                                 Take the Fast Fourier Transforms.
realft(wk2_t,1);
df=1.0/(xdif*ofac);
pmax = -1.0;
for (k=2,j=0;j<nout;j++,k+=2) {                  Compute the Lomb value for each fre-
    hypo=sqrt(wk2_t[k]*wk2_t[k]+wk2_t[k+1]*wk2_t[k+1]);      quency.
    hc2wt=0.5*wk2_t[k]/hypo;
    hs2wt=0.5*wk2_t[k+1]/hypo;
    cwt=sqrt(0.5+hc2wt);
    swt=SIGN(sqrt(0.5-hc2wt),hs2wt);
    den=0.5*n+hc2wt*wk2_t[k]+hs2wt*wk2_t[k+1];
    cterm=SQR(cwt*wk1_t[k]+swt*wk1_t[k+1])/den;
    sterm=SQR(cwt*wk1_t[k+1]-swt*wk1_t[k])/(n-den);
    wk1[j]=(j+1)*df;
    wk2[j]=(cterm+sterm)/(2.0*var);
    if (wk2[j] > pmax) pmax=wk2[jmax=j];
}
expy=exp(-pmax);                                 Estimate significance of largest peak value.
effm=2.0*nout/ofac;
prob=effm*expy;
if (prob > 0.01) prob=1.0-pow(1.0-expy,effm);
}
```

```
#include "nr.h"

void NR::spread(const DP y, Vec_IO_DP &yy, const DP x, const int m)
```
Given an array yy[0..n-1], extirpolate (spread) a value y into m actual array elements that
best approximate the "fictional" (i.e., possibly noninteger) array element number x. The weights
used are coefficients of the Lagrange interpolating polynomial.
```
{
    static int nfac[11]={0,1,1,2,6,24,120,720,5040,40320,362880};
    int ihi,ilo,ix,j,nden;
    DP fac;

    int n=yy.size();
    if (m > 10) nrerror("factorial table too small in spread");
    ix=int(x);
    if (x == DP(ix)) yy[ix-1] += y;
    else {
        ilo=MIN(MAX(int(x-0.5*m),0),int(n-m));
        ihi=ilo+m;
        nden=nfac[m];
        fac=x-ilo-1;
```

```
    for (j=ilo+1;j<ihi;j++) fac *= (x-j-1);
    yy[ihi-1] += y*fac/(nden*(x-ihi));
    for (j=ihi-1;j>ilo;j--) {
        nden=(nden/(j-ilo))*(j-ihi);
        yy[j-1] += y*fac/(nden*(x-j));
    }
  }
}
```

CITED REFERENCES AND FURTHER READING:

Lomb, N.R. 1976, *Astrophysics and Space Science*, vol. 39, pp. 447–462. [1]

Barning, F.J.M. 1963, *Bulletin of the Astronomical Institutes of the Netherlands*, vol. 17, pp. 22–28. [2]

Vaníček, P. 1971, *Astrophysics and Space Science*, vol. 12, pp. 10–33. [3]

Scargle, J.D. 1982, *Astrophysical Journal*, vol. 263, pp. 835–853. [4]

Horne, J.H., and Baliunas, S.L. 1986, *Astrophysical Journal*, vol. 302, pp. 757–763. [5]

Press, W.H. and Rybicki, G.B. 1989, *Astrophysical Journal*, vol. 338, pp. 277–280. [6]

13.9 Computing Fourier Integrals Using the FFT

Not uncommonly, one wants to calculate accurate numerical values for integrals of the form

$$I = \int_a^b e^{i\omega t} h(t)dt , \qquad (13.9.1)$$

or the equivalent real and imaginary parts

$$I_c = \int_a^b \cos(\omega t)h(t)dt \qquad I_s = \int_a^b \sin(\omega t)h(t)dt , \qquad (13.9.2)$$

and one wants to evaluate this integral for many different values of ω. In cases of interest, $h(t)$ is often a smooth function, but it is not necessarily periodic in $[a,b]$, nor does it necessarily go to zero at a or b. While it seems intuitively obvious that the *force majeure* of the FFT ought to be applicable to this problem, doing so turns out to be a surprisingly subtle matter, as we will now see.

Let us first approach the problem naively, to see where the difficulty lies. Divide the interval $[a,b]$ into M subintervals, where M is a large integer, and define

$$\Delta \equiv \frac{b-a}{M} , \quad t_j \equiv a + j\Delta , \quad h_j \equiv h(t_j) , \quad j = 0, \ldots, M \qquad (13.9.3)$$

Notice that $h_0 = h(a)$ and $h_M = h(b)$, and that there are $M+1$ values h_j. We can approximate the integral I by a sum,

$$I \approx \Delta \sum_{j=0}^{M-1} h_j \exp(i\omega t_j) \qquad (13.9.4)$$

which is at any rate first-order accurate. (If we centered the h_j's and the t_j's in the intervals, we could be accurate to second order.) Now for certain values of ω and M, the sum in equation (13.9.4) can be made into a discrete Fourier transform, or DFT, and evaluated by the fast Fourier transform (FFT) algorithm. In particular, we can choose M to be an integer power of 2, and define a set of special ω's by

$$\omega_m \Delta \equiv \frac{2\pi m}{M} \tag{13.9.5}$$

where m has the values $m = 0, 1, \ldots, M/2 - 1$. Then equation (13.9.4) becomes

$$I(\omega_m) \approx \Delta e^{i\omega_m a} \sum_{j=0}^{M-1} h_j e^{2\pi i m j / M} = \Delta e^{i\omega_m a} [\text{DFT}(h_0 \ldots h_{M-1})]_m \tag{13.9.6}$$

Equation (13.9.6), while simple and clear, is emphatically *not recommended* for use: It is likely to give wrong answers!

The problem lies in the oscillatory nature of the integral (13.9.1). If $h(t)$ is at all smooth, and if ω is large enough to imply several cycles in the interval $[a, b]$ — in fact, ω_m in equation (13.9.5) gives exactly m cycles — then the value of I is typically very small, so small that it is easily swamped by first-order, or even (with centered values) second-order, truncation error. Furthermore, the characteristic "small parameter" that occurs in the error term is not $\Delta/(b - a) = 1/M$, as it would be if the integrand were not oscillatory, but $\omega \Delta$, which can be as large as π for ω's within the Nyquist interval of the DFT (cf. equation 13.9.5). The result is that equation (13.9.6) becomes systematically inaccurate as ω increases.

It is a sobering exercise to implement equation (13.9.6) for an integral that can be done analytically, and to see just how bad it is. We recommend that you try it.

Let us therefore turn to a more sophisticated treatment. Given the sampled points h_j, we can approximate the function $h(t)$ everywhere in the interval $[a, b]$ by interpolation on nearby h_j's. The simplest case is linear interpolation, using the two nearest h_j's, one to the left and one to the right. A higher-order interpolation, e.g., would be cubic interpolation, using two points to the left and two to the right — except in the first and last subintervals, where we must interpolate with three h_j's on one side, one on the other.

The formulas for such interpolation schemes are (piecewise) polynomial in the independent variable t, but with coefficients that are of course linear in the function values h_j. Although one does not usually think of it in this way, interpolation can be viewed as approximating a function by a sum of kernel functions (which depend only on the interpolation scheme) times sample values (which depend only on the function). Let us write

$$h(t) \approx \sum_{j=0}^{M} h_j \, \psi\left(\frac{t - t_j}{\Delta}\right) + \sum_{j=\text{endpoints}} h_j \, \varphi_j\left(\frac{t - t_j}{\Delta}\right) \tag{13.9.7}$$

Here $\psi(s)$ is the kernel function of an interior point: It is zero for s sufficiently negative or sufficiently positive, and becomes nonzero only when s is in the range where the h_j multiplying it is actually used in the interpolation. We always have $\psi(0) = 1$ and $\psi(m) = 0$, $m = \pm 1, \pm 2, \ldots$, since interpolation right on a sample point should give the sampled function value. For linear interpolation $\psi(s)$ is piecewise linear, rises from 0 to 1 for s in $(-1, 0)$, and falls back to 0 for s in $(0, 1)$. For higher-order interpolation, $\psi(s)$ is made up piecewise of segments of Lagrange interpolation polynomials. It has discontinuous derivatives at integer values of s, where the pieces join, because the set of points used in the interpolation changes discretely.

As already remarked, the subintervals closest to a and b require different (noncentered) interpolation formulas. This is reflected in equation (13.9.7) by the second sum, with the special endpoint kernels $\varphi_j(s)$. Actually, for reasons that will become clearer below, we have included *all* the points in the *first* sum (with kernel ψ), so the φ_j's are actually differences between true endpoint kernels and the interior kernel ψ. It is a tedious, but straightforward, exercise to write down all the $\varphi_j(s)$'s for any particular order of interpolation, each one consisting of differences of Lagrange interpolating polynomials spliced together piecewise.

Now apply the integral operator $\int_a^b dt \exp(i\omega t)$ to both sides of equation (13.9.7), interchange the sums and integral, and make the changes of variable $s = (t - t_j)/\Delta$ in the first sum, $s = (t - a)/\Delta$ in the second sum. The result is

$$I \approx \Delta e^{i\omega a} \left[W(\theta) \sum_{j=0}^{M} h_j e^{ij\theta} + \sum_{j=\text{endpoints}} h_j \alpha_j(\theta) \right] \qquad (13.9.8)$$

Here $\theta \equiv \omega \Delta$, and the functions $W(\theta)$ and $\alpha_j(\theta)$ are defined by

$$W(\theta) \equiv \int_{-\infty}^{\infty} ds\, e^{i\theta s} \psi(s) \qquad (13.9.9)$$

$$\alpha_j(\theta) \equiv \int_{-\infty}^{\infty} ds\, e^{i\theta s} \varphi_j(s - j) \qquad (13.9.10)$$

The key point is that equations (13.9.9) and (13.9.10) can be evaluated, analytically, once and for all, for any given interpolation scheme. Then equation (13.9.8) is an algorithm for applying "endpoint corrections" to a sum which (as we will see) can be done using the FFT, giving a result with high-order accuracy.

We will consider only interpolations that are left-right symmetric. Then symmetry implies

$$\varphi_{M-j}(s) = \varphi_j(-s) \qquad \alpha_{M-j}(\theta) = e^{i\theta M} \alpha_j^*(\theta) = e^{i\omega(b-a)} \alpha_j^*(\theta) \qquad (13.9.11)$$

where * denotes complex conjugation. Also, $\psi(s) = \psi(-s)$ implies that $W(\theta)$ is real.

Turn now to the first sum in equation (13.9.8), which we want to do by FFT methods. To do so, choose some N that is an integer power of 2 with $N \geq M + 1$. (Note that M need not be a power of two, so $M = N - 1$ is allowed.) If $N > M + 1$, define $h_j \equiv 0$, $M + 1 < j \leq N - 1$, i.e., "zero pad" the array of h_j's so that j takes on the range $0 \leq j \leq N - 1$. Then the sum can be done as a DFT for the special values $\omega = \omega_n$ given by

$$\omega_n \Delta \equiv \frac{2\pi n}{N} \equiv \theta \qquad n = 0, 1, \ldots, \frac{N}{2} - 1 \qquad (13.9.12)$$

For fixed M, the larger N is chosen, the finer the sampling in frequency space. The value M, on the other hand, determines the *highest* frequency sampled, since Δ decreases with increasing M (equation 13.9.3), and the largest value of $\omega \Delta$ is always just under π (equation 13.9.12). In general it is advantageous to oversample by *at least* a factor of 4, i.e., $N > 4M$ (see below). We can now rewrite equation (13.9.8) in its final form as

$$I(\omega_n) = \Delta e^{i\omega_n a} \Big\{ W(\theta)[\text{DFT}(h_0 \ldots h_{N-1})]_n$$

$$+ \alpha_0(\theta)h_0 + \alpha_1(\theta)h_1 + \alpha_2(\theta)h_2 + \alpha_3(\theta)h_3 + \ldots$$

$$+ e^{i\omega(b-a)} \left[\alpha_0^*(\theta)h_M + \alpha_1^*(\theta)h_{M-1} + \alpha_2^*(\theta)h_{M-2} + \alpha_3^*(\theta)h_{M-3} + \ldots \right] \Big\}$$

$$(13.9.13)$$

For cubic (or lower) polynomial interpolation, at most the terms explicitly shown above are nonzero; the ellipses ($\ldots$) can therefore be ignored, and we need explicit forms only for the functions $W, \alpha_0, \alpha_1, \alpha_2, \alpha_3$, calculated with equations (13.9.9) and (13.9.10). We have worked these out for you, in the trapezoidal (second-order) and cubic (fourth-order) cases. Here are the results, along with the first few terms of their power series expansions for small θ:

Trapezoidal order:

$$W(\theta) = \frac{2(1 - \cos\theta)}{\theta^2} \approx 1 - \frac{1}{12}\theta^2 + \frac{1}{360}\theta^4 - \frac{1}{20160}\theta^6$$

$$\alpha_0(\theta) = -\frac{(1 - \cos\theta)}{\theta^2} + i\frac{(\theta - \sin\theta)}{\theta^2}$$

$$\approx -\frac{1}{2} + \frac{1}{24}\theta^2 - \frac{1}{720}\theta^4 + \frac{1}{40320}\theta^6 + i\theta\left(\frac{1}{6} - \frac{1}{120}\theta^2 + \frac{1}{5040}\theta^4 - \frac{1}{362880}\theta^6\right)$$

$$\alpha_1 = \alpha_2 = \alpha_3 = 0$$

Cubic order:

$$W(\theta) = \left(\frac{6 + \theta^2}{3\theta^4}\right)(3 - 4\cos\theta + \cos 2\theta) \approx 1 - \frac{11}{720}\theta^4 + \frac{23}{15120}\theta^6$$

$$\alpha_0(\theta) = \frac{(-42 + 5\theta^2) + (6 + \theta^2)(8\cos\theta - \cos 2\theta)}{6\theta^4} + i\frac{(-12\theta + 6\theta^3) + (6 + \theta^2)\sin 2\theta}{6\theta^4}$$

$$\approx -\frac{2}{3} + \frac{1}{45}\theta^2 + \frac{103}{15120}\theta^4 - \frac{169}{226800}\theta^6 + i\theta\left(\frac{2}{45} + \frac{2}{105}\theta^2 - \frac{8}{2835}\theta^4 + \frac{86}{467775}\theta^6\right)$$

$$\alpha_1(\theta) = \frac{14(3 - \theta^2) - 7(6 + \theta^2)\cos\theta}{6\theta^4} + i\frac{30\theta - 5(6 + \theta^2)\sin\theta}{6\theta^4}$$

$$\approx \frac{7}{24} - \frac{7}{180}\theta^2 + \frac{5}{3456}\theta^4 - \frac{7}{259200}\theta^6 + i\theta\left(\frac{7}{72} - \frac{1}{168}\theta^2 + \frac{11}{72576}\theta^4 - \frac{13}{5987520}\theta^6\right)$$

$$\alpha_2(\theta) = \frac{-4(3 - \theta^2) + 2(6 + \theta^2)\cos\theta}{3\theta^4} + i\frac{-12\theta + 2(6 + \theta^2)\sin\theta}{3\theta^4}$$

$$\approx -\frac{1}{6} + \frac{1}{45}\theta^2 - \frac{5}{6048}\theta^4 + \frac{1}{64800}\theta^6 + i\theta\left(-\frac{7}{90} + \frac{1}{210}\theta^2 - \frac{11}{90720}\theta^4 + \frac{13}{7484400}\theta^6\right)$$

$$\alpha_3(\theta) = \frac{2(3 - \theta^2) - (6 + \theta^2)\cos\theta}{6\theta^4} + i\frac{6\theta - (6 + \theta^2)\sin\theta}{6\theta^4}$$

$$\approx \frac{1}{24} - \frac{1}{180}\theta^2 + \frac{5}{24192}\theta^4 - \frac{1}{259200}\theta^6 + i\theta\left(\frac{7}{360} - \frac{1}{840}\theta^2 + \frac{11}{362880}\theta^4 - \frac{13}{29937600}\theta^6\right)$$

The program `dftcor`, below, implements the endpoint corrections for the cubic case. Given input values of ω, Δ, a, b, and an array with the eight values $h_0, \ldots, h_3, h_{M-3}, \ldots, h_M$, it returns the real and imaginary parts of the endpoint corrections in equation (13.9.13), and the factor $W(\theta)$. The code is turgid, but only because the formulas above are complicated. The formulas have cancellations to high powers of θ. It is therefore necessary to compute the right-hand sides in double precision, even when the corrections are desired only to single precision. It is also necessary to use the series expansion for small values of θ. The optimal cross-over value of θ depends on your machine's wordlength, but you can always find it experimentally as the largest value where the two methods give identical results to machine precision.

```
#include <cmath>
#include "nr.h"
using namespace std;

void NR::dftcor(const DP w, const DP delta, const DP a, const DP b,
    Vec_I_DP &endpts, DP &corre, DP &corim, DP &corfac)
```
For an integral approximated by a discrete Fourier transform, this routine computes the correction factor that multiplies the DFT and the endpoint correction to be added. Input is the angular frequency `w`, stepsize `delta`, lower and upper limits of the integral `a` and `b`, while the array `endpts` contains the first 4 and last 4 function values. The correction factor $W(\theta)$ is returned as `corfac`, while the real and imaginary parts of the endpoint correction are returned as `corre` and `corim`.
```
{
    DP aOi,aOr,a1i,a1r,a2i,a2r,a3i,a3r,arg,c,cl,cr,s,sl,sr,t,t2,t4,t6,
        cth,ctth,spth2,sth,sth4i,stth,th,th2,th4,tmth2,tth4i;

    th=w*delta;
    if (a >= b || th < 0.0e0 || th > 3.1416e0)
        nrerror("bad arguments to dftcor");
    if (fabs(th) < 5.0e-2) {        Use series.
        t=th;
        t2=t*t;
        t4=t2*t2;
        t6=t4*t2;
```

```
    corfac=1.0-(11.0/720.0)*t4+(23.0/15120.0)*t6;
    a0r=(-2.0/3.0)+t2/45.0+(103.0/15120.0)*t4-(169.0/226800.0)*t6;
    a1r=(7.0/24.0)-(7.0/180.0)*t2+(5.0/3456.0)*t4-(7.0/259200.0)*t6;
    a2r=(-1.0/6.0)+t2/45.0-(5.0/6048.0)*t4+t6/64800.0;
    a3r=(1.0/24.0)-t2/180.0+(5.0/24192.0)*t4-t6/259200.0;
    a0i=t*(2.0/45.0+(2.0/105.0)*t2-(8.0/2835.0)*t4+(86.0/467775.0)*t6);
    a1i=t*(7.0/72.0-t2/168.0+(11.0/72576.0)*t4-(13.0/5987520.0)*t6);
    a2i=t*(-7.0/90.0+t2/210.0-(11.0/90720.0)*t4+(13.0/7484400.0)*t6);
    a3i=t*(7.0/360.0-t2/840.0+(11.0/362880.0)*t4-(13.0/29937600.0)*t6);
} else {                     Use trigonometric formulas.
    cth=cos(th);
    sth=sin(th);
    ctth=cth*cth-sth*sth;
    stth=2.0e0*sth*cth;
    th2=th*th;
    th4=th2*th2;
    tmth2=3.0e0-th2;
    spth2=6.0e0+th2;
    sth4i=1.0/(6.0e0*th4);
    tth4i=2.0e0*sth4i;
    corfac=tth4i*spth2*(3.0e0-4.0e0*cth+ctth);
    a0r=sth4i*(-42.0e0+5.0e0*th2+spth2*(8.0e0*cth-ctth));
    a0i=sth4i*(th*(-12.0e0+6.0e0*th2)+spth2*stth);
    a1r=sth4i*(14.0e0*tmth2-7.0e0*spth2*cth);
    a1i=sth4i*(30.0e0*th-5.0e0*spth2*sth);
    a2r=tth4i*(-4.0e0*tmth2+2.0e0*spth2*cth);
    a2i=tth4i*(-12.0e0*th+2.0e0*spth2*sth);
    a3r=sth4i*(2.0e0*tmth2-spth2*cth);
    a3i=sth4i*(6.0e0*th-spth2*sth);
}
cl=a0r*endpts[0]+a1r*endpts[1]+a2r*endpts[2]+a3r*endpts[3];
sl=a0i*endpts[0]+a1i*endpts[1]+a2i*endpts[2]+a3i*endpts[3];
cr=a0r*endpts[7]+a1r*endpts[6]+a2r*endpts[5]+a3r*endpts[4];
sr= -a0i*endpts[7]-a1i*endpts[6]-a2i*endpts[5]-a3i*endpts[4];
arg=w*(b-a);
c=cos(arg);
s=sin(arg);
corre=cl+c*cr-s*sr;
corim=sl+s*cr+c*sr;
}
```

Since the use of dftcor can be confusing, we also give an illustrative program dftint which uses dftcor to compute equation (13.9.1) for general a, b, ω, and $h(t)$. Several points within this program bear mentioning: The constants M and NDFT correspond to M and N in the above discussion. On successive calls, we recompute the Fourier transform only if a or b or $h(t)$ has changed.

Since dftint is designed to work for any value of ω satisfying $\omega\Delta < \pi$, not just the special values returned by the DFT (equation 13.9.12), we do polynomial interpolation of degree MPOL on the DFT spectrum. You should be warned that a large factor of oversampling ($N \gg M$) is required for this interpolation to be accurate. After interpolation, we add the endpoint corrections from dftcor, which can be evaluated for any ω.

While dftcor is good at what it does, dftint is illustrative only. It is not a general purpose program, because it does not adapt its parameters M, NDFT, MPOL, or its interpolation scheme, to any particular function $h(t)$. You will have to experiment with your own application.

```
#include <cmath>
#include "nr.h"
using namespace std;

void NR::dftint(DP func(const DP), const DP a, const DP b, const DP w,
    DP &cosint, DP &sinint)
```

Example program illustrating how to use the routine `dftcor`. The user supplies an external function `func` that returns the quantity $h(t)$. The routine then returns $\int_a^b \cos(\omega t) h(t)\, dt$ as cosint and $\int_a^b \sin(\omega t) h(t)\, dt$ as sinint.

```
{
    static int init=0;
    static DP (*funcold)(const DP);
    static DP aold = -1.e30,bold = -1.e30,delta;
    const int M=64,NDFT=1024,MPOL=6;
```
The values of M, NDFT, and MPOL are merely illustrative and should be optimized for your particular application. M is the number of subintervals, NDFT is the length of the FFT (a power of 2), and MPOL is the degree of polynomial interpolation used to obtain the desired frequency from the FFT.
```
    const DP TWOPI=6.283185307179586476;
    int j,nn;
    DP c,cdft,cerr,corfac,corim,corre,en,s,sdft,serr;
    static Vec_DP data(NDFT),endpts(8);
    Vec_DP cpol(MPOL),spol(MPOL),xpol(MPOL);

    if (init != 1 || a != aold || b != bold || func != funcold) {
        Do we need to initialize, or is only ω changed?
        init=1;
        aold=a;
        bold=b;
        funcold=func;
        delta=(b-a)/M;
        for (j=0;j<M+1;j++)                    Load the function values into the data
            data[j]=func(a+j*delta);               array.
        for (j=M+1;j<NDFT;j++)                 Zero pad the rest of the data array.
            data[j]=0.0;
        for (j=0;j<4;j++) {                    Load the endpoints.
            endpts[j]=data[j];
            endpts[j+4]=data[M-3+j];
        }
        realft(data,1);
```
`realft` returns the unused value corresponding to $\omega_{N/2}$ in data[1]. We actually want this element to contain the imaginary part corresponding to ω_0, which is zero.
```
        data[1]=0.0;
    }
```
Now interpolate on the DFT result for the desired frequency. If the frequency is an ω_n, i.e., the quantity en is an integer, then cdft=data[2*en-2], sdft=data[2*en-1], and you could omit the interpolation.
```
    en=w*delta*NDFT/TWOPI+1.0;
    nn=MIN(MAX(int(en-0.5*MPOL+1.0),1),NDFT/2-MPOL+1);   Leftmost point for the
    for (j=0;j<MPOL;j++,nn++) {                              interpolation.
        cpol[j]=data[2*nn-2];
        spol[j]=data[2*nn-1];
        xpol[j]=nn;
    }
    polint(xpol,cpol,en,cdft,cerr);
    polint(xpol,spol,en,sdft,serr);
    dftcor(w,delta,a,b,endpts,corre,corim,corfac);    Now get the endpoint cor-
    cdft *= corfac;                                      rection and the mul-
    sdft *= corfac;                                      tiplicative factor W(θ).
    cdft += corre;
    sdft += corim;
    c=delta*cos(w*a);                        Finally multiply by Δ and exp(iωa).
    s=delta*sin(w*a);
    cosint=c*cdft-s*sdft;
    sinint=s*cdft+c*sdft;
}
```

Sometimes one is interested only in the discrete frequencies ω_m of equation (13.9.5), the ones that have integral numbers of periods in the interval $[a, b]$. For smooth $h(t)$, the value of I tends to be much smaller in magnitude at these ω's than at values in between, since the integral half-periods tend to cancel precisely. (That is why one must oversample for interpolation to be accurate: $I(\omega)$ is oscillatory with small magnitude near the ω_m's.) If you want these ω_m's without messy (and possibly inaccurate) interpolation, you have to set N to a multiple of M (compare equations 13.9.5 and 13.9.12). In the method implemented above, however, N must be at least $M + 1$, so the smallest such multiple is $2M$, resulting in a factor ~ 2 unnecessary computing. Alternatively, one can derive a formula like equation (13.9.13), but with the last function sample $h_M = h(b)$ omitted from the DFT, but included entirely in the endpoint correction for h_M. Then one can set $M = N$ (an integer power of 2) and get the special frequencies of equation (13.9.5) with no additional overhead. The modified formula is

$$
\begin{aligned}
I(\omega_m) = \Delta e^{i\omega_m a} \Big\{ & W(\theta)[\text{DFT}(h_0 \ldots h_{M-1})]_m \\
& + \alpha_0(\theta)h_0 + \alpha_1(\theta)h_1 + \alpha_2(\theta)h_2 + \alpha_3(\theta)h_3 \\
& + e^{i\omega(b-a)} \left[A(\theta)h_M + \alpha_1^*(\theta)h_{M-1} + \alpha_2^*(\theta)h_{M-2} + \alpha_3^*(\theta)h_{M-3} \right] \Big\}
\end{aligned}
\tag{13.9.14}
$$

where $\theta \equiv \omega_m \Delta$ and $A(\theta)$ is given by

$$
A(\theta) = -\alpha_0(\theta)
\tag{13.9.15}
$$

for the trapezoidal case, or

$$
\begin{aligned}
A(\theta) &= \frac{(-6 + 11\theta^2) + (6 + \theta^2)\cos 2\theta}{6\theta^4} - i\,\text{Im}[\alpha_0(\theta)] \\
&\approx \frac{1}{3} + \frac{1}{45}\theta^2 - \frac{8}{945}\theta^4 + \frac{11}{14175}\theta^6 - i\,\text{Im}[\alpha_0(\theta)]
\end{aligned}
\tag{13.9.16}
$$

for the cubic case.

Factors like $W(\theta)$ arise naturally whenever one calculates Fourier coefficients of smooth functions, and they are sometimes called attenuation factors [1]. However, the endpoint corrections are equally important in obtaining accurate values of integrals. Narasimhan and Karthikeyan [2] have given a formula that is algebraically equivalent to our trapezoidal formula. However, their formula requires the evaluation of *two* FFTs, which is unnecessary. The basic idea used here goes back at least to Filon [3] in 1928 (before the FFT!). He used Simpson's rule (quadratic interpolation). Since this interpolation is not left-right symmetric, two Fourier transforms are required. An alternative algorithm for equation (13.9.14) has been given by Lyness in [4]; for related references, see [5]. To our knowledge, the cubic-order formulas derived here have not previously appeared in the literature.

Calculating Fourier transforms when the range of integration is $(-\infty, \infty)$ can be tricky. If the function falls off reasonably quickly at infinity, you can split the integral at a large enough value of t. For example, the integration to $+\infty$ can be written

$$
\begin{aligned}
\int_a^\infty e^{i\omega t} h(t)\, dt &= \int_a^b e^{i\omega t} h(t)\, dt + \int_b^\infty e^{i\omega t} h(t)\, dt \\
&= \int_a^b e^{i\omega t} h(t)\, dt - \frac{h(b)e^{i\omega b}}{i\omega} + \frac{h'(b)e^{i\omega b}}{(i\omega)^2} - \cdots
\end{aligned}
\tag{13.9.17}
$$

The splitting point b must be chosen large enough that the remaining integral over (b, ∞) is small. Successive terms in its asymptotic expansion are found by integrating by parts. The integral over (a, b) can be done using `dftint`. You keep as many terms in the asymptotic expansion as you can easily compute. See [6] for some examples of this idea. More powerful methods, which work well for long-tailed functions but which do not use the FFT, are described in [7-9].

CITED REFERENCES AND FURTHER READING:

Stoer, J., and Bulirsch, R. 1993, *Introduction to Numerical Analysis*, 2nd ed. (New York: Springer-Verlag), §2.3.4. [1]

Narasimhan, M.S. and Karthikeyan, M. 1984, *IEEE Transactions on Antennas & Propagation*, vol. 32, pp. 404–408. [2]

Filon, L.N.G. 1928, *Proceedings of the Royal Society of Edinburgh*, vol. 49, pp. 38–47. [3]

Giunta, G. and Murli, A. 1987, *ACM Transactions on Mathematical Software*, vol. 13, pp. 97–107. [4]

Lyness, J.N. 1987, in *Numerical Integration*, P. Keast and G. Fairweather, eds. (Dordrecht: Reidel). [5]

Pantis, G. 1975, *Journal of Computational Physics*, vol. 17, pp. 229–233. [6]

Blakemore, M., Evans, G.A., and Hyslop, J. 1976, *Journal of Computational Physics*, vol. 22, pp. 352–376. [7]

Lyness, J.N., and Kaper, T.J. 1987, *SIAM Journal on Scientific and Statistical Computing*, vol. 8, pp. 1005–1011. [8]

Thakkar, A.J., and Smith, V.H. 1975, *Computer Physics Communications*, vol. 10, pp. 73–79. [9]

13.10 Wavelet Transforms

Like the fast Fourier transform (FFT), the discrete wavelet transform (DWT) is a fast, linear operation that operates on a data vector whose length is an integer power of two, transforming it into a numerically different vector of the same length. Also like the FFT, the wavelet transform is invertible and in fact orthogonal — the inverse transform, when viewed as a big matrix, is simply the transpose of the transform. Both FFT and DWT, therefore, can be viewed as a rotation in function space, from the input space (or time) domain, where the basis functions are the unit vectors e_i, or Dirac delta functions in the continuum limit, to a different domain. For the FFT, this new domain has basis functions that are the familiar sines and cosines. In the wavelet domain, the basis functions are somewhat more complicated and have the fanciful names "mother functions" and "wavelets."

Of course there are an infinity of possible bases for function space, almost all of them uninteresting! What makes the wavelet basis interesting is that, *unlike* sines and cosines, individual wavelet functions are quite localized in space; simultaneously, *like* sines and cosines, individual wavelet functions are quite localized in frequency or (more precisely) characteristic scale. As we will see below, the particular kind of dual localization achieved by wavelets renders large classes of functions and operators sparse, or sparse to some high accuracy, when transformed into the wavelet domain. Analogously with the Fourier domain, where a class of computations, like convolutions, become computationally fast, there is a large class of computations — those that can take advantage of sparsity — that become computationally fast in the wavelet domain [1].

Unlike sines and cosines, which define a unique Fourier transform, there is not one single unique set of wavelets; in fact, there are infinitely many possible sets. Roughly, the different sets of wavelets make different trade-offs between how compactly they are localized in space and how smooth they are. (There are further fine distinctions.)

Daubechies Wavelet Filter Coefficients

A particular set of wavelets is specified by a particular set of numbers, called *wavelet filter coefficients*. Here, we will largely restrict ourselves to wavelet filters in a class discovered by Daubechies [2]. This class includes members ranging from highly localized to highly smooth. The simplest (and most localized) member, often called *DAUB4*, has only four coefficients, $c_0, \ldots, c_3$. For the moment we specialize to this case for ease of notation.

Consider the following transformation matrix acting on a column vector of data to its right:

$$
\begin{bmatrix}
c_0 & c_1 & c_2 & c_3 & & & & & \\
c_3 & -c_2 & c_1 & -c_0 & & & & & \\
 & & c_0 & c_1 & c_2 & c_3 & & & \\
 & & c_3 & -c_2 & c_1 & -c_0 & & & \\
\vdots & \vdots & & & & & \ddots & & \\
 & & & & c_0 & c_1 & c_2 & c_3 \\
 & & & & c_3 & -c_2 & c_1 & -c_0 \\
c_2 & c_3 & & & & & c_0 & c_1 \\
c_1 & -c_0 & & & & & c_3 & -c_2
\end{bmatrix}
\tag{13.10.1}
$$

Here blank entries signify zeroes. Note the structure of this matrix. The first row generates one component of the data convolved with the filter coefficients $c_0 \ldots, c_3$. Likewise the third, fifth, and other odd rows. If the even rows followed this pattern, offset by one, then the matrix would be a circulant, that is, an ordinary convolution that could be done by FFT methods. (Note how the last two rows wrap around like convolutions with periodic boundary conditions.) Instead of convolving with $c_0, \ldots, c_3$, however, the even rows perform a different convolution, with coefficients $c_3, -c_2, c_1, -c_0$. The action of the matrix, overall, is thus to perform two related convolutions, then to decimate each of them by half (throw away half the values), and interleave the remaining halves.

It is useful to think of the filter $c_0, \ldots, c_3$ as being a smoothing filter, call it H, something like a moving average of four points. Then, because of the minus signs, the filter $c_3, -c_2, c_1, -c_0$, call it G, is *not* a smoothing filter. (In signal processing contexts, H and G are called *quadrature mirror filters* [3].) In fact, the c's are chosen so as to make G yield, insofar as possible, a *zero* response to a sufficiently smooth data vector. This is done by requiring the sequence $c_3, -c_2, c_1, -c_0$ to have a certain number of vanishing moments. When this is the case for p moments (starting with the zeroth), a set of wavelets is said to satisfy an "approximation condition of order p." This results in the output of H, decimated by half, accurately representing the data's "smooth" information. The output of G, also decimated, is referred to as the data's "detail" information [4].

For such a characterization to be useful, it must be possible to reconstruct the original data vector of length N from its $N/2$ smooth or s-components and its $N/2$ detail or d-components. That is effected by requiring the matrix (13.10.1) to be orthogonal, so that its inverse is just the transposed matrix

$$
\begin{bmatrix}
c_0 & c_3 & & & \cdots & & & & c_2 & c_1 \\
c_1 & -c_2 & & & \cdots & & & & c_3 & -c_0 \\
c_2 & c_1 & c_0 & c_3 & & & & & & \\
c_3 & -c_0 & c_1 & -c_2 & & & & & & \\
& & & & \ddots & & & & & \\
& & & & c_2 & c_1 & c_0 & c_3 & & \\
& & & & c_3 & -c_0 & c_1 & -c_2 & & \\
& & & & & & c_2 & c_1 & c_0 & c_3 \\
& & & & & & c_3 & -c_0 & c_1 & -c_2
\end{bmatrix}
\tag{13.10.2}
$$

One sees immediately that matrix (13.10.2) is inverse to matrix (13.10.1) if and only if these two equations hold,

$$
c_0^2 + c_1^2 + c_2^2 + c_3^2 = 1
$$
$$
c_2 c_0 + c_3 c_1 = 0
\tag{13.10.3}
$$

If additionally we require the approximation condition of order $p = 2$, then two additional relations are required,

$$
c_3 - c_2 + c_1 - c_0 = 0
$$
$$
0c_3 - 1c_2 + 2c_1 - 3c_0 = 0
\tag{13.10.4}
$$

Equations (13.10.3) and (13.10.4) are 4 equations for the 4 unknowns $c_0, \dots, c_3$, first recognized and solved by Daubechies. The unique solution (up to a left-right reversal) is

$$
c_0 = (1 + \sqrt{3})/4\sqrt{2} \qquad c_1 = (3 + \sqrt{3})/4\sqrt{2}
$$
$$
c_2 = (3 - \sqrt{3})/4\sqrt{2} \qquad c_3 = (1 - \sqrt{3})/4\sqrt{2}
\tag{13.10.5}
$$

In fact, DAUB4 is only the most compact of a sequence of wavelet sets: If we had six coefficients instead of four, there would be three orthogonality requirements in equation (13.10.3) (with offsets of zero, two, and four), and we could require the vanishing of $p = 3$ moments in equation (13.10.4). In this case, DAUB6, the solution coefficients can also be expressed in closed form,

$$
\begin{aligned}
c_0 &= (1 + \sqrt{10} + \sqrt{5 + 2\sqrt{10}})/16\sqrt{2} & c_1 &= (5 + \sqrt{10} + 3\sqrt{5 + 2\sqrt{10}})/16\sqrt{2} \\
c_2 &= (10 - 2\sqrt{10} + 2\sqrt{5 + 2\sqrt{10}})/16\sqrt{2} & c_3 &= (10 - 2\sqrt{10} - 2\sqrt{5 + 2\sqrt{10}})/16\sqrt{2} \\
c_4 &= (5 + \sqrt{10} - 3\sqrt{5 + 2\sqrt{10}})/16\sqrt{2} & c_5 &= (1 + \sqrt{10} - \sqrt{5 + 2\sqrt{10}})/16\sqrt{2}
\end{aligned}
\tag{13.10.6}
$$

For higher p, up to 10, Daubechies [2] has tabulated the coefficients numerically. The number of coefficients increases by two each time p is increased by one.

Discrete Wavelet Transform

We have not yet defined the discrete wavelet transform (DWT), but we are almost there: The DWT consists of applying a wavelet coefficient matrix like (13.10.1) *hierarchically*, first to the full data vector of length N, then to the "smooth" vector of length $N/2$, then to the "smooth-smooth" vector of length $N/4$, and so on until only a trivial number of "smooth-...-smooth" components (usually 2) remain. The procedure is sometimes called a *pyramidal algorithm* [4], for obvious reasons. The output of the DWT consists of these remaining components and all the "detail" components that were accumulated along the way. A diagram should make the procedure clear:

$$
\begin{bmatrix} y_0 \\ y_1 \\ y_2 \\ y_3 \\ y_4 \\ y_5 \\ y_6 \\ y_7 \\ y_8 \\ y_9 \\ y_{10} \\ y_{11} \\ y_{12} \\ y_{13} \\ y_{14} \\ y_{15} \end{bmatrix}
\xrightarrow{13.10.1}
\begin{bmatrix} s_0 \\ d_0 \\ s_1 \\ d_1 \\ s_2 \\ d_2 \\ s_3 \\ d_3 \\ s_4 \\ d_4 \\ s_5 \\ d_5 \\ s_6 \\ d_6 \\ s_7 \\ d_7 \end{bmatrix}
\xrightarrow{\text{permute}}
\begin{bmatrix} s_0 \\ s_1 \\ s_2 \\ s_3 \\ s_4 \\ s_5 \\ s_6 \\ s_7 \\ d_0 \\ d_1 \\ d_2 \\ d_3 \\ d_4 \\ d_5 \\ d_6 \\ d_7 \end{bmatrix}
\xrightarrow{13.10.1}
\begin{bmatrix} S_0 \\ D_0 \\ S_1 \\ D_1 \\ S_2 \\ D_2 \\ S_3 \\ D_3 \\ d_0 \\ d_1 \\ d_2 \\ d_3 \\ d_4 \\ d_5 \\ d_6 \\ d_7 \end{bmatrix}
\xrightarrow{\text{permute}}
\begin{bmatrix} S_0 \\ S_1 \\ S_2 \\ S_3 \\ D_0 \\ D_1 \\ D_2 \\ D_3 \\ d_0 \\ d_1 \\ d_2 \\ d_3 \\ d_4 \\ d_5 \\ d_6 \\ d_7 \end{bmatrix}
\xrightarrow{\text{etc.}}
\begin{bmatrix} S_0 \\ S_1 \\ \mathcal{D}_0 \\ \mathcal{D}_1 \\ D_0 \\ D_1 \\ D_2 \\ D_3 \\ d_0 \\ d_1 \\ d_2 \\ d_3 \\ d_4 \\ d_5 \\ d_6 \\ d_7 \end{bmatrix}
$$

$$(13.10.7)$$

If the length of the data vector were a higher power of two, there would be more stages of applying (13.10.1) (or any other wavelet coefficients) and permuting. The endpoint will always be a vector with two S's and a hierarchy of $\mathcal{D}$'s, D's, d's, etc. Notice that once d's are generated, they simply propagate through to all subsequent stages.

A value d_i of any level is termed a "wavelet coefficient" of the original data vector; the final values S_0, S_1 should strictly be called "mother-function coefficients," although the term "wavelet coefficients" is often used loosely for both d's and final S's. Since the full procedure is a composition of orthogonal linear operations, the whole DWT is itself an orthogonal linear operator.

To invert the DWT, one simply reverses the procedure, starting with the smallest level of the hierarchy and working (in equation 13.10.7) from right to left. The inverse matrix (13.10.2) is of course used instead of the matrix (13.10.1).

As already noted, the matrices (13.10.1) and (13.10.2) embody periodic ("wrap-around") boundary conditions on the data vector. One normally accepts this as a minor inconvenience: the last few wavelet coefficients at each level of the hierarchy are affected by data from both ends of the data vector. By circularly shifting the matrix (13.10.1) $N/2$ columns to the left, one can symmetrize the wrap-around; but this does not eliminate it. It is in fact possible to eliminate the wrap-around completely by altering the coefficients in the first and last N rows of (13.10.1), giving an orthogonal matrix that is purely band-diagonal [5]. This variant, beyond our scope here, is useful when, e.g., the data varies by many orders of magnitude from one end of the data vector to the other.

Here is a routine, wt1, that performs the pyramidal algorithm (or its inverse
if isign is negative) on some data vector a[0..n-1]. Successive applications of
the wavelet filter, and accompanying permutations, are done by an assumed routine
wtstep, which must be provided. (We give examples of several different wtstep
routines just below.)

```
#include "nr.h"

void NR::wt1(Vec_IO_DP &a, const int isign,
    void wtstep(Vec_IO_DP &, const int, const int))
```
One-dimensional discrete wavelet transform. This routine implements the pyramid algorithm,
replacing a[0..n-1] by its wavelet transform (for isign=1), or performing the inverse operation (for isign=-1). Note that n MUST be an integer power of 2. The routine wtstep,
whose actual name must be supplied in calling this routine, is the underlying wavelet filter.
Examples of wtstep are daub4 and (preceded by pwtset) pwt.
```
{
    int nn;

    int n=a.size();
    if (n < 4) return;
    if (isign >= 0) {                             Wavelet transform.
        for (nn=n;nn>=4;nn>>=1) wtstep(a,nn,isign);
        Start at largest hierarchy, and work towards smallest.
    } else {
        for (nn=4;nn<=n;nn<<=1) wtstep(a,nn,isign);
        Start at smallest hierarchy, and work towards largest.
    }
}
```

Here, as a specific instance of wtstep, is a routine for the DAUB4 wavelets:

```
#include "nr.h"

void NR::daub4(Vec_IO_DP &a, const int n, const int isign)
```
Applies the Daubechies 4-coefficient wavelet filter to data vector a[0..n-1] (for isign=1) or
applies its transpose (for isign=-1). Used hierarchically by routines wt1 and wtn.
```
{
    const DP C0=0.4829629131445341,C1=0.8365163037378079,
        C2=0.2241438680420134,C3=-0.1294095225512604;
    int nh,i,j;

    if (n < 4) return;
    Vec_DP wksp(n);
    nh=n >> 1;
    if (isign >= 0) {                             Apply filter.
        for (i=0,j=0;j<n-3;j+=2,i++) {
            wksp[i]=C0*a[j]+C1*a[j+1]+C2*a[j+2]+C3*a[j+3];
            wksp[i+nh]=C3*a[j]-C2*a[j+1]+C1*a[j+2]-C0*a[j+3];
        }
        wksp[i]=C0*a[n-2]+C1*a[n-1]+C2*a[0]+C3*a[1];
        wksp[i+nh]=C3*a[n-2]-C2*a[n-1]+C1*a[0]-C0*a[1];
    } else {                                      Apply transpose filter.
        wksp[0]=C2*a[nh-1]+C1*a[n-1]+C0*a[0]+C3*a[nh];
        wksp[1]=C3*a[nh-1]-C0*a[n-1]+C1*a[0]-C2*a[nh];
        for (i=0,j=2;i<nh-1;i++) {
            wksp[j++]=C2*a[i]+C1*a[i+nh]+C0*a[i+1]+C3*a[i+nh+1];
            wksp[j++]=C3*a[i]-C0*a[i+nh]+C1*a[i+1]-C2*a[i+nh+1];
        }
    }
    for (i=0;i<n;i++) a[i]=wksp[i];
}
```

For larger sets of wavelet coefficients, the wrap-around of the last rows or columns is a programming inconvenience. An efficient implementation would handle the wrap-arounds as special cases, outside of the main loop. Here, we will content ourselves with a more general scheme involving some extra arithmetic at run time. The following routine sets up any particular wavelet coefficients whose values you happen to know.

```
#include "nr.h"

wavefilt *wfilt_p;                    The wavefilt class is defined in nrutil.h.

void NR::pwtset(const int n)
```
Initializing routine for pwt, here implementing the Daubechies wavelet filters with 4, 12, and 20 coefficients, as selected by the input value n. Further wavelet filters can be included in the obvious manner. This routine must be called (once) before the first use of pwt. (For the case n=4, the specific routine daub4 is considerably faster than pwt.) Note that the default values of ioff and joff in the wavefilt class center the "support" of the wavelets at each level. Alternatively, the "peaks" of the wavelets can be approximately centered by the choices ioff=-2 and joff=-n+2. Also note that daub4 and pwtset with n=4 use different default centerings.
```
{
    const static DP c4_d[4]=
        {0.4829629131445341,0.8365163037378079,
        0.2241438680420134,-0.1294095225512604};
    const static DP c12_d[12]=
        {0.111540743350, 0.494623890398, 0.751133908021,
        0.315250351709,-0.226264693965,-0.129766867567,
        0.097501605587, 0.027522865530,-0.031582039318,
        0.000553842201, 0.004777257511,-0.001077301085};
    const static DP c20_d[20]=
        {0.026670057901, 0.188176800078, 0.527201188932,
        0.688459039454, 0.281172343661,-0.249846424327,
        -0.195946274377, 0.127369340336, 0.093057364604,
        -0.071394147166,-0.029457536822, 0.033212674059,
        0.003606553567,-0.010733175483, 0.001395351747,
        0.001992405295,-0.000685856695,-0.000116466855,
        0.000093588670,-0.000013264203};

    if (n == 4)
        wfilt_p=new wavefilt(c4_d,n);
    else if (n == 12)
        wfilt_p=new wavefilt(c12_d,n);
    else if (n == 20)
        wfilt_p=new wavefilt(c20_d,n);
    else nrerror("unimplemented value n in pwtset");
}
```

Once pwtset has been called, the following routine can be used as a specific instance of wtstep.

```
#include "nr.h"

extern wavefilt *wfilt_p;                    Defined in pwtset.

void NR::pwt(Vec_IO_DP &a, const int n, const int isign)
```
Partial wavelet transform: applies an arbitrary wavelet filter to data vector a[0..n-1] (for isign = 1) or applies its transpose (for isign = -1). Used hierarchically by routines wt1 and wtn. The actual filter is determined by a preceding (and required) call to pwtset, which initializes the struct wfilt.
```
{
    DP ai,ai1;
```

```
        int i,ii,j,jf,jr,k,n1,ni,nj,nh,nmod;

        if (n < 4) return;
        wavefilt &wfilt=*wfilt_p;                    Make alias to simplify coding.
        Vec_DP wksp(n);
        nmod=wfilt.ncof*n;                           A positive constant equal to zero mod n.
        n1=n-1;                                       Mask of all bits, since n a power of 2.
        nh=n >> 1;
        for (j=0;j<n;j++) wksp[j]=0.0;
        if (isign >= 0) {                             Apply filter.
            for (ii=0,i=0;i<n;i+=2,ii++) {
                ni=i+1+nmod+wfilt.ioff;               Pointer to be incremented and wrapped-around.
                nj=i+1+nmod+wfilt.joff;
                for (k=0;k<wfilt.ncof;k++) {
                    jf=n1 & (ni+k+1);                 We use "bitwise and" to wrap-around the
                    jr=n1 & (nj+k+1);                     pointers.
                    wksp[ii] += wfilt.cc[k]*a[jf];
                    wksp[ii+nh] += wfilt.cr[k]*a[jr];
                }
            }
        } else {                                      Apply transpose filter.
            for (ii=0,i=0;i<n;i+=2,ii++) {
                ai=a[ii];
                ai1=a[ii+nh];
                ni=i+1+nmod+wfilt.ioff;               See comments above.
                nj=i+1+nmod+wfilt.joff;
                for (k=0;k<wfilt.ncof;k++) {
                    jf=n1 & (ni+k+1);
                    jr=n1 & (nj+k+1);
                    wksp[jf] += wfilt.cc[k]*ai;
                    wksp[jr] += wfilt.cr[k]*ai1;
                }
            }
        }
        for (j=0;j<n;j++) a[j]=wksp[j];               Copy the results back from workspace.
    }
```

What Do Wavelets Look Like?

We are now in a position actually to see some wavelets. To do so, we simply run unit vectors through any of the above discrete wavelet transforms, with isign negative so that the inverse transform is performed. Figure 13.10.1 shows the DAUB4 wavelet that is the inverse DWT of a unit vector in component 4 of a vector of length 1024, and also the DAUB20 wavelet that is the inverse of component 21. (One needs to go to a later hierarchical level for DAUB20, to avoid a wavelet with a wrapped-around tail.) Other unit vectors would give wavelets with the same shapes, but different positions and scales.

One sees that both DAUB4 and DAUB20 have wavelets that are continuous. DAUB20 wavelets also have higher continuous derivatives. DAUB4 has the peculiar property that its derivative exists only *almost* everywhere. Examples of where it fails to exist are the points $p/2^n$, where p and n are integers; at such points, DAUB4 is left differentiable, but not right differentiable! This kind of discontinuity — at least in some derivative — is a necessary feature of wavelets with compact support, like the Daubechies series. For every increase in the number of wavelet coefficients by two, the Daubechies wavelets gain about *half* a derivative of continuity. (But not exactly half; the actual orders of regularity are irrational numbers!)

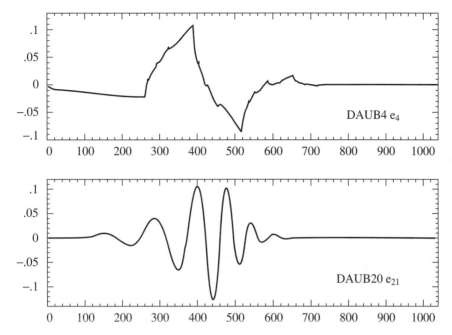

Figure 13.10.1.   Wavelet functions, that is, single basis functions from the wavelet families DAUB4 and DAUB20. A complete, orthonormal wavelet basis consists of scalings and translations of either one of these functions. DAUB4 has an infinite number of cusps; DAUB20 would show similar behavior in a higher derivative.

Note that the fact that wavelets are not smooth does not prevent their having exact representations for some smooth functions, as demanded by their approximation order p. The continuity of a wavelet is not the same as the continuity of functions that a set of wavelets can represent. For example, DAUB4 can represent (piecewise) linear functions of arbitrary slope: in the correct linear combinations, the cusps all cancel out. Every increase of two in the number of coefficients allows one higher order of polynomial to be exactly represented.

Figure 13.10.2 shows the result of performing the inverse DWT on the input vector $\mathbf{e}_9 + \mathbf{e}_{57}$, again for the two different particular wavelets. Since 9 lies early in the hierarchical range of $8 - 15$, that wavelet lies on the left side of the picture. Since 57 lies in a later (smaller-scale) hierarchy, it is a narrower wavelet; in the range of 32–63 it is towards the end, so it lies on the right side of the picture. Note that smaller-scale wavelets are taller, so as to have the same squared integral.

Wavelet Filters in the Fourier Domain

The Fourier transform of a set of filter coefficients c_j is given by

$$H(\omega) = \sum_j c_j e^{ij\omega} \tag{13.10.8}$$

Here H is a function periodic in 2π, and it has the same meaning as before: It is the wavelet filter, now written in the Fourier domain. A very useful fact is that the orthogonality conditions for the c's (e.g., equation 13.10.3 above) collapse to two

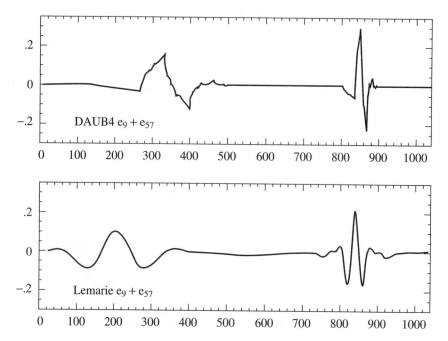

Figure 13.10.2.   More wavelets, here generated from the sum of two unit vectors, $e_9 + e_{57}$, which are in different hierarchical levels of scale, and also at different spatial positions. DAUB4 wavelets (a) are defined by a filter in coordinate space (equation 13.10.5), while Lemarie wavelets (b) are defined by a filter most easily written in Fourier space (equation 13.10.14).

simple relations in the Fourier domain,

$$\frac{1}{2}|H(0)|^2 = 1 \tag{13.10.9}$$

and

$$\frac{1}{2}\left[|H(\omega)|^2 + |H(\omega + \pi)|^2\right] = 1 \tag{13.10.10}$$

Likewise the approximation condition of order p (e.g., equation 13.10.4 above) has a simple formulation, requiring that $H(\omega)$ have a pth order zero at $\omega = \pi$, or (equivalently)

$$H^{(m)}(\pi) = 0 \qquad m = 0, 1, \ldots, p - 1 \tag{13.10.11}$$

It is thus relatively straightforward to invent wavelet sets in the Fourier domain. You simply invent a function $H(\omega)$ satisfying equations (13.10.9)–(13.10.11). To find the actual c_j's applicable to a data (or s-component) vector of length N, and with periodic wrap-around as in matrices (13.10.1) and (13.10.2), you invert equation (13.10.8) by the discrete Fourier transform

$$c_j = \frac{1}{N} \sum_{k=0}^{N-1} H\left(\frac{2\pi k}{N}\right) e^{-2\pi ijk/N} \tag{13.10.12}$$

The quadrature mirror filter G (reversed c_j's with alternating signs), incidentally, has the Fourier representation

$$G(\omega) = e^{-i\omega} H^*(\omega + \pi) \tag{13.10.13}$$

where asterisk denotes complex conjugation.

In general the above procedure will *not* produce wavelet filters with compact support. In other words, all N of the c_j's, $j = 0, \ldots, N - 1$ will in general be nonzero (though they may be rapidly decreasing in magnitude). The Daubechies wavelets, or other wavelets with compact support, are specially chosen so that $H(\omega)$ is a trigonometric polynomial with only a small number of Fourier components, guaranteeing that there will be only a small number of nonzero c_j's.

On the other hand, there is sometimes no particular reason to demand compact support. Giving it up in fact allows the ready construction of relatively smoother wavelets (higher values of p). Even without compact support, the convolutions implicit in the matrix (13.10.1) can be done efficiently by FFT methods.

Lemarie's wavelet (see [4]) has $p = 4$, does not have compact support, and is defined by the choice of $H(\omega)$,

$$H(\omega) = \left[2(1 - u)^4 \frac{315 - 420u + 126u^2 - 4u^3}{315 - 420v + 126v^2 - 4v^3} \right]^{1/2} \qquad (13.10.14)$$

where

$$u \equiv \sin^2 \frac{\omega}{2} \qquad v \equiv \sin^2 \omega \qquad (13.10.15)$$

It is beyond our scope to explain where equation (13.10.14) comes from. An informal description is that the quadrature mirror filter $G(\omega)$ deriving from equation (13.10.14) has the property that it gives identically zero when applied to any function whose odd-numbered samples are equal to the cubic spline interpolation of its even-numbered samples. Since this class of functions includes many very smooth members, it follows that $H(\omega)$ does a good job of truly selecting a function's smooth information content. Sample Lemarie wavelets are shown in Figure 13.10.2.

Truncated Wavelet Approximations

Most of the usefulness of wavelets rests on the fact that wavelet transforms can usefully be severely truncated, that is, turned into sparse expansions. The case of Fourier transforms is different: FFTs are ordinarily used without truncation, to compute fast convolutions, for example. This works because the convolution operator is particularly simple in the Fourier basis. There are not, however, any standard mathematical operations that are especially simple in the wavelet basis.

To see how truncation works, consider the simple example shown in Figure 13.10.3. The upper panel shows an arbitrarily chosen test function, smooth except for a square-root cusp, sampled onto a vector of length 2^{10}. The bottom panel (solid curve) shows, on a log scale, the absolute value of the vector's components after it has been run through the DAUB4 discrete wavelet transform. One notes, from right to left, the different levels of hierarchy, 512–1023, 256–511, 128–255, etc. Within each level, the wavelet coefficients are non-negligible only very near the location of the cusp, or very near the left and right boundaries of the hierarchical range (edge effects).

The dotted curve in the lower panel of Figure 13.10.3 plots the same amplitudes as the solid curve, but sorted into decreasing order of size. One can read off, for example, that the 130th largest wavelet coefficient has an amplitude less than 10^{-5}

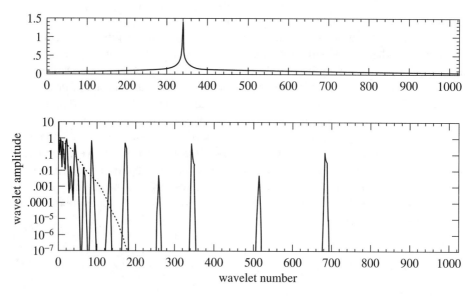

Figure 13.10.3.   (a) Arbitrary test function, with cusp, sampled on a vector of length 1024.   (b) Absolute value of the 1024 wavelet coefficients produced by the discrete wavelet transform of (a). Note log scale. The dotted curve plots the same amplitudes when sorted by decreasing size. One sees that only 130 out of 1024 coefficients are larger than 10^{-4} (or larger than about 10^{-5} times the largest coefficient, whose value is ~ 10).

of the largest coefficient, whose magnitude is ~ 10 (power or square integral ratio less than 10^{-10}). Thus, the example function can be represented quite accurately by only 130, rather than 1024, coefficients — the remaining ones being set to zero. Note that this kind of truncation makes the vector sparse, but not shorter than 1024. It is very important that vectors in wavelet space be truncated according to the *amplitude* of the components, not their position in the vector. Keeping the first 256 components of the vector (all levels of the hierarchy except the last two) would give an extremely poor, and jagged, approximation to the function. When you compress a function with wavelets, you have to record both the values *and the positions* of the nonzero coefficients.

Generally, compact (and therefore unsmooth) wavelets are better for lower accuracy approximation and for functions with discontinuities (like edges), while smooth (and therefore noncompact) wavelets are better for achieving high numerical accuracy. This makes compact wavelets a good choice for image compression, for example, while it makes smooth wavelets best for fast solution of integral equations.

Wavelet Transform in Multidimensions

A wavelet transform of a d-dimensional array is most easily obtained by transforming the array sequentially on its first index (for all values of its other indices), then on its second, and so on. Each transformation corresponds to multiplication by an orthogonal matrix. By matrix associativity, the result is independent of the order in which the indices were transformed. The situation is exactly like that for multidimensional FFTs. A routine for effecting the multidimensional DWT can thus be modeled on a multidimensional FFT routine like fourn:

```
#include "nr.h"

void NR::wtn(Vec_IO_DP &a, Vec_I_INT &nn, const int isign,
    void wtstep(Vec_IO_DP &, const int, const int))
```
Replaces a by its ndim-dimensional discrete wavelet transform, if isign is input as 1. Here
nn[0..ndim-1] is an integer array containing the lengths of each dimension (number of real
values), which MUST all be powers of 2. a is a real array of length equal to the product of
these lengths, in which the data are stored as in a multidimensional real array. If isign is input
as −1, a is replaced by its inverse wavelet transform. The routine wtstep, whose actual name
must be supplied in calling this routine, is the underlying wavelet filter. Examples of wtstep
are daub4 and (preceded by pwtset) pwt.
```
{
    int idim,i1,i2,i3,k,n,nnew,nprev=1,nt,ntot=1;

    int ndim=nn.size();
    for (idim=0;idim<ndim;idim++) ntot *= nn[idim];
    for (idim=0;idim<ndim;idim++) {            Main loop over the dimensions.
        n=nn[idim];
        Vec_DP wksp(n);
        nnew=n*nprev;
        if (n > 4) {
            for (i2=0;i2<ntot;i2+=nnew) {
                for (i1=0;i1<nprev;i1++) {
                    for (i3=i1+i2,k=0;k<n;k++,i3+=nprev) wksp[k]=a[i3];
                    Copy the relevant row or column or etc. into workspace.
                    if (isign >= 0) {          Do one-dimensional wavelet transform.
                        for(nt=n;nt>=4;nt >>= 1)
                            wtstep(wksp,nt,isign);
                    } else {                    Or inverse transform.
                        for(nt=4;nt<=n;nt <<= 1)
                            wtstep(wksp,nt,isign);
                    }
                    for (i3=i1+i2,k=0;k<n;k++,i3+=nprev) a[i3]=wksp[k];
                    Copy back from workspace.
                }
            }
        }
        nprev=nnew;
    }
}
```

Here, as before, wtstep is an individual wavelet step, either daub4 or pwt.

Compression of Images

An immediate application of the multidimensional transform wtn is to image
compression. The overall procedure is to take the wavelet transform of a digitized
image, and then to "allocate bits" among the wavelet coefficients in some highly
nonuniform, optimized, manner. In general, large wavelet coefficients get quantized
accurately, while small coefficients are quantized coarsely with only a bit or two
— or else are truncated completely. If the resulting quantization levels are still
statistically nonuniform, they may then be further compressed by a technique like
Huffman coding (§20.4).

While a more detailed description of the "back end" of this process, namely the
quantization and coding of the image, is beyond our scope, it is quite straightforward
to demonstrate the "front-end" wavelet encoding with a simple truncation: We keep
(with full accuracy) all wavelet coefficients larger than some threshold, and we delete

(a)   (b)

(c)   (d)

Figure 13.10.4.   (a) IEEE test image, 256×256 pixels with 8-bit grayscale. (b) The image is transformed into the wavelet basis; 77% of its wavelet components are set to zero (those of smallest magnitude); it is then reconstructed from the remaining 23%. (c) Same as (b), but 94.5% of the wavelet components are deleted. (d) Same as (c), but the Fourier transform is used instead of the wavelet transform. Wavelet coefficients are better than the Fourier coefficients at preserving relevant details.

(set to zero) all smaller wavelet coefficients. We can then adjust the threshold to vary the fraction of preserved coefficients.

Figure 13.10.4 shows a sequence of images that differ in the number of wavelet coefficients that have been kept. The original picture (a), which is an official IEEE test image, has 256 by 256 pixels with an 8-bit grayscale. The two reproductions following are reconstructed with 23% (b) and 5.5% (c) of the 65536 wavelet coefficients. The latter image illustrates the kind of compromises made by the truncated wavelet representation. High-contrast edges (the model's right cheek and hair highlights, e.g.) are maintained at a relatively high resolution, while low-contrast areas (the model's left eye and cheek, e.g.) are washed out into what amounts to large constant pixels. Figure 13.10.4 (d) is the result of performing the identical procedure with Fourier, instead of wavelet, transforms: The figure is reconstructed from the 5.5% of 65536 real Fourier components having the largest magnitudes. One sees that, since sines and cosines are nonlocal, the resolution is uniformly poor across the picture; also, the deletion of any components produces a mottled "ringing" everywhere. (Practical Fourier image compression schemes therefore break up an image into small blocks of pixels, 16×16, say, and do rather elaborate smoothing across block boundaries when the image is reconstructed.)

Fast Solution of Linear Systems

One of the most interesting, and promising, wavelet applications is linear algebra. The basic idea [1] is to think of an integral operator (that is, a large matrix) as a digital image. Suppose that the operator compresses well under a two-dimensional wavelet transform, i.e., that a large fraction of its wavelet coefficients are so small as to be negligible. Then any linear system involving the operator becomes a sparse system in the wavelet basis. In other words, to solve

$$\mathbf{A} \cdot \mathbf{x} = \mathbf{b} \qquad (13.10.16)$$

we first wavelet-transform the operator $\mathbf{A}$ and the right-hand side $\mathbf{b}$ by

$$\widetilde{\mathbf{A}} \equiv \mathbf{W} \cdot \mathbf{A} \cdot \mathbf{W}^T, \qquad \widetilde{\mathbf{b}} \equiv \mathbf{W} \cdot \mathbf{b} \qquad (13.10.17)$$

where $\mathbf{W}$ represents the one-dimensional wavelet transform, then solve

$$\widetilde{\mathbf{A}} \cdot \widetilde{\mathbf{x}} = \widetilde{\mathbf{b}} \qquad (13.10.18)$$

and finally transform to the answer by the inverse wavelet transform

$$\mathbf{x} = \mathbf{W}^T \cdot \widetilde{\mathbf{x}} \qquad (13.10.19)$$

(Note that the routine wtn does the complete transformation of $\mathbf{A}$ into $\widetilde{\mathbf{A}}$.)

A typical integral operator that compresses well into wavelets has arbitrary (or even nearly singular) elements near to its main diagonal, but becomes smooth away from the diagonal. An example might be

$$A_{ij} = \begin{cases} -1 & \text{if } i = j \\ |i - j|^{-1/2} & \text{otherwise} \end{cases} \qquad (13.10.20)$$

Figure 13.10.5 shows a graphical representation of the wavelet transform of this matrix, where i and j range over $0 \ldots 255$, using the DAUB12 wavelets. Elements larger in magnitude than 10^{-3} times the maximum element are shown as black pixels, while elements between 10^{-3} and 10^{-6} are shown in gray. White pixels are $< 10^{-6}$. The indices i and j each number from the lower left.

In the figure, one sees the hierarchical decomposition into power-of-two sized blocks. At the edges or corners of the various blocks, one sees edge effects caused by the wrap-around wavelet boundary conditions. Apart from edge effects, within each block, the nonnegligible elements are concentrated along the block diagonals. This is a statement that, for this type of linear operator, a wavelet is coupled mainly

Figure 13.10.5.   Wavelet transform of a 256×256 matrix, represented graphically. The original matrix has a discontinuous cusp along its diagonal, decaying smoothly away on both sides of the diagonal. In wavelet basis, the matrix becomes sparse: Components larger than 10^{-3} are shown as black, components larger than 10^{-6} as gray, and smaller-magnitude components are white. The matrix indices i and j number from the lower left.

to near neighbors in its own hierarchy (square blocks along the main diagonal) and near neighbors in other hierarchies (rectangular blocks off the diagonal).

The number of nonnegligible elements in a matrix like that in Figure 13.10.5 scales only as N, the linear size of the matrix; as a rough rule of thumb it is about $10N \log_{10}(1/\epsilon)$, where ϵ is the truncation level, e.g., 10^{-6}. For a 2000 by 2000 matrix, then, the matrix is sparse by a factor on the order of 30.

Various numerical schemes can be used to solve sparse linear systems of this "hierarchically band diagonal" form. Beylkin, Coifman, and Rokhlin [1] make the interesting observations that (1) the product of two such matrices is itself hierarchically band diagonal (truncating, of course, newly generated elements that are smaller than the predetermined threshold ϵ); and moreover that (2) the product can be formed in order N operations.

Fast matrix multiplication makes it possible to find the matrix inverse by Schultz's (or Hotelling's) method, see §2.5.

Other schemes are also possible for fast solution of hierarchically band diagonal forms. For example, one can use the conjugate gradient method, implemented in §2.7 as `linbcg`.

CITED REFERENCES AND FURTHER READING:

Daubechies, I. 1992, *Wavelets* (Philadelphia: S.I.A.M.).

Strang, G. 1989, *SIAM Review*, vol. 31, pp. 614–627.

Beylkin, G., Coifman, R., and Rokhlin, V. 1991, *Communications on Pure and Applied Mathematics*, vol. 44, pp. 141–183. [1]

Daubechies, I. 1988, *Communications on Pure and Applied Mathematics*, vol. 41, pp. 909–996. [2]

Vaidyanathan, P.P. 1990, *Proceedings of the IEEE*, vol. 78, pp. 56–93. [3]

Mallat, S.G. 1989, *IEEE Transactions on Pattern Analysis and Machine Intelligence*, vol. 11, pp. 674–693. [4]

Freedman, M.H., and Press, W.H. 1992, preprint. [5]

13.11 Numerical Use of the Sampling Theorem

In §6.10 we implemented an approximating formula for Dawson's integral due to Rybicki. Now that we have become Fourier sophisticates, we can learn that the formula derives from *numerical* application of the sampling theorem (§12.1), normally considered to be a purely analytic tool. Our discussion is identical to Rybicki [1].

For present purposes, the sampling theorem is most conveniently stated as follows: Consider an arbitrary function $g(t)$ and the grid of sampling points $t_n = \alpha + nh$, where n ranges over the integers and α is a constant that allows an arbitrary shift of the sampling grid. We then write

$$g(t) = \sum_{n=-\infty}^{\infty} g(t_n)\,\mathrm{sinc}\,\frac{\pi}{h}(t - t_n) + e(t) \tag{13.11.1}$$

where $\mathrm{sinc}\,x \equiv \sin x / x$. The summation over the sampling points is called the *sampling representation* of $g(t)$, and $e(t)$ is its error term. The sampling theorem asserts that the sampling representation is exact, that is, $e(t) \equiv 0$, if the Fourier transform of $g(t)$,

$$G(\omega) = \int_{-\infty}^{\infty} g(t)e^{i\omega t}\,dt \tag{13.11.2}$$

vanishes identically for $|\omega| \geq \pi/h$.

When can sampling representations be used to advantage for the approximate numerical computation of functions? In order that the error term be small, the Fourier transform $G(\omega)$ must be sufficiently small for $|\omega| \geq \pi/h$. On the other hand, in order for the summation in (13.11.1) to be approximated by a reasonably small number of terms, the function $g(t)$ itself should be very small outside of a fairly limited range of values of t. Thus we are led to two conditions to be satisfied in order that (13.11.1) be useful numerically: Both the function $g(t)$ and its Fourier transform $G(\omega)$ must rapidly approach zero for large values of their respective arguments.

Unfortunately, these two conditions are mutually antagonistic — the Uncertainty Principle in quantum mechanics. There exist strict limits on how rapidly the simultaneous approach to zero can be in both arguments. According to a theorem of Hardy [2], if $g(t) = O(e^{-t^2})$ as $|t| \to \infty$ and $G(\omega) = O(e^{-\omega^2/4})$ as $|\omega| \to \infty$, then $g(t) \equiv Ce^{-t^2}$, where C is a constant. This can be interpreted as saying that of all functions the Gaussian is the most rapidly decaying in both t and ω, and in this sense is the "best" function to be expressed numerically as a sampling representation.

Let us then write for the Gaussian $g(t) = e^{-t^2}$,

$$e^{-t^2} = \sum_{n=-\infty}^{\infty} e^{-t_n^2}\,\mathrm{sinc}\,\frac{\pi}{h}(t - t_n) + e(t) \tag{13.11.3}$$

The error $e(t)$ depends on the parameters h and α as well as on t, but it is sufficient for the present purposes to state the bound,

$$|e(t)| < e^{-(\pi/2h)^2} \tag{13.11.4}$$

which can be understood simply as the order of magnitude of the Fourier transform of the Gaussian at the point where it "spills over" into the region $|\omega| > \pi/h$.

When the summation in (13.11.3) is approximated by one with finite limits, say from $N_0 - N$ to $N_0 + N$, where N_0 is the integer nearest to $-\alpha/h$, there is a further truncation error. However, if N is chosen so that $N > \pi/(2h^2)$, the truncation error in the summation is less than the bound given by (13.11.4), and, since this bound is an overestimate, we shall continue to use it for (13.11.3) as well. The truncated summation gives a remarkably accurate representation for the Gaussian even for moderate values of N. For example, $|e(t)| < 5 \times 10^{-5}$ for $h = 1/2$ and $N = 7$; $|e(t)| < 2 \times 10^{-10}$ for $h = 1/3$ and $N = 15$; and $|e(t)| < 7 \times 10^{-18}$ for $h = 1/4$ and $N = 25$.

One may ask, what is the point of such a numerical representation for the Gaussian, which can be computed so easily and quickly as an exponential? The answer is that many transcendental functions can be expressed as an integral involving the Gaussian, and by substituting (13.11.3) one can often find excellent approximations to the integrals as a sum over elementary functions.

Let us consider as an example the function $w(z)$ of the complex variable $z = x + iy$, related to the complex error function by

$$w(z) = e^{-z^2} \operatorname{erfc}(-iz) \tag{13.11.5}$$

having the integral representation

$$w(z) = \frac{1}{\pi i} \int_C \frac{e^{-t^2} \, dt}{t - z} \tag{13.11.6}$$

where the contour C extends from $-\infty$ to ∞, passing below z (see, e.g., [3]). Many methods exist for the evaluation of this function (e.g., [4]). Substituting the sampling representation (13.11.3) into (13.11.6) and performing the resulting elementary contour integrals, we obtain

$$w(z) \approx \frac{1}{\pi i} \sum_{n=-\infty}^{\infty} h e^{-t_n^2} \frac{1 - (-1)^n e^{-\pi i(\alpha - z)/h}}{t_n - z} \tag{13.11.7}$$

where we now omit the error term. One should note that there is no singularity as $z \to t_m$ for some $n = m$, but a special treatment of the mth term will be required in this case (for example, by power series expansion).

An alternative form of equation (13.11.7) can be found by expressing the complex exponential in (13.11.7) in terms of trigonometric functions and using the sampling representation (13.11.3) with z replacing t. This yields

$$w(z) \approx e^{-z^2} + \frac{1}{\pi i} \sum_{n=-\infty}^{\infty} h e^{-t_n^2} \frac{1 - (-1)^n \cos \pi(\alpha - z)/h}{t_n - z} \tag{13.11.8}$$

This form is particularly useful in obtaining Re $w(z)$ when $|y| \ll 1$. Note that in evaluating (13.11.7) the complex exponential inside the summation is a constant and needs to be evaluated only once; a similar comment holds for the cosine in (13.11.8).

There are a variety of formulas that can now be derived from either equation (13.11.7) or (13.11.8) by choosing particular values of α. Eight interesting choices are: $\alpha = 0$, x, iy, or z, plus the values obtained by adding $h/2$ to each of these. Since the error bound (13.11.3) assumed a real value of α, the choices involving a complex α are useful only if the imaginary part of z is not too large. This is not the place to catalog all sixteen possible formulas, and we give only two particular cases that show some of the important features.

First of all let $\alpha = 0$ in equation (13.11.8), which yields,

$$w(z) \approx e^{-z^2} + \frac{1}{\pi i} \sum_{n=-\infty}^{\infty} h e^{-(nh)^2} \frac{1 - (-1)^n \cos(\pi z/h)}{nh - z} \tag{13.11.9}$$

This approximation is good over the entire z-plane. As stated previously, one has to treat the case where one denominator becomes small by expansion in a power series. Formulas for the case $\alpha = 0$ were discussed briefly in [5]. They are similar, but not identical, to formulas derived by Chiarella and Reichel [6], using the method of Goodwin [7].

Next, let $\alpha = z$ in (13.11.7), which yields

$$w(z) \approx e^{-z^2} - \frac{2}{\pi i} \sum_{n \text{ odd}} \frac{e^{-(z-nh)^2}}{n} \tag{13.11.10}$$

the sum being over all odd integers (positive and negative). Note that we have made the substitution $n \to -n$ in the summation. This formula is simpler than (13.11.9) and contains half the number of terms, but its error is worse if y is large. Equation (13.11.10) is the source of the approximation formula (6.10.3) for Dawson's integral, used in §6.10.

CITED REFERENCES AND FURTHER READING:

Rybicki, G.B. 1989, *Computers in Physics*, vol. 3, no. 2, pp. 85–87. [1]

Hardy, G.H. 1933, *Journal of the London Mathematical Society*, vol. 8, pp. 227–231. [2]

Abramowitz, M., and Stegun, I.A. 1964, *Handbook of Mathematical Functions*, Applied Mathematics Series, Volume 55 (Washington: National Bureau of Standards; reprinted 1968 by Dover Publications, New York). [3]

Gautschi, W. 1970, *SIAM Journal on Numerical Analysis*, vol. 7, pp. 187–198. [4]

Armstrong, B.H., and Nicholls, R.W. 1972, *Emission, Absorption and Transfer of Radiation in Heated Atmospheres* (New York: Pergamon). [5]

Chiarella, C., and Reichel, A. 1968, *Mathematics of Computation*, vol. 22, pp. 137–143. [6]

Goodwin, E.T. 1949, *Proceedings of the Cambridge Philosophical Society*, vol. 45, pp. 241–245. [7]

Chapter 14. Statistical Description of Data

14.0 Introduction

In this chapter and the next, the concept of *data* enters the discussion more prominently than before.

Data consist of numbers, of course. But these numbers are fed into the computer, not produced by it. These are numbers to be treated with considerable respect, neither to be tampered with, nor subjected to a numerical process whose character you do not completely understand. You are well advised to acquire a reverence for data that is rather different from the "sporty" attitude that is sometimes allowable, or even commendable, in other numerical tasks.

The analysis of data inevitably involves some trafficking with the field of *statistics*, that gray area which is not quite a branch of mathematics — and just as surely not quite a branch of science. In the following sections, you will repeatedly encounter the following paradigm:

- apply some formula to the data to compute "a statistic"
- compute where the value of that statistic falls in a probability distribution that is computed on the basis of some "null hypothesis"
- if it falls in a very unlikely spot, way out on a tail of the distribution, conclude that the null hypothesis is *false* for your data set

If a statistic falls in a *reasonable* part of the distribution, you must not make the mistake of concluding that the null hypothesis is "verified" or "proved." That is the curse of statistics, that it can never prove things, only disprove them! At best, you can substantiate a hypothesis by ruling out, statistically, a whole long list of competing hypotheses, every one that has ever been proposed. After a while your adversaries and competitors will give up trying to think of alternative hypotheses, or else they will grow old and die, and *then your hypothesis will become accepted.* Sounds crazy, we know, but that's how science works!

In this book we make a somewhat arbitrary distinction between data analysis procedures that are *model-independent* and those that are *model-dependent*. In the former category, we include so-called *descriptive statistics* that characterize a data set in general terms: its mean, variance, and so on. We also include statistical tests that seek to establish the "sameness" or "differentness" of two or more data sets, or that seek to establish and measure a degree of *correlation* between two data sets. These subjects are discussed in this chapter.

In the other category, model-dependent statistics, we lump the whole subject of fitting data to a theory, parameter estimation, least-squares fits, and so on. Those subjects are introduced in Chapter 15.

Section 14.1 deals with so-called *measures of central tendency*, the moments of a distribution, the median and mode. In §14.2 we learn to test whether different data sets are drawn from distributions with different values of these measures of central tendency. This leads naturally, in §14.3, to the more general question of whether two distributions can be shown to be (significantly) different.

In §14.4–§14.7, we deal with *measures of association* for two distributions. We want to determine whether two variables are "correlated" or "dependent" on one another. If they are, we want to characterize the degree of correlation in some simple ways. The distinction between parametric and nonparametric (rank) methods is emphasized.

Section 14.8 introduces the concept of data smoothing, and discusses the particular case of Savitzky-Golay smoothing filters.

This chapter draws mathematically on the material on special functions that was presented in Chapter 6, especially §6.1–§6.4. You may wish, at this point, to review those sections.

CITED REFERENCES AND FURTHER READING:

Bevington, P.R., and Robinson, D.K. 1992, *Data Reduction and Error Analysis for the Physical Sciences*, 2nd ed. (New York: McGraw-Hill).

Stuart, A., and Ord, J.K. 1994, *Kendall's Advanced Theory of Statistics*, 6th ed. (London: Edward Arnold) [previous eds. published as Kendall, M., and Stuart, A., *The Advanced Theory of Statistics*].

Norusis, M.J. 1999, *SPSS 9.0 Guide to Data Analysis* (Englewood Cliffs, NJ: Prentice-Hall).

Dunn, O.J., and Clark, V.A. 1974, *Applied Statistics: Analysis of Variance and Regression* (New York: Wiley).

14.1 Moments of a Distribution: Mean, Variance, Skewness, and So Forth

When a set of values has a sufficiently strong central tendency, that is, a tendency to cluster around some particular value, then it may be useful to characterize the set by a few numbers that are related to its *moments*, the sums of integer powers of the values.

Best known is the *mean* of the values $x_0, \ldots, x_{N-1}$,

$$\overline{x} = \frac{1}{N} \sum_{j=0}^{N-1} x_j \tag{14.1.1}$$

which estimates the value around which central clustering occurs. Note the use of an overbar to denote the mean; angle brackets are an equally common notation, e.g., $\langle x \rangle$. You should be aware that the mean is not the only available estimator of this quantity, nor is it necessarily the best one. For values drawn from a probability distribution with very broad "tails," the mean may converge poorly, or not at all, as the number of sampled points is increased. Alternative estimators, the *median* and the *mode*, are mentioned at the end of this section.

Having characterized a distribution's central value, one conventionally next characterizes its "width" or "variability" around that value. Here again, more than one measure is available. Most common is the *variance*,

$$\text{Var}(x_0 \ldots x_{N-1}) = \frac{1}{N-1} \sum_{j=0}^{N-1} (x_j - \overline{x})^2 \qquad (14.1.2)$$

or its square root, the *standard deviation*,

$$\sigma(x_0 \ldots x_{N-1}) = \sqrt{\text{Var}(x_0 \ldots x_{N-1})} \qquad (14.1.3)$$

Equation (14.1.2) estimates the mean squared deviation of x from its mean value. There is a long story about why the denominator of (14.1.2) is $N - 1$ instead of N. If you have never heard that story, you may consult any good statistics text. Here we will be content to note that the $N - 1$ *should* be changed to N if you are ever in the situation of measuring the variance of a distribution whose mean $\overline{x}$ is known *a priori* rather than being estimated from the data. (We might also comment that if the difference between N and $N - 1$ ever matters to you, then you are probably up to no good anyway — e.g., trying to substantiate a questionable hypothesis with marginal data.)

As the mean depends on the first moment of the data, so do the variance and standard deviation depend on the second moment. It is not uncommon, in real life, to be dealing with a distribution whose second moment does not exist (i.e., is infinite). In this case, the variance or standard deviation is useless as a measure of the data's width around its central value: The values obtained from equations (14.1.2) or (14.1.3) will not converge with increased numbers of points, nor show any consistency from data set to data set drawn from the same distribution. This can occur even when the width of the peak looks, by eye, perfectly finite. A more robust estimator of the width is the *average deviation* or *mean absolute deviation*, defined by

$$\text{ADev}(x_0 \ldots x_{N-1}) = \frac{1}{N} \sum_{j=0}^{N-1} |x_j - \overline{x}| \qquad (14.1.4)$$

One often substitutes the sample median x_{med} for $\overline{x}$ in equation (14.1.4). For any fixed sample, the median in fact minimizes the mean absolute deviation.

Statisticians have historically sniffed at the use of (14.1.4) instead of (14.1.2), since the absolute value brackets in (14.1.4) are "nonanalytic" and make theorem-proving difficult. In recent years, however, the fashion has changed, and the subject of *robust estimation* (meaning, estimation for broad distributions with significant numbers of "outlier" points) has become a popular and important one. Higher moments, or statistics involving higher powers of the input data, are almost always less robust than lower moments or statistics that involve only linear sums or (the lowest moment of all) counting.

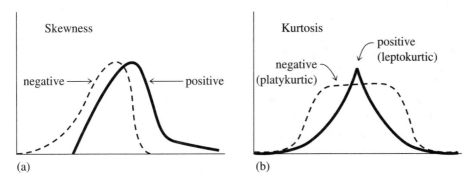

Figure 14.1.1.   Distributions whose third and fourth moments are significantly different from a normal (Gaussian) distribution. (a) Skewness or third moment. (b) Kurtosis or fourth moment.

That being the case, the *skewness* or *third moment*, and the *kurtosis* or *fourth moment* should be used with caution or, better yet, not at all.

The skewness characterizes the degree of asymmetry of a distribution around its mean. While the mean, standard deviation, and average deviation are *dimensional* quantities, that is, have the same units as the measured quantities x_j, the skewness is conventionally defined in such a way as to make it *nondimensional*. It is a pure number that characterizes only the shape of the distribution. The usual definition is

$$\text{Skew}(x_0 \ldots x_{N-1}) = \frac{1}{N} \sum_{j=0}^{N-1} \left[\frac{x_j - \overline{x}}{\sigma} \right]^3 \tag{14.1.5}$$

where $\sigma = \sigma(x_0 \ldots x_{N-1})$ is the distribution's standard deviation (14.1.3). A positive value of skewness signifies a distribution with an asymmetric tail extending out towards more positive x; a negative value signifies a distribution whose tail extends out towards more negative x (see Figure 14.1.1).

Of course, any set of N measured values is likely to give a nonzero value for (14.1.5), even if the underlying distribution is in fact symmetrical (has zero skewness). For (14.1.5) to be meaningful, we need to have some idea of *its* standard deviation as an estimator of the skewness of the underlying distribution. Unfortunately, that depends on the shape of the underlying distribution, and rather critically on its tails! For the idealized case of a normal (Gaussian) distribution, the standard deviation of (14.1.5) is approximately $\sqrt{15/N}$ when $\overline{x}$ is the true mean, and $\sqrt{6/N}$ when it is estimated by the sample mean, (14.1.1). In real life it is good practice to believe in skewnesses only when they are several or many times as large as this.

The kurtosis is also a nondimensional quantity. It measures the relative peakedness or flatness of a distribution. Relative to what? A normal distribution, what else! A distribution with positive kurtosis is termed *leptokurtic*; the outline of the Matterhorn is an example. A distribution with negative kurtosis is termed *platykurtic*; the outline of a loaf of bread is an example. (See Figure 14.1.1.) And, as you no doubt expect, an in-between distribution is termed *mesokurtic*.

The conventional definition of the kurtosis is

$$\text{Kurt}(x_0 \ldots x_{N-1}) = \left\{ \frac{1}{N} \sum_{j=0}^{N-1} \left[\frac{x_j - \overline{x}}{\sigma} \right]^4 \right\} - 3 \tag{14.1.6}$$

where the -3 term makes the value zero for a normal distribution.

The standard deviation of (14.1.6) as an estimator of the kurtosis of an underlying normal distribution is $\sqrt{96/N}$ when σ is the true standard deviation, and $\sqrt{24/N}$ when it is the sample estimate (14.1.3). However, the kurtosis depends on such a high moment that there are many real-life distributions for which the standard deviation of (14.1.6) as an estimator is effectively infinite.

Calculation of the quantities defined in this section is perfectly straightforward. Many textbooks use the binomial theorem to expand out the definitions into sums of various powers of the data, e.g., the familiar

$$\text{Var}(x_0 \dots x_{N-1}) = \frac{1}{N-1} \left[\left(\sum_{j=0}^{N-1} x_j^2 \right) - N\overline{x}^2 \right] \approx \overline{x^2} - \overline{x}^2 \qquad (14.1.7)$$

but this can magnify the roundoff error by a large factor and is generally unjustifiable in terms of computing speed. A clever way to minimize roundoff error, especially for large samples, is to use the *corrected two-pass algorithm* [1]: First calculate $\overline{x}$, then calculate $\text{Var}(x_0 \dots x_{N-1})$ by

$$\text{Var}(x_0 \dots x_{N-1}) = \frac{1}{N-1} \left\{ \sum_{j=0}^{N-1} (x_j - \overline{x})^2 - \frac{1}{N} \left[\sum_{j=0}^{N-1} (x_j - \overline{x}) \right]^2 \right\} \qquad (14.1.8)$$

The second sum would be zero if $\overline{x}$ were exact, but otherwise it does a good job of correcting the roundoff error in the first term.

```
#include <cmath>
#include "nr.h"
using namespace std;

void NR::moment(Vec_I_DP &data, DP &ave, DP &adev, DP &sdev, DP &var, DP &skew,
    DP &curt)
Given an array of data[0..n-1], this routine returns its mean ave, average deviation adev,
standard deviation sdev, variance var, skewness skew, and kurtosis curt.
{
    int j;
    DP ep=0.0,s,p;

    int n=data.size();
    if (n <= 1) nrerror("n must be at least 2 in moment");
    s=0.0;                           First pass to get the mean.
    for (j=0;j<n;j++) s += data[j];
    ave=s/n;
    adev=var=skew=curt=0.0;          Second pass to get the first (absolute), sec-
    for (j=0;j<n;j++) {                  ond, third, and fourth moments of the
        adev += fabs(s=data[j]-ave);     deviation from the mean.
        ep += s;
        var += (p=s*s);
        skew += (p *= s);
        curt += (p *= s);
    }
    adev /= n;
    var=(var-ep*ep/n)/(n-1);         Corrected two-pass formula.
    sdev=sqrt(var);                  Put the pieces together according to the con-
    if (var != 0.0) {                    ventional definitions.
```

```
      skew /= (n*var*sdev);
      curt=curt/(n*var*var)-3.0;
   } else nrerror("No skew/kurtosis when variance = 0 (in moment)");
}
```

Semi-Invariants

The mean and variance of independent random variables are additive: If x and y are drawn independently from two, possibly different, probability distributions, then

$$\overline{(x+y)} = \overline{x} + \overline{y} \qquad \mathrm{Var}(x+y) = \mathrm{Var}(x) + \mathrm{Var}(x) \tag{14.1.9}$$

Higher moments are not, in general, additive. However, certain combinations of them, called *semi-invariants*, are in fact additive. If the centered moments of a distribution are denoted M_k,

$$M_k \equiv \left\langle (x_i - \overline{x})^k \right\rangle \tag{14.1.10}$$

so that, e.g., $M_2 = \mathrm{Var}(x)$, then the first few semi-invariants, denoted I_k are given by

$$I_2 = M_2 \qquad I_3 = M_3 \qquad I_4 = M_4 - 3M_2^2$$
$$I_5 = M_5 - 10M_2M_3 \qquad I_6 = M_6 - 15M_2M_4 - 10M_3^2 + 30M_2^3 \tag{14.1.11}$$

Notice that the skewness and kurtosis, equations (14.1.5) and (14.1.6) are simple powers of the semi-invariants,

$$\mathrm{Skew}(x) = I_3/I_2^{3/2} \qquad \mathrm{Kurt}(x) = I_4/I_2^2 \tag{14.1.12}$$

A Gaussian distribution has all its semi-invariants higher than I_2 equal to zero. A Poisson distribution has all of its semi-invariants equal to its mean. For more details, see [2].

Median and Mode

The median of a probability distribution function $p(x)$ is the value x_{med} for which larger and smaller values of x are equally probable:

$$\int_{-\infty}^{x_{\mathrm{med}}} p(x)\, dx = \frac{1}{2} = \int_{x_{\mathrm{med}}}^{\infty} p(x)\, dx \tag{14.1.13}$$

The median of a distribution is estimated from a sample of values $x_0, \dots,$ x_{N-1} by finding that value x_i which has equal numbers of values above it and below it. Of course, this is not possible when N is even. In that case it is conventional to estimate the median as the mean of the unique *two* central values. If the values $x_j \; j = 0, \dots, N-1$ are sorted into ascending (or, for that matter, descending) order, then the formula for the median is

$$x_{\mathrm{med}} = \begin{cases} x_{(N-1)/2}, & N \text{ odd} \\ \frac{1}{2}(x_{(N/2)-1} + x_{N/2}), & N \text{ even} \end{cases} \tag{14.1.14}$$

If a distribution has a strong central tendency, so that most of its area is under a single peak, then the median is an estimator of the central value. It is a more robust estimator than the mean is: The median fails as an estimator only if the area in the tails is large, while the mean fails if the first moment of the tails is large; it is easy to construct examples where the first moment of the tails is large even though their area is negligible.

To find the median of a set of values, one can proceed by sorting the set and then applying (14.1.14). This is a process of order $N \log N$. You might rightly think that this is wasteful, since it yields much more information than just the median (e.g., the upper and lower quartile points, the deciles, etc.). In fact, we saw in §8.5 that the element $x_{(N-1)/2}$ can be located in of order N operations. Consult that section for routines.

The *mode* of a probability distribution function $p(x)$ is the value of x where it takes on a maximum value. The mode is useful primarily when there is a single, sharp maximum, in which case it estimates the central value. Occasionally, a distribution will be *bimodal*, with two relative maxima; then one may wish to know the two modes individually. Note that, in such cases, the mean and median are not very useful, since they will give only a "compromise" value between the two peaks.

CITED REFERENCES AND FURTHER READING:

Bevington, P.R., and Robinson, D.K. 1992, *Data Reduction and Error Analysis for the Physical Sciences*, 2nd ed. (New York: McGraw-Hill), Chapter 1.

Stuart, A., and Ord, J.K. 1994, *Kendall's Advanced Theory of Statistics*, 6th ed. (London: Edward Arnold) [previous eds. published as Kendall, M., and Stuart, A., *The Advanced Theory of Statistics*], vol. 1, §10.15

Norusis, M.J. 1999, *SPSS 9.0 Guide to Data Analysis* (Englewood Cliffs, NJ: Prentice-Hall).

Chan, T.F., Golub, G.H., and LeVeque, R.J. 1983, *American Statistician*, vol. 37, pp. 242–247. [1]

Cramér, H. 1946, *Mathematical Methods of Statistics* (Princeton: Princeton University Press), §15.10. [2]

14.2 Do Two Distributions Have the Same Means or Variances?

Not uncommonly we want to know whether two distributions have the same mean. For example, a first set of measured values may have been gathered before some event, a second set after it. We want to know whether the event, a "treatment" or a "change in a control parameter," made a difference.

Our first thought is to ask "how many standard deviations" one sample mean is from the other. That number may in fact be a useful thing to know. It does relate to the strength or "importance" of a difference of means *if that difference is genuine*. However, by itself, it says nothing about whether the difference *is* genuine, that is, statistically significant. A difference of means can be very small compared to the standard deviation, and yet very significant, if the number of data points is large. Conversely, a difference may be moderately large but not significant, if the data are sparse. We will be meeting these distinct concepts of *strength* and *significance* several times in the next few sections.

A quantity that measures the significance of a difference of means is not the number of standard deviations that they are apart, but the number of so-called *standard errors* that they are apart. The standard error of a set of values measures the accuracy with which the sample mean estimates the population (or "true") mean. Typically the standard error is equal to the sample's standard deviation divided by the square root of the number of points in the sample.

Student's t-test for Significantly Different Means

Applying the concept of standard error, the conventional statistic for measuring the significance of a difference of means is termed *Student's t*. When the two distributions are thought to have the same variance, but possibly different means, then Student's t is computed as follows: First, estimate the standard error of the difference of the means, s_D, from the "pooled variance" by the formula

$$s_D = \sqrt{\frac{\sum_{i \in A}(x_i - \overline{x_A})^2 + \sum_{i \in B}(x_i - \overline{x_B})^2}{N_A + N_B - 2} \left(\frac{1}{N_A} + \frac{1}{N_B} \right)} \qquad (14.2.1)$$

where each sum is over the points in one sample, the first or second, each mean likewise refers to one sample or the other, and N_A and N_B are the numbers of points in the first and second samples, respectively. Second, compute t by

$$t = \frac{\overline{x_A} - \overline{x_B}}{s_D} \qquad (14.2.2)$$

Third, evaluate the significance of this value of t for Student's distribution with $N_A + N_B - 2$ degrees of freedom, by equations (6.4.7) and (6.4.9), and by the routine betai (incomplete beta function) of §6.4.

The significance is a number between zero and one, and is the probability that $|t|$ could be this large or larger just by chance, for distributions with equal means. Therefore, a small numerical value of the significance (0.05 or 0.01) means that the observed difference is "very significant." The function $A(t|\nu)$ in equation (6.4.7) is one minus the significance.

As a routine, we have

```
#include <cmath>
#include "nr.h"
using namespace std;

void NR::ttest(Vec_I_DP &data1, Vec_I_DP &data2, DP &t, DP &prob)
Given the arrays data1[0..n1-1] and data2[0..n2-1], returns Student's t as t, and its
significance as prob, small values of prob indicating that the arrays have significantly different
means. The data arrays are assumed to be drawn from populations with the same true variance.
{
    DP var1,var2,svar,df,ave1,ave2;

    int n1=data1.size();
    int n2=data2.size();
    avevar(data1,ave1,var1);
    avevar(data2,ave2,var2);
    df=n1+n2-2;                            Degrees of freedom.
    svar=((n1-1)*var1+(n2-1)*var2)/df;     Pooled variance.
    t=(ave1-ave2)/sqrt(svar*(1.0/n1+1.0/n2));
    prob=betai(0.5*df,0.5,df/(df+t*t));    See equation (6.4.9).
}
```

which makes use of the following routine for computing the mean and variance of a set of numbers,

```
#include "nr.h"

void NR::avevar(Vec_I_DP &data, DP &ave, DP &var)
Given array data[0..n-1], returns its mean as ave and its variance as var.
{
    DP s,ep;
    int j;

    int n=data.size();
    ave=0.0;
    for (j=0;j<n;j++) ave += data[j];
    ave /= n;
    var=ep=0.0;
    for (j=0;j<n;j++) {
        s=data[j]-ave;
        ep += s;
        var += s*s;
    }
    var=(var-ep*ep/n)/(n-1);          Corrected two-pass formula (14.1.8).
}
```

The next case to consider is where the two distributions have significantly different variances, but we nevertheless want to know if their means are the same or different. (A treatment for baldness has caused some patients to *lose* all their hair and turned others into werewolves, but we want to know if it helps cure baldness *on the average*!) Be suspicious of the unequal-variance t-test: If two distributions have very different variances, then they may also be substantially different in shape; in that case, the difference of the means may not be a particularly useful thing to know.

To find out whether the two data sets have variances that are significantly different, you use the *F-test*, described later on in this section.

The relevant statistic for the unequal variance t-test is

$$t = \frac{\overline{x_A} - \overline{x_B}}{[\mathrm{Var}(x_A)/N_A + \mathrm{Var}(x_B)/N_B]^{1/2}} \qquad (14.2.3)$$

This statistic is distributed *approximately* as Student's t with a number of degrees of freedom equal to

$$\frac{\left[\dfrac{\mathrm{Var}(x_A)}{N_A} + \dfrac{\mathrm{Var}(x_B)}{N_B} \right]^2}{\dfrac{[\mathrm{Var}(x_A)/N_A]^2}{N_A - 1} + \dfrac{[\mathrm{Var}(x_B)/N_B]^2}{N_B - 1}} \qquad (14.2.4)$$

Expression (14.2.4) is in general not an integer, but equation (6.4.7) doesn't care.

The routine is

```
#include <cmath>
#include "nr.h"
using namespace std;

void NR::tutest(Vec_I_DP &data1, Vec_I_DP &data2, DP &t, DP &prob)
```

Given the arrays `data1[0..n1-1]` and `data2[0..n2-1]`, this routine returns Student's *t* as t, and its significance as `prob`, small values of `prob` indicating that the arrays have significantly different means. The data arrays are allowed to be drawn from populations with unequal variances.

```
{
    DP var1,var2,df,ave1,ave2;

    int n1=data1.size();
    int n2=data2.size();
    avevar(data1,ave1,var1);
    avevar(data2,ave2,var2);
    t=(ave1-ave2)/sqrt(var1/n1+var2/n2);
    df=SQR(var1/n1+var2/n2)/(SQR(var1/n1)/(n1-1)+SQR(var2/n2)/(n2-1));
    prob=betai(0.5*df,0.5,df/(df+SQR(t)));
}
```

Our final example of a Student's *t* test is the case of *paired samples*. Here we imagine that much of the variance in *both* samples is due to effects that are point-by-point identical in the two samples. For example, we might have two job candidates who have each been rated by the same ten members of a hiring committee. We want to know if the means of the ten scores differ significantly. We first try `ttest` above, and obtain a value of `prob` that is not especially significant (e.g., > 0.05). But perhaps the significance is being washed out by the tendency of some committee members always to give high scores, others always to give low scores, which increases the apparent variance and thus decreases the significance of any difference in the means. We thus try the paired-sample formulas,

$$\text{Cov}(x_A, x_B) \equiv \frac{1}{N-1} \sum_{i=0}^{N-1} (x_{Ai} - \overline{x_A})(x_{Bi} - \overline{x_B}) \tag{14.2.5}$$

$$s_D = \left[\frac{\text{Var}(x_A) + \text{Var}(x_B) - 2\text{Cov}(x_A, x_B)}{N} \right]^{1/2} \tag{14.2.6}$$

$$t = \frac{\overline{x_A} - \overline{x_B}}{s_D} \tag{14.2.7}$$

where N is the number in each sample (number of pairs). Notice that it is important that a particular value of i label the corresponding points in each sample, that is, the ones that are paired. The significance of the t statistic in (14.2.7) is evaluated for $N - 1$ degrees of freedom.

The routine is

```
#include <cmath>
#include "nr.h"
using namespace std;

void NR::tptest(Vec_I_DP &data1, Vec_I_DP &data2, DP &t, DP &prob)
```
Given the paired arrays `data1[0..n-1]` and `data2[0..n-1]`, this routine returns Student's *t* for paired data as t, and its significance as `prob`, small values of `prob` indicating a significant difference of means.
```
{
    int j;
    DP var1,var2,ave1,ave2,sd,df,cov=0.0;

    int n=data1.size();
    avevar(data1,ave1,var1);
```

```
      avevar(data2,ave2,var2);
      for (j=0;j<n;j++)
            cov += (data1[j]-ave1)*(data2[j]-ave2);
      cov /= df=n-1;
      sd=sqrt((var1+var2-2.0*cov)/n);
      t=(ave1-ave2)/sd;
      prob=betai(0.5*df,0.5,df/(df+t*t));
}
```

F-Test for Significantly Different Variances

The *F-test* tests the hypothesis that two samples have different variances by
trying to reject the null hypothesis that their variances are actually consistent. The
statistic F is the ratio of one variance to the other, so values either $\gg 1$ or $\ll 1$
will indicate very significant differences. The distribution of F in the null case is
given in equation (6.4.11), which is evaluated using the routine betai. In the most
common case, we are willing to disprove the null hypothesis (of equal variances) by
either very large or very small values of F, so the correct significance is *two-tailed*,
the sum of two incomplete beta functions. It turns out, by equation (6.4.3), that the
two tails are always equal; we need compute only one, and double it. Occasionally,
when the null hypothesis is strongly viable, the identity of the two tails can become
confused, giving an indicated probability greater than one. Changing the probability
to two minus itself correctly exchanges the tails. These considerations and equation
(6.4.3) give the routine

```
#include "nr.h"

void NR::ftest(Vec_I_DP &data1, Vec_I_DP &data2, DP &f, DP &prob)
Given the arrays data1[0..n1-1] and data2[0..n2-1], this routine returns the value of f,
and its significance as prob. Small values of prob indicate that the two arrays have significantly
different variances.
{
    DP var1,var2,ave1,ave2,df1,df2;

    int n1=data1.size();
    int n2=data2.size();
    avevar(data1,ave1,var1);
    avevar(data2,ave2,var2);
    if (var1 > var2) {                     Make F the ratio of the larger variance to the smaller
        f=var1/var2;                       one.
        df1=n1-1;
        df2=n2-1;
    } else {
        f=var2/var1;
        df1=n2-1;
        df2=n1-1;
    }
    prob = 2.0*betai(0.5*df2,0.5*df1,df2/(df2+df1*f));
    if (prob > 1.0) prob=2.0-prob;
}
```

CITED REFERENCES AND FURTHER READING:

von Mises, R. 1964, *Mathematical Theory of Probability and Statistics* (New York: Academic
      Press), Chapter IX(B).

Norusis, M.J. 1999, *SPSS 9.0 Guide to Data Analysis* (Englewood Cliffs, NJ: Prentice-Hall).

14.3 Are Two Distributions Different?

Given two sets of data, we can generalize the questions asked in the previous section and ask the single question: Are the two sets drawn from the same distribution function, or from different distribution functions? Equivalently, in proper statistical language, "Can we disprove, to a certain required level of significance, the null hypothesis that two data sets are drawn from the same population distribution function?" Disproving the null hypothesis in effect proves that the data sets are from different distributions. Failing to disprove the null hypothesis, on the other hand, only shows that the data sets can be *consistent* with a single distribution function. One can never *prove* that two data sets come from a single distribution, since (e.g.) no practical amount of data can distinguish between two distributions which differ only by one part in 10^{10}.

Proving that two distributions are different, or showing that they are consistent, is a task that comes up all the time in many areas of research: Are the visible stars distributed uniformly in the sky? (That is, is the distribution of stars as a function of declination — position in the sky — the same as the distribution of sky area as a function of declination?) Are educational patterns the same in Brooklyn as in the Bronx? (That is, are the distributions of people as a function of last-grade-attended the same?) Do two brands of fluorescent lights have the same distribution of burn-out times? Is the incidence of chicken pox the same for first-born, second-born, third-born children, etc.?

These four examples illustrate the four combinations arising from two different dichotomies: (1) The data are either continuous or binned. (2) Either we wish to compare one data set to a known distribution, or we wish to compare two equally unknown data sets. The data sets on fluorescent lights and on stars are continuous, since we can be given lists of individual burnout times or of stellar positions. The data sets on chicken pox and educational level are binned, since we are given tables of numbers of events in discrete categories: first-born, second-born, etc.; or 6th Grade, 7th Grade, etc. Stars and chicken pox, on the other hand, share the property that the null hypothesis is a known distribution (distribution of area in the sky, or incidence of chicken pox in the general population). Fluorescent lights and educational level involve the comparison of two equally unknown data sets (the two brands, or Brooklyn and the Bronx).

One can always turn continuous data into binned data, by grouping the events into specified ranges of the continuous variable(s): declinations between 0 and 10 degrees, 10 and 20, 20 and 30, etc. Binning involves a loss of information, however. Also, there is often considerable arbitrariness as to how the bins should be chosen. Along with many other investigators, we prefer to avoid unnecessary binning of data.

The accepted test for differences between binned distributions is the *chi-square test*. For continuous data as a function of a single variable, the most generally accepted test is the *Kolmogorov-Smirnov test*. We consider each in turn.

Chi-Square Test

Suppose that N_i is the number of events observed in the ith bin, and that n_i is the number expected according to some known distribution. Note that the N_i's are integers, while the n_i's may not be. Then the chi-square statistic is

$$\chi^2 = \sum_i \frac{(N_i - n_i)^2}{n_i} \qquad (14.3.1)$$

where the sum is over all bins. A large value of χ^2 indicates that the null hypothesis (that the N_i's are drawn from the population represented by the n_i's) is rather unlikely.

Any term j in (14.3.1) with $0 = n_j = N_j$ should be omitted from the sum. A term with $n_j = 0$, $N_j \neq 0$ gives an infinite χ^2, as it should, since in this case the N_i's cannot possibly be drawn from the n_i's!

The *chi-square probability function* $Q(\chi^2|\nu)$ is an incomplete gamma function, and was already discussed in §6.2 (see equation 6.2.18). Strictly speaking $Q(\chi^2|\nu)$ is the probability that the sum of the squares of ν random *normal* variables of unit variance (and zero mean) will be greater than χ^2. The terms in the sum (14.3.1) are not individually normal. However, if either the number of bins is large ($\gg 1$), or the number of events in each bin is large ($\gg 1$), then the chi-square probability function is a good approximation to the distribution of (14.3.1) in the case of the null hypothesis. Its use to estimate the significance of the chi-square test is standard.

The appropriate value of ν, the number of degrees of freedom, bears some additional discussion. If the data are collected with the model n_i's fixed — that is, not later renormalized to fit the total observed number of events ΣN_i — then ν equals the number of bins N_B. (Note that this is *not* the total number of *events*!) Much more commonly, the n_i's are normalized after the fact so that their sum equals the sum of the N_i's. In this case the correct value for ν is $N_B - 1$, and the model is said to have one constraint (knstrn=1 in the program below). If the model that gives the n_i's has additional free parameters that were adjusted after the fact to agree with the data, then each of these additional "fitted" parameters decreases ν (and increases knstrn) by one additional unit.

We have, then, the following program:

```
#include "nr.h"

void NR::chsone(Vec_I_DP &bins, Vec_I_DP &ebins, const int knstrn, DP &df,
    DP &chsq, DP &prob)
```
Given the array bins[0..nbins-1] containing the observed numbers of events, and an array ebins[0..nbins-1] containing the expected numbers of events, and given the number of constraints knstrn (normally one), this routine returns (trivially) the number of degrees of freedom df, and (nontrivially) the chi-square chsq and the significance prob. A small value of prob indicates a significant difference between the distributions bins and ebins. Note that bins and ebins are both double arrays, although bins will normally contain integer values.
```
{
    int j;
    DP temp;

    int nbins=bins.size();
    df=nbins-knstrn;
    chsq=0.0;
    for (j=0;j<nbins;j++) {
        if (ebins[j] <= 0.0) nrerror("Bad expected number in chsone");
        temp=bins[j]-ebins[j];
        chsq += temp*temp/ebins[j];
    }
    prob=gammq(0.5*df,0.5*chsq);        Chi-square probability function. See §6.2.
}
```

Next we consider the case of comparing *two* binned data sets. Let R_i be the number of events in bin i for the first data set, S_i the number of events in the same bin i for the second data set. Then the chi-square statistic is

$$\chi^2 = \sum_i \frac{(R_i - S_i)^2}{R_i + S_i} \tag{14.3.2}$$

Comparing (14.3.2) to (14.3.1), you should note that the denominator of (14.3.2) is *not* just the average of R_i and S_i (which would be an estimator of n_i in 14.3.1). Rather, it is twice the average, the sum. The reason is that each term in a chi-square sum is supposed to approximate the square of a normally distributed quantity with unit variance. The variance of the difference of two normal quantities is the sum of their individual variances, not the average.

If the data were collected in such a way that the sum of the R_i's is necessarily equal to the sum of S_i's, then the number of degrees of freedom is equal to one less than the number of bins, $N_B - 1$ (that is, knstrn $= 1$), the usual case. If this requirement were absent, then the number of degrees of freedom would be N_B. Example: A birdwatcher wants to know whether the distribution of sighted birds as a function of species is the same this year as last. Each bin corresponds to one species. If the birdwatcher takes his data to be the first 1000 birds that he saw in each year, then the number of degrees of freedom is $N_B - 1$. If he takes his data to be all the birds he saw on a random sample of days, the same days in each year, then the number of degrees of freedom is N_B (knstrn $= 0$). In this latter case, note that he is also testing whether the birds were more numerous overall in one year or the other: That is the extra degree of freedom. Of course, any additional constraints on the data set lower the number of degrees of freedom (i.e., increase knstrn to *more positive* values) in accordance with their number.

The program is

```
#include "nr.h"

void NR::chstwo(Vec_I_DP &bins1, Vec_I_DP &bins2, const int knstrn, DP &df,
    DP &chsq, DP &prob)
```
Given the arrays bins1[0..nbins-1] and bins2[0..nbins-1], containing two sets of binned data, and given the number of constraints knstrn (normally 1 or 0), this routine returns the number of degrees of freedom df, the chi-square chsq, and the significance prob. A small value of prob indicates a significant difference between the distributions bins1 and bins2. Note that bins1 and bins2 are both double arrays, although they will normally contain integer values.
```
{
    int j;
    DP temp;

    int nbins=bins1.size();
    df=nbins-knstrn;
    chsq=0.0;
    for (j=0;j<nbins;j++)
        if (bins1[j] == 0.0 && bins2[j] == 0.0)
            --df;                                No data means one less degree of free-
        else {                                      dom.
            temp=bins1[j]-bins2[j];
            chsq += temp*temp/(bins1[j]+bins2[j]);
        }
    prob=gammq(0.5*df,0.5*chsq);                 Chi-square probability function. See §6.2.
}
```

Equation (14.3.2) and the routine `chstwo` both apply to the case where the total number of data points is the same in the two binned sets. For unequal numbers of data points, the formula analogous to (14.3.2) is

$$\chi^2 = \sum_i \frac{(\sqrt{S/R}\,R_i - \sqrt{R/S}\,S_i)^2}{R_i + S_i} \tag{14.3.3}$$

where

$$R \equiv \sum_i R_i \qquad S \equiv \sum_i S_i \tag{14.3.4}$$

are the respective numbers of data points. It is straightforward to make the corresponding change in `chstwo`.

Kolmogorov-Smirnov Test

The Kolmogorov-Smirnov (or *K–S*) test is applicable to unbinned distributions that are functions of a single independent variable, that is, to data sets where each data point can be associated with a single number (lifetime of each lightbulb when it burns out, or declination of each star). In such cases, the list of data points can be easily converted to an unbiased estimator $S_N(x)$ of the *cumulative* distribution function of the probability distribution from which it was drawn: If the N events are located at values x_i, $i = 0, \ldots, N-1$, then $S_N(x)$ is the function giving the fraction of data points to the left of a given value x. This function is obviously constant between consecutive (i.e., sorted into ascending order) x_i's, and jumps by the same constant $1/N$ at each x_i. (See Figure 14.3.1.)

Different distribution functions, or sets of data, give different cumulative distribution function estimates by the above procedure. However, all cumulative distribution functions agree at the smallest allowable value of x (where they are zero), and at the largest allowable value of x (where they are unity). (The smallest and largest values might of course be $\pm\infty$.) So it is the behavior between the largest and smallest values that distinguishes distributions.

One can think of any number of statistics to measure the overall difference between two cumulative distribution functions: the absolute value of the area between them, for example. Or their integrated mean square difference. The Kolmogorov-Smirnov D is a particularly simple measure: It is defined as the *maximum value* of the absolute difference between two cumulative distribution functions. Thus, for comparing one data set's $S_N(x)$ to a known cumulative distribution function $P(x)$, the K–S statistic is

$$D = \max_{-\infty < x < \infty} |S_N(x) - P(x)| \tag{14.3.5}$$

while for comparing two different cumulative distribution functions $S_{N_1}(x)$ and $S_{N_2}(x)$, the K–S statistic is

$$D = \max_{-\infty < x < \infty} |S_{N_1}(x) - S_{N_2}(x)| \tag{14.3.6}$$

What makes the K–S statistic useful is that *its* distribution in the case of the null hypothesis (data sets drawn from the same distribution) can be calculated, at least to

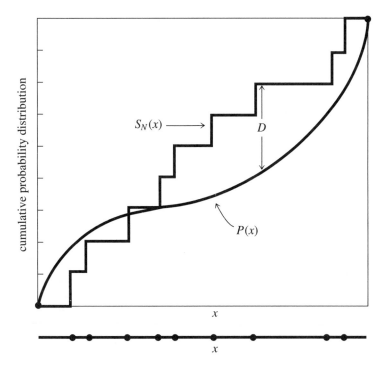

Figure 14.3.1.   Kolmogorov-Smirnov statistic D. A measured distribution of values in x (shown as N dots on the lower abscissa) is to be compared with a theoretical distribution whose cumulative probability distribution is plotted as $P(x)$. A step-function cumulative probability distribution $S_N(x)$ is constructed, one that rises an equal amount at each measured point. D is the greatest distance between the two cumulative distributions.

useful approximation, thus giving the significance of any observed nonzero value of D. A central feature of the K–S test is that it is invariant under reparametrization of x; in other words, you can locally slide or stretch the x axis in Figure 14.3.1, and the maximum distance D remains unchanged. For example, you will get the same significance using x as using $\log x$.

   The function that enters into the calculation of the significance can be written as the following sum:

$$Q_{KS}(\lambda) = 2 \sum_{j=1}^{\infty} (-1)^{j-1} \, e^{-2j^2 \lambda^2} \qquad (14.3.7)$$

which is a monotonic function with the limiting values

$$Q_{KS}(0) = 1 \qquad Q_{KS}(\infty) = 0 \qquad (14.3.8)$$

In terms of this function, the significance level of an observed value of D (as a disproof of the null hypothesis that the distributions are the same) is given approximately [1] by the formula

$$\text{Probability } (D > \text{observed }) = Q_{KS}\left(\left[\sqrt{N_e} + 0.12 + 0.11/\sqrt{N_e}\right] D\right) \qquad (14.3.9)$$

where N_e is the effective number of data points, $N_e = N$ for the case (14.3.5) of one distribution, and

$$N_e = \frac{N_1 N_2}{N_1 + N_2} \qquad (14.3.10)$$

for the case (14.3.6) of two distributions, where N_1 is the number of data points in the first distribution, N_2 the number in the second.

The nature of the approximation involved in (14.3.9) is that it becomes asymptotically accurate as the N_e becomes large, but is already quite good for $N_e \geq 4$, as small a number as one might ever actually use. (See [1].)

So, we have the following routines for the cases of one and two distributions:

```
#include <cmath>
#include "nr.h"
using namespace std;

void NR::ksone(Vec_IO_DP &data, DP func(const DP), DP &d, DP &prob)
```
Given an array data[0..n-1], and given a user-supplied function of a single variable func which is a cumulative distribution function ranging from 0 (for smallest values of its argument) to 1 (for largest values of its argument), this routine returns the K–S statistic d, and the significance level prob. Small values of prob show that the cumulative distribution function of data is significantly different from func. The array data is modified by being sorted into ascending order.
```
{
    int j;
    DP dt,en,ff,fn,fo=0.0;

    int n=data.size();
    sort(data);                                 If the data are already sorted into as-
    en=n;                                         cending order, then this call can be
    d=0.0;                                        omitted.
    for (j=0;j<n;j++) {                         Loop over the sorted data points.
        fn=(j+1)/en;                             Data's c.d.f. after this step.
        ff=func(data[j]);                        Compare to the user-supplied function.
        dt=MAX(fabs(fo-ff),fabs(fn-ff));         Maximum distance.
        if (dt > d) d=dt;
        fo=fn;
    }
    en=sqrt(en);
    prob=probks((en+0.12+0.11/en)*d);           Compute significance.
}
```

```
#include <cmath>
#include "nr.h"
using namespace std;

void NR::kstwo(Vec_IO_DP &data1, Vec_IO_DP &data2, DP &d, DP &prob)
```
Given an array data1[0..n1-1], and an array data2[0..n2-1], this routine returns the K–S statistic d, and the significance level prob for the null hypothesis that the data sets are drawn from the same distribution. Small values of prob show that the cumulative distribution function of data1 is significantly different from that of data2. The arrays data1 and data2 are modified by being sorted into ascending order.
```
{
    int j1=0,j2=0;
    DP d1,d2,dt,en1,en2,en,fn1=0.0,fn2=0.0;

    int n1=data1.size();
    int n2=data2.size();
```

```
        sort(data1);
        sort(data2);
        en1=n1;
        en2=n2;
        d=0.0;
        while (j1 < n1 && j2 < n2) {                            If we are not done...
            if ((d1=data1[j1]) <= (d2=data2[j2])) fn1=j1++/en1;  Next step is in data1.
            if (d2 <= d1) fn2=j2++/en2;                          Next step is in data2.
            if ((dt=fabs(fn2-fn1)) > d) d=dt;
        }
        en=sqrt(en1*en2/(en1+en2));
        prob=probks((en+0.12+0.11/en)*d);                      Compute significance.
    }
```

Both of the above routines use the following routine for calculating the function Q_{KS}:

```
#include <cmath>
#include "nr.h"
using namespace std;

DP NR::probks(const DP alam)
Kolmogorov-Smirnov probability function.
{
    const DP EPS1=1.0e-6,EPS2=1.0e-16;
    int j;
    DP a2,fac=2.0,sum=0.0,term,termbf=0.0;

    a2 = -2.0*alam*alam;
    for (j=1;j<=100;j++) {
        term=fac*exp(a2*j*j);
        sum += term;
        if (fabs(term) <= EPS1*termbf || fabs(term) <= EPS2*sum) return sum;
        fac = -fac;                      Alternating signs in sum.
        termbf=fabs(term);
    }
    return 1.0;                          Get here only by failing to converge.
}
```

Variants on the K–S Test

The sensitivity of the K–S test to deviations from a cumulative distribution function $P(x)$ is not independent of x. In fact, the K–S test tends to be most sensitive around the median value, where $P(x) = 0.5$, and less sensitive at the extreme ends of the distribution, where $P(x)$ is near 0 or 1. The reason is that the difference $|S_N(x) - P(x)|$ does not, in the null hypothesis, have a probability distribution that is independent of x. Rather, its variance is proportional to $P(x)[1 - P(x)]$, which is largest at $P = 0.5$. Since the K–S statistic (14.3.5) is the maximum difference over all x of two cumulative distribution functions, a deviation that might be statistically significant at *its own* value of x gets compared to the expected chance deviation at $P = 0.5$, and is thus discounted. A result is that, while the K–S test is good at finding *shifts* in a probability distribution, especially changes in the median value, it is not always so good at finding *spreads*, which more affect the tails of the probability distribution, and which may leave the median unchanged.

One way of increasing the power of the K–S statistic out on the tails is to replace D (equation 14.3.5) by a so-called *stabilized* or *weighted* statistic [2-4], for example the *Anderson-Darling statistic*,

$$D^* = \max_{-\infty < x < \infty} \frac{|S_N(x) - P(x)|}{\sqrt{P(x)[1 - P(x)]}} \qquad (14.3.11)$$

Unfortunately, there is no simple formula analogous to equations (14.3.7) and (14.3.9) for this statistic, although Noé [5] gives a computational method using a recursion relation and provides a graph of numerical results. There are many other possible similar statistics, for example

$$D^{**} = \int_{P=0}^{1} \frac{[S_N(x) - P(x)]^2}{P(x)[1 - P(x)]} dP(x) \qquad (14.3.12)$$

which is also discussed by Anderson and Darling (see [3]).

Another approach, which we prefer as simpler and more direct, is due to Kuiper [6,7]. We already mentioned that the standard K–S test is invariant under reparametrizations of the variable x. An even more general symmetry, which guarantees equal sensitivities at all values of x, is to wrap the x axis around into a circle (identifying the points at $\pm\infty$), and to look for a statistic that is now invariant under all shifts and parametrizations on the circle. This allows, for example, a probability distribution to be "cut" at some central value of x, and the left and right halves to be interchanged, without altering the statistic or its significance.

Kuiper's statistic, defined as

$$V = D_+ + D_- = \max_{-\infty < x < \infty} [S_N(x) - P(x)] + \max_{-\infty < x < \infty} [P(x) - S_N(x)] \quad (14.3.13)$$

is the sum of the maximum distance of $S_N(x)$ *above and below* $P(x)$. You should be able to convince yourself that this statistic has the desired invariance on the circle: Sketch the indefinite integral of two probability distributions defined on the circle as a function of angle around the circle, as the angle goes through several times 360°. If you change the starting point of the integration, D_+ and D_- change individually, but their sum is constant.

Furthermore, there is a simple formula for the asymptotic distribution of the statistic V, directly analogous to equations (14.3.7)–(14.3.10). Let

$$Q_{KP}(\lambda) = 2\sum_{j=1}^{\infty}(4j^2\lambda^2 - 1)e^{-2j^2\lambda^2} \qquad (14.3.14)$$

which is monotonic and satisfies

$$Q_{KP}(0) = 1 \qquad Q_{KP}(\infty) = 0 \qquad (14.3.15)$$

In terms of this function the significance level is [1]

$$\text{Probability } (V > \text{observed }) = Q_{KP}\left(\left[\sqrt{N_e} + 0.155 + 0.24/\sqrt{N_e}\right]V\right) \qquad (14.3.16)$$

Here N_e is N in the one-sample case, or is given by equation (14.3.10) in the case of two samples.

Of course, Kuiper's test is ideal for any problem originally defined on a circle, for example, to test whether the distribution in longitude of something agrees with some theory, or whether two somethings have different distributions in longitude. (See also [8].)

We will leave to you the coding of routines analogous to `ksone`, `kstwo`, and `probks`, above. (For $\lambda < 0.4$, don't try to do the sum 14.3.14. Its value is 1, to 7 figures, but the series can require many terms to converge, and loses accuracy to roundoff.)

Two final cautionary notes: First, we should mention that all varieties of K–S test lack the ability to discriminate some kinds of distributions. A simple example is a probability distribution with a narrow "notch" within which the probability falls to zero. Such a distribution is of course ruled out by the existence of even one data point within the notch, but, because of its cumulative nature, a K–S test would require many data points in the notch before signaling a discrepancy.

Second, we should note that, if you estimate any parameters from a data set (e.g., a mean and variance), then the distribution of the K–S statistic D for a cumulative distribution function $P(x)$ that *uses the estimated parameters* is no longer given by equation (14.3.9). In general, you will have to determine the new distribution yourself, e.g., by Monte Carlo methods.

CITED REFERENCES AND FURTHER READING:

von Mises, R. 1964, *Mathematical Theory of Probability and Statistics* (New York: Academic Press), Chapters IX(C) and IX(E).

Stephens, M.A. 1970, *Journal of the Royal Statistical Society*, ser. B, vol. 32, pp. 115–122. [1]

Anderson, T.W., and Darling, D.A. 1952, *Annals of Mathematical Statistics*, vol. 23, pp. 193–212. [2]

Darling, D.A. 1957, *Annals of Mathematical Statistics*, vol. 28, pp. 823–838. [3]

Michael, J.R. 1983, *Biometrika*, vol. 70, no. 1, pp. 11–17. [4]

Noé, M. 1972, *Annals of Mathematical Statistics*, vol. 43, pp. 58–64. [5]

Kuiper, N.H. 1962, *Proceedings of the Koninklijke Nederlandse Akademie van Wetenschappen*, ser. A., vol. 63, pp. 38–47. [6]

Stephens, M.A. 1965, *Biometrika*, vol. 52, pp. 309–321. [7]

Fisher, N.I., Lewis, T., and Embleton, B.J.J. 1987, *Statistical Analysis of Spherical Data* (New York: Cambridge University Press). [8]

14.4 Contingency Table Analysis of Two Distributions

In this section, and the next two sections, we deal with *measures of association* for two distributions. The situation is this: Each data point has two or more different quantities associated with it, and we want to know whether knowledge of one quantity gives us any demonstrable advantage in predicting the value of another quantity. In many cases, one variable will be an "independent" or "control" variable, and another will be a "dependent" or "measured" variable. Then, we want to know if the latter variable *is* in fact dependent on or *associated* with the former variable. If it is, we want to have some quantitative measure of the strength of the association. One often hears this loosely stated as the question of whether two variables are *correlated* or *uncorrelated*, but we will reserve those terms for a particular kind of association (linear, or at least monotonic), as discussed in §14.5 and §14.6.

Notice that, as in previous sections, the different concepts of significance and strength appear: The association between two distributions may be very significant even if that association is weak — if the quantity of data is large enough.

It is useful to distinguish among some different kinds of variables, with different categories forming a loose hierarchy.

- A variable is called *nominal* if its values are the members of some unordered set. For example, "state of residence" is a nominal variable that (in the U.S.) takes on one of 50 values; in astrophysics, "type of galaxy" is a nominal variable with the three values "spiral," "elliptical," and "irregular."

- A variable is termed *ordinal* if its values are the members of a discrete, but ordered, set. Examples are: grade in school, planetary order from the Sun (Mercury = 1, Venus = 2, ...), number of offspring. There need not be any concept of "equal metric distance" between the values of an ordinal variable, only that they be intrinsically ordered.

- We will call a variable *continuous* if its values are real numbers, as are times, distances, temperatures, etc. (Social scientists sometimes distinguish between *interval* and *ratio* continuous variables, but we do not find that distinction very compelling.)

A continuous variable can always be made into an ordinal one by binning it into ranges. If we choose to ignore the ordering of the bins, then we can turn it into

	0. red	1. green	. . .	
0. male	# of red males N_{00}	# of green males N_{01}	. . .	# of males $N_{0}.$
1. female	# of red females N_{10}	# of green females N_{11}	. . .	# of females $N_{1}.$
⋮	⋮	⋮	. . .	⋮
	# of red $N_{.0}$	# of green $N_{.1}$	. . .	total # N

Figure 14.4.1.   Example of a contingency table for two nominal variables, here sex and color. The row and column marginals (totals) are shown. The variables are "nominal," i.e., the order in which their values are listed is arbitrary and does not affect the result of the contingency table analysis. If the ordering of values has some intrinsic meaning, then the variables are "ordinal" or "continuous," and correlation techniques (§14.5–§14.6) can be utilized.

a nominal variable. Nominal variables constitute the lowest type of the hierarchy, and therefore the most general. For example, a set of *several* continuous or ordinal variables can be turned, if crudely, into a single nominal variable, by coarsely binning each variable and then taking each distinct combination of bin assignments as a single nominal value. When multidimensional data are sparse, this is often the only sensible way to proceed.

The remainder of this section will deal with measures of association between *nominal* variables. For any pair of nominal variables, the data can be displayed as a *contingency table*, a table whose rows are labeled by the values of one nominal variable, whose columns are labeled by the values of the other nominal variable, and whose entries are nonnegative integers giving the number of observed events for each combination of row and column (see Figure 14.4.1). The analysis of association between nominal variables is thus called *contingency table analysis* or *crosstabulation analysis*.

We will introduce two different approaches. The first approach, based on the chi-square statistic, does a good job of characterizing the significance of association, but is only so-so as a measure of the strength (principally because its numerical values have no very direct interpretations). The second approach, based on the information-theoretic concept of *entropy*, says nothing at all about the significance of association (use chi-square for that!), but is capable of very elegantly characterizing the strength of an association already known to be significant.

Measures of Association Based on Chi-Square

Some notation first: Let N_{ij} denote the number of events that occur with the first variable x taking on its ith value, and the second variable y taking on its jth value. Let N denote the total number of events, the sum of all the N_{ij}'s. Let $N_{i\cdot}$ denote the number of events for which the first variable x takes on its ith value regardless of the value of y; $N_{\cdot j}$ is the number of events with the jth value of y regardless of x. So we have

$$N_{i\cdot} = \sum_j N_{ij} \qquad N_{\cdot j} = \sum_i N_{ij}$$

$$N = \sum_i N_{i\cdot} = \sum_j N_{\cdot j} \qquad (14.4.1)$$

$N_{\cdot j}$ and $N_{i\cdot}$ are sometimes called the *row and column totals* or *marginals*, but we will use these terms cautiously since we can never keep straight which are the rows and which are the columns!

The null hypothesis is that the two variables x and y have no association. In this case, the probability of a particular value of x given a particular value of y should be the same as the probability of that value of x regardless of y. Therefore, in the null hypothesis, the expected number for any N_{ij}, which we will denote n_{ij}, can be calculated from only the row and column totals,

$$\frac{n_{ij}}{N_{\cdot j}} = \frac{N_{i\cdot}}{N} \qquad \text{which implies} \qquad n_{ij} = \frac{N_{i\cdot}.N_{\cdot j}}{N} \qquad (14.4.2)$$

Notice that if a column or row total is zero, then the expected number for all the entries in that column or row is also zero; in that case, the never-occurring bin of x or y should simply be removed from the analysis.

The chi-square statistic is now given by equation (14.3.1), which, in the present case, is summed over all entries in the table,

$$\chi^2 = \sum_{i,j} \frac{(N_{ij} - n_{ij})^2}{n_{ij}} \qquad (14.4.3)$$

The number of degrees of freedom is equal to the number of entries in the table (product of its row size and column size) minus the number of constraints that have arisen from our use of the data themselves to determine the n_{ij}. Each row total and column total is a constraint, except that this overcounts by one, since the total of the column totals and the total of the row totals both equal N, the total number of data points. Therefore, if the table is of size I by J, the number of degrees of freedom is $IJ - I - J + 1$. Equation (14.4.3), along with the chi-square probability function (§6.2), now give the significance of an association between the variables x and y.

Suppose there is a significant association. How do we quantify its strength, so that (e.g.) we can compare the strength of one association with another? The idea here is to find some reparametrization of χ^2 which maps it into some convenient interval, like 0 to 1, where the result is not dependent on the quantity of data that we happen to sample, but rather depends only on the underlying population from which the data were drawn. There are several different ways of doing this. Two of the more common are called *Cramer's V* and the *contingency coefficient C*.

The formula for Cramer's V is

$$V = \sqrt{\frac{\chi^2}{N \min (I - 1, J - 1)}} \qquad (14.4.4)$$

where I and J are again the numbers of rows and columns, and N is the total number of events. Cramer's V has the pleasant property that it lies between zero and one inclusive, equals zero when there is no association, and equals one only when the association is perfect: All the events in any row lie in one unique column, and vice versa. (In chess parlance, no two rooks, placed on a nonzero table entry, can capture each other.)

In the case of $I = J = 2$, Cramer's V is also referred to as the *phi* statistic.

The contingency coefficient C is defined as

$$C = \sqrt{\frac{\chi^2}{\chi^2 + N}} \qquad (14.4.5)$$

It also lies between zero and one, but (as is apparent from the formula) it can never achieve the upper limit. While it can be used to compare the strength of association of two tables with the same I and J, its upper limit depends on I and J. Therefore it can never be used to compare tables of different sizes.

The trouble with both Cramer's V and the contingency coefficient C is that, when they take on values in between their extremes, there is no very direct interpretation of what that value means. For example, you are in Las Vegas, and a friend tells you that there is a small, but significant, association between the color of a croupier's eyes and the occurrence of red and black on his roulette wheel. Cramer's V is about 0.028, your friend tells you. You know what the usual odds against you are (because of the green zero and double zero on the wheel). Is this association sufficient for you to make money? Don't ask us!

```
#include <cmath>
#include "nr.h"
using namespace std;

void NR::cntab1(Mat_I_INT &nn, DP &chisq, DP &df, DP &prob, DP &cramrv, DP &ccc)
Given a two-dimensional contingency table in the form of an array nn[0..ni-1][0..nj-1]
of integers, this routine returns the chi-square chisq, the number of degrees of freedom df,
the significance level prob (small values indicating a significant association), and two measures
of association, Cramer's V (cramrv) and the contingency coefficient C (ccc).
{
    const DP TINY=1.0e-30;              A small number.
    int i,j,nnj,nni,minij;
    DP sum=0.0,expctd,temp;

    int ni=nn.nrows();
    int nj=nn.ncols();
    Vec_DP sumi(ni),sumj(nj);
    nni=ni;                            Number of rows
    nnj=nj;                            and columns.
    for (i=0;i<ni;i++) {               Get the row totals.
        sumi[i]=0.0;
        for (j=0;j<nj;j++) {
            sumi[i] += nn[i][j];
            sum += nn[i][j];
```

```
        }
        if (sumi[i] == 0.0) --nni;          Eliminate any zero rows by reducing the num-
    }                                        ber.
    for (j=0;j<nj;j++) {                     Get the column totals.
        sumj[j]=0.0;
        for (i=0;i<ni;i++) sumj[j] += nn[i][j];
        if (sumj[j] == 0.0) --nnj;          Eliminate any zero columns.
    }
    df=nni*nnj-nni-nnj+1;                    Corrected number of degrees of freedom.
    chisq=0.0;
    for (i=0;i<ni;i++) {                     Do the chi-square sum.
        for (j=0;j<nj;j++) {
            expctd=sumj[j]*sumi[i]/sum;
            temp=nn[i][j]-expctd;
            chisq += temp*temp/(expctd+TINY);   Here TINY guarantees that any
        }                                        eliminated row or column will
    }                                            not contribute to the sum.
    prob=gammq(0.5*df,0.5*chisq);           Chi-square probability function.
    minij = nni < nnj ? nni-1 : nnj-1;
    cramrv=sqrt(chisq/(sum*minij));
    ccc=sqrt(chisq/(chisq+sum));
}
```

Measures of Association Based on Entropy

Consider the game of "twenty questions," where by repeated yes/no questions you try to eliminate all except one correct possibility for an unknown object. Better yet, consider a generalization of the game, where you are allowed to ask multiple choice questions as well as binary (yes/no) ones. The categories in your multiple choice questions are supposed to be mutually exclusive and exhaustive (as are "yes" and "no").

The value to you of an answer increases with the number of possibilities that it eliminates. More specifically, an answer that eliminates all except a fraction p of the remaining possibilities can be assigned a value $-\ln p$ (a positive number, since $p < 1$). The purpose of the logarithm is to make the value additive, since (e.g.) one question that eliminates all but 1/6 of the possibilities is considered as good as two questions that, in sequence, reduce the number by factors 1/2 and 1/3.

So that is the value of an answer; but what is the value of a question? If there are I possible answers to the question ($i = 0, \ldots, I - 1$) and the fraction of possibilities consistent with answer i is p_i (with the sum of the p_i's equal to one), then the value of the question is the expectation value of the value of the answer, denoted H,

$$H = -\sum_{i=0}^{I-1} p_i \ln p_i \tag{14.4.6}$$

In evaluating (14.4.6), note that

$$\lim_{p \to 0} p \ln p = 0 \tag{14.4.7}$$

The value H lies between 0 and $\ln I$. It is zero only when one of the p_i's is one, all the others zero: In this case, the question is valueless, since its answer is preordained.

H takes on its maximum value when all the p_i's are equal, in which case the question is sure to eliminate all but a fraction $1/I$ of the remaining possibilities.

The value H is conventionally termed the *entropy* of the distribution given by the p_i's, a terminology borrowed from statistical physics.

So far we have said nothing about the association of two variables; but suppose we are deciding what question to ask next in the game and have to choose between two candidates, or possibly want to ask both in one order or another. Suppose that one question, x, has I possible answers, labeled by i, and that the other question, y, as J possible answers, labeled by j. Then the possible outcomes of asking both questions form a contingency table whose entries N_{ij}, when normalized by dividing by the total number of remaining possibilities N, give all the information about the p's. In particular, we can make contact with the notation (14.4.1) by identifying

$$p_{ij} = \frac{N_{ij}}{N}$$

$$p_{i\cdot} = \frac{N_{i\cdot}}{N} \qquad \text{(outcomes of question } x \text{ alone)} \qquad (14.4.8)$$

$$p_{\cdot j} = \frac{N_{\cdot j}}{N} \qquad \text{(outcomes of question } y \text{ alone)}$$

The entropies of the questions x and y are, respectively,

$$H(x) = -\sum_i p_{i\cdot} \ln p_{i\cdot} \qquad H(y) = -\sum_j p_{\cdot j} \ln p_{\cdot j} \qquad (14.4.9)$$

The entropy of the two questions together is

$$H(x,y) = -\sum_{i,j} p_{ij} \ln p_{ij} \qquad (14.4.10)$$

Now what is the entropy of the question y *given* x (that is, if x is asked first)? It is the expectation value over the answers to x of the entropy of the restricted y distribution that lies in a single column of the contingency table (corresponding to the x answer):

$$H(y|x) = -\sum_i p_{i\cdot} \sum_j \frac{p_{ij}}{p_{i\cdot}} \ln \frac{p_{ij}}{p_{i\cdot}} = -\sum_{i,j} p_{ij} \ln \frac{p_{ij}}{p_{i\cdot}} \qquad (14.4.11)$$

Correspondingly, the entropy of x given y is

$$H(x|y) = -\sum_j p_{\cdot j} \sum_i \frac{p_{ij}}{p_{\cdot j}} \ln \frac{p_{ij}}{p_{\cdot j}} = -\sum_{i,j} p_{ij} \ln \frac{p_{ij}}{p_{\cdot j}} \qquad (14.4.12)$$

We can readily prove that the entropy of y given x is never more than the entropy of y alone, i.e., that asking x first can only reduce the usefulness of asking

y (in which case the two variables are *associated!*):

$$H(y|x) - H(y) = -\sum_{i,j} p_{ij} \ln \frac{p_{ij}/p_{i\cdot}}{p_{\cdot j}}$$

$$= \sum_{i,j} p_{ij} \ln \frac{p_{\cdot j} p_{i\cdot}}{p_{ij}}$$

$$\leq \sum_{i,j} p_{ij} \left(\frac{p_{\cdot j} p_{i\cdot}}{p_{ij}} - 1 \right) \qquad (14.4.13)$$

$$= \sum_{i,j} p_{i\cdot} p_{\cdot j} - \sum_{i,j} p_{ij}$$

$$= 1 - 1 = 0$$

where the inequality follows from the fact

$$\ln w \leq w - 1 \qquad (14.4.14)$$

We now have everything we need to define a measure of the "dependency" of y on x, that is to say a measure of association. This measure is sometimes called the *uncertainty coefficient* of y. We will denote it as $U(y|x)$,

$$U(y|x) \equiv \frac{H(y) - H(y|x)}{H(y)} \qquad (14.4.15)$$

This measure lies between zero and one, with the value 0 indicating that x and y have no association, the value 1 indicating that knowledge of x completely predicts y. For in-between values, $U(y|x)$ gives the fraction of y's entropy $H(y)$ that is lost if x is already known (i.e., that is redundant with the information in x). In our game of "twenty questions," $U(y|x)$ is the fractional loss in the utility of question y if question x is to be asked first.

If we wish to view x as the dependent variable, y as the independent one, then interchanging x and y we can of course define the dependency of x on y,

$$U(x|y) \equiv \frac{H(x) - H(x|y)}{H(x)} \qquad (14.4.16)$$

If we want to treat x and y symmetrically, then the useful combination turns out to be

$$U(x,y) \equiv 2 \left[\frac{H(y) + H(x) - H(x,y)}{H(x) + H(y)} \right] \qquad (14.4.17)$$

If the two variables are completely independent, then $H(x,y) = H(x) + H(y)$, so (14.4.17) vanishes. If the two variables are completely dependent, then $H(x) = H(y) = H(x,y)$, so (14.4.16) equals unity. In fact, you can use the identities (easily proved from equations 14.4.9–14.4.12)

$$H(x,y) = H(x) + H(y|x) = H(y) + H(x|y) \qquad (14.4.18)$$

to show that

$$U(x,y) = \frac{H(x)U(x|y) + H(y)U(y|x)}{H(x) + H(y)} \qquad (14.4.19)$$

i.e., that the symmetrical measure is just a weighted average of the two asymmetrical measures (14.4.15) and (14.4.16), weighted by the entropy of each variable separately.

Here is a program for computing all the quantities discussed, $H(x)$, $H(y)$, $H(x|y)$, $H(y|x)$, $H(x,y)$, $U(x|y)$, $U(y|x)$, and $U(x,y)$:

```
#include <cmath>
#include "nr.h"
using namespace std;
```

```
void NR::cntab2(Mat_I_INT &nn, DP &h, DP &hx, DP &hy, DP &hygx, DP &hxgy,
    DP &uygx, DP &uxgy, DP &uxy)
```
Given a two-dimensional contingency table in the form of an integer array nn[i][j], where i
labels the x variable and ranges from 0 to ni-1, j labels the y variable and ranges from 0 to
nj-1, this routine returns the entropy h of the whole table, the entropy hx of the x distribution,
the entropy hy of the y distribution, the entropy hygx of y given x, the entropy hxgy of x
given y, the dependency uygx of y on x (eq. 14.4.15), the dependency uxgy of x on y (eq.
14.4.16), and the symmetrical dependency uxy (eq. 14.4.17).

```
{
    const DP TINY=1.0e-30;                          A small number.
    int i,j;
    DP sum=0.0,p;

    int ni=nn.nrows();
    int nj=nn.ncols();
    Vec_DP sumi(ni),sumj(nj);
    for (i=0;i<ni;i++) {                            Get the row totals.
        sumi[i]=0.0;
        for (j=0;j<nj;j++) {
            sumi[i] += nn[i][j];
            sum += nn[i][j];
        }
    }
    for (j=0;j<nj;j++) {                            Get the column totals.
        sumj[j]=0.0;
        for (i=0;i<ni;i++)
            sumj[j] += nn[i][j];
    }
    hx=0.0;                                         Entropy of the x distribution,
    for (i=0;i<ni;i++)
        if (sumi[i] != 0.0) {
            p=sumi[i]/sum;
            hx -= p*log(p);
        }
    hy=0.0;                                         and of the y distribution.
    for (j=0;j<nj;j++)
        if (sumj[j] != 0.0) {
            p=sumj[j]/sum;
            hy -= p*log(p);
        }
    h=0.0;
    for (i=0;i<ni;i++)                              Total entropy: loop over both x
        for (j=0;j<nj;j++)                          and y.
            if (nn[i][j] != 0) {
                p=nn[i][j]/sum;
                h -= p*log(p);
            }
    hygx=h-hx;                                      Uses equation (14.4.18),
    hxgy=h-hy;                                      as does this.
    uygx=(hy-hygx)/(hy+TINY);                       Equation (14.4.15).
    uxgy=(hx-hxgy)/(hx+TINY);                       Equation (14.4.16).
    uxy=2.0*(hx+hy-h)/(hx+hy+TINY);                 Equation (14.4.17).
}
```

CITED REFERENCES AND FURTHER READING:

Dunn, O.J., and Clark, V.A. 1974, *Applied Statistics: Analysis of Variance and Regression* (New
    York: Wiley).

Norusis, M.J. 1999, *SPSS 9.0 Guide to Data Analysis* (Englewood Cliffs, NJ: Prentice-Hall).

Fano, R.M. 1961, *Transmission of Information* (New York: Wiley and MIT Press), Chapter 2.

14.5 Linear Correlation

We next turn to measures of association between variables that are ordinal or continuous, rather than nominal. Most widely used is the *linear correlation coefficient*. For pairs of quantities (x_i, y_i), $i = 0, \ldots, N-1$, the linear correlation coefficient r (also called the product-moment correlation coefficient, or *Pearson's r*) is given by the formula

$$r = \frac{\sum_i (x_i - \overline{x})(y_i - \overline{y})}{\sqrt{\sum_i (x_i - \overline{x})^2}\sqrt{\sum_i (y_i - \overline{y})^2}} \tag{14.5.1}$$

where, as usual, $\overline{x}$ is the mean of the x_i's, $\overline{y}$ is the mean of the y_i's.

The value of r lies between -1 and 1, inclusive. It takes on a value of 1, termed "complete positive correlation," when the data points lie on a perfect straight line with positive slope, with x and y increasing together. The value 1 holds independent of the magnitude of the slope. If the data points lie on a perfect straight line with negative slope, y decreasing as x increases, then r has the value -1; this is called "complete negative correlation." A value of r near zero indicates that the variables x and y are *uncorrelated*.

When a correlation is known to be significant, r is one conventional way of summarizing its strength. In fact, the value of r can be translated into a statement about what residuals (root mean square deviations) are to be expected if the data are fitted to a straight line by the least-squares method (see §15.2, especially equations 15.2.13 – 15.2.14). Unfortunately, r is a rather poor statistic for deciding *whether* an observed correlation is statistically significant, and/or whether one observed correlation is significantly stronger than another. The reason is that r is ignorant of the individual distributions of x and y, so there is no universal way to compute its distribution in the case of the null hypothesis.

About the only general statement that can be made is this: If the null hypothesis is that x and y are uncorrelated, and if the distributions for x and y each have enough convergent moments ("tails" die off sufficiently rapidly), and if N is large (typically > 500), then r is distributed approximately normally, with a mean of zero and a standard deviation of $1/\sqrt{N}$. In that case, the (double-sided) significance of the correlation, that is, the probability that $|r|$ should be larger than its observed value in the null hypothesis, is

$$\text{erfc}\left(\frac{|r|\sqrt{N}}{\sqrt{2}}\right) \tag{14.5.2}$$

where $\text{erfc}(x)$ is the complementary error function, equation (6.2.8), computed by the routines `erffc` or `erfcc` of §6.2. A small value of (14.5.2) indicates that the two distributions are significantly correlated. (See expression 14.5.9 below for a more accurate test.)

Most statistics books try to go beyond (14.5.2) and give additional statistical tests that can be made using r. In almost all cases, however, these tests are valid only for a very special class of hypotheses, namely that the distributions of x and y jointly form a *binormal* or *two-dimensional Gaussian* distribution around their mean values, with joint probability density

$$p(x, y)\, dxdy = \text{const.} \times \exp\left[-\frac{1}{2}(a_{00}x^2 - 2a_{01}xy + a_{11}y^2)\right] dxdy \quad (14.5.3)$$

where a_{00}, a_{01}, and a_{11} are arbitrary constants. For this distribution r has the value

$$r = -\frac{a_{01}}{\sqrt{a_{00}a_{11}}} \quad (14.5.4)$$

There are occasions when (14.5.3) may be known to be a good model of the data. There may be other occasions when we are willing to take (14.5.3) as at least a rough and ready guess, since many two-dimensional distributions do resemble a binormal distribution, at least not too far out on their tails. In either situation, we can use (14.5.3) to go beyond (14.5.2) in any of several directions:

First, we can allow for the possibility that the number N of data points is not large. Here, it turns out that the statistic

$$t = r\sqrt{\frac{N-2}{1-r^2}} \quad (14.5.5)$$

is distributed in the null case (of no correlation) like Student's t-distribution with $\nu = N - 2$ degrees of freedom, whose two-sided significance level is given by $1 - A(t|\nu)$ (equation 6.4.7). As N becomes large, this significance and (14.5.2) become asymptotically the same, so that one never does worse by using (14.5.5), even if the binormal assumption is not well substantiated.

Second, when N is only moderately large (≥ 10), we can compare whether the difference of two significantly nonzero r's, e.g., from different experiments, is itself significant. In other words, we can quantify whether a change in some control variable significantly alters an existing correlation between two other variables. This is done by using *Fisher's z-transformation* to associate each measured r with a corresponding z,

$$z = \frac{1}{2}\ln\left(\frac{1+r}{1-r}\right) \quad (14.5.6)$$

Then, each z is approximately normally distributed with a mean value

$$\bar{z} = \frac{1}{2}\left[\ln\left(\frac{1+r_{\text{true}}}{1-r_{\text{true}}}\right) + \frac{r_{\text{true}}}{N-1}\right] \quad (14.5.7)$$

where r_{true} is the actual or population value of the correlation coefficient, and with a standard deviation

$$\sigma(z) \approx \frac{1}{\sqrt{N-3}} \quad (14.5.8)$$

Equations (14.5.7) and (14.5.8), when they are valid, give several useful statistical tests. For example, the significance level at which a measured value of r differs from some hypothesized value r_{true} is given by

$$\text{erfc}\left(\frac{|z - \bar{z}|\sqrt{N - 3}}{\sqrt{2}}\right) \tag{14.5.9}$$

where z and $\bar{z}$ are given by (14.5.6) and (14.5.7), with small values of (14.5.9) indicating a significant difference. (Setting $\bar{z} = 0$ makes expression 14.5.9 a more accurate replacement for expression 14.5.2 above.) Similarly, the significance of a difference between two measured correlation coefficients r_1 and r_2 is

$$\text{erfc}\left(\frac{|z_1 - z_2|}{\sqrt{2}\sqrt{\frac{1}{N_1 - 3} + \frac{1}{N_2 - 3}}}\right) \tag{14.5.10}$$

where z_1 and z_2 are obtained from r_1 and r_2 using (14.5.6), and where N_1 and N_2 are, respectively, the number of data points in the measurement of r_1 and r_2.

All of the significances above are two-sided. If you wish to disprove the null hypothesis in favor of a one-sided hypothesis, such as that $r_1 > r_2$ (where the sense of the inequality was decided *a priori*), then (i) if your measured r_1 and r_2 have the *wrong* sense, you have failed to demonstrate your one-sided hypothesis, but (ii) if they have the right ordering, you can multiply the significances given above by 0.5, which makes them more significant.

But keep in mind: These interpretations of the r statistic can be completely meaningless if the joint probability distribution of your variables x and y is too different from a binormal distribution.

```
#include <cmath>
#include "nr.h"
using namespace std;

void NR::pearsn(Vec_I_DP &x, Vec_I_DP &y, DP &r, DP &prob, DP &z)
Given two arrays x[0..n-1] and y[0..n-1], this routine computes their correlation coefficient
r (returned as r), the significance level at which the null hypothesis of zero correlation is
disproved (prob whose small value indicates a significant correlation), and Fisher's z (returned
as z), whose value can be used in further statistical tests as described above.
{
    const DP TINY=1.0e-20;                    Will regularize the unusual case of
    int j;                                    complete correlation.
    DP yt,xt,t,df;
    DP syy=0.0,sxy=0.0,sxx=0.0,ay=0.0,ax=0.0;

    int n=x.size();
    for (j=0;j<n;j++) {                        Find the means.
        ax += x[j];
        ay += y[j];
    }
    ax /= n;
    ay /= n;
    for (j=0;j<n;j++) {                        Compute the correlation coefficient.
        xt=x[j]-ax;
        yt=y[j]-ay;
        sxx += xt*xt;
        syy += yt*yt;
```

```
      sxy += xt*yt;
   }
   r=sxy/(sqrt(sxx*syy)+TINY);
   z=0.5*log((1.0+r+TINY)/(1.0-r+TINY));          Fisher's z transformation.
   df=n-2;
   t=r*sqrt(df/((1.0-r+TINY)*(1.0+r+TINY)));       Equation (14.5.5).
   prob=betai(0.5*df,0.5,df/(df+t*t));             Student's t probability.
   // prob=erfcc(fabs(z*sqrt(n-1.0))/1.4142136);
   For large n, this easier computation of prob, using the short routine erfcc, would give
   approximately the same value.
}
```

CITED REFERENCES AND FURTHER READING:

Dunn, O.J., and Clark, V.A. 1974, *Applied Statistics: Analysis of Variance and Regression* (New York: Wiley).

Hoel, P.G. 1971, *Introduction to Mathematical Statistics*, 4th ed. (New York: Wiley), Chapter 7.

von Mises, R. 1964, *Mathematical Theory of Probability and Statistics* (New York: Academic Press), Chapters IX(A) and IX(B).

Korn, G.A., and Korn, T.M. 1968, *Mathematical Handbook for Scientists and Engineers*, 2nd ed. (New York: McGraw-Hill), §19.7.

Norusis, M.J. 1999, *SPSS 9.0 Guide to Data Analysis* (Englewood Cliffs, NJ: Prentice-Hall).

14.6 Nonparametric or Rank Correlation

It is precisely the uncertainty in interpreting the significance of the linear correlation coefficient r that leads us to the important concepts of *nonparametric* or *rank correlation*. As before, we are given N pairs of measurements (x_i, y_i). Before, difficulties arose because we did not necessarily know the probability distribution function from which the x_i's or y_i's were drawn.

The key concept of nonparametric correlation is this: If we replace the value of each x_i by the value of its *rank* among all the other x_i's in the sample, that is, $1, 2, 3, \ldots, N$, then the resulting list of numbers will be drawn from a perfectly known distribution function, namely uniformly from the integers between 1 and N, inclusive. Better than uniformly, in fact, since if the x_i's are all distinct, then each integer will occur precisely once. If some of the x_i's have identical values, it is conventional to assign to all these "ties" the mean of the ranks that they would have had if their values had been slightly different. This *midrank* will sometimes be an integer, sometimes a half-integer. In all cases the sum of all assigned ranks will be the same as the sum of the integers from 1 to N, namely $\frac{1}{2}N(N + 1)$.

Of course we do exactly the same procedure for the y_i's, replacing each value by its rank among the other y_i's in the sample.

Now we are free to invent statistics for detecting correlation between uniform sets of integers between 1 and N, keeping in mind the possibility of ties in the ranks. There is, of course, some loss of information in replacing the original numbers by ranks. We could construct some rather artificial examples where a correlation could be detected parametrically (e.g., in the linear correlation coefficient r), but could not be detected nonparametrically. Such examples are very rare in real life, however, and the slight loss of information in ranking is a small price to pay for a very major advantage: When a correlation is demonstrated to be present nonparametrically, then it is really there! (That is, to a certainty level that depends on the significance chosen.) Nonparametric correlation is more robust than linear correlation, more resistant to unplanned defects in the data, in the same sort of sense that the median is more robust than the mean. For more on the concept of robustness, see §15.7.

As always in statistics, some particular choices of a statistic have already been invented for us and consecrated, if not beatified, by popular use. We will discuss two, the *Spearman rank-order correlation coefficient* (r_s), and *Kendall's tau* (τ).

Spearman Rank-Order Correlation Coefficient

Let R_i be the rank of x_i among the other x's, S_i be the rank of y_i among the other y's, ties being assigned the appropriate midrank as described above. Then the rank-order correlation coefficient is defined to be the linear correlation coefficient of the ranks, namely,

$$r_s = \frac{\sum_i (R_i - \overline{R})(S_i - \overline{S})}{\sqrt{\sum_i (R_i - \overline{R})^2}\sqrt{\sum_i (S_i - \overline{S})^2}} \tag{14.6.1}$$

The significance of a nonzero value of r_s is tested by computing

$$t = r_s \sqrt{\frac{N-2}{1-r_s^2}} \tag{14.6.2}$$

which is distributed approximately as Student's distribution with $N - 2$ degrees of freedom. A key point is that this approximation does not depend on the original distribution of the x's and y's; it is always the same approximation, and always pretty good.

It turns out that r_s is closely related to another conventional measure of nonparametric correlation, the so-called *sum squared difference of ranks*, defined as

$$D = \sum_{i=0}^{N-1} (R_i - S_i)^2 \tag{14.6.3}$$

(This D is sometimes denoted D^{**}, where the asterisks are used to indicate that ties are treated by midranking.)

When there are no ties in the data, then the exact relation between D and r_s is

$$r_s = 1 - \frac{6D}{N^3 - N} \tag{14.6.4}$$

When there are ties, then the exact relation is slightly more complicated: Let f_k be the number of ties in the kth group of ties among the R_i's, and let g_m be the number of ties in the mth group of ties among the S_i's. Then it turns out that

$$r_s = \frac{1 - \frac{6}{N^3 - N}\left[D + \frac{1}{12}\sum_k (f_k^3 - f_k) + \frac{1}{12}\sum_m (g_m^3 - g_m)\right]}{\left[1 - \frac{\sum_k (f_k^3 - f_k)}{N^3 - N}\right]^{1/2}\left[1 - \frac{\sum_m (g_m^3 - g_m)}{N^3 - N}\right]^{1/2}} \tag{14.6.5}$$

holds exactly. Notice that if all the f_k's and all the g_m's are equal to one, meaning that there are no ties, then equation (14.6.5) reduces to equation (14.6.4).

In (14.6.2) we gave a t-statistic that tests the significance of a nonzero r_s. It is also possible to test the significance of D directly. The expectation value of D in the null hypothesis of uncorrelated data sets is

$$\overline{D} = \frac{1}{6}(N^3 - N) - \frac{1}{12}\sum_k (f_k^3 - f_k) - \frac{1}{12}\sum_m (g_m^3 - g_m) \qquad (14.6.6)$$

its variance is

$$\text{Var}(D) = \frac{(N-1)N^2(N+1)^2}{36} \\ \times \left[1 - \frac{\sum_k (f_k^3 - f_k)}{N^3 - N}\right]\left[1 - \frac{\sum_m (g_m^3 - g_m)}{N^3 - N}\right] \qquad (14.6.7)$$

and it is approximately normally distributed, so that the significance level is a complementary error function (cf. equation 14.5.2). Of course, (14.6.2) and (14.6.7) are not independent tests, but simply variants of the same test. In the program that follows, we calculate both the significance level obtained by using (14.6.2) and the significance level obtained by using (14.6.7); their discrepancy will give you an idea of how good the approximations are. You will also notice that we break off the task of assigning ranks (including tied midranks) into a separate function, `crank`.

```
#include <cmath>
#include "nr.h"
using namespace std;

void NR::spear(Vec_I_DP &data1, Vec_I_DP &data2, DP &d, DP &zd, DP &probd,
    DP &rs, DP &probrs)
```
Given two data arrays, `data1[0..n-1]` and `data2[0..n-1]`, this routine returns their sum-squared difference of ranks as D, the number of standard deviations by which D deviates from its null-hypothesis expected value as `zd`, the two-sided significance level of this deviation as `probd`, Spearman's rank correlation r_s as `rs`, and the two-sided significance level of its deviation from zero as `probrs`. The external routines `crank` (below) and `sort2` (§8.2) are used. A small value of either `probd` or `probrs` indicates a significant correlation (`rs` positive) or anticorrelation (`rs` negative).
```
{
    int j;
    DP vard,t,sg,sf,fac,en3n,en,df,aved;

    int n=data1.size();
    Vec_DP wksp1(n),wksp2(n);
    for (j=0;j<n;j++) {
        wksp1[j]=data1[j];
        wksp2[j]=data2[j];
    }
    sort2(wksp1,wksp2);              Sort each of the data arrays, and convert the entries to
    crank(wksp1,sf);                 ranks. The values sf and sg return the sums $\sum(f_k^3 - f_k)$
    sort2(wksp2,wksp1);              and $\sum(g_m^3 - g_m)$, respectively.
    crank(wksp2,sg);
    d=0.0;
    for (j=0;j<n;j++)                Sum the squared difference of ranks.
        d += SQR(wksp1[j]-wksp2[j]);
    en=n;
    en3n=en*en*en-en;
    aved=en3n/6.0-(sf+sg)/12.0;                              Expectation value of $D$,
```

```
    fac=(1.0-sf/en3n)*(1.0-sg/en3n);
    vard=((en-1.0)*en*en*SQR(en+1.0)/36.0)*fac;          and variance of D give
    zd=(d-aved)/sqrt(vard);                              number of standard devia-
    probd=erfcc(fabs(zd)/1.4142136);                     tions and significance.
    rs=(1.0-(6.0/en3n)*(d+(sf+sg)/12.0))/sqrt(fac);      Rank correlation coefficient,
    fac=(rs+1.0)*(1.0-rs);
    if (fac > 0.0) {
        t=rs*sqrt((en-2.0)/fac);                         and its t value,
        df=en-2.0;
        probrs=betai(0.5*df,0.5,df/(df+t*t));            give its significance.
    } else
        probrs=0.0;
}
```

```
#include "nr.h"

void NR::crank(Vec_IO_DP &w, DP &s)
```
Given a sorted array `w[0..n-1]`, replaces the elements by their rank, including midranking of ties, and returns as s the sum of $f^3 - f$, where f is the number of elements in each tie.
```
{
    int j=1,ji,jt;
    DP t,rank;

    int n=w.size();
    s=0.0;
    while (j < n) {
        if (w[j] != w[j-1]) {                   Not a tie.
            w[j-1]=j;
            ++j;
        } else {                                A tie:
            for (jt=j+1;jt<=n && w[jt-1]==w[j-1];jt++);   How far does it go?
            rank=0.5*(j+jt-1);                  This is the mean rank of the tie,
            for (ji=j;ji<=(jt-1);ji++)          so enter it into all the tied entries,
                w[ji-1]=rank;
            t=jt-j;
            s += (t*t*t-t);                     and update s.
            j=jt;
        }
    }
    if (j == n) w[n-1]=n;                       If the last element was not tied, this is its
}                                               rank.
```

Kendall's Tau

Kendall's τ is even more nonparametric than Spearman's r_s or D. Instead of using the numerical difference of ranks, it uses only the relative ordering of ranks: higher in rank, lower in rank, or the same in rank. But in that case we don't even have to rank the data! Ranks will be higher, lower, or the same if and only if the values are larger, smaller, or equal, respectively. On balance, we prefer r_s as being the more straightforward nonparametric test, but both statistics are in general use. In fact, τ and r_s are very strongly correlated and, in most applications, are effectively the same test.

To define τ, we start with the N data points (x_i, y_i). Now consider all $\frac{1}{2}N(N-1)$ *pairs* of data points, where a data point cannot be paired with itself, and where the points in either order count as one pair. We call a pair *concordant* if the relative ordering of the ranks of the two x's (or for that matter the two x's themselves) is the same as the relative ordering of the ranks of the two y's (or for that matter the two y's themselves). We call a pair *discordant* if the relative ordering of the ranks of the two x's is opposite from the relative ordering of the ranks of the two y's. If there is a tie in either the ranks of the two x's or the ranks of the two y's, then we don't call the pair either concordant or discordant. If the tie is in the x's, we will call the pair an "extra y pair." If the tie is in the y's, we will call the pair an "extra x pair." If the tie is in both the x's and the y's, we don't call the pair anything at all. Are you still with us?

Kendall's τ is now the following simple combination of these various counts:

$$\tau = \frac{\text{concordant} - \text{discordant}}{\sqrt{\text{concordant} + \text{discordant} + \text{extra-}y}\ \sqrt{\text{concordant} + \text{discordant} + \text{extra-}x}}$$

$$(14.6.8)$$

You can easily convince yourself that this must lie between 1 and -1, and that it takes on the extreme values only for complete rank agreement or complete rank reversal, respectively.

More important, Kendall has worked out, from the combinatorics, the approximate distribution of τ in the null hypothesis of no association between x and y. In this case τ is approximately normally distributed, with zero expectation value and a variance of

$$\text{Var}(\tau) = \frac{4N + 10}{9N(N - 1)} \tag{14.6.9}$$

The following program proceeds according to the above description, and therefore loops over all pairs of data points. Beware: This is an $O(N^2)$ algorithm, unlike the algorithm for r_s, whose dominant sort operations are of order $N \log N$. If you are routinely computing Kendall's τ for data sets of more than a few thousand points, you may be in for some serious computing. If, however, you are willing to bin your data into a moderate number of bins, then read on.

```
#include <cmath>
#include "nr.h"
using namespace std;

void NR::kendl1(Vec_I_DP &data1, Vec_I_DP &data2, DP &tau, DP &z, DP &prob)
Given data arrays data1[0..n-1] and data2[0..n-1], this program returns Kendall's τ as
tau, its number of standard deviations from zero as z, and its two-sided significance level as
prob. Small values of prob indicate a significant correlation (tau positive) or anticorrelation
(tau negative).
{
    int is=0,j,k,n2=0,n1=0;
    DP svar,aa,a2,a1;

    int n=data1.size();
    for (j=0;j<n-1;j++) {                        Loop over first member of pair,
        for (k=j+1;k<n;k++) {                    and second member.
            a1=data1[j]-data1[k];
            a2=data2[j]-data2[k];
            aa=a1*a2;
            if (aa != 0.0) {                     Neither array has a tie.
                ++n1;
```

```
            ++n2;
            aa > 0.0 ? ++is : --is;        One or both arrays have ties.
        } else {
            if (a1 != 0.0) ++n1;           An "extra x" event.
            if (a2 != 0.0) ++n2;           An "extra y" event.
        }
    }
}
tau=is/(sqrt(DP(n1))*sqrt(DP(n2)));        Equation (14.6.8).
svar=(4.0*n+10.0)/(9.0*n*(n-1.0));         Equation (14.6.9).
z=tau/sqrt(svar);
prob=erfcc(fabs(z)/1.4142136);             Significance.
}
```

Sometimes it happens that there are only a few possible values each for x and y. In that case, the data can be recorded as a contingency table (see §14.4) that gives the number of data points for each contingency of x and y.

Spearman's rank-order correlation coefficient is not a very natural statistic under these circumstances, since it assigns to each x and y bin a not-very-meaningful midrank value and then totals up vast numbers of identical rank differences. Kendall's tau, on the other hand, with its simple counting, remains quite natural. Furthermore, its $O(N^2)$ algorithm is no longer a problem, since we can arrange for it to loop over pairs of contingency table entries (each containing many data points) instead of over pairs of data points. This is implemented in the program that follows.

Note that Kendall's tau can be applied only to contingency tables where both variables are *ordinal*, i.e., well-ordered, and that it looks specifically for monotonic correlations, not for arbitrary associations. These two properties make it less general than the methods of §14.4, which applied to *nominal*, i.e., unordered, variables and arbitrary associations.

Comparing kendl1 above with kendl2 below, you will see that we have changed a number of variables from int to double. This is because the number of events in a contingency table might be sufficiently large as to cause overflows in some of the integer arithmetic, while the number of individual data points in a list could not possibly be that large [for an $O(N^2)$ routine!].

```
#include <cmath>
#include "nr.h"
using namespace std;

void NR::kendl2(Mat_I_DP &tab, DP &tau, DP &z, DP &prob)
Given a two-dimensional table tab[0..i-1][0..j-1], such that tab[k][l] contains the
number of events falling in bin k of one variable and bin l of another, this program returns
Kendall's τ as tau, its number of standard deviations from zero as z, and its two-sided signif-
icance level as prob. Small values of prob indicate a significant correlation (tau positive) or
anticorrelation (tau negative) between the two variables. Although tab is a double array, it
will normally contain integral values.
{
    int k,l,nn,mm,m2,m1,lj,li,kj,ki;
    DP svar,s=0.0,points,pairs,en2=0.0,en1=0.0;

    int i=tab.nrows();
    int j=tab.ncols();
    nn=i*j;                              Total number of entries in contingency table.
    points=tab[i-1][j-1];
    for (k=0;k<=nn-2;k++) {              Loop over entries in table,
        ki=(k/j);                        decoding a row,
```

```
    kj=k-j*ki;                              and a column.
    points += tab[ki][kj];                  Increment the total count of events.
    for (l=k+1;l<=nn-1;l++) {               Loop over other member of the pair,
        li=l/j;                             decoding its row
        lj=l-j*li;                          and column.
        mm=(m1=li-ki)*(m2=lj-kj);
        pairs=tab[ki][kj]*tab[li][lj];
        if (mm != 0) {                      Not a tie.
            en1 += pairs;
            en2 += pairs;
            s += (mm > 0 ? pairs : -pairs);         Concordant, or discordant.
        } else {
            if (m1 != 0) en1 += pairs;
            if (m2 != 0) en2 += pairs;
        }
    }
}
tau=s/sqrt(en1*en2);
svar=(4.0*points+10.0)/(9.0*points*(points-1.0));
z=tau/sqrt(svar);
prob=erfcc(fabs(z)/1.4142136);
}
```

CITED REFERENCES AND FURTHER READING:

Lehmann, E.L. 1975, *Nonparametrics: Statistical Methods Based on Ranks* (San Francisco: Holden-Day).

Downie, N.M., and Heath, R.W. 1965, *Basic Statistical Methods*, 2nd ed. (New York: Harper & Row), pp. 206–209.

Norusis, M.J. 1999, *SPSS 9.0 Guide to Data Analysis* (Englewood Cliffs, NJ: Prentice-Hall).

14.7 Do Two-Dimensional Distributions Differ?

We here discuss a useful generalization of the K–S test (§14.3) to *two-dimensional* distributions. This generalization is due to Fasano and Franceschini [1], a variant on an earlier idea due to Peacock [2].

In a two-dimensional distribution, each data point is characterized by an (x, y) pair of values. An example near to our hearts is that each of the 19 neutrinos that were detected from Supernova 1987A is characterized by a time t_i and by an energy E_i (see [3]). We might wish to know whether these measured pairs (t_i, E_i), $i = 0 \ldots 18$ are consistent with a theoretical model that predicts neutrino flux as a function of both time and energy — that is, a two-dimensional probability distribution in the (x, y) [here, (t, E)] plane. That would be a one-sample test. Or, given two sets of neutrino detections, from two comparable detectors, we might want to know whether they are compatible with each other, a two-sample test.

In the spirit of the tried-and-true, one-dimensional K–S test, we want to range over the (x, y) plane in search of some kind of maximum *cumulative* difference between two two-dimensional distributions. Unfortunately, cumulative probability distribution is not well-defined in more than one dimension! Peacock's insight was that a good surrogate is the *integrated probability in each of four natural quadrants* around a given point (x_i, y_i), namely the total probabilities (or fraction of data) in $(x > x_i, y > y_i)$, $(x < x_i, y > y_i)$, $(x < x_i, y < y_i)$, $(x > x_i, y < y_i)$. The two-dimensional K–S statistic D is now taken to be the maximum difference (ranging both over data points and over quadrants) of the corresponding integrated probabilities. When comparing two data sets, the value of D may depend on which data set is ranged over. In that case, define an effective D as the average

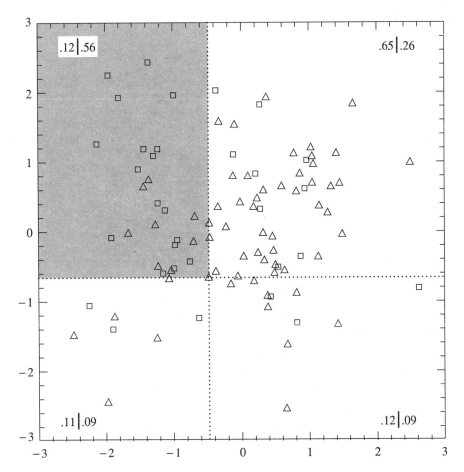

Figure 14.7.1.   Two-dimensional distributions of 65 triangles and 35 squares. The two-dimensional K–S test finds that point one of whose quadrants (shown by dotted lines) maximizes the difference between fraction of triangles and fraction of squares. Then, equation (14.7.1) indicates whether the difference is statistically significant, i.e., whether the triangles and squares must have different underlying distributions.

of the two values obtained. If you are confused at this point about the exact definition of D, don't fret; the accompanying computer routines amount to a precise algorithmic definition.

Figure 14.7.1 gives a feeling for what is going on. The 65 triangles and 35 squares seem to have somewhat different distributions in the plane. The dotted lines are centered on the triangle that maximizes the D statistic; the maximum occurs in the upper-left quadrant. That quadrant contains only 0.12 of all the triangles, but it contains 0.56 of all the squares. The value of D is thus 0.44. Is this statistically significant?

Even for fixed sample sizes, it is unfortunately not rigorously true that the distribution of D in the null hypothesis is independent of the shape of the two-dimensional distribution. In this respect the two-dimensional K–S test is not as natural as its one-dimensional parent. However, extensive Monte Carlo integrations have shown that the distribution of the two-dimensional D is *very nearly* identical for even quite different distributions, as long as they have the same coefficient of correlation r, defined in the usual way by equation (14.5.1). In their paper, Fasano and Franceschini tabulate Monte Carlo results for (what amounts to) the distribution of D as a function of (of course) D, sample size N, and coefficient of correlation r. Analyzing their results, one finds that the significance levels for the two-dimensional K–S test can be summarized by the simple, though approximate, formulas,

$$\text{Probability } (D > \text{observed }) = Q_{KS}\left(\frac{\sqrt{N}\,D}{1+\sqrt{1-r^2}(0.25-0.75/\sqrt{N})}\right) \quad (14.7.1)$$

for the one-sample case, and the same for the two-sample case, but with

$$N = \frac{N_1 N_2}{N_1 + N_2}.$$ (14.7.2)

The above formulas are accurate enough when $N \gtrsim 20$, and when the indicated probability (significance level) is less than (more significant than) 0.20 or so. When the indicated probability is > 0.20, its value may not be accurate, but the implication that the data and model (or two data sets) are not significantly different is certainly correct. Notice that in the limit of $r \to 1$ (perfect correlation), equations (14.7.1) and (14.7.2) reduce to equations (14.3.9) and (14.3.10): The two-dimensional data lie on a perfect straight line, and the two-dimensional K–S test becomes a one-dimensional K–S test.

The significance level for the data in Figure 14.7.1, by the way, is about 0.001. This establishes to a near-certainty that the triangles and squares were drawn from different distributions. (As in fact they were.)

Of course, if you do not want to rely on the Monte Carlo experiments embodied in equation (14.7.1), you can do your own: Generate a lot of synthetic data sets from your model, each one with the same number of points as the real data set. Compute D for each synthetic data set, using the accompanying computer routines (but ignoring their calculated probabilities), and count what fraction of the time these synthetic D's exceed the D from the real data. That fraction is your significance.

One disadvantage of the two-dimensional tests, by comparison with their one-dimensional progenitors, is that the two-dimensional tests require of order N^2 operations: Two nested loops of order N take the place of an $N \log N$ sort. For small computers, this restricts the usefulness of the tests to N less than several thousand.

We now give computer implementations. The one-sample case is embodied in the routine `ks2d1s` (that is, 2-dimensions, 1-sample). This routine calls a straightforward utility routine `quadct` to count points in the four quadrants, and it calls a user-supplied routine `quadvl` that must be capable of returning the integrated probability of an analytic model in each of four quadrants around an arbitrary (x, y) point. A trivial sample `quadvl` is shown; realistic `quadvl`s can be quite complicated, often incorporating numerical quadratures over analytic two-dimensional distributions.

```
#include <cmath>
#include "nr.h"
using namespace std;

void NR::ks2d1s(Vec_I_DP &x1, Vec_I_DP &y1, void quadvl(const DP, const DP,
    DP &, DP &, DP &, DP &), DP &d1, DP &prob)
```
Two-dimensional Kolmogorov-Smirnov test of one sample against a model. Given the x and y coordinates of n1 data points in arrays x1[0..n1-1] and y1[0..n1-1], and given a user-supplied function quadvl that exemplifies the model, this routine returns the two-dimensional K-S statistic as d1, and its significance level as prob. Small values of prob show that the sample is significantly different from the model. Note that the test is slightly distribution-dependent, so prob is only an estimate.
```
{
    int j;
    DP dum,dumm,fa,fb,fc,fd,ga,gb,gc,gd,r1,rr,sqen;

    int n1=x1.size();
    d1=0.0;
    for (j=0;j<n1;j++) {                     Loop over the data points.
        quadct(x1[j],y1[j],x1,y1,fa,fb,fc,fd);
        quadvl(x1[j],y1[j],ga,gb,gc,gd);
        d1=MAX(d1,fabs(fa-ga));
        d1=MAX(d1,fabs(fb-gb));
        d1=MAX(d1,fabs(fc-gc));
        d1=MAX(d1,fabs(fd-gd));
        For both the sample and the model, the distribution is integrated in each of four
        quadrants, and the maximum difference is saved.
    }
```

```
    pearsn(x1,y1,r1,dum,dumm);                    Get the linear correlation coefficient r1.
    sqen=sqrt(DP(n1));
    rr=sqrt(1.0-r1*r1);
    Estimate the probability using the K-S probability function probks.
    prob=probks(d1*sqen/(1.0+rr*(0.25-0.75/sqen)));
}
```

```
#include "nr.h"

void NR::quadct(const DP x, const DP y, Vec_I_DP &xx, Vec_I_DP &yy, DP &fa,
    DP &fb, DP &fc, DP &fd)
```
Given an origin (x, y), and an array of nn points with coordinates xx[0..nn-1] and yy[0..nn-1],
count how many of them are in each quadrant around the origin, and return the normalized
fractions. Quadrants are labeled alphabetically, counterclockwise from the upper right. Used
by ks2d1s and ks2d2s.
```
{
    int k,na,nb,nc,nd;
    DP ff;

    int nn=xx.size();
    na=nb=nc=nd=0;
    for (k=0;k<nn;k++) {
        if (yy[k] > y)
            xx[k] > x ? ++na : ++nb;
        else
            xx[k] > x ? ++nd : ++nc;
    }
    ff=1.0/nn;
    fa=ff*na;
    fb=ff*nb;
    fc=ff*nc;
    fd=ff*nd;
}
```

```
#include "nr.h"

void NR::quadvl(const DP x, const DP y, DP &fa, DP &fb, DP &fc, DP &fd)
```
This is a sample of a user-supplied routine to be used with ks2d1s. In this case, the model
distribution is uniform inside the square $-1 < x < 1$, $-1 < y < 1$. In general this routine
should return, for any point (x, y), the fraction of the total distribution in each of the four
quadrants around that point. The fractions, fa, fb, fc, and fd, must add up to 1. Quadrants
are alphabetical, counterclockwise from the upper right.
```
{
    DP qa,qb,qc,qd;

    qa=MIN(2.0,MAX(0.0,1.0-x));
    qb=MIN(2.0,MAX(0.0,1.0-y));
    qc=MIN(2.0,MAX(0.0,x+1.0));
    qd=MIN(2.0,MAX(0.0,y+1.0));
    fa=0.25*qa*qb;
    fb=0.25*qb*qc;
    fc=0.25*qc*qd;
    fd=0.25*qd*qa;
}
```

The routine ks2d2s is the two-sample case of the two-dimensional K–S test. It also calls
quadct, pearsn, and probks. Being a two-sample test, it does not need an analytic model.

```
#include <cmath>
#include "nr.h"
using namespace std;

void NR::ks2d2s(Vec_I_DP &x1, Vec_I_DP &y1, Vec_I_DP &x2, Vec_I_DP &y2, DP &d,
    DP &prob)
```

Two-dimensional Kolmogorov-Smirnov test on two samples. Given the x and y coordinates of the first sample as n1 values in arrays x1[0..n1-1] and y1[0..n1-1], and likewise for the second sample, n2 values in arrays x2 and y2, this routine returns the two-dimensional, two-sample K-S statistic as d, and its significance level as prob. Small values of prob show that the two samples are significantly different. Note that the test is slightly distribution-dependent, so prob is only an estimate.

```
{
    int j;
    DP d1,d2,dum,dumm,fa,fb,fc,fd,ga,gb,gc,gd,r1,r2,rr,sqen;

    int n1=x1.size();
    int n2=x2.size();
    d1=0.0;
    for (j=0;j<n1;j++) {                     First, use points in the first sample as ori-
        quadct(x1[j],y1[j],x1,y1,fa,fb,fc,fd);             gins.
        quadct(x1[j],y1[j],x2,y2,ga,gb,gc,gd);
        d1=MAX(d1,fabs(fa-ga));
        d1=MAX(d1,fabs(fb-gb));
        d1=MAX(d1,fabs(fc-gc));
        d1=MAX(d1,fabs(fd-gd));
    }
    d2=0.0;
    for (j=0;j<n2;j++) {                     Then, use points in the second sample as
        quadct(x2[j],y2[j],x1,y1,fa,fb,fc,fd);             origins.
        quadct(x2[j],y2[j],x2,y2,ga,gb,gc,gd);
        d2=MAX(d2,fabs(fa-ga));
        d2=MAX(d2,fabs(fb-gb));
        d2=MAX(d2,fabs(fc-gc));
        d2=MAX(d2,fabs(fd-gd));
    }
    d=0.5*(d1+d2);                           Average the K-S statistics.
    sqen=sqrt(n1*n2/DP(n1+n2));
    pearsn(x1,y1,r1,dum,dumm);               Get the linear correlation coefficient for each
    pearsn(x2,y2,r2,dum,dumm);                 sample.
    rr=sqrt(1.0-0.5*(r1*r1+r2*r2));
    Estimate the probability using the K-S probability function probks.
    prob=probks(d*sqen/(1.0+rr*(0.25-0.75/sqen)));
}
```

CITED REFERENCES AND FURTHER READING:

Fasano, G. and Franceschini, A. 1987, *Monthly Notices of the Royal Astronomical Society*, vol. 225, pp. 155–170. [1]

Peacock, J.A. 1983, *Monthly Notices of the Royal Astronomical Society*, vol. 202, pp. 615–627. [2]

Spergel, D.N., Piran, T., Loeb, A., Goodman, J., and Bahcall, J.N. 1987, *Science*, vol. 237, pp. 1471–1473. [3]

14.8 Savitzky-Golay Smoothing Filters

In §13.5 we learned something about the construction and application of digital filters, but little guidance was given on *which particular* filter to use. That, of course, depends on what you want to accomplish by filtering. One obvious use for *low-pass* filters is to smooth noisy data.

The premise of data smoothing is that one is measuring a variable that is both slowly varying and also corrupted by random noise. Then it can sometimes be useful to replace each data point by some kind of local average of surrounding data points. Since nearby points measure very nearly the same underlying value, averaging can reduce the level of noise without (much) biasing the value obtained.

We must comment editorially that the smoothing of data lies in a murky area, beyond the fringe of some better posed, and therefore more highly recommended, techniques that are discussed elsewhere in this book. If you are fitting data to a parametric model, for example (see Chapter 15), it is almost always better to use raw data than to use data that has been pre-processed by a smoothing procedure. Another alternative to blind smoothing is so-called "optimal" or Wiener filtering, as discussed in §13.3 and more generally in §13.6. Data smoothing is probably most justified when it is used simply as a graphical technique, to guide the eye through a forest of data points all with large error bars; or as a means of making initial *rough* estimates of simple parameters from a graph.

In this section we discuss a particular type of low-pass filter, well-adapted for data smoothing, and termed variously *Savitzky-Golay* [1], *least-squares* [2], or *DISPO* (Digital Smoothing Polynomial) [3] filters. Rather than having their properties defined in the Fourier domain, and then translated to the time domain, Savitzky-Golay filters derive directly from a particular formulation of the data smoothing problem in the time domain, as we will now see. Savitzky-Golay filters were initially (and are still often) used to render visible the relative widths and heights of spectral lines in noisy spectrometric data.

Recall that a digital filter is applied to a series of equally spaced data values $f_i \equiv f(t_i)$, where $t_i \equiv t_0 + i\Delta$ for some constant sample spacing Δ and $i = \ldots - 2, -1, 0, 1, 2, \ldots$. We have seen (§13.5) that the simplest type of digital filter (the nonrecursive or finite impulse response filter) replaces each data value f_i by a linear combination g_i of itself and some number of nearby neighbors,

$$g_i = \sum_{n=-n_L}^{n_R} c_n f_{i+n} \tag{14.8.1}$$

Here n_L is the number of points used "to the left" of a data point i, i.e., earlier than it, while n_R is the number used to the right, i.e., later. A so-called *causal* filter would have $n_R = 0$.

As a starting point for understanding Savitzky-Golay filters, consider the simplest possible averaging procedure: For some fixed $n_L = n_R$, compute each g_i as the average of the data points from f_{i-n_L} to f_{i+n_R}. This is sometimes called *moving window averaging* and corresponds to equation (14.8.1) with constant $c_n = 1/(n_L + n_R + 1)$. If the underlying function is constant, or is changing linearly with time (increasing or decreasing), then no bias is introduced into the result. Higher points at one end of the averaging interval are on the average balanced by lower points at the other end. A bias is introduced, however, if the underlying function has a nonzero second derivative. At a local maximum, for example, moving window averaging always reduces the function value. In the spectrometric application, a narrow spectral line has its height reduced and its width increased. Since these parameters are themselves of physical interest, the bias introduced is distinctly undesirable.

Note, however, that moving window averaging does preserve the area under a spectral line, which is its zeroth moment, and also (if the window is symmetric with $n_L = n_R$) its mean position in time, which is its first moment. What is violated is the second moment, equivalent to the line width.

The idea of Savitzky-Golay filtering is to find filter coefficients c_n that preserve higher moments. Equivalently, the idea is to approximate the underlying function within the moving window not by a constant (whose estimate is the average), but by a polynomial of higher order, typically quadratic or quartic: For each point f_i, we least-squares fit a polynomial to all

M	n_L	n_R	Sample Savitzky-Golay Coefficients
2	2	2	-0.086 0.343 0.486 0.343 -0.086
2	3	1	-0.143 0.171 0.343 0.371 0.257
2	4	0	0.086 -0.143 -0.086 0.257 0.886
2	5	5	-0.084 0.021 0.103 0.161 0.196 0.207 0.196 0.161 0.103 0.021 -0.084
4	4	4	0.035 -0.128 0.070 0.315 0.417 0.315 0.070 -0.128 0.035
4	5	5	0.042 -0.105 -0.023 0.140 0.280 0.333 0.280 0.140 -0.023 -0.105 0.042

$n_L + n_R + 1$ points in the moving window, and then set g_i to be the value of that polynomial at position i. (If you are not familiar with least-squares fitting, you might want to look ahead to Chapter 15.) We make no use of the value of the polynomial at any other point. When we move on to the next point f_{i+1}, we do a whole new least-squares fit using a shifted window.

All these least-squares fits would be laborious if done as described. Luckily, since the process of least-squares fitting involves only a linear matrix inversion, the coefficients of a fitted polynomial are themselves linear in the values of the data. That means that we can do all the fitting in advance, for fictitious data consisting of all zeros except for a single 1, and then do the fits on the real data just by taking linear combinations. This is the key point, then: There are particular sets of filter coefficients c_n for which equation (14.8.1) "automatically" accomplishes the process of polynomial least-squares fitting inside a moving window.

To derive such coefficients, consider how g_0 might be obtained: We want to fit a polynomial of degree M in i, namely $a_0 + a_1 i + \cdots + a_M i^M$ to the values $f_{-n_L}, \ldots, f_{n_R}$. Then g_0 will be the value of that polynomial at $i = 0$, namely a_0. The design matrix for this problem (§15.4) is

$$A_{ij} = i^j \qquad i = -n_L, \ldots, n_R, \quad j = 0, \ldots, M \tag{14.8.2}$$

and the normal equations for the vector of a_j's in terms of the vector of f_i's is in matrix notation

$$(\mathbf{A}^T \cdot \mathbf{A}) \cdot \mathbf{a} = \mathbf{A}^T \cdot \mathbf{f} \qquad \text{or} \qquad \mathbf{a} = (\mathbf{A}^T \cdot \mathbf{A})^{-1} \cdot (\mathbf{A}^T \cdot \mathbf{f}) \tag{14.8.3}$$

We also have the specific forms

$$\left\{\mathbf{A}^T \cdot \mathbf{A}\right\}_{ij} = \sum_{k=-n_L}^{n_R} A_{ki} A_{kj} = \sum_{k=-n_L}^{n_R} k^{i+j} \tag{14.8.4}$$

and

$$\left\{\mathbf{A}^T \cdot \mathbf{f}\right\}_j = \sum_{k=-n_L}^{n_R} A_{kj} f_k = \sum_{k=-n_L}^{n_R} k^j f_k \tag{14.8.5}$$

Since the coefficient c_n is the component a_0 when $\mathbf{f}$ is replaced by the unit vector $\mathbf{e}_n$, $-n_L \leq n < n_R$, we have

$$c_n = \left\{(\mathbf{A}^T \cdot \mathbf{A})^{-1} \cdot (\mathbf{A}^T \cdot \mathbf{e}_n)\right\}_0 = \sum_{m=0}^{M} \left\{(\mathbf{A}^T \cdot \mathbf{A})^{-1}\right\}_{0m} n^m \tag{14.8.6}$$

Note that equation (14.8.6) says that we need only one row of the inverse matrix. (Numerically we can get this by LU decomposition with only a single backsubstitution.)

The function savgol, below, implements equation (14.8.6). As input, it takes the parameters nl $= n_L$, nr $= n_R$, and m $= M$ (the desired order). Also input is np, the physical length of the output array c, and a parameter ld which for data fitting should be zero. In fact, ld specifies which coefficient among the a_i's should be returned, and we are here interested in a_0. For another purpose, namely the computation of numerical derivatives (already mentioned in §5.7) the useful choice is ld ≥ 1. With ld $= 1$, for example, the filtered first derivative is the convolution (14.8.1) divided by the stepsize Δ. For ld $= k > 1$, the array c must be multiplied by $k!$ to give derivative coefficients. For derivatives, one usually wants m $= 4$ or larger.

```
#include <cmath>
#include "nr.h"
using namespace std;

void NR::savgol(Vec_O_DP &c, const int np, const int nl, const int nr,
    const int ld, const int m)
```
Returns in c[0..np-1], in wrap-around order (N.B.!) consistent with the argument respns in routine convlv, a set of Savitzky-Golay filter coefficients. nl is the number of leftward (past) data points used, while nr is the number of rightward (future) data points, making the total number of data points used nl + nr + 1. ld is the order of the derivative desired (e.g., ld = 0 for smoothed function. For the derivative of order k, you must multiply the array c by $k!$.) m is the order of the smoothing polynomial, also equal to the highest conserved moment; usual values are m = 2 or m = 4.
```
{
    int j,k,imj,ipj,kk,mm;
    DP d,fac,sum;

    if (np < nl+nr+1 || nl < 0 || nr < 0 || ld > m || nl+nr < m)
        nrerror("bad args in savgol");
    Vec_INT indx(m+1);
    Mat_DP a(m+1,m+1);
    Vec_DP b(m+1);
    for (ipj=0;ipj<=(m << 1);ipj++) {          Set up the normal equations of the desired
        sum=(ipj ? 0.0 : 1.0);                            least-squares fit.
        for (k=1;k<=nr;k++) sum += pow(DP(k),DP(ipj));
        for (k=1;k<=nl;k++) sum += pow(DP(-k),DP(ipj));
        mm=MIN(ipj,2*m-ipj);
        for (imj = -mm;imj<=mm;imj+=2) a[(ipj+imj)/2][(ipj-imj)/2]=sum;
    }
    ludcmp(a,indx,d);                          Solve them: LU decomposition.
    for (j=0;j<m+1;j++) b[j]=0.0;
    b[ld]=1.0;
    Right-hand side vector is unit vector, depending on which derivative we want.
    lubksb(a,indx,b);                          Get one row of the inverse matrix.
    for (kk=0;kk<np;kk++) c[kk]=0.0;           Zero the output array (it may be bigger than
    for (k = -nl;k<=nr;k++) {                         number of coefficients).
        sum=b[0];                              Each Savitzky-Golay coefficient is the dot
        fac=1.0;                                     product of powers of an integer with the
        for (mm=1;mm<=m;mm++) sum += b[mm]*(fac *= k);     inverse matrix row.
        kk=(np-k) % np;                        Store in wrap-around order.
        c[kk]=sum;
    }
}
```

As output, savgol returns the coefficients c_n, for $-n_L \leq n \leq n_R$. These are stored in c in "wrap-around order"; that is, c_0 is in c[0], c_{-1} is in c[1], and so on for further negative indices. The value c_1 is stored in c[np-1], c_2 in c[np-2], and so on for positive indices. This order may seem arcane, but it is the natural one where causal filters have nonzero coefficients in low array elements of c. It is also the order required by the function convlv in §13.1, which can be used to apply the digital filter to a data set.

The accompanying table shows some typical output from savgol. For orders 2 and 4, the coefficients of Savitzky-Golay filters with several choices of n_L and n_R are shown. The central column is the coefficient applied to the data f_i in obtaining the smoothed g_i. Coefficients to the left are applied to earlier data; to the right, to later. The coefficients always add (within roundoff error) to unity. One sees that, as befits a smoothing operator, the coefficients always have a central positive lobe, but with smaller, outlying corrections of both positive and negative sign. In practice, the Savitzky-Golay filters are most useful for much larger values of n_L and n_R, since these few-point formulas can accomplish only a relatively small amount of smoothing.

Figure 14.8.1 shows a numerical experiment using a 33 point smoothing filter, that is,

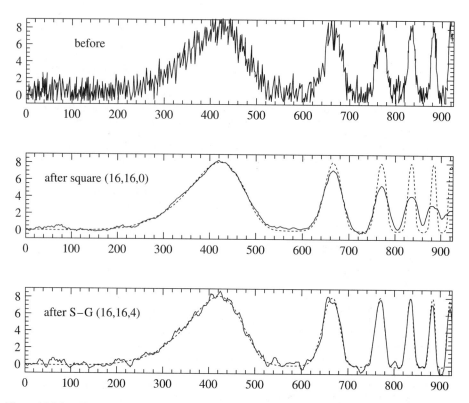

Figure 14.8.1.   Top: Synthetic noisy data consisting of a sequence of progressively narrower bumps, and additive Gaussian white noise. Center: Result of smoothing the data by a simple moving window average. The window extends 16 points leftward and rightward, for a total of 33 points. Note that narrow features are broadened and suffer corresponding loss of amplitude. The dotted curve is the underlying function used to generate the synthetic data. Bottom: Result of smoothing the data by a Savitzky-Golay smoothing filter (of degree 4) using the same 33 points. While there is less smoothing of the broadest feature, narrower features have their heights and widths preserved.

$n_L = n_R = 16$. The upper panel shows a test function, constructed to have six "bumps" of varying widths, all of height 8 units. To this function Gaussian white noise of unit variance has been added. (The test function without noise is shown as the dotted curves in the center and lower panels.) The widths of the bumps (full width at half of maximum, or FWHM) are 140, 43, 24, 17, 13, and 10, respectively.

The middle panel of Figure 14.8.1 shows the result of smoothing by a moving window average. One sees that the window of width 33 does quite a nice job of smoothing the broadest bump, but that the narrower bumps suffer considerable loss of height and increase of width. The underlying signal (dotted) is very badly represented.

The lower panel shows the result of smoothing with a Savitzky-Golay filter of the identical width, and degree $M = 4$. One sees that the heights and widths of the bumps are quite extraordinarily preserved. A trade-off is that the broadest bump is less smoothed. That is because the central positive lobe of the Savitzky-Golay filter coefficients fills only a fraction of the full 33 point width. As a rough guideline, best results are obtained when the full width of the degree 4 Savitzky-Golay filter is between 1 and 2 times the FWHM of desired features in the data. (References [3] and [4] give additional practical hints.)

Figure 14.8.2 shows the result of smoothing the same noisy "data" with broader Savitzky-Golay filters of 3 different orders. Here we have $n_L = n_R = 32$ (65 point filter) and $M = 2, 4, 6$. One sees that, when the bumps are too narrow with respect to the filter size, then even the Savitzky-Golay filter must at some point give out. The higher order filter manages to track narrower features, but at the cost of less smoothing on broad features.

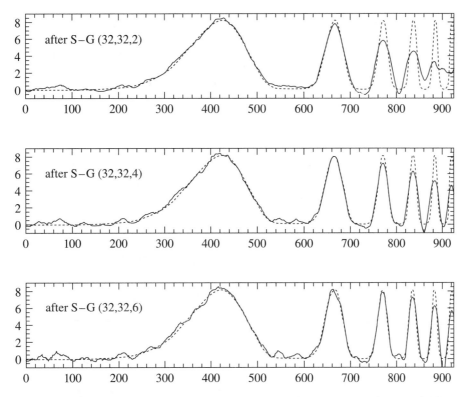

Figure 14.8.2. Result of applying wider 65 point Savitzky-Golay filters to the same data set as in Figure 14.8.1. Top: degree 2. Center: degree 4. Bottom: degree 6. All of these filters are inoptimally broad for the resolution of the narrow features. Higher-order filters do best at preserving feature heights and widths, but do less smoothing on broader features.

To summarize: Within limits, Savitzky-Golay filtering does manage to provide smoothing without loss of resolution. It does this by assuming that relatively distant data points have some significant redundancy that can be used to reduce the level of noise. The specific nature of the assumed redundancy is that the underlying function should be locally well-fitted by a polynomial. When this is true, as it is for smooth line profiles not too much narrower than the filter width, then the performance of Savitzky-Golay filters can be spectacular. When it is not true, then these filters have no compelling advantage over other classes of smoothing filter coefficients.

A last remark concerns irregularly sampled data, where the values f_i are not uniformly spaced in time. The obvious generalization of Savitzky-Golay filtering would be to do a least-squares fit within a moving window around each data point, one containing a fixed number of data points to the left (n_L) and right (n_R). Because of the irregular spacing, however, there is no way to obtain universal filter coefficients applicable to more than one data point. One must instead do the actual least-squares fits for each data point. This becomes computationally burdensome for larger n_L, n_R, and M.

As a cheap alternative, one can simply pretend that the data points *are* equally spaced. This amounts to virtually shifting, within each moving window, the data points to equally spaced positions. Such a shift introduces the equivalent of an additional source of noise into the function values. In those cases where smoothing is useful, this noise will often be much smaller than the noise already present. Specifically, if the location of the points is approximately random within the window, then a rough criterion is this: If the change in f across the full width of the $N = n_L + n_R + 1$ point window is less than $\sqrt{N/2}$ times the measurement noise on a single point, then the cheap method can be used.

CITED REFERENCES AND FURTHER READING:

Savitzky A., and Golay, M.J.E. 1964, *Analytical Chemistry*, vol. 36, pp. 1627–1639. [1]

Hamming, R.W. 1983, *Digital Filters*, 2nd ed. (Englewood Cliffs, NJ: Prentice-Hall). [2]

Ziegler, H. 1981, *Applied Spectroscopy*, vol. 35, pp. 88–92. [3]

Bromba, M.U.A., and Ziegler, H. 1981, *Analytical Chemistry*, vol. 53, pp. 1583–1586. [4]

Chapter 15.   Modeling of Data

15.0 Introduction

Given a set of observations, one often wants to condense and summarize the data by fitting it to a "model" that depends on adjustable parameters. Sometimes the model is simply a convenient class of functions, such as polynomials or Gaussians, and the fit supplies the appropriate coefficients. Other times, the model's parameters come from some underlying theory that the data are supposed to satisfy; examples are coefficients of rate equations in a complex network of chemical reactions, or orbital elements of a binary star. Modeling can also be used as a kind of constrained interpolation, where you want to extend a few data points into a continuous function, but with some underlying idea of what that function should look like.

The basic approach in all cases is usually the same: You choose or design a *figure-of-merit function* ("merit function," for short) that measures the agreement between the data and the model with a particular choice of parameters. The merit function is conventionally arranged so that small values represent close agreement. The parameters of the model are then adjusted to achieve a minimum in the merit function, yielding *best-fit parameters*. The adjustment process is thus a problem in minimization in many dimensions. This optimization was the subject of Chapter 10; however, there exist special, more efficient, methods that are specific to modeling, and we will discuss these in this chapter.

There are important issues that go beyond the mere finding of best-fit parameters. Data are generally not exact. They are subject to *measurement errors* (called *noise* in the context of signal-processing). Thus, typical data never exactly fit the model that is being used, even when that model is correct. We need the means to assess whether or not the model is appropriate, that is, we need to test the *goodness-of-fit* against some useful statistical standard.

We usually also need to know the accuracy with which parameters are determined by the data set. In other words, we need to know the likely errors of the best-fit parameters.

Finally, it is not uncommon in fitting data to discover that the merit function is not unimodal, with a single minimum. In some cases, we may be interested in global rather than local questions. Not, "how good is this fit?" but rather, "how sure am I that there is not a *very much better* fit in some corner of parameter space?" As we have seen in Chapter 10, especially §10.9, this kind of problem is generally quite difficult to solve.

The important message we want to deliver is that fitting of parameters is not the end-all of parameter estimation. To be genuinely useful, a fitting procedure should provide (i) parameters, (ii) error estimates on the parameters, and (iii) a statistical measure of goodness-of-fit. When the third item suggests that the model is an unlikely match to the data, then items (i) and (ii) are probably worthless. Unfortunately, many practitioners of parameter estimation never proceed beyond item (i). They deem a fit acceptable if a graph of data and model "looks good." This approach is known as *chi-by-eye*. Luckily, its practitioners get what they deserve.

CITED REFERENCES AND FURTHER READING:

Bevington, P.R., and Robinson, D.K. 1992, *Data Reduction and Error Analysis for the Physical Sciences*, 2nd ed. (New York: McGraw-Hill).

Brownlee, K.A. 1965, *Statistical Theory and Methodology*, 2nd ed. (New York: Wiley).

Martin, B.R. 1971, *Statistics for Physicists* (New York: Academic Press).

von Mises, R. 1964, *Mathematical Theory of Probability and Statistics* (New York: Academic Press), Chapter X.

Korn, G.A., and Korn, T.M. 1968, *Mathematical Handbook for Scientists and Engineers*, 2nd ed. (New York: McGraw-Hill), Chapters 18–19.

15.1 Least Squares as a Maximum Likelihood Estimator

Suppose that we are fitting N data points (x_i, y_i) $i = 0, \ldots, N - 1$, to a model that has M adjustable parameters a_j, $j = 0, \ldots, M - 1$. The model predicts a functional relationship between the measured independent and dependent variables,

$$y(x) = y(x; a_0 \ldots a_{M-1}) \tag{15.1.1}$$

where the dependence on the parameters is indicated explicitly on the right-hand side.

What, exactly, do we want to minimize to get fitted values for the a_j's? The first thing that comes to mind is the familiar least-squares fit,

$$\text{minimize over } a_0 \ldots a_{M-1}: \quad \sum_{i=0}^{N-1} [y_i - y(x_i; a_0 \ldots a_{M-1})]^2 \tag{15.1.2}$$

But where does this come from? What general principles is it based on? The answer to these questions takes us into the subject of *maximum likelihood estimators*.

Given a particular data set of x_i's and y_i's, we have the intuitive feeling that some parameter sets $a_0 \ldots a_{M-1}$ are very unlikely — those for which the model function $y(x)$ looks *nothing like* the data — while others may be very likely — those that closely resemble the data. How can we quantify this intuitive feeling? How can we select fitted parameters that are "most likely" to be correct? It is not meaningful to ask the question, "What is the probability that a particular set of fitted parameters $a_0 \ldots a_{M-1}$ is correct?" The reason is that there is no statistical universe of models from which the parameters are drawn. There is just one model, the correct one, and a statistical universe of data sets that are drawn from it!

That being the case, we can, however, turn the question around, and ask, "*Given a particular set of parameters*, what is the probability that this data set could have occurred?" If the y_i's take on continuous values, the probability will always be zero unless we add the phrase, "...plus or minus some fixed Δy on each data point." So let's always take this phrase as understood. If the probability of obtaining the data set is infinitesimally small, then we can conclude that the parameters under consideration are "unlikely" to be right. Conversely, our intuition tells us that the data set should not be too improbable for the correct choice of parameters.

In other words, we identify the probability of the data given the parameters (which is a mathematically computable number), as the *likelihood* of the parameters given the data. This identification is entirely based on intuition. It has no formal mathematical basis in and of itself; as we already remarked, statistics is *not* a branch of mathematics!

Once we make this intuitive identification, however, it is only a small further step to decide to fit for the parameters $a_0 \ldots a_{M-1}$ precisely by finding those values that *maximize* the likelihood defined in the above way. This form of parameter estimation is *maximum likelihood estimation*.

We are now ready to make the connection to (15.1.2). Suppose that each data point y_i has a measurement error that is independently random and distributed as a normal (Gaussian) distribution around the "true" model $y(x)$. And suppose that the standard deviations σ of these normal distributions are the same for all points. Then the probability of the data set is the product of the probabilities of each point,

$$P \propto \prod_{i=0}^{N-1} \left\{ \exp\left[-\frac{1}{2}\left(\frac{y_i - y(x_i)}{\sigma} \right)^2 \right] \Delta y \right\} \qquad (15.1.3)$$

Notice that there is a factor Δy in each term in the product. Maximizing (15.1.3) is equivalent to maximizing its logarithm, or minimizing the negative of its logarithm, namely,

$$\left[\sum_{i=0}^{N-1} \frac{[y_i - y(x_i)]^2}{2\sigma^2} \right] - N \log \Delta y \qquad (15.1.4)$$

Since N, σ, and Δy are all constants, minimizing this equation is equivalent to minimizing (15.1.2).

What we see is that least-squares fitting *is* a maximum likelihood estimation of the fitted parameters *if* the measurement errors are independent and normally distributed with constant standard deviation. Notice that we made no assumption about the linearity or nonlinearity of the model $y(x; a_0 \ldots)$ in its parameters $a_0 \ldots a_{M-1}$. Just below, we will relax our assumption of constant standard deviations and obtain the very similar formulas for what is called "chi-square fitting" or "weighted least-squares fitting." First, however, let us discuss further our very stringent assumption of a normal distribution.

For a hundred years or so, mathematical statisticians have been in love with the fact that the probability distribution of the sum of a very large number of very small random deviations almost always converges to a normal distribution. (For precise statements of this *central limit theorem*, consult [1] or other standard works on mathematical statistics.) This infatuation tended to focus interest away from the fact that, for real data, the normal distribution is often rather poorly realized, if it is realized at all. We are often taught, rather casually, that, on average, measurements will fall within $\pm\sigma$ of the true value 68 percent of the time, within $\pm2\sigma$ 95 percent of the time, and within $\pm3\sigma$ 99.7 percent of the time. Extending this, one would expect a measurement to be off by $\pm20\sigma$ only one time out of 2×10^{88}. We all know that "glitches" are much more likely than *that*!

In some instances, the deviations from a normal distribution are easy to understand and quantify. For example, in measurements obtained by counting events, the measurement errors are usually distributed as a Poisson distribution, whose cumulative probability function was already discussed in §6.2. When the number of counts going into one data point is large, the Poisson distribution converges towards a Gaussian. However, the convergence is not uniform when measured in fractional accuracy. The more standard deviations out on the tail of the distribution, the larger the number of counts must be before a value close to the Gaussian is realized. The sign of the effect is always the same: The Gaussian predicts that "tail" events are much less likely than they actually (by Poisson) are. This causes such events, when they occur, to skew a least-squares fit much more than they ought.

Other times, the deviations from a normal distribution are not so easy to understand in detail. Experimental points are occasionally just *way off*. Perhaps the power flickered during a point's measurement, or someone kicked the apparatus, or someone wrote down a wrong number. Points like this are called *outliers*. They can easily turn a least-squares fit on otherwise adequate data into nonsense. Their probability of occurrence in the assumed Gaussian model is so small that the maximum likelihood estimator is willing to distort the whole curve to try to bring them, mistakenly, into line.

The subject of *robust statistics* deals with cases where the normal or Gaussian model is a bad approximation, or cases where outliers are important. We will discuss robust methods briefly in §15.7. All the sections between this one and that one assume, one way or the other, a Gaussian model for the measurement errors in the data. It it quite important that you keep the limitations of that model in mind, even as you use the very useful methods that follow from assuming it.

Finally, note that our discussion of measurement errors has been limited to *statistical* errors, the kind that will average away if we only take enough data. Measurements are also susceptible to *systematic* errors that will not go away with any amount of averaging. For example, the calibration of a metal meter stick might depend on its temperature. If we take all our measurements at the same wrong temperature, then no amount of averaging or numerical processing will correct for this unrecognized systematic error.

Chi-Square Fitting

We considered the chi-square statistic once before, in §14.3. Here it arises in a slightly different context.

If each data point (x_i, y_i) has its own, known standard deviation σ_i, then equation (15.1.3) is modified only by putting a subscript i on the symbol σ. That subscript also propagates docilely into (15.1.4), so that the maximum likelihood estimate of the model parameters is obtained by minimizing the quantity

$$\chi^2 \equiv \sum_{i=0}^{N-1} \left(\frac{y_i - y(x_i; a_0 \dots a_{M-1})}{\sigma_i} \right)^2 \tag{15.1.5}$$

called the "chi-square."

To whatever extent the measurement errors actually *are* normally distributed, the quantity χ^2 is correspondingly a sum of N squares of normally distributed quantities, each normalized to unit variance. Once we have adjusted the $a_0 \dots a_{M-1}$ to minimize the value of χ^2, the terms in the sum are not all statistically independent. For models that are linear in the a's, however, it turns out that the probability distribution for different values of χ^2 at its minimum can nevertheless be derived analytically, and is the *chi-square distribution for $N - M$ degrees of freedom*. We learned how to compute this probability function using the incomplete gamma function gammq in §6.2. In particular, equation (6.2.18) gives the probability Q that the chi-square should exceed a particular value χ^2 by chance, where $\nu = N - M$ is the *number of degrees of freedom*. The quantity Q, or its complement $P \equiv 1 - Q$, is frequently tabulated in appendices to statistics books, but we generally find it easier to use gammq and compute our own values: $Q = \text{gammq}\,(0.5\nu, 0.5\chi^2)$. It is quite common, and usually not too wrong, to assume that the chi-square distribution holds even for models that are not strictly linear in the a's.

This computed probability gives a quantitative measure for the goodness-of-fit of the model. If Q is a very small probability for some particular data set, then the apparent discrepancies are unlikely to be chance fluctuations. Much more probably either (i) the model is wrong — can be statistically rejected, or (ii) someone has lied to you about the size of the measurement errors σ_i — they are really larger than stated.

It is an important point that the chi-square probability Q does not directly measure the credibility of the assumption that the measurement errors are normally distributed. It assumes they are. In most, but not all, cases, however, the effect of nonnormal errors is to create an abundance of outlier points. These decrease the probability Q, so that we can add another possible, though less definitive, conclusion to the above list: (iii) the measurement errors may not be normally distributed.

Possibility (iii) is fairly common, and also fairly benign. It is for this reason that reasonable experimenters are often rather tolerant of low probabilities Q. It is not uncommon to deem acceptable on equal terms any models with, say, $Q > 0.001$. This is not as sloppy as it sounds: Truly *wrong* models will often be rejected with vastly smaller values of Q, 10^{-18}, say. However, if day-in and day-out you find yourself accepting models with $Q \sim 10^{-3}$, you really should track down the cause.

If you happen to know the actual distribution law of your measurement errors, then you might wish to *Monte Carlo simulate* some data sets drawn from a particular model, cf. §7.2–§7.3. You can then subject these synthetic data sets to your actual fitting procedure, so as to determine both the probability distribution of the χ^2 statistic, and also the accuracy with which your model parameters are reproduced by the fit. We discuss this further in §15.6. The technique is very general, but it can also be very expensive.

At the opposite extreme, it sometimes happens that the probability Q is too large, too near to 1, literally too good to be true! Nonnormal measurement errors cannot in general produce this disease, since the normal distribution is about as "compact" as a distribution can be. Almost always, the cause of too good a chi-square fit is that the experimenter, in a "fit" of conservativism, has *overestimated* his or her measurement errors. Very rarely, too good a chi-square signals actual fraud, data that has been "fudged" to fit the model.

A rule of thumb is that a "typical" value of χ^2 for a "moderately" good fit is $\chi^2 \approx \nu$. More precise is the statement that the χ^2 statistic has a mean ν and a standard deviation $\sqrt{2\nu}$, and, asymptotically for large ν, becomes normally distributed.

In some cases the uncertainties associated with a set of measurements are not known in advance, and considerations related to χ^2 fitting are used to derive a value for σ. If we assume that all measurements have the same standard deviation, $\sigma_i = \sigma$, and that the model does fit well, then we can proceed by first assigning an arbitrary constant σ to all points, next fitting for the model parameters by minimizing χ^2, and finally recomputing

$$\sigma^2 = \sum_{i=0}^{N-1} [y_i - y(x_i)]^2/(N-M) \qquad (15.1.6)$$

Obviously, this approach prohibits an independent assessment of goodness-of-fit, a fact occasionally missed by its adherents. When, however, the measurement error is not known, this approach at least allows *some* kind of error bar to be assigned to the points.

If we take the derivative of equation (15.1.5) with respect to the parameters a_k, we obtain equations that must hold at the chi-square minimum,

$$0 = \sum_{i=0}^{N-1} \left(\frac{y_i - y(x_i)}{\sigma_i^2}\right) \left(\frac{\partial y(x_i; \ldots a_k \ldots)}{\partial a_k}\right) \qquad k = 0, \ldots, M-1 \quad (15.1.7)$$

Equation (15.1.7) is, in general, a set of M nonlinear equations for the M unknown a_k. Various of the procedures described subsequently in this chapter derive from (15.1.7) and its specializations.

CITED REFERENCES AND FURTHER READING:

Bevington, P.R., and Robinson, D.K. 1992, *Data Reduction and Error Analysis for the Physical Sciences*, 2nd ed. (New York: McGraw-Hill), Chapters 1–3.

von Mises, R. 1964, *Mathematical Theory of Probability and Statistics* (New York: Academic Press), §VI.C. [1]

15.2 Fitting Data to a Straight Line

A concrete example will make the considerations of the previous section more meaningful. We consider the problem of fitting a set of N data points (x_i, y_i) to a straight-line model

$$y(x) = y(x; a, b) = a + bx \qquad (15.2.1)$$

This problem is often called *linear regression*, a terminology that originated, long ago, in the social sciences. We assume that the uncertainty σ_i associated with each measurement y_i is known, and that the x_i's (values of the dependent variable) are known exactly.

To measure how well the model agrees with the data, we use the chi-square merit function (15.1.5), which in this case is

$$\chi^2(a, b) = \sum_{i=0}^{N-1} \left(\frac{y_i - a - bx_i}{\sigma_i} \right)^2 \tag{15.2.2}$$

If the measurement errors are normally distributed, then this merit function will give maximum likelihood parameter estimations of a and b; if the errors are not normally distributed, then the estimations are not maximum likelihood, but may still be useful in a practical sense. In §15.7, we will treat the case where outlier points are so numerous as to render the χ^2 merit function useless.

Equation (15.2.2) is minimized to determine a and b. At its minimum, derivatives of $\chi^2(a, b)$ with respect to a, b vanish.

$$0 = \frac{\partial \chi^2}{\partial a} = -2 \sum_{i=0}^{N-1} \frac{y_i - a - bx_i}{\sigma_i^2}$$

$$\tag{15.2.3}$$

$$0 = \frac{\partial \chi^2}{\partial b} = -2 \sum_{i=0}^{N-1} \frac{x_i(y_i - a - bx_i)}{\sigma_i^2}$$

These conditions can be rewritten in a convenient form if we define the following sums:

$$S \equiv \sum_{i=0}^{N-1} \frac{1}{\sigma_i^2} \quad S_x \equiv \sum_{i=0}^{N-1} \frac{x_i}{\sigma_i^2} \quad S_y \equiv \sum_{i=0}^{N-1} \frac{y_i}{\sigma_i^2}$$

$$\tag{15.2.4}$$

$$S_{xx} \equiv \sum_{i=0}^{N-1} \frac{x_i^2}{\sigma_i^2} \quad S_{xy} \equiv \sum_{i=0}^{N-1} \frac{x_i y_i}{\sigma_i^2}$$

With these definitions (15.2.3) becomes

$$aS + bS_x = S_y$$

$$\tag{15.2.5}$$

$$aS_x + bS_{xx} = S_{xy}$$

The solution of these two equations in two unknowns is calculated as

$$\Delta \equiv SS_{xx} - (S_x)^2$$

$$a = \frac{S_{xx}S_y - S_x S_{xy}}{\Delta} \tag{15.2.6}$$

$$b = \frac{SS_{xy} - S_x S_y}{\Delta}$$

Equation (15.2.6) gives the solution for the best-fit model parameters a and b.

We are not done, however. We must estimate the probable uncertainties in the estimates of a and b, since obviously the measurement errors in the data must introduce some uncertainty in the determination of those parameters. If the data are independent, then each contributes its own bit of uncertainty to the parameters. Consideration of propagation of errors shows that the variance σ_f^2 in the value of any function will be

$$\sigma_f^2 = \sum_{i=0}^{N-1} \sigma_i^2 \left(\frac{\partial f}{\partial y_i} \right)^2 \tag{15.2.7}$$

For the straight line, the derivatives of a and b with respect to y_i can be directly evaluated from the solution:

$$\begin{aligned} \frac{\partial a}{\partial y_i} &= \frac{S_{xx} - S_x x_i}{\sigma_i^2 \Delta} \\ \frac{\partial b}{\partial y_i} &= \frac{S x_i - S_x}{\sigma_i^2 \Delta} \end{aligned} \tag{15.2.8}$$

Summing over the points as in (15.2.7), we get

$$\begin{aligned} \sigma_a^2 &= S_{xx}/\Delta \\ \sigma_b^2 &= S/\Delta \end{aligned} \tag{15.2.9}$$

which are the variances in the estimates of a and b, respectively. We will see in §15.6 that an additional number is also needed to characterize properly the probable uncertainty of the parameter estimation. That number is the *covariance* of a and b, and (as we will see below) is given by

$$\mathrm{Cov}(a, b) = -S_x/\Delta \tag{15.2.10}$$

The coefficient of correlation between the uncertainty in a and the uncertainty in b, which is a number between -1 and 1, follows from (15.2.10) (compare equation 14.5.1),

$$r_{ab} = \frac{-S_x}{\sqrt{SS_{xx}}} \tag{15.2.11}$$

A positive value of r_{ab} indicates that the errors in a and b are likely to have the same sign, while negative value indicates the errors are anticorrelated, likely to have opposite signs.

We are *still* not done. We must estimate the goodness-of-fit of the data to the model. Absent this estimate, we have not the slightest indication that the parameters a and b in the model have any meaning at all! The probability Q that a value of chi-square as *poor* as the value (15.2.2) should occur by chance is

$$Q = \mathrm{gammq} \left(\frac{N-2}{2}, \frac{\chi^2}{2} \right) \tag{15.2.12}$$

Here gammq is our routine for the incomplete gamma function $Q(a, x)$, §6.2. If Q is larger than, say, 0.1, then the goodness-of-fit is believable. If it is larger than, say, 0.001, then the fit *may* be acceptable if the errors are nonnormal or have been moderately underestimated. If Q is less than 0.001 then the model and/or estimation procedure can rightly be called into question. In this latter case, turn to §15.7 to proceed further.

If you do not know the individual measurement errors of the points σ_i, and are proceeding (dangerously) to use equation (15.1.6) for estimating these errors, then here is the procedure for estimating the probable uncertainties of the parameters a and b: Set $\sigma_i \equiv 1$ in all equations through (15.2.6), and multiply σ_a and σ_b, as obtained from equation (15.2.9), by the additional factor $\sqrt{\chi^2/(N-2)}$, where χ^2 is computed by (15.2.2) using the fitted parameters a and b. As discussed above, this procedure is equivalent to *assuming* a good fit, so you get no independent goodness-of-fit probability Q.

In §14.5 we promised a relation between the linear correlation coefficient r (equation 14.5.1) and a goodness-of-fit measure, χ^2 (equation 15.2.2). For unweighted data (all $\sigma_i = 1$), that relation is

$$\chi^2 = (1 - r^2)\text{NVar}\,(y_0 \ldots y_{N-1}) \tag{15.2.13}$$

where

$$\text{NVar}\,(y_0 \ldots y_{N-1}) \equiv \sum_{i=0}^{N-1} (y_i - \bar{y})^2 \tag{15.2.14}$$

For data with varying weights σ_i, the above equations remain valid if the sums in equation (14.5.1) are weighted by $1/\sigma_i^2$.

The following function, fit, carries out exactly the operations that we have discussed. When the weights σ are known in advance, the calculations exactly correspond to the formulas above. However, when weights σ are unavailable, the routine *assumes* equal values of σ for each point and *assumes* a good fit, as discussed in §15.1.

The formulas (15.2.6) are susceptible to roundoff error. Accordingly, we rewrite them as follows: Define

$$t_i = \frac{1}{\sigma_i}\left(x_i - \frac{S_x}{S}\right), \qquad i = 0, 1, \ldots, N-1 \tag{15.2.15}$$

and

$$S_{tt} = \sum_{i=0}^{N-1} t_i^2 \tag{15.2.16}$$

Then, as you can verify by direct substitution,

$$b = \frac{1}{S_{tt}} \sum_{i=0}^{N-1} \frac{t_i y_i}{\sigma_i} \tag{15.2.17}$$

$$a = \frac{S_y - S_x b}{S} \tag{15.2.18}$$

$$\sigma_a^2 = \frac{1}{S}\left(1 + \frac{S_x^2}{SS_{tt}}\right) \tag{15.2.19}$$

$$\sigma_b^2 = \frac{1}{S_{tt}} \tag{15.2.20}$$

$$\mathrm{Cov}(a,b) = -\frac{S_x}{SS_{tt}} \tag{15.2.21}$$

$$r_{ab} = \frac{\mathrm{Cov}(a,b)}{\sigma_a\sigma_b} \tag{15.2.22}$$

```
#include <cmath>
#include "nr.h"
using namespace std;

void NR::fit(Vec_I_DP &x, Vec_I_DP &y, Vec_I_DP &sig, const bool mwt, DP &a,
    DP &b, DP &siga, DP &sigb, DP &chi2, DP &q)
```
Given a set of data points x[0..ndata-1], y[0..ndata-1] with individual standard devia-
tions sig[0..ndata-1], fit them to a straight line $y = a + bx$ by minimizing χ^2. Returned
are a,b and their respective probable uncertainties siga and sigb, the chi-square chi2, and
the goodness-of-fit probability q (that the fit would have χ^2 this large or larger). If mwt=false
on input, then the standard deviations are assumed to be unavailable: q is returned as 1.0 and
the normalization of chi2 is to unit standard deviation on all points.
```
{
    int i;
    DP wt,t,sxoss,sx=0.0,sy=0.0,st2=0.0,ss,sigdat;

    int ndata=x.size();
    b=0.0;
    if (mwt) {                                          Accumulate sums ...
        ss=0.0;
        for (i=0;i<ndata;i++) {                         ...with weights
            wt=1.0/SQR(sig[i]);
            ss += wt;
            sx += x[i]*wt;
            sy += y[i]*wt;
        }
    } else {
        for (i=0;i<ndata;i++) {                         ...or without weights.
            sx += x[i];
            sy += y[i];
        }
        ss=ndata;
    }
    sxoss=sx/ss;
    if (mwt) {
        for (i=0;i<ndata;i++) {
            t=(x[i]-sxoss)/sig[i];
            st2 += t*t;
            b += t*y[i]/sig[i];
        }
    } else {
        for (i=0;i<ndata;i++) {
            t=x[i]-sxoss;
            st2 += t*t;
            b += t*y[i];
        }
    }
    b /= st2;                                           Solve for a, b, σ_a, and σ_b.
    a=(sy-sx*b)/ss;
    siga=sqrt((1.0+sx*sx/(ss*st2))/ss);
```

```
    sigb=sqrt(1.0/st2);
    chi2=0.0;                                          Calculate χ².
    q=1.0;
    if (!mwt) {
        for (i=0;i<ndata;i++)
            chi2 += SQR(y[i]-a-b*x[i]);
        sigdat=sqrt(chi2/(ndata-2));                   For unweighted data evaluate typ-
        siga *= sigdat;                                ical sig using chi2, and ad-
        sigb *= sigdat;                                just the standard deviations.
    } else {
        for (i=0;i<ndata;i++)
            chi2 += SQR((y[i]-a-b*x[i])/sig[i]);
        if (ndata>2) q=gammq(0.5*(ndata-2),0.5*chi2);  Equation (15.2.12).
    }
}
```

CITED REFERENCES AND FURTHER READING:

Bevington, P.R., and Robinson, D.K. 1992, *Data Reduction and Error Analysis for the Physical Sciences*, 2nd ed. (New York: McGraw-Hill), Chapter 6.

15.3 Straight-Line Data with Errors in Both Coordinates

If experimental data are subject to measurement error not only in the y_i's, but also in the x_i's, then the task of fitting a straight-line model

$$y(x) = a + bx \tag{15.3.1}$$

is considerably harder. It is straightforward to write down the χ^2 merit function for this case,

$$\chi^2(a, b) = \sum_{i=0}^{N-1} \frac{(y_i - a - bx_i)^2}{\sigma_{y\,i}^2 + b^2 \sigma_{x\,i}^2} \tag{15.3.2}$$

where $\sigma_{x\,i}$ and $\sigma_{y\,i}$ are, respectively, the x and y standard deviations for the ith point. The weighted sum of variances in the denominator of equation (15.3.2) can be understood both as the variance in the direction of the smallest χ^2 between each data point and the line with slope b, and also as the variance of the linear combination $y_i - a - bx_i$ of two random variables x_i and y_i,

$$\text{Var}(y_i - a - bx_i) = \text{Var}(y_i) + b^2 \text{Var}(x_i) = \sigma_{y\,i}^2 + b^2 \sigma_{x\,i}^2 \equiv 1/w_i \tag{15.3.3}$$

The sum of the square of N random variables, each normalized by its variance, is thus χ^2-distributed.

We want to minimize equation (15.3.2) with respect to a and b. Unfortunately, the occurrence of b in the denominator of equation (15.3.2) makes the resulting equation for the slope $\partial \chi^2/\partial b = 0$ nonlinear. However, the corresponding condition for the intercept, $\partial \chi^2/\partial a = 0$, is still linear and yields

$$a = \left[\sum_i w_i(y_i - bx_i) \right] \bigg/ \sum_i w_i \tag{15.3.4}$$

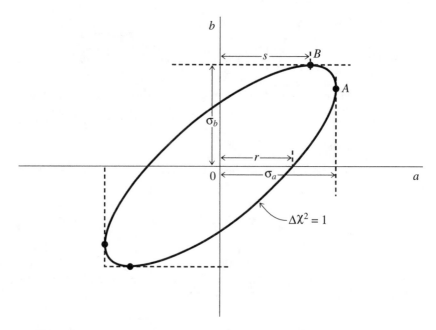

Figure 15.3.1.   Standard errors for the parameters a and b. The point B can be found by varying the slope b while simultaneously minimizing the intercept a. This gives the standard error σ_b, and also the value s. The standard error σ_a can then be found by the geometric relation $\sigma_a^2 = s^2 + r^2$.

where the w_i's are defined by equation (15.3.3). A reasonable strategy, now, is to use the machinery of Chapter 10 (e.g., the routine brent) for minimizing a general one-dimensional function to minimize with respect to b, while using equation (15.3.4) at each stage to ensure that the minimum with respect to b is also minimized with respect to a.

Because of the finite error bars on the x_i's, the minimum χ^2 as a function of b will be finite, though usually large, when b equals infinity (line of infinite slope). The angle $\theta \equiv \arctan b$ is thus more suitable as a parametrization of slope than b itself. The value of χ^2 will then be periodic in θ with period π (not 2π!). If any data points have very small σ_y's but moderate or large σ_x's, then it is also possible to have a maximum in χ^2 near zero slope, $\theta \approx 0$. In that case, there can sometimes be two χ^2 minima, one at positive slope and the other at negative. Only one of these is the correct global minimum. It is therefore important to have a good starting guess for b (or θ). Our strategy, implemented below, is to scale the y_i's so as to have variance equal to the x_i's, then to do a conventional (as in §15.2) linear fit with weights derived from the (scaled) sum $\sigma_{y\,i}^2 + \sigma_{x\,i}^2$. This yields a good starting guess for b if the data are even *plausibly* related to a straight-line model.

Finding the standard errors σ_a and σ_b on the parameters a and b is more complicated. We will see in §15.6 that, in appropriate circumstances, the standard errors in a and b are the respective projections onto the a and b axes of the "confidence region boundary" where χ^2 takes on a value one greater than its minimum, $\Delta\chi^2 = 1$. In the linear case of §15.2, these projections follow from the Taylor series expansion

$$\Delta\chi^2 \approx \frac{1}{2}\left[\frac{\partial^2\chi^2}{\partial a^2}(\Delta a)^2 + \frac{\partial^2\chi^2}{\partial b^2}(\Delta b)^2\right] + \frac{\partial^2\chi^2}{\partial a\partial b}\Delta a\Delta b \qquad (15.3.5)$$

Because of the present nonlinearity in b, however, analytic formulas for the second derivatives are quite unwieldy; more important, the lowest-order term frequently gives a poor approximation to $\Delta\chi^2$. Our strategy is therefore to find the roots of $\Delta\chi^2 = 1$ numerically, by adjusting the value of the slope b away from the minimum. In the program below the general root finder zbrent is used. It may occur that there are no roots at all — for example, if all error bars are so large that all the data points are compatible with each other. It is important, therefore, to make some effort at bracketing a putative root before refining it (cf. §9.1).

Because a is minimized at each stage of varying b, successful numerical root-finding leads to a value of Δa that minimizes χ^2 for the value of Δb that gives $\Delta\chi^2 = 1$. This (see Figure 15.3.1) directly gives the tangent projection of the confidence region onto the b axis, and thus σ_b. It does not, however, give the tangent projection of the confidence region onto the a axis. In the figure, we have found the point labeled B; to find σ_a we need to find the point A. Geometry to the rescue: To the extent that the confidence region is approximated by an ellipse, then you can prove (see figure) that $\sigma_a^2 = r^2 + s^2$. The value of s is known from having found the point B. The value of r follows from equations (15.3.2) and (15.3.3) applied at the χ^2 minimum (point O in the figure), giving

$$r^2 = 1 \left/ \sum_i w_i \right. \tag{15.3.6}$$

Actually, since b can go through infinity, this whole procedure makes more sense in (a, θ) space than in (a, b) space. That is in fact how the following program works. Since it is conventional, however, to return standard errors for a and b, not a and θ, we finally use the relation

$$\sigma_b = \sigma_\theta / \cos^2 \theta \tag{15.3.7}$$

We caution that if b and its standard error are both large, so that the confidence region actually includes infinite slope, then the standard error σ_b is not very meaningful. The function `chixy` is normally called only by the routine `fitexy`. However, if you want, you can yourself explore the confidence region by making repeated calls to `chixy` (whose argument is an angle θ, not a slope b), after a single initializing call to `fitexy`.

A final caution, repeated from §15.0, is that if the goodness-of-fit is not acceptable (returned probability is too small), the standard errors σ_a and σ_b are surely not believable. In dire circumstances, you might try scaling all your x and y error bars by a constant factor until the probability is acceptable (0.5, say), to get more plausible values for σ_a and σ_b.

```
#include <cmath>
#include "nr.h"
using namespace std;

Vec_DP *xx_p,*yy_p,*sx_p,*sy_p,*ww_p;            Global variables communicate with
DP aa,offs;                                           chixy.
```

```
void NR::fitexy(Vec_I_DP &x, Vec_I_DP &y, Vec_I_DP &sigx, Vec_I_DP &sigy,
    DP &a, DP &b, DP &siga, DP &sigb, DP &chi2, DP &q)
```
Straight-line fit to input data x[0..ndat-1] and y[0..ndat-1] with errors in both x and y, the respective standard deviations being the input quantities sigx[0..ndat-1] and sigy[0..ndat-1]. Output quantities are a and b such that $y = a + bx$ minimizes χ^2, whose value is returned as chi2. The χ^2 probability is returned as q, a small value indicating a poor fit (sometimes indicating underestimated errors). Standard errors on a and b are returned as siga and sigb. These are not meaningful if either (i) the fit is poor, or (ii) b is so large that the data are consistent with a vertical (infinite b) line. If siga and sigb are returned as BIG, then the data are consistent with *all* values of b.
```
{
    int j;
    const DP POTN=1.571000,BIG=1.0e30,ACC=1.0e-3;
    const DP PI=3.141592653589793238;
    DP amx,amn,varx,vary,ang[7],ch[7],scale,bmn,bmx,d1,d2,r2,
        dum1,dum2,dum3,dum4,dum5;

    int ndat=x.size();
    xx_p=new Vec_DP(ndat);
    yy_p=new Vec_DP(ndat);
    sx_p=new Vec_DP(ndat);
    sy_p=new Vec_DP(ndat);
```

```
        ww_p=new Vec_DP(ndat);
        Vec_DP &xx=*xx_p, &yy=*yy_p;                        Make aliases to simplify coding.
        Vec_DP &sx=*sx_p, &sy=*sy_p, &ww=*ww_p;
        avevar(x,dum1,varx);                               Find the x and y variances, and scale
        avevar(y,dum1,vary);                                   the data into the global variables
        scale=sqrt(varx/vary);                                 for communication with the func-
        for (j=0;j<ndat;j++) {                                 tion chixy.
            xx[j]=x[j];
            yy[j]=y[j]*scale;
            sx[j]=sigx[j];
            sy[j]=sigy[j]*scale;
            ww[j]=sqrt(SQR(sx[j])+SQR(sy[j]));             Use both x and y weights in first
        }                                                      trial fit.
        fit(xx,yy,ww,true,dum1,b,dum2,dum3,dum4,dum5);       Trial fit for b.
        offs=ang[0]=0.0;                                   Construct several angles for refer-
        ang[1]=atan(b);                                        ence points, and make b an an-
        ang[3]=0.0;                                            gle.
        ang[4]=ang[1];
        ang[5]=POTN;
        for (j=3;j<6;j++) ch[j]=chixy(ang[j]);
        mnbrak(ang[0],ang[1],ang[2],ch[0],ch[1],ch[2],chixy);
        Bracket the χ² minimum and then locate it with brent.
        chi2=brent(ang[0],ang[1],ang[2],chixy,ACC,b);
        chi2=chixy(b);
        a=aa;
        q=gammq(0.5*(ndat-2),chi2*0.5);                    Compute χ² probability.
        r2=0.0;
        for (j=0;j<ndat;j++) r2 += ww[j];                  Save the inverse sum of weights at
        r2=1.0/r2;                                             the minimum.
        bmx=BIG;                                           Now, find standard errors for b as
        bmn=BIG;                                               points where Δχ² = 1.
        offs=chi2+1.0;
        for (j=0;j<6;j++) {                                Go through saved values to bracket
            if (ch[j] > offs) {                                the desired roots. Note period-
                d1=fabs(ang[j]-b);                             icity in slope angles.
                while (d1 >= PI) d1 -= PI;
                d2=PI-d1;
                if (ang[j] < b)
                    SWAP(d1,d2);
                if (d1 < bmx) bmx=d1;
                if (d2 < bmn) bmn=d2;
            }
        }
        if (bmx < BIG) {                                   Call zbrent to find the roots.
            bmx=zbrent(chixy,b,b+bmx,ACC)-b;
            amx=aa-a;
            bmn=zbrent(chixy,b,b-bmn,ACC)-b;
            amn=aa-a;
            sigb=sqrt(0.5*(bmx*bmx+bmn*bmn))/(scale*SQR(cos(b)));
            siga=sqrt(0.5*(amx*amx+amn*amn)+r2)/scale;     Error in a has additional piece
        } else sigb=siga=BIG;                                  r2.
        a /= scale;                                        Unscale the answers.
        b=tan(b)/scale;
        delete ww_p; delete sy_p; delete sx_p; delete yy_p; delete xx_p;
    }
```

```
#include <cmath>
#include "nr.h"
using namespace std;

extern Vec_DP *xx_p,*yy_p,*sx_p,*sy_p,*ww_p;
extern DP aa,offs;
```

```
DP NR::chixy(const DP bang)
```
Captive function of `fitexy`, returns the value of $(\chi^2 - \text{offs})$ for the slope `b=tan(bang)`. Scaled data and `offs` are communicated via the global variables.

```
{
    const DP BIG=1.0e30;
    int j;
    DP ans,avex=0.0,avey=0.0,sumw=0.0,b;

    Vec_DP &xx=*xx_p, &yy=*yy_p;          Make aliases to simplify coding.
    Vec_DP &sx=*sx_p, &sy=*sy_p, &ww=*ww_p;
    int nn=xx.size();
    b=tan(bang);
    for (j=0;j<nn;j++) {
        ww[j] = SQR(b*sx[j])+SQR(sy[j]);
        sumw += (ww[j]=(ww[j] < 1.0/BIG ? BIG : 1.0/ww[j]));
        avex += ww[j]*xx[j];
        avey += ww[j]*yy[j];
    }
    avex /= sumw;
    avey /= sumw;
    aa=avey-b*avex;
    for (ans = -offs,j=0;j<nn;j++)
        ans += ww[j]*SQR(yy[j]-aa-b*xx[j]);
    return ans;
}
```

Be aware that the literature on the seemingly straightforward subject of this section is generally confusing and sometimes plain wrong. Deming's [1] early treatment is sound, but its reliance on Taylor expansions gives inaccurate error estimates. References [2-4] are reliable, more recent, general treatments with critiques of earlier work. York [5] and Reed [6] usefully discuss the simple case of a straight line as treated here, but the latter paper has some errors, corrected in [7]. All this commotion has attracted the Bayesians [8-10], who have still different points of view.

CITED REFERENCES AND FURTHER READING:

Deming, W.E. 1943, *Statistical Adjustment of Data* (New York: Wiley), reprinted 1964 (New York: Dover). [1]

Jefferys, W.H. 1980, *Astronomical Journal*, vol. 85, pp. 177–181; see also vol. 95, p. 1299 (1988). [2]

Jefferys, W.H. 1981, *Astronomical Journal*, vol. 86, pp. 149–155; see also vol. 95, p. 1300 (1988). [3]

Lybanon, M. 1984, *American Journal of Physics*, vol. 52, pp. 22–26. [4]

York, D. 1966, *Canadian Journal of Physics*, vol. 44, pp. 1079–1086. [5]

Reed, B.C. 1989, *American Journal of Physics*, vol. 57, pp. 642–646; see also vol. 58, p. 189, and vol. 58, p. 1209. [6]

Reed, B.C. 1992, *American Journal of Physics*, vol. 60, pp. 59–62. [7]

Zellner, A. 1971, *An Introduction to Bayesian Inference in Econometrics* (New York: Wiley); reprinted 1987 (Malabar, FL: R. E. Krieger Pub. Co.). [8]

Gull, S.F. 1989, in *Maximum Entropy and Bayesian Methods*, J. Skilling, ed. (Boston: Kluwer). [9]

Jaynes, E.T. 1991, in *Maximum-Entropy and Bayesian Methods, Proc. 10th Int. Workshop*, W.T. Grandy, Jr., and L.H. Schick, eds. (Boston: Kluwer). [10]

Macdonald, J.R., and Thompson, W.J. 1992, *American Journal of Physics*, vol. 60, pp. 66–73.

15.4 General Linear Least Squares

An immediate generalization of §15.2 is to fit a set of data points (x_i, y_i) to a model that is not just a linear combination of 1 and x (namely $a + bx$), but rather a linear combination of *any* M specified functions of x. For example, the functions could be $1, x, x^2, \ldots, x^{M-1}$, in which case their general linear combination,

$$y(x) = a_0 + a_1 x + a_2 x^2 + \cdots + a_{M-1} x^{M-1} \qquad (15.4.1)$$

is a polynomial of degree $M - 1$. Or, the functions could be sines and cosines, in which case their general linear combination is a harmonic series.

The general form of this kind of model is

$$y(x) = \sum_{k=0}^{M-1} a_k X_k(x) \qquad (15.4.2)$$

where $X_0(x), \ldots, X_{M-1}(x)$ are arbitrary fixed functions of x, called the *basis functions*.

Note that the functions $X_k(x)$ can be wildly nonlinear functions of x. In this discussion "linear" refers only to the model's dependence on its *parameters* a_k.

For these linear models we generalize the discussion of the previous section by defining a merit function

$$\chi^2 = \sum_{i=0}^{N-1} \left[\frac{y_i - \sum_{k=0}^{M-1} a_k X_k(x_i)}{\sigma_i} \right]^2 \qquad (15.4.3)$$

As before, σ_i is the measurement error (standard deviation) of the ith data point, presumed to be known. If the measurement errors are not known, they may all (as discussed at the end of §15.1) be set to the constant value $\sigma = 1$.

Once again, we will pick as best parameters those that minimize χ^2. There are several different techniques available for finding this minimum. Two are particularly useful, and we will discuss both in this section. To introduce them and elucidate their relationship, we need some notation.

Let $\mathbf{A}$ be a matrix whose $N \times M$ components are constructed from the M basis functions evaluated at the N abscissas x_i, and from the N measurement errors σ_i, by the prescription

$$A_{ij} = \frac{X_j(x_i)}{\sigma_i} \qquad (15.4.4)$$

The matrix $\mathbf{A}$ is called the *design matrix* of the fitting problem. Notice that in general $\mathbf{A}$ has more rows than columns, $N \geq M$, since there must be more data points than model parameters to be solved for. (You can fit a straight line to two points, but not a very meaningful quintic!) The design matrix is shown schematically in Figure 15.4.1.

Also define a vector $\mathbf{b}$ of length N by

$$b_i = \frac{y_i}{\sigma_i} \qquad (15.4.5)$$

and denote the M vector whose components are the parameters to be fitted, $a_0, \ldots, a_{M-1}$, by $\mathbf{a}$.

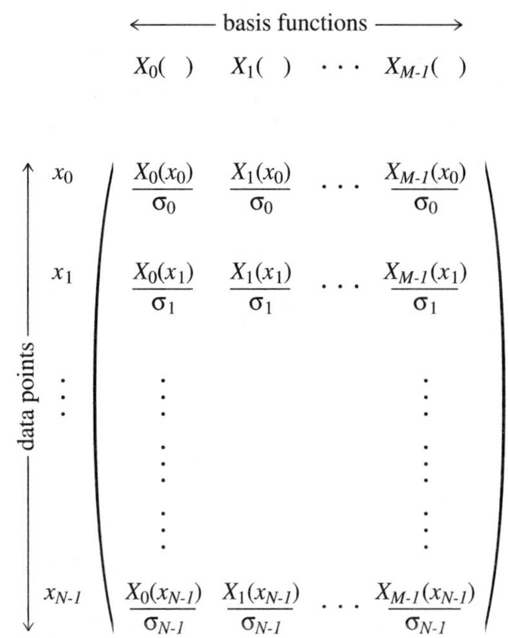

Figure 15.4.1. Design matrix for the least-squares fit of a linear combination of M basis functions to N data points. The matrix elements involve the basis functions evaluated at the values of the independent variable at which measurements are made, and the standard deviations of the measured dependent variable. The measured values of the dependent variable do not enter the design matrix.

Solution by Use of the Normal Equations

The minimum of (15.4.3) occurs where the derivative of χ^2 with respect to all M parameters a_k vanishes. Specializing equation (15.1.7) to the case of the model (15.4.2), this condition yields the M equations

$$0 = \sum_{i=0}^{N-1} \frac{1}{\sigma_i^2} \left[y_i - \sum_{j=0}^{M-1} a_j X_j(x_i) \right] X_k(x_i) \qquad k = 0, \ldots, M-1 \qquad (15.4.6)$$

Interchanging the order of summations, we can write (15.4.6) as the matrix equation

$$\sum_{j=0}^{M-1} \alpha_{kj} a_j = \beta_k \qquad (15.4.7)$$

where

$$\alpha_{kj} = \sum_{i=0}^{N-1} \frac{X_j(x_i) X_k(x_i)}{\sigma_i^2} \qquad \text{or equivalently} \qquad [\alpha] = \mathbf{A}^T \cdot \mathbf{A} \qquad (15.4.8)$$

an $M \times M$ matrix, and

$$\beta_k = \sum_{i=0}^{N-1} \frac{y_i X_k(x_i)}{\sigma_i^2} \qquad \text{or equivalently} \qquad [\beta] = \mathbf{A}^T \cdot \mathbf{b} \qquad (15.4.9)$$

a vector of length M.

The equations (15.4.6) or (15.4.7) are called the *normal equations* of the least-squares problem. They can be solved for the vector of parameters **a** by the standard methods of Chapter 2, notably LU decomposition and backsubstitution, Choleksy decomposition, or Gauss-Jordan elimination. In matrix form, the normal equations can be written as either

$$[\alpha] \cdot \mathbf{a} = [\beta] \qquad \text{or as} \qquad \left(\mathbf{A}^T \cdot \mathbf{A}\right) \cdot \mathbf{a} = \mathbf{A}^T \cdot \mathbf{b} \qquad (15.4.10)$$

The inverse matrix $C_{jk} \equiv [\alpha]_{jk}^{-1}$ is closely related to the probable (or, more precisely, *standard*) uncertainties of the estimated parameters **a**. To estimate these uncertainties, consider that

$$a_j = \sum_{k=0}^{M-1} [\alpha]_{jk}^{-1} \beta_k = \sum_{k=0}^{M-1} C_{jk} \left[\sum_{i=0}^{N-1} \frac{y_i X_k(x_i)}{\sigma_i^2} \right] \qquad (15.4.11)$$

and that the variance associated with the estimate a_j can be found as in (15.2.7) from

$$\sigma^2(a_j) = \sum_{i=0}^{N-1} \sigma_i^2 \left(\frac{\partial a_j}{\partial y_i} \right)^2 \qquad (15.4.12)$$

Note that α_{jk} is independent of y_i, so that

$$\frac{\partial a_j}{\partial y_i} = \sum_{k=0}^{M-1} C_{jk} X_k(x_i) / \sigma_i^2 \qquad (15.4.13)$$

Consequently, we find that

$$\sigma^2(a_j) = \sum_{k=0}^{M-1} \sum_{l=0}^{M-1} C_{jk} C_{jl} \left[\sum_{i=0}^{N-1} \frac{X_k(x_i) X_l(x_i)}{\sigma_i^2} \right] \qquad (15.4.14)$$

The final term in brackets is just the matrix $[\alpha]$. Since this is the matrix inverse of $[C]$, (15.4.14) reduces immediately to

$$\sigma^2(a_j) = C_{jj} \qquad (15.4.15)$$

In other words, the diagonal elements of $[C]$ are the variances (squared uncertainties) of the fitted parameters **a**. It should not surprise you to learn that the off-diagonal elements C_{jk} are the covariances between a_j and a_k (cf. 15.2.10); but we shall defer discussion of these to §15.6.

We will now give a routine that implements the above formulas for the general linear least-squares problem, by the method of normal equations. Since we wish to compute not only the solution vector **a** but also the covariance matrix $[C]$, it is most convenient to use Gauss-Jordan elimination (routine gaussj of §2.1) to perform the linear algebra. The operation count, in this application, is no larger than that for LU decomposition. If you have no need for the covariance matrix, however, you can save a factor of 3 on the linear algebra by switching to LU decomposition, without computation of the matrix inverse. In theory, since $\mathbf{A}^T \cdot \mathbf{A}$ is positive definite, Cholesky decomposition is the most efficient way to solve the normal equations. However, in practice most of the computing time is spent in looping over the data to form the equations, and Gauss-Jordan is quite adequate.

We need to warn you that the solution of a least-squares problem directly from the normal equations is rather susceptible to roundoff error. An alternative, and preferred, technique involves QR decomposition (§2.10, §11.3, and §11.6) of the design matrix $\mathbf{A}$. This is essentially what we did at the end of §15.2 for fitting data to a straight line, but without invoking all the machinery of QR to derive the necessary formulas. Later in this section, we will discuss other difficulties in the least-squares problem, for which the cure is *singular value decomposition* (SVD), of which we give an implementation. It turns out that SVD also fixes the roundoff problem, so it is our recommended technique for all but "easy" least-squares problems. It is for these easy problems that the following routine, which solves the normal equations, is intended.

The routine below introduces one bookkeeping trick that is quite useful in practical work. Frequently it is a matter of "art" to decide which parameters a_k in a model should be fit from the data set, and which should be held constant at fixed values, for example values predicted by a theory or measured in a previous experiment. One wants, therefore, to have a convenient means for "freezing" and "unfreezing" the parameters a_k. In the following routine the total number of parameters a_k is denoted ma (called M above). As input to the routine, you supply a boolean array ia[0..ma-1]. Components that are false indicate that you want the corresponding elements of the parameter vector a[0..ma-1] to be held fixed at their input values. Components that are true indicate parameters that should be fitted for. On output, any frozen parameters will have their variances, and all their covariances, set to zero in the covariance matrix.

```
#include "nr.h"

void NR::lfit(Vec_I_DP &x, Vec_I_DP &y, Vec_I_DP &sig, Vec_IO_DP &a,
    Vec_I_BOOL &ia, Mat_O_DP &covar, DP &chisq,
    void funcs(const DP, Vec_O_DP &))
```
Given a set of data points x[0..ndat-1], y[0..ndat-1] with individual standard deviations sig[0..ndat-1], use χ^2 minimization to fit for some or all of the coefficients a[0..ma-1] of a function that depends linearly on a, $y = \sum_i a_i \times \mathrm{afunc}_i(x)$. The boolean input array ia[0..ma-1] indicates by true entries those components of a that should be fitted for, and by false entries those components that should be held fixed at their input values. The program returns values for a[0..ma-1], χ^2 = chisq, and the covariance matrix covar[0..ma-1][0..ma-1]. (Parameters held fixed will return zero covariances.) The user supplies a routine funcs(x,afunc) that returns the ma basis functions evaluated at $x =$ x in the array afunc[0..ma-1].
```
{
    int i,j,k,l,m,mfit=0;
    DP ym,wt,sum,sig2i;

    int ndat=x.size();
    int ma=a.size();
    Vec_DP afunc(ma);
    Mat_DP beta(ma,1);
    for (j=0;j<ma;j++)
        if (ia[j]) mfit++;
    if (mfit == 0) nrerror("lfit: no parameters to be fitted");
    for (j=0;j<mfit;j++) {                    Initialize the (symmetric) matrix.
        for (k=0;k<mfit;k++) covar[j][k]=0.0;
        beta[j][0]=0.0;
```

```
        }
        for (i=0;i<ndat;i++) {                    Loop over data to accumulate coefficients of
            funcs(x[i],afunc);                        the normal equations.
            ym=y[i];
            if (mfit < ma) {                      Subtract off dependences on known pieces
                for (j=0;j<ma;j++)                    of the fitting function.
                    if (!ia[j]) ym -= a[j]*afunc[j];
            }
            sig2i=1.0/SQR(sig[i]);
            for (j=0,l=0;l<ma;l++) {
                if (ia[l]) {
                    wt=afunc[l]*sig2i;
                    for (k=0,m=0;m<=l;m++)
                        if (ia[m]) covar[j][k++] += wt*afunc[m];
                    beta[j++][0] += ym*wt;
                }
            }
        }
        for (j=1;j<mfit;j++)                      Fill in above the diagonal from symmetry.
            for (k=0;k<j;k++)
                covar[k][j]=covar[j][k];
        Mat_DP temp(mfit,mfit);
        for (j=0;j<mfit;j++)
            for (k=0;k<mfit;k++)
                temp[j][k]=covar[j][k];
        gaussj(temp,beta);                        Matrix solution.
        for (j=0;j<mfit;j++)
            for (k=0;k<mfit;k++)
                covar[j][k]=temp[j][k];
        for (j=0,l=0;l<ma;l++)
            if (ia[l]) a[l]=beta[j++][0];         Partition solution to appropriate coefficients
        chisq=0.0;                                    a.
        for (i=0;i<ndat;i++) {                    Evaluate χ² of the fit.
            funcs(x[i],afunc);
            sum=0.0;
            for (j=0;j<ma;j++) sum += a[j]*afunc[j];
            chisq += SQR((y[i]-sum)/sig[i]);
        }
        covsrt(covar,ia,mfit);                    Sort covariance matrix to true order of fitting
    }                                                 coefficients.
```

That last call to a function covsrt is only for the purpose of spreading the covariances back into the full ma × ma covariance matrix, in the proper rows and columns and with zero variances and covariances set for variables which were held frozen.

The function covsrt is as follows.

```
#include "nr.h"

void NR::covsrt(Mat_IO_DP &covar, Vec_I_BOOL &ia, const int mfit)
Expand in storage the covariance matrix covar, so as to take into account parameters that are
being held fixed. (For the latter, return zero covariances.)
{
    int i,j,k;

    int ma=ia.size();
    for (i=mfit;i<ma;i++)
        for (j=0;j<i+1;j++) covar[i][j]=covar[j][i]=0.0;
    k=mfit-1;
    for (j=ma-1;j>=0;j--) {
        if (ia[j]) {
```

```
        for (i=0;i<ma;i++) SWAP(covar[i][k],covar[i][j]);
        for (i=0;i<ma;i++) SWAP(covar[k][i],covar[j][i]);
        k--;
      }
    }
}
```

Solution by Use of Singular Value Decomposition

In some applications, the normal equations are perfectly adequate for linear least-squares problems. However, in many cases the normal equations are very close to singular. A zero pivot element may be encountered during the solution of the linear equations (e.g., in gaussj), in which case you get no solution at all. Or a very small pivot may occur, in which case you typically get fitted parameters a_k with very large magnitudes that are delicately (and unstably) balanced to cancel out almost precisely when the fitted function is evaluated.

Why does this commonly occur? The reason is that, more often than experimenters would like to admit, data do not clearly distinguish between two or more of the basis functions provided. If two such functions, or two different combinations of functions, happen to fit the data about equally well — or equally badly — then the matrix $[\alpha]$, unable to distinguish between them, neatly folds up its tent and becomes singular. There is a certain mathematical irony in the fact that least-squares problems are *both* overdetermined (number of data points greater than number of parameters) *and* underdetermined (ambiguous combinations of parameters exist); but that is how it frequently is. The ambiguities can be extremely hard to notice *a priori* in complicated problems.

Enter singular value decomposition (SVD). This would be a good time for you to review the material in §2.6, which we will not repeat here. In the case of an overdetermined system, SVD produces a solution that is the best approximation in the least-squares sense, cf. equation (2.6.10). That is exactly what we want. In the case of an underdetermined system, SVD produces a solution whose values (for us, the a_k's) are smallest in the least-squares sense, cf. equation (2.6.8). That is also what we want: When some combination of basis functions is irrelevant to the fit, that combination will be driven down to a small, innocuous, value, rather than pushed up to delicately canceling infinities.

In terms of the design matrix $\mathbf{A}$ (equation 15.4.4) and the vector $\mathbf{b}$ (equation 15.4.5), minimization of χ^2 in (15.4.3) can be written as

$$\text{find} \quad \mathbf{a} \quad \text{that minimizes} \quad \chi^2 = |\mathbf{A} \cdot \mathbf{a} - \mathbf{b}|^2 \qquad (15.4.16)$$

Comparing to equation (2.6.9), we see that this is precisely the problem that routines svdcmp and svbksb are designed to solve. The solution, which is given by equation (2.6.12), can be rewritten as follows: If $\mathbf{U}$ and $\mathbf{V}$ enter the SVD decomposition of $\mathbf{A}$ according to equation (2.6.1), as computed by svdcmp, then let the vectors $\mathbf{U}_{(i)}$ $i = 0, \ldots, M - 1$ denote the *columns* of $\mathbf{U}$ (each one a vector of length N); and let the vectors $\mathbf{V}_{(i)}$ $i = 0, \ldots, M - 1$ denote the *columns* of $\mathbf{V}$ (each one a vector of length M). Then the solution (2.6.12) of the least-squares problem

(15.4.16) can be written as

$$\mathbf{a} = \sum_{i=0}^{M-1} \left(\frac{\mathbf{U}_{(i)} \cdot \mathbf{b}}{w_i} \right) \mathbf{V}_{(i)} \tag{15.4.17}$$

where the w_i are, as in §2.6, the singular values calculated by svdcmp.

Equation (15.4.17) says that the fitted parameters $\mathbf{a}$ are linear combinations of the columns of $\mathbf{V}$, with coefficients obtained by forming dot products of the columns of $\mathbf{U}$ with the weighted data vector (15.4.5). Though it is beyond our scope to prove here, it turns out that the standard (loosely, "probable") errors in the fitted parameters are also linear combinations of the columns of $\mathbf{V}$. In fact, equation (15.4.17) can be written in a form displaying these errors as

$$\mathbf{a} = \left[\sum_{i=0}^{M-1} \left(\frac{\mathbf{U}_{(i)} \cdot \mathbf{b}}{w_i} \right) \mathbf{V}_{(i)} \right] \pm \frac{1}{w_0} \mathbf{V}_{(0)} \pm \cdots \pm \frac{1}{w_{M-1}} \mathbf{V}_{(M-1)} \tag{15.4.18}$$

Here each $\pm$ is followed by a standard deviation. The amazing fact is that, decomposed in this fashion, the standard deviations are all mutually independent (uncorrelated). Therefore they can be added together in root-mean-square fashion. What is going on is that the vectors $\mathbf{V}_{(i)}$ are the principal axes of the error ellipsoid of the fitted parameters $\mathbf{a}$ (see §15.6).

It follows that the variance in the estimate of a parameter a_j is given by

$$\sigma^2(a_j) = \sum_{i=0}^{M-1} \frac{1}{w_i^2} [\mathbf{V}_{(i)}]_j^2 = \sum_{i=0}^{M-1} \left(\frac{V_{ji}}{w_i} \right)^2 \tag{15.4.19}$$

whose result should be identical with (15.4.14). As before, you should not be surprised at the formula for the covariances, here given without proof,

$$\text{Cov}(a_j, a_k) = \sum_{i=0}^{M-1} \left(\frac{V_{ji} V_{ki}}{w_i^2} \right) \tag{15.4.20}$$

We introduced this subsection by noting that the normal equations can fail by encountering a zero pivot. We have not yet, however, mentioned how SVD overcomes this problem. The answer is: If any singular value w_i is zero, its reciprocal in equation (15.4.18) should be set to zero, not infinity. (Compare the discussion preceding equation 2.6.7.) This corresponds to adding to the fitted parameters $\mathbf{a}$ a *zero* multiple, rather than some random large multiple, of any linear combination of basis functions that are degenerate in the fit. It is a good thing to do!

Moreover, if a singular value w_i is nonzero but very small, you should also define *its* reciprocal to be zero, since its apparent value is probably an artifact of roundoff error, not a meaningful number. A plausible answer to the question "how small is small?" is to edit in this fashion all singular values whose ratio to the largest singular value is less than N times the machine precision ϵ. (You might argue for $\sqrt{N}$, or a constant, instead of N as the multiple; that starts getting into hardware-dependent questions.)

There is another reason for editing even *additional* singular values, ones large enough that roundoff error is not a question. Singular value decomposition allows you to identify linear combinations of variables that just happen not to contribute much to reducing the χ^2 of your data set. Editing these can sometimes reduce the probable error on your coefficients quite significantly, while increasing the minimum χ^2 only negligibly. We will learn more about identifying and treating such cases in §15.6. In the following routine, the point at which this kind of editing would occur is indicated.

Generally speaking, we recommend that you always use SVD techniques instead of using the normal equations. SVD's only significant disadvantage is that it requires an extra array of size $N \times M$ to store the whole design matrix. This storage is overwritten by the matrix **U**. Storage is also required for the $M \times M$ matrix **V**, but this is instead of the same-sized matrix for the coefficients of the normal equations. SVD can be significantly slower than solving the normal equations; however, its great advantage, that it (theoretically) *cannot fail*, more than makes up for the speed disadvantage.

In the routine that follows, the matrices u,v and the vector w are input as working space. The logical dimensions of the problem are ndata data points by ma basis functions (and fitted parameters). If you care only about the values a of the fitted parameters, then u,v,w contain no useful information on output. If you want probable errors for the fitted parameters, read on.

```
#include "nr.h"

void NR::svdfit(Vec_I_DP &x, Vec_I_DP &y, Vec_I_DP &sig, Vec_O_DP &a,
    Mat_O_DP &u, Mat_O_DP &v, Vec_O_DP &w, DP &chisq,
    void funcs(const DP, Vec_O_DP &))
```
Given a set of data points x[0..ndata-1],y[0..ndata-1] with individual standard devia-
tions sig[0..ndata-1], use χ^2 minimization to determine the coefficients a[0..ma-1] of the
fitting function $y = \sum_i a_i \times \text{afunc}_i(x)$. Here we solve the fitting equations using singular value
decomposition of the ndata by ma matrix, as in §2.6. Arrays u[0..ndata-1][0..ma-1],
v[0..ma-1][0..ma-1], and w[0..ma-1] provide workspace on input; on output they define the singular value decomposition, and can be used to obtain the covariance matrix. The
program returns values for the ma fit parameters a, and χ^2, chisq. The user supplies a routine funcs(x,afunc) that returns the ma basis functions evaluated at $x =$ x in the array
afunc[0..ma-1].
```
{
    int i,j;
    const DP TOL=1.0e-13;                   Default value for double precision and vari-
    DP wmax,tmp,thresh,sum;                     ables scaled to order unity.

    int ndata=x.size();
    int ma=a.size();
    Vec_DP b(ndata),afunc(ma);
    for (i=0;i<ndata;i++) {                  Accumulate coefficients of the fitting ma-
        funcs(x[i],afunc);                      trix.
        tmp=1.0/sig[i];
        for (j=0;j<ma;j++) u[i][j]=afunc[j]*tmp;
        b[i]=y[i]*tmp;
    }
    svdcmp(u,w,v);                           Singular value decomposition.
    wmax=0.0;                                Edit the singular values, given TOL defined
    for (j=0;j<ma;j++)                           above, between here ...
        if (w[j] > wmax) wmax=w[j];
    thresh=TOL*wmax;
    for (j=0;j<ma;j++)
        if (w[j] < thresh) w[j]=0.0;         ...and here.
```

```
        svbksb(u,w,v,b,a);
        chisq=0.0;                                    Evaluate chi-square.
        for (i=0;i<ndata;i++) {
            funcs(x[i],afunc);
            sum=0.0;
            for (j=0;j<ma;j++) sum += a[j]*afunc[j];
            chisq += (tmp=(y[i]-sum)/sig[i],tmp*tmp);
        }
    }
```

Feeding the matrix v and vector w output by the above program into the following short routine, you easily obtain variances and covariances of the fitted parameters a. The square roots of the variances are the standard deviations of the fitted parameters. The routine straightforwardly implements equation (15.4.20) above, with the convention that singular values equal to zero are recognized as having been edited out of the fit.

```
#include "nr.h"

void NR::svdvar(Mat_I_DP &v, Vec_I_DP &w, Mat_O_DP &cvm)
To evaluate the covariance matrix cvm[0..ma-1][0..ma-1] of the fit for ma parameters
obtained by svdfit, call this routine with matrices v[0..ma-1][0..ma-1], w[0..ma-1]
as returned from svdfit.
{
    int i,j,k;
    DP sum;

    int ma=w.size();
    Vec_DP wti(ma);
    for (i=0;i<ma;i++) {
        wti[i]=0.0;
        if (w[i] != 0.0) wti[i]=1.0/(w[i]*w[i]);
    }
    for (i=0;i<ma;i++) {          Sum contributions to covariance matrix (15.4.20).
        for (j=0;j<i+1;j++) {
            sum=0.0;
            for (k=0;k<ma;k++)
                sum += v[i][k]*v[j][k]*wti[k];
            cvm[j][i]=cvm[i][j]=sum;
        }
    }
}
```

Examples

Be aware that some apparently nonlinear problems can be expressed so that they are linear. For example, an exponential model with two parameters a and b,

$$y(x) = a \exp(-bx) \qquad (15.4.21)$$

can be rewritten as

$$\log[y(x)] = c - bx \qquad (15.4.22)$$

which is linear in its parameters c and b. (Of course you must be aware that such transformations do not exactly take Gaussian errors into Gaussian errors.)

Also watch out for "non-parameters," as in

$$y(x) = a \exp(-bx + d) \qquad (15.4.23)$$

Here the parameters a and d are, in fact, indistinguishable. This is a good example of where the normal equations will be exactly singular, and where SVD will find a zero singular value. SVD will then make a "least-squares" choice for setting a balance between a and d (or, rather, their equivalents in the linear model derived by taking the logarithms). However — and this is true whenever SVD gives back a zero singular value — you are better advised to figure out analytically where the degeneracy is among your basis functions, and then make appropriate deletions in the basis set.

Here are two examples for user-supplied routines `funcs`. The first one is trivial and fits a general polynomial to a set of data:

```
#include "nr.h"

void NR::fpoly(const DP x, Vec_O_DP &p)
Fitting routine for a polynomial of degree np-1, with coefficients in the array p[0..np-1].
{
    int j;

    int np=p.size();
    p[0]=1.0;
    for (j=1;j<np;j++) p[j]=p[j-1]*x;
}
```

The second example is slightly less trivial. It is used to fit Legendre polynomials up to some order `nl-1` to a data set.

```
#include "nr.h"

void NR::fleg(const DP x, Vec_O_DP &pl)
Fitting routine for an expansion with nl Legendre polynomials pl, evaluated using the recurrence
relation as in §5.5.
{
    int j;
    DP twox,f2,f1,d;

    int nl=pl.size();
    pl[0]=1.0;
    pl[1]=x;
    if (nl > 2) {
        twox=2.0*x;
        f2=x;
        d=1.0;
        for (j=2;j<nl;j++) {
            f1=d++;
            f2+=twox;
            pl[j]=(f2*pl[j-1]-f1*pl[j-2])/d;
        }
    }
}
```

Multidimensional Fits

If you are measuring a single variable y as a function of more than one variable — say, a *vector* of variables $\mathbf{x}$, then your basis functions will be functions of a vector, $X_0(\mathbf{x}), \ldots, X_{M-1}(\mathbf{x})$. The χ^2 merit function is now

$$\chi^2 = \sum_{i=0}^{N-1} \left[\frac{y_i - \sum_{k=0}^{M-1} a_k X_k(\mathbf{x}_i)}{\sigma_i} \right]^2 \qquad (15.4.24)$$

All of the preceding discussion goes through unchanged, with x replaced by $\mathbf{x}$. In fact, if you are willing to tolerate a bit of programming hack, you can use the above programs without any modification: In both lfit and svdfit, the only use made of the array elements x[i] is that each element is in turn passed to the user-supplied routine funcs, which duly gives back the values of the basis functions at that point. If you set x[i]=i before calling lfit or svdfit, and independently provide funcs with the true vector values of your data points (e.g., in global variables), then funcs can translate from the fictitious x[i]'s to the actual data points before doing its work.

CITED REFERENCES AND FURTHER READING:

Bevington, P.R., and Robinson, D.K. 1992, *Data Reduction and Error Analysis for the Physical Sciences*, 2nd ed. (New York: McGraw-Hill), Chapter 7.

Lawson, C.L., and Hanson, R. 1974, *Solving Least Squares Problems* (Englewood Cliffs, NJ: Prentice-Hall).

Forsythe, G.E., Malcolm, M.A., and Moler, C.B. 1977, *Computer Methods for Mathematical Computations* (Englewood Cliffs, NJ: Prentice-Hall), Chapter 9.

15.5 Nonlinear Models

We now consider fitting when the model depends *nonlinearly* on the set of M unknown parameters $a_k, k = 0, 1, \ldots, M - 1$. We use the same approach as in previous sections, namely to define a χ^2 merit function and determine best-fit parameters by its minimization. With nonlinear dependences, however, the minimization must proceed iteratively. Given trial values for the parameters, we develop a procedure that improves the trial solution. The procedure is then repeated until χ^2 stops (or effectively stops) decreasing.

How is this problem different from the general nonlinear function minimization problem already dealt with in Chapter 10? Superficially, not at all: Sufficiently close to the minimum, we expect the χ^2 function to be well approximated by a quadratic form, which we can write as

$$\chi^2(\mathbf{a}) \approx \gamma - \mathbf{d} \cdot \mathbf{a} + \frac{1}{2} \mathbf{a} \cdot \mathbf{D} \cdot \mathbf{a} \qquad (15.5.1)$$

where $\mathbf{d}$ is an M-vector and $\mathbf{D}$ is an $M \times M$ matrix. (Compare equation 10.6.1.) If the approximation is a good one, we know how to jump from the current trial parameters $\mathbf{a}_{\mathrm{cur}}$ to the minimizing ones $\mathbf{a}_{\mathrm{min}}$ in a single leap, namely

$$\mathbf{a}_{\mathrm{min}} = \mathbf{a}_{\mathrm{cur}} + \mathbf{D}^{-1} \cdot \left[-\nabla \chi^2(\mathbf{a}_{\mathrm{cur}}) \right] \qquad (15.5.2)$$

(Compare equation 10.7.4.)

On the other hand, (15.5.1) might be a poor local approximation to the shape of the function that we are trying to minimize at $\mathbf{a}_{\text{cur}}$. In that case, about all we can do is take a step down the gradient, as in the steepest descent method (§10.6). In other words,

$$\mathbf{a}_{\text{next}} = \mathbf{a}_{\text{cur}} - \text{constant} \times \nabla\chi^2(\mathbf{a}_{\text{cur}}) \tag{15.5.3}$$

where the constant is small enough not to exhaust the downhill direction.

To use (15.5.2) or (15.5.3), we must be able to compute the gradient of the χ^2 function at any set of parameters $\mathbf{a}$. To use (15.5.2) we also need the matrix $\mathbf{D}$, which is the second derivative matrix (Hessian matrix) of the χ^2 merit function, at any $\mathbf{a}$.

Now, this is the crucial difference from Chapter 10: There, we had no way of directly evaluating the Hessian matrix. We were given only the ability to evaluate the function to be minimized and (in some cases) its gradient. Therefore, we had to resort to iterative methods *not just* because our function was nonlinear, *but also* in order to build up information about the Hessian matrix. Sections 10.7 and 10.6 concerned themselves with two different techniques for building up this information.

Here, life is much simpler. We *know* exactly the form of χ^2, since it is based on a model function that we ourselves have specified. Therefore the Hessian matrix is known to us. Thus we are free to use (15.5.2) whenever we care to do so. The only reason to use (15.5.3) will be failure of (15.5.2) to improve the fit, signaling failure of (15.5.1) as a good local approximation.

Calculation of the Gradient and Hessian

The model to be fitted is

$$y = y(x; \mathbf{a}) \tag{15.5.4}$$

and the χ^2 merit function is

$$\chi^2(\mathbf{a}) = \sum_{i=0}^{N-1} \left[\frac{y_i - y(x_i; \mathbf{a})}{\sigma_i} \right]^2 \tag{15.5.5}$$

The gradient of χ^2 with respect to the parameters $\mathbf{a}$, which will be zero at the χ^2 minimum, has components

$$\frac{\partial\chi^2}{\partial a_k} = -2 \sum_{i=0}^{N-1} \frac{[y_i - y(x_i; \mathbf{a})]}{\sigma_i^2} \frac{\partial y(x_i; \mathbf{a})}{\partial a_k} \qquad k = 0, 1, \ldots, M-1 \tag{15.5.6}$$

Taking an additional partial derivative gives

$$\frac{\partial^2\chi^2}{\partial a_k \partial a_l} = 2 \sum_{i=0}^{N-1} \frac{1}{\sigma_i^2} \left[\frac{\partial y(x_i; \mathbf{a})}{\partial a_k} \frac{\partial y(x_i; \mathbf{a})}{\partial a_l} - [y_i - y(x_i; \mathbf{a})] \frac{\partial^2 y(x_i; \mathbf{a})}{\partial a_l \partial a_k} \right] \tag{15.5.7}$$

It is conventional to remove the factors of 2 by defining

$$\beta_k \equiv -\frac{1}{2} \frac{\partial\chi^2}{\partial a_k} \qquad \alpha_{kl} \equiv \frac{1}{2} \frac{\partial^2\chi^2}{\partial a_k \partial a_l} \tag{15.5.8}$$

making $[\alpha] = \frac{1}{2}\mathbf{D}$ in equation (15.5.2), in terms of which that equation can be rewritten as the set of linear equations

$$\sum_{l=0}^{M-1} \alpha_{kl}\, \delta a_l = \beta_k \qquad (15.5.9)$$

This set is solved for the increments δa_l that, added to the current approximation, give the next approximation. In the context of least-squares, the matrix $[\alpha]$, equal to one-half times the Hessian matrix, is usually called the *curvature matrix*.

Equation (15.5.3), the steepest descent formula, translates to

$$\delta a_l = \text{constant} \times \beta_l \qquad (15.5.10)$$

Note that the components α_{kl} of the Hessian matrix (15.5.7) depend both on the first derivatives and on the second derivatives of the basis functions with respect to their parameters. Some treatments proceed to ignore the second derivative without comment. We will ignore it also, but only *after* a few comments.

Second derivatives occur because the gradient (15.5.6) already has a dependence on $\partial y/\partial a_k$, so the next derivative simply must contain terms involving $\partial^2 y/\partial a_l \partial a_k$. The second derivative term can be dismissed when it is zero (as in the linear case of equation 15.4.8), or small enough to be negligible when compared to the term involving the first derivative. It also has an additional possibility of being ignorably small in practice: The term multiplying the second derivative in equation (15.5.7) is $[y_i - y(x_i; \mathbf{a})]$. For a successful model, this term should just be the random measurement error of each point. This error can have either sign, and should in general be uncorrelated with the model. Therefore, the second derivative terms tend to cancel out when summed over i.

Inclusion of the second-derivative term can in fact be destabilizing if the model fits badly or is contaminated by outlier points that are unlikely to be offset by compensating points of opposite sign. From this point on, we will always use as the definition of α_{kl} the formula

$$\alpha_{kl} = \sum_{i=0}^{N-1} \frac{1}{\sigma_i^2} \left[\frac{\partial y(x_i; \mathbf{a})}{\partial a_k} \frac{\partial y(x_i; \mathbf{a})}{\partial a_l} \right] \qquad (15.5.11)$$

This expression more closely resembles its linear cousin (15.4.8). You should understand that minor (or even major) fiddling with $[\alpha]$ has no effect at all on what final set of parameters $\mathbf{a}$ is reached, but affects only the iterative route that is taken in getting there. The condition at the χ^2 minimum, that $\beta_k = 0$ for all k, is independent of how $[\alpha]$ is defined.

Levenberg-Marquardt Method

Marquardt [1] has put forth an elegant method, related to an earlier suggestion of Levenberg, for varying smoothly between the extremes of the inverse-Hessian method (15.5.9) and the steepest descent method (15.5.10). The latter method is used far from the minimum, switching continuously to the former as the minimum is approached.

This *Levenberg-Marquardt method* (also called *Marquardt method*) works very well in practice and has become the standard of nonlinear least-squares routines.

The method is based on two elementary, but important, insights. Consider the "constant" in equation (15.5.10). What should it be, even in order of magnitude? What sets its scale? There is no information about the answer in the gradient. That tells only the slope, not how far that slope extends. Marquardt's first insight is that the components of the Hessian matrix, even if they are not usable in any precise fashion, give *some* information about the order-of-magnitude scale of the problem.

The quantity χ^2 is nondimensional, i.e., is a pure number; this is evident from its definition (15.5.5). On the other hand, β_k has the dimensions of $1/a_k$, which may well be dimensional, i.e., have units like cm^{-1}, or kilowatt-hours, or whatever. (In fact, each component of β_k can have different dimensions!) The constant of proportionality between β_k and δa_k must therefore have the dimensions of a_k^2. Scan the components of $[\alpha]$ and you see that there is only one obvious quantity with these dimensions, and that is $1/\alpha_{kk}$, the reciprocal of the diagonal element. So that must set the scale of the constant. But that scale might itself be too big. So let's divide the constant by some (nondimensional) fudge factor λ, with the possibility of setting $\lambda \gg 1$ to cut down the step. In other words, replace equation (15.5.10) by

$$\delta a_l = \frac{1}{\lambda \alpha_{ll}} \beta_l \quad \text{or} \quad \lambda \alpha_{ll} \, \delta a_l = \beta_l \qquad (15.5.12)$$

It is necessary that α_{ll} be positive, but this is guaranteed by definition (15.5.11) — another reason for adopting that equation.

Marquardt's second insight is that equations (15.5.12) and (15.5.9) can be combined if we define a new matrix α' by the following prescription

$$\begin{aligned}
\alpha'_{jj} &\equiv \alpha_{jj}(1 + \lambda) \\
\alpha'_{jk} &\equiv \alpha_{jk} \qquad (j \neq k)
\end{aligned} \qquad (15.5.13)$$

and then replace both (15.5.12) and (15.5.9) by

$$\sum_{l=0}^{M-1} \alpha'_{kl} \, \delta a_l = \beta_k \qquad (15.5.14)$$

When λ is very large, the matrix α' is forced into being *diagonally dominant*, so equation (15.5.14) goes over to be identical to (15.5.12). On the other hand, as λ approaches zero, equation (15.5.14) goes over to (15.5.9).

Given an initial guess for the set of fitted parameters **a**, the recommended Marquardt recipe is as follows:

- Compute $\chi^2(\mathbf{a})$.
- Pick a modest value for λ, say $\lambda = 0.001$.
- (†) Solve the linear equations (15.5.14) for $\delta \mathbf{a}$ and evaluate $\chi^2(\mathbf{a} + \delta \mathbf{a})$.
- If $\chi^2(\mathbf{a} + \delta \mathbf{a}) \geq \chi^2(\mathbf{a})$, *increase* λ by a factor of 10 (or any other substantial factor) and go back to (†).

- If $\chi^2(\mathbf{a} + \delta\mathbf{a}) < \chi^2(\mathbf{a})$, *decrease* λ by a factor of 10, update the trial solution $\mathbf{a} \leftarrow \mathbf{a} + \delta\mathbf{a}$, and go back to (†).

Also necessary is a condition for stopping. Iterating to convergence (to machine accuracy or to the roundoff limit) is generally wasteful and unnecessary since the minimum is at best only a statistical estimate of the parameters $\mathbf{a}$. As we will see in §15.6, a change in the parameters that changes χ^2 by an amount $\ll 1$ is *never* statistically meaningful.

Furthermore, it is not uncommon to find the parameters wandering around near the minimum in a flat valley of complicated topography. The reason is that Marquardt's method generalizes the method of normal equations (§15.4), hence has the same problem as that method with regard to near-degeneracy of the minimum. Outright failure by a zero pivot is possible, but unlikely. More often, a small pivot will generate a large correction which is then rejected, the value of λ being then increased. For sufficiently large λ the matrix $[\alpha']$ is positive definite and can have no small pivots. Thus the method does tend to stay away from zero pivots, but at the cost of a tendency to wander around doing steepest descent in very un-steep degenerate valleys.

These considerations suggest that, in practice, one might as well stop iterating on the first or second occasion that χ^2 decreases by a negligible amount, say either less than 0.01 absolutely or (in case roundoff prevents that being reached) some fractional amount like 10^{-3}. Don't stop after a step where χ^2 *increases*: That only shows that λ has not yet adjusted itself optimally.

Once the acceptable minimum has been found, one wants to set $\lambda = 0$ and compute the matrix

$$[C] \equiv [\alpha]^{-1} \tag{15.5.15}$$

which, as before, is the estimated covariance matrix of the standard errors in the fitted parameters $\mathbf{a}$ (see next section).

The following pair of functions encodes Marquardt's method for nonlinear parameter estimation. Much of the organization matches that used in lfit of §15.4. In particular the array ia[0..ma-1] must be input with components true or false corresponding to whether the respective parameter values a[0..ma-1] are to be fitted for or held fixed at their input values, respectively.

The routine mrqmin performs one iteration of Marquardt's method. It is first called (once) with alamda < 0, which signals the routine to initialize. alamda is set on the first and all subsequent calls to the suggested value of λ for the next iteration; a and chisq are always given back as the best parameters found so far and their χ^2. When convergence is deemed satisfactory, set alamda to zero before a final call. The matrices alpha and covar (which were used as workspace in all previous calls) will then be set to the curvature and covariance matrices for the converged parameter values. The arguments alpha, a, and chisq must not be modified between calls, nor should alamda be, except to set it to zero for the final call. When an uphill step is taken, chisq and a are given back with their input (best) values, but alamda is set to an increased value.

The routine mrqmin calls the routine mrqcof for the computation of the matrix $[\alpha]$ (equation 15.5.11) and vector β (equations 15.5.6 and 15.5.8). In turn mrqcof calls the user-supplied routine funcs(x,a,y,dyda), which for input values $\mathbf{x} \equiv x_i$

and a $\equiv$ **a** calculates the model function y $\equiv y(x_i; \mathbf{a})$ and the vector of derivatives dyda $\equiv \partial y/\partial a_k$.

```
#include "nr.h"

void NR::mrqmin(Vec_I_DP &x, Vec_I_DP &y, Vec_I_DP &sig, Vec_IO_DP &a,
    Vec_I_BOOL &ia, Mat_O_DP &covar, Mat_O_DP &alpha, DP &chisq,
    void funcs(const DP, Vec_I_DP &, DP &, Vec_O_DP &), DP &alamda)
```
Levenberg-Marquardt method: attempt to reduce the χ^2 of a fit between a set of data points x[0..ndata-1], y[0..ndata-1] with individual standard deviations sig[0..ndata-1], and a nonlinear function that depends on ma coefficients a[0..ma-1]. Entries that are true in the input array ia[0..ma-1] indicate those components of a that should be fitted for, and false entries indicate those components that should be held fixed at their input values. The program returns current best-fit values for the parameters a[0..ma-1], and χ^2 = chisq. The arrays covar[0..ma-1][0..ma-1], alpha[0..ma-1][0..ma-1] are used as working space during most iterations. Supply a routine funcs(x,a,yfit,dyda) that evaluates the fitting function yfit, and its derivatives dyda[0..ma-1] with respect to the fitting parameters a at x. On the first call provide an initial guess for the parameters a, and set alamda<0 for initialization (which then sets alamda=.001). If a step succeeds chisq becomes smaller and alamda decreases by a factor of 10. If a step fails alamda grows by a factor of 10. You must call this routine repeatedly until convergence is achieved. Then, make one final call with alamda=0, so that covar[0..ma-1][0..ma-1] returns the covariance matrix, and alpha the curvature matrix. (Parameters held fixed will return zero covariances.)
```
{
    static int mfit;
    static DP ochisq;
    int j,k,l;

    int ma=a.size();
    static Mat_DP oneda(ma,1);
    static Vec_DP atry(ma),beta(ma),da(ma);
    if (alamda < 0.0) {                     Initialization.
        mfit=0;
        for (j=0;j<ma;j++)
            if (ia[j]) mfit++;
        alamda=0.001;
        mrqcof(x,y,sig,a,ia,alpha,beta,chisq,funcs);
        ochisq=chisq;
        for (j=0;j<ma;j++) atry[j]=a[j];
    }
    Mat_DP temp(mfit,mfit);
    for (j=0;j<mfit;j++) {                   Alter linearized fitting matrix, by augmenting di-
        for (k=0;k<mfit;k++) covar[j][k]=alpha[j][k];     agonal elements.
        covar[j][j]=alpha[j][j]*(1.0+alamda);
        for (k=0;k<mfit;k++) temp[j][k]=covar[j][k];
        oneda[j][0]=beta[j];
    }
    gaussj(temp,oneda);                      Matrix solution.
    for (j=0;j<mfit;j++) {
        for (k=0;k<mfit;k++) covar[j][k]=temp[j][k];
        da[j]=oneda[j][0];
    }
    if (alamda == 0.0) {                     Once converged, evaluate covariance matrix.
        covsrt(covar,ia,mfit);
        covsrt(alpha,ia,mfit);               Spread out alpha to its full size too.
        return;
    }
    for (j=0,l=0;l<ma;l++)                    Did the trial succeed?
        if (ia[l]) atry[l]=a[l]+da[j++];
    mrqcof(x,y,sig,atry,ia,covar,da,chisq,funcs);
    if (chisq < ochisq) {                    Success, accept the new solution.
        alamda *= 0.1;
        ochisq=chisq;
```

```
        for (j=0;j<mfit;j++) {
            for (k=0;k<mfit;k++) alpha[j][k]=covar[j][k];
                beta[j]=da[j];
        }
        for (l=0;l<ma;l++) a[l]=atry[l];
    } else {                        Failure, increase alamda and return.
        alamda *= 10.0;
        chisq=ochisq;
    }
}
```

Notice the use of the routine covsrt from §15.4. This is merely for rearranging the covariance matrix covar into the order of all ma parameters. The above routine also makes use of

```
#include "nr.h"

void NR::mrqcof(Vec_I_DP &x, Vec_I_DP &y, Vec_I_DP &sig, Vec_I_DP &a,
    Vec_I_BOOL &ia, Mat_O_DP &alpha, Vec_O_DP &beta, DP &chisq,
    void funcs(const DP, Vec_I_DP &,DP &, Vec_O_DP &))
```
Used by mrqmin to evaluate the linearized fitting matrix alpha, and vector beta as in (15.5.8), and to calculate χ^2.
```
{
    int i,j,k,l,m,mfit=0;
    DP ymod,wt,sig2i,dy;

    int ndata=x.size();
    int ma=a.size();
    Vec_DP dyda(ma);
    for (j=0;j<ma;j++)
        if (ia[j]) mfit++;
    for (j=0;j<mfit;j++) {                  Initialize (symmetric) alpha, beta.
        for (k=0;k<=j;k++) alpha[j][k]=0.0;
        beta[j]=0.0;
    }
    chisq=0.0;
    for (i=0;i<ndata;i++) {                  Summation loop over all data.
        funcs(x[i],a,ymod,dyda);
        sig2i=1.0/(sig[i]*sig[i]);
        dy=y[i]-ymod;
        for (j=0,l=0;l<ma;l++) {
            if (ia[l]) {
                wt=dyda[l]*sig2i;
                for (k=0,m=0;m<l+1;m++)
                    if (ia[m]) alpha[j][k++] += wt*dyda[m];
                beta[j++] += dy*wt;
            }
        }
        chisq += dy*dy*sig2i;               And find $\chi^2$.
    }
    for (j=1;j<mfit;j++)                     Fill in the symmetric side.
        for (k=0;k<j;k++) alpha[k][j]=alpha[j][k];
}
```

Example

The following function fgauss is an example of a user-supplied function funcs. Used with the above routine mrqmin (in turn using mrqcof, covsrt, and gaussj), it fits for the model

$$y(x) = \sum_{k=0}^{K-1} B_k \exp\left[-\left(\frac{x-E_k}{G_k}\right)^2\right] \qquad (15.5.16)$$

which is a sum of K Gaussians, each having a variable position, amplitude, and width. We store the parameters in the order $B_0, E_0, G_0, B_1, E_1, G_1, \ldots, B_{K-1}, E_{K-1}, G_{K-1}$.

```
#include <cmath>
#include "nr.h"
using namespace std;

void NR::fgauss(const DP x, Vec_I_DP &a, DP &y, Vec_O_DP &dyda)
```
$y(x; a)$ is the sum of na/3 Gaussians (15.5.16). The amplitude, center, and width of the Gaussians are stored in consecutive locations of a: $a[3k] = B_k$, $a[3k+1] = E_k$, $a[3k+2] = G_k$, $k = 0, ..., na/3 - 1$. The dimensions of the arrays are $a[0..na-1]$, $dyda[0..na-1]$.
```
{
    int i;
    DP fac,ex,arg;

    int na=a.size();
    y=0.0;
    for (i=0;i<na-1;i+=3) {
        arg=(x-a[i+1])/a[i+2];
        ex=exp(-arg*arg);
        fac=a[i]*ex*2.0*arg;
        y += a[i]*ex;
        dyda[i]=ex;
        dyda[i+1]=fac/a[i+2];
        dyda[i+2]=fac*arg/a[i+2];
    }
}
```

More Advanced Methods for Nonlinear Least Squares

The Levenberg-Marquardt algorithm can be implemented as a model-trust region method for minimization (see §9.7 and ref. [2]) applied to the special case of a least squares function. A code of this kind due to Moré [3] can be found in MINPACK [4]. Another algorithm for nonlinear least-squares keeps the second-derivative term we dropped in the Levenberg-Marquardt method whenever it would be better to do so. These methods are called "full Newton-type" methods and are reputed to be more robust than Levenberg-Marquardt, but more complex. One implementation is the code NL2SOL [5].

CITED REFERENCES AND FURTHER READING:

Bevington, P.R., and Robinson, D.K. 1992, *Data Reduction and Error Analysis for the Physical Sciences*, 2nd ed. (New York: McGraw-Hill), Chapter 8.

Marquardt, D.W. 1963, *Journal of the Society for Industrial and Applied Mathematics*, vol. 11, pp. 431–441. [1]

Jacobs, D.A.H. (ed.) 1977, *The State of the Art in Numerical Analysis* (London: Academic Press), Chapter III.2 (by J.E. Dennis).

Dennis, J.E., and Schnabel, R.B. 1983, *Numerical Methods for Unconstrained Optimization and Nonlinear Equations*; reprinted 1996 (Philadelphia: S.I.A.M.). [2]

Moré, J.J. 1977, in *Numerical Analysis*, Lecture Notes in Mathematics, vol. 630, G.A. Watson, ed. (Berlin: Springer-Verlag), pp. 105–116. [3]

Moré, J.J., Garbow, B.S., and Hillstrom, K.E. 1980, *User Guide for MINPACK-1*, Argonne National Laboratory Report ANL-80-74. [4]

Dennis, J.E., Gay, D.M, and Welsch, R.E. 1981, *ACM Transactions on Mathematical Software*, vol. 7, pp. 348–368; *op. cit.*, pp. 369–383. [5].

15.6 Confidence Limits on Estimated Model Parameters

Several times already in this chapter we have made statements about the standard errors, or uncertainties, in a set of M estimated parameters **a**. We have given some formulas for computing standard deviations or variances of individual parameters (equations 15.2.9, 15.4.15, 15.4.19), as well as some formulas for covariances between pairs of parameters (equation 15.2.10; remark following equation 15.4.15; equation 15.4.20; equation 15.5.15).

In this section, we want to be more explicit regarding the precise meaning of these quantitative uncertainties, and to give further information about how quantitative confidence limits on fitted parameters can be estimated. The subject can get somewhat technical, and even somewhat confusing, so we will try to make precise statements, even when they must be offered without proof.

Figure 15.6.1 shows the conceptual scheme of an experiment that "measures" a set of parameters. There is some underlying true set of parameters $\mathbf{a}_{\text{true}}$ that are known to Mother Nature but hidden from the experimenter. These true parameters are statistically realized, along with random measurement errors, as a measured data set, which we will symbolize as $\mathcal{D}_{(0)}$. The data set $\mathcal{D}_{(0)}$ *is* known to the experimenter. He or she fits the data to a model by χ^2 minimization or some other technique, and obtains measured, i.e., fitted, values for the parameters, which we here denote $\mathbf{a}_{(0)}$.

Because measurement errors have a random component, $\mathcal{D}_{(0)}$ is not a unique realization of the true parameters $\mathbf{a}_{\text{true}}$. Rather, there are infinitely many other realizations of the true parameters as "hypothetical data sets" each of which *could* have been the one measured, but happened not to be. Let us symbolize these by $\mathcal{D}_{(1)}, \mathcal{D}_{(2)}, \ldots$. Each one, had it been realized, would have given a slightly different set of fitted parameters, $\mathbf{a}_{(1)}, \mathbf{a}_{(2)}, \ldots$, respectively. These parameter sets $\mathbf{a}_{(i)}$ therefore occur with some probability distribution in the M-dimensional space of all possible parameter sets **a**. The actual measured set $\mathbf{a}_{(0)}$ is one member drawn from this distribution.

Even more interesting than the probability distribution of $\mathbf{a}_{(i)}$ would be the distribution of the difference $\mathbf{a}_{(i)} - \mathbf{a}_{\text{true}}$. This distribution differs from the former one by a translation that puts Mother Nature's true value at the origin. If we knew *this* distribution, we would know everything that there is to know about the quantitative uncertainties in our experimental measurement $\mathbf{a}_{(0)}$.

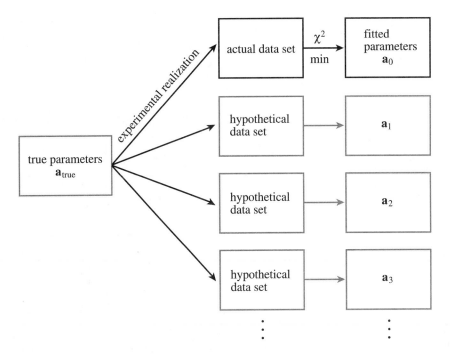

Figure 15.6.1.   A statistical universe of data sets from an underlying model. True parameters $\mathbf{a}_{\text{true}}$ are realized in a data set, from which fitted (observed) parameters $\mathbf{a}_0$ are obtained. If the experiment were repeated many times, new data sets and new values of the fitted parameters would be obtained.

So the name of the game is to find some way of estimating or approximating the probability distribution of $\mathbf{a}_{(i)} - \mathbf{a}_{\text{true}}$ without knowing $\mathbf{a}_{\text{true}}$ and without having available to us an infinite universe of hypothetical data sets.

Monte Carlo Simulation of Synthetic Data Sets

Although the measured parameter set $\mathbf{a}_{(0)}$ is not the true one, let us consider a fictitious world in which it *was* the true one. Since we hope that our measured parameters are not *too* wrong, we hope that that fictitious world is not too different from the actual world with parameters $\mathbf{a}_{\text{true}}$. In particular, let us hope — no, let us *assume* — that the shape of the probability distribution $\mathbf{a}_{(i)} - \mathbf{a}_{(0)}$ in the fictitious world is the same, or very nearly the same, as the shape of the probability distribution $\mathbf{a}_{(i)} - \mathbf{a}_{\text{true}}$ in the real world. Notice that we are not assuming that $\mathbf{a}_{(0)}$ and $\mathbf{a}_{\text{true}}$ are equal; they are certainly not. We are only assuming that the way in which random errors enter the experiment and data analysis does not vary rapidly as a function of $\mathbf{a}_{\text{true}}$, so that $\mathbf{a}_{(0)}$ can serve as a reasonable surrogate.

Now, often, the distribution of $\mathbf{a}_{(i)} - \mathbf{a}_{(0)}$ in the fictitious world *is* within our power to calculate (see Figure 15.6.2). If we know something about the process that generated our data, given an assumed set of parameters $\mathbf{a}_{(0)}$, then we can usually figure out how to *simulate* our own sets of "synthetic" realizations of these parameters as "synthetic data sets." The procedure is to draw random numbers from appropriate distributions (cf. §7.2–§7.3) so as to mimic our best understanding of the underlying process and measurement errors in our apparatus. With such random draws, we construct data sets with exactly the same numbers of measured points, and

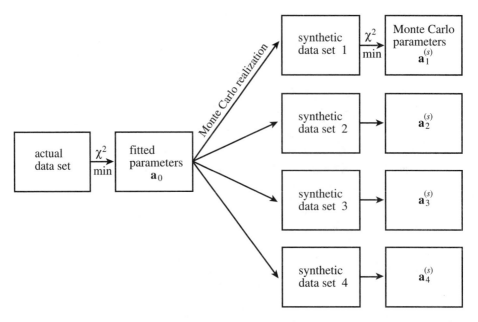

Figure 15.6.2. Monte Carlo simulation of an experiment. The fitted parameters from an actual experiment are used as surrogates for the true parameters. Computer-generated random numbers are used to simulate many synthetic data sets. Each of these is analyzed to obtain its fitted parameters. The distribution of these fitted parameters around the (known) surrogate true parameters is thus studied.

precisely the same values of all control (independent) variables, as our actual data set $\mathcal{D}_{(0)}$. Let us call these simulated data sets $\mathcal{D}^S_{(1)}, \mathcal{D}^S_{(2)}, \ldots$. By construction these are supposed to have exactly the same statistical relationship to $\mathbf{a}_{(0)}$ as the $\mathcal{D}_{(i)}$'s have to $\mathbf{a}_{\text{true}}$. (For the case where you don't know enough about what you are measuring to do a credible job of simulating it, see below.)

Next, for each $\mathcal{D}^S_{(j)}$, perform exactly the same procedure for estimation of parameters, e.g., χ^2 minimization, as was performed on the actual data to get the parameters $\mathbf{a}_{(0)}$, giving simulated measured parameters $\mathbf{a}^S_{(1)}, \mathbf{a}^S_{(2)}, \ldots$. Each simulated measured parameter set yields a point $\mathbf{a}^S_{(i)} - \mathbf{a}_{(0)}$. Simulate enough data sets and enough derived simulated measured parameters, and you map out the desired probability distribution in M dimensions.

In fact, the ability to do *Monte Carlo simulations* in this fashion has revolutionized many fields of modern experimental science. Not only is one able to characterize the errors of parameter estimation in a very precise way; one can also try out on the computer different methods of parameter estimation, or different data reduction techniques, and seek to minimize the uncertainty of the result according to any desired criteria. Offered the choice between mastery of a five-foot shelf of analytical statistics books and middling ability at performing statistical Monte Carlo simulations, we would surely choose to have the latter skill.

Quick-and-Dirty Monte Carlo: The Bootstrap Method

Here is a powerful technique that can often be used when you don't know enough about the underlying process, or the nature of your measurement errors, to do a credible Monte Carlo simulation. Suppose that your data set consists of

N *independent and identically distributed* (or *iid*) "data points." Each data point probably consists of several numbers, e.g., one or more control variables (uniformly distributed, say, in the range that you have decided to measure) and one or more associated measured values (each distributed however Mother Nature chooses). "Iid" means that the sequential order of the data points is not of consequence to the process that you are using to get the fitted parameters $\mathbf{a}$. For example, a χ^2 sum like (15.5.5) does not care in what order the points are added. Even simpler examples are the mean value of a measured quantity, or the mean of some function of the measured quantities.

The *bootstrap method* [1] uses the actual data set $\mathcal{D}_{(0)}^S$, with its N data points, to generate any number of synthetic data sets $\mathcal{D}_{(1)}^S, \mathcal{D}_{(2)}^S, \ldots$, also with N data points. The procedure is simply to draw N data points at a time *with replacement* from the set $\mathcal{D}_{(0)}^S$. Because of the replacement, you do not simply get back your original data set each time. You get sets in which a random fraction of the original points, typically $\sim 1/e \approx 37\%$, are replaced by *duplicated* original points. Now, exactly as in the previous discussion, you subject these data sets to the same estimation procedure as was performed on the actual data, giving a set of simulated measured parameters $\mathbf{a}_{(1)}^S, \mathbf{a}_{(2)}^S, \ldots$. These will be distributed around $\mathbf{a}_{(0)}$ in close to the same way that $\mathbf{a}_{(0)}$ is distributed around $\mathbf{a}_{\text{true}}$.

Sounds like getting something for nothing, doesn't it? In fact, it has taken more than a decade for the bootstrap method to become accepted by statisticians. By now, however, enough theorems have been proved to render the bootstrap reputable (see [2] for references). The basic idea behind the bootstrap is that the actual data set, viewed as a probability distribution consisting of delta functions at the measured values, is in most cases the best — or only — available estimator of the underlying probability distribution. It takes courage, but one can often simply use *that* distribution as the basis for Monte Carlo simulations.

Watch out for cases where the bootstrap's "iid" assumption is violated. For example, if you have made measurements at evenly spaced intervals of some control variable, then you can *usually* get away with pretending that these are "iid," uniformly distributed over the measured range. However, some estimators of $\mathbf{a}$ (e.g., ones involving Fourier methods) might be particularly sensitive to all the points on a grid being present. In that case, the bootstrap is going to give a wrong distribution. Also watch out for estimators that look at anything like small-scale clumpiness within the N data points, or estimators that sort the data and look at sequential differences. Obviously the bootstrap will fail on these, too. (The theorems justifying the method are still true, but some of their technical assumptions are violated by these examples.)

For a large class of problems, however, the bootstrap does yield easy, *very quick*, Monte Carlo estimates of the errors in an estimated parameter set.

Confidence Limits

Rather than present all details of the probability distribution of errors in parameter estimation, it is common practice to summarize the distribution in the form of *confidence limits*. The full probability distribution is a function defined on the M-dimensional space of parameters $\mathbf{a}$. A *confidence region* (or *confidence interval*) is just a region of that M-dimensional space (hopefully a small region) that contains a certain (hopefully large) percentage of the total probability distribution.

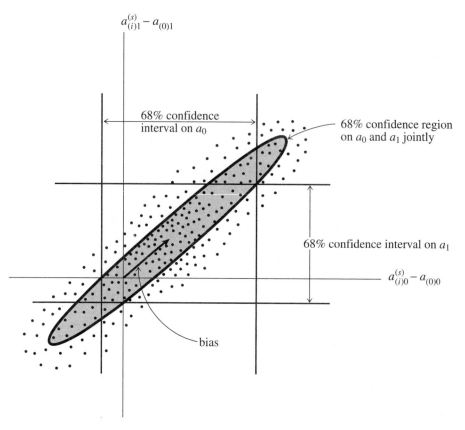

Figure 15.6.3.   Confidence intervals in 1 and 2 dimensions. The same fraction of measured points (here 68%) lies (i) between the two vertical lines, (ii) between the two horizontal lines, (iii) within the ellipse.

You point to a confidence region and say, e.g., "there is a 99 percent chance that the true parameter values fall within this region around the measured value."

It is worth emphasizing that you, the experimenter, get to pick both the *confidence level* (99 percent in the above example), and the shape of the confidence region. The only requirement is that your region does include the stated percentage of probability. Certain percentages are, however, customary in scientific usage: 68.3 percent (the lowest confidence worthy of quoting), 90 percent, 95.4 percent, 99 percent, and 99.73 percent. Higher confidence levels are conventionally "ninety-nine point nine ... nine." As for shape, obviously you want a region that is compact and reasonably centered on your measurement $\mathbf{a}_{(0)}$, since the whole purpose of a confidence limit is to inspire confidence in that measured value. In one dimension, the convention is to use a line segment centered on the measured value; in higher dimensions, ellipses or ellipsoids are most frequently used.

You might suspect, correctly, that the numbers 68.3 percent, 95.4 percent, and 99.73 percent, and the use of ellipsoids, have some connection with a normal distribution. That is true historically, but not always relevant nowadays. In general, the probability distribution of the parameters will not be normal, and the above numbers, used as levels of confidence, are purely matters of convention.

Figure 15.6.3 sketches a possible probability distribution for the case $M = 2$. Shown are three different confidence regions which might usefully be given, all at the same confidence level. The two vertical lines enclose a band (horizontal interval) which represents the 68 percent confidence interval for the variable a_0 without regard to the value of a_1. Similarly the horizontal lines enclose a 68 percent confidence interval for a_1. The ellipse shows a 68 percent confidence interval for a_0 and a_1 jointly. Notice that to enclose the same probability as the two bands, the ellipse must necessarily extend outside of both of them (a point we will return to below).

Constant Chi-Square Boundaries as Confidence Limits

When the method used to estimate the parameters $\mathbf{a}_{(0)}$ is chi-square minimization, as in the previous sections of this chapter, then there is a natural choice for the shape of confidence intervals, whose use is almost universal. For the observed data set $\mathcal{D}_{(0)}$, the value of χ^2 is a minimum at $\mathbf{a}_{(0)}$. Call this minimum value $\chi^2_{\min}$. If the vector $\mathbf{a}$ of parameter values is perturbed away from $\mathbf{a}_{(0)}$, then χ^2 increases. The region within which χ^2 increases by no more than a set amount $\Delta\chi^2$ defines some M-dimensional confidence region around $\mathbf{a}_{(0)}$. If $\Delta\chi^2$ is set to be a large number, this will be a big region; if it is small, it will be small. Somewhere in between there will be choices of $\Delta\chi^2$ that cause the region to contain, variously, 68 percent, 90 percent, etc. of probability distribution for $\mathbf{a}$'s, as defined above. These regions are taken as the confidence regions for the parameters $\mathbf{a}_{(0)}$.

Very frequently one is interested not in the full M-dimensional confidence region, but in individual confidence regions for some smaller number ν of parameters. For example, one might be interested in the confidence interval of each parameter taken separately (the bands in Figure 15.6.3), in which case $\nu = 1$. In that case, the natural confidence regions in the ν-dimensional subspace of the M-dimensional parameter space are the *projections* of the M-dimensional regions defined by fixed $\Delta\chi^2$ into the ν-dimensional spaces of interest. In Figure 15.6.4, for the case $M = 2$, we show regions corresponding to several values of $\Delta\chi^2$. The one-dimensional confidence interval in a_1 corresponding to the region bounded by $\Delta\chi^2 = 1$ lies between the lines A and A'.

Notice that the projection of the higher-dimensional region on the lower-dimension space is used, not the intersection. The intersection would be the band between Z and Z'. It is *never* used. It is shown in the figure only for the purpose of making this cautionary point, that it should not be confused with the projection.

Probability Distribution of Parameters in the Normal Case

You may be wondering why we have, in this section up to now, made no connection at all with the error estimates that come out of the χ^2 fitting procedure, most notably the covariance matrix C_{ij}. The reason is this: χ^2 minimization is a useful means for estimating parameters even if the measurement errors are not normally distributed. While normally distributed errors are required if the χ^2 parameter estimate is to be a maximum likelihood estimator (§15.1), one is often willing to give up that property in return for the relative convenience of the χ^2 procedure. Only in extreme cases, measurement error distributions with very large "tails," is χ^2 minimization abandoned in favor of more robust techniques, as will be discussed in §15.7.

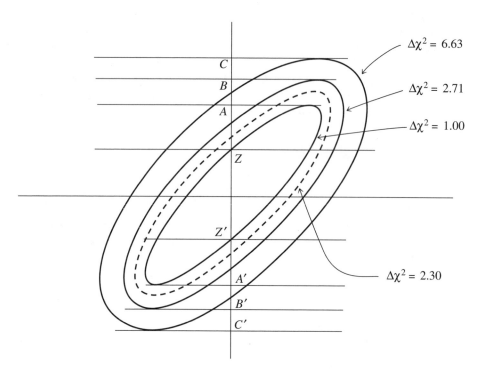

Figure 15.6.4.   Confidence region ellipses corresponding to values of chi-square larger than the fitted minimum. The solid curves, with $\Delta\chi^2 = 1.00, 2.71, 6.63$ project onto one-dimensional intervals AA', BB', CC'. These intervals — not the ellipses themselves — contain 68.3%, 90%, and 99% of normally distributed data. The ellipse that contains 68.3% of normally distributed data is shown dashed, and has $\Delta\chi^2 = 2.30$. For additional numerical values, see accompanying table.

However, the formal covariance matrix that comes out of a χ^2 minimization has a clear quantitative interpretation only if (or to the extent that) the measurement errors actually are normally distributed. In the case of *non*normal errors, you are "allowed"

- to fit for parameters by minimizing χ^2
- to use a contour of constant $\Delta\chi^2$ as the boundary of your confidence region
- to use Monte Carlo simulation or detailed analytic calculation in determining *which* contour $\Delta\chi^2$ is the correct one for your desired confidence level
- to give the covariance matrix C_{ij} as the "formal covariance matrix of the fit."

You are *not* allowed

- to use formulas that we now give for the case of normal errors, which establish quantitative relationships among $\Delta\chi^2$, C_{ij}, and the confidence level.

Here are the key theorems that hold when (i) the measurement errors are normally distributed, and either (ii) the model is linear in its parameters or (iii) the sample size is large enough that the uncertainties in the fitted parameters **a** do not extend outside a region in which the model could be replaced by a suitable linearized model. [Note that condition (iii) does not preclude your use of a nonlinear routine like mqrfit to *find* the fitted parameters.]

Theorem A. $\chi^2_{\min}$ is distributed as a chi-square distribution with $N - M$ degrees of freedom, where N is the number of data points and M is the number of fitted parameters. This is the basic theorem that lets you evaluate the goodness-of-fit of the model, as discussed above in §15.1. We list it first to remind you that unless the goodness-of-fit is credible, the whole estimation of parameters is suspect.

Theorem B.   If $\mathbf{a}_{(j)}^S$ is drawn from the universe of simulated data sets with actual parameters $\mathbf{a}_{(0)}$, then the probability distribution of $\delta\mathbf{a} \equiv \mathbf{a}_{(j)}^S - \mathbf{a}_{(0)}$ is the multivariate normal distribution

$$ P(\delta\mathbf{a})\, da_0 \ldots da_{M-1} = \text{const.} \times \exp\left(-\frac{1}{2}\delta\mathbf{a} \cdot [\alpha] \cdot \delta\mathbf{a}\right)\, da_0 \ldots da_{M-1} $$

where $[\alpha]$ is the curvature matrix defined in equation (15.5.8).

Theorem C.   If $\mathbf{a}_{(j)}^S$ is drawn from the universe of simulated data sets with actual parameters $\mathbf{a}_{(0)}$, then the quantity $\Delta\chi^2 \equiv \chi^2(\mathbf{a}_{(j)}) - \chi^2(\mathbf{a}_{(0)})$ is distributed as a chi-square distribution with M degrees of freedom. Here the χ^2's are all evaluated using the fixed (actual) data set $\mathcal{D}_{(0)}$. This theorem makes the connection between particular values of $\Delta\chi^2$ and the fraction of the probability distribution that they enclose as an M-dimensional region, i.e., the confidence level of the M-dimensional confidence region.

Theorem D.   Suppose that $\mathbf{a}_{(j)}^S$ is drawn from the universe of simulated data sets (as above), that its first ν components $a_0, \ldots, a_{\nu-1}$ are held fixed, and that its remaining $M - \nu$ components are varied so as to minimize χ^2. Call this minimum value χ^2_ν. Then $\Delta\chi^2_\nu \equiv \chi^2_\nu - \chi^2_{\min}$ is distributed as a chi-square distribution with ν degrees of freedom. If you consult Figure 15.6.4, you will see that this theorem connects the *projected* $\Delta\chi^2$ region with a confidence level. In the figure, a point that is held fixed in a_1 and allowed to vary in a_0 minimizing χ^2 will seek out the ellipse whose top or bottom edge is tangent to the line of constant a_1, and is therefore the line that projects it onto the smaller-dimensional space.

As a first example, let us consider the case $\nu = 1$, where we want to find the confidence interval of a single parameter, say a_0. Notice that the chi-square distribution with $\nu = 1$ degree of freedom is the same distribution as that of the square of a single normally distributed quantity. Thus $\Delta\chi^2_\nu < 1$ occurs 68.3 percent of the time (1-σ for the normal distribution), $\Delta\chi^2_\nu < 4$ occurs 95.4 percent of the time (2-σ for the normal distribution), $\Delta\chi^2_\nu < 9$ occurs 99.73 percent of the time (3-σ for the normal distribution), etc. In this manner you find the $\Delta\chi^2_\nu$ that corresponds to your desired confidence level. (Additional values are given in the accompanying table.)

Let $\delta\mathbf{a}$ be a change in the parameters whose first component is arbitrary, δa_0, but the rest of whose components are chosen to minimize the $\Delta\chi^2$. Then Theorem D applies. The value of $\Delta\chi^2$ is given in general by

$$ \Delta\chi^2 = \delta\mathbf{a} \cdot [\alpha] \cdot \delta\mathbf{a} \qquad (15.6.1) $$

which follows from equation (15.5.8) applied at $\chi^2_{\min}$ where $\beta_k = 0$. Since $\delta\mathbf{a}$ by hypothesis minimizes χ^2 in all but its zeroth component, components 1 through $M - 1$ of the normal equations (15.5.9) continue to hold. Therefore, the solution of (15.5.9) is

$$ \delta\mathbf{a} = [\alpha]^{-1} \cdot \begin{pmatrix} c \\ 0 \\ \vdots \\ 0 \end{pmatrix} = [C] \cdot \begin{pmatrix} c \\ 0 \\ \vdots \\ 0 \end{pmatrix} \qquad (15.6.2) $$

$\Delta\chi^2$ as a Function of Confidence Level and Degrees of Freedom						
			ν			
p	1	2	3	4	5	6
68.3%	1.00	2.30	3.53	4.72	5.89	7.04
90%	2.71	4.61	6.25	7.78	9.24	10.6
95.4%	4.00	6.17	8.02	9.70	11.3	12.8
99%	6.63	9.21	11.3	13.3	15.1	16.8
99.73%	9.00	11.8	14.2	16.3	18.2	20.1
99.99%	15.1	18.4	21.1	23.5	25.7	27.8

where c is one arbitrary constant that we get to adjust to make (15.6.1) give the desired left-hand value. Plugging (15.6.2) into (15.6.1) and using the fact that $[C]$ and $[\alpha]$ are inverse matrices of one another, we get

$$c = \delta a_0/C_{00} \qquad \text{and} \qquad \Delta\chi^2_\nu = (\delta a_0)^2/C_{00} \qquad (15.6.3)$$

or

$$\delta a_0 = \pm\sqrt{\Delta\chi^2_\nu}\,\sqrt{C_{00}} \qquad (15.6.4)$$

At last! A relation between the confidence interval $\pm\delta a_0$ and the formal standard error $\sigma_0 \equiv \sqrt{C_{00}}$. Not unreasonably, we find that the 68 percent confidence interval is $\pm\sigma_0$, the 95 percent confidence interval is $\pm2\sigma_0$, etc.

These considerations hold not just for the individual parameters a_i, but also for any linear combination of them: If

$$b \equiv \sum_{k=0}^{M-1} c_i a_i = \mathbf{c}\cdot\mathbf{a} \qquad (15.6.5)$$

then the 68 percent confidence interval on b is

$$\delta b = \pm\sqrt{\mathbf{c}\cdot[C]\cdot\mathbf{c}} \qquad (15.6.6)$$

However, these simple, normal-sounding numerical relationships do *not* hold in the case $\nu > 1$ [3]. In particular, $\Delta\chi^2 = 1$ is not the boundary, nor does it project onto the boundary, of a 68.3 percent confidence region when $\nu > 1$. If you want to calculate not confidence intervals in one parameter, but confidence ellipses in two parameters jointly, or ellipsoids in three, or higher, then you must follow the following prescription for implementing Theorems C and D above:

- Let ν be the number of fitted parameters whose joint confidence region you wish to display, $\nu \leq M$. Call these parameters the "parameters of interest."
- Let p be the confidence limit desired, e.g., $p = 0.68$ or $p = 0.95$.
- Find Δ (i.e., $\Delta\chi^2$) such that the probability of a chi-square variable with ν degrees of freedom being less than Δ is p. For some useful values of p and ν, Δ is given in the table. For other values, you can use the routine gammq and a simple root-finding routine (e.g., bisection) to find Δ such that gammq$(\nu/2, \Delta/2) = 1 - p$.

- Take the $M \times M$ covariance matrix $[C] = [\alpha]^{-1}$ of the chi-square fit. Copy the intersection of the ν rows and columns corresponding to the parameters of interest into a $\nu \times \nu$ matrix denoted $[C_{\mathrm{proj}}]$.
- Invert the matrix $[C_{\mathrm{proj}}]$. (In the one-dimensional case this was just taking the reciprocal of the element C_{00}.)
- The equation for the elliptical boundary of your desired confidence region in the ν-dimensional subspace of interest is

$$\Delta = \delta \mathbf{a}' \cdot [C_{\mathrm{proj}}]^{-1} \cdot \delta \mathbf{a}' \tag{15.6.7}$$

where $\delta \mathbf{a}'$ is the ν-dimensional vector of parameters of interest.

If you are confused at this point, you may find it helpful to compare Figure 15.6.4 and the accompanying table, considering the case $M = 2$ with $\nu = 1$ and $\nu = 2$. You should be able to verify the following statements: (i) The horizontal band between C and C' contains 99 percent of the probability distribution, so it is a confidence limit on a_1 alone at this level of confidence. (ii) Ditto the band between B and B' at the 90 percent confidence level. (iii) The dashed ellipse, labeled by $\Delta\chi^2 = 2.30$, contains 68.3 percent of the probability distribution, so it is a confidence region for a_0 and a_1 jointly, at this level of confidence.

Confidence Limits from Singular Value Decomposition

When you have obtained your χ^2 fit by singular value decomposition (§15.4), the information about the fit's formal errors comes packaged in a somewhat different, but generally more convenient, form. The columns of the matrix $\mathbf{V}$ are an orthonormal set of M vectors that are the principal axes of the $\Delta\chi^2 = $ constant ellipsoids. We denote the columns as $\mathbf{V}_{(0)} \ldots \mathbf{V}_{(M-1)}$. The lengths of those axes are inversely proportional to the corresponding singular values $w_0 \ldots w_{M-1}$; see Figure 15.6.5. The boundaries of the ellipsoids are thus given by

$$\Delta\chi^2 = w_0^2 (\mathbf{V}_{(0)} \cdot \delta \mathbf{a})^2 + \cdots + w_{M-1}^2 (\mathbf{V}_{(M-1)} \cdot \delta \mathbf{a})^2 \tag{15.6.8}$$

which is the justification for writing equation (15.4.18) above. Keep in mind that it is *much* easier to plot an ellipsoid given a list of its vector principal axes, than given its matrix quadratic form!

The formula for the covariance matrix $[C]$ in terms of the columns $\mathbf{V}_{(i)}$ is

$$[C] = \sum_{i=0}^{M-1} \frac{1}{w_i^2} \mathbf{V}_{(i)} \otimes \mathbf{V}_{(i)} \tag{15.6.9}$$

or, in components,

$$C_{jk} = \sum_{i=0}^{M-1} \frac{1}{w_i^2} V_{ji} V_{ki} \tag{15.6.10}$$

CITED REFERENCES AND FURTHER READING:

Efron, B. 1982, *The Jackknife, the Bootstrap, and Other Resampling Plans* (Philadelphia: S.I.A.M.). [1]

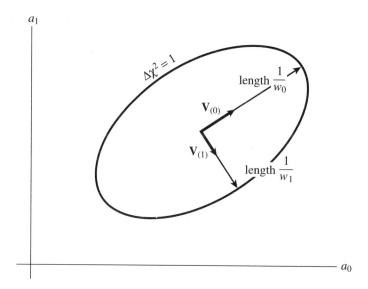

Figure 15.6.5.   Relation of the confidence region ellipse $\Delta\chi^2 = 1$ to quantities computed by singular value decomposition. The vectors $\mathbf{V}_{(i)}$ are unit vectors along the principal axes of the confidence region. The semi-axes have lengths equal to the reciprocal of the singular values w_i. If the axes are all scaled by some constant factor α, $\Delta\chi^2$ is scaled by the factor α^2.

Efron, B., and Tibshirani, R. 1986, *Statistical Science* vol. 1, pp. 54–77. [2]

Avni, Y. 1976, *Astrophysical Journal*, vol. 210, pp. 642–646. [3]

Lampton, M., Margon, M., and Bowyer, S. 1976, *Astrophysical Journal*, vol. 208, pp. 177–190.

Brownlee, K.A. 1965, *Statistical Theory and Methodology*, 2nd ed. (New York: Wiley).

Martin, B.R. 1971, *Statistics for Physicists* (New York: Academic Press).

15.7 Robust Estimation

The concept of *robustness* has been mentioned in passing several times already. In §14.1 we noted that the median was a more robust estimator of central value than the mean; in §14.6 it was mentioned that rank correlation is more robust than linear correlation. The concept of outlier points as exceptions to a Gaussian model for experimental error was discussed in §15.1.

The term "robust" was coined in statistics by G.E.P. Box in 1953. Various definitions of greater or lesser mathematical rigor are possible for the term, but in general, referring to a statistical estimator, it means "insensitive to small departures from the idealized assumptions for which the estimator is optimized." [1,2] The word "small" can have two different interpretations, both important: either fractionally small departures for all data points, or else fractionally large departures for a small number of data points. It is the latter interpretation, leading to the notion of outlier points, that is generally the most stressful for statistical procedures.

Statisticians have developed various sorts of robust statistical estimators. Many, if not most, can be grouped in one of three categories.

M-estimates follow from maximum-likelihood arguments very much as equations (15.1.5) and (15.1.7) followed from equation (15.1.3). M-estimates are usually the most relevant class for model-fitting, that is, estimation of parameters. We therefore consider these estimates in some detail below.

L-estimates are "linear combinations of order statistics." These are most applicable to estimations of central value and central tendency, though they can occasionally be applied to some problems in estimation of parameters. Two "typical" L-estimates will give you the general idea. They are (i) the median, and (ii) *Tukey's trimean*, defined as the weighted average of the first, second, and third quartile points in a distribution, with weights 1/4, 1/2, and 1/4, respectively.

R-estimates are estimates based on rank tests. For example, the equality or inequality of two distributions can be estimated by the *Wilcoxon test* of computing the mean rank of one distribution in a combined sample of both distributions. The Kolmogorov-Smirnov statistic (equation 14.3.6) and the Spearman rank-order correlation coefficient (14.6.1) are R-estimates in essence, if not always by formal definition.

Some other kinds of robust techniques, coming from the fields of optimal control and filtering rather than from the field of mathematical statistics, are mentioned at the end of this section. Some examples where robust statistical methods are desirable are shown in Figure 15.7.1.

Estimation of Parameters by Local M-Estimates

Suppose we know that our measurement errors are not normally distributed. Then, in deriving a maximum-likelihood formula for the estimated parameters **a** in a model $y(x; \mathbf{a})$, we would write instead of equation (15.1.3)

$$P = \prod_{i=0}^{N-1} \left\{ \exp\left[-\rho(y_i, y\{x_i; \mathbf{a}\})\right] \Delta y \right\} \tag{15.7.1}$$

where the function ρ is the negative logarithm of the probability density. Taking the logarithm of (15.7.1) analogously with (15.1.4), we find that we want to minimize the expression

$$\sum_{i=0}^{N-1} \rho(y_i, y\{x_i; \mathbf{a}\}) \tag{15.7.2}$$

Very often, it is the case that the function ρ depends not independently on its two arguments, measured y_i and predicted $y(x_i)$, but only on their difference, at least if scaled by some weight factors σ_i which we are able to assign to each point. In this case the M-estimate is said to be *local*, and we can replace (15.7.2) by the prescription

$$\text{minimize over } \mathbf{a} \quad \sum_{i=0}^{N-1} \rho\left(\frac{y_i - y(x_i; \mathbf{a})}{\sigma_i}\right) \tag{15.7.3}$$

where the function $\rho(z)$ is a function of a single variable $z \equiv [y_i - y(x_i)]/\sigma_i$.

If we now define the derivative of $\rho(z)$ to be a function $\psi(z)$,

$$\psi(z) \equiv \frac{d\rho(z)}{dz} \tag{15.7.4}$$

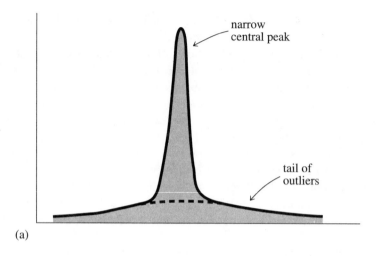

(a)

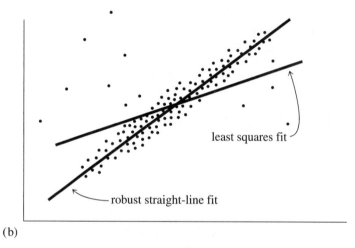

(b)

Figure 15.7.1.   Examples where robust statistical methods are desirable: (a) A one-dimensional distribution with a tail of outliers; statistical fluctuations in these outliers can prevent accurate determination of the position of the central peak. (b) A distribution in two dimensions fitted to a straight line; non-robust techniques such as least-squares fitting can have undesired sensitivity to outlying points.

then the generalization of (15.1.7) to the case of a general M-estimate is

$$0 = \sum_{i=0}^{N-1} \frac{1}{\sigma_i} \psi \left(\frac{y_i - y(x_i)}{\sigma_i} \right) \left(\frac{\partial y(x_i; \mathbf{a})}{\partial a_k} \right) \qquad k = 0, \ldots, M-1 \qquad (15.7.5)$$

If you compare (15.7.3) to (15.1.3), and (15.7.5) to (15.1.7), you see at once that the specialization for normally distributed errors is

$$\rho(z) = \frac{1}{2} z^2 \qquad \psi(z) = z \qquad \text{(normal)} \qquad (15.7.6)$$

If the errors are distributed as a *double* or *two-sided exponential*, namely

$$\text{Prob} \{ y_i - y(x_i) \} \sim \exp \left(- \left| \frac{y_i - y(x_i)}{\sigma_i} \right| \right) \qquad (15.7.7)$$

then, by contrast,

$$\rho(x) = |z| \qquad \psi(z) = \mathrm{sgn}(z) \qquad \text{(double exponential)} \qquad (15.7.8)$$

Comparing to equation (15.7.3), we see that in this case the maximum likelihood estimator is obtained by minimizing the *mean absolute deviation*, rather than the mean square deviation. Here the tails of the distribution, although exponentially decreasing, are asymptotically much larger than any corresponding Gaussian.

A distribution with even more extensive — therefore sometimes even more realistic — tails is the *Cauchy* or *Lorentzian* distribution,

$$\text{Prob} \, \{y_i - y(x_i)\} \sim \frac{1}{1 + \dfrac{1}{2}\left(\dfrac{y_i - y(x_i)}{\sigma_i}\right)^2} \qquad (15.7.9)$$

This implies

$$\rho(z) = \log\left(1 + \frac{1}{2}z^2\right) \qquad \psi(z) = \frac{z}{1 + \frac{1}{2}z^2} \qquad \text{(Lorentzian)} \qquad (15.7.10)$$

Notice that the ψ function occurs as a weighting function in the generalized normal equations (15.7.5). For normally distributed errors, equation (15.7.6) says that the more deviant the points, the greater the weight. By contrast, when tails are somewhat more prominent, as in (15.7.7), then (15.7.8) says that all deviant points get the same relative weight, with only the sign information used. Finally, when the tails are even larger, (15.7.10) says the ψ increases with deviation, then starts *decreasing*, so that very deviant points — the true outliers — are not counted at all in the estimation of the parameters.

This general idea, that the weight given individual points should first increase with deviation, then decrease, motivates some additional prescriptions for ψ which do not especially correspond to standard, textbook probability distributions. Two examples are

Andrew's sine

$$\psi(z) = \begin{cases} \sin(z/c) & |z| < c\pi \\ 0 & |z| > c\pi \end{cases} \qquad (15.7.11)$$

If the measurement errors happen to be normal after all, with standard deviations σ_i, then it can be shown that the optimal value for the constant c is $c = 2.1$.

Tukey's biweight

$$\psi(z) = \begin{cases} z(1 - z^2/c^2)^2 & |z| < c \\ 0 & |z| > c \end{cases} \qquad (15.7.12)$$

where the optimal value of c for normal errors is $c = 6.0$.

Numerical Calculation of M-Estimates

To fit a model by means of an M-estimate, you first decide which M-estimate you want, that is, which matching pair ρ, ψ you want to use. We rather like (15.7.8) or (15.7.10).

You then have to make an unpleasant choice between two fairly difficult problems. Either find the solution of the nonlinear set of M equations (15.7.5), or else minimize the single function in M variables (15.7.3).

Notice that the function (15.7.8) has a discontinuous ψ, and a discontinuous derivative for ρ. Such discontinuities frequently wreak havoc on both general nonlinear equation solvers and general function minimizing routines. You might now think of rejecting (15.7.8) in favor of (15.7.10), which is smoother. However, you will find that the latter choice is also bad news for many general equation solving or minimization routines: small changes in the fitted parameters can drive $\psi(z)$ off its peak into one or the other of its asymptotically small regimes. Therefore, different terms in the equation spring into or out of action (almost as bad as analytic discontinuities).

Don't despair. If your computer budget (or, for personal computers, patience) is up to it, this is an excellent application for the downhill simplex minimization algorithm exemplified in amoeba §10.4 or amebsa in §10.9. Those algorithms make no assumptions about continuity; they just ooze downhill and will work for virtually any sane choice of the function ρ.

It is very much to your (financial) advantage to find good starting values, however. Often this is done by first fitting the model by the standard χ^2 (nonrobust) techniques, e.g., as described in §15.4 or §15.5. The fitted parameters thus obtained are then used as starting values in amoeba, now using the robust choice of ρ and minimizing the expression (15.7.3).

Fitting a Line by Minimizing Absolute Deviation

Occasionally there is a special case that happens to be much easier than is suggested by the general strategy outlined above. The case of equations (15.7.7)–(15.7.8), when the model is a simple straight line

$$y(x; a, b) = a + bx \tag{15.7.13}$$

and where the weights σ_i are all equal, happens to be such a case. The problem is precisely the robust version of the problem posed in equation (15.2.1) above, namely fit a straight line through a set of data points. The merit function to be minimized is

$$\sum_{i=0}^{N-1} |y_i - a - bx_i| \tag{15.7.14}$$

rather than the χ^2 given by equation (15.2.2).

The key simplification is based on the following fact: The median c_M of a set of numbers c_i is also that value which minimizes the sum of the absolute deviations

$$\sum_i |c_i - c_M|$$

(Proof: Differentiate the above expression with respect to c_M and set it to zero.)
It follows that, for fixed b, the value of a that minimizes (15.7.14) is

$$a = \text{median}\{y_i - bx_i\} \tag{15.7.15}$$

Equation (15.7.5) for the parameter b is

$$0 = \sum_{i=0}^{N-1} x_i \, \text{sgn}(y_i - a - bx_i) \tag{15.7.16}$$

(where $\text{sgn}(0)$ is to be interpreted as zero). If we replace a in this equation by the implied function $a(b)$ of (15.7.15), then we are left with an equation in a single variable which can be solved by bracketing and bisection, as described in §9.1. (In fact, it is dangerous to use any fancier method of root-finding, because of the discontinuities in equation 15.7.16.)

Here is a routine that does all this. It calls select (§8.5) to find the median. The bracketing and bisection are built in to the routine, as is the χ^2 solution that generates the initial guesses for a and b. Notice that the evaluation of the right-hand side of (15.7.16) occurs in the function rofunc, with communication via global (top-level) variables.

```
#include <cmath>
#include "nr.h"
using namespace std;

DP aa,abdevt;
const Vec_DP *xt_p,*yt_p;

void NR::medfit(Vec_I_DP &x, Vec_I_DP &y, DP &a, DP &b, DP &abdev)
```
Fits $y = a + bx$ by the criterion of least absolute deviations. The arrays x[0..ndata-1] and y[0..ndata-1] are the input experimental points. The fitted parameters a and b are output, along with abdev, which is the mean absolute deviation (in y) of the experimental points from the fitted line. This routine uses the routine rofunc, with communication via global variables.
```
{
    int j;
    DP bb,b1,b2,del,f,f1,f2,sigb,temp;
    DP sx=0.0,sy=0.0,sxy=0.0,sxx=0.0,chisq=0.0;

    int ndata=x.size();
    xt_p= &x;
    yt_p= &y;
    for (j=0;j<ndata;j++) {            As a first guess for a and b, we will find the
        sx += x[j];                        least-squares fitting line.
        sy += y[j];
        sxy += x[j]*y[j];
        sxx += x[j]*x[j];
    }
    del=ndata*sxx-sx*sx;
    aa=(sxx*sy-sx*sxy)/del;            Least-squares solutions.
    bb=(ndata*sxy-sx*sy)/del;
    for (j=0;j<ndata;j++)
        chisq += (temp=y[j]-(aa+bb*x[j]),temp*temp);
    sigb=sqrt(chisq/del);              The standard deviation will give some idea of
    b1=bb;                                how big an iteration step to take.
    f1=rofunc(b1);
    if (sigb > 0.0) {
```

```
        b2=bb+SIGN(3.0*sigb,f1);          Guess bracket as 3-σ away, in the downhill direc-
        f2=rofunc(b2);                       tion known from f1.
        if (b2 == b1) {
            a=aa;
            b=bb;
            abdev=abdevt/ndata;
            return;
        }
        while (f1*f2 > 0.0) {             Bracketing.
            bb=b2+1.6*(b2-b1);
            b1=b2;
            f1=f2;
            b2=bb;
            f2=rofunc(b2);
        }
        sigb=0.01*sigb;
        while (fabs(b2-b1) > sigb) {
            bb=b1+0.5*(b2-b1);           Bisection.
            if (bb == b1 || bb == b2) break;
            f=rofunc(bb);
            if (f*f1 >= 0.0) {
                f1=f;
                b1=bb;
            } else {
                f2=f;
                b2=bb;
            }
        }
    }
    a=aa;
    b=bb;
    abdev=abdevt/ndata;
}



#include <cmath>
#include <limits>
#include "nr.h"
using namespace std;

extern DP aa,abdevt;                              Defined in medfit.
extern const Vec_DP *xt_p,*yt_p;

DP NR::rofunc(const DP b)
Evaluates the right-hand side of equation (15.7.16) for a given value of b. Communication with
the routine medfit is through global variables.
{
    const DP EPS=numeric_limits<DP>::epsilon();
    int j;
    DP d,sum=0.0;

    const Vec_DP &xt=*xt_p,&yt=*yt_p;           Make aliases to simplify coding.
    int ndatat=xt.size();
    Vec_DP arr(ndatat);
    for (j=0;j<ndatat;j++) arr[j]=yt[j]-b*xt[j];
    if (ndatat & 1 == 1) {
        aa=select((ndatat-1)>>1,arr);
    } else {
        j=ndatat >> 1;
        aa=0.5*(select(j-1,arr)+select(j,arr));
    }
    abdevt=0.0;
    for (j=0;j<ndatat;j++) {
```

```
        d=yt[j]-(b*xt[j]+aa);
        abdevt += fabs(d);
        if (yt[j] != 0.0) d /= fabs(yt[j]);
        if (fabs(d) > EPS) sum += (d >= 0.0 ? xt[j] : -xt[j]);
    }
    return sum;
}
```

Other Robust Techniques

Sometimes you may have *a priori* knowledge about the probable values and probable uncertainties of some parameters that you are trying to estimate from a data set. In such cases you may want to perform a fit that takes this advance information properly into account, neither completely freezing a parameter at a predetermined value (as in lfit §15.4) nor completely leaving it to be determined by the data set. The formalism for doing this is called "use of *a priori* covariances."

A related problem occurs in signal processing and control theory, where it is sometimes desired to "track" (i.e., maintain an estimate of) a time-varying signal in the presence of noise. If the signal is known to be characterized by some number of parameters that vary only slowly, then the formalism of *Kalman filtering* tells how the incoming, raw measurements of the signal should be processed to produce best parameter estimates as a function of time. For example, if the signal is a frequency-modulated sine wave, then the slowly varying parameter might be the instantaneous frequency. The Kalman filter for this case is called a *phase-locked loop* and is implemented in the circuitry of good radio receivers [3,4].

CITED REFERENCES AND FURTHER READING:

Huber, P.J. 1981, *Robust Statistics* (New York: Wiley). [1]

Launer, R.L., and Wilkinson, G.N. (eds.) 1979, *Robustness in Statistics* (New York: Academic Press). [2]

Bryson, A. E., and Ho, Y.C. 1969, *Applied Optimal Control* (Waltham, MA: Ginn). [3]

Jazwinski, A. H. 1970, *Stochastic Processes and Filtering Theory* (New York: Academic Press). [4]

Chapter 16. Integration of Ordinary Differential Equations

16.0 Introduction

Problems involving ordinary differential equations (ODEs) can always be reduced to the study of sets of first-order differential equations. For example the second-order equation

$$\frac{d^2 y}{dx^2} + q(x)\frac{dy}{dx} = r(x) \tag{16.0.1}$$

can be rewritten as two first-order equations

$$\begin{aligned} \frac{dy}{dx} &= z(x) \\ \frac{dz}{dx} &= r(x) - q(x)z(x) \end{aligned} \tag{16.0.2}$$

where z is a new variable. This exemplifies the procedure for an arbitrary ODE. The usual choice for the new variables is to let them be just derivatives of each other (and of the original variable). Occasionally, it is useful to incorporate into their definition some other factors in the equation, or some powers of the independent variable, for the purpose of mitigating singular behavior that could result in overflows or increased roundoff error. Let common sense be your guide: If you find that the original variables are smooth in a solution, while your auxiliary variables are doing crazy things, then figure out why and choose different auxiliary variables.

The generic problem in ordinary differential equations is thus reduced to the study of a set of N coupled *first-order* differential equations for the functions y_i, $i = 0, 1, \ldots, N-1$, having the general form

$$\frac{dy_i(x)}{dx} = f_i(x, y_0, \ldots, y_{N-1}), \qquad i = 0, \ldots, N-1 \tag{16.0.3}$$

where the functions f_i on the right-hand side are known.

A problem involving ODEs is not completely specified by its equations. Even more crucial in determining how to attack the problem numerically is the nature of the problem's boundary conditions. Boundary conditions are algebraic conditions on the values of the functions y_i in (16.0.3). In general they can be satisfied at discrete specified points, but do not hold between those points, i.e., are not preserved automatically by the differential equations. Boundary conditions can be as simple as requiring that certain variables have certain numerical values, or as complicated as a set of nonlinear algebraic equations among the variables.

Usually, it is the nature of the boundary conditions that determines which numerical methods will be feasible. Boundary conditions divide into two broad categories.

- In *initial value problems* all the y_i are given at some starting value x_s, and it is desired to find the y_i's at some final point x_f, or at some discrete list of points (for example, at tabulated intervals).

- In *two-point boundary value problems*, on the other hand, boundary conditions are specified at more than one x. Typically, some of the conditions will be specified at x_s and the remainder at x_f.

This chapter will consider exclusively the initial value problem, deferring two-point boundary value problems, which are generally more difficult, to Chapter 17.

The underlying idea of any routine for solving the initial value problem is always this: Rewrite the dy's and dx's in (16.0.3) as finite steps Δy and Δx, and multiply the equations by Δx. This gives algebraic formulas for the change in the functions when the independent variable x is "stepped" by one "stepsize" Δx. In the limit of making the stepsize very small, a good approximation to the underlying differential equation is achieved. Literal implementation of this procedure results in *Euler's method* (16.1.1, below), which is, however, *not* recommended for any practical use. Euler's method is conceptually important, however; one way or another, practical methods all come down to this same idea: Add small increments to your functions corresponding to derivatives (right-hand sides of the equations) multiplied by stepsizes.

In this chapter we consider three major types of practical numerical methods for solving initial value problems for ODEs:

- Runge-Kutta methods
- Richardson extrapolation and its particular implementation as the Bulirsch-Stoer method
- predictor-corrector methods.

A brief description of each of these types follows.

1. *Runge-Kutta* methods propagate a solution over an interval by combining the information from several Euler-style steps (each involving one evaluation of the right-hand f's), and then using the information obtained to match a Taylor series expansion up to some higher order.

2. *Richardson extrapolation* uses the powerful idea of extrapolating a computed result to the value that *would* have been obtained if the stepsize had been very much smaller than it actually was. In particular, extrapolation to zero stepsize is the desired goal. The first practical ODE integrator that implemented this idea was developed by Bulirsch and Stoer, and so extrapolation methods are often called Bulirsch-Stoer methods.

3. *Predictor-corrector* methods store the solution along the way, and use those results to extrapolate the solution one step advanced; they then correct the extrapolation using derivative information at the new point. These are best for very smooth functions.

Runge-Kutta is what you use when (i) you don't know any better, or (ii) you have an intransigent problem where Bulirsch-Stoer is failing, or (iii) you have a trivial problem where computational efficiency is of no concern. Runge-Kutta succeeds virtually always; but it is not usually fastest, except when evaluating f_i is cheap and moderate accuracy ($\lesssim 10^{-5}$) is required. Predictor-corrector methods, since they use past information, are somewhat more difficult to start up, but, for many smooth problems, they are computationally more efficient than Runge-Kutta. In recent years Bulirsch-Stoer has been replacing predictor-corrector in many applications, but it is too soon to say that predictor-corrector is dominated in all cases. However, it appears that only rather sophisticated predictor-corrector routines are competitive. Accordingly, we have chosen *not* to give an implementation of predictor-corrector in this book. We discuss predictor-corrector further in §16.7, so that you can use a canned routine should you encounter a suitable problem. In our experience, the relatively simple Runge-Kutta and Bulirsch-Stoer routines we give are adequate for most problems.

Each of the three types of methods can be organized to monitor internal consistency. This allows numerical errors which are inevitably introduced into the solution to be controlled by automatic, (*adaptive*) changing of the fundamental stepsize. We always recommend that adaptive stepsize control be implemented, and we will do so below.

In general, all three types of methods can be applied to any initial value problem. Each comes with its own set of debits and credits that must be understood before it is used.

We have organized the routines in this chapter into three nested levels. The lowest or "nitty-gritty" level is the piece we call the *algorithm* routine. This implements the basic formulas of the method, starts with dependent variables y_i at x, and calculates new values of the dependent variables at the value $x + h$. The algorithm routine also yields up some information about the quality of the solution after the step. The routine is dumb, however, and it is unable to make any adaptive decision about whether the solution is of acceptable quality or not.

That quality-control decision we encode in a *stepper* routine. The stepper routine calls the algorithm routine. It may reject the result, set a smaller stepsize, and call the algorithm routine again, until compatibility with a predetermined accuracy criterion has been achieved. The stepper's fundamental task is to take the largest stepsize consistent with specified performance. Only when this is accomplished does the true power of an algorithm come to light.

Above the stepper is the *driver* routine, which starts and stops the integration, stores intermediate results, and generally acts as an interface with the user. There is nothing at all canonical about our driver routines. You should consider them to be examples, and you can customize them for your particular application.

Of the routines that follow, rk4, rkck, mmid, stoerm, and simpr are algorithm routines; rkqs, bsstep, stiff, and stifbs are steppers; rkdumb and odeint are drivers.

Section 16.6 of this chapter treats the subject of *stiff equations*, relevant both to ordinary differential equations and also to partial differential equations (Chapter 19).

CITED REFERENCES AND FURTHER READING:

Gear, C.W. 1971, *Numerical Initial Value Problems in Ordinary Differential Equations* (Englewood Cliffs, NJ: Prentice-Hall).

Acton, F.S. 1970, *Numerical Methods That Work*; 1990, corrected edition (Washington: Mathematical Association of America), Chapter 5.

Stoer, J., and Bulirsch, R. 1993, *Introduction to Numerical Analysis*, 2nd ed. (New York: Springer-Verlag), Chapter 7.

Lambert, J. 1973, *Computational Methods in Ordinary Differential Equations* (New York: Wiley).

Lapidus, L., and Seinfeld, J. 1971, *Numerical Solution of Ordinary Differential Equations* (New York: Academic Press).

16.1 Runge-Kutta Method

The formula for the Euler method is

$$y_{n+1} = y_n + hf(x_n, y_n) \tag{16.1.1}$$

which advances a solution from x_n to $x_{n+1} \equiv x_n + h$. The formula is unsymmetrical: It advances the solution through an interval h, but uses derivative information only at the beginning of that interval (see Figure 16.1.1). That means (and you can verify by expansion in power series) that the step's error is only one power of h smaller than the correction, i.e $O(h^2)$ added to (16.1.1).

There are several reasons that Euler's method is not recommended for practical use, among them, (i) the method is not very accurate when compared to other, fancier, methods run at the equivalent stepsize, and (ii) neither is it very stable (see §16.6 below).

Consider, however, the use of a step like (16.1.1) to take a "trial" step to the midpoint of the interval. Then use the value of both x and y at that midpoint to compute the "real" step across the whole interval. Figure 16.1.2 illustrates the idea. In equations,

$$k_1 = hf(x_n, y_n)$$
$$k_2 = hf\left(x_n + \tfrac{1}{2}h, y_n + \tfrac{1}{2}k_1\right) \tag{16.1.2}$$
$$y_{n+1} = y_n + k_2 + O(h^3)$$

As indicated in the error term, this symmetrization cancels out the first-order error term, making the method *second order*. [A method is conventionally called nth order if its error term is $O(h^{n+1})$.] In fact, (16.1.2) is called the *second-order Runge-Kutta* or *midpoint* method.

We needn't stop there. There are many ways to evaluate the right-hand side $f(x, y)$ that all agree to first order, but that have different coefficients of higher-order error terms. Adding up the right combination of these, we can eliminate the error terms order by order. That is the basic idea of the Runge-Kutta method. Abramowitz and Stegun [1], and Gear [2], give various specific formulas that derive from this basic

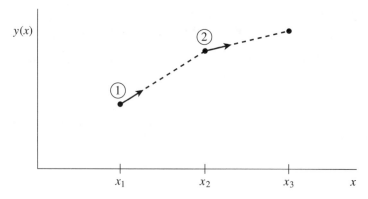

Figure 16.1.1.   Euler's method. In this simplest (and least accurate) method for integrating an ODE, the derivative at the starting point of each interval is extrapolated to find the next function value. The method has first-order accuracy.

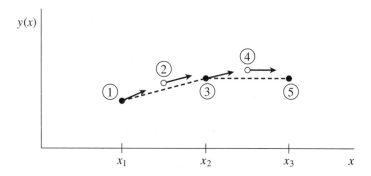

Figure 16.1.2.   Midpoint method. Second-order accuracy is obtained by using the initial derivative at each step to find a point halfway across the interval, then using the midpoint derivative across the full width of the interval. In the figure, filled dots represent final function values, while open dots represent function values that are discarded once their derivatives have been calculated and used.

idea. By far the most often used is the classical *fourth-order Runge-Kutta formula*, which has a certain sleekness of organization about it:

$$k_1 = hf(x_n, y_n)$$
$$k_2 = hf(x_n + \frac{h}{2}, y_n + \frac{k_1}{2})$$
$$k_3 = hf(x_n + \frac{h}{2}, y_n + \frac{k_2}{2})$$
$$k_4 = hf(x_n + h, y_n + k_3)$$
$$y_{n+1} = y_n + \frac{k_1}{6} + \frac{k_2}{3} + \frac{k_3}{3} + \frac{k_4}{6} + O(h^5) \qquad (16.1.3)$$

The fourth-order Runge-Kutta method requires four evaluations of the right-hand side per step h (see Figure 16.1.3). This will be superior to the midpoint method (16.1.2) *if* at least twice as large a step is possible with (16.1.3) for the same accuracy. Is that so? The answer is: often, perhaps even usually, but surely not always! This takes us back to a central theme, namely that *high order* does not always mean *high accuracy*. The statement "fourth-order Runge-Kutta is generally superior to second-order" is a true one, but you should recognize it as a statement about the

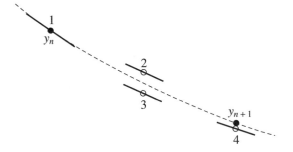

Figure 16.1.3.   Fourth-order Runge-Kutta method. In each step the derivative is evaluated four times: once at the initial point, twice at trial midpoints, and once at a trial endpoint. From these derivatives the final function value (shown as a filled dot) is calculated. (See text for details.)

contemporary practice of science rather than as a statement about strict mathematics. That is, it reflects the nature of the problems that contemporary scientists like to solve.

For many scientific users, fourth-order Runge-Kutta is not just the first word on ODE integrators, but the last word as well. In fact, you can get pretty far on this old workhorse, especially if you combine it with an adaptive stepsize algorithm. Keep in mind, however, that the old workhorse's last trip may well be to take you to the poorhouse: Bulirsch-Stoer or predictor-corrector methods can be very much more efficient for problems where very high accuracy is a requirement. Those methods are the high-strung racehorses. Runge-Kutta is for ploughing the fields. However, even the old workhorse is more nimble with new horseshoes. In §16.2 we will give a modern implementation of a Runge-Kutta method that is quite competitive as long as very high accuracy is not required. An excellent discussion of the pitfalls in constructing a good Runge-Kutta code is given in [3].

Here is the routine for carrying out one classical Runge-Kutta step on a set of n differential equations. You input the values of the independent variables, and you get out new values which are stepped by a stepsize h (which can be positive or negative). You will notice that the routine requires you to supply not only function derivs for calculating the right-hand side, but also values of the derivatives at the starting point. Why not let the routine call derivs for this first value? The answer will become clear only in the next section, but in brief is this: This call may not be your only one with these starting conditions. You may have taken a previous step with too large a stepsize, and this is your replacement. In that case, you do not want to call derivs unnecessarily at the start. Note that the routine that follows has, therefore, only three calls to derivs.

```
#include "nr.h"

void NR::rk4(Vec_I_DP &y, Vec_I_DP &dydx, const DP x, const DP h,
    Vec_O_DP &yout, void derivs(const DP, Vec_I_DP &, Vec_O_DP &))
```
Given values for the variables y[0..n-1] and their derivatives dydx[0..n-1] known at x, use the fourth-order Runge-Kutta method to advance the solution over an interval h and return the incremented variables as yout[0..n-1]. The user supplies the routine derivs(x,y,dydx), which returns derivatives dydx at x.
```
{
    int i;
    DP xh,hh,h6;

    int n=y.size();
    Vec_DP dym(n),dyt(n),yt(n);
```

```
    hh=h*0.5;
    h6=h/6.0;
    xh=x+hh;
    for (i=0;i<n;i++) yt[i]=y[i]+hh*dydx[i];          First step.
    derivs(xh,yt,dyt);                                Second step.
    for (i=0;i<n;i++) yt[i]=y[i]+hh*dyt[i];
    derivs(xh,yt,dym);                                Third step.
    for (i=0;i<n;i++) {
        yt[i]=y[i]+h*dym[i];
        dym[i] += dyt[i];
    }
    derivs(x+h,yt,dyt);                               Fourth step.
    for (i=0;i<n;i++)                                 Accumulate increments with proper
        yout[i]=y[i]+h6*(dydx[i]+dyt[i]+2.0*dym[i]);     weights.
}
```

The Runge-Kutta method treats every step in a sequence of steps in identical manner. Prior behavior of a solution is not used in its propagation. This is mathematically proper, since any point along the trajectory of an ordinary differential equation can serve as an initial point. The fact that all steps are treated identically also makes it easy to incorporate Runge-Kutta into relatively simple "driver" schemes.

We consider adaptive stepsize control, discussed in the next section, an essential for serious computing. Occasionally, however, you just want to tabulate a function at equally spaced intervals, and without particularly high accuracy. In the most common case, you want to produce a graph of the function. Then all you need may be a simple driver program that goes from an initial x_s to a final x_f in a specified number of steps. To check accuracy, double the number of steps, repeat the integration, and compare results. This approach surely does not minimize computer time, and it can fail for problems whose nature *requires* a variable stepsize, but it may well minimize user effort. On small problems, this may be the paramount consideration.

Here is such a driver, self-explanatory, which tabulates the integrated functions in the arrays pointed to by the global pointers xx_p and y_p; be sure to allocate memory for them with statements like the following in the calling routine:

```
    xx_p=new Vec_DP(nstep+1);
    y_p=new Mat_DP(nvar,nstep+1);
```

```
#include "nr.h"

extern Vec_DP *xx_p;                    For communication back to main.
extern Mat_DP *y_p;

void NR::rkdumb(Vec_I_DP &vstart, const DP x1, const DP x2,
    void derivs(const DP, Vec_I_DP &, Vec_O_DP &))
```
Starting from initial values vstart[0..nvar-1] known at x1 use fourth-order Runge-Kutta to advance nstep equal increments to x2. The user-supplied routine derivs(x,v,dvdx) evaluates derivatives. The results are stored in the arrays y[0..nvar-1][0..nstep] and xx[0..nstep] pointed to by y_p and xx_p. nstep and nvar are communicated by the dimensions of the global array *y_p.
```
{
    int i,k;
    DP x,h;

    Vec_DP &xx=*xx_p;                    Make aliases to simplify coding.
    Mat_DP &y=*y_p;
    int nvar=y.nrows();
    int nstep=y.ncols()-1;
```

```
Vec_DP v(nvar),vout(nvar),dv(nvar);
for (i=0;i<nvar;i++) {          Load starting values.
    v[i]=vstart[i];
    y[i][0]=v[i];
}
xx[0]=x1;
x=x1;
h=(x2-x1)/nstep;
for (k=0;k<nstep;k++) {          Take nstep steps.
    derivs(x,v,dv);
    rk4(v,dv,x,h,vout,derivs);
    if (x+h == x)
        nrerror("Step size too small in routine rkdumb");
    x += h;
    xx[k+1]=x;                    Store intermediate steps.
    for (i=0;i<nvar;i++) {
        v[i]=vout[i];
        y[i][k+1]=v[i];
    }
}
}
```

CITED REFERENCES AND FURTHER READING:

Abramowitz, M., and Stegun, I.A. 1964, *Handbook of Mathematical Functions*, Applied Mathematics Series, Volume 55 (Washington: National Bureau of Standards; reprinted 1968 by Dover Publications, New York), §25.5. [1]

Gear, C.W. 1971, *Numerical Initial Value Problems in Ordinary Differential Equations* (Englewood Cliffs, NJ: Prentice-Hall), Chapter 2. [2]

Shampine, L.F., and Watts, H.A. 1977, in *Mathematical Software III*, J.R. Rice, ed. (New York: Academic Press), pp. 257–275; 1979, *Applied Mathematics and Computation*, vol. 5, pp. 93–121. [3]

Rice, J.R. 1983, *Numerical Methods, Software, and Analysis* (New York: McGraw-Hill), §9.2.

16.2 Adaptive Stepsize Control for Runge-Kutta

A good ODE integrator should exert some adaptive control over its own progress, making frequent changes in its stepsize. Usually the purpose of this adaptive stepsize control is to achieve some predetermined accuracy in the solution with minimum computational effort. Many small steps should tiptoe through treacherous terrain, while a few great strides should speed through smooth uninteresting countryside. The resulting gains in efficiency are not mere tens of percents or factors of two; they can sometimes be factors of ten, a hundred, or more. Sometimes accuracy may be demanded not directly in the solution itself, but in some related conserved quantity that can be monitored.

Implementation of adaptive stepsize control requires that the stepping algorithm signal information about its performance, most important, an estimate of its truncation error. In this section we will learn how such information can be obtained. Obviously, the calculation of this information will add to the computational overhead, but the investment will generally be repaid handsomely.

With fourth-order Runge-Kutta, the most straightforward technique by far is *step doubling* (see, e.g., [1]). We take each step twice, once as a full step, then, independently, as two half steps (see Figure 16.2.1). How much overhead is this, say in terms of the number of evaluations of the right-hand sides? Each of the three separate Runge-Kutta steps in the procedure requires 4 evaluations, but the single and double sequences share a starting point, so the total is 11. This is to be compared not to 4, but to 8 (the two half-steps), since — stepsize control aside — we are achieving the accuracy of the smaller (half) stepsize. The overhead cost is therefore a factor 1.375. What does it buy us?

Let us denote the exact solution for an advance from x to $x + 2h$ by $y(x + 2h)$ and the two approximate solutions by y_1 (one step $2h$) and y_2 (2 steps each of size h). Since the basic method is fourth order, the true solution and the two numerical approximations are related by

$$
\begin{aligned}
y(x + 2h) &= y_1 + (2h)^5\phi + O(h^6) + \dots \\
y(x + 2h) &= y_2 + 2(h^5)\phi + O(h^6) + \dots
\end{aligned}
\tag{16.2.1}
$$

where, to order h^5, the value ϕ remains constant over the step. [Taylor series expansion tells us the ϕ is a number whose order of magnitude is $y^{(5)}(x)/5!$.] The first expression in (16.2.1) involves $(2h)^5$ since the stepsize is $2h$, while the second expression involves $2(h^5)$ since the error on each step is $h^5\phi$. The difference between the two numerical estimates is a convenient indicator of truncation error

$$
\Delta \equiv y_2 - y_1
\tag{16.2.2}
$$

It is this difference that we shall endeavor to keep to a desired degree of accuracy, neither too large nor too small. We do this by adjusting h.

It might also occur to you that, ignoring terms of order h^6 and higher, we can solve the two equations in (16.2.1) to improve our numerical estimate of the true solution $y(x + 2h)$, namely,

$$
y(x + 2h) = y_2 + \frac{\Delta}{15} + O(h^6)
\tag{16.2.3}
$$

This estimate is accurate to *fifth order*, one order higher than the original Runge-Kutta steps. However, we can't have our cake and eat it: (16.2.3) may be fifth-order accurate, but we have no way of monitoring *its* truncation error. Higher order is not always higher accuracy! Use of (16.2.3) rarely does harm, but we have no way of directly knowing whether it is doing any good. Therefore we should use Δ as the error estimate and take as "gravy" any additional accuracy gain derived from (16.2.3). In the technical literature, use of a procedure like (16.2.3) is called "local extrapolation."

An alternative stepsize adjustment algorithm is based on the *embedded Runge-Kutta formulas*, originally invented by Fehlberg. An interesting fact about Runge-Kutta formulas is that for orders M higher than four, more than M function evaluations (though never more than $M + 2$) are required. This accounts for the popularity of the classical fourth-order method: It seems to give the most bang for the buck. However, Fehlberg discovered a fifth-order method with six function

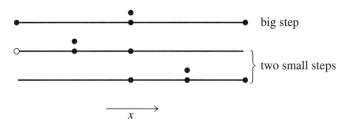

Figure 16.2.1.   Step-doubling as a means for adaptive stepsize control in fourth-order Runge-Kutta. Points where the derivative is evaluated are shown as filled circles. The open circle represents the same derivatives as the filled circle immediately above it, so the total number of evaluations is 11 per two steps. Comparing the accuracy of the big step with the two small steps gives a criterion for adjusting the stepsize on the next step, or for rejecting the current step as inaccurate.

evaluations where another combination of the six functions gives a fourth-order method. The difference between the two estimates of $y(x + h)$ can then be used as an estimate of the truncation error to adjust the stepsize. Since Fehlberg's original formula, several other embedded Runge-Kutta formulas have been found.

Many practitioners were at one time wary of the robustness of Runge-Kutta-Fehlberg methods. The feeling was that using the same evaluation points to advance the function and to estimate the error was riskier than step-doubling, where the error estimate is based on independent function evaluations. However, experience has shown that this concern is not a problem in practice. Accordingly, embedded Runge-Kutta formulas, which are roughly a factor of two more efficient, have superseded algorithms based on step-doubling.

The general form of a fifth-order Runge-Kutta formula is

$$k_1 = hf(x_n, y_n)$$
$$k_2 = hf(x_n + a_2h, y_n + b_{21}k_1)$$
$$\cdots \tag{16.2.4}$$
$$k_6 = hf(x_n + a_6h, y_n + b_{61}k_1 + \cdots + b_{65}k_5)$$
$$y_{n+1} = y_n + c_1k_1 + c_2k_2 + c_3k_3 + c_4k_4 + c_5k_5 + c_6k_6 + O(h^6)$$

The embedded fourth-order formula is

$$y^*_{n+1} = y_n + c^*_1k_1 + c^*_2k_2 + c^*_3k_3 + c^*_4k_4 + c^*_5k_5 + c^*_6k_6 + O(h^5) \tag{16.2.5}$$

and so the error estimate is

$$\Delta \equiv y_{n+1} - y^*_{n+1} = \sum_{i=1}^{6}(c_i - c^*_i)k_i \tag{16.2.6}$$

The particular values of the various constants that we favor are those found by Cash and Karp [2], and given in the accompanying table. These give a more efficient method than Fehlberg's original values, with somewhat better error properties.

Now that we know, at least approximately, what our error is, we need to consider how to keep it within desired bounds. What is the relation between Δ

Cash-Karp Parameters for Embedded Runga-Kutta Method							
i	a_i	b_{ij}					c_i	c_i^*
1							$\frac{37}{378}$	$\frac{2825}{27648}$
2	$\frac{1}{5}$	$\frac{1}{5}$					0	0
3	$\frac{3}{10}$	$\frac{3}{40}$	$\frac{9}{40}$				$\frac{250}{621}$	$\frac{18575}{48384}$
4	$\frac{3}{5}$	$\frac{3}{10}$	$-\frac{9}{10}$	$\frac{6}{5}$			$\frac{125}{594}$	$\frac{13525}{55296}$
5	1	$-\frac{11}{54}$	$\frac{5}{2}$	$-\frac{70}{27}$	$\frac{35}{27}$		0	$\frac{277}{14336}$
6	$\frac{7}{8}$	$\frac{1631}{55296}$	$\frac{175}{512}$	$\frac{575}{13824}$	$\frac{44275}{110592}$	$\frac{253}{4096}$	$\frac{512}{1771}$	$\frac{1}{4}$
$j =$		1	2	3	4	5		

and h? According to (16.2.4) – (16.2.5), Δ scales as h^5. If we take a step h_1 and produce an error Δ_1, therefore, the step h_0 that *would have given* some other value Δ_0 is readily estimated as

$$h_0 = h_1 \left| \frac{\Delta_0}{\Delta_1} \right|^{0.2} \tag{16.2.7}$$

Henceforth we will let Δ_0 denote the *desired* accuracy. Then equation (16.2.7) is used in two ways: If Δ_1 is larger than Δ_0 in magnitude, the equation tells how much to decrease the stepsize *when we retry the present (failed) step*. If Δ_1 is smaller than Δ_0, on the other hand, then the equation tells how much we can safely increase the stepsize *for the next step*. Local extrapolation consists in accepting the fifth order value y_{n+1}, even though the error estimate actually applies to the fourth order value y_{n+1}^*.

Our notation hides the fact that Δ_0 is actually a vector of desired accuracies, one for each equation in the set of ODEs. In general, our accuracy requirement will be that all equations are within their respective allowed errors. In other words, we will rescale the stepsize according to the needs of the "worst-offender" equation.

How is Δ_0, the desired accuracy, related to some looser prescription like "get a solution good to one part in 10^6"? That can be a subtle question, and it depends on exactly what your application is! You may be dealing with a set of equations whose dependent variables differ enormously in magnitude. In that case, you probably want to use fractional errors, $\Delta_0 = \epsilon y$, where ϵ is the number like 10^{-6} or whatever. On the other hand, you may have oscillatory functions that pass through zero but are bounded by some maximum values. In that case you probably want to set Δ_0 equal to ϵ times those maximum values.

A convenient way to fold these considerations into a generally useful stepper routine is this: One of the arguments of the routine will of course be the vector of dependent variables at the beginning of a proposed step. Call that y[0..n-1]. Let us require the user to specify for each step another, corresponding, vector argument yscal[0..n-1], and also an overall tolerance level eps. Then the desired accuracy for the ith equation will be taken to be

$$\Delta_0 = \text{eps} \times \text{yscal[i]} \tag{16.2.8}$$

If you desire constant fractional errors, plug y into the `yscal` calling slot (no need to copy the values into a different array). If you desire constant absolute errors relative to some maximum values, set the elements of `yscal` equal to those maximum values. A useful "trick" for getting constant fractional errors *except* "very" near zero crossings is to set `yscal[i]` equal to $|y[i]| + |h \times dydx[i]|$. (The routine `odeint`, below, does this.)

Here is a more technical point. We have to consider one additional possibility for `yscal`. The error criteria mentioned thus far are "local," in that they bound the error of each step individually. In some applications you may be unusually sensitive about a "global" accumulation of errors, from beginning to end of the integration and in the worst possible case where the errors all are presumed to add with the same sign. Then, the smaller the stepsize h, the smaller the value Δ_0 that you will need to impose. Why? Because there will be *more steps* between your starting and ending values of x. In such cases you will want to set `yscal` proportional to h, typically to something like

$$\Delta_0 = \epsilon h \times \text{dydx[i]} \qquad (16.2.9)$$

This enforces fractional accuracy ϵ not on the values of y but (much more stringently) on the *increments* to those values at each step. But now look back at (16.2.7). If Δ_0 has an implicit scaling with h, then the exponent 0.20 is no longer correct: When the stepsize is reduced from a too-large value, the new predicted value h_1 will fail to meet the desired accuracy when `yscal` is also altered to this new h_1 value. Instead of $0.20 = 1/5$, we must scale by the exponent $0.25 = 1/4$ for things to work out.

The exponents 0.20 and 0.25 are not really very different. This motivates us to adopt the following pragmatic approach, one that frees us from having to know in advance whether or not you, the user, plan to scale your `yscal`'s with stepsize. Whenever we decrease a stepsize, let us use the larger value of the exponent (whether we need it or not!), and whenever we increase a stepsize, let us use the smaller exponent. Furthermore, because our estimates of error are not exact, but only accurate to the leading order in h, we are advised to put in a safety factor S which is a few percent smaller than unity. Equation (16.2.7) is thus replaced by

$$h_0 = \begin{cases} Sh_1 \left| \dfrac{\Delta_0}{\Delta_1} \right|^{0.20} & \Delta_0 \geq \Delta_1 \\[2ex] Sh_1 \left| \dfrac{\Delta_0}{\Delta_1} \right|^{0.25} & \Delta_0 < \Delta_1 \end{cases} \qquad (16.2.10)$$

We have found this prescription to be a reliable one in practice.

Here, then, is a stepper program that takes one "quality-controlled" Runge-Kutta step.

```
#include <cmath>
#include "nr.h"
using namespace std;

void NR::rkqs(Vec_IO_DP &y, Vec_IO_DP &dydx, DP &x, const DP htry,
    const DP eps, Vec_I_DP &yscal, DP &hdid, DP &hnext,
    void derivs(const DP, Vec_I_DP &, Vec_O_DP &))
```
Fifth-order Runge-Kutta step with monitoring of local truncation error to ensure accuracy and adjust stepsize. Input are the dependent variable vector y[0..n-1] and its derivative

dydx[0..n-1] at the starting value of the independent variable x. Also input are the stepsize to be attempted htry, the required accuracy eps, and the vector yscal[0..n-1] against which the error is scaled. On output, y and x are replaced by their new values, hdid is the stepsize that was actually accomplished, and hnext is the estimated next stepsize. derivs is the user-supplied routine that computes the right-hand side derivatives.

```
{
    const DP SAFETY=0.9, PGROW=-0.2, PSHRNK=-0.25, ERRCON=1.89e-4;
    The value ERRCON equals (5/SAFETY) raised to the power (1/PGROW), see use below.
    int i;
    DP errmax,h,htemp,xnew;

    int n=y.size();
    h=htry;                                  Set stepsize to the initial trial value.
    Vec_DP yerr(n),ytemp(n);
    for (;;) {
        rkck(y,dydx,x,h,ytemp,yerr,derivs);        Take a step.
        errmax=0.0;                                Evaluate accuracy.
        for (i=0;i<n;i++) errmax=MAX(errmax,fabs(yerr[i]/yscal[i]));
        errmax /= eps;                             Scale relative to required tolerance.
        if (errmax <= 1.0) break;                  Step succeeded. Compute size of next step.
        htemp=SAFETY*h*pow(errmax,PSHRNK);
        Truncation error too large, reduce stepsize.
        h=(h >= 0.0 ? MAX(htemp,0.1*h) : MIN(htemp,0.1*h));
        No more than a factor of 10.
        xnew=x+h;
        if (xnew == x) nrerror("stepsize underflow in rkqs");
    }
    if (errmax > ERRCON) hnext=SAFETY*h*pow(errmax,PGROW);
    else hnext=5.0*h;                              No more than a factor of 5 increase.
    x += (hdid=h);
    for (i=0;i<n;i++) y[i]=ytemp[i];
}
```

The routine rkqs calls the routine rkck to take a Cash-Karp Runge-Kutta step:

```
#include "nr.h"

void NR::rkck(Vec_I_DP &y, Vec_I_DP &dydx, const DP x,
    const DP h, Vec_O_DP &yout, Vec_O_DP &yerr,
    void derivs(const DP, Vec_I_DP &, Vec_O_DP &))
```
Given values for n variables y[0..n-1] and their derivatives dydx[0..n-1] known at x, use the fifth-order Cash-Karp Runge-Kutta method to advance the solution over an interval h and return the incremented variables as yout[0..n-1]. Also return an estimate of the local truncation error in yout using the embedded fourth-order method. The user supplies the routine derivs(x,y,dydx), which returns derivatives dydx at x.
```
{
    static const DP a2=0.2, a3=0.3, a4=0.6, a5=1.0, a6=0.875,
        b21=0.2, b31=3.0/40.0, b32=9.0/40.0, b41=0.3, b42 = -0.9,
        b43=1.2, b51 = -11.0/54.0, b52=2.5, b53 = -70.0/27.0,
        b54=35.0/27.0, b61=1631.0/55296.0, b62=175.0/512.0,
        b63=575.0/13824.0, b64=44275.0/110592.0, b65=253.0/4096.0,
        c1=37.0/378.0, c3=250.0/621.0, c4=125.0/594.0, c6=512.0/1771.0,
        dc1=c1-2825.0/27648.0, dc3=c3-18575.0/48384.0,
        dc4=c4-13525.0/55296.0, dc5 = -277.00/14336.0, dc6=c6-0.25;
    int i;

    int n=y.size();
    Vec_DP ak2(n),ak3(n),ak4(n),ak5(n),ak6(n),ytemp(n);
    for (i=0;i<n;i++)                          First step.
        ytemp[i]=y[i]+b21*h*dydx[i];
    derivs(x+a2*h,ytemp,ak2);                  Second step.
    for (i=0;i<n;i++)
        ytemp[i]=y[i]+h*(b31*dydx[i]+b32*ak2[i]);
```

```
derivs(x+a3*h,ytemp,ak3);              Third step.
for (i=0;i<n;i++)
    ytemp[i]=y[i]+h*(b41*dydx[i]+b42*ak2[i]+b43*ak3[i]);
derivs(x+a4*h,ytemp,ak4);              Fourth step.
for (i=0;i<n;i++)
    ytemp[i]=y[i]+h*(b51*dydx[i]+b52*ak2[i]+b53*ak3[i]+b54*ak4[i]);
derivs(x+a5*h,ytemp,ak5);              Fifth step.
for (i=0;i<n;i++)
    ytemp[i]=y[i]+h*(b61*dydx[i]+b62*ak2[i]+b63*ak3[i]+b64*ak4[i]+b65*ak5[i]);
derivs(x+a6*h,ytemp,ak6);              Sixth step.
for (i=0;i<n;i++)                      Accumulate increments with proper weights.
    yout[i]=y[i]+h*(c1*dydx[i]+c3*ak3[i]+c4*ak4[i]+c6*ak6[i]);
for (i=0;i<n;i++)
    yerr[i]=h*(dc1*dydx[i]+dc3*ak3[i]+dc4*ak4[i]+dc5*ak5[i]+dc6*ak6[i]);
    Estimate error as difference between fourth and fifth order methods.
}
```

A warning: don't be too greedy in specifying eps. The punishment for excessive greediness is interesting and worthy of Gilbert and Sullivan's *Mikado*: The routine can always achieve an apparent *zero* error by making the stepsize so small that quantities of order hy' add to quantities of order y as if they were zero. Then the routine chugs happily along taking infinitely many infinitesimal steps and never changing the dependent variables one iota. (You guard against this catastrophic loss of your computer budget by signaling on abnormally small stepsizes or on the dependent variable vector remaining unchanged from step to step. On a personal workstation you guard against it by not taking too long a lunch hour while your program is running.)

Here is a full-fledged "driver" for Runge-Kutta with adaptive stepsize control. We warmly recommend this routine, or one like it, for a variety of problems, notably including garden-variety ODEs or sets of ODEs, and definite integrals (augmenting the methods of Chapter 4). For storage of intermediate results (if you desire to inspect them) we assume that the top-level pointer references *xp_p and *yp_p have been validly initialized by statements analogous to those shown before rkdumb in §16.1. Because steps occur at unequal intervals results are only stored at intervals greater than dxsav. The top-level variable kmax indicates the maximum number of steps that can be stored. If kmax=0 there is no intermediate storage, and the pointers xp_p and yp_p need not point to valid memory. Storage of steps stops if kmax is exceeded, except that the ending values are always stored. Again, these controls are merely indicative of what you might need. The routine odeint should be customized to the problem at hand.

```
#include <cmath>
#include "nr.h"
using namespace std;

extern DP dxsav;
extern int kmax,kount;
extern Vec_DP *xp_p;
extern Mat_DP *yp_p;
```
User storage for intermediate results. Preset kmax and dxsav in the calling program. If kmax $\neq 0$ results are stored at approximate intervals dxsav in the arrays xp[0..kount-1], yp[0..nvar-1] [0..kount-1], where kount is output by odeint. Defining declarations for these variables, with memory allocations xp[0..kmax-1] and yp[0..nvar-1][0..kmax-1] for the arrays, should be in the calling program.

```
void NR::odeint(Vec_IO_DP &ystart, const DP x1, const DP x2, const DP eps,
    const DP h1, const DP hmin, int &nok, int &nbad,
    void derivs(const DP, Vec_I_DP &, Vec_O_DP &),
    void rkqs(Vec_IO_DP &, Vec_IO_DP &, DP &, const DP, const DP,
    Vec_I_DP &, DP &, DP &, void (*)(const DP, Vec_I_DP &, Vec_O_DP &)))
```
Runge-Kutta driver with adaptive stepsize control. The routine integrates starting values
ystart[0..nvar-1] from x1 to x2 with accuracy eps, storing intermediate results in global
variables. h1 should be set as a guessed first stepsize, hmin as the minimum allowed stepsize
(can be zero). On output nok and nbad are the number of good and bad (but retried and
fixed) steps taken, and ystart is replaced by values at the end of the integration interval.
derivs is the user-supplied routine for calculating the right-hand side derivative, while rkqs
is the name of the stepper routine to be used.
```
{
    const int MAXSTP=10000;
    const DP TINY=1.0e-30;
    int i,nstp;
    DP xsav,x,hnext,hdid,h;

    int nvar=ystart.size();
    Vec_DP yscal(nvar),y(nvar),dydx(nvar);
    Vec_DP &xp=*xp_p;                            Make aliases to simplify coding.
    Mat_DP &yp=*yp_p;
    x=x1;
    h=SIGN(h1,x2-x1);
    nok = nbad = kount = 0;
    for (i=0;i<nvar;i++) y[i]=ystart[i];
    if (kmax > 0) xsav=x-dxsav*2.0;             Assures storage of first step.
    for (nstp=0;nstp<MAXSTP;nstp++) {
        derivs(x,y,dydx);
        for (i=0;i<nvar;i++)
            Scaling used to monitor accuracy. This general-purpose choice can be modified
            if need be.
            yscal[i]=fabs(y[i])+fabs(dydx[i]*h)+TINY;
        if (kmax > 0 && kount < kmax-1 && fabs(x-xsav) > fabs(dxsav)) {
            for (i=0;i<nvar;i++) yp[i][kount]=y[i];
            xp[kount++]=x;                       Store intermediate results.
            xsav=x;
        }
        if ((x+h-x2)*(x+h-x1) > 0.0) h=x2-x;    If stepsize can overshoot, decrease.
        rkqs(y,dydx,x,h,eps,yscal,hdid,hnext,derivs);
        if (hdid == h) ++nok; else ++nbad;
        if ((x-x2)*(x2-x1) >= 0.0) {            Are we done?
            for (i=0;i<nvar;i++) ystart[i]=y[i];
            if (kmax != 0) {
                for (i=0;i<nvar;i++) yp[i][kount]=y[i];
                xp[kount++]=x;                   Save final step.
            }
            return;                              Normal exit.
        }
        if (fabs(hnext) <= hmin) nrerror("Step size too small in odeint");
        h=hnext;
    }
    nrerror("Too many steps in routine odeint");
}
```

CITED REFERENCES AND FURTHER READING:

Gear, C.W. 1971, *Numerical Initial Value Problems in Ordinary Differential Equations* (Englewood Cliffs, NJ: Prentice-Hall). [1]

Cash, J.R., and Karp, A.H. 1990, *ACM Transactions on Mathematical Software*, vol. 16, pp. 201–222. [2]

Shampine, L.F., and Watts, H.A. 1977, in *Mathematical Software III*, J.R. Rice, ed. (New York: Academic Press), pp. 257–275; 1979, *Applied Mathematics and Computation*, vol. 5, pp. 93–121.

Forsythe, G.E., Malcolm, M.A., and Moler, C.B. 1977, *Computer Methods for Mathematical Computations* (Englewood Cliffs, NJ: Prentice-Hall).

16.3 Modified Midpoint Method

This section discusses the *modified midpoint method*, which advances a vector of dependent variables $y(x)$ from a point x to a point $x + H$ by a sequence of n substeps each of size h,

$$h = H/n \tag{16.3.1}$$

In principle, one could use the modified midpoint method in its own right as an ODE integrator. In practice, the method finds its most important application as a part of the more powerful Bulirsch-Stoer technique, treated in §16.4. You can therefore consider this section as a preamble to §16.4.

The number of right-hand side evaluations required by the modified midpoint method is $n + 1$. The formulas for the method are

$$z_0 \equiv y(x)$$

$$z_1 = z_0 + h f(x, z_0)$$

$$z_{m+1} = z_{m-1} + 2h f(x + mh, z_m) \qquad \text{for} \quad m = 1, 2, \ldots, n - 1$$

$$y(x + H) \approx y_n \equiv \frac{1}{2}[z_n + z_{n-1} + h f(x + H, z_n)] \tag{16.3.2}$$

Here the z's are intermediate approximations which march along in steps of h, while y_n is the final approximation to $y(x + H)$. The method is basically a "centered difference" or "midpoint" method (compare equation 16.1.2), except at the first and last points. Those give the qualifier "modified."

The modified midpoint method is a second-order method, like (16.1.2), but with the advantage of requiring (asymptotically for large n) only one derivative evaluation per step h instead of the two required by second-order Runge-Kutta. Perhaps there are applications where the simplicity of (16.3.2), easily coded in-line in some other program, recommends it. In general, however, use of the modified midpoint method by itself will be dominated by the embedded Runge-Kutta method with adaptive stepsize control, as implemented in the preceding section.

The usefulness of the modified midpoint method to the Bulirsch-Stoer technique (§16.4) derives from a "deep" result about equations (16.3.2), due to Gragg. It turns out that the error of (16.3.2), expressed as a power series in h, the stepsize, contains only *even* powers of h,

$$y_n - y(x + H) = \sum_{i=1}^{\infty} \alpha_i h^{2i} \tag{16.3.3}$$

where H is held constant, but h changes by varying n in (16.3.1). The importance of this even power series is that, if we play our usual tricks of combining steps to knock out higher-order error terms, we can gain *two* orders at a time!

For example, suppose n is even, and let $y_{n/2}$ denote the result of applying (16.3.1) and (16.3.2) with half as many steps, $n \to n/2$. Then the estimate

$$y(x + H) \approx \frac{4y_n - y_{n/2}}{3} \tag{16.3.4}$$

is *fourth-order* accurate, the same as fourth-order Runge-Kutta, but requires only about 1.5 derivative evaluations per step h instead of Runge-Kutta's 4 evaluations. Don't be too anxious to implement (16.3.4), since we will soon do even better.

Now would be a good time to look back at the routine qsimp in §4.2, and especially to compare equation (4.2.4) with equation (16.3.4) above. You will see that the transition in Chapter 4 to the idea of Richardson extrapolation, as embodied in Romberg integration of §4.3, is exactly analogous to the transition in going from this section to the next one.

Here is the routine that implements the modified midpoint method, which will be used below.

```
#include "nr.h"

void NR::mmid(Vec_I_DP &y, Vec_I_DP &dydx, const DP xs, const DP htot,
    const int nstep, Vec_O_DP &yout,
    void derivs(const DP, Vec_I_DP &, Vec_O_DP &))
Modified midpoint step. At xs, input the dependent variable vector y[0..nvar-1] and its
derivative vector dydx[0..nvar-1]. Also input is htot, the total step to be made, and
nstep, the number of substeps to be used. The output is returned as yout[0..nvar-1].
{
    int i,n;
    DP x,swap,h2,h;

    int nvar=y.size();
    Vec_DP ym(nvar),yn(nvar);
    h=htot/nstep;                        Stepsize this trip.
    for (i=0;i<nvar;i++) {
        ym[i]=y[i];
        yn[i]=y[i]+h*dydx[i];            First step.
    }
    x=xs+h;
    derivs(x,yn,yout);                   Will use yout for temporary storage of deriva-
    h2=2.0*h;                                tives.
    for (n=1;n<nstep;n++) {              General step.
        for (i=0;i<nvar;i++) {
            swap=ym[i]+h2*yout[i];
            ym[i]=yn[i];
            yn[i]=swap;
        }
        x += h;
        derivs(x,yn,yout);
    }
    for (i=0;i<nvar;i++)                 Last step.
        yout[i]=0.5*(ym[i]+yn[i]+h*yout[i]);
}
```

CITED REFERENCES AND FURTHER READING:

Gear, C.W. 1971, *Numerical Initial Value Problems in Ordinary Differential Equations* (Englewood Cliffs, NJ: Prentice-Hall), §6.1.4.

Stoer, J., and Bulirsch, R. 1993, *Introduction to Numerical Analysis*, 2nd ed. (New York: Springer-Verlag), §7.2.12.

16.4 Richardson Extrapolation and the Bulirsch-Stoer Method

The techniques described in this section are not for differential equations containing nonsmooth functions. For example, you might have a differential equation whose right-hand side involves a function that is evaluated by table look-up and interpolation. If so, go back to Runge-Kutta with adaptive stepsize choice: That method does an excellent job of feeling its way through rocky or discontinuous terrain. It is also an excellent choice for quick-and-dirty, low-accuracy solution of a set of equations. A second warning is that the techniques in this section are not particularly good for differential equations that have singular points *inside* the interval of integration. A regular solution must tiptoe very carefully across such points. Runge-Kutta with adaptive stepsize can sometimes effect this; more generally, there are special techniques available for such problems, beyond our scope here.

Apart from those two caveats, we believe that the Bulirsch-Stoer method, discussed in this section, is the best known way to obtain high-accuracy solutions to ordinary differential equations with minimal computational effort. (A possible exception, infrequently encountered in practice, is discussed in §16.7.)

Three key ideas are involved. The first is *Richardson's deferred approach to the limit*, which we already met in §4.3 on Romberg integration. The idea is to consider the final answer of a numerical calculation as itself being an analytic function (if a complicated one) of an adjustable parameter like the stepsize h. That analytic function can be probed by performing the calculation with various values of h, *none* of them being necessarily small enough to yield the accuracy that we desire. When we know enough about the function, we *fit* it to some analytic form, and then *evaluate* it at that mythical and golden point $h = 0$ (see Figure 16.4.1). Richardson extrapolation is a method for turning straw into gold! (Lead into gold for alchemist readers.)

The second idea has to do with what kind of fitting function is used. Bulirsch and Stoer first recognized the strength of *rational function extrapolation* in Richardson-type applications. That strength is to break the shackles of the power series and its limited radius of convergence, out only to the distance of the first pole in the complex plane. Rational function fits can remain good approximations to analytic functions even after the various terms in powers of h all have comparable magnitudes. In other words, h can be so large as to make the whole notion of the "order" of the method meaningless — and the method can still work superbly. Nevertheless, more recent experience suggests that for smooth problems straightforward polynomial extrapolation is slightly more efficient than rational function extrapolation. We will accordingly adopt polynomial extrapolation as the default, but the routine bsstep

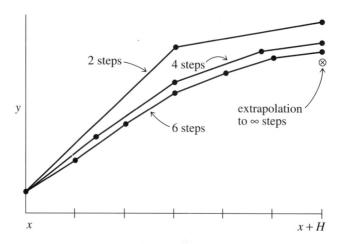

Figure 16.4.1.   Richardson extrapolation as used in the Bulirsch-Stoer method. A large interval H is spanned by different sequences of finer and finer substeps. Their results are extrapolated to an answer that is supposed to correspond to infinitely fine substeps. In the Bulirsch-Stoer method, the integrations are done by the modified midpoint method, and the extrapolation technique is rational function or polynomial extrapolation.

below allows easy substitution of one kind of extrapolation for the other. You might wish at this point to review §3.1–§3.2, where polynomial and rational function extrapolation were already discussed.

The third idea was discussed in the section before this one, namely to use a method whose error function is strictly even, allowing the rational function or polynomial approximation to be in terms of the variable h^2 instead of just h.

Put these ideas together and you have the *Bulirsch-Stoer method* [1]. A single Bulirsch-Stoer step takes us from x to $x + H$, where H is supposed to be quite a large — not at all infinitesimal — distance. That single step is a grand leap consisting of many (e.g., dozens to hundreds) substeps of modified midpoint method, which are then extrapolated to zero stepsize.

The sequence of separate attempts to cross the interval H is made with increasing values of n, the number of substeps. Bulirsch and Stoer originally proposed the sequence

$$n = 2, 4, 6, 8, 12, 16, 24, 32, 48, 64, 96, \ldots, [n_j = 2n_{j-2}], \ldots \qquad (16.4.1)$$

More recent work by Deuflhard [2,3] suggests that the sequence

$$n = 2, 4, 6, 8, 10, 12, 14, \ldots, [n_j = 2(j + 1)], \ldots \qquad (16.4.2)$$

is usually more efficient. For each step, we do not know in advance how far up this sequence we will go. After each successive n is tried, a polynomial extrapolation is attempted. That extrapolation gives both extrapolated values and error estimates. If the errors are not satisfactory, we go higher in n. If they are satisfactory, we go on to the next step and begin anew with $n = 2$.

Of course there must be some upper limit, beyond which we conclude that there is some obstacle in our path in the interval H, so that we must reduce H rather than just subdivide it more finely. In the implementations below, the maximum number of n's to be tried is called KMAXX. For reasons described below we usually take this equal to 8; the 8th value of the sequence (16.4.2) is 16, so this is the maximum number of subdivisions of H that we allow.

We enforce error control, as in the Runge-Kutta method, by monitoring internal consistency, and adapting stepsize to match a prescribed bound on the local truncation error. Each new result from the sequence of modified midpoint integrations allows a tableau like that in §3.1 to be extended by one additional set of diagonals. The size of the new correction added at each stage is taken as the (conservative) error estimate. How should we use this error estimate to adjust the stepsize? The best strategy now known is due to Deuflhard [2,3]. For completeness we describe it here:

Suppose the absolute value of the error estimate returned from column k (and hence row $k + 1$) of the extrapolation tableau is $\epsilon_{k+1,k}$. (We number from $k = 0$.) Error control is enforced by requiring

$$\epsilon_{k+1,k} < \epsilon \tag{16.4.3}$$

as the criterion for accepting the current step, where ϵ is the required tolerance. For the even sequence (16.4.2) the order of the method is $2k + 3$:

$$\epsilon_{k+1,k} \sim H^{2k+3} \tag{16.4.4}$$

Thus a simple estimate of a new stepsize H_k to obtain convergence in a fixed column k would be

$$H_k = H \left(\frac{\epsilon}{\epsilon_{k+1,k}} \right)^{1/(2k+3)} \tag{16.4.5}$$

Which column k should we aim to achieve convergence in? Let's compare the work required for different k. Suppose A_k is the work to obtain row k of the extrapolation tableau, so A_{k+1} is the work to obtain column k. We will assume the work is dominated by the cost of evaluating the functions defining the right-hand sides of the differential equations. For n_k subdivisions in H, the number of function evaluations can be found from the recurrence

$$A_0 = n_0 + 1$$
$$A_{k+1} = A_k + n_{k+1} \tag{16.4.6}$$

The work per unit step to get column k is A_{k+1}/H_k, which we nondimensionalize with a factor of H and write as

$$W_k = \frac{A_{k+1}}{H_k} H \tag{16.4.7}$$

$$= A_{k+1} \left(\frac{\epsilon_{k+1,k}}{\epsilon} \right)^{1/(2k+3)} \tag{16.4.8}$$

The quantities W_k can be calculated during the integration. The optimal column index q is then defined by

$$W_q = \min_{k=0,\dots,k_f} W_k \tag{16.4.9}$$

where k_f is the final column, in which the error criterion (16.4.3) was satisfied. The q determined from (16.4.9) defines the stepsize H_q to be used as the next basic stepsize, so that we can expect to get convergence in the optimal column q.

Two important refinements have to be made to the strategy outlined so far:
- If the current H is "too small," then k_f will be "too small," and so q remains "too small." It may be desirable to increase H and aim for convergence in a column $q > k_f$.

- If the current H is "too big," we may not converge at all on the current step and we will have to decrease H. We would like to detect this by monitoring the quantities $\epsilon_{k+1,k}$ for each k so we can stop the current step as soon as possible.

Deuflhard's prescription for dealing with these two problems uses ideas from communication theory to determine the "average expected convergence behavior" of the extrapolation. His model produces certain correction factors $\alpha(k,q)$ by which H_k is to be multiplied to try to get convergence in column q. The factors $\alpha(k,q)$ depend only on ϵ and the sequence $\{n_i\}$ and so can be computed once during initialization:

$$\alpha(k,q) = \epsilon^{\frac{A_{k+1}-A_{q+1}}{(2k+3)(A_{q+1}-A_0+1)}} \qquad \text{for} \quad k < q \qquad (16.4.10)$$

with $\alpha(q,q) = 1$.

Now to handle the first problem, suppose convergence occurs in column $q = k_f$. Then rather than taking H_q for the next step, we might aim to increase the stepsize to get convergence in column $q+1$. Since we don't have H_{q+1} available from the computation, we estimate it as

$$H_{q+1} = H_q \alpha(q, q+1) \qquad (16.4.11)$$

By equation (16.4.7) this replacement is efficient, i.e., reduces the work per unit step, if

$$\frac{A_{q+1}}{H_q} > \frac{A_{q+2}}{H_{q+1}} \qquad (16.4.12)$$

or

$$A_{q+1}\alpha(q, q+1) > A_{q+2} \qquad (16.4.13)$$

During initialization, this inequality can be checked for $q = 0, 1, \ldots$ to determine $k_{\max}$, the largest allowed column. Then when (16.4.12) is satisfied it will always be efficient to use H_{q+1}. (In practice we limit $k_{\max}$ to 7, i.e., 8 columns, even when ϵ is very small as there is very little further gain in efficiency whereas roundoff can become a problem.)

The problem of stepsize reduction is handled by computing stepsize estimates

$$\bar{H}_k \equiv H_k \alpha(k, q), \qquad k = 0, \ldots, q-1 \qquad (16.4.14)$$

during the current step. The $\bar{H}$'s are estimates of the stepsize to get convergence in the optimal column q. If any $\bar{H}_k$ is "too small," we abandon the current step and restart using $\bar{H}_k$. The criterion of being "too small" is taken to be

$$H_k \alpha(k, q+1) < H \qquad (16.4.15)$$

The α's satisfy $\alpha(k, q+1) > \alpha(k, q)$.

During the first step, when we have no information about the solution, the stepsize reduction check is made for all k. Afterwards, we test for convergence and for possible stepsize reduction only in an "order window"

$$\max(0, q-1) \le k \le \min(k_{\max}, q+1) \qquad (16.4.16)$$

The rationale for the order window is that if convergence appears to occur for $k < q - 1$ it is often spurious, resulting from some fortuitously small error estimate in the extrapolation. On the other hand, if you need to go beyond $k = q + 1$ to obtain convergence, your local model of the convergence behavior is obviously not very good and you need to cut the stepsize and reestablish it.

In the routine `bsstep`, these various tests are actually carried out using quantities

$$\epsilon(k) \equiv \frac{H}{H_k} = \left(\frac{\epsilon_{k+1,k}}{\epsilon}\right)^{1/(2k+3)} \qquad (16.4.17)$$

called `err[k]` in the code. As usual, we include a "safety factor" in the stepsize selection. This is implemented by replacing ϵ by 0.25ϵ. Other safety factors are explained in the program comments.

Note that while the optimal convergence column is restricted to increase by at most one on each step, a sudden drop in order is allowed by equation (16.4.9). This gives the method a degree of robustness for problems with discontinuities.

Let us remind you once again that *scaling* of the variables is often crucial for successful integration of differential equations. The scaling "trick" suggested in the discussion following equation (16.2.8) is a good general purpose choice, but not foolproof. Scaling by the maximum values of the variables is more robust, but requires you to have some prior information.

The following implementation of a Bulirsch-Stoer step has exactly the same calling sequence as the quality-controlled Runge-Kutta stepper rkqs. This means that the driver odeint in §16.2 can be used for Bulirsch-Stoer as well as Runge-Kutta: Just substitute bsstep for rkqs in odeint's argument list. The routine bsstep calls mmid to take the modified midpoint sequences, and calls pzextr, given below, to do the polynomial extrapolation.

```cpp
#include <cmath>
#include "nr.h"
using namespace std;

Vec_DP *x_p;
Mat_DP *d_p;
```
Pointers to matrix and vector used by pzextr or rzextr.

```cpp
void NR::bsstep(Vec_IO_DP &y, Vec_IO_DP &dydx, DP &xx, const DP htry,
    const DP eps, Vec_I_DP &yscal, DP &hdid, DP &hnext,
    void derivs(const DP, Vec_I_DP &, Vec_O_DP &))
```
Bulirsch-Stoer step with monitoring of local truncation error to ensure accuracy and adjust step-size. Input are the dependent variable vector y[0..nv-1] and its derivative dydx[0..nv-1] at the starting value of the independent variable x. Also input are the stepsize to be attempted htry, the required accuracy eps, and the vector yscal[0..nv-1] against which the error is scaled. On output, y and x are replaced by their new values, hdid is the stepsize that was actually accomplished, and hnext is the estimated next stepsize. derivs is the user-supplied routine that computes the right-hand side derivatives. Be sure to set htry on successive steps to the value of hnext returned from the previous step, as is the case if the routine is called by odeint.
```cpp
{
    const int KMAXX=8, IMAXX=(KMAXX+1);
    const DP SAFE1=0.25, SAFE2=0.7, REDMAX=1.0e-5, REDMIN=0.7;
    const DP TINY=1.0e-30, SCALMX=0.1;
```
Here KMAXX is the maximum number of rows used in the extrapolation; SAFE1 and SAFE2 are safety factors; REDMAX is the maximum factor for stepsize reduction; REDMIN is the minimum factor for stepsize reduction; TINY prevents division by zero; and 1/SCALMX is the maximum factor by which a stepsize can be increased.
```cpp
    static const int nseq_d[IMAXX]={2,4,6,8,10,12,14,16,18};
    static int first=1,kmax,kopt;
    static DP epsold = -1.0,xnew;
    static Vec_DP a(IMAXX);
    static Mat_DP alf(KMAXX,KMAXX);
    bool exitflag=false;
    int i,iq,k,kk,km,reduct;
    DP eps1,errmax,fact,h,red,scale,work,wrkmin,xest;
    Vec_INT nseq(nseq_d,IMAXX);
    Vec_DP err(KMAXX);

    int nv=y.size();
    Vec_DP yerr(nv),ysav(nv),yseq(nv);
    x_p=new Vec_DP(KMAXX);
    d_p=new Mat_DP(nv,KMAXX);
    if (eps != epsold) {                           A new tolerance, so reinitialize.
        hnext = xnew = -1.0e29;                     "Impossible" values.
        eps1=SAFE1*eps;
        a[0]=nseq[0]+1;                             Compute work coefficients $A_k$.
        for (k=0;k<KMAXX;k++) a[k+1]=a[k]+nseq[k+1];
```

```
    for (iq=1;iq<KMAXX;iq++) {                    Compute α(k, q).
        for (k=0;k<iq;k++)
            alf[k][iq]=pow(eps1,(a[k+1]-a[iq+1])/
                ((a[iq+1]-a[0]+1.0)*(2*k+3)));
    }
    epsold=eps;
    for (kopt=1;kopt<KMAXX-1;kopt++)              Determine optimal row number for
        if (a[kopt+1] > a[kopt]*alf[kopt-1][kopt]) break;     convergence.
    kmax=kopt;
}
h=htry;
for (i=0;i<nv;i++) ysav[i]=y[i];                 Save the starting values.
if (xx != xnew || h != hnext) {                  A new stepsize or a new integration:
    first=1;                                         re-establish the order window.
    kopt=kmax;
}
reduct=0;
for (;;) {
    for (k=0;k<=kmax;k++) {                       Evaluate the sequence of modified
        xnew=xx+h;                                   midpoint integrations.
        if (xnew == xx) nrerror("step size underflow in bsstep");
        mmid(ysav,dydx,xx,h,nseq[k],yseq,derivs);
        xest=SQR(h/nseq[k]);                      Squared, since error series is even.
        pzextr(k,xest,yseq,y,yerr);               Perform extrapolation.
        if (k != 0) {                             Compute normalized error estimate
            errmax=TINY;                              ε(k).
            for (i=0;i<nv;i++) errmax=MAX(errmax,fabs(yerr[i]/yscal[i]));
            errmax /= eps;                        Scale error relative to tolerance.
            km=k-1;
            err[km]=pow(errmax/SAFE1,1.0/(2*km+3));
        }
        if (k != 0 && (k >= kopt-1 || first)) {   In order window.
            if (errmax < 1.0) {                   Converged.
                exitflag=true;
                break;
            }
            if (k == kmax || k == kopt+1) {       Check for possible stepsize
                red=SAFE2/err[km];                    reduction.
                break;
            }
            else if (k == kopt && alf[kopt-1][kopt] < err[km]) {
                red=1.0/err[km];
                break;
            }
            else if (kopt == kmax && alf[km][kmax-1] < err[km]) {
                red=alf[km][kmax-1]*SAFE2/err[km];
                break;
            }
            else if (alf[km][kopt] < err[km]) {
                red=alf[km][kopt-1]/err[km];
                break;
            }
        }
    }
    if (exitflag) break;
    red=MIN(red,REDMIN);                          Reduce stepsize by at least REDMIN
    red=MAX(red,REDMAX);                              and at most REDMAX.
    h *= red;
    reduct=1;
}                                                 Try again.
xx=xnew;                                          Successful step taken.
hdid=h;
first=0;
wrkmin=1.0e35;                                    Compute optimal row for convergence
                                                      and corresponding stepsize.
```

```
for (kk=0;kk<=km;kk++) {
    fact=MAX(err[kk],SCALMX);
    work=fact*a[kk+1];
    if (work < wrkmin) {
        scale=fact;
        wrkmin=work;
        kopt=kk+1;
    }
}
hnext=h/scale;
if (kopt >= k && kopt != kmax && !reduct) {
Check for possible order increase, but not if stepsize was just reduced.
    fact=MAX(scale/alf[kopt-1][kopt],SCALMX);
    if (a[kopt+1]*fact <= wrkmin) {
        hnext=h/fact;
        kopt++;
    }
}
delete d_p;
delete x_p;
}
```

The polynomial extrapolation routine is based on the same algorithm as `polint`
§3.1. It is simpler in that it is always extrapolating to zero, rather than to an arbitrary
value. However, it is more complicated in that it must individually extrapolate each
component of a vector of quantities.

```
#include "nr.h"

extern Vec_DP *x_p;                          Defined in bsstep.
extern Mat_DP *d_p;

void NR::pzextr(const int iest, const DP xest, Vec_I_DP &yest, Vec_O_DP &yz,
    Vec_O_DP &dy)
```
Use polynomial extrapolation to evaluate `nv` functions at $x = 0$ by fitting a polynomial to a
sequence of estimates with progressively smaller values $x = $ `xest`, and corresponding function
vectors `yest[0..nv-1]`. This call is number `iest` in the sequence of calls. Extrapolated function values are output as `yz[0..nv-1]`, and their estimated error is output as `dy[0..nv-1]`.
```
{
    int j,k1;
    DP q,f2,f1,delta;

    int nv=yz.size();
    Vec_DP c(nv);
    Vec_DP &x=*x_p;                          Make aliases to simplify coding.
    Mat_DP &d=*d_p;
    x[iest]=xest;                            Save current independent variable.
    for (j=0;j<nv;j++) dy[j]=yz[j]=yest[j];
    if (iest == 0) {                         Store first estimate in first column.
        for (j=0;j<nv;j++) d[j][0]=yest[j];
    } else {
        for (j=0;j<nv;j++) c[j]=yest[j];
        for (k1=0;k1<iest;k1++) {
            delta=1.0/(x[iest-k1-1]-xest);
            f1=xest*delta;
            f2=x[iest-k1-1]*delta;
            for (j=0;j<nv;j++) {             Propagate tableau 1 diagonal more.
                q=d[j][k1];
                d[j][k1]=dy[j];
                delta=c[j]-q;
                dy[j]=f1*delta;
                c[j]=f2*delta;
```

```
                yz[j] += dy[j];
            }
        }
        for (j=0;j<nv;j++) d[j][iest]=dy[j];
    }
}
```

Current wisdom favors polynomial extrapolation over rational function extrapolation in the Bulirsch-Stoer method. However, our feeling is that this view is guided more by the kinds of problems used for tests than by one method being actually "better." Accordingly, we provide the optional routine `rzextr` for rational function extrapolation, an exact substitution for `pzextr` above.

```
#include "nr.h"

extern Vec_DP *x_p;                          Defined in bsstep.
extern Mat_DP *d_p;

void NR::rzextr(const int iest, const DP xest, Vec_I_DP &yest, Vec_O_DP &yz,
    Vec_O_DP &dy)
Exact substitute for pzextr, but uses diagonal rational function extrapolation instead of poly-
nomial extrapolation.
{
    int j,k,nv;
    DP yy,v,ddy,c,b1,b;

    nv=yz.size();
    Vec_DP fx(iest+1);
    Vec_DP &x=*x_p;                          Make aliases to simplify coding.
    Mat_DP &d=*d_p;
    x[iest]=xest;                            Save current independent variable.
    if (iest == 0)
        for (j=0;j<nv;j++) {
            yz[j]=yest[j];
            d[j][0]=yest[j];
            dy[j]=yest[j];
        }
    else {
        for (k=0;k<iest;k++)
            fx[k+1]=x[iest-(k+1)]/xest;
        for (j=0;j<nv;j++) {                 Evaluate next diagonal in tableau.
            v=d[j][0];
            d[j][0]=yy=c=yest[j];
            for (k=1;k<=iest;k++) {
                b1=fx[k]*v;
                b=b1-c;
                if (b != 0.0) {
                    b=(c-v)/b;
                    ddy=c*b;
                    c=b1*b;
                } else                       Care needed to avoid division by 0.
                    ddy=v;
                if (k != iest) v=d[j][k];
                d[j][k]=ddy;
                yy += ddy;
            }
            dy[j]=ddy;
            yz[j]=yy;
        }
    }
}
```

CITED REFERENCES AND FURTHER READING:

Stoer, J., and Bulirsch, R. 1993, *Introduction to Numerical Analysis*, 2nd ed. (New York: Springer-Verlag), §7.2.14. [1]

Gear, C.W. 1971, *Numerical Initial Value Problems in Ordinary Differential Equations* (Englewood Cliffs, NJ: Prentice-Hall), §6.2.

Deuflhard, P. 1983, *Numerische Mathematik*, vol. 41, pp. 399–422. [2]

Deuflhard, P. 1985, *SIAM Review*, vol. 27, pp. 505–535. [3]

16.5 Second-Order Conservative Equations

Usually when you have a system of high-order differential equations to solve it is best to reformulate them as a system of first-order equations, as discussed in §16.0. There is a particular class of equations that occurs quite frequently in practice where you can gain about a factor of two in efficiency by differencing the equations directly. The equations are second-order systems where the derivative does not appear on the right-hand side:

$$y'' = f(x, y), \qquad y(x_0) = y_0, \qquad y'(x_0) = z_0 \qquad (16.5.1)$$

As usual, y can denote a vector of values.

Stoermer's rule, dating back to 1907, has been a popular method for discretizing such systems. With $h = H/m$ we have

$$y_1 = y_0 + h[z_0 + \tfrac{1}{2}hf(x_0, y_0)]$$

$$y_{k+1} - 2y_k + y_{k-1} = h^2 f(x_0 + kh, y_k), \qquad k = 1, \ldots, m-1 \qquad (16.5.2)$$

$$z_m = (y_m - y_{m-1})/h + \tfrac{1}{2}hf(x_0 + H, y_m)$$

Here z_m is $y'(x_0 + H)$. Henrici showed how to rewrite equations (16.5.2) to reduce roundoff error by using the quantities $\Delta_k \equiv y_{k+1} - y_k$. Start with

$$\Delta_0 = h[z_0 + \tfrac{1}{2}hf(x_0, y_0)]$$
$$y_1 = y_0 + \Delta_0 \qquad (16.5.3)$$

Then for $k = 1, \ldots, m-1$, set

$$\Delta_k = \Delta_{k-1} + h^2 f(x_0 + kh, y_k)$$
$$y_{k+1} = y_k + \Delta_k \qquad (16.5.4)$$

Finally compute the derivative from

$$z_m = \Delta_{m-1}/h + \tfrac{1}{2}hf(x_0 + H, y_m) \qquad (16.5.5)$$

Gragg again showed that the error series for equations (16.5.3)–(16.5.5) contains only even powers of h, and so the method is a logical candidate for extrapolation à la Bulirsch-Stoer. We replace `mmid` by the following routine `stoerm`:

```
#include "nr.h"

void NR::stoerm(Vec_I_DP &y, Vec_I_DP &d2y, const DP xs,
    const DP htot, const int nstep, Vec_O_DP &yout,
    void derivs(const DP, Vec_I_DP &, Vec_O_DP &))
```

Stoermer's rule for integrating $y'' = f(x, y)$ for a system of $n = $ nv/2 equations. On input y[0..nv-1] contains y in its first n elements and y' in its second n elements, all evaluated at xs. d2y[0..nv-1] contains the right-hand side function f (also evaluated at xs) in its first n elements. Its second n elements are not referenced. Also input is htot, the total step to be taken, and nstep, the number of substeps to be used. The output is returned as yout[0..nv-1], with the same storage arrangement as y. derivs is the user-supplied routine that calculates f.

```
{
    int i,nn,n,neqns;
    DP h,h2,halfh,x;

    int nv=y.size();
    Vec_DP ytemp(nv);
    h=htot/nstep;                         Stepsize this trip.
    halfh=0.5*h;
    neqns=nv/2;                           Number of equations.
    for (i=0;i<neqns;i++) {               First step.
        n=neqns+i;
        ytemp[i]=y[i]+(ytemp[n]=h*(y[n]+halfh*d2y[i]));
    }
    x=xs+h;
    derivs(x,ytemp,yout);                 Use yout for temporary storage of derivatives.
    h2=h*h;
    for (nn=1;nn<nstep;nn++) {            General step.
        for (i=0;i<neqns;i++)
            ytemp[i] += (ytemp[neqns+i] += h2*yout[i]);
        x += h;
        derivs(x,ytemp,yout);
    }
    for (i=0;i<neqns;i++) {               Last step.
        n=neqns+i;
        yout[n]=ytemp[n]/h+halfh*yout[i];
        yout[i]=ytemp[i];
    }
}
```

Note that for compatibility with bsstep the arrays y and d2y are of length $2n$ for a system of n second-order equations. The values of y are stored in the first n elements of y, while the first derivatives are stored in the second n elements. The right-hand side f is stored in the first n elements of the array d2y; the second n elements are unused. With this storage arrangement you can use bsstep simply by replacing the call to mmid with one to stoerm using the same arguments; just be sure that the argument nv of bsstep is set to $2n$. You should also use the more efficient sequence of stepsizes suggested by Deuflhard:

$$n = 1, 2, 3, 4, 5, \ldots \tag{16.5.6}$$

and set KMAXX $= 12$ in bsstep.

CITED REFERENCES AND FURTHER READING:

Deuflhard, P. 1985, *SIAM Review*, vol. 27, pp. 505–535.

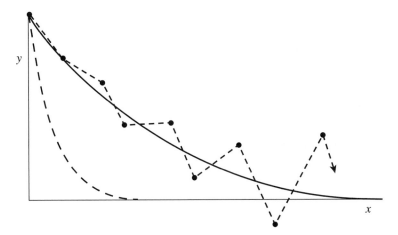

Figure 16.6.1.   Example of an instability encountered in integrating a stiff equation (schematic). Here it is supposed that the equation has two solutions, shown as solid and dashed lines. Although the initial conditions are such as to give the solid solution, the stability of the integration (shown as the unstable dotted sequence of segments) is determined by the more rapidly varying dashed solution, even after that solution has effectively died away to zero. Implicit integration methods are the cure.

16.6 Stiff Sets of Equations

As soon as one deals with more than one first-order differential equation, the possibility of a *stiff* set of equations arises. Stiffness occurs in a problem where there are two or more very different scales of the independent variable on which the dependent variables are changing. For example, consider the following set of equations [1]:

$$
\begin{aligned}
u' &= 998u + 1998v \\
v' &= -999u - 1999v
\end{aligned}
\tag{16.6.1}
$$

with boundary conditions

$$
u(0) = 1 \qquad v(0) = 0
\tag{16.6.2}
$$

By means of the transformation

$$
u = 2y - z \qquad v = -y + z
\tag{16.6.3}
$$

we find the solution

$$
\begin{aligned}
u &= 2e^{-x} - e^{-1000x} \\
v &= -e^{-x} + e^{-1000x}
\end{aligned}
\tag{16.6.4}
$$

If we integrated the system (16.6.1) with any of the methods given so far in this chapter, the presence of the e^{-1000x} term would require a stepsize $h \ll 1/1000$ for the method to be stable (the reason for this is explained below). This is so even though the e^{-1000x} term is completely negligible in determining the values of u and v as soon as one is away from the origin (see Figure 16.6.1).

This is the generic disease of stiff equations: we are required to follow the variation in the solution on the shortest length scale to maintain stability of the integration, even though accuracy requirements allow a much larger stepsize.

To see how we might cure this problem, consider the single equation

$$y' = -cy \tag{16.6.5}$$

where $c > 0$ is a constant. The explicit (or *forward*) Euler scheme for integrating this equation with stepsize h is

$$y_{n+1} = y_n + hy'_n = (1 - ch)y_n \tag{16.6.6}$$

The method is called explicit because the new value y_{n+1} is given explicitly in terms of the old value y_n. Clearly the method is unstable if $h > 2/c$, for then $|y_n| \to \infty$ as $n \to \infty$.

The simplest cure is to resort to *implicit* differencing, where the right-hand side is evaluated at the *new y* location. In this case, we get the *backward Euler* scheme:

$$y_{n+1} = y_n + hy'_{n+1} \tag{16.6.7}$$

or

$$y_{n+1} = \frac{y_n}{1 + ch} \tag{16.6.8}$$

The method is absolutely stable: even as $h \to \infty$, $y_{n+1} \to 0$, which is in fact the correct solution of the differential equation. If we think of x as representing time, then the implicit method converges to the true equilibrium solution (i.e., the solution at late times) for large stepsizes. This nice feature of implicit methods holds only for linear systems, but even in the general case implicit methods give better stability. Of course, we give up *accuracy* in following the evolution towards equilibrium if we use large stepsizes, but we maintain *stability*.

These considerations can easily be generalized to sets of linear equations with constant coefficients:

$$\mathbf{y}' = -\mathbf{C} \cdot \mathbf{y} \tag{16.6.9}$$

where $\mathbf{C}$ is a positive definite matrix. Explicit differencing gives

$$\mathbf{y}_{n+1} = (\mathbf{1} - \mathbf{C}h) \cdot \mathbf{y}_n \tag{16.6.10}$$

Now a matrix $\mathbf{A}^n$ tends to zero as $n \to \infty$ only if the largest eigenvalue of $\mathbf{A}$ has magnitude less than unity. Thus $\mathbf{y}_n$ is bounded as $n \to \infty$ only if the largest eigenvalue of $\mathbf{1} - \mathbf{C}h$ is less than 1, or in other words

$$h < \frac{2}{\lambda_{\max}} \tag{16.6.11}$$

where $\lambda_{\max}$ is the largest eigenvalue of $\mathbf{C}$.

On the other hand, implicit differencing gives

$$\mathbf{y}_{n+1} = \mathbf{y}_n + h\mathbf{y}'_{n+1} \tag{16.6.12}$$

or

$$\mathbf{y}_{n+1} = (\mathbf{1} + \mathbf{C}h)^{-1} \cdot \mathbf{y}_n \qquad (16.6.13)$$

If the eigenvalues of $\mathbf{C}$ are λ, then the eigenvalues of $(\mathbf{1} + \mathbf{C}h)^{-1}$ are $(1 + \lambda h)^{-1}$, which has magnitude less than one for all h. (Recall that all the eigenvalues of a positive definite matrix are nonnegative.) Thus the method is stable for all stepsizes h. The penalty we pay for this stability is that we are required to invert a matrix at each step.

Not all equations are linear with constant coefficients, unfortunately! For the system

$$\mathbf{y}' = \mathbf{f}(\mathbf{y}) \qquad (16.6.14)$$

implicit differencing gives

$$\mathbf{y}_{n+1} = \mathbf{y}_n + h\mathbf{f}(\mathbf{y}_{n+1}) \qquad (16.6.15)$$

In general this is some nasty set of nonlinear equations that has to be solved iteratively at each step. Suppose we try linearizing the equations, as in Newton's method:

$$\mathbf{y}_{n+1} = \mathbf{y}_n + h\left[\mathbf{f}(\mathbf{y}_n) + \frac{\partial \mathbf{f}}{\partial \mathbf{y}}\bigg|_{\mathbf{y}_n} \cdot (\mathbf{y}_{n+1} - \mathbf{y}_n)\right] \qquad (16.6.16)$$

Here $\partial \mathbf{f}/\partial \mathbf{y}$ is the matrix of the partial derivatives of the right-hand side (the Jacobian matrix). Rearrange equation (16.6.16) into the form

$$\mathbf{y}_{n+1} = \mathbf{y}_n + h\left[\mathbf{1} - h\frac{\partial \mathbf{f}}{\partial \mathbf{y}}\right]^{-1} \cdot \mathbf{f}(\mathbf{y}_n) \qquad (16.6.17)$$

If h is not too big, only one iteration of Newton's method may be accurate enough to solve equation (16.6.15) using equation (16.6.17). In other words, at each step we have to invert the matrix

$$\mathbf{1} - h\frac{\partial \mathbf{f}}{\partial \mathbf{y}} \qquad (16.6.18)$$

to find $\mathbf{y}_{n+1}$. Solving implicit methods by linearization is called a "semi-implicit" method, so equation (16.6.17) is the *semi-implicit Euler method*. It is not guaranteed to be stable, but it usually is, because the behavior is locally similar to the case of a constant matrix $\mathbf{C}$ described above.

So far we have dealt only with implicit methods that are first-order accurate. While these are very robust, most problems will benefit from higher-order methods. There are three important classes of higher-order methods for stiff systems:

- Generalizations of the Runge-Kutta method, of which the most useful are the Rosenbrock methods. The first practical implementation of these ideas was by Kaps and Rentrop, and so these methods are also called Kaps-Rentrop methods.
- Generalizations of the Bulirsch-Stoer method, in particular a semi-implicit extrapolation method due to Bader and Deuflhard.

- Predictor-corrector methods, most of which are descendants of Gear's backward differentiation method.

We shall give implementations of the first two methods. Note that systems where the right-hand side depends explicitly on x, $\mathbf{f}(\mathbf{y}, x)$, can be handled by adding x to the list of dependent variables so that the system to be solved is

$$\begin{pmatrix} \mathbf{y} \\ x \end{pmatrix}' = \begin{pmatrix} \mathbf{f} \\ 1 \end{pmatrix} \tag{16.6.19}$$

In both the routines to be given in this section, we have explicitly carried out this replacement for you, so the routines can handle right-hand sides of the form $\mathbf{f}(\mathbf{y}, x)$ without any special effort on your part.

We now mention an important point: *It is absolutely crucial to scale your variables properly when integrating stiff problems with automatic stepsize adjustment.* As in our nonstiff routines, you will be asked to supply a vector $\mathbf{y}_{\text{scal}}$ with which the error is to be scaled. For example, to get constant fractional errors, simply set $\mathbf{y}_{\text{scal}} = |\mathbf{y}|$. You can get constant absolute errors relative to some maximum values by setting $\mathbf{y}_{\text{scal}}$ equal to those maximum values. In stiff problems, there are often strongly decreasing pieces of the solution which you are not particularly interested in following once they are small. You can control the relative error above some threshold $\mathbf{C}$ and the absolute error below the threshold by setting

$$\mathbf{y}_{\text{scal}} = \max(\mathbf{C}, |\mathbf{y}|) \tag{16.6.20}$$

If you are using appropriate nondimensional units, then each component of $\mathbf{C}$ should be of order unity. If you are not sure what values to take for $\mathbf{C}$, simply try setting each component equal to unity. *We strongly advocate the choice (16.6.20) for stiff problems.*

One final warning: Solving stiff problems can sometimes lead to catastrophic precision loss. Double precision is often a requirement, not an option.

Rosenbrock Methods

These methods have the advantage of being relatively simple to understand and implement. For moderate accuracies ($\epsilon \lesssim 10^{-4} - 10^{-5}$ in the error criterion) and moderate-sized systems ($N \lesssim 10$), they are competitive with the more complicated algorithms. For more stringent parameters, Rosenbrock methods remain reliable; they merely become less efficient than competitors like the semi-implicit extrapolation method (see below).

A Rosenbrock method seeks a solution of the form

$$\mathbf{y}(x_0 + h) = \mathbf{y}_0 + \sum_{i=1}^{s} c_i \mathbf{k}_i \tag{16.6.21}$$

where the corrections $\mathbf{k}_i$ are found by solving s linear equations that generalize the structure in (16.6.17):

$$(\mathbf{1} - \gamma h \mathbf{f}') \cdot \mathbf{k}_i = h\mathbf{f}\left(\mathbf{y}_0 + \sum_{j=1}^{i-1} \alpha_{ij} \mathbf{k}_j\right) + h\mathbf{f}' \cdot \sum_{j=1}^{i-1} \gamma_{ij} \mathbf{k}_j, \qquad i = 1, \ldots, s \tag{16.6.22}$$

Here we denote the Jacobian matrix by $\mathbf{f}'$. The coefficients γ, c_i, α_{ij}, and γ_{ij} are fixed constants independent of the problem. If $\gamma = \gamma_{ij} = 0$, this is simply a Runge-Kutta scheme. Equations (16.6.22) can be solved successively for $\mathbf{k}_1, \mathbf{k}_2, \ldots$.

Crucial to the success of a stiff integration scheme is an automatic stepsize adjustment algorithm. Kaps and Rentrop [2] discovered an *embedded* or Runge-Kutta-Fehlberg method as described in §16.2: Two estimates of the form (16.6.21) are computed, the "real" one $\mathbf{y}$ and a lower-order estimate $\widehat{\mathbf{y}}$ with different coefficients $\hat{c}_i, i = 1, \ldots, \hat{s}$, where $\hat{s} < s$ but the $\mathbf{k}_i$ are the same. The difference between $\mathbf{y}$ and $\widehat{\mathbf{y}}$ leads to an estimate of the local truncation error, which can then be used for stepsize control. Kaps and Rentrop showed that the smallest value of s for which embedding is possible is $s = 4$, $\hat{s} = 3$, leading to a fourth-order method.

To minimize the matrix-vector multiplications on the right-hand side of (16.6.22), we rewrite the equations in terms of quantities

$$\mathbf{g}_i = \sum_{j=1}^{i-1} \gamma_{ij}\mathbf{k}_j + \gamma\mathbf{k}_i \tag{16.6.23}$$

The equations then take the form

$$(1/\gamma h - \mathbf{f}') \cdot \mathbf{g}_1 = \mathbf{f}(\mathbf{y}_0)$$

$$(1/\gamma h - \mathbf{f}') \cdot \mathbf{g}_2 = \mathbf{f}(\mathbf{y}_0 + a_{21}\mathbf{g}_1) + c_{21}\mathbf{g}_1/h$$

$$(1/\gamma h - \mathbf{f}') \cdot \mathbf{g}_3 = \mathbf{f}(\mathbf{y}_0 + a_{31}\mathbf{g}_1 + a_{32}\mathbf{g}_2) + (c_{31}\mathbf{g}_1 + c_{32}\mathbf{g}_2)/h$$

$$(1/\gamma h - \mathbf{f}') \cdot \mathbf{g}_4 = \mathbf{f}(\mathbf{y}_0 + a_{41}\mathbf{g}_1 + a_{42}\mathbf{g}_2 + a_{43}\mathbf{g}_3) + (c_{41}\mathbf{g}_1 + c_{42}\mathbf{g}_2 + c_{43}\mathbf{g}_3)/h$$

$$\tag{16.6.24}$$

In our implementation `stiff` of the Kaps-Rentrop algorithm, we have carried out the replacement (16.6.19) explicitly in equations (16.6.24), so you need not concern yourself about it. Simply provide a routine (called `derivs` in `stiff`) that returns $\mathbf{f}$ (called `dydx`) as a function of x and $\mathbf{y}$. Also supply a routine `jacobn_s` that returns $\mathbf{f}'$ (`dfdy`) and $\partial\mathbf{f}/\partial x$ (`dfdx`) as functions of x and $\mathbf{y}$. If x does not occur explicitly on the right-hand side, then `dfdx` will be zero. Usually the Jacobian matrix will be available to you by analytic differentiation of the right-hand side $\mathbf{f}$. If not, your routine will have to compute it by numerical differencing with appropriate increments $\Delta\mathbf{y}$.

Kaps and Rentrop gave two different sets of parameters, which have slightly different stability properties. Several other sets have been proposed. Our default choice is that of Shampine [3], but we also give you one of the Kaps-Rentrop sets as an option. Some proposed parameter sets require function evaluations outside the domain of integration; we prefer to avoid that complication.

The calling sequence of `stiff` is exactly the same as the nonstiff routines given earlier in this chapter. It is thus "plug-compatible" with them in the general ODE integrating routine `odeint`. This compatibility requires, unfortunately, one slight anomaly: While the user-supplied routine `derivs` is a dummy argument (which can therefore have any actual name), the other user-supplied routine is *not* an argument and must be named (exactly) `jacobn_s`.

`stiff` begins by saving the initial values, in case the step has to be repeated because the error tolerance is exceeded. The linear equations (16.6.24) are solved by first computing the LU decomposition of the matrix $1/\gamma h - \mathbf{f}'$ using the routine `ludcmp`. Then the four $\mathbf{g}_i$ are found by back-substitution of the four different right-hand sides using `lubksb`. Note that each step of the integration requires one call to `jacobn_s` and three calls to `derivs` (one call to get `dydx` before calling `stiff`, and two calls inside `stiff`). The reason only three calls are needed and not four is that the parameters have been chosen so that the last two calls in equation (16.6.24) are done with the same arguments. Counting the evaluation of the Jacobian matrix as roughly equivalent to N evaluations of the right-hand side $\mathbf{f}$, we see that the Kaps-Rentrop scheme involves about $N + 3$ function evaluations per step. Note that if N is large and the Jacobian matrix is sparse, you should replace the LU decomposition by a suitable sparse matrix procedure.

Stepsize control depends on the fact that

$$\mathbf{y}_{\text{exact}} = \mathbf{y} + O(h^5)$$

$$\mathbf{y}_{\text{exact}} = \widehat{\mathbf{y}} + O(h^4) \tag{16.6.25}$$

Thus

$$|\mathbf{y} - \widehat{\mathbf{y}}| = O(h^4) \qquad (16.6.26)$$

Referring back to the steps leading from equation (16.2.4) to equation (16.2.10), we see that the new stepsize should be chosen as in equation (16.2.10) but with the exponents 1/4 and 1/5 replaced by 1/3 and 1/4, respectively. Also, experience shows that it is wise to prevent too large a stepsize change in one step, otherwise we will probably have to undo the large change in the next step. We adopt 0.5 and 1.5 as the maximum allowed decrease and increase of h in one step.

```
#include <cmath>
#include "nr.h"
using namespace std;

void NR::stiff(Vec_IO_DP &y, Vec_IO_DP &dydx, DP &x, const DP htry,
    const DP eps, Vec_I_DP &yscal, DP &hdid, DP &hnext,
    void derivs(const DP, Vec_I_DP &, Vec_O_DP &))
```
Fourth-order Rosenbrock step for integrating stiff o.d.e.'s, with monitoring of local truncation error to adjust stepsize. Input are the dependent variable vector y[0..n-1] and its derivative dydx[0..n-1] at the starting value of the independent variable x. Also input are the stepsize to be attempted htry, the required accuracy eps, and the vector yscal[0..n-1] against which the error is scaled. On output, y and x are replaced by their new values, hdid is the stepsize that was actually accomplished, and hnext is the estimated next stepsize. derivs is a user-supplied routine that computes the derivatives of the right-hand side with respect to x, while jacobn_s (a fixed name) is a user-supplied routine that computes the Jacobi matrix of derivatives of the right-hand side with respect to the components of y.
```
{
    const DP SAFETY=0.9,GROW=1.5,PGROW= -0.25,SHRNK=0.5;
    const DP PSHRNK=(-1.0/3.0),ERRCON=0.1296;
```
Here GROW and SHRNK are the largest and smallest factors by which stepsize can change in one step; ERRCON equals (GROW/SAFETY) raised to the power (1/PGROW) and handles the case when errmax $\simeq$ 0.
```
    const int MAXTRY=40;                    Maximum number of stepsize changes attempted.
    const DP GAM=1.0/2.0,A21=2.0,A31=48.0/25.0,A32=6.0/25.0,C21= -8.0,
        C31=372.0/25.0,C32=12.0/5.0,C41=(-112.0/125.0),
        C42=(-54.0/125.0),C43=(-2.0/5.0),B1=19.0/9.0,B2=1.0/2.0,
        B3=25.0/108.0,B4=125.0/108.0,E1=17.0/54.0,E2=7.0/36.0,E3=0.0,
        E4=125.0/108.0,C1X=1.0/2.0,C2X=(-3.0/2.0),C3X=(121.0/50.0),
        C4X=(29.0/250.0),A2X=1.0,A3X=3.0/5.0;
    int i,j,jtry;
    DP d,errmax,h,xsav;

    int n=y.size();
    Mat_DP a(n,n),dfdy(n,n);
    Vec_INT indx(n);
    Vec_DP dfdx(n),dysav(n),err(n),ysav(n),g1(n),g2(n),g3(n),g4(n);
    xsav=x;                                 Save initial values.
    for (i=0;i<n;i++) {
        ysav[i]=y[i];
        dysav[i]=dydx[i];
    }
    jacobn_s(xsav,ysav,dfdx,dfdy);
```
The user must supply this routine to return the n-by-n matrix dfdy and the vector dfdx.
```
    h=htry;                                 Set stepsize to the initial trial value.
    for (jtry=0;jtry<MAXTRY;jtry++) {
        for (i=0;i<n;i++) {                 Set up the matrix 1 − γhf'.
            for (j=0;j<n;j++) a[i][j] = -dfdy[i][j];
            a[i][i] += 1.0/(GAM*h);
        }
        ludcmp(a,indx,d);                   LU decomposition of the matrix.
        for (i=0;i<n;i++)                   Set up right-hand side for g₁.
            g1[i]=dysav[i]+h*C1X*dfdx[i];
        lubksb(a,indx,g1);                  Solve for g₁.
        for (i=0;i<n;i++)                   Compute intermediate values of y and x.
```

```
        y[i]=ysav[i]+A21*g1[i];
    x=xsav+A2X*h;
    derivs(x,y,dydx);                 Compute dydx at the intermediate values.
    for (i=0;i<n;i++)                 Set up right-hand side for g₂.
        g2[i]=dydx[i]+h*C2X*dfdx[i]+C21*g1[i]/h;
    lubksb(a,indx,g2);                Solve for g₂.
    for (i=0;i<n;i++)                 Compute intermediate values of y and x.
        y[i]=ysav[i]+A31*g1[i]+A32*g2[i];
    x=xsav+A3X*h;
    derivs(x,y,dydx);                 Compute dydx at the intermediate values.
    for (i=0;i<n;i++)                 Set up right-hand side for g₃.
        g3[i]=dydx[i]+h*C3X*dfdx[i]+(C31*g1[i]+C32*g2[i])/h;
    lubksb(a,indx,g3);                Solve for g₃.
    for (i=0;i<n;i++)                 Set up right-hand side for g₄.
        g4[i]=dydx[i]+h*C4X*dfdx[i]+(C41*g1[i]+C42*g2[i]+C43*g3[i])/h;
    lubksb(a,indx,g4);                Solve for g₄.
    for (i=0;i<n;i++) {               Get fourth-order estimate of y and error estimate.
        y[i]=ysav[i]+B1*g1[i]+B2*g2[i]+B3*g3[i]+B4*g4[i];
        err[i]=E1*g1[i]+E2*g2[i]+E3*g3[i]+E4*g4[i];
    }
    x=xsav+h;
    if (x == xsav) nrerror("stepsize not significant in stiff");
    errmax=0.0;                       Evaluate accuracy.
    for (i=0;i<n;i++) errmax=MAX(errmax,fabs(err[i]/yscal[i]));
    errmax /= eps;                    Scale relative to required tolerance.
    if (errmax <= 1.0) {              Step succeeded. Compute size of next step and re-
        hdid=h;                       turn.
        hnext=(errmax > ERRCON ? SAFETY*h*pow(errmax,PGROW) : GROW*h);
        return;
    } else {                          Truncation error too large, reduce stepsize.
        hnext=SAFETY*h*pow(errmax,PSHRNK);
        h=(h >= 0.0 ? MAX(hnext,SHRNK*h) : MIN(hnext,SHRNK*h));
    }
}                                     Go back and re-try step.
nrerror("exceeded MAXTRY in stiff");
}
```

Here are the Kaps-Rentrop parameters, which can be substituted for those of Shampine simply by replacing the appropriate const DP statement:

```
const DP GAM=0.231,A21=2.0,A31=4.52470820736,A32=4.16352878860,
    C21=-5.07167533877,C31=6.02015272865,C32=0.159750684673,
    C41=-1.856343618677,C42=-8.50538085819,C43=-2.08407513602,
    B1=3.95750374663,B2=4.62489238836,B3=0.617477263873,
    B4=1.282612945268,E1=-2.30215540292,E2=-3.07363448539,
    E3=0.873280801802,E4=1.282612945268,C1X=GAM,
    C2X=-0.396296677520e-01,C3X=0.550778939579,
    C4X=-0.553509845700e-01,A2X=0.462,A3X=0.880208333333
```

As an example of how stiff is used, one can solve the system

$$y_0' = -.013y_0 - 1000y_0y_2$$

$$y_1' = -2500y_1y_2 \tag{16.6.27}$$

$$y_2' = -.013y_0 - 1000y_0y_2 - 2500y_1y_2$$

with initial conditions

$$y_0(0) = 1, \qquad y_1(0) = 1, \qquad y_2(0) = 0 \tag{16.6.28}$$

(This is test problem D4 in [4].) We integrate the system up to $x = 50$ with an initial stepsize of $h = 2.9 \times 10^{-4}$ using odeint. The components of $\mathbf{C}$ in (16.6.20) are all set to unity.

The routines `derivs` and `jacobn_s` for this problem are given below. Even though the ratio of largest to smallest decay constants for this problem is around 10^6, `stiff` succeeds in integrating this set in only 29 steps with $\epsilon = 10^{-4}$. By contrast, the Runge-Kutta routine `rkqs` requires 51,012 steps!

```
#include "nr.h"

void NR::jacobn_s(const DP x, Vec_I_DP &y, Vec_O_DP &dfdx, Mat_O_DP &dfdy)
{
    int i;

    int n=y.size();
    for (i=0;i<n;i++) dfdx[i]=0.0;
    dfdy[0][0] = -0.013-1000.0*y[2];
    dfdy[0][1] = 0.0;
    dfdy[0][2] = -1000.0*y[0];
    dfdy[1][0] = 0.0;
    dfdy[1][1] = -2500.0*y[2];
    dfdy[1][2] = -2500.0*y[1];
    dfdy[2][0] = -0.013-1000.0*y[2];
    dfdy[2][1] = -2500.0*y[2];
    dfdy[2][2] = -1000.0*y[0]-2500.0*y[1];
}

void NR::derivs_s(const DP x, Vec_I_DP &y, Vec_O_DP &dydx)
{
    dydx[0] = -0.013*y[0]-1000.0*y[0]*y[2];
    dydx[1] = -2500.0*y[1]*y[2];
    dydx[2] = -0.013*y[0]-1000.0*y[0]*y[2]-2500.0*y[1]*y[2];
}
```

Semi-implicit Extrapolation Method

The Bulirsch-Stoer method, which discretizes the differential equation using the modified midpoint rule, does not work for stiff problems. Bader and Deuflhard [5] discovered a semi-implicit discretization that works very well and that lends itself to extrapolation exactly as in the original Bulirsch-Stoer method.

The starting point is an implicit form of the midpoint rule:

$$\mathbf{y}_{n+1} - \mathbf{y}_{n-1} = 2h\mathbf{f}\left(\frac{\mathbf{y}_{n+1} + \mathbf{y}_{n-1}}{2}\right) \qquad (16.6.29)$$

Convert this equation into semi-implicit form by linearizing the right-hand side about $\mathbf{f}(\mathbf{y}_n)$. The result is the *semi-implicit midpoint rule*:

$$\left[1 - h\frac{\partial \mathbf{f}}{\partial \mathbf{y}}\right] \cdot \mathbf{y}_{n+1} = \left[1 + h\frac{\partial \mathbf{f}}{\partial \mathbf{y}}\right] \cdot \mathbf{y}_{n-1} + 2h\left[\mathbf{f}(\mathbf{y}_n) - \frac{\partial \mathbf{f}}{\partial \mathbf{y}} \cdot \mathbf{y}_n\right] \qquad (16.6.30)$$

It is used with a special first step, the semi-implicit Euler step (16.6.17), and a special "smoothing" last step in which the last $\mathbf{y}_n$ is replaced by

$$\bar{\mathbf{y}}_n \equiv \tfrac{1}{2}(\mathbf{y}_{n+1} + \mathbf{y}_{n-1}) \qquad (16.6.31)$$

Bader and Deuflhard showed that the error series for this method once again involves only even powers of h.

For practical implementation, it is better to rewrite the equations using $\Delta_k \equiv \mathbf{y}_{k+1} - \mathbf{y}_k$. With $h = H/m$, start by calculating

$$\Delta_0 = \left[1 - h\frac{\partial \mathbf{f}}{\partial \mathbf{y}}\right]^{-1} \cdot h\mathbf{f}(\mathbf{y}_0)$$
$$\mathbf{y}_1 = \mathbf{y}_0 + \Delta_0 \qquad (16.6.32)$$

Then for $k = 1, \ldots, m - 1$, set

$$\Delta_k = \Delta_{k-1} + 2 \left[1 - h \frac{\partial \mathbf{f}}{\partial \mathbf{y}} \right]^{-1} \cdot [h\mathbf{f}(\mathbf{y}_k) - \Delta_{k-1}]$$

$$\mathbf{y}_{k+1} = \mathbf{y}_k + \Delta_k$$

(16.6.33)

Finally compute

$$\Delta_m = \left[1 - h \frac{\partial \mathbf{f}}{\partial \mathbf{y}} \right]^{-1} \cdot [h\mathbf{f}(\mathbf{y}_m) - \Delta_{m-1}]$$

$$\bar{\mathbf{y}}_m = \mathbf{y}_m + \Delta_m$$

(16.6.34)

It is easy to incorporate the replacement (16.6.19) in the above formulas. The additional terms in the Jacobian that come from $\partial \mathbf{f}/\partial x$ all cancel out of the semi-implicit midpoint rule (16.6.30). In the special first step (16.6.17), and in the corresponding equation (16.6.32), the term $h\mathbf{f}$ becomes $h\mathbf{f} + h^2 \partial \mathbf{f}/\partial x$. The remaining equations are all unchanged.

This algorithm is implemented in the routine simpr:

```
#include "nr.h"

void NR::simpr(Vec_I_DP &y, Vec_I_DP &dydx, Vec_I_DP &dfdx, Mat_I_DP &dfdy,
    const DP xs, const DP htot, const int nstep, Vec_O_DP &yout,
    void derivs(const DP, Vec_I_DP &, Vec_O_DP &))
```
Performs one step of the semi-implicit midpoint rule. Input quantities are the dependent variable y[0..n-1], its derivative dydx[0..n-1], the derivative of the right-hand side with respect to x, dfdx[0..n-1], and the Jacobian dfdy[0..n-1][0..n-1] at xs. Also input are htot, the total step to be taken, and nstep, the number of substeps to be used. The output is returned as yout[0..n-1]. derivs is the user-supplied routine that calculates dydx.
```
{
    int i,j,nn;
    DP d,h,x;

    int n=y.size();
    Mat_DP a(n,n);
    Vec_INT indx(n);
    Vec_DP del(n),ytemp(n);
    h=htot/nstep;                       Stepsize this trip.
    for (i=0;i<n;i++) {                 Set up the matrix 1 − hf′.
        for (j=0;j<n;j++) a[i][j] = -h*dfdy[i][j];
        ++a[i][i];
    }
    ludcmp(a,indx,d);                   LU decomposition of the matrix.
    for (i=0;i<n;i++)                   Set up right-hand side for first step. Use yout
        yout[i]=h*(dydx[i]+h*dfdx[i]);      for temporary storage.
    lubksb(a,indx,yout);
    for (i=0;i<n;i++)                   First step.
        ytemp[i]=y[i]+(del[i]=yout[i]);
    x=xs+h;
    derivs(x,ytemp,yout);               Use yout for temporary storage of derivatives.
    for (nn=2;nn<=nstep;nn++) {         General step.
        for (i=0;i<n;i++)              Set up right-hand side for general step.
            yout[i]=h*yout[i]-del[i];
        lubksb(a,indx,yout);
        for (i=0;i<n;i++) ytemp[i] += (del[i] += 2.0*yout[i]);
        x += h;
        derivs(x,ytemp,yout);
    }
    for (i=0;i<n;i++)                   Set up right-hand side for last step.
        yout[i]=h*yout[i]-del[i];
    lubksb(a,indx,yout);
    for (i=0;i<n;i++)                   Take last step.
        yout[i] += ytemp[i];
}
```

The routine `simpr` is intended to be used in a routine `stifbs` that is almost exactly the same as `bsstep`. The only differences are:

- The stepsize sequence is

$$n = 2, 6, 10, 14, 22, 34, 50, \ldots, \qquad (16.6.35)$$

where each member differs from its predecessor by the smallest multiple of 4 that makes the ratio of successive terms be $\leq \frac{5}{7}$. The constant `KMAXX` is taken to be 7.

- The work per unit step now includes the cost of Jacobian evaluations as well as function evaluations. We count one Jacobian evaluation as equivalent to N function evaluations, where N is the number of equations.

- Once again the user-supplied routine `derivs` is a dummy argument and so can have any name. However, to maintain "plug-compatibility" with `rkqs`, `bsstep` and `stiff`, the routine `jacobn_s` is not an argument and *must* have exactly this name. It is called once per step to return $\mathbf{f}'$ (`dfdy`) and $\partial \mathbf{f}/\partial x$ (`dfdx`) as functions of x and $\mathbf{y}$.

Here is the routine, with comments pointing out only the differences from `bsstep`:

```
#include <cmath>
#include "nr.h"
using namespace std;

Vec_DP *x_p;
Mat_DP *d_p;

void NR::stifbs(Vec_IO_DP &y, Vec_IO_DP &dydx, DP &xx, const DP htry,
    const DP eps, Vec_I_DP &yscal, DP &hdid, DP &hnext,
    void derivs(const DP, Vec_I_DP &, Vec_O_DP &))
```
Semi-implicit extrapolation step for integrating stiff o.d.e.'s, with monitoring of local truncation error to adjust stepsize. Input are the dependent variable vector y[0..nv-1] and its derivative dydx[0..nv-1] at the starting value of the independent variable x. Also input are the stepsize to be attempted htry, the required accuracy eps, and the vector yscal[0..nv-1] against which the error is scaled. On output, y and x are replaced by their new values, hdid is the stepsize that was actually accomplished, and hnext is the estimated next stepsize. derivs is a user-supplied routine that computes the derivatives of the right-hand side with respect to x, while jacobn_s (a fixed name) is a user-supplied routine that computes the Jacobi matrix of derivatives of the right-hand side with respect to the components of y. Be sure to set htry on successive steps to the value of hnext returned from the previous step, as is the case if the routine is called by odeint.
```
{
    const int KMAXX=7,IMAXX=KMAXX+1;
    const DP SAFE1=0.25,SAFE2=0.7,REDMAX=1.0e-5,REDMIN=0.7;
    const DP TINY=1.0e-30,SCALMX=0.1;
    bool exitflag=false;
    int i,iq,k,kk,km,reduct;
    static int first=1,kmax,kopt,nvold = -1;
    DP eps1,errmax,fact,h,red,scale,work,wrkmin,xest;
    static DP epsold = -1.0,xnew;
    static Vec_DP a(IMAXX);
    static Mat_DP alf(KMAXX,KMAXX);
    static int nseq_d[IMAXX]={2,6,10,14,22,34,50,70};     Sequence is different from
    Vec_INT nseq(nseq_d,IMAXX);                                      bsstep.

    int nv=y.size();
    d_p=new Mat_DP(nv,KMAXX);
    x_p=new Vec_DP(KMAXX);
    Vec_DP dfdx(nv),err(KMAXX),yerr(nv),ysav(nv),yseq(nv);
    Mat_DP dfdy(nv,nv);
    if (eps != epsold || nv != nvold) {          Reinitialize also if nv has changed.
        hnext = xnew = -1.0e29;
        eps1=SAFE1*eps;
        a[0]=nseq[0]+1;
        for (k=0;k<KMAXX;k++) a[k+1]=a[k]+nseq[k+1];
```

```
        for (iq=1;iq<KMAXX;iq++) {
            for (k=0;k<iq;k++)
                alf[k][iq]=pow(eps1,(a[k+1]-a[iq+1])/
                    ((a[iq+1]-a[0]+1.0)*(2*k+3)));
        }
        epsold=eps;
        nvold=nv;                               Save nv.
        a[0] += nv;                             Add cost of Jacobian evaluations to work
        for (k=0;k<KMAXX;k++) a[k+1]=a[k]+nseq[k+1];         coefficients.
        for (kopt=1;kopt<KMAXX-1;kopt++)
            if (a[kopt+1] > a[kopt]*alf[kopt-1][kopt]) break;
        kmax=kopt;
    }
    h=htry;
    for (i=0;i<nv;i++) ysav[i]=y[i];
    jacobn_s(xx,y,dfdx,dfdy);                   Evaluate Jacobian.
    if (xx != xnew || h != hnext) {
        first=1;
        kopt=kmax;
    }
    reduct=0;
    for (;;) {
        for (k=0;k<=kmax;k++) {
            xnew=xx+h;
            if (xnew == xx) nrerror("step size underflow in stifbs");
            simpr(ysav,dydx,dfdx,dfdy,xx,h,nseq[k],yseq,derivs);
            Semi-implicit midpoint rule.
            xest=SQR(h/nseq[k]);                The rest of the routine is identical to
            pzextr(k,xest,yseq,y,yerr);             bsstep.
            if (k != 0) {
                errmax=TINY;
                for (i=0;i<nv;i++) errmax=MAX(errmax,fabs(yerr[i]/yscal[i]));
                errmax /= eps;
                km=k-1;
                err[km]=pow(errmax/SAFE1,1.0/(2*km+3));
            }
            if (k != 0 && (k >= kopt-1 || first)) {
                if (errmax < 1.0) {
                    exitflag=true;
                    break;
                }
                if (k == kmax || k == kopt+1) {
                    red=SAFE2/err[km];
                    break;
                }
                else if (k == kopt && alf[kopt-1][kopt] < err[km]) {
                    red=1.0/err[km];
                    break;
                }
                else if (kopt == kmax && alf[km][kmax-1] < err[km]) {
                    red=alf[km][kmax-1]*SAFE2/err[km];
                    break;
                }
                else if (alf[km][kopt] < err[km]) {
                    red=alf[km][kopt-1]/err[km];
                    break;
                }
            }
        }
        if (exitflag) break;
        red=MIN(red,REDMIN);
        red=MAX(red,REDMAX);
        h *= red;
        reduct=1;
```

```
    }
    xx=xnew;
    hdid=h;
    first=0;
    wrkmin=1.0e35;
    for (kk=0;kk<=km;kk++) {
        fact=MAX(err[kk],SCALMX);
        work=fact*a[kk+1];
        if (work < wrkmin) {
            scale=fact;
            wrkmin=work;
            kopt=kk+1;
        }
    }
    hnext=h/scale;
    if (kopt >= k && kopt != kmax && !reduct) {
        fact=MAX(scale/alf[kopt-1][kopt],SCALMX);
        if (a[kopt+1]*fact <= wrkmin) {
            hnext=h/fact;
            kopt++;
        }
    }
    delete d_p;
    delete x_p;
}
```

The routine stifbs is an excellent routine for all stiff problems, competitive with the best Gear-type routines. stiff is comparable in execution time for moderate N and $\epsilon \lesssim 10^{-4}$. By the time $\epsilon \sim 10^{-8}$, stifbs is roughly an order of magnitude faster. There are further improvements that could be applied to stifbs to make it even more robust. For example, very occasionally ludcmp in simpr will encounter a singular matrix. You could arrange for the stepsize to be reduced, say by a factor of the current nseq[k]. There are also certain stability restrictions on the stepsize that come into play on some problems. For a discussion of how to implement these automatically, see [6].

CITED REFERENCES AND FURTHER READING:

Gear, C.W. 1971, *Numerical Initial Value Problems in Ordinary Differential Equations* (Englewood Cliffs, NJ: Prentice-Hall). [1]

Kaps, P., and Rentrop, P. 1979, *Numerische Mathematik*, vol. 33, pp. 55–68. [2]

Shampine, L.F. 1982, *ACM Transactions on Mathematical Software*, vol. 8, pp. 93–113. [3]

Enright, W.H., and Pryce, J.D. 1987, *ACM Transactions on Mathematical Software*, vol. 13, pp. 1–27. [4]

Bader, G., and Deuflhard, P. 1983, *Numerische Mathematik*, vol. 41, pp. 373–398. [5]

Deuflhard, P. 1983, *Numerische Mathematik*, vol. 41, pp. 399–422.

Deuflhard, P. 1985, *SIAM Review*, vol. 27, pp. 505–535.

Deuflhard, P. 1987, "Uniqueness Theorems for Stiff ODE Initial Value Problems," *Preprint SC-87-3* (Berlin: Konrad Zuse Zentrum für Informationstechnik). [6]

Enright, W.H., Hull, T.E., and Lindberg, B. 1975, *BIT*, vol. 15, pp. 10–48.

Wanner, G. 1988, in *Numerical Analysis 1987*, Pitman Research Notes in Mathematics, vol. 170, D.F. Griffiths and G.A. Watson, eds. (Harlow, Essex, U.K.: Longman Scientific and Technical).

Stoer, J., and Bulirsch, R. 1993, *Introduction to Numerical Analysis*, 2nd ed. (New York: Springer-Verlag).

16.7 Multistep, Multivalue, and Predictor-Corrector Methods

The terms multistep and multivalue describe two different ways of implementing essentially the same integration technique for ODEs. Predictor-corrector is a particular subcategrory of these methods — in fact, the most widely used. Accordingly, the name predictor-corrector is often loosely used to denote all these methods.

We suspect that predictor-corrector integrators have had their day, and that they are no longer the method of choice for most problems in ODEs. For high-precision applications, or applications where evaluations of the right-hand sides are expensive, Bulirsch-Stoer dominates. For convenience, or for low precision, adaptive-stepsize Runge-Kutta dominates. Predictor-corrector methods have been, we think, squeezed out in the middle. There is possibly only one exceptional case: high-precision solution of very smooth equations with very complicated right-hand sides, as we will describe later.

Nevertheless, these methods have had a long historical run. Textbooks are full of information on them, and there are a lot of standard ODE programs around that are based on predictor-corrector methods. Many capable researchers have a lot of experience with predictor-corrector routines, and they see no reason to make a precipitous change of habit. It is not a bad idea for you to be familiar with the principles involved, and even with the sorts of bookkeeping details that are the bane of these methods. Otherwise there will be a big surprise in store when you first have to fix a problem in a predictor-corrector routine.

Let us first consider the multistep approach. Think about how integrating an ODE is different from finding the integral of a function: For a function, the integrand has a known dependence on the independent variable x, and can be evaluated at will. For an ODE, the "integrand" is the right-hand side, which depends both on x and on the dependent variables y. Thus to advance the solution of $y' = f(x, y)$ from x_n to x, we have

$$y(x) = y_n + \int_{x_n}^{x} f(x', y) \, dx' \qquad (16.7.1)$$

In a single-step method like Runge-Kutta or Bulirsch-Stoer, the value y_{n+1} at x_{n+1} depends only on y_n. In a multistep method, we approximate $f(x, y)$ by a polynomial passing through *several* previous points $x_n, x_{n-1}, \ldots$ and possibly also through x_{n+1}. The result of evaluating the integral (16.7.1) at $x = x_{n+1}$ is then of the form

$$y_{n+1} = y_n + h(\beta_0 y'_{n+1} + \beta_1 y'_n + \beta_2 y'_{n-1} + \beta_3 y'_{n-2} + \cdots) \qquad (16.7.2)$$

where y'_n denotes $f(x_n, y_n)$, and so on. If $\beta_0 = 0$, the method is explicit; otherwise it is implicit. The order of the method depends on how many previous steps we use to get each new value of y.

Consider how we might solve an implicit formula of the form (16.7.2) for y_{n+1}. Two methods suggest themselves: *functional iteration* and *Newton's method*. In functional iteration, we take some initial guess for y_{n+1}, insert it into the right-hand side of (16.7.2) to get an updated value of y_{n+1}, insert this updated value back into the right-hand side, and continue iterating. But how are we to get an initial guess for

y_{n+1}? Easy! Just use some *explicit* formula of the same form as (16.7.2). This is called the *predictor step*. In the predictor step we are essentially *extrapolating* the polynomial fit to the derivative from the previous points to the new point x_{n+1} and then doing the integral (16.7.1) in a Simpson-like manner from x_n to x_{n+1}. The subsequent Simpson-like integration, using the prediction step's value of y_{n+1} to *interpolate* the derivative, is called the *corrector step*. The difference between the predicted and corrected function values supplies information on the local truncation error that can be used to control accuracy and to adjust stepsize.

If one corrector step is good, aren't many better? Why not use each corrector as an improved predictor and iterate to convergence on each step? Answer: Even if you had a *perfect* predictor, the step would still be accurate only to the finite order of the corrector. This incurable error term is on the same order as that which your iteration is supposed to cure, so you are at best changing only the coefficient in front of the error term by a fractional amount. So dubious an improvement is certainly not worth the effort. Your extra effort would be better spent in taking a smaller stepsize.

As described so far, you might think it desirable or necessary to predict several intervals ahead at each step, then to use all these intervals, with various weights, in a Simpson-like corrector step. That is not a good idea. Extrapolation is the least stable part of the procedure, and it is desirable to minimize its effect. Therefore, the integration steps of a predictor-corrector method are overlapping, each one involving several stepsize intervals h, but extending just one such interval farther than the previous ones. Only that one extended interval is extrapolated by each predictor step.

The most popular predictor-corrector methods are probably the Adams-Bashforth-Moulton schemes, which have good stability properties. The Adams-Bashforth part is the predictor. For example, the third-order case is

$$\text{predictor:} \quad y_{n+1} = y_n + \frac{h}{12}(23y'_n - 16y'_{n-1} + 5y'_{n-2}) + O(h^4) \quad (16.7.3)$$

Here information at the current point x_n, together with the two previous points x_{n-1} and x_{n-2} (assumed equally spaced), is used to predict the value y_{n+1} at the next point, x_{n+1}. The Adams-Moulton part is the corrector. The third-order case is

$$\text{corrector:} \quad y_{n+1} = y_n + \frac{h}{12}(5y'_{n+1} + 8y'_n - y'_{n-1}) + O(h^4) \quad (16.7.4)$$

Without the trial value of y_{n+1} from the predictor step to insert on the right-hand side, the corrector would be a nasty implicit equation for y_{n+1}.

There are actually three separate processes occurring in a predictor-corrector method: the predictor step, which we call P, the evaluation of the derivative y'_{n+1} from the latest value of y, which we call E, and the corrector step, which we call C. In this notation, iterating m times with the corrector (a practice we inveighed against earlier) would be written $P(EC)^m$. One also has the choice of finishing with a C or an E step. The lore is that a final E is superior, so the strategy usually recommended is PECE.

Notice that a PC method with a fixed number of iterations (say, one) is an explicit method! When we fix the number of iterations in advance, then the final value of y_{n+1} can be written as some complicated function of known quantities. Thus fixed iteration PC methods lose the strong stability properties of implicit methods and *should only be used for nonstiff problems*.

For stiff problems we *must* use an implicit method if we want to avoid having tiny stepsizes. (Not all implicit methods are good for stiff problems, but fortunately some good ones such as the Gear formulas are known.) We then appear to have two choices for solving the implicit equations: functional iteration to convergence, or Newton iteration. However, it turns out that for stiff problems functional iteration will not even converge unless we use tiny stepsizes, no matter how close our prediction is! Thus Newton iteration is usually an essential part of a multistep stiff solver. For convergence, Newton's method doesn't particularly care what the stepsize is, as long as the prediction is accurate enough.

Multistep methods, as we have described them so far, suffer from two serious difficulties when one tries to implement them:

- Since the formulas require results from equally spaced steps, adjusting the stepsize is difficult.
- Starting and stopping present problems. For starting, we need the initial values plus several previous steps to prime the pump. Stopping is a problem because equal steps are unlikely to land directly on the desired termination point.

Older implementations of PC methods have various cumbersome ways of dealing with these problems. For example, they might use Runge-Kutta to start and stop. Changing the stepsize requires considerable bookkeeping to do some kind of interpolation procedure. Fortunately both these drawbacks disappear with the multivalue approach.

For multivalue methods the basic data available to the integrator are the first few terms of the Taylor series expansion of the solution at the current point x_n. The aim is to advance the solution and obtain the expansion coefficients at the next point x_{n+1}. This is in contrast to multistep methods, where the data are the values of the solution at $x_n, x_{n-1}, \ldots$. We'll illustrate the idea by considering a four-value method, for which the basic data are

$$\mathbf{y}_n \equiv \begin{pmatrix} y_n \\ hy'_n \\ (h^2/2)y''_n \\ (h^3/6)y'''_n \end{pmatrix} \tag{16.7.5}$$

It is also conventional to scale the derivatives with the powers of $h = x_{n+1} - x_n$ as shown. Note that here we use the vector notation $\mathbf{y}$ to denote the solution and its first few derivatives at a point, not the fact that we are solving a system of equations with many components y.

In terms of the data in (16.7.5), we can approximate the value of the solution y at some point x:

$$y(x) = y_n + (x - x_n)y'_n + \frac{(x - x_n)^2}{2}y''_n + \frac{(x - x_n)^3}{6}y'''_n \tag{16.7.6}$$

Set $x = x_{n+1}$ in equation (16.7.6) to get an approximation to y_{n+1}. Differentiate equation (16.7.6) and set $x = x_{n+1}$ to get an approximation to y'_{n+1}, and similarly for y''_{n+1} and y'''_{n+1}. Call the resulting approximation $\widetilde{\mathbf{y}}_{n+1}$, where the tilde is a reminder that all we have done so far is a polynomial extrapolation of the solution and its derivatives; we have not yet used the differential equation. You can easily verify that

$$\widetilde{\mathbf{y}}_{n+1} = \mathbf{B} \cdot \mathbf{y}_n \qquad (16.7.7)$$

where the matrix $\mathbf{B}$ is

$$\mathbf{B} = \begin{pmatrix} 1 & 1 & 1 & 1 \\ 0 & 1 & 2 & 3 \\ 0 & 0 & 1 & 3 \\ 0 & 0 & 0 & 1 \end{pmatrix} \qquad (16.7.8)$$

We now write the actual approximation to $\mathbf{y}_{n+1}$ that we will use by adding a correction to $\widetilde{\mathbf{y}}_{n+1}$:

$$\mathbf{y}_{n+1} = \widetilde{\mathbf{y}}_{n+1} + \alpha \mathbf{r} \qquad (16.7.9)$$

Here $\mathbf{r}$ will be a fixed vector of numbers, in the same way that $\mathbf{B}$ is a fixed matrix. We fix α by requiring that the differential equation

$$y'_{n+1} = f(x_{n+1}, y_{n+1}) \qquad (16.7.10)$$

be satisfied. The second of the equations in (16.7.9) is

$$hy'_{n+1} = h\widetilde{y}'_{n+1} + \alpha r_1 \qquad (16.7.11)$$

and this will be consistent with (16.7.10) provided

$$r_1 = 1, \qquad \alpha = hf(x_{n+1}, y_{n+1}) - h\widetilde{y}'_{n+1} \qquad (16.7.12)$$

The values of r_0, r_2, and r_3 are free for the inventor of a given four-value method to choose. Different choices give different orders of method (i.e., through what order in h the final expression 16.7.9 actually approximates the solution), and different stability properties.

    An interesting result, not obvious from our presentation, is that multivalue and multistep methods are entirely equivalent. In other words, the value y_{n+1} given by a multivalue method with given $\mathbf{B}$ and $\mathbf{r}$ is exactly the same value given by some multistep method with given β's in equation (16.7.2). For example, it turns out that the Adams-Bashforth formula (16.7.3) corresponds to a four-value method with $r_0 = 0$, $r_2 = 3/4$, and $r_3 = 1/6$. The method is explicit because $r_0 = 0$. The Adams-Moulton method (16.7.4) corresponds to the implicit four-value method with $r_0 = 5/12$, $r_2 = 3/4$, and $r_3 = 1/6$. Implicit multivalue methods are solved the same way as implicit multistep methods: either by a predictor-corrector approach using an explicit method for the predictor, or by Newton iteration for stiff systems.

    Why go to all the trouble of introducing a whole new method that turns out to be equivalent to a method you already knew? The reason is that multivalue methods allow an easy solution to the two difficulties we mentioned above in actually implementing multistep methods.

    Consider first the question of stepsize adjustment. To change stepsize from h to h' at some point x_n, simply multiply the components of $\mathbf{y}_n$ in (16.7.5) by the appropriate powers of h'/h, and you are ready to continue to $x_n + h'$.

Multivalue methods also allow a relatively easy change in the *order* of the method: Simply change **r**. The usual strategy for this is first to determine the new stepsize with the current order from the error estimate. Then check what stepsize would be predicted using an order one greater and one smaller than the current order. Choose the order that allows you to take the biggest next step. Being able to change order also allows an easy solution to the starting problem: Simply start with a first-order method and let the order automatically increase to the appropriate level.

For low accuracy requirements, a Runge-Kutta routine like rkqs is almost always the most efficient choice. For high accuracy, bsstep is both robust and efficient. For very smooth functions, a variable-order PC method can invoke very high orders. If the right-hand side of the equation is relatively complicated, so that the expense of evaluating it outweighs the bookkeeping expense, then the best PC packages can outperform Bulirsch-Stoer on such problems. As you can imagine, however, such a variable-stepsize, variable-order method is not trivial to program. If you suspect that your problem is suitable for this treatment, we recommend use of a canned PC package. For further details consult Gear [1] or Shampine and Gordon [2].

Our prediction, nevertheless, is that, as extrapolation methods like Bulirsch-Stoer continue to gain sophistication, they will eventually beat out PC methods in all applications. We are willing, however, to be corrected.

CITED REFERENCES AND FURTHER READING:

Gear, C.W. 1971, *Numerical Initial Value Problems in Ordinary Differential Equations* (Englewood Cliffs, NJ: Prentice-Hall), Chapter 9. [1]

Shampine, L.F., and Gordon, M.K. 1975, *Computer Solution of Ordinary Differential Equations. The Initial Value Problem.* (San Francisco: W.H Freeman). [2]

Acton, F.S. 1970, *Numerical Methods That Work*; 1990, corrected edition (Washington: Mathematical Association of America), Chapter 5.

Kahaner, D., Moler, C., and Nash, S. 1989, *Numerical Methods and Software* (Englewood Cliffs, NJ: Prentice Hall), Chapter 8.

Hamming, R.W. 1962, *Numerical Methods for Engineers and Scientists*; reprinted 1986 (New York: Dover), Chapters 14–15.

Stoer, J., and Bulirsch, R. 1993, *Introduction to Numerical Analysis*, 2nd ed. (New York: Springer-Verlag), Chapter 7.

Chapter 17.   Two Point Boundary Value Problems

17.0 Introduction

When ordinary differential equations are required to satisfy boundary conditions at more than one value of the independent variable, the resulting problem is called a *two point boundary value problem*. As the terminology indicates, the most common case by far is where boundary conditions are supposed to be satisfied at two points — usually the starting and ending values of the integration. However, the phrase "two point boundary value problem" is also used loosely to include more complicated cases, e.g., where some conditions are specified at endpoints, others at interior (usually singular) points.

The crucial distinction between initial value problems (Chapter 16) and two point boundary value problems (this chapter) is that in the former case we are able to start an acceptable solution at its beginning (initial values) and just march it along by numerical integration to its end (final values); while in the present case, the boundary conditions at the starting point do not determine a unique solution to start with — and a "random" choice among the solutions that satisfy these (incomplete) starting boundary conditions is almost certain *not* to satisfy the boundary conditions at the other specified point(s).

It should not surprise you that iteration is in general required to meld these spatially scattered boundary conditions into a single global solution of the differential equations. For this reason, two point boundary value problems require considerably more effort to solve than do initial value problems. You have to integrate your differential equations over the interval of interest, or perform an analogous "relaxation" procedure (see below), at least several, and sometimes very many, times. Only in the special case of linear differential equations can you say in advance just how many such iterations will be required.

The "standard" two point boundary value problem has the following form: We desire the solution to a set of N coupled first-order ordinary differential equations, satisfying n_1 boundary conditions at the starting point x_1, and a remaining set of $n_2 = N - n_1$ boundary conditions at the final point x_2. (Recall that all differential equations of order higher than first can be written as coupled sets of first-order equations, cf. §16.0.)

The differential equations are

$$\frac{dy_i(x)}{dx} = g_i(x, y_0, y_1, \dots, y_{N-1}) \qquad i = 0, 1, \dots, N - 1 \qquad (17.0.1)$$

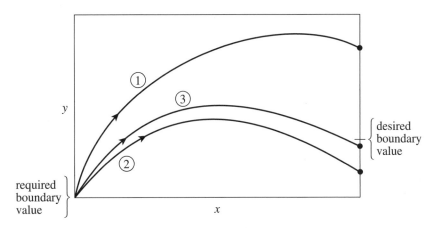

Figure 17.0.1. Shooting method (schematic). Trial integrations that satisfy the boundary condition at one endpoint are "launched." The discrepancies from the desired boundary condition at the other endpoint are used to adjust the starting conditions, until boundary conditions at both endpoints are ultimately satisfied.

At x_1, the solution is supposed to satisfy

$$B_{1j}(x_1, y_0, y_1, \ldots, y_{N-1}) = 0 \qquad j = 0, \ldots, n_1 - 1 \qquad (17.0.2)$$

while at x_2, it is supposed to satisfy

$$B_{2k}(x_2, y_0, y_1, \ldots, y_{N-1}) = 0 \qquad k = 0, \ldots, n_2 - 1 \qquad (17.0.3)$$

There are two distinct classes of numerical methods for solving two point boundary value problems. In the *shooting method* (§17.1) we choose values for all of the dependent variables at one boundary. These values must be consistent with any boundary conditions for *that* boundary, but otherwise are arranged to depend on arbitrary free parameters whose values we initially "randomly" guess. We then integrate the ODEs by initial value methods, arriving at the other boundary (and/or any interior points with boundary conditions specified). In general, we find discrepancies from the desired boundary values there. Now we have a multidimensional root-finding problem, as was treated in §9.6 and §9.7: Find the adjustment of the free parameters at the starting point that zeros the discrepancies at the other boundary point(s). If we liken integrating the differential equations to following the trajectory of a shot from gun to target, then picking the initial conditions corresponds to aiming (see Figure 17.0.1). The shooting method provides a systematic approach to taking a set of "ranging" shots that allow us to improve our "aim" systematically.

As another variant of the shooting method (§17.2), we can guess unknown free parameters at both ends of the domain, integrate the equations to a common midpoint, and seek to adjust the guessed parameters so that the solution joins "smoothly" at the fitting point. In all shooting methods, trial solutions satisfy the differential equations "exactly" (or as exactly as we care to make our numerical integration), but the trial solutions come to satisfy the required boundary conditions only after the iterations are finished.

Relaxation methods use a different approach. The differential equations are replaced by finite-difference equations on a mesh of points that covers the range of

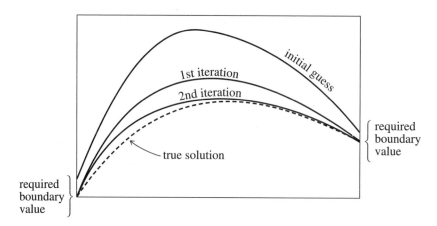

Figure 17.0.2. Relaxation method (schematic). An initial solution is guessed that approximately satisfies the differential equation and boundary conditions. An iterative process adjusts the function to bring it into close agreement with the true solution.

the integration. A trial solution consists of values for the dependent variables at each mesh point, *not* satisfying the desired finite-difference equations, nor necessarily even satisfying the required boundary conditions. The iteration, now called *relaxation*, consists of adjusting all the values on the mesh so as to bring them into successively closer agreement with the finite-difference equations and, simultaneously, with the boundary conditions (see Figure 17.0.2). For example, if the problem involves three coupled equations and a mesh of one hundred points, we must guess and improve three hundred variables representing the solution.

With all this adjustment, you may be surprised that relaxation is ever an efficient method, but (for the right problems) it really is! Relaxation works better than shooting when the boundary conditions are especially delicate or subtle, or where they involve complicated algebraic relations that cannot easily be solved in closed form. Relaxation works best when the solution is smooth and not highly oscillatory. Such oscillations would require many grid points for accurate representation. The number and position of required points may not be known *a priori*. Shooting methods are usually preferred in such cases, because their variable stepsize integrations adjust naturally to a solution's peculiarities.

Relaxation methods are often preferred when the ODEs have extraneous solutions which, while not appearing in the final solution satisfying all boundary conditions, may wreak havoc on the initial value integrations required by shooting. The typical case is that of trying to maintain a dying exponential in the presence of growing exponentials.

Good initial guesses are the secret of efficient relaxation methods. Often one has to solve a problem many times, each time with a slightly different value of some parameter. In that case, the previous solution is usually a good initial guess when the parameter is changed, and relaxation will work well.

Until you have enough experience to make your own judgment between the two methods, you might wish to follow the advice of your authors, who are notorious computer gunslingers: We always shoot first, and only then relax.

Problems Reducible to the Standard Boundary Problem

There are two important problems that can be reduced to the standard boundary value problem described by equations (17.0.1) – (17.0.3). The first is the *eigenvalue problem for differential equations.* Here the right-hand side of the system of differential equations depends on a parameter λ,

$$\frac{dy_i(x)}{dx} = g_i(x, y_0, \ldots, y_{N-1}, \lambda) \tag{17.0.4}$$

and one has to satisfy $N + 1$ boundary conditions instead of just N. The problem is overdetermined and in general there is no solution for arbitrary values of λ. For certain special values of λ, the eigenvalues, equation (17.0.4) does have a solution.

We reduce this problem to the standard case by introducing a new dependent variable

$$y_N \equiv \lambda \tag{17.0.5}$$

and another differential equation

$$\frac{dy_N}{dx} = 0 \tag{17.0.6}$$

An example of this trick is given in §17.4.

The other case that can be put in the standard form is a *free boundary problem.* Here only one boundary abscissa x_1 is specified, while the other boundary x_2 is to be determined so that the system (17.0.1) has a solution satisfying a total of $N + 1$ boundary conditions. Here we again add an extra constant dependent variable:

$$y_N \equiv x_2 - x_1 \tag{17.0.7}$$

$$\frac{dy_N}{dx} = 0 \tag{17.0.8}$$

We also define a new *independent* variable t by setting

$$x - x_1 \equiv t\, y_N, \qquad 0 \le t \le 1 \tag{17.0.9}$$

The system of $N + 1$ differential equations for dy_i/dt is now in the standard form, with t varying between the known limits 0 and 1.

CITED REFERENCES AND FURTHER READING:

Keller, H.B. 1968, *Numerical Methods for Two-Point Boundary-Value Problems*; reprinted 1991 (New York: Dover).

Kippenhan, R., Weigert, A., and Hofmeister, E. 1968, in *Methods in Computational Physics*, vol. 7 (New York: Academic Press), pp. 129ff.

Eggleton, P.P. 1971, *Monthly Notices of the Royal Astronomical Society*, vol. 151, pp. 351–364.

London, R.A., and Flannery, B.P. 1982, *Astrophysical Journal*, vol. 258, pp. 260–269.

Stoer, J., and Bulirsch, R. 1993, *Introduction to Numerical Analysis*, 2nd ed. (New York: Springer-Verlag), §§7.3–7.4.

17.1 The Shooting Method

In this section we discuss "pure" shooting, where the integration proceeds from x_1 to x_2, and we try to match boundary conditions at the end of the integration. In the next section, we describe shooting to an intermediate fitting point, where the solution to the equations and boundary conditions is found by launching "shots" from both sides of the interval and trying to match continuity conditions at some intermediate point.

Our implementation of the shooting method exactly implements multidimensional, globally convergent Newton-Raphson (§9.7). It seeks to zero n_2 functions of n_2 variables. The functions are obtained by integrating N differential equations from x_1 to x_2. Let us see how this works:

At the starting point x_1 there are N starting values y_i to be specified, but subject to n_1 conditions. Therefore there are $n_2 = N - n_1$ *freely specifiable* starting values. Let us imagine that these freely specifiable values are the components of a vector $\mathbf{V}$ that lives in a vector space of dimension n_2. Then you, the user, knowing the functional form of the boundary conditions (17.0.2), can write a function that generates a complete set of N starting values $\mathbf{y}$, satisfying the boundary conditions at x_1, from an arbitrary vector value of $\mathbf{V}$ in which there are no restrictions on the n_2 component values. In other words, (17.0.2) converts to a prescription

$$y_i(x_1) = y_i(x_1; V_0, \dots, V_{n_2-1}) \qquad i = 0, \dots, N-1 \qquad (17.1.1)$$

Below, the function that implements (17.1.1) will be called `load`.

Notice that the components of $\mathbf{V}$ might be exactly the values of certain "free" components of $\mathbf{y}$, with the other components of $\mathbf{y}$ determined by the boundary conditions. Alternatively, the components of $\mathbf{V}$ might parametrize the solutions that satisfy the starting boundary conditions in some other convenient way. Boundary conditions often impose algebraic relations among the y_i, rather than specific values for each of them. Using some auxiliary set of parameters often makes it easier to "solve" the boundary relations for a consistent set of y_i's. It makes no difference which way you go, as long as your vector space of $\mathbf{V}$'s generates (through 17.1.1) all allowed starting vectors $\mathbf{y}$.

Given a particular $\mathbf{V}$, a particular $\mathbf{y}(x_1)$ is thus generated. It can then be turned into a $\mathbf{y}(x_2)$ by integrating the ODEs to x_2 as an initial value problem (e.g., using Chapter 16's `odeint`). Now, at x_2, let us define a *discrepancy vector* $\mathbf{F}$, also of dimension n_2, whose components measure how far we are from satisfying the n_2 boundary conditions at x_2 (17.0.3). Simplest of all is just to use the right-hand sides of (17.0.3),

$$F_k = B_{2k}(x_2, \mathbf{y}) \qquad k = 0, \dots, n_2 - 1 \qquad (17.1.2)$$

As in the case of $\mathbf{V}$, however, you can use any other convenient parametrization, as long as your space of $\mathbf{F}$'s spans the space of possible discrepancies from the desired boundary conditions, with all components of $\mathbf{F}$ equal to zero if and only if the boundary conditions at x_2 are satisfied. Below, you will be asked to supply a user-written function `score` which uses (17.0.3) to convert an N-vector of ending values $\mathbf{y}(x_2)$ into an n_2-vector of discrepancies $\mathbf{F}$.

Now, as far as Newton-Raphson is concerned, we are nearly in business. We want to find a vector value of $\mathbf{V}$ that zeros the vector value of $\mathbf{F}$. We do this by invoking the globally convergent Newton's method implemented in the routine newt of §9.7. Recall that the heart of Newton's method involves solving the set of n_2 linear equations

$$\mathbf{J} \cdot \delta\mathbf{V} = -\mathbf{F} \qquad (17.1.3)$$

and then adding the correction back,

$$\mathbf{V}^{\text{new}} = \mathbf{V}^{\text{old}} + \delta\mathbf{V} \qquad (17.1.4)$$

In (17.1.3), the Jacobian matrix $\mathbf{J}$ has components given by

$$J_{ij} = \frac{\partial F_i}{\partial V_j} \qquad (17.1.5)$$

It is not feasible to compute these partial derivatives analytically. Rather, each requires a *separate* integration of the N ODEs, followed by the evaluation of

$$\frac{\partial F_i}{\partial V_j} \approx \frac{F_i(V_0, \ldots, V_j + \Delta V_j, \ldots) - F_i(V_0, \ldots, V_j, \ldots)}{\Delta V_j} \qquad (17.1.6)$$

This is done automatically for you in the routine fdjac that comes with newt. The only input to newt that you have to provide is the routine vecfunc that calculates $\mathbf{F}$ by integrating the ODEs. Here is the appropriate routine, called shoot, that is to be passed as the actual argument in newt:

```
#include "nr.h"

extern int nvar;                          Variables that you must define and set in your main pro-
extern DP x1,x2;                             gram.

int kmax,kount;                           Communicates with odeint.
DP dxsav;
Vec_DP *xp_p;
Mat_DP *yp_p;

void derivs(const DP x, Vec_I_DP &y, Vec_O_DP &dydx);
void load(const DP x1, Vec_I_DP &v, Vec_O_DP &y);
void score(const DP xf, Vec_I_DP &y, Vec_O_DP &f);

void NR::shoot(Vec_I_DP &v, Vec_O_DP &f)
```
Routine for use with newt to solve a two point boundary value problem for nvar coupled ODEs by shooting from x1 to x2. Initial values for the nvar ODEs at x1 are generated from the n2 input coefficients v[0..n2-1], using the user-supplied routine load. The routine integrates the ODEs to x2 using the Runge-Kutta method with tolerance EPS, initial stepsize h1, and minimum stepsize hmin. At x2 it calls the user-supplied routine score to evaluate the n2 functions f[0..n2-1] that ought to be zero to satisfy the boundary conditions at x2. The function values f are returned on output. newt uses a globally convergent Newton's method to adjust the values of v until the functions f are zero. The user-supplied routine derivs(x,y,dydx) supplies derivative information to the ODE integrator (see Chapter 16). The first set of global variables above receives its values from the main program so that shoot can have the syntax required for it to be the argument vecfunc of newt.
```
{
    const DP EPS=1.0e-14;
```

```
      int nbad,nok;
      DP h1,hmin=0.0;

      Vec_DP y(nvar);
      kmax=0;
      h1=(x2-x1)/100.0;
      load(x1,v,y);
      odeint(y,x1,x2,EPS,h1,hmin,nok,nbad,derivs,rkqs);
      score(x2,y,f);
  }
```

For some problems the initial stepsize ΔV might depend sensitively upon the initial conditions. It is straightforward to alter load to include a suggested stepsize h1 as another output variable and feed it to fdjac via a global variable.

A complete cycle of the shooting method thus requires $n_2 + 1$ integrations of the N coupled ODEs: one integration to evaluate the current degree of mismatch, and n_2 for the partial derivatives. Each new cycle requires a new round of $n_2 + 1$ integrations. This illustrates the enormous extra effort involved in solving two point boundary value problems compared with initial value problems.

If the differential equations are *linear*, then only one complete cycle is required, since (17.1.3)–(17.1.4) should take us right to the solution. A second round can be useful, however, in mopping up some (never all) of the roundoff error.

As given here, shoot uses the quality controlled Runge-Kutta method of §16.2 to integrate the ODEs, but any of the other methods of Chapter 16 could just as well be used.

You, the user, must supply shoot with: (i) a function load(x1,v,y) which calculates the n-vector y[0..n-1] (satisfying the starting boundary conditions, of course), given the freely specifiable variables of v[0..n2-1] at the initial point x1; (ii) a function score(x2,y,f) which calculates the discrepancy vector f[0..n2-1] of the ending boundary conditions, given the vector y[0..n-1] at the endpoint x2; (iii) a starting vector v[0..n2-1]; (iv) a function derivs for the ODE integration; and other obvious parameters as described in the header comment above.

In §17.4 we give a sample program illustrating how to use shoot.

CITED REFERENCES AND FURTHER READING:

Acton, F.S. 1970, *Numerical Methods That Work*; 1990, corrected edition (Washington: Mathematical Association of America).

Keller, H.B. 1968, *Numerical Methods for Two-Point Boundary-Value Problems*; reprinted 1991 (New York: Dover).

17.2 Shooting to a Fitting Point

The shooting method described in §17.1 tacitly assumed that the "shots" would be able to traverse the entire domain of integration, even at the early stages of convergence to a correct solution. In some problems it can happen that, for very wrong starting conditions, an initial solution can't even get from x_1 to x_2 without encountering some incalculable, or catastrophic, result. For example, the argument of a square root might go negative, causing the numerical code to crash. Simple shooting would be stymied.

A different, but related, case is where the endpoints are both singular points of the set of ODEs. One frequently needs to use special methods to integrate near the singular points, analytic asymptotic expansions, for example. In such cases it is feasible to integrate in the direction *away* from a singular point, using the special method to get through the first little bit and then reading off "initial" values for further numerical integration. However it is usually not feasible to integrate *into* a singular point, if only because one has not usually expended the same analytic effort to obtain expansions of "wrong" solutions near the singular point (those not satisfying the desired boundary condition).

The solution to the above mentioned difficulties is *shooting to a fitting point*. Instead of integrating from x_1 to x_2, we integrate first from x_1 to some point x_f that is *between* x_1 and x_2; and second from x_2 (in the opposite direction) to x_f.

If (as before) the number of boundary conditions imposed at x_1 is n_1, and the number imposed at x_2 is n_2, then there are n_2 freely specifiable starting values at x_1 and n_1 freely specifiable starting values at x_2. (If you are confused by this, go back to §17.1.) We can therefore define an n_2-vector $\mathbf{V}^{(1)}$ of starting parameters at x_1, and a prescription load1(x1,v1,y) for mapping $\mathbf{V}^{(1)}$ into a $\mathbf{y}$ that satisfies the boundary conditions at x_1,

$$y_i(x_1) = y_i(x_1; V_0^{(1)}, \ldots, V_{n_2-1}^{(1)}) \qquad i = 0, \ldots, N-1 \qquad (17.2.1)$$

Likewise we can define an n_1-vector $\mathbf{V}^{(2)}$ of starting parameters at x_2, and a prescription load2(x2,v2,y) for mapping $\mathbf{V}^{(2)}$ into a $\mathbf{y}$ that satisfies the boundary conditions at x_2,

$$y_i(x_2) = y_i(x_2; V_0^{(2)}, \ldots, V_{n_1-1}^{(2)}) \qquad i = 0, \ldots, N-1 \qquad (17.2.2)$$

We thus have a total of N freely adjustable parameters in the combination of $\mathbf{V}^{(1)}$ and $\mathbf{V}^{(2)}$. The N conditions that must be satisfied are that there be agreement in N components of $\mathbf{y}$ at x_f between the values obtained integrating from one side and from the other,

$$y_i(x_f; \mathbf{V}^{(1)}) = y_i(x_f; \mathbf{V}^{(2)}) \qquad i = 0, \ldots, N-1 \qquad (17.2.3)$$

In some problems, the N matching conditions can be better described (physically, mathematically, or numerically) by using N different functions F_i, $i = 0 \ldots N-1$, each possibly depending on the N components y_i. In those cases, (17.2.3) is replaced by

$$F_i[\mathbf{y}(x_f; \mathbf{V}^{(1)})] = F_i[\mathbf{y}(x_f; \mathbf{V}^{(2)})] \qquad i = 0, \ldots, N-1 \qquad (17.2.4)$$

In the program below, the user-supplied function score(xf,y,f) is supposed to map an input N-vector $\mathbf{y}$ into an output N-vector $\mathbf{F}$. In most cases, you can simply use the identity mapping $\mathbf{F} = \mathbf{y}$.

Shooting to a fitting point uses globally convergent Newton-Raphson exactly as in §17.1. Comparing closely with the routine shoot of the previous section, you should have no difficulty in understanding the following routine shootf. The main differences in use are that you have to supply both load1 and load2. Also, in the calling program you must supply initial guesses for v1[0..n2-1] and v2[0..n1-1]. Once again a sample program illustrating shooting to a fitting point is given in §17.4.

```
#include "nr.h"
```

`extern int n2;`	Variables that you must define and set in your main program.
`extern DP x1,x2,xf;`	

`int kmax,kount;`	Communicates with odeint.
`DP dxsav;`	
`Mat_DP *yp_p;`	
`Vec_DP *xp_p;`	

```
void derivs(const DP x, Vec_I_DP &y, Vec_O_DP &dydx);
void load1(const DP x1, Vec_I_DP &v1, Vec_O_DP &y);
void load2(const DP x2, Vec_I_DP &v2, Vec_O_DP &y);
void score(const DP xf, Vec_I_DP &y, Vec_O_DP &f);
```

```
void NR::shootf(Vec_I_DP &v, Vec_O_DP &f)
```
Routine for use with `newt` to solve a two point boundary value problem for `nvar` coupled ODEs by shooting from `x1` and `x2` to a fitting point `xf`. Initial values for the `nvar` ODEs at `x1` are generated from the `n2` coefficients `v1` and the user-supplied routine `load1`. Likewise, those at `x2` are from the `n1=nvar-n2` coefficients `v2`, using `load2`. The coefficients `v1` and `v2` should be stored in a single array `v[0..nvar-1]` in the main program with `v1` in `v[0..n2-1]` and `v2` in `v[n2..nvar-1]`. The routine integrates the ODEs to `xf` using the Runge-Kutta method with tolerance EPS, initial stepsize `h1`, and minimum stepsize `hmin`. At `xf` it calls the user-supplied routine `score` to evaluate the `nvar` functions `f1` and `f2` that ought to match at `xf`. The differences `f` are returned on output. `newt` uses a globally convergent Newton's method to adjust the values of `v` until the functions `f` are zero. The user-supplied routine `derivs(x,y,dydx)` supplies derivative information to the ODE integrator (see Chapter 16). The first set of global variables above receives its values from the main program so that `shoot` can have the syntax required for it to be the argument `vecfunc` of `newt`.
```
{
    const DP EPS=1.0e-14;
    int i,nbad,nok;
    DP h1,hmin=0.0;

    int nvar=v.size();
    Vec_DP f1(nvar),f2(nvar),y(nvar);
    Vec_DP v2(&v[n2],nvar-n2);
    kmax=0;
    h1=(x2-x1)/100.0;
    load1(x1,v,y);                    Path from x1 to xf with best trial values v1.
    odeint(y,x1,xf,EPS,h1,hmin,nok,nbad,derivs,rkqs);
    score(xf,y,f1);
    load2(x2,v2,y);                   Path from x2 to xf with best trial values v2.
    odeint(y,x2,xf,EPS,h1,hmin,nok,nbad,derivs,rkqs);
    score(xf,y,f2);
    for (i=0;i<nvar;i++) f[i]=f1[i]-f2[i];
}
```

There are boundary value problems where even shooting to a fitting point fails — the integration interval has to be partitioned by several fitting points with the solution being matched at each such point. For more details see [1].

CITED REFERENCES AND FURTHER READING:

Acton, F.S. 1970, *Numerical Methods That Work*; 1990, corrected edition (Washington: Mathematical Association of America).

Keller, H.B. 1968, *Numerical Methods for Two-Point Boundary-Value Problems*; reprinted 1991 (New York: Dover).

Stoer, J., and Bulirsch, R. 1993, *Introduction to Numerical Analysis*, 2nd ed. (New York: Springer-Verlag), §§7.3.5–7.3.6. [1]

17.3 Relaxation Methods

In *relaxation methods* we replace ODEs by approximate *finite-difference equations* (FDEs) on a grid or mesh of points that spans the domain of interest. As a typical example, we could replace a general first-order differential equation

$$\frac{dy}{dx} = g(x, y) \qquad (17.3.1)$$

with an algebraic equation relating function values at two points $k, k-1$:

$$y_k - y_{k-1} - (x_k - x_{k-1})\, g\left[\tfrac{1}{2}(x_k + x_{k-1}),\, \tfrac{1}{2}(y_k + y_{k-1})\right] = 0 \qquad (17.3.2)$$

The form of the FDE in (17.3.2) illustrates the idea, but not uniquely: There are many ways to turn the ODE into an FDE. When the problem involves N coupled first-order ODEs represented by FDEs on a mesh of M points, a solution consists of values for N dependent functions given at each of the M mesh points, or $N \times M$ variables in all. The relaxation method determines the solution by starting with a guess and improving it, iteratively. As the iterations improve the solution, the result is said to *relax* to the true solution.

While several iteration schemes are possible, for most problems our old standby, multi-dimensional Newton's method, works well. The method produces a matrix equation that must be solved, but the matrix takes a special, "block diagonal" form, that allows it to be inverted far more economically both in time and storage than would be possible for a general matrix of size $(MN) \times (MN)$. Since MN can easily be several thousand, this is crucial for the feasibility of the method.

Our implementation couples at most pairs of points, as in equation (17.3.2). More points can be coupled, but then the method becomes more complex. We will provide enough background so that you can write a more general scheme if you have the patience to do so.

Let us develop a general set of algebraic equations that represent the ODEs by FDEs. The ODE problem is exactly identical to that expressed in equations (17.0.1)–(17.0.3) where we had N coupled first-order equations that satisfy n_1 boundary conditions at one end of the interval and $n_2 = N - n_1$ boundary conditions at the other. We first define a mesh or grid by a set of $k = 0, 1, ..., M - 1$ points at which we supply values for the independent variable x_k. In particular, x_0 is the initial boundary, and x_{M-1} is the final boundary. We use the notation $\mathbf{y}_k$ to refer to the entire set of dependent variables $y_0, y_1, \ldots, y_{N-1}$ at point x_k. At an arbitrary point k in the middle of the mesh, we approximate the set of N first-order ODEs by algebraic relations of the form

$$0 = \mathbf{E}_k \equiv \mathbf{y}_k - \mathbf{y}_{k-1} - (x_k - x_{k-1})\mathbf{g}_k(x_k, x_{k-1}, \mathbf{y}_k, \mathbf{y}_{k-1}), \quad k = 1, 2, \ldots, M-1 \quad (17.3.3)$$

The notation signifies that $\mathbf{g}_k$ can be evaluated using information from both points $k, k-1$. The FDEs labeled by $\mathbf{E}_k$ provide N equations coupling $2N$ variables at points $k, k-1$. There are $M - 1$ points, $k = 1, 2, \ldots, M - 1$, at which difference equations of the form (17.3.3) apply. Thus the FDEs provide a total of $(M - 1)N$ equations for the MN unknowns. The remaining N equations come from the boundary conditions.

At the first boundary we have

$$0 = \mathbf{E}_0 \equiv \mathbf{B}(x_0, \mathbf{y}_0) \qquad (17.3.4)$$

while at the second boundary

$$0 = \mathbf{E}_M \equiv \mathbf{C}(x_{M-1}, \mathbf{y}_{M-1}) \qquad (17.3.5)$$

The vectors $\mathbf{E}_0$ and $\mathbf{B}$ have only n_1 nonzero components, corresponding to the n_1 boundary conditions at x_0. It will turn out to be useful to take these nonzero components to be the *last* n_1 components. In other words, $E_{j,0} \neq 0$ only for $j = n_2, n_2 + 1, \ldots, N - 1$. At the other boundary, only the first n_2 components of $\mathbf{E}_M$ and $\mathbf{C}$ are nonzero: $E_{j,M} \neq 0$ only for $j = 0, 1, \ldots, n_2 - 1$.

The "solution" of the FDE problem in (17.3.3)–(17.3.5) consists of a set of variables $y_{j,k}$, the values of the N variables y_j at the M points x_k. The algorithm we describe below requires an initial guess for the $y_{j,k}$. We then determine increments $\Delta y_{j,k}$ such that $y_{j,k} + \Delta y_{j,k}$ is an improved approximation to the solution.

Equations for the increments are developed by expanding the FDEs in first-order Taylor series with respect to small changes $\Delta \mathbf{y}_k$. At an interior point, $k = 1, 2, \ldots, M - 1$ this gives:

$$\mathbf{E}_k(\mathbf{y}_k + \Delta \mathbf{y}_k, \mathbf{y}_{k-1} + \Delta \mathbf{y}_{k-1}) \approx \mathbf{E}_k(\mathbf{y}_k, \mathbf{y}_{k-1})$$

$$+ \sum_{n=0}^{N-1} \frac{\partial \mathbf{E}_k}{\partial y_{n,k-1}} \Delta y_{n,k-1} + \sum_{n=0}^{N-1} \frac{\partial \mathbf{E}_k}{\partial y_{n,k}} \Delta y_{n,k} \qquad (17.3.6)$$

For a solution we want the updated value $\mathbf{E}(\mathbf{y} + \Delta \mathbf{y})$ to be zero, so the general set of equations at an interior point can be written in matrix form as

$$\sum_{n=0}^{N-1} S_{j,n} \Delta y_{n,k-1} + \sum_{n=N}^{2N-1} S_{j,n} \Delta y_{n-N,k} = -E_{j,k}, \quad j = 0, 1, \ldots, N-1 \qquad (17.3.7)$$

where

$$S_{j,n} = \frac{\partial E_{j,k}}{\partial y_{n,k-1}}, \quad S_{j,n+N} = \frac{\partial E_{j,k}}{\partial y_{n,k}}, \quad n = 0, 1, \ldots, N-1 \qquad (17.3.8)$$

The quantity $S_{j,n}$ is an $N \times 2N$ matrix at each point k. Each interior point thus supplies a block of N equations coupling $2N$ corrections to the solution variables at the points $k, k-1$.

Similarly, the algebraic relations at the boundaries can be expanded in a first-order Taylor series for increments that improve the solution. Since $\mathbf{E}_0$ depends only on $\mathbf{y}_0$, we find at the first boundary:

$$\sum_{n=0}^{N-1} S_{j,n} \Delta y_{n,0} = -E_{j,0}, \quad j = n_2, n_2 + 1, \ldots, N-1 \qquad (17.3.9)$$

where

$$S_{j,n} = \frac{\partial E_{j,0}}{\partial y_{n,0}}, \quad n = 0, 1, \ldots, N-1 \qquad (17.3.10)$$

At the second boundary,

$$\sum_{n=0}^{N-1} S_{j,n} \Delta y_{n,M-1} = -E_{j,M}, \quad j = 0, 1, \ldots, n_2 - 1 \qquad (17.3.11)$$

where

$$S_{j,n} = \frac{\partial E_{j,M}}{\partial y_{n,M-1}}, \quad n = 0, 1, \ldots, N-1 \qquad (17.3.12)$$

We thus have in equations (17.3.7)–(17.3.12) a set of linear equations to be solved for the corrections $\Delta \mathbf{y}$, iterating until the corrections are sufficiently small. The equations have a special structure, because each $S_{j,n}$ couples only points $k, k-1$. Figure 17.3.1 illustrates the typical structure of the complete matrix equation for the case of 5 variables and 4 mesh points, with 3 boundary conditions at the first boundary and 2 at the second. The 3×5 block of nonzero entries in the top left-hand corner of the matrix comes from the boundary condition $S_{j,n}$ at point $k = 0$. The next three 5×10 blocks are the $S_{j,n}$ at the interior points, coupling variables at mesh points (2,1), (3,2), and (4,3). Finally we have the block corresponding to the second boundary condition.

We can solve equations (17.3.7)–(17.3.12) for the increments $\Delta \mathbf{y}$ using a form of Gaussian elimination that exploits the special structure of the matrix to minimize the total number of operations, and that minimizes storage of matrix coefficients by packing the elements in a special blocked structure. (You might wish to review Chapter 2, especially §2.2, if you are unfamiliar with the steps involved in Gaussian elimination.) Recall that Gaussian elimination consists of manipulating the equations by elementary operations such as dividing rows of coefficients by a common factor to produce unity in diagonal elements,

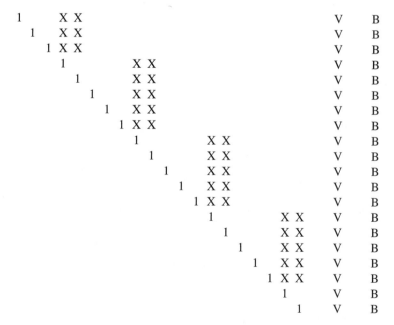

```
X X  X X                                                          V   B
X X  X X                                                          V   B
X X  X X                                                          V   B
X X  X X  X X X X X X                                             V   B
X X  X X  X X X X X X                                             V   B
X X  X X  X X X X X X                                             V   B
X X  X X  X X X X X X                                             V   B
X X  X X  X X X X X X                                             V   B
          X X X X X X X X X X                                     V   B
          X X X X X X X X X X                                     V   B
          X X X X X X X X X X                                     V   B
          X X X X X X X X X X                                     V   B
          X X X X X X X X X X                                     V   B
                    X X X X X X X X X X                           V   B
                    X X X X X X X X X X                           V   B
                    X X X X X X X X X X                           V   B
                    X X X X X X X X X X                           V   B
                    X X X X X X X X X X                           V   B
                              X X X X                             V   B
                              X X X X                             V   B
```

Figure 17.3.1.   Matrix structure of a set of linear finite-difference equations (FDEs) with boundary conditions imposed at both endpoints. Here X represents a coefficient of the FDEs, V represents a component of the unknown solution vector, and B is a component of the known right-hand side. Empty spaces represent zeros. The matrix equation is to be solved by a special form of Gaussian elimination. (See text for details.)

```
1     X X                                                        V   B
  1   X X                                                        V   B
    1 X X                                                        V   B
      1            X X                                           V   B
        1          X X                                           V   B
          1        X X                                           V   B
            1      X X                                           V   B
              1 X X                                              V   B
                1            X X                                 V   B
                  1          X X                                 V   B
                    1        X X                                 V   B
                      1      X X                                 V   B
                        1 X X                                    V   B
                          1            X X                       V   B
                            1          X X                       V   B
                              1        X X                       V   B
                                1      X X                       V   B
                                  1 X X                          V   B
                                    1                            V   B
                                      1                          V   B
```

Figure 17.3.2.   Target structure of the Gaussian elimination. Once the matrix of Figure 17.3.1 has been reduced to this form, the solution follows quickly by backsubstitution.

and adding appropriate multiples of other rows to produce zeros below the diagonal. Here we take advantage of the block structure by performing a bit more reduction than in pure Gaussian elimination, so that the storage of coefficients is minimized. Figure 17.3.2 shows the form that we wish to achieve by elimination, just prior to the backsubstitution step. Only a small subset of the reduced $MN \times MN$ matrix elements needs to be stored as the elimination progresses. Once the matrix elements reach the stage in Figure 17.3.2, the solution follows quickly by a backsubstitution procedure.

Furthermore, the entire procedure, except the backsubstitution step, operates only on one block of the matrix at a time. The procedure contains four types of operations: (1) partial reduction to zero of certain elements of a block using results from a previous step, (2) elimination of the square structure of the remaining block elements such that the square section contains unity along the diagonal, and zero in off-diagonal elements, (3) storage of the remaining nonzero coefficients for use in later steps, and (4) backsubstitution. We illustrate the steps schematically by figures.

Consider the block of equations describing corrections available from the initial boundary conditions. We have n_1 equations for N unknown corrections. We wish to transform the first block so that its left-hand $n_1 \times n_1$ square section becomes unity along the diagonal, and zero in off-diagonal elements. Figure 17.3.3 shows the original and final form of the first block of the matrix. In the figure we designate matrix elements that are subject to diagonalization by "D", and elements that will be altered by "A"; in the final block, elements that are stored are labeled by "S". We get from start to finish by selecting in turn n_1 "pivot" elements from among the first n_1 columns, normalizing the pivot row so that the value of the "pivot" element is unity, and adding appropriate multiples of this row to the remaining rows so that they contain zeros in the pivot column. In its final form, the reduced block expresses values for the corrections to the first n_1 variables at mesh point 0 in terms of values for the remaining n_2 unknown corrections at point 0, i.e., we now know what the first n_1 elements are in terms of the remaining n_2 elements. We store only the final set of n_2 nonzero columns from the initial block, plus the column for the altered right-hand side of the matrix equation.

We must emphasize here an important detail of the method. To exploit the reduced storage allowed by operating on blocks, it is essential that the ordering of columns in the s matrix of derivatives be such that pivot elements can be found among the first n_1 rows of the matrix. This means that the n_1 boundary conditions at the first point must contain some dependence on the first j=0,1,...,$n_1 - 1$ dependent variables, y[j][0]. If not, then the original square $n_1 \times n_1$ subsection of the first block will appear to be singular, and the method will fail. Alternatively, we would have to allow the search for pivot elements to involve all N columns of the block, and this would require column swapping and far more bookkeeping. The code provides a simple method of reordering the variables, i.e., the columns of the s matrix, so that this can be done easily. End of important detail.

Next consider the block of N equations representing the FDEs that describe the relation between the $2N$ corrections at points 1 and 0. The elements of that block, together with results from the previous step, are illustrated in Figure 17.3.4. Note that by adding suitable multiples of rows from the first block we can reduce to zero the first n_1 columns of the block (labeled by "Z"), and, to do so, we will need to alter only the columns from n_1 to $N - 1$ and the vector element on the right-hand side. Of the remaining columns we can diagonalize a square subsection of $N \times N$ elements, labeled by "D" in the figure. In the process we alter the final set of n_2 columns, denoted "A" in the figure. The second half of the figure shows the block when we finish operating on it, with the stored $n_2 \times N$ elements labeled by "S."

If we operate on the next set of equations corresponding to the FDEs coupling corrections at points 2 and 1, we see that the state of available results and new equations exactly reproduces the situation described in the previous paragraph. Thus, we can carry out those steps again for each block in turn through block $M - 1$. Finally on block M we encounter the remaining boundary conditions.

Figure 17.3.5 shows the final block of n_2 FDEs relating the N corrections for variables at mesh point $M - 1$, together with the result of reducing the previous block. Again, we can first use the prior results to zero the first n_1 columns of the block. Now, when we diagonalize the remaining square section, we strike gold: We get values for the final n_2 corrections at mesh point $M - 1$.

```
(a) D D D A A        V    A
    D D D A A        V    A
    D D D A A        V    A

(b) 1 0 0 S S        V    S
    0 1 0 S S        V    S
    0 0 1 S S        V    S
```

Figure 17.3.3.   Reduction process for the first (upper left) block of the matrix in Figure 17.3.1. (a) Original form of the block, (b) final form. (See text for explanation.)

```
(a)  1 0 0 S S                        V    S
     0 1 0 S S                        V    S
     0 0 1 S S                        V    S
     Z Z Z D D D D D A A              V    A
     Z Z Z D D D D D A A              V    A
     Z Z Z D D D D D A A              V    A
     Z Z Z D D D D D A A              V    A
     Z Z Z D D D D D A A              V    A

(b)  1 0 0 S S                        V    S
     0 1 0 S S                        V    S
     0 0 1 S S                        V    S
     0 0 0 1 0 0 0 0 S S              V    S
     0 0 0 0 1 0 0 0 S S              V    S
     0 0 0 0 0 1 0 0 S S              V    S
     0 0 0 0 0 0 1 0 S S              V    S
     0 0 0 0 0 0 0 1 S S              V    S
```

Figure 17.3.4.   Reduction process for intermediate blocks of the matrix in Figure 17.3.1. (a) Original form, (b) final form. (See text for explanation.)

```
(a)  0 0 0 1 0 0 0 0 S S              V    S
     0 0 0 0 1 0 0 0 S S              V    S
     0 0 0 0 0 1 0 0 S S              V    S
     0 0 0 0 0 0 1 0 S S              V    S
     0 0 0 0 0 0 0 1 S S              V    S
                 Z Z Z D D            V    A
                 Z Z Z D D            V    A

(b)  0 0 0 1 0 0 0 0 S S              V    S
     0 0 0 0 1 0 0 0 S S              V    S
     0 0 0 0 0 1 0 0 S S              V    S
     0 0 0 0 0 0 1 0 S S              V    S
     0 0 0 0 0 0 0 1 S S              V    S
                 0 0 0 1 0            V    S
                 0 0 0 0 1            V    S
```

Figure 17.3.5.   Reduction process for the last (lower right) block of the matrix in Figure 17.3.1. (a) Original form, (b) final form. (See text for explanation.)

With the final block reduced, the matrix has the desired form shown previously in Figure 17.3.2, and the matrix is ripe for backsubstitution. Starting with the bottom row and working up towards the top, at each stage we can simply determine one unknown correction in terms of known quantities.

The function solvde organizes the steps described above. The principal procedures used in the algorithm are performed by functions called internally by solvde. The function red eliminates leading columns of the s matrix using results from prior blocks. pinvs diagonalizes the square subsection of s and stores unreduced coefficients. bksub carries out the backsubstitution step. The user of solvde must understand the calling arguments, as described below, and supply a function difeq, called by solvde, that evaluates the s matrix for each block.

Most of the arguments in the call to solvde have already been described, but some require discussion. Array y[j][k] contains the initial guess for the solution, with j labeling the dependent variables at mesh points k. The problem involves ne FDEs spanning points k=0,..., m-1. nb boundary conditions apply at the first point k=0. The array indexv[j] establishes the correspondence between columns of the s matrix, equations (17.3.8), (17.3.10), and (17.3.12), and the dependent variables. As described above it is essential that the nb boundary conditions at k=0 involve the dependent variables referenced by the first nb columns of the s matrix. Thus, columns j of the s matrix can be ordered by the user in difeq to refer to derivatives with respect to the dependent variable indexv[j].

The function only attempts itmax correction cycles before returning, even if the solution has not converged. The parameters conv, slowc, scalv relate to convergence. Each inversion of the matrix produces corrections for ne variables at m mesh points. We want these to become vanishingly small as the iterations proceed, but we must define a measure for the size of corrections. This error "norm" is very problem specific, so the user might wish to rewrite this section of the code as appropriate. In the program below we compute a value for the average correction err by summing the absolute value of all corrections, weighted by a scale factor appropriate to each type of variable:

$$\text{err} = \frac{1}{m \times ne} \sum_{k=0}^{m-1} \sum_{j=0}^{ne-1} \frac{|\Delta Y[j][k]|}{\text{scalv}[j]} \tag{17.3.13}$$

When $\text{err} \leq \text{conv}$, the method has converged. Note that the user gets to supply an array scalv which measures the typical size of each variable.

Obviously, if err is large, we are far from a solution, and perhaps it is a bad idea to believe that the corrections generated from a first-order Taylor series are accurate. The number slowc modulates application of corrections. After each iteration we apply only a fraction of the corrections found by matrix inversion:

$$Y[j][k] \rightarrow Y[j][k] + \frac{\text{slowc}}{\max(\text{slowc},\text{err})} \Delta Y[j][k] \tag{17.3.14}$$

Thus, when $\text{err} > \text{slowc}$ only a fraction of the corrections are used, but when $\text{err} \leq \text{slowc}$ the entire correction gets applied.

The call statement also supplies solvde with the array y[0..nyj-1][0..nyk-1] containing the initial trial solution. Internally, workspace arrays c[0..ne-1][0..ne-nb][0..m], s[0..ne-1][0..2*ne] are allocated. The array c is the blockbuster: It stores the unreduced elements of the matrix built up for the backsubstitution step. If there are m mesh points, then there will be m+1 blocks, each requiring ne rows and ne-nb+1 columns. Although large, this is small compared with $(ne \times m)^2$ elements required for the whole matrix if we did not break it into blocks.

We now describe the workings of the user-supplied function difeq. The declaration of the function is

```
void difeq(const int k, const int k1, const int k2, const int jsf,
    const int is1, const int isf, Vec_I_INT &indexv,
    Mat_O_DP &s, Mat_I_DP &y);
```

The only information passed from difeq to solvde is the matrix of derivatives
s[0..ne-1][0..2*ne]; all other arguments are input to difeq and should not be altered.
k indicates the current mesh point, or block number. k1,k2 label the first and last point in
the mesh. If k=k1 or k>k2, the block involves the boundary conditions at the first or final
points; otherwise the block acts on FDEs coupling variables at points k-1, k.

The convention on storing information into the array s[i][j] follows that used in
equations (17.3.8), (17.3.10), and (17.3.12): Rows i label equations, columns j refer to
derivatives with respect to dependent variables in the solution. Recall that each equation will
depend on the ne dependent variables at either one or two points. Thus, j runs from 0 to
either ne-1 or 2*ne-1. The column ordering for dependent variables at each point must agree
with the list supplied in indexv[j]. Thus, for a block not at a boundary, the first column
multiplies ΔY(l=indexv[0],k-1), and the column ne multiplies ΔY(l=indexv[0],k).
is1,isf give the numbers of the starting and final *rows* that need to be filled in the s matrix
for this block. jsf labels the column in which the difference equations $E_{j,k}$ of equations
(17.3.3)–(17.3.5) are stored. Thus, −s[i][jsf] is the vector on the right-hand side of the
matrix. The reason for the minus sign is that difeq supplies the actual difference equation,
$E_{j,k}$, not its negative. Note that solvde supplies a value for jsf such that the difference
equation is put in the column *just after* all derivatives in the s matrix. Thus, difeq expects to
find values entered into s[i][j] for rows is1 $\leq$ i $\leq$ isf and 0 $\leq$ j $\leq$ jsf.

Finally, s[0..nsi-1][0..nsj-1] and y[0..nyj-1][0..nyk-1] supply difeq with
storage for s and the solution variables y for this iteration. An example of how to use this
routine is given in the next section.

Many ideas in the following code are due to Eggleton[1].

```
#include <iostream>
#include <iomanip>
#include <cmath>
#include "nr.h"
using namespace std;

void NR::solvde(const int itmax, const DP conv, const DP slowc,
    Vec_I_DP &scalv, Vec_I_INT &indexv, const int nb, Mat_IO_DP &y)
```
Driver routine for solution of two point boundary value problems by relaxation. itmax is the
maximum number of iterations. conv is the convergence criterion (see text). slowc controls
the fraction of corrections actually used after each iteration. scalv[0..ne-1] contains typical
sizes for each dependent variable, used to weight errors. indexv[0..ne-1] lists the column
ordering of variables used to construct the matrix s of derivatives, defined internally. (The nb
boundary conditions at the first mesh point must contain some dependence on the first nb
variables listed in indexv.) The problem involves ne equations for ne adjustable dependent
variables at each point. At the first mesh point there are nb boundary conditions. There are a
total of m mesh points. y[0..ne-1][0..m-1] is the two-dimensional array that contains the
initial guess for all the dependent variables at each mesh point. On each iteration, it is updated
by the calculated correction.
```
{
    int ic1,ic2,ic3,ic4,it,j,j1,j2,j3,j4,j5,j6,j7,j8,j9;
    int jc1,jcf,jv,k,k1,k2,km,kp,nvars;
    DP err,errj,fac,vmax,vz;

    int ne=y.nrows();
    int m=y.ncols();
    Vec_INT kmax(ne);
    Vec_DP ermax(ne);
    Mat3D_DP c(ne,ne-nb+1,m+1);
    Mat_DP s(ne,2*ne+1);
    k1=0; k2=m;                     Set up row and column markers.
    nvars=ne*m;
    j1=0,j2=nb,j3=nb,j4=ne,j5=j4+j1;
    j6=j4+j2,j7=j4+j3,j8=j4+j4,j9=j8+j1;
    ic1=0,ic2=ne-nb,ic3=ic2,ic4=ne;
    jc1=0,jcf=ic3;
    for (it=0;it<itmax;it++) {      Primary iteration loop.
```

```
        k=k1;                                 Boundary conditions at first point.
        difeq(k,k1,k2,j9,ic3,ic4,indexv,s,y);
        pinvs(ic3,ic4,j5,j9,jc1,k1,c,s);
        for (k=k1+1;k<k2;k++) {               Finite difference equations at all point pairs.
            kp=k;
            difeq(k,k1,k2,j9,ic1,ic4,indexv,s,y);
            red(ic1,ic4,j1,j2,j3,j4,j9,ic3,jc1,jcf,kp,c,s);
            pinvs(ic1,ic4,j3,j9,jc1,k,c,s);
        }
        k=k2;                                 Final boundary conditions.
        difeq(k,k1,k2,j9,ic1,ic2,indexv,s,y);
        red(ic1,ic2,j5,j6,j7,j8,j9,ic3,jc1,jcf,k2,c,s);
        pinvs(ic1,ic2,j7,j9,jcf,k2,c,s);
        bksub(ne,nb,jcf,k1,k2,c);             Backsubstitution.
        err=0.0;
        for (j=0;j<ne;j++) {                  Convergence check, accumulate average er-
            jv=indexv[j];                     ror.
            errj=vmax=0.0;
            km=0;
            for (k=k1;k<k2;k++) {             Find point with largest error, for each de-
                vz=fabs(c[jv][0][k]);         pendent variable.
                if (vz > vmax) {
                    vmax=vz;
                    km=k+1;
                }
                errj += vz;
            }
            err += errj/scalv[j];             Note weighting for each dependent variable.
            ermax[j]=c[jv][0][km-1]/scalv[j];
            kmax[j]=km;
        }
        err /= nvars;
        fac=(err > slowc ? slowc/err : 1.0);
        Reduce correction applied when error is large.
        for (j=0;j<ne;j++) {                  Apply corrections.
            jv=indexv[j];
            for (k=k1;k<k2;k++)
            y[j][k] -= fac*c[jv][0][k];
        }
        cout << setw(8) << "Iter.";           Summary of corrections for this
        cout << setw(10) << "Error" << setw(10) << "FAC" << endl;        step.
        cout << setw(6) << it;
        cout << fixed << setprecision(6) << setw(13) << err;
        cout << setw(12) << fac << endl;   Point with largest error for each variable can
        if (err < conv) return;               be monitored by writing out kmax and
    }                                         ermax arrays.
    nrerror("Too many iterations in solvde");         Convergence failed.
}



#include "nr.h"

void NR::bksub(const int ne, const int nb, const int jf, const int k1,
    const int k2, Mat3D_IO_DP &c)
Backsubstitution, used internally by solvde.
{
    int nbf,im,kp,k,j,i;
    DP xx;

    nbf=ne-nb;
    im=1;
    for (k=k2-1;k>=k1;k--) {               Use recurrence relations to eliminate remaining de-
        if (k == k1) im=nbf+1;             pendences.
```

```
            kp=k+1;
            for (j=0;j<nbf;j++) {
                xx=c[j][jf][kp];
                for (i=im-1;i<ne;i++)
                    c[i][jf][k] -= c[i][j][k]*xx;
            }
        }
        for (k=k1;k<k2;k++) {            Reorder corrections to be in column 1.
            kp=k+1;
            for (i=0;i<nb;i++) c[i][0][k]=c[i+nbf][jf][k];
            for (i=0;i<nbf;i++) c[i+nb][0][k]=c[i][jf][kp];
        }
    }
}



#include <cmath>
#include "nr.h"
using namespace std;

void NR::pinvs(const int ie1, const int ie2, const int je1, const int jsf,
    const int jc1, const int k, Mat3D_O_DP &c, Mat_IO_DP &s)
Diagonalize the square subsection of the s matrix, and store the recursion coefficients in c;
used internally by solvde.
{
    int jpiv,jp,je2,jcoff,j,irow,ipiv,id,icoff,i;
    DP pivinv,piv,dum,big;

    const int iesize=ie2-ie1;
    Vec_INT indxr(iesize);
    Vec_DP pscl(iesize);
    je2=je1+iesize;
    for (i=ie1;i<ie2;i++) {                   Implicit pivoting, as in §2.1.
        big=0.0;
        for (j=je1;j<je2;j++)
            if (fabs(s[i][j]) > big) big=fabs(s[i][j]);
        if (big == 0.0)
            nrerror("Singular matrix - row all 0, in pinvs");
        pscl[i-ie1]=1.0/big;
        indxr[i-ie1]=0;
    }
    for (id=0;id<iesize;id++) {
        piv=0.0;
        for (i=ie1;i<ie2;i++) {          Find pivot element.
            if (indxr[i-ie1] == 0) {
                big=0.0;
                for (j=je1;j<je2;j++) {
                    if (fabs(s[i][j]) > big) {
                        jp=j;
                        big=fabs(s[i][j]);
                    }
                }
                if (big*pscl[i-ie1] > piv) {
                    ipiv=i;
                    jpiv=jp;
                    piv=big*pscl[i-ie1];
                }
            }
        }
        if (s[ipiv][jpiv] == 0.0)
            nrerror("Singular matrix in routine pinvs");
        indxr[ipiv-ie1]=jpiv+1;           In place reduction. Save column ordering.
        pivinv=1.0/s[ipiv][jpiv];
        for (j=je1;j<=jsf;j++) s[ipiv][j] *= pivinv;    Normalize pivot row.
```

```
            s[ipiv][jpiv]=1.0;
            for (i=ie1;i<ie2;i++) {              Reduce nonpivot elements in column.
                if (indxr[i-ie1] != jpiv+1) {
                    if (s[i][jpiv] != 0.0) {
                        dum=s[i][jpiv];
                        for (j=je1;j<=jsf;j++)
                            s[i][j] -= dum*s[ipiv][j];
                        s[i][jpiv]=0.0;
                    }
                }
            }
        }
        jcoff=jc1-je2;                           Sort and store unreduced coefficients.
        icoff=ie1-je1;
        for (i=ie1;i<ie2;i++) {
            irow=indxr[i-ie1]+icoff;
            for (j=je2;j<=jsf;j++) c[irow-1][j+jcoff][k]=s[i][j];
        }
    }
```

```
#include "nr.h"

void NR::red(const int iz1, const int iz2, const int jz1, const int jz2,
    const int jm1, const int jm2, const int jmf, const int ic1,
    const int jc1, const int jcf, const int kc, Mat3D_I_DP &c,
    Mat_IO_DP &s)
```
Reduce columns jz1..jz2-1 of the s matrix, using previous results as stored in the c matrix.
Only columns jm1..jm2-1 and jmf are affected by the prior results. red is used internally
by solvde.
```
{
    int loff,l,j,ic,i;
    DP vx;

    loff=jc1-jm1;
    ic=ic1;
    for (j=jz1;j<jz2;j++) {                      Loop over columns to be zeroed.
        for (l=jm1;l<jm2;l++) {                  Loop over columns altered.
            vx=c[ic][l+loff][kc-1];
            for (i=iz1;i<iz2;i++) s[i][l] -= s[i][j]*vx;   Loop over rows.
        }
        vx=c[ic][jcf][kc-1];
        for (i=iz1;i<iz2;i++) s[i][jmf] -= s[i][j]*vx;     Plus final element.
        ic += 1;
    }
}
```

"Algebraically Difficult" Sets of Differential Equations

Relaxation methods allow you to take advantage of an additional opportunity that, while
not obvious, can speed up some calculations enormously. It is not necessary that the set
of variables $y_{j,k}$ correspond exactly with the dependent variables of the original differential
equations. They can be related to those variables through algebraic equations. Obviously, it
is necessary only that the solution variables allow us to *evaluate* the functions $y, g, \mathbf{B}, \mathbf{C}$ that
are used to construct the FDEs from the ODEs. In some problems g depends on functions of
y that are known only implicitly, so that iterative solutions are necessary to evaluate functions
in the ODEs. Often one can dispense with this "internal" nonlinear problem by defining
a new set of variables from which both y, g and the boundary conditions can be obtained
directly. A typical example occurs in physical problems where the equations require solution
of a complex equation of state that can be expressed in more convenient terms using variables

other than the original dependent variables in the ODE. While this approach is analogous to performing an *analytic* change of variables directly on the original ODEs, such an analytic transformation might be prohibitively complicated. The change of variables in the relaxation method is easy and requires no analytic manipulations.

CITED REFERENCES AND FURTHER READING:

Eggleton, P.P. 1971, *Monthly Notices of the Royal Astronomical Society*, vol. 151, pp. 351–364. [1]

Keller, H.B. 1968, *Numerical Methods for Two-Point Boundary-Value Problems*; reprinted 1991 (New York: Dover).

Kippenhan, R., Weigert, A., and Hofmeister, E. 1968, in *Methods in Computational Physics*, vol. 7 (New York: Academic Press), pp. 129ff.

17.4 A Worked Example: Spheroidal Harmonics

The best way to understand the algorithms of the previous sections is to see them employed to solve an actual problem. As a sample problem, we have selected the computation of spheroidal harmonics. (The more common name is spheroidal angle functions, but we prefer the explicit reminder of the kinship with spherical harmonics.) We will show how to find spheroidal harmonics, first by the method of relaxation (§17.3), and then by the methods of shooting (§17.1) and shooting to a fitting point (§17.2).

Spheroidal harmonics typically arise when certain partial differential equations are solved by separation of variables in spheroidal coordinates. They satisfy the following differential equation on the interval $-1 \le x \le 1$:

$$\frac{d}{dx}\left[(1-x^2)\frac{dS}{dx}\right] + \left(\lambda - c^2 x^2 - \frac{m^2}{1-x^2}\right)S = 0 \qquad (17.4.1)$$

Here m is an integer, c is the "oblateness parameter," and λ is the eigenvalue. Despite the notation, c^2 can be positive or negative. For $c^2 > 0$ the functions are called "prolate," while if $c^2 < 0$ they are called "oblate." The equation has singular points at $x = \pm 1$ and is to be solved subject to the boundary conditions that the solution be regular at $x = \pm 1$. Only for certain values of λ, the eigenvalues, will this be possible.

If we consider first the spherical case, where $c = 0$, we recognize the differential equation for Legendre functions $P_n^m(x)$. In this case the eigenvalues are $\lambda_{mn} = n(n+1)$, $n = m, m+1, \ldots$. The integer n labels successive eigenvalues for fixed m: When $n = m$ we have the lowest eigenvalue, and the corresponding eigenfunction has no nodes in the interval $-1 < x < 1$; when $n = m+1$ we have the next eigenvalue, and the eigenfunction has one node inside $(-1, 1)$; and so on.

A similar situation holds for the general case $c^2 \ne 0$. We write the eigenvalues of (17.4.1) as $\lambda_{mn}(c)$ and the eigenfunctions as $S_{mn}(x; c)$. For fixed m, $n = m, m+1, \ldots$ labels the successive eigenvalues.

The computation of $\lambda_{mn}(c)$ and $S_{mn}(x; c)$ traditionally has been quite difficult. Complicated recurrence relations, power series expansions, etc., can be found in references [1-3]. Cheap computing makes evaluation by direct solution of the differential equation quite feasible.

The first step is to investigate the behavior of the solution near the singular points $x = \pm 1$. Substituting a power series expansion of the form

$$S = (1 \pm x)^{\alpha} \sum_{k=0}^{\infty} a_k (1 \pm x)^k \qquad (17.4.2)$$

in equation (17.4.1), we find that the regular solution has $\alpha = m/2$. (Without loss of generality we can take $m \geq 0$ since $m \to -m$ is a symmetry of the equation.) We get an equation that is numerically more tractable if we factor out this behavior. Accordingly we set

$$S = (1 - x^2)^{m/2} y \qquad (17.4.3)$$

We then find from (17.4.1) that y satisfies the equation

$$(1 - x^2)\frac{d^2 y}{dx^2} - 2(m+1)x\frac{dy}{dx} + (\mu - c^2 x^2)y = 0 \qquad (17.4.4)$$

where

$$\mu \equiv \lambda - m(m+1) \qquad (17.4.5)$$

Both equations (17.4.1) and (17.4.4) are invariant under the replacement $x \to -x$. Thus the functions S and y must also be invariant, except possibly for an overall scale factor. (Since the equations are linear, a constant multiple of a solution is also a solution.) Because the solutions will be normalized, the scale factor can only be ± 1. If $n - m$ is odd, there are an odd number of zeros in the interval $(-1, 1)$. Thus we must choose the antisymmetric solution $y(-x) = -y(x)$ which has a zero at $x = 0$. Conversely, if $n - m$ is even we must have the symmetric solution. Thus

$$y_{mn}(-x) = (-1)^{n-m} y_{mn}(x) \qquad (17.4.6)$$

and similarly for S_{mn}.

The boundary conditions on (17.4.4) require that y be regular at $x = \pm 1$. In other words, near the endpoints the solution takes the form

$$y = a_0 + a_1(1 - x^2) + a_2(1 - x^2)^2 + \ldots \qquad (17.4.7)$$

Substituting this expansion in equation (17.4.4) and letting $x \to 1$, we find that

$$a_1 = -\frac{\mu - c^2}{4(m+1)} a_0 \qquad (17.4.8)$$

Equivalently,

$$y'(1) = \frac{\mu - c^2}{2(m+1)} y(1) \qquad (17.4.9)$$

A similar equation holds at $x = -1$ with a minus sign on the right-hand side. The irregular solution has a different relation between function and derivative at the endpoints.

Instead of integrating the equation from -1 to 1, we can exploit the symmetry (17.4.6) to integrate from 0 to 1. The boundary condition at $x = 0$ is

$$y(0) = 0, \quad n - m \text{ odd}$$
$$y'(0) = 0, \quad n - m \text{ even} \tag{17.4.10}$$

A third boundary condition comes from the fact that any constant multiple of a solution y is a solution. We can thus *normalize* the solution. We adopt the normalization that the function S_{mn} has the same limiting behavior as P_n^m at $x = 1$:

$$\lim_{x \to 1} (1 - x^2)^{-m/2} S_{mn}(x; c) = \lim_{x \to 1} (1 - x^2)^{-m/2} P_n^m(x) \tag{17.4.11}$$

Various normalization conventions in the literature are tabulated by Flammer [1].

Imposing three boundary conditions for the second-order equation (17.4.4) turns it into an eigenvalue problem for λ or equivalently for μ. We write it in the standard form by setting

$$y_0 = y \tag{17.4.12}$$
$$y_1 = y' \tag{17.4.13}$$
$$y_2 = \mu \tag{17.4.14}$$

Then

$$y_0' = y_1 \tag{17.4.15}$$
$$y_1' = \frac{1}{1 - x^2} \left[2x(m + 1)y_1 - (y_2 - c^2 x^2)y_0 \right] \tag{17.4.16}$$
$$y_2' = 0 \tag{17.4.17}$$

The boundary condition at $x = 0$ in this notation is

$$y_0 = 0, \quad n - m \text{ odd}$$
$$y_1 = 0, \quad n - m \text{ even} \tag{17.4.18}$$

At $x = 1$ we have two conditions:

$$y_1 = \frac{y_2 - c^2}{2(m + 1)} y_0 \tag{17.4.19}$$

$$y_0 = \lim_{x \to 1} (1 - x^2)^{-m/2} P_n^m(x) = \frac{(-1)^m (n + m)!}{2^m m!(n - m)!} \equiv \gamma \tag{17.4.20}$$

We are now ready to illustrate the use of the methods of previous sections on this problem.

Relaxation

If we just want a few isolated values of λ or S, shooting is probably the quickest method. However, if we want values for a large sequence of values of c, relaxation is better. Relaxation rewards a good initial guess with rapid convergence, and the previous solution should be a good initial guess if c is changed only slightly.

For simplicity, we choose a uniform grid on the interval $0 \le x \le 1$. For a total of M mesh points, we have

$$h = \frac{1}{M-1} \tag{17.4.21}$$

$$x_k = kh, \qquad k = 0, 1, \ldots, M-1 \tag{17.4.22}$$

At interior points $k = 1, 2, \ldots, M-1$, equation (17.4.15) gives

$$E_{0,k} = y_{0,k} - y_{0,k-1} - \frac{h}{2}(y_{1,k} + y_{1,k-1}) \tag{17.4.23}$$

Equation (17.4.16) gives

$$
\begin{aligned}
E_{1,k} = y_{1,k} - y_{1,k-1} - \beta_k \\
\times \left[\frac{(x_k + x_{k-1})(m+1)(y_{1,k} + y_{1,k-1})}{2} - \alpha_k \frac{(y_{0,k} + y_{0,k-1})}{2} \right]
\end{aligned}
\tag{17.4.24}
$$

where

$$\alpha_k = \frac{y_{2,k} + y_{2,k-1}}{2} - \frac{c^2(x_k + x_{k-1})^2}{4} \tag{17.4.25}$$

$$\beta_k = \frac{h}{1 - \frac{1}{4}(x_k + x_{k-1})^2} \tag{17.4.26}$$

Finally, equation (17.4.17) gives

$$E_{2,k} = y_{2,k} - y_{2,k-1} \tag{17.4.27}$$

Now recall that the matrix of partial derivatives $S_{i,j}$ of equation (17.3.8) is defined so that i labels the equation and j the variable. In our case, j runs from 0 to 2 for y_j at $k-1$ and from 3 to 5 for y_j at k. Thus equation (17.4.23) gives

$$
S_{0,0} = -1, \qquad S_{0,1} = -\frac{h}{2}, \qquad S_{0,2} = 0
$$
$$
S_{0,3} = 1, \qquad S_{0,4} = -\frac{h}{2}, \qquad S_{0,5} = 0
\tag{17.4.28}
$$

Similarly equation (17.4.24) yields

$$
\begin{aligned}
&S_{1,0} = \alpha_k \beta_k / 2, &&S_{1,1} = -1 - \beta_k(x_k + x_{k-1})(m+1)/2, \\
&S_{1,2} = \beta_k(y_{0,k} + y_{0,k-1})/4, &&S_{1,3} = S_{1,0}, \\
&S_{1,4} = 2 + S_{1,1}, &&S_{1,5} = S_{1,2}
\end{aligned}
\tag{17.4.29}
$$

while from equation (17.4.27) we find

$$S_{2,0} = 0, \qquad S_{2,1} = 0, \qquad S_{2,2} = -1$$
$$S_{2,3} = 0, \qquad S_{2,4} = 0, \qquad S_{2,5} = 1 \tag{17.4.30}$$

At $x = 0$ we have the boundary condition

$$E_{2,0} = \begin{cases} y_{0,0}, & n - m \text{ odd} \\ y_{1,0}, & n - m \text{ even} \end{cases} \tag{17.4.31}$$

Recall the convention adopted in the `solvde` routine that for one boundary condition at $k = 0$ only $S_{2,j}$ can be nonzero. Also, j takes on the values 3 to 5 since the boundary condition involves only y_k, not y_{k-1}. Accordingly, the only nonzero values of $S_{2,j}$ at $x = 0$ are

$$S_{2,3} = 1, \qquad n - m \text{ odd}$$
$$S_{2,4} = 1, \qquad n - m \text{ even} \tag{17.4.32}$$

At $x = 1$ we have

$$E_{0,M} = y_{1,M-1} - \frac{y_{2,M-1} - c^2}{2(m+1)} y_{0,M-1} \tag{17.4.33}$$

$$E_{1,M} = y_{0,M-1} - \gamma \tag{17.4.34}$$

Thus

$$S_{0,3} = -\frac{y_{2,M-1} - c^2}{2(m+1)}, \qquad S_{0,4} = 1, \qquad S_{0,5} = -\frac{y_{0,M-1}}{2(m+1)} \tag{17.4.35}$$
$$S_{1,3} = 1, \qquad\qquad S_{1,4} = 0, \qquad S_{1,5} = 0 \tag{17.4.36}$$

Here now is the sample program that implements the above algorithm. We need a `main` program, `sfroid`, that calls the routine `solvde`, and we must supply the function `difeq` called by `solvde`. For simplicity we choose an equally spaced mesh of m = 41 points, that is, $h = .025$. As we shall see, this gives good accuracy for the eigenvalues up to moderate values of $n - m$.

Since the boundary condition at $x = 0$ does not involve y_0 if $n - m$ is even, we have to use the `indexv` feature of `solvde`. Recall that the value of `indexv[j]` describes which column of `s[i][j]` the variable `y[j]` has been put in. If $n - m$ is even, we need to interchange the columns for y_0 and y_1 so that there is not a zero pivot element in `s[i][j]`.

The program prompts for values of m and n. It then computes an initial guess for y based on the Legendre function P_n^m. It next prompts for c^2, solves for y, prompts for c^2, solves for y using the previous values as an initial guess, and so on.

```
#include <iostream>
#include <iomanip>
#include <cmath>
#include "nr.h"
using namespace std;
```

```
const int M=40;                                 Global variables communicating
int mm,n,mpt=M+1;                               with difeq.
DP h,c2=0.0,anorm;
Vec_DP *x_p;
```

```
int main(void)     // Program sfroid
```
Sample program using solvde. Computes eigenvalues of spheroidal harmonics $S_{mn}(x;c)$ for $m \geq 0$ and $n \geq m$. In the program, m is mm, c^2 is c2, and γ of equation (17.4.20) is anorm.
```
{
    const int NE=3,NB=1,NYJ=NE,NYK=M+1;
    int i,itmax,k;
    DP conv,deriv,fac1,fac2,q1,slowc;
    Vec_INT indexv(NE);
    Vec_DP scalv(NE);
    Mat_DP y(NYJ,NYK);

    x_p=new Vec_DP(M+1);
    Vec_DP &x=*x_p;                             Make an alias to simplify cod-
    itmax=100;                                  ing.
    conv=1.0e-14;
    slowc=1.0;
    h=1.0/M;
    cout << endl << "Enter m n" << endl;
    cin >> mm >> n;
    if ((n+mm & 1) != 0) {                      No interchanges necessary.
        indexv[0]=0;
        indexv[1]=1;
        indexv[2]=2;
    } else {                                    Interchange $y_0$ and $y_1$.
        indexv[0]=1;
        indexv[1]=0;
        indexv[2]=2;
    }
    anorm=1.0;                                  Compute $\gamma$.
    if (mm != 0) {
        q1=n;
        for (i=1;i<=mm;i++) anorm = -0.5*anorm*(n+i)*(q1--/i);
    }
    for (k=0;k<M;k++) {                         Initial guess.
        x[k]=k*h;
        fac1=1.0-x[k]*x[k];
        fac2=exp((-mm/2.0)*log(fac1));
        y[0][k]=NR::plgndr(n,mm,x[k])*fac2;      $P_n^m$ from §6.8.
        deriv = -((n-mm+1)*NR::plgndr(n+1,mm,x[k])-   Derivative of $P_n^m$ from a recur-
            (n+1)*x[k]*NR::plgndr(n,mm,x[k]))/fac1;    rence relation.
        y[1][k]=mm*x[k]*y[0][k]/fac1+deriv*fac2;
        y[2][k]=n*(n+1)-mm*(mm+1);
    }
    x[M]=1.0;                                   Initial guess at $x = 1$ done sep-
    y[0][M]=anorm;                              arately.
    y[2][M]=n*(n+1)-mm*(mm+1);
    y[1][M]=(y[2][M]-c2)*y[0][M]/(2.0*(mm+1.0)));
    scalv[0]=fabs(anorm);
    scalv[1]=(y[1][M] > scalv[0] ? y[1][M] : scalv[0]);
    scalv[2]=(y[2][M] > 1.0 ? y[2][M] : 1.0);
    for (;;) {
        cout << endl << "Enter c**2 or 999 to end" << endl;
        cin >> c2;
```

```
        if (c2 == 999) {
            delete x_p;
            return 0;
        }
        NR::solvde(itmax,conv,slowc,scalv,indexv,NB,y);
        cout << endl << " m = " << setw(3) << mm;
        cout << " n = " << setw(3) << n << " c**2 = ";
        cout << fixed << setprecision(3) << setw(7) << c2;
        cout << " lamda = " << setprecision(6) << (y[2][0]+mm*(mm+1));
        cout << endl;                          Return for another value of c².
    }
}
```

```
#include "nr.h"

extern int mm,n,mpt;                           Defined in sfroid.
extern DP h,c2,anorm;
extern Vec_DP *x_p;

void NR::difeq(const int k, const int k1, const int k2, const int jsf,
    const int is1, const int isf, Vec_I_INT &indexv, Mat_O_DP &s,
    Mat_I_DP &y)
Returns matrix s for solvde.
{
    DP temp,temp1,temp2;

    Vec_DP &x=*x_p;                            Make an alias to simplify coding.
    if (k == k1) {                             Boundary condition at first point.
        if ((n+mm & 1) != 0) {
            s[2][3+indexv[0]]=1.0;             Equation (17.4.32).
            s[2][3+indexv[1]]=0.0;
            s[2][3+indexv[2]]=0.0;
            s[2][jsf]=y[0][0];                 Equation (17.4.31).
        } else {
            s[2][3+indexv[0]]=0.0;             Equation (17.4.32).
            s[2][3+indexv[1]]=1.0;
            s[2][3+indexv[2]]=0.0;
            s[2][jsf]=y[1][0];                 Equation (17.4.31).
        }
    } else if (k > k2-1) {                     Boundary conditions at last point.
        s[0][3+indexv[0]] = -(y[2][mpt-1]-c2)/(2.0*(mm+1.0));      (17.4.35).
        s[0][3+indexv[1]]=1.0;
        s[0][3+indexv[2]] = -y[0][mpt-1]/(2.0*(mm+1.0));
        s[0][jsf]=y[1][mpt-1]-(y[2][mpt-1]-c2)*y[0][mpt-1]/        (17.4.33).
            (2.0*(mm+1.0));
        s[1][3+indexv[0]]=1.0;                 Equation (17.4.36).
        s[1][3+indexv[1]]=0.0;
        s[1][3+indexv[2]]=0.0;
        s[1][jsf]=y[0][mpt-1]-anorm;           Equation (17.4.34).
    } else {                                   Interior point.
        s[0][indexv[0]] = -1.0;                Equation (17.4.28).
        s[0][indexv[1]] = -0.5*h;
        s[0][indexv[2]]=0.0;
        s[0][3+indexv[0]]=1.0;
        s[0][3+indexv[1]] = -0.5*h;
        s[0][3+indexv[2]]=0.0;
        temp1=x[k]+x[k-1];
        temp=h/(1.0-temp1*temp1*0.25);
        temp2=0.5*(y[2][k]+y[2][k-1])-c2*0.25*temp1*temp1;
        s[1][indexv[0]]=temp*temp2*0.5;        Equation (17.4.29).
        s[1][indexv[1]] = -1.0-0.5*temp*(mm+1.0)*temp1;
        s[1][indexv[2]]=0.25*temp*(y[0][k]+y[0][k-1]);
```

```
        s[1][3+indexv[0]]=s[1][indexv[0]];
        s[1][3+indexv[1]]=2.0+s[1][indexv[1]];
        s[1][3+indexv[2]]=s[1][indexv[2]];
        s[2][indexv[0]]=0.0;                          Equation (17.4.30).
        s[2][indexv[1]]=0.0;
        s[2][indexv[2]] = -1.0;
        s[2][3+indexv[0]]=0.0;
        s[2][3+indexv[1]]=0.0;
        s[2][3+indexv[2]]=1.0;
        s[0][jsf]=y[0][k]-y[0][k-1]-0.5*h*(y[1][k]+y[1][k-1]);    (17.4.23).
        s[1][jsf]=y[1][k]-y[1][k-1]-temp*((x[k]+x[k-1])          (17.4.24).
            *0.5*(mm+1.0)*(y[1][k]+y[1][k-1])-temp2
            *0.5*(y[0][k]+y[0][k-1]));
        s[2][jsf]=y[2][k]-y[2][k-1];                   Equation (17.4.27).
    }
}
```

You can run the program and check it against values of $\lambda_{mn}(c)$ given in the tables at the back of Flammer's book [1] or in Table 21.1 of Abramowitz and Stegun [2]. Typically it converges in about 3 iterations. The table below gives a few comparisons.

		Selected Output of sfroid		
m	n	c^2	λ_{exact}	λ_{sfroid}
2	2	0.1	6.01427	6.01427
		1.0	6.14095	6.14095
		4.0	6.54250	6.54253
2	5	1.0	30.4361	30.4372
		16.0	36.9963	37.0135
4	11	−1.0	131.560	131.554

Shooting

To solve the same problem via shooting (§17.1), we supply a function derivs that implements equations (17.4.15)–(17.4.17). We will integrate the equations over the range $-1 \le x \le 0$. We provide the function load which sets the eigenvalue y_2 to its current best estimate, v[0]. It also sets the boundary values of y_0 and y_1 using equations (17.4.20) and (17.4.19) (with a minus sign corresponding to $x = -1$). Note that the boundary condition is actually applied a distance dx from the boundary to avoid having to evaluate y_1' right on the boundary. The function score follows from equation (17.4.18).

```
#include <iostream>
#include <iomanip>
#include "nr.h"
using namespace std;

int m,n;                          Communicates with load, score, and derivs.
DP c2,dx,gmma;
```

```
int nvar;                          Communicates with shoot.
DP x1,x2;

int main(void)     // Program sphoot
```
Sample program using shoot. Computes eigenvalues of spheroidal harmonics $S_{mn}(x;c)$ for $m \geq 0$ and $n \geq m$. Note how the routine vecfunc for newt is provided by shoot (§17.1).
```
{
    const int N2=1;
    bool check;
    int i;
    DP q1;
    Vec_DP v(N2);

    dx=1.0e-8;                     Avoid evaluating derivatives exactly at x = −1.
    nvar=3;                        Number of equations.
    for (;;) {
        cout << endl << "input m,n,c-squared (999 to end)" << endl;
        cin >> m >> n >> c2;
        if (c2 == 999) break;
        if (n < m || m < 0) continue;
        gmma=1.0;                  Compute γ of equation (17.4.20).
        q1=n;
        for (i=1;i<=m;i++) gmma *= -0.5*(n+i)*(q1--/i);
        v[0]=n*(n+1)-m*(m+1)+c2/2.0;   Initial guess for eigenvalue.
        x1= -1.0+dx;               Set range of integration.
        x2=0.0;
        NR::newt(v,check,NR::shoot);   Find v that zeros function f in score.
        if (check) {
            cout << "shoot failed; bad initial guess" << endl;
        } else {
            cout << " " << "mu(m,n)" << endl;
            cout << fixed << setprecision(6);
            cout << setw(12) << v[0] << endl;
        }
    }
    return 0;
}


void load(const DP x1, Vec_I_DP &v, Vec_O_DP &y)
```
Supplies starting values for integration at $x = -1 + dx$.
```
{
    DP y1 = ((n-m & 1) != 0 ? -gmma : gmma);
    y[2]=v[0];
    y[1] = -(y[2]-c2)*y1/(2*(m+1));
    y[0]=y1+y[1]*dx;
}


void score(const DP xf, Vec_I_DP &y, Vec_O_DP &f)
```
Computes amount by which boundary condition at $x = 0$ is violated.
```
{
    f[0]=((n-m & 1) != 0 ? y[0] : y[1]);
}


void derivs(const DP x, Vec_I_DP &y, Vec_O_DP &dydx)
```
Evaluates derivatives for odeint.
```
{
    dydx[0]=y[1];
    dydx[1]=(2.0*x*(m+1.0)*y[1]-(y[2]-c2*x*x)*y[0])/(1.0-x*x);
    dydx[2]=0.0;
}
```

Shooting to a Fitting Point

For variety we illustrate shootf from §17.2 by integrating over the whole range $-1 + dx \le x \le 1 - dx$, with the fitting point chosen to be at $x = 0$. The routine derivs is identical to the one for shoot. Now, however, there are two load routines. The routine load1 for $x = -1$ is essentially identical to load above. At $x = 1$, load2 sets the function value y_0 and the eigenvalue y_2 to their best current estimates, v2[0] and v2[1], respectively. If you quite sensibly make your initial guess of the eigenvalue the same in the two intervals, then v1[0] will stay equal to v2[1] during the iteration. The function score computes the degree of mismatch of the three function values at the fitting point.

```
#include <iostream>
#include <iomanip>
#include <cmath>
#include "nr.h"
using namespace std;

int m,n;                                Communicates with load1, load2, score,
DP c2,dx,gmma;                             and derivs.

int n2;                                 Communicates with shootf.
DP x1,x2,xf;

int main(void)      // Program sphfpt
```
Sample program using shootf. Computes eigenvalues of spheroidal harmonics $S_{mn}(x; c)$ for $m \ge 0$ and $n \ge m$. Note how the routine vecfunc for newt is provided by shootf (§17.2). The routine derivs is the same as for sphoot.
```
{
    const int N1=2,N2=1,NTOT=N1+N2;        Number of equations.
    const DP DXX=1.0e-8;
    bool check;
    int i;
    DP q1;
    Vec_DP v(NTOT);

    n2=N2;
    dx=DXX;                                Avoid evaluating derivatives exactly at x =
    for (;;) {                                ±1.
        cout << endl << "input m,n,c-squared (n >= m, c=999 to end)" << endl;
        cin >> m >> n >> c2;
        if (c2 == 999) break;
        if (n < m || m < 0) continue;
        gmma=1.0;                          Compute γ of equation (17.4.20).
        q1=n;
        for (i=1;i<=m;i++) gmma *= -0.5*(n+i)*(q1--/i);
        v[0]=n*(n+1)-m*(m+1)+c2/2.0;       Initial guess for eigenvalue and function value.
        v[2]=v[0];
        v[1]=gmma*(1.0-(v[2]-c2)*dx/(2*(m+1)));
        x1= -1.0+dx;                       Set range of integration.
        x2=1.0-dx;
        xf=0.0;                            Fitting point.
        NR::newt(v,check,NR::shootf);      Find v that zeros function f in score.
        if (check) {
            cout << "shootf failed; bad initial guess" << endl;
        } else {
            cout << " " << "mu(m,n)" << endl;
            cout << fixed << setprecision(6);
            cout << setw(12) << v[0] << endl;
        }
    }
```

```
    }
    return 0;
}

void load1(const DP x1, Vec_I_DP &v1, Vec_O_DP &y)
```
Supplies starting values for integration at $x = -1 + dx$.
```
{
    DP y1 = ((n-m & 1) != 0 ? -gmma : gmma);
    y[2]=v1[0];
    y[1] = -(y[2]-c2)*y1/(2*(m+1));
    y[0]=y1+y[1]*dx;
}

void load2(const DP x2, Vec_I_DP &v2, Vec_O_DP &y)
```
Supplies starting values for integration at $x = 1 - dx$.
```
{
    y[2]=v2[1];
    y[0]=v2[0];
    y[1]=(y[2]-c2)*y[0]/(2*(m+1));
}

void score(const DP xf, Vec_I_DP &y, Vec_O_DP &f)
```
Computes the mismatch of the solutions at the fitting point $x = 0$.
```
{
    int i;

    for (i=0;i<3;i++) f[i]=y[i];
}

void derivs(const DP x, Vec_I_DP &y, Vec_O_DP &dydx)
{
    dydx[0]=y[1];
    dydx[1]=(2.0*x*(m+1.0)*y[1]-(y[2]-c2*x*x)*y[0])/(1.0-x*x);
    dydx[2]=0.0;
}
```

CITED REFERENCES AND FURTHER READING:

Flammer, C. 1957, *Spheroidal Wave Functions* (Stanford, CA: Stanford University Press). [1]

Abramowitz, M., and Stegun, I.A. 1964, *Handbook of Mathematical Functions*, Applied Mathematics Series, Volume 55 (Washington: National Bureau of Standards; reprinted 1968 by Dover Publications, New York), §21. [2]

Morse, P.M., and Feshbach, H. 1953, *Methods of Theoretical Physics*, Part II (New York: McGraw-Hill), pp. 1502ff. [3]

17.5 Automated Allocation of Mesh Points

In relaxation problems, you have to choose values for the independent variable at the mesh points. This is called *allocating* the grid or mesh. The usual procedure is to pick a plausible set of values and, if it works, to be content. If it doesn't work, increasing the number of points usually cures the problem.

If we know ahead of time where our solutions will be rapidly varying, we can put more grid points there and less elsewhere. Alternatively, we can solve the problem first on a uniform mesh and then examine the solution to see where we should add more points. We then repeat the solution with the improved grid. The object of the exercise is to allocate points in such a way as to represent the solution accurately.

It is also possible to automate the allocation of mesh points, so that it is done "dynamically" during the relaxation process. This powerful technique not only improves the accuracy of the relaxation method, but also (as we will see in the next section) allows internal singularities to be handled in quite a neat way. Here we learn how to accomplish the automatic allocation.

We want to focus attention on the independent variable x, and consider two alternative reparametrizations of it. The first, we term q; this is just the coordinate corresponding to the mesh points themselves, so that $q = 0$ at $k = 0$, $q = 1$ at $k = 1$, and so on. Between any two mesh points we have $\Delta q = 1$. In the change of independent variable in the ODEs from x to q,

$$\frac{d\mathbf{y}}{dx} = \mathbf{g} \qquad (17.5.1)$$

becomes

$$\frac{d\mathbf{y}}{dq} = \mathbf{g}\frac{dx}{dq} \qquad (17.5.2)$$

In terms of q, equation (17.5.2) as an FDE might be written

$$\mathbf{y}_k - \mathbf{y}_{k-1} - \tfrac{1}{2}\left[\left(\mathbf{g}\frac{dx}{dq}\right)_k + \left(\mathbf{g}\frac{dx}{dq}\right)_{k-1}\right] = 0 \qquad (17.5.3)$$

or some related version. Note that dx/dq should accompany $\mathbf{g}$. The transformation between x and q depends only on the *Jacobian* dx/dq. Its reciprocal dq/dx is proportional to the density of mesh points.

Now, given the function $\mathbf{y}(x)$, or its approximation at the current stage of relaxation, we are supposed to have some idea of how we want to specify the density of mesh points. For example, we might want dq/dx to be larger where $\mathbf{y}$ is changing rapidly, or near to the boundaries, or both. In fact, we can probably make up a formula for what we would like dq/dx to be proportional to. The problem is that we do not know the proportionality constant. That is, the formula that we might invent would not have the correct integral over the whole range of x so as to make q vary from 0 to $M - 1$, according to its definition. To solve this problem we introduce a second reparametrization $Q(q)$, where Q is a new independent variable. The relation between Q and q is taken to be *linear*, so that a mesh spacing formula for dQ/dx differs only in its unknown proportionality constant. A linear relation implies

$$\frac{d^2Q}{dq^2} = 0 \qquad (17.5.4)$$

or, expressed in the usual manner as coupled first-order equations,

$$\frac{dQ(x)}{dq} = \psi \qquad \frac{d\psi}{dq} = 0 \qquad (17.5.5)$$

where ψ is a new intermediate variable. We add these two equations to the set of ODEs being solved.

Completing the prescription, we add a third ODE that is just our desired mesh-density function, namely

$$\phi(x) = \frac{dQ}{dx} = \frac{dQ}{dq}\frac{dq}{dx} \qquad (17.5.6)$$

where $\phi(x)$ is chosen by us. Written in terms of the mesh variable q, this equation is

$$\frac{dx}{dq} = \frac{\psi}{\phi(x)} \qquad (17.5.7)$$

Notice that $\phi(x)$ should be chosen to be positive definite, so that the density of mesh points is everywhere positive. Otherwise (17.5.7) can have a zero in its denominator.

To use automated mesh spacing, you add the three ODEs (17.5.5) and (17.5.7) to your set of equations, i.e., to the array y[j][k]. Now x becomes a dependent variable! Q and ψ

also become new dependent variables. Normally, evaluating ϕ requires little extra work since it will be composed from pieces of the g's that exist anyway. The automated procedure allows one to investigate quickly how the numerical results might be affected by various strategies for mesh spacing. (A special case occurs if the desired mesh spacing function Q can be found analytically, i.e., dQ/dx is directly integrable. Then, you need to add only two equations, those in 17.5.5, and two new variables x, ψ.)

As an example of a typical strategy for implementing this scheme, consider a system with one dependent variable $y(x)$. We could set

$$dQ = \frac{dx}{\Delta} + \frac{|d \ln y|}{\delta} \qquad (17.5.8)$$

or

$$\phi(x) = \frac{dQ}{dx} = \frac{1}{\Delta} + \left| \frac{dy/dx}{y\delta} \right| \qquad (17.5.9)$$

where Δ and δ are constants that we choose. The first term would give a uniform spacing in x if it alone were present. The second term forces more grid points to be used where y is changing rapidly. The constants act to make every logarithmic change in y of an amount δ about as "attractive" to a grid point as a change in x of amount Δ. You adjust the constants according to taste. Other strategies are possible, such as a logarithmic spacing in x, replacing dx in the first term with $d \ln x$.

CITED REFERENCES AND FURTHER READING:

Eggleton, P. P. 1971, *Monthly Notices of the Royal Astronomical Society*, vol. 151, pp. 351–364.

Kippenhan, R., Weigert, A., and Hofmeister, E. 1968, in *Methods in Computational Physics*, vol. 7 (New York: Academic Press), pp. 129ff.

17.6 Handling Internal Boundary Conditions or Singular Points

Singularities can occur in the interiors of two point boundary value problems. Typically, there is a point x_s at which a derivative must be evaluated by an expression of the form

$$S(x_s) = \frac{N(x_s, \mathbf{y})}{D(x_s, \mathbf{y})} \qquad (17.6.1)$$

where the denominator $D(x_s, \mathbf{y}) = 0$. In physical problems with finite answers, singular points usually come with their own cure: Where $D \to 0$, there the physical solution $\mathbf{y}$ must be such as to make $N \to 0$ simultaneously, in such a way that the ratio takes on a meaningful value. This constraint on the solution $\mathbf{y}$ is often called a *regularity condition*. The condition that $D(x_s, \mathbf{y})$ satisfy some special constraint at x_s is entirely analogous to an extra boundary condition, an algebraic relation among the dependent variables that must hold at a point.

We discussed a related situation earlier, in §17.2, when we described the "fitting point method" to handle the task of integrating equations with singular behavior at the boundaries. In those problems you are unable to integrate from one side of the domain to the other. However, the ODEs do have well-behaved derivatives and solutions in the neighborhood of the singularity, so it is readily possible to integrate away from the point. Both the relaxation method and the method of "shooting" to a fitting point handle such problems easily. Also, in those problems the presence of singular behavior served to isolate some special boundary values that had to be satisfied to solve the equations.

The difference here is that we are concerned with singularities arising at intermediate points, where the location of the singular point depends on the solution, so is not known *a priori*. Consequently, we face a circular task: The singularity prevents us from finding a numerical solution, but we need a numerical solution to find its location. Such singularities are also associated with selecting a special value for some variable which allows the solution to satisfy the regularity condition at the singular point. Thus, internal singularities take on aspects of being internal boundary conditions.

One way of handling internal singularities is to treat the problem as a free boundary problem, as discussed at the end of §17.0. Suppose, as a simple example, we consider the equation

$$\frac{dy}{dx} = \frac{N(x, y)}{D(x, y)} \tag{17.6.2}$$

where N and D are required to pass through zero at some unknown point x_s. We add the equation

$$z \equiv x_s - x_1 \qquad \frac{dz}{dx} = 0 \tag{17.6.3}$$

where x_s is the unknown location of the singularity, and change the independent variable to t by setting

$$x - x_1 = tz, \qquad 0 \le t \le 1 \tag{17.6.4}$$

The boundary conditions at $t = 1$ become

$$N(x, y) = 0, \qquad D(x, y) = 0 \tag{17.6.5}$$

Use of an adaptive mesh as discussed in the previous section is another way to overcome the difficulties of an internal singularity. For the problem (17.6.2), we add the mesh spacing equations

$$\frac{dQ}{dq} = \psi \tag{17.6.6}$$

$$\frac{d\psi}{dq} = 0 \tag{17.6.7}$$

with a simple mesh spacing function that maps x uniformly into q, where q runs from 0 to $M - 1$, with M the number of mesh points:

$$Q(x) = x - x_1, \qquad \frac{dQ}{dx} = 1 \tag{17.6.8}$$

Having added three first-order differential equations, we must also add their corresponding boundary conditions. If there were no singularity, these could simply be

$$\text{at} \quad q = 0: \qquad x = x_1, \quad Q = 0 \tag{17.6.9}$$
$$\text{at} \quad q = M - 1: \qquad x = x_2 \tag{17.6.10}$$

and a total of N values y_i specified at $q = 0$. In this case the problem is essentially an initial value problem with all boundary conditions specified at x_1 and the mesh spacing function is superfluous.

However, in the actual case at hand we impose the conditions

$$\text{at} \quad q = 0: \qquad x = x_1, \qquad Q = 0 \tag{17.6.11}$$
$$\text{at} \quad q = M - 1: \quad N(x, y) = 0, \quad D(x, y) = 0 \tag{17.6.12}$$

and $N - 1$ values y_i at $q = 0$. The "missing" y_i is to be adjusted, in other words, so as to make the solution go through the singular point in a regular (zero-over-zero) rather than irregular (finite-over-zero) manner. Notice also that these boundary conditions do not directly impose a value for x_2, which becomes an adjustable parameter that the code varies in an attempt to match the regularity condition.

In this example the singularity occurred at a boundary, and the complication arose because the location of the boundary was unknown. In other problems we might wish to continue the integration beyond the internal singularity. For the example given above, we could simply integrate the ODEs to the singular point, then as a separate problem recommence the integration from the singular point on as far we care to go. However, in other cases the

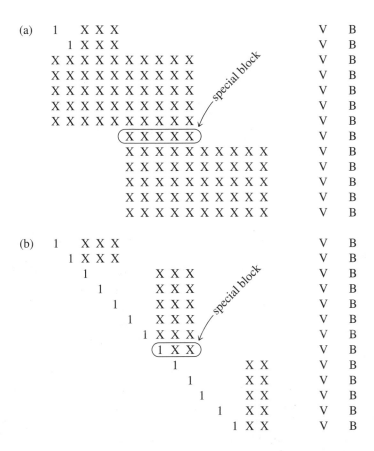

Figure 17.6.1. FDE matrix structure with an internal boundary condition. The internal condition introduces a special block. (a) Original form, compare with Figure 17.3.1; (b) final form, compare with Figure 17.3.2.

singularity occurs internally, but does not completely determine the problem: There are still some more boundary conditions to be satisfied further along in the mesh. Such cases present no difficulty in principle, but do require some adaptation of the relaxation code given in §17.3. In effect all you need to do is to add a "special" block of equations at the mesh point where the internal boundary conditions occur, and do the proper bookkeeping.

Figure 17.6.1 illustrates a concrete example where the overall problem contains 5 equations with 2 boundary conditions at the first point, one "internal" boundary condition, and two final boundary conditions. The figure shows the structure of the overall matrix equations along the diagonal in the vicinity of the special block. In the middle of the domain, blocks typically involve 5 equations (rows) in 10 unknowns (columns). For each block prior to the special block, the initial boundary conditions provided enough information to zero the first two columns of the blocks. The five FDEs eliminate five more columns, and the final three columns need to be stored for the backsubstitution step (as described in §17.3). To handle the extra condition we break the normal cycle and add a special block with only one equation: the internal boundary condition. This effectively reduces the required storage of unreduced coefficients by one column for the rest of the grid, and allows us to reduce to zero the first three columns of subsequent blocks. The functions red, pinvs, bksub can readily handle these cases with minor recoding, but each problem makes for a special case, and you will have to make the modifications as required.

CITED REFERENCES AND FURTHER READING:

London, R.A., and Flannery, B.P. 1982, *Astrophysical Journal*, vol. 258, pp. 260–269.

Chapter 18. Integral Equations and Inverse Theory

18.0 Introduction

Many people, otherwise numerically knowledgable, imagine that the numerical solution of integral equations must be an extremely arcane topic, since, until recently, it was almost never treated in numerical analysis textbooks. Actually there is a large and growing literature on the numerical solution of integral equations; several monographs have by now appeared [1-3]. One reason for the sheer volume of this activity is that there are many different kinds of equations, each with many different possible pitfalls; often many different algorithms have been proposed to deal with a single case.

There is a close correspondence between linear integral equations, which specify linear, integral relations among functions in an infinite-dimensional function space, and plain old linear equations, which specify analogous relations among vectors in a finite-dimensional vector space. Because this correspondence lies at the heart of most computational algorithms, it is worth making it explicit as we recall how integral equations are classified.

Fredholm equations involve definite integrals with fixed upper and lower limits. An *inhomogeneous Fredholm equation of the first kind* has the form

$$g(t) = \int_a^b K(t,s) f(s) \, ds \qquad (18.0.1)$$

Here $f(t)$ is the unknown function to be solved for, while $g(t)$ is a known "right-hand side." (In integral equations, for some odd reason, the familiar "right-hand side" is conventionally written on the left!) The function of two variables, $K(t,s)$ is called the *kernel*. Equation (18.0.1) is analogous to the matrix equation

$$\mathbf{K} \cdot \mathbf{f} = \mathbf{g} \qquad (18.0.2)$$

whose solution is $\mathbf{f} = \mathbf{K}^{-1} \cdot \mathbf{g}$, where $\mathbf{K}^{-1}$ is the matrix inverse. Like equation (18.0.2), equation (18.0.1) has a unique solution whenever g is nonzero (the homogeneous case with $g = 0$ is almost never useful) and K is invertible. However, as we shall see, this latter condition is as often the exception as the rule.

The analog of the finite-dimensional eigenvalue problem

$$(\mathbf{K} - \sigma \mathbf{1}) \cdot \mathbf{f} = \mathbf{g} \qquad (18.0.3)$$

is called a *Fredholm equation of the second kind*, usually written

$$f(t) = \lambda \int_a^b K(t,s)f(s)\,ds + g(t) \qquad (18.0.4)$$

Again, the notational conventions do not exactly correspond: λ in equation (18.0.4) is $1/\sigma$ in (18.0.3), while $\mathbf{g}$ is $-g/\lambda$. If g (or $\mathbf{g}$) is zero, then the equation is said to be *homogeneous*. If the kernel $K(t,s)$ is bounded, then, like equation (18.0.3), equation (18.0.4) has the property that its homogeneous form has solutions for at most a denumerably infinite set $\lambda = \lambda_n$, $n = 1, 2, \ldots$, the *eigenvalues*. The corresponding solutions $f_n(t)$ are the *eigenfunctions*. The eigenvalues are real if the kernel is symmetric.

In the *inhomogeneous* case of nonzero g (or $\mathbf{g}$), equations (18.0.3) and (18.0.4) are soluble *except* when λ (or σ) is an eigenvalue — because the integral operator (or matrix) is singular then. In integral equations this dichotomy is called *the Fredholm alternative*.

Fredholm equations of the first kind are often extremely ill-conditioned. Applying the kernel to a function is generally a smoothing operation, so the solution, which requires inverting the operator, will be extremely sensitive to small changes or errors in the input. Smoothing often actually loses information, and there is no way to get it back in an inverse operation. Specialized methods have been developed for such equations, which are often called *inverse problems*. In general, a method must augment the information given with some prior knowledge of the nature of the solution. This prior knowledge is then used, in one way or another, to restore lost information. We will introduce such techniques in §18.4.

Inhomogeneous Fredholm equations of the second kind are much less often ill-conditioned. Equation (18.0.4) can be rewritten as

$$\int_a^b [K(t,s) - \sigma\delta(t-s)]f(s)\,ds = -\sigma g(t) \qquad (18.0.5)$$

where $\delta(t-s)$ is a Dirac delta function (and where we have changed from λ to its reciprocal σ for clarity). If σ is large enough in magnitude, then equation (18.0.5) is, in effect, diagonally dominant and thus well-conditioned. Only if σ is small do we go back to the ill-conditioned case.

Homogeneous Fredholm equations of the second kind are likewise not particularly ill-posed. If K is a smoothing operator, then it will map many f's to zero, or near-zero; there will thus be a large number of degenerate or nearly degenerate eigenvalues around $\sigma = 0$ ($\lambda \to \infty$), but this will cause no particular computational difficulties. In fact, we can now see that the magnitude of σ needed to rescue the inhomogeneous equation (18.0.5) from an ill-conditioned fate is generally much *less* than that required for diagonal dominance. Since the σ term shifts all eigenvalues, it is enough that it be large enough to shift a smoothing operator's forest of near-zero eigenvalues away from zero, so that the resulting operator becomes invertible (except, of course, at the discrete eigenvalues).

Volterra equations are a special case of Fredholm equations with $K(t,s) = 0$ for $s > t$. Chopping off the unnecessary part of the integration, Volterra equations are written in a form where the upper limit of integration is the independent variable t.

The *Volterra equation of the first kind*

$$g(t) = \int_a^t K(t,s) f(s) \, ds \qquad (18.0.6)$$

has as its analog the matrix equation (now written out in components)

$$\sum_{j=0}^k K_{kj} f_j = g_k \qquad (18.0.7)$$

Comparing with equation (18.0.2), we see that the Volterra equation corresponds to a matrix $\mathbf{K}$ that is lower (i.e., left) triangular, with zero entries above the diagonal. As we know from Chapter 2, such matrix equations are trivially soluble by forward substitution. Techniques for solving Volterra equations are similarly straightforward. When experimental measurement noise does not dominate, Volterra equations of the first kind tend *not* to be ill-conditioned; the upper limit to the integral introduces a sharp step that conveniently spoils any smoothing properties of the kernel.

The Volterra equation of the second kind is written

$$f(t) = \int_a^t K(t,s) f(s) \, ds + g(t) \qquad (18.0.8)$$

whose matrix analog is the equation

$$(\mathbf{K} - \mathbf{1}) \cdot \mathbf{f} = \mathbf{g} \qquad (18.0.9)$$

with $\mathbf{K}$ lower triangular. The reason there is no λ in these equations is that (i) in the inhomogeneous case (nonzero g) it can be absorbed into K, while (ii) in the homogeneous case ($g = 0$), it is a theorem that Volterra equations of the second kind with bounded kernels have no eigenvalues with square-integrable eigenfunctions.

We have specialized our definitions to the case of linear integral equations. The integrand in a nonlinear version of equation (18.0.1) or (18.0.6) would be $K(t, s, f(s))$ instead of $K(t, s) f(s)$; a nonlinear version of equation (18.0.4) or (18.0.8) would have an integrand $K(t, s, f(t), f(s))$. Nonlinear Fredholm equations are considerably more complicated than their linear counterparts. Fortunately, they do not occur as frequently in practice and we shall by and large ignore them in this chapter. By contrast, solving nonlinear Volterra equations usually involves only a slight modification of the algorithm for linear equations, as we shall see.

Almost all methods for solving integral equations numerically make use of *quadrature rules*, frequently Gaussian quadratures. This would be a good time for you to go back and review §4.5, especially the advanced material towards the end of that section.

In the sections that follow, we first discuss Fredholm equations of the second kind with smooth kernels (§18.1). Nontrivial quadrature rules come into the discussion, but we will be dealing with well-conditioned systems of equations. We then return to Volterra equations (§18.2), and find that simple and straightforward methods are generally satisfactory for these equations.

In §18.3 we discuss how to proceed in the case of singular kernels, focusing largely on Fredholm equations (both first and second kinds). Singularities require special quadrature rules, but they are also sometimes blessings in disguise, since they can spoil a kernel's smoothing and make problems well-conditioned.

In §§18.4–18.7 we face up to the issues of inverse problems. §18.4 is an introduction to this large subject.

We should note here that wavelet transforms, already discussed in §13.10, are applicable not only to data compression and signal processing, but can also be used to transform some classes of integral equations into sparse linear problems that allow fast solution. You may wish to review §13.10 as part of reading this chapter.

Some subjects, such as *integro-differential equations*, we must simply declare to be beyond our scope. For a review of methods for integro-differential equations, see Brunner [4].

It should go without saying that this one short chapter can only barely touch on a few of the most basic methods involved in this complicated subject.

CITED REFERENCES AND FURTHER READING:

Delves, L.M., and Mohamed, J.L. 1985, *Computational Methods for Integral Equations* (Cambridge, U.K.: Cambridge University Press). [1]

Linz, P. 1985, *Analytical and Numerical Methods for Volterra Equations* (Philadelphia: S.I.A.M.). [2]

Atkinson, K.E. 1976, *A Survey of Numerical Methods for the Solution of Fredholm Integral Equations of the Second Kind* (Philadelphia: S.I.A.M.). [3]

Brunner, H. 1988, in *Numerical Analysis 1987*, Pitman Research Notes in Mathematics vol. 170, D.F. Griffiths and G.A. Watson, eds. (Harlow, Essex, U.K.: Longman Scientific and Technical), pp. 18–38. [4]

Smithies, F. 1958, *Integral Equations* (Cambridge, U.K.: Cambridge University Press).

Kanwal, R.P. 1971, *Linear Integral Equations* (New York: Academic Press).

Green, C.D. 1969, *Integral Equation Methods* (New York: Barnes & Noble).

18.1 Fredholm Equations of the Second Kind

We desire a numerical solution for $f(t)$ in the equation

$$f(t) = \lambda \int_a^b K(t,s)f(s)\,ds + g(t) \qquad (18.1.1)$$

The method we describe, a very basic one, is called the *Nystrom method*. It requires the choice of some approximate *quadrature rule*:

$$\int_a^b y(s)\,ds = \sum_{j=0}^{N-1} w_j y(s_j) \qquad (18.1.2)$$

Here the set $\{w_j\}$ are the weights of the quadrature rule, while the N points $\{s_j\}$ are the abscissas.

What quadrature rule should we use? It is certainly possible to solve integral equations with low-order quadrature rules like the extended trapezoidal or Simpson's rules. We will see, however, that the solution method involves $O(N^3)$ operations, and so the most efficient methods tend to use high-order quadrature rules to keep N as small as possible. For smooth, nonsingular problems, nothing beats Gaussian quadrature (e.g., Gauss-Legendre quadrature, §4.5). (For non-smooth or singular kernels, see §18.3.)

Delves and Mohamed [1] investigated methods more complicated than the Nystrom method. For straightforward Fredholm equations of the second kind, they concluded "... the clear winner of this contest has been the Nystrom routine ... with the N-point Gauss-Legendre rule. This routine is extremely simple.... Such results are enough to make a numerical analyst weep."

If we apply the quadrature rule (18.1.2) to equation (18.1.1), we get

$$f(t) = \lambda \sum_{j=0}^{N-1} w_j K(t, s_j) f(s_j) + g(t) \tag{18.1.3}$$

Evaluate equation (18.1.3) at the quadrature points:

$$f(t_i) = \lambda \sum_{j=0}^{N-1} w_j K(t_i, s_j) f(s_j) + g(t_i) \tag{18.1.4}$$

Let f_i be the vector $f(t_i)$, g_i the vector $g(t_i)$, K_{ij} the matrix $K(t_i, s_j)$, and define

$$\widetilde{K}_{ij} = K_{ij} w_j \tag{18.1.5}$$

Then in matrix notation equation (18.1.4) becomes

$$(\mathbf{1} - \lambda \widetilde{\mathbf{K}}) \cdot \mathbf{f} = \mathbf{g} \tag{18.1.6}$$

This is a set of N linear algebraic equations in N unknowns that can be solved by standard triangular decomposition techniques (§2.3) — that is where the $O(N^3)$ operations count comes in. The solution is usually well-conditioned, unless λ is very close to an eigenvalue.

Having obtained the solution at the quadrature points $\{t_i\}$, how do you get the solution at some other point t? You do *not* simply use polynomial interpolation. This destroys all the accuracy you have worked so hard to achieve. Nystrom's key observation was that you should use equation (18.1.3) as an interpolatory formula, maintaining the accuracy of the solution.

We here give two routines for use with linear Fredholm equations of the second kind. The routine `fred2` sets up equation (18.1.6) and then solves it by LU decomposition with calls to the routines `ludcmp` and `lubksb`. The Gauss-Legendre quadrature is implemented by first getting the weights and abscissas with a call to `gauleg`. Routine `fred2` requires that you provide an external function that returns $g(t)$ and another that returns λK_{ij}. It then returns the solution f at the quadrature points. It also returns the quadrature points and weights. These are used by the second routine `fredin` to carry out the Nystrom interpolation of equation (18.1.3) and return the value of f at any point in the interval $[a, b]$.

```
#include "nr.h"

void NR::fred2(const DP a, const DP b, Vec_O_DP &t, Vec_O_DP &f, Vec_O_DP &w,
    DP g(const DP), DP ak(const DP, const DP))
```
Solves a linear Fredholm equation of the second kind. Quantities a and b are input as the limits of integration. g and ak are user-supplied external functions that respectively return $g(t)$ and $\lambda K(t,s)$. The routine returns arrays t[0..n-1] and f[0..n-1] containing the abscissas t_i of the Gaussian quadrature and the solution f at these abscissas. Also returned is the array w[0..n-1] of Gaussian weights for use with the Nystrom interpolation routine fredin. The quantity n is the number of points to use in the Gaussian quadrature, and is determined from the dimension of the arrays you supply to this routine.
```
{
    int i,j;
    DP d;

    int n=t.size();
    Mat_DP omk(n,n);
    Vec_INT indx(n);
    gauleg(a,b,t,w);              Replace gauleg with another routine if not using
    for (i=0;i<n;i++) {           Gauss-Legendre quadrature.
        for (j=0;j<n;j++)         Form 1 − λK.
            omk[i][j]=DP(i == j)-ak(t[i],t[j])*w[j];
        f[i]=g(t[i]);
    }
    ludcmp(omk,indx,d);           Solve linear equations.
    lubksb(omk,indx,f);
}
```

```
#include "nr.h"

DP NR::fredin(const DP x, const DP a, const DP b, Vec_I_DP &t, Vec_I_DP &f,
    Vec_I_DP &w, DP g(const DP), DP ak(const DP, const DP))
```
Given arrays t[0..n-1] and w[0..n-1] containing the abscissas and weights of the Gaussian quadrature, and given the solution array f[0..n-1] from fred2, this function returns the value of f at x using the Nystrom interpolation formula. Quantities a and b are input as the limits of integration. g and ak are user-supplied external functions that respectively return $g(t)$ and $\lambda K(t,s)$.
```
{
    int i;
    DP sum=0.0;

    int n=t.size();
    for (i=0;i<n;i++) sum += ak(x,t[i])*w[i]*f[i];
    return g(x)+sum;
}
```

One disadvantage of a method based on Gaussian quadrature is that there is no simple way to obtain an estimate of the error in the result. The best practical method is to increase N by 50%, say, and treat the difference between the two estimates as a conservative estimate of the error in the result obtained with the larger value of N.

Turn now to solutions of the homogeneous equation. If we set $\lambda = 1/\sigma$ and $\mathbf{g} = 0$, then equation (18.1.6) becomes a standard eigenvalue equation

$$\widetilde{\mathbf{K}} \cdot \mathbf{f} = \sigma \mathbf{f} \tag{18.1.7}$$

which we can solve with any convenient matrix eigenvalue routine (see Chapter 11). Note that if our original problem had a symmetric kernel, then the matrix $\mathbf{K}$

is symmetric. However, since the weights w_j are not equal for most quadrature rules, the matrix $\widetilde{\mathbf{K}}$ (equation 18.1.5) is not symmetric. The matrix eigenvalue problem is much easier for symmetric matrices, and so we should restore the symmetry if possible. Provided the weights are positive (which they are for Gaussian quadrature), we can define the diagonal matrix $\mathbf{D} = \text{diag}(w_j)$ and its square root, $\mathbf{D}^{1/2} = \text{diag}(\sqrt{w_j})$. Then equation (18.1.7) becomes

$$\mathbf{K} \cdot \mathbf{D} \cdot \mathbf{f} = \sigma \mathbf{f}$$

Multiplying by $\mathbf{D}^{1/2}$, we get

$$\left(\mathbf{D}^{1/2} \cdot \mathbf{K} \cdot \mathbf{D}^{1/2} \right) \cdot \mathbf{h} = \sigma \mathbf{h} \tag{18.1.8}$$

where $\mathbf{h} = \mathbf{D}^{1/2} \cdot \mathbf{f}$. Equation (18.1.8) is now in the form of a symmetric eigenvalue problem.

Solution of equations (18.1.7) or (18.1.8) will in general give N eigenvalues, where N is the number of quadrature points used. For square-integrable kernels, these will provide good approximations to the lowest N eigenvalues of the integral equation. Kernels of *finite rank* (also called *degenerate* or *separable* kernels) have only a finite number of nonzero eigenvalues (possibly none). You can diagnose this situation by a cluster of eigenvalues σ that are zero to machine precision. The number of nonzero eigenvalues will stay constant as you increase N to improve their accuracy. Some care is required here: A nondegenerate kernel can have an infinite number of eigenvalues that have an accumulation point at $\sigma = 0$. You distinguish the two cases by the behavior of the solution as you increase N. If you suspect a degenerate kernel, you will usually be able to solve the problem by analytic techniques described in all the textbooks.

CITED REFERENCES AND FURTHER READING:

Delves, L.M., and Mohamed, J.L. 1985, *Computational Methods for Integral Equations* (Cambridge, U.K.: Cambridge University Press). [1]

Atkinson, K.E. 1976, *A Survey of Numerical Methods for the Solution of Fredholm Integral Equations of the Second Kind* (Philadelphia: S.I.A.M.).

18.2 Volterra Equations

Let us now turn to Volterra equations, of which our prototype is the Volterra equation of the second kind,

$$f(t) = \int_a^t K(t,s) f(s)\, ds + g(t) \tag{18.2.1}$$

Most algorithms for Volterra equations march out from $t = a$, building up the solution as they go. In this sense they resemble not only forward substitution (as discussed in §18.0), but also initial-value problems for ordinary differential equations. In fact, many algorithms for ODEs have counterparts for Volterra equations.

The simplest way to proceed is to solve the equation on a mesh with uniform spacing:

$$t_i = a + ih, \quad i = 0, 1, \ldots, N, \qquad h \equiv \frac{b-a}{N} \tag{18.2.2}$$

To do so, we must choose a quadrature rule. For a uniform mesh, the simplest scheme is the trapezoidal rule, equation (4.1.11):

$$\int_a^{t_i} K(t_i, s) f(s) \, ds = h \left(\tfrac{1}{2} K_{i0} f_0 + \sum_{j=1}^{i-1} K_{ij} f_j + \tfrac{1}{2} K_{ii} f_i \right) \tag{18.2.3}$$

Thus the trapezoidal method for equation (18.2.1) is:

$$f_0 = g_0$$

$$(1 - \tfrac{1}{2} h K_{ii}) f_i = h \left(\tfrac{1}{2} K_{i0} f_0 + \sum_{j=1}^{i-1} K_{ij} f_j \right) + g_i, \qquad i = 1, \ldots, N \tag{18.2.4}$$

(For a Volterra equation of the first kind, the leading 1 on the left would be absent, and g would have opposite sign, with corresponding straightforward changes in the rest of the discussion.)

Equation (18.2.4) is an explicit prescription that gives the solution in $O(N^2)$ operations. Unlike Fredholm equations, it is not necessary to solve a system of linear equations. Volterra equations thus usually involve less work than the corresponding Fredholm equations which, as we have seen, do involve the inversion of, sometimes large, linear systems.

The efficiency of solving Volterra equations is somewhat counterbalanced by the fact that *systems* of these equations occur more frequently in practice. If we interpret equation (18.2.1) as a *vector* equation for the vector of m functions $f(t)$, then the kernel $K(t, s)$ is an $m \times m$ matrix. Equation (18.2.4) must now also be understood as a vector equation. For each i, we have to solve the $m \times m$ set of linear algebraic equations by Gaussian elimination.

The routine `voltra` below implements this algorithm. You must supply an external function that returns the kth function of the vector $g(t)$ at the point t, and another that returns the (k, l) element of the matrix $K(t, s)$ at (t, s). The routine `voltra` then returns the vector $f(t)$ at the regularly spaced points t_i.

```
#include "nr.h"

void NR::voltra(const DP t0, const DP h, Vec_O_DP &t, Mat_O_DP &f,
    DP g(const int, const DP),
    DP ak(const int, const int, const DP, const DP))
```
Solves a set of m linear Volterra equations of the second kind using the extended trapezoidal rule. On input, t0 is the starting point of the integration and h is the stepsize. g(k,t) is a user-supplied external function that returns $g_k(t)$, while ak(k,l,t,s) is another user-supplied external function that returns the (k, l) element of the matrix $K(t, s)$. The solution is returned in f[0..m-1][0..n-1], with the corresponding abscissas in t[0..n-1], where n-1 is the

number of steps to be taken. The value of m is determined from the row-dimension of the solution matrix f.

```
{
    int i,j,k,l;
    DP d,sum;

    int m=f.nrows();
    int n=f.ncols();
    Vec_INT indx(m);
    Vec_DP b(m);
    Mat_DP a(m,m);
    t[0]=t0;
    for (k=0;k<m;k++) f[k][0]=g(k,t[0]);          Initialize.
    for (i=1;i<n;i++) {                           Take a step h.
        t[i]=t[i-1]+h;
        for (k=0;k<m;k++) {
            sum=g(k,t[i]);                        Accumulate right-hand side of linear
            for (l=0;l<m;l++) {                       equations in sum.
                sum += 0.5*h*ak(k,l,t[i],t[0])*f[l][0];
                for (j=1;j<i;j++)
                    sum += h*ak(k,l,t[i],t[j])*f[l][j];
                if (k == l)                       Left-hand side goes in matrix a.
                    a[k][l]=1.0-0.5*h*ak(k,l,t[i],t[i]);
                else
                    a[k][l] = -0.5*h*ak(k,l,t[i],t[i]);
            }
            b[k]=sum;
        }
        ludcmp(a,indx,d);                         Solve linear equations.
        lubksb(a,indx,b);
        for (k=0;k<m;k++) f[k][i]=b[k];
    }
}
```

For nonlinear Volterra equations, equation (18.2.4) holds with the product $K_{ii}f_i$ replaced by $K_{ii}(f_i)$, and similarly for the other two products of K's and f's. Thus for each i we solve a nonlinear equation for f_i with a known right-hand side. Newton's method (§9.4 or §9.6) with an initial guess of f_{i-1} usually works very well provided the stepsize is not too big.

Higher-order methods for solving Volterra equations are, in our opinion, not as important as for Fredholm equations, since Volterra equations are relatively easy to solve. However, there is an extensive literature on the subject. Several difficulties arise. First, any method that achieves higher order by operating on several quadrature points simultaneously will need a special method to get started, when values at the first few points are not yet known.

Second, stable quadrature rules can give rise to unexpected instabilities in integral equations. For example, suppose we try to replace the trapezoidal rule in the algorithm above with Simpson's rule. Simpson's rule naturally integrates over an interval $2h$, so we easily get the function values at the even mesh points. For the odd mesh points, we could try appending one panel of trapezoidal rule. But to which end of the integration should we append it? We could do one step of trapezoidal rule followed by all Simpson's rule, or Simpson's rule with one step of trapezoidal rule at the end. Surprisingly, the former scheme is unstable, while the latter is fine!

A simple approach that can be used with the trapezoidal method given above is Richardson extrapolation: Compute the solution with stepsize h and $h/2$. Then, assuming the error scales with h^2, compute

$$f_{\mathrm{E}} = \frac{4f(h/2) - f(h)}{3} \qquad (18.2.5)$$

This procedure can be repeated as with Romberg integration.

The general consensus is that the best of the higher order methods is the *block-by-block method* (see [1]). Another important topic is the use of variable stepsize methods, which are much more efficient if there are sharp features in K or f. Variable stepsize methods are quite a bit more complicated than their counterparts for differential equations; we refer you to the literature [1,2] for a discussion.

You should also be on the lookout for singularities in the integrand. If you find them, then look to §18.3 for additional ideas.

CITED REFERENCES AND FURTHER READING:

Linz, P. 1985, *Analytical and Numerical Methods for Volterra Equations* (Philadelphia: S.I.A.M.). [1]

Delves, L.M., and Mohamed, J.L. 1985, *Computational Methods for Integral Equations* (Cambridge, U.K.: Cambridge University Press). [2]

18.3 Integral Equations with Singular Kernels

Many integral equations have singularities in either the kernel or the solution or both. A simple quadrature method will show poor convergence with N if such singularities are ignored. There is sometimes art in how singularities are best handled.

We start with a few straightforward suggestions:

1. Integrable singularities can often be removed by a change of variable. For example, the singular behavior $K(t, s) \sim s^{1/2}$ or $s^{-1/2}$ near $s = 0$ can be removed by the transformation $z = s^{1/2}$. Note that we are assuming that the singular behavior is confined to K, whereas the quadrature actually involves the product $K(t, s)f(s)$, and it is this product that must be "fixed." Ideally, you must deduce the singular nature of the product before you try a numerical solution, and take the appropriate action. Commonly, however, a singular kernel does *not* produce a singular solution $f(t)$. (The highly singular kernel $K(t, s) = \delta(t - s)$ is simply the identity operator, for example.)

2. If $K(t, s)$ can be factored as $w(s)\overline{K}(t, s)$, where $w(s)$ is singular and $\overline{K}(t, s)$ is smooth, then a Gaussian quadrature based on $w(s)$ as a weight function will work well. Even if the factorization is only approximate, the convergence is often improved dramatically. All you have to do is replace `gauleg` in the routine `fred2` by another quadrature routine. Section 4.5 explained how to construct such quadratures; or you can find tabulated abscissas and weights in the standard references [1,2]. You must of course supply $\overline{K}$ instead of K.

This method is a special case of the *product Nystrom method* [3,4], where one factors out a singular term $p(t, s)$ depending on both t and s from K and constructs suitable weights for its Gaussian quadrature. The calculations in the general case are quite cumbersome, because the weights depend on the chosen $\{t_i\}$ as well as the form of $p(t, s)$.

We prefer to implement the product Nystrom method on a uniform grid, with a quadrature scheme that generalizes the extended Simpson's 3/8 rule (equation 4.1.5) to arbitrary weight functions. We discuss this in the subsections below.

3. Special quadrature formulas are also useful when the kernel is not strictly singular, but is "almost" so. One example is when the kernel is concentrated near $t = s$ on a scale much smaller than the scale on which the solution $f(t)$ varies. In that case, a quadrature formula

can be based on locally approximating $f(s)$ by a polynomial or spline, while calculating the first few *moments* of the kernel $K(t, s)$ at the tabulation points t_i. In such a scheme the narrow width of the kernel becomes an asset, rather than a liability: The quadrature becomes exact as the width of the kernel goes to zero.

4. An infinite range of integration is also a form of singularity. Truncating the range at a large finite value should be used only as a last resort. If the kernel goes rapidly to zero, then a Gauss-Laguerre $[w \sim \exp(-\alpha s)]$ or Gauss-Hermite $[w \sim \exp(-s^2)]$ quadrature should work well. Long-tailed functions often succumb to the transformation

$$s = \frac{2\alpha}{z+1} - \alpha \tag{18.3.1}$$

which maps $0 < s < \infty$ to $1 > z > -1$ so that Gauss-Legendre integration can be used. Here $\alpha > 0$ is a constant that you adjust to improve the convergence.

5. A common situation in practice is that $K(t, s)$ is singular along the diagonal line $t = s$. Here the Nystrom method fails completely because the kernel gets evaluated at (t_i, s_i). *Subtraction of the singularity* is one possible cure:

$$\int_a^b K(t, s) f(s)\, ds = \int_a^b K(t, s)[f(s) - f(t)]\, ds + \int_a^b K(t, s) f(t)\, ds$$
$$= \int_a^b K(t, s)[f(s) - f(t)]\, ds + r(t) f(t) \tag{18.3.2}$$

where $r(t) = \int_a^b K(t, s)\, ds$ is computed analytically or numerically. If the first term on the right-hand side is now regular, we can use the Nystrom method. Instead of equation (18.1.4), we get

$$f_i = \lambda \sum_{\substack{j=0 \\ j \neq i}}^{N-1} w_j K_{ij} [f_j - f_i] + \lambda r_i f_i + g_i \tag{18.3.3}$$

Sometimes the subtraction process must be repeated before the kernel is completely regularized. See [3] for details. (And read on for a different, we think better, way to handle diagonal singularities.)

Quadrature on a Uniform Mesh with Arbitrary Weight

It is possible in general to find n-point linear quadrature rules that approximate the integral of a function $f(x)$, times an arbitrary weight function $w(x)$, over an arbitrary range of integration (a, b), as the sum of weights times n evenly spaced values of the function $f(x)$, say at $x = kh, (k+1)h, \ldots, (k+n-1)h$. The general scheme for deriving such quadrature rules is to write down the n linear equations that must be satisfied if the quadrature rule is to be exact for the n functions $f(x) = \text{const}, x, x^2, \ldots, x^{n-1}$, and then solve these for the coefficients. This can be done analytically, once and for all, if the moments of the weight function over the same range of integration,

$$W_n \equiv \frac{1}{h^n} \int_a^b x^n w(x) dx \tag{18.3.4}$$

are assumed to be known. Here the prefactor h^{-n} is chosen to make W_n scale as h if (as in the usual case) $b - a$ is proportional to h.

Carrying out this prescription for the four-point case gives the result

$$\int_a^b w(x)f(x)dx =$$

$$\frac{1}{6}f(kh)\Big[(k+1)(k+2)(k+3)W_0 - (3k^2 + 12k + 11)W_1 + 3(k+2)W_2 - W_3\Big]$$

$$+\frac{1}{2}f([k+1]h)\Big[-k(k+2)(k+3)W_0 + (3k^2 + 10k + 6)W_1 - (3k+5)W_2 + W_3\Big]$$

$$+\frac{1}{2}f([k+2]h)\Big[k(k+1)(k+3)W_0 - (3k^2 + 8k + 3)W_1 + (3k+4)W_2 - W_3\Big]$$

$$+\frac{1}{6}f([k+3]h)\Big[-k(k+1)(k+2)W_0 + (3k^2 + 6k + 2)W_1 - 3(k+1)W_2 + W_3\Big]$$

$$(18.3.5)$$

While the terms in brackets superficially appear to scale as k^2, there is typically cancellation at both $O(k^2)$ and $O(k)$.

Equation (18.3.5) can be specialized to various choices of (a, b). The obvious choice is $a = kh$, $b = (k+3)h$, in which case we get a four-point quadrature rule that generalizes Simpson's 3/8 rule (equation 4.1.5). In fact, we can recover this special case by setting $w(x) = 1$, in which case (18.3.4) becomes

$$W_n = \frac{h}{n+1}[(k+3)^{n+1} - k^{n+1}] \tag{18.3.6}$$

The four terms in square brackets equation (18.3.5) each become independent of k, and (18.3.5) in fact reduces to

$$\int_{kh}^{(k+3)h} f(x)dx = \frac{3h}{8}f(kh) + \frac{9h}{8}f([k+1]h) + \frac{9h}{8}f([k+2]h) + \frac{3h}{8}f([k+3]h) \tag{18.3.7}$$

Back to the case of general $w(x)$, some other choices for a and b are also useful. For example, we may want to choose (a, b) to be $([k+1]h, [k+3]h)$ or $([k+2]h, [k+3]h)$, allowing us to finish off an extended rule whose number of intervals is not a multiple of three, without loss of accuracy: The integral will be estimated using the four values $f(kh), \dots, f([k+3]h)$. Even more useful is to choose (a, b) to be $([k+1]h, [k+2]h)$, thus using four points to integrate a centered single interval. These weights, when sewed together into an extended formula, give quadrature schemes that have smooth coefficients, i.e., without the Simpson-like $2, 4, 2, 4, 2$ alternation. (In fact, this was the technique that we used to derive equation 4.1.14, which you may now wish to reexamine.)

All these rules are of the same order as the extended Simpson's rule, that is, exact for $f(x)$ a cubic polynomial. Rules of lower order, if desired, are similarly obtained. The three point formula is

$$\int_a^b w(x)f(x)dx = \frac{1}{2}f(kh)\Big[(k+1)(k+2)W_0 - (2k+3)W_1 + W_2\Big]$$

$$+ f([k+1]h)\Big[-k(k+2)W_0 + 2(k+1)W_1 - W_2\Big] \tag{18.3.8}$$

$$+ \frac{1}{2}f([k+2]h)\Big[k(k+1)W_0 - (2k+1)W_1 + W_2\Big]$$

Here the simple special case is to take, $w(x) = 1$, so that

$$W_n = \frac{h}{n+1}[(k+2)^{n+1} - k^{n+1}] \tag{18.3.9}$$

Then equation (18.3.8) becomes Simpson's rule,

$$\int_{kh}^{(k+2)h} f(x)dx = \frac{h}{3}f(kh) + \frac{4h}{3}f([k+1]h) + \frac{h}{3}f([k+2]h) \tag{18.3.10}$$

For nonconstant weight functions $w(x)$, however, equation (18.3.8) gives rules of one order less than Simpson, since they do not benefit from the extra symmetry of the constant case.

The two point formula is simply

$$\int_{kh}^{(k+1)h} w(x)f(x)dx = f(kh)[(k+1)W_0 - W_1] + f([k+1]h)[-kW_0 + W_1] \quad (18.3.11)$$

Here is a routine `wwghts` that uses the above formulas to return an extended N-point quadrature rule for the interval $(a,b) = (0, [N-1]h)$. Input to `wwghts` is a user-supplied routine, `kermom`, that is called to get the first four *indefinite-integral* moments of $w(x)$, namely

$$F_m(y) \equiv \int^y s^m w(s)ds \qquad m = 0,1,2,3 \qquad (18.3.12)$$

(The lower limit is arbitrary and can be chosen for convenience.) Cautionary note: When called with $N < 4$, `wwghts` returns a rule of lower order than Simpson; you should structure your problem to avoid this.

```
#include "nr.h"

void NR::wwghts(Vec_O_DP &wghts, const DP h,
    void kermom(Vec_O_DP &w, const DP y))
```
Constructs in `wghts[0..n-1]` weights for the n-point equal-interval quadrature from 0 to $(n-1)h$ of a function $f(x)$ times an arbitrary (possibly singular) weight function $w(x)$ whose indefinite-integral moments $F_n(y)$ are provided by the user-supplied routine `kermom`.
```
{
    int j,k;
    Vec_DP wold(4);
    DP hh,hi,c,fac,a,b;

    int n=wghts.size();
    hh=h;
    hi=1.0/hh;
    for (j=0;j<n;j++) wghts[j]=0.0;
    Zero all the weights so we can sum into them.
    if (n >= 4) {                                   Use highest available order.
        Vec_DP wold(4),wnew(4),w(4);
        kermom(wold,0.0);                           Evaluate indefinite integrals at lower end.
        b=0.0;                                      For another problem, you might change
        for (j=0;j<n-3;j++) {                           this lower limit.
            c=j;                                    This is called k in equation (18.3.5).
            a=b;                                    Set upper and lower limits for this step.
            b=a+hh;
            if (j == n-4) b=(n-1)*hh;               Last interval: go all the way to end.
            kermom(wnew,b);
            for (fac=1.0,k=0;k<4;k++,fac*=hi)             Equation (18.3.4).
                w[k]=(wnew[k]-wold[k])*fac;
            wghts[j]   += (((c+1.0)*(c+2.0)*(c+3.0)*w[0]       Equation (18.3.5).
                -(11.0+c*(12.0+c*3.0))*w[1]+3.0*(c+2.0)*w[2]-w[3])/6.0);
            wghts[j+1] += ((-c*(c+2.0)*(c+3.0)*w[0]
                +(6.0+c*(10.0+c*3.0))*w[1]-(3.0*c+5.0)*w[2]+w[3])*0.5);
            wghts[j+2] += ((c*(c+1.0)*(c+3.0)*w[0]
                -(3.0+c*(8.0+c*3.0))*w[1]+(3.0*c+4.0)*w[2]-w[3])*0.5);
            wghts[j+3] += ((-c*(c+1.0)*(c+2.0)*w[0]
                +(2.0+c*(6.0+c*3.0))*w[1]-3.0*(c+1.0)*w[2]+w[3])/6.0);
            for (k=0;k<4;k++) wold[k]=wnew[k];      Reset lower limits for moments.
        }
    } else if (n == 3) {                            Lower-order cases; not recommended.
        Vec_DP wold(3),wnew(3),w(3);
        kermom(wold,0.0);
        kermom(wnew,hh+hh);
        w[0]=wnew[0]-wold[0];
        w[1]=hi*(wnew[1]-wold[1]);
        w[2]=hi*hi*(wnew[2]-wold[2]);
        wghts[0]=w[0]-1.5*w[1]+0.5*w[2];
        wghts[1]=2.0*w[1]-w[2];
```

```
        wghts[2]=0.5*(w[2]-w[1]);
    } else if (n == 2) {
        Vec_DP wold(2),wnew(2),w(2);
        kermom(wold,0.0);
        kermom(wnew,hh);
        wghts[0]=wnew[0]-wold[0]-(wghts[1]=hi*(wnew[1]-wold[1]));
    }
}
```

We will now give an example of how to apply wwghts to a singular integral equation.

Worked Example: A Diagonally Singular Kernel

As a particular example, consider the integral equation

$$f(x) + \int_0^\pi K(x,y)f(y)dy = \sin x \qquad (18.3.13)$$

with the (arbitrarily chosen) nasty kernel

$$K(x,y) = \cos x \cos y \times \begin{cases} -\ln(x-y) & y < x \\ \sqrt{y-x} & y \geq x \end{cases} \qquad (18.3.14)$$

which has a logarithmic singularity on the left of the diagonal, combined with a square-root discontinuity on the right.

The first step is to do (analytically, in this case) the required moment integrals over the singular part of the kernel, equation (18.3.12). Since these integrals are done at a fixed value of x, we can use x as the lower limit. For any specified value of y, the required indefinite integral is then either

$$F_m(y;x) = \int_x^y s^m(s-x)^{1/2}ds = \int_0^{y-x}(x+t)^m t^{1/2}dt \qquad \text{if } y > x \qquad (18.3.15)$$

or

$$F_m(y;x) = -\int_x^y s^m \ln(x-s)ds = \int_0^{x-y}(x-t)^m \ln t\, dt \qquad \text{if } y < x \qquad (18.3.16)$$

(where a change of variable has been made in the second equality in each case). Doing these integrals analytically (actually, we used a symbolic integration package!), we package the resulting formulas in the following routine. Note that $w(j+1)$ returns $F_j(y;x)$.

```
#include <cmath>
#include "nr.h"
using namespace std;

extern DP x;                    Defined in quadmx.

void NR::kermom(Vec_O_DP &w, const DP y)
```
Returns in w[0..m-1] the first m indefinite-integral moments of one row of the singular part of the kernel. (For this example, m is hard-wired to be 4.) The input variable y labels the column, while the global variable x is the row. We can take x as the lower limit of integration. Thus, we return the moment integrals either purely to the left or purely to the right of the diagonal.
```
{
    DP d,df,clog,x2,x3,x4,y2;

    int m=w.size();
    if (y >= x) {
        d=y-x;
        df=2.0*sqrt(d)*d;
        w[0]=df/3.0;
        w[1]=df*(x/3.0+d/5.0);
        w[2]=df*((x/3.0 + 0.4*d)*x + d*d/7.0);
        w[3]=df*(((x/3.0 + 0.6*d)*x + 3.0*d*d/7.0)*x+d*d*d/9.0);
```

```
    } else {
        x3=(x2=x*x)*x;
        x4=x2*x2;
        y2=y*y;
        d=x-y;
        w[0]=d*((clog=log(d))-1.0);
        w[1] = -0.25*(3.0*x+y-2.0*clog*(x+y))*d;
        w[2]=(-11.0*x3+y*(6.0*x2+y*(3.0*x+2.0*y))
            +6.0*clog*(x3-y*y2))/18.0;
        w[3]=(-25.0*x4+y*(12.0*x3+y*(6.0*x2+y*
            (4.0*x+3.0*y)))+12.0*clog*(x4-(y2*y2)))/48.0;
    }
}
```

Next, we write a routine that constructs the quadrature matrix.

```
#include <cmath>
#include "nr.h"
using namespace std;

DP x;                            Communicates with kermom.

void NR::quadmx(Mat_O_DP &a)
```
Constructs in a[0..n-1][0..n-1] the quadrature matrix for an example Fredholm equation of the second kind. The nonsingular part of the kernel is computed within this routine, while the quadrature weights that integrate the singular part of the kernel are obtained via calls to wwghts. An external routine kermom, which supplies indefinite-integral moments of the singular part of the kernel, is passed to wwghts.
```
{
    const DP PI=3.14159263589793238;
    int j,k;
    DP h,xx,cx;

    int n=a.nrows();
    Vec_DP wt(n);
    h=PI/(n-1);
    for (j=0;j<n;j++) {
        x=xx=j*h;               Put x in global variable for use by kermom.
        wwghts(wt,h,kermom);
        cx=cos(xx);             Part of nonsingular kernel.
        for (k=0;k<n;k++)       Put together all the pieces of the kernel.
            a[j][k]=wt[k]*cx*cos(k*h);
        ++a[j][j];              For equations of the second kind, there is a diagonal
    }                                   piece independent of h.
}
```

Finally, we solve the linear system for any particular right-hand side, here sin x.

```
#include <iostream>
#include <iomanip>
#include <cmath>
#include "nr.h"
using namespace std;

int main(void)     // Program fredex
```
This sample program shows how to solve a Fredholm equation of the second kind using the product Nystrom method and a quadrature rule especially constructed for a particular, singular, kernel.
```
{
    const int N=40;                 Here the size of the grid is specified.
    const DP PI=3.141592653589793238;
    int j;
```

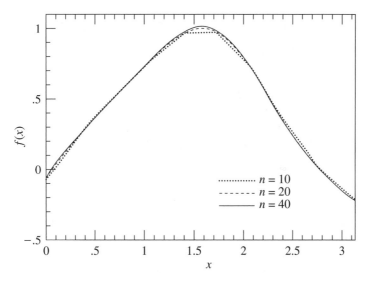

Figure 18.3.1. Solution of the example integral equation (18.3.14) with grid sizes $N = 10$, 20, and 40. The tabulated solution values have been connected by straight lines; in practice one would interpolate a small N solution more smoothly.

```
    DP d,x;
    Vec_INT indx(N);
    Vec_DP g(N);
    Mat_DP a(N,N);

    NR::quadmx(a);              Make the quadrature matrix; all the action is here.
    NR::ludcmp(a,indx,d);      Decompose the matrix.
    for (j=0;j<N;j++)          Construct the right hand side, here sin x.
        g[j]=sin(j*PI/(N-1));
    NR::lubksb(a,indx,g);      Backsubstitute.
    for (j=0;j<N;j++) {        Write out the solution.
        x=j*PI/(N-1);
        cout << fixed << setprecision(2) << setw(6) << (j+1);
        cout << setprecision(6) << setw(13) << x << setw(13) << g[j] << endl;
    }
    return 0;
}
```

With $N = 40$, this program gives accuracy at about the 10^{-5} level. The accuracy increases as N^4 (as it should for our Simpson-order quadrature scheme) *despite* the highly singular kernel. Figure 18.3.1 shows the solution obtained, also plotting the solution for smaller values of N, which are themselves seen to be remarkably faithful. Notice that the solution is smooth, even though the kernel is singular, a common occurrence.

CITED REFERENCES AND FURTHER READING:

Abramowitz, M., and Stegun, I.A. 1964, *Handbook of Mathematical Functions*, Applied Mathematics Series, Volume 55 (Washington: National Bureau of Standards; reprinted 1968 by Dover Publications, New York). [1]

Stroud, A.H., and Secrest, D. 1966, *Gaussian Quadrature Formulas* (Englewood Cliffs, NJ: Prentice-Hall). [2]

Delves, L.M., and Mohamed, J.L. 1985, *Computational Methods for Integral Equations* (Cambridge, U.K.: Cambridge University Press). [3]

Atkinson, K.E. 1976, *A Survey of Numerical Methods for the Solution of Fredholm Integral Equations of the Second Kind* (Philadelphia: S.I.A.M.). [4]

18.4 Inverse Problems and the Use of A Priori Information

Later discussion will be facilitated by some preliminary mention of a couple of mathematical points. Suppose that $\mathbf{u}$ is an "unknown" vector that we plan to determine by some minimization principle. Let $\mathcal{A}[\mathbf{u}] > 0$ and $\mathcal{B}[\mathbf{u}] > 0$ be two positive functionals of $\mathbf{u}$, so that we can try to determine $\mathbf{u}$ by either

$$\text{minimize:} \quad \mathcal{A}[\mathbf{u}] \qquad \text{or} \qquad \text{minimize:} \quad \mathcal{B}[\mathbf{u}] \tag{18.4.1}$$

(Of course these will generally give different answers for $\mathbf{u}$.) As another possibility, now suppose that we want to minimize $\mathcal{A}[\mathbf{u}]$ subject to the *constraint* that $\mathcal{B}[\mathbf{u}]$ have some particular value, say b. The method of Lagrange multipliers gives the variation

$$\frac{\delta}{\delta \mathbf{u}} \left\{ \mathcal{A}[\mathbf{u}] + \lambda_1 (\mathcal{B}[\mathbf{u}] - b) \right\} = \frac{\delta}{\delta \mathbf{u}} \left(\mathcal{A}[\mathbf{u}] + \lambda_1 \mathcal{B}[\mathbf{u}] \right) = 0 \tag{18.4.2}$$

where λ_1 is a Lagrange multiplier. Notice that b is absent in the second equality, since it doesn't depend on $\mathbf{u}$.

Next, suppose that we change our minds and decide to minimize $\mathcal{B}[\mathbf{u}]$ subject to the constraint that $\mathcal{A}[\mathbf{u}]$ have a particular value, a. Instead of equation (18.4.2) we have

$$\frac{\delta}{\delta \mathbf{u}} \left\{ \mathcal{B}[\mathbf{u}] + \lambda_2 (\mathcal{A}[\mathbf{u}] - a) \right\} = \frac{\delta}{\delta \mathbf{u}} \left(\mathcal{B}[\mathbf{u}] + \lambda_2 \mathcal{A}[\mathbf{u}] \right) = 0 \tag{18.4.3}$$

with, this time, λ_2 the Lagrange multiplier. Multiplying equation (18.4.3) by the constant $1/\lambda_2$, and identifying $1/\lambda_2$ with λ_1, we see that the actual variations are exactly the same in the two cases. Both cases will yield the same one-parameter family of solutions, say, $\mathbf{u}(\lambda_1)$. As λ_1 varies from 0 to ∞, the solution $\mathbf{u}(\lambda_1)$ varies along a so-called *trade-off curve* between the problem of minimizing $\mathcal{A}$ and the problem of minimizing $\mathcal{B}$. Any solution along this curve can equally well be thought of as either (i) a minimization of $\mathcal{A}$ for some constrained value of $\mathcal{B}$, or (ii) a minimization of $\mathcal{B}$ for some constrained value of $\mathcal{A}$, or (iii) a weighted minimization of the sum $\mathcal{A} + \lambda_1 \mathcal{B}$.

The second preliminary point has to do with *degenerate* minimization principles. In the example above, now suppose that $\mathcal{A}[\mathbf{u}]$ has the particular form

$$\mathcal{A}[\mathbf{u}] = |\mathbf{A} \cdot \mathbf{u} - \mathbf{c}|^2 \tag{18.4.4}$$

for some matrix $\mathbf{A}$ and vector $\mathbf{c}$. If $\mathbf{A}$ has fewer rows than columns, or if $\mathbf{A}$ is square but degenerate (has a nontrivial nullspace, see §2.6, especially Figure 2.6.1), then minimizing $\mathcal{A}[\mathbf{u}]$ will *not* give a unique solution for $\mathbf{u}$. (To see why, review §15.4, and note that for a "design matrix" $\mathbf{A}$ with fewer rows than columns, the matrix $\mathbf{A}^T \cdot \mathbf{A}$ in the normal equations 15.4.10 is degenerate.) *However*, if we add any multiple λ times a nondegenerate quadratic form $\mathcal{B}[\mathbf{u}]$, for example $\mathbf{u} \cdot \mathbf{H} \cdot \mathbf{u}$ with $\mathbf{H}$ a positive definite matrix, then minimization of $\mathcal{A}[\mathbf{u}] + \lambda \mathcal{B}[\mathbf{u}]$ *will* lead to a unique solution for $\mathbf{u}$. (The sum of two quadratic forms is itself a quadratic form, with the second piece guaranteeing nondegeneracy.)

We can combine these two points, for this conclusion: When a quadratic minimization principle is combined with a quadratic constraint, and both are positive, only *one* of the two need be nondegenerate for the overall problem to be well-posed. We are now equipped to face the subject of inverse problems.

The Inverse Problem with Zeroth-Order Regularization

Suppose that $u(x)$ is some unknown or underlying (u stands for both unknown and underlying!) physical process, which we hope to determine by a set of N measurements c_i, $i = 0, 1, \ldots, N - 1$. The relation between $u(x)$ and the c_i's is that each c_i measures a (hopefully distinct) aspect of $u(x)$ through its own linear response kernel r_i, and with its own measurement error n_i. In other words,

$$c_i \equiv s_i + n_i = \int r_i(x)u(x)dx + n_i \tag{18.4.5}$$

(compare this to equations 13.3.1 and 13.3.2). Within the assumption of linearity, this is quite a general formulation. The c_i's might approximate values of $u(x)$ at certain locations x_i, in which case $r_i(x)$ would have the form of a more or less narrow instrumental response centered around $x = x_i$. Or, the c_i's might "live" in an entirely different function space from $u(x)$, measuring different Fourier components of $u(x)$ for example.

The *inverse problem* is, given the c_i's, the $r_i(x)$'s, and perhaps some information about the errors n_i such as their covariance matrix

$$S_{ij} \equiv \text{Covar}[n_i, n_j] \tag{18.4.6}$$

how do we find a good statistical estimator of $u(x)$, call it $\widehat{u}(x)$?

It should be obvious that this is an ill-posed problem. After all, how can we reconstruct a whole function $\widehat{u}(x)$ from only a finite number of discrete values c_i? Yet, whether formally or informally, we do this all the time in science. We routinely measure "enough points" and then "draw a curve through them." In doing so, we are making some assumptions, either about the underlying function $u(x)$, or about the nature of the response functions $r_i(x)$, or both. Our purpose now is to formalize these assumptions, and to extend our abilities to cases where the measurements and underlying function live in quite different function spaces. (How do you "draw a curve" through a scattering of Fourier coefficients?)

We can't really want every point x of the function $\widehat{u}(x)$. We do want some large number M of discrete points x_μ, $\mu = 0, 1, \ldots, M - 1$, where M is sufficiently large, and the x_μ's are sufficiently evenly spaced, that neither $u(x)$ nor $r_i(x)$ varies much between any x_μ and $x_{\mu+1}$. (Here and following we will use Greek letters like μ to denote values in the space of the underlying process, and Roman letters like i to denote values of immediate observables.) For such a dense set of x_μ's, we can replace equation (18.4.5) by a quadrature like

$$c_i = \sum_\mu R_{i\mu} u(x_\mu) + n_i \tag{18.4.7}$$

where the $N \times M$ matrix $\mathbf{R}$ has components

$$R_{i\mu} \equiv r_i(x_\mu)(x_{\mu+1} - x_{\mu-1})/2 \tag{18.4.8}$$

(or any other simple quadrature — it rarely matters which). We will view equations (18.4.5) and (18.4.7) as being equivalent for practical purposes.

How do you solve a set of equations like equation (18.4.7) for the unknown $u(x_\mu)$'s? Here is a bad way, but one that contains the germ of some correct ideas: Form a χ^2 measure of how well a model $\widehat{u}(x)$ agrees with the measured data,

$$
\begin{aligned}
\chi^2 &= \sum_{i=0}^{N-1} \sum_{j=0}^{N-1} \left[c_i - \sum_{\mu=0}^{M-1} R_{i\mu}\widehat{u}(x_\mu) \right] S_{ij}^{-1} \left[c_j - \sum_{\mu=0}^{M-1} R_{j\mu}\widehat{u}(x_\mu) \right] \\
&\approx \sum_{i=0}^{N-1} \left[\frac{c_i - \sum_{\mu=0}^{M-1} R_{i\mu}\widehat{u}(x_\mu)}{\sigma_i} \right]^2
\end{aligned}
\tag{18.4.9}
$$

(compare with equation 15.1.5). Here $\mathbf{S}^{-1}$ is the inverse of the covariance matrix, and the approximate equality holds if you can neglect the off-diagonal covariances, with $\sigma_i \equiv (\mathrm{Covar}[i,i])^{1/2}$.

Now you can use the method of singular value decomposition (SVD) in §15.4 to find the vector $\widehat{\mathbf{u}}$ that minimizes equation (18.4.9). Don't try to use the method of normal equations; since M is greater than N they will be singular, as we already discussed. The SVD process will thus surely find a large number of zero singular values, indicative of a highly non-unique solution. Among the infinity of degenerate solutions (most of them badly behaved with arbitrarily large $\widehat{u}(x_\mu)$'s) SVD will select the one with smallest $|\widehat{\mathbf{u}}|$ in the sense of

$$
\sum_{\mu} [\widehat{u}(x_\mu)]^2 \quad \text{a minimum}
\tag{18.4.10}
$$

(look at Figure 2.6.1). This solution is often called the *principal solution*. It is a limiting case of what is called *zeroth-order regularization*, corresponding to minimizing the sum of the two positive functionals

$$
\text{minimize:} \quad \chi^2[\widehat{\mathbf{u}}] + \lambda(\widehat{\mathbf{u}} \cdot \widehat{\mathbf{u}})
\tag{18.4.11}
$$

in the limit of small λ. Below, we will learn how to do such minimizations, as well as more general ones, without the *ad hoc* use of SVD.

What happens if we determine $\widehat{\mathbf{u}}$ by equation (18.4.11) with a non-infinitesimal value of λ? First, note that if $M \gg N$ (many more unknowns than equations), then $\mathbf{u}$ will often have enough freedom to be able to make χ^2 (equation 18.4.9) quite unrealistically small, if not zero. In the language of §15.1, the number of degrees of freedom $\nu = N - M$, which is approximately the expected value of χ^2 when ν is large, is being driven down to zero (and, not meaningfully, beyond). Yet, we know that for the *true* underlying function $u(x)$, which has no adjustable parameters, the number of degrees of freedom and the expected value of χ^2 should be about $\nu \approx N$.

Increasing λ pulls the solution away from minimizing χ^2 in favor of minimizing $\widehat{\mathbf{u}} \cdot \widehat{\mathbf{u}}$. From the preliminary discussion above, we can view this as minimizing $\widehat{\mathbf{u}} \cdot \widehat{\mathbf{u}}$ subject to the *constraint* that χ^2 have some constant nonzero value. A popular choice, in fact, is to find that value of λ which yields $\chi^2 = N$, that is, to get about as much extra regularization as a plausible value of χ^2 dictates. The resulting $\widehat{u}(x)$ is called *the solution of the inverse problem with zeroth-order regularization*.

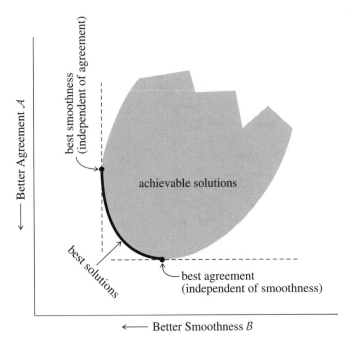

Figure 18.4.1.   Almost all inverse problem methods involve a trade-off between two optimizations: agreement between data and solution, or "sharpness" of mapping between true and estimated solution (here denoted $\mathcal{A}$), and smoothness or stability of the solution (here denoted $\mathcal{B}$). Among all possible solutions, shown here schematically as the shaded region, those on the boundary connecting the unconstrained minimum of $\mathcal{A}$ and the unconstrained minimum of $\mathcal{B}$ are the "best" solutions, in the sense that every other solution is dominated by at least one solution on the curve.

The value N is actually a surrogate for any value drawn from a Gaussian distribution with mean N and standard deviation $(2N)^{1/2}$ (the asymptotic χ^2 distribution). One might equally plausibly try two values of λ, one giving $\chi^2 = N + (2N)^{1/2}$, the other $N - (2N)^{1/2}$.

Zeroth-order regularization, though dominated by better methods, demonstrates most of the basic ideas that are used in inverse problem theory. In general, there are two positive functionals, call them $\mathcal{A}$ and $\mathcal{B}$. The first, $\mathcal{A}$, measures something like the agreement of a model to the data (e.g., χ^2), or sometimes a related quantity like the "sharpness" of the mapping between the solution and the underlying function. When $\mathcal{A}$ by itself is minimized, the agreement or sharpness becomes very good (often impossibly good), but the solution becomes unstable, wildly oscillating, or in other ways unrealistic, reflecting that $\mathcal{A}$ alone typically defines a highly degenerate minimization problem.

That is where $\mathcal{B}$ comes in. It measures something like the "smoothness" of the desired solution, or sometimes a related quantity that parametrizes the stability of the solution with respect to variations in the data, or sometimes a quantity reflecting *a priori* judgments about the likelihood of a solution. $\mathcal{B}$ is called the *stabilizing functional* or *regularizing operator*. In any case, minimizing $\mathcal{B}$ by itself is supposed to give a solution that is "smooth" or "stable" or "likely" — and that has nothing at all to do with the measured data.

The single central idea in inverse theory is the prescription

$$\text{minimize:} \quad \mathcal{A} + \lambda \mathcal{B} \quad\quad\quad (18.4.12)$$

for various values of $0 < \lambda < \infty$ along the so-called trade-off curve (see Figure 18.4.1), and then to settle on a "best" value of λ by one or another criterion, ranging from fairly objective (e.g., making $\chi^2 = N$) to entirely subjective. Successful methods, several of which we will now describe, differ as to their choices of $\mathcal{A}$ and $\mathcal{B}$, as to whether the prescription (18.4.12) yields linear or nonlinear equations, as to their recommended method for selecting a final λ, and as to their practicality for computer-intensive two-dimensional problems like image processing.

They also differ as to the philosophical baggage that they (or rather, their proponents) carry. We have thus far avoided the word "Bayesian." (Courts have consistently held that academic license does not extend to shouting "Bayesian" in a crowded lecture hall.) But it is hard, nor have we any wish, to disguise the fact that $\mathcal{B}$ has something to do with *a priori* expectation, or knowledge, of a solution, while $\mathcal{A}$ has something to do with *a posteriori* knowledge. The constant λ adjudicates a delicate compromise between the two. Some inverse methods have acquired a more Bayesian stamp than others, but we think that this is purely an accident of history. An outsider looking only at the equations that are actually solved, and not at the accompanying philosophical justifications, would have a difficult time separating the so-called Bayesian methods from the so-called empirical ones, we think.

The next three sections discuss three different approaches to the problem of inversion, which have had considerable success in different fields. All three fit within the general framework that we have outlined, but they are quite different in detail and in implementation.

CITED REFERENCES AND FURTHER READING:

Craig, I.J.D., and Brown, J.C. 1986, *Inverse Problems in Astronomy* (Bristol, U.K.: Adam Hilger).

Twomey, S. 1977, *Introduction to the Mathematics of Inversion in Remote Sensing and Indirect Measurements* (Amsterdam: Elsevier).

Tikhonov, A.N., and Arsenin, V.Y. 1977, *Solutions of Ill-Posed Problems* (New York: Wiley).

Tikhonov, A.N., and Goncharsky, A.V. (eds.) 1987, *Ill-Posed Problems in the Natural Sciences* (Moscow: MIR).

Parker, R.L. 1977, *Annual Review of Earth and Planetary Science*, vol. 5, pp. 35–64.

Frieden, B.R. 1975, in *Picture Processing and Digital Filtering*, T.S. Huang, ed. (New York: Springer-Verlag).

Tarantola, A. 1987, *Inverse Problem Theory* (Amsterdam: Elsevier).

Baumeister, J. 1987, *Stable Solution of Inverse Problems* (Braunschweig, Germany: Friedr. Vieweg & Sohn) [mathematically oriented].

Titterington, D.M. 1985, *Astronomy and Astrophysics*, vol. 144, pp. 381–387.

Jeffrey, W., and Rosner, R. 1986, *Astrophysical Journal*, vol. 310, pp. 463–472.

18.5 Linear Regularization Methods

What we will call *linear regularization* is also called the *Phillips-Twomey method* [1,2], the *constrained linear inversion method* [3], the *method of regularization* [4], and *Tikhonov-Miller regularization* [5-7]. (It probably has other names also, since it is so obviously a good idea.) In its simplest form, the method is an immediate generalization of zeroth-order regularization (equation 18.4.11, above). As before, the functional $\mathcal{A}$ is taken to be the χ^2 deviation, equation (18.4.9), but the functional $\mathcal{B}$ is replaced by more sophisticated measures of smoothness that derive from first or higher derivatives.

For example, suppose that your *a priori* belief is that a credible $u(x)$ is not too different from a constant. Then a reasonable functional to minimize is

$$\mathcal{B} \propto \int [\widehat{u}'(x)]^2 dx \propto \sum_{\mu=0}^{M-2} [\widehat{u}_\mu - \widehat{u}_{\mu+1}]^2 \qquad (18.5.1)$$

since it is nonnegative and equal to zero only when $\widehat{u}(x)$ is constant. Here $\widehat{u}_\mu \equiv \widehat{u}(x_\mu)$, and the second equality (proportionality) assumes that the x_μ's are uniformly spaced. We can write the second form of $\mathcal{B}$ as

$$\mathcal{B} = |\mathbf{B} \cdot \widehat{\mathbf{u}}|^2 = \widehat{\mathbf{u}} \cdot (\mathbf{B}^T \cdot \mathbf{B}) \cdot \widehat{\mathbf{u}} \equiv \widehat{\mathbf{u}} \cdot \mathbf{H} \cdot \widehat{\mathbf{u}} \qquad (18.5.2)$$

where $\widehat{\mathbf{u}}$ is the vector of components $\widehat{u}_\mu$, $\mu = 0, \ldots, M-1$, $\mathbf{B}$ is the $(M-1) \times M$ first difference matrix

$$\mathbf{B} = \begin{pmatrix} -1 & 1 & 0 & 0 & 0 & 0 & 0 & \cdots & 0 \\ 0 & -1 & 1 & 0 & 0 & 0 & 0 & \cdots & 0 \\ \vdots & & & & \ddots & & & & \vdots \\ 0 & \cdots & 0 & 0 & 0 & 0 & -1 & 1 & 0 \\ 0 & \cdots & 0 & 0 & 0 & 0 & 0 & -1 & 1 \end{pmatrix} \qquad (18.5.3)$$

and $\mathbf{H}$ is the $M \times M$ matrix

$$\mathbf{H} = \mathbf{B}^T \cdot \mathbf{B} = \begin{pmatrix} 1 & -1 & 0 & 0 & 0 & 0 & 0 & \cdots & 0 \\ -1 & 2 & -1 & 0 & 0 & 0 & 0 & \cdots & 0 \\ 0 & -1 & 2 & -1 & 0 & 0 & 0 & \cdots & 0 \\ \vdots & & & & \ddots & & & & \vdots \\ 0 & \cdots & 0 & 0 & 0 & -1 & 2 & -1 & 0 \\ 0 & \cdots & 0 & 0 & 0 & 0 & -1 & 2 & -1 \\ 0 & \cdots & 0 & 0 & 0 & 0 & 0 & -1 & 1 \end{pmatrix} \qquad (18.5.4)$$

Note that $\mathbf{B}$ has one fewer row than column. It follows that the symmetric $\mathbf{H}$ is degenerate; it has exactly one zero eigenvalue corresponding to the *value* of a constant function, any one of which makes $\mathcal{B}$ exactly zero.

If, just as in §15.4, we write

$$A_{i\mu} \equiv R_{i\mu}/\sigma_i \qquad b_i \equiv c_i/\sigma_i \qquad (18.5.5)$$

then, using equation (18.4.9), the minimization principle (18.4.12) is

$$\text{minimize:} \quad \mathcal{A} + \lambda\mathcal{B} = |\mathbf{A} \cdot \widehat{\mathbf{u}} - \mathbf{b}|^2 + \lambda\widehat{\mathbf{u}} \cdot \mathbf{H} \cdot \widehat{\mathbf{u}} \tag{18.5.6}$$

This can readily be reduced to a linear set of *normal equations*, just as in §15.4: The components $\widehat{u}_\mu$ of the solution satisfy the set of M equations in M unknowns,

$$\sum_\rho \left[\left(\sum_i A_{i\mu} A_{i\rho} \right) + \lambda H_{\mu\rho} \right] \widehat{u}_\rho = \sum_i A_{i\mu} b_i \qquad \mu = 0, 1, \ldots, M - 1 \tag{18.5.7}$$

or, in vector notation,

$$(\mathbf{A}^T \cdot \mathbf{A} + \lambda\mathbf{H}) \cdot \widehat{\mathbf{u}} = \mathbf{A}^T \cdot \mathbf{b} \tag{18.5.8}$$

Equations (18.5.7) or (18.5.8) can be solved by the standard techniques of Chapter 2, e.g., *LU* decomposition. The usual warnings about normal equations being ill-conditioned do not apply, since the whole purpose of the λ term is to cure that same ill-conditioning. Note, however, that the λ term *by itself* is ill-conditioned, since it does not select a preferred constant value. You hope your data can at least do *that*!

Although inversion of the matrix $(\mathbf{A}^T \cdot \mathbf{A} + \lambda\mathbf{H})$ is not generally the best way to solve for $\widehat{\mathbf{u}}$, let us digress to write the solution to equation (18.5.8) schematically as

$$\widehat{\mathbf{u}} = \left(\frac{1}{\mathbf{A}^T \cdot \mathbf{A} + \lambda\mathbf{H}} \cdot \mathbf{A}^T \cdot \mathbf{A} \right) \mathbf{A}^{-1} \cdot \mathbf{b} \qquad \text{(schematic only!)} \tag{18.5.9}$$

where the identity matrix in the form $\mathbf{A} \cdot \mathbf{A}^{-1}$ has been inserted. This is schematic not only because the matrix inverse is fancifully written as a denominator, but also because, in general, the inverse matrix $\mathbf{A}^{-1}$ does not exist. However, it is illuminating to compare equation (18.5.9) with equation (13.3.6) for optimal or Wiener filtering, or with equation (13.6.6) for general linear prediction. One sees that $\mathbf{A}^T \cdot \mathbf{A}$ plays the role of S^2, the signal power or autocorrelation, while $\lambda\mathbf{H}$ plays the role of N^2, the noise power or autocorrelation. The term in parentheses in equation (18.5.9) is something like an optimal filter, whose effect is to pass the ill-posed inverse $\mathbf{A}^{-1} \cdot \mathbf{b}$ through unmodified when $\mathbf{A}^T \cdot \mathbf{A}$ is sufficiently large, but to suppress it when $\mathbf{A}^T \cdot \mathbf{A}$ is small.

The above choices of $\mathbf{B}$ and $\mathbf{H}$ are only the simplest in an obvious sequence of derivatives. If your *a priori* belief is that a *linear* function is a good approximation to $u(x)$, then minimize

$$\mathcal{B} \propto \int [\widehat{u}''(x)]^2 dx \propto \sum_{\mu=0}^{M-3} [-\widehat{u}_\mu + 2\widehat{u}_{\mu+1} - \widehat{u}_{\mu+2}]^2 \tag{18.5.10}$$

implying

$$\mathbf{B} = \begin{pmatrix} -1 & 2 & -1 & 0 & 0 & 0 & 0 & \cdots & 0 \\ 0 & -1 & 2 & -1 & 0 & 0 & 0 & \cdots & 0 \\ \vdots & & & & \ddots & & & & \vdots \\ 0 & \cdots & 0 & 0 & 0 & -1 & 2 & -1 & 0 \\ 0 & \cdots & 0 & 0 & 0 & 0 & -1 & 2 & -1 \end{pmatrix} \tag{18.5.11}$$

and

$$
\mathbf{H} = \mathbf{B}^T \cdot \mathbf{B} =
\begin{pmatrix}
1 & -2 & 1 & 0 & 0 & 0 & 0 & \cdots & 0 \\
-2 & 5 & -4 & 1 & 0 & 0 & 0 & \cdots & 0 \\
1 & -4 & 6 & -4 & 1 & 0 & 0 & \cdots & 0 \\
0 & 1 & -4 & 6 & -4 & 1 & 0 & \cdots & 0 \\
\vdots & & & & \ddots & & & & \vdots \\
0 & \cdots & 0 & 1 & -4 & 6 & -4 & 1 & 0 \\
0 & \cdots & 0 & 0 & 1 & -4 & 6 & -4 & 1 \\
0 & \cdots & 0 & 0 & 0 & 1 & -4 & 5 & -2 \\
0 & \cdots & 0 & 0 & 0 & 0 & 1 & -2 & 1
\end{pmatrix}
\tag{18.5.12}
$$

This $\mathbf{H}$ has two zero eigenvalues, corresponding to the two undetermined parameters of a linear function.

If your *a priori* belief is that a *quadratic* function is preferable, then minimize

$$
\mathcal{B} \propto \int [\widehat{u}'''(x)]^2 dx \propto \sum_{\mu=0}^{M-4} [-\widehat{u}_\mu + 3\widehat{u}_{\mu+1} - 3\widehat{u}_{\mu+2} + \widehat{u}_{\mu+3}]^2
\tag{18.5.13}
$$

with

$$
\mathbf{B} =
\begin{pmatrix}
-1 & 3 & -3 & 1 & 0 & 0 & 0 & \cdots & 0 \\
0 & -1 & 3 & -3 & 1 & 0 & 0 & \cdots & 0 \\
\vdots & & & & \ddots & & & & \vdots \\
0 & \cdots & 0 & 0 & -1 & 3 & -3 & 1 & 0 \\
0 & \cdots & 0 & 0 & 0 & -1 & 3 & -3 & 1
\end{pmatrix}
\tag{18.5.14}
$$

and now

$$
\mathbf{H} =
\begin{pmatrix}
1 & -3 & 3 & -1 & 0 & 0 & 0 & 0 & 0 & \cdots & 0 \\
-3 & 10 & -12 & 6 & -1 & 0 & 0 & 0 & 0 & \cdots & 0 \\
3 & -12 & 19 & -15 & 6 & -1 & 0 & 0 & 0 & \cdots & 0 \\
-1 & 6 & -15 & 20 & -15 & 6 & -1 & 0 & 0 & \cdots & 0 \\
0 & -1 & 6 & -15 & 20 & -15 & 6 & -1 & 0 & \cdots & 0 \\
\vdots & & & & & \ddots & & & & & \vdots \\
0 & \cdots & 0 & -1 & 6 & -15 & 20 & -15 & 6 & -1 & 0 \\
0 & \cdots & 0 & 0 & -1 & 6 & -15 & 20 & -15 & 6 & -1 \\
0 & \cdots & 0 & 0 & 0 & -1 & 6 & -15 & 19 & -12 & 3 \\
0 & \cdots & 0 & 0 & 0 & 0 & -1 & 6 & -12 & 10 & -3 \\
0 & \cdots & 0 & 0 & 0 & 0 & 0 & -1 & 3 & -3 & 1
\end{pmatrix}
\tag{18.5.15}
$$

(We'll leave the calculation of cubics and above to the compulsive reader.)

Notice that you can regularize with "closeness to a differential equation," if you want. Just pick $\mathbf{B}$ to be the appropriate sum of finite-difference operators (the coefficients can depend on x), and calculate $\mathbf{H} = \mathbf{B}^T \cdot \mathbf{B}$. You don't need to know the values of your boundary conditions, since $\mathbf{B}$ can have fewer rows than columns, as above; hopefully, your data will determine them. Of course, if you do know some boundary conditions, you can build these into $\mathbf{B}$ too.

With all the proportionality signs above, you may have lost track of what actual value of λ to try first. A simple trick for at least getting "on the map" is to first try

$$\lambda = \mathrm{Tr}(\mathbf{A}^T \cdot \mathbf{A})/\mathrm{Tr}(\mathbf{H}) \qquad (18.5.16)$$

where Tr is the trace of the matrix (sum of diagonal components). This choice will tend to make the two parts of the minimization have comparable weights, and you can adjust from there.

As for what is the "correct" value of λ, an objective criterion, if you know your errors σ_i with reasonable accuracy, is to make χ^2 (that is, $|\mathbf{A} \cdot \widehat{\mathbf{u}} - \mathbf{b}|^2$) equal to N, the number of measurements. We remarked above on the twin acceptable choices $N \pm (2N)^{1/2}$. A subjective criterion is to pick any value that you like in the range $0 < \lambda < \infty$, depending on your relative degree of belief in the *a priori* and *a posteriori* evidence. (Yes, people actually do that. Don't blame us.)

Two-Dimensional Problems and Iterative Methods

Up to now our notation has been indicative of a one-dimensional problem, finding $\widehat{u}(x)$ or $\widehat{u}_\mu = \widehat{u}(x_\mu)$. However, all of the discussion easily generalizes to the problem of estimating a two-dimensional set of unknowns $\widehat{u}_{\mu\kappa}$, $\mu = 0, \ldots,$ $M - 1$, $\kappa = 0, \ldots, K - 1$, corresponding, say, to the pixel intensities of a measured image. In this case, equation (18.5.8) is still the one we want to solve.

In image processing, it is usual to have the same number of input pixels in a measured "raw" or "dirty" image as desired "clean" pixels in the processed output image, so the matrices $\mathbf{R}$ and $\mathbf{A}$ (equation 18.5.5) are square and of size $MK \times MK$. $\mathbf{A}$ is typically much too large to represent as a full matrix, but often it is either (i) sparse, with coefficients blurring an underlying pixel (i, j) only into measurements $(i \pm \text{few}, j \pm \text{few})$, or (ii) translationally invariant, so that $A_{(i,j)(\mu,\nu)} = A(i - \mu, j - \nu)$. Both of these situations lead to tractable problems.

In the case of translational invariance, fast Fourier transforms (FFTs) are the obvious method of choice. The general linear relation between underlying function and measured values (18.4.7) now becomes a discrete convolution like equation (13.1.1). If $\mathbf{k}$ denotes a two-dimensional wave-vector, then the two-dimensional FFT takes us back and forth between the transform pairs

$$A(i - \mu, j - \nu) \iff \widetilde{\mathbf{A}}(\mathbf{k}) \qquad b_{(i,j)} \iff \widetilde{b}(\mathbf{k}) \qquad \widehat{u}_{(i,j)} \iff \widetilde{u}(\mathbf{k}) \quad (18.5.17)$$

We also need a regularization or smoothing operator $\mathbf{B}$ and the derived $\mathbf{H} = \mathbf{B}^T \cdot \mathbf{B}$. One popular choice for $\mathbf{B}$ is the five-point finite-difference approximation of the Laplacian operator, that is, the difference between the value of each point and the average of its four Cartesian neighbors. In Fourier space, this choice implies,

$$\widetilde{B}(\mathbf{k}) \propto \sin^2(\pi k_1/M)\sin^2(\pi k_2/K)$$
$$\widetilde{H}(\mathbf{k}) \propto \sin^4(\pi k_1/M)\sin^4(\pi k_2/K) \qquad (18.5.18)$$

In Fourier space, equation (18.5.7) is merely algebraic, with solution

$$\widetilde{u}(\mathbf{k}) = \frac{\widetilde{A}^*(\mathbf{k})\widetilde{b}(\mathbf{k})}{|\widetilde{A}(\mathbf{k})|^2 + \lambda\widetilde{H}(\mathbf{k})} \qquad (18.5.19)$$

where asterisk denotes complex conjugation. You can make use of the FFT routines for real data in §12.5.

Turn now to the case where $\mathbf{A}$ is not translationally invariant. Direct solution of (18.5.8) is now hopeless, since the matrix $\mathbf{A}$ is just too large. We need some kind of iterative scheme.

One way to proceed is to use the full machinery of the conjugate gradient method in §10.6 to find the minimum of $\mathcal{A} + \lambda\mathcal{B}$, equation (18.5.6). Of the various methods in Chapter 10, conjugate gradient is the unique best choice because (i) it does not require storage of a Hessian matrix, which would be infeasible here, and (ii) it does exploit gradient information, which we can readily compute: The gradient of equation (18.5.6) is

$$\nabla(\mathcal{A} + \lambda\mathcal{B}) = 2[(\mathbf{A}^T \cdot \mathbf{A} + \lambda\mathbf{H}) \cdot \widehat{\mathbf{u}} - \mathbf{A}^T \cdot \mathbf{b}] \tag{18.5.20}$$

(cf. 18.5.8). Evaluation of both the function and the gradient should of course take advantage of the sparsity of $\mathbf{A}$, for example via the routines `sprsax` and `sprstx` in §2.7. We will discuss the conjugate gradient technique further in §18.7, in the context of the (nonlinear) maximum entropy method. Some of that discussion can apply here as well.

The conjugate gradient method notwithstanding, application of the unsophisticated steepest descent method (see §10.6) can sometimes produce useful results, particularly when combined with projections onto convex sets (see below). If the solution after k iterations is denoted $\widehat{\mathbf{u}}^{(k)}$, then after $k + 1$ iterations we have

$$\widehat{\mathbf{u}}^{(k+1)} = [\mathbf{1} - \epsilon(\mathbf{A}^T \cdot \mathbf{A} + \lambda\mathbf{H})] \cdot \widehat{\mathbf{u}}^{(k)} + \epsilon\mathbf{A}^T \cdot \mathbf{b} \tag{18.5.21}$$

Here ϵ is a parameter that dictates how far to move in the downhill gradient direction. The method converges when ϵ is small enough, in particular satisfying

$$0 < \epsilon < \frac{2}{\text{max eigenvalue } (\mathbf{A}^T \cdot \mathbf{A} + \lambda\mathbf{H})} \tag{18.5.22}$$

There exist complicated schemes for finding optimal values or sequences for ϵ, see [7]; or, one can adopt an experimental approach, evaluating (18.5.6) to be sure that downhill steps are in fact being taken.

In those image processing problems where the final measure of success is somewhat subjective (e.g., "how good does the picture look?"), iteration (18.5.21) sometimes produces significantly improved images long before convergence is achieved. This probably accounts for much of its use, since its mathematical convergence is extremely slow. In fact, (18.5.21) can be used with $\mathbf{H} = 0$, in which case the solution is not regularized at all, and full convergence would be disastrous! This is called *Van Cittert's method* and goes back to the 1930s. A number of iterations the order of 1000 is not uncommon [7].

Deterministic Constraints: Projections onto Convex Sets

A set of possible underlying functions (or images) $\{\widehat{\mathbf{u}}\}$ is said to be *convex* if, for any two elements $\widehat{\mathbf{u}}_a$ and $\widehat{\mathbf{u}}_b$ in the set, all the linearly interpolated combinations

$$(1 - \eta)\widehat{\mathbf{u}}_a + \eta\widehat{\mathbf{u}}_b \qquad 0 \le \eta \le 1 \tag{18.5.23}$$

are also in the set. Many *deterministic constraints* that one might want to impose on the solution $\widehat{\mathbf{u}}$ to an inverse problem in fact define convex sets, for example:

- positivity
- compact support (i.e., zero value outside of a certain region)
- known bounds (i.e., $u_L(x) \leq \widehat{u}(x) \leq u_U(x)$ for specified functions u_L and u_U).

(In this last case, the bounds might be related to an initial estimate and its error bars, e.g., $\widehat{u}_0(x) \pm \gamma\sigma(x)$, where γ is of order 1 or 2.) Notice that these, and similar, constraints can be either in the image space, or in the Fourier transform space, or (in fact) in the space of any linear transformation of $\widehat{u}$.

If C_i is a convex set, then $\mathcal{P}_i$ is called a *nonexpansive projection operator* onto that set if (i) $\mathcal{P}_i$ leaves unchanged any $\widehat{u}$ already in C_i, and (ii) $\mathcal{P}_i$ maps any $\widehat{u}$ outside C_i to the *closest* element of C_i, in the sense that

$$|\mathcal{P}_i\widehat{u} - \widehat{u}| \leq |\widehat{u}_a - \widehat{u}| \quad \text{for all } \widehat{u}_a \text{ in } C_i \qquad (18.5.24)$$

While this definition sounds complicated, examples are very simple: A nonexpansive projection onto the set of positive $\widehat{u}$'s is "set all negative components of $\widehat{u}$ equal to zero." A nonexpansive projection onto the set of $\widehat{u}(x)$'s bounded by $u_L(x) \leq \widehat{u}(x) \leq u_U(x)$ is "set all values less than the lower bound equal to that bound, and set all values greater than the upper bound equal to *that* bound." A nonexpansive projection onto functions with compact support is "zero the values outside of the region of support."

The usefulness of these definitions is the following remarkable theorem: Let C be the intersection of m convex sets $C_1, C_2, \ldots, C_m$. Then the iteration

$$\widehat{u}^{(k+1)} = (\mathcal{P}_1\mathcal{P}_2\cdots\mathcal{P}_m)\widehat{u}^{(k)} \qquad (18.5.25)$$

will converge to C from all starting points, as $k \to \infty$. Also, if C is empty (there is no intersection), then the iteration will have no limit point. Application of this theorem is called the *method of projections onto convex sets* or sometimes *POCS* [7].

A generalization of the POCS theorem is that the $\mathcal{P}_i$'s can be replaced by a set of $\mathcal{T}_i$'s,

$$\mathcal{T}_i \equiv \mathbf{1} + \beta_i(\mathcal{P}_i - \mathbf{1}) \qquad 0 < \beta_i < 2 \qquad (18.5.26)$$

A well-chosen set of β_i's can accelerate the convergence to the intersection set C.

Some inverse problems can be completely solved by iteration (18.5.25) alone! For example, a problem that occurs in both astronomical imaging and X-ray diffraction work is to recover an image given only the *modulus* of its Fourier transform (equivalent to its power spectrum or autocorrelation) and not the *phase*. Here three convex sets can be utilized: the set of all images whose Fourier transform has the specified modulus to within specified error bounds; the set of all positive images; and the set of all images with zero intensity outside of some specified region. In this case the POCS iteration (18.5.25) cycles among these three, imposing each constraint in turn; FFTs are used to get in and out of Fourier space each time the Fourier constraint is imposed.

The specific application of POCS to constraints alternately in the spatial and Fourier domains is also known as the *Gerchberg-Saxton* algorithm [8]. While this algorithm is non-expansive, and is frequently convergent in practice, it has not been proved to converge in all cases [9]. In the phase-retrieval problem mentioned above, the algorithm often "gets stuck" on a plateau for many iterations before making sudden, dramatic improvements. As many as 10^4 to 10^5 iterations are sometimes necessary. (For "unsticking" procedures, see [10].) The uniqueness of the solution is also not well understood, although for two-dimensional images of reasonable complexity it is believed to be unique.

Deterministic constraints can be incorporated, via projection operators, into iterative methods of linear regularization. In particular, rearranging terms somewhat, we can write the iteration (18.5.21) as

$$\widehat{\mathbf{u}}^{(k+1)} = (\mathbf{1} - \epsilon \lambda \mathbf{H}) \cdot \widehat{\mathbf{u}}^{(k)} + \epsilon \mathbf{A}^T \cdot (\mathbf{b} - \mathbf{A} \cdot \widehat{\mathbf{u}}^{(k)}) \qquad (18.5.27)$$

If the iteration is modified by the insertion of projection operators at each step

$$\widehat{\mathbf{u}}^{(k+1)} = (\mathcal{P}_1 \mathcal{P}_2 \cdots \mathcal{P}_m)[(\mathbf{1} - \epsilon \lambda \mathbf{H}) \cdot \widehat{\mathbf{u}}^{(k)} + \epsilon \mathbf{A}^T \cdot (\mathbf{b} - \mathbf{A} \cdot \widehat{\mathbf{u}}^{(k)})] \qquad (18.5.28)$$

(or, instead of $\mathcal{P}_i$'s, the $\mathcal{T}_i$ operators of equation 18.5.26), then it can be shown that the convergence condition (18.5.22) is unmodified, and the iteration will converge to minimize the quadratic functional (18.5.6) subject to the desired nonlinear deterministic constraints. See [7] for references to more sophisticated, and faster converging, iterations along these lines.

CITED REFERENCES AND FURTHER READING:

Phillips, D.L. 1962, *Journal of the Association for Computing Machinery*, vol. 9, pp. 84–97. [1]

Twomey, S. 1963, *Journal of the Association for Computing Machinery*, vol. 10, pp. 97–101. [2]

Twomey, S. 1977, *Introduction to the Mathematics of Inversion in Remote Sensing and Indirect Measurements* (Amsterdam: Elsevier). [3]

Craig, I.J.D., and Brown, J.C. 1986, *Inverse Problems in Astronomy* (Bristol, U.K.: Adam Hilger). [4]

Tikhonov, A.N., and Arsenin, V.Y. 1977, *Solutions of Ill-Posed Problems* (New York: Wiley). [5]

Tikhonov, A.N., and Goncharsky, A.V. (eds.) 1987, *Ill-Posed Problems in the Natural Sciences* (Moscow: MIR).

Miller, K. 1970, *SIAM Journal on Mathematical Analysis*, vol. 1, pp. 52–74. [6]

Schafer, R.W., Mersereau, R.M., and Richards, M.A. 1981, *Proceedings of the IEEE*, vol. 69, pp. 432–450.

Biemond, J., Lagendijk, R.L., and Mersereau, R.M. 1990, *Proceedings of the IEEE*, vol. 78, pp. 856–883. [7]

Gerchberg, R.W., and Saxton, W.O. 1972, *Optik*, vol. 35, pp. 237–246. [8]

Fienup, J.R. 1982, *Applied Optics*, vol. 15, pp. 2758–2769. [9]

Fienup, J.R., and Wackerman, C.C. 1986, *Journal of the Optical Society of America A*, vol. 3, pp. 1897–1907. [10]

18.6 Backus-Gilbert Method

The *Backus-Gilbert method* [1,2] (see, e.g., [3] or [4] for summaries) differs from other regularization methods in the nature of its functionals $\mathcal{A}$ and $\mathcal{B}$. For $\mathcal{B}$, the method seeks to maximize the *stability* of the solution $\widehat{u}(x)$ rather than, in the first instance, its smoothness. That is,

$$\mathcal{B} \equiv \text{Var}[\widehat{u}(x)] \tag{18.6.1}$$

is used as a measure of how much the solution $\widehat{u}(x)$ varies as the data vary within their measurement errors. Note that this variance is not the expected deviation of $\widehat{u}(x)$ from the true $u(x)$ — that will be constrained by $\mathcal{A}$ — but rather measures the expected experiment-to-experiment scatter among estimates $\widehat{u}(x)$ if the whole experiment were to be repeated many times.

For $\mathcal{A}$ the Backus-Gilbert method looks at the relationship between the solution $\widehat{u}(x)$ and the true function $u(x)$, and seeks to make the mapping between these as close to the identity map as possible in the limit of error-free data. The method is linear, so the relationship between $\widehat{u}(x)$ and $u(x)$ can be written as

$$\widehat{u}(x) = \int \widehat{\delta}(x, x') u(x') dx' \tag{18.6.2}$$

for some so-called *resolution function* or *averaging kernel* $\widehat{\delta}(x, x')$. The Backus-Gilbert method seeks to minimize the width or *spread* of $\widehat{\delta}$ (that is, maximize the resolving power). $\mathcal{A}$ is chosen to be some positive measure of the spread.

While Backus-Gilbert's philosophy is thus rather different from that of Phillips-Twomey and related methods, in practice the differences between the methods are less than one might think. A *stable* solution is almost inevitably bound to be *smooth*: The wild, unstable oscillations that result from an unregularized solution are always exquisitely sensitive to small changes in the data. Likewise, making $\widehat{u}(x)$ close to $u(x)$ inevitably will bring error-free data into agreement with the model. Thus $\mathcal{A}$ and $\mathcal{B}$ play roles closely analogous to their corresponding roles in the previous two sections.

The principal advantage of the Backus-Gilbert formulation is that it gives good control over just those properties that it seeks to measure, namely stability and resolving power. Moreover, in the Backus-Gilbert method, the choice of λ (playing its usual role of compromise between $\mathcal{A}$ and $\mathcal{B}$) is conventionally made, or at least can easily be made, *before* any actual data are processed. One's uneasiness at making a *post hoc*, and therefore potentially subjectively biased, choice of λ is thus removed. Backus-Gilbert is often recommended as the method of choice for designing, and predicting the performance of, experiments that require data inversion.

Let's see how this all works. Starting with equation (18.4.5),

$$c_i \equiv s_i + n_i = \int r_i(x) u(x) dx + n_i \tag{18.6.3}$$

and building in linearity from the start, we seek a set of *inverse response kernels* $q_i(x)$ such that

$$\widehat{u}(x) = \sum_i q_i(x) c_i \tag{18.6.4}$$

is the desired estimator of $u(x)$. It is useful to define the integrals of the response kernels for each data point,

$$R_i \equiv \int r_i(x)dx \tag{18.6.5}$$

Substituting equation (18.6.4) into equation (18.6.3), and comparing with equation (18.6.2), we see that

$$\widehat{\delta}(x, x') = \sum_i q_i(x)r_i(x') \tag{18.6.6}$$

We can require this averaging kernel to have unit area at every x, giving

$$1 = \int \widehat{\delta}(x, x')dx' = \sum_i q_i(x) \int r_i(x')dx' = \sum_i q_i(x)R_i \equiv \mathbf{q}(x) \cdot \mathbf{R} \quad (18.6.7)$$

where $\mathbf{q}(x)$ and $\mathbf{R}$ are each vectors of length N, the number of measurements.

Standard propagation of errors, and equation (18.6.1), give

$$\mathcal{B} = \mathrm{Var}[\widehat{u}(x)] = \sum_i \sum_j q_i(x)S_{ij}q_j(x) = \mathbf{q}(x) \cdot \mathbf{S} \cdot \mathbf{q}(x) \tag{18.6.8}$$

where S_{ij} is the covariance matrix (equation 18.4.6). If one can neglect off-diagonal covariances (as when the errors on the c_i's are independent), then $S_{ij} = \delta_{ij}\sigma_i^2$ is diagonal.

We now need to define a measure of the width or spread of $\widehat{\delta}(x, x')$ at each value of x. While many choices are possible, Backus and Gilbert choose the second moment of its square. This measure becomes the functional $\mathcal{A}$,

$$
\begin{aligned}
\mathcal{A} \equiv w(x) &= \int (x' - x)^2 [\widehat{\delta}(x, x')]^2 dx' \\
&= \sum_i \sum_j q_i(x)W_{ij}(x)q_j(x) \equiv \mathbf{q}(x) \cdot \mathbf{W}(x) \cdot \mathbf{q}(x)
\end{aligned}
\tag{18.6.9}
$$

where we have here used equation (18.6.6) and defined the *spread matrix* $\mathbf{W}(x)$ by

$$W_{ij}(x) \equiv \int (x' - x)^2 r_i(x')r_j(x')dx' \tag{18.6.10}$$

The functions $q_i(x)$ are now determined by the minimization principle

$$\text{minimize:} \quad \mathcal{A} + \lambda\mathcal{B} = \mathbf{q}(x) \cdot \big[\mathbf{W}(x) + \lambda\mathbf{S}\big] \cdot \mathbf{q}(x) \tag{18.6.11}$$

subject to the constraint (18.6.7) that $\mathbf{q}(x) \cdot \mathbf{R} = 1$.

The solution of equation (18.6.11) is

$$\mathbf{q}(x) = \frac{[\mathbf{W}(x) + \lambda\mathbf{S}]^{-1} \cdot \mathbf{R}}{\mathbf{R} \cdot [\mathbf{W}(x) + \lambda\mathbf{S}]^{-1} \cdot \mathbf{R}} \tag{18.6.12}$$

(Reference [4] gives an accessible proof.)   For any particular data set $\mathbf{c}$ (set of measurements c_i), the solution $\widehat{u}(x)$ is thus

$$\widehat{u}(x) = \frac{\mathbf{c} \cdot [\mathbf{W}(x) + \lambda \mathbf{S}]^{-1} \cdot \mathbf{R}}{\mathbf{R} \cdot [\mathbf{W}(x) + \lambda \mathbf{S}]^{-1} \cdot \mathbf{R}} \qquad (18.6.13)$$

(Don't let this notation mislead you into inverting the full matrix $\mathbf{W}(x) + \lambda \mathbf{S}$. You only need to solve the linear system $(\mathbf{W}(x) + \lambda \mathbf{S}) \cdot \mathbf{y} = \mathbf{R}$ for the vector $\mathbf{y}$, and then substitute $\mathbf{y}$ into both the numerators and denominators of 18.6.12 or 18.6.13.)

Equations (18.6.12) and (18.6.13) have a completely different character from the linearly regularized solutions to (18.5.7) and (18.5.8). The vectors and matrices in (18.6.12) all have size N, the number of measurements. There is no discretization of the underlying variable x, so M does not come into play at all. One solves a different $N \times N$ set of linear equations for each desired value of x. By contrast, in (18.5.8), one solves an $M \times M$ linear set, but only once. In general, the computational burden of repeatedly solving linear systems makes the Backus-Gilbert method unsuitable for other than one-dimensional problems.

How does one choose λ within the Backus-Gilbert scheme?   As already mentioned, you can (in some cases *should*) make the choice *before* you see any actual data. For a given trial value of λ, and for a sequence of x's, use equation (18.6.12) to calculate $\mathbf{q}(x)$; then use equation (18.6.6) to plot the resolution functions $\widehat{\delta}(x, x')$ as a function of x'. These plots will exhibit the amplitude with which different underlying values x' contribute to the point $\widehat{u}(x)$ of your estimate. For the same value of λ, also plot the function $\sqrt{\text{Var}[\widehat{u}(x)]}$ using equation (18.6.8). (You need an estimate of your measurement covariance matrix for this.)

As you change λ you will see very explicitly the trade-off between resolution and stability. Pick the value that meets your needs. You can even choose λ to be a function of x, $\lambda = \lambda(x)$, in equations (18.6.12) and (18.6.13), should you desire to do so. (This is one benefit of solving a separate set of equations for each x.) For the chosen value or values of λ, you now have a quantitative understanding of your inverse solution procedure. This can prove invaluable if — once you are processing real data — you need to judge whether a particular feature, a spike or jump for example, is genuine, and/or is actually resolved. The Backus-Gilbert method has found particular success among geophysicists, who use it to obtain information about the structure of the Earth (e.g., density run with depth) from seismic travel time data.

CITED REFERENCES AND FURTHER READING:

Backus, G.E., and Gilbert, F. 1968, *Geophysical Journal of the Royal Astronomical Society*, vol. 16, pp. 169–205. [1]

Backus, G.E., and Gilbert, F. 1970, *Philosophical Transactions of the Royal Society of London A*, vol. 266, pp. 123–192. [2]

Parker, R.L. 1977, *Annual Review of Earth and Planetary Science*, vol. 5, pp. 35–64. [3]

Loredo, T.J., and Epstein, R.I. 1989, *Astrophysical Journal*, vol. 336, pp. 896–919. [4]

18.7 Maximum Entropy Image Restoration

Above, we commented that the association of certain inversion methods with Bayesian arguments is more historical accident than intellectual imperative. *Maximum entropy methods*, so-called, are notorious in this regard; to summarize these methods without some, at least introductory, Bayesian invocations would be to serve a steak without the sizzle, or a sundae without the cherry. We should also comment in passing that the connection between maximum entropy inversion methods, considered here, and maximum entropy spectral estimation, discussed in §13.7, is rather abstract. For practical purposes the two techniques, though both named *maximum entropy method* or *MEM*, are unrelated.

Bayes' Theorem, which follows from the standard axioms of probability, relates the conditional probabilities of two events, say A and B:

$$\text{Prob}(A|B) = \text{Prob}(A)\frac{\text{Prob}(B|A)}{\text{Prob}(B)} \tag{18.7.1}$$

Here $\text{Prob}(A|B)$ is the probability of A *given* that B has occurred, and similarly for $\text{Prob}(B|A)$, while $\text{Prob}(A)$ and $\text{Prob}(B)$ are unconditional probabilities.

"Bayesians" (so-called) adopt a broader interpretation of probabilities than do so-called "frequentists." To a Bayesian, $P(A|B)$ is a measure of the degree of plausibility of A (given B) on a scale ranging from zero to one. In this broader view, A and B need not be repeatable events; they can be propositions or hypotheses. The equations of probability theory then become a set of consistent rules for conducting inference [1,2]. Since plausibility is itself always conditioned on some, perhaps unarticulated, set of assumptions, all Bayesian probabilities are viewed as conditional on some collective background information I.

Suppose H is some hypothesis. Even before there exist any explicit data, a Bayesian can assign to H some degree of plausibility $\text{Prob}(H|I)$, called the "Bayesian prior." Now, when some data D_1 comes along, Bayes theorem tells how to reassess the plausibility of H,

$$\text{Prob}(H|D_1 I) = \text{Prob}(H|I)\frac{\text{Prob}(D_1|HI)}{\text{Prob}(D_1|I)} \tag{18.7.2}$$

The factor in the numerator on the right of equation (18.7.2) is calculable as the probability of a data set *given* the hypothesis (compare with "likelihood" in §15.1). The denominator, called the "prior predictive probability" of the data, is in this case merely a normalization constant which can be calculated by the requirement that the probability of all hypotheses should sum to unity. (In other Bayesian contexts, the prior predictive probabilities of two qualitatively different models can be used to assess their relative plausibility.)

If some additional data D_2 comes along tomorrow, we can further refine our estimate of H's probability, as

$$\text{Prob}(H|D_2 D_1 I) = \text{Prob}(H|D_1 I)\frac{\text{Prob}(D_2|HD_1 I)}{\text{Prob}(D_2|D_1 I)} \tag{18.7.3}$$

Using the product rule for probabilities, $\text{Prob}(AB|C) = \text{Prob}(A|C)\text{Prob}(B|AC)$, we find that equations (18.7.2) and (18.7.3) imply

$$\text{Prob}(H|D_2 D_1 I) = \text{Prob}(H|I)\frac{\text{Prob}(D_2 D_1|HI)}{\text{Prob}(D_2 D_1|I)} \qquad (18.7.4)$$

which shows that we would have gotten the same answer if all the data $D_1 D_2$ had been taken together.

From a Bayesian perspective, inverse problems are inference problems [3,4]. The underlying parameter set $\mathbf{u}$ is a hypothesis whose probability, given the measured data values $\mathbf{c}$, and the Bayesian prior $\text{Prob}(\mathbf{u}|I)$ can be calculated. We might want to report a single "best" inverse $\mathbf{u}$, the one that maximizes

$$\text{Prob}(\mathbf{u}|\mathbf{c}I) = \text{Prob}(\mathbf{c}|\mathbf{u}I)\frac{\text{Prob}(\mathbf{u}|I)}{\text{Prob}(\mathbf{c}|I)} \qquad (18.7.5)$$

over all possible choices of $\mathbf{u}$. Bayesian analysis also admits the possibility of reporting additional information that characterizes the region of possible $\mathbf{u}$'s with high relative probability, the so-called "posterior bubble" in $\mathbf{u}$.

The calculation of the probability of the data $\mathbf{c}$, given the hypothesis $\mathbf{u}$ proceeds exactly as in the maximum likelihood method. For Gaussian errors, e.g., it is given by

$$\text{Prob}(\mathbf{c}|\mathbf{u}I) = \exp(-\frac{1}{2}\chi^2)\Delta u_0 \Delta u_1 \cdots \Delta u_{M-1} \qquad (18.7.6)$$

where χ^2 is calculated from $\mathbf{u}$ and $\mathbf{c}$ using equation (18.4.9), and the Δu_μ's are constant, small ranges of the components of $\mathbf{u}$ whose actual magnitude is irrelevant, because they do not depend on $\mathbf{u}$ (compare equations 15.1.3 and 15.1.4).

In maximum likelihood estimation we, in effect, chose the prior $\text{Prob}(\mathbf{u}|I)$ to be constant. That was a luxury that we could afford when estimating a small number of parameters from a large amount of data. Here, the number of "parameters" (components of $\mathbf{u}$) is comparable to or larger than the number of measured values (components of $\mathbf{c}$); we *need* to have a nontrivial prior, $\text{Prob}(\mathbf{u}|I)$, to resolve the degeneracy of the solution.

In maximum entropy image restoration, that is where *entropy* comes in. The entropy of a physical system in some macroscopic state, usually denoted S, is the logarithm of the number of microscopically distinct configurations that all have the same macroscopic observables (i.e., consistent with the observed macroscopic state). Actually, we will find it useful to denote the *negative* of the entropy, also called the *negentropy*, by $H \equiv -S$ (a notation that goes back to Boltzmann). In situations where there is reason to believe that the *a priori* probabilities of the *microscopic* configurations are all the same (these situations are called *ergodic*), then the Bayesian prior $\text{Prob}(\mathbf{u}|I)$ for a *macroscopic* state with entropy S is proportional to $\exp(S)$ or $\exp(-H)$.

MEM uses this concept to assign a prior probability to any given underlying function $\mathbf{u}$. For example [5-7], suppose that the measurement of luminance in each pixel is quantized to (in some units) an integer value. Let

$$U = \sum_{\mu=0}^{M-1} u_\mu \qquad (18.7.7)$$

be the total number of luminance quanta in the whole image. Then we can base our "prior" on the notion that each luminance quantum has an equal *a priori* chance of being in any pixel. (See [8] for a more abstract justification of this idea.) The number of ways of getting a particular configuration $\mathbf{u}$ is

$$\frac{U!}{u_0!u_1!\cdots u_{M-1}!} \propto \exp\left[-\sum_\mu u_\mu \ln(u_\mu/U) + \frac{1}{2}\left(\ln U - \sum_\mu \ln u_\mu\right)\right] \tag{18.7.8}$$

Here the left side can be understood as the number of distinct orderings of all the luminance quanta, divided by the numbers of equivalent reorderings within each pixel, while the right side follows by Stirling's approximation to the factorial function. Taking the negative of the logarithm, and neglecting terms of order $\log U$ in the presence of terms of order U, we get the negentropy

$$H(\mathbf{u}) = \sum_{\mu=0}^{M-1} u_\mu \ln(u_\mu/U) \tag{18.7.9}$$

From equations (18.7.5), (18.7.6), and (18.7.9) we now seek to maximize

$$\text{Prob}(\mathbf{u}|\mathbf{c}) \propto \exp\left[-\frac{1}{2}\chi^2\right]\exp[-H(\mathbf{u})] \tag{18.7.10}$$

or, equivalently,

$$\text{minimize:}\quad -\ln\left[\text{Prob}(\mathbf{u}|\mathbf{c})\right] = \frac{1}{2}\chi^2[\mathbf{u}] + H(\mathbf{u}) = \frac{1}{2}\chi^2[\mathbf{u}] + \sum_{\mu=0}^{M-1} u_\mu \ln(u_\mu/U) \tag{18.7.11}$$

This ought to remind you of equation (18.4.11), or equation (18.5.6), or in fact any of our previous minimization principles along the lines of $\mathcal{A} + \lambda\mathcal{B}$, where $\lambda\mathcal{B} = H(\mathbf{u})$ is a regularizing operator. Where is λ? We need to put it in for exactly the reason discussed following equation (18.4.11): Degenerate inversions are likely to be able to achieve unrealistically small values of χ^2. We need an adjustable parameter to bring χ^2 into its expected narrow statistical range of $N \pm (2N)^{1/2}$. The discussion at the beginning of §18.4 showed that it makes no difference which term we attach the λ to. For consistency in notation, we absorb a factor 2 into λ and put it on the entropy term. (Another way to see the necessity of an undetermined λ factor is to note that it is necessary if our minimization principle is to be invariant under changing the units in which $\mathbf{u}$ is quantized, e.g., if an 8-bit analog-to-digital converter is replaced by a 12-bit one.) We can now also put "hats" back to indicate that this is the procedure for obtaining our chosen statistical estimator:

$$\text{minimize:}\quad \mathcal{A} + \lambda\mathcal{B} = \chi^2[\widehat{\mathbf{u}}] + \lambda H(\widehat{\mathbf{u}}) = \chi^2[\widehat{\mathbf{u}}] + \lambda\sum_{\mu=0}^{M-1} \widehat{u}_\mu \ln(\widehat{u}_\mu) \tag{18.7.12}$$

(Formally, we might also add a second Lagrange multiplier $\lambda' U$, to constrain the total intensity U to be constant.)

It is not hard to see that the negentropy, $H(\widehat{\mathbf{u}})$, is in fact a regularizing operator, similar to $\widehat{\mathbf{u}} \cdot \widehat{\mathbf{u}}$ (equation 18.4.11) or $\widehat{\mathbf{u}} \cdot \mathbf{H} \cdot \widehat{\mathbf{u}}$ (equation 18.5.6). The following of its properties are noteworthy:

1. When U is held constant, $H(\widehat{\mathbf{u}})$ is minimized for $\widehat{u}_\mu = U/M = $ constant, so it smooths in the sense of trying to achieve a constant solution, similar to equation (18.5.4). The fact that the constant solution is a minimum follows from the fact that the second derivative of $u \ln u$ is positive.
2. Unlike equation (18.5.4), however, $H(\widehat{\mathbf{u}})$ is *local*, in the sense that it does not difference neighboring pixels. It simply sums some function f, here

$$f(u) = u \ln u \qquad (18.7.13)$$

over all pixels; it is invariant, in fact, under a complete scrambling of the pixels in an image. This form implies that $H(\widehat{\mathbf{u}})$ is not seriously increased by the occurrence of a small number of very bright pixels (point sources) embedded in a low-intensity smooth background.
3. $H(\widehat{\mathbf{u}})$ goes to infinite slope as any one pixel goes to zero. This causes it to enforce positivity of the image, without the necessity of additional deterministic constraints.
4. The biggest difference between $H(\widehat{\mathbf{u}})$ and the other regularizing operators that we have met is that $H(\widehat{\mathbf{u}})$ is not a quadratic functional of $\widehat{\mathbf{u}}$, so the equations obtained by varying equation (18.7.12) are *nonlinear*. This fact is itself worthy of some additional discussion.

Nonlinear equations are harder to solve than linear equations. For image processing, however, the large number of equations usually dictates an iterative solution procedure, even for linear equations, so the practical effect of the nonlinearity is somewhat mitigated. Below, we will summarize some of the methods that are successfully used for MEM inverse problems.

For some problems, notably the problem in radio-astronomy of image recovery from an incomplete set of Fourier coefficients, the superior performance of MEM inversion can be, in part, traced to the nonlinearity of $H(\widehat{\mathbf{u}})$. One way to see this [5] is to consider the limit of perfect measurements $\sigma_i \to 0$. In this case the χ^2 term in the minimization principle (18.7.12) gets replaced by a set of constraints, each with its own Lagrange multiplier, requiring agreement between model and data; that is,

$$\text{minimize:} \quad \sum_j \lambda_j \left[c_j - \sum_\mu R_{j\mu}\widehat{u}_\mu \right] + H(\widehat{\mathbf{u}}) \qquad (18.7.14)$$

(cf. equation 18.4.7). Setting the formal derivative with respect to $\widehat{u}_\mu$ to zero gives

$$\frac{\partial H}{\partial \widehat{u}_\mu} = f'(\widehat{u}_\mu) = \sum_j \lambda_j R_{j\mu} \qquad (18.7.15)$$

or defining a function G as the inverse function of f',

$$\widehat{u}_\mu = G\left(\sum_j \lambda_j R_{j\mu} \right) \qquad (18.7.16)$$

This solution is only formal, since the λ_j's must be found by requiring that equation (18.7.16) satisfy all the constraints built into equation (18.7.14). However, equation (18.7.16) does show the crucial fact that if G is *linear*, then the solution $\hat{\mathbf{u}}$ contains *only* a linear combination of basis functions $R_{j\mu}$ corresponding to actual measurements j. This is equivalent to setting unmeasured c_j's to zero. Notice that the principal solution obtained from equation (18.4.11) in fact has a linear G.

In the problem of incomplete Fourier image reconstruction, the typical $R_{j\mu}$ has the form $\exp(-2\pi i \mathbf{k}_j \cdot \mathbf{x}_\mu)$, where $\mathbf{x}_\mu$ is a two-dimensional vector in the image space and $\mathbf{k}_\mu$ is a two-dimensional wave-vector. If an image contains strong point sources, then the effect of setting unmeasured c_j's to zero is to produce sidelobe ripples throughout the image plane. These ripples can mask any actual extended, low-intensity image features lying between the point sources. If, however, the slope of G is smaller for small values of its argument, larger for large values, then ripples in low-intensity portions of the image are relatively suppressed, while strong point sources will be relatively sharpened ("superresolution"). This behavior on the slope of G is equivalent to requiring $f'''(u) < 0$. For $f(u) = u \ln u$, we in fact have $f'''(u) = -1/u^2 < 0$.

In more picturesque language, the nonlinearity acts to "create" nonzero values for the unmeasured c_i's, so as to suppress the low-intensity ripple and sharpen the point sources.

Is MEM Really Magical?

How unique is the negentropy functional (18.7.9)? Recall that that equation is based on the assumption that luminance elements are *a priori* distributed over the pixels uniformly. If we instead had some other preferred *a priori* image in mind, one with pixel intensities m_μ, then it is easy to show that the negentropy becomes

$$H(\mathbf{u}) = \sum_{\mu=0}^{M-1} u_\mu \ln(u_\mu/m_\mu) + \text{constant} \qquad (18.7.17)$$

(the constant can then be ignored). All the rest of the discussion then goes through.

More fundamentally, and despite statements by zealots to the contrary [7], there is actually nothing universal about the functional form $f(u) = u \ln u$. In some other physical situations (for example, the entropy of an electromagnetic field in the limit of many photons per mode, as in radio-astronomy) the physical negentropy functional is actually $f(u) = -\ln u$ (see [5] for other examples). In general, the question, "Entropy of what?" is not uniquely answerable in any particular situation. (See reference [9] for an attempt at articulating a more general principle that reduces to one or another entropy functional under appropriate circumstances.)

The four numbered properties summarized above, plus the desirable sign for nonlinearity, $f'''(u) < 0$, are all as true for $f(u) = -\ln u$ as for $f(u) = u \ln u$. In fact these properties are shared by a nonlinear function as simple as $f(u) = -\sqrt{u}$, which has no information theoretic justification at all (no logarithms!). MEM reconstructions of test images using any of these entropy forms are virtually indistinguishable [5].

By all available evidence, MEM seems to be neither more nor less than one usefully nonlinear version of the general regularization scheme $\mathcal{A} + \lambda\mathcal{B}$ that we have by now considered in many forms. Its peculiarities become strengths when applied to the reconstruction from incomplete Fourier data of images that are expected to be dominated by very bright point sources, but which also contain interesting low-intensity, extended sources. For images of some other character, there is no reason to suppose that MEM methods will generally dominate other regularization schemes, either ones already known or yet to be invented.

Algorithms for MEM

The goal is to find the vector $\widehat{\mathbf{u}}$ that minimizes $\mathcal{A} + \lambda\mathcal{B}$ where in the notation of equations (18.5.5), (18.5.6), and (18.7.13),

$$\mathcal{A} = |\mathbf{b} - \mathbf{A} \cdot \widehat{\mathbf{u}}|^2 \qquad \mathcal{B} = \sum_\mu f(\widehat{u}_\mu) \qquad (18.7.18)$$

Compared with a "general" minimization problem, we have the advantage that we can compute the gradients and the second partial derivative matrices (Hessian matrices) explicitly,

$$\nabla\mathcal{A} = 2(\mathbf{A}^T \cdot \mathbf{A} \cdot \widehat{\mathbf{u}} - \mathbf{A}^T \cdot \mathbf{b}) \qquad \frac{\partial^2 \mathcal{A}}{\partial \widehat{u}_\mu \partial \widehat{u}_\rho} = [2\mathbf{A}^T \cdot \mathbf{A}]_{\mu\rho}$$

$$[\nabla\mathcal{B}]_\mu = f'(\widehat{u}_\mu) \qquad \frac{\partial^2 \mathcal{B}}{\partial \widehat{u}_\mu \partial \widehat{u}_\rho} = \delta_{\mu\rho} f''(\widehat{u}_\mu) \qquad (18.7.19)$$

It is important to note that while $\mathcal{A}$'s second partial derivative matrix cannot be stored (its size is the square of the number of pixels), it can be applied to any vector by first applying $\mathbf{A}$, then $\mathbf{A}^T$. In the case of reconstruction from incomplete Fourier data, or in the case of convolution with a translation invariant point spread function, these applications will typically involve several FFTs. Likewise, the calculation of the gradient $\nabla\mathcal{A}$ will involve FFTs in the application of $\mathbf{A}$ and $\mathbf{A}^T$.

While some success has been achieved with the classical conjugate gradient method (§10.6), it is often found that the nonlinearity in $f(u) = u \ln u$ causes problems. Attempted steps that give $\widehat{\mathbf{u}}$ with even one negative value must be cut in magnitude, sometimes so severely as to slow the solution to a crawl. The underlying problem is that the conjugate gradient method develops its information about the inverse of the Hessian matrix a bit at a time, while changing its location in the search space. When a nonlinear function is quite different from a pure quadratic form, the old information becomes obsolete before it gets usefully exploited.

Skilling and collaborators [6,7,10,11] developed a complicated but highly successful scheme, wherein a minimum is repeatedly sought not along a single search direction, but in a small- (typically three-) dimensional subspace, spanned by vectors that are calculated anew at each landing point. The subspace basis vectors are chosen in such a way as to avoid directions leading to negative values. One of the most successful choices is the three-dimensional subspace spanned by the vectors with components given by

$$e_\mu^{(1)} = \widehat{u}_\mu [\nabla \mathcal{A}]_\mu$$

$$e_\mu^{(2)} = \widehat{u}_\mu [\nabla \mathcal{B}]_\mu$$

$$e_\mu^{(3)} = \frac{\widehat{u}_\mu \sum_\rho (\partial^2 \mathcal{A}/\partial \widehat{u}_\mu \partial \widehat{u}_\rho) \widehat{u}_\rho [\nabla \mathcal{B}]_\rho}{\sqrt{\sum_\rho \widehat{u}_\rho ([\nabla \mathcal{B}]_\rho)^2}} - \frac{\widehat{u}_\mu \sum_\rho (\partial^2 \mathcal{A}/\partial \widehat{u}_\mu \partial \widehat{u}_\rho) \widehat{u}_\rho [\nabla \mathcal{A}]_\rho}{\sqrt{\sum_\rho \widehat{u}_\rho ([\nabla \mathcal{A}]_\rho)^2}}$$

$$(18.7.20)$$

(In these equations there is no sum over μ.) The form of the $\mathbf{e}^{(3)}$ has some justification if one views dot products as occurring in a space with the metric $g_{\mu\nu} = \delta_{\mu\nu}/u_\mu$, chosen to make zero values "far away"; see [6].

Within the three-dimensional subspace, the three-component gradient and nine-component Hessian matrix are computed by projection from the large space, and the minimum in the subspace is estimated by (trivially) solving three simultaneous linear equations, as in §10.7, equation (10.7.4). The size of a step $\Delta\widehat{\mathbf{u}}$ is required to be limited by the inequality

$$\sum_\mu (\Delta \widehat{u}_\mu)^2 / \widehat{u}_\mu < (0.1 \text{ to } 0.5)U \qquad (18.7.21)$$

Because the gradient directions $\nabla \mathcal{A}$ and $\nabla \mathcal{B}$ are separately available, it is possible to combine the minimum search with a simultaneous adjustment of λ so as finally to satisfy the desired constraint. There are various further tricks employed.

A less general, but in practice often equally satisfactory, approach is due to Cornwell and Evans [12]. Here, noting that $\mathcal{B}$'s Hessian (second partial derivative) matrix is diagonal, one asks whether there is a useful diagonal approximation to $\mathcal{A}$'s Hessian, namely $2\mathbf{A}^T \cdot \mathbf{A}$. If Λ_μ denotes the diagonal components of such an approximation, then a useful step in $\widehat{\mathbf{u}}$ would be

$$\Delta \widehat{u}_\mu = -\frac{1}{\Lambda_\mu + \lambda f''(\widehat{u}_\mu)} (\nabla \mathcal{A} + \lambda \nabla \mathcal{B}) \qquad (18.7.22)$$

(again compare equation 10.7.4). Even more extreme, one might seek an approximation with constant diagonal elements, $\Lambda_\mu = \Lambda$, so that

$$\Delta \widehat{u}_\mu = -\frac{1}{\Lambda + \lambda f''(\widehat{u}_\mu)} (\nabla \mathcal{A} + \lambda \nabla \mathcal{B}) \qquad (18.7.23)$$

Since $\mathbf{A}^T \cdot \mathbf{A}$ has something of the nature of a doubly convolved point spread function, and since in real cases one often has a point spread function with a sharp central peak, even the more extreme of these approximations is often fruitful. One starts with a rough estimate of Λ obtained from the $A_{i\mu}$'s, e.g.,

$$\Lambda \sim \left\langle \sum_i [A_{i\mu}]^2 \right\rangle \qquad (18.7.24)$$

An accurate value is not important, since in practice Λ is adjusted adaptively: If Λ is too large, then equation (18.7.23)'s steps will be too small (that is, larger steps in the same direction will produce even greater decrease in $\mathcal{A} + \lambda\mathcal{B}$). If Λ is too small, then attempted steps will land in an unfeasible region (negative values of $\widehat{u}_\mu$), or will result in an increased $\mathcal{A} + \lambda\mathcal{B}$. There is an obvious similarity between the adjustment of Λ here and the Levenberg-Marquardt method of §15.5; this should not be too surprising, since MEM is closely akin to the problem of nonlinear least-squares fitting. Reference [12] also discusses how the value of $\Lambda + \lambda f''(\widehat{u}_\mu)$ can be used to adjust the Lagrange multiplier λ so as to converge to the desired value of χ^2.

All practical MEM algorithms are found to require on the order of 30 to 50 iterations to converge. This convergence behavior is not now understood in any fundamental way.

"Bayesian" versus "Historic" Maximum Entropy

Several more recent developments in maximum entropy image restoration go under the rubric "Bayesian" to distinguish them from the previous "historic" methods. See [13] for details and references.

- Better priors: We already noted that the entropy functional (equation 18.7.13) is invariant under scrambling all pixels and has no notion of smoothness. The so-called "intrinsic correlation function" (ICF) model (Ref. [13], where it is called "New MaxEnt") is similar enough to the entropy functional to allow similar algorithms, but it makes the values of neighboring pixels correlated, enforcing smoothness.

- Better estimation of λ: Above we chose λ to bring χ^2 into its expected narrow statistical range of $N \pm (2N)^{1/2}$. This in effect overestimates χ^2, however, since some effective number γ of parameters are being "fitted" in doing the reconstruction. A Bayesian approach leads to a self-consistent estimate of this γ and an objectively better choice for λ.

CITED REFERENCES AND FURTHER READING:

Jaynes, E.T. 1976, in *Foundations of Probability Theory, Statistical Inference, and Statistical Theories of Science*, W.L. Harper and C.A. Hooker, eds. (Dordrecht: Reidel). [1]

Jaynes, E.T. 1985, in *Maximum-Entropy and Bayesian Methods in Inverse Problems*, C.R. Smith and W.T. Grandy, Jr., eds. (Dordrecht: Reidel). [2]

Jaynes, E.T. 1984, in *SIAM-AMS Proceedings*, vol. 14, D.W. McLaughlin, ed. (Providence, RI: American Mathematical Society). [3]

Titterington, D.M. 1985, *Astronomy and Astrophysics*, vol. 144, 381–387. [4]

Narayan, R., and Nityananda, R. 1986, *Annual Review of Astronomy and Astrophysics*, vol. 24, pp. 127–170. [5]

Skilling, J., and Bryan, R.K. 1984, *Monthly Notices of the Royal Astronomical Society*, vol. 211, pp. 111–124. [6]

Burch, S.F., Gull, S.F., and Skilling, J. 1983, *Computer Vision, Graphics and Image Processing*, vol. 23, pp. 113–128. [7]

Skilling, J. 1989, in *Maximum Entropy and Bayesian Methods*, J. Skilling, ed. (Boston: Kluwer). [8]

Frieden, B.R. 1983, *Journal of the Optical Society of America*, vol. 73, pp. 927–938. [9]

Skilling, J., and Gull, S.F. 1985, in *Maximum-Entropy and Bayesian Methods in Inverse Problems*, C.R. Smith and W.T. Grandy, Jr., eds. (Dordrecht: Reidel). [10]

Skilling, J. 1986, in *Maximum Entropy and Bayesian Methods in Applied Statistics*, J.H. Justice, ed. (Cambridge: Cambridge University Press). [11]

Cornwell, T.J., and Evans, K.F. 1985, *Astronomy and Astrophysics*, vol. 143, pp. 77–83. [12]

Gull, S.F. 1989, in *Maximum Entropy and Bayesian Methods*, J. Skilling, ed. (Boston: Kluwer). [13]

Chapter 19.   Partial Differential Equations

19.0 Introduction

The numerical treatment of partial differential equations is, by itself, a vast subject. Partial differential equations are at the heart of many, if not most, computer analyses or simulations of continuous physical systems, such as fluids, electromagnetic fields, the human body, and so on. The intent of this chapter is to give the briefest possible useful introduction. Ideally, there would be an entire second volume of *Numerical Recipes* dealing with partial differential equations alone. (The references [1-4] provide, of course, available alternatives.)

In most mathematics books, partial differential equations (PDEs) are classified into the three categories, *hyperbolic, parabolic,* and *elliptic,* on the basis of their *characteristics,* or curves of information propagation. The prototypical example of a hyperbolic equation is the one-dimensional *wave* equation

$$\frac{\partial^2 u}{\partial t^2} = v^2 \frac{\partial^2 u}{\partial x^2} \tag{19.0.1}$$

where $v = $ constant is the velocity of wave propagation. The prototypical parabolic equation is the *diffusion* equation

$$\frac{\partial u}{\partial t} = \frac{\partial}{\partial x}\left(D\frac{\partial u}{\partial x}\right) \tag{19.0.2}$$

where D is the diffusion coefficient.   The prototypical elliptic equation is the *Poisson* equation

$$\frac{\partial^2 u}{\partial x^2} + \frac{\partial^2 u}{\partial y^2} = \rho(x,y) \tag{19.0.3}$$

where the source term ρ is given. If the source term is equal to zero, the equation is *Laplace's equation.*

From a computational point of view, the classification into these three canonical types is not very meaningful — or at least not as important as some other essential distinctions. Equations (19.0.1) and (19.0.2) both define *initial value* or *Cauchy* problems: If information on u (perhaps including time derivative information) is

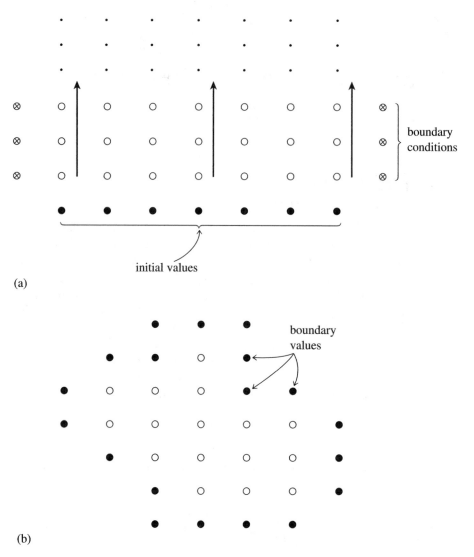

Figure 19.0.1.   Initial value problem (a) and boundary value problem (b) are contrasted. In (a) initial
values are given on one "time slice," and it is desired to advance the solution in time, computing successive
rows of open dots in the direction shown by the arrows. Boundary conditions at the left and right edges
of each row (⊗) must also be supplied, but only one row at a time. Only one, or a few, previous rows
need be maintained in memory. In (b), boundary values are specified around the edge of a grid, and an
iterative process is employed to find the values of all the internal points (open circles). All grid points
must be maintained in memory.

given at some initial time t_0 for all x, then the equations describe how $u(x,t)$
propagates itself forward in time. In other words, equations (19.0.1) and (19.0.2)
describe time evolution. The goal of a numerical code should be to track that time
evolution with some desired accuracy.

By contrast, equation (19.0.3) directs us to find a single "static" function $u(x,y)$
which satisfies the equation within some (x,y) region of interest, and which — one
must also specify — has some desired behavior on the boundary of the region of
interest. These problems are called *boundary value problems*. In general it is not

possible stably to just "integrate in from the boundary" in the same sense that an initial value problem can be "integrated forward in time." Therefore, the goal of a numerical code is somehow to converge on the correct solution everywhere at once.

This, then, is the most important classification from a computational point of view: Is the problem at hand an *initial value* (time evolution) problem? or is it a *boundary value* (static solution) problem? Figure 19.0.1 emphasizes the distinction. Notice that while the italicized terminology is standard, the terminology in parentheses is a much better description of the dichotomy from a computational perspective. The subclassification of initial value problems into parabolic and hyperbolic is much less important because (i) many actual problems are of a mixed type, and (ii) as we will see, most hyperbolic problems get parabolic pieces mixed into them by the time one is discussing practical computational schemes.

Initial Value Problems

An initial value problem is defined by answers to the following questions:
- What are the dependent variables to be propagated forward in time?
- What is the evolution equation for each variable? Usually the evolution equations will all be coupled, with more than one dependent variable appearing on the right-hand side of each equation.
- What is the highest time derivative that occurs in each variable's evolution equation? If possible, this time derivative should be put alone on the equation's left-hand side. Not only the value of a variable, but also the value of all its time derivatives — up to the highest one — must be specified to define the evolution.
- What special equations (boundary conditions) govern the evolution in time of points on the boundary of the spatial region of interest? Examples: *Dirichlet conditions* specify the values of the boundary points as a function of time; *Neumann conditions* specify the values of the normal gradients on the boundary; *outgoing-wave boundary conditions* are just what they say.

Sections 19.1–19.3 of this chapter deal with initial value problems of several different forms. We make no pretence of completeness, but rather hope to convey a certain amount of generalizable information through a few carefully chosen model examples. These examples will illustrate an important point: One's principal *computational* concern must be the *stability* of the algorithm. Many reasonable-looking algorithms for initial value problems just don't work — they are numerically unstable.

Boundary Value Problems

The questions that define a boundary value problem are:
- What are the variables?
- What equations are satisfied in the interior of the region of interest?
- What equations are satisfied by points on the boundary of the region of interest? (Here Dirichlet and Neumann conditions are possible choices for elliptic second-order equations, but more complicated boundary conditions can also be encountered.)

In contrast to initial value problems, stability is relatively easy to achieve for boundary value problems. Thus, the *efficiency* of the algorithms, both in computational load and storage requirements, becomes the principal concern.

Because all the conditions on a boundary value problem must be satisfied "simultaneously," these problems usually boil down, at least conceptually, to the solution of large numbers of simultaneous algebraic equations. When such equations are nonlinear, they are usually solved by linearization and iteration; so without much loss of generality we can view the problem as being the solution of special, large linear sets of equations.

As an example, one which we will refer to in §§19.4–19.6 as our "model problem," let us consider the solution of equation (19.0.3) by the *finite-difference method*. We represent the function $u(x, y)$ by its values at the discrete set of points

$$x_j = x_0 + j\Delta, \qquad j = 0, 1, ..., J$$
$$y_l = y_0 + l\Delta, \qquad l = 0, 1, ..., L \tag{19.0.4}$$

where Δ is the *grid spacing*. From now on, we will write $u_{j,l}$ for $u(x_j, y_l)$, and $\rho_{j,l}$ for $\rho(x_j, y_l)$. For (19.0.3) we substitute a finite-difference representation (see Figure 19.0.2),

$$\frac{u_{j+1,l} - 2u_{j,l} + u_{j-1,l}}{\Delta^2} + \frac{u_{j,l+1} - 2u_{j,l} + u_{j,l-1}}{\Delta^2} = \rho_{j,l} \tag{19.0.5}$$

or equivalently

$$u_{j+1,l} + u_{j-1,l} + u_{j,l+1} + u_{j,l-1} - 4u_{j,l} = \Delta^2 \rho_{j,l} \tag{19.0.6}$$

To write this system of linear equations in matrix form we need to make a vector out of u. Let us number the two dimensions of grid points in a single one-dimensional sequence by defining

$$i \equiv j(L + 1) + l \qquad \text{for} \qquad j = 0, 1, ..., J, \qquad l = 0, 1, ..., L \tag{19.0.7}$$

In other words, i increases most rapidly along the columns representing y values. Equation (19.0.6) now becomes

$$u_{i+L+1} + u_{i-(L+1)} + u_{i+1} + u_{i-1} - 4u_i = \Delta^2 \rho_i \tag{19.0.8}$$

This equation holds only at the interior points $j = 1, 2, ..., J - 1; l = 1, 2, ..., L - 1$.

The points where

$$
\begin{aligned}
j = 0 \qquad & [\text{i.e., } i = 0, ..., L] \\
j = J \qquad & [\text{i.e., } i = J(L + 1), ..., J(L + 1) + L] \\
l = 0 \qquad & [\text{i.e., } i = 0, L + 1, ..., J(L + 1)] \\
l = L \qquad & [\text{i.e., } i = L, L + 1 + L, ..., J(L + 1) + L]
\end{aligned}
\tag{19.0.9}
$$

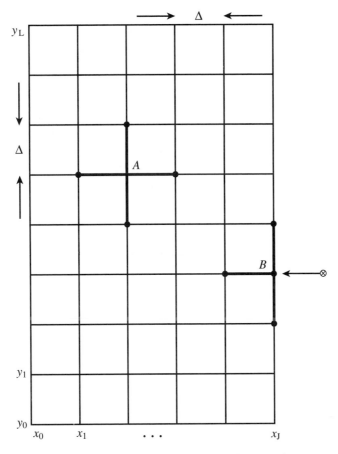

Figure 19.0.2. Finite-difference representation of a second-order elliptic equation on a two-dimensional grid. The second derivatives at the point A are evaluated using the points to which A is shown connected. The second derivatives at point B are evaluated using the connected points and also using "right-hand side" boundary information, shown schematically as $\otimes$.

are boundary points where either u or its derivative has been specified. If we pull all this "known" information over to the right-hand side of equation (19.0.8), then the equation takes the form

$$\mathbf{A} \cdot \mathbf{u} = \mathbf{b} \qquad (19.0.10)$$

where $\mathbf{A}$ has the form shown in Figure 19.0.3. The matrix $\mathbf{A}$ is called "tridiagonal with fringes." A general linear second-order elliptic equation

$$
\begin{aligned}
a(x,y)\frac{\partial^2 u}{\partial x^2} + b(x,y)\frac{\partial u}{\partial x} + c(x,y)\frac{\partial^2 u}{\partial y^2} + d(x,y)\frac{\partial u}{\partial y} \\
+ e(x,y)\frac{\partial^2 u}{\partial x \partial y} + f(x,y)u = g(x,y)
\end{aligned}
\qquad (19.0.11)
$$

will lead to a matrix of similar structure except that the nonzero entries will not be constants.

As a rough classification, there are three different approaches to the solution of equation (19.0.10), not all applicable in all cases: relaxation methods, "rapid" methods (e.g., Fourier methods), and direct matrix methods.

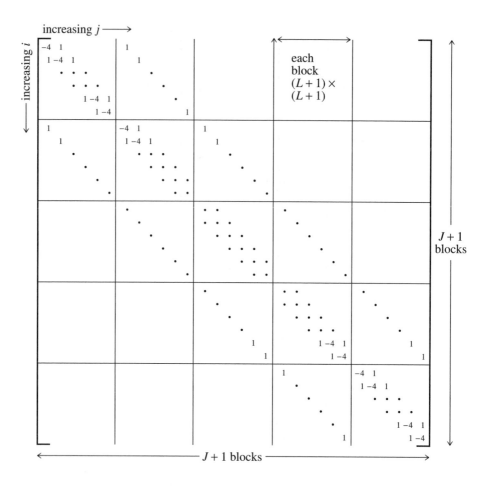

Figure 19.0.3.   Matrix structure derived from a second-order elliptic equation (here equation 19.0.6). All elements not shown are zero. The matrix has diagonal blocks that are themselves tridiagonal, and sub- and super-diagonal blocks that are diagonal. This form is called "tridiagonal with fringes." A matrix this sparse would never be stored in its full form as shown here.

Relaxation methods make immediate use of the structure of the sparse matrix **A**. The matrix is split into two parts

$$\mathbf{A} = \mathbf{E} - \mathbf{F} \tag{19.0.12}$$

where **E** is easily invertible and **F** is the remainder. Then (19.0.10) becomes

$$\mathbf{E} \cdot \mathbf{u} = \mathbf{F} \cdot \mathbf{u} + \mathbf{b} \tag{19.0.13}$$

The relaxation method involves choosing an initial guess $\mathbf{u}^{(0)}$ and then solving successively for iterates $\mathbf{u}^{(r)}$ from

$$\mathbf{E} \cdot \mathbf{u}^{(r)} = \mathbf{F} \cdot \mathbf{u}^{(r-1)} + \mathbf{b} \tag{19.0.14}$$

Since **E** is chosen to be easily invertible, each iteration is fast. We will discuss relaxation methods in some detail in §19.5 and §19.6.

So-called rapid methods [5] apply for only a rather special class of equations: those with constant coefficients, or, more generally, those that are separable in the chosen coordinates. In addition, the boundaries must coincide with coordinate lines. This special class of equations is met quite often in practice. We defer detailed discussion to §19.4. Note, however, that the multigrid relaxation methods discussed in §19.6 can be faster than "rapid" methods.

Matrix methods attempt to solve the equation

$$\mathbf{A} \cdot \mathbf{x} = \mathbf{b} \tag{19.0.15}$$

directly. The degree to which this is practical depends very strongly on the exact structure of the matrix $\mathbf{A}$ for the problem at hand, so our discussion can go no farther than a few remarks and references at this point.

Sparseness of the matrix *must* be the guiding force. Otherwise the matrix problem is prohibitively large. For example, the simplest problem on a 100×100 spatial grid would involve 10000 unknown $u_{j,l}$'s, implying a 10000×10000 matrix $\mathbf{A}$, containing 10^8 elements!

As we discussed at the end of §2.7, if $\mathbf{A}$ is symmetric and positive definite (as it usually is in elliptic problems), the conjugate-gradient algorithm can be used. In practice, rounding error often spoils the effectiveness of the conjugate gradient algorithm for solving finite-difference equations. However, it is useful when incorporated in methods that first rewrite the equations so that $\mathbf{A}$ is transformed to a matrix $\mathbf{A}'$ that is close to the identity matrix. The quadratic surface defined by the equations then has almost spherical contours, and the conjugate gradient algorithm works very well. In §2.7, in the routine `linbcg`, an analogous *preconditioner* was exploited for non-positive definite problems with the more general biconjugate gradient method. For the positive definite case that arises in PDEs, an example of a successful implementation is the *incomplete Cholesky conjugate gradient method (ICCG)* (see [6-8]).

Another method that relies on a transformation approach is the *strongly implicit procedure* of Stone [9]. A program called SIPSOL that implements this routine has been published [10].

A third class of matrix methods is the Analyze-Factorize-Operate approach as described in §2.7.

Generally speaking, when you have the storage available to implement these methods — not nearly as much as the 10^8 above, but usually much more than is required by relaxation methods — then you should consider doing so. Only multigrid relaxation methods (§19.6) are competitive with the best matrix methods. For grids larger than, say, 300×300, however, it is generally found that only relaxation methods, or "rapid" methods when they are applicable, are possible.

There Is More to Life than Finite Differencing

Besides finite differencing, there are other methods for solving PDEs. Most important are finite element, Monte Carlo, spectral, and variational methods. Unfortunately, we shall barely be able to do justice to finite differencing in this chapter, and so shall not be able to discuss these other methods in this book. Finite element methods [11-12] are often preferred by practitioners in solid mechanics and structural engineering; these methods allow considerable freedom in putting computational elements where you want them, important when dealing with highly irregular geometries. Spectral methods [13-15] are preferred for very regular geometries and smooth functions; they converge more rapidly than finite-difference methods (cf. §19.4), but they do not work well for problems with discontinuities.

CITED REFERENCES AND FURTHER READING:

Ames, W.F. 1977, *Numerical Methods for Partial Differential Equations*, 2nd ed. (New York: Academic Press). [1]

Richtmyer, R.D., and Morton, K.W. 1967, *Difference Methods for Initial Value Problems*, 2nd ed. (New York: Wiley-Interscience). [2]

Roache, P.J. 1976, *Computational Fluid Dynamics* (Albuquerque: Hermosa). [3]

Mitchell, A.R., and Griffiths, D.F. 1980, *The Finite Difference Method in Partial Differential Equations* (New York: Wiley) [includes discussion of finite element methods]. [4]

Dorr, F.W. 1970, *SIAM Review*, vol. 12, pp. 248–263. [5]

Meijerink, J.A., and van der Vorst, H.A. 1977, *Mathematics of Computation*, vol. 31, pp. 148–162. [6]

van der Vorst, H.A. 1981, *Journal of Computational Physics*, vol. 44, pp. 1–19 [review of sparse iterative methods]. [7]

Kershaw, D.S. 1970, *Journal of Computational Physics*, vol. 26, pp. 43–65. [8]

Stone, H.J. 1968, *SIAM Journal on Numerical Analysis*, vol. 5, pp. 530–558. [9]

Jesshope, C.R. 1979, *Computer Physics Communications*, vol. 17, pp. 383–391. [10]

Strang, G., and Fix, G. 1973, *An Analysis of the Finite Element Method* (Englewood Cliffs, NJ: Prentice-Hall). [11]

Burnett, D.S. 1987, *Finite Element Analysis: From Concepts to Applications* (Reading, MA: Addison-Wesley). [12]

Gottlieb, D. and Orszag, S.A. 1977, *Numerical Analysis of Spectral Methods: Theory and Applications* (Philadelphia: S.I.A.M.). [13]

Canuto, C., Hussaini, M.Y., Quarteroni, A., and Zang, T.A. 1988, *Spectral Methods in Fluid Dynamics* (New York: Springer-Verlag). [14]

Boyd, J.P. 1989, *Chebyshev and Fourier Spectral Methods* (New York: Springer-Verlag). [15]

19.1 Flux-Conservative Initial Value Problems

A large class of initial value (time-evolution) PDEs in one space dimension can be cast into the form of a *flux-conservative equation*,

$$\frac{\partial \mathbf{u}}{\partial t} = -\frac{\partial \mathbf{F}(\mathbf{u})}{\partial x} \tag{19.1.1}$$

where $\mathbf{u}$ and $\mathbf{F}$ are vectors, and where (in some cases) $\mathbf{F}$ may depend not only on $\mathbf{u}$ but also on spatial derivatives of $\mathbf{u}$. The vector $\mathbf{F}$ is called the *conserved flux*.

For example, the prototypical hyperbolic equation, the one-dimensional wave equation with constant velocity of propagation v

$$\frac{\partial^2 u}{\partial t^2} = v^2 \frac{\partial^2 u}{\partial x^2} \tag{19.1.2}$$

can be rewritten as a set of two first-order equations

$$\frac{\partial r}{\partial t} = v\frac{\partial s}{\partial x}$$
$$\frac{\partial s}{\partial t} = v\frac{\partial r}{\partial x} \qquad (19.1.3)$$

where

$$r \equiv v\frac{\partial u}{\partial x}$$
$$s \equiv \frac{\partial u}{\partial t} \qquad (19.1.4)$$

In this case r and s become the two components of $\mathbf{u}$, and the flux is given by the linear matrix relation

$$\mathbf{F}(\mathbf{u}) = \begin{pmatrix} 0 & -v \\ -v & 0 \end{pmatrix} \cdot \mathbf{u} \qquad (19.1.5)$$

(The physicist-reader may recognize equations (19.1.3) as analogous to Maxwell's equations for one-dimensional propagation of electromagnetic waves.)

We will consider, in this section, a prototypical example of the general flux-conservative equation (19.1.1), namely the equation for a scalar u,

$$\frac{\partial u}{\partial t} = -v\frac{\partial u}{\partial x} \qquad (19.1.6)$$

with v a constant. As it happens, we already know analytically that the general solution of this equation is a wave propagating in the positive x-direction,

$$u = f(x - vt) \qquad (19.1.7)$$

where f is an arbitrary function. However, the numerical strategies that we develop will be equally applicable to the more general equations represented by (19.1.1). In some contexts, equation (19.1.6) is called an *advective* equation, because the quantity u is transported by a "fluid flow" with a velocity v.

How do we go about finite differencing equation (19.1.6) (or, analogously, 19.1.1)? The straightforward approach is to choose equally spaced points along both the t- and x-axes. Thus denote

$$x_j = x_0 + j\Delta x, \qquad j = 0, 1, \ldots, J$$
$$t_n = t_0 + n\Delta t, \qquad n = 0, 1, \ldots, N \qquad (19.1.8)$$

Let u_j^n denote $u(t_n, x_j)$. We have several choices for representing the time derivative term. The obvious way is to set

$$\left.\frac{\partial u}{\partial t}\right|_{j,n} = \frac{u_j^{n+1} - u_j^n}{\Delta t} + O(\Delta t) \qquad (19.1.9)$$

This is called *forward Euler* differencing (cf. equation 16.1.1). While forward Euler is only first-order accurate in Δt, it has the advantage that one is able to calculate

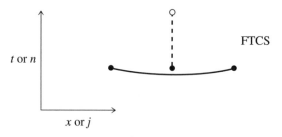

Figure 19.1.1. Representation of the Forward Time Centered Space (FTCS) differencing scheme. In this and subsequent figures, the open circle is the new point at which the solution is desired; filled circles are known points whose function values are used in calculating the new point; the solid lines connect points that are used to calculate spatial derivatives; the dashed lines connect points that are used to calculate time derivatives. The FTCS scheme is generally unstable for hyperbolic problems and cannot usually be used.

quantities at timestep $n + 1$ in terms of only quantities known at timestep n. For the space derivative, we can use a second-order representation still using only quantities known at timestep n:

$$\left.\frac{\partial u}{\partial x}\right|_{j,n} = \frac{u_{j+1}^n - u_{j-1}^n}{2\Delta x} + O(\Delta x^2) \qquad (19.1.10)$$

The resulting finite-difference approximation to equation (19.1.6) is called the FTCS representation (Forward Time Centered Space),

$$\frac{u_j^{n+1} - u_j^n}{\Delta t} = -v\left(\frac{u_{j+1}^n - u_{j-1}^n}{2\Delta x}\right) \qquad (19.1.11)$$

which can easily be rearranged to be a formula for u_j^{n+1} in terms of the other quantities. The FTCS scheme is illustrated in Figure 19.1.1. It's a fine example of an algorithm that is easy to derive, takes little storage, and executes quickly. Too bad it doesn't work! (See below.)

The FTCS representation is an *explicit* scheme. This means that u_j^{n+1} for each j can be calculated explicitly from the quantities that are already known. Later we shall meet *implicit* schemes, which require us to solve implicit equations coupling the u_j^{n+1} for various j. (Explicit and implicit methods for ordinary differential equations were discussed in §16.6.) The FTCS algorithm is also an example of a *single-level* scheme, since only values at time level n have to be stored to find values at time level $n + 1$.

von Neumann Stability Analysis

Unfortunately, equation (19.1.11) is of very limited usefulness. It is an *unstable* method, which can be used only (if at all) to study waves for a short fraction of one oscillation period. To find alternative methods with more general applicability, we must introduce the *von Neumann stability analysis*.

The von Neumann analysis is local: We imagine that the coefficients of the difference equations are so slowly varying as to be considered constant in space and time. In that case, the independent solutions, or *eigenmodes*, of the difference equations are all of the form

$$u_j^n = \xi^n e^{ikj\Delta x} \qquad (19.1.12)$$

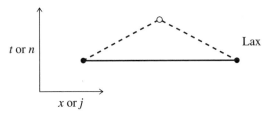

Figure 19.1.2.   Representation of the Lax differencing scheme, as in the previous figure. The stability criterion for this scheme is the Courant condition.

where k is a real spatial wave number (which can have any value) and $\xi = \xi(k)$ is a complex number that depends on k. The key fact is that the time dependence of a single eigenmode is nothing more than successive integer powers of the complex number ξ. Therefore, the difference equations are unstable (have exponentially growing modes) if $|\xi(k)| > 1$ for *some* k. The number ξ is called the *amplification factor* at a given wave number k.

To find $\xi(k)$, we simply substitute (19.1.12) back into (19.1.11). Dividing by ξ^n, we get

$$\xi(k) = 1 - i\frac{v\Delta t}{\Delta x}\sin k\Delta x \tag{19.1.13}$$

whose modulus is > 1 for *all* k; so the FTCS scheme is unconditionally unstable.

If the velocity v were a function of t and x, then we would write v_j^n in equation (19.1.11). In the von Neumann stability analysis we would still treat v as a constant, the idea being that for v slowly varying the analysis is local. In fact, even in the case of strictly constant v, the von Neumann analysis does not rigorously treat the end effects at $j = 0$ and $j = N$.

More generally, if the equation's right-hand side were nonlinear in u, then a von Neumann analysis would linearize by writing $u = u_0 + \delta u$, expanding to linear order in δu. Assuming that the u_0 quantities already satisfy the difference equation exactly, the analysis would look for an unstable eigenmode of δu.

Despite its lack of rigor, the von Neumann method generally gives valid answers and is much easier to apply than more careful methods. We accordingly adopt it exclusively. (See, for example, [1] for a discussion of other methods of stability analysis.)

Lax Method

The instability in the FTCS method can be cured by a simple change due to Lax. One replaces the term u_j^n in the time derivative term by its average (Figure 19.1.2):

$$u_j^n \to \frac{1}{2}\left(u_{j+1}^n + u_{j-1}^n\right) \tag{19.1.14}$$

This turns (19.1.11) into

$$u_j^{n+1} = \frac{1}{2}\left(u_{j+1}^n + u_{j-1}^n\right) - \frac{v\Delta t}{2\Delta x}\left(u_{j+1}^n - u_{j-1}^n\right) \tag{19.1.15}$$

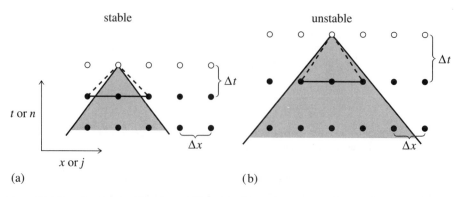

Figure 19.1.3.   Courant condition for stability of a differencing scheme. The solution of a hyperbolic problem at a point depends on information within some domain of dependency to the past, shown here shaded. The differencing scheme (19.1.15) has its own domain of dependency determined by the choice of points on one time slice (shown as connected solid dots) whose values are used in determining a new point (shown connected by dashed lines). A differencing scheme is Courant stable if the differencing domain of dependency is larger than that of the PDEs, as in (a), and unstable if the relationship is the reverse, as in (b). For more complicated differencing schemes, the domain of dependency might not be determined simply by the outermost points.

Substituting equation (19.1.12), we find for the amplification factor

$$\xi = \cos k\Delta x - i\frac{v\Delta t}{\Delta x}\sin k\Delta x \tag{19.1.16}$$

The stability condition $|\xi|^2 \le 1$ leads to the requirement

$$\frac{|v|\Delta t}{\Delta x} \le 1 \tag{19.1.17}$$

This is the famous Courant-Friedrichs-Lewy stability criterion, often called simply the *Courant condition*. Intuitively, the stability condition can be understood as follows (Figure 19.1.3): The quantity u_j^{n+1} in equation (19.1.15) is computed from information at points $j-1$ and $j+1$ at time n. In other words, x_{j-1} and x_{j+1} are the boundaries of the spatial region that is allowed to communicate information to u_j^{n+1}. Now recall that in the continuum wave equation, information actually propagates with a maximum velocity v. If the point u_j^{n+1} is outside of the shaded region in Figure 19.1.3, then it requires information from points more distant than the differencing scheme allows. Lack of that information gives rise to an instability. Therefore, Δt cannot be made too large.

The surprising result, that the simple replacement (19.1.14) stabilizes the FTCS scheme, is our first encounter with the fact that differencing PDEs is an art as much as a science. To see if we can demystify the art somewhat, let us compare the FTCS and Lax schemes by rewriting equation (19.1.15) so that it is in the form of equation (19.1.11) with a remainder term:

$$\frac{u_j^{n+1} - u_j^n}{\Delta t} = -v\left(\frac{u_{j+1}^n - u_{j-1}^n}{2\Delta x}\right) + \frac{1}{2}\left(\frac{u_{j+1}^n - 2u_j^n + u_{j-1}^n}{\Delta t}\right) \tag{19.1.18}$$

But this is exactly the FTCS representation of the equation

$$\frac{\partial u}{\partial t} = -v\frac{\partial u}{\partial x} + \frac{(\Delta x)^2}{2\Delta t}\nabla^2 u \tag{19.1.19}$$

where $\nabla^2 = \partial^2/\partial x^2$ in one dimension. We have, in effect, added a diffusion term to the equation, or, if you recall the form of the Navier-Stokes equation for viscous fluid flow, a dissipative term. The Lax scheme is thus said to have *numerical dissipation*, or *numerical viscosity*. We can see this also in the amplification factor. Unless $|v|\Delta t$ is exactly equal to Δx, $|\xi| < 1$ and the amplitude of the wave decreases spuriously.

Isn't a spurious decrease as bad as a spurious increase? No. The scales that we hope to study accurately are those that encompass many grid points, so that they have $k\Delta x \ll 1$. (The spatial wave number k is defined by equation 19.1.12.) For these scales, the amplification factor can be seen to be very close to one, in both the stable and unstable schemes. The stable and unstable schemes are therefore about equally accurate. For the unstable scheme, however, short scales with $k\Delta x \sim 1$, *which we are not interested in*, will blow up and swamp the interesting part of the solution. Much better to have a stable scheme in which these short wavelengths die away innocuously. Both the stable and the unstable schemes are *inaccurate* for these short wavelengths, but the inaccuracy is of a tolerable character when the scheme is stable.

When the independent variable **u** is a vector, then the von Neumann analysis is slightly more complicated. For example, we can consider equation (19.1.3), rewritten as

$$\frac{\partial}{\partial t}\begin{bmatrix} r \\ s \end{bmatrix} = \frac{\partial}{\partial x}\begin{bmatrix} vs \\ vr \end{bmatrix} \tag{19.1.20}$$

The Lax method for this equation is

$$
\begin{aligned}
r_j^{n+1} &= \frac{1}{2}(r_{j+1}^n + r_{j-1}^n) + \frac{v\Delta t}{2\Delta x}(s_{j+1}^n - s_{j-1}^n) \\
s_j^{n+1} &= \frac{1}{2}(s_{j+1}^n + s_{j-1}^n) + \frac{v\Delta t}{2\Delta x}(r_{j+1}^n - r_{j-1}^n)
\end{aligned}
\tag{19.1.21}
$$

The von Neumann stability analysis now proceeds by assuming that the eigenmode is of the following (vector) form,

$$\begin{bmatrix} r_j^n \\ s_j^n \end{bmatrix} = \xi^n e^{ikj\Delta x}\begin{bmatrix} r^0 \\ s^0 \end{bmatrix} \tag{19.1.22}$$

Here the vector on the right-hand side is a constant (both in space and in time) eigenvector, and ξ is a complex number, as before. Substituting (19.1.22) into (19.1.21), and dividing by the power ξ^n, gives the homogeneous vector equation

$$\begin{bmatrix} (\cos k\Delta x) - \xi & i\dfrac{v\Delta t}{\Delta x}\sin k\Delta x \\ i\dfrac{v\Delta t}{\Delta x}\sin k\Delta x & (\cos k\Delta x) - \xi \end{bmatrix} \cdot \begin{bmatrix} r^0 \\ s^0 \end{bmatrix} = \begin{bmatrix} 0 \\ 0 \end{bmatrix} \tag{19.1.23}$$

This admits a solution only if the determinant of the matrix on the left vanishes, a condition easily shown to yield the two roots ξ

$$\xi = \cos k\Delta x \pm i\frac{v\Delta t}{\Delta x}\sin k\Delta x \tag{19.1.24}$$

The stability condition is that both roots satisfy $|\xi| \le 1$. This again turns out to be simply the Courant condition (19.1.17).

Other Varieties of Error

Thus far we have been concerned with *amplitude error*, because of its intimate connection with the stability or instability of a differencing scheme. Other varieties of error are relevant when we shift our concern to accuracy, rather than stability.

Finite-difference schemes for hyperbolic equations can exhibit dispersion, or *phase errors*. For example, equation (19.1.16) can be rewritten as

$$\xi = e^{-ik\Delta x} + i\left(1 - \frac{v\Delta t}{\Delta x}\right)\sin k\Delta x \qquad (19.1.25)$$

An arbitrary initial wave packet is a superposition of modes with different k's. At each timestep the modes get multiplied by different phase factors (19.1.25), depending on their value of k. If $\Delta t = \Delta x/v$, then the exact solution for each mode of a wave packet $f(x - vt)$ is obtained if each mode gets multiplied by $\exp(-ik\Delta x)$. For this value of Δt, equation (19.1.25) shows that the finite-difference solution gives the exact analytic result. However, if $v\Delta t/\Delta x$ is not exactly 1, the phase relations of the modes can become hopelessly garbled and the wave packet disperses. Note from (19.1.25) that the dispersion becomes large as soon as the wavelength becomes comparable to the grid spacing Δx.

A third type of error is one associated with nonlinear hyperbolic equations and is therefore sometimes called *nonlinear instability*. For example, a piece of the Euler or Navier-Stokes equations for fluid flow looks like

$$\frac{\partial v}{\partial t} = -v\frac{\partial v}{\partial x} + \ldots \qquad (19.1.26)$$

The nonlinear term in v can cause a transfer of energy in Fourier space from long wavelengths to short wavelengths. This results in a wave profile steepening until a vertical profile or "shock" develops. Since the von Neumann analysis suggests that the stability can depend on $k\Delta x$, a scheme that was stable for shallow profiles can become unstable for steep profiles. This kind of difficulty arises in a differencing scheme where the cascade in Fourier space is halted at the shortest wavelength representable on the grid, that is, at $k \sim 1/\Delta x$. If energy simply accumulates in these modes, it eventually swamps the energy in the long wavelength modes of interest.

Nonlinear instability and shock formation is thus somewhat controlled by numerical viscosity such as that discussed in connection with equation (19.1.18) above. In some fluid problems, however, shock formation is not merely an annoyance, but an actual physical behavior of the fluid whose detailed study is a goal. Then, numerical viscosity alone may not be adequate or sufficiently controllable. This is a complicated subject which we discuss further in the subsection on fluid dynamics, below.

For wave equations, propagation errors (amplitude or phase) are usually most worrisome. For advective equations, on the other hand, *transport errors* are usually of greater concern. In the Lax scheme, equation (19.1.15), a disturbance in the advected quantity u at mesh point j propagates to mesh points $j + 1$ and $j - 1$ at the next timestep. In reality, however, if the velocity v is positive then only mesh point $j + 1$ should be affected.

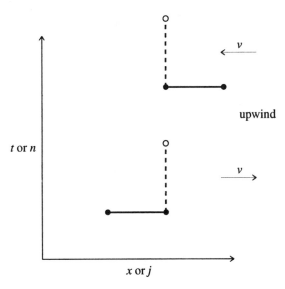

Figure 19.1.4.   Representation of upwind differencing schemes. The upper scheme is stable when the advection constant v is negative, as shown; the lower scheme is stable when the advection constant v is positive, also as shown. The Courant condition must, of course, also be satisfied.

The simplest way to model the transport properties "better" is to use *upwind differencing* (see Figure 19.1.4):

$$\frac{u_j^{n+1} - u_j^n}{\Delta t} = -v_j^n \begin{cases} \dfrac{u_j^n - u_{j-1}^n}{\Delta x}, & v_j^n > 0 \\[2mm] \dfrac{u_{j+1}^n - u_j^n}{\Delta x}, & v_j^n < 0 \end{cases} \tag{19.1.27}$$

Note that this scheme is only first-order, not second-order, accurate in the calculation of the spatial derivatives. How can it be "better"? The answer is one that annoys the mathematicians: The goal of numerical simulations is not always "accuracy" in a strictly mathematical sense, but sometimes "fidelity" to the underlying physics in a sense that is looser and more pragmatic. In such contexts, some kinds of error are much more tolerable than others. Upwind differencing generally adds fidelity to problems where the advected variables are liable to undergo sudden changes of state, e.g., as they pass through shocks or other discontinuities. You will have to be guided by the specific nature of your own problem.

For the differencing scheme (19.1.27), the amplification factor (for constant v) is

$$\xi = 1 - \left|\frac{v\Delta t}{\Delta x}\right|(1 - \cos k\Delta x) - i\frac{v\Delta t}{\Delta x}\sin k\Delta x \tag{19.1.28}$$

$$|\xi|^2 = 1 - 2\left|\frac{v\Delta t}{\Delta x}\right|\left(1 - \left|\frac{v\Delta t}{\Delta x}\right|\right)(1 - \cos k\Delta x) \tag{19.1.29}$$

So the stability criterion $|\xi|^2 \leq 1$ is (again) simply the Courant condition (19.1.17).

There are various ways of improving the accuracy of first-order upwind differencing. In the continuum equation, material originally a distance $v\Delta t$ away

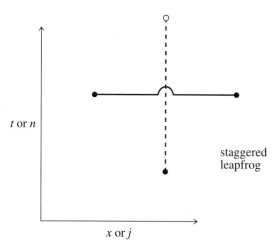

Figure 19.1.5.   Representation of the staggered leapfrog differencing scheme. Note that information from two previous time slices is used in obtaining the desired point. This scheme is second-order accurate in both space and time.

arrives at a given point after a time interval Δt. In the first-order method, the material always arrives from Δx away. If $v\Delta t \ll \Delta x$ (to insure accuracy), this can cause a large error. One way of reducing this error is to interpolate u between $j - 1$ and j before transporting it. This gives effectively a second-order method. Various schemes for second-order upwind differencing are discussed and compared in [2-3].

Second-Order Accuracy in Time

When using a method that is first-order accurate in time but second-order accurate in space, one generally has to take $v\Delta t$ significantly smaller than Δx to achieve desired accuracy, say, by at least a factor of 5. Thus the Courant condition is not actually the limiting factor with such schemes in practice. However, there are schemes that are second-order accurate in both space and time, and these can often be pushed right to their stability limit, with correspondingly smaller computation times.

For example, the *staggered leapfrog* method for the conservation equation (19.1.1) is defined as follows (Figure 19.1.5): Using the values of u^n at time t^n, compute the fluxes F_j^n. Then compute new values u^{n+1} using the time-centered values of the fluxes:

$$u_j^{n+1} - u_j^{n-1} = -\frac{\Delta t}{\Delta x}(F_{j+1}^n - F_{j-1}^n) \qquad (19.1.30)$$

The name comes from the fact that the time levels in the time derivative term "leapfrog" over the time levels in the space derivative term. The method requires that u^{n-1} and u^n be stored to compute u^{n+1}.

For our simple model equation (19.1.6), staggered leapfrog takes the form

$$u_j^{n+1} - u_j^{n-1} = -\frac{v\Delta t}{\Delta x}(u_{j+1}^n - u_{j-1}^n) \qquad (19.1.31)$$

The von Neumann stability analysis now gives a quadratic equation for ξ, rather than a linear one, because of the occurrence of three consecutive powers of ξ when the form (19.1.12) for an eigenmode is substituted into equation (19.1.31),

$$\xi^2 - 1 = -2i\xi \frac{v\Delta t}{\Delta x} \sin k\Delta x \tag{19.1.32}$$

whose solution is

$$\xi = -i \frac{v\Delta t}{\Delta x} \sin k\Delta x \pm \sqrt{1 - \left(\frac{v\Delta t}{\Delta x} \sin k\Delta x \right)^2} \tag{19.1.33}$$

Thus the Courant condition is again required for stability. In fact, in equation (19.1.33), $|\xi|^2 = 1$ for any $v\Delta t \leq \Delta x$. This is the great advantage of the staggered leapfrog method: There is no amplitude dissipation.

Staggered leapfrog differencing of equations like (19.1.20) is most transparent if the variables are centered on appropriate half-mesh points:

$$
\begin{aligned}
r^n_{j+1/2} &\equiv v \left. \frac{\partial u}{\partial x} \right|^n_{j+1/2} = v \frac{u^n_{j+1} - u^n_j}{\Delta x} \\
s^{n+1/2}_j &\equiv \left. \frac{\partial u}{\partial t} \right|^{n+1/2}_j = \frac{u^{n+1}_j - u^n_j}{\Delta t}
\end{aligned}
\tag{19.1.34}
$$

This is purely a notational convenience: we can think of the mesh on which r and s are defined as being twice as fine as the mesh on which the original variable u is defined. The leapfrog differencing of equation (19.1.20) is

$$
\begin{aligned}
\frac{r^{n+1}_{j+1/2} - r^n_{j+1/2}}{\Delta t} &= \frac{s^{n+1/2}_{j+1} - s^{n+1/2}_j}{\Delta x} \\
\frac{s^{n+1/2}_j - s^{n-1/2}_j}{\Delta t} &= v \frac{r^n_{j+1/2} - r^n_{j-1/2}}{\Delta x}
\end{aligned}
\tag{19.1.35}
$$

If you substitute equation (19.1.22) in equation (19.1.35), you will find that once again the Courant condition is required for stability, and that there is no amplitude dissipation when it is satisfied.

If we substitute equation (19.1.34) in equation (19.1.35), we find that equation (19.1.35) is equivalent to

$$\frac{u^{n+1}_j - 2u^n_j + u^{n-1}_j}{(\Delta t)^2} = v^2 \frac{u^n_{j+1} - 2u^n_j + u^n_{j-1}}{(\Delta x)^2} \tag{19.1.36}$$

This is just the "usual" second-order differencing of the wave equation (19.1.2). We see that it is a two-level scheme, requiring both u^n and u^{n-1} to obtain u^{n+1}. In equation (19.1.35) this shows up as both $s^{n-1/2}$ and r^n being needed to advance the solution.

For equations more complicated than our simple model equation, especially nonlinear equations, the leapfrog method usually becomes unstable when the gradients get large. The instability is related to the fact that odd and even mesh points are completely decoupled, like the black and white squares of a chess board, as shown

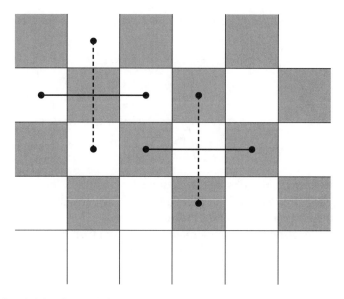

Figure 19.1.6.   Origin of mesh-drift instabilities in a staggered leapfrog scheme. If the mesh points are imagined to lie in the squares of a chess board, then white squares couple to themselves, black to themselves, but there is no coupling between white and black. The fix is to introduce a small diffusive mesh-coupling piece.

in Figure 19.1.6. This mesh drifting instability is cured by coupling the two meshes through a numerical viscosity term, e.g., adding to the right side of (19.1.31) a small coefficient ($\ll 1$) times $u_{j+1}^n - 2u_j^n + u_{j-1}^n$. For more on stabilizing difference schemes by adding numerical dissipation, see, e.g., [4].

The *Two-Step Lax-Wendroff* scheme is a second-order in time method that avoids large numerical dissipation and mesh drifting. One defines intermediate values $u_{j+1/2}$ at the half timesteps $t_{n+1/2}$ and the half mesh points $x_{j+1/2}$. These are calculated by the Lax scheme:

$$u_{j+1/2}^{n+1/2} = \frac{1}{2}(u_{j+1}^n + u_j^n) - \frac{\Delta t}{2\Delta x}(F_{j+1}^n - F_j^n) \tag{19.1.37}$$

Using these variables, one calculates the fluxes $F_{j+1/2}^{n+1/2}$. Then the updated values u_j^{n+1} are calculated by the properly centered expression

$$u_j^{n+1} = u_j^n - \frac{\Delta t}{\Delta x}\left(F_{j+1/2}^{n+1/2} - F_{j-1/2}^{n+1/2}\right) \tag{19.1.38}$$

The provisional values $u_{j+1/2}^{n+1/2}$ are now discarded. (See Figure 19.1.7.)

Let us investigate the stability of this method for our model advective equation, where $F = vu$. Substitute (19.1.37) in (19.1.38) to get

$$\begin{aligned} u_j^{n+1} = u_j^n - \alpha \Bigg[&\frac{1}{2}(u_{j+1}^n + u_j^n) - \frac{1}{2}\alpha(u_{j+1}^n - u_j^n) \\ &- \frac{1}{2}(u_j^n + u_{j-1}^n) + \frac{1}{2}\alpha(u_j^n - u_{j-1}^n)\Bigg] \end{aligned} \tag{19.1.39}$$

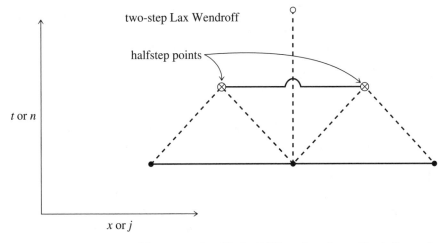

Figure 19.1.7. Representation of the two-step Lax-Wendroff differencing scheme. Two halfstep points (⊗) are calculated by the Lax method. These, plus one of the original points, produce the new point via staggered leapfrog. Halfstep points are used only temporarily and do not require storage allocation on the grid. This scheme is second-order accurate in both space and time.

where

$$\alpha \equiv \frac{v\Delta t}{\Delta x} \tag{19.1.40}$$

Then

$$\xi = 1 - i\alpha \sin k\Delta x - \alpha^2(1 - \cos k\Delta x) \tag{19.1.41}$$

so

$$|\xi|^2 = 1 - \alpha^2(1 - \alpha^2)(1 - \cos k\Delta x)^2 \tag{19.1.42}$$

The stability criterion $|\xi|^2 \leq 1$ is therefore $\alpha^2 \leq 1$, or $v\Delta t \leq \Delta x$ as usual. Incidentally, you should not think that the Courant condition is the only stability requirement that ever turns up in PDEs. It keeps doing so in our model examples just because those examples are so simple in form. The method of analysis is, however, general.

Except when $\alpha = 1$, $|\xi|^2 < 1$ in (19.1.42), so some amplitude damping does occur. The effect is relatively small, however, for wavelengths large compared with the mesh size Δx. If we expand (19.1.42) for small $k\Delta x$, we find

$$|\xi|^2 = 1 - \alpha^2(1 - \alpha^2)\frac{(k\Delta x)^4}{4} + \ldots \tag{19.1.43}$$

The departure from unity occurs only at fourth order in k. This should be contrasted with equation (19.1.16) for the Lax method, which shows that

$$|\xi|^2 = 1 - (1 - \alpha^2)(k\Delta x)^2 + \ldots \tag{19.1.44}$$

for small $k\Delta x$.

In summary, our recommendation for initial value problems that can be cast in flux-conservative form, and especially problems related to the wave equation, is to use the staggered leapfrog method when possible. We have personally had better success with it than with the Two-Step Lax-Wendroff method. For problems sensitive to transport errors, upwind differencing or one of its refinements should be considered.

Fluid Dynamics with Shocks

As we alluded to earlier, the treatment of fluid dynamics problems with shocks has become a very complicated and very sophisticated subject. All we can attempt to do here is to guide you to some starting points in the literature.

There are basically three important general methods for handling shocks. The oldest and simplest method, invented by von Neumann and Richtmyer, is to add *artificial viscosity* to the equations, modeling the way Nature uses real viscosity to smooth discontinuities. A good starting point for trying out this method is the differencing scheme in §12.11 of [1]. This scheme is excellent for nearly all problems in one spatial dimension.

The second method combines a high-order differencing scheme that is accurate for smooth flows with a low order scheme that is very dissipative and can smooth the shocks. Typically, various upwind differencing schemes are combined using weights chosen to zero the low order scheme unless steep gradients are present, and also chosen to enforce various "monotonicity" constraints that prevent nonphysical oscillations from appearing in the numerical solution. References [2-3,5] are a good place to start with these methods.

The third, and potentially most powerful method, is Godunov's approach. Here one gives up the simple linearization inherent in finite differencing based on Taylor series and includes the nonlinearity of the equations explicitly. There is an analytic solution for the evolution of two uniform states of a fluid separated by a discontinuity, the Riemann shock problem. Godunov's idea was to approximate the fluid by a large number of cells of uniform states, and piece them together using the Riemann solution. There have been many generalizations of Godunov's approach, of which the most powerful is probably the PPM method [6].

Readable reviews of all these methods, discussing the difficulties arising when one-dimensional methods are generalized to multidimensions, are given in [7-9].

CITED REFERENCES AND FURTHER READING:

Ames, W.F. 1977, *Numerical Methods for Partial Differential Equations*, 2nd ed. (New York: Academic Press), Chapter 4.

Richtmyer, R.D., and Morton, K.W. 1967, *Difference Methods for Initial Value Problems*, 2nd ed. (New York: Wiley-Interscience). [1]

Centrella, J., and Wilson, J.R. 1984, *Astrophysical Journal Supplement*, vol. 54, pp. 229–249, Appendix B. [2]

Hawley, J.F., Smarr, L.L., and Wilson, J.R. 1984, *Astrophysical Journal Supplement*, vol. 55, pp. 211–246, §2c. [3]

Kreiss, H.-O. 1978, *Numerical Methods for Solving Time-Dependent Problems for Partial Differential Equations* (Montreal: University of Montreal Press), pp. 66ff. [4]

Harten, A., Lax, P.D., and Van Leer, B. 1983, *SIAM Review*, vol. 25, pp. 36–61. [5]

Woodward, P., and Colella, P. 1984, *Journal of Computational Physics*, vol. 54, pp. 174–201. [6]

Roache, P.J. 1976, *Computational Fluid Dynamics* (Albuquerque: Hermosa). [7]

Woodward, P., and Colella, P. 1984, *Journal of Computational Physics*, vol. 54, pp. 115–173. [8]

Rizzi, A., and Engquist, B. 1987, *Journal of Computational Physics*, vol. 72, pp. 1–69. [9]

19.2 Diffusive Initial Value Problems

Recall the model parabolic equation, the diffusion equation in one space dimension,

$$\frac{\partial u}{\partial t} = \frac{\partial}{\partial x}\left(D\frac{\partial u}{\partial x}\right) \qquad (19.2.1)$$

where D is the diffusion coefficient. Actually, this equation is a flux-conservative equation of the form considered in the previous section, with

$$F = -D\frac{\partial u}{\partial x} \qquad (19.2.2)$$

the flux in the x-direction. We will assume $D \geq 0$, otherwise equation (19.2.1) has physically unstable solutions: A small disturbance evolves to become more and more concentrated instead of dispersing. (Don't make the mistake of trying to find a stable differencing scheme for a problem whose underlying PDEs are themselves unstable!)

Even though (19.2.1) is of the form already considered, it is useful to consider it as a model in its own right. The particular form of flux (19.2.2), and its direct generalizations, occur quite frequently in practice. Moreover, we have already seen that numerical viscosity and artificial viscosity can introduce diffusive pieces like the right-hand side of (19.2.1) in many other situations.

Consider first the case when D is a constant. Then the equation

$$\frac{\partial u}{\partial t} = D\frac{\partial^2 u}{\partial x^2} \qquad (19.2.3)$$

can be differenced in the obvious way:

$$\frac{u_j^{n+1} - u_j^n}{\Delta t} = D\left[\frac{u_{j+1}^n - 2u_j^n + u_{j-1}^n}{(\Delta x)^2}\right] \qquad (19.2.4)$$

This is the FTCS scheme again, except that it is a second derivative that has been differenced on the right-hand side. But this makes a world of difference! The FTCS scheme was unstable for the hyperbolic equation; however, a quick calculation shows that the amplification factor for equation (19.2.4) is

$$\xi = 1 - \frac{4D\Delta t}{(\Delta x)^2}\sin^2\left(\frac{k\Delta x}{2}\right) \qquad (19.2.5)$$

The requirement $|\xi| \leq 1$ leads to the stability criterion

$$\frac{2D\Delta t}{(\Delta x)^2} \leq 1 \qquad (19.2.6)$$

The physical interpretation of the restriction (19.2.6) is that the maximum allowed timestep is, up to a numerical factor, the diffusion time across a cell of width Δx.

More generally, the diffusion time τ across a spatial scale of size λ is of order

$$\tau \sim \frac{\lambda^2}{D} \qquad (19.2.7)$$

Usually we are interested in modeling accurately the evolution of features with spatial scales $\lambda \gg \Delta x$. If we are limited to timesteps satisfying (19.2.6), we will need to evolve through of order $\lambda^2/(\Delta x)^2$ steps before things start to happen on the scale of interest. This number of steps is usually prohibitive. We must therefore find a stable way of taking timesteps comparable to, or perhaps — for accuracy — somewhat smaller than, the time scale of (19.2.7).

This goal poses an immediate "philosophical" question. Obviously the large timesteps that we propose to take are going to be woefully inaccurate for the small scales that we have decided not to be interested in. We want those scales to do something stable, "innocuous," and perhaps not too physically unreasonable. We want to build this innocuous behavior into our differencing scheme. What should it be?

There are two different answers, each of which has its pros and cons. The first answer is to seek a differencing scheme that drives small-scale features to their *equilibrium* forms, e.g., satisfying equation (19.2.3) with the left-hand side set to zero. This answer generally makes the best physical sense; but, as we will see, it leads to a differencing scheme ("fully implicit") that is only *first-order* accurate in time for the scales that we are interested in. The second answer is to let small-scale features *maintain* their initial amplitudes, so that the evolution of the larger-scale features of interest takes place superposed with a kind of "frozen in" (though fluctuating) background of small-scale stuff. This answer gives a differencing scheme ("Crank-Nicolson") that is *second-order* accurate in time. Toward the end of an evolution calculation, however, one might want to switch over to some steps of the other kind, to drive the small-scale stuff into equilibrium. Let us now see where these distinct differencing schemes come from:

Consider the following differencing of (19.2.3),

$$\frac{u_j^{n+1} - u_j^n}{\Delta t} = D \left[\frac{u_{j+1}^{n+1} - 2u_j^{n+1} + u_{j-1}^{n+1}}{(\Delta x)^2} \right] \qquad (19.2.8)$$

This is exactly like the FTCS scheme (19.2.4), except that the spatial derivatives on the right-hand side are evaluated at timestep $n+1$. Schemes with this character are called *fully implicit* or *backward time*, by contrast with FTCS (which is called *fully explicit*). To solve equation (19.2.8) one has to solve a set of simultaneous linear equations at each timestep for the u_j^{n+1}. Fortunately, this is a simple problem because the system is tridiagonal: Just group the terms in equation (19.2.8) appropriately:

$$-\alpha u_{j-1}^{n+1} + (1 + 2\alpha)u_j^{n+1} - \alpha u_{j+1}^{n+1} = u_j^n, \qquad j = 1, 2 \ldots J - 1 \qquad (19.2.9)$$

where

$$\alpha \equiv \frac{D \Delta t}{(\Delta x)^2} \qquad (19.2.10)$$

Supplemented by Dirichlet or Neumann boundary conditions at $j = 0$ and $j = J$, equation (19.2.9) is clearly a tridiagonal system, which can easily be solved at each timestep by the method of §2.4.

What is the behavior of (19.2.8) for very large timesteps? The answer is seen most clearly in (19.2.9), in the limit $\alpha \to \infty$ $(\Delta t \to \infty)$. Dividing by α, we see that the difference equations are just the finite-difference form of the equilibrium equation

$$\frac{\partial^2 u}{\partial x^2} = 0 \qquad (19.2.11)$$

What about stability? The amplification factor for equation (19.2.8) is

$$\xi = \frac{1}{1 + 4\alpha \sin^2\left(\dfrac{k\Delta x}{2}\right)} \qquad (19.2.12)$$

Clearly $|\xi| < 1$ for any stepsize Δt. The scheme is unconditionally stable. The details of the small-scale evolution from the initial conditions are obviously inaccurate for large Δt. But, as advertised, the correct equilibrium solution is obtained. This is the characteristic feature of implicit methods.

Here, on the other hand, is how one gets to the second of our above philosophical answers, combining the stability of an implicit method with the accuracy of a method that is second-order in both space and time. Simply form the average of the explicit and implicit FTCS schemes:

$$\frac{u_j^{n+1} - u_j^n}{\Delta t} = \frac{D}{2} \left[\frac{(u_{j+1}^{n+1} - 2u_j^{n+1} + u_{j-1}^{n+1}) + (u_{j+1}^n - 2u_j^n + u_{j-1}^n)}{(\Delta x)^2} \right]$$

$$(19.2.13)$$

Here both the left- and right-hand sides are centered at timestep $n + \frac{1}{2}$, so the method is second-order accurate in time as claimed. The amplification factor is

$$\xi = \frac{1 - 2\alpha \sin^2\left(\dfrac{k\Delta x}{2}\right)}{1 + 2\alpha \sin^2\left(\dfrac{k\Delta x}{2}\right)} \qquad (19.2.14)$$

so the method is stable for any size Δt. This scheme is called the *Crank-Nicolson* scheme, and is our recommended method for any simple diffusion problem (perhaps supplemented by a few fully implicit steps at the end). (See Figure 19.2.1.)

Now turn to some generalizations of the simple diffusion equation (19.2.3). Suppose first that the diffusion coefficient D is not constant, say $D = D(x)$. We can adopt either of two strategies. First, we can make an analytic change of variable

$$y = \int \frac{dx}{D(x)} \qquad (19.2.15)$$

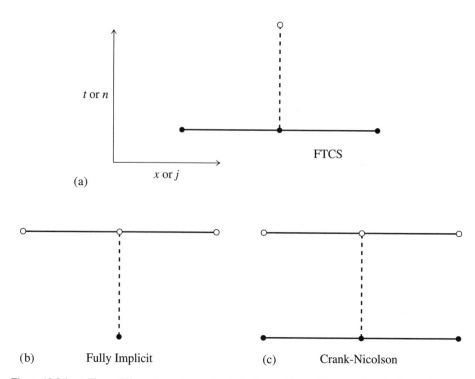

Figure 19.2.1.   Three differencing schemes for diffusive problems (shown as in Figure 19.1.2). (a) Forward Time Center Space is first-order accurate, but stable only for sufficiently small timesteps. (b) Fully Implicit is stable for arbitrarily large timesteps, but is still only first-order accurate. (c) Crank-Nicolson is second-order accurate, and is usually stable for large timesteps.

Then

$$\frac{\partial u}{\partial t} = \frac{\partial}{\partial x} D(x) \frac{\partial u}{\partial x} \tag{19.2.16}$$

becomes

$$\frac{\partial u}{\partial t} = \frac{1}{D(y)} \frac{\partial^2 u}{\partial y^2} \tag{19.2.17}$$

and we evaluate D at the appropriate y_j. Heuristically, the stability criterion (19.2.6) in an explicit scheme becomes

$$\Delta t \leq \min_j \left[\frac{(\Delta y)^2}{2D_j^{-1}} \right] \tag{19.2.18}$$

Note that constant spacing Δy in y does not imply constant spacing in x.

An alternative method that does not require analytically tractable forms for D is simply to difference equation (19.2.16) as it stands, centering everything appropriately. Thus the FTCS method becomes

$$\frac{u_j^{n+1} - u_j^n}{\Delta t} = \frac{D_{j+1/2}(u_{j+1}^n - u_j^n) - D_{j-1/2}(u_j^n - u_{j-1}^n)}{(\Delta x)^2} \tag{19.2.19}$$

where

$$D_{j+1/2} \equiv D(x_{j+1/2}) \tag{19.2.20}$$

and the heuristic stability criterion is

$$\Delta t \le \min_j \left[\frac{(\Delta x)^2}{2D_{j+1/2}} \right] \tag{19.2.21}$$

The Crank-Nicolson method can be generalized similarly.

The second complication one can consider is a nonlinear diffusion problem, for example where $D = D(u)$. Explicit schemes can be generalized in the obvious way. For example, in equation (19.2.19) write

$$D_{j+1/2} = \frac{1}{2} \left[D(u_{j+1}^n) + D(u_j^n) \right] \tag{19.2.22}$$

Implicit schemes are not as easy. The replacement (19.2.22) with $n \to n+1$ leaves us with a nasty set of coupled nonlinear equations to solve at each timestep. Often there is an easier way: If the form of $D(u)$ allows us to integrate

$$dz = D(u)du \tag{19.2.23}$$

analytically for $z(u)$, then the right-hand side of (19.2.1) becomes $\partial^2 z / \partial x^2$, which we difference implicitly as

$$\frac{z_{j+1}^{n+1} - 2z_j^{n+1} + z_{j-1}^{n+1}}{(\Delta x)^2} \tag{19.2.24}$$

Now linearize each term on the right-hand side of equation (19.2.24), for example

$$\begin{aligned} z_j^{n+1} \equiv z(u_j^{n+1}) &= z(u_j^n) + (u_j^{n+1} - u_j^n) \left. \frac{\partial z}{\partial u} \right|_{j,n} \\ &= z(u_j^n) + (u_j^{n+1} - u_j^n)D(u_j^n) \end{aligned} \tag{19.2.25}$$

This reduces the problem to tridiagonal form again and in practice usually retains the stability advantages of fully implicit differencing.

Schrödinger Equation

Sometimes the physical problem being solved imposes constraints on the differencing scheme that we have not yet taken into account. For example, consider the time-dependent Schrödinger equation of quantum mechanics. This is basically a parabolic equation for the evolution of a complex quantity ψ. For the scattering of a wavepacket by a one-dimensional potential $V(x)$, the equation has the form

$$i\frac{\partial \psi}{\partial t} = -\frac{\partial^2 \psi}{\partial x^2} + V(x)\psi \tag{19.2.26}$$

(Here we have chosen units so that Planck's constant $\hbar = 1$ and the particle mass $m = 1/2$.) One is given the initial wavepacket, $\psi(x, t = 0)$, together with boundary conditions that $\psi \to 0$ at $x \to \pm\infty$. Suppose we content ourselves with first-order accuracy in time, but want to use an implicit scheme, for stability. A slight generalization of (19.2.8) leads to

$$i\left[\frac{\psi_j^{n+1} - \psi_j^n}{\Delta t}\right] = -\left[\frac{\psi_{j+1}^{n+1} - 2\psi_j^{n+1} + \psi_{j-1}^{n+1}}{(\Delta x)^2}\right] + V_j \psi_j^{n+1} \qquad (19.2.27)$$

for which

$$\xi = \frac{1}{1 + i\left[\frac{4\Delta t}{(\Delta x)^2} \sin^2\left(\frac{k\Delta x}{2}\right) + V_j\Delta t\right]} \qquad (19.2.28)$$

This is unconditionally stable, but unfortunately is not *unitary*. The underlying physical problem requires that the total probability of finding the particle somewhere remains unity. This is represented formally by the modulus-square norm of ψ remaining unity:

$$\int_{-\infty}^{\infty} |\psi|^2 dx = 1 \qquad (19.2.29)$$

The initial wave function $\psi(x, 0)$ is normalized to satisfy (19.2.29). The Schrödinger equation (19.2.26) then guarantees that this condition is satisfied at all later times.

Let us write equation (19.2.26) in the form

$$i\frac{\partial\psi}{\partial t} = H\psi \qquad (19.2.30)$$

where the operator H is

$$H = -\frac{\partial^2}{\partial x^2} + V(x) \qquad (19.2.31)$$

The formal solution of equation (19.2.30) is

$$\psi(x, t) = e^{-iHt}\psi(x, 0) \qquad (19.2.32)$$

where the exponential of the operator is defined by its power series expansion.

The unstable explicit FTCS scheme approximates (19.2.32) as

$$\psi_j^{n+1} = (1 - iH\Delta t)\psi_j^n \qquad (19.2.33)$$

where H is represented by a centered finite-difference approximation in x. The stable implicit scheme (19.2.27) is, by contrast,

$$\psi_j^{n+1} = (1 + iH\Delta t)^{-1}\psi_j^n \qquad (19.2.34)$$

These are both first-order accurate in time, as can be seen by expanding equation (19.2.32). However, neither operator in (19.2.33) or (19.2.34) is unitary.

The correct way to difference Schrödinger's equation [1,2] is to use *Cayley's form* for the finite-difference representation of e^{-iHt}, which is second-order accurate *and* unitary:

$$e^{-iHt} \simeq \frac{1 - \frac{1}{2}iH\Delta t}{1 + \frac{1}{2}iH\Delta t} \qquad (19.2.35)$$

In other words,

$$\left(1 + \tfrac{1}{2}iH\Delta t\right)\psi_j^{n+1} = \left(1 - \tfrac{1}{2}iH\Delta t\right)\psi_j^n \qquad (19.2.36)$$

On replacing H by its finite-difference approximation in x, we have a complex tridiagonal system to solve. The method is stable, unitary, and second-order accurate in space and time. In fact, it is simply the Crank-Nicolson method once again!

CITED REFERENCES AND FURTHER READING:

Ames, W.F. 1977, *Numerical Methods for Partial Differential Equations*, 2nd ed. (New York: Academic Press), Chapter 2.

Goldberg, A., Schey, H.M., and Schwartz, J.L. 1967, *American Journal of Physics*, vol. 35, pp. 177–186. [1]

Galbraith, I., Ching, Y.S., and Abraham, E. 1984, *American Journal of Physics*, vol. 52, pp. 60–68. [2]

19.3 Initial Value Problems in Multidimensions

The methods described in §19.1 and §19.2 for problems in $1+1$ dimension (one space and one time dimension) can easily be generalized to $N+1$ dimensions. However, the computing power necessary to solve the resulting equations is enormous. If you have solved a one-dimensional problem with 100 spatial grid points, solving the two-dimensional version with 100×100 mesh points requires *at least* 100 times as much computing. You generally have to be content with very modest spatial resolution in multidimensional problems.

Indulge us in offering a bit of advice about the development and testing of multidimensional PDE codes: You should always first run your programs on *very small* grids, e.g., 8×8, even though the resulting accuracy is so poor as to be useless. When your program is all debugged and demonstrably stable, *then* you can increase the grid size to a reasonable one and start looking at the results. We have actually heard someone protest, "my program would be unstable for a crude grid, but I am sure the instability will go away on a larger grid." That is nonsense of a most pernicious sort, evidencing total confusion between accuracy and stability. In fact, new instabilities sometimes do show up on *larger* grids; but old instabilities never (in our experience) just go away.

Forced to live with modest grid sizes, some people recommend going to higher-order methods in an attempt to improve accuracy. This is very dangerous. Unless the solution you are looking for is known to be smooth, and the high-order method you are using is known to be extremely stable, we do not recommend anything higher than second-order in time (for sets of first-order equations). For spatial differencing, we recommend the order of the underlying PDEs, perhaps allowing second-order spatial differencing for first-order-in-space PDEs. When you increase the order of a differencing method to greater than the order of the original PDEs, you introduce spurious solutions to the difference equations. This does not create a problem if they all happen to decay exponentially; otherwise you are going to see all hell break loose!

Lax Method for a Flux-Conservative Equation

As an example, we show how to generalize the Lax method (19.1.15) to two dimensions for the conservation equation

$$\frac{\partial u}{\partial t} = -\nabla \cdot \mathbf{F} = -\left(\frac{\partial F_x}{\partial x} + \frac{\partial F_y}{\partial y} \right) \tag{19.3.1}$$

Use a spatial grid with

$$
\begin{aligned}
x_j &= x_0 + j\Delta \\
y_l &= y_0 + l\Delta
\end{aligned}
\tag{19.3.2}
$$

We have chosen $\Delta x = \Delta y \equiv \Delta$ for simplicity. Then the Lax scheme is

$$
\begin{aligned}
u_{j,l}^{n+1} = \frac{1}{4}&(u_{j+1,l}^n + u_{j-1,l}^n + u_{j,l+1}^n + u_{j,l-1}^n) \\
&- \frac{\Delta t}{2\Delta}(F_{j+1,l}^n - F_{j-1,l}^n + F_{j,l+1}^n - F_{j,l-1}^n)
\end{aligned}
\tag{19.3.3}
$$

Note that as an abbreviated notation F_{j+1} and F_{j-1} refer to F_x, while F_{l+1} and F_{l-1} refer to F_y.

Let us carry out a stability analysis for the model advective equation (analog of 19.1.6) with

$$F_x = v_x u, \qquad F_y = v_y u \tag{19.3.4}$$

This requires an eigenmode with two dimensions in space, though still only a simple dependence on powers of ξ in time,

$$u_{j,l}^n = \xi^n e^{ik_x j\Delta} e^{ik_y l\Delta} \tag{19.3.5}$$

Substituting in equation (19.3.3), we find

$$\xi = \frac{1}{2}(\cos k_x\Delta + \cos k_y\Delta) - i\alpha_x \sin k_x\Delta - i\alpha_y \sin k_y\Delta \tag{19.3.6}$$

where

$$\alpha_x = \frac{v_x \Delta t}{\Delta}, \qquad \alpha_y = \frac{v_y \Delta t}{\Delta} \tag{19.3.7}$$

The expression for $|\xi|^2$ can be manipulated into the form

$$
\begin{aligned}
|\xi|^2 = 1 - (\sin^2 k_x \Delta + \sin^2 k_y \Delta) \left[\frac{1}{2} - (\alpha_x^2 + \alpha_y^2) \right] \\
- \frac{1}{4}(\cos k_x \Delta - \cos k_y \Delta)^2 - (\alpha_y \sin k_x \Delta - \alpha_x \sin k_y \Delta)^2
\end{aligned}
\tag{19.3.8}
$$

The last two terms are negative, and so the stability requirement $|\xi|^2 \leq 1$ becomes

$$
\frac{1}{2} - (\alpha_x^2 + \alpha_y^2) \geq 0
\tag{19.3.9}
$$

or

$$
\Delta t \leq \frac{\Delta}{\sqrt{2}(v_x^2 + v_y^2)^{1/2}}
\tag{19.3.10}
$$

This is an example of the general result for the N-dimensional Courant condition: If $|v|$ is the maximum propagation velocity in the problem, then

$$
\Delta t \leq \frac{\Delta}{\sqrt{N}|v|}
\tag{19.3.11}
$$

is the Courant condition.

Diffusion Equation in Multidimensions

Let us consider the two-dimensional diffusion equation,

$$
\frac{\partial u}{\partial t} = D \left(\frac{\partial^2 u}{\partial x^2} + \frac{\partial^2 u}{\partial y^2} \right)
\tag{19.3.12}
$$

An explicit method, such as FTCS, can be generalized from the one-dimensional case in the obvious way. However, we have seen that diffusive problems are usually best treated implicitly. Suppose we try to implement the Crank-Nicolson scheme in two dimensions. This would give us

$$
u_{j,l}^{n+1} = u_{j,l}^n + \frac{1}{2}\alpha \left(\delta_x^2 u_{j,l}^{n+1} + \delta_x^2 u_{j,l}^n + \delta_y^2 u_{j,l}^{n+1} + \delta_y^2 u_{j,l}^n \right)
\tag{19.3.13}
$$

Here

$$
\alpha \equiv \frac{D\Delta t}{\Delta^2} \qquad \Delta \equiv \Delta x = \Delta y
\tag{19.3.14}
$$

$$
\delta_x^2 u_{j,l}^n \equiv u_{j+1,l}^n - 2u_{j,l}^n + u_{j-1,l}^n
\tag{19.3.15}
$$

and similarly for $\delta_y^2 u_{j,l}^n$. This is certainly a viable scheme; the problem arises in solving the coupled linear equations. Whereas in one space dimension the system was tridiagonal, that is no longer true, though the matrix is still very sparse. One possibility is to use a suitable sparse matrix technique (see §2.7 and §19.0).

Another possibility, which we generally prefer, is a slightly different way of generalizing the Crank-Nicolson algorithm. It is still second-order accurate in time and space, and unconditionally stable, but the equations are easier to solve than

(19.3.13). Called the *alternating-direction implicit method (ADI)*, this embodies the powerful concept of *operator splitting* or *time splitting*, about which we will say more below. Here, the idea is to divide each timestep into two steps of size $\Delta t/2$. In each substep, a different dimension is treated implicitly:

$$
\begin{aligned}
u_{j,l}^{n+1/2} &= u_{j,l}^{n} + \frac{1}{2}\alpha \left(\delta_x^2 u_{j,l}^{n+1/2} + \delta_y^2 u_{j,l}^{n} \right) \\
u_{j,l}^{n+1} &= u_{j,l}^{n+1/2} + \frac{1}{2}\alpha \left(\delta_x^2 u_{j,l}^{n+1/2} + \delta_y^2 u_{j,l}^{n+1} \right)
\end{aligned}
\tag{19.3.16}
$$

The advantage of this method is that each substep requires only the solution of a simple tridiagonal system.

Operator Splitting Methods Generally

The basic idea of operator splitting, which is also called *time splitting* or *the method of fractional steps*, is this: Suppose you have an initial value equation of the form

$$
\frac{\partial u}{\partial t} = \mathcal{L}u
\tag{19.3.17}
$$

where $\mathcal{L}$ is some operator. While $\mathcal{L}$ is not necessarily linear, suppose that it can at least be written as a linear sum of m pieces, which act additively on u,

$$
\mathcal{L}u = \mathcal{L}_1 u + \mathcal{L}_2 u + \cdots + \mathcal{L}_m u
\tag{19.3.18}
$$

Finally, suppose that for *each* of the pieces, you already know a differencing scheme for updating the variable u from timestep n to timestep $n+1$, valid if that piece of the operator were the *only* one on the right-hand side. We will write these updatings symbolically as

$$
\begin{aligned}
u^{n+1} &= \mathcal{U}_1(u^n, \Delta t) \\
u^{n+1} &= \mathcal{U}_2(u^n, \Delta t) \\
&\cdots \\
u^{n+1} &= \mathcal{U}_m(u^n, \Delta t)
\end{aligned}
\tag{19.3.19}
$$

Now, one form of operator splitting would be to get from n to $n+1$ by the following sequence of updatings:

$$
\begin{aligned}
u^{n+(1/m)} &= \mathcal{U}_1(u^n, \Delta t) \\
u^{n+(2/m)} &= \mathcal{U}_2(u^{n+(1/m)}, \Delta t) \\
&\cdots \\
u^{n+1} &= \mathcal{U}_m(u^{n+(m-1)/m}, \Delta t)
\end{aligned}
\tag{19.3.20}
$$

For example, a combined advective-diffusion equation, such as

$$\frac{\partial u}{\partial t} = -v\frac{\partial u}{\partial x} + D\frac{\partial^2 u}{\partial x^2} \tag{19.3.21}$$

might profitably use an explicit scheme for the advective term combined with a Crank-Nicolson or other implicit scheme for the diffusion term.

The alternating-direction implicit (ADI) method, equation (19.3.16), is an example of operator splitting with a slightly different twist. Let us reinterpret (19.3.19) to have a different meaning: Let $\mathcal{U}_1$ now denote an updating method that includes algebraically *all* the pieces of the total operator $\mathcal{L}$, but which is desirably *stable* only for the $\mathcal{L}_1$ piece; likewise $\mathcal{U}_2, \ldots \mathcal{U}_m$. Then a method of getting from u^n to u^{n+1} is

$$u^{n+1/m} = \mathcal{U}_1(u^n, \Delta t/m)$$

$$u^{n+2/m} = \mathcal{U}_2(u^{n+1/m}, \Delta t/m)$$

$$\cdots \tag{19.3.22}$$

$$u^{n+1} = \mathcal{U}_m(u^{n+(m-1)/m}, \Delta t/m)$$

The timestep for each fractional step in (19.3.22) is now only $1/m$ of the full timestep, because each partial operation acts with all the terms of the original operator.

Equation (19.3.22) is usually, though not always, stable as a differencing scheme for the operator $\mathcal{L}$. In fact, as a rule of thumb, it is often sufficient to have stable $\mathcal{U}_i$'s only for the operator pieces having the highest number of spatial derivatives — the other $\mathcal{U}_i$'s can be *unstable* — to make the overall scheme stable!

It is at this point that we turn our attention from initial value problems to boundary value problems. These will occupy us for the remainder of the chapter.

CITED REFERENCES AND FURTHER READING:

Ames, W.F. 1977, *Numerical Methods for Partial Differential Equations*, 2nd ed. (New York: Academic Press).

19.4 Fourier and Cyclic Reduction Methods for Boundary Value Problems

As discussed in §19.0, most boundary value problems (elliptic equations, for example) reduce to solving large sparse linear systems of the form

$$\mathbf{A} \cdot \mathbf{u} = \mathbf{b} \tag{19.4.1}$$

either once, for boundary value equations that are linear, or iteratively, for boundary value equations that are nonlinear.

Two important techniques lead to "rapid" solution of equation (19.4.1) when the sparse matrix is of certain frequently occurring forms. The *Fourier transform method* is directly applicable when the equations have coefficients that are constant in space. The *cyclic reduction* method is somewhat more general; its applicability is related to the question of whether the equations are separable (in the sense of "separation of variables"). Both methods require the boundaries to coincide with the coordinate lines. Finally, for some problems, there is a powerful combination of these two methods called *FACR (Fourier Analysis and Cyclic Reduction)*. We now consider each method in turn, using equation (19.0.3), with finite-difference representation (19.0.6), as a model example. Generally speaking, the methods in this section are faster, when they apply, than the simpler relaxation methods discussed in §19.5; but they are not necessarily faster than the more complicated multigrid methods discussed in §19.6.

Fourier Transform Method

The discrete inverse Fourier transform in both x and y is

$$u_{jl} = \frac{1}{JL} \sum_{m=0}^{J-1} \sum_{n=0}^{L-1} \widehat{u}_{mn} e^{-2\pi i jm/J} e^{-2\pi i ln/L} \tag{19.4.2}$$

This can be computed using the FFT independently in each dimension, or else all at once via the routine fourn of §12.4 or the routine rlft3 of §12.5. Similarly,

$$\rho_{jl} = \frac{1}{JL} \sum_{m=0}^{J-1} \sum_{n=0}^{L-1} \widehat{\rho}_{mn} e^{-2\pi i jm/J} e^{-2\pi i ln/L} \tag{19.4.3}$$

If we substitute expressions (19.4.2) and (19.4.3) in our model problem (19.0.6), we find

$$\widehat{u}_{mn} \left(e^{2\pi im/J} + e^{-2\pi im/J} + e^{2\pi in/L} + e^{-2\pi in/L} - 4 \right) = \widehat{\rho}_{mn} \Delta^2 \tag{19.4.4}$$

or

$$\widehat{u}_{mn} = \frac{\widehat{\rho}_{mn} \Delta^2}{2 \left(\cos \dfrac{2\pi m}{J} + \cos \dfrac{2\pi n}{L} - 2 \right)} \tag{19.4.5}$$

Thus the strategy for solving equation (19.0.6) by FFT techniques is:
• Compute $\widehat{\rho}_{mn}$ as the Fourier transform

$$\widehat{\rho}_{mn} = \sum_{j=0}^{J-1} \sum_{l=0}^{L-1} \rho_{jl}\, e^{2\pi imj/J} e^{2\pi inl/L} \tag{19.4.6}$$

• Compute $\widehat{u}_{mn}$ from equation (19.4.5).

- Compute u_{jl} by the inverse Fourier transform (19.4.2).

The above procedure is valid for periodic boundary conditions. In other words, the solution satisfies

$$u_{jl} = u_{j+J,l} = u_{j,l+L} \tag{19.4.7}$$

Next consider a Dirichlet boundary condition $u = 0$ on the rectangular boundary. Instead of the expansion (19.4.2), we now need an expansion in sine waves:

$$u_{jl} = \frac{2}{J}\frac{2}{L} \sum_{m=1}^{J-1} \sum_{n=1}^{L-1} \widehat{u}_{mn} \sin\frac{\pi jm}{J} \sin\frac{\pi ln}{L} \tag{19.4.8}$$

This satisfies the boundary conditions that $u = 0$ at $j = 0, J$ and at $l = 0, L$. If we substitute this expansion and the analogous one for ρ_{jl} into equation (19.0.6), we find that the solution procedure parallels that for periodic boundary conditions:

- Compute $\widehat{\rho}_{mn}$ by the sine transform

$$\widehat{\rho}_{mn} = \sum_{j=1}^{J-1} \sum_{l=1}^{L-1} \rho_{jl} \sin\frac{\pi jm}{J} \sin\frac{\pi ln}{L} \tag{19.4.9}$$

(A fast sine transform algorithm was given in §12.3.)
- Compute $\widehat{u}_{mn}$ from the expression analogous to (19.4.5),

$$\widehat{u}_{mn} = \frac{\Delta^2 \widehat{\rho}_{mn}}{2\left(\cos\dfrac{\pi m}{J} + \cos\dfrac{\pi n}{L} - 2\right)} \tag{19.4.10}$$

- Compute u_{jl} by the inverse sine transform (19.4.8).

If we have inhomogeneous boundary conditions, for example $u = 0$ on all boundaries except $u = f(y)$ on the boundary $x = J\Delta$, we have to add to the above solution a solution u^H of the homogeneous equation

$$\frac{\partial^2 u}{\partial x^2} + \frac{\partial^2 u}{\partial y^2} = 0 \tag{19.4.11}$$

that satisfies the required boundary conditions. In the continuum case, this would be an expression of the form

$$u^H = \sum_n A_n \sinh\frac{n\pi x}{L\Delta} \sin\frac{n\pi y}{L\Delta} \tag{19.4.12}$$

where A_n would be found by requiring that $u = f(y)$ at $x = J\Delta$. In the discrete case, we have

$$u_{jl}^H = \frac{2}{L} \sum_{n=1}^{L-1} A_n \sinh\frac{\pi nj}{L} \sin\frac{\pi nl}{L} \tag{19.4.13}$$

If $f(y = l\Delta) \equiv f_l$, then we get A_n from the inverse formula

$$A_n = \frac{1}{\sinh(\pi n J/L)} \sum_{l=1}^{L-1} f_l \sin \frac{\pi nl}{L} \tag{19.4.14}$$

The complete solution to the problem is

$$u = u_{jl} + u_{jl}^H \tag{19.4.15}$$

By adding appropriate terms of the form (19.4.12), we can handle inhomogeneous terms on any boundary surface.

A much simpler procedure for handling inhomogeneous terms is to note that whenever boundary terms appear on the left-hand side of (19.0.6), they can be taken over to the right-hand side since they are known. The effective source term is therefore ρ_{jl} plus a contribution from the boundary terms. To implement this idea formally, write the solution as

$$u = u' + u^B \tag{19.4.16}$$

where $u' = 0$ on the boundary, while u^B vanishes everywhere *except* on the boundary. There it takes on the given boundary value. In the above example, the only nonzero values of u^B would be

$$u_{J,l}^B = f_l \tag{19.4.17}$$

The model equation (19.0.3) becomes

$$\nabla^2 u' = -\nabla^2 u^B + \rho \tag{19.4.18}$$

or, in finite-difference form,

$$u'_{j+1,l} + u'_{j-1,l} + u'_{j,l+1} + u'_{j,l-1} - 4u'_{j,l} = \\ - (u^B_{j+1,l} + u^B_{j-1,l} + u^B_{j,l+1} + u^B_{j,l-1} - 4u^B_{j,l}) + \Delta^2 \rho_{j,l} \tag{19.4.19}$$

All the u^B terms in equation (19.4.19) vanish except when the equation is evaluated at $j = J - 1$, where

$$u'_{J,l} + u'_{J-2,l} + u'_{J-1,l+1} + u'_{J-1,l-1} - 4u'_{J-1,l} = -f_l + \Delta^2 \rho_{J-1,l} \tag{19.4.20}$$

Thus the problem is now equivalent to the case of zero boundary conditions, except that one row of the source term is modified by the replacement

$$\Delta^2 \rho_{J-1,l} \rightarrow \Delta^2 \rho_{J-1,l} - f_l \tag{19.4.21}$$

The case of Neumann boundary conditions $\nabla u = 0$ is handled by the cosine expansion (12.3.17):

$$u_{jl} = \frac{2}{J}\frac{2}{L} \sum_{m=0}^{J}{}'' \sum_{n=0}^{L}{}'' \hat{u}_{mn} \cos\frac{\pi jm}{J} \cos\frac{\pi ln}{L} \tag{19.4.22}$$

Here the double prime notation means that the terms for $m = 0$ and $m = J$ should be multiplied by $\frac{1}{2}$, and similarly for $n = 0$ and $n = L$. Inhomogeneous terms $\nabla u = g$ can be again included by adding a suitable solution of the homogeneous equation, or more simply by taking boundary terms over to the right-hand side. For example, the condition

$$\frac{\partial u}{\partial x} = g(y) \qquad \text{at} \quad x = 0 \tag{19.4.23}$$

becomes

$$\frac{u_{1,l} - u_{-1,l}}{2\Delta} = g_l \tag{19.4.24}$$

where $g_l \equiv g(y = l\Delta)$. Once again we write the solution in the form (19.4.16), where now $\nabla u' = 0$ on the boundary. This time ∇u^B takes on the prescribed value on the boundary, but u^B vanishes everywhere except just *outside* the boundary. Thus equation (19.4.24) gives

$$u^B_{-1,l} = -2\Delta g_l \tag{19.4.25}$$

All the u^B terms in equation (19.4.19) vanish except when $j = 0$:

$$u'_{1,l} + u'_{-1,l} + u'_{0,l+1} + u'_{0,l-1} - 4u'_{0,l} = 2\Delta g_l + \Delta^2 \rho_{0,l} \tag{19.4.26}$$

Thus u' is the solution of a zero-gradient problem, with the source term modified by the replacement

$$\Delta^2 \rho_{0,l} \to \Delta^2 \rho_{0,l} + 2\Delta g_l \tag{19.4.27}$$

Sometimes Neumann boundary conditions are handled by using a staggered grid, with the u's defined midway between zone boundaries so that first derivatives are centered on the mesh points. You can solve such problems using similar techniques to those described above if you use the alternative form of the cosine transform, equation (12.3.23).

Cyclic Reduction

Evidently the FFT method works only when the original PDE has constant coefficients, and boundaries that coincide with the coordinate lines. An alternative algorithm, which can be used on somewhat more general equations, is called *cyclic reduction (CR)*.

We illustrate cyclic reduction on the equation

$$\frac{\partial^2 u}{\partial x^2} + \frac{\partial^2 u}{\partial y^2} + b(y)\frac{\partial u}{\partial y} + c(y)u = g(x, y) \tag{19.4.28}$$

This form arises very often in practice from the Helmholtz or Poisson equations in polar, cylindrical, or spherical coordinate systems. More general separable equations are treated in [1].

The finite-difference form of equation (19.4.28) can be written as a set of vector equations

$$\mathbf{u}_{j-1} + \mathbf{T} \cdot \mathbf{u}_j + \mathbf{u}_{j+1} = \mathbf{g}_j \Delta^2 \qquad (19.4.29)$$

Here the index j comes from differencing in the x-direction, while the y-differencing (denoted by the index l previously) has been left in vector form. The matrix $\mathbf{T}$ has the form

$$\mathbf{T} = \mathbf{B} - 2\mathbf{1} \qquad (19.4.30)$$

where the $2\mathbf{1}$ comes from the x-differencing and the matrix $\mathbf{B}$ from the y-differencing. The matrix $\mathbf{B}$, and hence $\mathbf{T}$, is tridiagonal with variable coefficients.

The CR method is derived by writing down three successive equations like (19.4.29):

$$\mathbf{u}_{j-2} + \mathbf{T} \cdot \mathbf{u}_{j-1} + \mathbf{u}_j = \mathbf{g}_{j-1} \Delta^2$$
$$\mathbf{u}_{j-1} + \mathbf{T} \cdot \mathbf{u}_j + \mathbf{u}_{j+1} = \mathbf{g}_j \Delta^2 \qquad (19.4.31)$$
$$\mathbf{u}_j + \mathbf{T} \cdot \mathbf{u}_{j+1} + \mathbf{u}_{j+2} = \mathbf{g}_{j+1} \Delta^2$$

Matrix-multiplying the middle equation by $-\mathbf{T}$ and then adding the three equations, we get

$$\mathbf{u}_{j-2} + \mathbf{T}^{(1)} \cdot \mathbf{u}_j + \mathbf{u}_{j+2} = \mathbf{g}_j^{(1)} \Delta^2 \qquad (19.4.32)$$

This is an equation of the same form as (19.4.29), with

$$\mathbf{T}^{(1)} = 2\mathbf{1} - \mathbf{T}^2$$
$$\mathbf{g}_j^{(1)} = \Delta^2 (\mathbf{g}_{j-1} - \mathbf{T} \cdot \mathbf{g}_j + \mathbf{g}_{j+1}) \qquad (19.4.33)$$

After one level of CR, we have reduced the number of equations by a factor of two. Since the resulting equations are of the same form as the original equation, we can repeat the process. Taking the number of mesh points to be a power of 2 for simplicity, we finally end up with a single equation for the central line of variables:

$$\mathbf{T}^{(f)} \cdot \mathbf{u}_{J/2} = \Delta^2 \mathbf{g}_{J/2}^{(f)} - \mathbf{u}_0 - \mathbf{u}_J \qquad (19.4.34)$$

Here we have moved $\mathbf{u}_0$ and $\mathbf{u}_J$ to the right-hand side because they are known boundary values. Equation (19.4.34) can be solved for $\mathbf{u}_{J/2}$ by the standard tridiagonal algorithm. The two equations at level $f - 1$ involve $\mathbf{u}_{J/4}$ and $\mathbf{u}_{3J/4}$. The equation for $\mathbf{u}_{J/4}$ involves $\mathbf{u}_0$ and $\mathbf{u}_{J/2}$, both of which are known, and hence can be solved by the usual tridiagonal routine. A similar result holds true at every stage, so we end up solving $J - 1$ tridiagonal systems.

In practice, equations (19.4.33) should be rewritten to avoid numerical instability. For these and other practical details, refer to [2].

FACR Method

The *best* way to solve equations of the form (19.4.28), including the constant coefficient problem (19.0.3), is a combination of Fourier analysis and cyclic reduction, the FACR method [3-6]. If at the rth stage of CR we Fourier analyze the equations of the form (19.4.32) along y, that is, with respect to the suppressed vector index, we will have a tridiagonal system in the x-direction for each y-Fourier mode:

$$\widehat{u}^k_{j-2^r} + \lambda^{(r)}_k \widehat{u}^k_j + \widehat{u}^k_{j+2^r} = \Delta^2 g^{(r)k}_j \qquad (19.4.35)$$

Here $\lambda^{(r)}_k$ is the eigenvalue of $\mathbf{T}^{(r)}$ corresponding to the kth Fourier mode. For the equation (19.0.3), equation (19.4.5) shows that $\lambda^{(r)}_k$ will involve terms like $\cos(2\pi k/L) - 2$ raised to a power. Solve the tridiagonal systems for $\widehat{u}^k_j$ at the levels $j = 2^r, 2 \times 2^r, 4 \times 2^r, ..., J - 2^r$. Fourier synthesize to get the y-values on these x-lines. Then fill in the intermediate x-lines as in the original CR algorithm.

The trick is to choose the number of levels of CR so as to minimize the total number of arithmetic operations. One can show that for a typical case of a 128×128 mesh, the optimal level is $r = 2$; asymptotically, $r \rightarrow \log_2(\log_2 J)$.

A rough estimate of running times for these algorithms for equation (19.0.3) is as follows: The FFT method (in both x and y) and the CR method are roughly comparable. FACR with $r = 0$ (that is, FFT in one dimension and solve the tridiagonal equations by the usual algorithm in the other dimension) gives about a factor of two gain in speed. The optimal FACR with $r = 2$ gives another factor of two gain in speed.

CITED REFERENCES AND FURTHER READING:

Swartzrauber, P.N. 1977, *SIAM Review*, vol. 19, pp. 490–501. [1]

Buzbee, B.L, Golub, G.H., and Nielson, C.W. 1970, *SIAM Journal on Numerical Analysis*, vol. 7, pp. 627–656; see also *op. cit.* vol. 11, pp. 753–763. [2]

Hockney, R.W. 1965, *Journal of the Association for Computing Machinery*, vol. 12, pp. 95–113. [3]

Hockney, R.W. 1970, in *Methods of Computational Physics*, vol. 9 (New York: Academic Press), pp. 135–211. [4]

Hockney, R.W., and Eastwood, J.W. 1981, *Computer Simulation Using Particles* (New York: McGraw-Hill), Chapter 6. [5]

Temperton, C. 1980, *Journal of Computational Physics*, vol. 34, pp. 314–329. [6]

19.5 Relaxation Methods for Boundary Value Problems

As we mentioned in §19.0, relaxation methods involve splitting the sparse matrix that arises from finite differencing and then iterating until a solution is found.

There is another way of thinking about relaxation methods that is somewhat more physical. Suppose we wish to solve the elliptic equation

$$\mathcal{L}u = \rho \qquad (19.5.1)$$

where $\mathcal{L}$ represents some elliptic operator and ρ is the source term. Rewrite the equation as a diffusion equation,

$$\frac{\partial u}{\partial t} = \mathcal{L}u - \rho \qquad (19.5.2)$$

An initial distribution u *relaxes* to an equilibrium solution as $t \to \infty$. This equilibrium has all time derivatives vanishing. Therefore it is the solution of the original elliptic problem (19.5.1). We see that all the machinery of §19.2, on diffusive initial value equations, can be brought to bear on the solution of boundary value problems by relaxation methods.

Let us apply this idea to our model problem (19.0.3). The diffusion equation is

$$\frac{\partial u}{\partial t} = \frac{\partial^2 u}{\partial x^2} + \frac{\partial^2 u}{\partial y^2} - \rho \qquad (19.5.3)$$

If we use FTCS differencing (cf. equation 19.2.4), we get

$$u_{j,l}^{n+1} = u_{j,l}^n + \frac{\Delta t}{\Delta^2} \left(u_{j+1,l}^n + u_{j-1,l}^n + u_{j,l+1}^n + u_{j,l-1}^n - 4u_{j,l}^n \right) - \rho_{j,l}\Delta t \quad (19.5.4)$$

Recall from (19.2.6) that FTCS differencing is stable in one spatial dimension only if $\Delta t/\Delta^2 \leq \frac{1}{2}$. In two dimensions this becomes $\Delta t/\Delta^2 \leq \frac{1}{4}$. Suppose we try to take the largest possible timestep, and set $\Delta t = \Delta^2/4$. Then equation (19.5.4) becomes

$$u_{j,l}^{n+1} = \frac{1}{4} \left(u_{j+1,l}^n + u_{j-1,l}^n + u_{j,l+1}^n + u_{j,l-1}^n \right) - \frac{\Delta^2}{4}\rho_{j,l} \qquad (19.5.5)$$

Thus the algorithm consists of using the average of u at its four nearest-neighbor points on the grid (plus the contribution from the source). This procedure is then iterated until convergence.

This method is in fact a classical method with origins dating back to the last century, called *Jacobi's method* (not to be confused with the Jacobi method for eigenvalues). The method is not practical because it converges too slowly. However, it is the basis for understanding the modern methods, which are always compared with it.

Another classical method is the *Gauss-Seidel* method, which turns out to be important in multigrid methods (§19.6). Here we make use of updated values of u on the right-hand side of (19.5.5) as soon as they become available. In other words, the averaging is done "in place" instead of being "copied" from an earlier timestep to a later one. If we are proceeding along the rows, incrementing j for fixed l, we have

$$u_{j,l}^{n+1} = \frac{1}{4} \left(u_{j+1,l}^n + u_{j-1,l}^{n+1} + u_{j,l+1}^n + u_{j,l-1}^{n+1} \right) - \frac{\Delta^2}{4}\rho_{j,l} \qquad (19.5.6)$$

This method is also slowly converging and only of theoretical interest when used by itself, but some analysis of it will be instructive.

Let us look at the Jacobi and Gauss-Seidel methods in terms of the matrix splitting concept. We change notation and call $\mathbf{u}$ "$\mathbf{x}$," to conform to standard matrix notation. To solve

$$\mathbf{A} \cdot \mathbf{x} = \mathbf{b} \qquad (19.5.7)$$

we can consider splitting $\mathbf{A}$ as

$$\mathbf{A} = \mathbf{L} + \mathbf{D} + \mathbf{U} \tag{19.5.8}$$

where $\mathbf{D}$ is the diagonal part of $\mathbf{A}$, $\mathbf{L}$ is the lower triangle of $\mathbf{A}$ with zeros on the diagonal, and $\mathbf{U}$ is the upper triangle of $\mathbf{A}$ with zeros on the diagonal.

In the Jacobi method we write for the rth step of iteration

$$\mathbf{D} \cdot \mathbf{x}^{(r)} = -(\mathbf{L} + \mathbf{U}) \cdot \mathbf{x}^{(r-1)} + \mathbf{b} \tag{19.5.9}$$

For our model problem (19.5.5), $\mathbf{D}$ is simply the identity matrix. The Jacobi method converges for matrices $\mathbf{A}$ that are "diagonally dominant" in a sense that can be made mathematically precise. For matrices arising from finite differencing, this condition is usually met.

What is the rate of convergence of the Jacobi method? A detailed analysis is beyond our scope, but here is some of the flavor: The matrix $-\mathbf{D}^{-1} \cdot (\mathbf{L} + \mathbf{U})$ is the *iteration matrix* which, apart from an additive term, maps one set of $\mathbf{x}$'s into the next. The iteration matrix has eigenvalues, each one of which reflects the factor by which the amplitude of a particular eigenmode of undesired residual is suppressed during one iteration. Evidently those factors had better all have modulus < 1 for the relaxation to work at all! The rate of convergence of the method is set by the rate for the slowest-decaying eigenmode, i.e., the factor with largest modulus. The modulus of this largest factor, therefore lying between 0 and 1, is called the *spectral radius* of the relaxation operator, denoted ρ_s.

The number of iterations r required to reduce the overall error by a factor 10^{-p} is thus estimated by

$$r \approx \frac{p \ln 10}{(-\ln \rho_s)} \tag{19.5.10}$$

In general, the spectral radius ρ_s goes asymptotically to the value 1 as the grid size J is increased, so that more iterations are required. For any given equation, grid geometry, *and boundary condition*, the spectral radius can, in principle, be computed analytically. For example, for equation (19.5.5) on a $J \times J$ grid with Dirichlet boundary conditions on all four sides, the asymptotic formula for large J turns out to be

$$\rho_s \simeq 1 - \frac{\pi^2}{2J^2} \tag{19.5.11}$$

The number of iterations r required to reduce the error by a factor of 10^{-p} is thus

$$r \simeq \frac{2pJ^2 \ln 10}{\pi^2} \simeq \frac{1}{2} pJ^2 \tag{19.5.12}$$

In other words, the number of iterations is proportional to the number of mesh points, J^2. Since 100×100 and larger problems are common, it is clear that the Jacobi method is only of theoretical interest.

The Gauss-Seidel method, equation (19.5.6), corresponds to the matrix decomposition

$$(\mathbf{L} + \mathbf{D}) \cdot \mathbf{x}^{(r)} = -\mathbf{U} \cdot \mathbf{x}^{(r-1)} + \mathbf{b} \tag{19.5.13}$$

The fact that $\mathbf{L}$ is on the left-hand side of the equation follows from the updating in place, as you can easily check if you write out (19.5.13) in components. One can show [1-3] that the spectral radius is just the square of the spectral radius of the Jacobi method. For our model problem, therefore,

$$\rho_s \simeq 1 - \frac{\pi^2}{J^2} \tag{19.5.14}$$

$$r \simeq \frac{pJ^2 \ln 10}{\pi^2} \simeq \frac{1}{4} pJ^2 \tag{19.5.15}$$

The factor of two improvement in the number of iterations over the Jacobi method still leaves the method impractical.

Successive Overrelaxation (SOR)

We get a better algorithm — one that was the standard algorithm until the 1970s — if we make an *overcorrection* to the value of $\mathbf{x}^{(r)}$ at the rth stage of Gauss-Seidel iteration, thus anticipating future corrections. Solve (19.5.13) for $\mathbf{x}^{(r)}$, add and subtract $\mathbf{x}^{(r-1)}$ on the right-hand side, and hence write the Gauss-Seidel method as

$$\mathbf{x}^{(r)} = \mathbf{x}^{(r-1)} - (\mathbf{L} + \mathbf{D})^{-1} \cdot [(\mathbf{L} + \mathbf{D} + \mathbf{U}) \cdot \mathbf{x}^{(r-1)} - \mathbf{b}] \tag{19.5.16}$$

The term in square brackets is just the residual vector $\xi^{(r-1)}$, so

$$\mathbf{x}^{(r)} = \mathbf{x}^{(r-1)} - (\mathbf{L} + \mathbf{D})^{-1} \cdot \xi^{(r-1)} \tag{19.5.17}$$

Now *overcorrect*, defining

$$\mathbf{x}^{(r)} = \mathbf{x}^{(r-1)} - \omega(\mathbf{L} + \mathbf{D})^{-1} \cdot \xi^{(r-1)} \tag{19.5.18}$$

Here ω is called the *overrelaxation parameter*, and the method is called *successive overrelaxation* (SOR).

The following theorems can be proved [1-3]:

- The method is convergent only for $0 < \omega < 2$. If $0 < \omega < 1$, we speak of *underrelaxation*.
- Under certain mathematical restrictions generally satisfied by matrices arising from finite differencing, only overrelaxation ($1 < \omega < 2$) can give faster convergence than the Gauss-Seidel method.
- If ρ_{Jacobi} is the spectral radius of the Jacobi iteration (so that the square of it is the spectral radius of the Gauss-Seidel iteration), then the *optimal* choice for ω is given by

$$\omega = \frac{2}{1 + \sqrt{1 - \rho_{\text{Jacobi}}^2}} \tag{19.5.19}$$

- For this optimal choice, the spectral radius for SOR is

$$\rho_{\text{SOR}} = \left(\frac{\rho_{\text{Jacobi}}}{1 + \sqrt{1 - \rho_{\text{Jacobi}}^2}} \right)^2 \tag{19.5.20}$$

As an application of the above results, consider our model problem for which ρ_{Jacobi} is given by equation (19.5.11). Then equations (19.5.19) and (19.5.20) give

$$\omega \simeq \frac{2}{1 + \pi/J} \tag{19.5.21}$$

$$\rho_{\text{SOR}} \simeq 1 - \frac{2\pi}{J} \qquad \text{for large} \quad J \tag{19.5.22}$$

Equation (19.5.10) gives for the number of iterations to reduce the initial error by a factor of 10^{-p},

$$r \simeq \frac{pJ \ln 10}{2\pi} \simeq \frac{1}{3} pJ \tag{19.5.23}$$

Comparing with equation (19.5.12) or (19.5.15), we see that optimal SOR requires of order J iterations, as opposed to of order J^2. Since J is typically 100 or larger, this makes a tremendous difference! Equation (19.5.23) leads to the mnemonic that 3-figure accuracy ($p = 3$) requires a number of iterations equal to the number of mesh points along a side of the grid. For 6-figure accuracy, we require about twice as many iterations.

How do we choose ω for a problem for which the answer is not known analytically? That is just the weak point of SOR! The advantages of SOR obtain only in a fairly narrow window around the correct value of ω. It is better to take ω slightly too large, rather than slightly too small, but best to get it right.

One way to choose ω is to map your problem approximately onto a known problem, replacing the coefficients in the equation by average values. Note, however, that the known problem must have the same grid size and boundary conditions as the actual problem. We give for reference purposes the value of ρ_{Jacobi} for our model problem on a rectangular $J \times L$ grid, allowing for the possibility that $\Delta x \neq \Delta y$:

$$\rho_{\text{Jacobi}} = \frac{\cos \dfrac{\pi}{J} + \left(\dfrac{\Delta x}{\Delta y} \right)^2 \cos \dfrac{\pi}{L}}{1 + \left(\dfrac{\Delta x}{\Delta y} \right)^2} \tag{19.5.24}$$

Equation (19.5.24) holds for homogeneous Dirichlet or Neumann boundary conditions. For periodic boundary conditions, make the replacement $\pi \to 2\pi$.

A second way, which is especially useful if you plan to solve many similar elliptic equations each time with slightly different coefficients, is to determine the optimum value ω empirically on the first equation and then use that value for the remaining equations. Various automated schemes for doing this and for "seeking out" the best values of ω are described in the literature.

While the matrix notation introduced earlier is useful for theoretical analyses, for practical implementation of the SOR algorithm we need explicit formulas.

Consider a general second-order elliptic equation in x and y, finite differenced on a square as for our model equation. Corresponding to each row of the matrix $\mathbf{A}$ is an equation of the form

$$a_{j,l}u_{j+1,l} + b_{j,l}u_{j-1,l} + c_{j,l}u_{j,l+1} + d_{j,l}u_{j,l-1} + e_{j,l}u_{j,l} = f_{j,l} \qquad (19.5.25)$$

For our model equation, we had $a = b = c = d = 1, e = -4$. The quantity f is proportional to the source term. The iterative procedure is defined by solving (19.5.25) for $u_{j,l}$:

$$u^*{}_{j,l} = \frac{1}{e_{j,l}}\left(f_{j,l} - a_{j,l}u_{j+1,l} - b_{j,l}u_{j-1,l} - c_{j,l}u_{j,l+1} - d_{j,l}u_{j,l-1}\right) \quad (19.5.26)$$

Then $u_{j,l}^{\text{new}}$ is a weighted average

$$u_{j,l}^{\text{new}} = \omega u^*{}_{j,l} + (1-\omega)u_{j,l}^{\text{old}} \qquad (19.5.27)$$

We calculate it as follows: The residual at any stage is

$$\xi_{j,l} = a_{j,l}u_{j+1,l} + b_{j,l}u_{j-1,l} + c_{j,l}u_{j,l+1} + d_{j,l}u_{j,l-1} + e_{j,l}u_{j,l} - f_{j,l} \quad (19.5.28)$$

and the SOR algorithm (19.5.18) or (19.5.27) is

$$u_{j,l}^{\text{new}} = u_{j,l}^{\text{old}} - \omega\frac{\xi_{j,l}}{e_{j,l}} \qquad (19.5.29)$$

This formulation is very easy to program, and the norm of the residual vector $\xi_{j,l}$ can be used as a criterion for terminating the iteration.

  Another practical point concerns the order in which mesh points are processed. The obvious strategy is simply to proceed in order down the rows (or columns). Alternatively, suppose we divide the mesh into "odd" and "even" meshes, like the red and black squares of a checkerboard. Then equation (19.5.26) shows that the odd points depend only on the even mesh values and vice versa. Accordingly, we can carry out one half-sweep updating the odd points, say, and then another half-sweep updating the even points with the new odd values. For the version of SOR implemented below, we shall adopt odd-even ordering.

  The last practical point is that in practice the asymptotic rate of convergence in SOR is not attained until of order J iterations. The error often grows by a factor of 20 before convergence sets in. A trivial modification to SOR resolves this problem. It is based on the observation that, while ω is the optimum *asymptotic* relaxation parameter, it is not necessarily a good initial choice. In SOR with *Chebyshev acceleration*, one uses odd-even ordering and changes ω at each half-sweep according to the following prescription:

$$\begin{aligned}
\omega^{(0)} &= 1 \\
\omega^{(1/2)} &= 1/(1 - \rho_{\text{Jacobi}}^2/2) \\
\omega^{(n+1/2)} &= 1/(1 - \rho_{\text{Jacobi}}^2\omega^{(n)}/4), \qquad n = 1/2, 1, ..., \infty \\
\omega^{(\infty)} &\to \omega_{\text{optimal}}
\end{aligned} \qquad (19.5.30)$$

The beauty of Chebyshev acceleration is that the norm of the error always decreases with each iteration. (This is the norm of the actual error in $u_{j,l}$. The norm of the residual $\xi_{j,l}$ need not decrease monotonically.) While the asymptotic rate of convergence is the same as ordinary SOR, there is never any excuse for not using Chebyshev acceleration to reduce the total number of iterations required.

Here we give a routine for SOR with Chebyshev acceleration.

```
#include <cmath>
#include "nr.h"
using namespace std;

void NR::sor(Mat_I_DP &a, Mat_I_DP &b, Mat_I_DP &c, Mat_I_DP &d, Mat_I_DP &e,
    Mat_I_DP &f, Mat_IO_DP &u, const DP rjac)
```
Successive overrelaxation solution of equation (19.5.25) with Chebyshev acceleration. a, b, c, d, e, and f are input as the coefficients of the equation, each dimensioned to the grid size [0..jmax-1][0..jmax-1]. u is input as the initial guess to the solution, usually zero, and returns with the final value. rjac is input as the spectral radius of the Jacobi iteration, or an estimate of it.
```
{
    const int MAXITS=1000;
    const DP EPS=1.0e-13;
    int j,l,n,ipass,jsw,lsw;
    DP anorm,anormf=0.0,omega=1.0,resid;

    int jmax=a.nrows();
    for (j=1;j<jmax-1;j++)
```
Compute initial norm of residual and terminate iteration when norm has been reduced by a factor EPS.
```
        for (l=1;l<jmax-1;l++)
            anormf += fabs(f[j][l]);          Assumes initial u is zero.
    for (n=0;n<MAXITS;n++) {
        anorm=0.0;
        jsw=1;
        for (ipass=0;ipass<2;ipass++) {       Odd-even ordering.
            lsw=jsw;
            for (j=1;j<jmax-1;j++) {
                for (l=lsw;l<jmax-1;l+=2) {
                    resid=a[j][l]*u[j+1][l]+b[j][l]*u[j-1][l]
                        +c[j][l]*u[j][l+1]+d[j][l]*u[j][l-1]
                        +e[j][l]*u[j][l]-f[j][l];
                    anorm += fabs(resid);
                    u[j][l] -= omega*resid/e[j][l];
                }
                lsw=3-lsw;
            }
            jsw=3-jsw;
            omega=(n == 0 && ipass == 0 ? 1.0/(1.0-0.5*rjac*rjac) :
                1.0/(1.0-0.25*rjac*rjac*omega));
        }
        if (anorm < EPS*anormf) return;
    }
    nrerror("MAXITS exceeded");
}
```

The main advantage of SOR is that it is very easy to program. Its main disadvantage is that it is still very inefficient on large problems.

ADI (Alternating-Direction Implicit) Method

The ADI method of §19.3 for diffusion equations can be turned into a relaxation method for elliptic equations [1-4]. In §19.3, we discussed ADI as a method for solving the time-dependent heat-flow equation

$$\frac{\partial u}{\partial t} = \nabla^2 u - \rho \qquad (19.5.31)$$

By letting $t \to \infty$ one also gets an iterative method for solving the elliptic equation

$$\nabla^2 u = \rho \qquad (19.5.32)$$

In either case, the operator splitting is of the form

$$\mathcal{L} = \mathcal{L}_x + \mathcal{L}_y \qquad (19.5.33)$$

where $\mathcal{L}_x$ represents the differencing in x and $\mathcal{L}_y$ that in y.

For example, in our model problem (19.0.6) with $\Delta x = \Delta y = \Delta$, we have

$$\begin{aligned} \mathcal{L}_x u &= 2u_{j,l} - u_{j+1,l} - u_{j-1,l} \\ \mathcal{L}_y u &= 2u_{j,l} - u_{j,l+1} - u_{j,l-1} \end{aligned} \qquad (19.5.34)$$

More complicated operators may be similarly split, but there is some art involved. A bad choice of splitting can lead to an algorithm that fails to converge. Usually one tries to base the splitting on the physical nature of the problem. We know for our model problem that an initial transient diffuses away, and we set up the x and y splitting to mimic diffusion in each dimension.

Having chosen a splitting, we difference the time-dependent equation (19.5.31) implicitly in two half-steps:

$$\begin{aligned} \frac{u^{n+1/2} - u^n}{\Delta t/2} &= -\frac{\mathcal{L}_x u^{n+1/2} + \mathcal{L}_y u^n}{\Delta^2} - \rho \\ \frac{u^{n+1} - u^{n+1/2}}{\Delta t/2} &= -\frac{\mathcal{L}_x u^{n+1/2} + \mathcal{L}_y u^{n+1}}{\Delta^2} - \rho \end{aligned} \qquad (19.5.35)$$

(cf. equation 19.3.16). Here we have suppressed the spatial indices (j, l). In matrix notation, equations (19.5.35) are

$$(\mathbf{L}_x + r\mathbf{1}) \cdot \mathbf{u}^{n+1/2} = (r\mathbf{1} - \mathbf{L}_y) \cdot \mathbf{u}^n - \Delta^2 \rho \qquad (19.5.36)$$

$$(\mathbf{L}_y + r\mathbf{1}) \cdot \mathbf{u}^{n+1} = (r\mathbf{1} - \mathbf{L}_x) \cdot \mathbf{u}^{n+1/2} - \Delta^2 \rho \qquad (19.5.37)$$

where

$$r \equiv \frac{2\Delta^2}{\Delta t} \qquad (19.5.38)$$

The matrices on the left-hand sides of equations (19.5.36) and (19.5.37) are tridiagonal (and usually positive definite), so the equations can be solved by the standard tridiagonal algorithm. Given $\mathbf{u}^n$, one solves (19.5.36) for $\mathbf{u}^{n+1/2}$, substitutes on the right-hand side of (19.5.37), and then solves for $\mathbf{u}^{n+1}$. The key question is how to choose the iteration parameter r, the analog of a choice of timestep for an initial value problem.

As usual, the goal is to minimize the spectral radius of the iteration matrix. Although it is beyond our scope to go into details here, it turns out that, for the optimal choice of r, the ADI method has the same rate of convergence as SOR. The individual iteration steps in the ADI method are much more complicated than in SOR, so the ADI method would appear to be inferior. This is in fact true if we choose the same parameter r for every iteration step. However, it is possible to choose a *different* r for each step. If this is done optimally, then ADI is generally more efficient than SOR. We refer you to the literature [1-4] for details.

Our reason for not fully implementing ADI here is that, in most applications, it has been superseded by the multigrid methods described in the next section. Our advice is to use SOR for trivial problems (e.g., 20×20), or for solving a larger problem once only, where ease of programming outweighs expense of computer time. Occasionally, the sparse matrix methods of §2.7 are useful for solving a set of difference equations directly. For production solution of large elliptic problems, however, multigrid is now almost always the method of choice.

CITED REFERENCES AND FURTHER READING:

Hockney, R.W., and Eastwood, J.W. 1981, *Computer Simulation Using Particles* (New York: McGraw-Hill), Chapter 6.

Young, D.M. 1971, *Iterative Solution of Large Linear Systems* (New York: Academic Press). [1]

Stoer, J., and Bulirsch, R. 1993, *Introduction to Numerical Analysis*, 2nd ed. (New York: Springer-Verlag), §§8.3–8.6. [2]

Varga, R.S. 1962, *Matrix Iterative Analysis* (Englewood Cliffs, NJ: Prentice-Hall). [3]

Spanier, J. 1967, in *Mathematical Methods for Digital Computers, Volume 2* (New York: Wiley), Chapter 11. [4]

19.6 Multigrid Methods for Boundary Value Problems

Practical multigrid methods were first introduced in the 1970s by Brandt. These methods can solve elliptic PDEs discretized on N grid points in $O(N)$ operations. The "rapid" direct elliptic solvers discussed in §19.4 solve special kinds of elliptic equations in $O(N \log N)$ operations. The numerical coefficients in these estimates are such that multigrid methods are comparable to the rapid methods in execution speed. Unlike the rapid methods, however, the multigrid methods can solve general elliptic equations with nonconstant coefficients with hardly any loss in efficiency. Even nonlinear equations can be solved with comparable speed.

Unfortunately there is not a single multigrid algorithm that solves all elliptic problems. Rather there is a multigrid technique that provides the framework for solving these problems. You have to adjust the various components of the algorithm within this framework to solve your specific problem. We can only give a brief introduction to the subject here. In particular, we will give two sample multigrid routines, one linear and one nonlinear. By following these prototypes and by perusing the references [1-4], you should be able to develop routines to solve your own problems.

There are two related, but distinct, approaches to the use of multigrid techniques. The first, termed "the multigrid method," is a means for speeding up the convergence of a traditional relaxation method, as defined by you on a grid of pre-specified fineness. In this case, you need define your problem (e.g., evaluate its source terms) only on this grid. Other, coarser, grids defined by the method can be viewed as temporary computational adjuncts.

The second approach, termed (perhaps confusingly) "the full multigrid (FMG) method," requires you to be able to define your problem on grids of various sizes (generally by discretizing the same underlying PDE into different-sized sets of finite-difference equations). In this approach, the method obtains successive solutions on finer and finer grids. You can stop the solution either at a pre-specified fineness, or you can monitor the truncation error due to the discretization, quitting only when it is tolerably small.

In this section we will first discuss the "multigrid method," then use the concepts developed to introduce the FMG method. The latter algorithm is the one that we implement in the accompanying programs.

From One-Grid, through Two-Grid, to Multigrid

The key idea of the multigrid method can be understood by considering the simplest case of a two-grid method. Suppose we are trying to solve the linear elliptic problem

$$\mathcal{L}u = f \tag{19.6.1}$$

where $\mathcal{L}$ is some linear elliptic operator and f is the source term. Discretize equation (19.6.1) on a uniform grid with mesh size h. Write the resulting set of linear algebraic equations as

$$\mathcal{L}_h u_h = f_h \tag{19.6.2}$$

Let $\tilde{u}_h$ denote some approximate solution to equation (19.6.2). We will use the symbol u_h to denote the exact solution to the difference equations (19.6.2). Then the *error* in $\tilde{u}_h$ or the *correction* is

$$v_h = u_h - \tilde{u}_h \tag{19.6.3}$$

The *residual* or *defect* is

$$d_h = \mathcal{L}_h \tilde{u}_h - f_h \tag{19.6.4}$$

(Beware: some authors define residual as minus the defect, and there is not universal agreement about which of these two quantities 19.6.4 defines.) Since $\mathcal{L}_h$ is linear, the error satisfies

$$\mathcal{L}_h v_h = -d_h \tag{19.6.5}$$

At this point we need to make an approximation to $\mathcal{L}_h$ in order to find v_h. The classical iteration methods, such as Jacobi or Gauss-Seidel, do this by finding, at each stage, an approximate solution of the equation

$$\widehat{\mathcal{L}}_h \widehat{v}_h = -d_h \tag{19.6.6}$$

where $\widehat{\mathcal{L}}_h$ is a "simpler" operator than $\mathcal{L}_h$. For example, $\widehat{\mathcal{L}}_h$ is the diagonal part of $\mathcal{L}_h$ for Jacobi iteration, or the lower triangle for Gauss-Seidel iteration. The next approximation is generated by

$$\widetilde{u}_h^{\mathrm{new}} = \widetilde{u}_h + \widehat{v}_h \tag{19.6.7}$$

Now consider, as an alternative, a completely different type of approximation for $\mathcal{L}_h$, one in which we "coarsify" rather than "simplify." That is, we form some appropriate approximation $\mathcal{L}_H$ of $\mathcal{L}_h$ on a coarser grid with mesh size H (we will always take $H = 2h$, but other choices are possible). The residual equation (19.6.5) is now approximated by

$$\mathcal{L}_H v_H = -d_H \tag{19.6.8}$$

Since $\mathcal{L}_H$ has smaller dimension, this equation will be easier to solve than equation (19.6.5). To define the defect d_H on the coarse grid, we need a *restriction operator* $\mathcal{R}$ that restricts d_h to the coarse grid:

$$d_H = \mathcal{R}d_h \tag{19.6.9}$$

The restriction operator is also called the *fine-to-coarse operator* or the *injection operator*. Once we have a solution $\widetilde{v}_H$ to equation (19.6.8), we need a *prolongation operator* $\mathcal{P}$ that prolongates or interpolates the correction to the fine grid:

$$\widetilde{v}_h = \mathcal{P}\widetilde{v}_H \tag{19.6.10}$$

The prolongation operator is also called the *coarse-to-fine operator* or the *interpolation operator*. Both $\mathcal{R}$ and $\mathcal{P}$ are chosen to be linear operators. Finally the approximation $\widetilde{u}_h$ can be updated:

$$\widetilde{u}_h^{\mathrm{new}} = \widetilde{u}_h + \widetilde{v}_h \tag{19.6.11}$$

One step of this *coarse-grid correction scheme* is thus:

Coarse-Grid Correction

- Compute the defect on the fine grid from (19.6.4).
- Restrict the defect by (19.6.9).
- Solve (19.6.8) exactly on the coarse grid for the correction.
- Interpolate the correction to the fine grid by (19.6.10).

- Compute the next approximation by (19.6.11).

Let's contrast the advantages and disadvantages of relaxation and the coarse-grid correction scheme. Consider the error v_h expanded into a discrete Fourier series. Call the components in the lower half of the frequency spectrum the *smooth components* and the high-frequency components the *nonsmooth components*. We have seen that relaxation becomes very slowly convergent in the limit $h \to 0$, i.e., when there are a large number of mesh points. The reason turns out to be that the smooth components are only slightly reduced in amplitude on each iteration. However, many relaxation methods reduce the amplitude of the nonsmooth components by large factors on each iteration: They are good *smoothing operators*.

For the two-grid iteration, on the other hand, components of the error with wavelengths $\lesssim 2H$ are not even representable on the coarse grid and so cannot be reduced to zero on this grid. But it is exactly these high-frequency components that can be reduced by relaxation on the fine grid! This leads us to combine the ideas of relaxation and coarse-grid correction:

Two-Grid Iteration

- Pre-smoothing: Compute $\bar{u}_h$ by applying $\nu_1 \geq 0$ steps of a relaxation method to $\widetilde{u}_h$.
- Coarse-grid correction: As above, using $\bar{u}_h$ to give $\bar{u}_h^{\text{new}}$.
- Post-smoothing: Compute $\widetilde{u}_h^{\text{new}}$ by applying $\nu_2 \geq 0$ steps of the relaxation method to $\bar{u}_h^{\text{new}}$.

It is only a short step from the above two-grid method to a multigrid method. Instead of solving the coarse-grid defect equation (19.6.8) exactly, we can get an approximate solution of it by introducing an even coarser grid and using the two-grid iteration method. If the convergence factor of the two-grid method is small enough, we will need only a few steps of this iteration to get a good enough approximate solution. We denote the number of such iterations by γ. Obviously we can apply this idea recursively down to some coarsest grid. There the solution is found easily, for example by direct matrix inversion or by iterating the relaxation scheme to convergence.

One iteration of a multigrid method, from finest grid to coarser grids and back to finest grid again, is called a *cycle*. The exact structure of a cycle depends on the value of γ, the number of two-grid iterations at each intermediate stage. The case $\gamma = 1$ is called a V-cycle, while $\gamma = 2$ is called a W-cycle (see Figure 19.6.1). These are the most important cases in practice.

Note that once more than two grids are involved, the pre-smoothing steps after the first one on the finest grid need an initial approximation for the error v. This should be taken to be zero.

Smoothing, Restriction, and Prolongation Operators

The most popular smoothing method, and the one you should try first, is Gauss-Seidel, since it usually leads to a good convergence rate. If we order the mesh points from 0 to $N - 1$, then the Gauss-Seidel scheme is

$$u_i = -\left(\sum_{\substack{j=0 \\ j\neq i}}^{N-1} L_{ij}u_j - f_i \right) \frac{1}{L_{ii}} \qquad i = 0, \dots, N-1 \qquad (19.6.12)$$

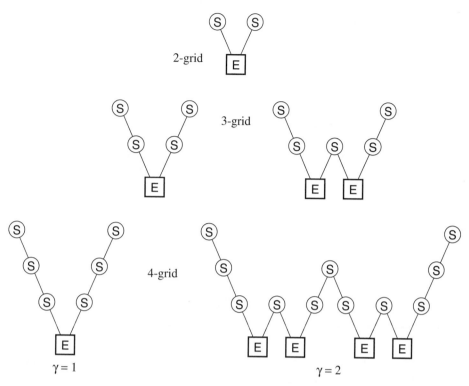

Figure 19.6.1.   Structure of multigrid cycles. S denotes smoothing, while E denotes exact solution on the coarsest grid. Each descending line \ denotes restriction ($\mathcal{R}$) and each ascending line / denotes prolongation ($\mathcal{P}$). The finest grid is at the top level of each diagram. For the V-cycles ($\gamma = 1$) the E step is replaced by one 2-grid iteration each time the number of grid levels is increased by one. For the W-cycles ($\gamma = 2$), each E step gets replaced by two 2-grid iterations.

where new values of u are used on the right-hand side as they become available. The exact form of the Gauss-Seidel method depends on the ordering chosen for the mesh points. For typical second-order elliptic equations like our model problem equation (19.0.3), as differenced in equation (19.0.8), it is usually best to use red-black ordering, making one pass through the mesh updating the "even" points (like the red squares of a checkerboard) and another pass updating the "odd" points (the black squares). When quantities are more strongly coupled along one dimension than another, one should relax a whole line along that dimension simultaneously. Line relaxation for nearest-neighbor coupling involves solving a tridiagonal system, and so is still efficient. Relaxing odd and even lines on successive passes is called zebra relaxation and is usually preferred over simple line relaxation.

Note that SOR should *not* be used as a smoothing operator. The overrelaxation destroys the high-frequency smoothing that is so crucial for the multigrid method.

A succint notation for the prolongation and restriction operators is to give their *symbol*. The symbol of $\mathcal{P}$ is found by considering v_H to be 1 at some mesh point (x, y), zero elsewhere, and then asking for the values of $\mathcal{P}v_H$. The most popular prolongation operator is simple bilinear interpolation. It gives nonzero values at the 9 points (x, y), $(x + h, y)$, . . . , $(x - h, y - h)$, where the values are $1, \frac{1}{2}, . . . , \frac{1}{4}$.

Its symbol is therefore

$$
\begin{bmatrix}
\frac{1}{4} & \frac{1}{2} & \frac{1}{4} \\
\frac{1}{2} & 1 & \frac{1}{2} \\
\frac{1}{4} & \frac{1}{2} & \frac{1}{4}
\end{bmatrix}
\tag{19.6.13}
$$

The symbol of $\mathcal{R}$ is defined by considering v_h to be defined everywhere on the fine grid, and then asking what is $\mathcal{R}v_h$ at (x, y) as a linear combination of these values. The simplest possible choice for $\mathcal{R}$ is *straight injection*, which means simply filling each coarse-grid point with the value from the corresponding fine-grid point. Its symbol is "[1]." However, difficulties can arise in practice with this choice. It turns out that a safe choice for $\mathcal{R}$ is to make it the adjoint operator to $\mathcal{P}$. To define the adjoint, define the scalar product of two grid functions u_h and v_h for mesh size h as

$$
\langle u_h | v_h \rangle_h \equiv h^2 \sum_{x,y} u_h(x, y) v_h(x, y)
\tag{19.6.14}
$$

Then the adjoint of $\mathcal{P}$, denoted $\mathcal{P}^\dagger$, is defined by

$$
\langle u_H | \mathcal{P}^\dagger v_h \rangle_H = \langle \mathcal{P}u_H | v_h \rangle_h
\tag{19.6.15}
$$

Now take $\mathcal{P}$ to be bilinear interpolation, and choose $u_H = 1$ at (x, y), zero elsewhere. Set $\mathcal{P}^\dagger = \mathcal{R}$ in (19.6.15) and $H = 2h$. You will find that

$$
(\mathcal{R}v_h)_{(x,y)} = \tfrac{1}{4}v_h(x, y) + \tfrac{1}{8}v_h(x + h, y) + \tfrac{1}{16}v_h(x + h, y + h) + \cdots
\tag{19.6.16}
$$

so that the symbol of $\mathcal{R}$ is

$$
\begin{bmatrix}
\frac{1}{16} & \frac{1}{8} & \frac{1}{16} \\
\frac{1}{8} & \frac{1}{4} & \frac{1}{8} \\
\frac{1}{16} & \frac{1}{8} & \frac{1}{16}
\end{bmatrix}
\tag{19.6.17}
$$

Note the simple rule: The symbol of $\mathcal{R}$ is $\frac{1}{4}$ the transpose of the matrix defining the symbol of $\mathcal{P}$, equation (19.6.13). This rule is general whenever $\mathcal{R} = \mathcal{P}^\dagger$ and $H = 2h$.

The particular choice of $\mathcal{R}$ in (19.6.17) is called *full weighting*. Another popular choice for $\mathcal{R}$ is *half weighting*, "halfway" between full weighting and straight injection. Its symbol is

$$
\begin{bmatrix}
0 & \frac{1}{8} & 0 \\
\frac{1}{8} & \frac{1}{2} & \frac{1}{8} \\
0 & \frac{1}{8} & 0
\end{bmatrix}
\tag{19.6.18}
$$

A similar notation can be used to describe the difference operator $\mathcal{L}_h$. For example, the standard differencing of the model problem, equation (19.0.6), is represented by the *five-point difference star*

$$
\mathcal{L}_h = \frac{1}{h^2}
\begin{bmatrix}
0 & 1 & 0 \\
1 & -4 & 1 \\
0 & 1 & 0
\end{bmatrix}
\tag{19.6.19}
$$

If you are confronted with a new problem and you are not sure what $\mathcal{P}$ and $\mathcal{R}$ choices are likely to work well, here is a safe rule: Suppose m_p is the order of the interpolation $\mathcal{P}$ (i.e., it interpolates polynomials of degree $m_p - 1$ exactly). Suppose m_r is the order of $\mathcal{R}$, and that $\mathcal{R}$ is the adjoint of some $\mathcal{P}$ (not necessarily the $\mathcal{P}$ you intend to use). Then if m is the order of the differential operator $\mathcal{L}_h$, you should satisfy the inequality $m_p + m_r > m$. For example, bilinear interpolation and its adjoint, full weighting, for Poisson's equation satisfy $m_p + m_r = 4 > m = 2$.

Of course the $\mathcal{P}$ and $\mathcal{R}$ operators should enforce the boundary conditions for your problem. The easiest way to do this is to rewrite the difference equation to have homogeneous boundary conditions by modifying the source term if necessary (cf. §19.4). Enforcing homogeneous boundary conditions simply requires the $\mathcal{P}$ operator to produce zeros at the appropriate boundary points. The corresponding $\mathcal{R}$ is then found by $\mathcal{R} = \mathcal{P}^{\dagger}$.

Full Multigrid Algorithm

So far we have described multigrid as an iterative scheme, where one starts with some initial guess on the finest grid and carries out enough cycles (V-cycles, W-cycles,...) to achieve convergence. This is the simplest way to use multigrid: Simply apply enough cycles until some appropriate convergence criterion is met. However, efficiency can be improved by using the *Full Multigrid Algorithm* (FMG), also known as *nested iteration*.

Instead of starting with an arbitrary approximation on the finest grid (e.g., $u_h = 0$), the first approximation is obtained by interpolating from a coarse-grid solution:

$$u_h = \mathcal{P}u_H \qquad (19.6.20)$$

The coarse-grid solution itself is found by a similar FMG process from even coarser grids. At the coarsest level, you start with the exact solution. Rather than proceed as in Figure 19.6.1, then, FMG gets to its solution by a series of increasingly tall "N's," each taller one probing a finer grid (see Figure 19.6.2).

Note that $\mathcal{P}$ in (19.6.20) need not be the same $\mathcal{P}$ used in the multigrid cycles. It should be at least of the same order as the discretization $\mathcal{L}_h$, but sometimes a higher-order operator leads to greater efficiency.

It turns out that you usually need one or at most two multigrid cycles at each level before proceeding down to the next finer grid. While there is theoretical guidance on the required number of cycles (e.g., [2]), you can easily determine it empirically. Fix the finest level and study the solution values as you increase the number of cycles per level. The asymptotic value of the solution is the exact solution of the difference equations. The difference between this exact solution and the solution for a small number of cycles is the iteration error. Now fix the number of cycles to be large, and vary the number of levels, i.e., the smallest value of h used. In this way you can estimate the truncation error for a given h. In your final production code, there is no point in using more cycles than you need to get the iteration error down to the size of the truncation error.

The simple multigrid iteration (cycle) needs the right-hand side f only at the finest level. FMG needs f at all levels. If the boundary conditions are homogeneous,

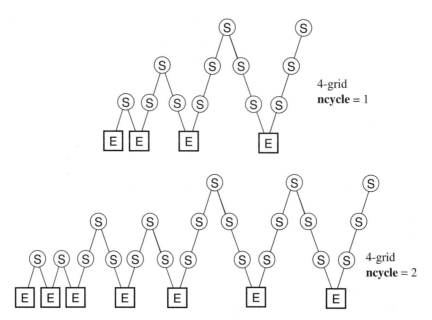

Figure 19.6.2.   Structure of cycles for the full multigrid (FMG) method (notation as in Fig. 19.6.1). This method starts on the coarsest grid, interpolates, and then refines (by "V's"), the solution onto grids of increasing fineness.

you can use $f_H = \mathcal{R} f_h$. This prescription is not always safe for inhomogeneous boundary conditions. In that case it is better to discretize f on each coarse grid.

Note that the FMG algorithm produces the solution on all levels. It can therefore be combined with techniques like Richardson extrapolation.

We now give a routine `mglin` that implements the Full Multigrid Algorithm for a linear equation, the model problem (19.0.6). It uses red-black Gauss-Seidel as the smoothing operator, bilinear interpolation for $\mathcal{P}$, and half-weighting for $\mathcal{R}$. To change the routine to handle another linear problem, all you need do is modify the functions `relax`, `resid`, and `slvsml` appropriately. A feature of the routine is the dynamical allocation of storage for variables defined on the various grids.

```
#include "nr.h"

namespace {
    void mg(int j, Mat_IO_DP &u, Mat_I_DP &rhs)
    Recursive multigrid iteration. On input, j is the current level, u is the current value of the
    solution, and rhs is the right-hand side. On output u contains the improved solution at
    the current level.
    {
        using namespace NR;
        const int NPRE=1,NPOST=1;             Number of relaxation sweeps before and af-
        int jpost,jpre,nc,nf;                 ter the coarse-grid correction is computed.

        nf=u.nrows();
        nc=(nf+1)/2;
        if (j == 0)                           Bottom of V: solve on coarsest grid.
            slvsml(u,rhs);
        else {                                On downward stoke of the V.
            Mat_DP res(nc,nc),v(0.0,nc,nc),temp(nf,nf);
            v is zero for initial guess in each relaxation.
            for (jpre=0;jpre<NPRE;jpre++)
```

```
                relax(u,rhs);                    Pre-smoothing.
             resid(temp,u,rhs);
             rstrct(res,temp);                   Restriction of the residual is the next r.h.s.
             mg(j-1,v,res);                      Recursive call for the coarse grid correction.
             addint(u,v,temp);                   On upward stroke of V.
             for (jpost=0;jpost<NPOST;jpost++)
                relax(u,rhs);                    Post-smoothing.
        }
    }
}
```

```
void NR::mglin(Mat_IO_DP &u, const int ncycle)
```
Full Multigrid Algorithm for solution of linear elliptic equation, here the model problem (19.0.6).
On input u[0..n-1][0..n-1] contains the right-hand side ρ, while on output it returns the
solution. The dimension n must be of the form $2^j + 1$ for some integer j. (j is actually the
number of grid levels used in the solution, called ng below.) ncycle is the number of V-cycles
to be used at each level.
```
{
    int j,jcycle,ng=0,ngrid,nn;
    Mat_DP *uj,*uj1;

    int n=u.nrows();

    nn=n;
    while (nn >>= 1) ng++;
    if ((n-1) != (1 << ng))
        nrerror("n-1 must be a power of 2 in mglin.");
    Vec_Mat_DP_p rho(ng);                        Vector of pointers to ρ on each level.
    nn=n;
    ngrid=ng-1;
    rho[ngrid] = new Mat_DP(nn,nn);              Allocate storage for r.h.s. on grid ng − 1,
    copy(*rho[ngrid],u);                         and fill it with the input r.h.s.
    while (nn > 3) {                             Similarly allocate storage and fill r.h.s. on
        nn=nn/2+1;                                   all coarse grids by restricting from finer
        rho[--ngrid]=new Mat_DP(nn,nn);          grids.
        rstrct(*rho[ngrid],*rho[ngrid+1]);
    }
    nn=3;
    uj=new Mat_DP(nn,nn);
    slvsml(*uj,*rho[0]);                         Initial solution on coarsest grid.
    for (j=1;j<ng;j++) {                         Nested iteration loop.
        nn=2*nn-1;
        uj1=uj;
        uj=new Mat_DP(nn,nn);
        interp(*uj,*uj1);                        Interpolate from grid j-1 to next finer grid
        delete uj1;                                  j.
        for (jcycle=0;jcycle<ncycle;jcycle++)    V-cycle loop.
            mg(j,*uj,*rho[j]);
    }
    copy(u,*uj);                                 Return solution in u.
    delete uj;
    for (j=0;j<ng;j++)
        delete rho[j];
}
```

```
#include "nr.h"

void NR::rstrct(Mat_O_DP &uc, Mat_I_DP &uf)
```
Half-weighting restriction. If nc is the coarse-grid dimension, the fine-grid solution is input in
uf[0..2*nc-2][0..2*nc-2], the coarse-grid solution is returned in uc[0..nc-1][0..nc-1].
```
{
    int ic,iif,jc,jf,ncc;
```

```
    int nc=uc.nrows();
    ncc=2*nc-2;
    for (jf=2,jc=1;jc<nc-1;jc++,jf+=2) {              Interior points.
        for (iif=2,ic=1;ic<nc-1;ic++,iif+=2) {
            uc[ic][jc]=0.5*uf[iif][jf]+0.125*(uf[iif+1][jf]+uf[iif-1][jf]
                +uf[iif][jf+1]+uf[iif][jf-1]);
        }
    }
    for (jc=0,ic=0;ic<nc;ic++,jc+=2) {              Boundary points.
        uc[ic][0]=uf[jc][0];
        uc[ic][nc-1]=uf[jc][ncc];
    }
    for (jc=0,ic=0;ic<nc;ic++,jc+=2) {
        uc[0][ic]=uf[0][jc];
        uc[nc-1][ic]=uf[ncc][jc];
    }
}
```

```
#include "nr.h"
```

```
void NR::interp(Mat_O_DP &uf, Mat_I_DP &uc)
```
Coarse-to-fine prolongation by bilinear interpolation. If nf is the fine-grid dimension, the coarse-grid solution is input as uc[0..nc-1][0..nc-1], where $nc = nf/2+1$. The fine-grid solution is returned in uf[0..nf-1][0..nf-1].
```
{
    int ic,iif,jc,jf,nc;

    int nf=uf.nrows();
    nc=nf/2+1;
    for (jc=0;jc<nc;jc++)                    Do elements that are copies.
        for (ic=0;ic<nc;ic++) uf[2*ic][2*jc]=uc[ic][jc];
    for (jf=0;jf<nf;jf+=2)                   Do even-numbered columns, interpolating ver-
        for (iif=1;iif<nf-1;iif+=2)                tically.
            uf[iif][jf]=0.5*(uf[iif+1][jf]+uf[iif-1][jf]);
    for (jf=1;jf<nf-1;jf+=2)                 Do odd-numbered columns, interpolating hor-
        for (iif=0;iif<nf;iif++)                   izontally.
            uf[iif][jf]=0.5*(uf[iif][jf+1]+uf[iif][jf-1]);
}
```

```
#include "nr.h"
```

```
void NR::addint(Mat_O_DP &uf, Mat_I_DP &uc, Mat_O_DP &res)
```
Does coarse-to-fine interpolation and adds result to uf. If nf is the fine-grid dimension, the coarse-grid solution is input as uc[0..nc-1][0..nc-1], where $nc = nf/2 + 1$. The fine-grid solution is returned in uf[0..nf-1][0..nf-1]. res[0..nf-1][0..nf-1] is used for temporary storage.
```
{
    int i,j;

    int nf=uf.nrows();
    interp(res,uc);
    for (j=0;j<nf;j++)
        for (i=0;i<nf;i++)
            uf[i][j] += res[i][j];
}
```

```
#include "nr.h"
```

```
void NR::slvsml(Mat_O_DP &u, Mat_I_DP &rhs)
```
Solution of the model problem on the coarsest grid, where $h = \frac{1}{2}$. The right-hand side is input in rhs[0..2][0..2] and the solution is returned in u[0..2][0..2].
```
{
```

```
    int i,j;
    DP h=0.5;

    for (i=0;i<3;i++)
        for (j=0;j<3;j++)
            u[i][j]=0.0;
    u[1][1] = -h*h*rhs[1][1]/4.0;
}
```

```
#include "nr.h"

void NR::relax(Mat_IO_DP &u, Mat_I_DP &rhs)
```
Red-black Gauss-Seidel relaxation for model problem. Updates the current value of the solution
u[0..n-1][0..n-1], using the right-hand side function rhs[0..n-1][0..n-1].
```
{
    int i,ipass,isw,j,jsw=1;
    DP h,h2;

    int n=u.nrows();
    h=1.0/(n-1);
    h2=h*h;
    for (ipass=0;ipass<2;ipass++,jsw=3-jsw) {          Red and black sweeps.
        isw=jsw;
        for (j=1;j<n-1;j++,isw=3-isw)
            for (i=isw;i<n-1;i+=2)                      Gauss-Seidel formula.
                u[i][j]=0.25*(u[i+1][j]+u[i-1][j]+u[i][j+1]
                    +u[i][j-1]-h2*rhs[i][j]);
    }
}
```

```
#include "nr.h"

void NR::resid(Mat_O_DP &res, Mat_I_DP &u, Mat_I_DP &rhs)
```
Returns *minus* the residual for the model problem. Input quantities are u[0..n-1][0..n-1]
and rhs[0..n-1][0..n-1], while res[0..n-1][0..n-1] is returned.
```
{
    int i,j;
    DP h,h2i;

    int n=u.nrows();
    h=1.0/(n-1);
    h2i=1.0/(h*h);
    for (j=1;j<n-1;j++)              Interior points.
        for (i=1;i<n-1;i++)
            res[i][j] = -h2i*(u[i+1][j]+u[i-1][j]+u[i][j+1]
                +u[i][j-1]-4.0*u[i][j])+rhs[i][j];
    for (i=0;i<n;i++)                Boundary points.
        res[i][0]=res[i][n-1]=res[0][i]=res[n-1][i]=0.0;
}
```

```
#include "nr.h"

void NR::copy(Mat_O_DP &aout, Mat_I_DP &ain)
```
Copies ain[0..n-1][0..n-1] to aout[0..n-1][0..n-1].
```
{
    int i,j;

    int n=ain.nrows();
    for (i=0;i<n;i++)
        for (j=0;j<n;j++)
            aout[j][i]=ain[j][i];

}
```

The routine `mglin` is written for clarity, not maximum efficiency, so that it is easy to modify. Several simple changes will speed up the execution time:

- The defect d_h vanishes identically at all black mesh points after a red-black Gauss-Seidel step. Thus $d_H = \mathcal{R}d_h$ for half-weighting reduces to simply copying half the defect from the fine grid to the corresponding coarse-grid point. The calls to `resid` followed by `rstrct` in the first part of the V-cycle can be replaced by a routine that loops only over the coarse grid, filling it with half the defect.
- Similarly, the quantity $\tilde{u}_h^{\text{new}} = \tilde{u}_h + \mathcal{P}\tilde{v}_H$ need not be computed at red mesh points, since they will immediately be redefined in the subsequent Gauss-Seidel sweep. This means that `addint` need only loop over black points.
- You can speed up `relax` in several ways. First, you can have a special form when the initial guess is zero. Next, you can store $h^2 f_h$ on the various grids and save a multiplication. Finally, it is possible to save an addition in the Gauss-Seidel formula by rewriting it with intermediate variables.
- On typical problems, `mglin` with `ncycle` = 1 will return a solution with the iteration error bigger than the truncation error for the given size of h. To knock the error down to the size of the truncation error, you have to set `ncycle` = 2 or, more cheaply, `NPRE` = 2. A more efficient way turns out to be to use a higher-order $\mathcal{P}$ in (19.6.20) than the linear interpolation used in the V-cycle.

Implementing all the above features typically gives up to a factor of two improvement in execution time and is certainly worthwhile in a production code.

Nonlinear Multigrid: The FAS Algorithm

Now turn to solving a nonlinear elliptic equation, which we write symbolically as

$$\mathcal{L}(u) = 0 \tag{19.6.21}$$

Any explicit source term has been moved to the left-hand side. Suppose equation (19.6.21) is suitably discretized:

$$\mathcal{L}_h(u_h) = 0 \tag{19.6.22}$$

We will see below that in the multigrid algorithm we will have to consider equations where a nonzero right-hand side is generated during the course of the solution:

$$\mathcal{L}_h(u_h) = f_h \tag{19.6.23}$$

One way of solving nonlinear problems with multigrid is to use Newton's method, which produces linear equations for the correction term at each iteration. We can then use linear multigrid to solve these equations. A great strength of the multigrid idea, however, is that it can be applied *directly* to nonlinear problems. All we need is a suitable *nonlinear* relaxation method to smooth the errors, plus a procedure for approximating corrections on coarser grids. This direct approach is Brandt's Full Approximation Storage Algorithm (FAS). No nonlinear equations need be solved, except perhaps on the coarsest grid.

To develop the nonlinear algorithm, suppose we have a relaxation procedure that can smooth the residual vector as we did in the linear case. Then we can seek a smooth correction v_h to solve (19.6.23):

$$\mathcal{L}_h(\tilde{u}_h + v_h) = f_h \tag{19.6.24}$$

To find v_h, note that

$$\mathcal{L}_h(\widetilde{u}_h + v_h) - \mathcal{L}_h(\widetilde{u}_h) = f_h - \mathcal{L}_h(\widetilde{u}_h)$$
$$= -d_h \tag{19.6.25}$$

The right-hand side is smooth after a few nonlinear relaxation sweeps. Thus we can transfer the left-hand side to a coarse grid:

$$\mathcal{L}_H(u_H) - \mathcal{L}_H(\mathcal{R}\widetilde{u}_h) = -\mathcal{R}d_h \tag{19.6.26}$$

that is, we solve

$$\mathcal{L}_H(u_H) = \mathcal{L}_H(\mathcal{R}\widetilde{u}_h) - \mathcal{R}d_h \tag{19.6.27}$$

on the coarse grid. (This is how nonzero right-hand sides appear.) Suppose the approximate solution is $\widetilde{u}_H$. Then the coarse-grid correction is

$$\widetilde{v}_H = \widetilde{u}_H - \mathcal{R}\widetilde{u}_h \tag{19.6.28}$$

and

$$\widetilde{u}_h^{\text{new}} = \widetilde{u}_h + \mathcal{P}(\widetilde{u}_H - \mathcal{R}\widetilde{u}_h) \tag{19.6.29}$$

Note that $\mathcal{PR} \neq 1$ in general, so $\widetilde{u}_h^{\text{new}} \neq \mathcal{P}\widetilde{u}_H$. This is a key point: In equation (19.6.29) the interpolation error comes only from the correction, not from the full solution $\widetilde{u}_H$.

Equation (19.6.27) shows that one is solving for the full approximation u_H, not just the error as in the linear algorithm. This is the origin of the name FAS.

The FAS multigrid algorithm thus looks very similar to the linear multigrid algorithm. The only differences are that both the defect d_h and the relaxed approximation u_h have to be restricted to the coarse grid, where now it is equation (19.6.27) that is solved by recursive invocation of the algorithm. However, instead of implementing the algorithm this way, we will first describe the so-called *dual viewpoint*, which leads to a powerful alternative way of looking at the multigrid idea.

The dual viewpoint considers the *local truncation error*, defined as

$$\tau \equiv \mathcal{L}_h(u) - f_h \tag{19.6.30}$$

where u is the exact solution of the original continuum equation. If we rewrite this as

$$\mathcal{L}_h(u) = f_h + \tau \tag{19.6.31}$$

we see that τ can be regarded as the correction to f_h so that the solution of the fine-grid equation will be the exact solution u.

Now consider the *relative truncation error* τ_h, which is defined on the H-grid relative to the h-grid:

$$\tau_h \equiv \mathcal{L}_H(\mathcal{R}u_h) - \mathcal{R}\mathcal{L}_h(u_h) \tag{19.6.32}$$

Since $\mathcal{L}_h(u_h) = f_h$, this can be rewritten as

$$\mathcal{L}_H(u_H) = f_H + \tau_h \tag{19.6.33}$$

In other words, we can think of τ_h as the correction to f_H that makes the solution of the coarse-grid equation equal to the fine-grid solution. Of course we cannot compute τ_h, but we do have an approximation to it from using $\widetilde{u}_h$ in equation (19.6.32):

$$\tau_h \simeq \widetilde{\tau}_h \equiv \mathcal{L}_H(\mathcal{R}\widetilde{u}_h) - \mathcal{R}\mathcal{L}_h(\widetilde{u}_h) \tag{19.6.34}$$

Replacing τ_h by $\widetilde{\tau}_h$ in equation (19.6.33) gives

$$\mathcal{L}_H(u_H) = \mathcal{L}_H(\mathcal{R}\widetilde{u}_h) - \mathcal{R}d_h \tag{19.6.35}$$

which is just the coarse-grid equation (19.6.27)!

Thus we see that there are two complementary viewpoints for the relation between coarse and fine grids:

- Coarse grids are used to accelerate the convergence of the smooth components of the fine-grid residuals.

- Fine grids are used to compute correction terms to the coarse-grid equations, yielding fine-grid accuracy on the coarse grids.

One benefit of this new viewpoint is that it allows us to derive a natural stopping criterion for a multigrid iteration. Normally the criterion would be

$$\|d_h\| \le \epsilon \tag{19.6.36}$$

and the question is how to choose ϵ. There is clearly no benefit in iterating beyond the point when the remaining error is dominated by the local truncation error τ. The computable quantity is $\widetilde{\tau}_h$. What is the relation between τ and $\widetilde{\tau}_h$? For the typical case of a second-order accurate differencing scheme,

$$\tau = \mathcal{L}_h(u) - \mathcal{L}_h(u_h) = h^2 \tau_2(x, y) + \cdots \tag{19.6.37}$$

Assume the solution satisfies $u_h = u + h^2 u_2(x, y) + \cdots$. Then, assuming $\mathcal{R}$ is of high enough order that we can neglect its effect, equation (19.6.32) gives

$$\tau_h \simeq \mathcal{L}_H(u + h^2 u_2) - \mathcal{L}_h(u + h^2 u_2)$$

$$= \mathcal{L}_H(u) - \mathcal{L}_h(u) + h^2[\mathcal{L}'_H(u_2) - \mathcal{L}'_h(u_2)] + \cdots \tag{19.6.38}$$

$$= (H^2 - h^2)\tau_2 + O(h^4)$$

For the usual case of $H = 2h$ we therefore have

$$\tau \simeq \tfrac{1}{3}\tau_h \simeq \tfrac{1}{3}\widetilde{\tau}_h \tag{19.6.39}$$

The stopping criterion is thus equation (19.6.36) with

$$\epsilon = \alpha\|\widetilde{\tau}_h\|, \qquad \alpha \sim \tfrac{1}{3} \tag{19.6.40}$$

We have one remaining task before implementing our nonlinear multigrid algorithm: choosing a nonlinear relaxation scheme. Once again, your first choice should probably be the nonlinear Gauss-Seidel scheme. If the discretized equation (19.6.23) is written with some choice of ordering as

$$L_i(u_0, \ldots, u_{N-1}) = f_i, \qquad i = 0, \ldots, N-1 \tag{19.6.41}$$

then the nonlinear Gauss-Seidel schemes solves

$$L_i(u_0, \ldots, u_{i-1}, u_i^{\text{new}}, u_{i+1}, \ldots, u_{N-1}) = f_i \tag{19.6.42}$$

for u_i^{new}. As usual new u's replace old u's as soon as they have been computed. Often equation (19.6.42) is linear in u_i^{new}, since the nonlinear terms are discretized by means of its neighbors. If this is not the case, we replace equation (19.6.42) by one step of a Newton iteration:

$$u_i^{\text{new}} = u_i^{\text{old}} - \frac{L_i(u_i^{\text{old}}) - f_i}{\partial L_i(u_i^{\text{old}})/\partial u_i} \tag{19.6.43}$$

For example, consider the simple nonlinear equation

$$\nabla^2 u + u^2 = \rho \tag{19.6.44}$$

In two-dimensional notation, we have

$$\mathcal{L}(u_{i,j}) = (u_{i+1,j} + u_{i-1,j} + u_{i,j+1} + u_{i,j-1} - 4u_{i,j})/h^2 + u_{i,j}^2 - \rho_{i,j} = 0 \tag{19.6.45}$$

Since

$$\frac{\partial \mathcal{L}}{\partial u_{i,j}} = -4/h^2 + 2u_{i,j} \tag{19.6.46}$$

the Newton Gauss-Seidel iteration is

$$u_{i,j}^{\text{new}} = u_{i,j} - \frac{\mathcal{L}(u_{i,j})}{-4/h^2 + 2u_{i,j}} \tag{19.6.47}$$

Here is a routine mgfas that solves equation (19.6.44) using the Full Multigrid Algorithm and the FAS scheme. Restriction and prolongation are done as in mglin. We have included the convergence test based on equation (19.6.40). A successful multigrid solution of a problem should aim to satisfy this condition with the maximum number of V-cycles, maxcyc, equal to 1 or 2. The routine mgfas uses the same functions copy, interp, and rstrct as mglin.

```
#include "nr.h"

namespace {
    void mg(const int j, Mat_IO_DP &u, Mat_I_DP &rhs, Vec_Mat_DP_p &rho,
        DP &trerr)
```
Recursive multigrid iteration. On input, j is the current level and u is the current value of the
solution. For the first call on a given level, the right-hand side is zero, and the argument rhs
is dummy. This is signaled by inputting trerr positive. Subsequent recursive calls supply
a nonzero rhs as in equation (19.6.33). This is signaled by inputting trerr negative. rho
is the vector of pointers to ρ on each level. On output u contains the improved solution
at the current level. When the first call on a given level is made, the relative truncation
error τ is returned in trerr.

```
{
    using namespace NR;
    const int NPRE=1,NPOST=1;
```
Number of relaxation sweeps before and after the coarse-grid correction is computed.
```
    const DP ALPHA=0.33;                    Relates the estimated truncation error to the
    int jpost,jpre,nc,nf;                        norm of the residual.
    DP dum=-1.0;

    nf=u.nrows();
    nc=(nf+1)/2;
    Mat_DP temp(nf,nf);
    if (j == 0) {                           Bottom of V: Solve on coarsest grid.
        matadd(rhs,*rho[j],temp);
        slvsm2(u,temp);
    } else {                                On downward stoke of the V.
        Mat_DP v(nc,nc),ut(nc,nc),tau(nc,nc),tempc(nc,nc);
        for (jpre=0;jpre<NPRE;jpre++) {            Pre-smoothing.
            if (trerr < 0.0) {
                matadd(rhs,*rho[j],temp);
                relax2(u,temp);
            }
            else
                relax2(u,*rho[j]);
        }
        rstrct(ut,u);                       $\mathcal{R}\tilde{u}_h$.
        copy(v,ut);                         Make a copy in v.
        lop(tau,ut);                        $\mathcal{L}_H(\mathcal{R}\tilde{u}_h)$ stored temporarily in $\tilde{\tau}_h$.
        lop(temp,u);                        $\mathcal{L}_h(\tilde{u}_h)$.
        if (trerr < 0.0)                    $\mathcal{L}_h(\tilde{u}_h) - f_h$.
            matsub(temp,rhs,temp);
        rstrct(tempc,temp);                 $\mathcal{R}\mathcal{L}_h(\tilde{u}_h) - f_H$.
        matsub(tau,tempc,tau);              $\tilde{\tau}_h + f_H = \mathcal{L}_H(\mathcal{R}\tilde{u}_h) - \mathcal{R}\mathcal{L}_h(\tilde{u}_h) + f_H$.
        if (trerr > 0.0)
            trerr=ALPHA*anorm2(tau);        Estimate truncation error $\tau$.
        mg(j-1,v,tau,rho,dum);              Recursive call for the coarse-grid correction.
        matsub(v,ut,tempc);                 On upward stroke of V: form $\tilde{u}_h^{\text{new}} = \tilde{u}_h +$
        interp(temp,tempc);                     $\mathcal{P}(\tilde{u}_H - \mathcal{R}\tilde{u}_h)$.
        matadd(u,temp,u);
        for (jpost=0;jpost<NPOST;jpost++) {        Post-smoothing.
            if (trerr < 0.0) {
                matadd(rhs,*rho[j],temp);
                relax2(u,temp);
            }
            else
                relax2(u,*rho[j]);
        }
    }
}
}

void NR::mgfas(Mat_IO_DP &u, const int maxcyc)
```
Full Multigrid Algorithm for FAS solution of nonlinear elliptic equation, here equation (19.6.44).

On input u[0..n-1][0..n-1] contains the right-hand side ρ, while on output it returns the solution. The dimension n must be of the form $2^j + 1$ for some integer j. (j is actually the number of grid levels used in the solution, called ng below.) maxcyc is the maximum number of V-cycles to be used at each level.

```
{
    int j,jcycle,ng=0,ngrid,nn;
    DP res,trerr;
    Mat_DP *uj,*uj1;

    int n=u.nrows();
    nn=n;
    while (nn >>= 1) ng++;
    if ((n-1) != (1 << ng))
        nrerror("n-1 must be a power of 2 in mgfas.");
    Vec_Mat_DP_p rho(ng);                          Vector of pointers to ρ on each level.
    nn=n;
    ngrid=ng-1;
    rho[ngrid]=new Mat_DP(nn,nn);                  Allocate storage for r.h.s. on grid ng − 1,
    copy(*rho[ngrid],u);                              and fill it with the input r.h.s.
    while (nn > 3) {                               Similarly allocate storage and fill r.h.s. by
        nn=nn/2+1;                                     restriction on all coarse grids.
        rho[--ngrid]=new Mat_DP(nn,nn);
        rstrct(*rho[ngrid],*rho[ngrid+1]);
    }
    nn=3;
    uj=new Mat_DP(nn,nn);
    slvsm2(*uj,*rho[0]);                           Initial solution on coarsest grid.
    for (j=1;j<ng;j++) {                           Nested iteration loop.
        nn=2*nn-1;
        uj1=uj;
        uj=new Mat_DP(nn,nn);
        Mat_DP temp(nn,nn);
        interp(*uj,*uj1);                          Interpolate from grid j−1 to next finer grid
        delete uj1;                                    j.
        for (jcycle=0;jcycle<maxcyc;jcycle++) {            V-cycle loop.
            trerr=1.0;                             R.h.s. is dummy.
            mg(j,*uj,temp,rho,trerr);
            lop(temp,*uj);                         Form residual ∥d_h∥.
            matsub(temp,*rho[j],temp);
            res=anorm2(temp);
            if (res < trerr) break;                No more V-cycles needed if residual small
        }                                              enough.
    }
    copy(u,*uj);                                   Return solution in u.
    delete uj;
    for (j=0;j<ng;j++)
        delete rho[j];
}
```

```
#include "nr.h"

void NR::relax2(Mat_IO_DP &u, Mat_I_DP &rhs)
```
Red-black Gauss-Seidel relaxation for equation (19.6.44). The current value of the solution u[0..n-1][0..n-1] is updated, using the right-hand side function rhs[0..n-1][0..n-1].
```
{
    int i,ipass,isw,j,jsw=1;
    DP foh2,h,h2i,res;

    int n=u.nrows();
    h=1.0/(n-1);
    h2i=1.0/(h*h);
    foh2 = -4.0*h2i;
    for (ipass=0;ipass<2;ipass++,jsw=3-jsw) {                 Red and black sweeps.
```

```
        isw=jsw;
        for (j=1;j<n-1;j++,isw=3-isw) {
            for (i=isw;i<n-1;i+=2) {
                res=h2i*(u[i+1][j]+u[i-1][j]+u[i][j+1]+u[i][j-1]-
                    4.0*u[i][j])+u[i][j]*u[i][j]-rhs[i][j];
                u[i][j] -= res/(foh2+2.0*u[i][j]);        Newton Gauss-Seidel formula.
            }
        }
    }
}
```

```
#include <cmath>
#include "nr.h"
using namespace std;

void NR::slvsm2(Mat_O_DP &u, Mat_I_DP &rhs)
```
Solution of equation (19.6.44) on the coarsest grid, where $h = \frac{1}{2}$. The right-hand side is input
in rhs[0..2][0..2] and the solution is returned in u[0..2][0..2].
```
{
    int i,j;
    DP disc,fact,h=0.5;

    for (i=0;i<3;i++)
        for (j=0;j<3;j++)
            u[i][j]=0.0;
    fact=2.0/(h*h);
    disc=sqrt(fact*fact+rhs[1][1]);
    u[1][1]= -rhs[1][1]/(fact+disc);
}
```

```
#include "nr.h"

void NR::lop(Mat_O_DP &out, Mat_I_DP &u)
```
Given u[0..n-1][0..n-1], returns $\mathcal{L}_h(\tilde{u}_h)$ for equation (19.6.44) in out[0..n-1][0..n-1].
```
{
    int i,j;
    DP h,h2i;

    int n=u.nrows();
    h=1.0/(n-1);
    h2i=1.0/(h*h);
    for (j=1;j<n-1;j++)            Interior points.
        for (i=1;i<n-1;i++)
            out[i][j]=h2i*(u[i+1][j]+u[i-1][j]+u[i][j+1]+u[i][j-1]-
                4.0*u[i][j])+u[i][j]*u[i][j];
    for (i=0;i<n;i++)             Boundary points.
        out[i][0]=out[i][n-1]=out[0][i]=out[n-1][i]=0.0;
}
```

```
#include "nr.h"

void NR::matadd(Mat_I_DP &a, Mat_I_DP &b, Mat_O_DP &c)
```
Adds a[0..n-1][0..n-1] to b[0..n-1][0..n-1] and returns result in c[0..n-1][0..n-1].
```
{
    int i,j;

    int n=a.nrows();
    for (j=0;j<n;j++)
        for (i=0;i<n;i++)
            c[i][j]=a[i][j]+b[i][j];
}
```

```
#include "nr.h"

void NR::matsub(Mat_I_DP &a, Mat_I_DP &b, Mat_O_DP &c)
Subtracts b[0..n-1][0..n-1] from a[0..n-1][0..n-1], returns result in c[0..n-1][0..n-1].
{
    int i,j;

    int n=a.nrows();
    for (j=0;j<n;j++)
        for (i=0;i<n;i++)
            c[i][j]=a[i][j]-b[i][j];
}
```

```
#include <cmath>
#include "nr.h"
using namespace std;

DP NR::anorm2(Mat_I_DP &a)
Returns the Euclidean norm of the matrix a[0..n-1][0..n-1].
{
    int i,j;
    DP sum=0.0;

    int n=a.nrows();
    for (j=0;j<n;j++)
        for (i=0;i<n;i++)
            sum += a[i][j]*a[i][j];
    return sqrt(sum)/n;
}
```

CITED REFERENCES AND FURTHER READING:

Brandt, A. 1977, *Mathematics of Computation*, vol. 31, pp. 333–390. [1]

Hackbusch, W. 1985, *Multi-Grid Methods and Applications* (New York: Springer-Verlag). [2]

Stuben, K., and Trottenberg, U. 1982, in *Multigrid Methods*, W. Hackbusch and U. Trottenberg, eds. (Springer Lecture Notes in Mathematics No. 960) (New York: Springer-Verlag), pp. 1–176. [3]

Brandt, A. 1982, in *Multigrid Methods*, W. Hackbusch and U. Trottenberg, eds. (Springer Lecture Notes in Mathematics No. 960) (New York: Springer-Verlag). [4]

Baker, L. 1991, *More C Tools for Scientists and Engineers* (New York: McGraw-Hill).

Briggs, W.L. 1987, *A Multigrid Tutorial* (Philadelphia: S.I.A.M.).

Jespersen, D. 1984, *Multigrid Methods for Partial Differential Equations* (Washington: Mathematical Association of America).

McCormick, S.F. (ed.) 1988, *Multigrid Methods: Theory, Applications, and Supercomputing* (New York: Marcel Dekker).

Hackbusch, W., and Trottenberg, U. (eds.) 1991, *Multigrid Methods III* (Boston: Birkhauser).

Wesseling, P. 1992, *An Introduction to Multigrid Methods* (New York: Wiley).

Chapter 20.   Less-Numerical Algorithms

20.0 Introduction

You can stop reading now. You are done with *Numerical Recipes*, as such. This final chapter is an idiosyncratic collection of "*less*-numerical recipes" which, for one reason or another, we have decided to include between the covers of an otherwise *more*-numerically oriented book. Authors of computer science texts, we've noticed, like to throw in a token numerical subject (usually quite a dull one — quadrature, for example). We find that we are not free of the reverse tendency.

Our selection of material is not completely arbitrary. One topic, Gray codes, was already used in the construction of quasi-random sequences (§7.7), and here needs only some additional explication. Two other topics, on diagnosing a computer's floating-point parameters, and on arbitrary precision arithmetic, give additional insight into the machinery behind the casual assumption that computers are useful for doing things with numbers (as opposed to bits or characters). The latter of these topics also shows a very different use for Chapter 12's fast Fourier transform.

The three other topics (checksums, Huffman and arithmetic coding) involve different aspects of data coding, compression, and validation. If you handle a large amount of data — numerical data, even — then a passing familiarity with these subjects might at some point come in handy. In §13.6, for example, we already encountered a good use for Huffman coding.

But again, you don't have to read this chapter. (And you should learn about quadrature from Chapters 4 and 16, not from a computer science text!)

20.1 Diagnosing Machine Parameters

A convenient fiction is that a computer's floating-point arithmetic is "accurate enough." If you believe this fiction, then numerical analysis becomes a very clean subject. Roundoff error disappears from view; many finite algorithms become "exact"; only docile truncation error (§1.4) stands between you and a perfect calculation. Sounds rather naive, doesn't it?

Yes, it is naive. Notwithstanding, it is a fiction necessarily adopted throughout most of this book. To do a good job of answering the question of how roundoff error propagates, or can be bounded, for every algorithm that we have discussed would be impractical. In fact, it would not be possible: Rigorous analysis of many practical algorithms has never been made, by us or anyone.

Proper numerical analysts cringe when they hear a user say, "I was getting roundoff errors with single precision, so I switched to double." The actual meaning is, "for this particular algorithm, and my particular data, double precision *seemed* able to restore my erroneous belief in the 'convenient fiction'." We admit that most of the mentions of precision or roundoff in *Numerical Recipes* are only slightly more quantitative in character. That comes along with our trying to be "practical."

It is important to know what the limitations of your machine's floating-point arithmetic actually are — the more so when your treatment of floating-point roundoff error is going to be intuitive, experimental, or casual. Methods for determining useful floating-point parameters experimentally have been developed by Cody [1], Malcolm [2], and others, and are embodied in the routine machar, below, which follows Cody's implementation. Many floating-point parameters are available by calling functions like numeric_limits<double>::epsilon() in the standard library limits, so a routine like machar is not of much practical importance any more.

All of machar's arguments are returned values. Here is what they mean:
- ibeta (called B in §1.4) is the radix in which numbers are represented, almost always 2, but occasionally 16, or even 10.
- it is the number of base-ibeta digits in the floating-point mantissa M (see Figure 1.4.1).
- machep is the exponent of the smallest (most negative) power of ibeta that, added to 1.0, gives something different from 1.0.
- eps is the floating-point number ibeta^{machep}, loosely referred to as the "floating-point precision."
- negep is the exponent of the smallest power of ibeta that, subtracted from 1.0, gives something different from 1.0.
- epsneg is ibeta^{negep}, another way of defining floating-point precision. Not infrequently epsneg is 0.5 times eps; occasionally eps and epsneg are equal.
- iexp is the number of bits in the exponent (including its sign or bias).
- minexp is the smallest (most negative) power of ibeta consistent with there being no leading zeros in the mantissa.
- xmin is the floating-point number ibeta^{minexp}, generally the smallest (in magnitude) useable floating value.
- maxexp is the smallest (positive) power of ibeta that causes overflow.
- xmax is $(1-\text{epsneg}) \times \text{ibeta}^{\text{maxexp}}$, generally the largest (in magnitude) useable floating value.
- irnd returns a code in the range 0 . . . 5, giving information on what kind of rounding is done in addition, and on how underflow is handled. See below.
- ngrd is the number of "guard digits" used when truncating the product of two mantissas to fit the representation.

There is a lot of subtlety in a program like machar, whose purpose is to ferret out machine properties that are supposed to be transparent to the user. Further, it must do so avoiding error conditions, like overflow and underflow, that might interrupt its execution. In some cases the program is able to do this only by recognizing certain characteristics of "standard" representations. For example, it recognizes

Sample Results Returned by `machar`			
	typical IEEE-compliant machine		DEC VAX
precision	single	double	single
`ibeta`	2	2	2
`it`	24	53	24
`machep`	-23	-52	-24
`eps`	1.19×10^{-7}	2.22×10^{-16}	5.96×10^{-8}
`negep`	-24	-53	-24
`epsneg`	5.96×10^{-8}	1.11×10^{-16}	5.96×10^{-8}
`iexp`	8	11	8
`minexp`	-126	-1022	-128
`xmin`	1.18×10^{-38}	2.23×10^{-308}	2.94×10^{-39}
`maxexp`	128	1024	127
`xmax`	3.40×10^{38}	1.79×10^{308}	1.70×10^{38}
`irnd`	5	5	1
`ngrd`	0	0	0

the IEEE standard representation [3] by its rounding behavior, and assumes certain features of its exponent representation as a consequence. We refer you to [1] and references therein for details. Be aware that `machar` can give incorrect results on some nonstandard machines.

The parameter `irnd` needs some additional explanation. In the IEEE standard, bit patterns correspond to exact, "representable" numbers. The specified method for rounding an addition is to add two representable numbers "exactly," and then round the sum to the closest representable number. If the sum is precisely halfway between two representable numbers, it should be rounded to the even one (low-order bit zero). The same behavior should hold for all the other arithmetic operations, that is, they should be done in a manner equivalent to infinite precision, and then rounded to the closest representable number.

If `irnd` returns 2 or 5, then your computer is compliant with this standard. If it returns 1 or 4, then it is doing some kind of rounding, but not the IEEE standard. If `irnd` returns 0 or 3, then it is truncating the result, not rounding it — not desirable.

The other issue addressed by `irnd` concerns underflow. If a floating value is less than `xmin`, many computers underflow its value to zero. Values `irnd` $= 0, 1$, or 2 indicate this behavior. The IEEE standard specifies a more graceful kind of underflow: As a value becomes smaller than `xmin`, its exponent is frozen at the smallest allowed value, while its mantissa is decreased, acquiring leading zeros and "gracefully" losing precision. This is indicated by `irnd` $= 3, 4$, or 5.

```
#include <cmath>
#include "nr.h"
using namespace std;

void NR::machar(int &ibeta, int &it, int &irnd, int &ngrd, int &machep,
    int &negep, int &iexp, int &minexp, int &maxexp, DP &eps, DP &epsneg,
    DP &xmin, DP &xmax)
```

Determines and returns machine-specific parameters affecting floating-point arithmetic. Returned values include ibeta, the floating-point radix; it, the number of base-ibeta digits in the floating-point mantissa; eps, the smallest positive number that, added to 1.0, is not equal to 1.0; epsneg, the smallest positive number that, subtracted from 1.0, is not equal to 1.0; xmin, the smallest representable positive number; and xmax, the largest representable positive number. See text for description of other returned parameters. Change DP to float to find single precision parameters.

```
{
    int i,itemp,iz,j,k,mx,nxres;
    DP a,b,beta,betah,betain,one,t,temp,temp1,tempa,two,y,z,zero;

    one=DP(1);
    two=one+one;
    zero=one-one;
    a=one;                                    Determine ibeta and beta by the method of M.
    do {                                         Malcolm.
        a += a;
        temp=a+one;
        temp1=temp-a;
    } while (temp1-one == zero);
    b=one;
    do {
        b += b;
        temp=a+b;
        itemp=int(temp-a);
    } while (itemp == 0);
    ibeta=itemp;
    beta=DP(ibeta);
    it=0;                                     Determine it and irnd.
    b=one;
    do {
        ++it;
        b *= beta;
        temp=b+one;
        temp1=temp-b;
    } while (temp1-one == zero);
    irnd=0;
    betah=beta/two;
    temp=a+betah;
    if (temp-a != zero) irnd=1;
    tempa=a+beta;
    temp=tempa+betah;
    if (irnd == 0 && temp-tempa != zero) irnd=2;
    negep=it+3;                               Determine negep and epsneg.
    betain=one/beta;
    a=one;
    for (i=1;i<=negep;i++) a *= betain;
    b=a;
    for (;;) {
        temp=one-a;
        if (temp-one != zero) break;
        a *= beta;
        --negep;
    }
    negep = -negep;
    epsneg=a;
    machep = -it-3;                           Determine machep and eps.
```

```
a=b;
for (;;) {
    temp=one+a;
    if (temp-one != zero) break;
    a *= beta;
    ++machep;
}
eps=a;
ngrd=0;                             Determine ngrd.
temp=one+eps;
if (irnd == 0 && temp*one-one != zero) ngrd=1;
i=0;                                Determine iexp.
k=1;
z=betain;
t=one+eps;
nxres=0;
for (;;) {                          Loop until an underflow occurs, then exit.
    y=z;
    z=y*y;
    a=z*one;                        Check here for the underflow.
    temp=z*t;
    if (a+a == zero || fabs(z) >= y) break;
    temp1=temp*betain;
    if (temp1*beta == z) break;
    ++i;
    k += k;
}
if (ibeta != 10) {
    iexp=i+1;
    mx=k+k;
} else {                            For decimal machines only.
    iexp=2;
    iz=ibeta;
    while (k >= iz) {
        iz *= ibeta;
        ++iexp;
    }
    mx=iz+iz-1;
}
for (;;) {                          To determine minexp and xmin, loop until an
    xmin=y;                             underflow occurs, then exit.
    y *= betain;
    a=y*one;                        Check here for the underflow.
    temp=y*t;
    if (a+a != zero && fabs(y) < xmin) {
        ++k;
        temp1=temp*betain;
        if (temp1*beta == y && temp != y) {
            nxres=3;
            xmin=y;
            break;
        }
    }
    else break;
}
minexp = -k;                        Determine maxexp, xmax.
if (mx <= k+k-3 && ibeta != 10) {
    mx += mx;
    ++iexp;
}
maxexp=mx+minexp;
irnd += nxres;                      Adjust irnd to reflect partial underflow.
if (irnd >= 2) maxexp -= 2;         Adjust for IEEE-style machines.
i=maxexp+minexp;
```

Adjust for machines with implicit leading bit in binary mantissa, and machines with radix
point at extreme right of mantissa.

```
if (ibeta == 2 && !i) --maxexp;
if (i > 20) --maxexp;
if (a != y) maxexp -= 2;
xmax=one-epsneg;
if (xmax*one != xmax) xmax=one-beta*epsneg;
xmax /= (xmin*beta*beta*beta);
i=maxexp+minexp+3;
for (j=1;j<=i;j++) {
    if (ibeta == 2) xmax += xmax;
    else xmax *= beta;
}
}
```

Some typical values returned by machar are given in the table, above. IEEE-compliant machines referred to in the table include most UNIX workstations (SUN, DEC, MIPS), and Apple Macintoshes. Pentium and similar chips with either on-chip or separate floating point co-processors are generally IEEE-compliant, except that some compilers underflow intermediate results ungracefully, yielding $irnd = 2$ rather than 5. Notice, as in the case of a VAX (fourth column), that representations with a "phantom" leading 1 bit in the mantissa achieve a smaller eps for the same wordlength, but cannot underflow gracefully.

CITED REFERENCES AND FURTHER READING:

Goldberg, D. 1991, *ACM Computing Surveys*, vol. 23, pp. 5–48.

Cody, W.J. 1988, *ACM Transactions on Mathematical Software*, vol. 14, pp. 303–311. [1]

Malcolm, M.A. 1972, *Communications of the ACM*, vol. 15, pp. 949–951. [2]

IEEE Standard for Binary Floating-Point Numbers, ANSI/IEEE Std 754–1985 (New York: IEEE, 1985). [3]

20.2 Gray Codes

A Gray code is a function $G(i)$ of the integers i, that for each integer $N \geq 0$ is one-to-one for $0 \leq i \leq 2^N - 1$, and that has the following remarkable property: The binary representation of $G(i)$ and $G(i+1)$ differ in *exactly one bit*. An example of a Gray code (in fact, the most commonly used one) is the sequence 0000, 0001, 0011, 0010, 0110, 0111, 0101, 0100, 1100, 1101, 1111, 1110, 1010, 1011, 1001, and 1000, for $i = 0, \ldots, 15$. The algorithm for generating this code is simply to form the bitwise exclusive-or (XOR) of i with $i/2$ (integer part). Think about how the carries work when you add one to a number in binary, and you will be able to see why this works. You will also see that $G(i)$ and $G(i+1)$ differ in the bit position of the rightmost zero bit of i (prefixing a leading zero if necessary).

The spelling is "Gray," not "gray": The codes are named after one Frank Gray, who first patented the idea for use in shaft encoders. A shaft encoder is a wheel with concentric coded stripes each of which is "read" by a fixed optical sensor or conducting brush. The idea is to generate a binary code describing the angle of the wheel. The obvious, but wrong, way to build a shaft encoder is to have one stripe

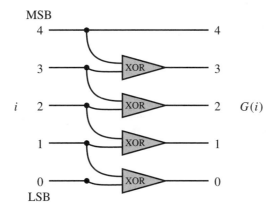

(a)

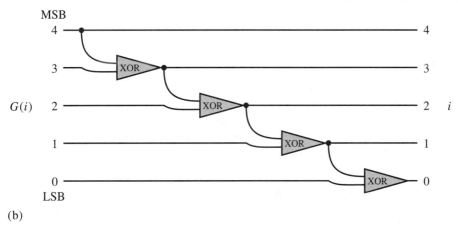

(b)

Figure 20.2.1.   Single-bit operations for calculating the Gray code $G(i)$ from i (a), or the inverse (b). LSB and MSB indicate the least and most significant bits, respectively. XOR denotes exclusive-or.

(the innermost, say) present on half the wheel, but absent on the other half; the next stripe is present in quadrants 1 and 3; the next stripe is present in octants 1, 3, 5, and 7; and so on. The optical or electrical sensors together then read a direct binary code for the position of the wheel.

The reason this method is bad, is that there is no way to guarantee that all the brushes will make or break contact *exactly* simultaneously as the wheel turns. Going from position 7 (0111) to 8 (1000), one might pass spuriously and transiently through 6 (0110), 14 (1110), and 10 (1010), as the different brushes make or break contact. Use of a Gray code on the encoding stripes guarantees that there is no transient state between 7 (0100 in the sequence above) and 8 (1100).

Of course we then need circuitry, or algorithmics, to translate from $G(i)$ to i. Figure 20.2.1 (b) shows how this is done by a cascade of XOR gates. The idea is that each output bit should be the XOR of all more significant input bits. To do N bits of Gray code inversion requires $N-1$ steps (or gate delays) in the circuit. (Nevertheless, this is typically very fast in circuitry.) In a register with word-wide binary operations, we don't have to do N consecutive operations, but only $\ln_2 N$.

The trick is to use the associativity of XOR and group the operations hierarchically. This involves sequential right-shifts by $1, 2, 4, 8, \ldots$ bits until the wordlength is exhausted. Here is a piece of code for doing both $G(i)$ and its inverse.

```
#include "nr.h"

unsigned long NR::igray(const unsigned long n, const int is)
For zero or positive values of is, return the Gray code of n; if is is negative, return the inverse
Gray code of n.
{
    int ish;
    unsigned long ans,idiv;

    if (is >= 0)                    This is the easy direction!
        return n ^ (n >> 1);
    ish=1;                          This is the more complicated direction: In hierarchical
    ans=n;                          stages, starting with a one-bit right shift, cause each
    for (;;) {                      bit to be XORed with all more significant bits.
        ans ^= (idiv=ans >> ish);
        if (idiv <= 1 || ish == 16) return ans;
        ish <<= 1;                  Double the amount of shift on the next cycle.
    }
}
```

In numerical work, Gray codes can be useful when you need to do some task that depends intimately on the bits of i, looping over many values of i. Then, if there are economies in repeating the task for values differing by only one bit, it makes sense to do things in Gray code order rather than consecutive order. We saw an example of this in §7.7, for the generation of quasi-random sequences.

CITED REFERENCES AND FURTHER READING:

Horowitz, P., and Hill, W. 1989, *The Art of Electronics*, 2nd ed. (New York: Cambridge University Press), §8.02.

Knuth, D.E. *Combinatorial Algorithms*, vol. 4 of *The Art of Computer Programming* (Reading, MA: Addison-Wesley), §7.2.1. [Unpublished. Will it be always so?]

20.3 Cyclic Redundancy and Other Checksums

When you send a sequence of bits from point A to point B, you want to know that it will arrive without error. A common form of insurance is the "parity bit," attached to 7-bit ASCII characters to put them into 8-bit format. The parity bit is chosen so as to make the total number of one-bits (versus zero-bits) either always even ("even parity") or always odd ("odd parity"). Any *single bit* error in a character will thereby be detected. When errors are sufficiently rare, and do not occur closely bunched in time, use of parity provides sufficient error detection.

Unfortunately, in real situations, a single noise "event" is likely to disrupt more than one bit. Since the parity bit has two possible values (0 and 1), it gives, on average, only a 50% chance of detecting an erroneous character with more than one wrong bit. That probability, 50%, is not nearly good enough for most applications.

Most communications protocols [1] use a multibit generalization of the parity bit called a "cyclic redundancy check" or CRC. In typical applications the CRC is 16 bits long (two bytes or two characters), so that the chance of a random error going undetected is 1 in $2^{16} = 65536$. Moreover, M-bit CRCs have the mathematical property of detecting *all* errors that occur in M or fewer *consecutive* bits, for any length of message. (We prove this below.) Since noise in communication channels tends to be "bursty," with short sequences of adjacent bits getting corrupted, this consecutive-bit property is highly desirable.

Normally CRCs lie in the province of communications software experts and chip-level hardware designers — people with bits under their fingernails. However, there are at least two kinds of situations where some understanding of CRCs can be useful to the rest of us. First, we sometimes need to be able to communicate with a lower-level piece of hardware or software that expects a valid CRC as part of its input. For example, it can be convenient to have a program generate XMODEM or Kermit [2] packets directly into the communications line rather than having to store the data in a local file.

Second, in the manipulation of large quantities of (e.g., experimental) data, it is useful to be able to tag aggregates of data (whether numbers, records, lines, or whole files) with a statistically unique "key," its CRC. Aggregates of any size can then be compared for identity by comparing only their short CRC keys. Differing keys imply nonidentical records. Identical keys imply, to high statistical certainty, identical records. If you can't tolerate the very small probability of being wrong, you can do a full comparison of the records when the keys are identical. When there is a possibility of files or data records being inadvertently or irresponsibly modified (for example, by a computer virus), it is useful to have their prior CRCs stored externally on a physically secure medium, like a floppy disk.

Sometimes CRCs can be used to compress data as it is recorded. If identical data records occur frequently, one can keep sorted in memory the CRCs of previously encountered records. A new record is archived in full if its CRC is different, otherwise only a pointer to a previous record need be archived. In this application one might desire a 4- or 8-byte CRC, to make the odds of mistakenly discarding a different data record tolerably small; or, if previous records can be randomly accessed, a full comparison can be made to decide whether records with identical CRCs are in fact identical.

Now let us briefly discuss the theory of CRCs. After that, we will give implementations of various (related) CRCs that are used by the official or de facto standard protocols [1-3] listed in the accompanying table.

The mathematics underlying CRCs is "polynomials over the integers modulo 2." Any binary message can be thought of as a polynomial with coefficients 0 and 1. For example, the message "1100001101" is the polynomial $x^9 + x^8 + x^3 + x^2 + 1$. Since 0 and 1 are the only integers modulo 2, a power of x in the polynomial is either present (1) or absent (0). A polynomial over the integers modulo 2 may be irreducible, meaning that it can't be factored. A subset of the irreducible polynomials are the "primitive" polynomials. These generate maximum length sequences when used in shift registers, as described in §7.4. The polynomial $x^2 + 1$ is not irreducible: $x^2 + 1 = (x+1)(x+1)$, so it is also not primitive. The polynomial $x^4 + x^3 + x^2 + x + 1$ is irreducible, but it turns out not to be primitive. The polynomial $x^4 + x + 1$ is both irreducible and primitive.

Conventions and Test Values for Various CRC Protocols						
	icrc args		Test Values (C_2C_1 in hex)		Packet	
Protocol	jinit	jrev	T	CatMouse987654321	*Format*	*CRC*
XMODEM	0	1	1A71	E556	$S_1S_2\ldots S_NC_2C_1$	0
X.25	255	-1	1B26	F56E	$S_1S_2\ldots S_N\overline{C_1C_2}$	F0B8
(no name)	255	-1	1B26	F56E	$S_1S_2\ldots S_NC_1C_2$	0
SDLC (IBM)	same as X.25					
HDLC (ISO)	same as X.25					
CRC-CCITT	0	-1	14A1	C28D	$S_1S_2\ldots S_NC_1C_2$	0
(no name)	0	-1	14A1	C28D	$S_1S_2\ldots S_N\overline{C_1C_2}$	F0B8
Kermit	same as CRC-CCITT				see Notes	

Notes: Overbar denotes bit complement. $S_1\ldots S_N$ are character data. C_1 is CRC's least significant 8 bits, C_2 is its most significant 8 bits, so $CRC = 256\,C_2 + C_1$ (shown in hex). Kermit (block check level 3) sends the CRC as 3 printable ASCII characters (sends value $+32$). These contain, respectively, 4 most significant bits, 6 middle bits, 6 least significant bits.

An M-bit long CRC is based on a primitive polynomial of degree M, called the generator polynomial. Alternatively, the generator is chosen to be a primitive polynomial times $(1 + x)$ (this finds all parity errors). For 16-bit CRC's, the CCITT (Comité Consultatif International Télégraphique et Téléphonique) has anointed the "CCITT polynomial," which is $x^{16} + x^{12} + x^5 + 1$. This polynomial is used by all of the protocols listed in the table. Another common choice is the "CRC-16" polynomial $x^{16} + x^{15} + x^2 + 1$, which is used for EBCDIC messages in IBM's BISYNCH [1]. A common 12-bit choice, "CRC-12," is $x^{12} + x^{11} + x^3 + x + 1$. A common 32-bit choice, "AUTODIN-II," is $x^{32} + x^{26} + x^{23} + x^{22} + x^{16} + x^{12} + x^{11} + x^{10} + x^8 + x^7 + x^5 + x^4 + x^2 + x + 1$. For a table of some other primitive polynomials, see §7.4.

Given the generator polynomial G of degree M (which can be written either in polynomial form or as a bit-string, e.g., 10001000000100001 for CCITT), here is how you compute the CRC for a sequence of bits S: First, multiply S by x^M, that is, append M zero bits to it. Second divide — by long division — G into Sx^M. Keep in mind that the subtractions in the long division are done modulo 2, so that there are never any "borrows": Modulo 2 subtraction is the same as logical exclusive-or (XOR). Third, ignore the quotient you get. Fourth, when you eventually get to a remainder, it is the CRC, call it C. C will be a polynomial of degree $M - 1$ or less, otherwise you would not have finished the long division. Therefore, in bit string form, it has M bits, which may include leading zeros. (C might even be all zeros, see below.) See [3] for a worked example.

If you work through the above steps in an example, you will see that most of what you write down in the long-division tableau is superfluous. You are actually just left-shifting sequential bits of S, from the right, into an M-bit register. Every time a 1 bit gets shifted off the left end of this register, you zap the register by an XOR with the M low order bits of G (that is, all the bits of G except its leading 1). When a 0 bit is shifted off the left end you don't zap the register. When the last bit that was originally part of S gets shifted off the left end of the register, what remains is the CRC.

You can immediately recognize how efficiently this procedure can be implemented in hardware. It requires only a shift register with a few hard-wired XOR taps into it. That is how CRCs are computed in communications devices, by a single chip (or small part of one). In software, the implementation is not so elegant, since bit-shifting is not generally very efficient. One therefore typically finds (as in our implementation below) table-driven routines that pre-calculate the result of a bunch of shifts and XORs, say for each of 256 possible 8-bit inputs [4].

We can now see how the CRC gets its ability to detect all errors in M consecutive bits. Suppose two messages, S and T, differ only within a frame of M bits. Then their CRCs differ by an amount that is the remainder when G is divided into $(S - T)x^M \equiv D$. Now D has the form of leading zeros (which can be ignored), followed by some 1's in an M-bit frame, followed by trailing zeros (which are just multiplicative factors of x): $D = x^n F$ where F is a polynomial of degree at most $M - 1$ and $n > 0$. Since G is always primitive or primitive times $(1 + x)$, it is not divisible by x. So G cannot divide D. Therefore S and T must have different CRCs.

In most protocols, a transmitted block of data consists of some N data bits, directly followed by the M bits of their CRC (or the CRC XORed with a constant, see below). There are two equivalent ways of validating a block at the receiving end. Most obviously, the receiver can compute the CRC of the data bits, and compare it to the transmitted CRC bits. Less obviously, but more elegantly, the receiver can simply compute the CRC of the total block, with $N + M$ bits, and verify that a result of zero is obtained. Proof: The total block is the polynomial $Sx^M + C$ (data left-shifted to make room for the CRC bits). The definition of C is that $Sx^m = QG + C$, where Q is the discarded quotient. But then $Sx^M + C = QG + C + C = QG$ (remember modulo 2), which is a perfect multiple of G. It remains a multiple of G when it gets multiplied by an additional x^M on the receiving end, so it has a zero CRC, q.e.d.

A couple of small variations on the basic procedure need to be mentioned [1,3]: First, when the CRC is computed, the M-bit register need not be initialized to zero. Initializing it to some other M-bit value (e.g., all 1's) in effect prefaces all blocks by a phantom message that would have given the initialization value as its remainder. It is advantageous to do this, since the CRC described thus far otherwise cannot detect the addition or removal of any number of initial zero bits. (Loss of an initial bit, or insertion of zero bits, are common "clocking errors.") Second, one can add (XOR) any M-bit constant K to the CRC before it is transmitted. This constant can either be XORed away at the receiving end, or else it just changes the expected CRC of the whole block by a known amount, namely the remainder of dividing G into Kx^M. The constant K is frequently "all bits," changing the CRC into its ones complement. This has the advantage of detecting another kind of error that the CRC would otherwise not find: deletion of an initial 1 bit in the message with spurious insertion of a 1 bit at the end of the block.

The accompanying function `icrc` implements the above CRC calculation, including the possibility of the mentioned variations. Input to the function is a pointer to an array of characters, and the length of that array. `icrc` has two "switch" arguments that specify variations in the CRC calculation. A zero or positive value of `jinit` causes the 16-bit register to have each byte initialized with the value `jinit`. A negative value of `jrev` causes each input character to be interpreted as

its bit-reverse image, and a similar bit reversal to be done on the output CRC. You
do not have to understand this; just use the values of jinit and jrev specified in
the table. (If you *insist* on knowing, the explanation is that serial data ports send
characters *least-significant bit first* (!), and many protocols shift bits into the CRC
register in exactly the order received.) The table shows how to construct a block of
characters from the input array and output CRC of icrc. You should not need to
do any additional bit-reversal outside of icrc.

The switch jinit has one additional use: When negative it causes the input
value of the array crc to be used as initialization of the register. If you set crc to the
result of the last call to icrc, this in effect appends the current input array to that of
the previous call or calls. Use this feature, for example, to build up the CRC of a
whole file a line at a time, without keeping the whole file in memory.

The routine icrc is loosely based on the function in [4]. Here is how to
understand its operation: First look at the function icrc1. This incorporates one
input character into a 16-bit CRC register. The only trick used is that character bits
are XORed into the most significant bits, eight at a time, instead of being fed into
the least significant bit, one bit at a time, at the time of the register shift. This works
because XOR is associative and commutative — we can feed in character bits *any*
time before they will determine whether to zap with the generator polynomial. (The
decimal constant 4129 has the generator's bits in it.)

```
#include "nr.h"

unsigned short NR::icrc1(const unsigned short crc, const unsigned char onech)
Given a remainder up to now, return the new CRC after one character is added. This routine is
functionally equivalent to icrc(,,-1,1), but slower. It is used by icrc to initialize its table.
{
    int i;
    unsigned short ans=(crc ^ onech << 8);

    for (i=0;i<8;i++) {                    Here is where 8 one-bit shifts, and some XORs with
        if (ans & 0x8000)                      the generator polynomial, are done.
            ans = (ans <<= 1) ^ 4129;
        else
            ans <<= 1;
    }
    return ans;
}
```

Now look at icrc. There are two parts to understand, how it builds a table
when it initializes, and how it uses that table later on. Go back to thinking about a
character's bits being shifted into the CRC register from the least significant end. The
key observation is that while 8 bits are being shifted into the register's low end, all
the generator zapping is being determined by the bits already in the high end. Since
XOR is commutative and associative, all we need is a table of the result of all this
zapping, for each of 256 possible high-bit configurations. Then we can play catch-up
and XOR an input character into the result of a lookup into this table. The only other
content to icrc is the construction at initialization time of an 8-bit bit-reverse table
from the 4-bit table stored in it, and the logic associated with doing the bit reversals.
References [4-6] give further details on table-driven CRC computations.

```
#include "nr.h"

namespace {
    inline unsigned char lobyte(const unsigned short x)
    {
        return (unsigned char)((x) & 0xff);
    }

    inline unsigned char hibyte(const unsigned short x)
    {
        return (unsigned char)((x >> 8) & 0xff);
    }
}
```

unsigned short NR::icrc(const unsigned short crc, const string &bufptr,
    const short jinit, const int jrev)
Computes a 16-bit Cyclic Redundancy Check for a string bufptr, using any of several conventions as determined by the settings of jinit and jrev (see accompanying table). If jinit is negative, then crc is used on input to initialize the remainder register, in effect (for crc set to the last returned value) concatenating bufptr to the previous call.

```
{
    static unsigned short icrctb[256],init=0;
    static unsigned char rchr[256];
    unsigned short j,cword=crc;
    static unsigned char it[16]={0,8,4,12,2,10,6,14,1,9,5,13,3,11,7,15};
    Table of 4-bit bit-reverses.

    unsigned long len=bufptr.length();
    if (init == 0) {                        Do we need to initialize tables?
        init=1;
        for (j=0;j<256;j++) {
        The two tables are: CRCs of all characters, and bit-reverses of all characters.
            icrctb[j]=icrc1(j << 8,0);
            rchr[j]=(unsigned char)((it[j & 0xf] << 4) | (it[j >> 4]));
        }
    }
    if (jinit >= 0)                         Initialize the remainder register.
        cword=(jinit | (jinit << 8));
    else if (jrev < 0)                      If not initializing, do we reverse the register?
        cword=(rchr[hibyte(cword)] | (rchr[lobyte(cword)] << 8));
    for (j=0;j<len;j++) {                   Main loop over the characters in the array.
        cword=icrctb[(jrev < 0 ? rchr[(unsigned char) bufptr[j]] :
            (unsigned char) bufptr[j]) ^ hibyte(cword)] ^ (lobyte(cword) << 8);
    }
    return (jrev >= 0 ? cword :             Do we need to reverse the output?
        rchr[hibyte(cword)] | (rchr[lobyte(cword)] << 8));
}
```

What if you need a 32-bit checksum? For a true 32-bit CRC, you will need to rewrite the routines given to work with a longer generating polynomial. For example, $x^{32} + x^7 + x^5 + x^3 + x^2 + x + 1$ is primitive modulo 2, and has nonleading, nonzero bits only in its least significant byte (which makes for some simplification). The idea of table lookup on only the most significant byte of the CRC register goes through unchanged.

If you do not care about the M-consecutive bit property of the checksum, but rather only need a statistically random 32 bits, then you can use icrc as given here: Call it once with jrev $= 1$ to get 16 bits, and *again* with jrev $= -1$ to get another 16 bits. The internal bit reversals make these two 16-bit CRCs in effect totally independent of each other.

Other Kinds of Checksums

Quite different from CRCs are the various techniques used to append a decimal "check digit" to numbers that are handled by human beings (e.g., typed into a computer). Check digits need to be proof against the kinds of highly structured errors that humans tend to make, such as transposing consecutive digits. Wagner and Putter [7] give an interesting introduction to this subject, including specific algorithms.

Checksums now in widespread use vary from fair to poor. The 10-digit ISBN (International Standard Book Number) that you find on most books, including this one, uses the check equation

$$10d_1 + 9d_2 + 8d_3 + \cdots + 2d_9 + d_{10} = 0 \quad (\text{mod } 11) \qquad (20.3.1)$$

where d_{10} is the right-hand check digit. The character "X" is used to represent a check digit value of 10. Another popular scheme is the so-called "IBM check," often used for account numbers (including, e.g., MasterCard). Here, the check equation is

$$2\#d_1 + d_2 + 2\#d_3 + d_4 + \cdots = 0 \quad (\text{mod } 10) \qquad (20.3.2)$$

where $2\#d$ means, "multiply d by two and add the resulting decimal digits." United States banks code checks with a 9-digit processing number whose check equation is

$$3a_1 + 7a_2 + a_3 + 3a_4 + 7a_5 + a_6 + 3a_7 + 7a_8 + a_9 = 0 \quad (\text{mod } 10) \quad (20.3.3)$$

The bar code put on many envelopes by the U.S. Postal Service is decoded by removing the single tall marker bars at each end, and breaking the remaining bars into 6 or 10 groups of five. In each group the five bars signify (from left to right) the values 7,4,2,1,0. Exactly two of them will be tall. Their sum is the represented digit, except that zero is represented as $7 + 4$. The 5- or 9-digit Zip Code is followed by a check digit, with the check equation

$$\sum d_i = 0 \quad (\text{mod } 10) \qquad (20.3.4)$$

None of these schemes is close to optimal. An elegant scheme due to Verhoeff is described in [7]. The underlying idea is to use the ten-element *dihedral group* D_5, which corresponds to the symmetries of a pentagon, instead of the cyclic group of the integers modulo 10. The check equation is

$$a_1 {}^* f(a_2) {}^* f^2(a_3) {}^* \cdots {}^* f^{n-1}(a_n) = 0 \qquad (20.3.5)$$

where $*$ is (noncommutative) multiplication in D_5, and f^i denotes the ith iteration of a certain fixed permutation. Verhoeff's method finds *all* single errors in a string, and *all* adjacent transpositions. It also finds about 95% of twin errors ($aa \rightarrow bb$), jump transpositions ($acb \rightarrow bca$), and jump twin errors ($aca \rightarrow bcb$). Here is an implementation:

```
#include "nr.h"
```

```
bool NR::decchk(string str, char &ch)
```
Decimal check digit computation or verification. Returns as `ch` a check digit for appending to `string[0..n-1]`, that is, for storing into `string[n]`. In this mode, ignore the returned boolean value. If `string[0..n-1]` already ends with a check digit (`string[n-1]`), returns the function value `true` if the check digit is valid, otherwise `false`. In this mode, ignore the returned value of `ch`. Note that `string` and `ch` contain ASCII characters corresponding to the digits 0-9, *not* byte values in that range. Other ASCII characters are allowed in `string`, and are ignored in calculating the check digit.

```
{
    char c;
    int j,k=0,m=0;
    static int ip[10][8]={{0,1,5,8,9,4,2,7},{1,5,8,9,4,2,7,0},
        {2,7,0,1,5,8,9,4},{3,6,3,6,3,6,3,6},{4,2,7,0,1,5,8,9},
        {5,8,9,4,2,7,0,1},{6,3,6,3,6,3,6,3},{7,0,1,5,8,9,4,2},
        {8,9,4,2,7,0,1,5},{9,4,2,7,0,1,5,8}};
    static int ij[10][10]={{0,1,2,3,4,5,6,7,8,9},{1,2,3,4,0,6,7,8,9,5},
        {2,3,4,0,1,7,8,9,5,6},{3,4,0,1,2,8,9,5,6,7},{4,0,1,2,3,9,5,6,7,8},
        {5,9,8,7,6,0,4,3,2,1},{6,5,9,8,7,1,0,4,3,2},{7,6,5,9,8,2,1,0,4,3},
        {8,7,6,5,9,3,2,1,0,4},{9,8,7,6,5,4,3,2,1,0}};
    Group multiplication and permutation tables.

    int n=str.length();
    for (j=0;j<n;j++) {                     Look at successive characters.
        c=str[j];
        if (c >= 48 && c <= 57)             Ignore everything except digits.
            k=ij[k][ip[(c+2) % 10][7 & m++]];
    }
    for (j=0;j<10;j++)                      Find which appended digit will check properly.
        if (ij[k][ip[j][m & 7]] == 0) break;
    ch=char(j+48);                          Convert to ASCII.
    return k==0;
}
```

CITED REFERENCES AND FURTHER READING:

McNamara, J.E. 1982, *Technical Aspects of Data Communication*, 2nd ed. (Bedford, MA: Digital Press). [1]

da Cruz, F. 1987, *Kermit, A File Transfer Protocol* (Bedford, MA: Digital Press). [2]

Morse, G. 1986, *Byte*, vol. 11, pp. 115–124 (September). [3]

LeVan, J. 1987, *Byte*, vol. 12, pp. 339–341 (November). [4]

Sarwate, D.V. 1988, *Communications of the ACM*, vol. 31, pp. 1008–1013. [5]

Griffiths, G., and Stones, G.C. 1987, *Communications of the ACM*, vol. 30, pp. 617–620. [6]

Wagner, N.R., and Putter, P.S. 1989, *Communications of the ACM*, vol. 32, pp. 106–110. [7]

20.4 Huffman Coding and Compression of Data

A lossless data compression algorithm takes a string of symbols (typically ASCII characters or bytes) and translates it *reversibly* into another string, one that is *on the average* of shorter length. The words "on the average" are crucial; it is obvious that no reversible algorithm can make all strings shorter — there just aren't enough short strings to be in one-to-one correspondence with longer strings. Compression algorithms are possible only when, on the input side, some strings, or some input symbols, are more common than others. These can then be encoded in fewer bits than rarer input strings or symbols, giving a net average gain.

There exist many, quite different, compression techniques, corresponding to different ways of detecting and using departures from equiprobability in input strings. In this section and the next we shall consider only *variable length codes* with *defined word* inputs. In these, the input is sliced into fixed units, for example ASCII characters, while the corresponding output comes in chunks of variable size. The simplest such method is Huffman coding [1], discussed in this section. Another example, *arithmetic compression*, is discussed in §20.5.

At the opposite extreme from defined-word, variable length codes are schemes that divide up the *input* into units of variable length (words or phrases of English text, for example) and then transmit these, often with a fixed-length output code. The most widely used code of this type is the Ziv-Lempel code [2]. References [3-6] give the flavor of some other compression techniques, with references to the large literature.

The idea behind Huffman coding is simply to use shorter bit patterns for more common characters. We can make this idea quantitative by considering the concept of *entropy*. Suppose the input alphabet has N_{ch} characters, and that these occur in the input string with respective probabilities p_i, $i = 1, \ldots, N_{ch}$, so that $\sum p_i = 1$. Then the fundamental theorem of information theory says that strings consisting of independently random sequences of these characters (a conservative, but not always realistic assumption) require, on the average, at least

$$H = -\sum p_i \log_2 p_i \qquad (20.4.1)$$

bits per character. Here H is the entropy of the probability distribution. Moreover, coding schemes exist which approach the bound arbitrarily closely. For the case of equiprobable characters, with all $p_i = 1/N_{ch}$, one easily sees that $H = \log_2 N_{ch}$, which is the case of no compression at all. Any other set of p_i's gives a smaller entropy, allowing some useful compression.

Notice that the bound of (20.4.1) would be achieved if we could encode character i with a code of length $L_i = -\log_2 p_i$ bits: Equation (20.4.1) would then be the average $\sum p_i L_i$. The trouble with such a scheme is that $-\log_2 p_i$ is not generally an integer. How can we encode the letter "Q" in 5.32 bits? Huffman coding makes a stab at this by, in effect, approximating all the probabilities p_i by integer powers of 1/2, so that all the L_i's are integral. If all the p_i's are in fact of this form, then a Huffman code does achieve the entropy bound H.

The construction of a Huffman code is best illustrated by example. Imagine a language, Vowellish, with the $N_{ch} = 5$ character alphabet A, E, I, O, and U, occurring with the respective probabilities 0.12, 0.42, 0.09, 0.30, and 0.07. Then the construction of a Huffman code for Vowellish is accomplished in the following table:

Node	Stage:	1	2	3	4	5
1	A:	0.12	0.12 ∎			
2	E:	0.42	0.42	0.42	0.42 ∎	
3	I:	0.09 ∎				
4	O:	0.30	0.30	0.30 ∎		
5	U:	0.07 ∎				
6		UI:	0.16 ∎			
7			AUI:	0.28 ∎		
8				AUIO:	0.58 ∎	
9					EAUIO:	1.00

Here is how it works, proceeding in sequence through N_{ch} stages, represented by the columns of the table. The first stage starts with N_{ch} nodes, one for each letter of the alphabet, containing their respective relative frequencies. At each stage, the two smallest probabilities are found, summed to make a new node, and then dropped from the list of active nodes. (A "block" denotes the stage where a node is dropped.) All active nodes (including the new composite) are then carried over to the next stage (column). In the table, the names assigned to new nodes (e.g., AUI) are inconsequential. In the example shown, it happens that (after stage 1) the two smallest nodes are always an original node and a composite one; this need not be true in general: The two smallest probabilities might be both original nodes, or both composites, or one of each. At the last stage, all nodes will have been collected into one grand composite of total probability 1.

Now, to see the code, you redraw the data in the above table as a tree (Figure 20.4.1). As shown, each node of the tree corresponds to a node (row) in the table, indicated by the integer to its left and probability value to its right. Terminal nodes, so called, are shown as circles; these are single alphabetic characters. The branches of the tree are labeled 0 and 1. The code for a character is the sequence of zeros and ones that lead to it, from the top down. For example, E is simply 0, while U is 1010.

Any string of zeros and ones can now be decoded into an alphabetic sequence. Consider, for example, the string 1011111010. Starting at the top of the tree we descend through 1011 to I, the first character. Since we have reached a terminal node, we reset to the top of the tree, next descending through 11 to O. Finally 1010 gives U. The string thus decodes to IOU.

These ideas are embodied in the following routines. Input to the first routine hufmak is an integer vector of the frequency of occurrence of the nchin $\equiv N_{ch}$ alphabetic characters, i.e., a set of integers proportional to the p_i's. hufmak, along with hufapp, which it calls, performs the construction of the above table, and also the tree of Figure 20.4.1. The routine utilizes a heap structure (see §8.3) for efficiency; for a detailed description, see Sedgewick [7].

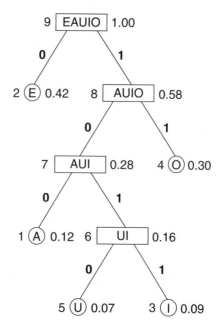

Figure 20.4.1.   Huffman code for the fictitious language Vowellish, in tree form. A letter (A, E, I, O, or U) is encoded or decoded by traversing the tree from the top down; the code is the sequence of 0's and 1's on the branches. The value to the right of each node is its probability; to the left, its node number in the accompanying table.

```
#include "nr.h"
```

```
void NR::hufmak(Vec_I_ULNG &nfreq, const unsigned long nchin,
      unsigned long &ilong, unsigned long &nlong, huffcode &hcode)
```
Given the frequency of occurrence table nfreq[0..nchin-1] for nchin characters, construct the Huffman code in the class object hcode. Returned values ilong and nlong are the character number that produced the longest code symbol, and the length of that symbol. You should check that nlong is not larger than your machine's word length.
```
{
    int ibit,j,node;
    unsigned long k,n,nused;
    static unsigned long setbit[32]={0x1L,0x2L,0x4L,0x8L,0x10L,0x20L,
        0x40L,0x80L,0x100L,0x200L,0x400L,0x800L,0x1000L,0x2000L,
        0x4000L,0x8000L,0x10000L,0x20000L,0x40000L,0x80000L,0x100000L,
        0x200000L,0x400000L,0x800000L,0x1000000L,0x2000000L,0x4000000L,
        0x8000000L,0x10000000L,0x20000000L,0x40000000L,0x80000000L};
    hcode.nch=nchin;                          Initialization.
    Vec_ULNG index(2*hcode.nch-1);
    Vec_ULNG nprob(2*hcode.nch-1);
    Vec_INT up(2*hcode.nch-1);                Vector that will keep track of heap.
    for (nused=0,j=0;j<hcode.nch;j++) {
        nprob[j]=nfreq[j];
        hcode.icod[j]=hcode.ncod[j]=0;
        if (nfreq[j] != 0) index[nused++]=j;
    }
    for (j=nused-1;j>=0;j--)                   Sort nprob into a heap structure in index.
        hufapp(index,nprob,nused,j);
    k=hcode.nch;
    while (nused > 1) {                        Combine heap nodes, remaking the heap at
        node=index[0];                              each stage.
        index[0]=index[(nused--)-1];
        hufapp(index,nprob,nused,0);
```

```
            nprob[k]=nprob[index[0]]+nprob[node];
            hcode.left[k]=node;                    Store left and right children of a node.
            hcode.right[k++]=index[0];
            up[index[0]] = -int(k);               Indicate whether a node is a left or right child
            index[0]=k-1;                              of its parent.
            up[node]=k;
            hufapp(index,nprob,nused,0);
    }
    up[(hcode.nodemax=k)-1]=0;
    for (j=0;j<hcode.nch;j++) {                    Make the Huffman code from the tree.
        if (nprob[j] != 0) {
            for (n=0,ibit=0,node=up[j];node;node=up[node-1],ibit++) {
                if (node < 0) {
                    n |= setbit[ibit];
                    node = -node;
                }
            }
            hcode.icod[j]=n;
            hcode.ncod[j]=ibit;
        }
    }
    nlong=0;
    for (j=0;j<hcode.nch;j++) {
        if (hcode.ncod[j] > nlong) {
            nlong=hcode.ncod[j];
            ilong=j;
        }
    }
}
```

```
#include "nr.h"

void NR::hufapp(Vec_IO_ULNG &index, Vec_I_ULNG &nprob, const unsigned long n,
    const unsigned long m)
Used by hufmak to maintain a heap structure in the array index[0..m-1].
{
    unsigned long i=m,j,k;

    k=index[i];
    while (i < (n >> 1)) {
        if ((j = 2*i+1) < n-1
            && nprob[index[j]] > nprob[index[j+1]]) j++;
        if (nprob[k] <= nprob[index[j]]) break;
        index[i]=index[j];
        i=j;
    }
    index[i]=k;
}
```

Note that the class huffcode is defined in nrutil.h. The variable hcode of type huffcode must be set up in your main program with statements like this:

```
#include "nr.h"
    ...
    const unsigned int MC=512;           Maximum anticipated value of nchin in hufmak.
    const unsigned int MQ=2*MC-1;
    huffcode hcode(MQ,MQ,MQ,MQ);
    for (i=0;i<MQ;i++)
        hcode.icod[i]=hcode.ncod[i]=0;
```

Once the code is constructed, one encodes a string of characters by repeated calls to hufenc, which simply does a table lookup of the code and appends it to the output message.

```
#include <string>
#include "nr.h"
using namespace std;

void NR::hufenc(const unsigned long ich, string &code, unsigned long &nb,
    huffcode &hcode)
```
Huffman encode the single character ich (in the range 0..nch−1) using the code in the class object hcode, write the result to the string code starting at bit nb (whose smallest valid value is zero), and increment nb appropriately. This routine is called repeatedly to encode consecutive characters in a message, but must be preceded by a single initializing call to hufmak, which constructs hcode.
```
{
    int m,n;
    unsigned long k,nc;
    static unsigned long setbit[32]={0x1L,0x2L,0x4L,0x8L,0x10L,0x20L,
        0x40L,0x80L,0x100L,0x200L,0x400L,0x800L,0x1000L,0x2000L,
        0x4000L,0x8000L,0x10000L,0x20000L,0x40000L,0x80000L,0x100000L,
        0x200000L,0x400000L,0x800000L,0x1000000L,0x2000000L,0x4000000L,
        0x8000000L,0x10000000L,0x20000000L,0x40000000L,0x80000000L};

    k=ich;
    if (k >= hcode.nch)
        nrerror("ich out of range in hufenc.");
    for (n=hcode.ncod[k]-1;n >= 0;n--,++nb) {       Loop over the bits in the stored
        nc=nb >> 3;                                     Huffman code for ich.
        if (code.length() < nc+1)                   Increase the string length.
            code.resize(2*(nc+1));
        m=nb & 7;
        if (m == 0) code[nc]=0;                      Set appropriate bits in code.
        if ((hcode.icod[k] & setbit[n]) != 0) code[nc] |= setbit[m];
    }
}
```

Decoding a Huffman-encoded message is slightly more complicated. The coding tree must be traversed from the top down, using up a variable number of bits:

```
#include <string>
#include "nr.h"
using namespace std;

void NR::hufdec(unsigned long &ich, string &code, const unsigned long lcode,
    unsigned long &nb, huffcode &hcode)
```
Starting at bit number nb in the string code, use the Huffman code stored in the class object hcode to decode a single character (returned as ich in the range 0..nch−1) and increment nb appropriately. code has valid data in code[0..lcode−1]. Repeated calls, starting with nb = 0, will return successive characters in a compressed message. The returned value ich=nch indicates end-of-message. The class object hcode must already have been declared and allocated in your main program, and also filled by a call to hufmak.
```
{
    unsigned long nc;
    static unsigned char setbit[8]={0x1,0x2,0x4,0x8,0x10,0x20,0x40,0x80};

    int node=hcode.nodemax-1;
    for (;;) {                          Set node to the top of the decoding tree, and loop
        nc=nb >> 3;                         until a valid character is obtained.
        if (nc >= lcode) {              Ran out of input; with ich=nch indicating end of
            ich=hcode.nch;                  message.
```

```
            return;
        }
    node=((code[nc] & setbit[7 & nb++]) != 0 ?
        hcode.right[node] : hcode.left[node]);
        Branch left or right in tree, depending on its value.
    if (node < hcode.nch) {       If we reach a terminal node, we have a complete
        ich=node;                            character and can return.
        return;
    }
    }
}
```

For simplicity, `hufdec` quits when it runs out of code bytes; if your coded message is not an integral number of bytes, and if N_{ch} is less than 256, `hufdec` can return a spurious final character or two, decoded from the spurious trailing bits in your last code byte. If you have independent knowledge of the number of characters sent, you can readily discard these. Otherwise, you can fix this behavior by providing a bit, not byte, count, and modifying the routine accordingly. (When N_{ch} is 256 or larger, `hufdec` will normally run out of code in the middle of a spurious character, and it will be discarded.)

Run-Length Encoding

For the compression of highly correlated bit-streams (for example the black or white values along a facsimile scan line), Huffman compression is often combined with *run-length encoding*: Instead of sending each bit, the input stream is converted to a series of integers indicating how many consecutive bits have the same value. These integers are then Huffman-compressed. The Group 3 CCITT facsimile standard functions in this manner, with a fixed, immutable, Huffman code, optimized for a set of eight standard documents [8,9].

CITED REFERENCES AND FURTHER READING:

Gallager, R.G. 1968, *Information Theory and Reliable Communication* (New York: Wiley).

Hamming, R.W. 1980, *Coding and Information Theory* (Englewood Cliffs, NJ: Prentice-Hall).

Storer, J.A. 1988, *Data Compression: Methods and Theory* (Rockville, MD: Computer Science Press).

Nelson, M. 1991, *The Data Compression Book* (Redwood City, CA: M&T Books).

Huffman, D.A. 1952, *Proceedings of the Institute of Radio Engineers*, vol. 40, pp. 1098–1101. [1]

Ziv, J., and Lempel, A. 1978, *IEEE Transactions on Information Theory*, vol. IT-24, pp. 530–536. [2]

Cleary, J.G., and Witten, I.H. 1984, *IEEE Transactions on Communications*, vol. COM-32, pp. 396–402. [3]

Welch, T.A. 1984, *Computer*, vol. 17, no. 6, pp. 8–19. [4]

Bentley, J.L., Sleator, D.D., Tarjan, R.E., and Wei, V.K. 1986, *Communications of the ACM*, vol. 29, pp. 320–330. [5]

Jones, D.W. 1988, *Communications of the ACM*, vol. 31, pp. 996–1007. [6]

Sedgewick, R. 1998, *Algorithms in C*, 3rd ed. (Reading, MA: Addison-Wesley), Chapter 22. [7]

Hunter, R., and Robinson, A.H. 1980, *Proceedings of the IEEE*, vol. 68, pp. 854–867. [8]

Marking, M.P. 1990, *The C Users' Journal*, vol. 8, no. 6, pp. 45–54. [9]

20.5 Arithmetic Coding

We saw in the previous section that a perfect (entropy-bounded) coding scheme would use $L_i = -\log_2 p_i$ bits to encode character i (in the range $1 \le i \le N_{ch}$), if p_i is its probability of occurrence. Huffman coding gives a way of rounding the L_i's to close integer values and constructing a code with those lengths. *Arithmetic coding* [1], which we now discuss, actually does manage to encode characters using noninteger numbers of bits! It also provides a convenient way to output the result not as a stream of bits, but as a stream of symbols in any desired radix. This latter property is particularly useful if you want, e.g., to convert data from bytes (radix 256) to printable ASCII characters (radix 94), or to case-independent alphanumeric sequences containing only A-Z and 0-9 (radix 36).

In arithmetic coding, an input message of any length is represented as a real number R in the range $0 \le R < 1$. The longer the message, the more precision required of R. This is best illustrated by an example, so let us return to the fictitious language, Vowellish, of the previous section. Recall that Vowellish has a 5 character alphabet (A, E, I, O, U), with occurrence probabilities 0.12, 0.42, 0.09, 0.30, and 0.07, respectively. Figure 20.5.1 shows how a message beginning "IOU" is encoded: The interval $[0, 1)$ is divided into segments corresponding to the 5 alphabetical characters; the length of a segment is the corresponding p_i. We see that the first message character, "I", narrows the range of R to $0.37 \le R < 0.46$. This interval is now subdivided into five subintervals, again with lengths proportional to the p_i's. The second message character, "O", narrows the range of R to $0.3763 \le R < 0.4033$. The "U" character further narrows the range to $0.37630 \le R < 0.37819$. *Any* value of R in this range can be sent as encoding "IOU". In particular, the binary fraction .011000001 is in this range, so "IOU" can be sent in 9 bits. (Huffman coding took 10 bits for this example, see §20.4.)

Of course there is the problem of knowing when to stop decoding. The fraction .011000001 represents not simply "IOU," but "IOU...," where the ellipses represent an infinite string of successor characters. To resolve this ambiguity, arithmetic coding generally assumes the existence of a special $N_{ch} + 1$th character, EOM (end of message), which occurs only once at the end of the input. Since EOM has a low probability of occurrence, it gets allocated only a very tiny piece of the number line.

In the above example, we gave R as a binary fraction. We could just as well have output it in any other radix, e.g., base 94 or base 36, whatever is convenient for the anticipated storage or communication channel.

You might wonder how one deals with the seemingly incredible precision required of R for a long message. The answer is that R is never actually represented all at once. At any give stage we have upper and lower bounds for R represented as a finite number of digits in the output radix. As digits of the upper and lower bounds become identical, we can left-shift them away and bring in new digits at the low-significance end. The routines below have a parameter NWK for the number of working digits to keep around. This must be large enough to make the chance of an accidental degeneracy vanishingly small. (The routines signal if a degeneracy ever occurs.) Since the process of discarding old digits and bringing in new ones is performed identically on encoding and decoding, everything stays synchronized.

The routine arcmak constructs the cumulative frequency distribution table used

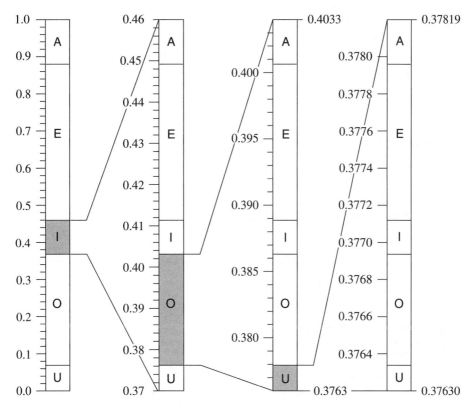

Figure 20.5.1.   Arithmetic coding of the message "IOU..." in the fictitious language Vowellish. Successive characters give successively finer subdivisions of the initial interval between 0 and 1. The final value can be output as the digits of a fraction in any desired radix. Note how the subinterval allocated to a character is proportional to its probability of occurrence.

to partition the interval at each stage. In the principal routine `arcode`, when an interval of size `jdif` is to be partitioned in the proportions of some `n` to some `ntot`, say, then we must compute `(n*jdif)/ntot`. With integer arithmetic, the numerator is likely to overflow; and, unfortunately, an expression like `jdif/(ntot/n)` is not equivalent. In the implementation below, we resort to double precision floating arithmetic for this calculation. Not only is this inefficient, but different roundoff errors can (albeit very rarely) make different machines encode differently, though any one type of machine will decode exactly what it encoded, since identical roundoff errors occur in the two processes. For serious use, one needs to replace this floating calculation with an integer computation in a double register (not available to the C++ programmer).

The internally set variable `minint`, which is the minimum allowed number of discrete steps between the upper and lower bounds, determines when new low-significance digits are added. `minint` must be large enough to provide resolution of all the input characters. That is, we must have $p_i \times \mathtt{minint} > 1$ for all i. A value of $100N_{ch}$, or $1.1/\min p_i$, whichever is larger, is generally adequate. However, for safety, the routine below takes `minint` to be as large as possible, with the product `minint*nradd` just smaller than overflow. This results in some time inefficiency, and in a few unnecessary characters being output at the end of a message. You can decrease `minint` if you want to live closer to the edge.

A final safety feature in `arcmak` is its refusal to believe zero values in the table `nfreq`; a 0 is treated as if it were a 1. If this were not done, the occurrence in a message of a single character whose `nfreq` entry is zero would result in scrambling the entire rest of the message. If you want to live dangerously, with a very slightly more efficient coding, you can delete the `MAX(  ,(unsigned long) 1)` operation.

```
#include <limits>
#include "nr.h"
using namespace std;

void NR::arcmak(Vec_I_ULNG &nfreq, unsigned long nchh, unsigned long nradd,
    arithcode &acode)
```
Given a table `nfreq[0..nchh-1]` of the frequency of occurrence of `nchh` symbols, and given a desired output radix `nradd`, initialize the cumulative frequency table and other variables for arithmetic compression in the class object `acode`.
```
{
    const unsigned long MAXULNG=numeric_limits<unsigned long>::max();
    unsigned long j;

    unsigned long MC=acode.ncumfq.size()-2;
    if (nchh > MC) nrerror("input radix may not exceed MC in arcmak.");
    if (nradd > 256) nrerror("output radix may not exceed 256 in arcmak.");

    acode.minint=MAXULNG/nradd;
    acode.nch=nchh;
    acode.nrad=nradd;
    acode.ncumfq[0]=0;
    for (j=1;j<=acode.nch;j++)
        acode.ncumfq[j]=acode.ncumfq[j-1]+MAX(nfreq[j-1],(unsigned long) 1);
    acode.ncum=acode.ncumfq[acode.nch+1]=acode.ncumfq[acode.nch]+1;
}
```

The class `arithcode` is defined in `nrutil.h`. The variable `acode` of type `arithcode` must be set up in your main program with statements like this:

```
#include "nr.h"
    ...
    const unsigned long MC=512;          Maximum anticipated value of nchh in arcmak.
    const unsigned long NWK=20;          arcode returns an error if this is too small.
    arithcode acode(NWK,NWK,MC+2);
```

Individual characters in a message are coded or decoded by the routine `arcode`, which in turn uses the utility `arcsum`.

```
#include "nr.h"

namespace {
    inline unsigned long JTRY(const unsigned long j, const unsigned long k,
        const unsigned long m)
        This function is used to calculate (k*j)/m without overflow. Program efficiency can
        be improved by substituting an assembly language routine that does integer multiply
        to a double register.
    {
        return (unsigned long) (DP(j)*DP(k)/DP (m));
    }
}

void NR::arcode(unsigned long &ich, string &code, unsigned long &lcd,
    const int isign, arithcode &acode)
```

Compress (isign = 1) or decompress (isign = −1) the single character ich into or out of the string code, starting with byte code[lcd] and (if necessary) incrementing lcd so that, on return, lcd points to the first unused byte in code. Note that the class object acode contains both information on the code, and also state information on the particular output being written into the string code. An initializing call with isign=0 is required before beginning any code string, whether for encoding or decoding. This is in addition to the initializing call to arcmak that is required to initialize the code itself. A call with ich=nch (as set in arcmak) has the reserved meaning "end of message."

```
{
    int j,k;
    unsigned long ihi,ja,jh,jl,m;

    int NWK=acode.ilob.size();
    if (isign == 0) {                Initialize enough digits of the upper and lower bounds.
        acode.jdif=acode.nrad-1;
        for (j=NWK-1;j>=0;j--) {
            acode.iupb[j]=acode.nrad-1;
            acode.ilob[j]=0;
            acode.nc=j;
            if (acode.jdif > acode.minint) return;        Initialization complete.
            acode.jdif=(acode.jdif+1)*acode.nrad-1;
        }
        nrerror("NWK too small in arcode.");
    } else {
        if (isign > 0) {          If encoding, check for valid input character.
            if (ich > acode.nch) nrerror("bad ich in arcode.");
        } else {                  If decoding, locate the character ich by bisection.
            ja=(unsigned char) code[lcd]-acode.ilob[acode.nc];
            for (j=acode.nc+1;j<NWK;j++) {
                ja *= acode.nrad;
                ja += ((unsigned char) code[lcd+j-acode.nc]-acode.ilob[j]);
            }
            ihi=acode.nch+1;
            ich=0;
            while (ihi-ich > 1) {
                m=(ich+ihi)>>1;
                if (ja >= JTRY(acode.jdif,acode.ncumfq[m],acode.ncum))
                    ich=m;
                else ihi=m;
            }
            if (ich == acode.nch) return;             Detected end of message.
        }
        Following code is common for encoding and decoding. Convert character ich to a new
        subrange [ilob,iupb).
        jh=JTRY(acode.jdif,acode.ncumfq[ich+1],acode.ncum);
        jl=JTRY(acode.jdif,acode.ncumfq[ich],acode.ncum);
        acode.jdif=jh-jl;
        arcsum(acode.ilob,acode.iupb,jh,NWK,acode.nrad,acode.nc);
        arcsum(acode.ilob,acode.ilob,jl,NWK,acode.nrad,acode.nc);
        How many leading digits to output (if encoding) or skip over?
        for (j=acode.nc;j<NWK;j++) {
            if (ich != acode.nch && acode.iupb[j] != acode.ilob[j]) break;
            if (isign > 0) code += (unsigned char)acode.ilob[j];
            lcd++;
        }
        if (j+1 > NWK) return;   Ran out of message. Did someone forget to encode a
        acode.nc=j;                      terminating ncd?
        for(j=0;acode.jdif<acode.minint;j++)      How many digits to shift?
            acode.jdif *= acode.nrad;
        if (j > acode.nc) nrerror("NWK too small in arcode.");
        if (j != 0) {              Shift them.
            for (k=acode.nc;k<NWK;k++) {
                acode.iupb[k-j]=acode.iupb[k];
                acode.ilob[k-j]=acode.ilob[k];
```

```
        }
    }
    acode.nc -= j;
    for (k=NWK-j;k<NWK;k++) acode.iupb[k]=acode.ilob[k]=0;
    }
    return;                          Normal return.
}
```

```
#include "nr.h"

void NR::arcsum(Vec_I_ULNG &iin, Vec_O_ULNG &iout, unsigned long ja,
    const int nwk, const unsigned long nrad, const unsigned long nc)
```
Used by arcode. Add the integer ja to the radix nrad multiple-precision integer iin[nc..nwk-1].
Return the result in iout[nc..nwk-1].
```
{
    int karry=0;
    unsigned long j,jtmp;

    for (j=nwk-1;j>nc;j--) {
        jtmp=ja;
        ja /= nrad;
        iout[j]=iin[j]+(jtmp-ja*nrad)+karry;
        if (iout[j] >= nrad) {
            iout[j] -= nrad;
            karry=1;
        } else karry=0;
    }
    iout[nc]=iin[nc]+ja+karry;
}
```

If radix-changing, rather than compression, is your primary aim (for example to convert an arbitrary file into printable characters) then you are of course free to set all the components of nfreq equal, say, to 1.

CITED REFERENCES AND FURTHER READING:

Bell, T.C., Cleary, J.G., and Witten, I.H. 1990, *Text Compression* (Englewood Cliffs, NJ: Prentice-Hall).

Nelson, M. 1991, *The Data Compression Book* (Redwood City, CA: M&T Books).

Witten, I.H., Neal, R.M., and Cleary, J.G. 1987, *Communications of the ACM*, vol. 30, pp. 520–540. [1]

20.6 Arithmetic at Arbitrary Precision

Let's compute the number π to a couple of thousand decimal places. In doing so, we'll learn some things about multiple precision arithmetic on computers and meet quite an unusual application of the fast Fourier transform (FFT). We'll also develop a set of routines that you can use for other calculations at any desired level of arithmetic precision.

To start with, we need an analytic algorithm for π.   Useful algorithms are quadratically convergent, i.e., they double the number of significant digits at each iteration. Quadratically convergent algorithms for π are based on the *AGM (arithmetic geometric mean) method*, which also finds application to the calculation of elliptic integrals (cf. §6.11) and in advanced implementations of the ADI method for elliptic partial differential equations (§19.5). Borwein and Borwein [1] treat this subject, which is beyond our scope here. One of their algorithms for π starts with the initializations

$$X_0 = \sqrt{2}$$

$$\pi_0 = 2 + \sqrt{2} \tag{20.6.1}$$

$$Y_0 = \sqrt[4]{2}$$

and then, for $i = 0, 1, \ldots$, repeats the iteration

$$X_{i+1} = \frac{1}{2}\left(\sqrt{X_i} + \frac{1}{\sqrt{X_i}}\right)$$

$$\pi_{i+1} = \pi_i\left(\frac{X_{i+1} + 1}{Y_i + 1}\right) \tag{20.6.2}$$

$$Y_{i+1} = \frac{Y_i\sqrt{X_{i+1}} + \dfrac{1}{\sqrt{X_{i+1}}}}{Y_i + 1}$$

The value π emerges as the limit π_∞.

Now, to the question of how to do arithmetic to arbitrary precision: In a high-level language like C++, a natural choice is to work in radix (base) 256, so that character arrays can be directly interpreted as strings of digits. At the very end of our calculation, we will want to convert our answer to radix 10, but that is essentially a frill for the benefit of human ears, accustomed to the familiar chant, "three point one four one five nine...." For any less frivolous calculation, we would likely never leave base 256 (or the thence trivially reachable hexadecimal, octal, or binary bases).

We will adopt the convention of storing digit strings in the "human" ordering, that is, with the first stored digit in an array being most significant, the last stored digit being least significant. The opposite convention would, of course, also be possible. "Carries," where we need to partition a number larger than 255 into a low-order byte and a high-order carry, present a minor programming annoyance, solved, in the routines below, by the use of the functions `lobyte` and `hibyte`. It will be our usual convention to assume that the digit strings represent floating point numbers with the radix point falling after the the first digit. When an operation results in a number that requires more digits in front of the decimal point, it is the responsibility of the user to shift the digits to the right and keep track of any excess factors of 256 that this implies.

It is easy at this point, following Knuth [2], to write a routine for the "fast" arithmetic operations: short addition (adding a single byte to a string), addition, subtraction, short multiplication (multiplying a string by a single byte), short division, ones-complement negation; and a couple of utility operations, copying and left-shifting strings. (In the machine-readable distribution, these functions are all in the single file `mpops.cpp`.)

```
#include "nr.h"
```

Multiple precision arithmetic operations done on character strings, interpreted as radix 256 numbers with the radix point after the first digit. This set of routines collects the simpler operations.

```
namespace {
    unsigned char lobyte(unsigned short x)
    {return (x & 0xff);}

    unsigned char hibyte(unsigned short x)
    {return ((x >> 8) & 0xff);}
}
```

void NR::mpadd(Vec_O_UCHR &w, Vec_I_UCHR &u, Vec_I_UCHR &v)
Adds the unsigned radix 256 numbers u and v, yielding the unsigned result w. To achieve the full available accuracy, the array w must be longer, by one element, than the shorter of the two arrays u and v.

```
{
    int j,n_min,p_min;
    unsigned short ireg=0;

    int n=u.size();
    int m=v.size();
    int p=w.size();
    n_min=MIN(n,m);
    p_min=MIN(n_min,p-1);
    for (j=p_min-1;j>=0;j--) {
        ireg=u[j]+v[j]+hibyte(ireg);
        w[j+1]=lobyte(ireg);
    }
    w[0]=hibyte(ireg);
    if (p > p_min+1)
        for (j=p_min+1;j<p;j++) w[j]=0;
}
```

void NR::mpsub(int &is, Vec_O_UCHR &w, Vec_I_UCHR &u, Vec_I_UCHR &v)
Subtracts the unsigned radix 256 number v from u yielding the unsigned result w. If the result is negative (wraps around), is is returned as −1; otherwise it is returned as 0. To achieve the full available accuracy, the array w must be as long as the shorter of the two arrays u and v.

```
{
    int j,n_min,p_min;
    unsigned short ireg=256;

    int n=u.size();
    int m=v.size();
    int p=w.size();
    n_min=MIN(n,m);
    p_min=MIN(n_min,p);
    for (j=p_min-1;j>=0;j--) {
        ireg=255+u[j]-v[j]+hibyte(ireg);
        w[j]=lobyte(ireg);
    }
    is=hibyte(ireg)-1;
    if (p > p_min)
        for (j=p_min;j<p;j++) w[j]=0;
}
```

void NR::mpsad(Vec_O_UCHR &w, Vec_I_UCHR &u, const int iv)
Short addition: the integer iv (in the range $0 \le iv \le 255$) is added to the least significant radix position of unsigned radix 256 number u, yielding result w. To ensure that the result does not require two digits before the radix point, one may first right-shift the operand u so that the first digit is 0, and keep track of multiples of 256 separately.

```
{
```

```
    int j;
    unsigned short ireg;

    int n=u.size();
    int p=w.size();
    ireg=256*iv;
    for (j=n-1;j>=0;j--) {
        ireg=u[j]+hibyte(ireg);
        if (j+1 < p) w[j+1]=lobyte(ireg);
    }
    w[0]=hibyte(ireg);
    for (j=n+1;j<p;j++) w[j]=0;
}
```

void NR::mpsmu(Vec_O_UCHR &w, Vec_I_UCHR &u, const int iv)
Short multiplication: the unsigned radix 256 number u is multiplied by the integer iv (in the range $0 \leq iv \leq 255$), yielding result w. To ensure that the result does not require two digits before the radix point, one may first right-shift the operand u so that the first digit is 0, and keep track of multiples of 256 separately.

```
{
    int j;
    unsigned short ireg=0;

    int n=u.size();
    int p=w.size();
    for (j=n-1;j>=0;j--) {
        ireg=u[j]*iv+hibyte(ireg);
        if (j < p-1) w[j+1]=lobyte(ireg);
    }
    w[0]=hibyte(ireg);
    for (j=n+1;j<p;j++) w[j]=0;
}
```

void NR::mpsdv(Vec_O_UCHR &w, Vec_I_UCHR &u, const int iv, int &ir)
Short division: the unsigned radix 256 number u is divided by the integer iv (in the range $0 \leq iv \leq 255$), yielding a quotient w and a remainder ir (with $0 \leq ir \leq 255$). To achieve the full available accuracy, the array w must be as long as the array u.

```
{
    int i,j,p_min;

    int n=u.size();
    int p=w.size();
    p_min=MIN(n,p);
    ir=0;
    for (j=0;j<p_min;j++) {
        i=256*ir+u[j];
        w[j]=(unsigned char) (i/iv);
        ir=i % iv;
    }
    if (p > p_min)
        for (j=p_min;j<p;j++) w[j]=0;
}
```

void NR::mpneg(Vec_IO_UCHR &u)
Ones-complement negate the unsigned radix 256 number u.

```
{
    int j;
    unsigned short ireg=256;

    int n=u.size();
    for (j=n-1;j>=0;j--) {
        ireg=255-u[j]+hibyte(ireg);
        u[j]=lobyte(ireg);
    }
```

```
}

void NR::mpmov(Vec_O_UCHR &u, Vec_I_UCHR &v)
```
Move the unsigned radix 256 number v into u. To achieve full accuracy, the array v must be
as long as the array u.
```
{
    int j,n_min;

    int n=u.size();
    int m=v.size();
    n_min=MIN(n,m);
    for (j=0;j<n_min;j++) u[j]=v[j];
    if (n > n_min)
        for(j=n_min;j<n-1;j++) u[j]=0;
}

void NR::mplsh(Vec_IO_UCHR &u)
```
Left shift digits of unsigned radix 256 number u. The final element of the array is set to 0.
```
{
    int j;

    int n=u.size();
    for (j=0;j<n-1;j++) u[j]=u[j+1];
    u[n-1]=0;
}
```

Full multiplication of two digit strings, if done by the traditional hand method, is not a fast operation: In multiplying two strings of length N, the multiplicand would be short-multiplied in turn by each byte of the multiplier, requiring $O(N^2)$ operations in all. We will see, however, that *all* the arithmetic operations on numbers of length N can in fact be done in $O(N \times \log N \times \log \log N)$ operations.

The trick is to recognize that multiplication is essentially a *convolution* (§13.1) of the digits of the multiplicand and multiplier, followed by some kind of carry operation. Consider, for example, two ways of writing the calculation 456×789:

$$
\begin{array}{r}
456 \\
\times\ 789 \\
\hline
4104 \\
3648 \\
3192 \\
\hline
359784
\end{array}
\qquad\qquad
\begin{array}{rrrrrr}
 & & 4 & 5 & 6 \\
 & \times & 7 & 8 & 9 \\
\hline
 & & 36 & 45 & 54 \\
 & 32 & 40 & 48 \\
28 & 35 & 42 \\
\hline
28 & 67 & 118 & 93 & 54 \\
\hline
3 & 5 & 9 & 7 & 8 & 4
\end{array}
$$

The tableau on the left shows the conventional method of multiplication, in which three separate short multiplications of the full multiplicand (by 9, 8, and 7) are added to obtain the final result. The tableau on the right shows a different method (sometimes taught for mental arithmetic), where the single-digit cross products are all computed (e.g. $8 \times 6 = 48$), then added in columns to obtain an incompletely carried result (here, the list $28, 67, 118, 93, 54$). The final step is a single pass from right to left, recording the single least-significant digit and carrying the higher digit or digits into the total to the left (e.g. $93 + 5 = 98$, record the 8, carry 9).

You can see immediately that the column sums in the right-hand method are components of the convolution of the digit strings, for example $118 = 4 \times 9 + 5 \times 8 + 6 \times 7$. In §13.1 we learned how to compute the convolution of two vectors by the fast Fourier transform (FFT): Each vector is FFT'd, the two complex transforms are multiplied, and the result is inverse-FFT'd. Since the transforms are done with floating arithmetic, we need sufficient precision so that the exact integer value of each component of the result is discernible in the presence of roundoff error. We should therefore allow a (conservative) few times $\log_2(\log_2 N)$ bits for roundoff in the FFT. A number of length N bytes in radix 256 can generate convolution components as large as the order of $(256)^2 N$, thus requiring $16 + \log_2 N$ bits of precision for exact storage. If it is the number of bits in the floating mantissa (cf. §20.1), we obtain the condition

$$16 + \log_2 N + \text{few} \times \log_2 \log_2 N < \text{it} \qquad (20.6.3)$$

We see that single precision, say with it $= 24$, is inadequate for any interesting value of N, while double precision, say with it $= 53$, allows N to be greater than 10^6, corresponding to some millions of decimal digits. The double precision used in the routines realft (§12.3) and four1 (§12.2) is therefore a necessity, not an extravagance, for this application.

```
#include "nr.h"

void NR::mpmul(Vec_O_UCHR &w, Vec_I_UCHR &u, Vec_I_UCHR &v)
Uses Fast Fourier Transform to multiply the unsigned radix 256 integers u[0..n-1] and
v[0..m-1], yielding a product w[0..n+m-1].
{
    const DP RX=256.0;
    int j,n_max,nn=1;
    DP cy,t;

    int n=u.size();
    int m=v.size();
    int p=w.size();
    n_max=MAX(m,n);
    while (nn < n_max) nn <<= 1;       Find the smallest usable power of two for the trans-
    nn <<= 1;                              form.
    Vec_DP a(0.0,nn),b(0.0,nn);
    for (j=0;j<n;j++) a[j]=u[j];        Move U and V to double precision floating arrays.
    for (j=0;j<m;j++) b[j]=v[j];
    realft(a,1);                        Perform the convolution: First, the two Fourier trans-
    realft(b,1);                            forms.
    b[0] *= a[0];                       Then multiply the complex results (real and imagi-
    b[1] *= a[1];                           nary parts).
    for (j=2;j<nn;j+=2) {
        b[j]=(t=b[j])*a[j]-b[j+1]*a[j+1];
        b[j+1]=t*a[j+1]+b[j+1]*a[j];
    }
    realft(b,-1);                       Then do the inverse Fourier transform.
    cy=0.0;                             Make a final pass to do all the carries.
    for (j=nn-1;j>=0;j--) {
        t=b[j]/(nn >> 1)+cy+0.5;        The 0.5 allows for roundoff error.
        cy=(unsigned long) (t/RX);
        b[j]=t-cy*RX;
    }
    if (cy >= RX) nrerror("cannot happen in mpmul");
    for (j=0;j<p;j++) w[j]=0;
    w[0]=(unsigned char) cy;            Copy answer to output.
    for (j=1;j<MIN(n+m,p);j++)
        w[j]=(unsigned char) b[j-1];
}
```

With multiplication thus a "fast" operation, division is best performed by multiplying the dividend by the reciprocal of the divisor. The reciprocal of a value V is calculated by iteration of Newton's rule,

$$U_{i+1} = U_i(2 - VU_i) \qquad (20.6.4)$$

which results in the quadratic convergence of U_∞ to $1/V$, as you can easily prove. (Many supercomputers and RISC machines actually use this iteration to perform divisions.) We can now see where the operations count $N \log N \log \log N$, mentioned above, originates: $N \log N$ is in the Fourier transform, with the iteration to converge Newton's rule giving an additional factor of $\log \log N$.

```
#include "nr.h"

void NR::mpinv(Vec_O_UCHR &u, Vec_I_UCHR &v)
Character string v[0..m-1] is interpreted as a radix 256 number with the radix point after
(nonzero) v[0]; u[0..n-1] is set to the most significant digits of its reciprocal, with the
radix point after u[0].
{
    const int MF=4;
    const DP BI=1.0/256.0;
    int i,j,mm;
    DP fu,fv;

    int n=u.size();
    int m=v.size();
    Vec_UCHR s(n+m),r(2*n+m);
    mm=MIN(MF,m);
    fv=DP(v[mm-1]);                          Use ordinary floating arithmetic to get an initial
    for (j=mm-2;j>=0;j--) {                      approximation.
        fv *= BI;
        fv += v[j];
    }
    fu=1.0/fv;
    for (j=0;j<n;j++) {
        i=int(fu);
        u[j]=(unsigned char) i;
        fu=256.0*(fu-i);
    }
    for (;;) {                               Iterate Newton's rule to convergence.
        mpmul(s,u,v);                        Construct $2 - UV$ in $S$.
        mplsh(s);
        mpneg(s);
        s[0] += (unsigned char) 2;           Multiply $SU$ into $U$.
        mpmul(r,s,u);
        mplsh(r);
        mpmov(u,r);
        for (j=1;j<n-1;j++)                   If fractional part of $S$ is not zero, it has not
            if (s[j] != 0) break;               converged to 1.
        if (j==n-1) return;
    }
}
```

Division now follows as a simple corollary, with only the necessity of calculating the reciprocal to sufficient accuracy to get an exact quotient and remainder.

```
#include "nr.h"

void NR::mpdiv(Vec_O_UCHR &q, Vec_O_UCHR &r, Vec_I_UCHR &u, Vec_I_UCHR &v)
```
Divides unsigned radix 256 integers u[0..n-1] by v[0..m-1] (with m ≤ n required), yielding
a quotient q[0..n-m] and a remainder r[0..m-1].
```
{
    const int MACC=1;
    int i,is,mm;

    int n=u.size();
    int m=v.size();
    int p=r.size();
    int n_min=MIN(m,p);
    if (m > n) nrerror("Divisor longer than dividend in mpdiv");
    mm=m+MACC;
    Vec_UCHR s(mm),rr(mm),ss(mm+1),qq(n-m+1),t(n);
    mpinv(s,v);                         Set S = 1/V.
    mpmul(rr,s,u);                      Set Q = SU.
    mpsad(ss,rr,1);
    mplsh(ss);
    mplsh(ss);
    mpmov(qq,ss);
    mpmov(q,qq);
    mpmul(t,qq,v);                      Multiply and subtract to get the remainder.
    mplsh(t);
    mpsub(is,t,u,t);
    if (is != 0) nrerror("MACC too small in mpdiv");
    for (i=0;i<n_min;i++) r[i]=t[i+n-m];
    if (p>m)
        for (i=m;i<p;i++) r[i]=0;
}
```

Square roots are calculated by a Newton's rule much like division. If

$$U_{i+1} = \frac{1}{2}U_i(3 - VU_i^2) \qquad (20.6.5)$$

then U_∞ converges quadratically to $1/\sqrt{V}$. A final multiplication by V gives $\sqrt{V}$.

```
#include <cmath>
#include "nr.h"
using namespace std;

void NR::mpsqrt(Vec_O_UCHR &w, Vec_O_UCHR &u, Vec_I_UCHR &v)
```
Character string v[0..m-1] is interpreted as a radix 256 number with the radix point after
v[0]; w[0..n-1] is set to its square root (radix point after w[0]), and u[0..n-1] is set to
the reciprocal thereof (radix point before u[0]). w and u need not be distinct, in which case
they are set to the square root.
```
{
    const int MF=3;
    const DP BI=1.0/256.0;
    int i,ir,j,mm;
    DP fu,fv;

    int n=u.size();
    int m=v.size();
    Vec_UCHR r(2*n),x(n+m),s(2*n+m),t(3*n+m);
    mm=MIN(m,MF);
    fv=DP(v[mm-1]);                     Use ordinary floating arithmetic to get an initial ap-
    for (j=mm-2;j>=0;j--) {                 proximation.
        fv *= BI;
        fv += v[j];
    }
    fu=1.0/sqrt(fv);
```

```
for (j=0;j<n;j++) {
    i=int(fu);
    u[j]=(unsigned char) i;
    fu=256.0*(fu-i);
}
for (;;) {                              Iterate Newton's rule to convergence.
    mpmul(r,u,u);                       Construct $S = (3 - VU^2)/2$.
    mplsh(r);
    mpmul(s,r,v);
    mplsh(s);
    mpneg(s);
    s[0] += (unsigned char) 3;
    mpsdv(s,s,2,ir);
    for (j=1;j<n-1;j++) {               If fractional part of $S$ is not zero, it has not converged
        if (s[j] != 0) {                   to 1.
            mpmul(t,s,u);               Replace $U$ by $SU$.
            mplsh(t);
            mpmov(u,t);
            break;
        }
    }
    if (j<n-1) continue;
    mpmul(x,u,v);                       Get square root from reciprocal and return.
    mplsh(x);
    mpmov(w,x);
    return;
}
}
```

We already mentioned that radix conversion to decimal is a merely cosmetic operation that should normally be omitted. The simplest way to convert a fraction to decimal is to multiply it repeatedly by 10, picking off (and subtracting) the resulting integer part. This, has an operations count of $O(N^2)$, however, since each liberated decimal digit takes an $O(N)$ operation. It *is* possible to do the radix conversion as a fast operation by a "divide and conquer" strategy, in which the fraction is (fast) multiplied by a large power of 10, enough to move about half the desired digits to the left of the radix point. The integer and fractional pieces are now processed independently, each further subdivided. If our goal were a few billion digits of π, instead of a few thousand, we would need to implement this scheme. For present purposes, the following lazy routine is adequate:

```
#include "nr.h"

void NR::mp2dfr(Vec_IO_UCHR &a, string &s)
```
Converts a radix 256 fraction a[0..n-1] (radix point before a[0]) to a decimal fraction represented as an ascii string s[0..m-1], where m is a returned value. The input array a[0..n-1] is destroyed. NOTE: For simplicity, this routine implements a slow ($\propto N^2$) algorithm. Fast ($\propto N \ln N$), more complicated, radix conversion algorithms do exist.
```
{
    const unsigned int IAZ=48;
    int j,m;

    int n=a.size();
    m=int(2.408*n);
    mplsh(a);
    for (j=0;j<m;j++) {
        mpsmu(a,a,10);
        s += a[0]+IAZ;
        mplsh(a);
    }
}
```

Finally, then, we arrive at a routine implementing equations (20.6.1) and (20.6.2):

```
#include <iostream>
#include <iomanip>
#include "nr.h"
using namespace std;

void NR::mppi(const int np)
Demonstrate multiple precision routines by calculating and printing the first n bytes of π.
{
    const unsigned int IAOFF=48,MACC=2;
    int ir,j,n;
    unsigned char mm;
    string s;

    n=np+MACC;
    Vec_UCHR x(n),y(n),sx(n),sxi(n);
    Vec_UCHR z(n),t(n),pi(n);
    Vec_UCHR ss(2*n),tt(2*n);
    t[0]=2;                              Set T = 2.
    for (j=1;j<n;j++) t[j]=0;
    mpsqrt(x,x,t);                       Set X₀ = √2.
    mpadd(pi,t,x);                       Set π₀ = 2 + √2.
    mplsh(pi);
    mpsqrt(sx,sxi,x);                    Set Y₀ = 2^{1/4}.
    mpmov(y,sx);
    for (;;) {
        mpadd(z,sx,sxi);                 Set X_{i+1} = (X_i^{1/2} + X_i^{-1/2})/2.
        mplsh(z);
        mpsdv(x,z,2,ir);
        mpsqrt(sx,sxi,x);                Form the temporary T = Y_i X_{i+1}^{1/2} + X_{i+1}^{-1/2}.
        mpmul(tt,y,sx);
        mplsh(tt);
        mpadd(tt,tt,sxi);
        mplsh(tt);
        x[0]++;                          Increment X_{i+1} and Y_i by 1.
        y[0]++;
        mpinv(ss,y);                     Set Y_{i+1} = T/(Y_i + 1).
        mpmul(y,tt,ss);
        mplsh(y);
        mpmul(tt,x,ss);                  Form temporary T = (X_{i+1} + 1)/(Y_i + 1).
        mplsh(tt);
        mpmul(ss,pi,tt);                 Set π_{i+1} = Tπ_i.
        mplsh(ss);
        mpmov(pi,ss);
        mm=tt[0]-1;                      If T = 1 then we have converged.
        for (j=1;j < n-1;j++)
            if (tt[j] != mm) break;
        if (j == n-1) {
            cout << endl << "pi = ";
            s=pi[0]+IAOFF;
            s += '.';
            mp2dfr(pi,s);
            Convert to decimal for printing. NOTE: The conversion routine, for this demon-
            stration only, is a slow (∝ N²) algorithm. Fast (∝ N ln N), more complicated,
            radix conversion algorithms do exist.
            s.erase(2.408*np,s.length());
            cout << setw(64) << left << s << endl;
            return;
        }
    }
}
```

3.1415926535897932384626433832795028841971693993751058209749445923078164062
8620899862803482534211706798214808651328230664709384460955058223172535940813
2848111745028410270193852110555964462294895493038196442881097566593344612843
7564823337867831652712019091456485669234603486104543266482133936072602491412
7372458700660631558817488152092096282925409171536436789259036001133053054883
2046652138414695194151160943305727036575959195309218611738193261179310511853
4807446237996274956735188575272489122793818301194912983367336244065664308603
2139494639522473719070217986094370277053921717629317675253846748184676694051
3200056812714526356082778577134275778960917363717872146844090122495343014654
9958537105079227968925892354201995611212902196086403441815981362977477130993
6051870721134999999837297804995105973173281609631859502445945534690830264253
2230825334468503526193118817101000313783875288658753320838142061717766914733
0359825349042875546873115956286388235378759375195778185778053217122680661303
0192787661119590921642019893809525720106548586327886593615338182796823030195
2035301852968995773622599413891249721775283479131515574857242454150695950833
2953311686172785588907509838175463746493931925506040092770167113900984882403
1285836160356370766010471018194295559619894676783744944825537977472684710403
4753464620804668425906949129331367702898915210475216205696602405803815019353
1125338243003558764024749647326391419927260426992279678235478163600934172163
4121999245863150302861829745557067498385054945885869269956909272107975093029
5532116534498720275596023648066549911988183479775356636980742654252786255183
1841757467289097777279380008164706001614524919217321721477235014144197356853
4816136115735255213347574184946843852332390739414333454477624168625189835693
8556209922192221842725502542568876717904946016534668049886272327917860857843
8382796797668145410095388378636095068006422512520511739298489608412848862693
4560424196528502221066118630674427862203919494504712371378696095636437191723
8746776465757396241389086583264599581339047802759009946576407895126946838933
5259570982582262052248940772671947826848260147699090264013639443745530506823
0349625245174939965143142980919065925093722169646151570985838741059578859593
7297549893016175392846813826868386894277415599185592524595395943104997252463
8084598727364469584865383670622626099124608051243884390451244136549762780773
9771569143599770012961608944169486855584840635342207222582848864815845602852

Figure 20.6.1.   The first 2398 decimal digits of π, computed by the routines in this section.

Figure 20.6.1 gives the result, computed with $n = 1000$. As an exercise, you might enjoy checking the first hundred digits of the figure against the first 12 terms of Ramanujan's celebrated identity [3]

$$\frac{1}{\pi} = \frac{\sqrt{8}}{9801} \sum_{n=0}^{\infty} \frac{(4n)!\,(1103 + 26390n)}{(n!\,396^n)^4} \qquad (20.6.6)$$

using the above routines.   You might also use the routines to verify that the number $2^{512} + 1$ is not a prime, but has factors 2,424,833 and 7,455,602,825,647,884,208,337,395,736,200,454,918,783,366,342,657 (which are in fact prime; the remaining prime factor being about 7.416×10^{98}) [4].

CITED REFERENCES AND FURTHER READING:

Borwein, J.M., and Borwein, P.B. 1987, *Pi and the AGM: A Study in Analytic Number Theory and Computational Complexity* (New York: Wiley). [1]

Knuth, D.E. 1997, *Seminumerical Algorithms*, 3rd ed., vol. 2 of *The Art of Computer Programming* (Reading, MA: Addison-Wesley), §4.3. [2]

Ramanujan, S. 1927, *Collected Papers of Srinivasa Ramanujan*, G.H. Hardy, P.V. Seshu Aiyar, and B.M. Wilson, eds. (Cambridge, U.K.: Cambridge University Press), pp. 23–39. [3]

Kolata, G. 1990, June 20, *The New York Times*. [4]

Kronsjö, L. 1987, *Algorithms: Their Complexity and Efficiency*, 2nd ed. (New York: Wiley).

References

The references collected here are those of general usefulness, usually cited in more than one section of this book. More specialized sources, usually cited in a single section, are not repeated here.

We first list a small number of books that form the nucleus of a recommended personal reference collection on numerical methods, numerical analysis, and closely related subjects. These are the books that we like to have within easy reach.

Abramowitz, M., and Stegun, I.A. 1964, *Handbook of Mathematical Functions*, Applied Mathematics Series, Volume 55 (Washington: National Bureau of Standards; reprinted 1968 by Dover Publications, New York)

Acton, F.S. 1970, *Numerical Methods That Work*; 1990, corrected edition (Washington: Mathematical Association of America)

Ames, W.F. 1977, *Numerical Methods for Partial Differential Equations*, 2nd ed. (New York: Academic Press)

Bratley, P., Fox, B.L., and Schrage, E.L. 1983, *A Guide to Simulation*, 2nd ed. (New York: Springer-Verlag)

Dahlquist, G., and Bjorck, A. 1974, *Numerical Methods* (Englewood Cliffs, NJ: Prentice-Hall)

Delves, L.M., and Mohamed, J.L. 1985, *Computational Methods for Integral Equations* (Cambridge, U.K.: Cambridge University Press)

Dennis, J.E., and Schnabel, R.B. 1983, *Numerical Methods for Unconstrained Optimization and Nonlinear Equations*; reprinted 1996 (Philadelphia: S.I.A.M.)

Gill, P.E., Murray, W., and Wright, M.H. 1991, *Numerical Linear Algebra and Optimization*, vol. 1 (Redwood City, CA: Addison-Wesley)

Golub, G.H., and Van Loan, C.F. 1996, *Matrix Computations*, 3rd ed. (Baltimore: Johns Hopkins University Press)

Oppenheim, A.V., and Schafer, R.W. 1989, *Discrete-Time Signal Processing* (Englewood Cliffs, NJ: Prentice-Hall)

Ralston, A., and Rabinowitz, P. 1978, *A First Course in Numerical Analysis*, 2nd ed.; reprinted 2001 (New York: Dover)

Sedgewick, R. 1998, *Algorithms in C*, 3rd ed. (Reading, MA: Addison-Wesley)

Stoer, J., and Bulirsch, R. 1993, *Introduction to Numerical Analysis*, 2nd ed. (New York: Springer-Verlag)

Wilkinson, J.H., and Reinsch, C. 1971, *Linear Algebra*, vol. II of *Handbook for Automatic Computation* (New York: Springer-Verlag)

We next list the larger collection of books, which, in our view, should be included in any serious research library on computing, numerical methods, or analysis.

Anderson, E., et al. 2000, LAPACK User's Guide, 3rd ed. (Philadelphia: S.I.A.M.)

Bevington, P.R., and Robinson, D.K. 1992, *Data Reduction and Error Analysis for the Physical Sciences*, 2nd ed. (New York: McGraw-Hill)

Bloomfield, P. 1976, *Fourier Analysis of Time Series – An Introduction* (New York: Wiley)

Bowers, R.L., and Wilson, J.R. 1991, *Numerical Modeling in Applied Physics and Astrophysics* (Boston: Jones & Bartlett)

Brent, R.P. 1973, *Algorithms for Minimization without Derivatives* (Englewood Cliffs, NJ: Prentice-Hall)

Brigham, E.O. 1974, *The Fast Fourier Transform* (Englewood Cliffs, NJ: Prentice-Hall)

Brownlee, K.A. 1965, *Statistical Theory and Methodology*, 2nd ed. (New York: Wiley)

Bunch, J.R., and Rose, D.J. (eds.) 1976, *Sparse Matrix Computations* (New York: Academic Press)

Canuto, C., Hussaini, M.Y., Quarteroni, A., and Zang, T.A. 1988, *Spectral Methods in Fluid Dynamics* (New York: Springer-Verlag)

Carnahan, B., Luther, H.A., and Wilkes, J.O. 1969, *Applied Numerical Methods* (New York: Wiley)

Champeney, D.C. 1973, *Fourier Transforms and Their Physical Applications* (New York: Academic Press)

Childers, D.G. (ed.) 1978, *Modern Spectrum Analysis* (New York: IEEE Press)

Cooper, L., and Steinberg, D. 1970, *Introduction to Methods of Optimization* (Philadelphia: Saunders)

Dantzig, G.B. 1963, *Linear Programming and Extensions* (Princeton, NJ: Princeton University Press)

Devroye, L. 1986, *Non-Uniform Random Variate Generation* (New York: Springer-Verlag)

Downie, N.M., and Heath, R.W. 1965, *Basic Statistical Methods*, 2nd ed. (New York: Harper & Row)

Duff, I.S., and Stewart, G.W. (eds.) 1979, *Sparse Matrix Proceedings 1978* (Philadelphia: S.I.A.M.)

Elliott, D.F., and Rao, K.R. 1982, *Fast Transforms: Algorithms, Analyses, Applications* (New York: Academic Press)

Fike, C.T. 1968, *Computer Evaluation of Mathematical Functions* (Englewood Cliffs, NJ: Prentice-Hall)

Forsythe, G.E., Malcolm, M.A., and Moler, C.B. 1977, *Computer Methods for Mathematical Computations* (Englewood Cliffs, NJ: Prentice-Hall)

Forsythe, G.E., and Moler, C.B. 1967, *Computer Solution of Linear Algebraic Systems* (Englewood Cliffs, NJ: Prentice-Hall)

Gass, S.I. 1985, *Linear Programming*, 5th ed. (New York: McGraw-Hill)

Gear, C.W. 1971, *Numerical Initial Value Problems in Ordinary Differential Equations* (Englewood Cliffs, NJ: Prentice-Hall)

Goodwin, E.T. (ed.) 1961, *Modern Computing Methods*, 2nd ed. (New York: Philosophical Library)

Gottlieb, D. and Orszag, S.A. 1977, *Numerical Analysis of Spectral Methods: Theory and Applications* (Philadelphia: S.I.A.M.)

Hackbusch, W. 1985, *Multi-Grid Methods and Applications* (New York: Springer-Verlag)

Hamming, R.W. 1962, *Numerical Methods for Engineers and Scientists*; reprinted 1986 (New York: Dover)

Hart, J.F., et al. 1968, *Computer Approximations* (New York: Wiley)

Hastings, C. 1955, *Approximations for Digital Computers* (Princeton: Princeton University Press)

Hildebrand, F.B. 1974, *Introduction to Numerical Analysis*, 2nd ed.; reprinted 1987 (New York: Dover)

Hoel, P.G. 1971, *Introduction to Mathematical Statistics*, 4th ed. (New York: Wiley)

Horn, R.A., and Johnson, C.R. 1985, *Matrix Analysis* (Cambridge: Cambridge University Press)

Householder, A.S. 1970, *The Numerical Treatment of a Single Nonlinear Equation* (New York: McGraw-Hill)

Huber, P.J. 1981, *Robust Statistics* (New York: Wiley)

Isaacson, E., and Keller, H.B. 1966, *Analysis of Numerical Methods*; reprinted 1994 (New York: Dover)

Jacobs, D.A.H. (ed.) 1977, *The State of the Art in Numerical Analysis* (London: Academic Press)

Johnson, L.W., and Riess, R.D. 1982, *Numerical Analysis*, 2nd ed. (Reading, MA: Addison-Wesley)

Kahaner, D., Moler, C., and Nash, S. 1989, *Numerical Methods and Software* (Englewood Cliffs, NJ: Prentice Hall)

Keller, H.B. 1968, *Numerical Methods for Two-Point Boundary-Value Problems*; reprinted 1991 (New York: Dover)

Knuth, D.E. 1997, *Fundamental Algorithms*, 3rd ed., vol. 1 of *The Art of Computer Programming* (Reading, MA: Addison-Wesley)

Knuth, D.E. 1997, *Seminumerical Algorithms*, 3rd ed., vol. 2 of *The Art of Computer Programming* (Reading, MA: Addison-Wesley)

Knuth, D.E. 1997, *Sorting and Searching*, 3rd ed., vol. 3 of *The Art of Computer Programming* (Reading, MA: Addison-Wesley)

Koonin, S.E., and Meredith, D.C. 1990, *Computational Physics, Fortran Version* (Redwood City, CA: Addison-Wesley)

Kuenzi, H.P., Tzschach, H.G., and Zehnder, C.A. 1971, *Numerical Methods of Mathematical Optimization* (New York: Academic Press)

Lanczos, C. 1956, *Applied Analysis*; reprinted 1988 (New York: Dover)

Land, A.H., and Powell, S. 1973, *Fortran Codes for Mathematical Programming* (London: Wiley-Interscience)

Lawson, C.L., and Hanson, R. 1974, *Solving Least Squares Problems* (Englewood Cliffs, NJ: Prentice-Hall)

Lehmann, E.L. 1975, *Nonparametrics: Statistical Methods Based on Ranks* (San Francisco: Holden-Day)

Luke, Y.L. 1975, *Mathematical Functions and Their Approximations* (New York: Academic Press)

Magnus, W., and Oberhettinger, F. 1949, *Formulas and Theorems for the Functions of Mathematical Physics* (New York: Chelsea)

Martin, B.R. 1971, *Statistics for Physicists* (New York: Academic Press)

Mathews, J., and Walker, R.L. 1970, *Mathematical Methods of Physics*, 2nd ed. (Reading, MA: W.A. Benjamin/Addison-Wesley)

von Mises, R. 1964, *Mathematical Theory of Probability and Statistics* (New York: Academic Press)

Murty, K.G. 1976, *Linear and Combinatorial Programming* (New York: Wiley)

Norusis, M.J. 1999, *SPSS 9.0 Guide to Data Analysis* (Englewood Cliffs, NJ: Prentice-Hall)

Nussbaumer, H.J. 1982, *Fast Fourier Transform and Convolution Algorithms* (New York: Springer-Verlag)

Ortega, J., and Rheinboldt, W. 1970, *Iterative Solution of Nonlinear Equations in Several Variables* (New York: Academic Press)

Ostrowski, A.M. 1966, *Solutions of Equations and Systems of Equations*, 2nd ed. (New York: Academic Press)

Polak, E. 1971, *Computational Methods in Optimization* (New York: Academic Press)

Rice, J.R. 1983, *Numerical Methods, Software, and Analysis* (New York: McGraw-Hill)

Richtmyer, R.D., and Morton, K.W. 1967, *Difference Methods for Initial Value Problems*, 2nd ed. (New York: Wiley-Interscience)

Roache, P.J. 1976, *Computational Fluid Dynamics* (Albuquerque: Hermosa)

Robinson, E.A., and Treitel, S. 1980, *Geophysical Signal Analysis* (Englewood Cliffs, NJ: Prentice-Hall)

Smith, B.T., et al. 1976, *Matrix Eigensystem Routines — EISPACK Guide*, 2nd ed., vol. 6 of Lecture Notes in Computer Science (New York: Springer-Verlag)

Strang, G. 1988, *Linear Algebra and its Applications*, 3rd ed. (New York: Harcourt Brace and Co.)

Stuart, A., and Ord, J.K. 1994, *Kendall's Advanced Theory of Statistics*, 6th ed. (London: Edward Arnold) [previous eds. published as Kendall, M., and Stuart, A., *The Advanced Theory of Statistics*]

Stroustrup, B. 1997, *The C++ Programming Language*, 3rd ed. (Reading, MA: Addison-Wesley)

Tewarson, R.P. 1973, *Sparse Matrices* (New York: Academic Press)

Westlake, J.R. 1968, *A Handbook of Numerical Matrix Inversion and Solution of Linear Equations* (New York: Wiley)

Wilkinson, J.H. 1965, *The Algebraic Eigenvalue Problem* (New York: Oxford University Press)

Young, D.M., and Gregory, R.T. 1973, *A Survey of Numerical Mathematics*, 2 vols.; reprinted 1988 (New York: Dover)

Appendix A: Table of Function Declarations

We here list the function declarations for all the routines in *Numerical Recipes in C++*. You should #include this listing in each separately compiled source file that contains or references any routine from this book. You have to apply the scope resolution operator NR:: either explicitly or implicitly to access a routine from the NR namespace (see §1.2).

In the machine-readable distribution of *Numerical Recipes* (e.g., CDROM or Internet download), this Appendix is in the file nr.h. Note that it automatically includes the files nrutil.h and nrtypes.h that are detailed in Appendix B, making those facilities available to your program.

Here is a listing of the file nr.h:

```
#ifndef _NR_H_
#define _NR_H_
#include <fstream>
#include <complex>
#include "nrutil.h"
#include "nrtypes.h"
using namespace std;

namespace NR {

void addint(Mat_O_DP &uf, Mat_I_DP &uc, Mat_O_DP &res);
void airy(const DP x, DP &ai, DP &bi, DP &aip, DP &bip);
void amebsa(Mat_IO_DP &p, Vec_IO_DP &y, Vec_O_DP &pb, DP &yb, const DP ftol,
    DP funk(Vec_I_DP &), int &iter, const DP temptr);
void amoeba(Mat_IO_DP &p, Vec_IO_DP &y, const DP ftol, DP funk(Vec_I_DP &),
    int &nfunk);
DP amotry(Mat_IO_DP &p, Vec_O_DP &y, Vec_IO_DP &psum, DP funk(Vec_I_DP &),
    const int ihi, const DP fac);
DP amotsa(Mat_IO_DP &p, Vec_O_DP &y, Vec_IO_DP &psum, Vec_O_DP &pb, DP &yb,
    DP funk(Vec_I_DP &), const int ihi, DP &yhi, const DP fac);
void anneal(Vec_I_DP &x, Vec_I_DP &y, Vec_IO_INT &iorder);
DP anorm2(Mat_I_DP &a);
void arcmak(Vec_I_ULNG &nfreq, unsigned long nchh, unsigned long nradd,
    arithcode &acode);
void arcode(unsigned long &ich, string &code, unsigned long &lcd,
    const int isign, arithcode &acode);
void arcsum(Vec_I_ULNG &iin, Vec_O_ULNG &iout, unsigned long ja,
    const int nwk, const unsigned long nrad, const unsigned long nc);
void asolve(Vec_I_DP &b, Vec_O_DP &x, const int itrnsp);
void atimes(Vec_I_DP &x, Vec_O_DP &r, const int itrnsp);
void avevar(Vec_I_DP &data, DP &ave, DP &var);
```

```
void balanc(Mat_IO_DP &a);
void banbks(Mat_I_DP &a, const int m1, const int m2, Mat_I_DP &al,
    Vec_I_INT &indx, Vec_IO_DP &b);
void bandec(Mat_IO_DP &a, const int m1, const int m2, Mat_O_DP &al,
    Vec_O_INT &indx, DP &d);
void banmul(Mat_I_DP &a, const int m1, const int m2, Vec_I_DP &x,
    Vec_O_DP &b);
void bcucof(Vec_I_DP &y, Vec_I_DP &y1, Vec_I_DP &y2, Vec_I_DP &y12,
    const DP d1, const DP d2, Mat_O_DP &c);
void bcuint(Vec_I_DP &y, Vec_I_DP &y1, Vec_I_DP &y2, Vec_I_DP &y12,
    const DP x1l, const DP x2l, const DP x2u,
    const DP x1, const DP x2, DP &ansy, DP &ansy1, DP &ansy2);
void beschb(const DP x, DP &gam1, DP &gam2, DP &gampl, DP &gammi);
DP bessi(const int n, const DP x);
DP bessi0(const DP x);
DP bessi1(const DP x);
void bessik(const DP x, const DP xnu, DP &ri, DP &rk, DP &rip, DP &rkp);
DP bessj(const int n, const DP x);
DP bessj0(const DP x);
DP bessj1(const DP x);
void bessjy(const DP x, const DP xnu, DP &rj, DP &ry, DP &rjp, DP &ryp);
DP bessk(const int n, const DP x);
DP bessk0(const DP x);
DP bessk1(const DP x);
DP bessy(const int n, const DP x);
DP bessy0(const DP x);
DP bessy1(const DP x);
DP beta(const DP z, const DP w);
DP betacf(const DP a, const DP b, const DP x);
DP betai(const DP a, const DP b, const DP x);
DP bico(const int n, const int k);
void bksub(const int ne, const int nb, const int jf, const int k1,
    const int k2, Mat3D_IO_DP &c);
DP bnldev(const DP pp, const int n, int &idum);
DP brent(const DP ax, const DP bx, const DP cx, DP f(const DP),
    const DP tol, DP &xmin);
void broydn(Vec_IO_DP &x, bool &check, void vecfunc(Vec_I_DP &, Vec_O_DP &));
void bsstep(Vec_IO_DP &y, Vec_IO_DP &dydx, DP &xx, const DP htry,
    const DP eps, Vec_I_DP &yscal, DP &hdid, DP &hnext,
    void derivs(const DP, Vec_I_DP &, Vec_O_DP &));
void caldat(const int julian, int &mm, int &id, int &iyyy);
void chder(const DP a, const DP b, Vec_I_DP &c, Vec_O_DP &cder, const int n);
DP chebev(const DP a, const DP b, Vec_I_DP &c, const int m, const DP x);
void chebft(const DP a, const DP b, Vec_O_DP &c, DP func(const DP));
void chebpc(Vec_I_DP &c, Vec_O_DP &d);
void chint(const DP a, const DP b, Vec_I_DP &c, Vec_O_DP &cint, const int n);
DP chixy(const DP bang);
void choldc(Mat_IO_DP &a, Vec_O_DP &p);
void cholsl(Mat_I_DP &a, Vec_I_DP &p, Vec_I_DP &b, Vec_O_DP &x);
void chsone(Vec_I_DP &bins, Vec_I_DP &ebins, const int knstrn, DP &df,
    DP &chsq, DP &prob);
void chstwo(Vec_I_DP &bins1, Vec_I_DP &bins2, const int knstrn, DP &df,
    DP &chsq, DP &prob);
void cisi(const DP x, complex<DP> &cs);
void cntab1(Mat_I_INT &nn, DP &chisq, DP &df, DP &prob, DP &cramrv, DP &ccc);
void cntab2(Mat_I_INT &nn, DP &h, DP &hx, DP &hy, DP &hygx, DP &hxgy,
    DP &uygx, DP &uxgy, DP &uxy);
void convlv(Vec_I_DP &data, Vec_I_DP &respns, const int isign,
    Vec_O_DP &ans);
void copy(Mat_O_DP &aout, Mat_I_DP &ain);
void correl(Vec_I_DP &data1, Vec_I_DP &data2, Vec_O_DP &ans);
void cosft1(Vec_IO_DP &y);
void cosft2(Vec_IO_DP &y, const int isign);
void covsrt(Mat_IO_DP &covar, Vec_I_BOOL &ia, const int mfit);
```

```
void crank(Vec_IO_DP &w, DP &s);
void cyclic(Vec_I_DP &a, Vec_I_DP &b, Vec_I_DP &c, const DP alpha,
    const DP beta, Vec_I_DP &r, Vec_O_DP &x);
void daub4(Vec_IO_DP &a, const int n, const int isign);
DP dawson(const DP x);
DP dbrent(const DP ax, const DP bx, const DP cx, DP f(const DP),
    DP df(const DP), const DP tol, DP &xmin);
void ddpoly(Vec_I_DP &c, const DP x, Vec_O_DP &pd);
bool decchk(string str, char &ch);
void derivs_s(const DP x, Vec_I_DP &y, Vec_O_DP &dydx);
DP df1dim(const DP x);
void dfpmin(Vec_IO_DP &p, const DP gtol, int &iter, DP &fret,
    DP func(Vec_I_DP &), void dfunc(Vec_I_DP &, Vec_O_DP &));
DP dfridr(DP func(const DP), const DP x, const DP h, DP &err);
void dftcor(const DP w, const DP delta, const DP a, const DP b,
    Vec_I_DP &endpts, DP &corre, DP &corim, DP &corfac);
void dftint(DP func(const DP), const DP a, const DP b, const DP w,
    DP &cosint, DP &sinint);
void difeq(const int k, const int k1, const int k2, const int jsf,
    const int is1, const int isf, Vec_I_INT &indexv, Mat_O_DP &s,
    Mat_I_DP &y);
void dlinmin(Vec_IO_DP &p, Vec_IO_DP &xi, DP &fret, DP func(Vec_I_DP &),
    void dfunc(Vec_I_DP &, Vec_O_DP &));
void eclass(Vec_O_INT &nf, Vec_I_INT &lista, Vec_I_INT &listb);
void eclazz(Vec_O_INT &nf, bool equiv(const int, const int));
DP ei(const DP x);
void eigsrt(Vec_IO_DP &d, Mat_IO_DP &v);
DP elle(const DP phi, const DP ak);
DP ellf(const DP phi, const DP ak);
DP ellpi(const DP phi, const DP en, const DP ak);
void elmhes(Mat_IO_DP &a);
DP erfcc(const DP x);
DP erff(const DP x);
DP erffc(const DP x);
void eulsum(DP &sum, const DP term, const int jterm, Vec_IO_DP &wksp);
DP evlmem(const DP fdt, Vec_I_DP &d, const DP xms);
DP expdev(int &idum);
DP expint(const int n, const DP x);
DP f1dim(const DP x);
DP factln(const int n);
DP factrl(const int n);
void fasper(Vec_I_DP &x, Vec_I_DP &y, const DP ofac, const DP hifac,
    Vec_O_DP &wk1, Vec_O_DP &wk2, int &nout, int &jmax, DP &prob);
void fdjac(Vec_IO_DP &x, Vec_I_DP &fvec, Mat_O_DP &df,
    void vecfunc(Vec_I_DP &, Vec_O_DP &));
void fgauss(const DP x, Vec_I_DP &a, DP &y, Vec_O_DP &dyda);
void fit(Vec_I_DP &x, Vec_I_DP &y, Vec_I_DP &sig, const bool mwt, DP &a,
    DP &b, DP &siga, DP &sigb, DP &chi2, DP &q);
void fitexy(Vec_I_DP &x, Vec_I_DP &y, Vec_I_DP &sigx, Vec_I_DP &sigy,
    DP &a, DP &b, DP &siga, DP &sigb, DP &chi2, DP &q);
void fixrts(Vec_IO_DP &d);
void fleg(const DP x, Vec_O_DP &pl);
void flmoon(const int n, const int nph, int &jd, DP &frac);
DP fmin(Vec_I_DP &x);
void four1(Vec_IO_DP &data, const int isign);
void fourew(Vec_FSTREAM_p &file, int &na, int &nb, int &nc, int &nd);
void fourfs(Vec_FSTREAM_p &file, Vec_I_INT &nn, const int isign);
void fourn(Vec_IO_DP &data, Vec_I_INT &nn, const int isign);
void fpoly(const DP x, Vec_O_DP &p);
void fred2(const DP a, const DP b, Vec_O_DP &t, Vec_O_DP &f, Vec_O_DP &w,
    DP g(const DP), DP ak(const DP, const DP));
DP fredin(const DP x, const DP a, const DP b, Vec_I_DP &t, Vec_I_DP &f,
    Vec_I_DP &w, DP g(const DP), DP ak(const DP, const DP));
void frenel(const DP x, complex<DP> &cs);
```

```
void frprmn(Vec_IO_DP &p, const DP ftol, int &iter, DP &fret,
    DP func(Vec_I_DP &), void dfunc(Vec_I_DP &, Vec_O_DP &));
void ftest(Vec_I_DP &data1, Vec_I_DP &data2, DP &f, DP &prob);
DP gamdev(const int ia, int &idum);
DP gammln(const DP xx);
DP gammp(const DP a, const DP x);
DP gammq(const DP a, const DP x);
DP gasdev(int &idum);
void gaucof(Vec_IO_DP &a, Vec_IO_DP &b, const DP amu0, Vec_O_DP &x,
    Vec_O_DP &w);
void gauher(Vec_O_DP &x, Vec_O_DP &w);
void gaujac(Vec_O_DP &x, Vec_O_DP &w, const DP alf, const DP bet);
void gaulag(Vec_O_DP &x, Vec_O_DP &w, const DP alf);
void gauleg(const DP x1, const DP x2, Vec_O_DP &x, Vec_O_DP &w);
void gaussj(Mat_IO_DP &a, Mat_IO_DP &b);
void gcf(DP &gammcf, const DP a, const DP x, DP &gln);
DP golden(const DP ax, const DP bx, const DP cx, DP f(const DP),
    const DP tol, DP &xmin);
void gser(DP &gamser, const DP a, const DP x, DP &gln);
void hpsel(Vec_I_DP &arr, Vec_O_DP &heap);
void hpsort(Vec_IO_DP &ra);
void hqr(Mat_IO_DP &a, Vec_O_CPLX_DP &wri);
void hufapp(Vec_IO_ULNG &index, Vec_I_ULNG &nprob, const unsigned long n,
    const unsigned long m);
void hufdec(unsigned long &ich, string &code, const unsigned long lcode,
    unsigned long &nb, huffcode &hcode);
void hufenc(const unsigned long ich, string &code, unsigned long &nb,
    huffcode &hcode);
void hufmak(Vec_I_ULNG &nfreq, const unsigned long nchin,
    unsigned long &ilong, unsigned long &nlong, huffcode &hcode);
void hunt(Vec_I_DP &xx, const DP x, int &jlo);
void hypdrv(const DP s, Vec_I_DP &yy, Vec_O_DP &dyyds);
complex<DP> hypgeo(const complex<DP> &a, const complex<DP> &b,
    const complex<DP> &c, const complex<DP> &z);
void hypser(const complex<DP> &a, const complex<DP> &b,
    const complex<DP> &c, const complex<DP> &z,
    complex<DP> &series, complex<DP> &deriv);
unsigned short icrc(const unsigned short crc, const string &bufptr,
    const short jinit, const int jrev);
unsigned short icrc1(const unsigned short crc, const unsigned char onech);
unsigned long igray(const unsigned long n, const int is);
void indexx(Vec_I_DP &arr, Vec_O_INT &indx);
void indexx(Vec_I_INT &arr, Vec_O_INT &indx);
void interp(Mat_O_DP &uf, Mat_I_DP &uc);
int irbit1(unsigned long &iseed);
int irbit2(unsigned long &iseed);
void jacobi(Mat_IO_DP &a, Vec_O_DP &d, Mat_O_DP &v, int &nrot);
void jacobn_s(const DP x, Vec_I_DP &y, Vec_O_DP &dfdx, Mat_O_DP &dfdy);
int julday(const int mm, const int id, const int iyyy);
void kendl1(Vec_I_DP &data1, Vec_I_DP &data2, DP &tau, DP &z, DP &prob);
void kendl2(Mat_I_DP &tab, DP &tau, DP &z, DP &prob);
void kermom(Vec_O_DP &w, const DP y);
void ks2d1s(Vec_I_DP &x1, Vec_I_DP &y1, void quadvl(const DP, const DP,
    DP &, DP &, DP &, DP &), DP &d1, DP &prob);
void ks2d2s(Vec_I_DP &x1, Vec_I_DP &y1, Vec_I_DP &x2, Vec_I_DP &y2, DP &d,
    DP &prob);
void ksone(Vec_IO_DP &data, DP func(const DP), DP &d, DP &prob);
void kstwo(Vec_IO_DP &data1, Vec_IO_DP &data2, DP &d, DP &prob);
void laguer(Vec_I_CPLX_DP &a, complex<DP> &x, int &its);
void lfit(Vec_I_DP &x, Vec_I_DP &y, Vec_I_DP &sig, Vec_IO_DP &a,
    Vec_I_BOOL &ia, Mat_O_DP &covar, DP &chisq,
    void funcs(const DP, Vec_O_DP &));
void linbcg(Vec_I_DP &b, Vec_IO_DP &x, const int itol, const DP tol,
    const int itmax, int &iter, DP &err);
```

```
void linmin(Vec_IO_DP &p, Vec_IO_DP &xi, DP &fret, DP func(Vec_I_DP &));
void lnsrch(Vec_I_DP &xold, const DP fold, Vec_I_DP &g, Vec_IO_DP &p,
    Vec_O_DP &x, DP &f, const DP stpmax, bool &check, DP func(Vec_I_DP &));
void locate(Vec_I_DP &xx, const DP x, int &j);
void lop(Mat_O_DP &out, Mat_I_DP &u);
void lubksb(Mat_I_DP &a, Vec_I_INT &indx, Vec_IO_DP &b);
void ludcmp(Mat_IO_DP &a, Vec_O_INT &indx, DP &d);
void machar(int &ibeta, int &it, int &irnd, int &ngrd, int &machep,
    int &negep, int &iexp, int &minexp, int &maxexp, DP &eps, DP &epsneg,
    DP &xmin, DP &xmax);
void matadd(Mat_I_DP &a, Mat_I_DP &b, Mat_O_DP &c);
void matsub(Mat_I_DP &a, Mat_I_DP &b, Mat_O_DP &c);
void medfit(Vec_I_DP &x, Vec_I_DP &y, DP &a, DP &b, DP &abdev);
void memcof(Vec_I_DP &data, DP &xms, Vec_O_DP &d);
bool metrop(const DP de, const DP t);
void mgfas(Mat_IO_DP &u, const int maxcyc);
void mglin(Mat_IO_DP &u, const int ncycle);
DP midexp(DP funk(const DP), const DP aa, const DP bb, const int n);
DP midinf(DP funk(const DP), const DP aa, const DP bb, const int n);
DP midpnt(DP func(const DP), const DP a, const DP b, const int n);
DP midsql(DP funk(const DP), const DP aa, const DP bb, const int n);
DP midsqu(DP funk(const DP), const DP aa, const DP bb, const int n);
void miser(DP func(Vec_I_DP &), Vec_I_DP &regn, const int npts,
    const DP dith, DP &ave, DP &var);
void mmid(Vec_I_DP &y, Vec_I_DP &dydx, const DP xs, const DP htot,
    const int nstep, Vec_O_DP &yout,
    void derivs(const DP, Vec_I_DP &, Vec_O_DP &));
void mnbrak(DP &ax, DP &bx, DP &cx, DP &fa, DP &fb, DP &fc,
    DP func(const DP));
void mnewt(const int ntrial, Vec_IO_DP &x, const DP tolx, const DP tolf);
void moment(Vec_I_DP &data, DP &ave, DP &adev, DP &sdev, DP &var, DP &skew,
    DP &curt);
void mp2dfr(Vec_IO_UCHR &a, string &s);
void mpadd(Vec_O_UCHR &w, Vec_I_UCHR &u, Vec_I_UCHR &v);
void mpdiv(Vec_O_UCHR &q, Vec_O_UCHR &r, Vec_I_UCHR &u, Vec_I_UCHR &v);
void mpinv(Vec_O_UCHR &u, Vec_I_UCHR &v);
void mplsh(Vec_IO_UCHR &u);
void mpmov(Vec_O_UCHR &u, Vec_I_UCHR &v);
void mpmul(Vec_O_UCHR &w, Vec_I_UCHR &u, Vec_I_UCHR &v);
void mpneg(Vec_IO_UCHR &u);
void mppi(const int np);
void mprove(Mat_I_DP &a, Mat_I_DP &alud, Vec_I_INT &indx, Vec_I_DP &b,
    Vec_IO_DP &x);
void mpsad(Vec_O_UCHR &w, Vec_I_UCHR &u, const int iv);
void mpsdv(Vec_O_UCHR &w, Vec_I_UCHR &u, const int iv, int &ir);
void mpsmu(Vec_O_UCHR &w, Vec_I_UCHR &u, const int iv);
void mpsqrt(Vec_O_UCHR &w, Vec_O_UCHR &u, Vec_I_UCHR &v);
void mpsub(int &is, Vec_O_UCHR &w, Vec_I_UCHR &u, Vec_I_UCHR &v);
void mrqcof(Vec_I_DP &x, Vec_I_DP &y, Vec_I_DP &sig, Vec_I_DP &a,
    Vec_I_BOOL &ia, Mat_O_DP &alpha, Vec_O_DP &beta, DP &chisq,
    void funcs(const DP, Vec_I_DP &,DP &, Vec_O_DP &));
void mrqmin(Vec_I_DP &x, Vec_I_DP &y, Vec_I_DP &sig, Vec_IO_DP &a,
    Vec_I_BOOL &ia, Mat_O_DP &covar, Mat_O_DP &alpha, DP &chisq,
    void funcs(const DP, Vec_I_DP &, DP &, Vec_O_DP &), DP &alamda);
void newt(Vec_IO_DP &x, bool &check, void vecfunc(Vec_I_DP &, Vec_O_DP &));
void odeint(Vec_IO_DP &ystart, const DP x1, const DP x2, const DP eps,
    const DP h1, const DP hmin, int &nok, int &nbad,
    void derivs(const DP, Vec_I_DP &, Vec_O_DP &),
    void rkqs(Vec_IO_DP &, Vec_IO_DP &, DP &, const DP, const DP,
    Vec_I_DP &, DP &, DP &, void (*)(const DP, Vec_I_DP &, Vec_O_DP &)));
void orthog(Vec_I_DP &anu, Vec_I_DP &alpha, Vec_I_DP &beta, Vec_O_DP &a,
    Vec_O_DP &b);
void pade(Vec_IO_DP &cof, DP &resid);
void pccheb(Vec_I_DP &d, Vec_O_DP &c);
```

```
void pcshft(const DP a, const DP b, Vec_IO_DP &d);
void pearsn(Vec_I_DP &x, Vec_I_DP &y, DP &r, DP &prob, DP &z);
void period(Vec_I_DP &x, Vec_I_DP &y, const DP ofac, const DP hifac,
    Vec_O_DP &px, Vec_O_DP &py, int &nout, int &jmax, DP &prob);
void piksr2(Vec_IO_DP &arr, Vec_IO_DP &brr);
void piksrt(Vec_IO_DP &arr);
void pinvs(const int ie1, const int ie2, const int je1, const int jsf,
    const int jc1, const int k, Mat3D_O_DP &c, Mat_IO_DP &s);
DP plgndr(const int l, const int m, const DP x);
DP poidev(const DP xm, int &idum);
void polcoe(Vec_I_DP &x, Vec_I_DP &y, Vec_O_DP &cof);
void polcof(Vec_I_DP &xa, Vec_I_DP &ya, Vec_O_DP &cof);
void poldiv(Vec_I_DP &u, Vec_I_DP &v, Vec_O_DP &q, Vec_O_DP &r);
void polin2(Vec_I_DP &x1a, Vec_I_DP &x2a, Mat_I_DP &ya, const DP x1,
    const DP x2, DP &y, DP &dy);
void polint(Vec_I_DP &xa, Vec_I_DP &ya, const DP x, DP &y, DP &dy);
void powell(Vec_IO_DP &p, Mat_IO_DP &xi, const DP ftol, int &iter,
    DP &fret, DP func(Vec_I_DP &));
void predic(Vec_I_DP &data, Vec_I_DP &d, Vec_O_DP &future);
DP probks(const DP alam);
void psdes(unsigned long &lword, unsigned long &irword);
void pwt(Vec_IO_DP &a, const int n, const int isign);
void pwtset(const int n);
DP pythag(const DP a, const DP b);
void pzextr(const int iest, const DP xest, Vec_I_DP &yest, Vec_O_DP &yz,
    Vec_O_DP &dy);
DP qgaus(DP func(const DP), const DP a, const DP b);
void qrdcmp(Mat_IO_DP &a, Vec_O_DP &c, Vec_O_DP &d, bool &sing);
DP qromb(DP func(const DP), DP a, DP b);
DP qromo(DP func(const DP), const DP a, const DP b,
    DP choose(DP (*)(const DP), const DP, const DP, const int));
void qroot(Vec_I_DP &p, DP &b, DP &c, const DP eps);
void qrsolv(Mat_I_DP &a, Vec_I_DP &c, Vec_I_DP &d, Vec_IO_DP &b);
void qrupdt(Mat_IO_DP &r, Mat_IO_DP &qt, Vec_IO_DP &u, Vec_I_DP &v);
DP qsimp(DP func(const DP), const DP a, const DP b);
DP qtrap(DP func(const DP), const DP a, const DP b);
DP quad3d(DP func(const DP, const DP, const DP), const DP x1, const DP x2);
void quadct(const DP x, const DP y, Vec_I_DP &xx, Vec_I_DP &yy, DP &fa,
    DP &fb, DP &fc, DP &fd);
void quadmx(Mat_O_DP &a);
void quadvl(const DP x, const DP y, DP &fa, DP &fb, DP &fc, DP &fd);
DP ran0(int &idum);
DP ran1(int &idum);
DP ran2(int &idum);
DP ran3(int &idum);
DP ran4(int &idum);
void rank(Vec_I_INT &indx, Vec_O_INT &irank);
void ranpt(Vec_O_DP &pt, Vec_I_DP &regn);
void ratint(Vec_I_DP &xa, Vec_I_DP &ya, const DP x, DP &y, DP &dy);
void ratlsq(DP fn(const DP), const DP a, const DP b, const int mm,
    const int kk, Vec_O_DP &cof, DP &dev);
DP ratval(const DP x, Vec_I_DP &cof, const int mm, const int kk);
DP rc(const DP x, const DP y);
DP rd(const DP x, const DP y, const DP z);
void realft(Vec_IO_DP &data, const int isign);
void rebin(const DP rc, const int nd, Vec_I_DP &r, Vec_O_DP &xin,
    Mat_IO_DP &xi, const int j);
void red(const int iz1, const int iz2, const int jz1, const int jz2,
    const int jm1, const int jm2, const int jmf, const int ic1,
    const int jc1, const int jcf, const int kc, Mat3D_I_DP &c,
    Mat_IO_DP &s);
void relax(Mat_IO_DP &u, Mat_I_DP &rhs);
void relax2(Mat_IO_DP &u, Mat_I_DP &rhs);
void resid(Mat_O_DP &res, Mat_I_DP &u, Mat_I_DP &rhs);
```

```
DP revcst(Vec_I_DP &x, Vec_I_DP &y, Vec_I_INT &iorder, Vec_IO_INT &n);
void reverse(Vec_IO_INT &iorder, Vec_I_INT &n);
DP rf(const DP x, const DP y, const DP z);
DP rj(const DP x, const DP y, const DP z, const DP p);
void rk4(Vec_I_DP &y, Vec_I_DP &dydx, const DP x, const DP h,
    Vec_O_DP &yout, void derivs(const DP, Vec_I_DP &, Vec_O_DP &));
void rkck(Vec_I_DP &y, Vec_I_DP &dydx, const DP x,
    const DP h, Vec_O_DP &yout, Vec_O_DP &yerr,
    void derivs(const DP, Vec_I_DP &, Vec_O_DP &));
void rkdumb(Vec_I_DP &vstart, const DP x1, const DP x2,
    void derivs(const DP, Vec_I_DP &, Vec_O_DP &));
void rkqs(Vec_IO_DP &y, Vec_IO_DP &dydx, DP &x, const DP htry,
    const DP eps, Vec_I_DP &yscal, DP &hdid, DP &hnext,
    void derivs(const DP, Vec_I_DP &, Vec_O_DP &));
void rlft3(Mat3D_IO_DP &data, Mat_IO_DP &speq, const int isign);
DP rofunc(const DP b);
void rotate(Mat_IO_DP &r, Mat_IO_DP &qt, const int i, const DP a,
    const DP b);
void rsolv(Mat_I_DP &a, Vec_I_DP &d, Vec_IO_DP &b);
void rstrct(Mat_O_DP &uc, Mat_I_DP &uf);
DP rtbis(DP func(const DP), const DP x1, const DP x2, const DP xacc);
DP rtflsp(DP func(const DP), const DP x1, const DP x2, const DP xacc);
DP rtnewt(void funcd(const DP, DP &, DP &), const DP x1, const DP x2,
    const DP xacc);
DP rtsafe(void funcd(const DP, DP &, DP &), const DP x1, const DP x2,
    const DP xacc);
DP rtsec(DP func(const DP), const DP x1, const DP x2, const DP xacc);
void rzextr(const int iest, const DP xest, Vec_I_DP &yest, Vec_O_DP &yz,
    Vec_O_DP &dy);
void savgol(Vec_O_DP &c, const int np, const int nl, const int nr,
    const int ld, const int m);
void scrsho(DP fx(const DP));
DP select(const int k, Vec_IO_DP &arr);
DP selip(const int k, Vec_I_DP &arr);
void shell(const int n, Vec_IO_DP &a);
void shoot(Vec_I_DP &v, Vec_O_DP &f);
void shootf(Vec_I_DP &v, Vec_O_DP &f);
void simp1(Mat_I_DP &a, const int mm, Vec_I_INT &ll, const int nll,
    const int iabf, int &kp, DP &bmax);
void simp2(Mat_I_DP &a, const int m, const int n, int &ip, const int kp);
void simp3(Mat_IO_DP &a, const int i1, const int k1, const int ip,
    const int kp);
void simplx(Mat_IO_DP &a, const int m1, const int m2, const int m3,
    int &icase, Vec_O_INT &izrov, Vec_O_INT &iposv);
void simpr(Vec_I_DP &y, Vec_I_DP &dydx, Vec_I_DP &dfdx, Mat_I_DP &dfdy,
    const DP xs, const DP htot, const int nstep, Vec_O_DP &yout,
    void derivs(const DP, Vec_I_DP &, Vec_O_DP &));
void sinft(Vec_IO_DP &y);
void slvsm2(Mat_O_DP &u, Mat_I_DP &rhs);
void slvsml(Mat_O_DP &u, Mat_I_DP &rhs);
void sncndn(const DP uu, const DP emmc, DP &sn, DP &cn, DP &dn);
DP snrm(Vec_I_DP &sx, const int itol);
void sobseq(const int n, Vec_O_DP &x);
void solvde(const int itmax, const DP conv, const DP slowc,
    Vec_I_DP &scalv, Vec_I_INT &indexv, const int nb, Mat_IO_DP &y);
void sor(Mat_I_DP &a, Mat_I_DP &b, Mat_I_DP &c, Mat_I_DP &d, Mat_I_DP &e,
    Mat_I_DP &f, Mat_IO_DP &u, const DP rjac);
void sort(Vec_IO_DP &arr);
void sort2(Vec_IO_DP &arr, Vec_IO_DP &brr);
void sort3(Vec_IO_DP &ra, Vec_IO_DP &rb, Vec_IO_DP &rc);
void spctrm(ifstream &fp, Vec_O_DP &p, const int k, const bool ovrlap);
void spear(Vec_I_DP &data1, Vec_I_DP &data2, DP &d, DP &zd, DP &probd,
    DP &rs, DP &probrs);
void sphbes(const int n, const DP x, DP &sj, DP &sy, DP &sjp, DP &syp);
```

```
void splie2(Vec_I_DP &x1a, Vec_I_DP &x2a, Mat_I_DP &ya, Mat_O_DP &y2a);
void splin2(Vec_I_DP &x1a, Vec_I_DP &x2a, Mat_I_DP &ya, Mat_I_DP &y2a,
    const DP x1, const DP x2, DP &y);
void spline(Vec_I_DP &x, Vec_I_DP &y, const DP yp1, const DP ypn,
    Vec_O_DP &y2);
void splint(Vec_I_DP &xa, Vec_I_DP &ya, Vec_I_DP &y2a, const DP x, DP &y);
void spread(const DP y, Vec_IO_DP &yy, const DP x, const int m);
void sprsax(Vec_I_DP &sa, Vec_I_INT &ija, Vec_I_DP &x, Vec_O_DP &b);
void sprsin(Mat_I_DP &a, const DP thresh, Vec_O_DP &sa, Vec_O_INT &ija);
void sprspm(Vec_I_DP &sa, Vec_I_INT &ija, Vec_I_DP &sb, Vec_I_INT &ijb,
    Vec_O_DP &sc, Vec_I_INT &ijc);
void sprstm(Vec_I_DP &sa, Vec_I_INT &ija, Vec_I_DP &sb, Vec_I_INT &ijb,
    const DP thresh, Vec_O_DP &sc, Vec_O_INT &ijc);
void sprstp(Vec_I_DP &sa, Vec_I_INT &ija, Vec_O_DP &sb, Vec_O_INT &ijb);
void sprstx(Vec_I_DP &sa, Vec_I_INT &ija, Vec_I_DP &x, Vec_O_DP &b);
void stifbs(Vec_IO_DP &y, Vec_IO_DP &dydx, DP &xx, const DP htry,
    const DP eps, Vec_I_DP &yscal, DP &hdid, DP &hnext,
    void derivs(const DP, Vec_I_DP &, Vec_O_DP &));
void stiff(Vec_IO_DP &y, Vec_IO_DP &dydx, DP &x, const DP htry,
    const DP eps, Vec_I_DP &yscal, DP &hdid, DP &hnext,
    void derivs(const DP, Vec_I_DP &, Vec_O_DP &));
void stoerm(Vec_I_DP &y, Vec_I_DP &d2y, const DP xs,
    const DP htot, const int nstep, Vec_O_DP &yout,
    void derivs(const DP, Vec_I_DP &, Vec_O_DP &));
void svbksb(Mat_I_DP &u, Vec_I_DP &w, Mat_I_DP &v, Vec_I_DP &b, Vec_O_DP &x);
void svdcmp(Mat_IO_DP &a, Vec_O_DP &w, Mat_O_DP &v);
void svdfit(Vec_I_DP &x, Vec_I_DP &y, Vec_I_DP &sig, Vec_O_DP &a,
    Mat_O_DP &u, Mat_O_DP &v, Vec_O_DP &w, DP &chisq,
    void funcs(const DP, Vec_O_DP &));
void svdvar(Mat_I_DP &v, Vec_I_DP &w, Mat_O_DP &cvm);
void toeplz(Vec_I_DP &r, Vec_O_DP &x, Vec_I_DP &y);
void tptest(Vec_I_DP &data1, Vec_I_DP &data2, DP &t, DP &prob);
void tqli(Vec_IO_DP &d, Vec_IO_DP &e, Mat_IO_DP &z);
DP trapzd(DP func(const DP), const DP a, const DP b, const int n);
void tred2(Mat_IO_DP &a, Vec_O_DP &d, Vec_O_DP &e);
void tridag(Vec_I_DP &a, Vec_I_DP &b, Vec_I_DP &c, Vec_I_DP &r, Vec_O_DP &u);
DP trncst(Vec_I_DP &x, Vec_I_DP &y, Vec_I_INT &iorder, Vec_IO_INT &n);
void trnspt(Vec_IO_INT &iorder, Vec_I_INT &n);
void ttest(Vec_I_DP &data1, Vec_I_DP &data2, DP &t, DP &prob);
void tutest(Vec_I_DP &data1, Vec_I_DP &data2, DP &t, DP &prob);
void twofft(Vec_I_DP &data1, Vec_I_DP &data2, Vec_O_DP &fft1,
    Vec_O_DP &fft2);
void vander(Vec_I_DP &x, Vec_O_DP &w, Vec_I_DP &q);
void vegas(Vec_I_DP &regn, DP fxn(Vec_I_DP &, const DP), const int init,
    const int ncall, const int itmx, const int nprn, DP &tgral, DP &sd,
    DP &chi2a);
void voltra(const DP t0, const DP h, Vec_O_DP &t, Mat_O_DP &f,
    DP g(const int, const DP),
    DP ak(const int, const int, const DP, const DP));
void wt1(Vec_IO_DP &a, const int isign,
    void wtstep(Vec_IO_DP &, const int, const int));
void wtn(Vec_IO_DP &a, Vec_I_INT &nn, const int isign,
    void wtstep(Vec_IO_DP &, const int, const int));
void wwghts(Vec_O_DP &wghts, const DP h,
    void kermom(Vec_O_DP &w, const DP y));
bool zbrac(DP func(const DP), DP &x1, DP &x2);
void zbrak(DP fx(const DP), const DP x1, const DP x2, const int n,
    Vec_O_DP &xb1, Vec_O_DP &xb2, int &nroot);
DP zbrent(DP func(const DP), const DP x1, const DP x2, const DP tol);
void zrhqr(Vec_I_DP &a, Vec_O_CPLX_DP &rt);
DP zriddr(DP func(const DP), const DP x1, const DP x2, const DP xacc);
void zroots(Vec_I_CPLX_DP &a, Vec_O_CPLX_DP &roots, const bool &polish);
}
#endif /* _NR_H_ */
```

Appendix B: Utility Routines and Classes

Non-Copyright Notice: This Appendix and its utility routines are herewith placed into the public domain. Anyone may copy them freely for any purpose. We of course accept no liability whatsoever for any such use.

The routines and classes listed below are used by essentially all of the Recipes in this book. First we give the listing of the default set of typedefs (see §1.2). In the machine-readable distribution of *Numerical Recipes*, this is the file nrtypes_nr.h. In practice, however we include these definitions with the statement

```
#include "nrtypes.h"
```

The file nrtypes.h is an intermediate file that, itself, contains a single include statement, calling into play an "nrtypes" file that is specific to the matrix/vector class library you use. The default content of nrtypes.h is

```
#include "nrtypes_nr.h"
```

which has the effect of including the following file:

```
#ifndef _NR_TYPES_H_
#define _NR_TYPES_H_

#include <complex>
#include <fstream>
#include "nrutil.h"
using namespace std;

typedef double DP;

// Vector Types

typedef const NRVec<bool> Vec_I_BOOL;
typedef NRVec<bool> Vec_BOOL, Vec_O_BOOL, Vec_IO_BOOL;

typedef const NRVec<char> Vec_I_CHR;
typedef NRVec<char> Vec_CHR, Vec_O_CHR, Vec_IO_CHR;

typedef const NRVec<unsigned char> Vec_I_UCHR;
typedef NRVec<unsigned char> Vec_UCHR, Vec_O_UCHR, Vec_IO_UCHR;

typedef const NRVec<int> Vec_I_INT;
```

939

```
typedef NRVec<int> Vec_INT, Vec_O_INT, Vec_IO_INT;

typedef const NRVec<unsigned int> Vec_I_UINT;
typedef NRVec<unsigned int> Vec_UINT, Vec_O_UINT, Vec_IO_UINT;

typedef const NRVec<long> Vec_I_LNG;
typedef NRVec<long> Vec_LNG, Vec_O_LNG, Vec_IO_LNG;

typedef const NRVec<unsigned long> Vec_I_ULNG;
typedef NRVec<unsigned long> Vec_ULNG, Vec_O_ULNG, Vec_IO_ULNG;

typedef const NRVec<float> Vec_I_SP;
typedef NRVec<float> Vec_SP, Vec_O_SP, Vec_IO_SP;

typedef const NRVec<DP> Vec_I_DP;
typedef NRVec<DP> Vec_DP, Vec_O_DP, Vec_IO_DP;

typedef const NRVec<complex<float> > Vec_I_CPLX_SP;
typedef NRVec<complex<float> > Vec_CPLX_SP, Vec_O_CPLX_SP, Vec_IO_CPLX_SP;

typedef const NRVec<complex<DP> > Vec_I_CPLX_DP;
typedef NRVec<complex<DP> > Vec_CPLX_DP, Vec_O_CPLX_DP, Vec_IO_CPLX_DP;

// Matrix Types

typedef const NRMat<bool> Mat_I_BOOL;
typedef NRMat<bool> Mat_BOOL, Mat_O_BOOL, Mat_IO_BOOL;

typedef const NRMat<char> Mat_I_CHR;
typedef NRMat<char> Mat_CHR, Mat_O_CHR, Mat_IO_CHR;

typedef const NRMat<unsigned char> Mat_I_UCHR;
typedef NRMat<unsigned char> Mat_UCHR, Mat_O_UCHR, Mat_IO_UCHR;

typedef const NRMat<int> Mat_I_INT;
typedef NRMat<int> Mat_INT, Mat_O_INT, Mat_IO_INT;

typedef const NRMat<unsigned int> Mat_I_UINT;
typedef NRMat<unsigned int> Mat_UINT, Mat_O_UINT, Mat_IO_UINT;

typedef const NRMat<long> Mat_I_LNG;
typedef NRMat<long> Mat_LNG, Mat_O_LNG, Mat_IO_LNG;

typedef const NRVec<unsigned long> Mat_I_ULNG;
typedef NRMat<unsigned long> Mat_ULNG, Mat_O_ULNG, Mat_IO_ULNG;

typedef const NRMat<float> Mat_I_SP;
typedef NRMat<float> Mat_SP, Mat_O_SP, Mat_IO_SP;

typedef const NRMat<DP> Mat_I_DP;
typedef NRMat<DP> Mat_DP, Mat_O_DP, Mat_IO_DP;

typedef const NRMat<complex<float> > Mat_I_CPLX_SP;
typedef NRMat<complex<float> > Mat_CPLX_SP, Mat_O_CPLX_SP, Mat_IO_CPLX_SP;

typedef const NRMat<complex<DP> > Mat_I_CPLX_DP;
typedef NRMat<complex<DP> > Mat_CPLX_DP, Mat_O_CPLX_DP, Mat_IO_CPLX_DP;

// 3D Matrix Types

typedef const NRMat3d<DP> Mat3D_I_DP;
typedef NRMat3d<DP> Mat3D_DP, Mat3D_O_DP, Mat3D_IO_DP;

// Miscellaneous Types
```

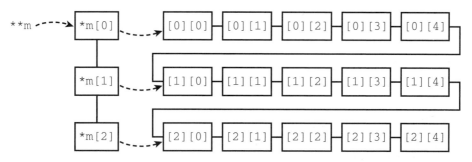

Figure B.1. Storage scheme for a matrix m, implemented as a pointer to an array of pointers to rows. Dotted lines denote address reference, while solid lines connect sequential memory locations.

```
typedef NRVec<unsigned long *> Vec_ULNG_p;
typedef NRVec<NRMat<DP> *> Vec_Mat_DP_p;
typedef NRVec<fstream *> Vec_FSTREAM_p;

#endif /* _NR_TYPES_H_ */
```

Next is a listing of the default vector and matrix classes, NRVec and NRMat, including the implementation (see §1.3).

Note that we do not use C-style arrays as the private data variables for vectors or matrices. For vectors we use a pointer to a block of storage that we explicitly allocate with new. For two-dimensional arrays, we use a pointer to an array of pointers, with the array elements pointing to the first element in the rows of the matrix (see Figure B.1). This scheme is much more flexible than fixed-size built-in arrays.

Also included in the listing are the utility routines SQR, MAX, MIN, SIGN, and SWAP, and the error routine nrerror, which is invoked to terminate program execution — with an appropriate message — when a fatal error is encountered.

In the machine-readable distribution of *Numerical Recipes*, this is the file nrutil_nr.h. In practice, however we include these definitions with the statement

```
#include "nrutil.h"
```

Like nrtypes.h, the file nrutil.h is an intermediate file that, itself, contains a single include statement, calling into play an "nrutil" file that is specific to the matrix/vector class library you use. The default content of nrutil.h is

```
#include "nrutil_nr.h"
```

which has the effect of including the following file:

```
#ifndef _NR_UTIL_H_
#define _NR_UTIL_H_

#include <string>
#include <cmath>
#include <complex>
#include <iostream>
using namespace std;

typedef double DP;
```

```
template<class T>
inline const T SQR(const T a) {return a*a;}

template<class T>
inline const T MAX(const T &a, const T &b)
{return b > a ? (b) : (a);}

inline float MAX(const double &a, const float &b)
{return b > a ? (b) : float(a);}

inline float MAX(const float &a, const double &b)
{return b > a ? float(b) : (a);}

template<class T>
inline const T MIN(const T &a, const T &b)
{return b < a ? (b) : (a);}

inline float MIN(const double &a, const float &b)
{return b < a ? (b) : float(a);}

inline float MIN(const float &a, const double &b)
{return b < a ? float(b) : (a);}

template<class T>
inline const T SIGN(const T &a, const T &b)
    {return b >= 0 ? (a >= 0 ? a : -a) : (a >= 0 ? -a : a);}

inline float SIGN(const float &a, const double &b)
    {return b >= 0 ? (a >= 0 ? a : -a) : (a >= 0 ? -a : a);}

inline float SIGN(const double &a, const float &b)
    {return b >= 0 ? (a >= 0 ? a : -a) : (a >= 0 ? -a : a);}

template<class T>
inline void SWAP(T &a, T &b)
    {T dum=a; a=b; b=dum;}

namespace NR {
    inline void nrerror(const string error_text)
    // Numerical Recipes standard error handler
    {
        cerr << "Numerical Recipes run-time error..." << endl;
        cerr << error_text << endl;
        cerr << "...now exiting to system..." << endl;
        exit(1);
    }
}

template <class T>
class NRVec {
private:
    int nn;     // size of array. upper index is nn-1
    T *v;
public:
    NRVec();
    explicit NRVec(int n);       // Zero-based array
    NRVec(const T &a, int n);    //initialize to constant value
    NRVec(const T *a, int n);    // Initialize to array
    NRVec(const NRVec &rhs);     // Copy constructor
    NRVec & operator=(const NRVec &rhs);    //assignment
    NRVec & operator=(const T &a);     //assign a to every element
    inline T & operator[](const int i);     //i'th element
    inline const T & operator[](const int i) const;
    inline int size() const;
```

```
         ~NRVec();
};

template <class T>
NRVec<T>::NRVec() : nn(0), v(0) {}

template <class T>
NRVec<T>::NRVec(int n) : nn(n), v(new T[n]) {}

template <class T>
NRVec<T>::NRVec(const T& a, int n) : nn(n), v(new T[n])
{
    for(int i=0; i<n; i++)
        v[i] = a;
}

template <class T>
NRVec<T>::NRVec(const T *a, int n) : nn(n), v(new T[n])
{
    for(int i=0; i<n; i++)
        v[i] = *a++;
}

template <class T>
NRVec<T>::NRVec(const NRVec<T> &rhs) : nn(rhs.nn), v(new T[nn])
{
    for(int i=0; i<nn; i++)
        v[i] = rhs[i];
}

template <class T>
NRVec<T> & NRVec<T>::operator=(const NRVec<T> &rhs)
// postcondition: normal assignment via copying has been performed;
//        if vector and rhs were different sizes, vector
//        has been resized to match the size of rhs
{
    if (this != &rhs)
    {
        if (nn != rhs.nn) {
            if (v != 0) delete [] (v);
            nn=rhs.nn;
            v= new T[nn];
        }
        for (int i=0; i<nn; i++)
            v[i]=rhs[i];
    }
    return *this;
}

template <class T>
NRVec<T> & NRVec<T>::operator=(const T &a)   //assign a to every element
{
    for (int i=0; i<nn; i++)
        v[i]=a;
    return *this;
}

template <class T>
inline T & NRVec<T>::operator[](const int i)    //subscripting
{
    return v[i];
}

template <class T>
```

```
inline const T & NRVec<T>::operator[](const int i) const    //subscripting
{
    return v[i];
}

template <class T>
inline int NRVec<T>::size() const
{
    return nn;
}

template <class T>
NRVec<T>::~NRVec()
{
    if (v != 0)
        delete[] (v);
}

template <class T>
class NRMat {
private:
    int nn;
    int mm;
    T **v;
public:
    NRMat();
    NRMat(int n, int m);                 // Zero-based array
    NRMat(const T &a, int n, int m);     //Initialize to constant
    NRMat(const T *a, int n, int m);     // Initialize to array
    NRMat(const NRMat &rhs);             // Copy constructor
    NRMat & operator=(const NRMat &rhs);    //assignment
    NRMat & operator=(const T &a);          //assign a to every element
    inline T* operator[](const int i);    //subscripting: pointer to row i
    inline const T* operator[](const int i) const;
    inline int nrows() const;
    inline int ncols() const;
    ~NRMat();
};

template <class T>
NRMat<T>::NRMat() : nn(0), mm(0), v(0) {}

template <class T>
NRMat<T>::NRMat(int n, int m) : nn(n), mm(m), v(new T*[n])
{
    v[0] = new T[m*n];
    for (int i=1; i< n; i++)
        v[i] = v[i-1] + m;
}

template <class T>
NRMat<T>::NRMat(const T &a, int n, int m) : nn(n), mm(m), v(new T*[n])
{
    int i,j;
    v[0] = new T[m*n];
    for (i=1; i< n; i++)
        v[i] = v[i-1] + m;
    for (i=0; i< n; i++)
        for (j=0; j<m; j++)
            v[i][j] = a;
}

template <class T>
NRMat<T>::NRMat(const T *a, int n, int m) : nn(n), mm(m), v(new T*[n])
```

```
{
    int i,j;
    v[0] = new T[m*n];
    for (i=1; i< n; i++)
        v[i] = v[i-1] + m;
    for (i=0; i< n; i++)
        for (j=0; j<m; j++)
            v[i][j] = *a++;
}

template <class T>
NRMat<T>::NRMat(const NRMat &rhs) : nn(rhs.nn), mm(rhs.mm), v(new T*[nn])
{
    int i,j;
    v[0] = new T[mm*nn];
    for (i=1; i< nn; i++)
        v[i] = v[i-1] + mm;
    for (i=0; i< nn; i++)
        for (j=0; j<mm; j++)
            v[i][j] = rhs[i][j];
}

template <class T>
NRMat<T> & NRMat<T>::operator=(const NRMat<T> &rhs)
// postcondition: normal assignment via copying has been performed;
//         if matrix and rhs were different sizes, matrix
//         has been resized to match the size of rhs
{
    if (this != &rhs) {
        int i,j;
        if (nn != rhs.nn || mm != rhs.mm) {
            if (v != 0) {
                delete[] (v[0]);
                delete[] (v);
            }
            nn=rhs.nn;
            mm=rhs.mm;
            v = new T*[nn];
            v[0] = new T[mm*nn];
        }
        for (i=1; i< nn; i++)
            v[i] = v[i-1] + mm;
        for (i=0; i< nn; i++)
            for (j=0; j<mm; j++)
                v[i][j] = rhs[i][j];
    }
    return *this;
}

template <class T>
NRMat<T> & NRMat<T>::operator=(const T &a)      //assign a to every element
{
    for (int i=0; i< nn; i++)
        for (int j=0; j<mm; j++)
            v[i][j] = a;
    return *this;
}

template <class T>
inline T* NRMat<T>::operator[](const int i)     //subscripting: pointer to row i
{
    return v[i];
}
```

```
template <class T>
inline const T* NRMat<T>::operator[](const int i) const
{
    return v[i];
}

template <class T>
inline int NRMat<T>::nrows() const
{
    return nn;
}

template <class T>
inline int NRMat<T>::ncols() const
{
    return mm;
}

template <class T>
NRMat<T>::~NRMat()
{
    if (v != 0) {
        delete[] (v[0]);
        delete[] (v);
    }
}

template <class T>
class NRMat3d {
private:
    int nn;
    int mm;
    int kk;
    T ***v;
public:
    NRMat3d();
    NRMat3d(int n, int m, int k);
    inline T** operator[](const int i);       //subscripting: pointer to row i
    inline const T* const * operator[](const int i) const;
    inline int dim1() const;
    inline int dim2() const;
    inline int dim3() const;
    ~NRMat3d();
};

template <class T>
NRMat3d<T>::NRMat3d(): nn(0), mm(0), kk(0), v(0) {}

template <class T>
NRMat3d<T>::NRMat3d(int n, int m, int k) : nn(n), mm(m), kk(k), v(new T**[n])
{
    int i,j;
    v[0] = new T*[n*m];
    v[0][0] = new T[n*m*k];
    for(j=1; j<m; j++)
        v[0][j] = v[0][j-1] + k;
    for(i=1; i<n; i++) {
        v[i] = v[i-1] + m;
        v[i][0] = v[i-1][0] + m*k;
        for(j=1; j<m; j++)
            v[i][j] = v[i][j-1] + k;
    }
}
```

```
template <class T>
inline T** NRMat3d<T>::operator[](const int i) //subscripting: pointer to row i
{
    return v[i];
}

template <class T>
inline const T* const * NRMat3d<T>::operator[](const int i) const
{
    return v[i];
}

template <class T>
inline int NRMat3d<T>::dim1() const
{
    return nn;
}

template <class T>
inline int NRMat3d<T>::dim2() const
{
    return mm;
}

template <class T>
inline int NRMat3d<T>::dim3() const
{
    return kk;
}

template <class T>
NRMat3d<T>::~NRMat3d()
{
    if (v != 0) {
        delete[] (v[0][0]);
        delete[] (v[0]);
        delete[] (v);
    }
}

//The next 3 classes are used in artihmetic coding, Huffman coding, and
//wavelet transforms respectively. This is as good a place as any to put them!

class arithcode {
private:
    NRVec<unsigned long> *ilob_p,*iupb_p,*ncumfq_p;
public:
    NRVec<unsigned long> &ilob,&iupb,&ncumfq;
    unsigned long jdif,nc,minint,nch,ncum,nrad;
    arithcode(unsigned long n1, unsigned long n2, unsigned long n3)
        : ilob_p(new NRVec<unsigned long>(n1)),
        iupb_p(new NRVec<unsigned long>(n2)),
        ncumfq_p(new NRVec<unsigned long>(n3)),
        ilob(*ilob_p),iupb(*iupb_p),ncumfq(*ncumfq_p) {}
    ~arithcode() {
        if (ilob_p != 0) delete ilob_p;
        if (iupb_p != 0) delete iupb_p;
        if (ncumfq_p != 0) delete ncumfq_p;
    }
};

class huffcode {
private:
    NRVec<unsigned long> *icod_p,*ncod_p,*left_p,*right_p;
```

```
public:
    NRVec<unsigned long> &icod,&ncod,&left,&right;
    int nch,nodemax;
    huffcode(unsigned long n1, unsigned long n2, unsigned long n3,
        unsigned long n4) :
        icod_p(new NRVec<unsigned long>(n1)),
        ncod_p(new NRVec<unsigned long>(n2)),
        left_p(new NRVec<unsigned long>(n3)),
        right_p(new NRVec<unsigned long>(n4)),
        icod(*icod_p),ncod(*ncod_p),left(*left_p),right(*right_p) {}
    ~huffcode() {
        if (icod_p != 0) delete icod_p;
        if (ncod_p != 0) delete ncod_p;
        if (left_p != 0) delete left_p;
        if (right_p != 0) delete right_p;
    }
};

class wavefilt {
private:
    NRVec<DP> *cc_p,*cr_p;
public:
    int ncof,ioff,joff;
    NRVec<DP> &cc,&cr;
    wavefilt() : cc(*cc_p),cr(*cr_p) {}
    wavefilt(const DP *a, const int n) : //initialize to array
        cc_p(new NRVec<DP>(n)),cr_p(new NRVec<DP>(n)),
        ncof(n),ioff(-(n >> 1)),joff(-(n >> 1)),cc(*cc_p),cr(*cr_p) {
            int i;
            for (i=0; i<n; i++)
                cc[i] = *a++;
            DP sig = -1.0;
            for (i=0; i<n; i++) {
                cr[n-1-i]=sig*cc[i];
                sig = -sig;
            }
        }
    ~wavefilt() {
        if (cc_p != 0) delete cc_p;
        if (cr_p != 0) delete cr_p;
    }
};

//Overloaded complex operations to handle mixed float and double
//This takes care of e.g. 1.0/z, z complex<float>

inline const complex<float> operator+(const double &a,
    const complex<float> &b) { return float(a)+b; }

inline const complex<float> operator+(const complex<float> &a,
    const double &b) { return a+float(b); }

inline const complex<float> operator-(const double &a,
    const complex<float> &b) { return float(a)-b; }

inline const complex<float> operator-(const complex<float> &a,
    const double &b) { return a-float(b); }

inline const complex<float> operator*(const double &a,
    const complex<float> &b) { return float(a)*b; }

inline const complex<float> operator*(const complex<float> &a,
    const double &b) { return a*float(b); }
```

```
inline const complex<float> operator/(const double &a,
    const complex<float> &b) { return float(a)/b; }

inline const complex<float> operator/(const complex<float> &a,
    const double &b) { return a/float(b); }

//some compilers choke on pow(float,double) in single precision. also atan2

inline float pow (float x, double y) {return pow(double(x),y);}
inline float pow (double x, float y) {return pow(x,double(y));}
inline float atan2 (float x, double y) {return atan2(double(x),y);}
inline float atan2 (double x, float y) {return atan2(x,double(y));}
#endif /* _NR_UTIL_H_ */
```

If you want to use a different matrix/vector class library than the above default, you need to do two things. The first is to change the constness of the type declarations, as discussed in §1.3. Here is a listing of the file nrtypes_lib.h. You use it by changing the file that is included by nrtypes.h from nrtypes_nr.h to nrtypes_lib.h.

```
#ifndef _NR_TYPES_H_
#define _NR_TYPES_H_

#include <complex>
#include <fstream>
#include "nrutil.h"
using namespace std;

typedef double DP;

// Vector Types

typedef const NRVec<bool> Vec_I_BOOL;
typedef const NRVec<bool> Vec_BOOL, Vec_O_BOOL, Vec_IO_BOOL;

typedef const NRVec<char> Vec_I_CHR;
typedef const NRVec<char> Vec_CHR, Vec_O_CHR, Vec_IO_CHR;

typedef const NRVec<unsigned char> Vec_I_UCHR;
typedef const NRVec<unsigned char> Vec_UCHR, Vec_O_UCHR, Vec_IO_UCHR;

typedef const NRVec<int> Vec_I_INT;
typedef const NRVec<int> Vec_INT, Vec_O_INT, Vec_IO_INT;

typedef const NRVec<unsigned int> Vec_I_UINT;
typedef const NRVec<unsigned int> Vec_UINT, Vec_O_UINT, Vec_IO_UINT;

typedef const NRVec<long> Vec_I_LNG;
typedef const NRVec<long> Vec_LNG, Vec_O_LNG, Vec_IO_LNG;

typedef const NRVec<unsigned long> Vec_I_ULNG;
typedef const NRVec<unsigned long> Vec_ULNG, Vec_O_ULNG, Vec_IO_ULNG;

typedef const NRVec<float> Vec_I_SP;
typedef const NRVec<float> Vec_SP, Vec_O_SP, Vec_IO_SP;

typedef const NRVec<DP> Vec_I_DP;
typedef const NRVec<DP> Vec_DP, Vec_O_DP, Vec_IO_DP;

typedef const NRVec<complex<float> > Vec_I_CPLX_SP;
typedef const NRVec<complex<float> > Vec_CPLX_SP, Vec_O_CPLX_SP, Vec_IO_CPLX_SP;
```

```
typedef const NRVec<complex<DP> > Vec_I_CPLX_DP;
typedef const NRVec<complex<DP> > Vec_CPLX_DP, Vec_O_CPLX_DP, Vec_IO_CPLX_DP;

// Matrix Types

typedef const NRMat<bool> Mat_I_BOOL;
typedef const NRMat<bool> Mat_BOOL, Mat_O_BOOL, Mat_IO_BOOL;

typedef const NRMat<char> Mat_I_CHR;
typedef const NRMat<char> Mat_CHR, Mat_O_CHR, Mat_IO_CHR;

typedef const NRMat<unsigned char> Mat_I_UCHR;
typedef const NRMat<unsigned char> Mat_UCHR, Mat_O_UCHR, Mat_IO_UCHR;

typedef const NRMat<int> Mat_I_INT;
typedef const NRMat<int> Mat_INT, Mat_O_INT, Mat_IO_INT;

typedef const NRMat<unsigned int> Mat_I_UINT;
typedef const NRMat<unsigned int> Mat_UINT, Mat_O_UINT, Mat_IO_UINT;

typedef const NRMat<long> Mat_I_LNG;
typedef const NRMat<long> Mat_LNG, Mat_O_LNG, Mat_IO_LNG;

typedef const NRVec<unsigned long> Mat_I_ULNG;
typedef const NRMat<unsigned long> Mat_ULNG, Mat_O_ULNG, Mat_IO_ULNG;

typedef const NRMat<float> Mat_I_SP;
typedef const NRMat<float> Mat_SP, Mat_O_SP, Mat_IO_SP;

typedef const NRMat<DP> Mat_I_DP;
typedef const NRMat<DP> Mat_DP, Mat_O_DP, Mat_IO_DP;

typedef const NRMat<complex<float> > Mat_I_CPLX_SP;
typedef const NRMat<complex<float> > Mat_CPLX_SP, Mat_O_CPLX_SP, Mat_IO_CPLX_SP;

typedef const NRMat<complex<DP> > Mat_I_CPLX_DP;
typedef const NRMat<complex<DP> > Mat_CPLX_DP, Mat_O_CPLX_DP, Mat_IO_CPLX_DP;

// 3D Matrix Types

typedef const NRMat3d<DP> Mat3D_I_DP;
typedef NRMat3d<DP> Mat3D_DP, Mat3D_O_DP, Mat3D_IO_DP;

// Miscellaneous Types

typedef NRVec<unsigned long *> Vec_ULNG_p;
typedef const NRVec<const NRMat<DP> *> Vec_Mat_DP_p;
typedef NRVec<fstream *> Vec_FSTREAM_p;

#endif /* _NR_TYPES_H_ */
```

The second thing you need to do to use a different matrix/vector library is to supply a wrapper-class replacement for `nrutil.h`. Here, for example, we give the listing of the file `nrutil_tnt.h` that enables you to use the TNT matrix/vector library [1] instead of our default one. You use it by changing the file that is included by `nrutil.h` from `nrutil_nr.h` to `nrutil_tnt.h`.

```
#ifndef _NR_UTIL_H_
#define _NR_UTIL_H_

#include <string>
#include <cmath>
#include <complex>
```

```cpp
#include <iostream>
using namespace std;

typedef double DP;

template<class T>
inline const T SQR(const T a) {return a*a;}

template<class T>
inline const T MAX(const T &a, const T &b)
{return b > a ? (b) : (a);}

inline float MAX(const double &a, const float &b)
{return b > a ? (b) : float(a);}

inline float MAX(const float &a, const double &b)
{return b > a ? float(b) : (a);}

template<class T>
inline const T MIN(const T &a, const T &b)
{return b < a ? (b) : (a);}

inline float MIN(const double &a, const float &b)
{return b < a ? (b) : float(a);}

inline float MIN(const float &a, const double &b)
{return b < a ? float(b) : (a);}

template<class T>
inline const T SIGN(const T &a, const T &b)
    {return b >= 0 ? (a >= 0 ? a : -a) : (a >= 0 ? -a : a);}

inline float SIGN(const float &a, const double &b)
    {return b >= 0 ? (a >= 0 ? a : -a) : (a >= 0 ? -a : a);}

inline float SIGN(const double &a, const float &b)
    {return b >= 0 ? (a >= 0 ? a : -a) : (a >= 0 ? -a : a);}

template<class T>
inline void SWAP(T &a, T &b)
    {T dum=a; a=b; b=dum;}

namespace NR {
    inline void nrerror(const string error_text)
    // Numerical Recipes standard error handler
    {
        cerr << "Numerical Recipes run-time error..." << endl;
        cerr << error_text << endl;
        cerr << "...now exiting to system..." << endl;
        exit(1);
    }
}

#include "tnt/tnt.h"
#include "tnt/vec.h"
#include "tnt/cmat.h"

// TNT Wrapper File
// This is the file that "joins" the TNT Vector<> and Matrix<> classes
// to the NRVec and NRMat classes by the Wrapper Class Method

// NRVec contains a Vector and a &Vector. All its constructors, except the
// conversion constructor, create the Vector and point the &Vector to it.
// The conversion constructor only points the &Vector. All operations
```

```
// (size, subscript) are through the &Vector, which as a reference
// (not pointer) has no indirection overhead.

template<class T>
class NRVec {
private:
    TNT::Vector<T> myvec;
    TNT::Vector<T> &myref;
public:
    NRVec<T>() : myvec(), myref(myvec) {}
    explicit NRVec<T>(const int n) : myvec(n), myref(myvec) {}
    NRVec<T>(const T &a, int n) : myvec(n,a), myref(myvec) {}
    NRVec<T>(const T *a, int n) : myvec(n,a), myref(myvec) {}
    NRVec<T>(TNT::Vector<T> &rhs) : myref(rhs) {}
    // conversion constructor makes a special NRVec pointing to Vector's data
    // this handles Vector actual args sent to NRVec formal args in functions
    NRVec(const NRVec<T>& rhs) : myvec(rhs.myref), myref(myvec) {}
    // copy constructor calls Vector copy constructor
    inline NRVec& operator=(const NRVec& rhs) { myref=rhs.myref; return *this;}
    // assignment operator calls Vector assignment operator
    inline NRVec& operator=(const T& rhs) { myvec=rhs; return *this;}
    // scalar assignment calls Vector assignment operator
    inline int size() const {return myref.size();}
    inline T & operator[](const int i) const {return myref[i];}
    // return element i
    inline operator TNT::Vector<T>() const {return myref;}
    // conversion operator to Vector
    // this handles NRVec function return types when used in Vector expressions
    ~NRVec() {}
};

template <class T>
class NRMat {
private:
    TNT::Matrix<T> mymat;
    TNT::Matrix<T> &myref;
public:
    NRMat() : mymat(), myref(mymat) {}
    NRMat(int n, int m) : mymat(n,m), myref(mymat) {}
    NRMat(const T& a, int n, int m) : mymat(n,m,a), myref(mymat) {}
    //Initialize to constant
    NRMat(const T* a, int n, int m) : mymat(n,m,a), myref(mymat) {}
    //Initialize to array
    NRMat<T>(TNT::Matrix<T> &rhs) : myref(rhs) {}
    // conversion constructor from Matrix
    NRMat(const NRMat& rhs) : mymat(rhs.myref), myref(mymat) {}
    // copy constructor
    inline NRMat& operator=(const NRMat& rhs) { myref=rhs.myref; return *this;}
    // assignment operator
    inline NRMat& operator=(const T& rhs) { mymat=rhs; return *this;}
    // scalar assignment calls Matrix assignment operator
    inline T* operator[](const int i) const {return myref[i];}
    //subscripting: pointer to row i
    //return type is whatever Matrix returns for a single [] dereference
    inline int nrows() const {return myref.num_rows();}
    inline int ncols() const {return myref.num_cols();}
    inline operator TNT::Matrix<T>() const {return myref;}
    // conversion operator to Matrix
    ~NRMat() {}
};

template <class T>
class NRMat3d {
The rest of the code is identical to nrutil_nr.h and is omitted.
```

Finally we give the listing of the file `nrutil_mtl.h` that enables you to use the MTL matrix/vector library [2]. Again, simply change the file that is included by `nrutil.h` from `nrutil_nr.h` to `nrutil_mtl.h`.

Note that the MTL library uses the `std::vector` class to define its basic vector type. This class has a specialization for `vector<bool>` that doesn't work with our wrapper class scheme. So we implement `NRVec<bool>` as our own specialization of `NRVec` by making it a derived class of `NRVec<int>`. (Recall that a `bool` variable is automatically converted to `int` when necessary.) The only potential problem with doing this occurs if you blindly try to mix objects of type `mtl::Vector<bool>` with objects of type `NRVec<bool>`!

```
#ifndef _NR_UTIL_H_
#define _NR_UTIL_H_

#include <string>
#include <cmath>
#include <complex>
#include <iostream>
using namespace std;

typedef double DP;

template<class T>
inline const T SQR(const T a) {return a*a;}

template<class T>
inline const T MAX(const T &a, const T &b)
{return b > a ? (b) : (a);}

inline float MAX(const double &a, const float &b)
{return b > a ? (b) : float(a);}

inline float MAX(const float &a, const double &b)
{return b > a ? float(b) : (a);}

template<class T>
inline const T MIN(const T &a, const T &b)
{return b < a ? (b) : (a);}

inline float MIN(const double &a, const float &b)
{return b < a ? (b) : float(a);}

inline float MIN(const float &a, const double &b)
{return b < a ? float(b) : (a);}

template<class T>
inline const T SIGN(const T &a, const T &b)
    {return b >= 0 ? (a >= 0 ? a : -a) : (a >= 0 ? -a : a);}

inline float SIGN(const float &a, const double &b)
    {return b >= 0 ? (a >= 0 ? a : -a) : (a >= 0 ? -a : a);}

inline float SIGN(const double &a, const float &b)
    {return b >= 0 ? (a >= 0 ? a : -a) : (a >= 0 ? -a : a);}

template<class T>
inline void SWAP(T &a, T &b)
    {T dum=a; a=b; b=dum;}

namespace NR {
    inline void nrerror(const string error_text)
```

```
    // Numerical Recipes standard error handler
    {
        cerr << "Numerical Recipes run-time error..." << endl;
        cerr << error_text << endl;
        cerr << "...now exiting to system..." << endl;
        exit(1);
    }
}


#include "mtl/mtl.h"
using namespace mtl;

// mtl Wrapper File
// This is the file that "joins" the mtl Vector<> and Matrix<> classes
// to the NRVec and NRMat classes by the Wrapper Class Method

// NRVec contains a Vector and a &Vector. All its constructors, except the
// conversion constructor, create the Vector and point the &Vector to it.
// The conversion constructor only points the &Vector. All operations
// (size, subscript) are through the &Vector, which as a reference
// (not pointer) has no indirection overhead.

template<class T>
class NRVec {
// Use the std::vector based dense1D for our Vector type
typedef dense1D<T> Vector;
protected:          //access required in NRVec<bool> below
    Vector myvec;
    Vector &myref;
public:
    NRVec<T>() : myvec(), myref(myvec) {}
    explicit NRVec<T>(const int n) : myvec(n), myref(myvec) {}
    NRVec<T>(const T &a, int n) : myvec(n), myref(myvec) {
        for (int i=0; i<n; i++) myvec[i] = a;}
    NRVec<T>(const T *a, int n) : myvec(n), myref(myvec) {
        for (int i=0; i<n; i++) myvec[i] = *a++;}
    NRVec<T>(Vector &rhs) : myref(rhs) {}
    // conversion constructor makes a special NRVec pointing to Vector's data
    // this handles Vector actual args sent to NRVec formal args in functions
    NRVec(const NRVec<T>& rhs) : myvec(rhs.myref.size()), myref(myvec)
        {copy(rhs.myref,myref);}
    // copy constructor. mtl copy constructor
    // does shallow copy only. so use copy() instead
    inline NRVec& operator=(const NRVec& rhs) {
        if (myref.size() != rhs.myref.size())
            myref.resize(rhs.myref.size());
        copy(rhs.myref,myref); return *this;}
    inline int size() const {return myref.size();}
    inline T & operator[](const int i) const {return myref[i];}
    inline operator Vector() const {return myref;}
    // conversion operator to Vector
    // handles NRVec function return types when used in Vector expressions
    ~NRVec() {}
};

//The std:vector class has a specialization for vector<bool> that doesn't
//work with the above wrapper class scheme. So implement our own
//specialization as a derived class of NRVec<int>. This could cause
//problems if you mix mtl::Vector<bool> with NRVec<bool>!

template <> class NRVec<bool> : public NRVec<int> {
public:
    NRVec() : NRVec<int>() {}
    explicit NRVec(const int n) : NRVec<int>(n) {}
```

```
        NRVec(const bool &a, int n) : NRVec<int>(int(a),n) {}
        NRVec(const bool *a, int n) : NRVec<int>(n) {
            for (int i=0; i<n; i++) myvec[i] = *a++;}
//note: defaults OK for copy constructor and assignment
};

template <class T>
class NRMat {
// Use the matrix generator to select a matrix type
typedef matrix< T,
    rectangle<>,
    dense<>,
    row_major>::type Matrix;
protected:
    Matrix mymat;
    Matrix &myref;
public:
    NRMat() : mymat(), myref(mymat) {}
    NRMat(int n, int m) : mymat(n,m), myref(mymat) {}
    NRMat(const T& a, int n, int m) : mymat(n,m), myref(mymat) {
        for (int i=0; i< n; i++)
            for (int j=0; j<m; j++)
                mymat[i][j] = a;}
    NRMat(const T* a, int n, int m) : mymat(n,m), myref(mymat) {
        for (int i=0; i< n; i++)
            for (int j=0; j<m; j++)
                mymat[i][j] = *a++;}
    NRMat<T>(Matrix &rhs) : myref(rhs) {}
    NRMat(const NRMat& rhs) :
        mymat(rhs.myref.nrows(),rhs.myref.ncols()), myref(mymat)
        {copy(rhs.myref,myref);}
    inline NRMat& operator=(const NRMat& rhs) {
        if (myref.nrows() != rhs.myref.nrows() && myref.ncols() !=
            rhs.myref.ncols()) {
                cerr << "assignment with incompatible matrix sizes\n";
                abort();
        }
        copy(rhs.myref,myref); return *this;
    }
    typename Matrix::OneD operator[](const int i) const {return myref[i];}
    //return type is whatever Matrix returns for a single [] dereference
    inline int nrows() const {return myref.nrows();}
    inline int ncols() const {return myref.ncols();}
    inline operator Matrix() const {return myref;}
    ~NRMat() {}
};

template <> class NRMat<bool> : public NRMat<int> {
public:
    NRMat() : NRMat<int>() {}
    explicit NRMat(int n, int m) : NRMat<int>(n,m) {}
    NRMat(const bool &a, int n, int m) : NRMat<int>(int(a),n,m) {}
    NRMat(const bool *a, int n, int m) : NRMat<int>(n,m) {
        for (int i=0; i< n; i++)
            for (int j=0; j<m; j++)
                mymat[i][j] = *a++;}
//note: defaults OK for copy constructor and assignment
};

template <class T>
class NRMat3d {
The rest of the code is identical to nrutil_nr.h and is omitted.
```

CITED REFERENCES AND FURTHER READING:

Pozo, R., *Template Numerical Toolkit*, `http://math.nist.gov/tnt`. [1]

Lumsdaine, A., and Siek, J. 1998, *The Matrix Template Library*, `http://www.lsc.nd.edu/research/mtl`. [2]

Appendix C: Converting to Single Precision

As discussed in §1.2, there is probably no good reason not to use the Recipes in the default mode supplied, namely in double precision. However, if you want to change them all to single precision, proceed as follows:

- Change the definition

```
typedef double DP;
```

   to

```
typedef float DP;
```

   in both `nrtypes.h` and `nrutil.h`. (Recall that these routines typically just include the default routines `nrtypes_nr.h` and `nrutil_nr.h`, which is where you will actually make the changes.)
- Change the "accuracy parameters" in the routines listed in the accompanying table as shown. (Note that these are only suggested values. Some experimentation may be required, depending on your problem.) To facilitate making these changes, we have provided with the machine-readable distribution a Perl script `nrtosp.pl` that will automatically do this for you. It reads in the information in the table from a file `nrtosp.dat`, which you can easily edit if there are any other changes you want to make.

A final warning: some of the Recipes are particularly susceptible to roundoff error, and are liable to give inaccurate results in single precision. We have mentioned this explicitly in the comments for the affected routines.

In routine...	*change...*	*to...*
beschb	NUSE1=7,NUSE2=8	NUSE1=5,NUSE2=5
bessi	ACC=200	ACC=40
bessj	ACC=160	ACC=40
broydn	TOLF=1.0e-8,TOLMIN=1.0e-12	TOLF=1.0e-4,TOLMIN=1.0e-6
dlinmin	TOL=2.0e-8	TOL=1.0e-4
fdjac	EPS=1.0e-8	EPS=1.0e-4
frprmn	EPS=1.0e-18	EPS=1.0e-10
gauher	EPS=1.0e-14	EPS=4.0e-7
gaujac	EPS=1.0e-14	EPS=1.0e-7
gaulag	EPS=1.0e-14	EPS=1.0e-6
gauleg	EPS=1.0e-14	EPS=1.0e-7
hypgeo	EPS=1.0e-14	EPS=1.0e-6
linmin	TOL=1.0e-8	TOL=1.0e-4
newt	TOLF=1.0e-8,TOLMIN=1.0e-12	TOLF=1.0e-4,TOLMIN=1.0e-6
probks	EPS1=1.0e-6,EPS2=1.0e-16	EPS1=0.001,EPS2=1.0e-8
qromb	EPS=1.0e-10	EPS=1.0e-6
qromo	EPS=3.0e-9	EPS=1.0e-6
qroot	TINY=1.0e-14	TINY=1.0e-6
qsimp	EPS=1.0e-10	EPS=1.0e-6
qtrap	EPS=1.0e-10	EPS=1.0e-6
ran1	EPS=3.0e-16	EPS=1.2e-7
ran2	EPS=3.0e-16	EPS=1.2e-7
rc	ERRTOL=0.0012	ERRTOL=0.04
rd	ERRTOL=0.0015	ERRTOL=0.05
rf	ERRTOL=0.0025	ERRTOL=0.08
rj	ERRTOL=0.0015	ERRTOL=0.05
sfroid	conv=1.0e-14	conv=5.0e-6
shoot	EPS=1.0e-14	EPS=1.0e-6
shootf	EPS=1.0e-14	EPS=1.0e-6
simp2	EPS=1.0e-14	EPS=1.0e-6
simplx	EPS=1.0e-14	EPS=1.0e-6
sncndn	CA=1.0e-8	CA=.0003
sor	EPS=1.0e-13	EPS=1.0e-4
sphfpt	DXX=1.0e-8	DXX=1.0e-4
sphoot	dx=1.0e-8	dx=1.0e-4
svdfit	TOL=1.0e-13	TOL=1.0e-5
zroots	EPS=1.0e-14	EPS=1.0e-6

Index of Programs and Dependencies

The following table lists, in alphabetical order, all the routines in *Numerical Recipes*. When a routine requires subsidiary routines, either from this book or else user-supplied, the full dependency tree is shown: A routine calls directly all routines to which it is connected by a solid line in the column immediately to its right; it calls indirectly the connected routines in all columns to its right. Typographical conventions: Routines from this book are in typewriter font (e.g., eulsum, *gammln*). The smaller, slanted font is used for the second and subsequent occurrences of a routine in a single dependency tree. (When you are getting routines from the *Numerical Recipes* diskettes, or their archive files, you need only specify names in the larger, upright font.) User-supplied routines are indicated by the use of text font and square brackets, e.g., [funcv]. Consult the text for individual specifications of these routines. The right-hand side of the table lists section and page numbers for each program.

General Index

Snyden
N.D. ILLUSTRATION.

Comics, Comix & Graphic Novels

Phaidon Press Limited
Regent's Wharf
All Saints Street
London N1 9PA

First published 1996
© 1996 Phaidon Press Limited
ISBN 0 7148 3008 9

Printed in Hong Kong